Welding Handbook

Ninth Edition

Volume 2

WELDING PROCESSES, PART 1

American Welding Society

Welding Handbook, Ninth Edition

Volume 1 *Welding Science and Technology*

Volume 2 *Welding Processes, Part 1*

Volume 3 *Welding Processes, Part 2*

Volume 4 *Materials and Applications, Part 1*

Volume 5 *Materials and Applications, Part 2*

Welding Handbook

Ninth Edition

Volume 2

WELDING PROCESSES, PART 1

Prepared under the direction of the
Welding Handbook Committee

Annette O'Brien
Editor

American Welding Society
550 N.W. LeJeune Road
Miami, FL 33126

Library of Congress Control Number: 2001089999
ISBN: 0-87171-729-8

The *Welding Handbook* is the result of the collective effort of many volunteer technical specialists who provide information to assist with the design and application of welding and allied processes.

The information and data presented in the *Welding Handbook* are intended for informational purposes only. Reasonable care is exercised in the compilation and publication of the *Welding Handbook* to ensure the authenticity of the contents. However, no representation is made as to the accuracy, reliability, or completeness of this information, and an independent substantiating investigation of the information should be undertaken by the user.

The information contained in the *Welding Handbook* shall not be construed as a grant of any right of manufacture, sale, use, or reproduction in connection with any method, process, apparatus, product, composition, or system, which is covered by patent, copyright, or trademark. Also, it shall not be construed as a defense against any liability for such infringement. Whether the use of any information in the *Welding Handbook* would result in an infringement of any patent, copyright, or trademark is a determination to be made by the user.

Printed in the United States of America

DEDICATION

Howard B. Cary
1920–2001

In recognition of his contributions to arc welding and the dissemination of arc welding technology.

Howard B. Cary's career in the welding industry spanned almost 60 years. His achievements ranged from original development work leading to the commercial introduction of gas-shielded metal arc welding to the eventual integration of arc welding processes into advanced robotic systems. He was internationally known and respected for the scope of his technical knowledge. He held five patents and wrote more than 100 technical papers and articles and two books, *Arc Welding Automation*, and the 5th edition of *Modern Welding Technology.*

Howard Cary worked with the Hobart Brothers Company for 30 years, advancing to Vice President of Welding Systems and simultaneously serving as President of the Hobart Institute of Welding Technology. His enthusiasm for welding technology, his skills in organizing information, teaching and training, and his rapport with colleagues and students were well known in the welding community. He served as a mentor to many as they built careers in welding.

Throughout his career Howard Cary was a volunteer member of various American Welding Society committees responsible for writing standards, welding procedures and performance qualifications for arc welding processes, welders, and welding educators.

The American Welding Society recognized his achievements with the Howard E. Atkins Instructor Membership award, the A. F. Davis Silver Metal Award, the Samuel Wylie Miller Memorial Award, the National Meritorious Certificate, the Honorary Membership Award and the R. D. Thomas Memorial Award. He was president of AWS in 1980 and 1981 and was elected a Fellow of the American Welding Society in 1991.

He was selected by AWS to present the Plummer Memorial Educational Lecture in 1994. True to his professional passion, the subject of his lecture was "The Importance of Being a Welder."

CONTENTS

ACKNOWLEDGMENTS

The American Welding Society, the Welding Handbook Committee and the Editor recognize with gratitude the volunteers who contributed to this volume of the 9th edition of the *Welding Handbook* and those who contributed to past editions from which the current volumes evolved. The Welding Handbook Committee appreciates the work of Robert L. O'Brien, Editor of the 8th edition of *Welding Processes.*

Contributors to the *Welding Handbook* represent many facets of the Welding industry—university, government, and private welding research and development institutions, manufacturers of welding equipment and materials, standards writing organizations, and fabricators and manufacturers who use welding technology. The Welding Handbook Committee recognizes and appreciates their contributions.

The Welding Handbook Committee extends its appreciation to the AWS technical committees responsible for creating the consensus standards that are discussed and referenced throughout this book, and to the AWS technical staff for the engineering expertise and support they contributed.

Continuing the work that began prior to 1938 when the first edition of the *Welding Handbook* was published, volunteers who contributed to this volume have once again demonstrated the enthusiasm for welding technology and dedication to the industry that are traditional with the *Welding Handbook.*

Welding Handbook Committee Chairs, 1938–2004

1938–1942	*D. S. Jacobus*
Circa 1950	*H. L. Boardman*
1956–1958	*F. L. Plummer*
1958–1960	*R. D. Stout*
1960–1962	*J. F. Randall*
1962–1965	*G. E. Claussen*
1965–1966	*H. Schwartzbart*
1966–1967	*A. Lesnewich*
1967–1968	*W. L. Burch*
1968–1969	*L. F. Lockwood*
1969–1970	*P. W. Ramsey*
1970–1971	*D. V. Wilcox*
1971–1972	*C. E. Jackson*
1972–1975	*S. Weiss*
1975–1978	*A. W. Pense*
1978–1981	*W. L. Wilcox*
1981–1984	*J. R. Condra*
1984–1987	*J. R. Hannahs*
1987–1990	*M. J. Tomsic*
1990–1992	*C. W. Case*
1992–1996	*B. R. Somers*
1996–1999	*P. I. Temple*
1999–2004	*H. R. Castner*

PREFACE

Welding Processes, Part 1 is the second of the five volumes of the 9th edition of the *Welding Handbook*. The fifteen chapters of this volume provide updated information on the arc welding and cutting processes, oxyfuel gas welding and cutting, brazing, and soldering. Volume 3, *Welding Processes, Part 2* will cover resistance, solid state, and other welding and cutting processes. Volumes 4 and 5 of the *Welding Handbook* will address welding materials and applications. These volumes represent the practical application of the principles discussed in the chapters of Volume 1, *Welding Science and Technology,* published in 2001.

This peer-reviewed volume of the *Welding Handbook* reflects a tremendous leap forward in welding technology. While many basics of the welding processes have remained substantially the same, the precise control of welding parameters, advanced techniques, complex applications and new materials discussed in this updated volume are dramatically changed from those described in previous editions. In particular, advancements in digital or computerized control of welding parameters have resulted in consistently high weld quality for manual and mechanized welding and the repeatability necessary for successful automated operations.

Chapter 1 of *Welding Processes, Part 1* is a compilation of information on arc welding power sources. Subsequent chapters present specific information on shielded metal arc welding, gas tungsten arc welding, gas metal arc welding, flux cored arc welding, submerged arc welding, plasma arc welding, electrogas welding, arc stud welding, electroslag welding, oxyfuel gas welding, brazing, soldering, oxygen cutting, and arc cutting and gouging.

Appendix A and B address safety issues. Appendix A reproduces the American Welding Society Lens Shade Selector. Appendix B is a list of national and international safety standards applicable to welding, cutting, and allied processes. Although each chapter in this volume has a section on safe practices as they pertain to the specific process, readers should refer to Chapter 17, "Safe Practices," of Volume 1 and to the appropriate standards listed in Appendix B. Appendix C is a list of American Welding Society filler metal specifications and related documents.

An index of this volume and a major subject index of previous volumes are included.

This volume was compiled by the members the Welding Handbook Volume 2 Committee and the Chapter Committees, with oversight by the Welding Handbook Committee. Chapter committee chairs, chapter committee members, and oversight persons are recognized on the title pages of the chapters. An important contribution to this volume is the review of each chapter provided by members of the Technical Activities Committee and the Safety and Health Committee of the American Welding Society.

The Welding Handbook Committee welcomes your comments and suggestions. Please address them to the Editor, *Welding Handbook*, American Welding Society, 550 N.W. LeJeune Road, Miami, Florida 33126.

Harvey R. Castner, Chair
Welding Handbook Committee

Ian D. Harris, Chair
Volume 2 Committee

Annette O'Brien, Editor
Welding Handbook

REVIEWERS
AMERICAN WELDING SOCIETY
SAFETY AND HEALTH COMMITTEE AND
TECHNICAL ACTIVITIES COMMITTEE

S. R. Fiore	Edison Welding Institute
R. L. Holdren	Edison Welding Institute
J. D. Karrow	Consultant
D. J. Landon	Vermeer Manufacturing Company
M. J. Lucas	GE Aircraft Engines
K. Lyttle	Consultant
A. F. Manz	A. F. Manz Associates
D. L. McQuaid	D. L. McQuaid and Associates, Incorporated
D.D. Rager	Rager Consulting
A. T. Sheppard	Consultant
A. W. Sindel	Sindel and Associates
W. J. Sperko	Sperko Engineering Associates
M. Wisbrock	Lockheed Martin Missiles and Fire Control

AMERICAN WELDING SOCIETY TECHNICAL STAFF ADVISORS

L. P. Connor
A. R. Davis
H. P. Ellison
J. L. Gayler
R. K. Gupta
S. P. Hedrick
P. Howe
C. L. Jenney

CONTRIBUTORS

WELDING HANDBOOK COMMITTEE

H. R. Castner, Chair	Edison Welding Institute
B. J. Bastian, First Vice-Chair	Benmar Associates
J. H. Myers, Second Vice-Chair	Weld Inspection and Consulting Services
A. O'Brien, Secretary	American Welding Society
I.D. Harris	Edison Welding Institute
C. E. Pepper	ENGlobal Engineering
W. L. Roth	Procter and Gamble
P. I. Temple	Detroit Edison

WELDING HANDBOOK VOLUME 2 COMMITTEE

I. D. Harris, Chair	Edison Welding Institute
D. R. Amos, First Vice-Chair	Siemens Westinghouse Power Corporation
A. O'Brien, Secretary	American Welding Society
D. W. Dickinson	The Ohio State University
D. B. Holliday	Northrop Grumman Marine Systems
B. R. Somers	Lucius Pitkin, Incorporated

CHAPTER CHAIRS

Chapter 1	*S. P. Moran*	Miller Electric Manufacturing Company
Chapter 2	*M. A. Amata*	ESAB Welding and Cutting Products
Chapter 3	*J. T. Salkin*	Arc Applications, Incorporated
Chapter 4	*D. B. Holliday*	Northrop Grumman Marine Systems
Chapter 5	*D. B. Arthur*	J.W. Harris
Chapter 6	*R. A. Swain*	Euroweld, Limited
Chapter 7	*W. L. Roth*	Procter and Gamble
Chapter 8	*D. A. Fink*	The Lincoln Electric Company
Chapter 9	*H. A. Chambers*	Nelson Stud Welding
Chapter 10	*J. R. Hannahs*	Edison Community College
Chapter 11	*G. R. Meyer*	Consultant
Chapter 12	*N. C. Cole*	NCC Engineering
Chapter 13	*F. M. Hosking*	Sandia National Laboratories
	P. T. Vianco	Sandia National Laboratories
Chapter 14	*G. R. Meyer*	Consultant
Chapter 15	*I. D. Harris*	Edison Welding Institute

CHAPTER 1

ARC WELDING POWER SOURCES

Photograph courtesy of NASA

Prepared by the Welding Handbook Chapter Committee on Arc Welding Power Sources:

S. P. Moran, Chair
Miller Electric Manufacturing Company

D. J. Erbe
Panasonic Factory Automation

W. E. Herwig
Miller Electric Manufacturing Company

W. E. Hoffman
ESAB Welding and Cutting Products

C. Hsu
The Lincoln Electric Company

J. O. Reynolds
Miller Electric Manufacturing Company

Welding Handbook Committee Member:

C. E. Pepper
ENGlobal Engineering

Contents

CHAPTER 1

ARC WELDING POWER SOURCES

INTRODUCTION

This chapter presents a general overview of the electrical power sources used for arc welding. It explores the many types of welding power sources available to meet the electrical requirements of the various arc welding processes.

Welding has a long and rich history. Commercial arc welding is over a hundred years old, and scores of processes and variations have been developed. Over the years, power sources have been developed or modified by equipment manufacturers in response to the changes and improvements in these processes. As welding processes continue to evolve, power sources continue to provide the means of controlling the welding current, voltage, and power. This chapter provides updated information on the basic electrical technologies, circuits, and functions designed into frequently used welding power sources. Topics covered in this chapter include the following:

1. The volt-ampere (V-A) characteristics required for common welding processes,
2. Basic electrical technologies and terminology used in power sources,
3. Simplified explanations of commonly used power source circuits, and
4. An introduction to useful national and international standards.

A basic knowledge of electrical power sources will provide the background for a more complete understanding of the welding processes presented in the other chapters of this book.

FUNDAMENTALS

This section introduces the fundamental functions of welding power sources and the concepts of constant-voltage (CV) and constant-current (CC) characteristics required for welding processes.

The voltage supplied by power companies for industrial purposes—120 volts (V), 230 V, 380 V, or 480 V—is too high for use in arc welding. Therefore, the first function of an arc welding power source is to reduce the high input or line voltage to a suitable output voltage range, 20 V to 80 V. A transformer, a solid-state inverter, or an electric motor-generator can be used to reduce the utility power to terminal or open-circuit voltage appropriate for arc welding.

Alternatively, a power source for arc welding may derive its power from a prime mover such as an internal combustion engine. The rotating power from an internal combustion engine is used to rotate a generator or an alternator for the source of electrical current.

Welding transformers, inverters, or generator/alternators provide high-amperage welding current, generally ranging from 30 amperes (A) to 1500 A. The output of a power source may be alternating current (ac), direct current (dc) or both. It may be constant current, constant voltage, or both. Welding power sources may also provide pulsed output of voltage or current.

Some power source configurations deliver only certain types of current. For example, transformer power sources deliver ac only. Transformer-rectifier power sources can deliver either alternating or direct current, as selected by the operator. Electric motor-generator power sources usually deliver dc output. A motor-alternator delivers ac, or when equipped with rectifiers, dc.

Power sources can also be classified into subcategories. For example, a gas tungsten arc welding power source might be identified as transformer-rectifier, constant-current, ac/dc. A complete description of any power source should include welding current rating, duty cycle rating, service classification, and input power

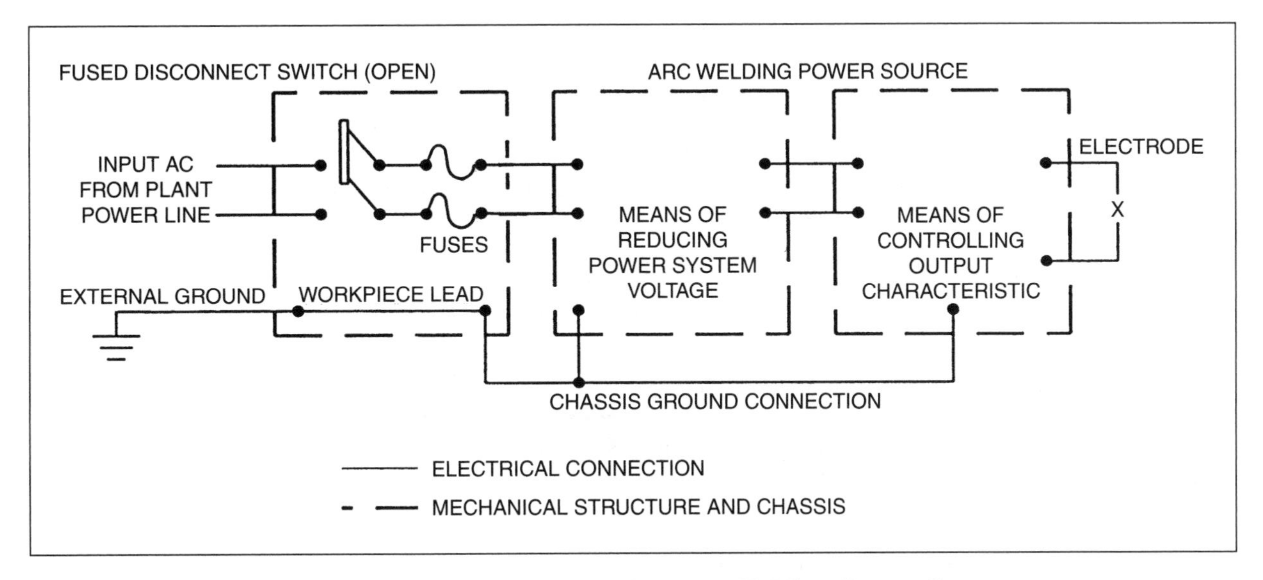

Figure 1.1—Basic Elements of an Arc Welding Power Source

requirements. Special features can also be included such as remote control, high-frequency stabilization, current-pulsing capability, starting and finishing current versus time programming, wave balancing capabilities, and line-voltage compensation. Conventional magnetic controls include movable shunts, saturable reactors, magnetic amplifiers, series impedance, or tapped windings. Solid-state electronic controls may be phase-controlled silicon-controlled rectifiers (SCRs) or inverter-controlled semiconductors. Electronic logic or microprocessor circuits may control these elements.

Figure 1.1 shows the basic elements of a welding power source with power supplied from utility lines. The arc welding power source itself does not usually include the fused disconnect switch; however, this is a necessary protective and safety element.

An engine-driven power source would require elements different from those shown in Figure 1.1. It would require an internal combustion engine, an engine speed regulator, and an alternator, with or without a rectifier, or a generator and an output control.

Before the advent of pulsed current welding processes in the 1970s, welding power sources were commonly classified as constant current or constant voltage. These classifications are based on the static volt-ampere characteristics of the power source, not the dynamic characteristic or arc characteristics. The term *constant* is true only in a general sense. A constant-voltage output actually reduces or droops slightly as the arc current increases, whereas a constant-current output gradually increases as the arc length and arc voltage decrease. In either case, specialized power sources are available that can hold output voltage or current truly constant. Constant-current power sources are also known as *variable-voltage* power sources, and constant-voltage power sources are often referred to as *constant-potential* power sources. These fast-response, solid-state power sources can provide power in pulses over a broad range of frequencies.

CONSTANT-CURRENT ARC WELDING POWER SOURCES

The National Electrical Manufacturers Association (NEMA) standard *Electric Arc-Welding Power Sources*, EW-1: 1988 (R1999), defines a constant-current arc power source as one "which has means for adjusting the load current and which has a static volt-ampere curve that tends to produce a relatively constant load current. At a given load current, the load voltage is responsive to the rate at which a consumable metal electrode is fed into the arc. When a tungsten electrode is used, the load voltage is responsive to the electrode-to-workpiece distance."[1,2] These characteristics are

1. National Electrical Manufacturers Association (NEMA), 1988 (R1999), *Electric Arc-Welding Power Sources*, EW-1: 1988, Washington, D.C.: National Electrical Manufacturers Association, p. 2.

2. At the time this chapter was prepared, the referenced codes and other standards were valid. If a code or other standard is cited without a date of publication, it is understood that the latest edition of the document referred to applies. If a code or other standard is cited with the date of publication, the citation refers to that edition only, and it is understood that any future revisions or amendments to the code or standard are not included; however, as codes and standards undergo frequent revision, the reader is advised to consult the most recent edition.

such that if the arc length varies because of external influences that result in slight changes in arc voltage, the welding current remains substantially constant. Each current setting yields a separate volt-ampere curve when tested under steady conditions with a resistive load. In the vicinity of the operating point, the percentage of change in current is lower than the percentage of change in voltage.

The no-load, or open-circuit, voltage of constant-current arc welding power sources is considerably higher than the arc voltage.

Constant-current power sources are generally used for manual welding processes such as shielded metal arc welding (SMAW), gas tungsten arc welding (GTAW), plasma arc welding (PAW), or plasma arc cutting (PAC), where variations in arc length are unavoidable because of the human element.

When used in a semiautomatic or automated application in which constant arc length is required, external control devices are necessary. For example, an arc-voltage-sensing wire feeder can be used to maintain constant arc length for gas metal arc welding (GMAW) or flux cored arc welding (FCAW). In GTAW, the arc voltage is monitored, and via a closed-loop feedback, the voltage is used to regulate a motorized slide that positions the torch to maintain a constant arc length (voltage).

CONSTANT-VOLTAGE ARC WELDING POWER SOURCES

The NEMA EW-1 standard defines a constant-voltage power source as follows: "A constant-voltage arc welding power source is a power source which has means for adjusting the load voltage and which has a static volt-ampere curve that tends to produce a relatively constant load voltage. The load current, at a given load voltage, is responsive to the rate at which a consumable electrode is fed into the arc."[3] Constant-voltage arc welding is generally used with welding processes that include a continuously fed consumable electrode, usually in the form of wire.

A welding arc powered by a constant-voltage source using a consumable electrode and a constant-speed wire feed is essentially a self-regulating system. It tends to stabilize the arc length despite momentary changes in the torch position. The arc current is approximately proportional to wire feed for all wire sizes.

CONSTANT-CURRENT/CONSTANT-VOLTAGE POWER SOURCES

A power source that provides both constant current and constant voltage is defined by NEMA as follows: "A constant-current/constant-voltage arc welding power source is a power source which has the selectable characteristics of a constant-current arc welding power source and a constant-voltage arc welding power source."[4]

Additionally, some power sources feature an automatic change from constant current to constant voltage (arc force control for SMAW) or constant voltage to constant current (current limit control for constant-voltage power sources).

3. See Reference 1, p. 3.

4. See Reference 1, p. 2.

PRINCIPLES OF OPERATION

The basic components of welding power sources—transformers, series inductors, generators/alternators, diodes, silicon-controlled rectifiers, and transistors—are introduced in this section. Simple circuits of reactance-controlled, phase-controlled, and inverter power sources are discussed as examples.

Most arc welding involves low-voltage, high-current arcs between an electrode and the workpiece. The means of reducing power-system voltage, as shown in Figure 1.1, may be a transformer or an electric generator or alternator driven by an electric motor.

Electric generators built for arc welding are usually designed for direct-current welding only. In these generators, the electromagnetic means of controlling the volt-ampere characteristic of the arc welding power source is usually an integral part of the generator and not a separate element. Unlike generators, alternators provide ac output that must be rectified to provide a dc output. Various configurations are employed in the construction of direct-current generators. They may use a separate exciter and either differential or cumulative compound winding for selecting and controlling volt-ampere output characteristics.

WELDING TRANSFORMER

A transformer is a magnetic device that operates on alternating current. As shown in Figure 1.2, a simple transformer is composed of three parts: a primary winding, a magnetic core, and a secondary winding. The primary winding, with N_1 turns of wire (in Equation 1.1), is energized by an alternating-current input voltage, thereby magnetizing the core. The core couples the alternating magnetic field into the secondary winding, with N_2 turns of wire, producing an output voltage.

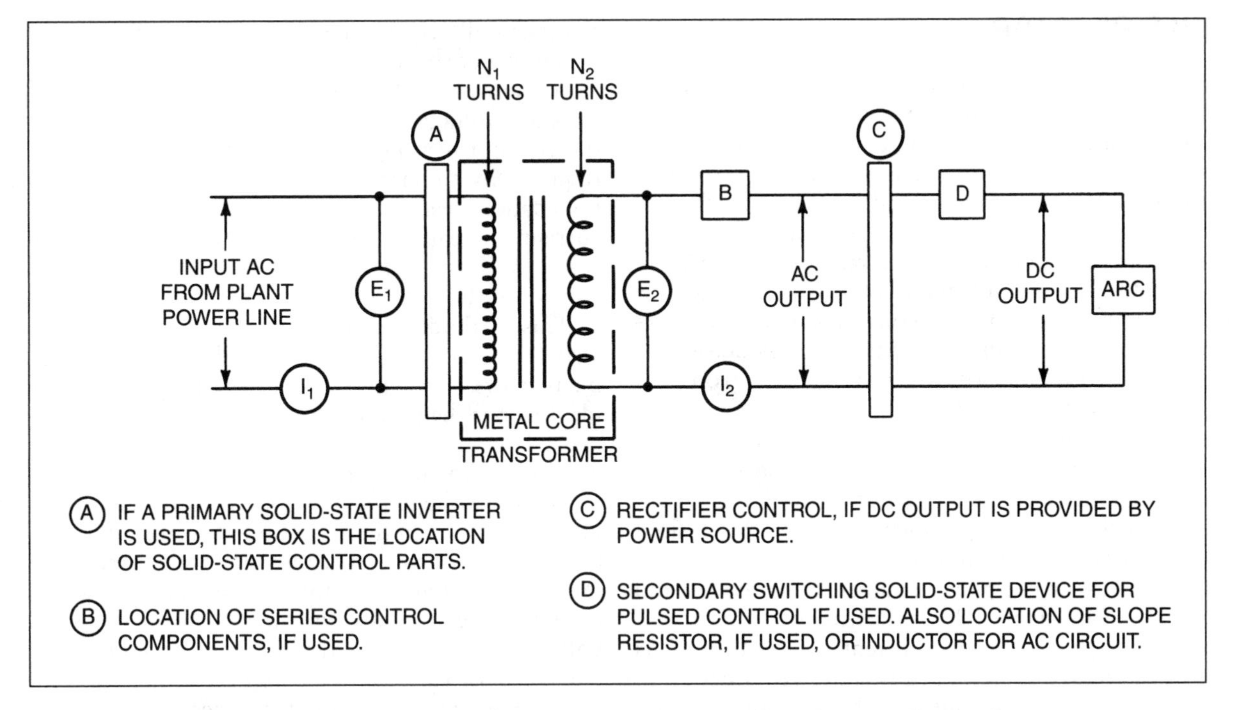

Figure 1.2—Principal Electrical Elements of a Transformer Power Source

Figure 1.2 also illustrates the principal elements of a welding transformer, with associated components. For a transformer, the significant relationships between voltages and currents and the turns in the primary and secondary windings are as follows:

$$\frac{N_1}{N_2} = \frac{E_1}{E_2} = \frac{I_2}{I_1} \tag{1.1}$$

where

N_1 = Number of turns on the primary winding of the transformer;
N_2 = Number of turns on the secondary winding;
E_1 = Input voltage, V;
E_2 = Output voltage, V;
I_1 = Input current, A; and
I_2 = Output (load) current, A.

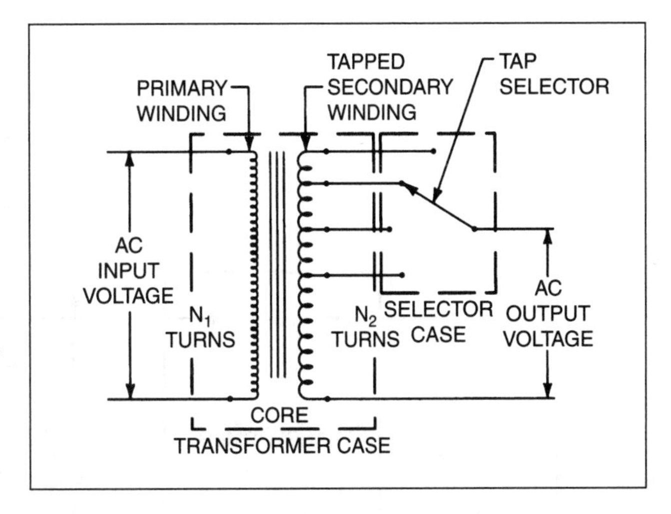

Figure 1.3—Welding Transformer with Tapped Secondary Winding

Taps in a transformer secondary winding may be used to change the number of turns in the secondary winding, as shown in Figure 1.3, to vary the open-circuit (no-load) output voltage. In this case, the tapped transformer permits the selection of the number of turns, N_2, in the secondary winding of the transformer. When the number of turns decreases on the secondary winding, output voltage is lowered because a smaller proportion of the transformer secondary winding is in use. The tap selection, therefore, controls the ac output voltage. As shown in Equation 1.1, the primary-secondary current ratio is inversely proportional to the primary-secondary voltage ratio. Thus, large secondary welding currents can be obtained from relatively low line input currents.

SERIES REACTOR

A transformer may be designed so that the tap selection directly adjusts the output volt-ampere slope characteristics for a specific welding condition. More often, however, an impedance source is inserted in series with the transformer secondary windings to provide this characteristic, as shown in Figure 1.4. The impedance is usually a magnetic device called a *reactor* when used in an ac welding circuit and an *inductor* when used in a dc welding circuit. Reactors are constructed with an electrical coil wound around a magnet core; inductors are constructed with an electrical coil wound around a magnet core with an air gap.

Some types of power sources use a combination of these arrangements, with the taps adjusting the open-circuit (or no-load) voltage, E_O, of the welding power source and the series impedance providing the desired volt-ampere slope characteristics.

In constant-current power sources, the voltage drop across the impedance, E_X (shown in Figure 1.4) increases greatly as the load current is increased. This increase in voltage drop, E_X, causes a large reduction in the arc voltage, E_A. Adjustment of the value of the series impedance controls the E_X voltage drop and the relation of load current to load voltage. This is called *current control*, or in some cases, *slope control.* Voltage E_O essentially equals the no-load (open-circuit) voltage of the power source.

As shown in Figure 1.5, the series impedance in constant-voltage power sources is typically small, and the transformer output voltage is very similar to that required by the arc. The voltage drop, E_X, across the impedance (reactor) increases only slightly as the load current increases. The reduction in load voltage is small. Adjustment in the value of reactance gives slight control of the relation of load current to load voltage.

This method of slope control, with simple reactors, also serves as a method to control voltage with saturable reactors or magnetic amplifiers. Figure 1.5 shows an ideal vector diagram of the relationship of the alternating voltages for the circuit of Figure 1.4, when a reactor is used as an impedance device. The no-load voltage equals the voltage drop across the impedance plus the load voltage when these are added vectorially. Vectorial addition is necessary because the alternating load and impedance voltages are not in time phase. In Figure 1.5, the open-circuit voltage of the transformer is 80 V, the voltage drop across the reactor is 69 V and the arc load voltage is 40 V.

The voltage drop across the series impedance, E_X, in an ac circuit is added vectorially to the load voltage, E_A, to equal the transformer secondary voltage, E_O. By varying the voltage drop across the impedance, the load or

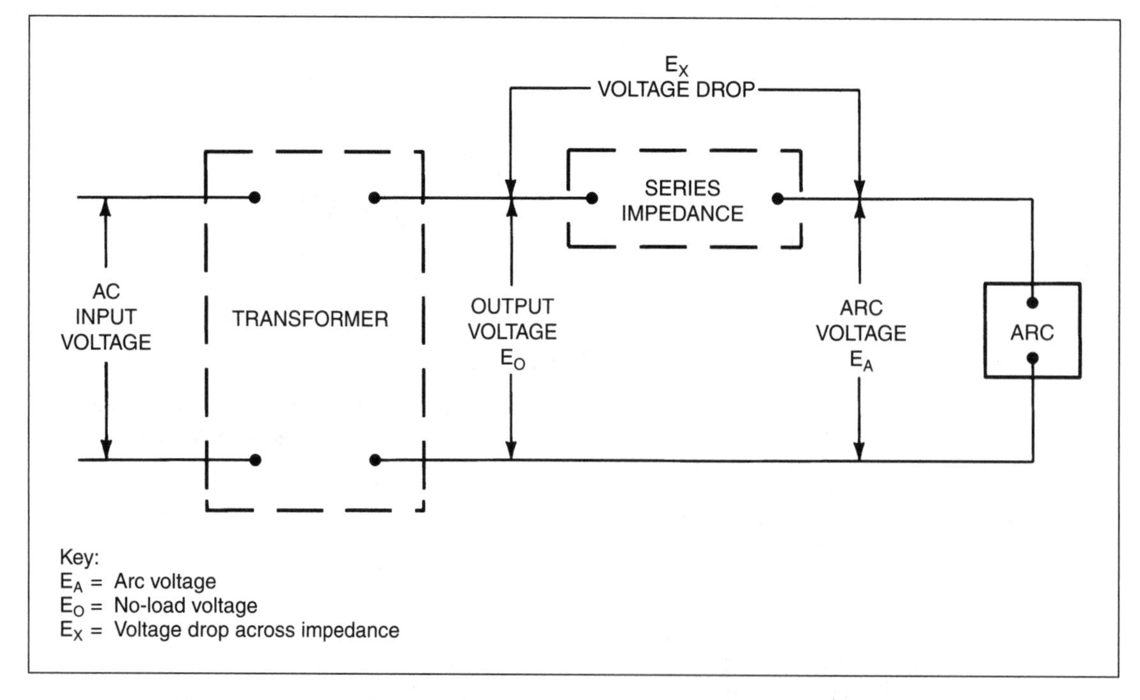

Figure 1.4—Typical Series Impedance Control of Output Current

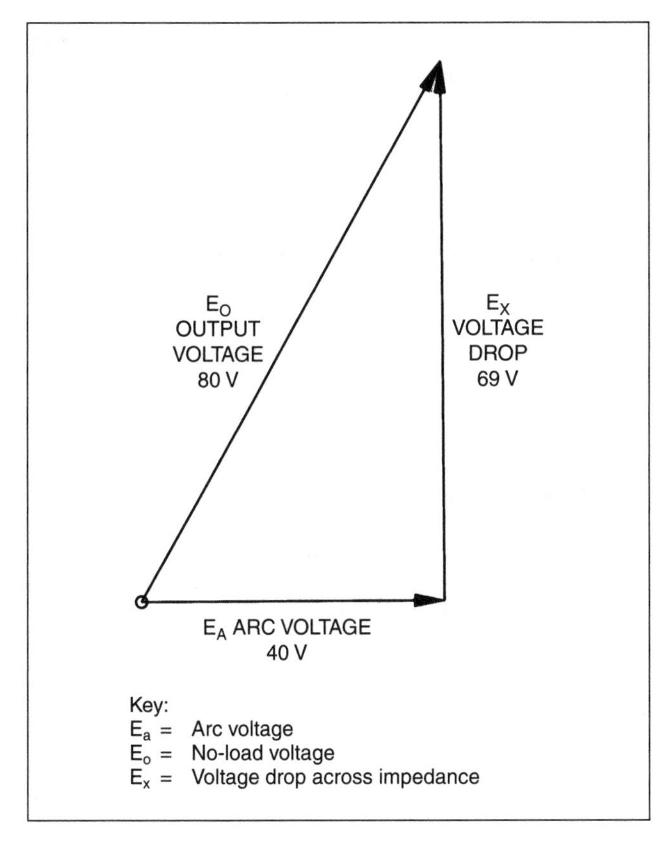

Figure 1.5—Ideal Vector Relationship of the Alternating-Voltage Output Using Reactor Control

arc voltage may be changed. This distinctive characteristic of vectorial addition for impedance voltages in ac circuits is related directly to the fact that both reactance and resistances may be used to produce a drooping voltage characteristic. An advantage of a reactor is that it consumes little or no power, even though a current flows through it and a voltage is developed across it.

When series resistors are used, power is lost and the temperature of the resistor rises. Theoretically, in a purely resistive circuit (no reactance), the voltage drop across the resistor could be added arithmetically to the load voltage to equal the output voltage of the transformer. For example, a power source with an approximately constant-current characteristic, an 80-V open circuit, and powering a 25-V, 200-A arc would need to dissipate 55 V × 200 A, or 11,000 watts (W), in the resistor to supply 5000 W to the arc. The reason is that the voltage and current are in phase in the resistive circuit. A resistance and reactance circuit phase shift accounts for the greatly reduced power loss.

Another major advantage of inductive reactance is that the phase shift produced in the alternating current by the reactor improves ac arc stability for a given open-circuit voltage. This is an advantage with the GTAW and SMAW processes.

The inductive reactance of a reactor can be varied by several means. One way is by changing taps on a coil or by other electrical or mechanical schemes. Varying the reactance alters the voltage drop across the reactor. Thus, for any given value of inductive reactance, a specific volt-ampere curve can be plotted. This creates the required control feature of these power sources.

In addition to adjusting series reactance, the mutual inductance between the primary and secondary coils of a transformer can also be adjusted. This can be done by moving the coils relative to one another or by using a movable magnetic shunt that can be inserted or withdrawn from between the primary and secondary windings. These methods change the magnetic coupling of the coils to produce adjustable mutual inductance, which is similar to series inductance.

In ac/dc welding power sources incorporating a rectifier, the rectifier is located between the magnetic control devices and the output terminal. In addition, transformer-rectifier arc welding power sources usually include a stabilizing inductance, or choke, located in the dc welding circuit to improve arc stability.

GENERATOR AND ALTERNATOR

Rotating machinery is also used as a source of power for arc welding. These machines are divided into two types—generators that produce direct current and alternators that produce alternating current.

The no-load output voltage of a direct-current generator can be controlled with a relatively small variable current in the winding of the main or shunt field. This current controls the output of the direct-current generator winding that supplies the welding current. The output polarity can be reversed by changing the interconnection between the exciter and the main field. An inductor or filter reactor is not usually needed to improve arc stability with this type of welding equipment. Instead, the several turns of series winding on the field poles of the rotating generator provide more than enough inductance to ensure satisfactory arc stability. These generators are described in greater detail in following sections of this chapter.

An alternator power source produces alternating current that is either used in that form or rectified into direct current. It can use a combination of the means of adjustment previously mentioned. A tapped reactor can be employed for gross adjustment of the welding output, and the field strength can be controlled for fine adjustment.

SOLID-STATE DIODES

The term *solid-state* is related to solid-state physics and the study of crystalline solids. Methods have been developed for treating crystalline materials to modify their electrical properties. The most important of these materials is silicon.

Transformer-rectifier and alternator-rectifier power sources rely on rectifiers, or groups of diodes, to convert alternating current to direct current. In earlier times, welding circuits relied on vacuum tube and selenium rectifiers, but most modern rectifiers are made of silicon for reasons of economy, current-carrying capacity, reliability, and efficiency.

A single rectifying element is called a *diode*, which is a one-way electrical valve. When placed in an electrical circuit, a diode allows current to flow in one direction only, when the anode of the diode is positive with respect to the cathode. Using a proper arrangement of diodes, it is possible to convert alternating current to direct current. An example of a diode symbol and a stud diode is shown in Figure 1.6.

As current flows through a diode, a voltage drop across the component develops and heat is produced within the diode. Unless this heat is dissipated, the diode temperature can increase enough to cause failure. Therefore, diodes are normally mounted on heat sinks (aluminum plates, many with fins) to remove the heat.

Diodes have limits as to the amount of voltage they can block in the reverse direction (anode negative and cathode positive). This is expressed as the voltage rating of the device. Welding power-source diodes are usually selected with a blocking rating at least twice the open-circuit voltage in order to provide a safe operating margin.

A diode can accommodate repetitive current peaks well beyond its normal steady-state rating, but a single high reverse-voltage transient will damage it. Most rectifier power sources have a resistor, capacitor, or other electronic devices, commonly called *snubber networks*, to suppress voltage transients that could damage the rectifiers.

SILICON-CONTROLLED RECTIFIER (THYRISTOR)

Solid-state devices with special characteristics can also be used to control welding power directly by altering the welding current or voltage wave form. These solid-state devices have replaced saturable reactors, moving shunts, moving coils, and other systems as control elements in large industrial power sources. One of the most important of these devices is the silicon-controlled rectifier (SCR), sometimes called a *thyristor.*

The SCR is a diode variation with a trigger, called a *gate*, as shown in Figure 1.7. An SCR is non-conducting until a positive electrical signal is applied to the gate. When this happens, the device becomes a diode and conducts current as long as the anode is positive with respect to the cathode. However, once it conducts, the current cannot be turned off by a signal to the gate. Conduction ceases only if the voltage applied to the anode becomes negative with respect to the cathode. Conduction will not take place again until a positive voltage is applied to the anode and another gate signal is received.

Silicon-controlled rectifiers are used principally in the phase-control mode with isolation transformers and in some inverter configurations. The output of a welding power source can be controlled by using the action of a gate signal to selectively turn on the SCR. A typical single-phase SCR circuit is shown in Figure 1.8.

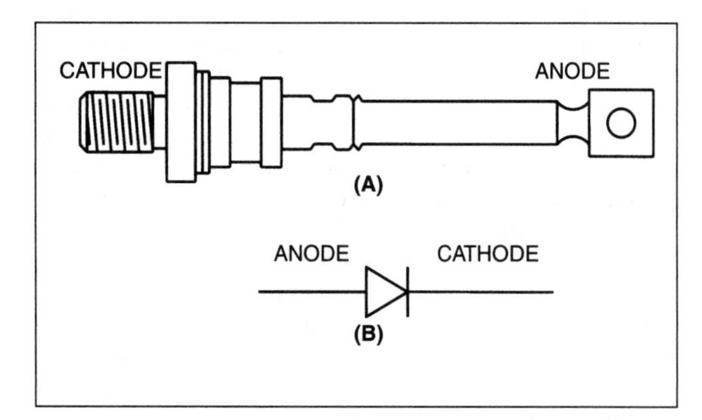

Figure 1.6—Stud Diode (A) and Diode Symbol (B)

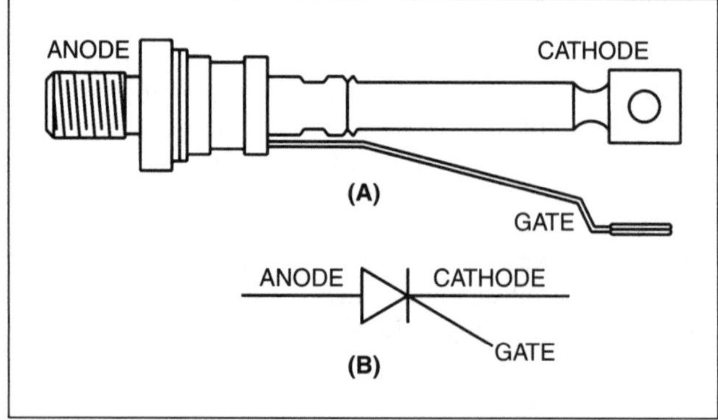

Figure 1.7—Silicon-Controlled Rectifier (A) and Silicon-Controlled Rectifier Symbol (B)

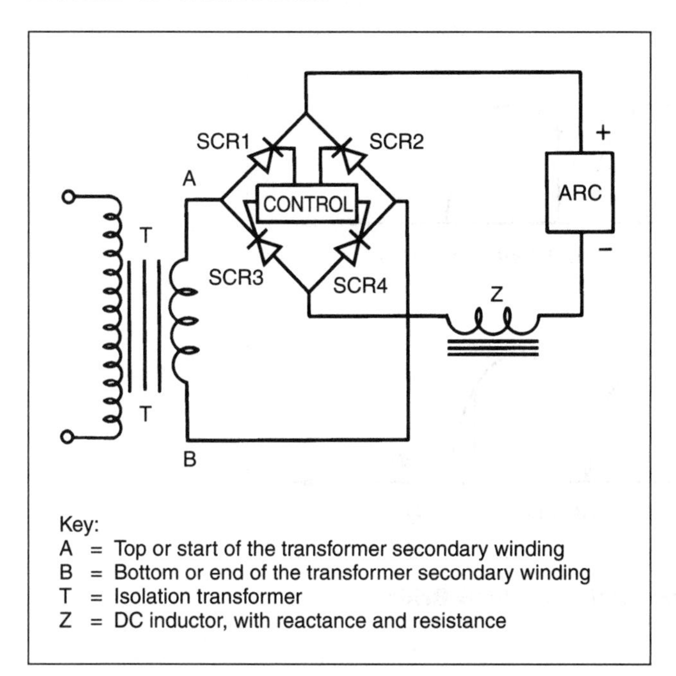

Figure 1.8—Single-Phase Direct-Current Power Source Using an SCR Bridge for Control

In Figure 1.8, during the time that Point A is positive with respective to Point B, no current will flow until both SCR 1 and SCR 4 receive gate signals to turn on. At that instant, current will flow through the load. At the end of that half-cycle, when the polarity of A and B reverses, a negative voltage will be impressed across SCR 1 and SCR 4, and they will turn off. With Point B positive relative to Point A, gate signals applied to SCR 2 and SCR 3 by the control will cause these two to conduct, again applying power to the load circuit. To adjust power in the load, it is necessary to precisely time when, in any given half-cycle, the gate triggers the SCR into conduction. With a 60-hertz (Hz) line frequency, this arrangement produces direct current with a 120-Hz ripple frequency at the arc or load.

The timing of the gate signals must be precisely controlled. This is a function of the control block shown in Figure 1.8. To adapt the system satisfactorily for welding service, another feature, feedback, is necessary. The nature of the feedback depends on the welding parameter to be controlled and the degree of control required. To provide constant-voltage characteristics, the feedback (not shown) must consist of a signal that is proportional to arc voltage. This signal controls the precise arc voltage at any instant so that the control can properly time and sequence the initiation of the SCR to hold a voltage pre-selected by the operator. The same effect is achieved with constant current by using feedback and an operator-selected current.

Figure 1.8 shows a large inductance, Z, in the load circuit. For a single-phase circuit to operate over a significant range of control, Z must be a large inductance to smooth out the voltage and current pulses. However, if SCRs were used in a three-phase circuit, the nonconducting intervals would be reduced significantly. Since three times as many output pulses are present in any time period, the inductance would also be significantly reduced.

When high power is required, conduction is started early in the half-cycle, as shown in Figure 1.9(A). If low power is required, conduction is delayed until later in a half-cycle, as shown in Figure 1.9(B). This is known as *phase control.* The resulting power is supplied in pulses to the load and is proportional to the shaded areas in Figure 1.9 under the wave form envelopes. Figure 1.9 illustrates that significant intervals may exist when no power is supplied to the load. This can cause arc outages, especially at low power levels. Therefore, wave filtering is required.

Most intermediate-sized or commercial SCR phase-controlled welding power sources are single-phase. Larger industrial SCR phase-controlled power sources are three-phase. Single- and three-phase power sources are the constant-current or constant-voltage type. Both constant-current and constant-voltage types have distinct features because the output characteristics are controlled electronically. For example, automatic line-voltage compensation is very easily accomplished, allowing welding power to be held precisely as set, even if the input line voltage varies. Volt-ampere curves can also be shaped and adapted for a particular welding process or its application. These power sources can adapt their static characteristic to any welding process, from one approaching a truly constant voltage to one having a relatively constant current. They are also capable of producing a controlled pulsed arc voltage and a high initial current or voltage pulse at the start of the weld.

An SCR can also serve as a secondary contactor, allowing welding current to flow only when the control allows the SCRs to conduct. This is a useful feature in rapid cycling operations, such as spot welding and tack welding. However, an SCR contactor does not provide the electrical isolation that a mechanical contactor or switch provides. Therefore, a primary circuit breaker or some other device is required to provide isolation for electrical safety.

Several SCR configurations can be used for arc welding. Figure 1.10 depicts a three-phase bridge with six SCR devices. With a 60-Hz line frequency, this arrangement produces direct current, with a 360-Hz ripple frequency at the load. It also provides precise control and quick response; in fact, each half-cycle of

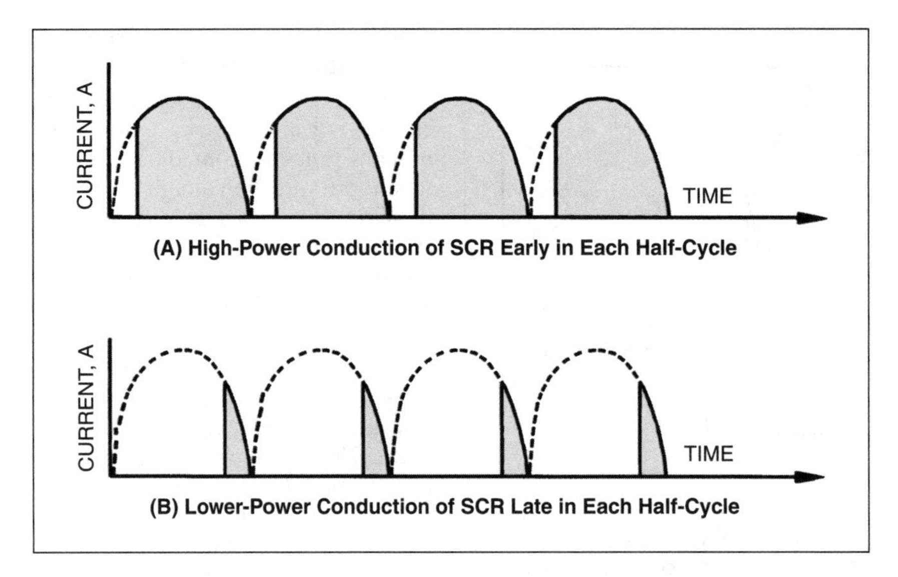

Figure 1.9—Phase Control Using an SCR Bridge

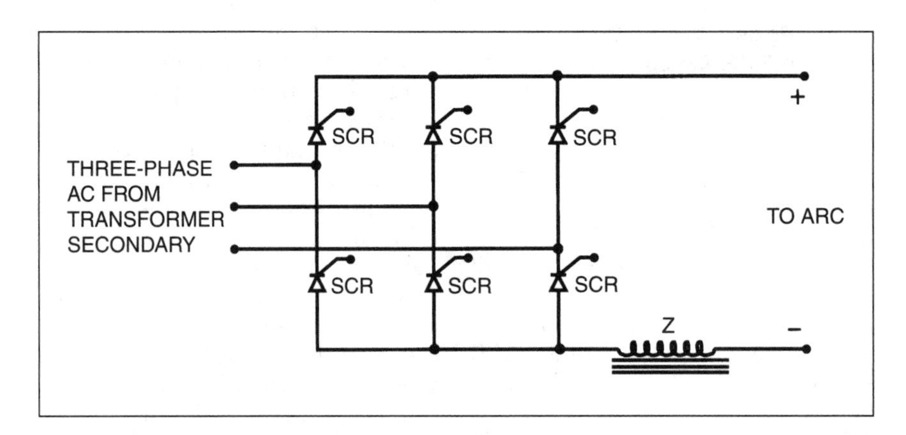

Figure 1.10—Three-Phase Bridge Using Six SCRs (Full-Wave Control)

each of the three-phase output is controlled separately. Dynamic response is enhanced because of the reduced size of the inductor needed to smooth out the welding current.

Figure 1.11 is a diagram of a three-phase bridge rectifier with three diodes and three SCRs. Because of greater current ripple, this configuration requires a larger inductor than the six-SCR unit. For that reason it has a slower dynamic response. A fourth diode, termed a *freewheeling diode*, can be added to recirculate the inductive currents from the inductor so that the SCRs will turn off, or commutate. This offers some economic advantage over the six-SCR unit because it uses fewer SCRs and a lower-cost control unit.

TRANSISTORS

The transistor is another solid-state device used in welding power sources. Transistors differ from SCRs in several ways. First, conduction through the device is proportional to the control signal applied. With no signal, no conduction occurs. The application of a small signal from base to emitter produces a correspondingly small conduction; likewise, a large signal results in a correspondingly large conduction. Unlike the SCR, the control can turn the device off without waiting for polarity reversal or an off time. Since transistors lack the current-carrying capacity of SCRs, several may be required to yield the output of one SCR.

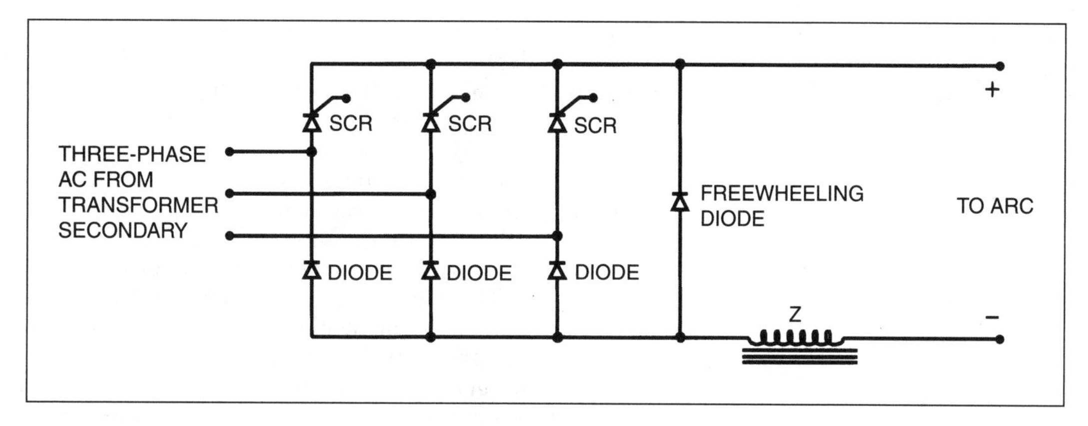

Figure 1.11—Three-Phase Hybrid Bridge Using Three SCRs and Four Diodes (Half-Wave Control)

Several methods can be used to take advantage of transistors in welding power sources. These include frequency modulation or pulse-width modulation. With frequency modulation, the welding current is controlled by varying the frequency supplied to a high-frequency transformer. Since the frequency is changing, the response time varies also. The size of the transformer and inductor must be optimized for the lowest operating frequency. With pulse-width modulation, varying the conduction time of the switching device controls welding current output. Since the frequency is constant, the response time is constant and the magnetic components can be optimized for one operating frequency.

Inverter circuits control the output power using the principle of time-ratio control (TRC) also referred to as *pulse-width modulation* (PWM). The solid-state devices (semiconductors) in an inverter act as switches; they are either switched on and conducting, or switched off and blocking. The function of switching on and off is sometimes referred to as *switch-mode operation. Time-ratio control* is the regulation of the on and off times of the switches to control the output. Figure 1.12 illustrates a simplified TRC circuit that controls the output to a load such as a welding arc. It should be noted that conditioning circuits include components such as a transformer, a rectifier, and an inductor, as represented previously in Figure 1.8.

SOLID-STATE INVERTER

An inverter is a circuit that uses solid-state devices called *metal oxide semiconductor field effect transistors* (MOSFETs), or *integrated gate bi-polar transistors* (IGBTs), to convert direct current into high-frequency ac, usually in the range of 20 kHz to 100 kHz. Conventional welding power sources use transformers operating from a line frequency of 50 Hz or 60 Hz.

Since transformer size is inversely proportional to line or applied frequency, reductions of up to 75% in power source size and weight is possible using inverter circuits. Inverter power sources are smaller and more compact than conventional welding power sources. They offer a faster response time and less electrical loss.

The primary contributors to weight or mass in any power source are the magnetic components, consisting of the main transformer and the filter inductor. Various efforts have been made by manufacturers to reduce the size and weight of power sources, for example, substituting aluminum windings for copper.

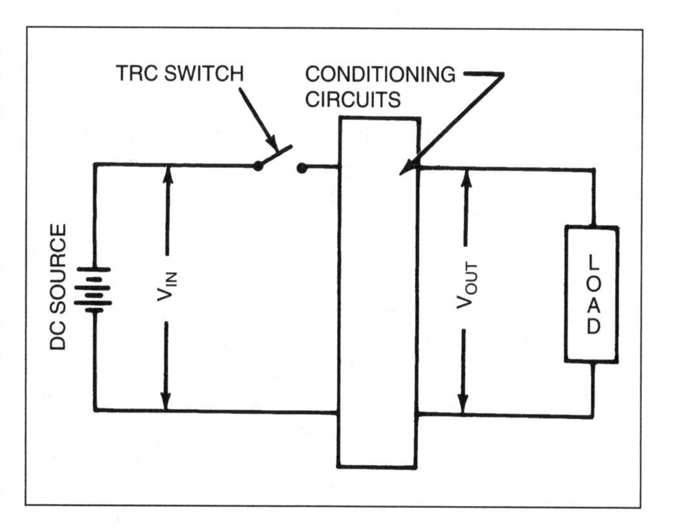

Figure 1.12—Simplified Diagram of an Inverter Circuit Used to Demonstrate the Principle of Time-Ratio Control (Pulse Width Modulation)

When the TRC switch is on, the voltage out (V_{OUT}) equals voltage in (V_{IN}). When the switch is off, V_{OUT} equals zero. The average value of V_{OUT} is calculated as follows:

$$V_{OUT} = \frac{V_{IN} \times t_{ON}}{t_{ON} + t_{OFF}} \tag{1.2}$$

where

V_{OUT} = Voltage out, V;
t_{ON} = On time (conducting), seconds (s);
V_{IN} = Voltage in, V;
t_{OFF} = Off time (blocking), s;

thus,

$$V_{OUT} = V_{IN} \times \frac{t_{ON}}{T_p} \tag{1.3}$$

where
T_p = $t_{ON} + t_{OFF}$ = Time period total, s.

Variable V_{OUT} is controlled by regulating the ratio of on time to off time for each alternation t_{ON}/T_p. Since the on/off cycle is repeated for every T_p interval, the frequency (f) of the on/off cycles is defined as follows:

$$f = \frac{1}{T_p} \tag{1.4}$$

where
f = Frequency, Hz

thus, the TRC formula can now be written as:

$$V_{OUT} = V_{IN} \times t_{ON} \times f \tag{1.5}$$

The TRC formula written in this manner points to two methods of controlling an inverter welding power source. By varying t_{ON}, the inverter uses pulse-width modulated TRC.

Another method of inverter control, *frequency-modulation TRC*, varies the frequency, f. Both frequency modulation and pulse-width modulation are used in commercially available welding inverters.

Figure 1.13 presents a block diagram of an inverter used for direct-current welding. A full-wave rectifier converts incoming three-phase or single-phase 50-Hz or 60-Hz power to direct current. This direct current is applied to the inverter, which inverts it into high-frequency square-wave alternating current using semiconductor switches. In another variation used for welding, the inverter produces sine waves in a resonant technology with frequency-modulation control. The switching of the semiconductors takes place between 1 kHz and 50 kHz, depending on the component used and method of control.

This high-frequency voltage allows the use of a smaller step-down transformer. After being transformed, the alternating current is rectified to direct current for welding. Solid-state controls enable the operator to select either constant-current or constant-voltage output, and with appropriate options these sources can also provide pulsed outputs.

The capabilities of the semiconductors and the particular circuit switching determine the response time and switching frequency. Faster output response times are generally associated with the higher switching and control frequencies, resulting in more stable arcs and superior arc performance. However, other variables, such as the length of the weld cables, must be considered because they may affect the performance of the power source. Table 1.1 compares inverter switching devices and the frequency applied to the transformer.

Inverter technology can be used to enhance the performance in ac welding power sources and can also be applied to dc constant-current power sources used for plasma arc cutting.

VOLT-AMPERE CHARACTERISTICS

The effectiveness of all welding power sources is determined by two kinds of operating characteristics, *static* and *dynamic*. Each has a different effect on welding performance. Both affect arc stability, but they do so in different ways depending on the welding process.

Static output characteristics are readily measured under steady-state conditions by conventional testing procedures using resistive loads. A set of output-voltage curves versus output-current characteristic curves (volt-ampere curves) is normally used to describe the static characteristics.

The dynamic characteristic of an arc welding power source is determined by measuring the transient variations in output current and voltage that appear in the arc. Dynamic characteristics describe instantaneous variations, or those that occur during very short intervals, such as 0.001 second.

Most welding arcs operate in continually changing conditions. Transient variations occur at specific times, such as the following:

1. During the striking of the arc,
2. During rapid changes in arc length,
3. During the transfer of metal across the arc, and
4. In alternating current welding, during arc extinction and reignition at each half-cycle.

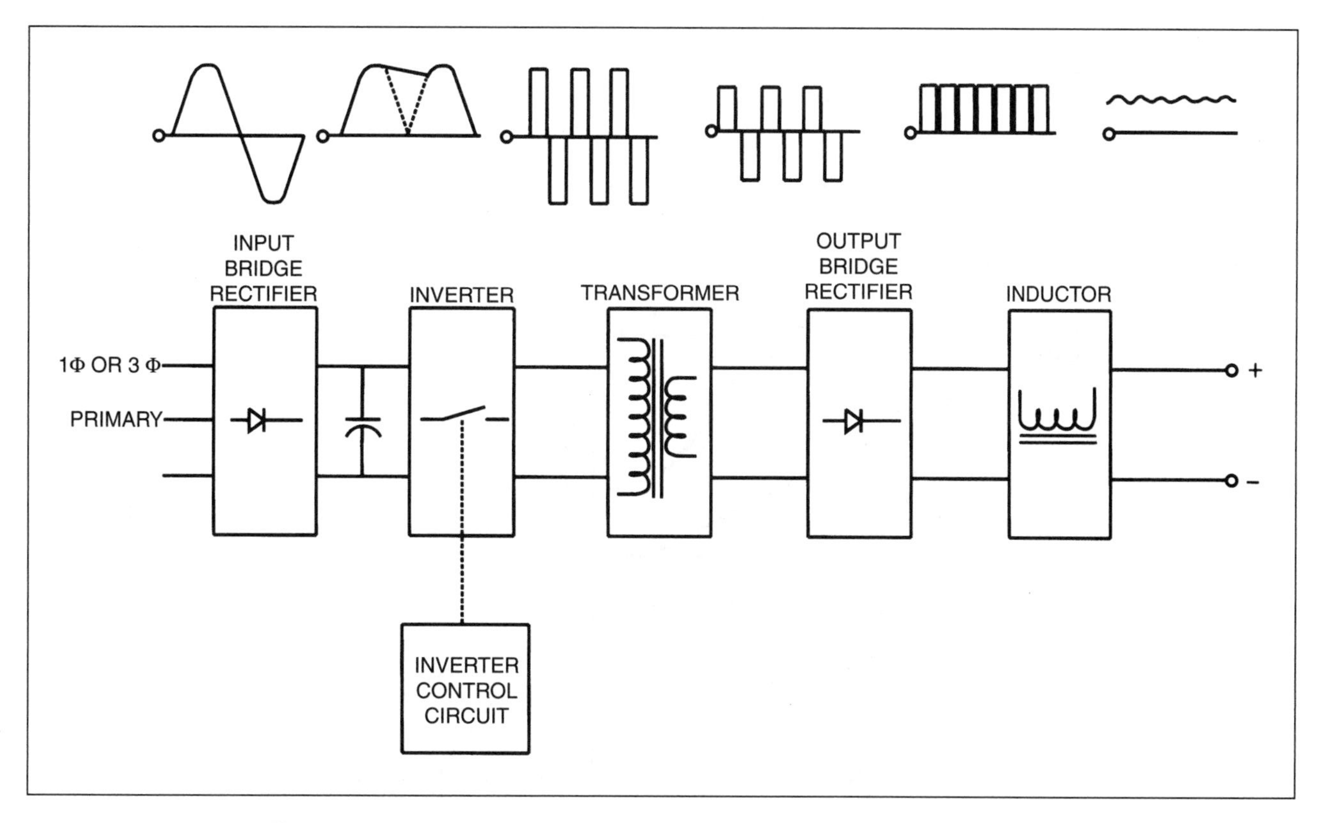

Figure 1.13—Inverter Diagram Showing Power Source Sections and Voltage Wave Forms with Pulse-Width Modulation Control

Table 1.1
Types of Inverter Switching Devices and Frequency Ranges Applied to the Transformer

Switching Device	Frequency Range
SCR devices	1 kHz to 10 kHz
Transistor devices	10 kHz to 100 kHz

The short arc-transient time of 0.001 second is the time interval during which a significant change in ionization of the arc column occurs. The power source must respond rapidly to these demands, and for this reason it is important to control the dynamic characteristics of an arc welding power source. The steady-state or static volt-ampere characteristics have little significance in determining the dynamic characteristics of an arc welding system.

Among the arc welding power source design features that do have an effect on dynamic characteristics are those that provide local transient energy storage such as parallel capacitance circuits or direct-current series inductance, feedback controls in automatically regulated systems, and modifications of wave form or circuit-operating frequencies.

Improving arc stability is typically the reason for modifying or controlling these characteristics. Beneficial results include improvement in the uniformity of metal transfer, reduction in metal spatter, and reduction in weld-pool turbulence.

Static volt-ampere characteristics are generally published by power source manufacturers. No universally recognized method exists by which dynamic characteristics are specified. The user should obtain assurance from the manufacturer that both the static and dynamic characteristics of the power source are acceptable for the intended application.

CONSTANT-CURRENT

Volt-ampere curves show graphically how welding current changes when arc voltage changes and power source settings remain unchanged, as illustrated in Figure 1.14 for a *drooper* power source. Constant-current

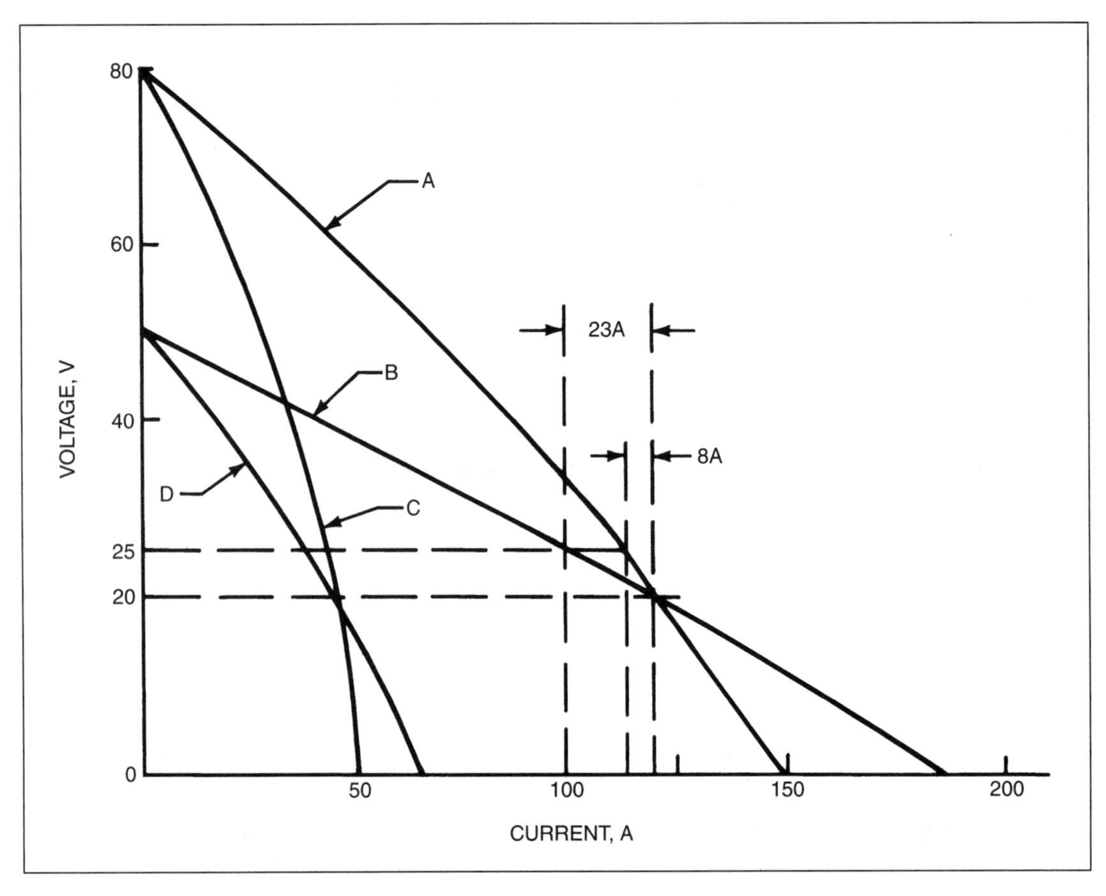

Figure 1.14—Typical Volt-Ampere Characteristics of a "Drooper" Power Source with Adjustable Open-Circuit Voltage

welding power sources are sometimes called *droopers* because of the substantial downward (negative) slope of the volt-ampere curves they produce. A constant-current V-A characteristic is suitable for shielded metal arc welding, gas tungsten arc welding, and other processes that use voltage-sensing wire feed systems.

The conventional constant-current output characteristic describes a power source that will produce a relatively small change in output current when a relatively large change in arc voltage occurs. Arc voltage is affected by arc length and process parameters such as electrode type, shielding gas, and arc current. Reducing the slope or the droop of a constant-current power source gives the operator a degree of real-time control over arc current or electrode melting rate. The power source might have open-circuit voltage adjustment in addition to output current control. A change in either control will change the slope of the volt-ampere curve.

The effect of the slope of the V-A curve on power output is shown in Figure 1.14. With Curve A, which has an 80-V open circuit, a steady increase in arc voltage from 20 V to 25 V (25%) would result in a decrease in current from 123 A to 115 A (6.5%). The change in current is relatively small. Therefore, with a consumable electrode welding process, the electrode melting rate would remain relatively constant with a slight change in arc length.

By setting the power source to Slope Curve B in Figure 1.14 the open circuit voltage is reduced from 80 volts to 50 volts. Curve B shows a shallower or flatter slope intercepting the same 20-V, 123-A output. In this case, the same increase in arc voltage from 20 V to 25 V would decrease the current from 123 to 100 A (19%), a significantly greater change. In manual welding, the flatter V-A curve would give a skilled welder the opportunity to substantially vary the output current by changing the arc length. This is useful for out-of-position welding because a welder can control the electrode melting rate and weld pool size in real time by simply changing the arc length. A flatter slope also provides increased short-circuit current. This helps reduce the tendency of some electrodes to stick to the workpiece during arc starts or times when the arc length is reduced to control penetration. Generally, however, less skilled welders would prefer the current to stay constant if the arc length should change. The higher open-circuit voltage of constant-current or drooping output curves also helps reduce arc outages with certain types of fast-freezing electrodes at longer arc lengths or when weaving the arc across a root opening.

Output current control is also used to provide lower output current. This results in volt-ampere curves with greater slope, as illustrated by Curves C and D in Figure 1.14. They offer the advantage of more nearly constant current output, allowing greater changes in voltage with minor changes in current.

CONSTANT-VOLTAGE CHARACTERISTICS

The volt-ampere curve in Figure 1.15 shows graphically how the output current is affected by changes in the arc voltage (arc length). It illustrates that this power source does not have true constant-voltage output. It has a slightly downward (negative) slope because internal electrical impedance in the welding circuit causes a minor voltage droop in the output. Changing that impedance will alter the slope of the volt-ampere curve.

Starting at Point B in Figure 1.15, the diagram shows that an increase or decrease in voltage to Points A or C (5 V or 25%), produces a large change in amperage (100 A or 50%), respectively. This V-A characteristic is suitable for maintaining a constant arc length in constant-speed electrode processes, such as GMAW, SAW, and FCAW. A slight change in arc length (voltage) causes a relatively large change in welding current. This automatically increases or decreases the electrode melting rate to regain the desired arc length (voltage). This effect is called *self-regulation*. Adjustments are sometimes provided with constant-voltage power sources to change or modify the slope or shape of the V-A curve. Typical adjustments involve changing the power source reactance, output inductance, or internal resistance. If adjustments are made with inductive devices, the dynamic characteristics will also change.

The curve shown in Figure 1.16 can also be used to explain the difference between static and dynamic characteristics of the power source. For example, during gas metal arc welding short-circuiting transfer (GMAW-S), the welding electrode tip touches the weld pool, causing a short-circuit. At this point, the arc voltage approaches zero, and only the circuit resistance and inductance limits the rapid increase of current. If the power source responded instantly, very high current would immediately flow through the welding circuit, quickly melting the short-circuited electrode and freeing it with an explosive force, expelling the weld metal as spatter. Dynamic characteristics designed into this power source compensate for this action by limiting the rate of current change, thereby decreasing the explosive force.

COMBINED CONSTANT-CURRENT AND CONSTANT-VOLTAGE CHARACTERISTICS

Electronic controls can be designed to provide either constant-voltage or constant-current outputs from single

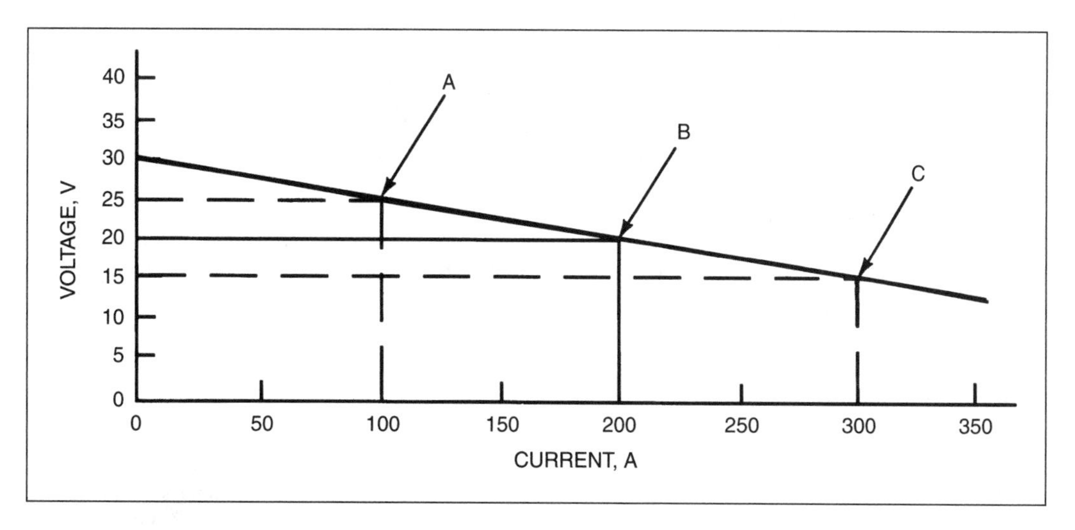

Figure 1.15—Volt-Ampere Output Relationship for a Constant-Voltage Power Source

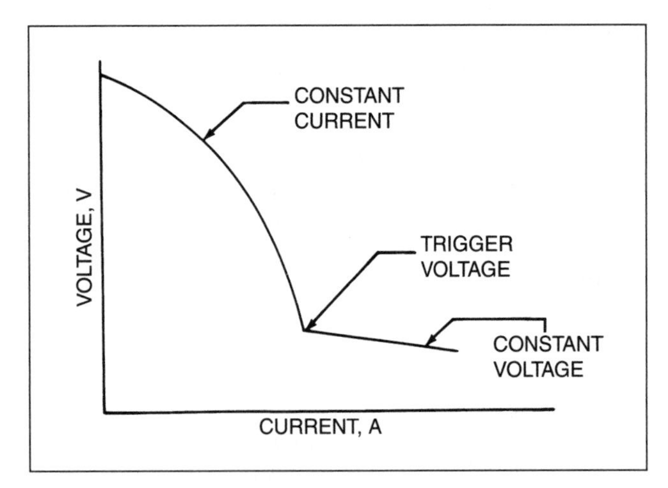

Figure 1.16—Combination Volt-Ampere Curve

power sources, making them useful for a variety of welding processes.

Electronically controlled outputs can also provide output curves that are a combination of constant-current and constant-voltage, as shown in Figure 1.16. The top part of the curve is essentially constant-current; below a certain trigger voltage, however, the curve switches to constant voltage. This type of curve is beneficial for shielded metal arc welding to assist starting and to avoid electrode stubbing (sticking in the weld pool) if the welder uses an arc length that is too short.

DUTY CYCLE

Internal components of a welding power source tend to heat up as welding current flows through. The amount of heat tolerated is determined by the breakdown temperature of the electrical components and the media used to insulate the transformer windings and other components. These maximum temperatures are specified by component manufacturers and organizations involved with standards in the field of electrical insulation.

Fundamentally, the *duty cycle* is a ratio of the load-on time allowed in a specified test interval time. Observing this ratio is important in preventing the internal windings and components and their electrical insulation system from heating above their rated temperature. These maximum temperature criteria do not change with the duty cycle or current rating of the power source.

Duty cycle is expressed as a percentage of the maximum time that the power source can deliver its rated output during each of a number of successive test intervals without exceeding a predetermined temperature limit. In the United States, for example, the National Electrical Manufacturers Association (NEMA) specifies duty cycles based on a test interval of 10 minutes in an ambient temperature of 40°C (104°F). Some agencies and manufacturers in other countries use shorter test intervals, such as 5 minutes. Thus, a 60% NEMA duty cycle (a standard industrial rating) means that the power source can deliver its rated output for 6 out of every 10 minutes without overheating.[5] A 100% duty-cycle power source is designed to produce its rated output continuously without exceeding the prescribed temperature limits of its components.

Duty cycle is a major factor in determining the type of service for which a power source is designed. Industrial units designed for manual welding are normally rated at a 60% duty cycle. For automatic and semiautomatic processes, the rating is usually a 100% duty cycle. Light-duty power sources usually have a 20% duty cycle. Power sources with ratings at other duty cycle values are available from the manufacturers.

An important point is that the duty cycle of a power source is based on the output current and not on a kilovolt-ampere load or kilowatt rating. Manufacturers perform duty-cycle tests under what NEMA defines as usual service conditions. Caution should be observed in basing operation on service conditions other than usual. Unusual service conditions such as high ambient temperatures, insufficient cooling air, and low line voltage are among the factors that contribute to performance that is lower than tested or calculated.

Equation 1.6 presents the formula for estimating the duty cycle at other than rated outputs, as follows:

$$T_a = \left(\frac{I}{I_a}\right)^2 \times T \tag{1.6}$$

where

T_a = Required duty cycle, %;
I = Rated current at the rated duty cycle, A;
I_a = Maximum current at the required duty cycle, A; and
T = Rated duty cycle, %.

Equation 1.7 presents the expression for estimating other than rated output current at a specified duty cycle, as follows:

$$I_a = I \times \left(\frac{T}{T_a}\right)^{1/2} \tag{1.7}$$

5. It should be noted that a power source specified for uninterrupted operation at a rated load for 36 minutes out of one hour would have a 100% duty cycle, rather than a 60% duty cycle, because it could operate continuously for the test-interval of 10 minutes.

where

I_a = Maximum current at the required duty cycle, A;
I = Rated current at the rated duty cycle, A;
T = Rated duty cycle, %; and
T_a = Required duty cycle, %.

The power source should never be operated above its rated current or duty cycle unless approved by the manufacturer. For example, Equation 1.8 applies Equation 1.6 to determine the duty cycle of a 200-A power source rated at a 60% duty cycle if operated at 250 A output (provided 250 A is permitted by the manufacturer), as follows:

$$T_a = \left(\frac{200}{250}\right)^2 \times 60\% = (0.8)^2 \times 0.6 = 38\% \quad (1.8)$$

Therefore, this unit must not be operated more than 3.8 minutes out of each 10-minute period at 250 A. If used in this way, welding at 250 A will not exceed the current rating of any power source component.

The output current that must not be exceeded when operating this power source continuously (100% duty cycle) can be determined by applying Equation 1.7, as shown in Equation 1.9:

$$I_a = 200 \times \left(\frac{60}{100}\right)^{1/2} = 200 \times 0.775 = 155 \text{ A} \quad (1.9)$$

If operated continuously, the current should be limited to an output of 155 A.

OPEN-CIRCUIT VOLTAGE

Open-circuit voltage is the voltage at the output terminals of a welding power source when it is energized but current is not being drawn. Open-circuit voltage is one of the design factors influencing the performance of all welding power sources. In a transformer, open-circuit voltage is a function of the primary input voltage and the ratio of primary-to-secondary coils. Although a high open-circuit voltage may be desirable from the standpoint of arc initiation and stability, the electrical hazard precludes the use of higher voltages.

The open-circuit voltage of generators or alternators is related to design features such as the strength of the magnetic field, the speed of rotation, the number of turns in the load coils, and so forth. These power sources generally have controls with which the open-circuit voltage can be varied.

Table 1.2
Maximum Open-Circuit Voltages for Various Types of Arc Welding Power Sources

Manual and Semiautomatic Applications	
Alternating current	80 V root mean square (rms)
Direct current—over 10% ripple voltage*	80 V rms
Direct current—10% or less ripple voltage	100 V average
Automatic Applications	
Alternating current	100 V rms
Direct current—over 10% ripple voltage	100 V rms
Direct current—10% or less ripple voltage	100 V average

$$\text{*Ripple voltage, \%} = \frac{\text{Ripple voltage, rms}}{\text{Average total voltage, V}}$$

NEMA EW-1, *Electric Arc-Welding Power Sources*,[6] contains specific requirements for maximum open-circuit voltage. When the rated line voltage is applied to the primary winding of a transformer or when a generator arc welding power source is operating at maximum-rated no-load speed, the open-circuit voltages are limited to the levels shown in Table 1.2.

NEMA Class I and Class II power sources normally have open-circuit voltages at or close to the maximum specified. Class III power sources frequently provide two or more open-circuit voltages. One arrangement is to have a high and low range of amperage output from the power source. The low range normally has approximately 80-V open circuit, with the high range somewhat lower. Another arrangement is the tapped secondary coil method, described previously, in which the open-circuit voltage changes approximately 2 V to 4 V at each current setting.

In the United States, 60-Hz power produces reversals in the direction of current flow each 1/120 second (60 Hz). Typical sine wave forms of a dual-range power source with open-circuit voltages of 80 V and 55 V root mean square (rms) are diagrammed in Figure 1.17. (The rms of alternating current or voltage is the effective current or voltage applied that produces the same heat as that produced by an equal value of direct current or voltage).

The current must change direction after each half-cycle. In order for it to do so, the current flow in the arc ceases for an instant at the point at which the current wave form crosses the zero line. An instant later, the current must reverse its direction of flow. However,

6. See Reference 1, p 91.

during the period in which current decreases and reaches zero, the arc plasma cools, reducing ionization of the arc stream.

Welding current cannot be reestablished in the opposite direction unless ionization within the arc length is either maintained or quickly reinitiated. With conventional power sources, ionization may not be sustained depending on the welding process and electrode being used. Reinitiating is improved by providing an appropriately high voltage across the arc, called a *recovery voltage*. The greater this recovery voltage, the shorter is the period during which the arc is extinguished. If recovery voltage is insufficient, the arc cannot be reestablished without shorting the electrode.

Figure 1.17 shows the phase relations between open-circuit voltage and equal currents and current for two different open-circuit voltages, assuming the same arc voltage (not shown) in each case. As shown, the available peak voltage of 113 V is greater with 80 V (rms) open-circuit voltage. The peak voltage of 78 V available with 55 V (rms) open circuit may not be sufficient to sustain a stable arc. The greater phase shift shown for the low-range condition causes a current reversal at a higher recovery voltage because it is near the peak of the open-circuit voltage wave form, which is the best condition for reignition. Adjustable resistance is not used to regulate alternating welding current because the power source voltage and current would be in phase. Since the recovery voltage would be zero during current reversal, it would be difficult to maintain a stable arc.

For shielded metal arc welding with low-voltage, open-circuit power sources, it is necessary to use electrodes with ingredients incorporated in the electrode coverings that help maintain ionization and provide favorable metal-transfer characteristics to prevent sudden, gross increases in the arc length.

In a direct-current system, once the arc is established, the welding current does not pass through zero. Thus, rapid voltage increase is not critical; resistors are suitable current controls for direct-current power sources. However, with some processes, direct-current power sources must function in much the same way with respect to the need to provide open-circuit voltage when the arc length changes abruptly. Often reactance or inductance is built into these power sources to enhance this effect.

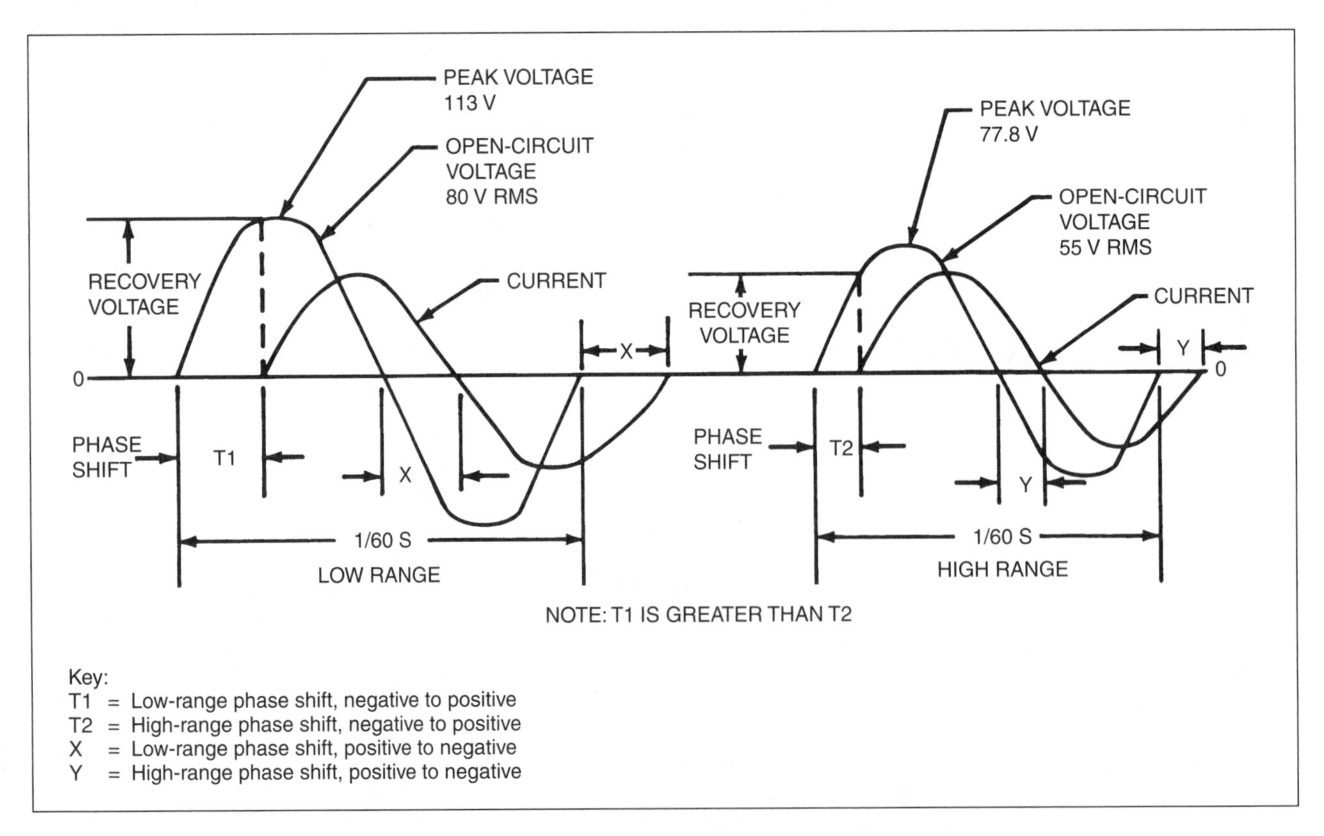

Figure 1.17—Typical Voltage and Current Waveforms of a Dual-Range Alternating-Current Power Source

NEMA POWER SOURCE REQUIREMENTS

The National Electrical Manufacturers Association (NEMA) power source requirements represent the technical judgment of the organization's Arc Welding Section concerning the performance and construction of electric arc welding power sources.[7] These requirements are based on sound engineering principles, research, records of tests, and field experience. The requirements cover both installation and manufacturing criteria obtained from manufacturers and users.

The reader should consult *Electric Arc-Welding Power Sources*, NEMA EW-1, for the requirements for electric arc welding apparatus, including power sources.[8]

NEMA CLASSIFICATIONS

NEMA categorizes arc welding power sources into the following three classes, primarily on the basis of duty cycle:

1. A NEMA Class I arc welding power source is characterized by its ability to deliver rated output at duty cycles of 60%, 80%, or 100%. If a power source is manufactured in accordance with the applicable standards for Class I power sources, it shall be marked "NEMA Class I (60)," "NEMA Class I (80)," or "NEMA Class I (100);"
2. A NEMA Class II arc welding power source is characterized by its ability to deliver rated output at duty cycles of 30%, 40%, or 50%. If a power source is manufactured in accordance with the applicable standards for Class II power sources, it shall be marked "NEMA Class II (30)," "NEMA Class II (40)," or "NEMA Class II (50)."
3. A NEMA Class III arc welding power source is characterized by its ability to deliver rated output at a duty cycle of 20%. If a power source is manufactured in accordance with the applicable standards for Class III power sources, it shall be marked "NEMA Class III (20)."

NEMA Class I and Class II power sources are further defined as completely assembled arc welding power sources in one of the following forms:

1. A constant-current, a constant-voltage, or a constant-current/constant-voltage power source;
2. A single-operator power source; or
3. One of the following:
   a. Direct-current generator arc welding power source,
   b. Alternating-current generator arc welding power source,
   c. Generator-rectifier arc welding power source,
   d. Alternating-current/direct-current generator-rectifier arc welding power source,
   e. Alternating-current transformer arc welding power source,
   f. Direct-current transformer-rectifier arc welding power source, or
   g. Alternating-current/direct-current transformer-rectifier arc welding power source.

OUTPUT AND INPUT REQUIREMENTS

In addition to duty cycle, the output ratings and performance capabilities of power sources of each class are specified by NEMA. Table 1.3 presents the output current ratings for NEMA Class I, Class II, and Class III arc welding power sources. The NEMA-rated load voltage for Class I and Class II power sources under 500 A can be calculated using the following formula:

$$E = 20 + 0.04 \times I \qquad (1.10)$$

where

E = Rated load voltage, V; and

I = Rated load current.

The NEMA-rated load voltage is 44 for output current ratings of 600 A and higher. The output ratings in amperes and load voltage and the minimum and maximum output currents and load voltage for power sources are given in NEMA publication EW-1.

The electrical input requirements of NEMA Class I, Class II, and Class III transformer arc welding power sources are 220 V, 380 V, and 440 V for 50 Hz; for 60 Hz they are 200 V, 230 V, 460 V, and 575 V.[9]

The voltage and frequency standards for welding generator drive motors are the same as for NEMA Class I and II transformer primaries.

7. The term *power source* is synonymous with *arc welding machine*.
8. See Reference 1, p. 4.
9. See Reference 1, p. 10.

Table 1.3
NEMA-Rated Output Currents for Arc Welding Power Sources (Amperes)

Class I	Class II	Class III
200	150	180–230
250	175	235–295
300	200	
400	225	
500	250	
600	300	
800	350	
1000		
1200		
1500		

Source: Adapted with permission from National Electrical Manufacturers Association (NEMA), *Electric Arc-Welding Power Sources*, EW-1:1988 (R1999), Washington, D.C.: National Electrical Manufacturers Association, Tables 5.1, 5.2, 5.3.

NAMEPLATE DATA

The minimum data on the nameplate of an arc welding power source specified in NEMA publication EW-1 are the following:[10]

1. Manufacturer's type designation or identification number, or both;
2. NEMA class designation;
3. Maximum open-circuit voltage;
4. Rated load voltage (V);
5. Rated load current (A);
6. Duty cycle at rated load;
7. Maximum speed in revolutions per minute (rpm) at no-load (generator or alternator);
8. Frequency of power source (Hz);
9. Number of phases of power source;
10. Input voltage(s) of power source;
11. Current (A) input at rated load output.;
12. No-load RPM; and
13. Power factor

The instruction book or owner's manual supplied with each power source is the prime source of data concerning electrical input requirements. General data is also stated on the power source nameplate, usually in tabular form along with other pertinent data that might apply to the particular unit. Table 1.4 shows typical information for a NEMA 300-A-rated constant-current power source. The welding current ranges are given with respect to welding process. The power source may use one of two input voltages with the corresponding input current listed for each voltage when the power source is producing its rated load. The kilovolt-ampere (kVA) and kilowatt (kW) input data are also listed. The power factor[11] can be calculated as follows:

$$pf = \frac{kW}{kVA} \tag{1.11}$$

where

pf = Power factor
kW = Kilowatts
kVA = Kilovolt-amperes

The manufacturer also provides other useful data concerning input requirements such as primary conductor size and recommended fuse size. Power sources cannot be protected with fuses of equal value to their primary current demand. If this is done, nuisance blowing of the fuses or tripping of circuit breakers will result. Table 1.5 presents typical input wire and fuse sizes for the 300-A power source specified in Table 1.4. All pertinent codes should be consulted in addition to these recommendations.

ALTERNATING-CURRENT POWER SOURCES

Except for the power produced by engine-driven dc welding generators and batteries, all welding power typically begins as alternating current. The two main reasons for this are that ac can be transformed to higher or lower voltages, and it can be economically transmitted over long distances. When ac is required for welding, the high-voltage power delivered by the utility company is converted to the proper welding voltage by transformers. Alternating current is not as simple to understand as dc because the voltage and current reverse at regular intervals. This section discusses the principles of how alternating current is used in a welding power source.

TRANSFORMERS

Alternating-current power sources can utilize single-phase or three-phase transformers that connect to alternating-current utility power lines. The ac power

10. See Reference 1, p. 24, 25.

11. The power factor is the ratio of circuit power (watts) to circuit volt-amperes.

Table 1.4
Typical Nameplate Specifications for AC-DC Arc Welding Power Sources

Rated Output Current, A				Open-Circuit Voltage, AC and DC	Input Current, A, at Rated 300 A Output			
Alternating Current		Direct Current			60-Hz Single-Phase			
GTAW	SMAW	GTAW	SMAW		230 V	460 V	kVA	kW
300 A	300 A	300 A	300 A	80 V	104 A	52 A	23.9	21.8

Source: Adapted with permission from National Electrical Manufacturers Association (NEMA), *Electric Arc-Welding Power Sources*, EW-1:1988 (R1999), Washington, D.C.: National Electrical Manufacturers Association, p. 20.

Table 1.5
Typical Primary Conductor and Fuse Size Recommendations

	Input Wire Size, AWG[a]				Fuse Size in Amperes			
Model	200 V	230 V	460 V	575 V	200 V	230 V	460 V	575 V
300 A	No. 2	No. 2	No. 8	No. 8				
	(No. 6)[b]	(No. 6)[b]	(No. 8)[b]	(No. 8)[b]	200	75	90	70

a. American Wire Gauge.
b. Indicates ground conductor size.

Source: Adapted with permission from Electric Arc-Welding Power Sources, EW-1:1988 (R1999), National Electrical Manufacturers Association (NEMA), Washington, D.C.: National Electrical Manufacturers Association, Table 4.2.

source transforms the input voltage and amperage to levels suitable for arc welding. The transformers also serve to isolate the welding circuits from the utility power lines. Because various welding applications have different welding power requirements, the means for the control of welding current or arc voltage, or both, must be incorporated within the welding transformer power source. The methods commonly used in transformers to control the welding circuit output are described in the following sections.

Movable-Coil Control

A movable-coil transformer consists of an elongated core on which are located primary and secondary coils. Either the primary coil or the secondary coil may be movable, while the other one is fixed in position. Most alternating-current transformers of this design have a fixed-position secondary coil. As shown in Figure 1.18, the primary coil is normally attached to a lead screw, and as the screw is turned, the coil moves closer to the secondary coil or farther from it.

The varying distance between the two coils regulates the inductive coupling of the magnetic lines of force between them. The farther apart the two coils are, the more vertical is the volt-ampere output curve and the lower is the maximum short-circuit current value. Conversely, when the two coils are closer together, the maximum short-circuit current is higher and the slope of the volt-ampere output curve is less steep.

Figure 1.18(A) shows one form of a movable-coil transformer with the coils far apart for minimum output and a steep slope of the volt-ampere curve. Figure 1.18(B) shows the coils as close together as possible. The volt-ampere curve is indicated at maximum output with less slope than the curve of Figure 1.18(A).

Another form of movable coil employs a pivot motion. When the two coils are at a right angle to one another, output is at minimum. When the coils are aligned with one coil nested inside the other, output is at maximum.

Movable-Shunt Control

In the movable shunt design, the primary coils and the secondary coils are fixed in position. Control is obtained with a laminated iron core shunt that is moved between the primary and secondary coils. The movable core is made of the same material as that used for the transformer core.

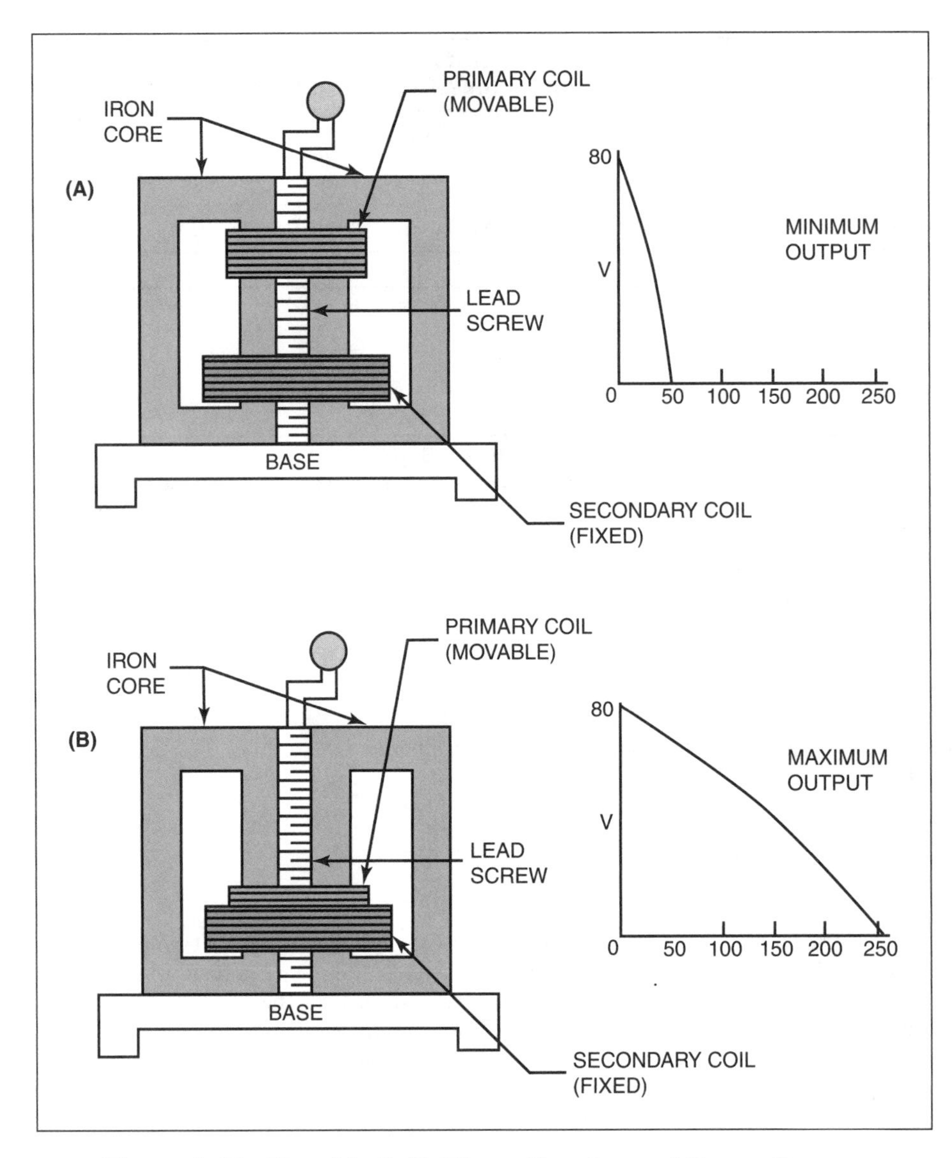

Figure 1.18—Movable-Coil Alternating-Current Power Source

As the shunt is moved into position between the primary and secondary coils, as shown in Figure 1.19(A), some magnetic lines of force are diverted through the iron shunt rather than to the secondary coils. With the iron shunt between the primary and secondary coils, the slope of the volt-ampere curve increases and the available welding current is decreased. Minimum current output is obtained when the shunt is fully in place.

As illustrated in Figure 1.19(B), the arrangement of the magnetic lines of force, or magnetic flux, is unobstructed when the iron shunt is separated from the primary and secondary coils. In this configuration, the output current is at its maximum.

Tapped Secondary Coil Control

A tapped secondary coil (refer to Figure 1.3) may be used for control of the volt-ampere output of a transformer. This method of adjustment is often used with NEMA Class III power sources. Basic construction is somewhat similar to the movable-shunt type, except that the shunt is permanently located inside the main core and the secondary coils are tapped to permit adjustment of the number of turns. Decreasing the secondary turns reduces open-circuit voltage as well as the inductance of the transformer, causing the welding current to increase.

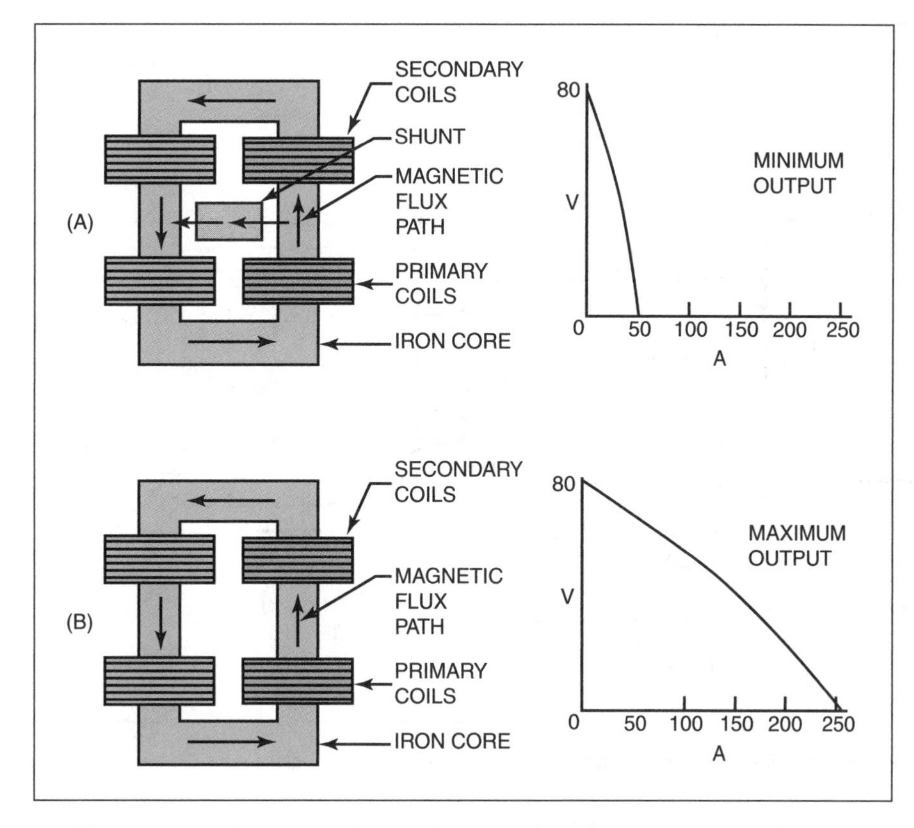

Figure 1.19—Movable-Shunt Alternating-Current Power Source

Movable-Core Reactor

The movable-core reactor type of alternating-current welding power source consists of a constant-voltage transformer and a reactor in series. The inductance of the reactor is varied by mechanically moving a section of its iron core. The power source is diagramed in Figure 1.20.

When the movable section of the core is in a withdrawn position, the permeability of the magnetic path is very low because of the air gap. The result is a low inductive reactance that permits a high welding current to flow. When the movable-core section is advanced into the stationary core, as shown by the broken-line rectangle in Figure 1.20, the increase in permeability causes an increase in inductive reactance, which reduces the welding current.

Saturable-Reactor Control

A saturable-reactor control is an electrical control that uses a low-voltage, low-amperage direct-current circuit to change the effective magnetic characteristics of reactor cores. Remote control of output from the power source is relatively easy with this type of control circuit, and it normally requires less maintenance than do mechanical controls. With this construction, the main transformer has no moving parts. The volt-ampere characteristics are determined by the transformer and the saturable-reactor configurations. The direct-current control circuit to the reactor system allows the adjustment of the output volt-ampere curve from minimum to maximum.

A simple, saturable-reactor power source is diagramed in Figure 1.21. The reactor coils are connected in opposition to the direct-current control coils. If this were not done, transformer action would cause high circulating currents to be present in the control circuit. With the opposing connection, the instantaneous voltages and currents tend to cancel out. Saturable reactors tend to cause severe distortion of the sine wave supplied by the transformer. This is not desirable for gas tungsten arc welding (GTAW) because the wave form for that process is important. One method of reducing this distortion is by introducing an air gap in the reactor core. Another is to insert a large choke in the direct-current control circuit. Either method, or a combination of both, will produce he desired results.

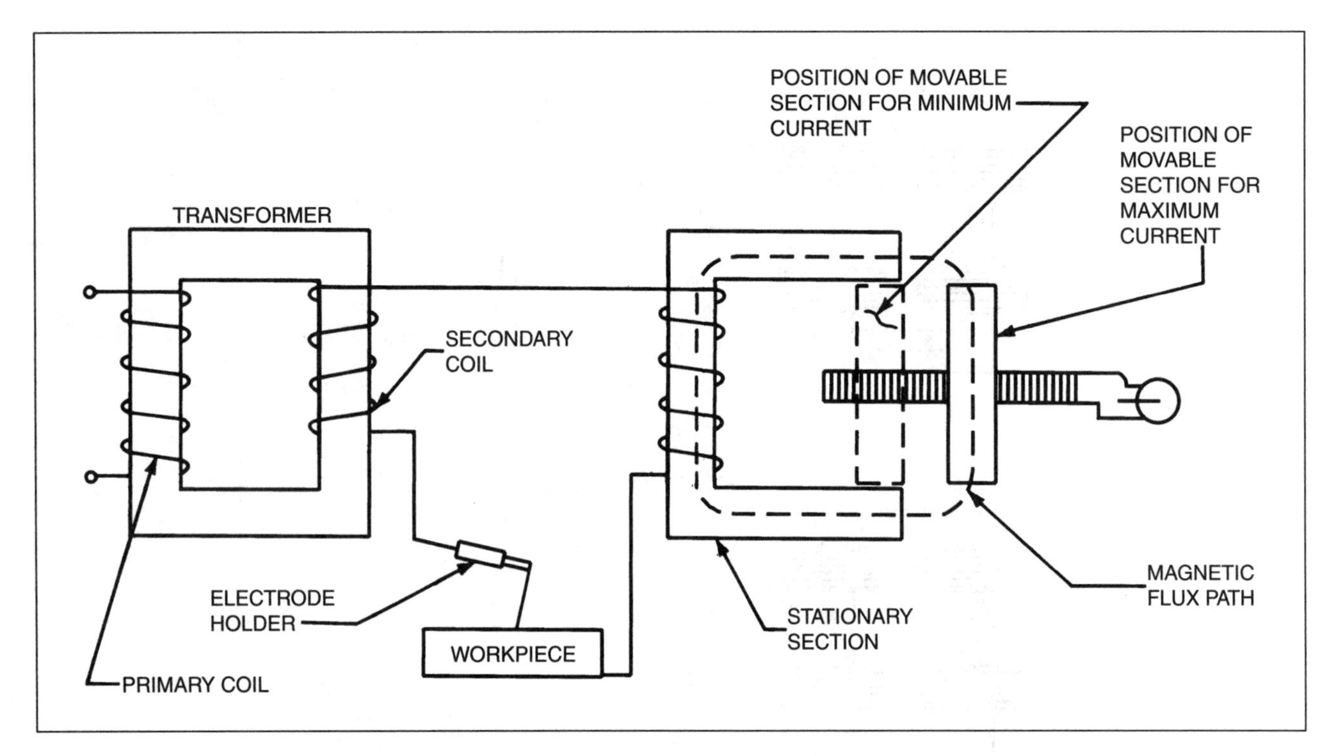

Figure 1.20—Movable-Core Reactor Alternating-Current Power Source

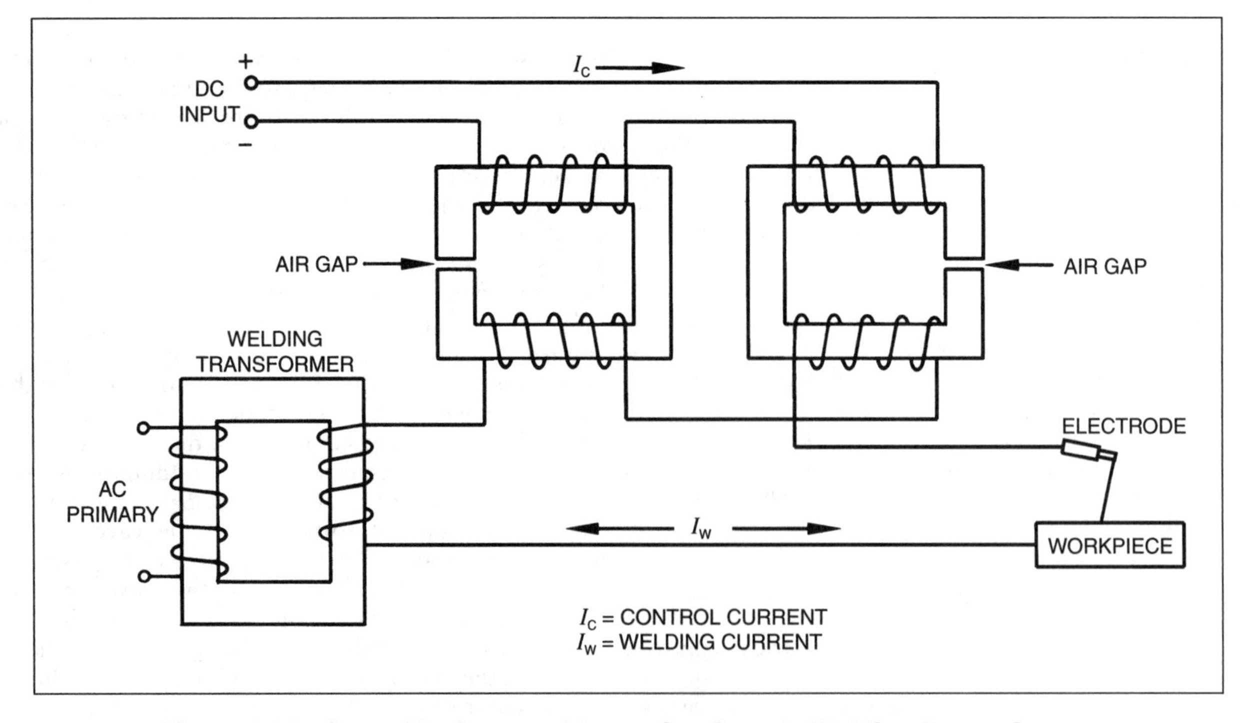

Figure 1.21—Saturable Reactor Alternating-Current Welding Power Source

The amount of current adjustment in a saturable reactor is based on the ampere-turns of the various coils. The term *ampere-turns* is defined as the number of turns in the coil multiplied by the current in amperes flowing through the coil.

In the basic saturable reactor, the law of equal ampere-turns applies. To increase output in the welding circuit, a current must be made to flow in the control circuit. The amount of change can be approximated with the following equation:

$$I_w = \frac{I_c N_c}{N_w} \tag{1.12}$$

where

I_w = Change in welding current, A;
I_c = Change in current, A, in the control circuit;
N_c = Number of turns in the control circuit; and
N_w = Number of turns in the welding current circuit.

The minimum current of the power source is established by the number of turns in the welding current reactor coils and the amount of iron in the reactor core. For a low minimum current, a large amount of iron or a relatively large number of turns, or both, are required. If a large number of turns are used, a large number of control turns or a high control current, or both, are necessary. The saturable reactors often employ taps on the welding current coils to reduce the requirement for large control coils, large amounts of iron, or high control currents, creating multiple-range power sources. The higher ranges would have fewer turns in these windings and thus correspondingly higher minimum currents.

Magnetic Amplifier Control

Technically, the magnetic amplifier is a self-saturating saturable reactor. It is called a *magnetic amplifier* because it uses the output current of the power source to provide additional magnetization of the reactors. In this way, the control currents can be reduced and control coils can be smaller. While magnetic amplifier power sources are often multiple-range, the ranges of control can be much broader than those possible with an ordinary saturable-reactor control.

Figure 1.22 illustrates that by using a different connection for the welding current coils and rectifying diodes in series with the coils, the load ampere-turns are used to assist the control ampere-turns in magnetizing the cores. A smaller number of control ampere-turns will cause a correspondingly higher welding current to flow because the welding current will essentially turn itself on. The control windings are polarity-sensitive.

Power Factor

The power factor (pf) of a welding power source is the ratio of circuit power (in watts) to the circuit volt amperes and can be measured and calculated as follows:

For a single-phase power source:

$$\text{pf} = \frac{W}{V_{\text{L to L}} \times A} \tag{1.13}$$

For a three-phase power source:

$$\text{pf} = \frac{W}{\sqrt{3} \times V_{\text{L to L}} \times A} \tag{1.14}$$

where

W = Watts as measured by a wattmeter connected to the single-phase or three-phase input circuit of the power source;
$V_{\text{L to L}}$ = Line-to-line voltage connected to the input line of the power source, as measured by a voltmeter; and
A = Amperes in an input line to the power source, as measured by an ammeter.

Wattmeters contain multiple electrical coils that detect the phase difference between the line voltage and line currents and display the actual power (wattage) consumed by the power source.

Constant-current alternating-current power sources are characterized by low power factors because of the large inductive reactance. This is often objectionable because the line currents can be high, and power utility rates can penalize industrial users for low power factors. Power factor may be improved by adding capacitors to the primary circuit of inductive loads such as welding transformers. This reduces the primary current from power lines while welding is being performed. Unfortunately, the current draw under light or no-load conditions will increase.

Large alternating-current-transformer power sources may be equipped with capacitors for power-factor correction to approximately 75% at rated load. At load-current settings lower than rated, the power factor may have a leading characteristic. When the transformer is operating at no-load or very light loads, the capacitors are drawing their full corrective kVA, thus contributing power-factor correction to the remainder of the load on the total electrical system.

When several transformer welding power sources are operating at light loads, it should be carefully ensured that the combined power-factor correction capacitance does not upset the voltage stability of the line. If three-phase primary power is used, the load on each phase of

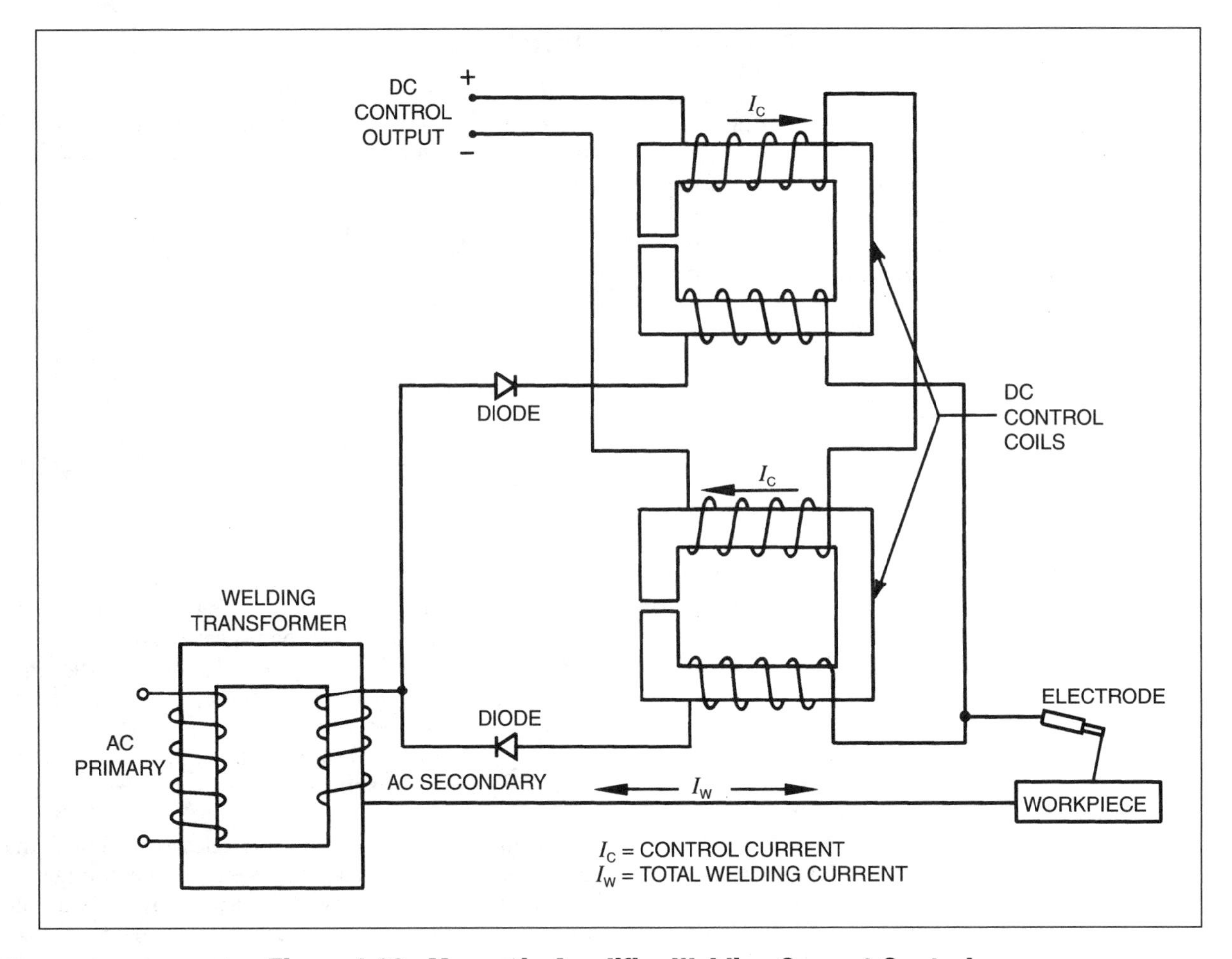

Figure 1.22—Magnetic Amplifier Welding Current Control

the primary system should be balanced for best performance. Power-factor correction, under normal conditions, has no bearing on welding performance.

Auxiliary Features

Constant-current alternating-current power sources are available in many configurations and with many auxiliary features. Generally, these features are incorporated to better adapt the unit to a specific process or application, or to make it more convenient to operate. The manufacturer should be consulted for available features when considering these power sources.

Primary contactors or manually operated power switches to turn the unit on and off are usually included in alternating-current power sources. Most NEMA Class I and Class II units are furnished with a terminal board or other means for the connection of various rated primary-line voltages. Input supply cables are not normally supplied with NEMA Class I and Class II welding power sources. The smaller NEMA Class III power sources are generally equipped with a manually operated primary switch and an input supply cable.

Some alternating-current power sources incorporate a system for supplying a higher-than-normal current to the arc for a fraction of a second at the start of a weld. This "hot start" feature provides starting surge characteristics similar to those of motor-generator units. The hot start assists in initiating the arc, particularly at current levels under 100 A. Other power sources, for example those used for GTAW, may be equipped with a start control to provide an adjustable "soft" start or reduced-current start to minimize the transfer of tungsten from the electrode.

Equipment designed for the GTAW process usually incorporates an additional valve and timer to control the flow of shielding gas to the electrode holder. A high-frequency, high-voltage unit may be added to assist in starting and stabilizing the alternating-current arc.

NEMA Class I and Class II power sources may be provided with a means for the remote adjustment of output power. This may consist of a motor-driven device for use with crank-adjusted units or a hand control at the workstation when an electrically adjusted power source is being used. When a weldment requires frequent changes of amperage or when welding must be performed in an inconvenient location, remote control adjustments can be very helpful. Foot-operated remote controls free the operator's hands and permit the gradual increase or decrease of welding current. This is of great assistance in crater filling for GTAW.

Safety controls are available on some power sources to reduce the open-circuit voltage of alternating-current arc welding power sources. They reduce the open-circuit voltage at the electrode holder to less than 30 V. Voltage reducers may consist of relays and contactors that either reconnect the secondary winding of the main transformer for a lower voltage or switch the welding load from the main transformer to an auxiliary transformer with a lower no-load voltage.

ALTERNATORS

Another source of alternating-current welding power is an alternator (often referred to as an *alternating-current generator)*, which converts mechanical energy into electrical power suitable for arc welding. The mechanical power may be obtained from various sources such as an internal combustion engine or an electric motor. As illustrated in Figure 1.23, ac generators differ from standard dc generators in that the alternator rotor assembly contains the magnetic field coils instead of the stator coils. Slip rings are used to conduct low direct-current power into the rotating member to produce a rotating magnetic field. This configuration precludes the need for the commutator and the brushes used with direct-current output generators. The stator (stationary portion), shown in Figure 1.24, has the welding current coils wound in slots in the iron core. The rotation of the field generates ac welding power in these coils.

The frequency of the output welding current is controlled by the rotation speed of the rotor assembly and by the number of poles in the alternator design. A two-pole alternator must operate at 3600 rpm to produce 60-Hz current, whereas a four-pole alternator design must operate at 1800 rpm to produce 60-Hz current.

Saturable reactors and moving-core reactors can be used for output control of alternators. However, the normal method is to provide a tapped reactor for broad control of current ranges in combination with control of the alternator magnetic field to produce fine control within these ranges. These controls are shown in Figure 1.25.

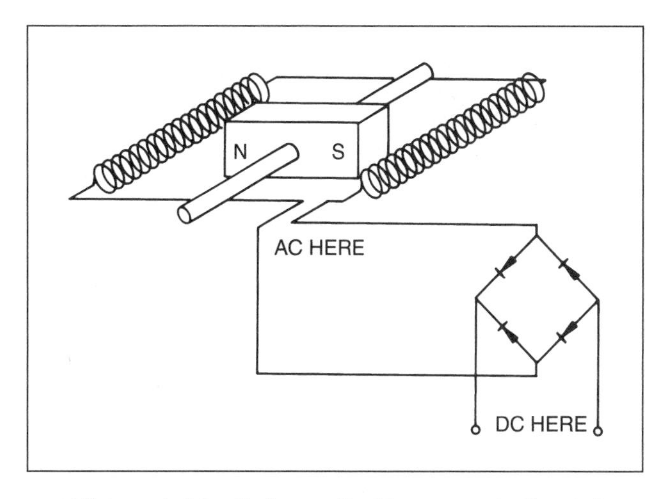

Figure 1.23—Schematic Representation of an Alternator Showing the Magnetic Field Contained in the Rotor Assembly

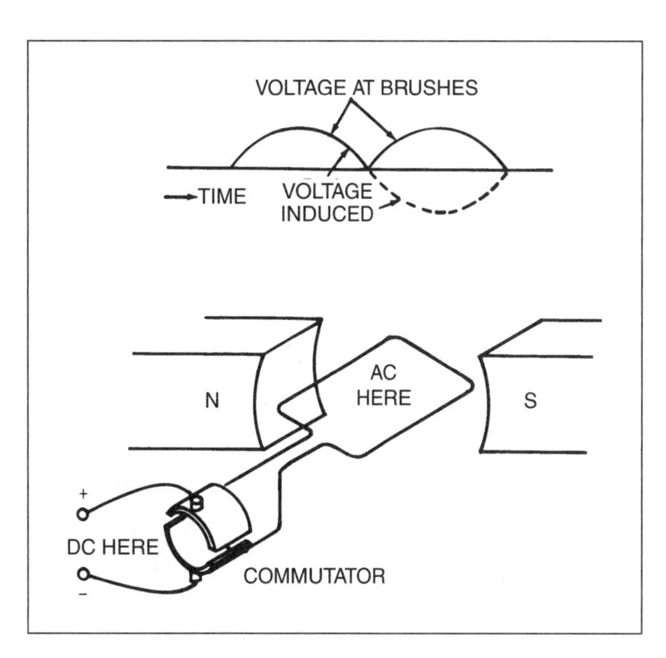

Figure 1.24—Schematic of a Generator Showing the Magnetic Field Contained in the Stator Assembly

SQUARE-WAVE ALTERNATING-CURRENT POWER SOURCES

Alternating-current welding power sources for the SMAW, GTAW, PAW, and submerged arc welding (SAW) processes were traditionally based on three methods of regulating their fields: moving coils, moving shunt, and saturable reactors. The need for wider

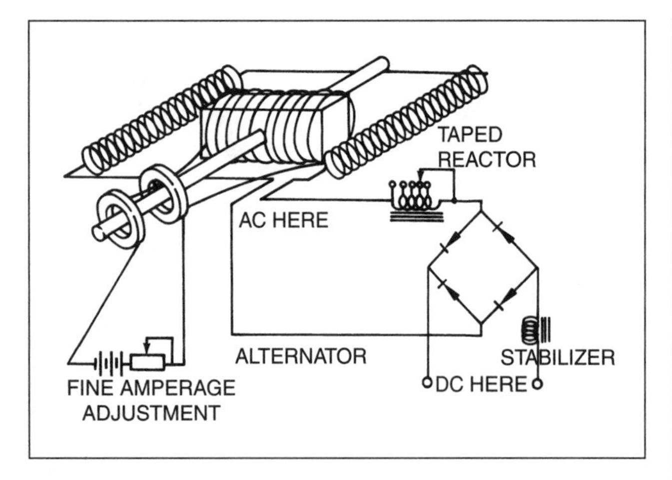

Figure 1.25—Schematic of an Alternator Power Source Showing a Tapped Reactor for Coarse Control of Current and Adjustable Magnetic Field Amperage for Fine Control of Output Current

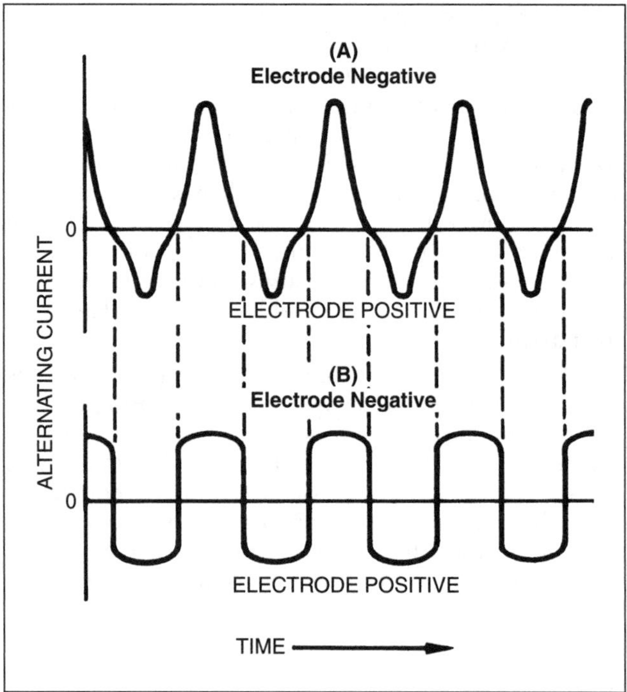

Figure 1.26—Comparison of Arc Current Wave Forms of (A) Magnetic Amplifier and (B) Square-Wave Power Source at Same Average Current Level

current ranges and remote current control led to the development of magnetic amplifiers with silicon diodes. While these technologies served the welding industry well, the need existed for power sources that would produce welds of higher quality and improved reliability. The development of power semiconductors has provided a new generation of welding power sources that meet these needs.

With 60-Hz alternating current, the welding current is reversed 120 times per second. With magnetic power sources, the current reversal occurs rather slowly, hampering reignition of the next half-cycle. Even though auxiliary methods can be used to provide a high ionizing voltage, such as superimposed high-frequency energy for GTAW and PAW, often the instantaneously available voltage is too low to assure reliable arc reignition.

The reignition problem can be minimized by using a current with a square wave form as diagramed in Figure 1.26. With its rapid zero crossing, deionization may not occur or, at the very least, arc reignition may be enhanced to the extent that the need for high-frequency reignition is reduced.

The trailing edge of the square wave form keeps the shielding gas ionized in preparation for reignition at the opposite polarity. These features are important in installations where it becomes desirable to eliminate high frequency (HF) for one or more of the following reasons:

1. High-frequency radiation may cause radio or television interference;
2. High-frequency may cause etching of the workpiece in the immediate vicinity of the weld, which may be cosmetically undesirable;
3. High-frequency leakage may bother the operator, and
4. Peripheral equipment may be damaged by HF.

Various design approaches have been used to produce square alternating-current wave forms. Some power sources use single-phase input and others use three-phase input. Two common approaches are the use of a memory core and inverter circuits.

Memory Core

A memory core is a magnetic device, such as an output inductor (arc stabilizer) that keeps the current flowing at a constant value (a kind of electric flywheel). In conjunction with a set of four power SCRs, it can be used to develop a square-wave alternating current. The memory core stores energy in proportion to the previous half-cycle of current, and then pumps that same amount of current to the arc at the beginning of each new opposite polarity half-cycle. The values of the reignition current and the extinguishing current of the half cycle are the same. The value is the "remembered" multi-cycle average-current value maintained by the memory core device. The transition time from one polarity to the other is very short, in the range of 80 microseconds.

A sensor placed in the memory core current path produces a voltage signal that is proportional to the alternating-current output. That current signal is compared with the desired current reference signal at a regulator amplifier. The resulting actuating error signal is processed to phase-fire four SCRs in the proper sequence to bring the output to the proper level. Consequently, the welding current is held within 1% for line voltage variations of 10%. Response time is fast, thus lending itself to pulsed alternating-current GTAW operations.

Another feature designed into this type of power source is a variable asymmetric wave shape. This enables the operator to obtain balanced current or various degrees of controlled imbalance of direct current electrode negative (DCEN) or *straight polarity,* versus direct current electrode positive (DCEP) or *reverse polarity.* This capability provides a powerful tool for arc control. The main reason for using alternating current with the GTAW process is that it provides a cleaning action. This is especially important when welding aluminum and magnesium. During DCEP cycles, the oxides on the surface of the workpiece are removed, exposing clean metal to be welded. Tests with various asymmetrical power sources established that only a small amount of DCEP current is required. Amounts as low as 10% would be adequate, with the exception of cases in which hydrocarbons may be introduced by the filler wire.

Balance is set with a single knob to control the split of positive and negative portions in a square-wave ac wave form cycle period, as illustrated in Figure 1.27. Trace A in Figure 1.27 has 50% of the ac cycle period in electrode positive, and 50% in electrode negative, hence a balanced ac wave form. Trace B favors electrode positive (55%) and trace C favors electrode negative (70%). In effect, the balance control adjusts the width of each polarity without changing current amplitude or frequency. The regulating system holds the selected balance ratio constant as other amperage values are selected.

This balance control is very useful. With a reasonably clean workpiece, the operator can adjust conditions for a low percentage of cleaning action (using DCEP). With the resulting high percentage of DCEN wave form, the heat balance approaches that of DCEN, providing more heat into the workpiece, less arc wander, and a narrower bead width. Considering that gas tungsten arc welding is often selected because of its concentrated arc, this balance control allows the greatest utilization of the best characteristic of this process.

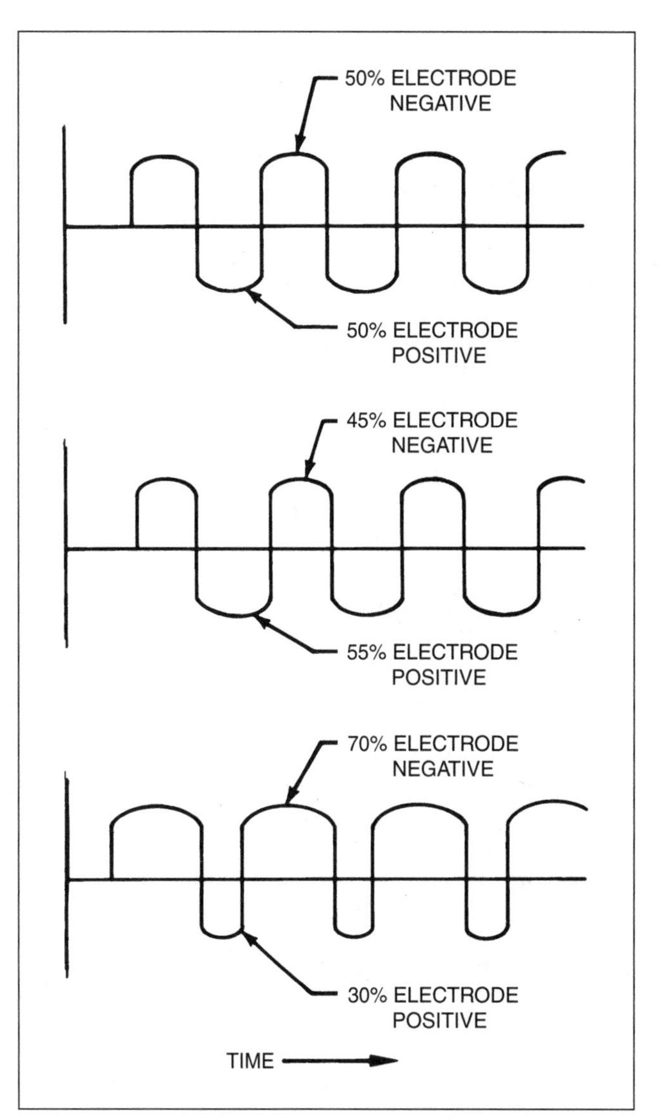

Figure 1.27—Typical Wave Forms Produced by Square-Wave Power Balance Control

The asymmetrical wave with less DCEP time allows the operator to use smaller-diameter electrodes without the risk of high temperatures eroding the tip. In effect, it allows a higher current density. This results in a smaller-diameter arc cone and better heat concentration. The smaller gas nozzle often allows the operator to get into tighter joint configurations.

Inverter with Alternating-Current Output

Another approach to achieving a square-wave alternating-current output is to use inverter circuits. Several systems are used with the inverter-circuit approach to achieve a square-wave alternating-current output with rapid zero crossover. These systems are dual sources with inverter switching, single sources with inverter switching, and synchronous rectifier inverters.

The dual source with inverter switching uses solid-state SCR technology. It combines two three-phase, adjustable-current, direct-current power sources. The power source that provides the main weld current is SCR-controlled and typically rated for 300-A, 50-V direct-current output. It supplies current during both DCEN and DCEP phases of operation. The other power source is a conventional reactor-controlled power source typically rated at 5 A to 100 A for 50-V direct-current output. Its function is to provide higher current during the DCEP phases of operation so that cleaning is improved. Tests have shown that the most effective etching is obtained when DCEP current is higher but applied for a much shorter time than the DCEN current. Both power sources must provide 50-V output to ensure good current regulation when welding.

The switching and combining of current from the two power sources is controlled by five SCRs, as shown in Figure 1.28. Four of these SCRs form part of an inverter circuit that switches the polarity of the current supplied to the arc. The four SCRs are arranged to operate in pairs.

One pair (SCR 1 and SCR 4) is switched on to provide current from the main power source during the DCEN portions of the square wave. The other pair (SCR 2 and SCR 3) is switched on to provide current from both power sources during DCEP portions of the square wave. A shorting SCR (SCR 5) is used with a blocking diode to bypass current from the second power source around the inverter circuit during the DCEN portion of the cycle, thereby preventing its addition to the welding current.

The SCRs are turned on by a gating circuit, which includes timing provisions for adjustment of the DCEN and DCEP portions of the square-wave output. The DCEN time can typically be adjusted from 5 milliseconds to 100 milliseconds, and the DCEP time from 1 millisecond to 100 milliseconds. For example, a typical time setting for welding thick aluminum might be 19 milliseconds DCEN and 3 milliseconds DCEP. The SCRs are turned off by individual commutation circuits. The current wave form is shown in Figure 1.29.

The single source with inverter switching is a much simpler and less bulky approach than the dual-source system. With the single-source approach, one dc constant-current power source is used. Figure 1.30 shows a single-source, ac, square-wave inverter using transistors instead of SCRs. The operation of this source is very similar to that of the dual source. The four transistors are arranged to operate in pairs. In the absence of an additional reverse-current source, a fifth transistor and blocking diode are not necessary. Alternating-current balance can be controlled like the memory-core source and the dual-source inverter. However, DCEP current must be the same in amplitude as the DCEN current and cannot be increased, as with the dual-source inverter. Both the single-source and dual-source inverters can vary the frequency of the ac square-wave output, whereas the memory-core source must operate at line frequency (50 Hz or 60 Hz).

A third approach uses a device called a *synchronous rectifier*. This method starts with a power source with an inverter in the primary that produces a high-frequency ac output. The high-frequency alternating current is applied to the synchronous rectifier circuit, which, on command, rectifies the high-frequency alternating current into either DCEN or DCEP output. By switching the synchronous rectifier alternately between DCEN and DCEP, a synthesized lower frequency alternating-current output can be created.

DIRECT-CURRENT POWER SOURCES

Direct-current power sources are the most commonly used. They can be used in a variety of arc welding processes, including GMAW, FCAW, GTAW, SMAW, SAW, and PAW. They are essentially power converters that convert high-voltage ac utility power or mechanical energy into low-voltage, high-current dc output suitable for welding. The load of a dc power source consists of a cable, electrode holder, and electrical arc in series. The output characteristics are process-dependent, which can be constant current (for GTAW and SMAW), constant voltage (for GMAW or FCAW), or a more dynamic volt-ampere curve, such as pulsed current for improved process control. The design can be a dc generator-engine drive, transformer type, thyrister-controlled, or a high-frequency switching topology such as an inverter.

CONSTANT-VOLTAGE POWER SOURCES

Constant-voltage, or constant potential, power sources are commonly used for GMAW, FCAW, and SAW. These power sources are rotating, transformer-rectifier, or inverter power sources. Generators that can supply constant-voltage welding power are normally the separately excited, modified compound-wound type. The compounding of constant-voltage units differs from that of constant-current units to produce flat

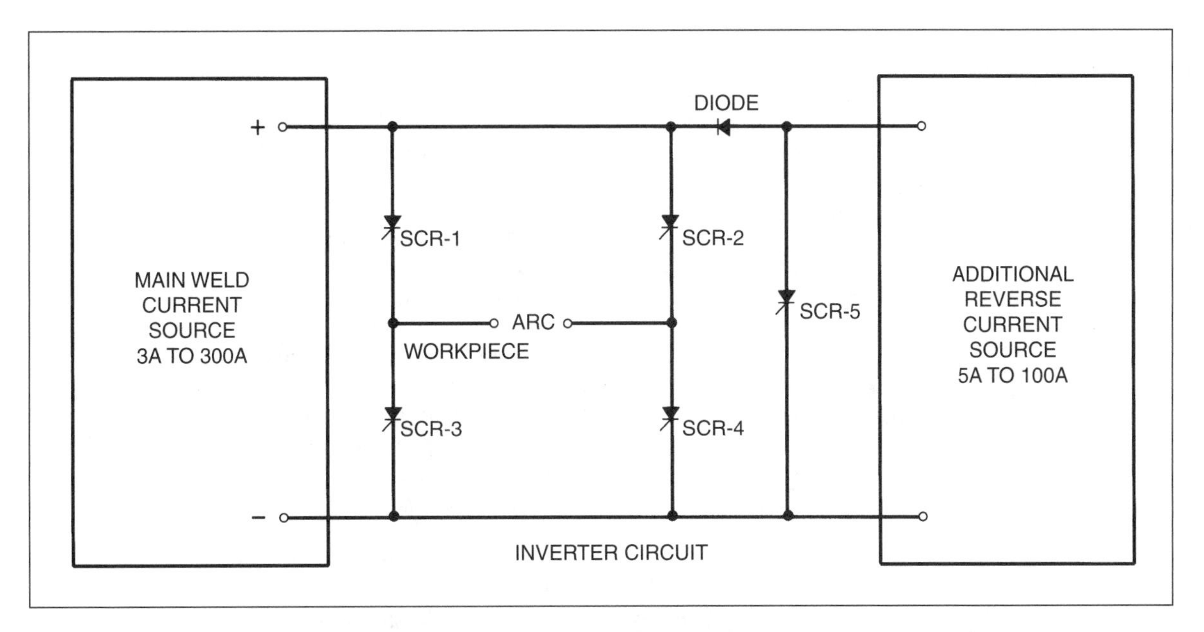

Figure 1.28—Inverter Circuit Used With Dual Direct-Current Power Sources for Controlling Heat Balance in Gas Tungsten Arc Welding

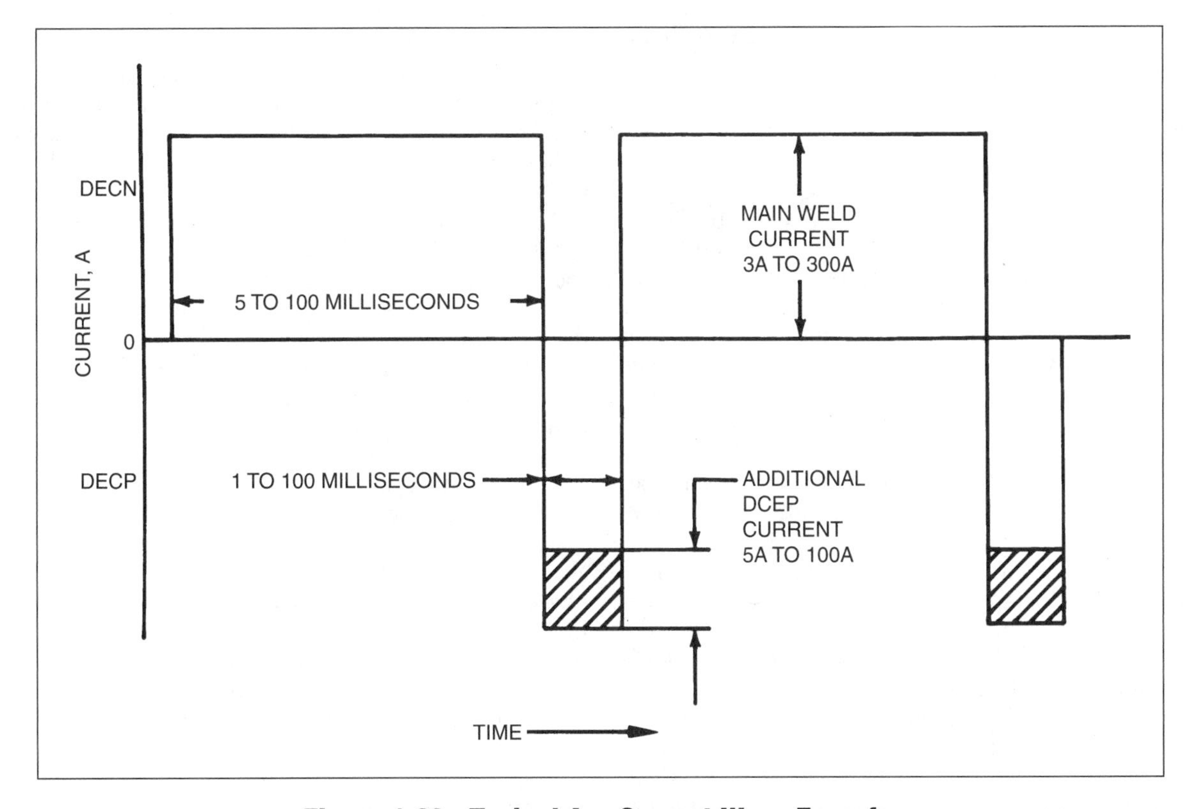

Figure 1.29—Typical Arc Current Wave Form for Dual-Adjustable Balance Inverter Power Source

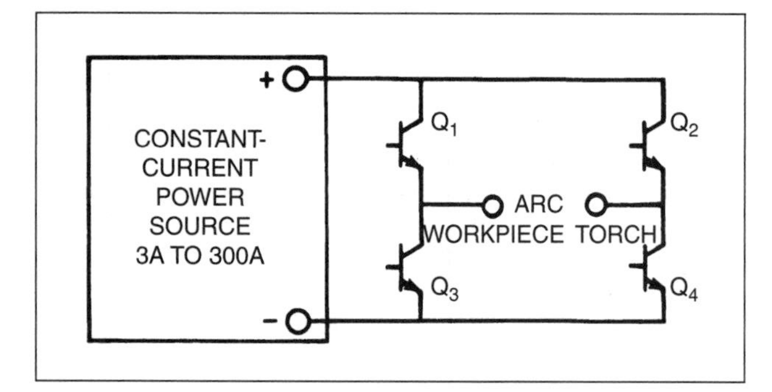

Figure 1.30—Alternating-Current Inverter Circuit (Single Source)

volt-ampere output characteristics. These power sources may have solid-state devices in the excitation circuit to optimize performance and to provide remote-control capability. Various types of electronic circuits such as phase-angle controlled SCRs and inverter circuits are used for this purpose.

Transformer-rectifier and inverter constant-voltage power sources for industrial applications are normally three-phase. Small single-phase power sources, usually rated 200 A or below, are typically designed for light-duty applications.

Electrical Characteristics

Constant-voltage power sources are characterized by their typically flat volt-ampere curves. A negative slope of 1 V to 5 V per 100 A is common. As a result, the maximum short-circuiting current is usually very high, sometimes in the range of thousands of amperes. Power sources with volt-ampere curves having slopes of up to 8 V per 100 A are still categorized as constant-voltage power sources.

There are many varieties and combinations of constant-voltage power sources. A fixed slope may be built into the power source, or the unit may have an adjustment to adapt the slope of the volt-ampere curve to the welding process.

The dynamic characteristics of these power sources are of prime importance. If inductance is used to adjust the slope, it will change not only the static but also the dynamic characteristics of the power source. In some cases, adjustable inductors are used in the direct-current portion of the circuit to obtain separate control of the static and dynamic features. The direct-current inductor will not alter the static characteristics, but will affect the dynamic characteristics. Direct-current inductors are very important for short-circuiting transfer in GMAW.

General Design

Many designs of constant-voltage power sources are available. The advantage of any particular type is related to the application and to the expectations of the user.

Open-Circuit Voltage. The open-circuit voltage of some transformer-rectifier power sources is adjusted by changing taps on the transformer. Another type of power source controls the open-circuit voltage with secondary coils, carbon brushes driven by a lead screw to slide along the secondary coil conductors. A second control is often included to adjust the volt-ampere characteristics to provide the requirements of the welding process. This is called *slope control* because of its additional effect on the volt-ampere output curve.

Constant-voltage power sources have a wide range of open-circuit voltages. Electrically controlled power sources may have an open circuit as high as 80 V. Tapped or adjustable transformers have open-circuit voltage that can be varied from 30 V to 50 V maximum to 10 V minimum.

Slope. Slope control is generally obtained by changing taps on reactors in series with the alternating-current portion of the circuit. Slope control can be provided by carbon brushes attached to a lead screw contacting the reactor turns. This variable reactor provides continuous adjustment of slope. Another method of control uses magnetic amplifiers or solid-state devices to electrically regulate output voltage. These power sources may have either voltage taps or slope taps in addition to electrical control.

Some of the advantages of electrical controls are easy adjustment, the capability to use remote control, and the absence of moving parts. Some electrically controlled power sources permit adjustment of output during welding. This is helpful for tasks such as crater filling or changing welding conditions. The combination of taps with electrical control to give fine output adjustment between taps is a suitable arrangement in an application in which the power source requires little attention during welding. Power sources that are fully controlled electrically are easier to set up and readjust when welding requirements change rapidly. Slope can also be controlled electronically by circuitry in most phase-angle controlled SCR and inverter power sources. Electrically controlled power sources frequently have fixed, all-purpose slopes designed into them.

Slope control on constant-voltage generators is usually provided by a tapped resistor in the welding circuit. This is desirable because of the inherent slow dynamic response of the generator to changing arc conditions. Resistance slope controls limit maximum short-circuiting current. Reactor slope control also limits maximum short-circuiting current. However, it slows the rate of response of the power source to changing arc conditions more than resistive slope control.

While no fixed rule exists for volt-ampere slope in the welding range, most power sources have slopes ranging from 1 V to 3 V per 100 A.

Inductance. Gas metal arc welding power sources designed for short-circuiting transfer generally incorporate additional direct-current inductance to improve performance by providing the dynamic characteristics required. This inductance can be variable or fixed.

Ripple. Single-phase power sources generally require some type of ripple filter arrangement in the welding circuit. This filter is usually a bank of electrolytic capacitors across the rectifier output. The purpose is to provide a smooth direct-current output capable of clearing a short circuit. An inductor is used to control the output of the capacitors. Without some inductance, the discharge of the capacitors through a short circuit would cause excessive spatter.

Control Devices

Constant-voltage power sources are usually equipped with primary contactors. Electrically controlled models typically have remote voltage-control capabilities. Other features available on certain power sources are line-voltage compensation and accessories to interface with wire feed equipment to change feed rate and welding current.

Electrical Rating

Primary ratings of constant-voltage power sources are similar to those discussed previously. Constant-voltage power sources typically have a more favorable power factor than constant-current power sources and do not require power-factor correction. Open-circuit voltage can be well below the established maximum of 80 V dc. Current ratings of NEMA Class I power sources range from 200 A to 1500 A.

Constant-voltage power sources are normally classified as NEMA Class I or Class II. It is the usual practice to rate them at 100% duty cycle, except for some of the light-duty units of 200 A and under, which may be rated as low as 20% duty cycle.

CONSTANT-CURRENT POWER SOURCES

Welding power sources that are called *constant-current power sources* can be used for SMAW, GTAW, SAW, and plasma arc welding and cutting. These power sources can be inverters, transformer-rectifiers, or generators. The transformer-rectifier and inverter power sources are static, transforming input alternating current to output direct-current power. Generators convert the mechanical energy of rotation to electrical power.

The open-circuit voltages of constant-current rectifier power sources vary as required by the specific welding application, ranging from 50 V to 100 V. Most NEMA Class I and Class II power sources are fixed in the 70 V to 80 V range.

Electrical Characteristics

The relationship of the output current to the output voltage is an important electrical characteristic. Both a static (steady-state) relationship and a dynamic (transient) relationship are of special interest. The static relationship is usually shown by volt-ampere curves such as those shown in Figure 1.14. The curves usually represent the maximum and minimum for each current-range setting. As discussed in a previous section, the dynamic relationship is difficult to define and measure for all load conditions. The dynamic characteristics determine the stability of the arc under actual welding conditions. Dynamic characteristics are influenced by circuit design and control.

General Design

The usual voltages of the alternating current supplied by electric utility companies in the United States are nominally 208 V, 240 V, and 480 V with a frequency of 60 Hz. The line frequency is 50 Hz in Europe, Asia, and Australia, and variations of 50 and 60 Hz are supplied in Central and South America. Line voltages vary in different countries. The 230/400 V lines are common in Western Europe, but 220/380, 100/500, 400/690, and 127/220 lines also exist. Voltages in Asia are more diverse: China uses 220/380 V lines; Japan uses 100/200 V lines, and Korea uses 110/220/380 V lines. The voltage tolerance is also country-specific. Transformers are designed to work on all these voltages. This is accomplished by arranging the primary coils in sections with taps. In that way, the leads from each section can be connected in series or parallel with other sections to suitably match the incoming line voltage. On three-phase power sources, the primary can be connected in the delta or "Y" configuration. The secondary is frequently connected in the delta configuration, as this connection is preferred for low voltage and high current.

The current is usually controlled in the section of the power source between the transformer and the rectifiers. Current control employs the principle of variable inductance or impedance. The following systems are used for varying the impedance for current control:

1. Moving coil,
2. Moving shunt,

3. Saturable reactor or magnetic amplifiers,
4. Tapped reactor,
5. Moving reactor core, and
6. Solid-state controls.

In addition to these six control systems, another configuration employs resistors in series with the direct-current portion of the welding circuit. Systems 1, 2, and 5 in the list above are classed as mechanical controls. Systems 3 and 6 are classed as electric controls, and System 4 and the external resistor types are classed as tap controls. These same systems are also used for controlling constant-current transformer sources.

An inductance is usually used in the direct-current welding circuit to control excessive surges in load current. These current surges may occur because of dynamic changes in arc load. Inductance is also used to reduce the inherent ripple found after rectifying the alternating current. A three-phase rectifier produces relatively little ripple; therefore, the size of its inductor is determined primarily by the need to control arc load surges. A high ripple is associated with single-phase rectification. The size of inductors for single-phase power sources is determined by the need to reduce ripple. Therefore, they are larger than those used in three-phase power sources of the same rating. Power sources of this type frequently have a switch on the direct-current output so that the polarity of the voltage at the power source terminals can be reversed without reversing the welding cables.

Auxiliary Features

The auxiliary features are similar to those available for constant-current ac power sources, although not all features are available on all power sources. The manufacturer can supply complete information.

In addition to the features listed previously, current-pulsing capabilities are available with many direct-current power sources as standard or optional equipment. Pulsed-power sources are capable of alternately switching from high to low welding current repetitively. Normally, high- and low-current values, pulse duration, and the pulse repetition rate are independently adjustable. This feature is useful for out-of-position GMAW and critical GTAW applications.

INVERTER POWER SOURCE

An inverter power source is different from a transformer-rectifier type in that the inverter rectifies ac line current, uses an inverter circuit to produce a high-frequency ac, reduces that voltage with an ac transformer, and rectifies that to obtain the required dc output. Changing the ac frequency to a much higher frequency allows a greatly reduced size of transformer and reduced transformer losses.

MOTOR-DRIVEN AND ENGINE-DRIVEN GENERATORS AND ALTERNATORS

Generator power sources and alternators convert mechanical energy into electrical power suitable for arc welding. Rotating mechanical power can be obtained from an internal combustion engine or an electric motor. For welding, two basic types of rotating power sources are used: the generator and the alternator. Both have a rotating component, called a *rotor* or an *armature*, and a stationary component, called a *stator.* A system of excitation is needed for both types.

In a rotating power source, voltage is induced in electrical conductors when the conductors are moved through a magnetic field. Physically, it makes no difference whether the magnetic field moves or the conductor moves, just so the coil experiences a changing magnetic field. In practice, a generator has a stationary field and moving conductors, while an alternator has a moving field and stationary conductors.

The dc generator has a commutator-brush arrangement for changing alternating current to direct-current welding power. Normally, the direct-current generator is a three-phase electrical device. Three-phase systems provide the smoothest welding power of any of the electro-mechanical welding power sources.

A direct-current generator consists of a rotor and stator. The rotor assembly is comprised of a through shaft; two end bearings to support the rotor and shaft load; an armature, which includes the laminated armature iron core and the current-carrying armature coils; and a commutator. It is in the armature or rotating coils that welding power is generated.

The stator is the stationary portion of the generator within which the rotor assembly turns. It holds the magnetic field coils of the generator. The magnetic field coils conduct a small amount of direct current to maintain the necessary continuous magnetic field required for power generation. The direct-current amperage is normally no more than 10 A to 15 A and very often is less.

In electric power generation, there must be relative motion between a magnetic field and a current-carrying conductor. In the direct-current generator, the rotating armature is the current-carrying conductor. The stationary magnetic field coils are located in the stator. The armature turns within the stator and generates the welding current.

The armature conductors of a welding generator are relatively heavy because they carry the welding current. The commutator, located at one end of the armature, is a group of conducting bars arranged parallel to the rotating shaft to make switching contact with a set of stationary carbon brushes. These bars are connected to the armature conductors. The whole arrangement is constructed in proper synchronization with the magnetic field so that, as the armature rotates, the commutator performs the function of mechanical rectification.

An alternator power source is very similar, except that the magnetic field coils are generally wound on the rotor and the heavy welding current winding is wound into the stator. These power sources are also referred to as *revolving* or *rotating field power sources.*

The alternating-current voltage produced by the armature coils moving through the magnetic field of the stator is carried to copper commutator bars through electrical conductors from the armature coils. The conductors are soft-soldered to the individual commutator bars. The latter may be considered as terminals, or collector bars, for the alternating current generated from the armature.

The commutator is a system of copper bars mounted on the rotor shaft. Each copper bar has a machined and polished top surface. Contact brushes ride on the top surface to pick up each half-cycle of the generated alternating current. The purpose of the commutator is to carry both half-cycles of the generated alternating current, but on separate copper commutator bars. Each of the copper commutator bars is insulated from the other copper bars.

The carbon contact brushes pick up each half-cycle of generated alternating current and direct it into a conductor as direct current. It may be said that the brush-commutator arrangement is a type of mechanical rectifier, since it changes the generated alternating current to direct current. Most of the brushes are an alloy of carbon, graphite, and small copper flakes. The direct-current characteristics are similar to those of single- and three-phase transformer-rectifier power sources.

Alternator construction (shown in Figure 1.23) places the heavy conductors in the stator, eliminating the need for carbon brushes and a commutator to carry high current. The output, however, is alternating current, which requires external rectification for direct-current application. Rectification is usually accomplished with a bridge using silicon diodes. An alternator usually has brushes and slip rings (shown in Figure 1.25) to provide the low direct current to the field coils. Both single- and three-phase alternators are available to supply alternating current to the necessary rectifier system.

An alternator or generator may be self-excited or separately excited depending on the source of the field power. Both may use a small auxiliary alternator or generator, with the rotor on the same shaft as the main rotor, to provide exciting power. In many engine-driven units, a portion of exciter field power is available to operate tools or lights necessary to the welding operation. In generators, this auxiliary power is usually 115-V direct current. With an alternator power source, 120-V or 120/240-V alternating current is usually available. Voltage frequency depends on the engine speed.

Output Characteristics

Both generator and alternator power sources generally provide welding current adjustment in broad steps called *ranges.* A rheostat or other control is usually placed in the field circuit to adjust the internal magnetic field strength for fine adjustment of power output. The fine adjustment, because it regulates the strength of the magnetic field, typically changes the open-circuit voltage. When adjusted near the bottom of the range, the open-circuit voltage can be substantially lower than at the high end of the range.

Figure 1.31 shows a "family" of volt-ampere curve characteristics for either an alternator or generator power source.

With many alternator power sources, broad ranges are obtained from taps on a reactor in the ac portion of the circuit. As such, the dynamic response of an alternator is limited for shielded metal arc welding, and thus a suitable inductor is generally inserted in series with one leg of the dc output from the rectifier. Direct-current welding typically requires an inductor.

A limited overlap of ranges is normally provided by rotating equipment so that the desired welding current can be obtained over a range of open-circuit voltages. If welding is performed in this range, welders have a better opportunity to adapt the power source to the job. With lower open-circuit voltage, the slope of the curve is less. This allows the welder to regulate the welding current to some degree by varying the arc length. This can assist in weld-pool control, particularly for out-of-position work.

Some dc welding generators carry this feature beyond the basic design described above. Generators that are compound-wound with separate current and voltage controls can provide the operator with a selection of volt-ampere curves at nearly any amperage, as shown in Figure 1.31. Thus, the welder can set the desired arc voltage with one control and the arc current with another. This adjusts the generator power source to provide a static volt-ampere characteristic that can be customized to the job throughout most of its range. The ranges of volt-ampere curves available as each control is changed independently are shown in Figures 1.32 and 1.33.

Welding power sources are available that produce both constant current and constant voltage. These units are used for field applications where both constant current and constant voltage are needed at the job site and utility electric power is not available. Many designs use solid-state circuitry to expand a variety of volt-ampere characteristics.

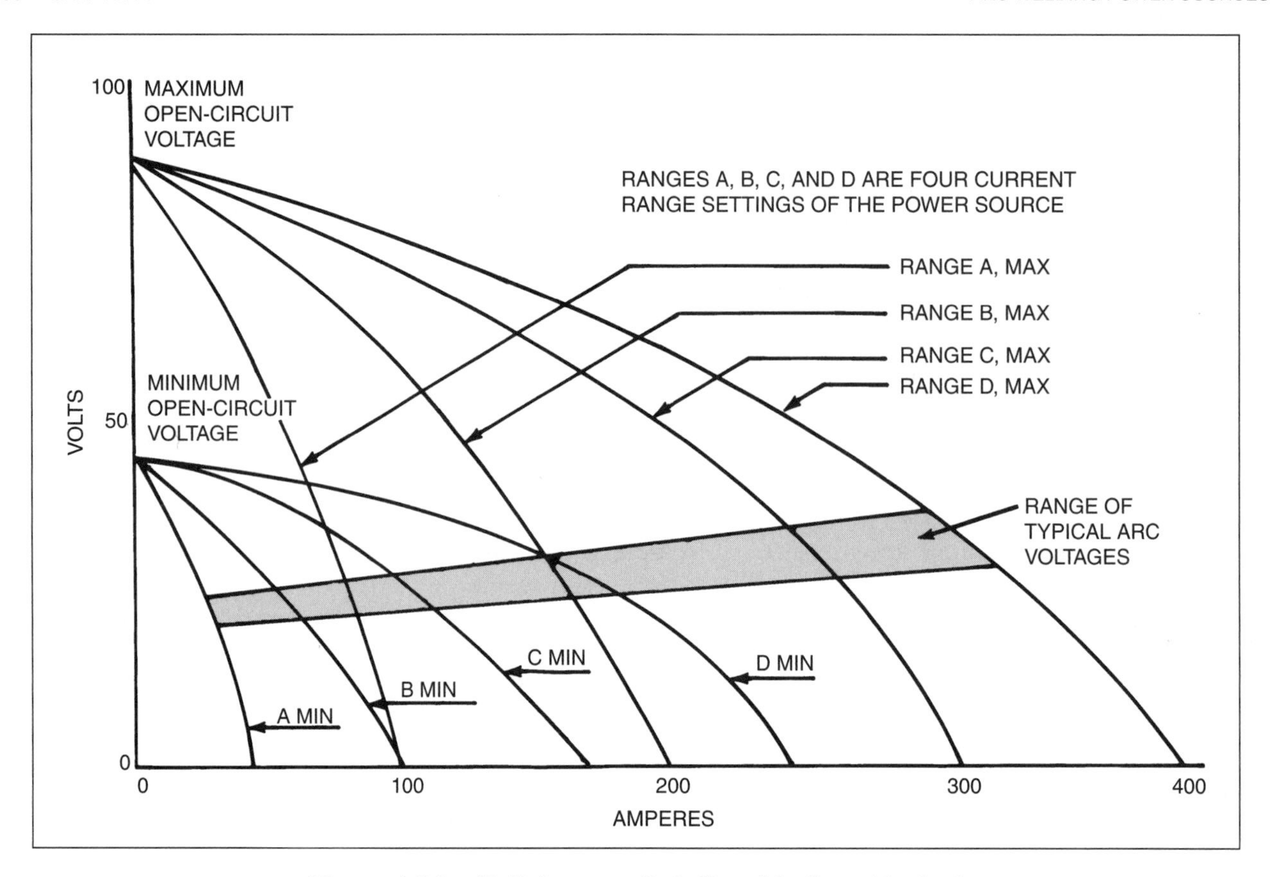

Figure 1.31— Volt-Ampere Relationship for a Typical Constant-Current Alternator or Generator Power Source

Sources of Mechanical Power

Generators are available with alternating-current drive motors of several voltage and frequency ratings and with direct-current motors. Welding generators are usually single units with the drive motor and generator assembled on the same shaft.

Induction motor-driven welding generators are normally available for 200-V, 240-V, 480-V, and 600-V three-phase, 60-Hz input. Other standard input requirements are 220 V, 380 V, and 440 V with 50-Hz input. Few are made with single-phase motors, since transformer welding power sources usually fill the need for single-phase operation. The most commonly used driving motor is the 230/460-V, three-phase, 60-Hz induction motor.

Figure 1.34 summarizes some of the electrical characteristics of a typical 230/460-V, three-phase, 60-Hz induction motor-generator set—overall efficiency, power factor, and current input. The motors of direct-current welding generators usually have a good power factor (80% to 90%) when under load, and from 30% to 40% lagging power factor at no-load. No-load power input can range from 2 kW to 5 kW, depending on the rating of the motor-generator set. The power factor of induction motor-driven welding generators may be improved by the use of static capacitors similar to those used on welding transformers. Some welding generators are built with synchronous motor drives to correct the low power factor.

Rotating-type power sources are used for field or on-site welding when utility-supplied electric power is not available. For this use, a wide variety of internal combustion engines are available. Both liquid-cooled and air-cooled engines are used depending on specific power source applications. In the United States, gasoline is the most popular fuel because of price and availability. Diesel fuel can lower operating costs and usage is therefore increased. However, local, state, and federal regulations should be consulted, as some codes permit the use of diesel fuel only for engines used in specific applications. An example is the use of diesel engines for welding power sources on offshore drilling rigs.

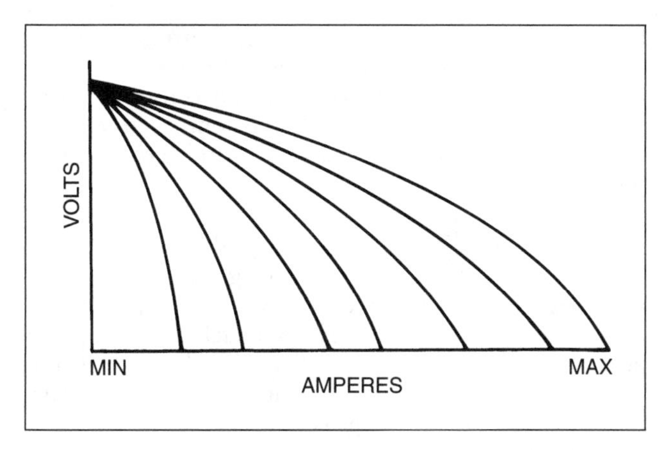

Figure 1.32—Effect of Current-Control Variations on Generator Output

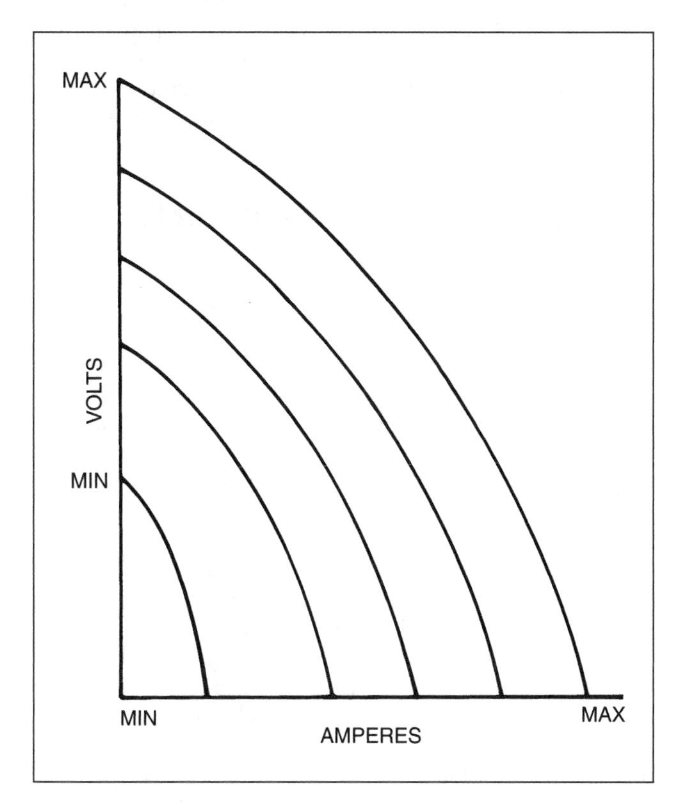

Figure 1.33—Effect of Voltage-Control Variations on Generator Output

Diesel fuel is more popular than gasoline in Western Europe and the Far East, but globally the consumption statistics of diesel and gasoline are about equal. In general, larger engine drives are designed for diesel fuel and smaller machines are designed for gasoline.

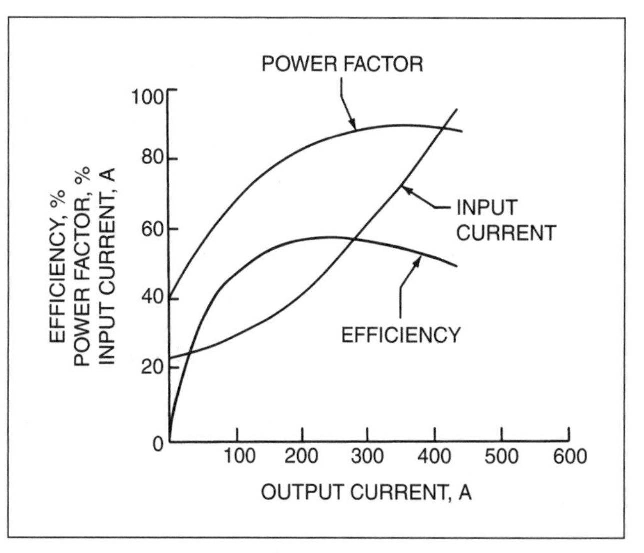

Figure 1.34—Typical Characteristic Curves of a 300-Ampere Direct-Current Motor-Generator Power Source

Propane is used in some applications because it burns cleaner than gasoline. However, it requires a special carburetion system.

Parallel Operation

Increased current output can be obtained by connecting engine-driven power sources in parallel. However, parallel connection is not advised unless the manufacturer's specific instructions are followed. Caution is necessary because successful parallel connections depend on matching the output voltage-ampere characteristic and polarity of each power source. In the case of self-excited generators, the problem is further complicated by the necessity to equalize the excitation between the generators.

The blocking nature of the rectifiers makes direct-current alternators easy to operate in parallel. Care should be taken to ensure that connections are of the same polarity. Rotating-type power sources operated in parallel require close operator attention to adjustment so that the welding load is equally shared.

Auxiliary Features

Rotating-type power sources are available with many auxiliary features such as attachments for hand or foot controls that the operator takes to the workstation to make power source adjustments while welding.

Gas engines are often equipped with idling devices to save fuel. These devices are automatic in that the engine will run at an idle speed until the electrode is touched to the work. Under idle, the open-circuit voltage of the generator is low. Touching the electrode to the workpiece energizes a sensing circuit that automatically accelerates the engine to operating speed. When the arc is broken, the engine will return to idle after a set time.

Engine-driven generators and alternators are often equipped with a provision for auxiliary electric power. Auxiliary power (typically 120 V to 240 V ac) is available when the engine is operating at rated speed and may be limited when welding power is used. Other auxiliary features that can often be obtained on engine-driven welding power sources are polarity switches (to change easily from DCEN to DCEP), running-hour meters, fuel gauges, battery chargers, high-frequency arc starters, tachometers, and output meters. Some larger engine-driven units are equipped with air compressors.

PULSED AND SYNERGIC POWER SOURCES

Pulsed-current power sources are used for GMAW, GTAW, SMAW, FCAW, and SAW. Of these processes, GMAW and GTAW are the most commonly used. Early versions of pulsed-current power sources appeared on the market in the 1970s.

Pulsed-Spray Transfer Power Sources

Pulsed-mode welding power sources are used with the GMAW process to reduce heat input, decrease workpiece distortion, and minimize fumes and spatter. The pulse spray transfer process is an advantage when welding thin-gauge and non-ferrous metals, such as aluminum and nickel alloys, and when performing all-position welding using projected-drop spray transfer. In a shielding gas environment in which argon is predominant, ejection of liquid metal from the tip of the electrode can be achieved when the peak instantaneous current exceeds a critical level, called the *transition current*. Instantaneously raising the current above the transition current causes the liquid metal to be propelled across the arc. This peak current period is then followed by a lower current period. It is thereby possible to obtain the desirable qualities of spray transfer while reducing the average current significantly, allowing the GMAW process to be used in all positions and for welding sheet metal.

The pulsed current level during the low interval is kept sufficiently low to prevent any metal transfer but high enough to prevent arc loss. Peak current is raised above the transition current for a sufficient time to allow at least one drop to form and transfer. Power sources have been designed with controlled pulsed-current outputs for pulsed gas metal arc welding (GMAW-P).

Independent settings can be made for all aspects of the pulse wave form, including peak current, background current, peak time, and background time. These variables are shown in Figure 1.35. The slope-up rate defines how fast current rises from background level to peak level, in amperes per millisecond. Because of the inductive component in the circuit, current decays exponentially from the peak current to an optional intermediate level called step-off current, at slope-down speed (or exponent), before it reaches the background current level.

By designing a wave form for specific welding conditions, the objectives of flat bead shape, penetration profile, high travel speed, low spatter level, and low tip wear can be optimized. Conditions may include the type of gas, wire type and size, workpiece thickness, surface mill scale, joint type, fitup, and even cable length.

Synergic Adjustments and Controls

Pulse welding involves controlling more parameters than constant-voltage spray or constant voltage short-circuiting processes. Instead of requiring the user to adjust numerous wave form control parameters, manufacturers preprogram the wave forms for commonly used welding wires and gases. The operator controls usually consist of two knobs—one to adjust wire feed speed (or current) and the other to refine or trim the voltage adjustment. The user simply turns the wire speed knob to pick one of the preprogrammed wave forms, hence the term *one-knob control*. When the wire-speed knob is turned, an entire new wave form is selected along with all the parameters defining the wave form, hence the term *synergic*. The preprogrammed wave forms assume a desired arc length (or arc voltage) for a particular wire speed, gas, stickout, travel speed, and cable length. A second knob is provided to allow the user to change the preprogrammed arc voltage, which basically trims the arc voltage up and down around the preprogrammed set point. A diagram of typical circuit elements for this type of power source is shown in Figure 1.36.

The top section of Figure 1.36 is a converter, although inverter circuitry is also commonly used. The bottom section of the control circuit can be implemented by discrete analog and logic electronics, but microprocessor designs are more often used. The wave form generator selects preprogrammed wave form parameters by wire-speed setting. At higher wire speed, pulse frequency is increased to maintain one-droplet-

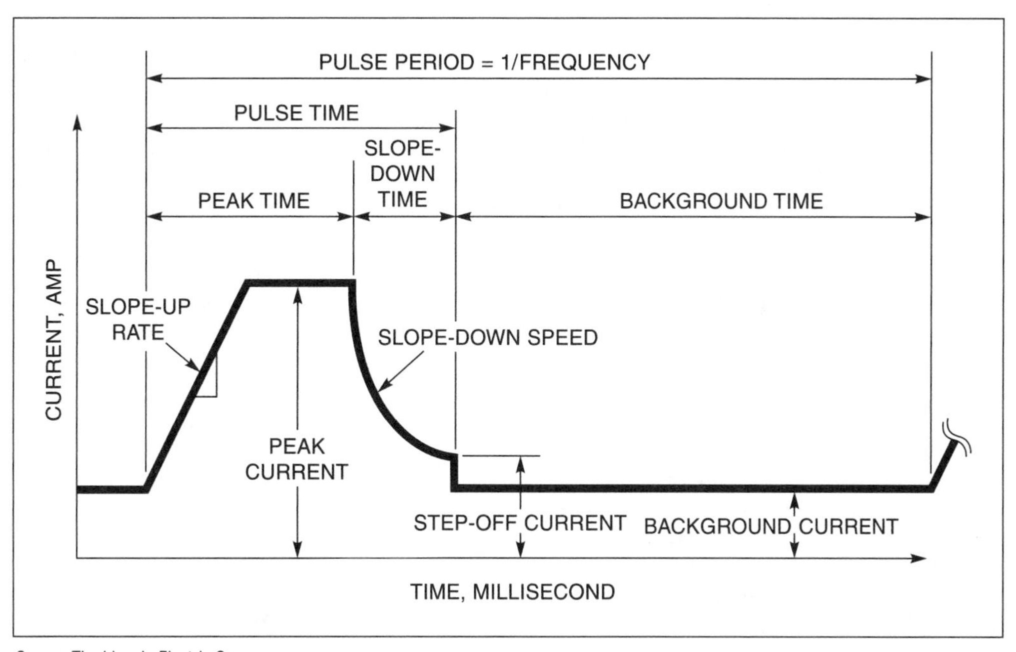

Source: The Lincoln Electric Company

Figure 1.35—Pulsed-Spray Metal Transfer Wave Form Variables

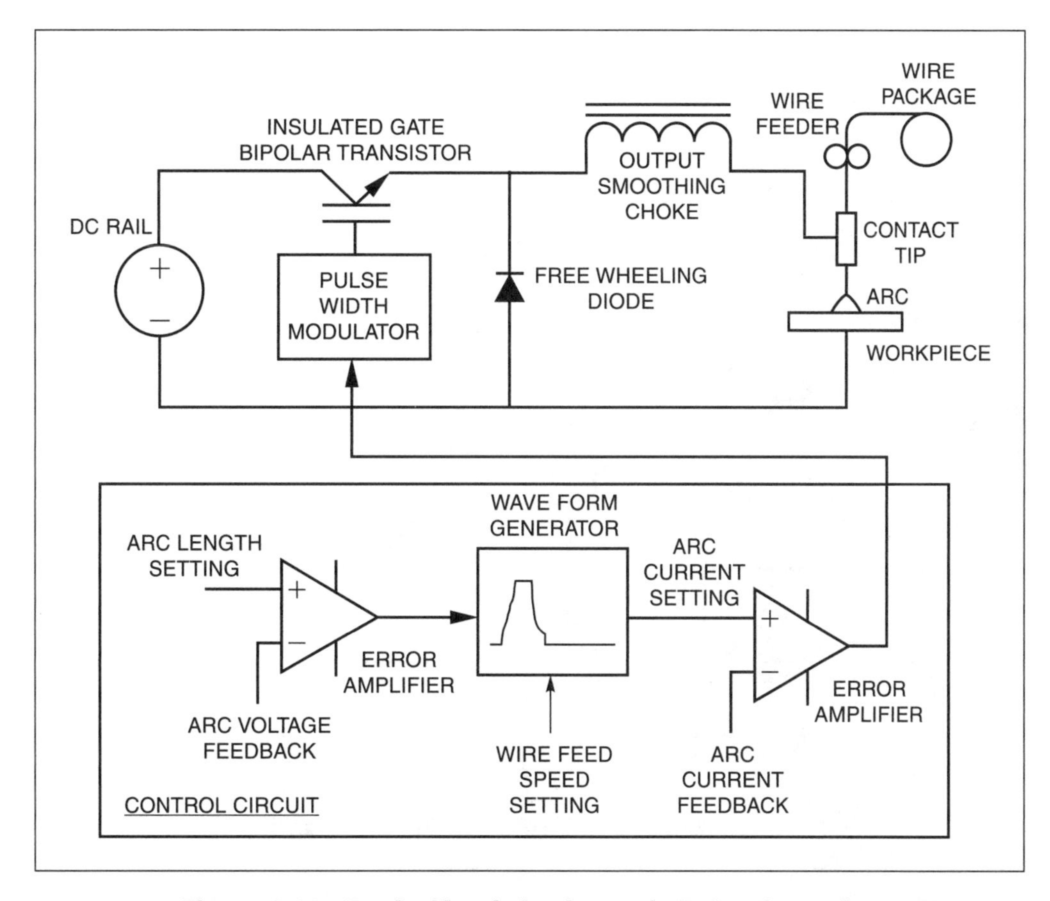

Figure 1.36—Basic Circuit for Synergic Pulse Operation

per-pulse, which proves to be a stable droplet transfer. The control regulates arc voltage by adjusting at least one aspect of the selected current wave form to adapt to contact-tip-to-work distance variations. The strategies of designing the wave form for a given wire speed and adjusting the wave form to control arc length are highly specific to the application and vary with different power source manufacturers.

Short Circuiting Gas Metal Arc Welding Power Sources

When pulsed gas metal arc welding (GMAW-P) is performed using a controlled short-circuiting transfer process, the wave form adapts to the physics of the welding arc and metal transfer, as illustrated in Figure 1.37. It reduces spatter by reducing the pinch force (current) when the liquid-metal bridge of the short is about to break and establish an arc. Without the energy of a high current, the liquid bridge is then broken by surface tension and fluid-mechanical inertia.

Figure 1.37 shows that a background current between 50 A and 100 A typically maintains the arc and contributes to workpiece heating. After the electrode initially shorts to the weld pool, the current is quickly reduced to below background current to ensure that the molten ball at the end of the electrode wets into the weld pool. A pinch current then establishes a magnetic field, which squeezes the molten metal down into the pool, while the necking of the liquid bridge is monitored from electrical signals. When the liquid bridge is about to break, the power source reacts by reducing current before the arc is re-established. Following the arc establishment, a peak current is applied to produce a plasma force pushing down on the weld pool to prevent accidental shorts to the melt of the wire while forming the next ball for transfer to the weld pool. Finally, the slope-out and background currents are adjusted to regulate input heat and fusion.

The controlled short-circuit process has advantages when welding thin-gauge materials, open root passes on pipe, single-sided heavy plate, and thin sheet metallic lining (called *wallpapering*). Notable advantages of the process are the following:

1. Less skill is required of the operator to control fusion when stickout is varied;

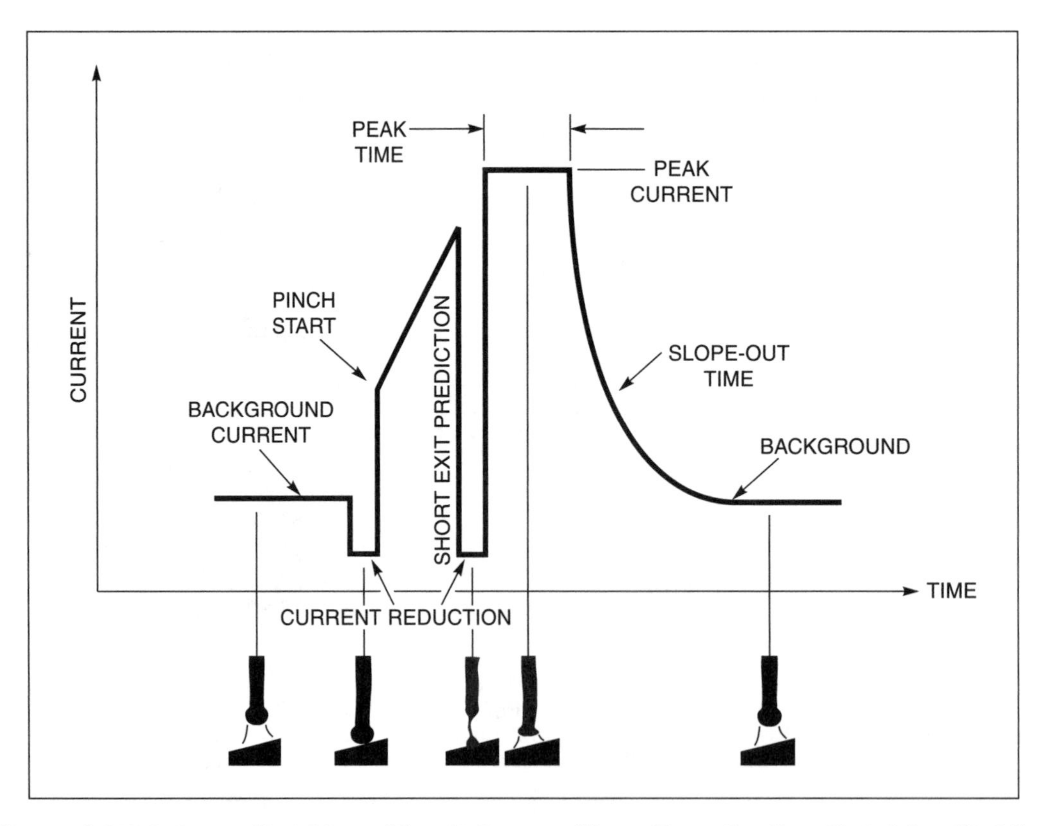

Figure 1.37–A Controlled Short-Circuit Current Wave Form for Gas Metal Arc Welding

2. Excellent backbead profile can be achieved without backing material;
3. Precisely controlled input heat helps to control joint distortion;
4. Fewer fumes are generated; and
5. Spatter is reduced, resulting in less cleanup and savings in labor cost.

Pulsed Gas Tungsten Arc Power Sources

The purpose of pulsing is to alternately heat and cool the molten weld metal. The heating cycle (peak current) is based on achieving a suitable weld pool size during the peak pulse without excessive groove face fusion or melt-through, depending on the joint being welded. The background current and duration are based on achieving the desired rate of cooling of the weld pool. The purpose of the cooling (background-current) portion of the cycle is to speed up the solidification rate and reduce the size of the weld pool without interrupting the arc. Thus, pulsing allows alternately increasing and decreasing the size of the weld pool.

The fluctuation in weld pool size and penetration is related to the pulsing variables: travel speed; the type, thickness, and mass of the base metal; filler metal size; and welding position. Because the size of the weld pool is partially controlled by the current pulsing action, the need for arc manipulation to control the molten metal is reduced or eliminated. Thus, pulsed current is a useful tool for manual out-of-position gas tungsten arc welding.

Pulsed current has also been applied with considerable effectiveness in the gas tungsten arc welding process (GTAW-P). The GTAW process frequency differs from that of the GMAW process and generally ranges from 2 seconds per pulse to 10 pulses per second, with the lower frequencies most commonly used. The complexity of control required for pulsing makes it advantageous to provide other appropriate control adjustments such as starting circuits and controlled current-time slopes. A diagram showing the current wave form possibilities when programming a GTAW power source is illustrated in Figure 1.38.

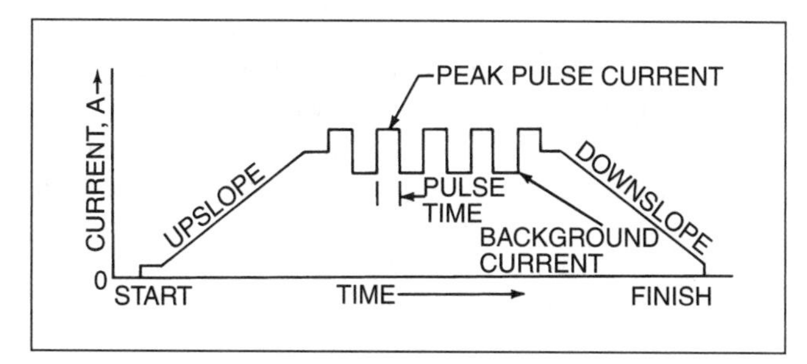

Figure 1.38—Typical Pulsed Gas Tungsten Arc Welding Program Showing Upslope and Downslope

Pulsed gas tungsten arc welding is characterized by a repetitive variation in arc current from a background (low) value to a peak value. Both peak and background current levels are adjustable over a wide range. The peak current time added to the background current time is equal to the duration of one pulse cycle. The peak current duration and background current duration are independently adjustable.

MULTIPLE-OPERATOR POWER SOURCES

Multiple-operator welding equipment is economical for shops that have a number of welding stations in a small area. This equipment has also been used to advantage in shipbuilding, in water tower erection, on oil platforms, at construction sites, and in welding training facilities.

Multiple-operator equipment is commercially available in three configurations: a single power source with multiple remote resistor or electronic control modules, a multiple output power source, and the single rack mounting of multiple conventional or inverter power sources. A brief description of each type of multiple-operator configuration follows, with a list of equipment advantages.

Multiple Modules

The grid resistor module and electronic module configurations are frequently used in shipbuilding and training facilities. One large-capacity constant-voltage power source can be connected with multiple weld cables to as many as eight to twelve operator stations. When a grid resistor module is connected to the work end of the weld cable, the operator station becomes a constant-current power source for shielded metal arc welding. Depending on the type of electronic modules connected to the weld cables, SMAW or GMAW is possible at the individual operator stations. All operator stations share a common workpiece connection. Since the 5% to 20% duty cycle of the SMAW process is factored into sizing the equipment, the cost of equipment for eight to twelve operator stations is significantly less than eight to twelve individual power sources.

Multiple Control Panels

A single, large-capacity constant voltage power source with six or eight control panels, each individually connected by weld cable to an operator station, is commonly used on construction sites, for example, in water tower erection. Control panels are available with constant-current, constant-voltage, and constant-current-constant-voltage outputs. All operator stations can share a common workpiece connection if the work site conditions allow. Equipment configurations will permit SMAW, GMAW, and GTAW processes on a single power source with appropriately selected control panels. Since the welding duty cycles increase for GMAW, fewer arcs or operator stations are typically included in this configuration. The equipment cost of control panel configurations for six or eight operator stations is less than the cost of six or eight individual power sources.

Multiple-Operator Facilities

The third multiple-operator configuration typically consists of from four to six conventional or inverter power sources mounted in a common rack or enclosure. Each power source has an individual electrode cable, and multiple power sources can share a common workpiece connection cable. Job site relocation is more convenient when one electric utility line and only one piece of equipment are involved. A rack of inverter power sources with SMAW, GMAW, GTAW, and pulsed/synergic processes can be very effective when applied to complex work-site demands. Capital equipment costs are typically higher for this configuration because the rack is required, but the higher welding duty cycles of the individual power sources are also available and may compensate for the additional cost.

Advantages of Multiple-Operator Equipment. Capital equipment selection and expenditures often present a trade-off when low (5% to 25%) duty cycle welding and frequent workstation moves are construction site requirements. Some advantages of multiple-operator equipment are the following:

1. One power source location provides welding current from multiple cables to multiple weld stations located in one general area;
2. Utility power electric service is only required at a single connection or drop point for all weld stations, which minimizes reconnect time and effort;
3. Only one power source (or rack) and the utility power service must be relocated on large construction sites or shipbuilding projects;
4. Capital equipment costs are reduced, except for the rack-mounted equipment, because low duty cycle welding allows an increase in the number of operator stations per power source; and
5. Parallel connections of multiple-operator modules, panels, or power sources permits increased output currents.

ECONOMICS

The economics of welding processes have cost factors similar to many manufacturing operations. The four major economic factors to be considered in choosing a welding process are the following:

1. Equipment cost required by the selected welding process,
2. Energy costs for operating the welding process,
3. Labor costs to staff the welding process, and
4. Material costs of metals, gases, and consumables used in the process.

The selection of the welding process is the single most important factor in determining the cost of a welding operation. The chapters in this volume on shielded metal arc welding, gas metal arc welding, gas tungsten arc welding, plasma arc cutting and gouging, and arc cutting and gouging provide an overview of welding processes and guidance in selecting the appropriate process and the best methods of preparing and joining metals.

Once a welding process has been determined, the selection of equipment and the required type of power become identifiable costs. This section focuses on the economic factors to be considered when selecting a welding power source and the type of energy required for operation.

CAPITAL INVESTMENT

The capital investment for a welding power source is determined by the type of welding process it is capable of performing, the rated output current, and the type of power or fuel used. Power sources for shielded metal arc welding are generally less expensive than those for gas metal arc welding. A gas tungsten arc welding power source tends to be more expensive, and a multiple-process power source, a unit that provides for all three of these processes, is most expensive.

The cost of a power source increases as the output current increases; for example, an 85-A GMAW power source is considerably less expensive than a 450-A unit. Generally, a power source that connects to utility-line power is less expensive than an engine-driven power source of the same output amperage. Within these classes, a single-phase power source is less expensive than a three-phase unit; a 60-Hz power source is less expensive than a 50 Hz unit; and a gasoline engine is less expensive than a diesel power source.

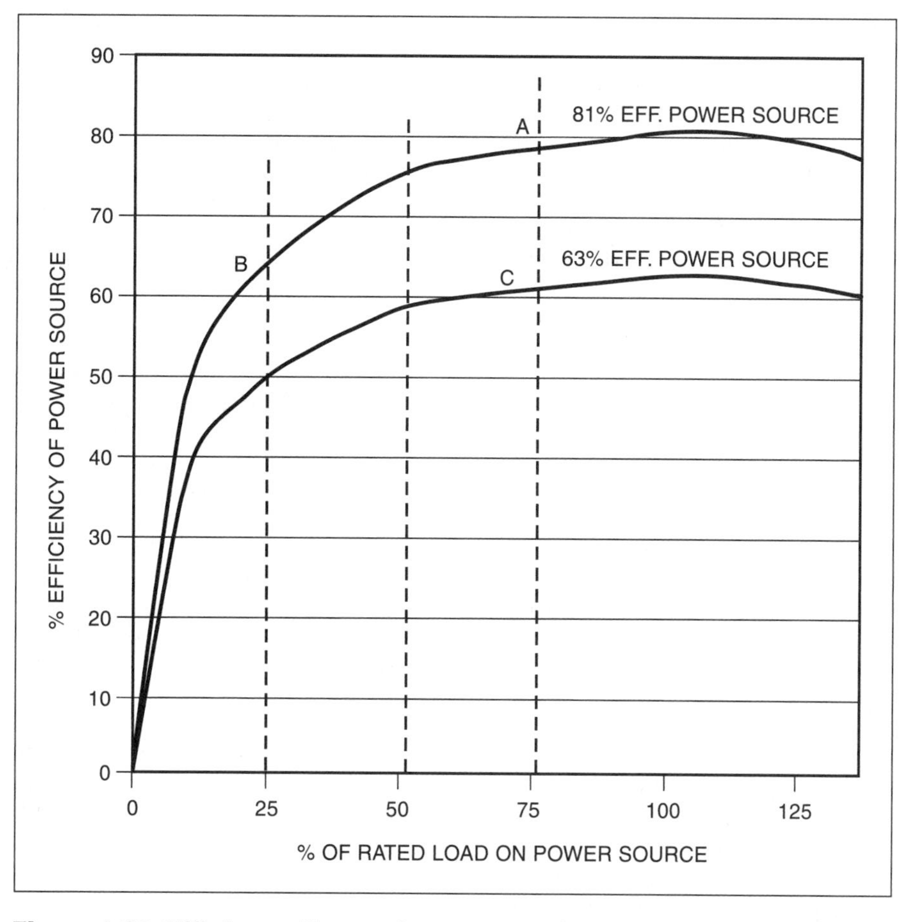

Figure 1.39–Efficiency Versus Output Load for Welding Power Sources

OPERATING EFFICIENCY

When specifying or purchasing a welding power source, the effect of operating efficiency on utility power or fuel costs should be considered. Figure 1.39 shows a comparison of two power sources, one with 81% efficiency and the second with 62% efficiency at rated output current. Note that when operated at 75% of load, the power source at Point C requires 28% more utility power or fuel over its operating life than the higher-efficiency power source operating at Point A. The impact of reducing fuel costs by 28% is significant, particularly when high fuel and fuel transport costs are involved. Older welding power sources tend to have lower efficiencies. High-efficiency power sources are readily available, and frequently a new capital equipment purchase can be justified when lower operating costs will return the purchase price in three years or less.

Oversized welding power sources are frequently purchased for an immediate small job for a number of justifiable reasons. A larger power source allows bigger jobs to be done in the future, and using a power source at lower currents may promote longer equipment life. However, the continuous use of a power source at only a fraction of the rated output current requires more electric power or fuel, and therefore costs more to operate. (For example, Figure 1.39, which charts a power source capable of operating at 81% efficiency, at 100% rated load. If this power source operates at Point A or 75% of rated current, the efficiency still remains high at 78%; but operation at Point B or 25% of rated current yields a much lower efficiency of 65%). Choosing a lower-cost, appropriately sized power source in this case also reduces operating costs. A higher-capacity power source also requires larger and more expensive electric services.

Low Power Factors

The cost of using power sources with low power factors (pf) should be evaluated at the time of purchase, because the cost of installing higher-capacity electric utility service to the building in which they are to be used is a consideration. The kVA rating or electric service entrance size is increased by the effect of low PF welding power sources. The service entrance size is determined by the total kilowatts of the connected power sources divided by the average pf, as follows;

$$kVA = \frac{kW}{PF} \quad (1.15)$$

Higher power factors are in the 0.8 to 1.0 (sometimes expressed as 80% to 100%) range, while lower power factors are 0.7 or less. Utility power companies may also charge a higher rate per kilowatt hour for the power used by customers with low-power-factor equipment.

Power and Fuel Costs

Utility power and fuel costs become more significant as the cost of energy increases. Electricity generated and distributed by large centralized power utilities are typically the least expensive sources of welding power. Small gasoline-driven welding power sources operating in remote areas and requiring fuel transportation over long distances are usually the most expensive to operate. Business conditions, such as the ability to rapidly move welding equipment for emergency repairs, may justify the specification of higher-priced power and equipment even though lower cost alternatives exist. Generally, welding power from gasoline and diesel engines is more expensive because of the additional energy conversions required. Gasoline and diesel fuel first must be converted to a rotating motion, which is then used to generate electricity, and finally must be converted to welding current. Calculations of the cost of power at local rates, whether it is electric utility power, diesel, or gasoline, should be a factor in equipment purchasing decisions.

Power costs are typically determined by geographical location and local policy. Therefore, a power source operating at idle or no-load may be more expensive to operate in some countries than the same power source when welding at the rated output current in a different geographic location.

SPECIAL EQUIPMENT FEATURES

Special equipment features that reduce operating costs should be considered when purchasing welding power sources. For example, when cooling is not required, some power sources automatically turn off the cooling fan, which reduces energy consumption and requires less cleaning to remove dust and dirt accumulations from the enclosure. When the output current is not present for a selected length of time, some power sources can be set to turn off automatically. Engine-driven power sources are also available that have an automatic return-to-idle speed feature when no arc conditions have existed for a period of time.

SAFE PRACTICES

Although electricity is a phenomenal force that has facilitated the extraordinary progress of industry, commerce, and lifestyle, it can be a destructive force that must be recognized and respected. The hazards of electric shock and fire are always present. Everyone who works with electricity, welding power sources, and the machines they drive must know how to use electricity and electric equipment safely.

Only qualified persons with knowledge and understanding of the principles of electricity and certified training in safe practices should test, repair, or perform maintenance on electrical equipment. They must read the manufacturers' manuals before working on power sources or accessory equipment and carefully follow all instructions. National, state, and local codes must be followed. The standard *Safety in Welding, Cutting, and Allied Processes*, ANSI Z.1-49, should be consulted.[12] Appendix B of this volume, "Safety and Health Codes and Other Standards" lists health and safety standards, codes, specifications, and other publications used in the welding industry. Workers should be trained in safe practices and should be proficient in cardiopulmonary resuscitation (CPR).

This section provides an overview of electrical safety hazards and preventive measures that must be followed when working with power sources and associated welding equipment.[13]

ELECTRIC SHOCK

Electrical shock can occur when the body becomes part of the electric circuit, either when an individual comes in contact with both wires of an electrical circuit, with one wire of an energized circuit and the ground, or

12. American National Standards Institute (ANSI) Accredited Standards Committee Z49, 1999, *Safety in Welding, Cutting, and Allied Processes*, ANSI Z49.1:1999, Miami: American Welding Society.
13. Adapted with permission from Miller Electric Manufacturing Company, R. Jennings, M. Sherman, and Training Group, 2003, *Safety Quick-Guide*, Appleton, Wis.: Miller Electric Manufacturing Company.

with a metallic part that has become energized by contact with an electrical conductor.

The severity and effects of electrical shock depend on factors such as the path of the current through the body, the amount of current, the length of time of the exposure, and whether the skin is wet or dry. Water readily conducts electricity, which flows more easily in wet conditions and through wet skin. The effect of shock may range from a slight tingle to severe burns and cardiac arrest. Table 1.6 shows body responses of an electric shock victim to various amounts of current. Values are in milliamperes (mA), where 1,000 mA equal 1 A. A typical welding process uses from 50 A to 300 A.

Table 1.6
Body Responses to Electric Shock

0.5–3 mA	Start to feel the energy, tingling sensation
3–10 mA	Experience pain, muscle contraction
10–40 mA	Grip paralysis threshold (brain says let go, but physically unable to do so)
30–75 mA	Respiratory system shuts down
100–200 mA	Experience heart fibrillation
200–500 mA	Heart clamps tight
Over 1,500 mA	Tissue and organs burn

Source: Adapted with permission from Electrical Safety Foundation International (ESFI), Rosslyn, Virginia.

GENERAL PRECAUTIONS

As used in this section, the word *technician* refers to all persons working with or near electrical equipment, including welders, fitters, electricians, repair technicians, engineers, supervisors, and helpers. Technicians should wear safety glasses with side shields. They should wear appropriate clothing, avoid loose-fitting or damp clothes, and remove jewelry. The technician should stand or work on a dry insulating mat large enough to provide insulation and prevent contact with the ground. Work should be performed carefully and precisely and should take place in an area where the technician will not be startled or distracted by other workers or visitors. An energized unit must never be left unattended. If a welding power source is to be moved, the supply line should be deenergized and the input power conductors disconnected before moving it.

Electric shock from the power source, wiring, or cables can be fatal. It is important for technicians to know the primary source and status of electrical power. During repair or replacement of parts, the power source should be turned off, the input power should be disconnected, and the supply circuit locked out and tagged out. A testing device, such as a meter or test lamp, should be used to verify that the circuit is deenergized. Engine-driven equipment should be turned off before work begins. Electrolytic capacitors should be discharged using a resistor, such as a 1000-ohm, 25-watt resistor.

If it is necessary to work on an energized unit, a good technique is to work with one hand (if it is safe to do so), keeping the other hand at the side or in a pocket to avoid contact with conductive material and to reduce the possibility of current passing through the chest cavity. If a person comes in contact with a live electrical conductor, a coworker or attendant should not touch the equipment, input power cable, or the person. The rescuer should immediately disconnect the input power by turning off the disconnect switch at the supply fuse box or circuit breaker panel and when applicable, pull out the plug. If necessary, cardiopulmonary resuscitation should be administered until emergency medical personnel arrive.

Preliminary Examination

When troubleshooting, the first steps are to disconnect the input power using the line disconnect switch or circuit breakers, or by removing the plug from the receptacle, and then to check for simple problems such as bare or shorted wiring, loose connectors, and corroded, burned, or pitted switch contacts.

Inverter power sources require special safety considerations, and recommendations in manufacturers' technical manuals must be followed. Failed parts or parts installed incorrectly can explode or cause other parts to explode when power is applied to inverters. The technician should always wear a face shield and long sleeves when servicing inverters.

Inspection items for both inverter and conventional power sources that should be performed prior to energizing the units for operation are the following:

1. Inspect primary power cables and input power cables for accessories, such as wire feeders and control systems.
2. Ensure that the wiring and the condition of the plug, cord, jumper links, and connections inside the unit are in good condition.
3. Check the internal safety ground circuit to be sure the input ground conductor is securely connected to the power source grounding terminal (see the input voltage/connection label, where applicable, or the equipment manual, or both).
4. Inspect all interior and exterior safety labels and replace damaged or unreadable labels.
5. Check hoses and cables, especially at connection points and stress areas of bending and flexing, and replace if cracked, damaged, or poorly spliced.

6. Clean the inside of the power source (vacuum or blow it out with compressed air), check for loose connections, and look for signs of overheating.
7. Clean the outside of the unit and check the condition and movement of knobs and switches.
8. Inspect the power source case for damage and check for possible affected parts inside.
9. Clean and tighten weld terminals.

These items should be also checked before repairing and replacing parts. The power must be off and electrolytic capacitors must be discharged before starting the inspection.

Testing

A digital multimeter (DMM) is commonly used for testing and troubleshooting welding power sources and associated equipment. The technician should read the meter manufacturer's technical manual and understand the proper use of DMMs to avoid possible injury before and during the taking of electrical measurements. The technician must avoid shorting metal tools, parts, or wires during testing and servicing.

The meter should be calibrated before use to assure correct readings. A DMM rated for the anticipated maximum voltage should be used. For example, to test 480 volts, the meter should be rated for 600 or 1000 volts to prevent injury or damage to the meter. The selector switch on the meter must be turned to the correct function and range. For example, the meter can become a hazard if an attempt is made to test voltage while the meter is set for current (amperes). For high-voltage circuits, the best practice is to turn the power off, connect the meter leads, and then turn the power on again. When using a meter that is not an auto-ranging model, the technician should start with the highest range when measuring unknown values and work down to prevent meter damage and injury.

When practical, the technician should perform "one-handed testing" by using at least a one-meter lead that has a self-retaining spring clip such as an alligator clip. Test leads must be in good condition; they should be replaced if the insulation is frayed or broken. The test leads must be properly connected to the meter.

Lockout and Tagout Procedures

Sometimes work must be performed on equipment and machinery that may contain electrical energy or other hazards. Contact with these hazards may result in injury or death. The term *lockout* means to install a locking device that keeps the switch or other mechanism from being turned on. *Tagout* means to put a tag on the locking device. The tag indicates "danger" or "warning" and carries a brief descriptive message. The tag has a space to put the date and the name of the person who locked out the equipment so that he or she may be easily found or notified. The job supervisor should be informed about the proposed work, and permission should be obtained to lockout and tagout the equipment. Employees should be trained in the purpose and procedures of the lockout and tagout modes and informed when it is taking place.

The equipment is shut down and locks and tags are placed on the switches and valves to prevent their use. If more than one person is performing work on the equipment, it is recommended that they have their own locks and tags on the lockout point.

STATIC ELECTRICITY

A static charge is an imbalance of electrons that can build up on objects and then transfer to another object, such as a person's body, producing a static electric shock. A static shock may be more surprising than painful. However, secondary injuries can occur, for example, when a person rapidly pulls his or her hand away from a static source and hits a hand or elbow against a wall, a fan, or another object. It should be noted that static electricity can produce a spark that could ignite flammable materials or gases in the work area.

Static electricity can cause costly damage to static-sensitive electrical parts and items such as printed-circuit boards, metal oxide semiconductor field effect transistors (MOFSETs), insulated gate bipolar transistors (IGBTs), erasable programmable read-only memory (E-Proms), complementary metal-oxide semiconductor (CMOS) devices, operational amplifiers (Op Amps), transistors, microprocessors, and inverter power modules. To prevent damage to the equipment, the technician should wear a grounded wrist strap. The wrist strap must make good contact with the skin and the clip end should be grounded to an earth ground, not a machine frame. The technician should refer to the manufacturer's equipment manual for guidance in when to wear the wrist strap. The wrist strap must **never** be worn with the equipment power turned on.

Static-sensitive items should be stored, moved, or transported in static-shielding bags or packages.

GENERAL RECOMMENDATIONS

In addition to observing the precautions for the safe use of electricity, workers must also follow the recommendations for safety in welding, cutting, and related processes. The hazards include electric shock, fumes and gases, radiation, noise, fire and explosion, mechanical hazards, falls, and falling objects. Special precautions must be followed by persons who wear pacemakers and those who wear contact lenses. Safety standards commonly used in the welding industry are listed in Appendix B.

Precautionary labels are often used with graphic symbols to supplement written messages with visual images. These symbols provide quick interpretation of possible hazards. Standardized symbols have been endorsed by the National Electric Manufacturing Association,[14] the American Welding Society, the American National Standards Institute, the FMC Corporation, and the International Organization for Standardization. Contact information for these organizations and others is provided in Appendix B of this volume, "Safety and Health Codes and Other Standards."

Examples of precautionary symbols from the National Electrical Manufacturers Association document *Guidelines for Precautionary Labeling for Arc Welding and Cutting Products*, NEMA EW6,[15] and endorsed by the American Welding Society are shown in Table 1.7.

14. National Electrical Manufacturers' Association (NEMA) *Guidelines for Precautionary Labeling for Arc Welding and Cutting Products*, Arc Welding Section, NEMA EW6, Washington, D.C.: National Electrical Manufacturers Association.

15. American Welding Society (AWS) Safety and Health Committee, *Safety and Health Fact Sheets*, Miami: American Welding Society, Fact Sheet 14.

Table 1.7
Sources of Hazards and Standard Symbols

HAZARD	SOURCE OF HAZARD	SYMBOL	SOURCE
Electric Shock	Welding Electrode		ISO, FMC, NEMA
Electric Shock	Wiring		ISO, FMC
Electric Shock	Welding Electrode and Wiring		ISO, FMC, NEMA
Fumes and Gases	Any Source		FMC, NEMA
Fumes and Gases	Welding Fumes and Gases		ISO, FMC, NEMA
Arc Rays	Welding Arc		ISO, FMC, NEMA

(Continued)

Table 1.7 (Continued)
Sources of Hazards and Standard Symbols

HAZARD	SOURCE OF HAZARD	SYMBOL	SOURCE
Fire	Engine Fuel		FMC, NEMA
Fumes and Gases	Engine Exhaust		ISO, FMC, NEMA
Fumes and Gases	Engine Exhaust and Welding Arc		ISO, FMC, NEMA
Moving Parts Causing Bodily Injury	Moving Parts Such as Fans and Rotors		FMC, NEMA

CONCLUSION

This chapter has provided basic information on arc welding power sources intended to enhance the readers' understanding of the principles of the arc welding processes presented in the subsequent chapters of this volume. The electrical basics of power sources were introduced in the following areas:

1. Electrical terminology—*duty cycle, open-circuit voltage, inductance, ripple, slope*, and other terms necessary to the understanding of arc welding power sources;
2. Electrical components, such as transformers, inductors, silicon-controlled rectifiers, and transistors;
3. Arc-welding electrical circuits, including generators, alternators, single-phase controls, three-phase bridges, and inverters; and
4. Types of power sources such as constant-current versus constant-voltage; alternating current, direct current, and pulsed welding wave forms;
5. Rotating generators and static transformers; and
6. Magnetic silicon-controlled rectifier and inverter power source controls.

This information will help in the selection of a power source compatible with the selected arc welding, cutting, or allied process, including shielded metal arc welding, submerged arc welding, gas metal arc welding, gas tungsten arc welding, flux cored arc welding, plasma arc welding, arc cutting and gouging, and thermal plasma spraying.

BIBLIOGRAPHY[16]

American National Standards Institute (ANSI) Accredited Standards Committee Z49. 1999. *Safety in welding, cutting, and allied processes,* ANSI Z49.1:1999. Miami: American Welding Society.

American Welding Society (AWS) Safety and Health Committee. 1998. *Safety and health fact sheets.* Miami: American Welding Society.

Electrical Safety Foundation International (ESFI). www.electrical-safety.org. Rosslyn, Virginia: Electrical Safety Foundation International.

16. The dates of publication given for the codes and other standards listed here were current at the time this chapter was prepared. The reader is advised to consult the latest editions.

Miller Electric Manufacturing Company. R. Jennings and M. Sherman. 2003. *Safety quick-guide*. Appleton, Wisconsin: Miller Electric Manufacturing Company.

National Electrical Manufacturers Association (NEMA). 1988 (R1999). *Electric arc-welding power sources*. EW-1:1988 (R1999). Washington, D.C.: National Electrical Manufacturers Association.

National Electrical Manufacturers Association (NEMA). *Guidelines for precautionary labeling for arc welding and cutting products,* EW-6. Washington, D.C.: National Electrical Manufacturers Association.

SUPPLEMENTARY READING LIST

Amin, M. 1981. *Microcomputer control of synergic pulsed MIG welding*. Report 166/1981. Abington, Cambridge, United Kingdom: The Welding Institute.

Amin, M. 1986. Microcomputer control of synergic pulsed MIG welding. *Metal Construction* 18(4): 216–221.

Amin, M., and P. V. C. Watkins. 1977. *Synergic pulse MIG welding*. Report 46/1977. Abington, Cambridge, United Kingdom: The Welding Institute.

Amin, M. 1981. Synergic pulse MIG welding. *Metal Construction* 13(6): 349–353.

Bailey, K., and R. Richardson. 1983. *A microprocessor-based SCR type arc welding power supply.* Technical Report 529501-83-14. Columbus, Ohio: Center for Welding Research.

Brosilow, R. 1987. The new GMAW [gas metal arc welding] power sources. *Welding Design and Fabrication* 60(6): 22–28.

Correy, T. B., D. G. Atteridge, R. E. Page, and M. C. Wismer. 1986. Radio frequency-free arc starting in gas tungsten arc welding. *Welding Journal* 65(2): 33–41.

Cullison, A., and Montiel B. Newton. 1990. Changes are coming for welding power sources. *Welding Journal* 69(5): 37–42.

Frederick, J. E., R. A. Morgan, and L. F. Stringer. 1978. Solid-state remote controllable welding power supplies. *Welding Journal* 57(8): 32–39.

Ogasawara, T., T. Maruyama, T. Saito, M. Sato, and Y. Hida. 1987. A power source for gas shielded arc welding with new current wave forms. *Welding Journal* 66(3): 57–63.

Grist, F. J. 1975. Improved, lower cost aluminum welding with solid-state power sources. *Welding Journal* 54(5): 348–357.

Grist, F. J., and F. W. Armstrong. 1980. A new ac constant-potential power source for heavy plate, deep groove welding. *Welding Journal* 59(6): 30–35.

Hackman, R. and A. F. Manz. 1964. *DC welding power sources for gas shielded metal arc welding.* Welding Research Council, Bulletin 97. New York: Welding Research Council.

Kashima, T. Development of the inverter controlled dc TIG arc welding power source. Document IIW-XII-878-85. London: International Institute of Welding.

Kyselica, S. 1987. High-frequency reversing arc switch for plasma arc welding of aluminum. *Welding Journal* 66(1): 31–35.

Lesnewich, A. 1972. *MIG welding with pulsed power.* Bulletin 170. New York: Welding Research Council.

Lucas, W. 1982. *A review of recent advancements in arc welding power sources and welding processes in Japan.* Document 199/1982. Abington, Cambridge, United Kingdom: The Welding Institute.

Malinowski-Brodnicka, M., G. den Ouden, and W. J. P. Vink. 1990. Effect of electromagnetic stirring on GTA welds in austenitic stainless steel. *Welding Journal* 69(2): 52-s to 59-s.

Manz, A. F. 1973. *Welding power handbook*. New York: Union Carbide Corporation.

Needham, J. C. 1980. *Review of new designs of power sources for arc welding processes.* Document XII-F-217-80. London, England: International Institute of Welding.

Pierre, E. R. 1985. *Welding processes and power sources.* 3rd ed. Minneapolis: Burgess Publishing Company.

Rankin, T. 1990. New power source design breaks with tradition. *Welding Journal* 69(5): 30–34.

Schiedermayer, M. 1987. The inverter [welding] power source. *Welding Design and Fabrication* 60(6): 30–33.

Shira, C. 1985. Converter power supplies—more options for arc welding. *Welding Design and Fabrication* 58(6): 52–55.

Spicer, R. A., W. A. Baeslack III, and T. J. Kelly. 1990. Elemental effects on GTA spot weld penetration in cast alloy 718. *Welding Journal* 69(8): 285-s to 288-s.

Tomsic, M. J., S. E. Barhorst, and H. B. Cary. 1984. *Welding of aluminum with variable polarity power.* Document No. XII–839-84. London: International Institute of Welding.

Tomsic, M. and S. E. Barhorst. 1984. Keyhole plasma arc welding of aluminum with variable polarity power. *Welding Journal* 63(2): 25–32.

Villafuerte, J. C., and H. W. Kerr. 1990. Electromagnetic stirring and grain refinement in stainless steel GTA welds. *Welding Journal* 69(1): 1-s to 13-s.

Xiao, Y. H., and G. den Ouden. 1990. A study of GTA weld pool oscillation. *Welding Journal* 69(8): 289-s to 293-s.

CHAPTER 2

SHIELDED METAL ARC WELDING

Photograph courtesy of The Lincoln Electric Company

Prepared by the Welding Handbook Chapter Committee on Shielded Metal Arc Welding:

M. A. Amata, Chair
ESAB Welding & Cutting Products

Y. I. Adonyi
LeTourneau University, Welding and Materials Joining Program

T. E. Brothers
ESAB Welding & Cutting Products

S. R. Fiore
Edison Welding Institute

D. Hatfield
Tulsa Welding School

K. Y. Lee
The Lincoln Electric Company

J. A. Luck
Miller Electric Manufacturing Company

J. M. Rolnick
Consultant

Welding Handbook Volume 2 Committee Member:

D. B. Holliday
Northrop Grumman Marine Systems

Contents

CHAPTER 2

SHIELDED METAL ARC WELDING

INTRODUCTION

Shielded metal arc welding (SMAW) is a process that uses an arc between a covered electrode and a weld pool to accomplish the weld. As the welder steadily feeds the covered electrode into the weld pool, the decomposition of the covering evolves gases that shield the pool. The process is used without the application of pressure, and with filler metal from the covered electrode.[1, 2] The sound weld metal deposited by the process is used not only for joining, but also for applying a functional surface to metal products. In welding booths and shops, the linear metal rod with a covering is commonly referred to as a *stick* and the shielded metal arc welding process is popularly referred to as *stick electrode welding*.

Because of the many possible variations in the composition of the electrode covering and the large selection of core wire chemistry, the process can produce an extensive range of weld metal deposits with desirable mechanical and physical properties, while providing for a smooth arc, uniform metal transfer characteristics, and ease of operation. These features combine to make shielded metal arc welding the favorite process of many welders and fabricators. It is one of the oldest and simplest welding processes and continues to be widely used. The simplicity of the process extends to the number and nature of the circuit components required—a power source of adequate current rating and duty cycle, a SMAW electrode compatible with the output of the power source, suitably sized welding cable, an electrode holder, and a workpiece lead.

This chapter covers the principles of operation, introduces the techniques that are applicable to the making of a weld, explains some of the metallurgical dynamics of the shielded metal arc weld, and describes the equipment and materials used in the process. Because they are of essential importance to the process and are definitive to the success of the weld, covered electrodes are discussed in detail. For the convenience of the reader, the text is arranged for easy reference to specific classifications of electrodes.

Other topics discussed are welding variables, joint design and preparation, welding techniques, procedures, weld quality, economics and safe practices. References to pertinent standards and specifications are included throughout the chapter and listed in the bibliography at the end of the chapter.

FUNDAMENTALS OF THE PROCESS

The phrase *shielded metal arc welding* refers to a method of joining two pieces of metal or adding metal to an existing metal surface. Each word in the phrase conveys an attribute of the process: *shielded* refers to its ability to displace the air surrounding the weld to avoid the harmful effects of the gases in air; *metal* denotes the core of the electrode, which is a conducting rod that contributes a substantial portion of liquid metal to the weld pool; *arc* refers to the plasma discharge that converts the electrical energy into heat. The term *welding*, in this case, denotes that the metals are joined by

1. American Welding Society (AWS) Committee on Definitions and Symbols, 2001, *Standard Welding Terms and Definitions,* AWS A3.0: 2001, Miami: American Welding Society.
2. At the time of the preparation of this chapter, the referenced codes and other standards were valid. If a code or other standard is cited without a date of publication, it is understood that the latest edition of the document referred to applies. If a code or other standard is cited with the date of publication, the citation refers to that edition only, and it is understood that any future revisions or amendments to the code or standard are not included. As codes and standards undergo frequent revision, the reader is advised to consult the most recent edition.

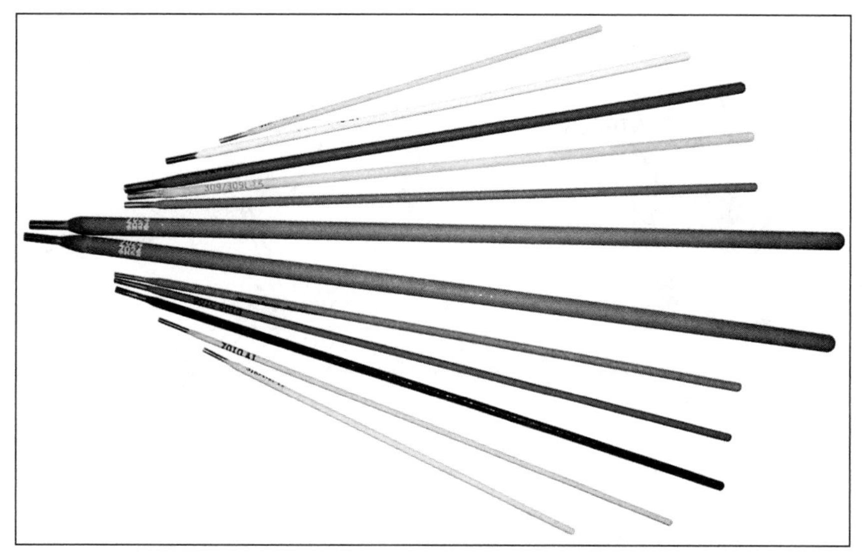

Photograph courtesy of ESAB Welding and Cutting Products

Figure 2.1—Typical Covered Electrodes Used in Shielded Metal Arc Welding

fusion. These attributes distinguish this process from other welding processes; however, the distinctive feature of shielded metal arc welding is the physical presence of the covering (coating) that surrounds the core wire of the consumable electrode. The covered rod is called an electrode because it functions as the terminal from which the electric flow changes from the conducting solid to the conducting plasma of the welding arc.

The electrode for any given application must meet three criteria. It must shield the arc and the weld metal, add metal to the weld, and sustain a welding arc. These functions are accomplished by the constituents of the covering on the metal core of the electrode. The covering contains ingredients that, when sufficiently heated, (1) decompose into gases and displace the air at the weld site, thus providing a shield for the arc and the weld metal, (2) ionize to support the arc plasma, and (3) flux the molten metal, and on cooling, form a protective slag cover on the weld bead. The electrode covering may also contain metal powders that enhance the metal contribution of the electrode to the weld pool. As a rule, the covering is nonconductive; therefore, to facilitate the electrical circuit, the covering is removed from one end (the grip end) of the electrode and is sharpened at the other end (the strike end). The core normally consists of a wire, but for some applications may consist of a cored wire with a powdered metal filler. Because of the numerous electrode coverings and core wire combinations, a large variety of SMAW electrodes are available for joining ferrous and nonferrous metals. Several typical SMAW electrodes are shown in Figure 2.1.

PRINCIPLES OF OPERATION

The shielded metal arc welding process uses an electric circuit that supports a welding arc to convert electric line power or fuel into heat. The heat from the welding arc is intense and extremely concentrated and immediately melts a portion of the workpiece and the end of the electrode. The welder maintains the arc length by holding a consistent space or "gap" between the electrode and the weld pool that forms on the workpiece. As the arc is removed, the liquid fuses and the melt solidifies into continuous metal.

As shown in the schematic of the process in Figure 2.2, the power source is connected into a circuit with the electrode and the workpiece in series. The welding cable used in the circuit, the electrode holder, and the connection between the cable and the workpiece are also important elements of the circuit.[3] The power source has two distinct output terminals. From one terminal, a connection is made to the workpiece; from the other terminal, a connection is made to the electrode. When using direct current (dc), the proper terminal for the electrode connection is dictated by the required polarity for that type electrode. When using alternating current (ac), the electrode may be connected to either terminal. The circuit is open between the workpiece and the electrode. As long as the SMAW electrode is held

3. Welders should not touch live electrical parts such as the conductive surface in the jaws of the electrode holder and the workpiece while the power source is running.

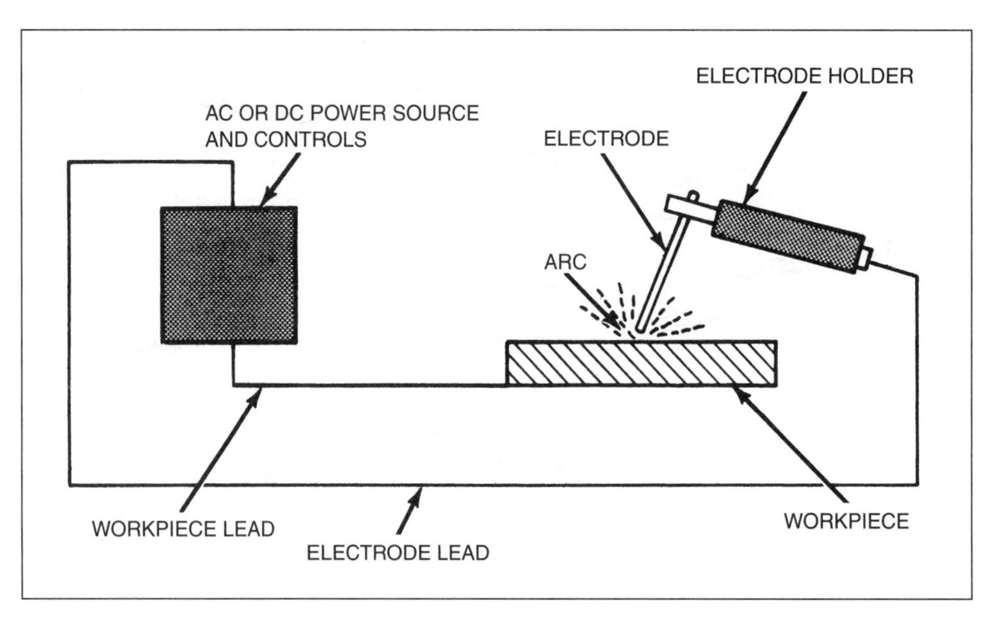

Figure 2.2—Elements of a Typical Welding Circuit for Shielded Metal Arc Welding

away from the workpiece, the circuit remains open and a voltmeter can be used to measure the voltage drop between the electrode holder and the workpiece for this open-circuit (pre-welding) condition.

The welder grasps the handle of the electrode holder and after lowering his or her welding helmet, initiates the arc by striking the tip of the electrode onto the workpiece and retracting it slightly. This momentary contact and retraction provides a path for current flow. As long as the tip of the electrode remains in proximity to the workpiece, the voltage drop across the narrow gap induces the flow of current through the air, resulting in an arc.

The current in the arc is carried by a plasma, which is the ionized state of a gas. In electrical terminology, the current flows out of the positive terminal of the arc (the anode) and to the negative terminal (the cathode), while electrons flow in the opposite direction. If the welding machine circuit is set for direct current electrode positive (DCEP), the cathode is on the workpiece and the anode is on the tip of the SMAW electrode. The amount of energy that is converted into heat by the arc is a function of the ease with which a gas ionizes and of the amount of current transmitted.[4] The temperature distribution is a function of the heat generated, the heat dissipated, and the dimensions of the arc. For the shielded metal arc welding process, the electrode coverings contain an abundant amount of stabilizers such as salts of sodium and potassium, which produce a relatively cooler arc that is nonetheless extremely effective for joining metals.[5]

The intense heat of the arc instantly melts the core wire adjacent to it and burns off the concentric covering. Some of the ingredients of the covering vaporize and decompose, producing a large volume of gases. Some ingredients may persist and begin to shape a protective cone inward to the core wire; other ingredients melt and combine with the core wire in the form of drops that are propelled across the arc. Simultaneously, a pool of molten metal begins to form at the surface of the workpiece in proximity to the arc. A quasi-steady state is instantly established, in which a distinct weld pool becomes visible and the welder is poised to manipulate the electrode. It is at this instant—at the onset of making the weld—that the weld is most vulnerable to porosity because the shielding has not fully evolved and the air at the weld site has not yet been totally expelled.

When the arc becomes established and the weld pool defined, the welder begins to feed the electrode into the arc and manipulate it within the weld joint while maintaining a constant arc length between the electrode

4. For shielded metal arc welding, the maximum temperature is reported as ~6000°K.

5. The characteristics of the welding arc are described in "Physics of Welding and Cutting," Chapter 2 of American Welding Society (AWS), Welding Handbook Committee, C. L. Jenney and A. O'Brien, eds., 2001, *Welding Science and Technology,* Vol. 1 of *Welding Handbook,* 9th ed., Miami: American Welding Society.

tip and the weld pool. (For best mechanical properties and weld metal soundness, the arc length should be held as short as practicable). As the electrode is consumed into the arc, the metal constituents of the electrode melt and transfer into the weld pool in droplets. In some cases, the droplets gather into a large globular mass that periodically transfers across the arc without disruption to the current flow. In some instances, the droplets are forcefully detached without regard to direction, for example, when using large electrode sizes of the E6010 type.[6] In other cases, the transfer is totally hidden and confined within the cone from the outer edge of the covering to the core wire (also referred to as the *crucible*). The exact nature of the forces acting on the droplets have not been extensively studied and quantified; however, the effects of gravity, explosive pressure from gas formation and expansion, electromagnetism, and surface tension are known to be important influences.[7]

As the arc is advanced, the quenching effect of the workpiece anchors some of the atoms in the melt onto the workpiece and starts to grow dendrites into the liquid of the weld pool. From the fusion with the workpiece, the solidification proceeds normal (perpendicular) to the temperature gradient in the weld pool and toward its geometric center. Various factors, such as weld travel speed, the steepness of the temperature gradient, the degree of undercooling and the composition and shape of the weld pool combine to determine whether the solidified structure will be planar, columnar, or equiaxed. The structure and composition of the weld will determine its mechanical properties. The flow of heat through the workpiece may cause structural changes within it. The region of change, an unmelted area called the *heat-affected zone* (see Figure 2.3), will have mechanical properties that are different from the balance of the workpiece.

The center of the weld pool may be stirred by magnetic forces, and some solidifying slag or oxidation products, or both, may be seen floating toward the outer rim of the weld pool. At the trailing edge of the weld pool, slag is seen advancing with the movement of the electrode. The slag consists of solidified fluxing compounds, oxidized metals, decomposition products, and high-melting oxides. These materials tend to congeal sooner than the weld metal solidifies, and because they have a lower density, will float to the top and form a protective cover on the weld bead. The desirable amount of slag is the minimum level that uniformly covers the weld bead. The bead-on-plate weld usually presents the greatest surface-to-weld ratio and the greatest challenge to fully cover.

The slag shields the high-temperature surface from the atmosphere. Premature removal or fracturing of the slag may cause discoloration of the weld or, at times, multi-coloration (i.e., stainless steel welds). The slag also serves as a heat barrier that lowers the cooling rate of the weld metal. Slag detachability is an important aspect of weld cleanup. For the most part, slag removal with shielded metal arc welding is easily accomplished, which contributes to the efficiency of the process. This removal becomes more difficult, however, in deep, narrow weld grooves.

The welder continually evaluates the flow of the solidifying slag, the wetting of the weld bead, the stability of the arc, the quietness of the metal transfer, and the direction of the arc. The welder intuitively guides and advances the electrode to maintain a consistent weld in the intended joint. With some electrodes, the welder merely drags the electrode covering along the joint. With others, the welder must quickly remove the arc from the weld pool and return it in a whipping motion to avoid overheating.

The welder must be sensitive to changes caused by arc blow (the deflection of the arc from its normal path due to magnetism) and react by reorienting the electrode. When the weld is completed, the welder withdraws the electrode, and the arc fades and extinguishes. However, in some joints merely withdrawing the electrode may leave an unacceptable void or crack in the crater (a depression in the weld face at the very end of the weld bead). In those instances, the welder backfills by returning past the center of the weld pool and then withdrawing the electrode. In this manner, a skilled welder uses the additional filler metal to achieve a sound weld joint termination free of porosity, cracks, and so forth.

COVERED ELECTRODES

All SMAW electrodes have a covering with constituents that facilitate the welding process and add alloying elements that impart useful properties to the weld. Without the covering, the arc would be very difficult to maintain, the weld deposit would be brittle with dissolved oxygen and nitrogen, the weld bead would be dull and irregularly shaped, and the workpiece would be undercut.

The manufacturer applies the covering on shielded metal arc electrodes by either the extrusion process or the dipping process. Extrusion is much more widely used and is achieved by mixing the dry components with liquid silicates. The dipping process is employed

6. E6010 electrodes and others are described in American Welding Society (AWS) Committee on Filler Metals and Allied Materials, 1991, *Specification for Carbon Steel Electrodes for Shielded Metal Arc Welding*, AWS A5.1, Miami: American Welding Society.

7. Metal transfer across the arc is described in "Physics of Welding and Cutting," Chapter 2 of American Welding Society (AWS) Welding Handbook Committee, Jenney, C. L. and A. O'Brien, eds., 2001, *Welding Science and Technology*, Vol. 1 of *Welding Handbook*, 9th ed., Miami: American Welding Society.

primarily for SMAW electrodes used to weld cast iron and for some specialty electrodes that have a complex core wire.

The covering contains most of the stabilizing, shielding, fluxing, deoxidizing and slag-forming materials essential to the process. In addition to sustaining the arc and supplying filler metal for the weld deposit, the decomposition of the electrode covering introduces other key materials into or around the arc, or both. Depending on the type of electrode being used, the electrode covering provides the following:

1. A gas to shield the arc and prevent excessive atmospheric contamination of the molten metal;
2. Deoxidizers to react with and deplete the level of dissolved gaseous elements that can cause porosity;
3. Fluxing agents to accelerate chemical reactions and cleanse the weld pool;
4. A slag blanket to protect the hot weld metal from the air and to enhance the mechanical properties, bead shape, and surface cleanliness of the weld metal;
5. Alloying elements to achieve the desired microstructure;
6. Elements and compounds to control grain growth;
7. Alloying materials to improve the mechanical properties of the weld metal;
8. Elements to affect the shape of the weld pool;
9. Elements that affect the wetting of the workpiece and the viscosity of the liquid weld metal; and
10. Stabilizers to help establish the desirable electrical characteristics of the electrode and minimize spattering.

The chemical compounds in the covering, in combination with the core wire composition, create unique mechanical properties in the weld and enhance welding characteristics such as arc stability, metal transfer type, and slag. The different types of electrodes are formulated not only to weld different metals but also to optimize certain characteristics of the process and gain an advantage in a particular area of application. For example, for farm or repair shop welding, the economical ac transformer welder is very popular; consequently an electrode covering designed for welding with ac would be ideally suited for non-industrial ferrous sheet metal and plate welding. With ac, the welding arc extinguishes and reestablishes each time the current reverses its direction. For good arc stability, it is necessary to have a gas in the gap between the electrode and the weld pool that will remain ionized during each reversal of the current. This ionized gas makes possible the smooth re-ignition of the arc. Gases that readily ionize are produced by a variety of compounds, including those that contain potassium. The incorporation of these compounds into the covering enables the electrode to operate on alternating current.

For commercial flat-position and horizontal welding, the joint completion speed is an important consideration. Therefore, electrodes with very high deposition rates and process efficiency are preferred, for example, ferrous SMAW electrodes with iron powder included in the coverings such as E7018, E7024, and E7028. The iron powder provides another source of metal available for deposition, supplementing that obtained from the core of the electrode. The presence of iron powder in the covering also makes more efficient use of the arc energy. Metal powders other than iron are frequently added to electrode coverings, and although they increase the deposition rate, their primary purpose is to alter the mechanical properties of the weld metal or to deoxidize the weld pool.

ARC SHIELDING

The shielding action of the shielded metal arc welding process illustrated in Figure 2.3 is essentially the same for all SMAW electrodes, but the specific method of shielding and the volume of slag produced vary from one electrode type to another. As depicted in Figure 2.3, two mechanisms are at work to prevent the detrimental effect on the weld pool caused by the gases contained in the air. The first is the forceful displacement of the air by gases produced by the burning and decomposition of the electrode covering. The second is the blanketing action of the flux or slag, which prevents diffusion of the air constituents into the liquid metal. Electrode coverings vary in their reliance on these two mechanisms to provide the most advantageous shielding action for a specific weld.

When electrodes that rely on air displacement are used, the bulk of the covering is converted to gas by the heat of the arc and only a small amount of slag is produced. This type of electrode depends largely on a gaseous shield to prevent atmospheric contamination. The weld bead produced with these electrodes will characteristically have a very light layer of slag that may not completely cover the surface.

When electrodes that rely on a blanketing action are used, the bulk of the covering is converted to slag by the heat of the arc, and only a small volume of shielding gas is produced. As the small globules of metal are transferred across the arc they are entirely coated with a thin film of molten slag. This slag floats to the surface of the weld pool because it has a lower density than that of the molten metal of the weld pool. As the temperature of the weld pool drops, the slag solidifies and continues to shield the weld pool and later shields the hot solidified weld metal from oxidation. Welds made with these

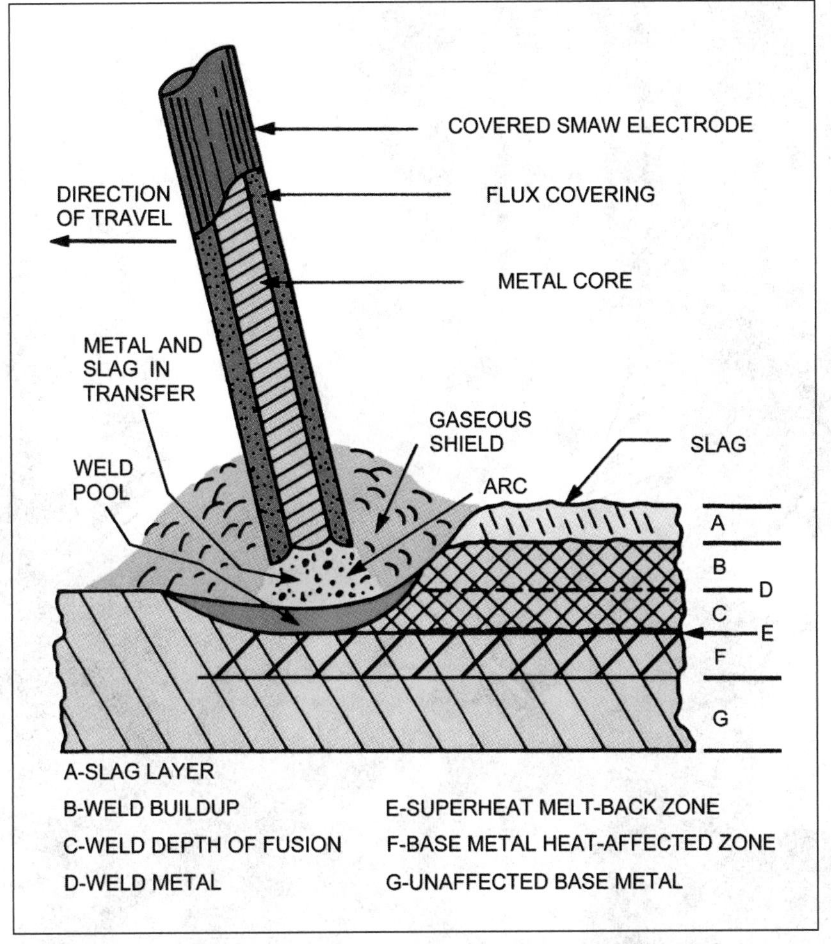

Source: Adapted from Linnert, G. E., 1994, *Welding Metallurgy,* 4th ed., Miami: American Welding Society

Figure 2.3—Schematic of Shielded Metal Arc Welding

electrodes can be identified by the heavy slag deposits that completely cover the weld beads. Various electrode types use different combinations of these two mechanisms, each with different combinations of gas and slag shielding.

Variations in the amount of slag and gas shielding also influence the welding characteristics of covered electrodes. Electrodes that produce a heavy slag can carry high amperage and provide high deposition rates, making them ideal for heavy weldments in the flat position. Electrodes that produce a light slag layer are used with lower amperage and provide lower deposition rates. These electrodes produce a smaller weld pool and are suitable for making welds in all positions. When the differences in welding characteristics are compared, one type of covered electrode usually emerges as the best selection for a given application.

PROCESS ADVANTAGES

A valuable advantage of shielded metal arc welding is the large variety of metals and alloys the process is capable of welding. Procedures and electrodes are available to weld carbon and low-alloy steels, high-alloy steels, coated steels, tool and die steels, stainless and heat resisting steels, cast irons, copper and copper alloys, nickel and cobalt alloys. Figures 2.4 and 2.5 illustrate two typical applications: pipe welding and structural steel welding. The process can also be used for some aluminum applications.

Short welds common to the production of components or finished products, maintenance and repair work, and field construction are important areas of application for the shielded metal arc welding process.

Photograph courtesy of The Lincoln Electric Company

Figure 2.4—Pipe Welding with Shielded Metal Arc Welding

Following are other advantages of the process:

1. The equipment is relatively simple, inexpensive, and portable;
2. The SMAW electrode provides both the shielding and the filler metal to make a sound weld;
3. Auxiliary gas shielding or granular flux is not required;
4. The process is less sensitive to wind and draft than the gas shielded arc welding processes;
5. The dimensions of the SMAW electrodes are ideal for reaching into areas of limited access (electrodes can be bent, and with the aid of mirrors, applied in blind spots;
6. The process is suitable for most of the commonly used metals and alloys;
7. The process is flexible and can be applied to a variety of joint configurations and welding positions; and
8. Optimum results can be readily and reliably obtained.

PROCESS LIMITATIONS

Metals with low melting temperatures, such as lead, tin, and zinc and their alloys are not welded with shielded metal arc welding. These metals also have relatively low boiling points and the intense heat of the SMAW arc immediately causes them to vaporize from the solid state. Shielded metal arc welding is likewise

Figure 2.5—Structural Steel Welding with the Shielded Metal Arc Process

not suitable for reactive metals such as titanium, zirconium, tantalum and niobium (columbium) because the shielding provided is not sufficiently inert to prevent contamination of the weld.

The shielded metal arc welding process yields lower deposition rates than the gas metal arc welding (GMAW) and flux cored arc welding (FCAW) processes. The deposition rate is lower because the maximum useful current is limited. Because covered electrodes are produced and used in discrete lengths (230 mm to 460 mm [9 to 18 in.]) that conduct current from the moment the arc is initiated until the electrode is practically consumed, they are subject to resistance heating. The amount of heat converted by the SMAW electrode is a function of the amount of current, the resistance of the core wire, and the welding time.

If the electrode is too long or if the current is too high, the amount of heat generated within the SMAW electrode will be excessive. After welding has started, the temperature of the covering will eventually rise to a range that will cause the premature breakdown of the covering. The breakdown of the covering, in turn, triggers a deterioration of the arc characteristics and reduces the level of shielding. Consequently, welding must stop before the electrode has been fully consumed. Hence, the amount of current that can be used is limited within a range that prevents the overheating of the electrode and the breakdown of the covering. The limited useful current results in generally lower deposition rates than those obtainable with the gas metal arc or flux cored arc welding processes.

Another inherent drawback of the shielded metal arc welding process is stub loss. The stub is the grip end of the SMAW electrode that is discarded. It consists of the core wire within the grip of the electrode holder and a small portion of the covered length. The stub loss affects the deposition efficiency, not the deposition rate. Longer stub losses translate directly into lower deposition efficiency.

The operator factor (i.e., arc time as a percentage of the welder's total labor time) for the shielded metal arc welding process is usually lower than that obtained with a continuous electrode process such as gas metal arc welding or flux cored arc welding.[8] Inherent in the shielded metal arc welding process is the need to reload a new electrode when the previous one has been consumed. A lengthy SMAW weld consists of a series of connected weld beads, each made with an individual electrode. The weld beads are tied together by slightly overlapping the previous weld bead, which requires the immediate removal of the slag from the crater of that bead. This necessary step adds to the total cleaning time. The reloading of the electrodes and the cleaning of craters are eliminated with a continuous electrode process.

When the weldment requires a large volume of filler metal, the combination of low deposition rates and a lower operator factor detracts from the use of the shielded metal arc welding process. In these instances, the weld completion rate may be too slow and the weld cost relatively high.

EQUIPMENT

Shielded metal arc welding equipment falls into four categories: (1) the equipment that is essential to establish the electric circuit that sustains a welding arc—the power source, welding cable, workpiece connector and electrode holder; (2) the devices and gear necessary to ensure the safety and well being of the welder and others in the welding area; (3) the accessories used to complete a weld according to a specified welding procedure; and (4) items of expedience, such as those needed to properly isolate the welding booth within the work environment or a fixture to position the workpiece for welding. Some of the more generally used equipment within the first three categories is considered in this chapter.

POWER SOURCES

Power sources deliver output current and voltage with the electrical characteristics that make arc welding possible.[9] Two types of power sources are used for shielded metal arc welding: static machines that convert the energy supplied by electric power companies and motor generators that combust fuel (chemical energy). The selection of a power source is the most important and most expensive element of the welding electrical circuit and is the first consideration in implementing the shielded metal arc welding process.

Power Source Selection

Several factors should be considered when selecting a power source for shielded metal arc welding:

1. The type of welding current required,
2. The output characteristics of the power source,
3. The amperage range required,
4. The positions in which welding will be done, and
5. The primary type of power available at the workstation.

The selection of the type of current—alternating current, direct current, or both—is based largely on the kind of welds to be made and on the types of electrodes that are suitable for use. For ac welding, a transformer

8. The operator factor and its use in the calculation of welding costs are discussed in "Economics of Welding and Cutting," Chapter 12 of American Welding Society (AWS), Welding Handbook Committee, Jenney, C. L. and A. O'Brien, eds., 2001, *Welding Science and Technology,* Vol. 1 of *Welding Handbook,* 9th ed., Miami: American Welding Society.

9. For additional information on power sources, see Chapter 1 of this volume. For information on power sources manufactured before 1991, see "Arc Welding Power Sources," Chapter 1 of American Welding Society (AWS) Welding Handbook Committee, O'Brien, R. L., ed. *Welding Processes,* Vol. 2 of *Welding Handbook,* 8th. ed., Miami: American Welding Society.

or an alternator-type power source can be used. For dc welding, the choice is between a transformer-rectifier or motor-generator power source. When both ac and dc are to be used, a single-phase transformer-rectifier or an alternator-rectifier power source can be used. Otherwise, two welding machines are required, one for ac and one for dc.

As explained in the section "Significance of the Volt-Ampere Curve," the output characteristics of the power source refer to the response of the power source to changes in arc length as caused by the welder. A response that maintains a nearly constant current is preferred for high-deposition (high-amperage) welding. For low-amperage applications, a substantial change in current with varying arc length can improve arc starting and can help the welder to avoid "snuffing" or "stubbing," a situation in which the electrode fuses in the weld pool and sticks to the workpiece.

The amperage requirements are determined by the sizes and types of electrodes to be used. When a variety of electrodes will be used, the power source must be capable of providing the amperage range needed. The power source must have the appropriate duty cycle for the maximum current that will be used. Because of the low operating factor for the shielded metal arc welding process, a satisfactory requirement is a 60% duty cycle at the maximum current.[10]

The positions in which welding will be done should also be considered. If vertical or overhead welding is planned, the ability to adjust the slope of the V-A curve is desirable (see Figure 2.6) and the power source must provide this feature. A machine with this capability usually requires controls for both the open-circuit voltage and the current.

A welding power source must be fueled. Therefore, a determination must be made as to whether electricity is available at the work site and whether it is single-phase or three-phase. The welding power source must be designed for the type of electricity available. If electricity is not available, an engine-driven generator or alternator can be used.

Type of Output Current

Either ac or dc can be employed for shielded metal arc welding. The specific type of current and the output of the power source influence the performance of the electrode. Each current type has its advantages and limitations, and these must be considered when selecting the type of current for a specific application. The distinctive performance in each of the following areas must be considered.

10. For an explanation of the duty cycle, see Chapter 1 of this volume.

Instances of Excessive Voltage Drop. When the work is located a long distance from the power source, long electrical cables will be required. The resistance to current flow in long cables may affect the output of the welding power source and the welding characteristics of the process. The voltage drop in the welding cables is lower with ac; therefore, if the welding is to be done at a long distance from the power supply ac is more efficient. It should be noted that long cables that carry ac should not be coiled because the inductive losses encountered in such cases can be substantial.

Welding with Low Current. With small-diameter electrodes and low welding currents, dc provides better operating characteristics and a more stable arc.

Arc Initiation. Striking the arc is generally easier with dc, particularly if small-diameter electrodes are used. With ac, the welding current passes through zero during each half cycle, requiring periodic re-ignition of the arc. This presents problems for arc starting and arc stability.

Maintaining Constant Arc Length. Welding with a short arc length (low arc voltage) is easier with dc than with ac. The shortest practical arc length is preferred for optimum physical properties of the weld metal. Therefore, the ability to weld with a short arc length is an important consideration, except when electrodes with a high iron powder content are to be used. With these electrodes, the deep crucible formed by the heavy covering automatically maintains the proper arc length when the electrode tip is dragged on the surface of the joint.

Arc Blow. Alternating current rarely presents a problem with arc blow because the magnetic field is constantly reversing (120 times per second). Arc blow can be a significant problem with dc when welding ferritic steel. Unbalanced magnetic fields that arise can deflect the arc and eject the transferring metal droplets.

Out-of-Position Welding. For vertical and overhead welds, dc is somewhat better than ac because the welding performance is better at lower amperages. However, with suitable electrodes, satisfactory welds can be made in all positions with ac.

Metal Thickness. Both sheet metal and heavy sections can be welded using dc. The welding of sheet metal with ac produces a less desirable weld than with dc. Arc stability is better with dc at the low current levels required for thin materials. However, properly sized and designed ac electrodes, such as the E6013 types, are used for sheet metal applications.

Mechanical Properties. As a rule, welds made with DCEP prove to have better mechanical properties, especially weld metal toughness. Skilled welders can exercise a greater degree of control of the average arc length with DCEP. The shorter arc results in lower traces of dissolved elements that originate from the air.

Power Source Preferences

Power sources are readily available for ac, dc, and combination ac/dc service. A review of the welding application will generally determine whether ac or dc is most suitable. The preferred power source for the shielded metal arc welding process is a constant-current type with at least 60% duty cycle at the maximum application current. Constant-voltage power sources are not used with shielded metal arc welding because the output electrical characteristics require the welder to maintain a constant arc length, which is nearly impossible.

Significance of the Volt-Ampere Curve

The electrical output characteristics of the power source are usually best presented by graphing the voltage as a function of current at varying resistances.[11] The resulting volt-ampere (V-A) curve spans from maximum voltage at zero current (analogous to infinite resistance or open circuit) and droops to zero voltage, maximum current (zero resistance, electrode shorted to the workpiece, zero arc length). The response of the power source can be gauged by noting the change in the voltage and in the current under varying resistances. During welding, changes in arc length vary the resistance of the arc. A shorter arc length produces lower resistance; a longer arc length produces higher resistance. By charting the curves with the machine set at minimum amperage and again with the machine set at maximum amperage, the response region of the power source can be fully depicted. For the shielded metal arc welding process, the preferred response is a minimal change in current with varying resistances. This type of response produces a steeply drooping volt-amperage curve, which is the trait of constant-current machines (called *droopers*). If a nearly steady current is maintained, the melting rate of the electrode will remain uniform and will provide the constancy required to make the weld.

There are occasions when the welder can benefit from a flatter response to a change of the arc length. A flatter (but still steep) response means that as the arc length is changed to produce a different arc voltage, the magnitude of the current change is substantially higher than the response of a drooper. At low amperage settings, the welder, with a small change in the arc length, can momentarily achieve a desired change to the deposition rate. At these low amperage settings, the larger increase in the current is beneficial because it acts as a built-in control of the arc length. As the arc length shortens, the arc voltage diminishes and the current substantially increases. The higher current accelerates the melting of the SMAW electrode, which tends to correct the arc to the intended length. As the arc length increases, there is a substantial decrease in the current that slows the melting of the electrode, which again helps to achieve the desired arc length. In this manner, a flatter response at low current settings on the power source can assist the welder in several ways: (1) maintaining a constant arc length, (2) making an instantaneous adjustment to deposition rate, and (3) improving arc starting by preventing the electrode from sticking to the workpiece in the weld pool. A constant-current, constant-voltage machine is capable of delivering the desired steep droop for normal welding and the flatter response at low amperage settings. Power sources with merely a less steep droop—a flatter response—are still considered constant-current machines; a constant-voltage machine is characterized by an almost flat response.

Figure 2.6 shows curves of typical volt-ampere output characteristics for both ac and dc power sources. These curves can be plotted by measuring the current flow and the voltage at various conditions. The starting point for plotting the curves is at zero current and open-circuit voltage, when the arc length between the electrode and the workpiece is too great for the voltage to induce current flow and the power source is not under load. The ending point is zero voltage and maximum current for a given setting on the amperage dial of the power source. In this condition, the electrode is shorted to the workpiece (arc length is zero) and the power source is under the maximum load for that setting. The functional relationship between these two points is obtained by holding an arc between the electrode and the workpiece and measuring the amperage and the voltage. As the arc diminishes in length, the current and the voltage approach the value at the shorting condition. At all times, the amperage and the voltage will fall on one curve. Although the arc length is difficult to measure absolutely, the functional relationship between the voltage and the current for the power source can be accurately graphed.

Two related sets of constant-current curves are illustrated in Figure 2.6. Each set consists of four volt-ampere curves plotted by measuring at four unique amperage settings on the power source. One group of curves was produced at a high open-circuit voltage (OCV) and another at a lower OCV. The OCV is common to each curve in its group because at this point it is

11. See "Volt-Ampere Characteristics" in Chapter 1 of this volume.

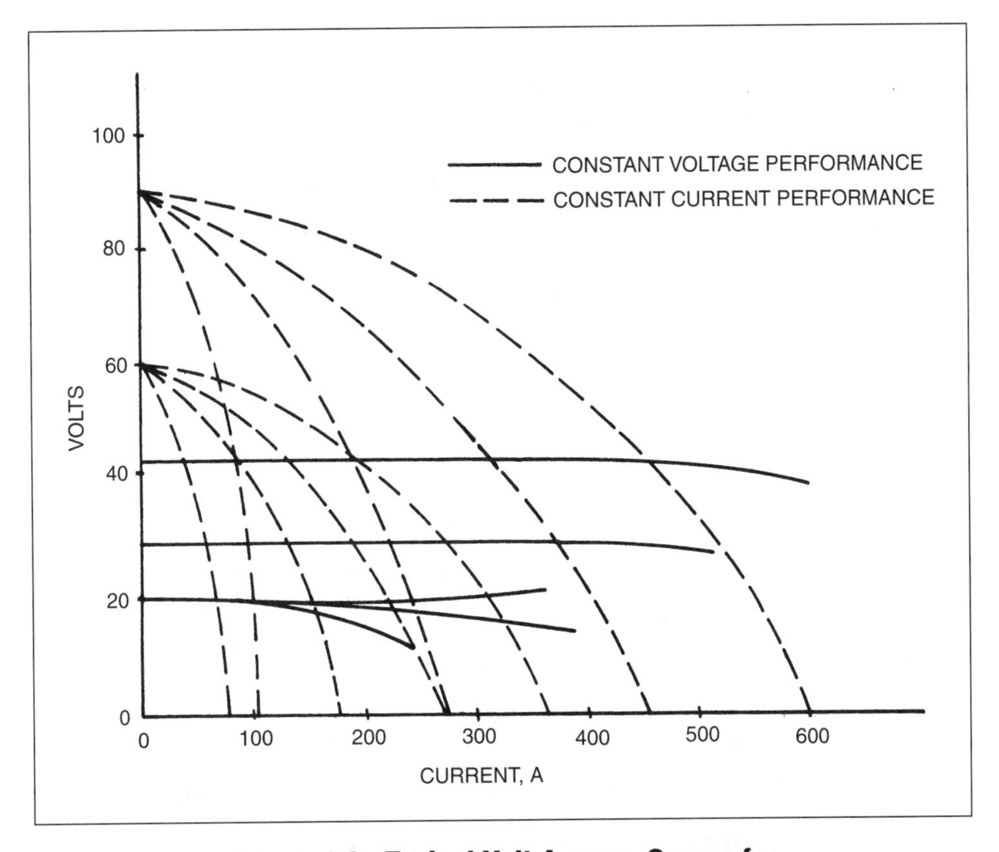

Figure 2.6—Typical Volt-Ampere Curves for Constant-Current and Constant-Voltage Power Sources

the characteristic response of the power source; the resistance (load) from the welding circuit is zero. The curves droop downward toward the maximum current, which occurs when the electrode is shorted to the workpiece. Figure 2.6 also depicts the typical flat response of a constant-voltage power source.

The graph can be used to assess the voltage and the current that the welder will encounter as the arc length is varied while making the weld. With a power source that produces a constant-voltage output, i.e., a flat volt-ampere curve, even a slight change in arc length produces a relatively large change in amperage. The large change in current translates into large changes in resistance heating of the electrode and hence in a variable rate of melting of the electrode. Therefore, a power source that produces a constant-voltage response will not produce a steady electrode melting rate and is not suitable for shielded metal arc welding. A power source that produces a constant-current response is preferred for manual welding because the steeper slope of the volt-ampere curve (within the welding range) results in a small change in the current, even with a substantial change in arc length.

For applications that involve high welding currents with large-diameter electrodes, a steeper volt-ampere curve is desirable. When more precise control of the size of the weld pool is required—for example, for out-of-position welds and root passes of joints with varying fitup configurations—a flatter volt-ampere curve is desirable. The flatter response enables the welder to substantially change the welding current within a specific range simply by changing the arc length. A flat response gives the welder greater control over the amount of filler metal that is being deposited. Figure 2.7 portrays these different volt-ampere curves for a typical welding power source. Even though there are substantial differences in the slope of the various curves, the power source is still considered a constant-current power source. The changes shown in the volt-ampere curve are accomplished by adjusting both the open-circuit voltage and the current settings on the power source.

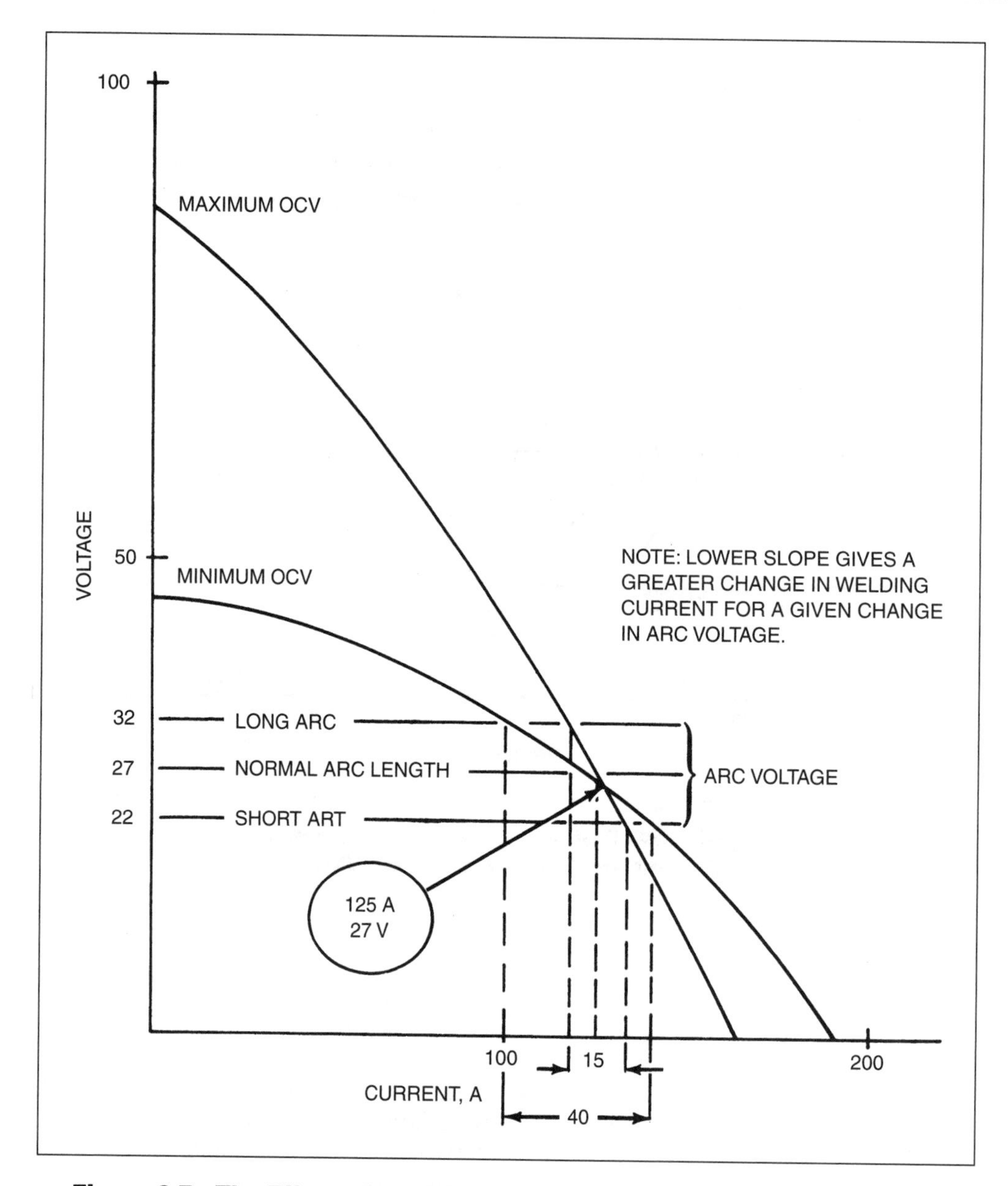

Figure 2.7—The Effect of the Volt-Ampere Curve on the Welding Current and Arc Voltage with a Change in Arc Length

Open-Circuit Voltage, Welding Current, and Arc Voltage

The term *open-circuit voltage*, which is a characteristic of the power source, must be distinguished from the term *arc voltage*. Open-circuit voltage is defined as the voltage between the output terminals generated by the welding machine when no welding is being done. Arc voltage is the electrical potential between the electrode and the workpiece during welding. For any given electrode, arc voltage is mostly determined by arc length. Open-circuit voltages generally range between 40 V and 100 V; arc voltages measure between 17 V and 40 V. If the circuit voltage is monitored, the reading starts with the open-circuit voltage and then drops to the arc voltage as the arc is struck and the welding machine begins to energize the welding circuit.

The term *welding current* denotes the amperage during welding. The welding current rises from zero to an ampere reading that is consistent with the arc length and the setting on the machine. The electrode being used and the amperage set at the machine generate a unique volt-ampere response curve that falls within the possible volt-ampere response range of the welding power source and that reflects the range of arc lengths held by the welder.

For the selected SMAW electrode and setting on the power source, the arc length determines the value of the arc voltage and the welding current. If the arc is lengthened, the arc voltage increases and the welding current decreases. The change in amperage produced by a change in arc length is determined by the volt-ampere curve. If the volt-ampere curve slopes steeply at the welding set point, then the change in amperage will be minimal. If the slope is only slightly less steep, the change will be greater; if it is extremely flat, the current will change radically and the process will not be stable.

Control of the open-circuit voltage is a useful feature for some shielded metal arc welding applications, but it is not a necessary item. Hence, many power sources do not provide for control of the open-circuit voltage; instead, the output is optimized at one set value.

ELECTRODE HOLDER

The electrode holder is the physical link in the process that allows the welder to take control of the SMAW covered electrode. It places the covered electrode in series in the welding circuit. Typically, the electrode holder has an electrically insulated handle and a clamping device that mechanically holds and conducts current to the electrode. The electrode holder should be designed for optimum welder comfort and safety.

The current is transferred to the electrode through the jaws of the holder, which may be grooved to maximize the contact area and prevent overheating. If the jaws are grooved, the orientation of the grooves will dictate the relative angle that the electrode will project out of the jaws. In order to provide multiple orientations, multiple grooves are machined onto the jaw plates. The jaws of the electrode holder must be kept in good condition to assure minimum contact resistance and avoid overheating the holder. The overheating of the holder makes grasping the handle uncomfortable for the welder, and the energy lost from the circuit impairs the welding performance of the electrode. Both conditions reduce the quality of the weld.

The electrode holder must be designed to grip the electrode securely and hold it in position with good electrical contact. Insertion of the electrode and ejection of the stub must be quick and easy. The holder should be light in weight and easy to manipulate, yet it must be sturdy enough to withstand rugged use. The handle on the holder must be insulated to protect the welder's hand from the welding circuit. The jaws must be recessed or caged to prevent incidental grounding. The design of the holder should allow it to be easily and safely set aside while the power source is running. The electrode holder should not be set aside with an electrode or stub extending from the jaws.

Electrode holders are sized or rated according to the maximum current that can be conducted by the holder. The jaws of the holder should accommodate the range of standard electrode diameters that are designed to weld within the electrical current range of the holder. The best electrode holder is the one that provides the highest level of operator safety and comfort. Usually it is the lightest in weight for the rated current.

Workpiece Connection

The workpiece connection fastens the cable lead to the workpiece. For light duty, a spring-loaded clamp may be suitable. For high currents, however, a screw clamp may be needed to provide a good connection without overheating the clamp. The clamp should produce a strong connection and attach easily to the workpiece.

WELDING CABLES

Welding cables are used to connect the power source to the electrode holder and to the workpiece connection. They are part of the welding circuit, as shown in Figure 2.2. Welding cable is constructed for maximum flexibility to permit easy access to the workpiece without unduly restricting the manipulation of the electrode holder. Welding cables must have an insulating cover that is resistant to wear and abrasion.

Suitable welding cable consists of many fine copper or aluminum strands of wire enclosed in an insulated cover. The cable cover is made of synthetic rubber or a type of plastic that has good toughness, high electrical resistance, and excellent heat endurance. Flexibility, a desired feature of welding cable, is achieved by a protective wrap placed between the cable and the insulating covering; the wrapping allows for slip between the cable and the plastic.

Welding cable is produced in a range of sizes standardized by American Wire Gauge (AWG) numbers and is available in sizes from about AWG 6 to 4/0. The size of the cable required for a particular welding circuit depends on the maximum welding current, the total length of welding cable required (roughly twice the distance between the power source and the workpiece), and the duty cycle of the welding machine. Table 2.1 shows the recommended size of copper welding cable for various power sources and circuit lengths. When

Table 2.1
Recommended Copper Welding Cable Sizes

Power Source		AWG Cable Size for Combined Length of Electrode and Ground Cables				
Size in Amperes	Duty Cycle, %	0 to 15 m (0 to 50 ft)	15 to 30 m (50 to 100 ft)	30 to 46 m (100 to 150 ft)	46 to 61 m (150 to 200 ft)	61 to 76 m (200 to 250 ft)
100	20	6	4	3	2	1
180	20–30	4	4	3	2	1
200	60	2	2	2	1	1/0
200	50	3	3	2	1	1/0
250	30	3	3	2	1	1/0
300	60	1/0	1/0	1/0	2/0	3/0
400	60	2/0	2/0	2/0	3/0	4/0
500	60	2/0	2/0	3/0	3/0	4/0
600	60	2/0	2/0	3/0	4/0	*

*Use two 3/0 cables in parallel.

aluminum cable is used, it should be two AWG sizes larger than designated copper cable for the application. Cable sizes are increased as the length of the welding circuit increases to keep the voltage drop and the resulting power loss in the cable at acceptable levels.

If long cables are necessary, several sections can be joined by suitable cable connectors. The connectors must provide good electrical contact with low resistance, and the insulation of the connectors must be equivalent to that of the cable. Soldered joints and mechanical connections are also used. The connection between the cable and a connector or lug must be strong with low electrical resistance. Aluminum cable requires a good mechanical connection to avoid overheating. Oxidation of the aluminum significantly increases the electrical resistance of the connection. The higher resistance can lead to overheating, excessive power loss, and cable failure.

The ends of the welding cables that attach to the power source are normally lugged to connect to the stud at the machine's output terminals. It is a common practice to locate the output connections safely on the power source and to secure the connections with nut-and-bolt assemblies. This protective connection on the welding machine does not require insulation.

Welding cables should be carefully handled to avoid damage to the covering of the cable, particularly the electrode cable. If hot metal burns the cable covering or if sharp edges cut it, the welder and others in the area will be at serious risk for electric shock. The cables should be routinely inspected and maintained.

PROTECTIVE EQUIPMENT

The welder must protect the eyes and skin from radiation from the arc by using a welding helmet with the correct filter lens and by wearing leather gloves and suitable clothing to protect against burns from arc spatter. Figure 2.8 shows appropriate protective clothing.

Welding Helmet

The welding helmet is a protective mechanical apparatus worn by the welder to protect the eyes, face and neck from arc radiation, radiated heat, spatter or other harmful matter that may be expelled during the welding process. Helmets are usually constructed of pressed fiber insulating material molded to the general contour of the face. A rectangular viewing portal fitted with a shaded filter lens provides protected vision. A helmet should be light in weight and should be designed to give the welder the greatest possible comfort.

The welding helmet is held in place by an adjustable band encircling the welder's head at the forehead. The front shielding surface is initially positioned above the head, and with a slight nod of the welder's head, the helmet pivots into the welding position before the welder strikes an arc.

Coordinating the position of the welding helmet with the strike of the arc is not an easy chore. In some instances, a confined welding zone may not provide sufficient height to execute the routine helmet start. For this reason, helmets with photosensitive millisecond-response filter lenses that darken to a target shade are

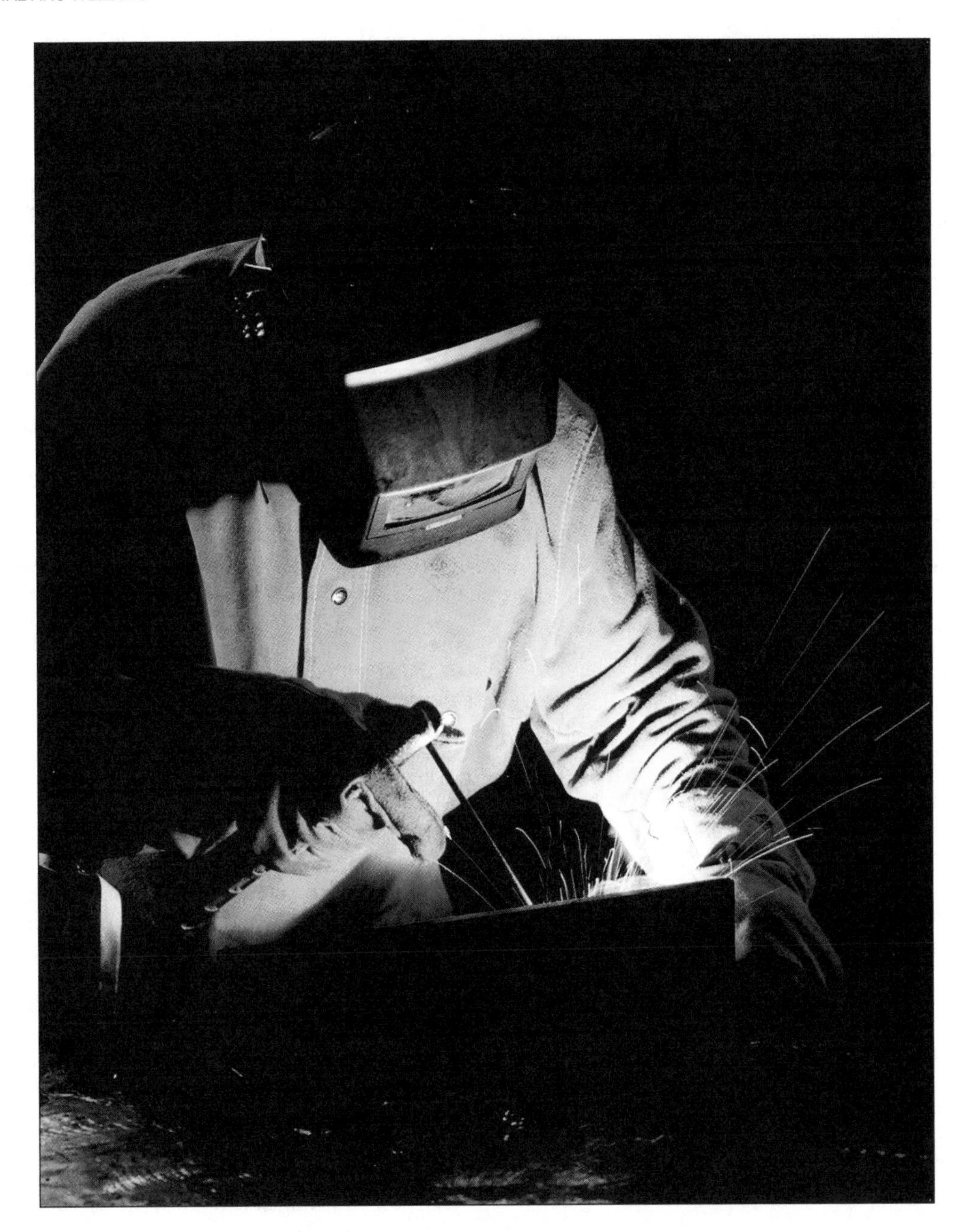

Figure 2.8–Welding Helmet and Protective Clothing for Shielded Metal Arc Welding

popular. Some helmets have an optional "flip lid" that permits the dark filter plate in the shield to be flipped up so the welder has normal vision for tasks such as chipping the slag from the weld after the arc is extinguished.

The welding helmet must be equipped with a lens that fully blocks the ultraviolet (UV) and infrared (IR) rays from the arc. The lens shade number is a proportionate measure of the effectiveness of the lens in blocking light—the higher the number, the darker the shade. Each application requires a minimum shade number. The "Safe Practices" section at the end of this chapter provides information on additional lens shade numbers. For shielded metal arc welding, a Number 7 shade is usually the minimum required. The minimum shade is determined by the intensity of the arc (welding current). Any darker shade can be used; the actual shade selected from the permissible range is a matter of welder preference for the application. Typically, a range of shades will provide the level of protection required before viewing becomes inadequate. A glass cover plate is frequently used to protect the shaded lens. The filter lens and the glass cover plate must remain clean. A dirty lens or cover plate will irregularly darken the view and interfere with the crisp image required for good electrode manipulation.

The welder and others in the area must not directly view the arc without proper protection. The rays emitted by the arc will cause burns and temporary loss of sight. Even with due diligence to protect the eyes, on occasion, especially in multiple arc welding zones, eye burn (flashing) can occur. After several hours, a "flashed" person will feel as if he or she has sand in the eyes. The irritation may last for several days. Consequently, the welder must use a welding helmet to view the arc and block the rays; others in the weld area can use a hand shield. The hand shield performs the same function as the helmet, but because the viewer is not as close to the arc and can choose a position that avoids direct exposure, the level of protection required is not as great. All persons in a welding area must wear goggles or safety glasses with side shields.[12]

Gloves

Gloves must be worn during welding to protect the hands from the rays of the arc and from the heat produced by the arc. The weld, the slag, the workpiece and the electrode stub are very hot immediately after welding and unfortunately, these items often give no physical sign of being at elevated temperatures. Consequently, gloves must also be worn for such tasks as cleaning the weld, handling the workpiece, and pinching the electrode covering to help restrike, thus rendering harmless any contact with these hot surfaces. Leather gloves in good condition (pliable) are highly recommended; they also provide an additional insulating layer to prevent electric shock.

MISCELLANEOUS EQUIPMENT

Cleanliness is important in welding. The surfaces of the workpieces and the previously deposited weld metal must be cleaned of dirt, slag, fume and any other foreign matter that would interfere with welding. To accomplish this task, the welder should have a steel wire brush, a hammer, a chisel and a chipping hammer. These tools are used to remove dirt and rust from the base metal, cut tack welds, and chip slag from the weld bead.

Welding is often performed in accordance with a welding procedure specification (WPS). The specified procedure usually requires control of the interpass and preheat temperatures and of the heat input. For these applications, temperature-indicating sticks, a timer, a metal ruler and an ammeter (or tong meter) are required to measure the pertinent variables to assure that the weld is in conformance with the WPS.

12. See American National Standards Institute (ANSI), *Practice for Occupational and Educational Eye and Face Protection*, ANSI Z87.1, Des Plaines, Illinois: American Society of Safety Engineers (ASSE).

MATERIALS

The essential materials for the shielded metal arc welding process are the base plate and the covered electrode. The materials for the base plate include the many metals and alloys that can be welded or surfaced by the process. Covered electrodes are materials that include numerous classified or unclassified types of solid and cored rods with a covering that is specially formulated for the shielded metal arc process to weld or to apply a surface.

BASE METALS

Shielded metal arc welding is used in joining and surfacing applications on a variety of metals. Whether a base metal (also referred to as *base plate* or *workpiece*) is a suitable material for the process depends on the intrinsic physical properties of the base metal and the availability of a covered electrode that will produce weld metal with compatible composition and mechanical properties. Shielded metal arc welding electrodes are available for joining the following base metals:

1. Carbon steels,
2. Low-alloy steels,
3. High-alloy steels,
4. High-strength low-alloy steels,
5. Coated steels,
6. Tool and die steel,
7. Stainless and heat-resisting steels,
8. Precipitation-hardening steels,
9. Cast irons (ductile and gray),
10. Aluminum and aluminum alloys,
11. Copper and copper alloys, and
12. Nickel and cobalt alloys.

Electrodes are also available for applying surfacing to these same base metals to make them resistant to wear, impact or corrosion. Surfacing may also be applied to enhance the appearance of the product.

Table 2.2
AWS Specifications for Covered Electrodes

Type of Electrode	AWS Specification
Carbon steel	*Specification for Carbon Steel Electrodes for Shielded Metal Arc Welding,* AWS A5.1
Low-alloy steel	*Specification for Low-Alloy Steel Electrodes for Shielded Metal Arc Welding,* AWS A5.5
Corrosion resistant steel	*Specification for Stainless Steel Electrodes for Shielded Metal Arc Welding, AWS A5.4*
Cast iron	*Specification for Welding Electrodes and Rods for Cast Iron,* AWS A5.15
Aluminum and aluminum alloys	*Specification for Aluminum and Aluminum Alloy Electrodes for Shielded Metal Arc Welding,* AWS A5.3/A5.3M
Copper alloys	*Specification for Covered Copper and Copper Alloy Welding Electrodes,* AWS A5.6
Nickel alloys	*Specification for Nickel and Nickel Alloy Welding Rods for Shielded Metal Arc Welding,* AWS A5.11/A5.11M
Surfacing	*Specification for Solid Surfacing Welding Rods and Electrodes,* AWS A5.13
	Specification for Composite Surfacing Welding Rods and Electrodes, AWS A5.21

COVERED ELECTRODES

In the United States, electrodes are categorized by the American Welding Society (AWS) and by the U.S. Department of Defense. The categorizing documents are referred to as *AWS specifications* and Mil/*Military specifications*. The content of the specification establishes classifications (or types) of SMAW covered electrodes for use on groups or categories of base material, as shown in Table 2.2. The specification defines the classifications on the basis of one or more of the following attributes:

1. The type of welding current,
2. The type of electrode covering,
3. Welding position,
4. Chemical composition of the undiluted weld metal, and
5. Mechanical properties of the undiluted weld metal.

Both AWS specifications and military specifications are periodically updated. An AWS revision of a specification is identified by the last two digits of the year of publication, which is appended to the specification number. Revisions to the military specifications are indicated by the appendage of a capital letter in alphabetical sequence beginning with "A."

In addition to the information specific to SMAW provided in Table 2.2, AWS filler metal specifications and related documents are listed in Appendix C.

Carbon Steel Electrodes. Carbon steels are alloys of iron and carbon in which the carbon does not exceed 1.0%; manganese does not exceed 1.65%; copper and silicon each do not exceed 0.60%, and any other element is merely a residual amount.[13] Shielded metal arc welding electrodes for welding carbon steels are described in *Specification for Carbon Steel Electrodes for Shielded Metal Arc Welding*, AWS A5.1.[14] This is a prime defining document that facilitates commercial applications of the shielded metal arc welding process.

In this standard, an alphanumeric system is used to title each electrode classification. For example, in E6010, the "E" designates an electrode (a material that functions as an electrical terminal or juncture). The first two digits, 60, specify the minimum tensile strength of the undiluted weld metal (thousands of pounds-force per square inch of area [ksi]).[15] The third digit, 1, represents the welding positions in which the electrode can be applied. (In this case, 1 refers to all positions). The last digit in conjunction with the prior digit, 10, designates the type of covering on the electrode and the type of current with which the electrode can be used.

The specification provides for optional designators to be appended to the primary classification title to accommodate electrodes with a low-hydrogen covering. The letter "R" is appended if the moisture content of

13. For guidance on electrodes for carbon steel plate, see "Carbon and Low-Alloy Steels," Chapter 1 of American Welding Society (AWS) Welding Handbook Committee, Oates, W. R., and A. M. Saitta, eds., 1998, *Materials and Applications—Part 2*, Vol. 4 of *Welding Handbook*, 8th ed., Miami: American Welding Society.
14. See Reference 6.
15. The reader familiar with base plate designators must not confuse the practice of designating yield strength with the AWS system of designating tensile strength in the type identifier. In the AWS system, the yield strength for SMAW electrodes intended for multiple-pass groove and fillet welding is also specified. Thus, the referenced specification must be researched to obtain the actual requirement.

the covering is equal to or less than the specified limits for *as-manufactured* electrodes even after atmospheric exposure of the electrodes. For testing purposes, exposure is simulated by placing the electrodes into a chamber maintained with a very high water vapor content (controlled humidity and temperature) for a minimum time.

The letter "H" immediately followed by a number indicates the level of the hydrogen content, (2, 4, 8, or 16) if a hydrogen test is successfully performed. The E7018H4R is an example of the designator for a low-hydrogen electrode that is moisture resistant and when hydrogen tested, gives results that are less than 4 milliliters (mL) per 100 grams of deposited metal.

For historical reasons, the designation for the E7018M class is somewhat of an exception. This classification title and the corresponding requirements were adopted from a military specification before a cohesive system for communicating the level of hydrogen evolution from the weld metal and the moisture content of the covering were incorporated in AWS A5.1.[16] Consequently, for E7018M electrodes, the tests for hydrogen evolution and for the moisture content of the covering are mandated by the specification, and therefore optional designators would be redundant and are not used with the E7018M designator. The only difference between an AWS A5.1-91 E7018M and a MIL-E-0022200/10A(SH) MIL 7018-M is that the AWS classification has a more lenient limit on the exposed moisture content of the covering. The final anomaly of the E7018M class is the lack of a minimum tensile requirement. Therefore, undiluted weld metal made with E7018M electrodes can have less than 480 kPa (70,000 psi) tensile strength.

The AWS A5.1 classification also uses a -1 designator to distinguish closely linked electrode types. The -1 is the first appendage to either the E7016, E7018 or the E7024 classifications, which indicates that the deposited weld metal meets more stringent toughness requirements as measured by the Charpy V-notch impact test.

For covered electrodes manufactured to meet the AWS A5.1 specification, at least one printing of the type designator must appear on the covering. However, the "E" common to all covered electrode types is omitted from the printed identification on each electrode. The remaining numeric class designator consists of one of two choices of tensile strength, 60 or 70, followed by a single digit that indicates the welding position: 1, all positions; 2, flat and horizontal; and 3, downhill (often called *vertical down*). A final digit is added, which is to be read in combination with the third digit. These last two digits taken together (i.e. Exx18) designate the type of covering on the electrode and the type of welding current used. Although the classification system can accommodate a large number of designators, only seventeen primary types are defined in the latest edition of AWS A5.1.

The A5.1 specification also specifies other important properties of the undiluted weld metal. These properties include yield strength, ductility, toughness, soundness, and chemical composition. In addition, physical properties of the electrodes are also specified. These include the length and diameter of the core wire, the concentricity of the covering, the configuration of the strike end and the holder end, and so forth. The specification also provides requirements for the manufacture, storage, and drying of the electrodes.

Of the types defined in AWS A5.1, some are closely related in design and intended function but they differ in the type of welding current to be used. For example, the Exx10 and Exx11 classifications are ideal for use on rusty or oily carbon steel, galvanized steel, and open root welding. The Exx10 and Exx11 classes are often the first choice for farm and repair shop use. However, the Exx10 class is designed to operate on dc and the Exx11 class on ac. As a result, the Exx10 class is widely used in the field when welding pipe with diesel-powered welding machines and the Exx11 class is often used for farm and repair shop applications with economical ac power sources. The primary polarity for dc is usually electrode positive, although a few of the electrodes are intended for direct current electrode negative. Some electrodes can be used with either polarity. The E6010s can be used with electrode negative to handle large root openings.

Low-Hydrogen Carbon Steel Electrodes. Another example of related electrode types are the basic slag (low-hydrogen) electrodes: the Exx15, Exx16, Exx18, Exx28, and Exx48 classes. These electrodes not only have coverings with a very low hydrogen content but they also contain minerals that are deemed basic in the welding operation. The term *basic* implies that the chemicals in the electrode cover function to reduce the oxygen content in the weld and diminish the number of harmful oxide inclusions at the grain boundaries of the microstructure. As a result, the weld metal deposited is cleaner and hence has the potential to attain very high toughness or Charpy V-notch values. The low-hydrogen content in the covering results in very low levels of diffusible hydrogen in the weld. Hydrogen in the weld metal can be extremely detrimental and is often the cause of brittle failures and delayed cracking. Hydrogen cracks occur when these four factors are simultaneously present:

1. Diffusible hydrogen is in the weld metal or the heat-affected zone of the base metal, or both;
2. The weld metal has high residual stresses caused by the constraint of the joint members and the heat input from the arc;

16. See Reference 6.

3. The weld metal has a susceptible microstructure or the heat-affected zone of the base plate develops a susceptible microstructure, or both; and
4. The temperature of the weldment is in the range of –101°C to 204°C] (–214°F to 400°F).

The crack may be parallel to the fusion boundary and within the heat-affected zone or may originate in the weld deposit. The crack tends to form after the completion of the weld. Because of the location of the crack and the time lag for forming, the cracking mechanism has been identified as *underbead cracking* and also as *delayed cracking*. The loss of weld metal ductility due to hydrogen is apparent in tensile testing. A hydrogen-affected tensile specimen contains a slight but noticeable level of hydrogen at the time of testing. The hydrogen affect manifests itself as voids on the fractured surface and a slight loss in the elongation at failure. The fracture surface of the hydrogen-damaged tensile specimens are partially or fully brittle (shiny in appearance) and measure very low in elongation. The risk of hydrogen-induced cracking is higher when welding hardenable carbon steels and rises proportionately to the thickness of the plate steel. Consequently, the low-hydrogen electrode types are excellent candidates for use in these applications.

Although there are many similarities in the performance of the weld metal, each classification of the low-hydrogen group provides different characteristics. The E7015 has a basic covering and is designed for dc; The E7016 is similar to the E7015 but operates with ac. These types are sometimes preferred for circumferential pipe welding because the covering is thin, which improves accessibility to the root of the joint.

The E7018 has a large amount of iron powder in the covering and performs best when applied with DCEP, although ac is possible. The iron powder quiets the metal transfer and increases the deposition rate. The E7018 class is a popular choice for many applications because it can be used in all positions with high deposition rates. Some E7018 electrodes are specifically designed for ac application and are sold as E7018AC. These ac E7018 electrodes are usually not as resistant to moisture as the more popular versions.

The E7028 class is designed with a thick covering and a high iron powder content. They achieve the highest deposition rate among the low-hydrogen electrode types. However, the thick covering and fluid nature of the molten weld limit the application versatility and hence they are the economical low-hydrogen choice for horizontal- and flat-position welding only.

The E7048 class is similar to the E7018 type, but is optimized for downhill applications. The E7048 electrodes are not very popular because multiple-pass, X-ray-quality welds are difficult to achieve consistently.

Of these types, the E7018 class is the most popular because it can be used in all positions, achieves high deposition rates, and has excellent welder appeal. Basic SMAW low-hydrogen electrodes are also used on carbon steel plate. The electrodes tend to desulfurize the weld pool and diminish the tendency for hot cracking.

Other related electrode types available for use on carbon steels are the titania (rutile) and the high-iron oxide groups. A good guide for the areas of application for these types is presented in the appendix to *Specification for Carbon Steel Electrodes for Shielded Metal Arc Welding*, AWS A5.1.[17]

Low-Alloy Steel Electrodes

Low-alloy steel SMAW electrodes are designed for the welding of carbon steels and low-alloy steels of matching composition. The low-alloy designation implies that no one element in the undiluted weld metal exceeds 10.5% and at least one element exceeds the chemical limits of the AWS A5.1 specification. As a general rule, the "matching" composition for the undiluted weld metal will be leaner in alloy content than the base material. The reduced alloy content offsets the increase in tensile strength that results from the high cooling rates inherent in arc welding.

The standard *Specification for Low-Alloy Steel Electrodes for Shielded Metal Arc Welding*, AWS A5.5,[18] classifies low-alloy steel electrodes according to the same numbering system described for carbon steel electrodes, except that the low-alloy steel classification uses an additional suffix consisting of one or more letters to indicate the dominant alloying elements, followed by a number to identify each type of alloy. For example, in the electrode classification E7010-A1, the letter "A" in the suffix A1 indicates that carbon and molybdenum are the primary alloying elements that this electrode adds to the undiluted weld metal, and the 1 means it is the first listed of the carbon-molybdenum type. Another example is E8016-C2, where the "C" signifies that nickel is the primary alloying element. The following alloy systems are defined in the specification AWS A5.5:

A = carbon-molybdenum;
B = chromium-molybdenum;
C = nickel;
NM = nickel-molybdenum; and
D = manganese-molybdenum.

In the low-alloy specification, the required minimum undiluted weld metal tensile strengths range from

17. See Reference 6.
18. American Welding Society (AWS) Committee on Filler Metals and Allied Materials, *Specification for Low-Alloy Steel Electrodes for Shielded Metal Arc Welding*, AWS A5.5, Miami: American Welding Society.

480 MPa to 830 MPa (70 ksi to 120 ksi) in 70 MPa (10 ksi) increments. The condition of the tensile specimen for testing the strength levels is either *as welded* or *stress relieved*, depending on the anticipated use of the alloy system. When stress relieving is required, the weld test assembly is subjected to postweld heat treatment (PWHT) for one hour at the temperature specified for that electrode classification. Fabricators using holding times that are significantly different from one hour at temperature or holding temperatures that are substantially higher may have to be more selective in the electrodes they use. They may be required to run tests to demonstrate that the mechanical properties of the weld metal from the selected electrode will be adequate after the specific heat treatment temperature and holding time. The AWS A5.5 specification also sets radiographic quality standards for deposited weld metal and notch toughness requirements for the SMAW electrode classifications.

There are some SMAW electrodes that are not classified in AWS specifications. These electrodes may not be listed because they are designed for very specific base material not commonly used or they broadly match a steel composition listed in a standard published by the American Iron and Steel Institute (AISI)[19] for low-alloy steel base metal compositions, such as 4130. These unclassified electrodes are usually imprinted with the alloy type and are made with a low-hydrogen covering.

Military specifications are also published that govern the use of SMAW electrodes for joining low-alloy and higher-strength steels. Fortunately, these specifications use designators similar to those in the AWS specification. A harmonization effort is in progress between the federal government and the American Welding Society in which military specifications will be discontinued and the requirements of the AWS classifications will be changed to better reflect the needs of military applications.

Controlling and maintaining a very low moisture content in SMAW electrodes is more important for the welding of low-alloy and higher-strength steels than for the welding of mild steel. The risk for hydrogen-induced cracking in welding high-strength and low-alloy steels is heightened because the residual stresses of the joint are higher and because the propensity for forming a susceptible microstructure is significantly increased. For this reason, the AWS A5.5 specification sets limits on the moisture content of low-hydrogen electrodes in the *as-received* or *reconditioned* state and in the *as-exposed* state.[20] These two conditions bracket the moisture content of the electrode between the start of welding and at the finish of a work shift on a day in which the air had a high water content (125 grains of moisture at 27°C (80°F). The limit of moisture content of the as-received electrodes ranges from 0.1% to 0.4% by weight, depending on the classification of the electrode. The higher the strength level of the classification, the lower the limit is on the as-received moisture content. The limit for moisture of electrodes with a low-moisture-absorbing covering (with the "R" suffix designator) in the exposed condition is 0.4% by weight. Unfortunately, the amount of moisture in the exposed condition is not specified for SMAW electrodes not bearing the "R" designator.

The first consideration in avoiding hydrogen-induced cracking when welding low-alloy steels is to select SMAW electrodes with the lowest moisture content and with coverings that have very little tendency to absorb moisture. It is equally important to follow the recommendations for preheat and interpass temperature, heat input, and cooling rate. Higher preheat temperature, higher heat input, and slow cooling rates are usually preferred. Finally, the code requirements for handling the electrodes should not be compromised. Exposure to high humidity (in the range of 70% relative humidity or higher) may increase the moisture content of the electrode to an unacceptable level in only a few hours.

High-Alloy Steel Electrodes

High-alloy steels generally possess high tensile strength with good toughness at the recommended service temperature. In these steels, the quantity of alloying elements exceeds 10% by weight. The objective of alloying is not to create exceptional corrosion resistance (stainless steels), improve high-temperature service (chrome-molybdenum steels), achieve very high hardness (tool-steels) or attain cryogenic toughness (austenitic stainless steels), but to achieve three valued properties: high strength, good ductility, and toughness. In this context, the high-alloy steels are similar to the low-alloy steels except for the amount of alloying elements they contain.

The specific high-alloy steels tend to be proprietary and fall into three categories: nickel-cobalt steels, maraging steels, and austenitic manganese (Hadfield) steels. Of the high-alloy steels, SMAW electrodes are practicable only for the Hadfield steels. The level of impurities resulting from the use of the shielded metal arc welding process is too high to obtain the desired toughness and avert the possibility of hot cracking in nickel-cobalt steels and maraging steels.

The Hadfield steels are alloyed with manganese (11% to 14%) and carbon (0.7% to 1.4%). During its manufacture, the steel is quenched from 982°C to 1038°C (1800°F to 1900°F) to room temperature with

19. American Iron and Steel Institute (AISI), 1101 17th St. N.W., Suite 1300, Washington, DC 20036-4700.
20. See Reference 18.

agitated water. The rapid cooling preserves the austenitic structure by suppressing the formation of carbides and the eventual transformation to ferrite and carbides. Although the room-temperature austenitic structure thus obtained is very stable, heating the steel to a temperature in excess of 316°C (600°F) will allow it to transform to ferrite and carbides. The formation of carbides embrittles the steel, and in service, may cause steel sections to fail by spalling. Consequently, the reheating of the austenitic manganese steels should be avoided.

The properties of austenitic manganese steel that make it commercially popular are moderate strength, good ductility and toughness, and most advantageous of all, the rapid work-hardening of the surface in response to compressive loads or impacts. As a result of these characteristics, a product fabricated with austenitic manganese steel has a very hard surface with a very tough underlying structure—an ideal combination for a variety of applications.

Some of the electrodes used for the welding of austenitic manganese steels are classified in *Specification for Solid Surfacing Welding Rods and Electrodes,* AWS A5.13.[21] This standard categorizes SMAW electrodes solely on the basis of the chemical composition of the undiluted weld metal. The class name is derived by appending the chemical symbols of the principal alloying elements to the letter "E." If the basic group requires subdivision, a hyphen (-) followed by a letter in alphabetical order starting with "A" is added. Further subdivision is achieved by adding a number to the letter. For example, two electrodes are used for welding austenitic manganese steel; one, an alloy of iron and manganese with some nickel, is designated EFeMn-A; the other, containing similar primary alloys but with molybdenum instead of nickel, is designated EFeMn-B. Both classifications can be used to join manganese steel. The EFeMn-A weld metal is tougher, whereas the EFeMn-B weld metal has higher strength.

For applications in which abrasive wear or high-temperature exposure is a dominant concern, proprietary and recently developed chromium-nickel austenitic SMAW electrodes offer superior performance. Austenitic stainless steel electrodes are often used for welding austenitic manganese steel to mild and low-alloy steels; however, these electrodes may develop a brittle interface that will cause the weld to have very low resistance to wear. Austenitic manganese steel electrodes, manganese-chromium electrodes, and proprietary electrodes, when properly applied, are better choices for the welding of these dissimilar metals.

21. American Welding Society (AWS) Committee on Filler Metals and Allied Materials, *Specification for Solid Surfacing Welding Rods and Electrodes,* AWS A5.13, Miami: American Welding Society.

Electrodes for Coated Steel

Coatings are applied to carbon steels to achieve a processing advantage, to protect it from corrosive environments, or to achieve a decorative finish. The coatings are either metallic or organic paint. The SMAW electrodes for joining these coated steels must possess unique characteristics to cope with the constituents of the coating and must deposit weld metal compatible with the base steel. SMAW electrodes that are successfully used with coated steels are classified under the appropriate steel specification (i.e., AWS A5.1).[22]

The most popular coatings applied to carbon steels are zinc metal and zinc alloys. The zinc is applied in a batch hot-dip process (general galvanizing), in a continuous hot-dip process, or in an electrolytic bath. The amount and type of the coating varies depending on the process. The welding procedure and techniques are likewise type-specific. Heavier coatings, characteristic of general galvanizing, require specific welding procedures and preparation to achieve sound welds. Steel with light coatings applied with a continuous hot-dip or an electrolytic process is typically welded with the qualified procedure normally applied to the underlying steel.

With heavily coated galvanized plate, the presence of liquid zinc at the weld root must be avoided or minimized, because the liquid zinc may penetrate into the alloy-rich dendrites of the base metal microstructure and cause cracking. Liquid zinc can similarly embrittle the weld metal. The liquid zinc is more severe in attacking weld deposits that contain more than 0.40% silicon. Consequently, cellulose (EXX10 and EXX11) electrodes and rutile (EXX12 and EXX13) electrodes are preferred when the level of zinc is high. Sometimes a combination of electrode types can be used to achieve the best mechanical properties. A low-silicon type can be used in the root and a basic type (EXX16 or EXX18) can be used to fill the joint.

The detrimental effects of the zinc can be minimized by removing it from the joint with grit blasting or by beveling the edges of the base plate. Alternatively, providing a slight gap at the root or between the backing strip and the plate will provide an exit path for the liquefied and vaporized zinc and will reduce the volume of zinc that can interact with the weld pool. In some instances, a slight whip of the electrode helps to volatilize some of the zinc ahead of the arc, thereby improving the probability for sound welds. Nonetheless, a reduction of joint speed is common for successful welding on general galvanized plate.

Galvanized sheet produced in a continuous hot dip or electrolytic galvanizing process has a lighter coating weight and is typically welded with the cellulosic or

22. See Reference 6.

rutile electrodes using settings that are normally used for uncoated mild steel. The cellulosic electrodes produce a highly penetrating arc and therefore may not be desirable for use on thin sheet steel. However, cellulosic electrodes are the lowest in silicon content and are best suited for minimizing the risk of liquid zinc embrittlement. A galvannealed coating is composed of iron-zinc alloys and is usually easily weldable because the amount of zinc per unit surface is tolerable.

A coating of aluminum on steel generally does not cause porosity or slag inclusions, but the wetting of the weld metal is detrimentally affected. Two reasons are cited for this behavior: (1) the aluminum oxide that forms around the weld pool is not easily fluxed, and (2) any aluminum dissolved into the weld pool causes it to be more viscous and sluggish, achieving less wetting of the base plate during solidification. When choosing a SMAW electrode for this application, it is necessary to consider the mechanical property requirements for the weld metal and whether protection of the weld from the corrosive environment must be employed. A choice is usually made between mild steel and stainless steel SMAW electrodes that would be proper if the steel were bare. The welds made with mild steel electrodes must be protected with a coating.

Steels that are coated with zinc-base paints can be readily welded using procedures similar to those used on uncoated steel. The coating may cause more spatter and arc instability, but it does not substantially affect weld integrity. The basic SMAW electrodes are most apt to produce occasional porosity. Organic paints should not be present on the joint surface because the paint will not survive the heat from the welding arc and will disrupt the welding. Organic paints may also cause porosity. If the paint cannot be removed, a cellulosic SMAW electrode is the best choice for achieving a sound weld.

Stainless And Heat-Resisting Steel Electrodes

Covered electrodes for welding stainless and heat-resisting steels are classified in *Specification for Stainless Steel Welding Electrodes for Shielded Metal Arc Welding*, AWS A5.4.[23] Each classification in this specification is based on the chemical composition of the undiluted weld metal, the position of welding, and the type of welding current for which the electrodes are suitable. The classification system uses the familiar numeric representations for the chemical composition and uses appendages to communicate the unique traits of each type.

Taking E310-15 and E310-16 as examples, the prefix "E" indicates an electrode. The first three digits refer to the alloy type, with respect to chemical composition. These may be followed by the symbol of a chemical element and either the letters "L" or "H." The letter "R" may be added to indicate a modification to the all-weld-metal chemistry of the parent class. For example, in E310Mo-15, Mo (the symbol for molybdenum) signifies that the molybdenum content is different from the E310-15 weld metal. The letters "L" and "H" inform the user that the carbon content is controlled within a desired range. When an electrode type bears the "L" appendage, it means that the carbon limit is lower. All electrodes of the "L" type also meet the requirements of the parent class. For example, the E308MoL-XX electrodes also meet the chemical requirements of the E308Mo-XX class. The "H" designator indicates that the minimum required carbon content is higher or that the carbon range exceeds the level of the parent class. The letter "R" signifies that detrimental residuals are restricted to low levels. Finally, a hyphen (-) links the chemistry designation to two digits that individually refer to the position of welding and the type of current for which the electrodes are suitable. The first number immediately following the hyphen is either 1 or 2. The number 1 indicates that electrode diameters 4 mm (5/32 in.) and smaller are usable in all positions and larger sizes are usable only in the flat and horizontal positions. The number 2 indicates that all sizes are meant to be used only in the flat and horizontal positions. If the second digit is a 5, the electrodes are suitable for use with DCEP only, but if the second digit is a 6 or a 7, the electrodes are suitable for either ac or DCEP.

The last two digits taken together describe a unique electrode covering that produces distinguishable weld properties and welding characteristics. Limestone powder, which consists primarily of calcium carbonate, is a dominant chemical in the covering of the E310-15 class. For this reason, this variety is also known as *lime electrodes*.[24] The E310-15 produces a sharp, forceful arc that promotes good weld penetration into the base plate, generally better weld metal soundness, and higher quality welds when subjected to radiographic inspection. The weld pool is somewhat agitated and at times this causes good-sized globules to be expelled from the joint as spatter. The liquid slag is very stiff (viscous) and crowds the arc by failing to clear the

23. American Welding Society (AWS) Committee on Filler Metals and Allied Materials, *Specification for Stainless Steel Welding Electrodes for Shielded Metal Arc Welding*, AWS A5.4, Miami: American Welding Society.

24. The E310-15, E310-16, and E310-17 varieties of SMAW stainless electrodes have familiar common names. Some of the most frequently used names are E31015: lime, lime-basic, dc; E310-16: rutile, titania, ac/dc, basic-rutile; E310-17: acid-rutile, European. The E310-17, E310-26 and E310-26 varieties were first listed in the AWS A5.4.92 revision.

surface of the weld pool about the hot spot. The slag solidifies rapidly as the electrode is advanced. Hence, the weld beads tend to be convex, with more irregularity and coarse ripples. However, the fast-freezing slag and good penetration allows the welder greater control during out-of-position work, such as pipe welding. The friable slag requires chipping and scraping for complete removal.

The covering of the E310-16 ac/dc variety contains high levels of titanium dioxide and may be bonded with potassium silicate or enhanced with potassium minerals to achieve a smooth arc, less spatter, and good performance when using alternating current. With this electrode class, the slag readily clears away from the weld pool, which allows the welder to have greater control when welding in the horizontal and flat positions. The finished beads are finely rippled, slightly convex, and very uniform. In out-of-position welding, the fluidity of the slag requires the welder to work the electrode with a wider weaving motion to achieve a reasonably good bead profile. The slag is dense and strong but easily detaches from the weld. The vigorous use of a chipping hammer is generally not required, unless, for example in the root pass of a flat butt weld, the slag locks between the joint faces.

The E310-17 type is a popular and distinct version of an electrode once falling within the E310-16 class. The E310-17 improvements were achieved by enlarging the size of the covering and by substituting silica (SiO_2) for some of the titania (TiO_2). These changes improved the welding performance in horizontal and flat applications. The E310-17 coverings achieve a smoother arc and a spray metal transfer mode. The wider arc "feathers" the bead edges into the base plate and enhances the wetting of the weld pool to achieve a concave bead cross section. The slag freezes slowly, resulting in a uniform, finely rippled weld surface. The slag often self detaches from the surface when the weld is completed. However, the slag is more fluid, which compromises uphill welding and generally requires the vertical beads to be made with a weaving motion. As a result, the smallest sized fillet deposit made with the E310-17 is larger than that achieved by the same diameter E310-15 or E310-16 varieties.

The EXXX-25 and EXXX-26 varieties are similar in composition to the E310-15 and E310-16, respectively. However, the coatings are purposely alloyed to obtain higher deposition rates. In some designs of EXXX-25 and EXXX-26 electrodes, the total alloy content of the core wire is very low relative to the alloy content of the undiluted weld metal, for example, when mild steel core wire is used to make an E308L-25. In these cases, large amounts of metallic powders must be added to the covering to achieve the required analysis of the weld deposit. Hence, the diameter of the covering is typically much larger than the corresponding all-position variety.

Stainless steels can be classified into three general types: austenitic, martensitic, and ferritic, and four specialty types: precipitation hardening, superaustenitic, superferritic, and duplex. The names of these steels emphasize the composition of the alloy; the prefix *super* conveys that there is an abundance of alloying. The resistance of these steels to corrosion is attributed to a passive oxide film that seals the surface. In iron alloys, this film is first achieved when the chromium content in solid solution exceeds 10.5%. Iron alloys, by definition, must have iron as the largest portion by weight. For this reason, the austenitic stainless steel electrodes are not unlike some electrodes designated as nickel types that use the same alloying elements. In nickel-alloy electrodes, however, nickel is the largest weight fraction.

The great variety of commercially available stainless steels arises from the need for products that resist diverse corrosive chemical solutions or atmospheres and that attain longevity in service. Each composition and microstructure has unique properties that must be considered for the intended use and for its suitability for joining by the shielded metal arc welding process. The ideal stainless alloy with the desired solid phase(s) would be stable throughout the temperature cycles experienced during welding and in service. Unfortunately, during welding and other exposure to high temperature, changes can occur in the constitution and structure of stainless steel plate and weld metal. Chemical reactions can generate new, unwanted phases and may deplete a needed element in the solid solution. The microstructure may become very coarse grained, which raises the ductile-to-brittle transition temperature to an unacceptable level. A hard, brittle martensite can form; or certain elements can segregate or diffuse to the grain boundaries. These changes can have a devastating effect on the performance of a welded stainless steel structure. Therefore, the joining of each stainless steel assembly should be well planned.[25]

Of the different categories of stainless steel electrodes, the austenitic group (2XX and 3XX) has by far the largest number of types. For this group, the required undiluted weld metal may differ from the composition of the base metal in order to produce a weld deposit that is not fully austenitic but which contains ferrite. The ferrite prevents the solidification cracking of the weld metal. The amount of ferrite common to the various welding electrodes is discussed in detail in the standard *Specification for Stainless Steel Welding Electrodes*

25. For more detailed information, see "Stainless and Heat-Resisting Steels," Chapter 5 of American Welding Society (AWS) Welding Handbook Committee, Oates, W. R. and A. M. Saitta, eds., *Materials and Applications—Part 2,* Vol. 4 of Welding Handbook, Miami: American Welding Society.

for Shielded Metal Arc Welding, AWS A5.4.[26] In general, a minimum ferrite content in the range of ferrite number (FN) 3 to 5 is sufficient. The WRC-1992 diagram[27, 28] can be used to predict the ferrite number of the 300 Series stainless steels and the Espy diagram[29] can be used to predict the ferrite number of the 200 Series. The ferrite number can also be directly measured with magnetic instruments.[30]

Certain austenitic stainless steel weld metals (types 310, 320, 320LR, and 330, for example) do not form ferrite because their nickel content is too high. For these materials, the phosphorus, sulfur, and silicon content of the weld metal should be limited or the carbon content should be increased as a means of minimizing the tendency for solidification cracking.

Using appropriate welding procedures can also reduce the solidification cracking tendency. Low heat input, for example, is beneficial. During welding, the welder can utilize a small amount of weaving as a means of promoting cellular grain growth, which is more resistant to cracking. Moving the electrode back over the weld pool before breaking the arc will help avoid deep craters and diminish the possibility of crater cracks.

The publication *Specification for Stainless Steel Welding Electrodes for Shielded Metal Arc Welding,* AWS A5.4 contains two classifications of electrodes for joining some of the 4XX Series stainless steels, E410-XX and E410NiMo-XX.[31] The weld metal produced with both types of electrodes is susceptible to the formation of martensite during the cooling phase of the welding cycle. The resulting martensitic structure may be excessively hard, brittle, and sensitive to hydrogen-induced cracking. Consequently, preheating the base plate before welding and heat-treating the welded structure immediately after is essential to achieving a weld deposit fit for its intended service.

The specification contains one ferritic type, E430. Although the use of the various ferritic stainless steels has grown, the need for a matching shielded metal arc welding electrode for these applications has not followed, and probably will not materialize. Most of the applications are light-gauge sheet, which is more economically welded with flux-cored or metal-cored electrodes. In addition, when welding the ferritic stainless steels, austenitic stainless steel electrodes and nickel electrodes are often more easily applied than a corresponding electrode of matching composition. For these reasons, the development of a ferritic stainless steel SMAW electrode with matching composition may not be warranted.

The weld metal of the E430-XX electrode is balanced to provide adequate strength and ductility. However, at elevated temperature the weld metal and the heat-affected zone of the plate can transform to austenite, and on air-cooling, into brittle martensite. During welding, chromium carbides can form near the grain boundaries, depleting the chromium in solid solution and leaving the grain boundaries of the welded structure susceptible to intergranular corrosion. Thus, to obtain optimum mechanical properties and corrosion resistance, the weldment should be heat-treated after welding.

The E630 electrodes are designed for welding the martensitic and semiaustenitic precipitation-hardening stainless steels. Optimum results are obtained when the dilution of the weld metal composition from the base plate is kept to a minimum. The welded structure must be postweld heat treated to attain the nominal or required strength and to relieve welding stresses. If high strength is not required (if under-matching is allowed,) austenitic electrodes are more commonly used.

The superferritic stainless steels are produced in a process that minimizes the content of the interstitial elements and adds elements that further stabilize the ferrite. The shielded metal arc welding process cannot achieve the precise control over the carbon-plus-nitrogen content and cannot efficiently transfer reactive elements such as aluminum and titanium. Therefore, no superferritic matching composition SMAW electrodes are available.

The term *superaustenitic* applies to austenitic compositions of stainless steel that have a higher molybdenum content and greater resistance to pitting corrosion. The pitting resistance equivalent (PRE) number is computed from the chemical composition of the stainless steel. The usual practice is to classify as superaustenitic those austenitic stainless steel compositions that meet or exceed a value of 40. The linchpin for the successful development of the various grades of superaustenitic steels was the introduction of nitrogen as an alloying element. Nitrogen, a strong austenite stabilizer, retards the formation of brittle phases at higher temperatures when a high amount of molybdenum is present. The E383, E385, and nickel-base electrodes are used for

26. See Reference 23.

27. Kotecki, D. J., and T. A. Siewert, 1992, WRC 1992 Constitution Diagram for Stainless Steel Weld Metal: a Modification of the WRC-1988 Diagram, *Welding Journal,* 71(5), 171s—178-s. Miami: American Welding Society.

28. See "Stainless and Heat-Resisting Steels," Chapter 5 of American Welding Society (AWS) Welding Handbook Committee, Oates, W. R., and A. M. Saitta, eds. *Materials and Applications—Part 2,* Vol. 4 of *Welding Handbook,* 8th ed., P 261, "Ferrite in Austenitic Stainless Steel Weld Metal," Miami: American Welding Society.

29. Espy, R. H., 1982, Weldability of Nitrogen-Strengthened Stainless Steels, *Welding Journal,* 61(5), 149-s—155-s, Miami: American Welding Society.

30. American Welding Society (AWS) Committee on Filler Metals and Allied Materials, 1997, *Standard Procedures for Calibrating Magnetic Instruments to Measure the Delta Ferrite Content of Austenitic Stainless Weld Metal,* AWS A4.2/A4.2M, Miami: American Welding Society.

31. See Reference 23.

joining these steels. During the welding process, the heat input, the loss of nitrogen, and the segregation of molybdenum must be considered.

The term *duplex* is a fitting name for the stainless steel compositions that have a dual structure of ferrite and austenite. The composition is balanced to solidify primarily as ferrite, and on further cooling, to partially transform to austenite. In welding, the cooling rate and the heat input are critical to the ratio of the phases that develop. For a given application, the acceptable heat input must be bracketed. The E2209 and E2553 electrodes are used to weld the corresponding 22% and 25% chromium duplex stainless steel alloys. Austenitic stainless steel electrodes and nickel electrodes are generally not used to join duplex stainless steel.

Electrodes for Cast Iron

The successful welding of cast iron is challenging. Success is generally inversely proportional to the strength of the material and directly proportional to its ductility, and only achieved by the conscientious application of a proven procedure with the correct SMAW electrode.

The standard *Specification for Welding Electrodes and Rods for Cast Iron*, AWS A5.15 classifies the SMAW electrodes for welding cast iron.[32] The designator for the electrodes classified in A5.15 consists of the letter "E" followed by the atomic symbol of the major elements in descending order. The SMAW filler metal designations are ENi for nickel, ENiFe for nickel-iron, ENiFeMn for nickel-iron-manganese, ENiCu for nickel-copper alloys, and ESt for mild steel compositions ("St" in the specification stands for mild steel). In addition, in those instances where the type designations may be confused with those from another specification, "-CI" (cast iron) must be appended to the designator. The designation is completed by adding a letter in alphabetical order beginning with "A" to the first and subsequent variations in a given class, e.g., ENiFe-CI-A.

Once an electrode with the required strength and machining characteristics has been selected, a welding procedure that addresses the amperage range (heat input), joint design, surface cleanliness, the need for preheat, preheat temperature, welding sequence, length of the stringer welds, interpass temperature, cooling, postweld heat treatment, peening, anchoring or grooving and so forth, must be worked out. The final values are a function of the casting grade, the mass and the complexity of the cast part, and the choice of electrode.

Small pits and cracks can be repair welded without excessive preparation of the casting. A "-CI" nickel electrode can be used with low amperage and no preheat. If the ENiFe-CI-X electrode is selected, a choice exists between electrodes made with a core wire that is a composite of nickel surrounded by iron or those made with a nickel-alloy core wire. The latter overheat because of the high resistivity of the nickel-alloy core wire, causing very high stub losses. Nonetheless, even with the high stub loss, the nickel alloy core wire electrodes are usually less expensive than the composite electrodes.

The tin-bronze and the aluminum-bronze electrodes described in *Specification for Covered Copper and Copper Alloy Arc Welding Electrodes,* AWS A5.6 are also used to braze weld cast iron or to apply surfacing.[33] The braze welding process is accomplished by preheating the casting to about 400°F (200°C), and using the lowest amperage (direct current electrode positive) that will produce a good bond between the weld metal and the base metal. Care must be exercised to avoid melting the surface of the casting.

Aluminum and Aluminum-Alloy Electrodes

The standard *Specification for Aluminum and Aluminum-Alloy Electrodes for Shielded Metal Arc Welding*, AWS A5.3, contains three classifications of covered electrodes for the welding of aluminum base metals.[34] The electrodes are classified according to the mechanical properties they produce in a defined groove weld and according to the chemical composition of the core wire. The chemical requirements for each type of core wire are identical to the corresponding wrought aluminum alloy classification. Consequently, the electrode type is designated by appending the four-digit number in common use to identify the wrought alloy to the familiar "E" for electrode.

The E1100 and the E3003 classifications are intended for the welding of alloys of similar composition. The E4043 electrodes have broader application; the silicon content increases the fluidity and reduces the melting point. When welding aluminum, it is imperative that the weld pool solidifies at a temperature that is lower than the melting point of the base plate. The E4043 electrodes are used to weld the 6XXX series, some 5XXX (up to 2.5% magnesium), aluminum-silicon casting alloys, and aluminum base metals 1100, 1350(EC), and 3003.

---

32. American Welding Society (AWS) Committee on Filler Metals and Allied Materials, *Specification for Welding Electrodes and Rods for Cast Iron,* AWS A5.15, Miami: American Welding Society.

33. American Welding Society (AWS) Committee on Filler Metals and Allied Materials, *Specification for Covered Copper and Copper Alloy Arc Welding Electrodes,* AWS A5.6 (R 2000), Miami: American Welding Society.

34. American Welding Society (AWS) Committee on Filler Metals and Allied Materials, *Specification for Aluminum and Aluminum-Alloy Electrodes for Shielded Metal Arc Welding,* AWS A5.3/A5.3M, Miami: American Welding Society.

The low melting point of aluminum and the tenacious oxide film that readily forms on the surface require the use of alkali (low-melting) halides in the electrode covering. These chemical compounds dissolve the oxide film to promote wetting and fusion. However, they are very corrosive to aluminum and must be completely removed after welding. Also, these halides may contain or acquire moisture if the electrodes are not properly stored. Moisture readily leads to porosity within the weld metal. Aluminum and its alloys have a very low solubility for hydrogen in the solid state; the liquid has a higher solubility. If hydrogen from moisture or other sources contaminates the weld pool, it is rejected as a gas during solidification, forming porosity in the weld.

Aluminum electrodes are used primarily for shielded metal arc welding and repair applications that are not critical. The standard *Structural Welding Code—Aluminum,* AWS D1.2, makes no provision for the use of the SMAW process.[35]

The SMAW aluminum electrodes are used with DCEP. To avoid porosity, the electrodes should be stored in a heated cabinet until they are used. Electrodes that have been exposed to moisture should be reconditioned (baked) before use, or discarded.

If the arc is broken during welding, the molten slag may fuse over the end of the electrode. To use the remaining stub, the slag must be removed before restriking an arc. Because slag on the weld bead can be very corrosive to aluminum, it is important that all of it be removed upon completion of the weld.

Copper and Copper-Alloy Electrodes

The standard *Specification for Covered Copper and Copper Alloy Arc Welding Electrodes*, AWS A5.6,[36] classifies copper and copper-alloy SMAW electrodes on the basis of the chemical composition and mechanical properties of the undiluted weld metal. The designation system, as first conceived and published in 1948, consists of the letter "E" followed by the chemical symbols of the major alloying elements. If more than one type contains the same major alloying elements, each is identified by the letters A, B, C, and so on. Later, the convention of following this letter with a number was adopted to further subdivide the types, for example, ECuAl-A2. The primary types are CuSi for silicon bronze; CuSn for phosphor (tin) bronze; CuNi for copper-nickel; CuAl for aluminum-bronze; and CuMnNiAl for manganese-nickel-aluminum-bronze.

The electrodes classified in the specification A5.6 are generally used with DCEP. The welder holds a short arc and often uses a weaving motion equal to three times the electrode's nominal diameter to minimize slag entrapment.

Copper electrodes are used to weld unalloyed copper and to repair copper cladding on steel or cast iron. Silicon-bronze electrodes are used to weld copper-zinc alloys, copper, and some iron-base materials. They are also used for surfacing to provide corrosion resistance.

Phosphor-bronze and brass base metals are welded with phosphor-bronze electrodes. These electrodes are also used to braze weld copper alloys to steel and cast iron. The phosphor-bronzes are rather viscous when molten, but fluidity is improved by preheating to about 200°C (400°F). The electrodes and the base plate must be dry.

Copper-nickel electrodes are used to weld a wide range of copper-nickel alloys and to apply copper-nickel cladding on steel. In general, preheat is not necessary for these materials.

Aluminum-bronze electrodes have broad use for welding copper-base alloys and some dissimilar metal combinations. They are used to braze weld many ferrous metals and to apply overlays that provide a bearing surface or surfaces that are resistant to wear and corrosion. Welding is usually done in the flat position with some preheat. Aluminum-bronze electrodes are particularly suitable for welding galvanized and aluminized steel. Welds on these coated steels have good corrosion resistance, which eliminates the need for painting or galvanizing.

Manganese-nickel-aluminum-bronze electrodes are used to weld and repair bronzes of similar composition. These bronzes have excellent resistance to corrosion, erosion, and cavitation. They are widely used for the casting of ship propellers.

Electrodes for Nickel and Cobalt Alloys

Nickel and nickel alloys are generally welded with SMAW electrodes that are of similar composition. However, the electrodes may also contain specific additions of elements such as titanium, manganese, magnesium, and niobium. These additions deoxidize the weld metal and neutralize the deleterious effects of some harmful impurities. Ultimately, the addition of these reactive metals improves the soundness of the weld and helps to prevent cracking.

The standard *Specification for Nickel and Nickel-Alloy Welding Electrodes for Shielded Metal Arc Welding*, AWS A5.11/A5.11M,[37] establishes five related

35. American Welding Society (AWS) Committee on Structural Welding, *Structural Welding Code—Aluminum*, AWS D1.2, Miami: American Welding Society.

36. See Reference 33.

37. American Welding Society (AWS) Committee on Filler Metals and Allied Materials, *Specification for Nickel and Nickel-Alloy Welding Electrodes for Shielded Metal Arc Welding*, AWS A5.11, Miami: American Welding Society.

groups of alloys and makes provision for the requirements of the electrodes and the naming of each classification. The type designation focuses on the composition of the alloy. The familiar "E" for electrode is followed by the symbol for nickel and then the symbols for the principal elements in the weld metal. The symbols are indicated sequentially in proportion to the weight fraction. Successive numbers are then added to identify each type within its classification. For example, ENiCrFe-1 contains nickel and significant amounts of chromium and iron.

Because nickel alloys have a face-centered cubic structure throughout the temperature range of the solid phase, the hardening of these alloys is accomplished by solid solution reactions or cold work. Nickel- and cobalt-base weld metal and alloys have excellent resistance to various corrosive environments, even at elevated temperatures. Therefore, nickel-base electrodes are widely used for surfacing, cladding, joining dissimilar metals and joining austenitic and superaustenitic stainless steels. Precipitation-hardenable nickel alloys are not welded with the shielded metal arc welding process.

Most of the electrodes are intended for use with DCEP. Some can also be used with ac to overcome problems that may be encountered with arc blow (for example, when 9% nickel steel is welded). The specification assumes that the manufacturer of the electrodes will specify the usable polarity and current range. It is extremely important to use a welding current within the recommended range. Because the core wire will typically have a high electrical resistance, the use of a current setting that exceeds the recommended maximum will overheat the electrode and damage the covering, causing arc instability and unacceptable amounts of spatter.

The weld pool during welding is sluggish (stiff), so a small weave is required to wet the edges of the weld. However, the width of the weave must be carefully controlled to avoid slag inclusions and porosity. Although most nickel-base electrodes can be used in all positions, it is best to plan for welding in the flat position. As a rule, when welding the Ni-Mo alloys with ENi-Mo electrodes, welding in other than the flat position is not recommended. The best results for out-of-position welds are achieved using electrodes of 3.2 mm (1/8 in.) diameter and smaller.

Electrodes for Surfacing

Surfacing with SMAW electrodes involves the application of a layer or layers of metal to a base metal by means of arc braze welding or arc welding. The reasons for applying the layer are reflected in the several terms that are often used interchangeably with surfacing. These terms, *cladding, buttering, buildup,* and *hardfacing*, however, are not synonymous and are best used to describe the particular function of the surfacing. The term *cladding* denotes that the surfacing will provide corrosion or heat resistance. The term *buttering* means that the layer will provide an underlay of compatible weld metal for the subsequent completion of the weld. When the term *buildup is used, it means* that the layer is added to achieve a required dimension. The term *hardfacing* is used to describe surfaces that are applied to withstand impact or abrasive wear.

Covered electrodes for a particular surfacing application should be selected after a careful review of the required properties of the weld metal when it is applied to a specific base metal. Many of the nickel- and copper-base SMAW electrodes previously described are used in cladding and buttering. The SMAW electrodes for buildup and hardfacing are classified in *Specification for Solid Surfacing Welding Rods and Electrodes*, AWS A5.13[38] or *Specification for Composite Surfacing Welding Rods and Electrodes*, AWS A5.21.[39] A wide range of SMAW electrodes are described in these and other AWS filler metal specifications to provide resistance to wear, impact, heat, or corrosion on a variety of base metals.

The covered electrodes specified in AWS A5.13 have a solid core wire; those specified in AWS A5.21 may have a composite core wire. The electrode designation system in both specifications is the same, and is similar to that used for copper-alloy electrodes.

The designation system for tungsten carbide SMAW electrodes is unique. All tungsten carbide electrodes have a core that consists of a sealed steel tube filled with tungsten carbide granules that comprise a substantial portion of the total weight of the electrode. Because the surfacing performance of tungsten carbide electrodes is also dependent on the size of the granules, the mesh range (U.S. Standard Sieve) is included in the classification designator. The complete designator for the tungsten carbide SMAW electrodes consists of the "E" followed by "WC," the chemical symbol for tungsten carbide, and the mesh size limits for the tungsten carbide granules in the core, i.e. EWC-30/40.

ELECTRODE STORAGE AND CONDITIONING

Shielded metal arc welding electrode coverings interact with the atmosphere. The coverings absorb moisture and carbon dioxide. The amounts absorbed vary, depending on the type of covering and the particular composition of the covering. The amounts absorbed are

38. See Reference 21.
39. American Welding Society (AWS) Committee on Filler Metals and Allied Materials, *Specification for Composite Surfacing Welding Rods and Electrodes*, AWS A5.15, Miami: American Welding Society.

also a function of the temperature and atmospheric content of these vapors in the storage environment. The absorption of carbon dioxide may appear as a white dusting of the surface of the covering. The extent of moisture absorption can be determined by laboratory measurement. The absorption of these vapors causes changes in the welding operation and in the quality of the deposit. A wet covering may increase the spatter rate, destabilize the arc, and cause porosity. The absorption of carbon dioxide has been primarily associated with a deterioration of the appearance and toughness of the electrode covering.

Control of moisture pickup in the covering is extremely important for the low-hydrogen electrodes used to join high-strength steels with the shielded metal arc welding process. The moisture these electrodes acquire on exposure to a humid atmosphere dissociates to form hydrogen and oxygen during welding. The atoms of hydrogen dissolve into the weld and diffuse into the heat-affected zone and may cause cold cracking. This type of crack is more prevalent in highly restrained joints of high-strength steel that form a susceptible microstructure in the heat-affected zone. Consequently, the moisture content of SMAW low-hydrogen electrodes in critical applications must be strictly controlled.

To minimize moisture problems, particularly for the low-hydrogen types, the electrodes must be properly packaged, stored, and handled. Electrodes shipped and stored in hermetically sealed steel containers offer the most resistance to moisture pickup. After the electrodes are removed from the container, a heated holding oven provides the best storage conditions. Once the electrodes are issued for use, it is a good practice to keep them in a rod warmer and to minimize the time interval in which the electrodes are exposed to the prevailing atmosphere. For critical work governed by construction codes, the maximum exposure time permitted is clearly specified. This may vary from as little as a half an hour to as much as eight hours, depending on the strength designation of the electrode, the humidity during exposure, and even the specific covering on the electrode. The time that an electrode can be kept out of an oven or rod warmer is reduced as the humidity increases. The temperature of the holding oven should be in the range of 110°C to 150°C (225°F to 300°F).

Low-hydrogen SMAW electrodes that have been exposed too long require baking at a substantially higher temperature to remove the absorbed moisture. The specific recommendations of the manufacturer of the electrode must be followed because the time and temperature limitations can vary from one manufacturer to another, even for electrodes within a given classification. It should be noted that excessive heating could damage the covering of the electrode.

APPLICATIONS

The application of the shielded metal arc welding process for a joining project requires a review of the nature of the base materials, the availability and selection of the appropriate electrode, the welding position, the dimensions of the workpieces, the orientation of the weld seam and the environmental conditions at the site of the project. Typical applications are shown in Figures 2.9 and 2.10.

MATERIALS

Shielded metal arc welding can be used to join most of the common metals and alloys, including carbon steels, low-alloy steels, stainless steels and cast iron, as well as copper, nickel, and aluminum, and some alloys of these metals. The process is also used to weld many chemically dissimilar metals.

The process is not used for metals that react unsatisfactorily to the gaseous products that the electrode cov-

Photograph courtesy of The Lincoln Electric Company

Figure 2.9—On-Site Welding with the Shielded Metal Arc Process

Figure 2.10—Repair Welding with the Shielded Metal Arc Process

ering normally produces to shield the arc. Among these are reactive metals such as titanium and zirconium and refractory metals such as niobium and tantalum. The shielded metal arc welding process is also not suited for the structural joining of aluminum and its alloys, or for some high-strength steels that require very clean, undiluted weld metal.

Base Metal Thickness

The shielded metal arc process is adaptable to any material thickness within certain practical and economic limitations. For thin materials less than about 1.6 mm (1/16 in.), the base metal will melt through and the molten metal will fall away before a weld pool can be established, unless the joint is supported in a proper fixture and a special welding procedure is employed. There is no upper limit on thickness, but other processes such as submerged arc welding (SAW) or FCAW are capable of providing higher deposition rates and better economies for most applications involving material thickness exceeding 38 mm (1-1/2 in.). Therefore, most of the applications of the shielded metal arc welding process are on thicknesses between 3 mm and 38 mm (1/8 and 1-1/2 in.). However, when irregular configurations are encountered or when the welding is a one-time event, automated welding processes may be an economic disadvantage. In such instances, the shielded metal arc process is used to weld materials as thick as 250 mm (10 in.).

WELDING POSITION

One of the major advantages of shielded metal arc welding is the ability to weld in any position. This versatility makes the process useful for joints that cannot be placed in the flat position. Despite this advantage, welding should be performed in the flat position whenever practical because larger electrodes with correspondingly higher deposition rates can be used and less skill is required of the welder. Vertical and overhead joints are welded with smaller-diameter electrodes with corresponding lower deposition rates and require more skill on the welder's part.

The position of welding is defined by the orientation of the joint, the base metal, the SMAW electrode and the weld pool relative to the earth. The welding positions are identified as *flat*, *horizontal*, *vertical*, and *overhead*.[40] The position of welding is an important variable that can affect the weld metal quality. Consequently, welding positions are strictly defined in welding codes, such as the American Society of Mechanical Engineers' (ASME) *Boiler and Pressure Vessel Code*[41] and *Structural Welding Code—Steel,* AWS D1.1.[42]

LOCATION OF WELDING

The simplicity and portability of shielded metal arc welding equipment makes the process easily adaptable to varied locations and environments. Welding can be done indoors or outdoors, on a production line, a ship, a bridge, a building framework, an oil refinery, a cross-country pipeline, or almost any other conceivable location. No gas or water hoses are needed and the welding cables can be extended quite some distance from the power source. In remote areas, gasoline- or diesel-powered units can be used. Despite this versatility, for outdoor applications the process should always be used in an environment that shelters it from the wind.

JOINT DESIGN AND PREPARATION

The weld joint is the junction of members or the edges of members of a weldment that are to be joined or have been joined. It is the spatial volume occupied or to be occupied by the weld metal. The configuration of the volume is determined by edge preparation and the orientations of the abutting base plates. The weld will achieve the desired mechanical properties and intended function only if the joint is properly designed and prepared.

TYPES OF WELDS

Welds are designed primarily on the basis of the strength and the safety required of the weldment under the service conditions imposed on it. The manner in which the stresses will be applied and the temperature of the weldment in the service environment must always be considered. A welded joint required to sustain dynamic loading may be quite different from one that is permitted by code to operate under static loading conditions. Dynamic loading requires consideration of fatigue strength and resistance to brittle fracture. These properties, among others, require that the joints be designed to reduce or eliminate points of stress concentration in the weld. The joint design and the welding procedure, particularly the bead sequence, should also balance the residual stresses resulting from welding to obtain as low a net residual stress level as possible. The weld must have the required strength, ductility, and toughness.[43]

In addition to service requirements, weld joints should be designed to achieve the lowest cost and to provide accessibility for the welder during fabrication. Good accessibility to the weld joint improves the ability of the welder to meet good workmanship and quality requirements, and can assist in the control of distortion and in the reduction of welding costs.

Groove Welds

Groove welds are joint designs in which at least part of the weld occurs within the thickness of the thinner base plate. The groove configuration that is most appropriate for a specific application is influenced by the following:

1. Suitability for the structure under consideration,
2. Accessibility to the joint for welding,
3. Position in which welding is to be done, and
4. Cost of welding.

A square groove weld is the most economical to prepare. It requires only the squaring-off of the edge of

40. American Welding Society (AWS) Committee on Definitions and Symbols, *Standard Welding Terms and Definitions*, AWS A3.0, Miami: American Welding Society, p. 84.

41. American Society of Mechanical Engineers' (ASME) Boiler and Pressure Vessel Committee, *Boiler and Pressure Vessel Code* New York: American Society of Mechanical Engineers.

42. American Welding Society (AWS) Committee on Structural Welding, *Structural Welding Code—Steel*, AWS D1.1, Miami: American Welding Society.

43. For further information on weld design, see Chapter 5 of American Welding Society (AWS) Welding Handbook Committee, Jenney, C. L. and A. O'Brien, eds., 2001, *Welding Science and Technology*, Vol. 1 of *Welding Handbook*, 9th ed., Miami: American Welding Society.

each member. However, this design greatly increases the potential for slag entrapment, and if the joint members are sufficiently spaced (gapped), the economic advantage is lost. Therefore, the square groove is limited to thicknesses with which satisfactory strength and soundness can be obtained. For shielded metal arc welding, that thickness is usually not greater than about 6 mm (1/4 in.) and then only when the joint can be welded in the flat position from both sides. The type of material to be welded is also a factor to be considered when evaluating whether a square groove weld will be satisfactory.

When thicker members are to be welded, the edge of each member must be prepared with a contour that will permit the arc to be directed to the point in the joint where the weld metal is to be deposited. The contour is necessary to provide the required depth of fusion of the base metals.

For economy as well as to reduce distortion and residual stresses, the joint design should have a root opening and a groove angle that will provide adequate strength and soundness with the deposition of the least amount of filler metal. The key to soundness is accessibility to the root and groove faces (sidewalls) of the joint. V-groove joints are typical, but may have to be replaced by the J-groove or U-groove joint on thick sections. In very thick sections, the savings in filler metal and welding time alone are sufficient to offset the added cost of preparing the base material to achieve one of these joint configurations. For each weld layer deposited, the angle of the groove faces must be large enough to prevent slag entrapment.

Fillet Welds

A fillet weld is used to join two base materials in a nearly perpendicular configuration. The joint profile appears as a "T," a lapped surface, or a corner. Fillet welds made on T-joints, lap joints and corner joints are approximately triangular in cross section. Unlike groove welds, fillet welds require little or no joint preparation. When the service requirements of the weldment permit, fillet welds are frequently used in preference to groove welds. In instances when a continuous fillet weld would provide more strength than is required to carry the intended load, greater economy may be achieved by substituting intermittent fillet welds.

A fillet weld is often combined with a groove weld to provide the required strength. The fillet weld eliminates abrupt junctures and hence reduces the stress concentration at the weld. To minimize the stress concentration at the toe of the fillet weld the weld metal should feather into the base material. A weld exhibiting overlap onto the base material is undesirable.

WELD BACKING

When full penetration welds are required and welding is done from one side of the joint, weld backing may be necessary. The backing supports the weld pool during deposition of the first layer of weld metal and prevents the molten metal from escaping through the root of the joint. Backing strips or inserts, or both, are used to accomplish this purpose.

Four types of backing are commonly used:

1. Backing strip,
2. Copper backing bar,
3. Nonmetallic backing, and
4. Backing weld.

Backing Strip

A backing strip is a narrow piece of metal placed on the back of the joint, as shown in Figure 2.11(A). The first weld layer ties both members of the joint together and to the backing strip. The strip can be left in place if it will not interfere with the serviceability of the weld or the function of the fabricated product. Otherwise, it must be removed, although the back side of the joint must be accessible to accomplish this. If the back side is not accessible, another means of obtaining a proper root pass must be used.

The backing strip must always be made of a material that is metallurgically compatible with the base metal and the welding electrode to be used. When design permits, another member of the structure may serve as backing for the weld. Figure 2.11(B) provides an example of this situation. In all cases, it is important that the backing strip and the surfaces of the joint are cleaned before welding to avoid porosity and inclusions in the weld. It is also important that the backing strip fits properly and is of sufficient thickness. Otherwise, the molten weld metal can run out through any gap between the strip and the base metal at the root of the joint.

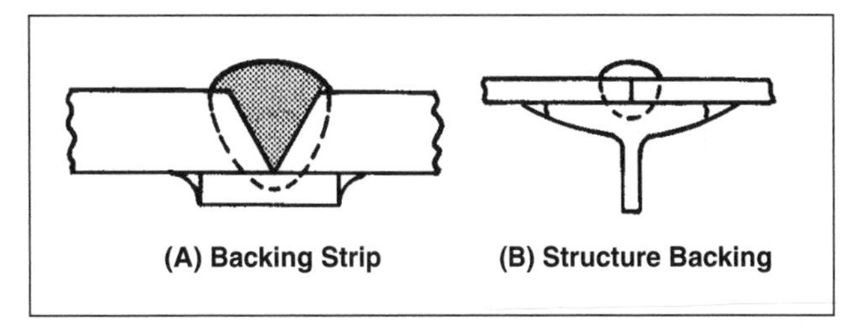

Figure 2.11—Fusible Metal Backing for a Weld

Copper Backing Bar

A copper backing bar is sometimes used as a means of supporting the weld pool at the root of the joint. Copper is used because of its high thermal conductivity, which helps prevent the weld metal from melting the surface of the copper and fusing to the backing bar. In addition to the rapid conduction of heat, the copper bar must have sufficient mass to avoid melting during the deposition of the first weld pass. In high-production applications, water can be passed through holes in the bar to remove the heat that accumulates during continuous welding. Regardless of the method of cooling, the arc should not be allowed to impinge on the copper backing bar, for if any copper melts, the weld metal can become contaminated with copper. The copper bar may be grooved to obtain a desired root surface contour and reinforcement.

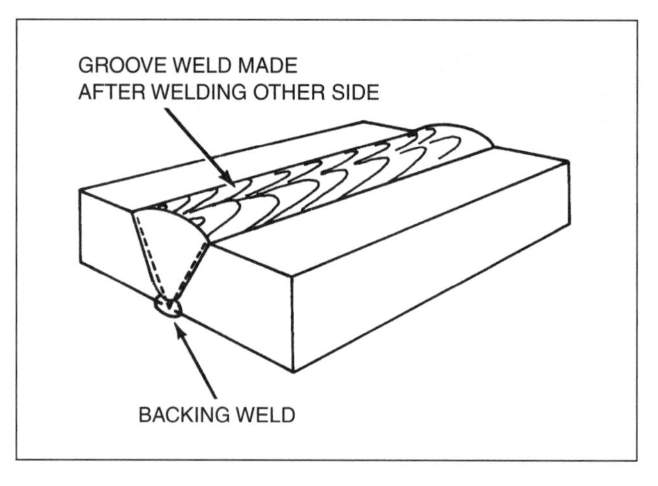

Figure 2.12—A Typical Backing Weld

Nonmetallic Backing

Nonmetallic backing of either granular flux or a refractory material can also be used to produce a sound first pass. These materials are used primarily to support the weld metal and to shape the root surface. The granular flux layer or backing made with a refractory material, which is usually supplied in individual tiles, are supported against the underside of the joint with aluminum adhesive tape. The tiles are shaped to function in a variety of joint configurations and often are grooved to obtain a properly shaped root bead. Refractory backing may also consist of a flexible, shaped form that is held on the back side of the joint by clamps or pressure-sensitive tape.

A method for supporting the flux against the back side of the weld involves the use of a pressurized fire hose. In this method, an inflatable rubberized canvas fire hose is placed in a trough under the weld; granular flux is arranged on a thin piece of flexible sheet material and placed over the hose. The workpieces are positioned over the trough containing the hose and flux so that the flux is in contact with the back side of the weld. The hose is inflated to no more than 35 kPa to 70 kPa (5 psi to 10 psi) to apply moderate pressure to the flux on the back side of the weld. A system of this type is generally used for production line work and is not widely used for shielded metal arc welding. An illustration of this method appears in Chapter 6, Figure 6.20.

Backing Weld

A backing weld is applied with one or more backing passes in a single-groove weld joint. The backing weld is deposited on the back side of the joint before the first pass is deposited on the face side, as illustrated in Figure 2.12. After the first pass of the backing weld is completed, all subsequent passes are made in the groove from the face side. After the backing weld is made, the root of the joint can be ground or gouged to produce sound, clean metal on which to deposit the first pass on the face side of the joint.

The backing weld can be made with the same process or with a different process from that to be used for welding the groove. If the same process is used, the electrodes should be of the same classification as those to be used for welding the groove. If a different process is used, such as gas tungsten arc welding (GTAW), the welding rods should deposit weld metal with composition and properties similar to those of the SMAW weld metal. The backing weld must be large enough to support any load that is temporarily placed on it during the welding operation. This requirement is especially important when the weldment must be repositioned after the backing weld has been deposited and before the groove weld is made.

JOINT FITUP

Joint fitup refers to the positioning of the members of the joint to provide the specified groove dimensions and alignment for welding. The points of concern are the size of the root opening and the misalignment of the members along the root of the weld. Both the root opening and the alignment of the joint have an important influence on the quality of the weld and the economics of the process. After the joint has been properly aligned throughout its length, the fitup of the workpieces should be maintained by clamps or tack welds. Finger bars or U-shaped bridges (strongbacks) can be placed across the joint and tack welded to each member of the weldment.

If the root opening is not uniform, the amount of weld metal will vary from location to location along the joint. Consequently, the shrinkage and the resulting distortion will not be uniform. This can cause problems when the finished dimensions have been predicated on the basis of uniform, controlled shrinkage.

Misalignment along the root of the weld may result in lack of penetration in some areas or poor root surface contour, or both. An inadequate root opening can also cause incomplete joint penetration. Too wide a root opening makes welding difficult and requires more weld metal to fill the joint. More weld metal means more welding time, and of course, additional cost. In thin members, an excessive root opening may cause excessive melt-through on the back side. It may even cause the edge of one or both members to melt away.

Typical Joint Geometries

The weld grooves shown in Figure 2.13 illustrate typical designs and dimensions of joints for the shielded metal arc welding of steel. These joints are generally suitable for economically achieving sound welds. Other joint designs or changes in the suggested dimensions of the noted joints may be required for special applications.

Runoff Weld Tabs. In some applications, it is necessary to completely fill the groove to the very ends of the joint. However, with the SMAW process, weld starts tend to be overfilled and weld craters underfilled. These features create an unfavorable shape at the start and at the end of multiple-pass groove and fillet welds. Therefore, runoff tabs are used to completely fill out the groove and achieve a uniform weld. The runoff tabs extend the groove beyond the ends of the workpieces. The welding is started on a runoff tab and carried over into the finishing runoff tab. This technique assures that the entire length of the joint is uniformly filled with sound weld metal to the necessary depth. Runoff tabs provide an excellent method for starting and stopping welding because all defects typically associated with starts and stops are located in areas that later will be discarded. A typical runoff tab is shown in Figure 2.14.

The selection of material for runoff tabs is important. The composition of the tabs should not be allowed to adversely affect the properties of the weld metal. For example, for stainless steel that is intended for corrosive service, the runoff tabs should be of a compatible grade of stainless steel. Carbon steel tabs would be less expensive, but fusion with the stainless steel filler metal would change the composition of the weld metal at the junction of the carbon steel tab and the stainless steel members of the joint. The weld metal at this location probably would not have adequate corrosion resistance.

WELDING VARIABLES

The application of the shielded metal arc welding process to a given joining project requires a review of several welding variables. The welding variables that must be considered are the type and size of the electrode and its orientation during welding, the type and magnitude of current, arc length, and travel speed. The welder must implement welding techniques that are fundamental for good workmanship, and also must exercise good judgment as to whether to grind starts and stops, clean the residues, adjust the workpiece temperature, and so forth. Control of the important variables in the shielded metal arc welding process will permit the trained welder to achieve a quality weld.

ELECTRODE DIAMETER

The diameter of the core wire used to manufacture the SMAW electrode defines its nominal size or electrode diameter. In a given application, the correct electrode diameter is one that, when used with the proper amperage and travel speed, produces a weld of the required size in the least amount of time. The electrode diameter ultimately selected for an application will depend largely on the thickness of the material to be welded, the position in which welding is to be performed, and the type of joint to be welded. In general, larger electrodes will be selected for applications involving thicker materials and for welding in the flat position in order to take advantage of the higher deposition rates of these electrodes.

For welding in the horizontal, uphill, and overhead positions, the bead shape is an important consideration. In these positions, the weld pool tends to flow out of the joint due to gravitational forces. Consequently, when it solidifies into weld metal, it has a convex profile. The severity of the convex profile of the bead shape can be controlled by the selection of small-diameter electrodes, typically 4.8 mm (1/8 in.) and smaller. The small-diameter electrodes have lower heat input and create a smaller weld pool. Electrode manipulation and increased travel speed along the joint also aid in reducing the weld pool size.

Weld groove design must also be considered when electrode size is selected. The electrode used in the first few passes must be small enough for easy manipulation in the root of the joint. In V-grooves, small-diameter electrodes are frequently used for the initial pass to control melt-through and bead shape. Larger electrodes can be used to complete the weld to take advantage of the deeper penetration and higher deposition rates they provide.

The expertise of the welder often has a bearing on the electrode size to be used. This consideration is

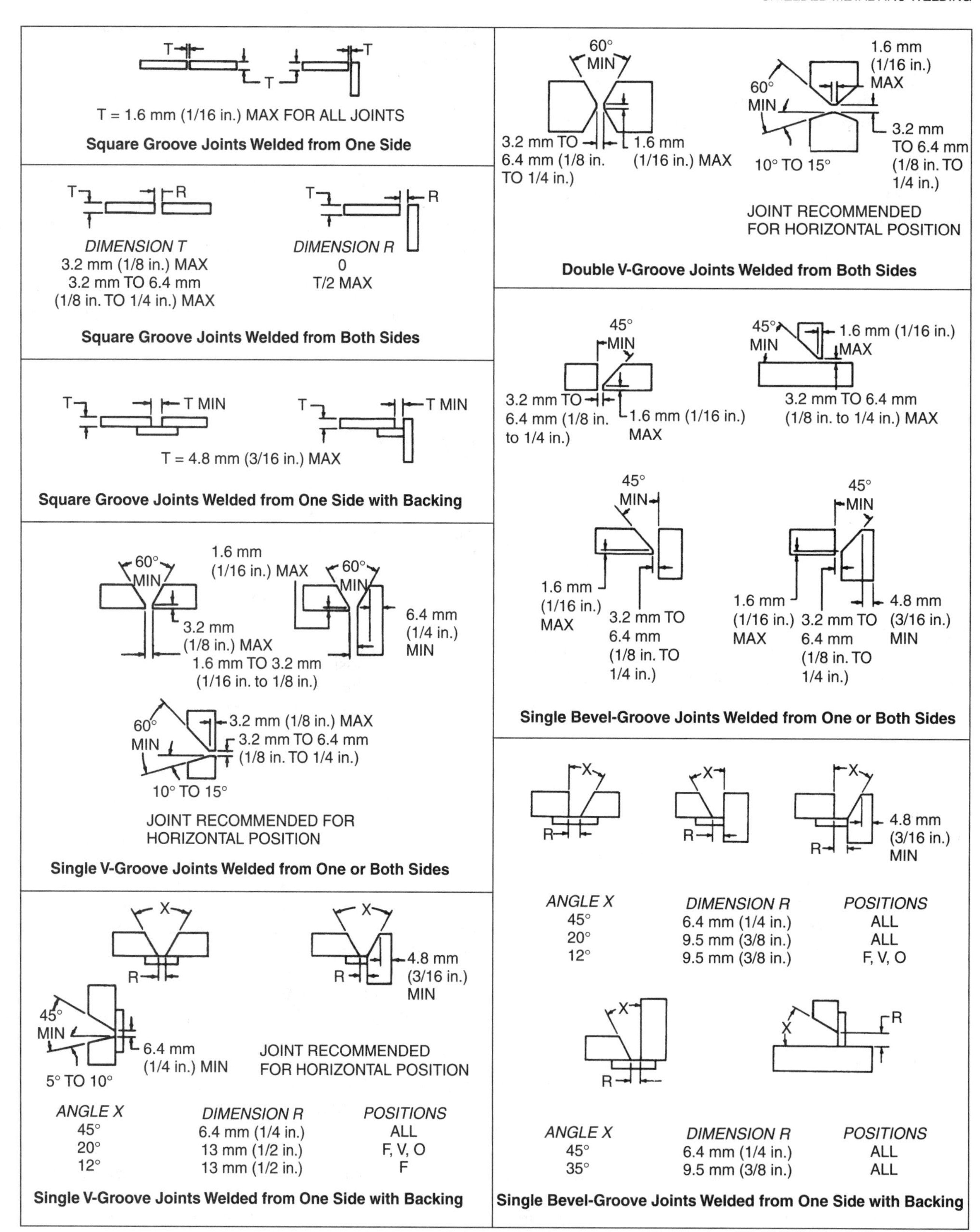

Figure 2.13—Typical Joint Geometries for the Shielded Metal Arc Welding of Steel

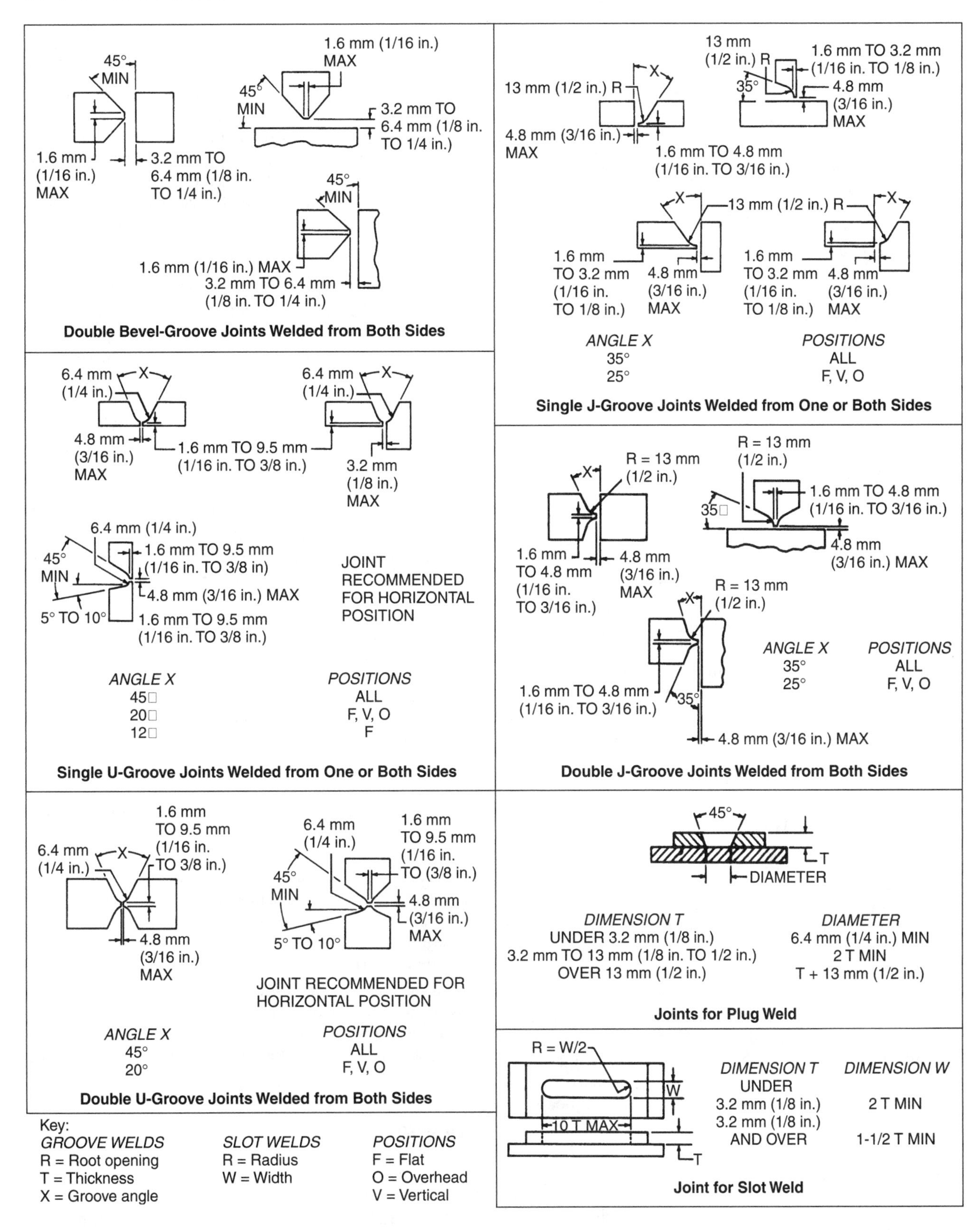

Figure 2.13 (Continued)—Typical Joint Geometries for the Shielded Metal Arc Welding of Steel

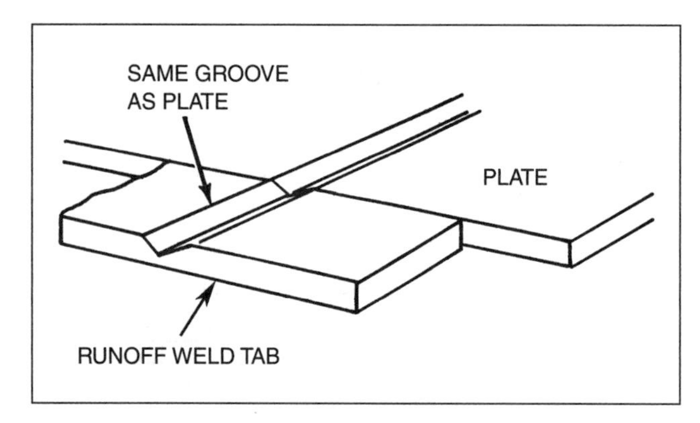

Figure 2.14—Runoff Weld Tab at End of a Joint

particularly true for out-of-position welding, since the welder's skill governs the size of the weld pool that can be successfully controlled and the maximum deposition rate that can be obtained.

The electrode used should be as large as possible, provided that it does not exceed any pertinent heat input limitations or layer thickness limitation or deposit too large a weld. Welds that are larger than necessary are more expensive, and in some instances, they may be detrimental in service. Any sudden change in section size or in the contour of a weld, such as might be caused by inconsistent welding, should be avoided because it creates stress concentrations. Choosing the correct electrode diameter and using it with the proper amperage and travel speed will produce a weld of the required size for the least cost.

WELDING CURRENT

Electric current is the transfer of energy in a circuit and amperage is a measure of the intensity of the current in units of coulombs per second, or amperes. Shielded metal arc welding can be accomplished with either alternating or direct electrical current, provided that the appropriate electrode has been selected. The magnitude of the welding current, the polarity, and the constituents in the electrode covering influence the melting rate of all covered electrodes. For any given electrode, the melting rate is directly related to the electrical energy supplied to the arc, but only a fraction of this energy is used to melt the electrode. The remaining energy melts and heats the base metal or is lost to the environment.

Direct Current

Direct current always provides a steadier arc and smoother metal transfer than alternating current because the amperage and direction of current flow is not constantly changing as it is with ac. Most covered electrodes operate better on direct current electrode positive, also referred to as *reverse polarity*, although some electrodes are designed for use with direct current electrode negative, also referred to as *straight polarity*. Direct current electrode positive produces deeper penetration, but direct current electrode negative produces a higher electrode melting rate. The dc arc produces good wetting action by the molten weld metal and uniform weld bead size, even at low amperage. For these reasons, dc is particularly suited to welding thin sections. Most electrodes designed to operate with either type of current actually operate better on dc than on ac.

Direct current is preferred for the vertical and overhead welding positions and for welding with a short arc. The dc arc has less tendency to short out and to eject the globules of molten metal as they transfer across the arc and into the weld pool. The sinusoidal profile of the ac current requires easily ionized elements in the electrode covering to sustain the arc.

Arc blow may be a problem when magnetic metals (iron and nickel) are welded with dc. The magnetic field that is induced by the current may deflect the arc and eject metal droplets from the arc area. One way to overcome this interference with dc is to change to ac.

Alternating Current

For shielded metal arc welding, alternating current offers two advantages over direct current. One is the absence of arc blow and the other is the cost of the power source.

Without arc blow, larger electrodes and higher welding currents can be used. Certain electrodes (specifically, those with iron powder in the covering) are designed for use at higher amperages with ac. The highest welding speeds for the shielded metal arc welding process can be obtained using these electrodes on ac with the drag technique. The absence of arc blow with ac also minimizes any impact of the material used in the construction of a welding fixture, the design of the fixture, and the location of the workpiece connection.

An ac transformer costs less than an equivalent dc power source. However, the cost of the equipment alone should not be the sole criterion in the selection of the power source. It is best to consider all of the pertinent factors.

Amperage

Covered electrodes of a specific size and classification will operate satisfactorily at various amperages within a given range. This range will vary somewhat with the formulation and thickness of the covering.

Deposition rates increase as the amperage increases. For a given size of electrode, the amperage ranges and the resulting deposition rates will vary from one electrode classification to another. The relationship between deposition rates and amperage for several classifications of carbon steel electrodes of one size is shown in Figure 2.15.

With a specific type and size of electrode, the optimum amperage depends on several factors, such as the position of welding and the type of joint. The amperage must be sufficient to obtain good fusion and penetration, yet permit proper control of the weld pool. For welding in the vertical and overhead positions, the optimum amperages will likely fall on the low end of the allowable range.

Amperage beyond the recommended maximum should not be used. This can overheat the electrode and cause excessive spatter, arc blow, undercut, and weld metal cracking. Figure 2.16 illustrates the effect of amperage, arc length, and travel speed on the resulting bead shape.

ARC LENGTH

Arc length is the distance from the molten tip of the electrode core wire to the surface of the weld pool. Although arc length is easily defined, it is rather difficult to visually perceive and measure. The welder must visualize the relative distance between the electrode tip and the weld pool while feeding the electrode and maintaining this distance as consistent as possible. The assessment of the proper arc length is important in

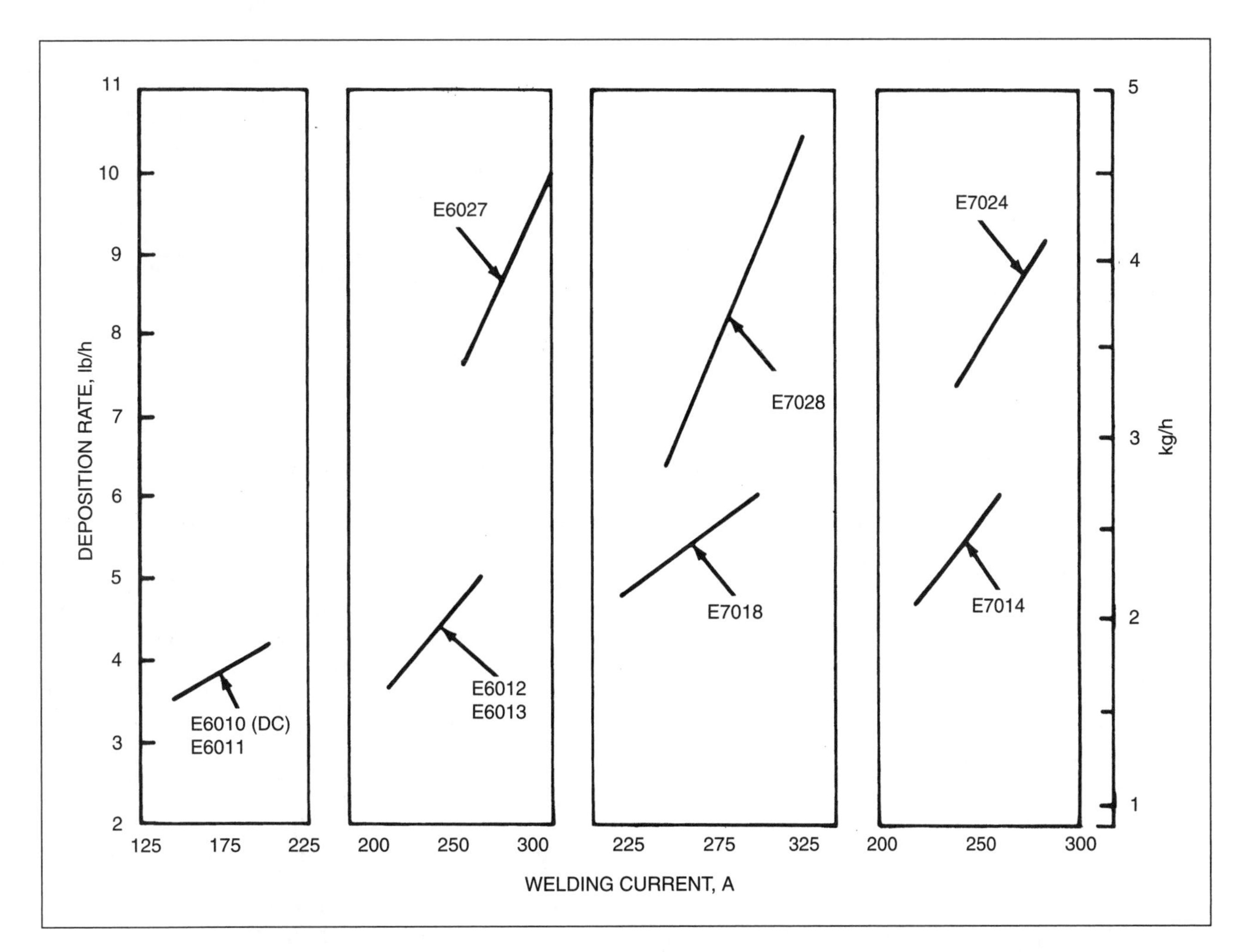

Figure 2.15—The Relationship between Deposition Rate and Welding Current for Various Types of 4.8 mm (3/16 in.)-Diameter Carbon Steel Electrodes

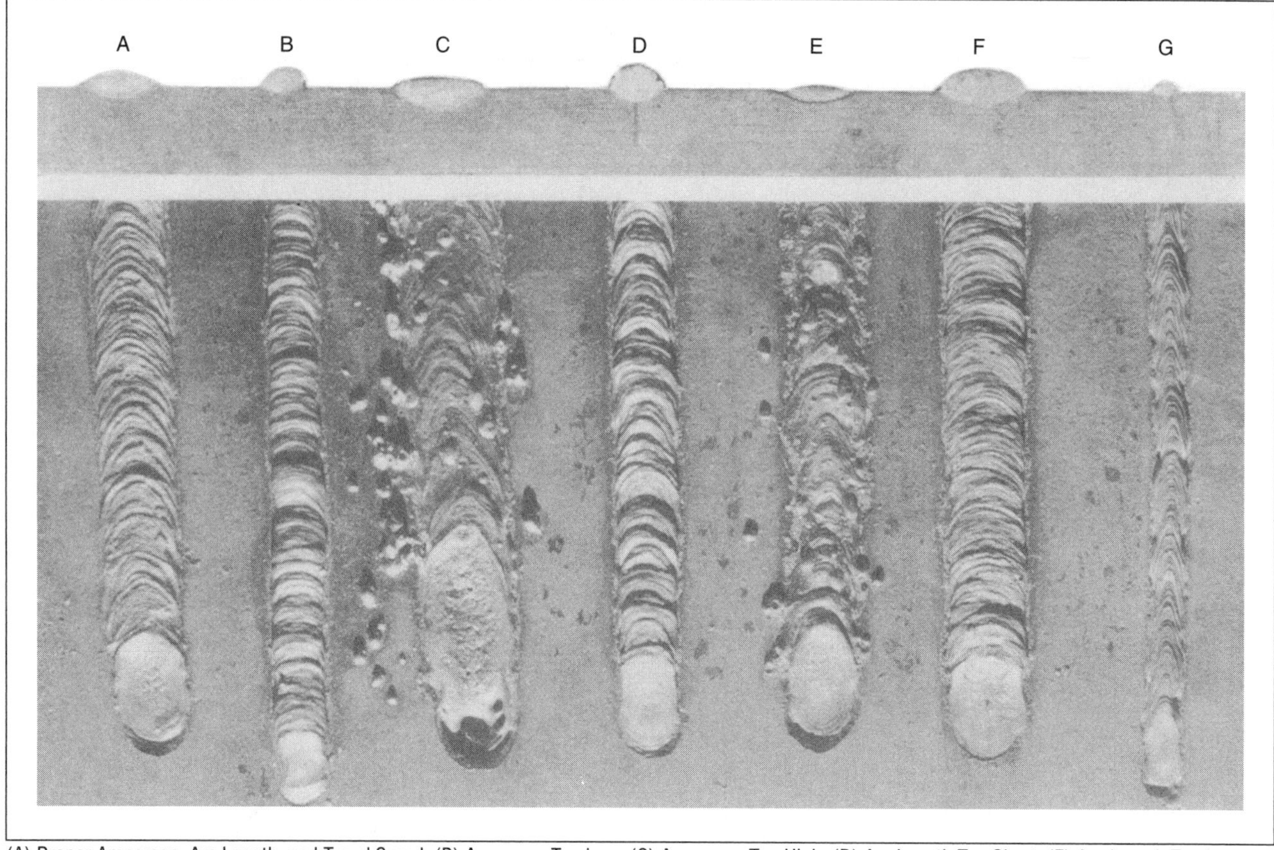

(A) Proper Amperage, Arc Length, and Travel Speed; (B) Amperage Too Low; (C) Amperage Too High; (D) Arc Length Too Short; (E) Arc Length Too Long; (F) Travel Speed Too Slow; (G) Travel Speed Too Fast

Figure 2.16—The Effect of Welding Amperage, Arc Length, and Travel Speed

obtaining a sound welded joint with optimum mechanical properties. The arc voltage is an indicator of the arc length. The metal transfer from the tip of the electrode to the weld pool is not a smooth, uniform action. As droplets of molten metal stream across the arc they cause the instantaneous arc voltage to vary, even when a constant arc length is maintained. However, the variation in arc voltage is usually minimal during welding when the proper amperage and arc length is maintained. In general, the welder has the highest probability for maintaining a consistent arc length when the electrode is run at optimum amperage. The best measure of arc voltage is obtained as an average value during a period in which all welding variables are held relatively constant.

The correct arc length varies according to the electrode classification, diameter, and covering composition as well as the amperage and welding position. The optimum arc length increases with increasing electrode diameter and amperage. As a rule, the arc length should not exceed the diameter of the core wire of the electrode. The arc length is usually shorter than this value for electrodes with thick coverings, such as iron powder (drag) electrodes.

Too short an arc length will lead to erratic operation as the metal droplets short-circuit during metal transfer, causing increased spatter. With some electrodes, shielding gases may not be generated and porosity will form. Too long an arc will lack direction and drive, which will tend to scatter the molten metal as it moves from the electrode toward the weld pool. Penetration is sacrificed because the base plate is not fully melted. With a long arc, the spatter may be heavy and hence the deposition efficiency will be low. In addition, the gas and flux generated by the electrode covering are not as concentrated, so they may not effectively shield the arc and the weld metal. The dissipation of the shielding gases can result in porosity and contamination of the weld metal by oxygen or nitrogen, or both. Control of the arc length is largely a matter of welder skill, involving the welder's knowledge, experience, visual perception, and manual dexterity. Although the optimum arc length does change to some extent with changing conditions, certain fundamental principles can serve as a guide to the proper arc length for a given set of conditions.

During flat or horizontal welding, particularly with large-diameter heavily covered electrodes, the electrode can be dragged lightly along the joint. The arc length, in this case, is automatically determined by the covering thickness and the melting rate of the electrode. Thus, the arc length is uniform. For vertical or overhead welding, the arc length is gauged by the welder. In such cases, the proper arc length is the one that permits the welder to control the size and motion of the weld pool.

For fillet welds, the electrode is pushed as deeply into the weld pool as the arc from the electrode allows without causing shorting or excessive penetration. This provides for the highest deposition rate and best penetration. The same is true of the root passes in groove welds in pipe where the electrode coating touches the groove faces of the joint.

When arc blow is encountered, the arc length should be reduced as much as possible without shorting the electrode. The various classifications of electrodes have widely different operating characteristics, including the proper arc length. It is important, therefore, for the welder to be familiar with the operating characteristics of the various types of electrodes in order to recognize the proper arc length and to know the effect of different arc lengths. The effect of a long and a short arc on bead appearance with a mild steel electrode is illustrated in Figures 2.16(D) and 2.16(E).

TRAVEL SPEED

Travel speed is the rate at which the electrode moves along the joint. The correct travel speed is the one that produces a weld bead of proper contour and appearance, as shown in Figure 2.16(A). Travel speed is influenced by several factors:

1. Type of welding current, amperage, and polarity;
2. Position of welding;
3. Melting rate of the electrode;
4. Thickness of base metal;
5. Surface condition of the base metal;
6. Type of joint;
7. Joint fit-up; and
8. Electrode manipulation.

The welding travel speed should be adjusted so that the arc is slightly ahead of the weld pool. Up to a point, increasing the travel speed will narrow the weld bead and increase penetration. Beyond this point, higher travel speeds can decrease penetration, cause the surface of the bead to deteriorate and cause undercut at the edges of the weld. It will also make slag removal difficult and entrap gas (causing porosity) in the weld metal. The effect of high travel speed on bead appearance is shown in Figure 2.16(G). With low travel speed, the weld bead will be wide and tend to be convex with shallow penetration, as illustrated in Figure 2.16(F). The shallow penetration is caused by allowing the arc to dwell on the weld pool instead of leading it. If the arc leads the weld pool, more heat is directed into the base metal. The depth of penetration, in turn, affects dilution of the weld metal. When dilution must be kept low, as in cladding operations, the travel speed must also be kept low.

Travel speed also influences heat input, which affects the metallurgical structure of the weld metal and the heat-affected zone. Low travel speed increases heat input and this, in turn, increases the size of the heat-affected zone and reduces the cooling rate of the weld. The travel speed is necessarily reduced with a weave bead as opposed to the higher travel speed that can be obtained with a stringer bead. Higher travel speed reduces the size of the heat-affected zone and increases the cooling rate of the weld. The increase in the cooling rate can increase the strength and hardness of a weld in a heat-hardenable steel, unless preheat of a level sufficient to prevent hardening is used. The cooling rate is also a function of the temperature of the base plate.

ELECTRODE ORIENTATION

The orientation of the electrode in relation to the workpiece and the weld groove controls the direction and location of the arc and is an important factor in the quality of a weld. Improper positioning of the electrode can result in slag entrapment, porosity, and weld undercut. The proper orientation in the joint depends on the type and size of electrode, the welding position, and the geometry of the joint. A skilled welder automatically evaluates these factors when deciding the orientation to be used for a specific joint. The positioning of the electrode relative to the joint and the workpiece is described by the travel angle and the work angle.

The term *travel angle* denotes the angle (less than 90°) between the electrode axis and a line perpendicular to the weld axis, in a plane determined by the electrode axis and the weld axis. The term *work angle* denotes the angle (less than 90°) between a line perpendicular to the major workpiece surface and a plane determined by the electrode axis and the weld axis.

When the electrode is pointed in the direction of welding, the technique is termed *forehand welding*. The travel angle, then, is known as the *push angle*. When the electrode is pointed in the opposite direction to that of welding, the technique is termed *backhand welding*. The travel angle in backhand welding is called the *drag angle*. These angles are shown in Figure 2.17.

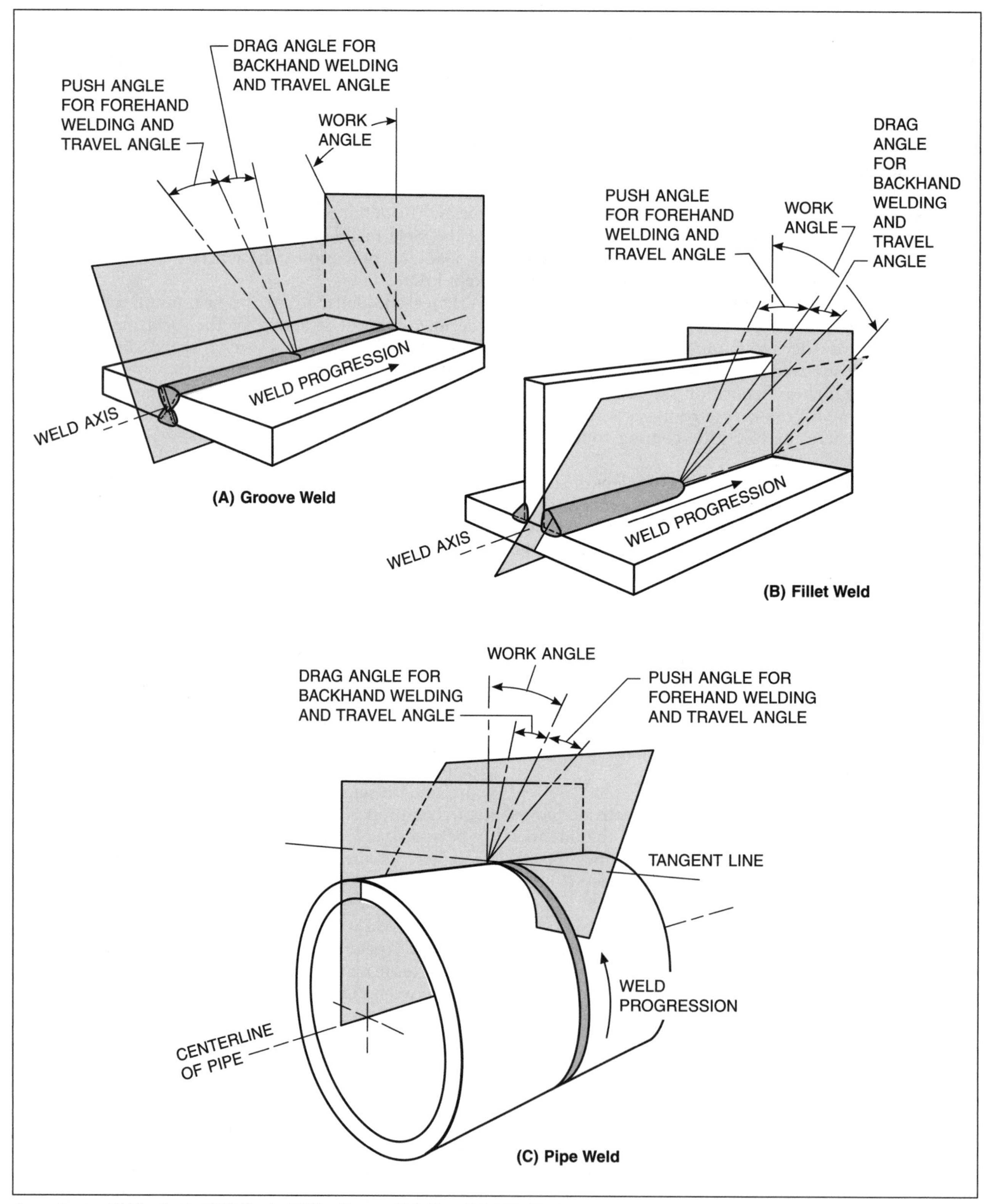

Source: American Welding Society (AWS) *Standard Welding Terms and Definitions*, A3.0: 2001, Miami: American Welding Society, Figure 21, p. 74.

Figure 2.17—Orientation of the Electrode

Table 2.3
Typical Shielded Metal Arc Electrode Positioning and Welding Technique for Carbon Steel Electrodes

Joint Type	Welding Position	Work Angle (Degree)	Travel Angle (Degree)	Welding Technique
Groove	Flat	90	5–10*	Backhand
Groove	Horizontal	80–100	5–10	Backhand
Groove	Uphill	90	5–10	Forehand
Groove	Overhead	90	5–10	Backhand
Fillet	Horizontal	45	5–10*	Backhand
Fillet	Uphill	35–55	5–10	Forehand
Fillet	Overhead	30–45	5–10	Backhand

*Travel angle may be 10° to 30° for electrodes with heavy iron powder coverings.

Correct placement of the electrode helps to achieve control of the weld pool, attain the desired penetration, and assure complete fusion to the base plate. Typical electrode orientation and welding technique for groove and fillet welds for use on carbon steel with carbon steel electrodes are listed in Table 2.3. These values may be different for other electrodes and materials.

A large travel angle may cause a convex, poorly shaped bead with inadequate penetration, whereas a small travel angle may cause slag entrapment. A large work angle can cause undercutting, while a small work angle can result in incomplete fusion.

WELDING TECHNIQUE

The first step in shielded metal arc welding is to assemble the proper equipment, materials, and tools for the job. The next steps are to determine the type (and polarity, if dc is used) of welding current needed, and to assure that the electrode and ground cables are connected to the proper terminals of the power source. If the power source permits, the open-circuit voltage must also be set to give the proper volt-ampere characteristic for the size and type of electrode to be used. After this, the workpiece is positioned for welding and clamped in place if necessary.

After lowering the welding helmet, the welder strikes an arc by tapping the end of the electrode on the workpiece near the point where welding is to begin, then quickly withdraws it to produce an arc of the proper length. Another technique is to use a scratching motion similar to that used in striking a match. The motion of the electrode helps to initiate a discharge across the narrow gap between the electrode and the workpiece.

If the electrode touches the workpiece, it may stick to the workpiece. When the electrode sticks, it must be broken free very quickly or it will overheat and fuse to the workpiece. When a conducting electrode fuses, any attempt to remove it is hazardous and usually only succeeds in bending the overheated electrode. At this point, the electrode should be released from the holder, the holder safely stored, and the electrode removed with a hammer and chisel.

The initiation of an arc with a stub (a partially consumed electrode) is referred to as *restriking*. The technique of restriking an arc varies somewhat, depending on the type of electrode and the timing of the restrike. Generally, the covering at the tip of the electrode becomes conductive when it is heated during welding. If the stub is sufficiently hot, the arc is easily initiated. If the stub has had time to cool, generally in excess of 20 seconds, the fused glassy material on the tip of the core wire may no longer be conductive, and arc initiation may be difficult. If the electrode is a type that has large amounts of metal powders or graphite in the covering, then it is possible that the covering is electrically conductive when cold. In these cases, arc initiation with the stub, as with a new electrode, is much easier. However, when using heavily covered electrodes that do not have conductive coverings, such as the low-hydrogen and the stainless steel electrodes, it may be necessary to break off (pinch) the covering of the stub to expose the core wire at the tip.

Initiating the arc with low-hydrogen electrodes requires a special backwashing technique to avoid porosity in the weld at the point where the arc is started. This technique consists of initiating the arc a few electrode diameters ahead of the place where welding is to begin. The arc is then quickly moved back to the starting point, where welding proceeds in the normal manner. As the arc approaches the original initiation point, it melts any small globules of weld metal that may have remained in the weld pool.

During welding, the welder maintains a uniform arc length by moving the electrode toward the workpiece as the electrode melts. At the same time, the welder steadily moves the electrode along the joint in the direction of welding to form a uniform bead. A uniform arc length is maintained by skillful manipulation in response to visual input; for most seasoned welders the action has become instinctive.

Any of a variety of techniques may be employed to extinguish (break) the arc. One method is to rapidly shorten the arc, then quickly remove the electrode sideways out of the crater. This technique is used when replacing a spent electrode, in which case welding will continue from the crater. Another technique is to stop the forward motion of the electrode and allow the crater to fill. Then the electrode is gradually withdrawn. When continuing a weld from a crater, the arc should be struck at the forward end of the crater. It should then be quickly moved to the back of the crater and slowly brought forward to continue the weld. In this manner, the crater is filled and porosity and entrapped slag are avoided. This technique is particularly important for low-hydrogen electrodes.

Slag Removal

Welds deposited with SMAW electrodes have a slag that covers and protects the weld during the welding cycle. After the weld is completed, it is customary and at times essential to remove the slag from the weld surface. The detachability of the slag depends on the composition of the covering and often on the condition of the covering. For a multiple-pass weld, the extent to which the slag is removed from each weld bead before welding over the bead has a direct bearing on weld quality. Failure to clean each bead thoroughly increases the probability of trapping slag and thus producing a defective weld. Complete and efficient slag removal requires that each bead be properly contoured and blended smoothly into the adjacent bead or base metal.

Small beads cool more rapidly than large beads. This fast quench tends to make slag removal from small beads easier. Concave or flat beads that wash or feather smoothly into the base metal or any adjoining beads tend to minimize undercutting and avoid a sharp notch along the edge of the bead where slag can become trapped in a subsequent layer. It is imperative that welders recognize the areas where slag entrapment is likely to occur and intervene to prevent it. Skilled welders understand that complete removal of slag is necessary in making quality multiple-pass welds. They also recognize that after the weld is made, complete cleanup is a very important step in the final inspection and preparation of the weldment for service.

Workpiece Connection

The workpiece connection, the attachment of the workpiece lead to the workpiece, must be properly executed because it is a juncture in the electrical circuit. The location of the workpiece lead is especially important with dc welding. Improper location may promote arc blow, making it difficult to control the arc. Moreover, the method of attaching the lead is important. A poorly attached workpiece lead will not provide consistent electrical contact, and the workpiece connection will heat up. A heated connection can interrupt the circuit and momentarily extinguish the arc. A copper contact shoe secured to the base metal with a C-clamp is the best method of connection. If copper pickup from this attachment is detrimental, the copper shoe should be fastened to a plate made of material compatible with the base metal and this assembly should be used to attach the workpiece lead. For rotating work, contact should be made by sliding shoes on the workpiece or through roller bearings on the spindle on which the workpiece is mounted. If sliding shoes are used, at least two shoes should be employed to guard against a break in the circuit if a single shoe temporarily loses contact.

Arc Stability

A stable arc is required if high-quality welds are to be produced. Discontinuities, such as incomplete fusion, entrapped slag and porosity (blowholes), can be the result of an unstable arc. The following are important factors influencing arc stability:

1. The open circuit voltage of the power source,
2. Transient voltage response characteristics of the power source,
3. Size of the molten drops of filler metal and slag in the arc,
4. Ionization of the arc path from the electrode to the workpiece, and
5. Manipulation of the electrode.

The first two factors are related to the design and operating characteristics of the power source. The next two are dependent on the type of welding electrode. The last one represents the skill of the welder.

The arc of a covered electrode is a transient element in the circuit even when the welder maintains a fairly constant arc length. The welding machine must be able to respond rapidly when the arc tends to extinguish because the resistance of the arc momentarily increases, or when a large droplet partially or totally short-circuits the gap. In these instances, a surge of current is needed to restore the arc. With ac, it is imperative that the voltage cycle leads the current cycle to sustain the arc. If the voltage and current were in phase, the arc would be very unstable. This phase shift must be designed into the welding machine.

Some ingredients of the electrode covering tend to stabilize the arc. These are necessary ingredients for an electrode to operate well on ac. A few of these ingredients are titanium dioxide, feldspar, and various potassium compounds (including the binder, potassium silicate). The inclusion of one or more of these arc-stabilizing compounds in the covering provides arc plasma that readily ionizes and achieves a quiet arc. Thus, the electrode, the power source, and the welder all contribute to arc stability.

Arc Blow

The term *arc blow* is used to describe the undesirable deflection of the arc by magnetic forces. Arc blow is encountered principally with the dc welding of magnetic materials (iron and nickel). It may be encountered with ac under some conditions, but those cases are rare and the intensity of the blow is always much less severe. Direct current flowing through the electrode and the base metal sets up magnetic fields around the electrical path that tend to deflect the arc from its intended course. The arc may be deflected to the side at times, but it is usually deflected either forward or backward along the joint. Back blow is encountered when welding away from the workpiece connection near the end of a joint or into a corner. Forward blow is encountered when welding away from the workpiece connection at the start of the joint, as shown in Figure 2.18.

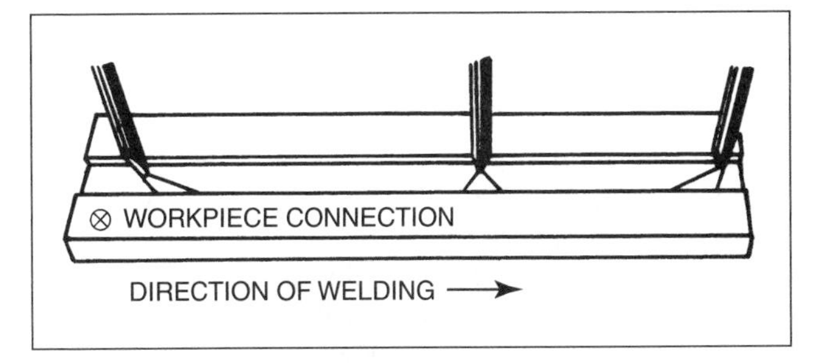

Figure 2.18—The Effect of the Location of the Workpiece Connection on Magnetic Arc Blow

Arc blow can result in incomplete fusion, excessive weld spatter, and distorted bead shape, and at times, may be so severe that a satisfactory weld cannot be made. When welding with high-iron electrodes and electrodes that produce a heavy slag, forward blow can be especially troublesome. When it is encountered, it pushes the molten slag, which is normally confined to the edge of the crater, forward under the arc, multiplying the risk of weld defects.

The deflection or "bending" of the arc is caused by an unbalanced magnetic field. A magnetic field naturally arises and encircles an electrical conductor that carries current. The intensity of the magnetic field is proportional to the current and to the reciprocal of the radius. When there is a substantial imbalance in the concentration of magnetic flux in the region of the arc, the arc will always bend away from the greatest concentration toward the least. The strength of the magnetic flux can be visualized as a closely packed series of concentric circles emanating from the center of the conductor. When the conductor makes a sharp turn, the circles thin out on one side and concentrate on the opposite side. Therefore, a change in the direction of current flow, for example, from the electrode into the workpiece, is one cause for an unbalanced magnetic field.

Another characteristic of magnetic flux is that it passes through a magnetic material more readily than it passes through air. In welding, the magnetic flux is superimposed on the steel and across the joint to be welded. The difference in permeability between the steel and the air causes the equal flux lines to assume elliptical shapes. At the beginning of current flow, if welding were started at the center of the plate, the lines would be closely packed on the steel sides and elongated along the joint. Moreover, because the magnetic flux passes through (permeates) steel more readily than it passes through air, the path of the flux concentrates within the steel plates. For this reason, when the electrode is near either end of the joint, the magnetic flux encircling the electrode is highly concentrated between the weld pool and the end of the plate and very diffuse on the opposite side. This higher concentration of magnetic flux on one side of the arc, at the start or at the finish of the weld, deflects the arc away from the end of the workpiece.

Forward blow exists for a short time at the start of a weld, and then diminishes. This calming effect occurs because the magnetic flux finds an easy path through the weld metal. Once the magnetic flux behind the arc is concentrated in the plate and the weld, the arc is influenced mainly by the flux in front, particularly by the imbalance as the magnetic field crosses the root opening. At this point, back blow may be encountered. Back blow can occur right up to the end of the joint, where it can become extremely severe. As the welding approaches the end, the magnetic flux ahead of the arc becomes more concentrated, increasing back blow.

The passing of the welding current through the cable, the workpiece, the arc, and the electrode creates a magnetic field around each conductor. The field is perpendicular to the path of current, and the strength of the field is proportional to the amperage. As the current passes from the electrode through the arc to the workpiece and through the workpiece, the current path bends. The change in the direction of the path always concentrates the magnetic flux on the side that is the shortest distance from the weld pool to the workpiece connection. This magnetic field imbalance tends to push the arc away from the location of the workpiece connection.

Arc blow is caused by an imbalance in the magnetic field in the region of the arc. The magnetic field can become unbalanced because the conducting path makes a turn. In addition, the vastly different permeability of the base plate and of the air may lead to substantial differences in magnetic flux. These two mechanisms cause the arc to deflect and the welding to become erratic.

Unless the arc blow is unusually severe, certain steps can be taken to eliminate it or reduce its severity. Corrective actions may include some or all of the following:

1. Placing workpiece lead connections as far as possible from the joints to be welded;
2. If back blow is the problem, placing the workpiece connection at the start of welding, and welding toward a heavy tack weld;
3. If forward blow is causing trouble, placing the workpiece connection at the end of the joint to be welded;
4. Positioning the electrode so that the arc force counteracts the arc blow;
5. Using the shortest possible arc consistent with good welding practice to help the arc force counteract the arc blow;
6. Reducing the welding current;
7. Welding toward a heavy tack or runoff tab;
8. Using the backstep sequence of welding;
9. If backing is used, welding the entire length of the backing to each base plate;
10. Changing to alternating current, which may require a change in electrode classification; and
11. Wrapping the workpiece lead around the workpiece in a direction so that the magnetic field it sets up will counteract the magnetic field causing the arc blow.

WELD QUALITY

A successfully welded joint must have the required physical and mechanical properties that will enable it to perform its function in service. The weld metal may also need a specific microstructure and chemical composition. The weldment may require abrasion or corrosion resistance. All of these objectives are influenced by the choice of the base metals, the welding materials, and the manner in which the weld is made.

The size and shape of the weld and the soundness of the joint affect its performance. Shielded metal arc welding is a manual welding process, and the quality of the welded joint largely depends on the skill of the welder who makes it. For this reason, the materials to be used must be selected with care, the welding procedure must be correct, and the welder must be proficient.

DISCONTINUITIES

Welded joints, by their nature, contain discontinuities of various types and sizes. The discontinuities are not considered harmful if they are below an applicable acceptance level. Above that level, they are considered defects. The acceptance level can vary with the severity of the service to which the weldment will be subjected. The acceptable limit should be considered at the design stage and should be established within the fabrication contract, usually by incorporating an applicable code or specification. The following discontinuities are sometimes encountered in welds made by the shielded metal arc welding process:

1. Porosity,
2. Slag inclusions,
3. Incomplete fusion,
4. Undercut, and
5. Cracks.

Porosity

The term *porosity* is used to describe gas pockets or voids in the weld metal. These voids result from gas that forms when certain chemical reactions take place during welding or that is expelled from the liquid but captured within the crystalline structure of the solidified weld. These discontinuities are usually round and contain gas rather than solids, and in this respect they differ from slag inclusions.

The welder can usually prevent porosity by using the proper amperage and holding a proper arc length. Assuring that the electrodes are dry is also helpful in many cases. The deoxidizers in a covered electrode that are needed to produce sound weld metal are more extensively reacted or lost during deposition when high amperage or a long arc is used. The remaining deoxidizers are then insufficient for the proper deoxidation of the molten metal.

Slag Inclusions

Inclusions are discontinuities caused by entrapped solids within the weld metal that can be detected with the unaided eye or with low magnification. An inclusion can occur if a solid with a high melting point or a dense solid, or both, is immersed in the weld pool and is trapped within the weld during solidification. Slag inclusions are oxides and nonmetallic solids that are intended to form the protective slag but are sometimes entrapped in the weld metal. This can be in the bead itself, between adjacent beads or between the weld and the base metal.

In shielded metal arc welding, during deposition and subsequent solidification of the weld metal, many chemical reactions occur. Some of the products of these reactions consist of solid, nonmetallic compounds that are insoluble in the weld pool. Because of their lower specific gravity, these compounds will rise to the surface of the molten metal and form a part of the slag unless they become entrapped in the solidified weld metal.

Slag formed from the covering on shielded metal arc electrodes may be forced below the surface of the molten metal by the stirring action caused by the arc within the pool. Slag may also flow ahead of the arc if the welder is not careful. This kind of slag entrapment can easily happen when welding over the crevice between two parallel but convex beads, or between one convex bead and a sidewall of the groove. It can also occur when welding in the downhill position. In such cases, the molten metal may flow over the slag and the slag becomes entrapped beneath the bead. The risk of slag entrapment is increased by factors such as highly viscous or rapidly solidifying slag or insufficient welding current.

Most slag inclusions can be prevented by good welding practice. In problem areas, the proper preparation of the groove before depositing the next bead of weld metal can reduce the likelihood of inclusions. In these cases, care must be taken to correct contours that are difficult to adequately penetrate with the arc.

Incomplete Fusion

The term *incomplete fusion* describes a weld discontinuity that occurs between the weld metal and the base metal or between the weld metal and an adjacent weld bead. This discontinuity may be localized or it may be extensive. It can occur at any point in the weld groove and may even occur at the root of the joint.

Incomplete fusion may be caused by failure to raise the base metal (or the previously deposited bead of weld metal) to the melting temperature. It may also be caused by improper fluxing, which fails to dissolve any oxides on the surface of the base metal, or by the improper removal of a coating on the surfaces of the workpieces. Incomplete fusion can also result from slag entrapment.

Incomplete fusion can be avoided by making certain that the surfaces to be welded are smooth, clean, and properly prepared and fitted. In the case of incomplete root fusion, the corrections involve assuring that (1) the root face is not too large, (2) the root opening is not too small, (3) the electrode is not too large, (4) the welding current is not too low, and (5) the travel speed is not excessive.

Undercut

The term *undercut* is used to describe an observable groove in the base metal adjacent to the weld toe or weld root that is left unfilled. Undercut in groove welds can occur adjacent to the back side of the root weld and on the beads in the final layer. In fillet welds, undercut can occur on either side of the weld. Undercut can result from either of two situations. One is the melting away of the face of a joint by the arc ahead of the weld pool, which forms a sharp recess in the face of the joint and is not filled by the weld. The other is the gradual melting and flowing of base plate material into the weld, which leaves a depression at the line where the weld metal ties into the surface of the base metal (i.e., at the toe of the weld).

Both types of undercut are usually due to the specific welding technique used by the welder and the type of electrode selected. High amperage and a long arc increase the tendency to undercut. Incorrect electrode position and travel speed are also causes, as is insufficient dwell time in a weave bead. The various classifications of electrodes show widely different characteristics in this respect. With some electrodes, even the most skilled welder may be unable to avoid undercutting completely in certain welding positions, particularly on joints with restricted access.

Undercut of the groove face has no effect on the completed weld if the undercut is removed and a bead is subsequently deposited at that location. A sharpened chipping tool or a grinding wheel is required to remove the undercut. If the undercut is slight, however, an experienced welder who knows just how deep the arc will penetrate may not need to remove the undercut.

The amount of undercut permitted in a completed weld is usually dictated by the governing fabrication code. The requirements specified should be strictly followed because excessive undercut can significantly reduce the strength of the joint. The reduced strength is particularly severe in applications subject to fatigue loading. Fortunately, undercut can be detected by visual examination of the completed weld, and it can be corrected by grinding it to blend in or by depositing an additional bead.

Cracks

Cracking is a fracture in the weld metal, the heat-affected zone of the base metal, or the surrounding base metal that occurs as a result of the welding cycle. A crack in a weldment is a three-dimensional discontinuity with a sharp tip and high ratio of length and width (depth) to the opening displacement, not unlike an exaggerated tall, narrow "V." If cracking is observed during welding, the cracks must be removed prior to further welding. A crack at the base of a joint is very likely to propagate into the newly deposited weld metal. Cracking in welded joints can be classified as either hot or cold cracking.

Hot Cracking. Hot cracking occurs during solidification. As the weld metal and base plate cool, stresses arise from shrinkage during transformation into the solid phase and from the contraction during the cooling of the metal. These stresses lead to failure if a liquid persists in the microstructure among the dendrites or at the edges of grains, or both. The stresses simply draw apart the solidifying dendrites or the grains touching the liquid. The rupture proceeds along the grain boundaries or in between dendrites. The main cause of hot cracking is the presence of constituents in the weld metal that have a relatively low melting temperature and that accumulate at the grain boundaries during solidification. A typical example is iron sulfide in steel. Coarse-grained, single-phase structures have a marked propensity for this type of cracking. Solutions to hot cracking problems include:

1. Changing the base metal, for example, using a steel with manganese additions, or another addition such as rare earths for sulfide shape control;
2. Changing filler metal, for example, using a filler metal with sufficient ferrite when welding austenitic stainless steel; and
3. Changing the welding technique or procedure, or both, by lowering the preheat and interpass temperatures and reducing the welding current.

Cold Cracking. Cold cracking encompasses restraint cracking and hydrogen-induced cracking. Restraint cracking is the result of inadequate ductility or strength in the base plate or in the weld metal. In hardenable steels, most cold cracking is hydrogen-induced cracking that occurs when these conditions are present:

1. Diffusible hydrogen is in the weld metal;
2. The weld metal has high residual tensile stresses caused by the constraint of the joint members and the heat input from the arc;
3. The weld metal has a susceptible microstructure (martensitic); and
4. The temperature of the weldment is in the range of –240°C to 750°C (–150°F to 400°F).

The use of dry low-hydrogen electrodes and proper preheat is required to prevent cold cracking in hardenable steels.

A crack will develop if the base metal or the weld metal has inadequate toughness in the presence of a mechanical or metallurgical notch and if the welding generates stresses of sufficient magnitude. These stresses do not have to be very high in some materials, for instance, large-grained ferritic stainless steel. Cold cracking will also occur if the weld is too small or too concave to provide sufficient strength to overcome the stresses. Preheat is also required for materials that are naturally low in ductility or toughness. Materials subject to extreme grain growth, for example, 28% chromium steel, must be welded with low heat input and low interpass temperatures. In all cases, notches must be avoided.

More information on the quality of welded joints can be found in Chapter 13, "Weld Quality" and Chapter 14, "Welding Inspection and Nondestructive Examination," *Welding Science and Technology,* Volume 1 of the *Welding Handbook*, 9th edition.[44]

ECONOMICS

Shielded metal arc welding is often the first choice for a great variety of welding applications because the equipment is inexpensive, portable, and versatile. The process provides a wide range of electrodes, parameters, and techniques that can have a positive influence on the cost of welding.

In some instances, choosing a welding process other than shielded metal arc welding is the best economic decision. Other processes may be used to advantage in the following situations:

1. When the weldment requires a large volume of filler metal and results in increased labor and material costs, unless only a few joints will be welded.
2. When welding light-gauge sheet, which is more economically welded with flux-cored or metal-cored electrodes.

One way to evaluate the economic performance of a joining process for a weldment is to determine the cost per unit length of the weld must be computed.[45] This calculation requires approximating the cost of filler metal, equipment and labor. In addition, consideration must be given to special circumstances, for example, if an accelerated completion date impacts the cost of financing.

To compute the labor cost, the operating factor (defined as the arc time divided by the total working time; for SMAW it is typically 30%) and the deposition rate (the amount of weld metal added to the joint per unit of arc time) must be known or approximated. To compute the electrode material cost the price must be known, the amount (total weight) of electrode approxi-

44. American Welding Society (AWS) Welding Handbook Committee, Jenney, C. L. and A. O'Brien, eds., 2001, *Welding Science and Technology*, Vol. 1 of Welding *Handbook*, 9th ed., Miami: American Welding Society, Chapters 13 and 14.

45. For information on welding costs refer to Chapter 12 of American Welding Society (AWS) Welding Handbook Committee, Jenney, C. L. and A. O'Brien, eds., 2001, *Welding Science and Technology*, Vol. 1 of Welding *Handbook*, 9th ed., Miami: American Welding Society.

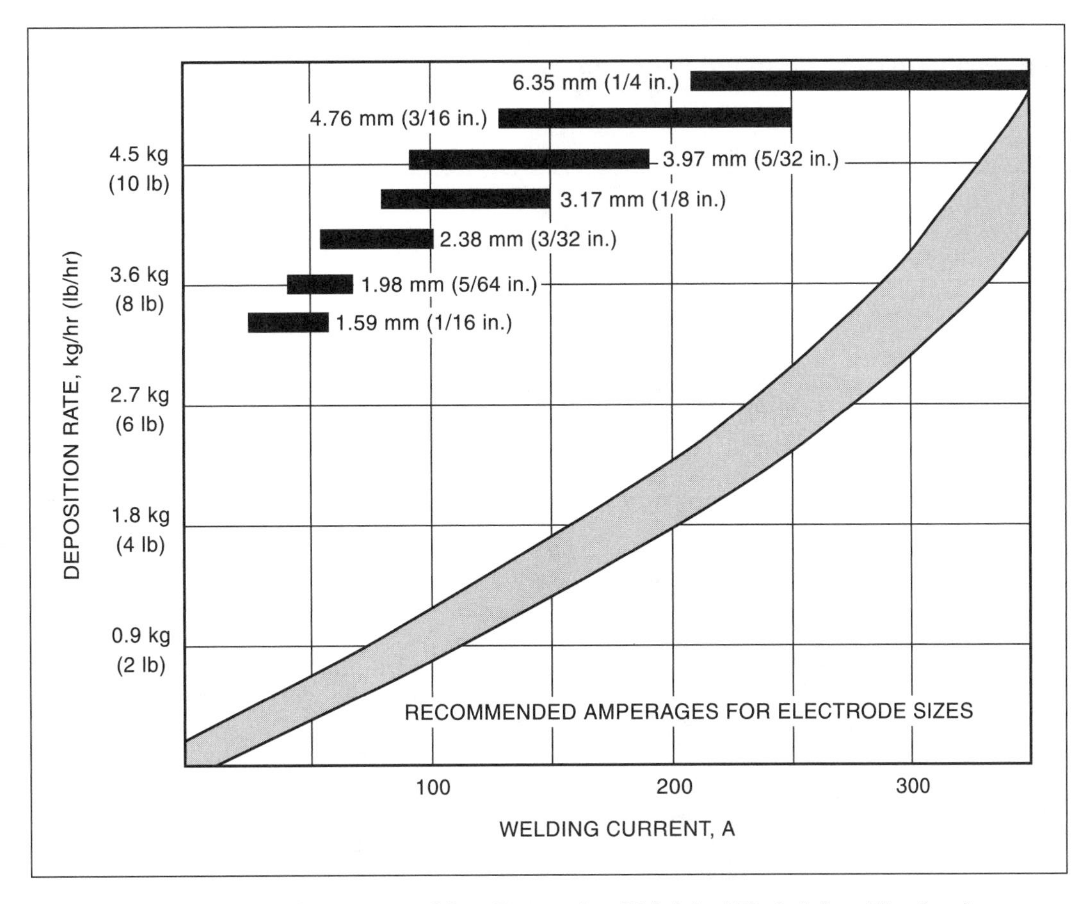

Figure 2.19–Typical Deposition Rates for Shielded Metal Arc Electrodes

mated from the joint volume, and the deposition efficiency (the ratio of the weight of metal deposited to the weight of electrode used, typically 0.60 for SMAW). Deposition rates for the particular electrodes, if not printed on the packaging, are readily available from the manufacturer. Figure 2.19 shows some typical values.

Once these values are known, the labor and material cost can be calculated as follows:

Labor cost, \$/ft =

$$\frac{\text{lb of weld/ft}}{\text{deposition rate, lb/hr} \times 0.30} \times \text{direct labor, \$/hr}$$

Material cost, \$/ft =

$$\text{Cost of electrode, \$/lb} \times \frac{\text{lbs of weld/ft}}{0.60}$$

The convention within industry for the shielded metal arc welding process is to report the deposition efficiency with no adjustment for stub loss, that is, simply state the amount of metal deposited divided by the amount of electrode consumed. This approximate deposition efficiency must then be adjusted by using a measure for the expected stub loss. The computation to adjust the deposition efficiency for stub loss is based on the original length of the electrode; the efficiency increases with increased length. Table 2.4 shows the approximate deposition efficiency for the various standard electrode lengths when adjusted for stub loss.

SAFE PRACTICES

Before undertaking the welding operation, the welder and all persons who will have access to the welding station must clearly understand the hazards and the potential harm that may result from the use of welding materials and equipment. They must know and

Table 2.4
Approximate Deposition Efficiency, in Percent, at Various Lengths of Stub Loss

		Stub Loss		
Electrode Length	**Zero Stub Loss**	**50 mm (2 in.)**	**75 mm (3 in.)**	**125 mm (5 in.)**
300 mm (12 in.)	60	50	45	35
	70	58	52	40
	80	67	60	47
350 mm (14 in.)	60	51	47	38
	70	60	55	45
	80	68	62	51
450 mm (18 in.)	60	53	50	43
	70	62	56	50
	80	71	66	57

practice the appropriate safety measures. A copy of *Safety in Welding, Cutting, and Allied Processes*, ANSI Z49.1[46] should be consulted and kept in the vicinity for reference. Manufacturers' safety data sheets (MSDS) detailing the proper handling and use of the materials should be consulted before using the materials. Operating manuals provided by the equipment manufacturer for the safe operation of the welding equipment should be consulted.

Other sources of safety information should be consulted. These include documents from such organizations as the National Electric Manufacturers Association,[47] Underwriters Laboratories (UL)[48] and the National Fire Protection Association (NFPA).[49] Appendix B of this volume, "Safety and Health Codes and Other Standards" lists health and safety standards, codes, specifications, and other publications. The publishers, the letter designations of the organizations, and the facts of publication are also listed.

A brief discussion of some of the safety considerations follows.

The operator must protect eyes and skin from radiation from the arc. A welding helmet with a suitable filter lens should be used, as well as dark clothing, preferably wool, to protect the skin. Leather gloves and clothing should be worn to protect against burns from arc spatter.

Welding helmets are available with filter plate windows that absorb 99% or more of the infrared and ultraviolet rays from the arc. The filter plate selected by the welder should be capable of absorbing infrared rays, ultraviolet rays, and most of the visible rays emanating from the arc. The standard filter plate size is 51 mm by 130 mm (2 in. by 4-1/8 in.), but larger openings are available.

The selection of the appropriate filter plate shade depends on the electrode size. For electrodes up to 4 mm (5/32 in.) diameter, a Number 10 filter plate shade should be used. For 4.8 mm to 6.4 mm (3/16 in. to 1/4 in.) electrodes, a Number 12 should be used. For electrodes over 6.4 mm (1/4 in.), a Number 14 filter plate shade should be used. The filter plate should be protected from molten spatter and from breakage. Placing a plate of clear glass or other suitable material on both sides of the filter plate can provide this protection.

Persons who are not welding but are working near the arc also need protection. Permanent or portable screens can be used to provide this protection. Failure to use adequate protection can result in eye burn (sometimes called *flashing*) to the welder or those working around the arc. Eye burn, which is similar to sunburn, is extremely painful for a period of 24 to 48 hours. Unprotected skin exposed to the arc may also be burned. A physician should be consulted in the case of severe arc burn, whether it involves the skin or the eyes.

If welding is performed in confined spaces with poor ventilation, auxiliary air should be supplied to the welder. This should be supplied through an attachment

46. American National Standards Institute (ANSI) Accredited Standards Committee Z49, *Safety in Welding, Cutting, and Allied Processes*, ANSI Z49.1, Miami: American Welding Society.
47. National Electric Manufacturers Association (NEMA), *Electric Arc Welding Power Sources*, EW1, Rosslyn, Virginia: National Electric Manufacturers Association.
48. Underwriters Laboratories (UL), *Transformer-Type Arc Welding Machines*, UL551, Northbrook, Illinois: Underwriters Laboratories.
49. National Fire Protection Association (NFPA). *National Electric Code*, (NEC®), ANSI/NFPA 70, Quincy, Massachusetts: National Fire Protection Association.

to the helmet. The method used must not restrict the welder's manipulation of the helmet, interfere with the field of vision, or make welding difficult. Additional information on eye protection and ventilation is published in *Safety in Welding, Cutting, and Allied Processes*, ANSI Z49.1.[50] Documents covering these safety issues are also available from the United States Occupational Safety and Health Administration.[51] Exposure limits for fumes and gases are detailed in documents from the American Conference of Governmental Industrial Hygienists.[52] From time to time during welding, sparks or globules of molten metal are thrown out from the arc. This is always a point of concern, but it becomes more serious when welding is performed out of position or when extremely high welding currents are used. To ensure protection from burns under these conditions, the welder should wear flame-resistant gloves, a protective apron, and a jacket. The welder's ankles and feet may also need protection from slag and spatter. Pants without cuffs and high-top work shoes or boots are recommended.

To avoid electric shock, the operator should follow recommendations provided by the equipment manufacturer. The welder should not weld while standing on a wet surface. Welding and auxiliary equipment should be examined periodically to make sure there are no cracks or worn spots on the electrode holder or cable insulation. Welders should not touch live electrical parts such as the conductive surface in the jaws of the electrode holder and the workpiece while the power source is running. A stub should never be left in the jaw of the electrode holder when the holder is stowed and the power source is on.

CONCLUSION

Shielded metal arc welding is a manual arc welding process that has a proven historical record of success in the field of metal joining and surfacing. In its special niche, the process will continue to serve welders and fabricators for generations to come.

No one process can fit all applications, but considering its worldwide use and its numerous types of electrodes, some have expressed the opinion that the SMAW process is one of the most important welding processes developed in the history of welding.[53]

Although many have predicted the demise of shielded metal arc welding because other arc welding processes appear to have tremendous economic advantages, the reliability and simplicity of the process continue to ensure its classic role in the industry.

BIBLIOGRAPHY[54]

American Conference of Governmental Industrial Hygienists (ACGIH). *Threshold limit values for chemical substances and physical agents in the workroom environment.* Cincinnati, Ohio: American Conference of Governmental Industrial Hygienists.

American National Standards Institute (ANSI) Accredited Standards Committee Z49. 1999. *Safety in welding, cutting, and allied processes.* ANSI Z49.1:1999. Miami: American Welding Society.

American National Standards Institute (ANSI). 1989. *Practice for occupational and educational eye and face protection.* ANSI Z87.1-1989. Des Plaines, Illinois: American Society of Safety Engineers (ASSE).

American Welding Society (AWS) Committee on Structural Welding. 2002. *Structural Welding Code—Steel*, AWS D1.1. Miami: American Welding Society.

American Welding Society Committee on Structural Welding. 1997. *Structural welding code—Aluminum,* AWS D1.2-97. Miami: American Welding Society.

American Welding Society Committee on Filler Metals and Allied Materials. 1997. *Specification for nickel and nickel-alloy welding electrodes for shielded metal arc welding.* AWS A5.11/A5.11M-97. Miami: American Welding Society.

American Welding Society (AWS) Committee on Filler Metals and Allied Materials and Welding Research Council (WRC) Subcommittee on Stainless Steels. 1997. *Standard procedures for calibrating magnetic instruments to measure the delta ferrite content of austenitic stainless weld metal.* AWS A4.2/A4.2M: 1997. Miami: American Welding Society.

American Welding Society Committee on Filler Metals and Allied Materials. 1996. *Specification for low-alloy steel electrodes for shielded metal arc welding.* AWS A5.5-96. Miami: American Welding Society.

American Welding Society (AWS) Committee on Filler Metals and Allied Materials. 1992. *Specification for*

50. See Reference 45.
51. United States Occupational Safety and Health Administration, *Code of Federal Regulations, Title 29 Labor*, Part 1910, Washington, D.C.: U.S. Government Printing Office.
52. American Conference of Governmental Industrial Hygienists (ACGIH), *Threshold Limit Values for Chemical Substances and Physical Agents in the Workroom Environment,* Cincinnati, Ohio: American Conference of Governmental Industrial Hygienists.
53. Welding Journal Reader Forum, *Welding Journal* 73(6), p. 19, Miami: American Welding Society.
54. The dates of publication for the codes and other standards listed here were current at the time this chapter was prepared. The reader is advised to consult the latest edition.

stainless steel welding electrodes for shielded metal arc welding. AWS A5.4-92 (R 2000). Miami: American Welding Society.

American Welding Society (AWS) Committee on Filler Metals and Allied Materials. 1990. *Specification for welding electrodes and rods for cast iron.* AWS A5.15-90. Miami: American Welding Society.

American Welding Society (AWS) Committee on Filler Metals and Allied Materials. 1984. *Specification for covered copper and copper alloy arc welding electrodes.* AWS A5.6-84. (R 2000), Miami: American Welding Society.

American Welding Society (AWS) Committee on Filler Metals and Allied Materials. 1980. *Specification for solid surfacing welding rods and electrodes.* AWS A5.13-80. Miami: American Welding Society.

American Welding Society Committee on Filler Metals and Allied Materials. 1980. *Specification for composite surfacing welding rods and electrodes.* AWS A5.21-80. Miami: American Welding Society.

National Electric Manufacturers Association (NEMA*), Electric arc welding power* sources, *EW1*. Rosslyn, Virginia: National Electric Manufacturers Association.

National Fire Protection Association (NFPA). *National electric code (NEC®),* ANSI/NFPA 70. Quincy, Massachusetts: National Fire Protection Association.

American Welding Society (AWS) Welding Handbook Committee. Jenney, C. L., and A. O'Brien, eds. 2001. *Welding science and technology.* Vol. 1 of *Welding handbook.* 9th ed. Miami: American Welding Society.

American Welding Society (AWS) Welding Handbook Committee. O'Brien, R. L., ed. 1991. *Welding Processes.* Vol. 2 of *Welding handbook.* 8th ed. Miami: American Welding Society.

American Welding Society (AWS) Welding Handbook Committee. Oates, W. R., ed. 1991. *Materials and Applications—Part 1.* Vol. 3 of *Welding handbook.* 8th ed. Miami: American Welding Society.

American Welding Society (AWS) Welding Handbook Committee. Oates, W. R., ed. 1996. *Materials and Applications—Part 2.* Vol. 4 of *Welding handbook.* 8th ed. Miami: American Welding Society.

Espy, R. H. 1982, Weldability of nitrogen-strengthened steels. *Welding Journal.* 61(5) 149-s–155-s.

Kotecki, D. J. 1992. Constitution diagram for stainless steel weld metal: a modification of the WRC-1988 diagram. *Welding Journal.* 71(5). 171-s–178-s.

Occupational Safety and Health Administration (OSHA). 1999. *Code of federal regulations, Title 29 Labor,* Part 1910. Washington, D.C.: U.S. Government Printing Office.

Underwriters Laboratories (UL). *Transformer-type arc welding machines, UL551.* Northbrook, Illinois: Underwriters Laboratories.

SUPPLEMENTARY READING LIST

ASM International 1993. *Welding, brazing, and soldering.* Vol. 6 of *Metals handbook.* 1st ed. Materials Park, Ohio: ASM International.

Bhadeshia H, K. D. H. 1992. *Bainite in steels: transformations, microstructure and properties.* London SY1Y 5DB UK: Institute of Metals.

Barbin, L. M. 1977. The new moisture-resistant electrodes. *Welding Journal* 56(7): 15–18.

Chew, B. 1976. Moisture loss and gain by some basic flux covered electrodes. *Welding Journal* 55(5): 127-s–134-s.

Cranyon, H., 1991. *Fundamentals of welding metallurgy.* New York: Welding Research Council.

Gregory, E. N. 1969. Shielded metal arc welding of galvanized steel. *Welding Journal* 48(8): 631-s–638-s.

Jackson, C. E. 1973. *Fluxes and slags in welding.* Welding Research Council Bulletin 190. New York: Welding Research Council.

Silva, E. A., and T. H. Hazlett. 1971. Shielded metal arc welding underwater with iron powder electrodes. *Welding Journal* 50(6): 406-s–415-s.

Stout, R. D., C. W. Ott, A. W. Pense, D. J. Snyder, B. R. Somers, and R. E. Somers. 1987. *Weldability of steels.* New York: Welding Research Council.

The Lincoln Electric Company. 1994. *The procedure handbook of arc welding.* 13th ed. Cleveland: The Lincoln Electric Company.

CHAPTER 3

GAS TUNGSTEN ARC WELDING

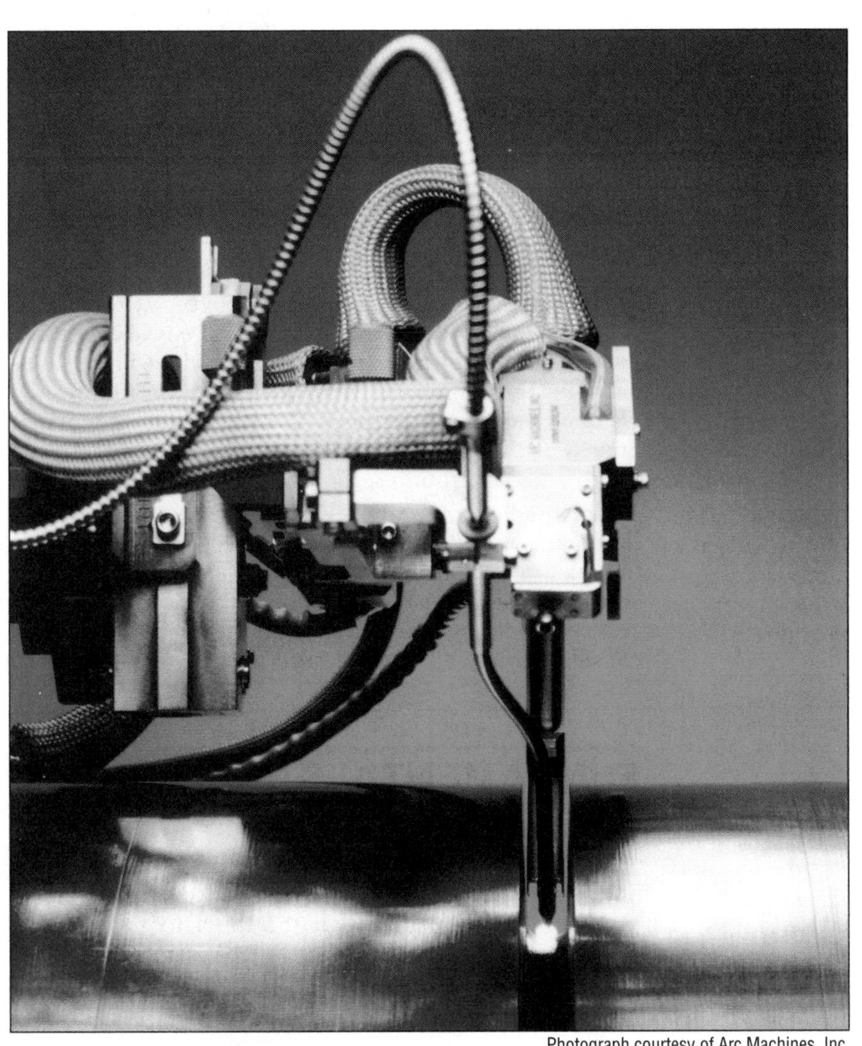

Photograph courtesy of Arc Machines, Inc.

Prepared by the Welding Handbook Chapter Committee on Gas Tungsten Arc Welding:

J. T. Salkin, Chair
Arc Applications, Inc.

K. W. Beedon
Elliott Turbomachinery, Inc.

B. K. Henon
Arc Machines, Inc.

K. R. Jelonek
Arc Applications, Inc.

D. B. O'Donnell
Raft Engineering, Inc.

Welding Handbook Volume 2 Committee Member:

D. R. Amos
Siemens Westinghouse Power Corporation

Contents

CHAPTER 3

GAS TUNGSTEN ARC WELDING

INTRODUCTION

Gas tungsten arc welding (GTAW) is an arc welding process that uses an arc between a nonconsumable tungsten electrode and the workpiece to establish a weld pool. The process is used with shielding gas and without the application of pressure, and may be used with or without the addition of filler metal.[1,2] Because of the high quality of welds that can be produced by gas tungsten arc welding, the process has become an indispensable tool for many manufacturers, including those in the aerospace, nuclear, marine, petrochemical and semiconductor industries.

The possibility of using helium to shield a welding arc and weld pool was first investigated in the 1920s.[3] However, there was no incentive for further development or use of this process until the beginning of World War II, when a great need emerged in the aircraft industry to replace riveting as the method for joining reactive materials, such as aluminum and magnesium. The welding industry responded by producing a stable, efficient heat source with which excellent welds could be made using a tungsten electrode and direct current arc power with the electrode negative. Helium was selected to provide the necessary shielding because it was the only inert gas readily available at the time. Tungsten electrode inert gas torches typical of that period are shown in Figure 3.1.

1. American Welding Society (AWS) Committee on Definitions and Symbols, 2001, *Standard Welding Terms and Definitions*, 2001, Standard Welding Terms and Definitions, AWS A3.0:2001, Miami: American Welding Society.
2. At the time this chapter was prepared, the referenced codes and other standards were valid. If a code or other standard is cited without a date of publication, it is understood that the latest edition of the document referred to applies. If a code or other standard is cited with the date of publication, the citation refers to that edition only, and it is understood that any future revisions or amendments to the code or other standard are not included. As codes and standards undergo frequent revision, the reader is advised to consult the most recent edition.
3. H. M. Hobart, U.S. Patent 1,746,081; 1926, and P. K. Devers, U.S. Patent 2,274,361; 1926.

The process has been called *nonconsumable electrode welding* and is very often referred to as *TIG* (tungsten inert gas) welding. However, because shielding gas mixtures that are not inert can be used for certain applications, the American Welding Society (AWS) adopted *gas tungsten arc welding* (GTAW) as the standard terminology for the process.

Numerous improvements have been made to the process and equipment since the early days of the invention. Welding power sources were developed specifically for the process, some providing pulsed direct current and variable-polarity alternating current. Water-cooled and gas-cooled torches were developed. The tungsten electrodes were alloyed with small amounts of active elements to increase emissivity, thus improving arc starting, arc stability, and electrode life. Shielding gas mixtures were identified for improved welding performance. Researchers continue to pursue improvements in such areas as automatic controls, vision and penetration sensors, and arc length controls.

The fundamentals of the GTAW process and a variation that uses pulsed current (GTAW-P) are discussed in this chapter, along with applications of the process, equipment and consumables used, techniques and procedures, welding variables, weld quality, and safety considerations.

FUNDAMENTALS

Gas tungsten arc welding uses a nonconsumable tungsten or tungsten alloy electrode held in a torch. Shielding gas is fed through the torch to provide an inert atmosphere that protects the electrode and the weld pool while the weld metal is solidifying. The electric arc, produced by the passing of current through the

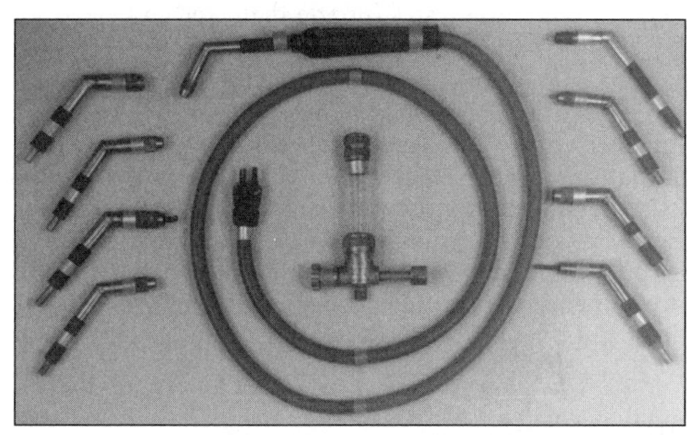

Photograph courtesy of Miller Electric Manufacturing Company

Figure 3.1—Early Gas Tungsten Arc Welding Heads with a Torch Body and Flow Meter, Circa 1943

conductive ionized shielding gas, is established between the tip of the electrode and the workpiece. The weld starts as heat generated by the arc melts the base metal and establishes a weld pool. The torch is moved along the workpiece and the arc progressively melts the surfaces of the joint. If specified, filler metal, usually in the form of wire, is added to the leading edge of the weld pool to fill the joint. The gas tungsten arc welding process is illustrated in Figure 3.2.

The four basic components common to all GTAW setups are a torch, the electrode, a welding power source, and shielding gas. A typical setup is shown in Figure 3.3.

ADVANTAGES OF GTAW

Gas tungsten arc welding offers advantages for an extensive range of applications, from the high-quality welds required in the aerospace and nuclear industries and the high-speed autogenous welds required in tube and sheet metal manufacturing to the welds typical of fabricating and repair shops, where the ease of operation and the flexibility of the process are welcomed.

The process can be automated and is readily programmable to provide precise control of the welding variables with remote welding control capability. Flexibility is gained when using gas tungsten arc welding because the process allows the heat source and filler metal additions to be controlled independently. Excellent control of root pass weld penetration can be maintained.

Welds can be made in any position, and applications are almost unlimited. The process is capable of producing consistent autogenous welds of superior quality at high speeds, spatter-free, and generally with few defects. Almost all metals, including dissimilar metals, can be welded with the GTAW process. The process can be used with or without filler metal, as required by the specific application. A further advantage is that relatively inexpensive power sources can be used.

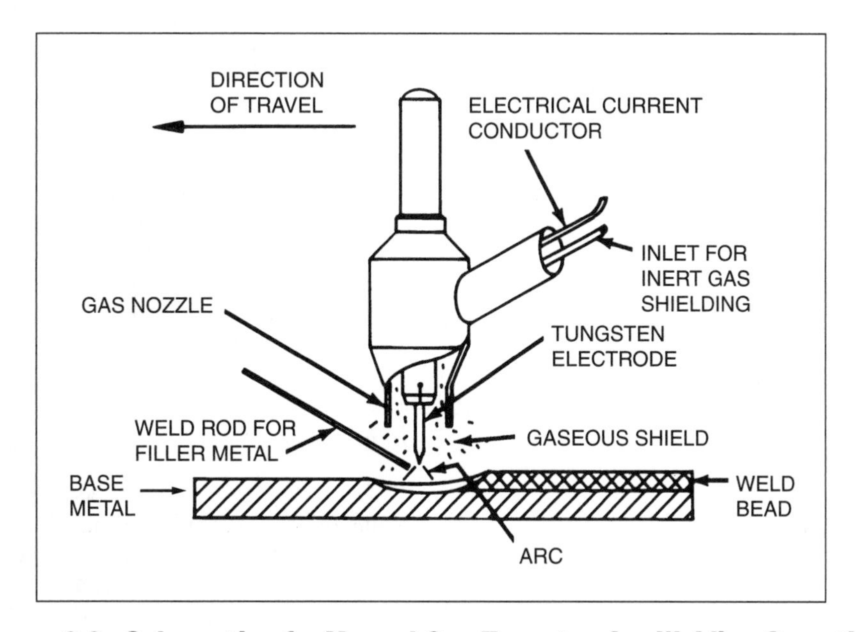

Figure 3.2—Schematic of a Manual Gas Tungsten Arc Welding Operation

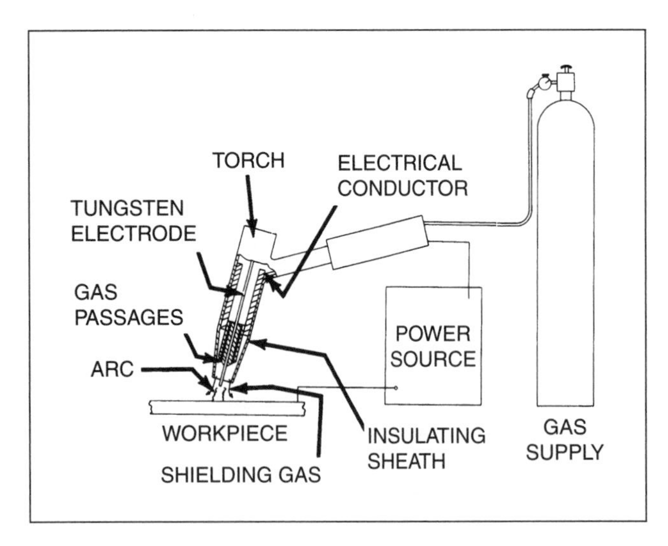

Figure 3.3—Gas Tungsten Arc Welding Equipment Arrangement

LIMITATIONS

The following limitations of the GTAW process should be considered when selecting a process for a specific application:

1. Deposition rates are generally lower than the rates possible with consumable electrode arc welding processes;
2. Slightly more dexterity and coordination is required of the welder using GTAW than with gas metal arc welding (GMAW) or shielded metal arc welding (SMAW) manual welding;
3. There is a low tolerance for contaminants on filler or base metals;
4. For welding thicknesses under 10 millimeters (mm) (3/8 inch [in.]), the GTAW process produces weld quality comparable to or better than the consumable arc welding processes, but it is more expensive;
5. Magnetic fields leading to arc blow or arc deflection, as with other arc processes, can make gas tungsten arc welding difficult to control; and
6. If welding takes place in windy or drafty environments, it can be difficult to shield the weld zone properly.

PROCESS VARIABLES

The principal variables in GTAW are arc voltage (arc length), welding current, travel speed, wire feed speed and shielding gas. Other variables may include shielding gas type, flow rates, nozzle diameter and electrode extension.

The amount of energy produced by the arc is proportional to the current and voltage. The amount of energy transferred per unit length of weld is inversely proportional to the travel speed. The arc shielded with helium is generally hotter and more penetrating than the argon-shielded arc. However, because all of these variables strongly interact with one another, it is impossible to treat them as truly independent variables when establishing welding procedures for fabricating specific joints.

Welding Current

In general, welding current controls weld penetration in gas tungsten arc welding—the effect of current on penetration is directly proportional, if not somewhat exponential. Welding current also affects the voltage. Voltage at a fixed arc length increases in proportion to the current. For this reason, it is necessary to adjust the voltage setting when the current is adjusted to keep a fixed arc length.

Gas tungsten arc welding can be used with direct current (dc) or alternating current (ac); the choice depends largely on the metal to be welded. Direct current with the electrode negative (DCEN) offers the advantages of deep penetration and fast welding speed, especially when helium or a helium mixture is used as the shielding gas. Alternating current provides a cathodic cleaning action during the electrode positive portion of the ac cycle, which removes refractory oxides from the metal surfaces. This cleaning action, known as *sputtering*, eliminates nearly all refractory oxide film from the weld pool surface and facilitates superior welds. In this case, argon must be used for the shielding because sputtering does not occur with helium. Argon is generally the gas of choice because it produces stable arc characteristics, whether used with direct current or alternating current, and it is lower in cost.

A third power option is also available, that of using direct current with the electrode positive (DCEP). However, this polarity is rarely used because it causes electrode overheating and reduced heating of the materials to be welded. The effects of polarity are explained in more detail in the section, "Direct Current."

Arc Voltage

The voltage measured between the tungsten electrode and the workpiece (with the arc initiated) is commonly referred to as the *arc voltage*. Arc voltage, which is generally proportional to arc length, is a dependent variable that is affected in decreasing amounts by the following:

1. The distance between the tungsten electrode and the workpiece,
2. Welding current,
3. Type of shielding gas,
4. The shape of the tungsten electrode tip, and
5. Ambient air pressure.

Arc voltage can also be affected by contaminants inadvertently entering the plasma from the workpiece or filler metal surfaces. The effects of these factors are particularly noticeable at low welding currents (under 75 amperes (A). Adjusting the arc voltage becomes a way to control the arc length, since other variables, such as current, shielding gas, nozzle diameter and electrode extension have been predetermined. Although arc voltage is changed by the effects of other variables, it is listed as a variable in welding procedures to assure similar results from weld to weld. Systems for mechanized welding can maintain consistent control of arc voltage. In manual welding, however, arc voltage is difficult to monitor and control.

Arc length is an important variable with GTAW because it affects the width of the weld pool (pool width is proportional to arc length) and to a lesser extent, penetration and shielding coverage. However, in most applications it is preferable to maintain minimum arc lengths. Consideration must also be given to the possibility of short-circuiting the electrode to the weld pool or filler wire if the arc is not long enough.

When using DCEN in mechanized welding with a relatively high current, it is possible to use the arc to depress the weld pool and submerge the electrode tip below the surface of the workpiece to produce deeply penetrating but narrow welds at high speeds without shorting out the tungsten electrode. This technique, which requires good control of arc voltage and arc length, is sometimes referred to as *buried arc*.

When arc voltage is used to control arc length in critical applications, the other variables that affect arc voltage must be carefully observed and controlled. Among these variables are electrode and shielding gas contaminants, improperly fed filler wire, temperature changes in the electrode, and electrode erosion. If any of the variables change enough to affect the arc voltage during mechanized welding, the arc length must be adjusted to restore the desired voltage.

Travel Speed

Travel speed affects both the width and penetration of a gas tungsten arc weld. However, the effect of travel speed on width is more pronounced than its effect on penetration. In some applications, travel speed is defined as an objective, with the other variables selected to achieve the desired weld configuration at that speed. In other cases, travel speed might be a dependent variable, selected to obtain the weld quality and uniformity needed under the best conditions possible with the other combination of variables. Regardless of the objectives, travel speed is generally fixed in mechanized welding while other variables, such as current or voltage, are varied to maintain control of the weld.

Wire Feed Speed

In manual welding, the welder controls the amount of filler metal to be used and the technique for depositing it in the weld pool. By adjusting the wire feed speed, the welder carefully controls the amount deposited to reduce the potential for incomplete fusion. Careful control also influences the number of passes required and the appearance of the finished weld.

In mechanized and automatic welding, the wire feed speed determines the amount of filler deposited per unit length of weld. Decreasing wire feed speed increases penetration and flattens the bead contour. Feeding the wire too slowly may produce weld beads that are more concave, which leads to undercut, incomplete joint fill, and the potential for centerline cracking. Increasing wire feed speed produces a weld bead that is more convex, but it may decrease weld penetration.

APPLICATIONS

Gas tungsten arc welding is often selected when critical weld specifications must be met. Because gas tungsten arc welding provides the best control of heat input, it is the preferred process for joining thin-gauge metals, for spot welding in sheet metal applications, and for making welds close to heat-sensitive components. The process is easy to control and offers the option of adding filler metal as necessary to produce high-quality welds with smooth, uniform bead shapes or deposit surfaces.

Gas tungsten arc welding can be used to weld almost all metals. It is especially useful for joining aluminum and magnesium, which form refractory oxides, and for reactive metals like titanium and zirconium, which can become embrittled if exposed to air while molten.

Gas tungsten arc welding can be used to weld many joint geometries and overlays in plate, sheet, pipe, tubing, and other structural shapes. It is particularly appropriate for welding sections less than 10 mm (3/8 in.) thick. Pipe welding is often accomplished using gas tungsten arc welding for the root pass and either shielded metal arc welding or gas metal arc welding for the fill passes.

Photograph courtesy of Arc Machines, Inc.

Figure 3.4—Gas Tungsten Arc Welding in a Semiconductor Plant

Photograph courtesy of Arc Machines, Inc.

Figure 3.5—Fabricating a Piping System with Orbital Gas Tungsten Arc Welding in a Dairy Processing Plant

Millions of orbital gas tungsten arc fusion welds have been applied in semiconductor fabrication plants worldwide to join small-diameter tubing, fittings and other components used in gas distribution systems, where weld quality is especially critical. Figure 3.4 shows a typical GTAW application in a semiconductor fabrication plant. In a similar application, shown in Figure 3.5, orbital GTAW fusion welds are used to fabricate the piping system in a dairy processing plant.

Innovations applied to the gas tungsten arc welding process have resulted in a range of equipment modifications for specific applications, such as narrow groove welding. Manufacturers have had increasing success in managing arc energy to achieve consistent bevel or groove face fusion, shielding gas coverage, and effective deposition rates in production. Typical narrow groove GTAW applications include piping for steam generating plants, superconductor magnet casings and large-diameter forged shafts. An example of this narrow groove modification of GTAW is a system that incorporates an oscillating angled tungsten arc synchronized with the side-to-side movement of the filler wire guide, as shown at left in Figure 3.6. This type of equipment has been used successfully for out-of-position welds in materials over 254 mm (10 in.) thick with included angles of 1° to 2°. An enlarged trailing view of the system making a weld within a narrow groove is shown at right.

The aerospace and other industries makes extensive use of gas tungsten arc welding to take advantage of the precise control and weld quality characteristics of the process. An aerospace application is shown in Figure 3.7. Pulsed current is used in this application for the autogenous welding of a flanged joint between two 17-4PH stainless steel machined castings, 180 mm (7 in.) in diameter. Argon shielding was used. Figure 3.8 shows a marine application, in which autogenous welds of copper alloy components were made using a mixed argon-helium shielding gas.

Photographs courtesy of Arc Applications, Inc.

Figure 3.6—Typical Application of Narrow Groove Orbital Welding of Thick-Wall Piping in a Power-Generating Plant

EQUIPMENT

Equipment for gas tungsten arc welding includes torches, electrodes, shielding gas delivery systems, and power sources. Mechanized GTAW systems may additionally incorporate a combination of arc voltage controls, travel motion controls, arc oscillators, and wire feeders.

WELDING TORCHES

The gas tungsten arc welding torch holds the tungsten electrode, which conducts welding current to the arc, and provides the means for conveying shielding gas to the arc zone. Most torches are designed to accommodate a range of electrode sizes and various types and sizes of gas nozzles.

Torches are rated in accordance with the maximum welding current that can be used without overheating. Typical current ranges are listed in Table 3.1. Most torches for manual applications are designed with a head angle (the angle between the electrode and handle) of approximately 120°. Torches are also available with adjustable-angle heads, 90° heads, or straight-line (pencil type) heads. Manual GTAW torches are often equipped with auxiliary switches and valves attached to the torch handles or a foot-controlled rheostat for controlling current and gas flow.

Torches for mechanized or automatic gas tungsten arc welding are typically mounted on a weld head or carriage that centers the torch over the joint. During welding, the joint can rotate under the arc or the torch can move along the joint, and can automatically change or maintain the arc length (torch-to-workpiece distance). Electrodes for gas tungsten arc orbital fusion welding are installed in a rotor in the weld head that rotates around the joint circumference, maintaining a fixed arc length while the tube remains in place.

Figure 3.7—Autogenous Gas Tungsten Arc Weld on Two 17-4PH Stainless Steel Machined Castings for an Aerospace Application

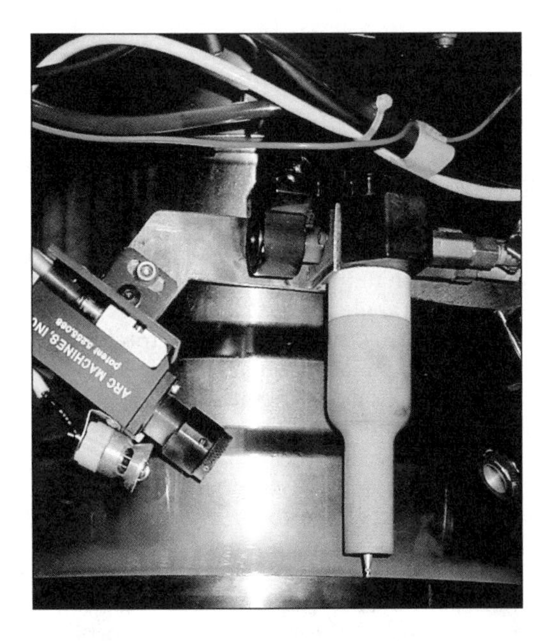

Figure 3.8—Autogenous Gas Tungsten Arc Welding of Copper Alloy Components for a Marine Application

Gas-Cooled Torches

The heat generated in the torch during welding is removed either by gas cooling or water cooling. Gas-cooled torches (sometimes called *air-cooled*) are cooled by the flow of the relatively cool shielding gas through the torch, as shown in Figure 3.3. Gas-cooled torches are limited to a maximum welding current of about 200 A.

Water-Cooled Torches

Water-cooled torches are cooled by the continuous flow of water through passageways in the holder. As illustrated in Figure 3.9, cooling water enters the torch through the inlet hose, circulates through the torch, and exits through an outlet hose. The power cable from the power source to the torch is typically enclosed within the cooling water outlet hose to assure that cool water reaches the torch prior to cooling the power cable.

Water-cooled torches are designed for use at higher welding currents on a continuous duty cycle than similar sizes of gas-cooled torches. Typical welding currents of 300 A to 500 A can be used, although some torches

Table 3.1
Typical Current Ratings for Gas- and Water-Cooled GTAW Torches

Torch Characteristic	Torch Size		
	Small	Medium	Large
Maximum current (continuous duty), amperes	200	200–300	500
Cooling method	Gas	Water	Water
Electrode diameters accommodated, mm (in.)	0.5–3.18 mm (0.020–1/8 in.)	1.0–3.96 mm (0.040–5/32 in.)	1.0–6.35 mm (0.040–1/4 in.)
Gas nozzle diameters (ID) accommodated, mm (in.)	6.35–19.05 mm (1/4–3/4 in.)	6.35–25.40 mm (1/4–1 in.)	9.53–43.45 mm (3/8–1-3/4 in.)
Gas Lens Nozzles (ID), mm (in.)	6.35–11.11 mm 1/4–7/16 in.)	7.94–19.05 mm (5/16–3/4 in.)	15.88–23.81 mm (5/8–15/16 in.)

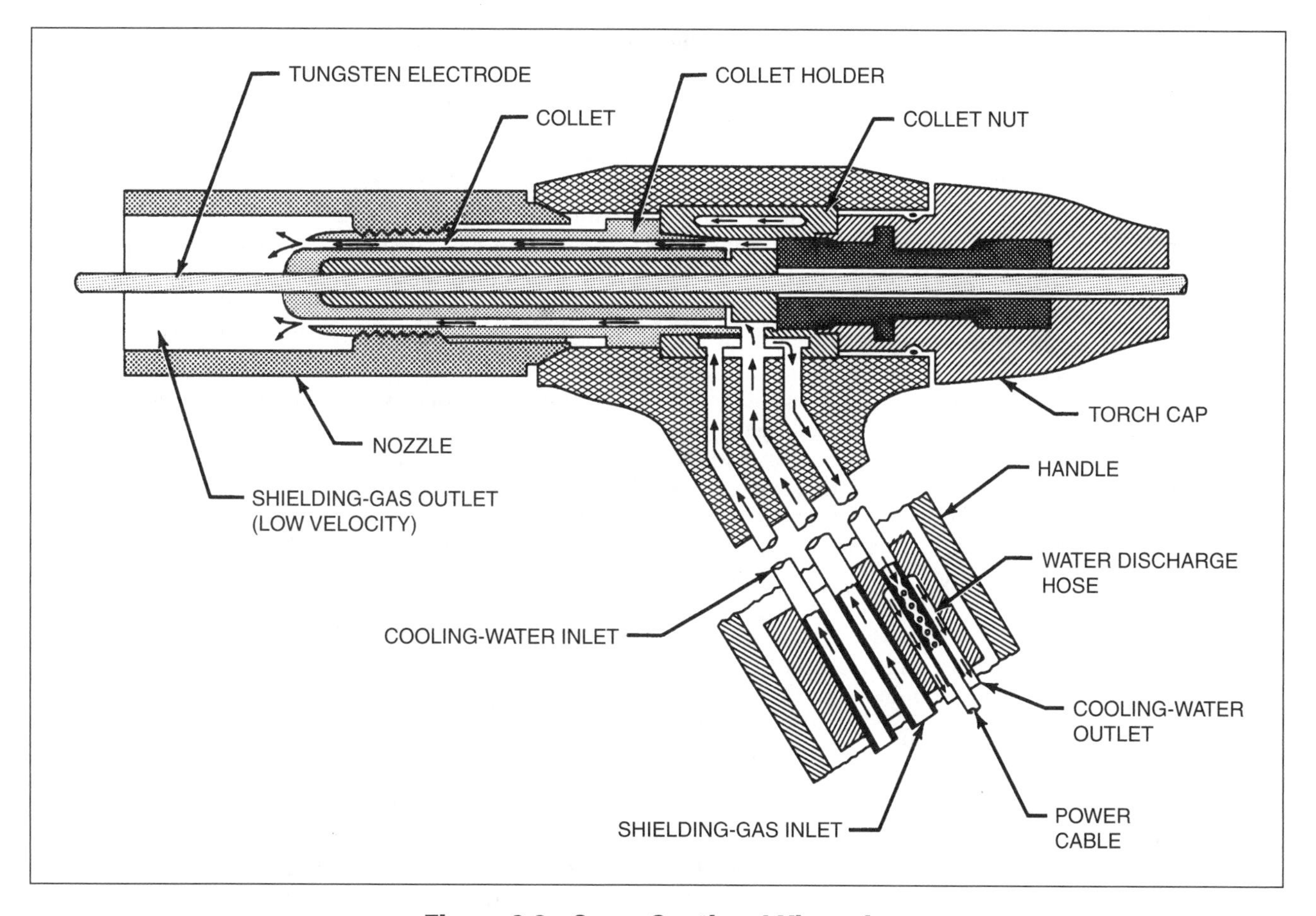

Figure 3.9—Cross-Sectional View of a Typical Water-Cooled Torch for Manual GTAW

are designed to handle welding currents up to 1000 A. Most mechanized or automatic welding applications use water-cooled torches.

Water-cooled torches can be cooled by tap water directed through the torch and then down a drain. However, to conserve water, a closed system involving a reservoir, pump, and radiator or water chiller to extract heat from the system is more common. The capacity of these systems ranges from one to fifty gallons. An automotive-type antifreeze coolant can be mixed with water and added to the cooling system to prevent freezing and corrosion and to provide lubrication for the water pump. Closed-system coolers, unless properly maintained, will eventually be contaminated by the proliferation of various bacteria, algae or fungal growth.

Collets

A collet is a clamping device that holds the electrode in the torch, as shown in Figure 3.9. Electrodes of various diameters are secured in position within the torch by appropriately sized collets, also called *chucks*. The collet body centers the tungsten in the torch and allows gas to flow around the collet and through holes in the collet body and out the nozzle. Collets and collet bodies are typically made of a copper alloy. The collet is forced against the collet body when the torch cap is tightened in place. Good contact between the electrode and the inside diameter of the collet and collet body is essential for proper current transfer and electrode cooling.

Nozzles

Shielding gas is directed to the weld zone by gas nozzles that fit onto the head of the torch as illustrated in Figure 3.2. Also incorporated in the torch body are diffusers or carefully patterned jets that feed the shielding gas to the nozzle. The purpose of the diffusers is to assist in producing a laminar flow of the exiting gas shield. Gas nozzles are made of various heat-resistant materials in different shapes, diameters, and lengths. These nozzles are either threaded to the torch or held by a friction fit.

Nozzle Materials. Nozzles are made of various ceramic materials, usually of high-strength alumina, metal, metal-jacketed ceramic, fused quartz or other materials. Ceramic nozzles are the least expensive and most popular, but are brittle and must be replaced if cracked or broken. Fused quartz nozzles are transparent and allow a better view of the arc and electrode. However, contamination from metal vapors from the weld can cause quartz nozzles to become opaque, and they are also brittle. Water-cooled metal nozzles last longer and are used mostly for mechanized and automatic welding applications in which high welding currents are employed.

Nozzle Sizes and Shapes. The gas nozzle must be large enough to provide shielding gas coverage of the weld pool area and the surrounding hot base metal. The nozzle diameter must be appropriate for the volume of shielding gas needed to provide protection and the stiffness needed to sustain shielding coverage in drafts. A delicate balance exists between the nozzle diameter and the flow rate. If the flow rate for a given diameter is excessive, the effectiveness of the shielding is reduced because of turbulence and the aspiration of the surrounding air. High flow rates, which are essential at high currents, require large-diameter nozzles to reduce turbulence. Nozzle size selection depends on the electrode size, the type of weld joint, the weld area to be effectively shielded, and access to the weld joint. Large nozzles (inside dimensions) offer the best shielding of the weld, therefore, the largest size possible should be used. Suggested gas nozzle sizes for various electrode diameters are listed in Table 3.2.

The smallest nozzle listed in Table 3.2 permits welding in restricted areas and offers a better view of the weld. However, if too small a nozzle is used, it may cause shielding gas turbulence and may cause melting of the lip of the nozzle. Larger nozzles provide better shielding gas coverage, as required for critical applications, and should be selected in conjunction with alternate shield sources for applications such as welding titanium or other reactive metals.

Nozzles are available in a variety of lengths to accommodate various joint geometries and the required clearance between the nozzle and the workpiece. Longer nozzles generally produce stiffer, less turbulent gas shields.

The majority of gas nozzles are cylindrical in shape with either straight or tapered ends. To minimize shielding gas turbulence, nozzles with internal streamlining are available. Nozzles are also available with elongated trailing sections or flared ends that provide additional coverage to the trailing portion of the weld pool. This allows hot solidified metal to cool in an inert atmosphere, providing better shielding for welding reactive metals such as titanium. Titanium is highly susceptible to contamination at elevated temperatures. Trailing shields are usually constructed of metal with a series of drilled holes or screening to act as a gas diffuser. Trailing shields, attached externally to gas nozzles, can provide additional gas coverage over the extended length of the weld.

Gas Lenses. One device used for assuring a laminar flow of shielding gas is an attachment called a *gas lens*. Gas lenses contain a porous barrier diffuser and are designed to fit around the electrode or collet. Gas lenses

Table 3.2
Recommended Tungsten Electrodes and Gas Nozzles for Various Welding Currents[a]

Electrode Diameter		Gas Nozzle (ID)		Direct Current, A		Alternating Current, A	
mm	in.	mm	in.	Polarity DCEN (Straight)[b]	Polarity DCEP (Reverse)[b]	Unbalanced Wave[c]	Balanced Wave[c]
0.25	0.010	6.4	1/4	up to 15		up to 15	up to 15
0.50	0.020	6.4	1/4	5–20		5–15	10–20
1.00	0.040	9.5	3/8	15–80		10–60	20–30
1.6	1/16	9.5	3/8	70–150	10–20	50–100	30–80
2.4	3/32	12.7	1/2	150–250	15–30	100–160	60–130
3.2	1/8	12.7	1/2	250–400	25–40	150–210	100–180
4.0	5/32	12.7	1/2	400–500	40–55	200–275	160–240
4.8	3/16	15.9	5/8	500–750	55–80	250–350	190–300
6.4	1/4	19.1	3/4	750–1100	80–125	325–450	325–450

a. All values are based on the use of argon as the shielding gas.
b. Use EWTh-2 electrodes.
c. Use EWP electrodes.

produce a longer, undisturbed flow of shielding gas. They enable operators to weld with the nozzle 25.4 mm (1 in.) or more from the workpiece, allowing greater tungsten electrode extension beyond the nozzle. This improves the welder's view of the weld pool and allows the welder to reach places with limited access, such as inside corners. Gas lenses allow more severely angled torch tilting for access to corners while maintaining more effective shielding than standard gas nozzles.

ELECTRODES

In gas tungsten arc welding, the word *tungsten* refers to the pure element, tungsten, and its various alloys used as electrodes. Tungsten electrodes are not consumed if the process is properly used, because they do not melt or transfer to the weld. In other welding processes, such as shielded metal arc welding, gas metal arc welding, and submerged arc welding, the electrode melts and becomes the filler metal. In gas tungsten arc welding, the function of a tungsten electrode is to serve as one of the electrical terminals of the arc, which supplies the heat required for welding. Its melting point is 3410°C (6170°F). Tungsten becomes thermionic when it approaches a temperature this high; therefore it is a ready source of electrons. The tungsten electrode reaches the melting temperature by resistance heating. Without the significant cooling effect of electrons emanating from the electrode tip, the resistance heating would cause the tip to melt. Because of this cooling effect, the electrode tip is much cooler than the immediate area adjacent to the tip.

Effect of Current on Electrodes

Current levels in excess of those recommended for a given electrode size and tip configuration will cause the tungsten to erode or melt. Tungsten particles may fall into the weld pool and be considered defects in the weld joint. Current that is too low for a specific electrode diameter can cause arc instability.

Direct current with the electrode positive requires a much larger diameter electrode to support a given level of current because the tip is not cooled by the evaporation of electrons but is heated by their impact. In general, a given electrode diameter on DCEP would be expected to handle only 10% of the current possible with DCEN. With alternating current, the tip is cooled during the electrode-negative cycle and heated when positive. Therefore, the current-carrying capacity of an electrode with ac is between that of DCEN and DCEP. In general, it is about 50% less than that of DCEN.

Classification of Electrodes

Tungsten electrode classifications are based on the chemical composition of the electrode, as noted in Table 3.3. Table 3.3 also shows the color identification system for the various classes of tungsten electrodes. Details of the requirements for tungsten electrodes are presented in the latest edition of AWS A5.12, *Specification for*

Table 3.3
Color Code and Alloying Elements for Various Tungsten Electrode Alloys

AWS Classification	Color[a]	Alloying Element	Alloying Oxide	Alloying Oxide%
EWP	Green	—	—	—
EWCe-2	Orange	Cerium	CeO_2	2
EWLa-1	Black	Lanthanum	La_2O_3	1
EWLa-1.5	Gold	Lanthanum	La_2O_3	1.5
EWLa-2	Blue	Lanthanum	La_2O_3	2
EWTh-1	Yellow	Thorium	ThO_2	1
EWTh-2	Red	Thorium	ThO_2	2
EWZr-1	Brown	Zirconium	ZrO_2	0.25
EWG	Gray	Not Specified[b]	—	—

a. Color may be applied in the form of bands, dots, or other, at any point on the surface of the electrode.
b. The manufacturer must identify the type and nominal content of the rare earth or other oxide additions.

Tungsten and Tungsten Alloy Electrodes for Arc Welding and Cutting.[4]

Because electrodes must be free of surface impurities or imperfections, they are produced with either a chemically cleaned finish, in which surface impurities are removed after the forming operation, or a centerless ground finish, in which surface imperfections are removed by grinding.

EWP Electrode Classification. Pure tungsten electrodes (EWP) contain a minimum of 99.5% tungsten, with no intentional alloying elements. The current-carrying capacity of pure tungsten electrodes is lower than that of the alloyed electrodes. Pure tungsten provides good arc stability when used with alternating current, either balanced-wave or continuous high frequency. The tip of the EWP electrode maintains a clean, balled end that promotes good arc stability. EWP electrodes may also be used with dc, but they do not provide the arc initiation and arc stability characteristics offered by thoriated, ceriated, or lanthanated electrodes. Pure tungsten electrodes are generally considered low-cost electrodes and are normally used for the welding of aluminum and magnesium alloys.

EWTh Electrode Classification. In the EWTh electrode classification, the thermionic emission of tungsten can be improved by alloying the tungsten with metal oxides that have very low work functions (low amounts of energy required to cause electrons to emit from the surface). As a result, these electrodes can be used with higher welding currents. Thorium oxide (ThO_2), called *thoria*, is one such additive. To prevent identification problems with these and other types of tungsten electrodes, they are color-coded as shown in Table 3.3. Two types of thoriated tungsten electrodes are available. The EWTh-1 and EWTh-2 electrodes contain 1% and 2% thoria, respectively, evenly dispersed through the entire length of the electrodes.

Thoriated tungsten electrodes are superior to pure tungsten (EWP) electrodes in several ways. The thoriated tungsten electrodes provide a 20% higher current-carrying capacity. They generally have a longer life and provide greater resistance to contamination of the weld. With these electrodes, arc starting is easier and the arc is more stable than with pure tungsten or zirconiated tungsten electrodes.

The EWTh-1 and EWTh-2 electrodes were designed for DCEN applications. They maintain a sharpened tip configuration during welding, which is the desired geometry for DCEN welding operations. They are seldom used with ac because it is difficult to maintain the balled end that is preferred for ac welding without splitting the electrode. The higher thoria content in the EWTh-2 electrode produces more pronounced improvements to the operating characteristics than the EWTh-1 electrodes with lower thoria content.

Thorium is a very low-level radioactive material. However, if welding is to be performed in confined spaces for prolonged periods of time, or if dust from grinding the electrode might be ingested, special ventilation precautions should be considered. The user should consult the appropriate safety personnel, and should consult the AWS standard *Specification for Tungsten and Tungsten Alloy Electrodes for Arc Welding and Cutting*, A5.12, for additional statements concerning safety.[5]

EWCe Electrode Classification. Ceriated tungsten electrodes were first introduced into the United States market in the early 1980s. These electrodes were developed as possible replacements for thoriated electrodes because cerium, unlike thorium, is not a radioactive element. The EWCe-2 electrodes are tungsten electrodes containing 2% cerium oxide (CeO_2), referred to as *ceria*. Compared with pure tungsten, the ceriated electrodes facilitate arc starting, improve arc stability, and reduce the rate of vaporization or burn-off. These advantages of ceriated electrodes improve in proportion to increased ceria content. EWCe-2 electrodes will operate successfully with ac or dc of either polarity.

4. American Welding Society (AWS) Committee on Filler Metals and Allied Materials, *Specification for Tungsten and Tungsten Alloy Electrodes for Arc Welding and Cutting*, ANSI/AWS A5.12, Miami: American Welding Society.

5. See Reference 4.

EWLa Electrode Classification. Electrodes with lanthanum oxide additions were developed about the same time as the ceriated electrodes and for the same reason—lanthanum is not radioactive. The advantages and operating characteristics of these electrodes are similar to that of the EWCe-2 electrodes.

Three types of lanthanted tungsten electrodes are available: EWLa-1, EWLa-1.5 and EWLa-2. The EWLa-1 electrodes contain 1% lanthanum oxide (La_2O_3), referred to as *lanthana*. The advantages and operating characteristics of these electrodes are very similar to the ceriated tungsten electrodes. The EWLa-1.5 electrodes contain 1.5% of dispersed lanthanum oxide, which enhances arc starting and stability, reduces the tip erosion rate, and extends the operating current range. These electrodes can be used as non-radioactive substitutes for 2% thoriated tungsten, as the operating characteristics are very similar. The EWLa-2 electrodes contain 2% dispersed lanthanum oxide. The EWLa-2 electrode has the highest volume of oxide of any of the specific single-additive, AWS-specified electrode types. The high oxide content enhances arc starting and stability, reduces the tip erosion rate, and extends the operating current range. In general, the user will experience very little operating difference between tungsten electrodes with varying element concentrations.

EWZr Electrode Classification. Zirconiated tungsten electrodes (EWZr) contain 0.25% zirconium oxide (ZrO_2). Zirconiated tungsten electrodes have welding characteristics that generally fall between those of pure tungsten and thoriated tungsten electrodes. With ac welding, EWZr combines desirable arc stability characteristics and a balled end typical of pure tungsten, with the current capacity and starting characteristics of thoriated tungsten. They have higher resistance to contamination than pure tungsten and are preferred for radiographic-quality welding applications where the tungsten contamination of the weld must be minimized.

EWG Electrode Classification. The EWG electrode classification is assigned for alloys not covered in the above classes. These electrodes contain an unspecified addition of an unspecified oxide or combination of oxides (rare earth or others). The purpose of the addition is to improve the nature or characteristics of the arc, as defined by the manufacturer. The manufacturer must identify the specific addition or additions and the nominal quantity or quantities added.

Several EWG electrodes are either commercially available or are being developed. These include electrodes with additions of yttrium oxide or magnesium oxide. This classification also includes ceriated and lanthanated electrodes that contain these oxides in amounts other than those listed above, or in combination with other oxides. Tri-mix electrodes are also being used.

Electrode Tip Configurations

The shape of the tungsten electrode tip is an important variable in the GTAW process. Tungsten electrodes can be used with a variety of tip preparations. With ac welding, pure or zirconiated tungsten electrodes form a hemispherical (balled) end. For dc welding, thoriated, ceriated, or lanthanated tungsten electrodes are usually used, and the electrode end is typically ground to a specific included angle and is often truncated. As shown in Figure 3.10, various geometries of the electrode tip affect the weld bead shape and size. In general, as the included angle increases, the weld penetration increases and the width of the weld bead decreases. Although small-diameter electrodes can be used with a square-end preparation for DCEN welding, conical tips provide improved welding performance.

Regardless of the electrode tip geometry selected, it is important that the electrode geometry remains the same once a welding procedure is established. Changes in electrode geometry can significantly influence the arc characteristics, penetration, and the size and shape of the weld bead; therefore, the electrode tip configuration is a welding variable that should be studied during the development of the welding procedure. While tungsten electrode geometry is a factor in manual welding, its effects on the arc characteristics and weld deposit can be more significant in mechanized welding.

Tungsten tips are generally prepared by grinding, balling, and to a much lesser extent, chemical sharpening. A tapered electrode tip is usually prepared on all but the smallest electrodes, even when the end later will be balled for ac welding.

Balling. With ac welding (usually performed with pure or zirconiated tungsten electrodes), a hemispherical tip is most desirable. Before use in welding, the electrode tip can be balled by striking an arc on a water-cooled copper block or another suitable material using ac or DCEP. Welding current is increased until the end of the electrode turns white hot and the tungsten begins to melt, causing a small ball to form at the end of the electrode. The current is down-sloped and extinguished, leaving a hemispherical ball on the end of the tungsten electrode. The size of the hemisphere should not exceed 1-1/2 times the electrode diameter; otherwise it may fall off before it solidifies.

Thoriated, ceriated, and lanthanated tungsten electrodes do not ball as readily as pure tungsten or zirconiated tungsten electrodes. They maintain a ground tip shape much better. These electrodes are not used with ac.

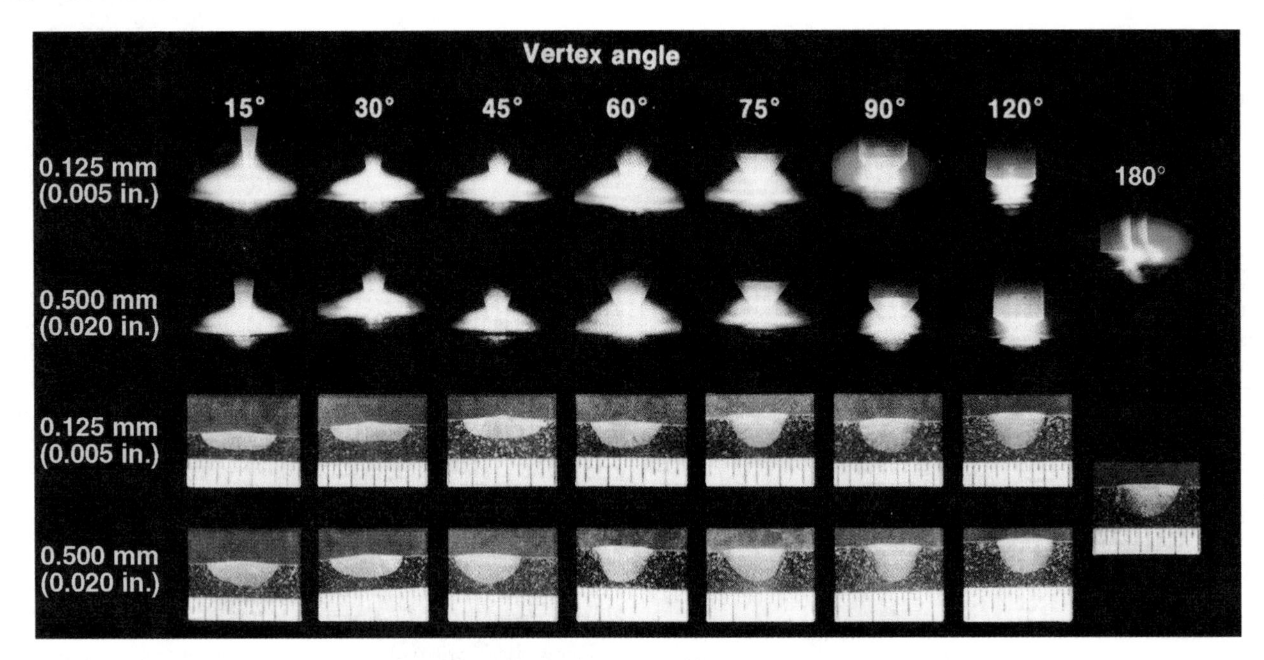

Figure 3.10—Arc Shape and Fusion Zone Profile as a Function of Electrode Tip Geometry in a Pure Argon Shield (150 A, 2.0 Seconds, Spot-on-Plate) with DCEN

Grinding. To produce optimum arc stability, the grinding of tungsten electrodes should be done with the axis of the electrode perpendicular to the axis of the grinding wheel. Tungsten should be ground in the linear direction, not circumferentially. The grinding wheel should be reserved for grinding only tungsten to eliminate possible contamination of the tungsten tip with foreign matter during the grinding operation. Exhaust hoods or dust collectors should be used when grinding electrodes to remove the grinding dust from the air in the work area.

Chemical Sharpening. Chemical sharpening is performed by submerging the red-hot end of a tungsten electrode into a container of sodium nitrate. The chemical reaction between the hot tungsten and the sodium nitrate will cause the tungsten to erode at a uniform rate all around the circumference and the end of the electrode. Repeated heating and dipping of the tungsten into the sodium nitrate will form a tapered tip.

Electrode Contamination

Contamination of the tungsten electrode is most likely to occur when a welder accidentally dips the tungsten into the weld pool or touches the tungsten with the filler metal. The tungsten electrode may also become oxidized by an improper shielding gas or insufficient gas flow during welding or after the arc has been extinguished. Other sources of contamination include metal vapors from the welding arc weld pool eruptions or spatter caused by gas entrapment and evaporated surface impurities.

The contaminated end of the tungsten electrode will adversely affect the arc characteristics and may cause tungsten inclusions in the weld metal. If this occurs, the welding operation should be stopped and the contaminated portion of the electrode removed. Contaminated tungsten electrodes must be properly repaired by breaking off the contaminated section and re-establishing the desired tip geometry according to the manufacturer's suggested procedure.

WIRE FEEDERS

Wire feeders are devices designed to add filler metal from various-sized spools of electrode wire during automatic and mechanized welding. Either cold wire (room temperature) or hot wire (electrically preheated) can be fed into the weld pool. Cold wire is fed into the leading edge and hot wire is normally fed into the trailing edge of the weld pool.

Cold Wire

The wire feed system for cold wire has three components:

1. The wire drive mechanism,

2. The speed control, and
3. A wire guide attachment to introduce the wire into the weld pool.

The drive consists of a motor and gear train to power a set of drive rolls that push the wire. The control is essentially a constant-speed governor that can be either a mechanical or an electronic device. The wire is fed to the wire guide through a flexible conduit.

An adjustable wire guide is attached to the welding torch body. It maintains the position at which the wire enters the weld and the angle of approach relative to the electrode, work surface, and the joint. In heavy-duty applications, the wire guide may be water-cooled. Wire diameters ranging from 0.4 to 2.4 mm (0.015 to 3/32 in.) are used. Special wire feeders are available to provide continuous, pulsed, or intermittent wire feed.

Hot Wire

The process for hot wire addition is similar to the cold wire process, except that the wire is resistance heated nearly to its melting temperature at the point of contact with the weld pool. When using the preheated wire in mechanized and automatic GTAW, the wire is fed mechanically to the weld pool through a holder to which inert gas may be added to protect the hot wire from oxidation. This system is illustrated in Figure 3.11. The hot wire process increases arc energy available to melt base material, thus in addition to increasing the deposition rate, the hot wire process is usually run at higher welding speeds than cold wire systems. The hot wire process is normally used in the flat welding position because of its higher deposition rates. Out-of-position welding has been done, but with limited success.

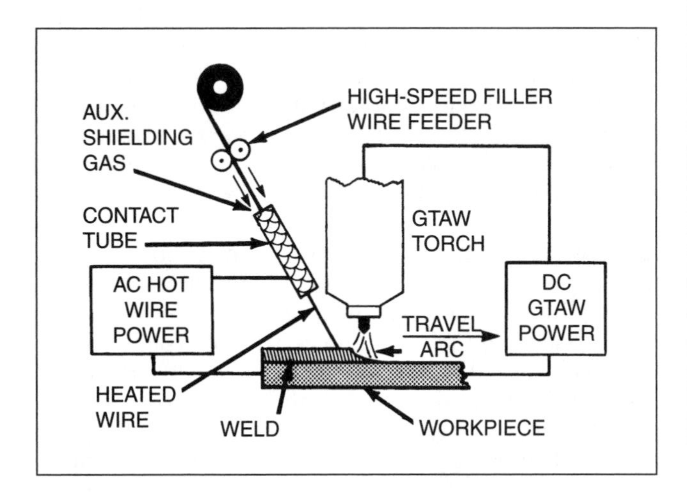

Figure 3.11—Gas Tungsten Arc Hot Wire System

While hot wire has not been used as extensively as cold wire, the hot wire system also produces high-quality weld deposits characteristic of the gas tungsten arc process. To accommodate increased deposition rates and weld pool sizes and to promote more desirable weld geometry, a mixture of shielding gases, such as 75% helium and 25% argon, is used in many applications. When welding at higher deposition rates and welding speeds, additional shielding gas can be supplied by trailing the gas shields to aid in protecting the solidifying weld metal.

The deposition rate is greater with hot wire than with cold wire, as shown in Figure 3.12. This rate can approach those of gas metal arc welding when welding at high currents. The current flow in the wire is initiated when the wire contacts the weld surface. The wire is usually fed into the weld pool directly behind the arc at a 40° to 60° angle relative to the tungsten electrode.

The wire is normally resistance heated by alternating current from a constant-voltage power source, although some systems apply direct current. Alternating current is used for heating the wire to avoid arc blow (deflection toward or away from the filler wire). When the heating current does not exceed 60% of the welding current, the arc oscillates 30° in the longitudinal direction. The oscillation increases to 120° when the heating and welding currents are equal. The amplitude of arc

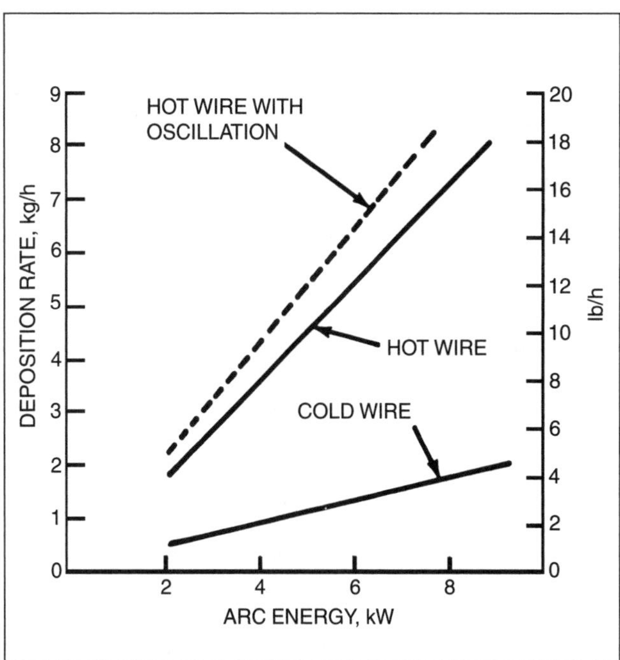

Figure 3.12—Deposition Rates for Gas Tungsten Arc Welding with Cold and Hot Steel Filler Wire

Photograph courtesy of Liburdi Dimetrics

Figure 3.13—Trial Welds for Nuclear Waste Storage Canisters Using Hot Wire Gas Tungsten Arc Welding

oscillation can be controlled by limiting the wire diameter to 1.2 mm (0.045 in.) and reducing the heating current below 60% of the welding current.

Hot wire filler has been used successfully for cladding applications and for welding a range of steels, stainless steels, nickel and copper alloys, and reactive metal such as titanium. Hot wire additions are not recommended for aluminum and copper because the low resistance of these filler wires requires high heating current, which results in excessive arc deflection and uneven melting. Figure 3.13 shows a GTAW application in which hot wire additions are used in the fabrication of stainless steel nuclear waste storage canisters.

POWER SOURCES

Constant-current power sources are used for gas tungsten arc welding. The power required for both ac and dc welding can be supplied by transformer-rectifier power sources or from rotating ac or dc generators. Advances in semiconductor electronics have made transformer-rectifier power sources popular for both shop and field applications of GTAW, but rotating-type power sources continue to be widely used in the field.

Gas tungsten arc welding power sources typically have either drooping or nearly true constant-current static output characteristics, such as those illustrated in Figure 3.14. The static output characteristic is a function of the type of welding current control used in the power source design.

Current Control

A drooping volt-ampere characteristic is typical of magnetically controlled power source models, including the moving coil, moving shunt, moving core reactor, saturable reactor, or magnetic amplifier models and also rotating power source designs. A true constant-current output can be achieved with electronically controlled power sources. The drooping characteristic is an advantage for manual welding when a remote foot-pedal current control is not available at the site of welding. With a drooping characteristic, the welder can vary the current level slightly by changing the arc length. The degree of current control possible by changing arc length can be inferred from Figure 3.14.

Magnetic Control. In most of the magnetically controlled power sources, control of the current level is accomplished with the ac function of the power source. As a result, these power sources are not typically used to provide pulsed current because of their slow dynamic response. The addition of a rectifier bridge allows these power sources to provide both alternating and direct current. The power sources that use a moving component for current control cannot readily be remotely controlled with a foot pedal, while the others typically can.

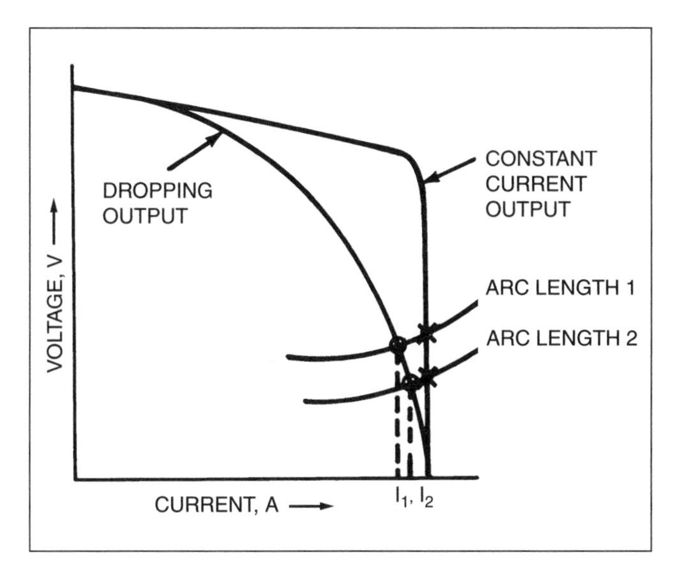

Figure 3.14—Static Volt-Ampere Characteristics for Drooping and Constant Current Power Sources

Most magnetically controlled power sources are considered to be open-loop controlled, in that the actual welding current for a given current setting depends on the welding conditions and may vary with them. Single-phase power sources can provide both ac and dc, while three-phase sources normally provide only dc. The direct current of a three-phase power source is typically smoother than that of a single-phase source because of reduced ripple-current amplitude.

The advantages of magnetically controlled power sources are that they are simple to operate, require little maintenance in adverse industrial environments, and are relatively inexpensive. The disadvantages are that they are large in size and weight and have a lower efficiency compared to electronically controlled power sources. Also, most magnetic-control techniques are open loop, which limits response, accuracy, and repeatability.

Electronic Control. An essentially constant-current volt-ampere characteristic can be provided by various models of electronically controlled power sources, such as the series linear regulator, silicon-controlled rectifier, secondary switcher, and inverter.

The essentially constant-current characteristic is typically an advantage for mechanized and automatic welding because it can maintain a current level that provides sufficient accuracy and repeatability. Most truly constant-current power sources are closed-loop controlled. The actual current is measured and compared to the desired current setting and adjustments are made electronically within the power source to maintain the specified current as welding conditions change.

Most electronically controlled power sources offer rapid dynamic response, a function that can be used to provide pulsed welding current. The series linear regulator and switched secondary models provide only dc welding current from single or three-phase input power. Silicon-controlled rectifier models can provide both ac and dc from single-phase power and dc from three-phase power. Depending on the design, inverters can provide ac and dc output from single- or three-phase input power. Inverter power sources are the most versatile, with many of these offering multi-process capabilities and variable welding current output. Inverters are also lighter and more compact than other power source designs of equivalent current rating.

The advantages of electronically controlled power sources are that they offer rapid dynamic response, provide variable current wave-form output, have excellent repeatability, and offer remote control. The disadvantages are that they are more complex to operate and maintain and are relatively expensive. It is important to select a GTAW power source based on the type of welding current required for a particular application. The types of welding current include ac sine wave, ac square wave, variable polarity ac, dc, and pulsed dc. Many types of power sources are available, providing a large selection of controls and functions such as cooling water and shielding gas control, wire feeder and travel mechanism sequencing, current up-slope and down-slope, and multiple-current sequences. Chapter 1 of this volume, "Arc Welding Power Sources," provides more detailed information.

Direct Current

Using direct current, the tungsten electrode can be connected to either the negative or the positive terminal of the power source. In almost all cases, electrode negative (cathode) is chosen. With that polarity, direct current electrode negative (DCEN), electrons flow from the electrode to the workpiece and positive ions are transferred from the workpiece to the electrode. When electrode positive (anode) is chosen, the directions of electron and positive ion flow are reversed for direct current electrode positive (DCEP). The welding characteristics of DCEN, DCEP, and balanced alternating current are illustrated in Figure 3.15.

When DCEN is used with a thermionic electrode such as tungsten, approximately 70% of the heat is generated at the anode and 30% at the cathode. Since DCEN produces the greatest amount of heat at the workpiece for a given welding current, DCEN will provide deeper weld penetration than DCEP (see Figure 3.15). Direct current electrode negative is the most common configuration used in GTAW and is used with

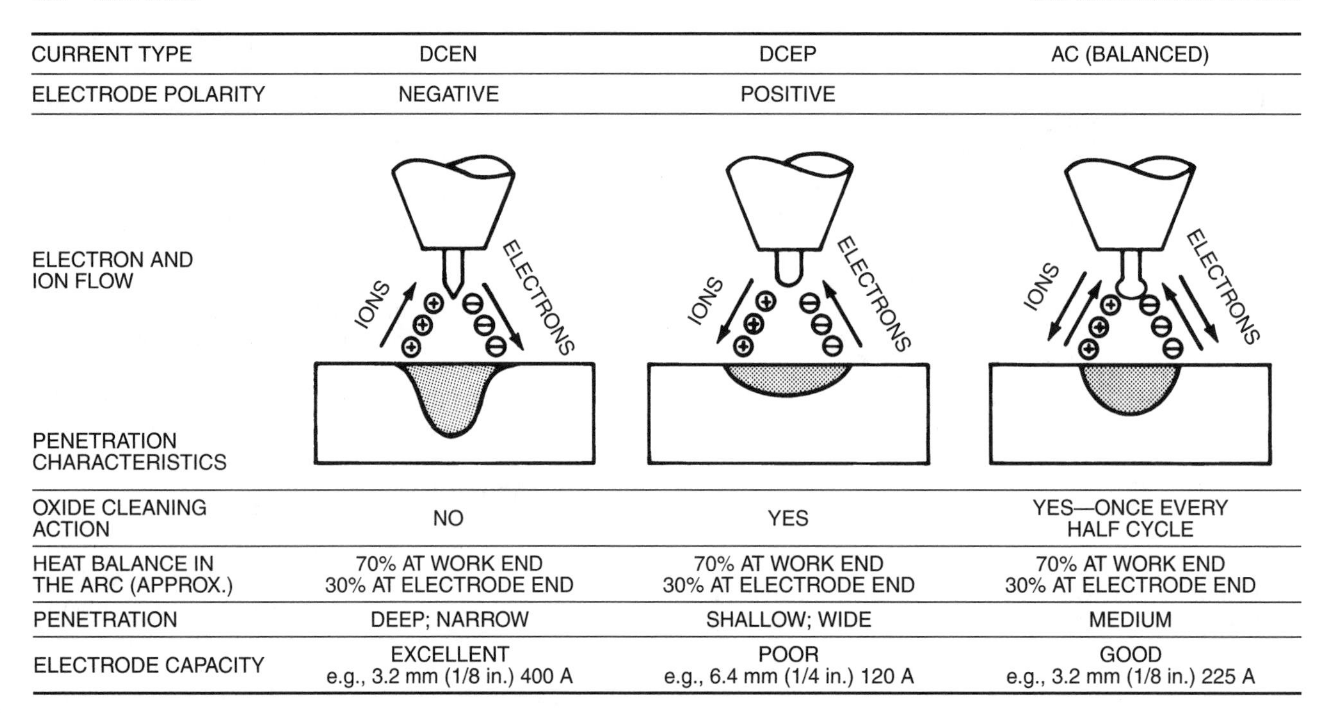

CURRENT TYPE	DCEN	DCEP	AC (BALANCED)
ELECTRODE POLARITY	NEGATIVE	POSITIVE	
ELECTRON AND ION FLOW PENETRATION CHARACTERISTICS			
OXIDE CLEANING ACTION	NO	YES	YES—ONCE EVERY HALF CYCLE
HEAT BALANCE IN THE ARC (APPROX.)	70% AT WORK END 30% AT ELECTRODE END	70% AT WORK END 30% AT ELECTRODE END	70% AT WORK END 30% AT ELECTRODE END
PENETRATION	DEEP; NARROW	SHALLOW; WIDE	MEDIUM
ELECTRODE CAPACITY	EXCELLENT e.g., 3.2 mm (1/8 in.) 400 A	POOR e.g., 6.4 mm (1/4 in.) 120 A	GOOD e.g., 3.2 mm (1/8 in.) 225 A

Figure 3.15—Characteristics of Current Types for Gas Tungsten Arc Welding

argon or helium, or an argon-helium mixture, to weld most metals.

When the tungsten electrode is connected to the positive terminal (DCEP), a cathodic cleaning action is created at the surface of the workpiece. This action occurs when welding most metals, but it is most important when welding aluminum and magnesium because the cleaning action removes the refractory oxide surface that inhibits the wetting of the base metal surfaces of the weldment by the weld metal.

Unlike DCEN, which cools the electrode tip by the evaporation of electrons, DCEP (with the electrode used as the positive pole) heats the electrode tip by the bombardment of electrons and by its resistance to the passage of the electrons through the electrode. When DCEP is used, a larger-diameter electrode is required for a given welding current to reduce resistance heating and increase thermal conduction into the electrode collet. The current-carrying capacity of an electrode connected to the positive terminal is approximately one-tenth that of an electrode connected to the negative terminal. DCEP is generally limited to welding sheet metal.

Pulsed DC Welding

Pulsed gas tungsten arc welding (GTAW-P) involves the repetitive variation in direct welding current from a low background value to a high peak value of current and is usually applied with the electrode negative. Pulsed dc power sources typically allow adjustments of the pulse current time, background current time, peak current level, and background current level to provide a current output wave form suited to a particular application. Figure 3.16 shows a typical pulsed-current wave form. Generally, the background and pulse duration times are adjustable so that the current will change in the range of one pulse every 2 seconds up to 20 pulses per second.

In pulsed dc welding, the pulse current level is usually set at 2 to 10 times the background current level. This combines the driving, forceful arc characteristics of high current with the low heat input of low current. The pulse current achieves good fusion and penetration, while the background current maintains the arc and allows the weld area to cool.

There are several advantages of pulsed current. For a given average current level, greater penetration (useful for metals sensitive to heat input) can be obtained with pulsed current than with continuous current. Pulsed current also minimizes distortion. Because there is insufficient time for significant heat flow during the short duration of a pulse, metals of dissimilar thicknesses usually respond equally, and equal penetration can be achieved. For a similar reason, very thin metals

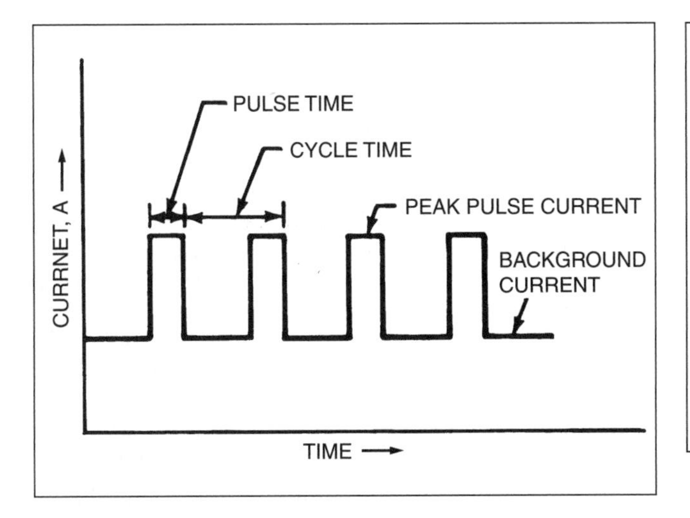

Figure 3.16—Pulsed DC Wave Form

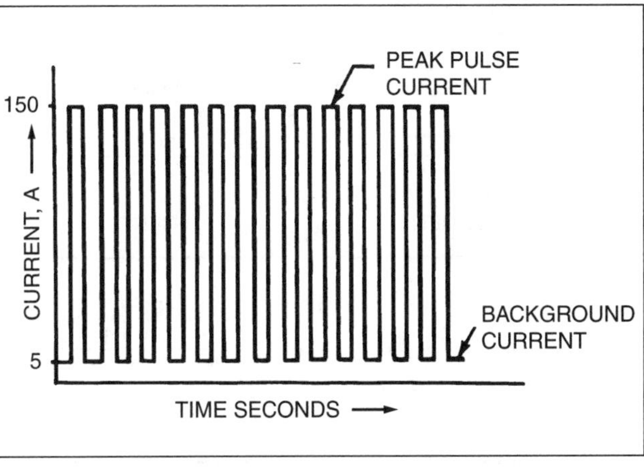

Figure 3.17—Current Wave Form for Switched DC Pulsed Welding

can be joined with pulsed dc. In addition, one set of welding variables can be used on a joint in all positions, such as those required by a circumferential weld in a horizontal pipe. Pulsed dc is also used for bridging gaps in open root joints.

Although most often used for mechanized and automatic GTAW, pulsed current offers advantages for manual welding. Inexperienced welders find that they can improve their proficiency by counting the pulses (from 1/2 to 2 pulses per second) and using them to time the movement of the torch and the cold wire. Experienced welders are able to weld thinner materials, dissimilar alloys and thicknesses with less difficulty.

High-Frequency Pulsed Welding. Switched direct current involves switching from a low current level to a high level at a rapid, fixed frequency, as shown in Figure 3.17. The peak current-on time is varied to change the average current level. A fixed frequency of approximately 20 kilohertz (kHz) can be obtained with switching.

The main reason for switching is to produce a constricted or "stiff" welding arc. A constricted arc provides a more concentrated heat source. Arc pressure is a measure of arc stiffness. As shown in Figure 3.18, when the switching frequency nears 10 kHz, arc pressure increases to nearly four times that of a steady dc arc. As arc pressure increases, lateral displacement of the arc, such as that produced by magnetic fields (arc blow) and shielding gas movement caused by wind, is reduced.

High-frequency switched dc is useful in precision mechanized and automatic welding applications that require an arc with exceptional directional properties and stability. It is also used where a stable arc is needed at very low average currents. The disadvantage of high-frequency switched dc is that the welding power sources are expensive. Also, if the switching frequency is in the audible range, the arc sound can be very annoying.

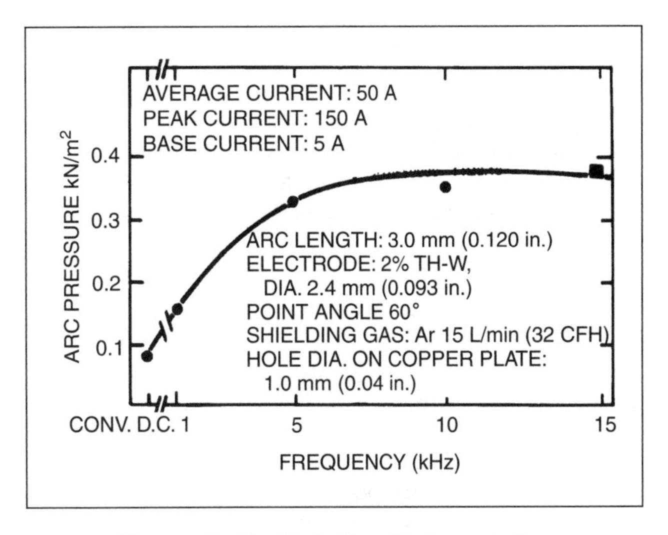

Figure 3.18—Relation Between Arc Pressure and Pulse Frequency

Alternating Current

Alternating current undergoes periodic reversal in polarity from electrode positive to electrode negative. Thus, ac can combine the workpiece cleaning action of DCEP with the deep penetration characteristic of DCEN. Alternating current welding is compared with DCEN and DCEP welding in Figure 3.19. Conventional

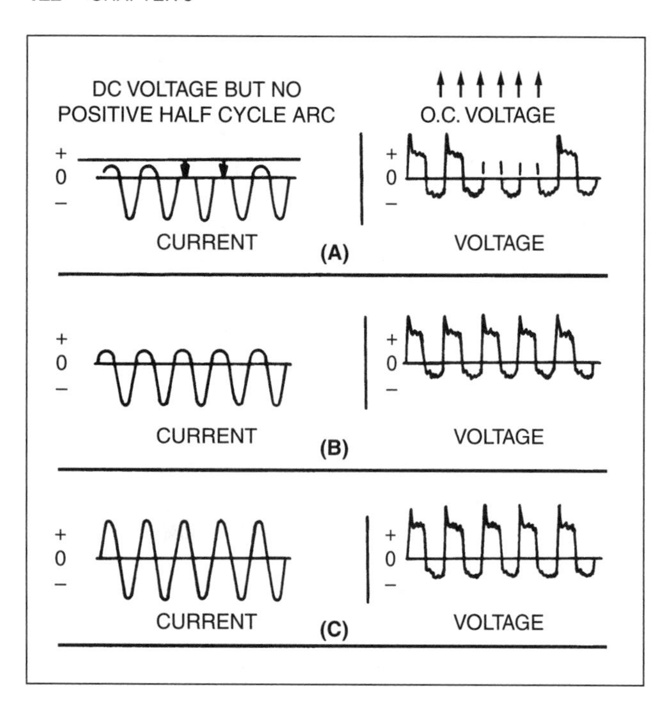

Figure 3.19—Voltage and Current Wave Form for AC Welding: (A) Partial and Complete Rectification; (B) With Arc Stabilization; (C) With Current Balancing

ac welding power sources produce a sinusoidal open-circuit voltage output, which is out of phase with the current by about 90°. The frequency of voltage reversal is typically fixed at the standard 60 hertz (Hz) frequency of the primary power. The actual arc voltage is in phase with the welding current. The voltage measured is the sum of voltage drops in the electrode and the plasma and also at the anode and cathode, all of which are the result of current flow.

When the current decays to zero, various effects will occur depending on the polarity. When the thermionic tungsten electrode becomes negative, it supplies electrons immediately to reignite the arc. However, when the weld pool becomes negative, it cannot supply electrons until the voltage is raised sufficiently to initiate cold-cathode emission. Without this voltage, the arc becomes unstable or fails to re-ignite. This is shown in Figure 3.19(A).

Some means of stabilizing the arc during voltage reversal is required with conventional sinusoidal welding power sources. This is accomplished by using a high-frequency circuit in the power sources, by discharging capacitors at the appropriate time in the cycle, by using high-voltage high-frequency sparks in parallel with the arc, and by using power sources with a square wave output. The results of this stabilization are shown in Figure 3.19(B).

To improve arc stability, the open-circuit voltage of the transformer can be increased. An open-circuit voltage of about 100 volts (V) (root mean square) is needed with helium shielding. The necessary voltage can also be obtained by adding a high-frequency voltage supply in series with the transformer. Generally, the high-frequency voltage should be on the order of several thousand volts, and its frequency can be as high as several megahertz. The current should be very low. The high-frequency voltage can be applied continuously or periodically during welding. In the latter case, a burst of high-frequency voltage should be timed to occur during the time when the welding current passes through zero.

Square-wave ac welding power sources can change the direction of the welding current in a short period of time. The presence of high voltage, coupled with high electrode and base-metal temperatures at current reversals, allows the arc to be reignited without the need for an arc stabilizer. Also, the lower effective "peak" of the square waveform tends to increase the usable current range of the electrode.

Since the electrons needed to sustain an arc are more readily provided when the electrode is negative, less voltage is required. The result is a higher welding current during the DCEN interval than during the DCEP time. In effect, the power source produces both direct current and alternating current. This type of rectification can cause damage to the power source due to overheating or, with some machines, decay in the output, but these can be eliminated by wave balancing as shown in Figure 3.19C.

The original technology of balanced-current power sources involved either series-connected capacitors or a dc voltage source (such as a battery) in the welding circuit. Modern power source circuits use electronic wave balancing. Although balanced current flow is not essential for most manual welding operations, it is beneficial for high-speed mechanized or automatic welding. The advantages of balanced current flow are the following:

1. Better oxide removal;
2. Smoother, better welding; and
3. No requirement for reduction in output rating of a given size of conventional welding transformer. (The unbalanced core magnetization that is produced by the dc component of an unbalanced current flow is minimized.)

The following are disadvantages of balanced current flow:

1. Larger tungsten electrodes are needed.

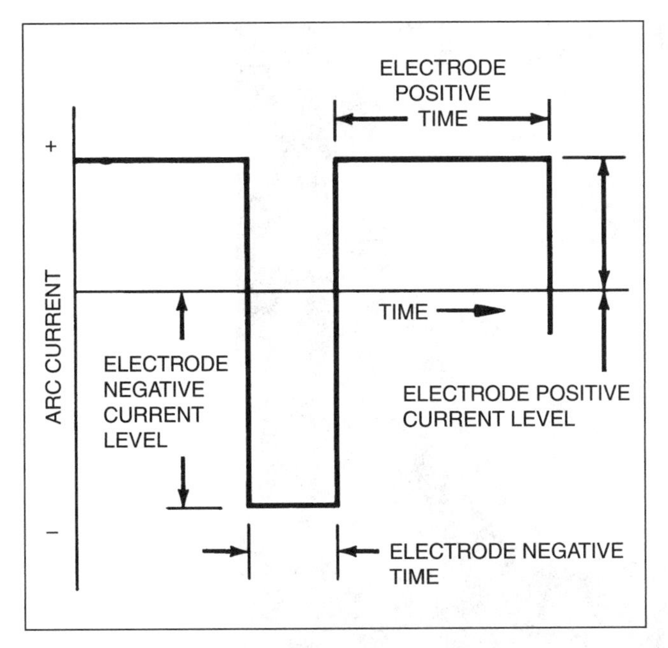

Figure 3.20—Characteristics of Variable Square-Wave AC

2. Higher open-circuit voltages generally associated with some methods of wave balancing may constitute a safety hazard.
3. Balanced-wave welding power sources are more expensive.

Some square-wave ac power sources adjust the current level during the electrode positive and electrode negative cycles at the standard 60 Hz frequency. Functions provided by the more sophisticated power sources adjust the time of each polarity half cycle as well as the current level during that half cycle. These variable wave forms will adjust the welding current to suit a particular application. The characteristics of variable square-wave alternating current are shown in Figure 3.20.

Power sources with variable polarity capabilities developed during the 1990s have provided the means to adjust the DCEN interval up to 90-95%. This greatly increases weld penetration while reducing the heating of the tungsten electrode, thus allowing the pointed tungsten electrode geometry to be maintained.

ARC VOLTAGE CONTROL

Arc voltage controllers (AVC) are used in mechanized and automatic gas tungsten arc welding to maintain arc length. In this case, the arc itself is a sensor, since it converts a measurement of arc length into an electrical signal (arc voltage).

The AVC measures the arc voltage and compares it to the specified arc voltage to determine the direction and speed the welding electrode should be moved. This determination, expressed as a voltage error signal, is amplified to drive motors in a slide that supports the torch. The changing voltage that results from the motion of the welding electrode is detected and the cycle repeats to maintain the desired arc voltage.

Arc Oscillation

Mechanical arc oscillation, an alternating (side-to-side) motion relative to the direction of travel, can be used to increase the width of gas tungsten arc welds. Mechanical oscillation can be achieved by mounting a GTAW torch on a cross slide that provides automatically controlled movement of the torch and wire feed transverse to the line of travel. This equipment provides adjustable cross-feed speed, amplitude of oscillation, and dwell time on each side of the oscillation cycle.

Magnetic oscillation can be used to increase the width of welds. Both mechanical and magnetic arc oscillation can result in better fusion of the bevel or groove faces of the joint. Magnetic control can be used to reduce the disruptive effects of arc blow. Figure 3.21 shows a four-pole magnetic probe used to reduce arc deflection and maintain the arc position during high-speed pipe seam welding. Oscillators deflect the arc longitudinally or laterally over the weld pool without moving the welding electrode or wire feed. The oscillators consist of electromagnets located close to the arc, powered by a variable-polarity, variable-amplitude power source. Control features include adjustable oscillation frequency and amplitude, and separately adjustable dwell times.

SHIELDING GASES

Shielding gas is directed by the torch to the arc and weld pool to protect the electrode and the molten weld metal from atmospheric contamination. Backup purge gas can also be used on the underside of the weld and the adjacent base metal surfaces to protect the weldment from oxidation during welding. The use of a gas backup under controlled conditions can achieve uniformity of root bead contour, elimination of undercutting, and the desired amount of root bead reinforcement. In some materials, gas backup reduces root cracking and porosity in the weld.

Argon and helium (used separately or together) are the two most common types of inert gas used for shielding, although argon-hydrogen mixtures are used for special applications. These gases can be supplied in

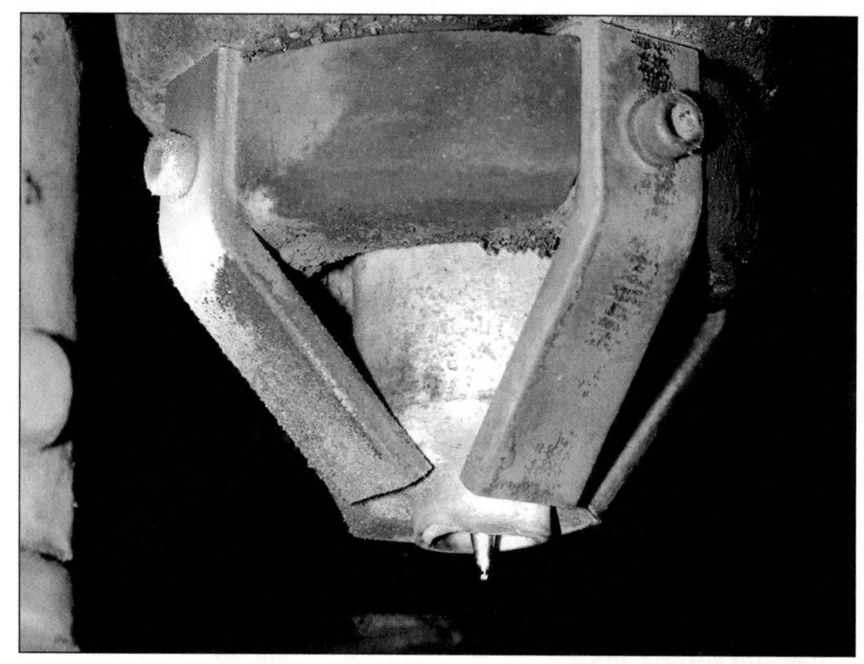

Photograph courtesy of Raft Engineering, Inc.

Figure 3.21—Four-Pole Magnet Used to Reduce Arc Deflection

cylinders or as a liquid in insulated tanks, depending on the volume of usage. For high-volume users, the liquid is vaporized and piped to various points within the fabricating plant, thus eliminating cylinder handling.

Argon

Argon (Ar) is an inert monatomic gas with a molecular weight of 40. It is obtained from the atmosphere by the separation of liquefied air. Welding-grade argon is refined to a minimum purity of 99.99%. This is acceptable for the gas tungsten arc welding of most metals except the reactive and refractory metals, for which a minimum purity of 99.997% is required. For this reason, weldments in reactive and refractory metals are often fabricated in purge chambers from which all traces of air have been removed prior to initiating the welding operation.

Argon is used more extensively than helium for shielding because argon offers the following advantages:

1. Smoother, quieter arc action;
2. Reduced penetration;
3. Cleaning action when welding materials such as aluminum and magnesium;
4. Lower cost and greater availability;
5. Lower flow rates for good shielding;
6. Better cross-draft resistance; and
7. Easier arc starting.

The reduced penetration of an argon-shielded arc is particularly helpful in the manual welding of thin material, because the tendency for excessive melt-through is lessened. This same characteristic is an advantage in vertical or overhead welding because it decreases the tendency for the base metal to sag or run.

Helium

Helium (He) is an inert, very light monatomic gas with the atomic weight of four. It is obtained by separation from natural gas. Unlike argon, welding-grade helium is seldom refined to a purity of 99.99%. Contaminants that may be present in welding-grade helium can interfere with its shielding function in gas tungsten arc welding, so the purity value may be specified to control weld quality.

Helium transfers more heat into the workpiece than argon for given values of welding current and arc length. The greater heating power of the helium arc can be an advantage when joining metals with high thermal conductivity and for high-speed mechanized applications. Helium is used more often than argon for the welding of heavy plate. Mixtures of argon and helium are used when some balance between the characteristics of both is desired.

Characteristics of Argon And Helium

The chief influence on the effectiveness of a shielding gas is the density of the gas. Argon is approximately one and one-third times heavier than air and ten times heavier than helium. Argon, after leaving the torch nozzle, forms a blanket over the weld area. Helium, because it is lighter, tends to rise around the nozzle. To produce equivalent shielding effectiveness, the flow of helium must be two to three times that of argon. The same general relationship is true for mixtures of argon and helium, particularly those high in helium content. The important characteristics of these gases are the relationships of voltage and current to the tungsten arc in argon and in helium, illustrated in Figure 3.22. For equivalent arc lengths at all current levels, the arc voltage obtained with helium is appreciably higher than the arc voltage obtained with argon.

Heat in the arc is roughly measured as the product of current and voltage (arc power). This measurement shows that helium produces more available heat than argon. The higher heat available with helium favors its selection for welding thick materials and metals that have high thermal conductivity or relatively high melting temperatures. However, it should be noted that at lower currents, the volt-ampere curves pass through a minimum voltage, at current levels of approximately 90 A, after which the voltage increases as the current decreases. When helium is used, this increase in voltage occurs in the range of 50 to 150 A, where much of the welding of thin materials is done. Since the voltage increase for argon occurs below 50 A, the use of argon in the range if 50 to 150 A provides the operator with more latitude in arc length to control the welding operation. It is apparent that to obtain equal arc power, appreciably higher current must be used with argon than with helium. Since undercutting will occur at about equal currents with either of these gases, helium can be selected to reduce undercutting at much higher welding speeds.

The other influential characteristic is arc stability. Both argon and helium provide excellent stability with direct current, but with alternating current, which is used extensively for welding aluminum and magnesium, argon yields much better arc stability. This stability and the highly desirable cleaning action of argon and argon-helium mixes make it superior to helium for ac welding applications.

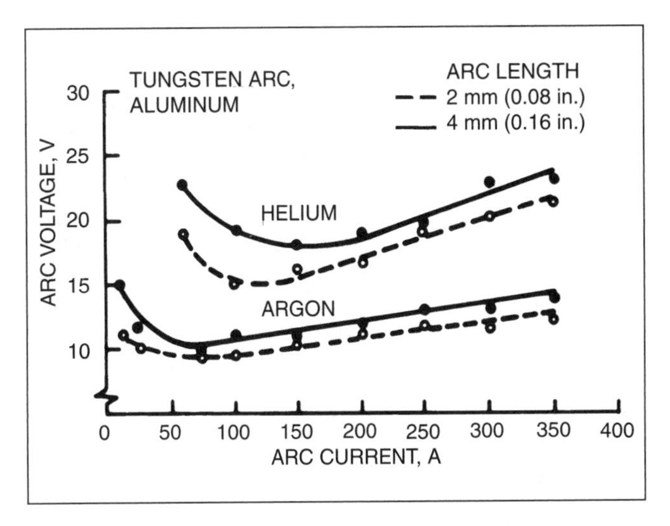

Figure 3.22—Voltage-Current Relationship With Argon and Helium Shielding

Argon-Hydrogen Mixtures

The addition of hydrogen to argon shielding gas increases the arc energy, for a given current, into the material being welded. Hydrogen also acts as a reducing agent, inhibiting the formation of oxides and thus producing cleaner weld surfaces. Argon-hydrogen mixtures are employed in special applications, such as the mechanized welding of light-gauge stainless steel tubing, where the hydrogen does not cause adverse metallurgical effects such as porosity and hydrogen-induced cracking. Increased welding speeds can be achieved in almost direct proportion to the amount of hydrogen added to argon because of the increased arc voltage. However, the amount of hydrogen that can be added varies with the metal thickness and type of joint for each particular application. Excessive hydrogen causes porosity. Hydrogen concentrations of up to 35% have been used, however, argon-hydrogen mixtures normally have from 1% to 8% hydrogen. These mixtures are usually limited to use on most grades of stainless steel, nickel-copper, and nickel-base alloys.

The most commonly used argon-hydrogen shielding gas mixture contains 5% hydrogen. The benefits gained with this mixture make it a widely used choice for welding thin as well as thick-section materials. With argon-hydrogen shielding, close-tolerance mechanized butt welds in stainless steel up to 1.6 mm (0.062 in.) thick can be made at speeds comparable to those made with helium and 50% faster than welds made with argon alone. The argon-hydrogen shielding mixture offers advantages when welding thick plate or vessel wall materials because of increased arc penetration, faster welding speeds, and weld surface cleanliness. Tube-to-tubesheet joints in a variety of stainless steels and nickel alloys can also be welded using this shielding mixture. For manual welding, a hydrogen content of 5% is sometimes preferred to obtain cleaner welds.

SHIELDING GAS SELECTION

No set rule governs the choice of shielding gas for any particular application. Argon, helium, or a mixture of argon and helium can be used successfully for most applications, with the possible exception of manual welding on extremely thin material, for which argon shielding is essential. Argon generally provides an arc that operates more smoothly and quietly, is handled more easily, and is less penetrating than an arc shielded by helium. In addition, the lower unit cost and the lower flow rate requirements of argon make argon preferable from an economic point of view. Argon is preferred for most applications, except where the higher heat penetration of helium is required for welding thick sections of metals with high heat conductivity, such as aluminum and copper. A guide to the selection of gases is provided in Table 3.4.

Recommended Gas Flow Rates

Shielding gas flow requirements are based on gas nozzle size, weld pool size, and air movement. In general, the flow rate increases in proportion to the cross-sectional area at the nozzle (considering the obstruction caused by the collet). The gas nozzle diameter is selected to suit the size of the weld pool and the reactivity of the metal to be welded. The minimum flow rate is determined by the need for a stiff gas stream to overcome the heating effects of the arc and crossdrafts. With the more commonly used torches, typical shielding gas flow rates are 7 to 16 liters per minute (L/min) (15 to 35 cubic feet per hour [cfh]) for argon and 14 to 24 L/min. (30 to 50 cfh) for helium. Excessive flow rates cause turbulence in the gas stream that may aspirate atmospheric contamination into the weld pool.

A crosswind or draft moving at five or more miles per hour can disrupt the shielding gas coverage. The stiffest, nonturbulent gas streams (with high-stream velocities) are obtained by incorporating gas lenses in the nozzle and by using helium as the shielding gas. However, in deference to cost, protective screens that block airflow are preferred over increasing the shielding gas flow rate.

Backup Purge

Exposure to air at the back side of the weldment can contaminate the weld when making the root pass. To avoid this problem, the air must be purged from this region. Argon and helium are satisfactory for use as the backup purge when welding all materials. Nitrogen can be used satisfactorily for purging the back side of welds in austenitic stainless steel, copper, and copper alloys. The backup purge gas purity may be specified for high- and ultra-high-purity welding applications, such as semiconductor tubing.

Table 3.4
Recommended Types of Current, Tungsten Electrodes and Shielding Gases for Welding Various Metals

Type of Metal	Thickness	Type of Current	Electrode*	Shielding Gas
Aluminum	All	ac	Pure or zirconium	Argon or argon-helium
	over 3.18 mm (1/8 in.)	DCEN	Thoriated	Argon-helium or helium
	under 3.18 mm (1/8 in.)	DCEP (Rare)	Thoriated or zirconium	Argon
Copper, copper alloys	All	DCEN	Thoriated	Helium, argon-helium
	under 3.18 mm (1/8 in.)	ac	Pure or zirconium	Argon, argon-helium
Magnesium alloys	All	ac	Pure or zirconium	Argon, argon-helium
	under 3.18 mm (1/8 in.)	DCEP (Rare)	Zirconium or thoriated	Argon
Nickel, nickel alloys	All	DCEN	Thoriated	Argon, argon-helium, argon-hydrogen
Plain carbon, low-alloy steels	All	DCEN	Thoriated	Argon, argon-helium
	under 3.18 mm (1/8 in.)	ac (Rare)	Pure or zirconium	Argon, argon-helium
Stainless steel	All	DCEN	Thoriated	Argon, argon-helium, argon-hydrogen
	under 3.18 mm (1/8 in.)	ac	Pure or zirconium	Argon, argon-helium
Titanium	All	DCEN	Thoriated	Argon, argon-helium

*If thoriated electrodes are recommended, ceriated or lanthanated electrodes may also be used.

Gas flow requirements for the backup purge may range from 0.5 to 42 L/min (1 to 90 cfh), based on the volume of air to be purged. As a general rule, a relatively inert atmosphere can be obtained by flushing the area with four times the volume to be purged. However, more volume changes are required for high-quality applications. After purging is completed, the flow of backup gas during welding should be reduced until only a slight positive pressure exists in the purged area. After the root and first filler passes are completed, the backup purge may be discontinued, depending on quality requirements.

Several devices are available to contain shielding gas on the back side of plate and piping weldments. One of these is shown in Figure 3.23. More information is presented in the latest edition of ANSI/AWS C5.5, *Recommended Practices for Gas Tungsten Arc Welding.*[6]

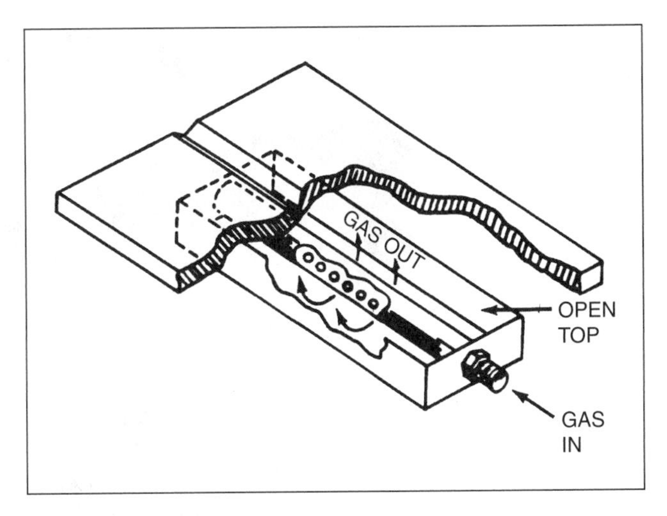

Figure 3.23—Backup Purge Gas Channel

When purging piping systems, provisions for an adequate vent or exhaust with baffles to contain the purge gas (shown in Figure 3.24) are important to prevent excessive pressure buildup during welding. The dimensions of the vents through which the backup gas is exhausted to the atmosphere should be at least equal to the dimensions of the opening through which the gas is admitted to the system. Extra care is required to ensure that the backup purge pressure is not excessive when welding the last inch or two of the root pass to prevent weld-pool blowout or root concavity.

When using argon or nitrogen, the backup gas should preferably enter the system at a low point to displace the atmosphere upward and should be vented at points beyond the joint to be welded, as illustrated in Figure 3.24. In piping systems that have several joints, all joints except the one being welded should be taped to prevent gas loss.

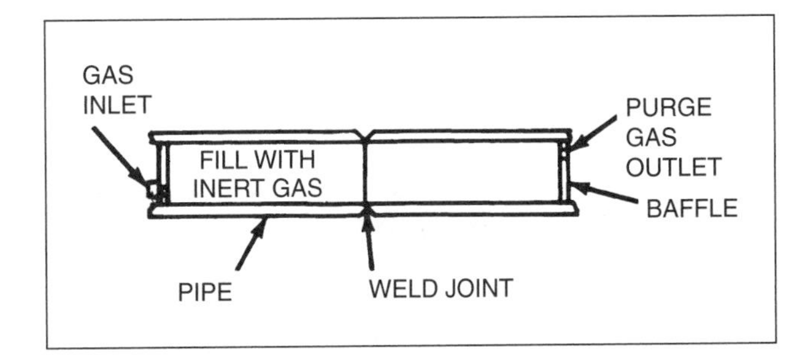

Figure 3.24—Backup Gas Purge Arrangement for Circumferential Pipe Joint

Controlled Atmosphere Welding Chamber

Maximum benefits can be obtained by using argon as a purge gas for the manual welding of reactive metals if the entire weldment can be placed in a controlled-atmosphere chamber. Such a chamber, shown in Figure 3.25, contains the workpieces, the shielding gas, and welding equipment. Purging is started after the assembly has been placed in the chamber. Readings are taken from instruments that analyze oxygen, nitrogen, and water vapor to ensure that welding is not started until contaminants are at a suitably low level, usually less than 50 parts per million (ppm).

6. American Welding Society (AWS) Committee on Filler Metals and Allied Materials, ANSI/AWS C5.5, *Recommended Practices for Gas Tungsten Arc Welding,* Miami: American Welding Society.

Trailing Shields

For some metals, such as titanium, a trailing shield is necessary if chambers or other shielding techniques are not available or practical. A trailing shield is a device that directs the flow of shielding gas so that it provides inert gas coverage over the weld area until the molten weld metal has cooled to the point that it will not react with the atmosphere. A trailing shield for manual welding is shown in Figure 3.26. Fixed barriers, as illustrated in Figure 3.27, also aid in containing shielding gas within the area immediately surrounding the electrode. Figure 3.28 shows a trailing shield used during orbital welding of titanium piping. A heat-resistant blanket and tape is used to assure a good seal around the torch and shielding gas.

Figure 3.25— Gas Tungsten Arc Welding of Reactive Metals in a Controlled-Atmosphere Chamber, Viewed Through a Transparent Window

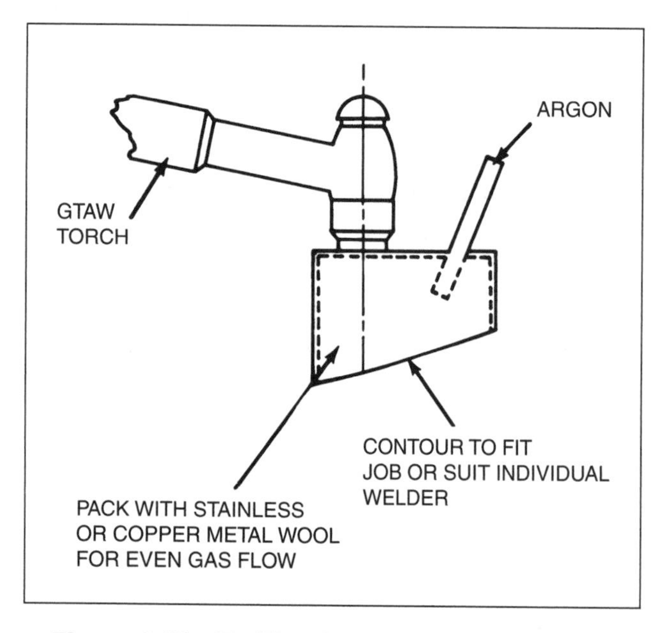

Figure 3.26—Trailing Shield for Manual Torch

TECHNIQUES

The start of the GTAW process is the initiation of an arc between the workpiece and the tungsten electrode. The method of starting the arc is based on equipment capability, production requirements, and ease of application.

ARC INITIATION METHODS

One of several techniques can be used to initiate the arc for gas tungsten arc welding, including the touch start (also called *scratch start* or *lift start*), the high-frequency start, the pulse start, and the pilot arc start.

Touch Start

A touch start is made with the power source energized and the shielding gas flowing from the nozzle. The torch is lowered toward the workpiece until the tungsten electrode makes contact with the workpiece, then is withdrawn a short distance to establish the arc. The advantage of this method of arc initiation for both manual and mechanized welding is its simplicity. The disadvantage of touch starting is the potential for the electrode to stick to the workpiece, causing electrode contamination and the transfer of tungsten to the workpiece.

High-Frequency Start

High-frequency starting can be used with dc or ac power sources for both manual and automatic applications. High-frequency generators usually have a spark-gap oscillator that superimposes a high-voltage ac output at radio frequencies in series with the welding circuit. The circuit is shown in Figure 3.29. The high voltage ionizes the gas between the electrode and the workpiece, and the ionized gas will then conduct the welding current that initiates the arc.

Because radiation from a high-frequency generator may disturb radio, electronic, and computer equipment, the use of this type of arc-starting equipment is governed by regulations of the Federal Communications Commission. The user should follow the instructions of the manufacturer for the proper installation and operation of high-frequency arc starting equipment.

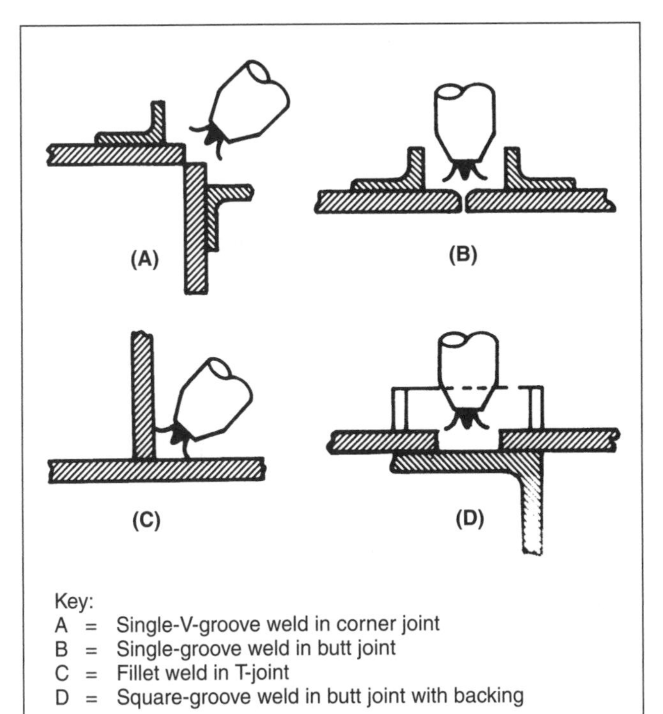

Figure 3.27—Barriers Used to Contain the Shielding Gas Near the Joint to be Welded

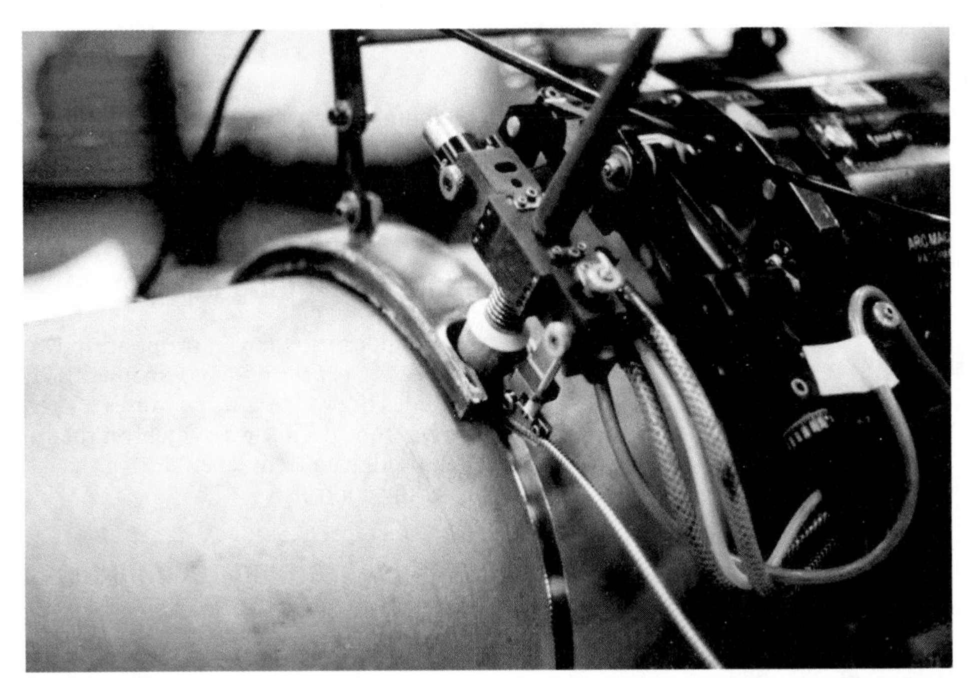

Photograph courtesy of RMI Titanium Company

Figure 3.28—Trailing Shield Using Argon Protects Titanium Piping during Orbital Gas Tungsten Arc Welding

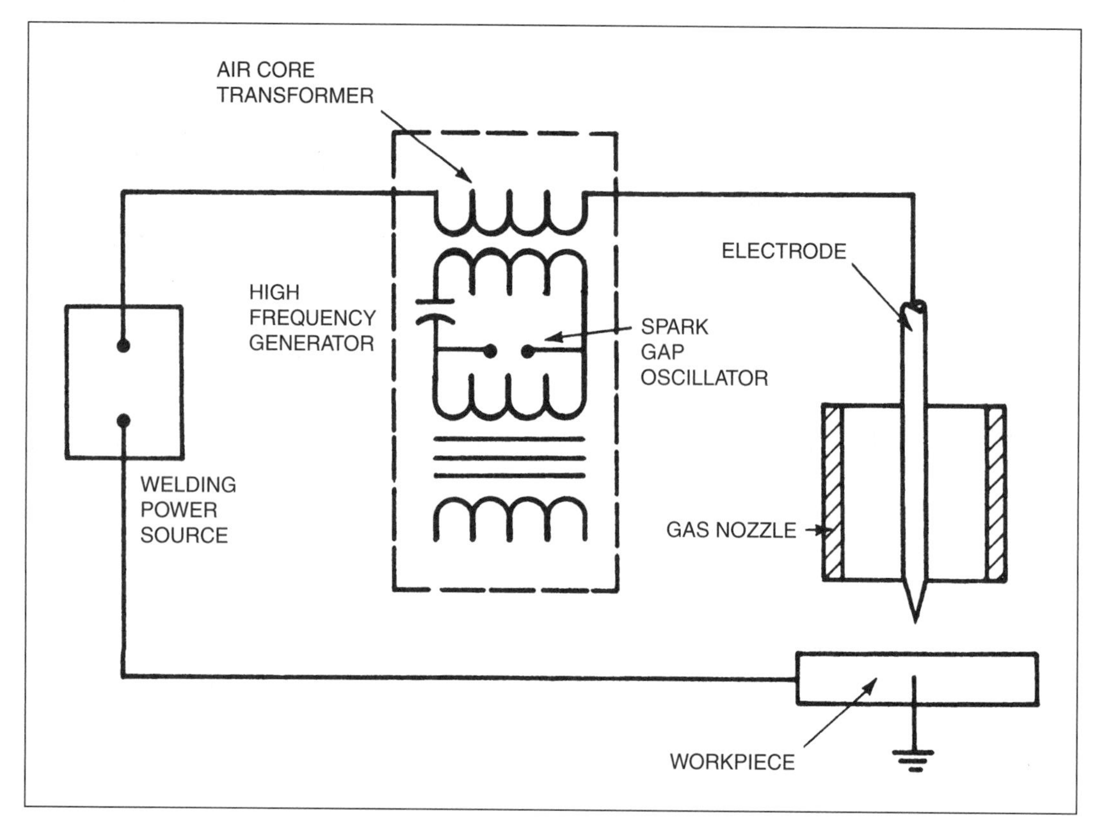

Figure 3.29—High-Frequency Arc Starting

Pulse Start

The application of a high-voltage pulse between the tungsten electrode and the workpiece will ionize the shielding gas and establish the welding arc. This method is generally used with dc power sources in mechanized welding applications.

Pilot Arc Start

Pilot arc starting can be used with dc welding power sources. The pilot arc is maintained between the welding electrode and the torch nozzle. The pilot arc supplies the ionized gas required to establish the main welding arc as shown in Figure 3.30. The pilot arc is powered by a small auxiliary power source and is started by high-frequency initiation.

MANUAL WELDING

The word *manual* in the gas tungsten arc process implies that a person controls all the functions of the welding process. The functions include manipulation of the electrode and control of filler metal additions, welding current, travel speed, and arc length. An application of the process is shown in Figure 3.31.

Manual Welding Equipment

In addition to an appropriate power source and a source of shielding gas, manual GTAW equipment includes the welding torch, hoses and electrical conductors, a foot pedal or a switch on the torch for controlling welding current levels during the welding cycle, and gas flow controls.

Manual Welding Techniques

The technique for manual welding is illustrated in Figure 3.32. Once the arc is started, the electrode is moved in a small circular motion until the desired weld pool is established. The torch is then held at an angle of 15° from the vertical and is moved along the joint to progressively melt the faying surfaces of the workpiece.

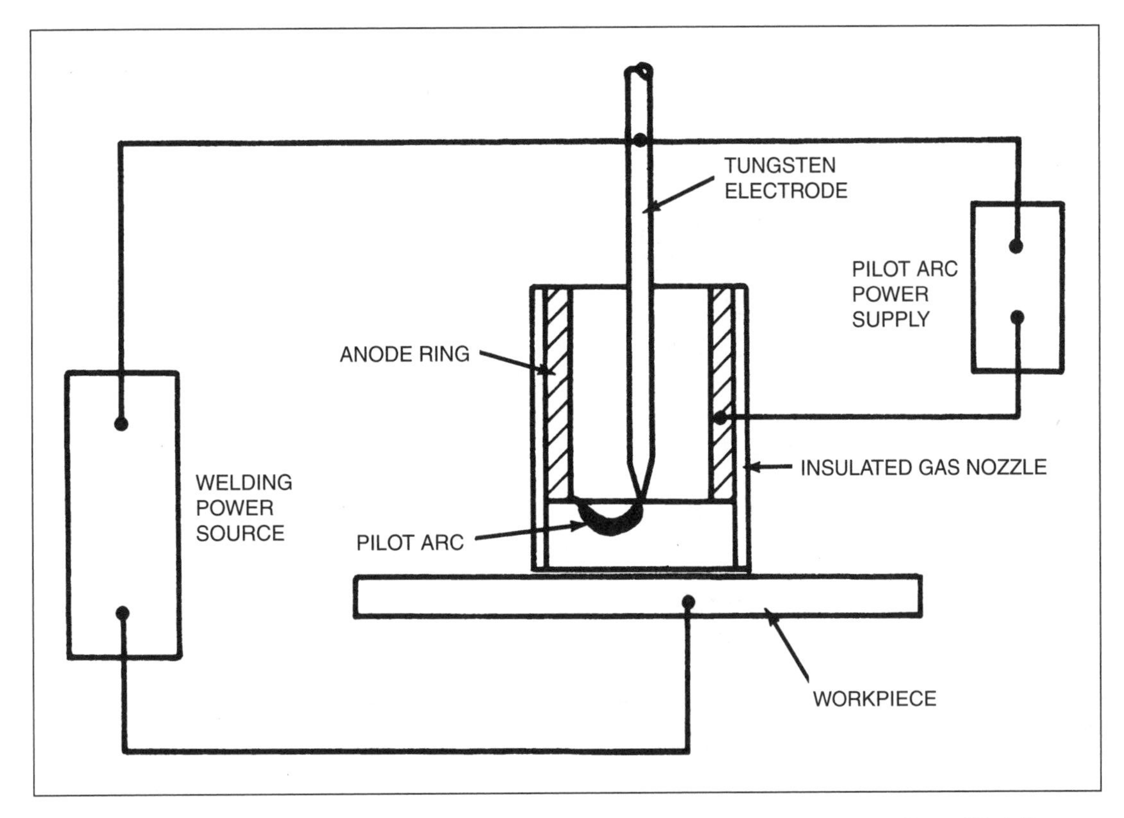

Figure 3.30—Pilot Arc Starting Circuit Used for Gas Tungsten Arc Spot Welding

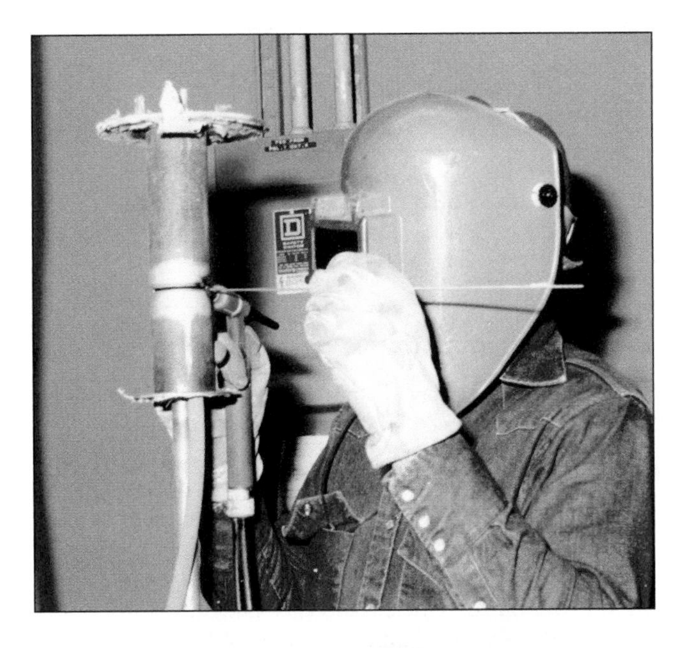

Figure 3.31—Manual Gas Tungsten Arc Welding of a Pipe Joint

If specified, filler metal in the form of a welding rod is added to the leading edge of the weld pool.

The torch and filler metal must be moved progressively and smoothly together so the weld pool, the hot welding rod end, and the hot solidified weld are not exposed to air that might contaminate the weld metal area or heat-affected zone. Generally, a large shielding envelope will prevent exposure to air.

The welding rod is usually held at an angle of about 15° to the surface of the workpiece and slowly fed into the weld pool. During welding, the hot end of the welding rod must not be removed from the protection of the inert gas shield. Upon completion of welding, the rod should remain in the protective gas shield until the rod has cooled sufficiently to prevent contamination. If the end of the rod becomes contaminated, the contaminated portion should be cut off.

MECHANIZED AND ORBITAL WELDING

Mechanized GTAW differs from orbital GTAW in that it is typically done in the flat position, in which the

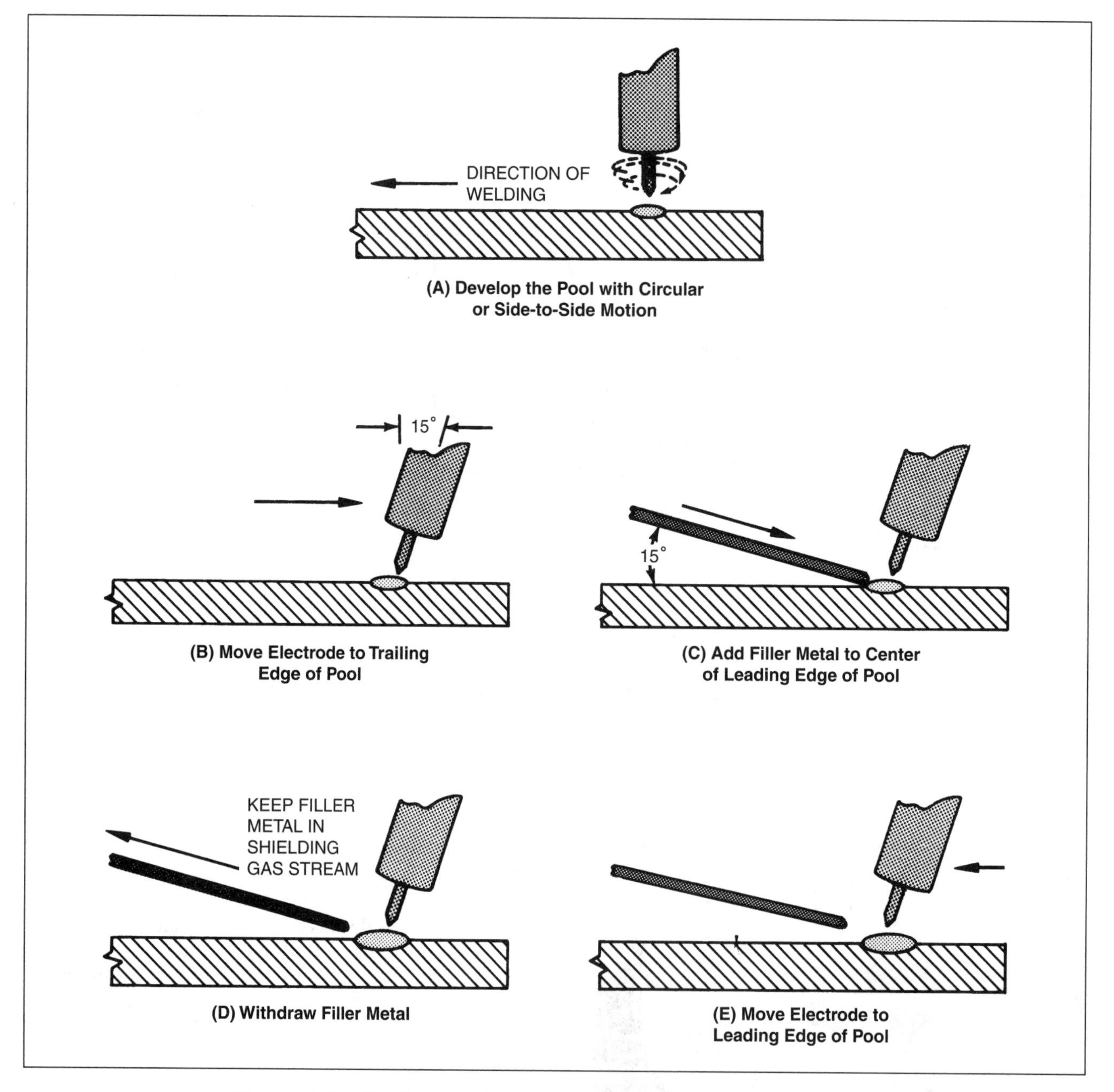

Figure 3.32—Technique for Manual Gas Tungsten Arc Welding

torch travels along the weld path or the workpiece moves or rotates under a fixed torch. Orbital welding is performed with the tube, pipe, fittings, or tubesheet remaining in place while the torch rotates around the joint to complete the weld. Both mechanized and orbital welding are implemented with equipment that performs the welding operation under the constant observation and control of a welding operator. Some equipment is designed to load and unload the workpieces.

Mechanized GTAW provides greater control over travel speed and heat input to the workpiece. The higher cost of equipment to provide these benefits must be justified by production and quality requirements.

Mechanized GTAW equipment ranges from simple weld program sequencers and mechanical manipulators to orbital tube and pipe welding systems and those with more complex motor controls. The controls vary from modular equipment, shown in Figure 3.33, to

Photograph courtesy of Jetline Engineering, Inc.

Figure 3.33—Typical Mechanized Welding Modular Panel for Control of Multiple Weld Process Variables

computer-controlled single-unit, multi-function power sources, shown in Figure 3.34.

Weld sequencers operate in an open-loop control mode: variables are maintained at preset levels and no attempt is made to adjust them in response to changing weld quality. Whereas orbital fusion tube welding is virtually automatic once the tube is loaded into the weld head, mechanized orbital welding requires the operator to exercise skill and judgment during the welding process. Some adjustment to torch steering, oscillation, wire feed rate or amperage may be required to compensate for observed welding conditions.

The sequencer automatically starts and completes the weld, switching from one variable setting to other settings at predetermined times or locations along the weld joint. A typical mechanized orbital pipe welding operation is shown in Figure 3.35. Backup gas is used to make the weld in this illustration, as evidenced by the baffle taped on a nozzle of the workpiece to contain the backup gas. Joint tolerances must be closely controlled, and fixturing must be strong, because the sequencer cannot compensate for unexpected movement of the components during welding. The preparation of high-precision weldments and the purchase or construction of sturdy fixturing increase production costs, but welding sequencers usually cost less than the more sophisticated automatic controllers.

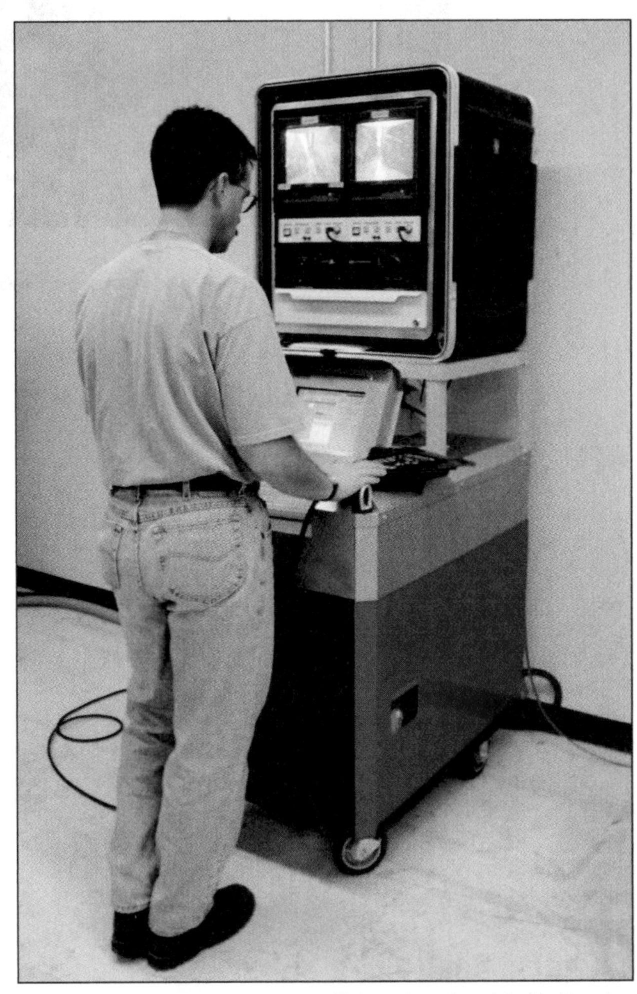

Photograph courtesy of Arc Machines, Inc.

Figure 3.34—Operator Using a Computer-Controlled Multi-Function GTAW Power Source with Touch Screen and Vision System for Remote Process Control

Figure 3.35—Mechanized Orbital Welding of Pipe Assembly

SEMIAUTOMATIC WELDING

Semiautomatic gas tungsten arc welding is defined as manual welding with equipment that automatically controls only the filler-metal feed. The advance of the welding torch is controlled manually. Semiautomatic systems for GTAW were introduced in the early 1950s but have been used only for special applications.

AUTOMATIC WELDING

Welding with equipment that performs the welding operation without adjustment of the controls by a welding operator is categorized as *automatic welding*. The equipment may or may not be designed to load and unload the workpieces.

Adaptive Control

Adaptive control welding systems (also called *feedback control systems*) make corrections to welding variables based on information gathered during welding. The objective is to maintain weld quality at a constant level in the presence of changing welding conditions. Automatic adjustment of individual welding variables, such as welding current or arc voltage, is made by monitoring a weld characteristic, such as pool width. Other adaptive control systems are available that provide electrode guidance and constant joint fill. Additional information on mechanized, automatic and robotic welding systems and techniques for monitoring and control of welding processes is presented in Chapters 10 and 11 of the *Welding Handbook*, 9th edition, Volume 1, *Welding Science and Technology*.[7]

ARC SPOT WELDING

Gas tungsten arc spot welding is often performed manually with a pistol-like holder that has a vented, water-cooled gas nozzle, a tungsten electrode that is concentrically positioned with respect to the gas nozzle,

7. American Welding Society (AWS) Welding Handbook Committee, Jenney, C. L. and A. O'Brien, eds., 2001, *Welding Science and Technology*, Vol. 1 of *Welding Handbook*, 9th ed., Chapter 10, Monitoring and Control of Welding and Joining Processes, p. 421-449, and Chapter 11, Mechanized, Automated and Robotic Welding, p. 452–481, Miami: American Welding Society.

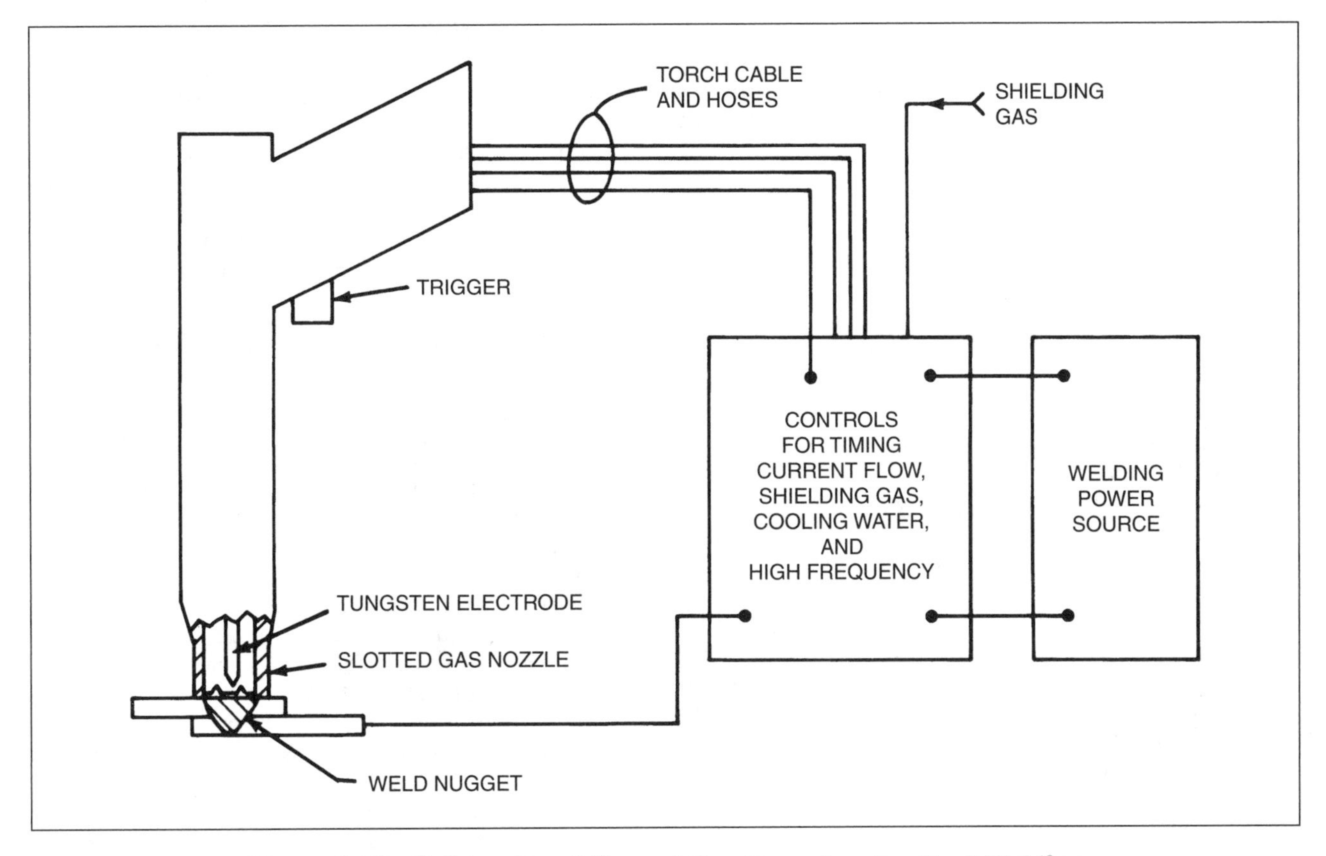

Figure 3.36—Schematic of Manual Gas Tungsten Arc Spot Welding

and a trigger switch for controlling the operation. Figure 3.36 illustrates such an arrangement. Gas tungsten arc spot welding electrode holders are also available for automatic applications.

The configuration of the nozzle is varied to fit the contour of the weldment. Edge-finding devices can be used to prevent variations in the distance of spot weld locations from the workpiece edge. The nozzle is pressed against the workpiece to assure tight fitup between the surface to be welded and the nozzle. The torch-gas nozzle assembly maintains a fixed electrode-to-workpiece distance.

Either ac or DCEN can be used for spot welding. Automatic sequencing controls are generally used because of the relatively complex cycles involved. The controls automatically establish the preweld gas and water flow, start the arc, time the arc duration, and provide the required postweld gas and water flow.

Penetration is controlled by adjusting the current and the length of time it flows. In some applications, multiple pulses of current are preferred over one long sustained pulse. Variations in the shear strength, nugget diameter, and penetration of the spot weld can be minimized with accurate timers, current monitors, and tungsten electrodes that have precision-ground tips.

When thinner materials are welded, a melted spot on the bottom of the lower workpiece is a positive indication of a good spot weld.

MATERIALS

This section describes the metals weldable with the GTAW process. Autogenous welds are made by melting base metal only. Filler metal, if used, can be in the form of welding wire or preplaced consumable inserts.

BASE METALS

Most metals can be welded by the gas tungsten arc welding process. Among these are various grades of carbon steels, steel alloys, stainless steels and other ferrous alloys; heat-resistant alloys of various types; aluminum;

magnesium; copper and its alloys (copper-nickel, bronzes, and brasses); and nickel. Certain metals must be welded with the GTAW process because it provides the greatest protection from contamination by the atmosphere. The process is especially useful for welding reactive and refractory metals and some nonferrous alloys. However, GTAW is not used to weld metals that have low vapor pressure when in the liquid state, such as cadmium, tin, or zinc.

Volume 3 and Volume 4 of the *Welding Handbook*, 8th edition, *Materials and Applications, Part 1* and *Materials and Applications, Part 2* provide details of the welding characteristics of specific metals and alloys, along with suggested types of welding current, electrode compositions, and shielding gas compositions for optimum weld quality.[8, 9]

Chapter 4 of the *Welding Handbook*, 9th edition, Volume 1, *Welding Science and Technology*, describes the metallurgical responses of metals and alloys to the heat of welding.[10]

The section below presents information on potential metallurgical problems unique to the gas tungsten arc welding process and addresses special concerns when welding certain metals and alloys with this process. In general, best welding results are obtained with DCEN for almost all metals, unless otherwise specified. The typical tungsten electrode composition is 2% thoriated, unless otherwise specified. (Refer to Table 3.4 for a listing of typical metals weldable by the gas tungsten arc welding process and the appropriate GTAW electrode).

Carbon and Alloy Steels

The quality of gas tungsten arc welds in carbon and alloy steels is more influenced by the base metal impurity content (e.g. sulfur, phosphorus, and oxygen) than is the quality of welds made with shielded metal arc welding or submerged arc welding. This is because gas tungsten arc welding does not provide fluxes to remove or tie up these impurities.

High-strength low-alloy (HSLA) steels are readily welded by the GTAW process. However, combined levels of phosphorus and sulfur in the base metal exceeding 0.03% can cause cracking in the fusion and heat-affected zones. Hydrogen embrittlement of these alloys is a problem if hydrocarbon or water vapor contamination is present. Hydrogen-induced cracking can be minimized by the application of preheat or a postweld heat treatment or, in high-humidity areas, with trailing gas shields.

Argon shielding is generally used for welding carbon and alloy steels because the weld pool is easier to control with argon than with helium. Argon or argon-helium mixtures can be used to obtain increased weld penetration, weld speed, and wettability.

Stainless Steels and Heat-Resistant Alloys

Stainless steels and the heat-resistant superalloys (alloys with an iron, nickel, or cobalt base) are extensively welded with the GTAW process because the welds are protected from the atmosphere by the inert gas. Weld-metal composition is essentially identical to the base-metal composition because the same alloys are used as filler metal and because the filler enters the liquid weld pool without passing through the arc, where losses of volatile alloys might be expected.

The successful welding of alloys that have mechanical and corrosion-resistance properties, i.e., duplex stainless steels, depends to a large extent on welding parameters. Mechanized or orbital GTAW is used to advantage for these alloys because of the repeatability of the process. Once successful welding procedures have been established for these materials, the procedures can be repeated with a high degree of precision.

Argon is the recommended shielding gas for the manual welding of thicknesses up to approximately 12 mm (1/2 in.) because it provides better control of the weld pool. For thick sections and for many mechanized and automatic applications, argon-helium mixtures or pure helium can be used to obtain increased weld penetration. Argon-hydrogen mixtures are used for some stainless steel welding applications to improve bead shape and wettability.

Alternating current can be used for automatic welding of the heat-resistant alloys when close control of arc length is possible.

Aluminum Alloys

Gas tungsten arc welding is ideally suited for welding aluminum alloys. All thicknesses of aluminum can be welded with this process, with or without filler metal.

Aluminum alloys form refractory surface oxides, which make joining more difficult. For this reason, most welding of aluminum is performed with alternating current (using high-frequency arc stabilization) because ac provides the surface cleaning action of DCEP in addition to the deeper penetration characteristics of DCEN. Sometimes DCEP is used for welding thin aluminum sections. DCEN with helium as the

8. American Welding Society (AWS) Welding Handbook Committee, 1994, Oates, W. R., ed., *Materials and Applications—Part 1*, Vol. 3, of *Welding Handbook*, 8th ed., Miami: American Welding Society.
9. American Welding Society (AWS) Welding Handbook Committee, 1996, Oates, W. R., and A. M. Saitta, eds., *Materials and Applications—Part 2*, Vol. 4 of *Welding Handbook*, 8th ed., Miami: American Welding Society.
10. American Welding Society (AWS) Welding Handbook Committee, 2001, Jenney, C. L., and A. O'Brien, eds., *Welding Science and Technology*, Vol. 1 of *Welding Handbook*, 9th ed., Miami: American Welding Society.

shielding gas is used for the high-current automatic welding of sections over 6.35 mm (1/4 in.) thick. Since DCEN produces no cleaning action, the aluminum workpieces must be thoroughly cleaned immediately prior to welding.

For welding with ac, electrodes of pure tungsten, ceriated tungsten, and zirconiated tungsten are recommended. Only thoriated tungsten electrodes are used on aluminum with dc.

Argon shielding is generally used for welding aluminum with ac because it provides better arc starting, better cleaning action, and superior weld quality than helium. When DCEN is used, helium shielding allows faster travel speeds and achieves deeper penetration. However, the poor surface cleaning action of this combination may result in porosity.

Magnesium Alloys

Magnesium alloys form refractory surface oxides similar to aluminum alloys. Alternating current GTAW is typically used for the welding of magnesium alloys because of the oxide cleaning action ac provides. Thicknesses less than 5 mm (3/16 in.) can be welded using DCEP, but ac provides better penetration for thicker sections. Argon shielding produces the best quality welds, but helium or argon–helium mixtures are also used. Pure tungsten, ceriated, and zirconiated electrodes can be used to weld magnesium alloys.

Beryllium

Beryllium is a light metal that is difficult to weld because of a tendency toward hot cracking and embrittlement. The gas tungsten arc welding of beryllium is performed in an inert-atmosphere chamber, generally using a shielding gas mixture of five parts helium to one part argon. Beryllium oxide fumes are toxic; thus an appropriate fume removal system must be used.

Copper Alloys

Gas tungsten arc welding is well suited for joining copper and its alloys because of the intense heat generated by the arc, which can produce melting with minimum heating of the surrounding highly conductive base metal. Most copper alloys are welded with DCEN and helium shielding because copper has high thermal conductivity. Alternating current is sometimes used to weld beryllium coppers and aluminum bronzes because it helps break up the surface oxides that are present during welding.

Nickel Alloys

Nickel alloys are often welded with the GTAW process, typically with filler metal additions. DCEN is recommended for all welding of nickel alloys. However, some applications, for example, those using nickel alloys with large quantities of surface oxides, may use ac with high-frequency stabilization. Argon, argon-helium, and helium are the most common shielding gases. Argon with additions of hydrogen is also used.

High-purity nickel alloys can exhibit variable weld penetration caused by differences in surface-active elements and may produce effects similar to sulfur variations in stainless steels.

Refractory and Reactive Metals

Gas tungsten arc welding is the most extensively used process for joining refractory and reactive metals. Refractory metals (notably tungsten, molybdenum, tantalum, niobium, and chromium) have extremely high melting temperatures, and like the reactive metals (titanium alloys, zirconium alloys, and hafnium), readily oxidize at elevated temperatures unless protected by an inert gas cover. The absorption of impurities, such as oxygen, nitrogen, hydrogen and carbon, decreases the toughness and ductility of the weld metal.

For these metals and alloys, GTAW produces a high concentration of heat and allows the greatest control over heat input while providing the best inert gas shielding of any welding process. Welding these metals is typically performed in purged chambers containing high-purity inert gases. Occasionally GTAW can be performed without special purge chambers; the necessary inert gas atmosphere is created with torch trailing and backup shielding.

Argon is most frequently used for shielding, but helium and mixtures of the two gases can be used. Argon flow rates of 7 L/min (15 cfh) or helium flow rates of 18.5 L/min (40 cfh) are sufficient, even with the large-diameter gas nozzles that are recommended.

Cast Irons

Cast irons can be welded with the GTAW process because dilution of the base metal can be minimized with independent control of heat input and filler metal placement. A high level of operator skill is required to minimize dilution while maintaining acceptable penetration and fusion.

The gas tungsten arc welding of cast irons is usually limited to the repair of small parts. Nickel-based and austenitic stainless steel filler metals are recommended; they minimize cracking because of their ductility and tolerance for hydrogen. Cracking can also be minimized by preheating and postweld heat treatment. DCEN is recommended, although ac may be used.

Table 3.5
AWS Specifications for Filler Metals Suitable for Gas Tungsten Arc Welding and Surfacing

Specification Number	Title
A5.2	Specification for Carbon and Low Alloy Steel Rods for Oxyfuel Gas Welding
A5.7	Specification for Copper and Copper Alloy Bare Welding Rods and Electrodes
A5.9	Specification for Bare Stainless Steel Electrodes and Rods
A5.10	Specification for Bare Aluminum and Aluminum Alloy Welding Electrodes and Rods
A5.13	Specification for Solid Surfacing Welding Rods and Electrodes
A5.14	Specification for Nickel and Nickel Alloy Bare Welding Electrodes and Rods
A5.16	Specification for Titanium and Titanium Alloy Electrodes and Welding Rods
A5.18	Specification for Carbon Steel Filler Metals for Gas Shielded Arc Welding
A5.19	Specification for Magnesium Alloy Welding Electrodes and Rods
A5.21	Specification for Composite Surfacing Welding Rods and Electrodes
A5.24	Specification for Zirconium and Zirconium Alloy Welding Electrodes and Rods
A5.28	Specification for Low-Alloy Steel Filler Metals for Gas Shielded Arc Welding
A5.30	Specifications for Consumable Inserts

FILLER METALS

Filler metals for use with gas tungsten arc welding are available for joining a wide variety of metals and alloys. If used, the filler metals should be similar, although not necessarily identical, to the metal that is being welded. For example, when welding dissimilar metals, the filler metals will be different from one or both of the base metals.

Generally, the filler metal composition is adjusted to match the properties of the base metal in its welded (cast) condition. Filler metals are produced with closer control of chemistry, purity, and quality than base metals. Deoxidizers are frequently added to ensure weld soundness. Further modifications are made to some filler metal compositions to improve response to postweld heat treatments.

The choice of filler metal for any application is a compromise involving metallurgical compatibility, suitability for the intended service, and cost. The tensile, impact, and hardness properties, corrosion resistance, and electrical or thermal conductivities required for a particular weldment must also be considered. Thus, the filler metal must suit both the alloy to be welded and the intended service.

Table 3.5 lists the American Welding Society's filler metal specification documents that are applicable to gas tungsten arc welding. These standards establish filler metal classifications based on the mechanical properties or chemical compositions, or both, of each filler metal. They also set forth the conditions under which the filler metals must be tested.

Several sources of information, specifically the appendices in the AWS filler metal specifications, provide background information on the properties and uses of the filler metals within the various classifications. Filler metal manufacturers' catalogs provide information on the proper use of their products. Brand name listings and addresses of vendors are published in the latest edition of AWS *Filler Metal Comparison Charts*.[11]

Filler metals for GTAW are available in most alloys in the form of cut lengths (rods), usually 0.91 meter (m) (36 in.) long for manual welding, and spooled or coiled continuous wire for mechanized or automatic welding. The filler metal rod diameters range from about 0.5 mm (0.020 in.) for delicate work on very thin metal to about 5 mm (3/16 in.) for high-current manual welding or for surfacing applications.

Extra care must be exercised to keep the filler metals clean and free of all contamination while in storage as well as while in use. The hot end of the wire or rod should not be removed from the protection of the inert gas shield during the welding operation.

Consumable inserts provide filler metal additions for root pass welds in certain pipe and plate applications. The advantages of this type of filler include broader fitup tolerances, more consistent weld bead fusion and smooth, uniform underbeads. Less operator skill is required.

11. American Welding Society (AWS) and Committee on Filler Metals and Allied Materials, 2000, *Filler Metal Comparison Charts*, Miami: American Welding Society.

JOINT DESIGN

Factors that affect joint design include base metal composition and thickness, weld penetration, joint restraint, and joint efficiency requirements. Because of the variety of base metals and their individual characteristics (such as melting temperature, fluidity, and surface tension), as well as joint geometries and weldment designs, these factors should be considered if optimum welding conditions are to be achieved.

BASIC JOINT CONFIGURATIONS

The five basic joints—the butt joint, lap joint, T-joint, edge joint and corner joint—are illustrated in Figure 3.37. They can be used for virtually all metals. Many variations can be derived from these basic joints. In all instances, the primary objective is to maximize the desired weld quality and performance level for the design while maintaining costs at a minimum. Factors that affect cost are preparation time, weld joint area to be filled, and setup time. While there are no fixed rules governing the use of a particular joint design for any one metal, certain designs were developed for specific purposes.

The primary variables of joint design are root opening, thickness of root face, and angle of bevel. All variables must be considered prior to joint preparation. The amount of root opening and thickness of the root face depend on whether the GTAW process is to be manual or automatic, whether filler metal is to be added during the root pass, or if a consumable insert will be used. Backing strips are generally not used due to additional costs of material and fitup, cost of removal (if required), as well as difficulty in interpretation of radiographs.

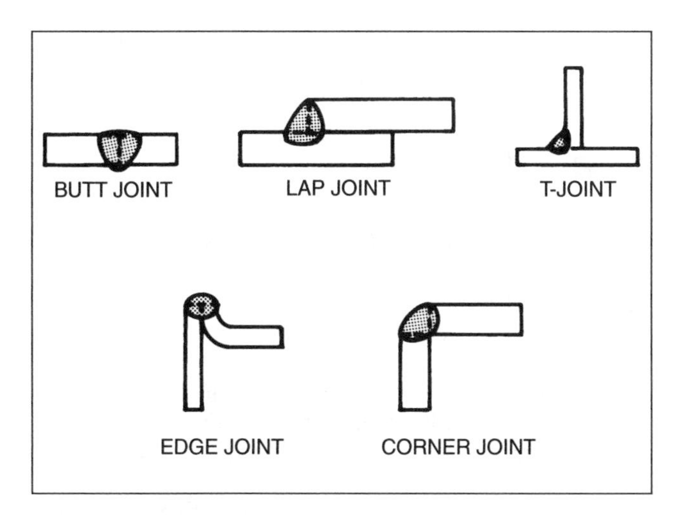

Figure 3.37—Five Basic Weld Joints

The amount of bevel angle depends on the thickness of the metal and the clearance needed for arc movement to assure adequate fusion on both sides of the joint. These variables are generally determined by producing sample joints welded in a variety of setups.

A major consideration in GTAW joint design is proper provision for weld accessibility. The groove angle must permit adequate manipulation of the electrode holder to obtain fusion of the groove face. Characteristics of the weld metal must also be considered. For example, alloys high in nickel content are very sluggish when molten, and the weld metal does not wet the groove face well. Therefore, groove angles for high-nickel alloys should be more open than those for carbon and alloy steel to provide space for manipulation. However, opening the groove angle increases distortion, weld time, and cost, and should be limited as far as possible.

Specific information on joint design is presented in Chapter 5 of the American Welding Society *Welding Handbook*, 9th ed., Vol.1, *Welding Science and Technology*,[12] and in metal supplier literature.

JOINT PREPARATION

After a particular joint design has been selected, the most important item for consideration is the method of joint preparation. There are many ways to remove metal to prepare a given joint configuration. However, many GTAW problems, or potential problems, are a direct result of using improper methods to prepare the joint. Chief among these is the improper use of grinding wheels to prepare joints. Soft materials such as aluminum become impregnated with abrasive particles and, unless subsequently removed, the particles will contaminate the weld and can result in excessive porosity. Grinding wheels should be scrupulously cleaned and dedicated exclusively to the material being welded.

The ideal joint preparation is obtained with machine cutting tools, such as a lathe or mill. Care must be exercised in the choice of cutting fluid, if used. Sulfur-bearing cutting fluids can create cracking in nickel alloys. Safety-approved solvents that are free of residues should be used for cleaning after cutting or turning the workpiece. Oxyfuel cutting and plasma arc cutting are also acceptable provided any slag is removed by careful grinding.

12. See Reference 9.

Joint Tolerance

The allowable tolerance of joint dimensions depends on whether GTAW is to be done manually or by mechanized means. Manual welding applications can tolerate greater irregularities in joint fitup than mechanized welding. The particular tolerance for a given application can be determined only by actual testing, and this tolerance should be specified for future work. Joint design and fitup are particularly important for mechanized and orbital welding. Consistent edge preparation is the key to maintaining the repeatability of the process.

Cleaning

Cleanliness of both the weld joint areas and the filler metal is an important consideration when welding with the GTAW process. Oil, grease, shop dirt, paint, marking crayon, and rust or corrosion deposits must be removed from the joint edges and metal surfaces to a distance beyond the heat-affected zone. The presence of these impurities during welding may lead to arc instability and contaminated welds. Depending on the metallurgical response to these contaminants, welds may contain pores, cracks, and inclusions. The cleaning of the weld joint areas can be accomplished by mechanical means, by the use of vapor or liquid cleaners, or by a combination of these.

FIXTURING

Fixturing devices or equipment to hold an assembly in place for welding may be required if the workpieces are not self-supported during welding, or if any distortion resulting from welding cannot be tolerated or corrected by straightening. The fixturing should be massive enough to support the weight of the weldment and to withstand welding stresses caused by thermal expansion and contraction. The fixtures must also handle the normal wear and tear that occurs during production and must not limit access to the weld locations.

The decision to use fixturing for the fabrication of a weldment is governed by quality requirements and economics. The use of appropriate fixturing, including heat sinks, can reduce welding time. The one-time fabrication of an assembly may not justify the use of fixturing; however, the fabrication of a large number of assemblies could justify even complex fixtures. Also, high-quality precision work may dictate that fixtures be used to maintain the close tolerances required by the design, or for nondestructive examination requirements. Fixturing performs the following primary functions:

1. Positions components precisely within the assembly,
2. Maintains stability and alignment during welding,
3. Minimizes distortion in the weldment, and
4. Controls heat buildup.

WELD QUALITY

The ability to produce high-quality welds with the GTAW process is based on the knowledge of process capabilities and limitations. The proper selection of the process for a given application must take into account the need for control of adequate shielding, potential contamination problems, and operator skills.

The solutions to some of the potential problems of gas tungsten arc welding begin with the recognition that the process has a low tolerance for contaminants on base metal and filler metals. The solutions become evident once the problems are identified. The following are typical issues in GTAW:

1. Tungsten inclusions can occur if the electrode is allowed to contact the weld pool or melt;
2. Contamination of the weld metal can occur if proper shielding of the filler metal by the gas stream is not maintained;
3. Contamination or porosity in the weld metal can occur because of poor gas quality, reduced shielding, or coolant leakage from water-cooled torches;
4. Faulty or loose torch connections can aspirate air into the gas stream, resulting in weld contamination.

DISCONTINUITIES AND DEFECTS

Discontinuities are interruptions in the typical structure of a weldment, and they may occur in the base metal, weld metal, or heat-affected zones. Discontinuities in a weldment that do not satisfy the requirements of the applicable fabrication code or specification are classified as defects, and they are required to be removed because they could impair the performance of the weldment in service. Various discontinuities, problems, and corrections are discussed in this section.

Tungsten Inclusions

A tungsten inclusion is a discontinuity found only in gas tungsten arc welds. Particles of tungsten from the electrode can become embedded in a weld when improper welding techniques are used with the GTAW process. Typical causes are the following:

1. Contact of electrode tip with weld pool;
2. Contact of filler metal with hot tip of electrode;
3. Contamination of the electrode tip by spatter from the weld pool;
4. Exceeding the current limit for a given electrode size or type;

5. Extension of electrodes beyond their normal distances from the collet (as with long nozzles), resulting in overheating of the electrode;
6. Inadequate tightening of the collet or electrode chuck;
7. Inadequate shielding gas flow rates or excessive wind drafts resulting in oxidation of the electrode tip;
8. Splits or cracks in the electrode; and
9. Improper shielding gases, such as argon-oxygen or argon-carbon dioxide mixtures that are used for gas metal arc welding.

Corrective steps are obvious once the causes are recognized and the welder is adequately trained.

Incomplete Shielding

Discontinuities related to the loss of inert gas shielding are porosity, oxide films (oxidization) and inclusions, incomplete fusion, cracking, and tungsten inclusions, as described above. The extent to which they occur is strongly related to the characteristics of the metal being welded. In addition, the mechanical properties of titanium, aluminum, nickel, and high-strength steel alloys can be seriously impaired with the loss of inert gas shielding. Gas shielding effectiveness can often be evaluated prior to production welding by making a spot weld and continuing gas flow until the weld has cooled to a low temperature. A bright, silvery spot will be evident if shielding is effective.

WELDING PROBLEMS AND REMEDIES

Numerous welding problems can develop while setting up or implementing a GTAW operation. The solutions will require careful evaluation of the material, fixturing, welding equipment, and procedures in use. Some problems that may be encountered and possible remedies are listed in Table 3.6.

Table 3.6
Troubleshooting Guide for Gas Tungsten Arc Welding

Problem	Cause	Remedy
Excessive electrode consumption	1. Inadequate gas flow	1. Increase gas flow.
	2. Operating on DCEP	2. Use larger electrode or change to DCEN.
	3. Improper size electrode for current required	3. Use larger electrode.
	4. Excessive heating in holder	4. Check for proper collet contact.
	5. Contaminated electrode	5. Remove contaminated portion. Erratic results will continue as long as contamination exists.
	6. Electrode oxidation during cooling	6. Keep gas flowing after stopping arc for at least 10 to 15 seconds.
	7. Using gas containing oxygen or carbon dioxide	7. Change to proper gas.
Erratic arc	1. Base metal is dirty, greasy	1. Use appropriate chemical cleaners, wire brush, or abrasives.
	2. Joint is too narrow	2. Open joint groove; bring electrode closer to work; decrease voltage.
	3. Electrode is contaminated	3. Remove contaminated portion of electrode.
	4. Arc is too long	4. Bring electrode holder closer to workpiece to shorten arc.
Porosity	1. Entrapped gas impurities (hydrogen, nitrogen, air, water vapor)	1. Blow out air from all lines before striking arc; remove condensed moisture from lines; use welding-grade (99.99%) inert gas.
	2. Defective gas hose or loose hose connections	2. Check hose and connections for leaks.
	3. Oil film on base metal	3. Clean with chemical cleaner not prone to break up in arc. Do not weld while metal is wet.
Tungsten contamination of workpiece	1. Contact starting with electrode	1. Use high-frequency starter; use copper striker plate.
	2. Electrode melting and alloying with base metal	2. Use less current or larger electrode; use proper electrode for the material being welded.
	3. Touching tungsten to molten pool	3. Keep tungsten out of weld pool.

ECONOMICS

When planning the economics of any welding project, the primary objective is to maximize the desired weld quality and ease of controlling the weld while maintaining costs at a minimum. To achieve these goals several factors must be considered.

The choice of gas tungsten arc welding as the preferred process for a given project is based on a combination of the service requirements of the weldments and the cost of GTAW compared to the cost of other processes. While weld quality and the properties required for the intended service of the weldment are often the driving factors in the selection of the process, other factors that impact the cost of welding are the skill levels required, accessibility to the weld, the level of control required, the equipment and consumables needed, and production rates.

Manual gas tungsten arc welding usually requires higher skill levels than other manual processes. Mechanized GTAW requires less operator skill and provides greater control over the process, but equipment costs are higher, and the benefits must be justified by production and quality requirements.

The cost of such items as shielding gases, gas nozzles and lenses or collets, the tungsten electrodes, and filler metals (if used) should be compared with the cost of materials typical of other processes. Filler metal for any application is a compromise involving metallurgical compatibility, suitability for the intended service, and cost. The ability of GTAW to produce certain welds without filler material results in decreased costs compared to other arc processes.

When intricate or repetitive welds are involved, the preparation requirements of the weldments and the purchase or construction of sturdy fixturing equipment increase production costs. Welding sequencers usually cost less than the more sophisticated automatic controllers.

Weld joint design can vary significantly depending on the in-service requirements of the weld, accessibility, and the type of materials being welded. Increasing the joint size can cause distortion and increase the weld time, and thus increase the cost, and should be limited as far as possible. The amount of root opening and the thickness of the root face will affect whether the GTAW process is to be manual or automated. Additional cost must be anticipated for machining tolerances, inserts or backing material, and the cost of removal of the backing (if required). As in most welding processes, accessibility for inspection and the interpretation of the results should be considered when calculating costs.

While relatively inexpensive power sources can be used, the complexity of the weld and the required control may mandate the purchase of more sophisticated power sources with ancillary equipment for voltage control, wire feed, oscillation, mechanical travel, remote vision, and other needs. Maintenance of equipment in industrial environments must also be addressed.

Applicable information on the economics of welding, including manual, mechanized and automatic, is presented in Chapter 12 of the *Welding Handbook*, 9th edition, Volume 1, *Welding Science and Technology.*[13]

SAFE PRACTICES

Safe practices should always be the foremost concern of the welder or welding operator. The general subject of safety and safe practices in welding, cutting, and allied processes is covered in ANSI Z49.1, *Safety in Welding, Cutting, and Allied Processes*.[14] All welding personnel should be familiar with the safe practices discussed in that document. Equipment manufacturers' operating instructions and material suppliers' Material Safety Data Sheets should be consulted.

Using safe practices in welding and cutting will ensure the protection of persons from injury and illness and the protection of property from damage. The potential hazard areas in arc welding, cutting, and surfacing include but are not limited to the handling of cylinders and regulators, gases, fumes, radiant energy, and electrical shock. The safety concerns associated with gas tungsten arc welding are briefly discussed in this section. Appendix B, "Safety and Health Codes and Other Standards," lists safety standards and other publications, and facts of publication.

SAFE HANDLING OF GAS CYLINDERS AND REGULATORS

Compressed gas cylinders should be handled carefully. Knocks, falls, or rough handling may damage cylinders, valves, or safety devices and cause leakage or explosive rupture accidents. Cylinder caps that protect the valves should be kept in place (hand-tight) except when cylinders are in use or connected for use. When in use, cylinders should be securely fastened to prevent accidental tipping. For further information, refer to

13. See Reference 9.
14. American National Standards Institute (ANSI) Accredited Standards Committee Z49, 1999, *Safety in Welding, Cutting, and Allied Processes*, ANSI Z49.1:1999, Miami: American Welding Society.

CGA Pamphlet P-1, *Safe Handling of Compressed Gases in Containers.*[15]

GAS HAZARDS

The major toxic gases associated with GTAW are ozone, nitrogen dioxide, and phosgene gas. Inert shielding gases, while not toxic, can be hazardous if they displace air in the welder's breathing area.

Ozone

The ultraviolet light emitted by the welding arc acts on the oxygen in the surrounding atmosphere to produce ozone. The amount of ozone produced will depend on the intensity of the ultraviolet energy, the humidity, the amount of screening afforded by the welding fume, and other factors. Test results based on present sampling methods indicate that the average concentration of ozone generated in the GTAW process does not constitute a hazard under conditions of good welding practices and good ventilation, but these conditions should be confirmed at the location of each welding application. (See ANSI Z49.1 for welding conditions requiring ventilation, particularly when welding is done in confined spaces.)[16]

Nitrogen Dioxide

High concentrations of nitrogen dioxide are found only within 150 mm (6 in.) of the arc. Natural ventilation quickly reduces these concentrations to safe levels in the welder's breathing zone. As long as the welder's head is kept out of the fumes, nitrogen dioxide is not considered to be a hazard in GTAW.

Phosgene Gas

Phosgene gas could be present as a result of thermal or ultraviolet decomposition of chlorinated hydrocarbon cleaning agents, such as trichlorethylene and perchlorethylene, that may be located in the vicinity of welding operations. Degreasing or other cleaning operations involving chlorinated hydrocarbons should be performed in an area where vapors from these operations are not exposed to radiation from the welding arc.

Inert Shielding Gases

Provision for adequate ventilation should be made when inert gas shielding and purging gases are used. Accumulation of these gases that would possibly displace the air in work areas could cause suffocation of welding and inspection personnel.

METAL FUMES

The welding fumes generated by the gas tungsten arc welding process can be controlled by natural ventilation, general ventilation, local exhaust ventilation, or by respiratory protective equipment, as described in ANSI Z49.1.[17] The method of ventilation required to keep the level of toxic substances within acceptable concentrations in the welder's breathing zone is directly dependent on a number of factors, among which are the material being welded, the size of the work area, and the degree of confinement or obstruction to normal air movement where the welding is being done. Each operation should be evaluated on an individual basis to determine which methods of ventilation will be required.

Acceptable levels of toxic substances associated with welding and designated as time-weighted average threshold limit values (TLVs®) and ceiling values are stated in the document *TLVs and BEIs—Threshold Limit Values for Chemical Substances and Biological Exposure Indices*, published by the American Conference of Governmental Industrial Hygienists (ACGIH).[18] The United States government Occupational Safety and Health Administration (OSHA) publishes standards for general industry that can be obtained through the U.S. Government Printing Office.[19] Compliance with these acceptable levels can be checked by sampling the atmosphere inside the welder's helmet or in the immediate vicinity of the helper's breathing zone. Sampling should be in accordance with ANSI/AWS F1.1, *Methods for Sampling Airborne Particulates Generated by Welding and Allied Processes.*[20]

RADIANT ENERGY

Radiant energy is a hazard that can cause injury to the welder (or others exposed to the welding arc) in two areas: eyes and skin. The general subject of eye protection is covered in ANSI Z49.1, *Safety in Welding and Cutting,*[21] and ANSI Z87.1, *Practice for Occupational and*

15. Compressed Gas Association (CGA), *Safe Handling of Compressed Gas in Containers*, CGA P-1, Arlington, Virginia: Compressed Gas Association.
16. See Reference 14.

17. See Reference 14.
18. American Conference of Governmental Industrial Hygienists, *TLVs and BEIs—Threshold Limit Values for Chemical Substances and Physical Agents and Biological Exposure Indices*, Cincinnati, Ohio: American Conference of Governmental Industrial Hygienists.
19. Occupational Safety and Health Administration (OSHA), 1999, *Occupational Safety and Health Standards for General Industry* in *Title 29, Code of Federal Regulations* (CFR), Chapter XVII, Parts 1901.l to 1910.1450, Washington D.C.: Superintendent of Documents, U.S. Government Printing Office.
20. American Welding Society (AWS) Committee on Fumes and Gases, *Methods for Sampling Airborne Particulates Generated by Welding and Allied Processes*, ANSI/AWS F1.1, Miami: American Welding Society.
21. See Reference 14.

Educational Eye and Face Protection.[22] All personnel within the immediate vicinity of a welding operation should have adequate protection from the radiation produced by the welding arc. Generally, the highest ultraviolet radiant energy intensities are produced when argon shielding gas is used and when aluminum or stainless steel is welded.

For the protection of eyes, filter glass or curtains should be used. As a guide, the filter glass shades recommended for GTAW and other processes are presented in Appendix A of this volume. The welder should use the darkest shade that is comfortable, but not lighter than the recommended shade.[23]

For the protection of skin, leather or dark protective clothing is recommended to reduce reflection which could cause ultraviolet burns to the face and neck underneath the helmet. High-intensity ultraviolet radiation will cause rapid disintegration of cotton and some synthetic materials.

In addition, when an area or room is set aside for GTAW, the walls should be coated with pigments such as titanium dioxide or zinc oxide because these will reduce ultraviolet reflection. Additional information is presented in *Ultraviolet Reflection of Paint*,[24] published by the American Welding Society.

ELECTRICAL SHOCK

Training in the avoidance of electrical shock is especially important for the welder. Safe procedures should be established according to applicable standards and observed at all times when working with equipment that uses the high voltages necessary for arc welding. Unexplained shocks should be reported to the supervisor for investigation and correction before welding is continued.

Even mild electrical shocks can cause involuntary muscular contraction, leading to injurious falls from high places. The severity of shock is dependent on voltage and contact resistance of the area of skin involved, and these are determined largely by the path, duration, and amount of current flowing through the body. Clothing that is damp from perspiration or wet working conditions may reduce contact resistance and increase current to a value high enough to cause such violent muscular contraction that the welder cannot release contact with the energized (live) workpiece.

22. American National Standards Institute (ANSI), *Practice for Occupational and Educational Eye and Face Protection*, ANSI Z87.1, Des Plaines, Illinois: American Society of Safety Engineers (ASSE).
23. American Welding Society (AWS) Committee on Safety and Health, *Lens Shade Selector,* ANSI/AWS F2.2, Miami: American Welding Society.
24. Ullrich, O. A., and R. M. Evans, 1976, *Ultraviolet Reflectance of Paint*, Miami: American Welding Society.

WELDING EQUIPMENT SAFETY

As a first requisite for the safe operation of welding equipment, the manufacturers' operating procedures should be strictly followed. All welding equipment should be on an approved list from a testing agency recognized by the National Fire Protection Agency (NFPA), such as Factory Mutual or Underwriters Laboratory. Damaged equipment should be repaired properly before use.

No welding should take place until all electrical connections, power source, welding leads, welding machines, and workpiece clamps are secure, and the welding power source frame is well grounded. The workpiece clamp must be secure and the cable connecting it to the power supply must be in good condition. The power supply should be turned off any time it is left unattended. The line supply disconnect switch should also be placed in the "off" position. Electrical lockouts should be used when performing maintenance on equipment.

CONCLUSION

Gas tungsten arc welding produces weld quality equal or superior to other arc welding processes. The flexibility of GTAW and its ability to produce high-quality welds in most materials in any position result in nearly unlimited applications of the process. This makes GTAW an excellent choice for fabricators, especially for applications requiring precision joints and those using thin materials. Fabricators value the versatility of the process.

While gas tungsten arc welding compares favorably to other processes, the required precision and quantity often specified for welds can increase equipment and production costs. Thus, precise control over the process and the ability to achieve consistent results promote the extensive use of the gas tungsten arc welding process in industry.

BIBLIOGRAPHY[25]

American Conference of Governmental Industrial Hygienists. *TLVs and BEIs—Threshold limit values for chemical substances and physical agents and biologi-*

25. The dates of publication given for the codes and other standards listed here were current at the time this chapter was prepared. The reader is advised to consult the latest edition.

cal indices. Cincinnati, Ohio: American Conference of Governmental Industrial Hygienists.

American National Standards Institute (ANSI) Accredited Standards Committee Z49. 1999. *Safety in welding, cutting and allied processes*. ANSI Z41.l. 1999. Miami: American Welding Society.

American National Standards Institute (ANSI). *Practice for occupational and educational eye and face protection*. ANSI Z87.1. Des Plaines, Illinois: American Society of Safety Engineers.

American Welding Society (AWS) Committee on Definitions and Symbols. 2001. *Standard welding terms and definitions*. AWS A3.0:2001. Miami: American Welding Society.

American Welding Society (AWS) Committee on Filler Metals and Allied Materials. 1998. *Specification for tungsten and tungsten alloy electrodes for arc welding and cutting*. ANSI/AWS A5.12/A5.12M–1998. Miami: American Welding Society.

American Welding Society (AWS) Committee on Filler Metals and Allied Materials. 1980. *Recommended practices for gas tungsten arc welding*. AWS C5.5-80R. Miami: American Welding Society.

American Welding Society (AWS) Committee on Filler Metals and Allied Materials. 2000. *Filler metal comparison charts*. Miami: American Welding Society.

American Welding Society (AWS) Committee on Fumes and Gases. *Methods for sampling airborne particulates generated by welding and allied processes*. ANSI/AWS F1.1. Miami: American Welding Society.

American Welding Society (AWS) Committee on Safety and Health. *Lens shade selector,* ANSI/AWS F2.2. Miami: American Welding Society.

American Welding Society (AWS) Welding Handbook Committee. 2001. Jenney, C. L., and A. O'Brien, eds. *Welding science and technology.* Vol. 1 of *Welding handbook*. 9th ed. Miami: American Welding Society.

American Welding Society (AWS) Welding Handbook Committee. 1994. Oates, W. R., ed. *Materials and applications—Part 1*. Vol. 3 of *Welding handbook*. 8th ed. Miami: American Welding Society.

American Welding Society (AWS) Welding Handbook Committee. 1996. Oates, W. R., and A. M. Saitta, eds. *Materials and applications —Part 2*, Volume 4 of *Welding handbook*. 8th ed. Miami: American Welding Society.

Compressed Gas Association (CGA). *Safe handling of compressed gas in containers*. CGA P-1. Arlington, Virginia: Compressed Gas Association.

Occupational Safety and Health Administration (OSHA). 1999. *Occupational safety and health standards for general industry, Title 29 (Labor), Code of Federal Regulations*. (CFR). Chapter XVII, Parts 1901.1 to 1910.1450. Washington D.C.: Superintendent of Documents, U.S. Government Printing Office.

Ullrich, O. A., and R. M. Evans. 1976. *Ultraviolet Reflectance of Paint*. Miami: American Welding Society.

SUPPLEMENTARY READING LIST

Baeslack, W. A., III, and C. M. Banas. 1981. A comparative evaluation of laser and gas tungsten arc weldments in high temperature titanium alloys. *Welding Journal* 60(7): 121-s to 130-s.

Burgardt, P., and C. R Heiple. 1986. Interaction between impurities and welding variables in determining GTA weld shape. *Welding Journal* 65(6): 150-s to 156-s.

Correy, T. B., D. G. Atteridge, R. E. Page, and M. C. Wismer. 1986. Radio frequency-free arc starting in gas tungsten arc welding. *Welding Journal* 64(2): 33–37.

Crement, D. J. 1993. Narrow groove welding of titanium using the hot-wire gas tungsten arc process. *Welding Journal*, 76(4): 71–76.

Geidt, W. H., L. N. Tallerico, and P. W. Fuersbach. 1989. GTA welding efficiency: calorimetric and temperature field measurements. *Welding Journal* 68(1): 28-s to 34-s.

Haberman, R. 1987. GTAW torch performance relies on component materials. *Welding Journal* 66(12): 55–60.

Heiple, C. R., J. R. Roper, R. T. Stagner, and R. J. Aden. 1983. Surface active elements effects on the shape of GTA, laser, and electron beam welds. *Welding Journal* 62(3): 72-s to 77-s.

Kanne, W. R. 1988. Remote reactor repair: GTA weld cracking caused by entrapped helium. *Welding Journal* 67(8): 33–38.

Katoh, M., and H. W. Ken. 1987. Investigation of heat-affected zone cracking of GTA welds of Al-Mg-Si alloys using the varestraint test. *Welding Journal* 66(12): 360-s.

Key, J. F. 1980. Anode/cathode geometry and shielding gas interrelationships in GTAW. *Welding Journal* 59(12): 364-s to 370-s.

Kraus, H. G. 1989. Experimental measurement of stationary SS 304, SS 316L and 8630 GTA weld pool surface temperatures. *Welding Journal* 68(7): 269-s to 279-s.

Kujanpaa, V. P., L. P. Karjalainen, and H. A. V. Sikanen. 1984. Role of shielding gases in flaw formation in GTAW of stainless steel strips. *Welding Journal* 63(5): 151-s to 155-s.

Lu, M., and S. Kou. 1988. Power and current distributions in gas tungsten arcs. *Welding Journal* 67(2): 29-s to 36-s.

Malinowski-Brodnicka, M., G. den Ouden, and W. J. P. Vink. 1990. Effect of electromagnetic stirring on GTA welds in austenitic stainless steel. *Welding Journal* 69(2): 52-s to 59-s.

Metcalfe, J. C., and M. C. B. Quigley. 1977. Arc and pool instability in GTA welding. *Welding Journal* 56(5): 133-s to 139-s.

International Institute of Welding. 1984. *The Physics of Welding*. Ed. J. F. Lancaster. Oxford, U.K.: Pergamon Press.

Oomen, W. J., and P. A. Verbeek. 1984. A real-time optical profile sensor for robot arc welding. *Robotic Welding*. Bedford, U.K.: IFS Publications Ltd.

Patterson, R. A., R. B. Nemec, and R. D. Reiswig. 1987. Discontinuities formed in Inconel GTA welds. *Welding Journal* 65(1): 19-s to 25-s.

Pearce, C. H., et al. 1986. Development and applications of microprocessor controlled systems for mechanized TIG welding. *Computer Technology in Welding*. Cambridge, U. K.: The Welding Institute.

Saede, H. R., and W. Unkel. 1988. Arc and weld pool behavior for pulsed current GTAW. *Welding Journal* 67(11): 247-s.

Sicard, P., and M. D. Levine. 1988. *IEEE Transactions on Systems, Man, and Cybernetics*. 18(2). Piscataway, N.J.: Institute of Electrical and Electronic Engineers.

Salkin, J. T. 1997. Rotating tungsten narrow groove GTAW—a summary of process development, capabilities, and applications. *Proceedings of the EWI International Conference on Advances in Welding Technology.* Columbus, Ohio: Edison Welding Institute.

Smith, J. S., et al. 1986. A vision-based seam tracker for TIG welding. *Computer Technology in Welding*. Cambridge, U. K.: The Welding Institute.

Troyer, W., M. Tomsik, and R. Barhorst. 1977. Investigation of pulsed wave shapes for gas tungsten arc welding. *Welding Journal* 56(1): 26–32.

Voigt, R. C., and Loper, C. R., Jr. 1980. Tungsten contamination during gas tungsten arc welding. *Welding Journal* 59(4): 99-s to 103-s.

Villafuerte, J. C., and H. W. Kerr. 1990. Electromagnetic stirring and grain refinement in stainless steel GTA welds. *Welding Journal* 69(1): 1-s.

Walsh, D. W., and W. F. Savage. 1985. Autogenous GTA weldments-bead geometry variations due to minor elements. *Welding Journal* 64(2): 59-s to 62-s.

Walsh, D. W., and W. F. Savage. 1985. Bead shape variance in AISI 8630 steel GTAW weldments. *Welding Journal* 64(5): 137-s to 139-s.

Zacharia, T., S. A. David, J. M. Vitek, and T. DebRoy. 1989. Weld pool development during GTA and laser beam welding of type 3 or 4 stainless steel—part 1 and part 2. *Welding Journal* 68(12): 499-s to 510-s.

GAS METAL ARC WELDING

Prepared by the Welding Handbook Chapter Committee on Gas Metal Arc Welding:

D. B. Holliday, Chair
Northrop Grumman Marine Systems

R. M. Dull
Edison Welding Institute

D. K. Hartman
The Lincoln Electric Company

D. A. Wright
Zephyr Products, Incorporated

Welding Handbook Volume 2 Committee Member:

B. R. Somers
Lucius Pitkin, Incorporated

Contents

Photograph courtesy of The Lincoln Electric Company

CHAPTER 4

GAS METAL ARC WELDING

INTRODUCTION

Gas metal arc welding (GMAW) is an arc welding process that uses an arc between a continuous filler metal electrode and the weld pool. The process incorporates shielding from an externally supplied gas and is used without the application of pressure.

The basic concept of gas metal arc welding was introduced in the 1920s; however, the process did not become commercially available until 1948. It was initially implemented as a high-current-density, small-diameter, bare-metal-electrode process using an inert gas for arc shielding. As a result, the process was called *MIG* (from "metal inert gas") *welding*, still a designation today, although colloquial.

Gas metal arc welding was primarily used for welding aluminum, but subsequent process developments included operation at low current densities and pulsed current, application to a broader range of materials, and the use of reactive gases (particularly carbon dioxide) and gas mixtures. The latter development led to the formal acceptance of the term *gas metal arc welding* for the process because both inert and reactive gases can be used.

Gas metal arc welding is used for a wide variety of applications in the fields of industrial manufacturing, agriculture, construction, shipbuilding and mining. The process is used to weld pipe, pressure vessels, structural steel components, furniture, automotive components and numerous other products.

Gas metal arc welding is described in detail following an examination of the fundamentals of the process. Topics include variations of the process, equipment, materials and consumables, welding variables, and process capabilities. The chapter concludes with a discussion of weld quality, troubleshooting, economics, and safe practices.

FUNDAMENTALS

Gas metal arc welding can be implemented in semiautomatic and automated operations. All commercially important metals—including carbon steel, high-strength low-alloy steel, stainless steel, aluminum, copper, titanium, and nickel alloys—can be welded in all positions with this process by choosing the appropriate combination of shielding gas, electrodes, and welding variables.

ADVANTAGES

Because of its versatility, gas metal arc welding has become more widely used and has replaced shielded metal arc welding for many applications. This increased usage can be attributed to its many advantages, the most important of which include the following:

1. It is an efficient consumable-electrode process that can be used to weld all commercial metals and alloys;
2. It overcomes the restriction of limited electrode length encountered with shielded metal arc welding;
3. Welding can be performed in all positions, a capability submerged arc welding does not have;
4. Deposition rates are significantly higher than those obtained with shielded metal arc welding;
5. Welding speeds are higher than those attained with shielded metal arc welding because of the continuous electrode feed and higher filler metal deposition rates;
6. Because the electrode (wire feed) is continuous, long welds can be deposited without intermediate stops and starts;
7. When spray transfer is used, deeper penetration is possible than with shielded metal arc welding, often permitting the use of smaller-sized fillet welds for equivalent joint strengths;
8. Minimal postweld cleaning is required due to the absence of heavy slag;
9. It is a low-hydrogen process, making it a good choice for welding materials that are susceptible to hydrogen embrittlement; and
10. Process skills are readily taught and acquired.

These advantages make the process particularly well suited to high-production and automated welding applications. This has become evident with the increased use of robotics, for which gas metal arc welding is the predominant process choice. A typical robotic GMAW welding installation is pictured in Figure 4.1.

LIMITATIONS

As with any welding process, certain limitations restrict the use of gas metal arc welding. The welding equipment required for gas metal arc welding is more complex, more costly, and less portable than that for shielded metal arc welding, for example. Gas metal arc welding is difficult to use in hard-to-reach places because the welding gun is larger than a shielded metal arc electrode holder and the welding gun must be close to the joint, (i.e., between 10 millimeters (mm) and 19 mm (3/8 inch [in.] and 3/4 in.) to ensure that the weld metal is properly shielded. The welding arc must be protected against air drafts in excess of 5 miles per hour (mph), which may disperse the shielding gas. This limits outdoor applications unless protective shields are

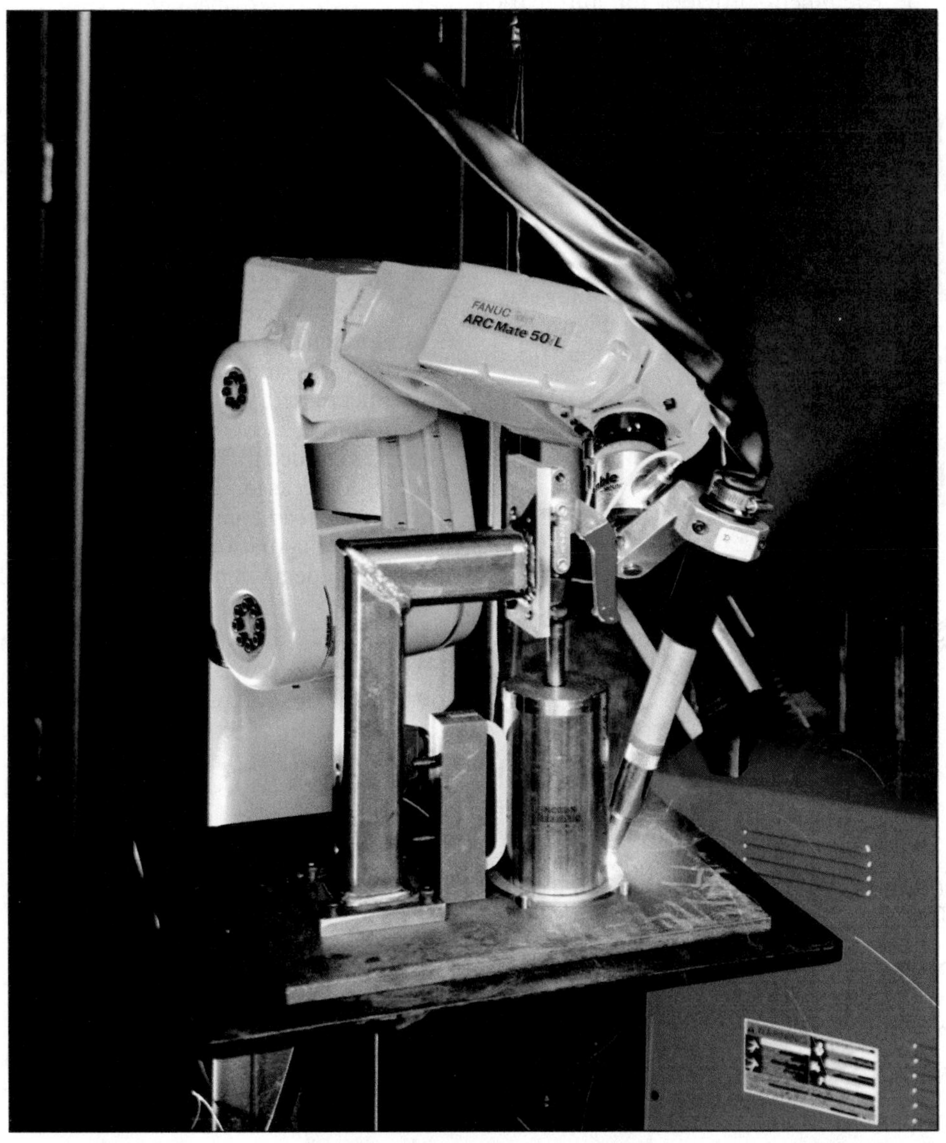

Photograph courtesy of Fanuc Robotics North America and The Lincoln Electric Company

Figure 4.1—Typical Robotic Gas Metal Arc Welding Installation

placed around the welding area. Another limitation involves the relatively high levels of radiated heat and arc intensity, which can contribute to resistance on the part of operators to accept the process.

PRINCIPLES OF OPERATION

The gas metal arc welding process utilizes the automated feeding of a continuous, consumable electrode that is shielded by an externally supplied gas. The process is illustrated schematically in Figure 4.2. After the initial settings on the welding machine are established by the welder, the equipment provides for automatic self-regulation of the electrical characteristics of the arc. Therefore, the only actions controlled by the welder for semiautomatic operations are the travel speed and direction and the gun positioning. Given the proper equipment and settings, the arc voltage and the wire feed speed (current) are automatically maintained.

The basic equipment components required for gas metal arc welding are a welding gun and cable assembly, an electrode (wire) feed unit, a welding power source, and the apparatus for delivery of shielding gas. This equipment is shown schematically in Figure 4.3 and described in more detail in the section titled "Equipment."

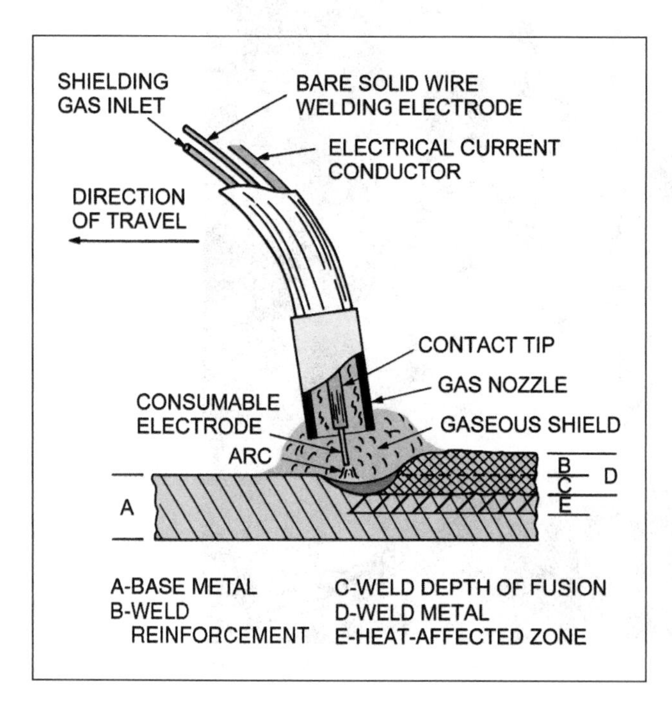

Figure 4.2—Gas Metal Arc Welding Process

The gun guides the consumable electrode, conducts the electrical current and directs shielding gas to the workpiece. This provides the means to establish and maintain the arc, melt the electrode, and provide the needed protection from the ambient atmosphere. One of two combinations of electrode feed units and power sources may be used to achieve the desirable self-regulation of arc length. Most commonly, a constant-voltage power source (characteristically providing an essentially flat volt-ampere [V-A] curve) is used in conjunction with a constant-speed electrode feed unit to regulate arc length. Alternatively, a constant-current power source (providing a drooping V-A curve) is coupled with an electrode feed unit that is arc-voltage controlled.

With the constant-voltage, constant wire-feed combination, changes in the gun position cause a change in the welding current due to a change in the electrode extension. For example, when the gun-to-workpiece distance is suddenly increased, the arc length momentarily becomes longer. The longer arc length causes a reduction in current, momentarily reducing the electrode melt-off rate. Since the feed rate remains the same, the arc length decreases and the current increases until the melt-off rate again equals the feed rate. The final current at the longer extension is lower than that resulting from the shorter extension with a slightly longer arc length. In effect, the resistance heating of the electrode extension has increased and arc heating at the tip has decreased. As a consequence, the arc heating of the workpiece, as well as weld penetration, is decreased. The opposite occurs if the gun-to-workpiece distance is suddenly decreased. The constant-current, arc-voltage-control wire feed combination results in self-regulation: arc voltage fluctuations readjust the control circuits of the wire feeder, which appropriately adjusts the wire feed speed.

In some cases (e.g., when welding aluminum), it may be preferable to deviate from these standard combinations and couple a constant-current power source with a constant-speed electrode feed unit. This combination provides only a small degree of automatic self-regulation and therefore requires more operator skill in semiautomatic welding. However, some users suggest that this combination affords a range of control over the arc energy (current) that may be important in coping with the high thermal conductivity of aluminum base metals.

METAL TRANSFER MODES

The characteristics of the gas metal arc welding process are best described in terms of the basic means by which metal is transferred from the electrode to the workpiece. The modes of metal transfer for gas metal arc welding are short-circuiting transfer, globular transfer, and spray transfer. The mode of transfer is deter-

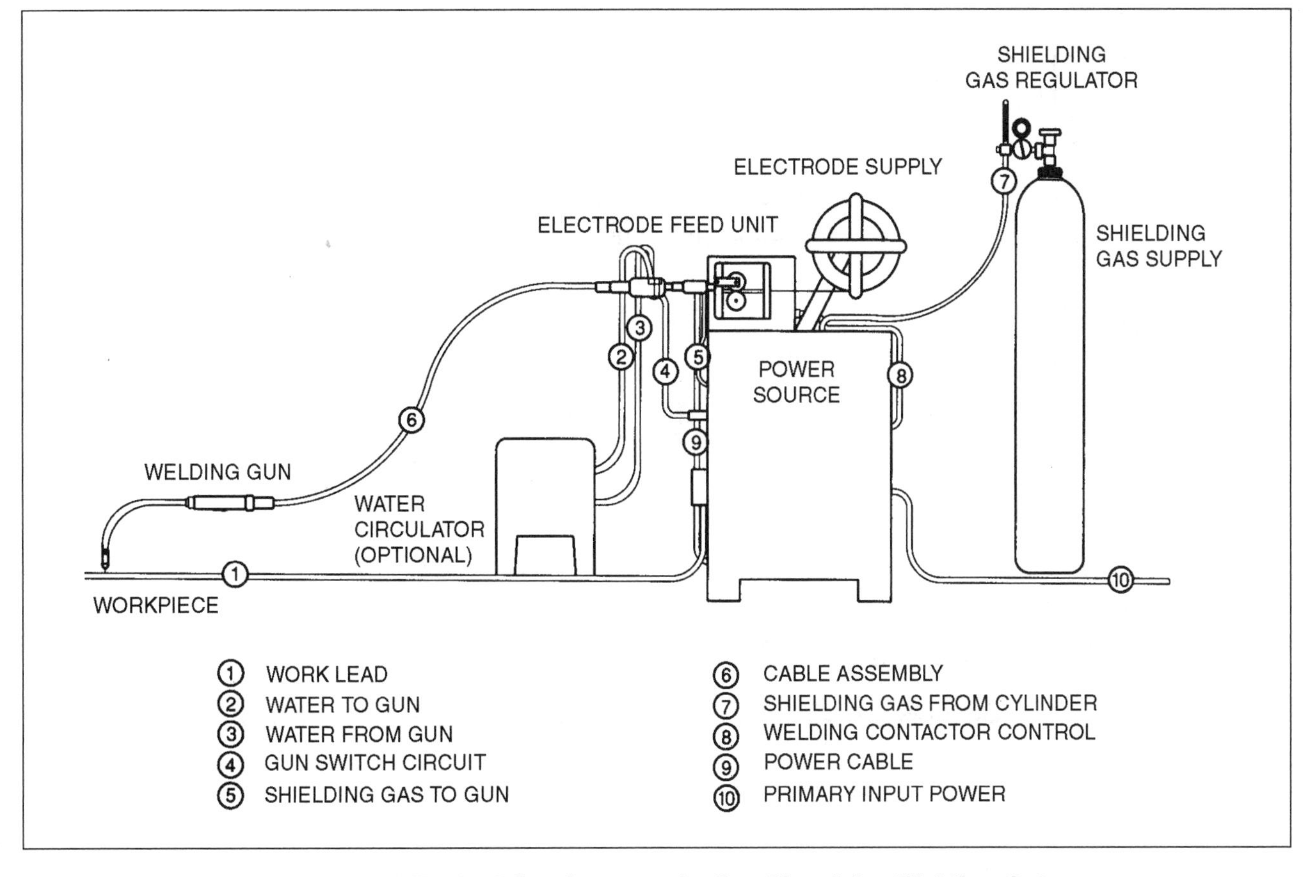

Figure 4.3—A Typical Semiautomatic Gas Metal Arc Welding Setup

mined by a number of factors, the most influential of which are the following:

1. Magnitude, type, and polarity of welding current,
2. Electrode diameter,
3. Electrode composition,
4. Electrode extension, and
5. Shielding gas composition.

Short-Circuiting Transfer

Short-circuiting transfer, employed in short-circuit gas metal arc welding (GMAW-S), encompasses the lowest range of welding currents and electrode diameters associated with gas metal arc welding. Metal transfer results when the molten metal from a consumable electrode is deposited during repeated short- circuits. This mode of transfer produces a small, fast-freezing weld pool that is generally suited for the joining of thin sections, for out-of-position welding, and for bridging large root openings.

In short-circuiting transfer, metal is transferred from the electrode to the workpiece only during the period in which the electrode is in contact with the weld pool. No metal is transferred across the arc. The electrode contacts the weld pool in a range of 20 to over 200 times per second.

The sequence of events in the transfer of metal and the corresponding current and voltage levels are shown in Figure 4.4. As the wire touches the weld metal, the current increases, as shown in Figure 4.4(A, B, C, and D). The molten metal at the wire tip pinches off at (D) and (E), initiating an arc as shown in (E) and (F). The rate of current increase must be high enough to heat the electrode and promote metal transfer, yet low enough to minimize the spatter caused by the violent separation of the drop of metal. This rate of current increase is controlled by adjusting the inductance in the power source.

The optimum inductance setting depends on both the electrical resistance of the welding circuit and the melting temperature of the electrode. When the arc is established, the wire melts at the tip as the wire is fed forward toward the next short-circuit, as indicated by

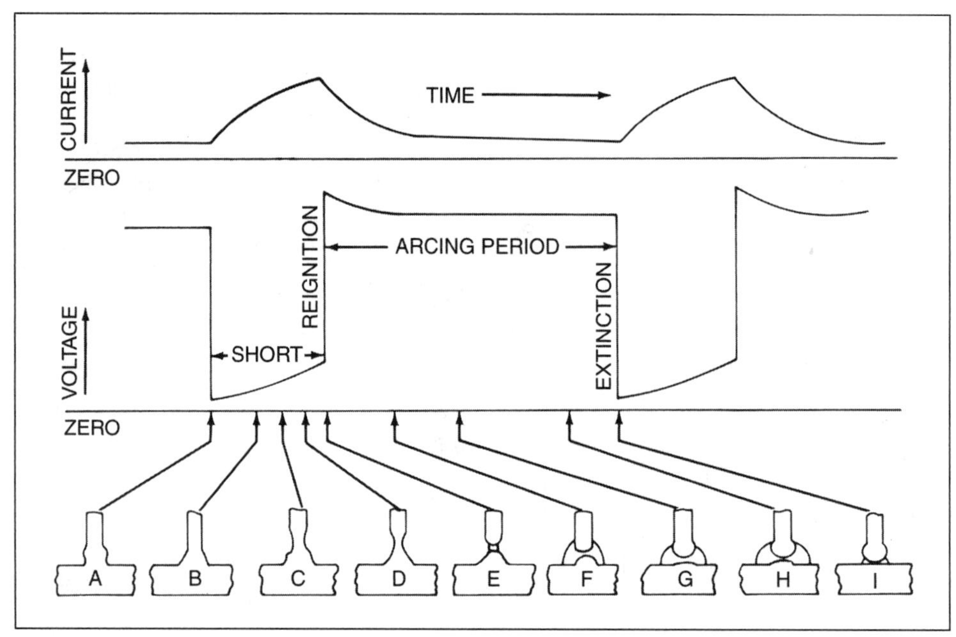

Source: Adapted from American Welding Society (AWS) Committee on Arc Welding and Cutting, 1994, *Recommended Practices for Gas Metal Arc Welding*, ANSI/AWS C5.6-94R, Miami: American Welding Society, Figure 6.

Figure 4.4–Schematic Representation of Short-Circuiting Metal Transfer

Figure 4.4(G, H, and I). The open-circuit voltage of the power source must be low enough that the drop of molten metal at the wire tip cannot transfer until it touches the base metal. The energy for arc maintenance is provided partly by the energy stored in the inductor during the period of short-circuiting.

Even though metal transfer occurs only during short-circuiting, the composition of the shielding gas used has a dramatic effect on the surface tension of the molten metal. Changes in the composition of the shielding gas may significantly affect the drop size and the duration of the short-circuit. In addition, the type of gas used influences the operating characteristics of the arc and the penetration into the base metal. For example, carbon dioxide generally produces high spatter levels compared to argon and helium, but it also promotes deeper penetration. To achieve a good compromise between spatter and penetration when welding carbon and low-alloy steels, mixtures of carbon dioxide and argon are often used. Additions of helium to argon may increase penetration in nonferrous metals and may promote the wetting of the base metal by the filler metal.

Globular Transfer

The globular transfer mode involves the transfer of molten metal in the form of large drops from the consumable electrode across the arc. This transfer mode is characterized by a drop size with a diameter greater than that of the electrode. This large drop is easily acted upon by gravity, generally limiting the successful application of this mode of transfer to the flat position.

At average current ranges that are only slightly higher than those used in short-circuiting transfer, axially directed globular transfer can be achieved in a substantially inert gas shield. If the arc length is too short (indicating low voltage), the enlarging drop may short to the workpiece, become superheated, and disintegrate, producing considerable spatter. The arc must therefore be long enough to ensure the detachment of the drop before it contacts the weld pool. However, a weld made with a higher voltage is likely to be unacceptable because of incomplete fusion, incomplete joint penetration, and excessive weld reinforcement. This characteristic greatly limits the use of the globular transfer mode in production applications.

Carbon dioxide shielding results in nonaxially directed globular transfer when the welding current and voltage are significantly above the range required for short-circuiting transfer. The departure from axial transfer is governed by electromagnetic forces that are generated by the welding current acting upon the molten tip, as shown in Figure 4.5. The most important of these are (1) the electromagnetic pinch force (P), that

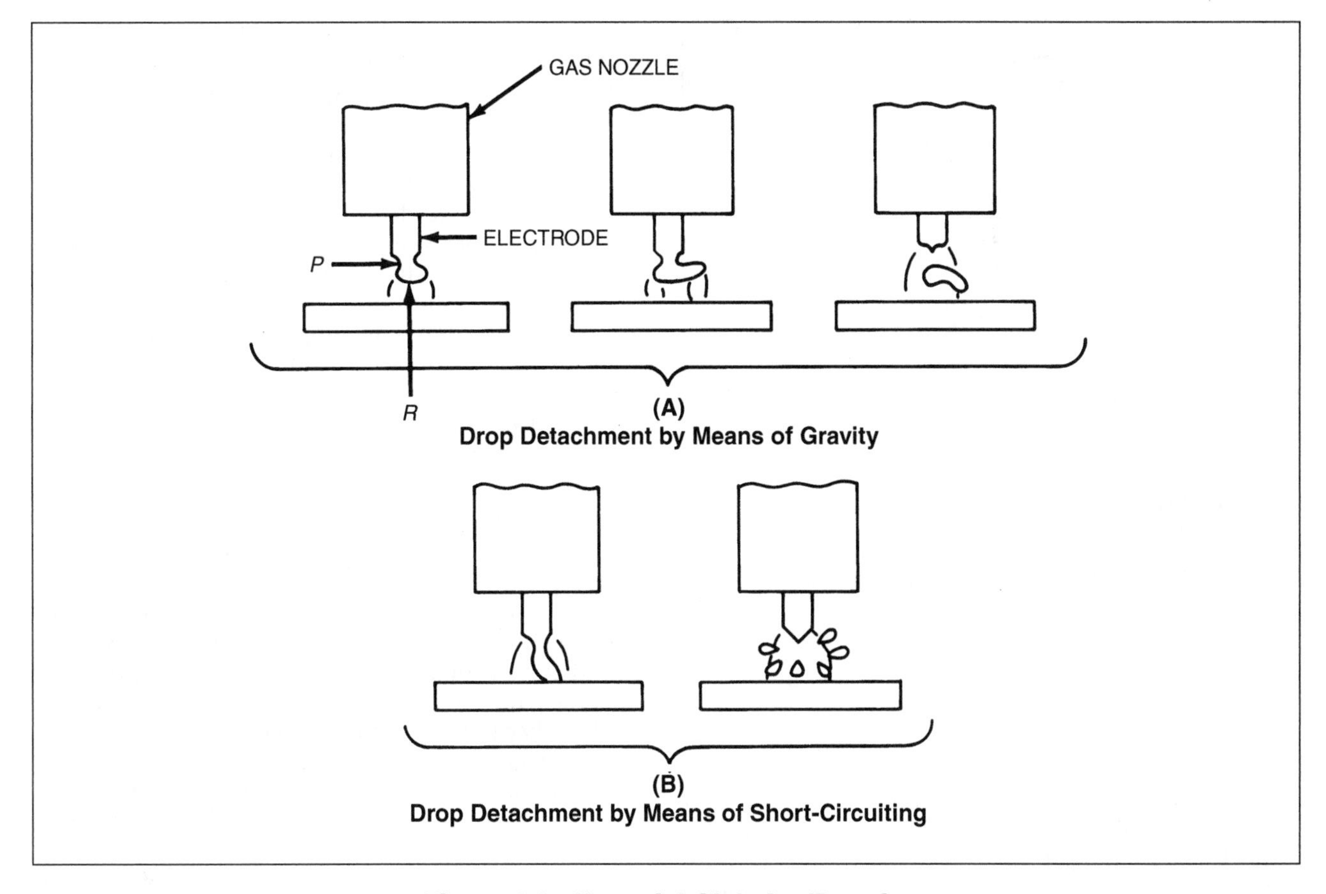

Figure 4.5—Nonaxial Globular Transfer

results in the momentary necking of the drop from the electrode because of the electromagnetic effects of the current, and (2) the anode reaction force (*R*).

The magnitude of the pinch force, which is a direct function of welding current and wire diameter (i.e., current density), is usually responsible for drop detachment. With carbon dioxide shielding, the welding current is conducted through the molten drop, and the arc plasma does not envelop the electrode tip. High-speed photography has revealed that the arc moves over the surface of the molten drop and the workpiece because the anode reaction force, *R*, tends to support the drop.

The molten drop grows until it detaches by means of short-circuiting, as shown in Figure 4.5(B), or gravity, as depicted in Figure 4.5(A), because the anode reaction force, *R*, is never overcome by the electromagnetic pinch force, *P*, alone. As shown in Figure 4.5(A), it is possible for the drop to become detached and transfer to the weld pool without disruption. The most likely situation is shown in Figure 4.5(B), which depicts the drop short-circuiting the arc column and exploding. Spatter can therefore be severe, limiting the use of carbon dioxide shielding for many commercial applications.

Nevertheless, carbon dioxide is commonly used for the welding of mild steels, since the spatter problem can be reduced significantly by "burying" the arc beneath the surface of the material being welded. In so doing, the arc's atmosphere becomes a mixture of the gas and iron vapor, allowing the transfer to become almost spray-like. The arc forces are sufficient to maintain a depressed cavity that traps much of the spatter. This technique requires a higher welding current and results in deep penetration. However, unless the arc voltage and travel speed are carefully controlled, poor wetting action may result in excessive weld reinforcement and a weld bead with a rope-like appearance.

Spray Transfer

The spray transfer mode occurs when the molten metal from a consumable electrode is propelled axially across the arc in the form of minute droplets, as illustrated in Figure 4.6. With argon-rich (at least 80%) gas

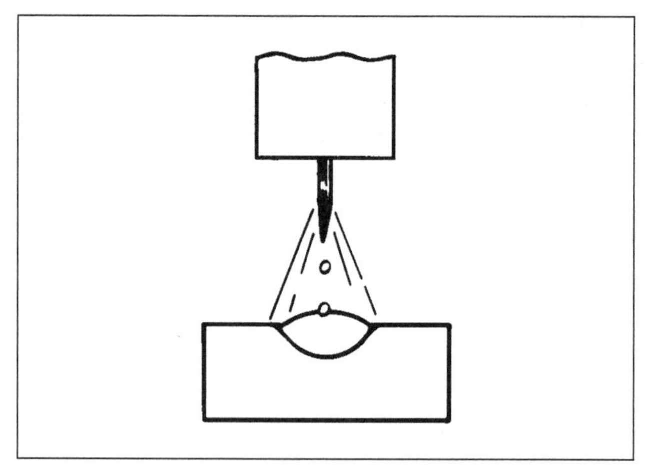

Source: Adapted from American Welding Society (AWS) Committee on Arc Welding and Cutting, 1994, *Recommended Practices for Gas Metal Arc Welding,* ANSI/AWS C5.6-94R, Miami: American Welding Society.

Figure 4.6—Axial Spray Transfer

shielding, it is possible to produce a very stable, spatter-free axial spray transfer mode.

This mode requires the use of direct current with a positive electrode (DCEP) and a current level above a critical value, termed the *spray transition current.* Below this current level, transfer occurs in the globular mode (described previously) at the rate of a few drops per second. Above the transition current, transfer occurs in the form of very small drops that are formed and detached at the rate of hundreds per second. They are accelerated axially across the arc. The relationship between transfer rate and current is plotted in Figure 4.7.

The transition current, which is dependent on the surface tension of the liquid metal, is inversely proportional to the electrode diameter and, to a lesser degree, to the electrode extension. It varies with the melting temperature of the filler metal and the composition of the shielding gas. Typical transition currents for some of the more common metals are shown in Table 4.1.

The spray transfer mode results in a highly directed stream of discrete drops that are accelerated by arc forces at velocities that overcome the effects of gravity. As a result, this process can be used in any welding position under certain conditions, although it is generally limited to the flat and horizontal positions. Because the drops are smaller than the arc length, short-circuits do not occur, and spatter is negligible if not totally eliminated. Another characteristic of the spray transfer mode is the narrow, deep finger-like penetration that it produces. Although this "finger" can be deep, it is affected by magnetic fields, which must be controlled to keep it located at the center of the weld joint profile.

The spray-arc transfer mode can be used to weld most metals and alloys because of the inert characteristics of the argon shield. However, applying this process variation to thin sheet metal may be difficult because of the high currents needed to produce the spray arc. The resultant arc forces may cut through relatively thin sheets instead of welding them. In addition, the characteristically high deposition rate may produce a weld pool that is too large to be supported by surface tension in the vertical or overhead positions.

The limitations of the spray arc transfer mode that are related to the thickness of the workpiece and the welding position have been largely overcome with the application of specially designed power sources. These machines produce precisely controlled waveforms and frequencies that "pulse" the welding current and voltage.

As shown in Figure 4.8, the power sources provide two levels of current. One level is a constant, low background current that sustains the arc without providing enough energy to cause drops to form on the electrode tip; the other is a superimposed pulsing current with an amplitude greater than that of the transition current necessary for spray transfer. During this pulse, one or more drops are formed and transferred. The frequency and amplitude of the pulses control the energy level of the arc and therefore the rate at which the wire melts. This allows spray transfer to occur at lower average current levels (see Table 4.1). By reducing the average arc energy and the wire melting rate, pulsing makes the desirable features of spray transfer available for the welding of sheet metals and thick metals in all positions.

Many variations of these power sources are available. The simplest of these provide a single frequency of pulsing (60 pulses per second [pps] or 120 pps) with independent control of the background and pulsing current levels. More sophisticated power sources, sometimes referred to as *synergic,* automatically provide the optimum combination of background and pulse currents for any given setting of wire feed speed for selected combinations of material type, shielding gas, and electrode diameter.

When pulsed power sources are used, the process is designated pulsed gas metal arc welding (GMAW-P). When operated within carefully controlled parameter limits, this process variation has wide application because of its many advantages, including the following:

1. It offers reduced spatter levels compared to short-circuiting and globular transfer, thereby increasing the deposition rate and minimizing postweld cleanup;

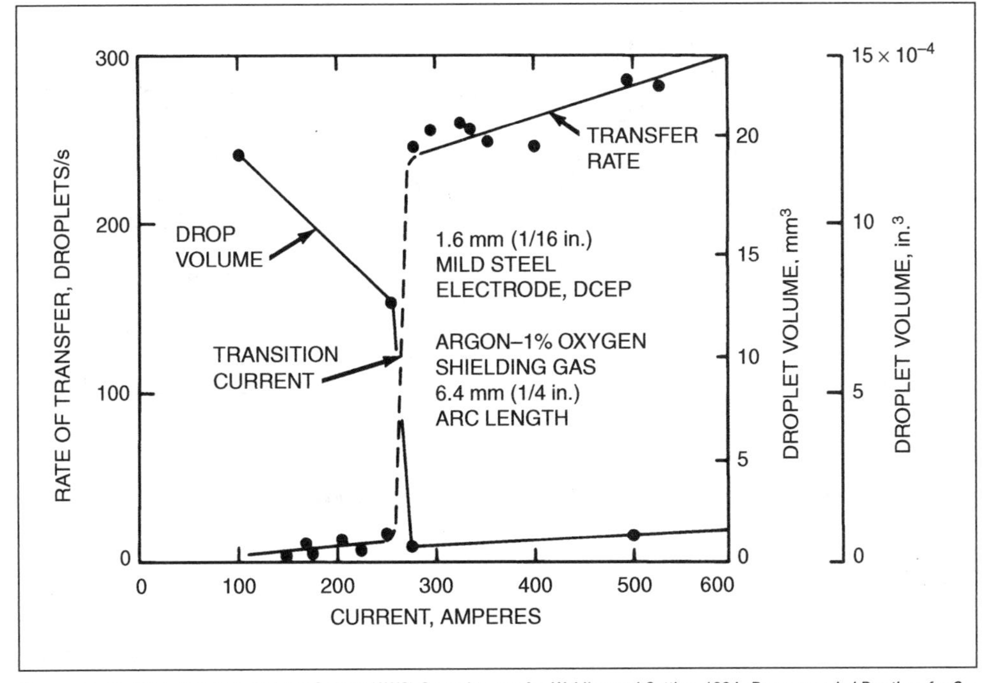

Source: Adapted from American Welding Society (AWS) Committee on Arc Welding and Cutting, 1994, *Recommended Practices for Gas Metal Arc Welding*, ANSI/AWS C5.6-94R, Miami: American Welding Society, Figure 4.

Figure 4.7—Variation in Volume and Transfer Rate of Drops with Welding Current (Steel Electrode)

Table 4.1
Globular-to-Spray Transition Currents for a Variety of Electrodes

Electrode Type	Wire Electrode Diameter		Shielding Gas	Minimum Spray Arc Current, A	Minimum Average Pulse Spray Current, A
	mm	in.			
Mild steel	0.8	0.030	98% argon–2% oxygen	150	—
Mild steel	0.9	0.035	98% argon–2% oxygen	165	48
Mild steel	1.1	0.045	98% argon–2% oxygen	220	68
Mild steel	1.6	0.062	98% argon–2% oxygen	275	—
Stainless steel	0.9	0.035	98% argon–2% oxygen	170	57
Stainless steel	1.1	0.045	98% argon–2% oxygen	225	104
Stainless steel	1.6	0.062	98% argon–2% oxygen	285	—
Aluminum	0.8	0.030	Argon	95	—
Aluminum	1.1	0.045	Argon	135	44
Aluminum	1.6	0.062	Argon	180	84
Deoxidized copper	0.9	0.035	Argon	180	—
Deoxidized copper	1.1	0.045	Argon	210	—
Deoxidized copper	1.6	0.062	Argon	310	—
Silicon bronze	0.9	0.035	Argon	165	107
Silicon bronze	1.1	0.045	Argon	205	133
Silicon bronze	1.6	0.062	Argon	270	—

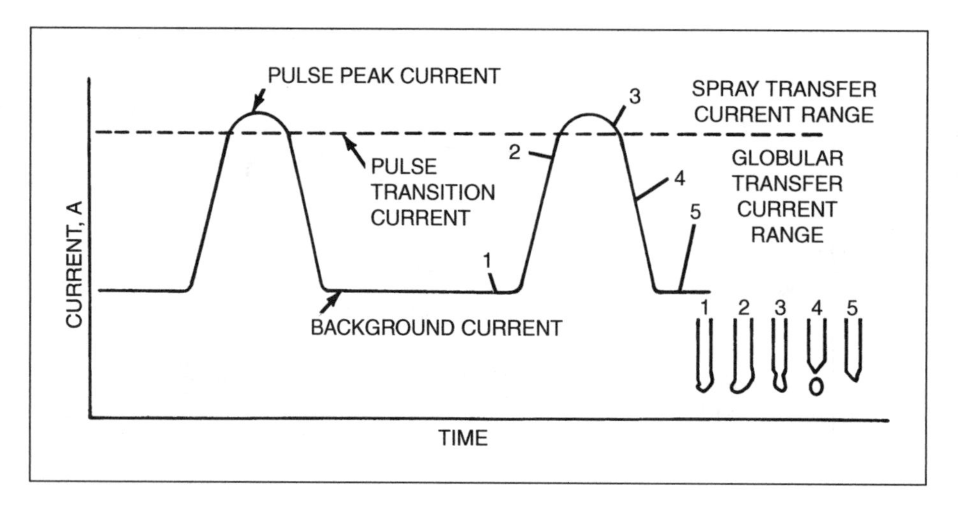

Figure 4.8—Pulsed Spray Arc Welding Current Characteristics

2. It generates lower fume levels, providing a healthier environment for the welder;
3. It provides a more controlled heat input, resulting in less distortion and improved weld quality for materials sensitive to heat input (e.g., high-strength, low-alloy steels, stainless steel, and nickel alloys); and
4. It can be used in place of GMAW-S for applications requiring low heat input with less propensity for incomplete fusion.

PROCESS VARIATIONS

The gas metal arc process has been adapted to provide specific characteristics for a wide range of applications. Some of these variations are discussed in the following sections.

Gas Metal Arc Spot Welding

Gas metal arc spot welding is a technique in which two overlapping workpieces are fused together by the penetration of the arc through one piece into the one lying behind it. The melted welding electrode and the base metals form a single nugget of solidified weld metal. The technique has been used for the joining of light-gauge materials up to approximately 5 mm (3/16 in.) thick in the production of automobile bodies, appliances, and electrical enclosures. This thickness limitation only applies to the member through which the melting occurs.

No joint preparation is required other than cleaning the overlapping areas. Heavier sections can also be welded with a similar technique by drilling or punching a hole in the upper workpiece. The arc is directed through this hole to fuse the upper workpiece to the underlying member. The type of weld produced with this technique is referred to as a *plug weld.*

A comparison between a gas metal arc spot weld and a resistance spot weld is shown in Figure 4.9. Resistance spot welds are made by applying resistance heating and electrode pressure, which melt the two components at their interface and fuse them together. In the gas metal arc spot weld, the arc penetrates through the top member and fuses the bottom component into its weld pool. An important advantage of the gas metal arc spot weld is that access to only one side of the joint is necessary.

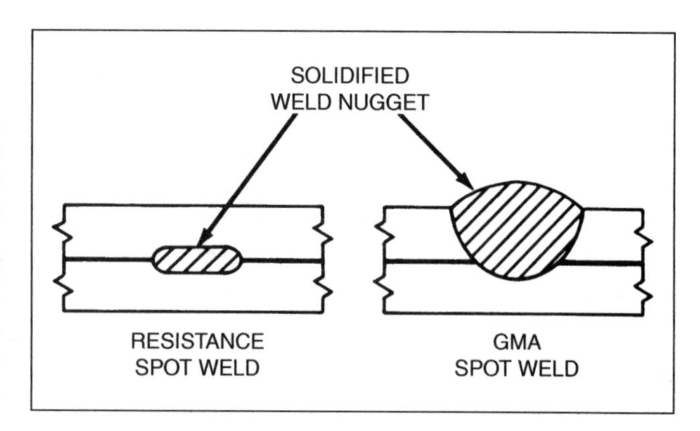

Figure 4.9—Comparison of Gas Metal Arc and Resistance Spot Welds

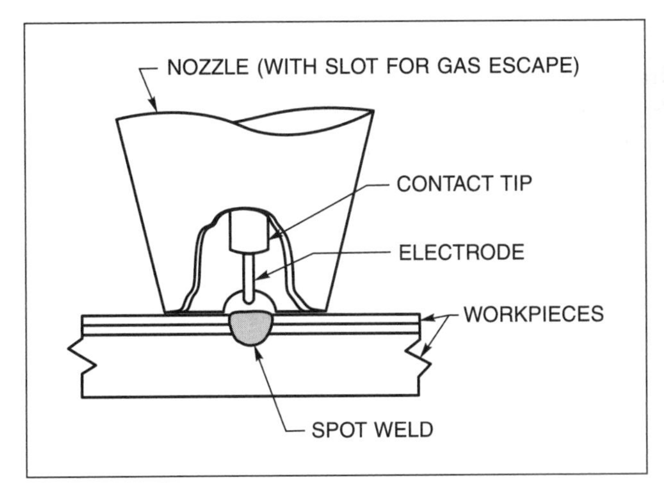

Figure 4.10—Gun Nozzle in Position for Gas Metal Arc Spot Welding

Equipment and Principles of Operation

The gas metal arc spot welding technique requires some modifications to conventional GMAW equipment. Special nozzles that have ports to allow the shield gas to escape as the gun is pressed to the workpiece are used. Also necessary are timers and wire feed speed controls to provide regulation of the actual welding time and a current decay period to fill the weld crater, leaving a desirable reinforcement contour.

The nozzle of the gas metal arc spot welding gun is brought into contact with the workpieces and a light pressure is applied to bring the two components together. The position of the nozzle relative to the workpieces is shown in Figure 4.10. The trigger of the gun is depressed to initiate the arc. The arc timer is started by a device that senses flow of welding current. The arc is maintained by the continuously fed consumable electrode until it melts through the top sheet and fuses into the bottom sheet without gun travel. The time cycle is set to maintain an arc until the melt-through and fusing sequence are complete, that is, until a spot weld has been formed. The electrode, which continues to feed during the arc cycle, should produce weld reinforcement on the upper surface of the top sheet.

Process Variables and Weld Quality

The weld diameter at the interface and the reinforcement are the two characteristics of a GMAW spot weld that determine whether the weld satisfies the intended service requirements. Three major process variables—weld current, voltage, and arc time—affect one or both of these characteristics.

Current. The welding current has the greatest effect on penetration. Penetration is increased by using higher currents with a corresponding increase in wire feed speed. Increased penetration typically results in a larger weld diameter at the interface.

Arc Voltage. The arc voltage has the greatest effect on spot weld shape. In general, when the current is held constant, an increase in the arc voltage augments the diameter of the fusion zone. However, it also causes a slight decrease in the weld reinforcement height and penetration. Welds made with arc voltages that are too low show a depression in the center of the reinforcement. Arc voltages that are too high create heavy spatter conditions.

Weld Time. The selected welding conditions should be those that produce a suitable weld within a time of 20 cycles to 100 cycles of 60 hertz (Hz) current (0.3 seconds [s] to 1.7 s) to join base metal up to 3.2 mm (0.125 in.) thick. An arc time up to 300 cycles (5 s) may be necessary on thicker materials to achieve adequate strength. The penetration, weld diameter, and weld reinforcement height generally increase with increased weld time.

As with conventional gas metal arc welding, the parameters for spot welding are very interdependent. Changing one usually requires the modification of one or more of the others. Some trial and error is needed to find a set or sets of conditions for a particular application. The starting parameters for the gas metal arc spot welding of carbon steel are shown in Table 4.2.

Process Capabilities

Gas metal arc spot welding can be used to weld lap joints in carbon steel, aluminum, magnesium, stainless steel, and copper alloys. Metals of the same or different thicknesses can be welded, but the thinner sheet should always be the top member when different thicknesses are welded. Gas metal arc spot welding is normally restricted to the flat position. By modifying the nozzle design, it may be adapted to spot weld lap-fillet, fillet, and corner joints in the horizontal position.

NARROW GROOVE GAS METAL ARC WELDING

Narrow groove gas metal arc welding is a multipass gas metal arc welding technique used to join heavy-section materials when the weld joint has a nearly square butt configuration with a minimal groove width (approximately 13 mm [1/2 in.]). A typical narrow groove joint configuration is shown in Figure 4.11. This technique is an efficient method for joining

Table 4.2
Variable Settings for GMAW Spot Welding of Carbon Steel in the Flat Position
(Carbon Dioxide Shielding Gas—6.4 mm [1/4 in.] Diameter Nugget)

Electrode Size			Material Thickness		Arc Spot Time	Current*	Voltage*
mm	in.	Gauge	mm	in.	s	A	V
0.8	0.030	24	0.56	0.022	1	90	24
		22	0.81	0.032	1.2	120	27
		20	0.94	0.037	1.2	120	27
0.9	0.035	18	0.99	0.039	1	190	27
		16	1.50	0.059	2	190	28
		14	1.83	0.072	5	190	28
1.2	0.045	14	1.83	0.072	1.5	300	30
		12	2.79	0.110	3.5	300	30
		11	3.15	0.124	4.2	300	30
1.6	0.063	11	3.15	0.124	1	490	32
			4.0	0.156	1.5	490	32

*Direct current electrode positive.

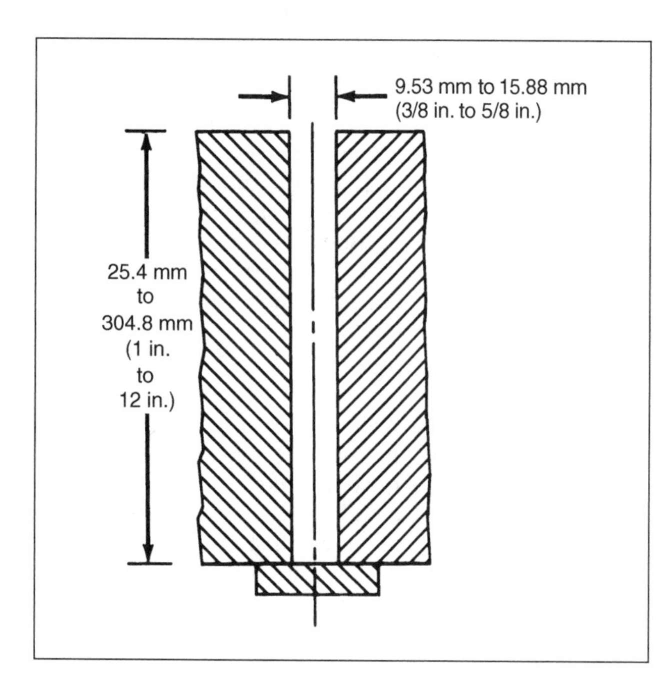

Figure 4.11—Typical Joint Configuration for Narrow Groove Gas Metal Arc Welding

heavy-section carbon and low-alloy steels with minimal distortion.

The use of gas metal arc welding to weld joints with the narrow groove technique requires special precautions to assure that the tip of the electrode is positioned accurately for proper fusion into the groove faces. Numerous wire feeding methods have been devised and successfully used in production environments. Examples of some of these are shown in Figure 4.12.

As shown in Figure 4.12(A), two wires with controlled cast and two contact tips can be used in tandem. The arcs are directed toward each groove face, producing a series of overlapping fillet welds. The same effect can be achieved with one wire by means of a weaving technique, which involves oscillating the arc across the groove in the course of welding. This oscillation can be created mechanically by moving the contact tip across the groove, as depicted in Figure 4.12(B). However, because of the small contact tip-to-groove face distance, this technique is impractical and seldom used.

Another mechanical technique uses a contact tip bent to an angle of about 15°, as illustrated in Figure 4.12(C). Along with a forward motion during welding, the contact tip twists to the right and left, giving the arc a weaving motion.

A more sophisticated technique is illustrated in Figure 4.12(D). During feeding, this electrode is formed into a waved shape by the bending action of a flapper plate and feed rollers as they rotate. The wire is continuously deformed plastically into this waved shape as the feed rollers press it against the bending plate. The electrode is almost straightened while passing through the contact tip, but it recovers its waviness after having passed through the tip. The continuous consumption of the waved electrode oscillates the arc from one side of the groove to the other. This technique produces an oscillating arc even in a very narrow groove with the contact tip remaining centered in the joint.

The twist electrode technique, shown in Figure 4.12(E), is another method that was developed to improve groove face penetration without moving the

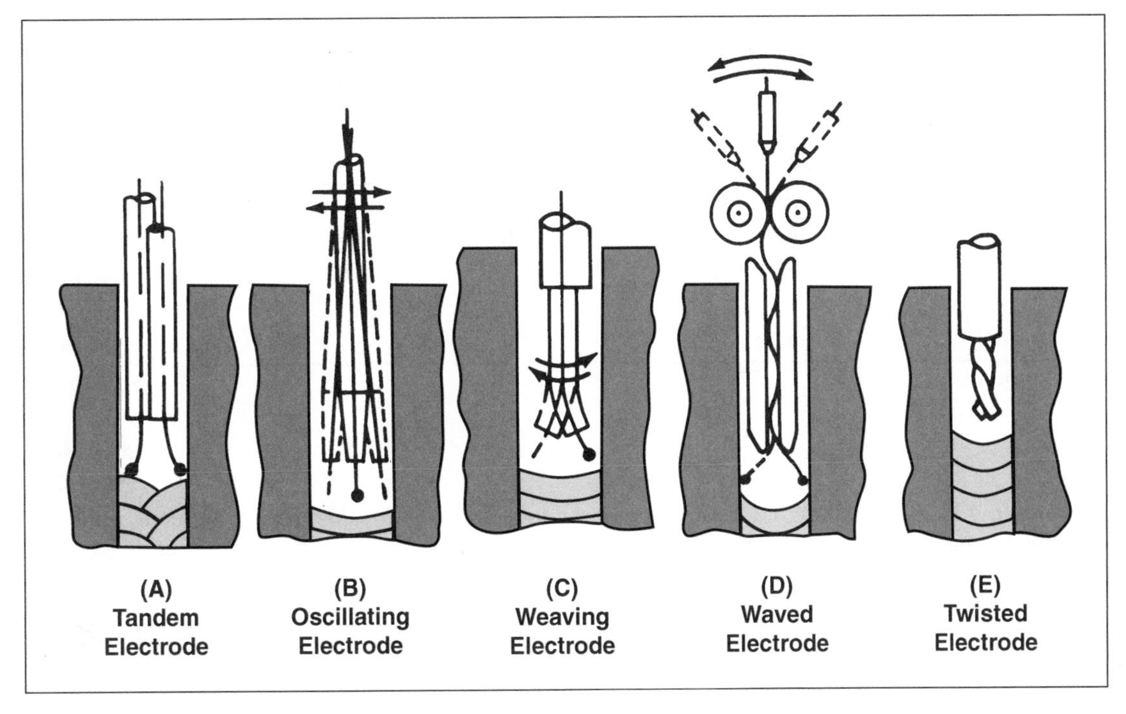

Figure 4.12—Typical Wire Feeding Techniques for Narrow Groove Gas Metal Arc Welding

contact tip. The twist electrode consists of two intertwined wires which, when fed into the groove, generate arcs from the tips of the two wires. Due to the twist, the arcs describe a continuous rotational movement that increases penetration into the groove face without any special weaving device.

Because these arc oscillation techniques often require special feeding equipment, an alternate method has been developed in which a larger electrode (e.g., 2.4 mm to 3.2 mm [0.093 in. to 0.125 in.] in diameter) is fed directly into the center of the groove from a contact tip situated above the plate surface. With this technique, the wire placement is still critical but there is less chance of arcing between the contact tip and the workpiece, and standard welding equipment can be used. However, it has a more limited thickness potential and is normally restricted to the flat position.

Materials and Consumables

Various shielding gases are used with the narrow groove technique, as with conventional gas metal arc welding. A gas mixture consisting of argon and 20% to 25% carbon dioxide is widely used because it provides a good combination of arc characteristics, bead profile, and groove face fusion. Delivering the shielding gas to the weld area is a challenge in the narrow groove configuration, and numerous nozzle designs have been developed to accommodate the varied weld areas.

Process Variables. The parameters for narrow groove gas metal arc welding are very similar to those used for conventional gas metal arc welding. For the narrow groove application, however, the quality of the weld is sensitive to slight changes in these parameters, voltage being particularly important. An excessive arc voltage (arc length) can cause undercut of the groove face, resulting in oxide entrapment or incomplete fusion in subsequent passes. High voltage may cause the arc to climb the groove face and damage the contact tip. For this reason, pulsed power sources have become widely used in this application. They can maintain a stable spray arc at low arc voltages. A summary of some typical welding conditions is shown in Table 4.3.

Gas Metal Arc Braze Welding

In gas metal arc braze welding, a copper-based electrode (e.g., aluminum bronze or silicon bronze) is used instead of a steel electrode to join steel. As the copper alloy has a lower melting temperature than steel, less heating of the base metal is required in order to deposit a weld bead, and little or no melting of the base metal occurs. (See "Braze Welding" in Chapter 12.)

Table 4.3
Typical Welding Conditions for the Narrow-Groove GMAW Technique

Technique and Weld Position	Groove Width		Current*	Voltage*	Travel Speed		Gas Shield
	mm	in.	A	V	mm/s	in./min.	
Narrow groove weld (NGW)-I; horizontal	9.5	0.375	260–270	25–26	17	40	Argon-carbon dioxide
NGW-I; horizontal	10–12	0.4–0.5	220–240	24–28†	6	13	Argon-carbon dioxide
NGW-I; flat	9.5	0.375	280–300	29†	4	9	Argon-carbon dioxide
NGW-II; flat	12.5	0.50	450	30–37.5	6	15	Argon-carbon dioxide
NGW-II; flat	12–14	0.47–0.55	450–550	38–42	8	20	Argon-carbon dioxide

*Direct current electrode positive.
† Pulsed power at 120 pulses per second.

Because of its low heat input, braze welding is sometimes used to join heat-sensitive materials such as cast iron, and for welding thin sheet steel to help prevent melt-through. Gas metal arc braze welding is also used to join galvanized steels. The lower heat input reduces the amount of coating that is melted away, and the copper-based weld bead furnishes better corrosion resistance than that provided by a carbon steel weld bead.

Twin, Two-Wire, and Tandem Gas Metal Arc Welding

The concept of a two-wire gas metal arc welding system was first examined and marketed in the 1960s. These systems found limited application, however, due to problems with arc interactions between the two electrodes. Recent developments in welding power source technology with computer controls have made possible the application of two-wire systems for production, while minimizing the arc interactions between the two electrodes. The two electrodes may be at the same electrical potential, or they may be at separate electrical potential. In either case, both electrodes feed into the same weld pool. While standard terminology has yet to be designated by the American Welding Society for these two variations, they have been referred to as *twin* or *twin-wire gas metal arc welding* and *tandem gas metal arc welding*, respectively.

Two-wire gas metal arc welding has the potential for a two- to three-fold increase in deposition rates for plate applications and a two- to six-fold increase in travel speed for sheet applications.

EQUIPMENT

The gas metal arc welding process can be implemented in semiautomated, mechanized, automated, and robotic installations. The equipment required in these installations is discussed in this section.

SEMIAUTOMATIC INSTALLATIONS

A semiautomatic gas metal arc welding installation is comprised of equipment components that must be integrated and positioned to provide the welder with an efficient workstation. A typical semiautomatic GMAW installation is depicted schematically in Figure 4.3. The components used are described below.

Arc Welding Guns and Accessories

Different types of welding guns have been designed to provide maximum efficiency for a wide variety of applications. These range from heavy-duty guns for high-current, high-production work to lightweight guns for lower-current, out-of-position welding. Gun designs also vary for use in semiautomatic and automated welding. Water or air cooling and curved or straight nozzles are available for heavy-duty and lightweight guns. An air-cooled gun is typically heavier than a water-cooled gun at the same rated amperage and duty cycle because the air-cooled gun requires more mass to overcome its less efficient cooling.

The following are basic components of arc welding guns:

1. Contact tip,
2. Gas shield nozzle,
3. Electrode conduit and liner,
4. Gas hose,
5. Water hose,
6. Power cable, and
7. Control switch.

These components are illustrated schematically in Figure 4.13.

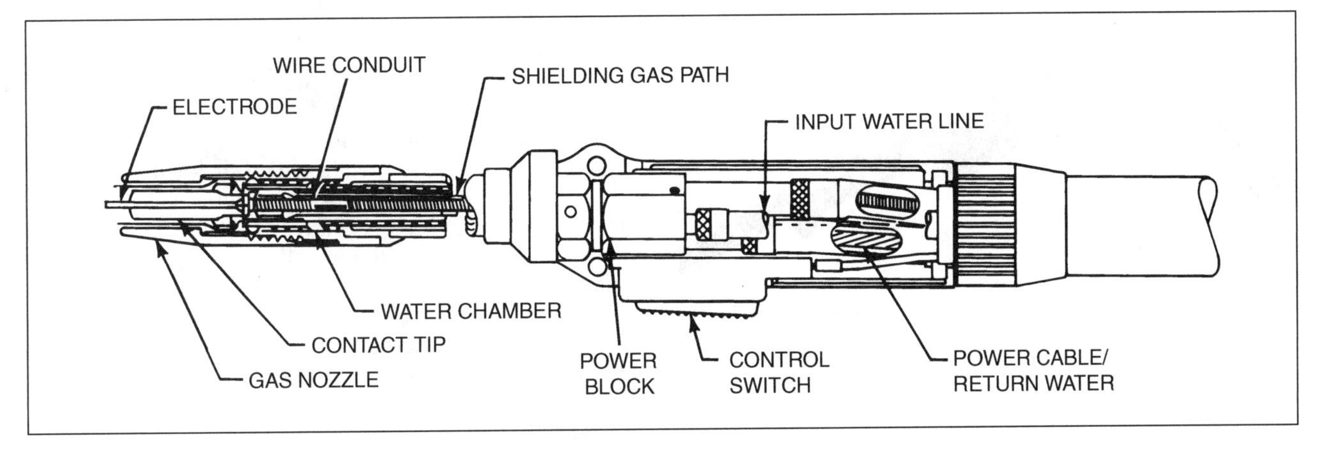

Figure 4.13—Cross-Sectional View of A Typical Gas Metal Arc Welding Gun

In the gas metal arc welding gun, the contact tip, a tube that is usually made of copper or a copper alloy, transfers welding current to the electrode and directs the electrode toward the work. The contact tip is connected electrically to the welding power source by the power cable. The inner surface of the contact tip should be smooth so that the electrode feeds easily and makes good electrical contact. The manufacturer's instruction booklet supplied with every gun specifies the correct contact tip size to use for each electrode size and material. The hole in the contact tip should generally be 0.13 mm to 0.25 mm (0.005 in. to 0.010 in.) larger than the wire being used, although larger hole sizes may be required for aluminum. The contact tip must be held firmly in the gun and centered in the gas shielding nozzle.

The positioning of the contact tip in relation to the end of the nozzle may be varied depending on the mode of transfer being used. For short-circuiting transfer, the tip is usually flush or extended beyond the nozzle, while for spray arc it is recessed approximately 3.2 mm (1/8 in.). The tip should be checked periodically during welding and replaced if the hole has become elongated due to excessive wear or clogged with spatter. Using a worn or clogged tip can result in poor electrical contact and erratic arc characteristics.

The nozzle directs an evenly flowing column of shielding gas into the welding zone. An even flow is extremely important to assure adequate protection of the molten weld metal from atmospheric contamination. Different sizes of nozzles are available and should be chosen according to the application. Larger nozzles are used for high-current work involving a large weld pool, and smaller nozzles are used for low-current and short-circuit gas metal arc welding. For gas metal arc spot welding applications, the nozzles are fabricated with ports that allow the gas to escape when the nozzle is pressed onto the workpiece.

The electrode conduit and its liner are connected to a bracket adjacent to the feed rolls on the electrode feed motor. The conduit supports, protects, and directs the electrode from the feed rolls to the gun and contact tip. Uninterrupted electrode feeding is necessary to ensure good arc stability. Buckling or kinking of the electrode must be prevented. The electrode will tend to jam anywhere between the drive rolls and the contact tip if not properly supported.

The liner may be an integral part of the conduit or may be supplied separately. In either case, the liner material and inner diameter are important. A helical steel liner is recommended when using hard electrode materials such as steel and nickel alloys. Nylon liners should be used for soft electrode materials such as aluminum and magnesium. Liners require periodic maintenance to ensure that they are clean and in good condition as required for consistent feeding of the wire.

Care must be taken not to crimp or bend the conduit excessively even though its outer surface is usually steel-supported. The instruction manual supplied with the unit typically specifies the recommended conduits and liners for each electrode size and material.

The remaining accessories deliver the shielding gas, cooling water, and welding power to the gun. These hoses and cables can be connected directly to the source of these facilities or to the welding control. Trailing gas shields are available and may be required to protect the weld pool during high-speed welding.

The basic gun, shown in Figure 4.14, is connected to an electrode feed unit that pushes the electrode from a remote location through the conduit. Other designs are also available, including a unit with a small electrode feed mechanism built into the gun, as shown in Figure 4.15.

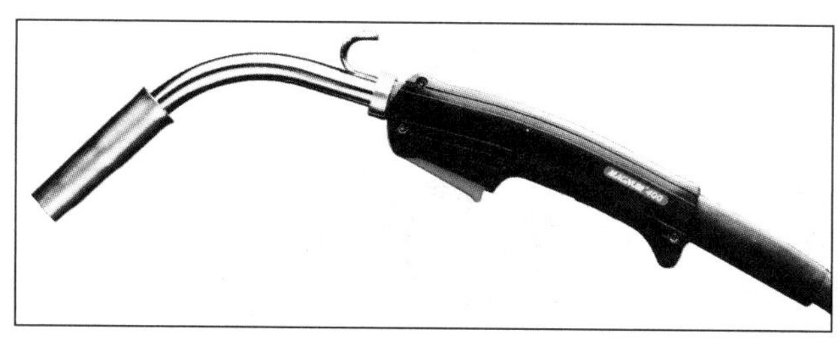

Photograph courtesy of The Lincoln Electric Company

Figure 4.14—Gas Metal Arc Welding Gun

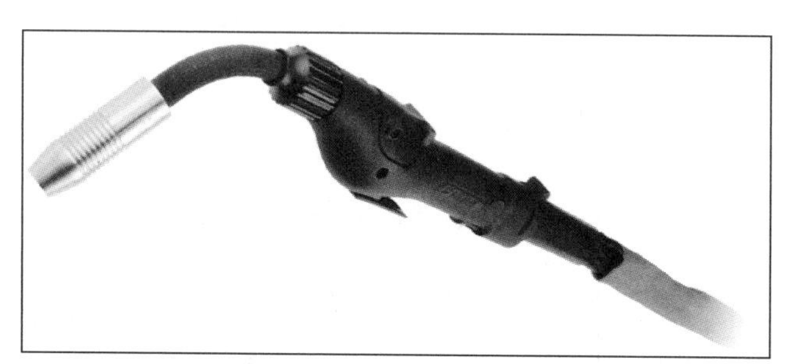

Photograph courtesy of M K Products

Figure 4.15—Push-Pull Gas Metal Arc Welding Gun

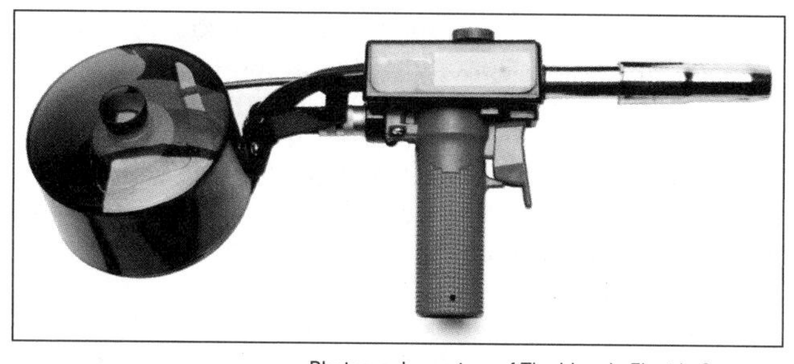

Photograph courtesy of The Lincoln Electric Company

Figure 4.16—Spool-on-Gun Welding Head

This gun pulls the electrode from the source, where an additional drive may also be located to push the electrode simultaneously into the conduit (i.e., a push-pull system). This type of gun is also useful for feeding small-diameter or soft electrodes (e.g., aluminum), when pushing might cause the electrode to buckle. Another variation is the spool-on-gun type (Figure 4.16), in which the electrode feed mechanism and the electrode source are self-contained.

Electrode Feed Unit

The electrode feed unit, or wire feeder, consists of an electric motor, drive rolls, and accessories for maintaining electrode alignment and pressure. These units can be integrated with the speed control or located at a distance from it. The electrode feed motor is usually a direct-current type. It pushes the electrode through the gun to the workpiece. It typically has a control circuit to vary the motor speed over a broad range.

Constant-speed wire feeders are normally used in combination with constant-voltage power sources. They may be used with constant-current power sources if a slow electrode run-in circuit is added. When a constant-current power source is used, an automatic voltage sensing control is necessary. This control detects changes in the arc voltage and adjusts the wire feed speed to maintain a constant arc length. The use of this combination of variable speed wire feeder and constant-current power source is limited to larger-diameter wires (over 1.6 mm [1/16 in.]), for which the feed speeds are lower. At high wire-feed speeds, the adjustments to motor speed cannot normally be made quickly enough to maintain arc stability.

The feed motor is connected to a drive-roll assembly. The drive rolls, in turn, transmit the force to the electrode, pulling it from the electrode source and pushing it through the welding gun. Wire feed units may use a two-roll or a four-roll arrangement. A four-roll wire feeding unit is shown in Figure 4.17. The pressure adjustment of the drive roll allows variable force to be applied to the wire, depending on its characteristics (e.g., solid or cored, hard or soft). The inlet and outlet guides provide for proper alignment of the wire to the drive rolls and support the wire to prevent buckling.

The type of feed rolls generally used with solid wires is shown in Figure 4.18. In this arrangement, a grooved roll is combined with a flat or grooved back-up roll. A V-groove is used for solid hard wires such as carbon and stainless steels. A U-groove is used for soft wires such as aluminum.

Serrated or knurled feed rolls with a knurled backup roll, shown in Figure 4.19, are generally used with tubular wires. The knurled design allows maximum drive force to be transmitted to the wire with a minimum of drive roll pressure. These types of rolls are not recommended for softer wire, such as aluminum, because the knurling tends to cause a flaking of the wire, which can eventually clog the gun or liner.

Welding Control

The welding control and electrode feed motor required for semiautomatic welding are available in one integrated package. The main function of the welding control is to regulate the speed of the electrode feed

Figure 4.17—Typical Four-Drive-Roll Wire Feeding Unit

(A) Grooved Feed Rolls Used to Feed Solid Wire

(B) Grooved Feed Roll with Flat Backup Used to Feed Solid Wire

Figure 4.18—Two Types of Grooved Feed Rolls for Solid Wire

Figure 4.19—Knurled Feed Rolls for use with Cored Wires

motor, usually with an electronic governor. By increasing the wire feed speed, the operator increases the welding current. Decreases in wire feed speed result in lower welding currents. The control also regulates the initiation and termination of the electrode feed by means of a signal received from the gun switch.

Also available are electrode feed control features that permit the use of a touch-start mechanism, in which the electrode feed is initiated when the electrode touches the work. With the slow run-in unit, the initial feed rate is slow until the arc is initiated and then increases to the rate required for welding. These two features, which are employed primarily in conjunction with constant-current power sources, are particularly useful for the gas metal arc welding of aluminum.

Shielding gas, cooling water, and welding power are also delivered to the gun through the welding control, requiring direct connection of the control to the gas, water, and the power source. Gas and water flows are regulated to coincide with the weld start and stop by use of solenoid valves. The control can also sequence the initiation and termination of gas flow and energize the power source contactor. The control may allow some gas to flow before the welding starts, referred to as *preflow* or *purging*, and after the welding stops, referred to as *post flow*, to protect the weld pool. The control is usually independently powered by 115 V alternating current.

Power Source

The welding power source delivers electrical power to the electrode and the workpiece to produce the arc. For the vast majority of gas metal arc welding applications, direct current electrode positive (DCEP) is used; therefore, the positive lead is connected to the gun and the negative lead to the workpiece. The common types of direct-current power sources are engine-driven generators, which are rotating, and transformer-rectifiers, which are static. Inverters are included in the static category. The static type is usually preferred for in-shop fabrication where a source of either 230 V or 460 V is available. The engine-driven generator is used at locations such as construction sites where no other source of electrical energy is available.

Power sources can be designed and built to provide either constant current or constant voltage. Early applications of the gas metal arc welding process used constant-current power sources (often referred to as a *droopers*). As illustrated in Figure 4.20, droopers maintain a relatively fixed current level during welding, regardless of variations in arc length (voltage). These machines are characterized by high open-circuit voltages and limited short-circuit current levels. Since they supply a virtually constant current output, the arc is maintained at a fixed length only if the contact-tip-to-work distance remains constant, along with a constant electrode feed rate.

In practice, since this distance varies, the arc tends to burn back to the contact tip or "stub" into the workpiece. This can be avoided by using a voltage-controlled electrode feed system. When the voltage (arc length) increases or decreases, the motor speeds up or slows down to hold the arc length constant. The electrode feed rate is adjusted automatically by the control system. This type of power source is generally used for spray transfer welding since the limited duration of the arc in short-circuiting transfer makes control by voltage regulation impractical.

As gas metal arc welding applications increased, it was found that constant-voltage (CV) power sources provided improvements in operations. Used in conjunction with a constant-speed wire feeder, it maintains a nearly constant voltage during the welding operation. The V-A curve of this type of power source is illustrated in Figure 4.21. The CV system compensates for the variations in the contact-tip-to-workpiece distance that occur during normal welding operations by instantaneously

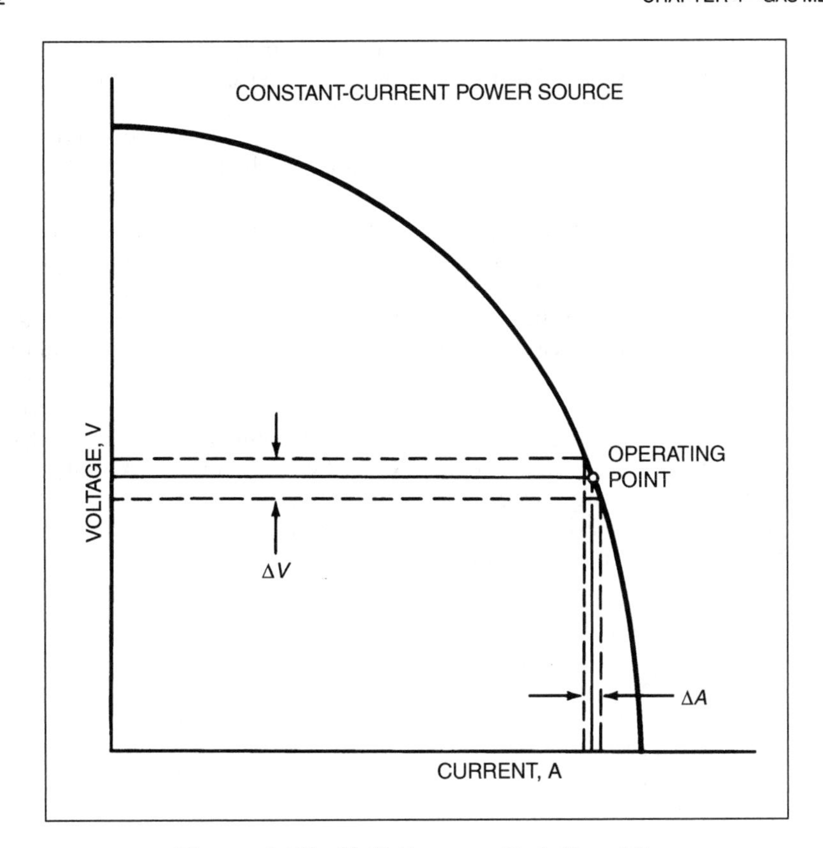

Figure 4.20—Volt-Ampere Relationship for a Constant-Current Power Source

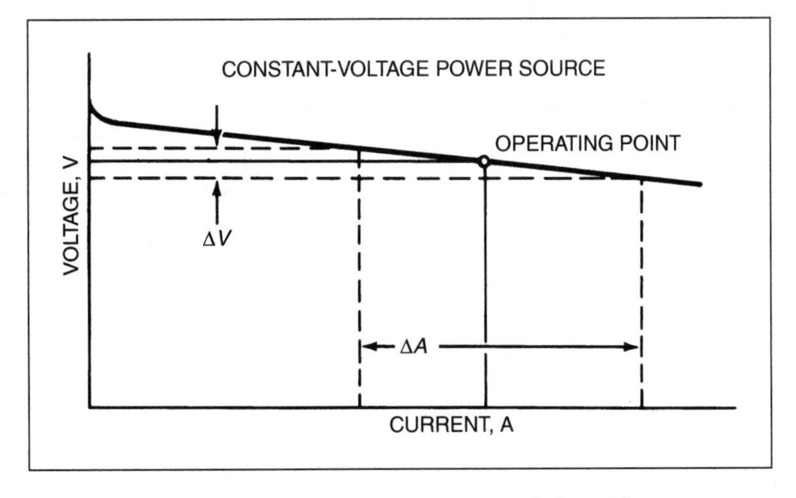

Figure 4.21—Volt-Ampere Relationship for a Constant-Voltage (CV) Power Supply

increasing or decreasing the welding current to compensate for the changes in electrode extension due to the changes in gun-to-workpiece distance.

The arc length is established by adjusting the welding voltage at the power source. Once this is set, no other changes are required during welding. The wire feed speed, which also becomes the current control, is preset by the welder or welding operator prior to welding. It can be adjusted over a considerable range before stubbing to the workpiece or burning back into the contact tip occurs. Welders and welding operators easily learn to adjust the wire feed and voltage controls with only minimum instruction.

The self-correction mechanism of a constant-voltage power source is illustrated in Figure 4.22. As the contact tip-to-work distance increases, the arc voltage and arc length would tend to increase. However, the welding current decreases with this slight increase in voltage, thus consuming less electrode and compensating for the increase in electrode extension. Conversely, if the contact tip-to-workpiece distance were shortened, the lower voltage would be accompanied by an increase in current, thus consuming more electrode to compensate for the shorter electrode extension.

The self-correcting arc length feature of the CV power source is important in producing stable welding conditions. However, several other variables contribute to optimum welding performance, particularly for short-circuiting transfer. In addition to the control of the output voltage, some degree of slope and inductance control may be desirable. The welder or welding operator should understand the effect of these variables on the welding arc and its stability.

Voltage. Arc voltage is the electrical potential between the electrode and the workpiece. Arc voltage is lower

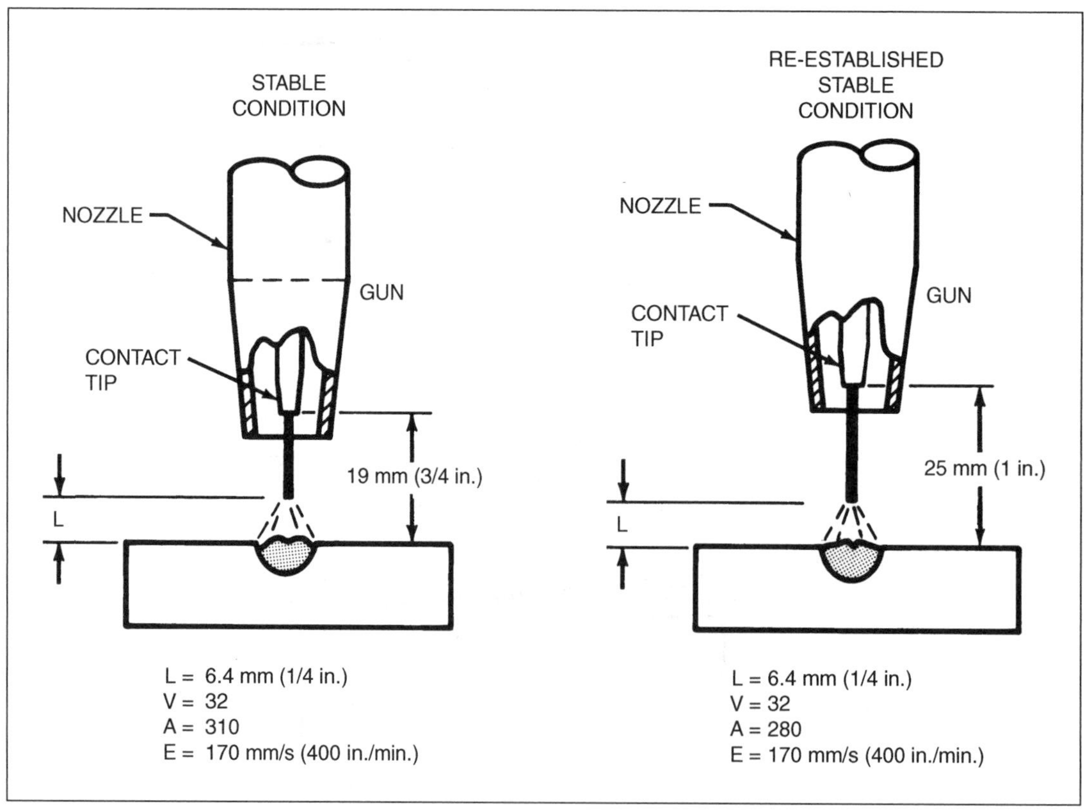

Key:
L = Arc length
V = Voltage
A = Arc current
E = Electrode feed speed

Figure 4.22—Automatic Regulation of Arc Length in the Gas Metal Arc Welding Process

than the voltage measured directly at the power source because of voltage drops at connections and along the length of the welding cable. As previously mentioned, arc voltage is directly related to arc length; therefore, an increase or a decrease in the output voltage at the power source will result in a corresponding change in the arc length.

Slope. The term *slope* refers to the relationship of the voltage change to the current change in the output of a power source. This slope, as specified by the manufacturer, is measured at the output terminals and is not the total slope of the arc welding system. Anything that adds resistance to the welding system (e.g., power cables, poor connections, loose terminals, dirty contacts, and the like) adds to the slope. Therefore, slope is best measured at the arc in a given welding system.

Two operating points are needed to calculate the slope of a constant-voltage welding system, as shown in Figure 4.23. With a change of 10 volts (V), the current changes 100 amperes (A). The slope would therefore be 10 V/100 A.

It should be noted that it is not accurate to use the open-circuit voltage as one of the operating points because of the sharp voltage drop that may occur with some machines at low currents. Two stable arc conditions should be chosen at currents that encompass the range likely to be used.

Slope has a major function in the short-circuiting transfer mode of gas metal arc welding. It controls the magnitude of the short-circuit current, which is the amperage that flows when the electrode is shorted to the workpiece. In gas metal arc welding, the separation of molten drops of metal from the electrode is controlled by an electrical phenomenon called the *electromagnetic pinch effect.* The term *pinch* refers to the magnetic "squeezing" force on a conductor produced by the current flowing through it. The pinch effect for short-circuiting transfer is illustrated in Figure 4.24.

The short-circuit current, and therefore the pinch effect force, is a function of the slope of the V-A curve of the power source, as illustrated in Figure 4.25. The operating voltage and the amperage of the two power sources are identical, but the short-circuiting current of Curve A is less than that of Curve B. Curve A has the steeper slope, or a greater voltage drop per 100 amperes, as compared to Curve B, resulting in a lower short-circuiting current and a lower pinch effect.

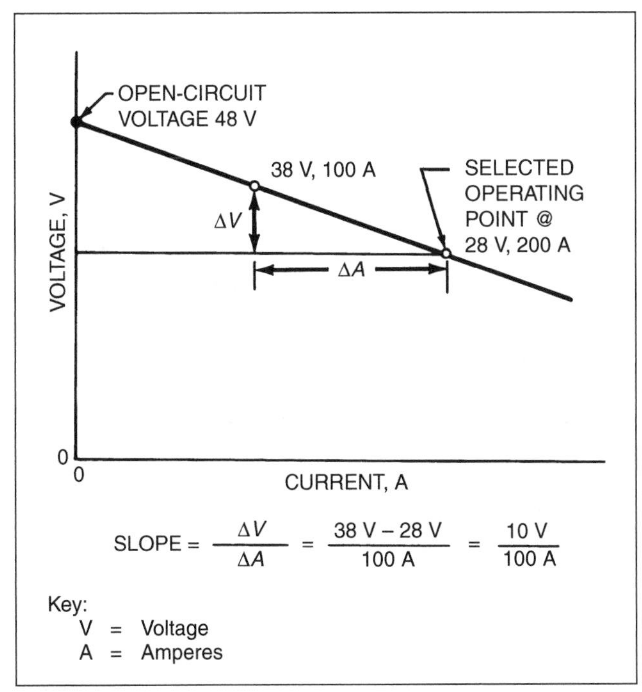

Source: Adapted from American Welding Society (AWS) Committee on Arc Welding and Cutting, 1994, *Recommended Practices for Gas Metal Arc Welding,* ANSI/AWS C5.6-94R, Miami: American Welding Society, Figure 13.

Figure 4.23—Calculation of the Slope for a Power Source

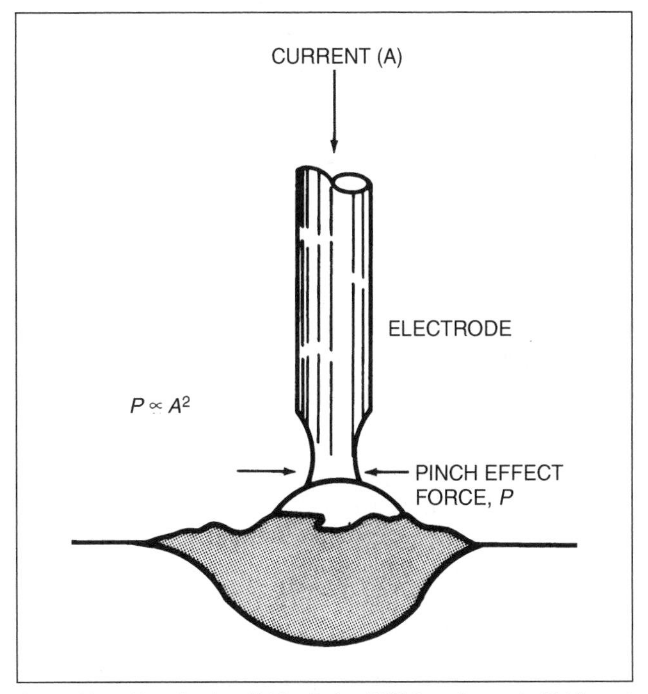

Source: Adapted from American Welding Society (AWS) Committee on Arc Welding and Cutting, 1994, *Recommended Practices for Gas Metal Arc Welding,* ANSI/AWS C5.6-94R, Miami: American Welding Society, Figure 14.

Figure 4.24—Illustration of Pinch Effect during Short-Circuiting Transfer

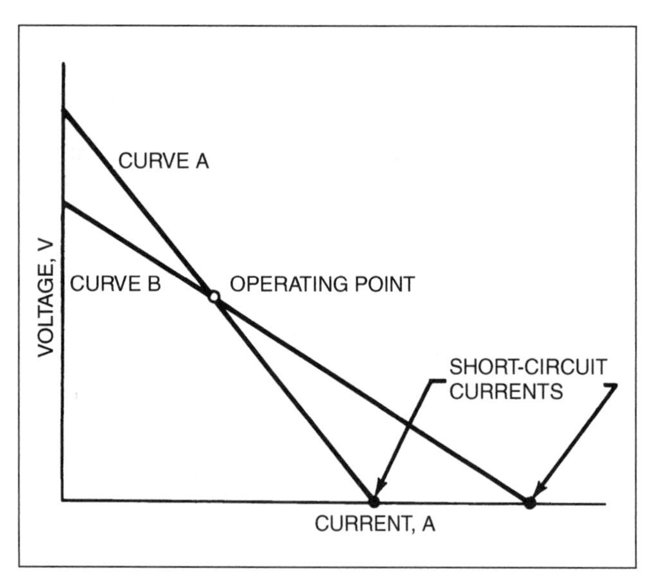

Source: Adapted from American Welding Society (AWS) Committee on Arc Welding and Cutting, 1994, *Recommended Practices for Gas Metal Arc Welding*, ANSI/AWS C5.6-94R, Miami: American Welding Society, Figure 15.

Figure 4.25—Effect of Changing Slope

In short-circuiting transfer, the amount of short-circuit current is important since the resultant pinch effect determines the way a molten drop detaches from the electrode. This, in turn, affects the stability of the arc. When little or no slope is present in the power source circuit, the short-circuit current rises rapidly to a high level. The pinch effect will also be high, and the molten drop will separate violently from the wire. Excessive pinch effect will abruptly squeeze the metal, clear the short-circuit, and create excessive spatter.

When the short-circuit current available from the power source is limited to a low value by a steep slope, the electrode carries the full current, but the pinch effect may be too low to separate the drop and reestablish the arc. Under these conditions, the electrode either piles up on the workpiece or freezes to the pool. When the short-circuit current is at an acceptable value, the parting of the molten drop from the electrode is smooth and creates very little spatter. Typical short-circuit currents required for metal transfer with the best arc stability are shown in Table 4.4.

In the past, many constant-voltage power sources were equipped with a slope adjustment mechanism. However, with the change in basic equipment designed to accommodate electronic controls, only a limited number of machines currently available have an adjustable slope control. Most power sources have a fixed slope that has been preset for the most common welding conditions.

Table 4.4
Typical Short-Circuit Currents Required for Metal Transfer in the Short-Circuiting Mode

Electrode Material	Electrode Diameter		Short-Circuit Current
	mm	in.	A (DCEP)
Carbon steel	0.8	0.030	300
Carbon steel	0.9	0.035	320
Aluminum	0.8	0.9	0.030
Aluminum	0.035	175	195

Inductance. When the electrode shorts with the work, the current increases rapidly to a higher level. The circuit characteristic affecting the time rate of this increase in current is inductance, which is usually measured in henrys (H).

The effect of inductance is illustrated by the curves plotted in Figure 4.26. Curve A is an example of a current-time curve immediately after a short-circuit when some inductance is in the circuit. Curve B illustrates the path the current would have taken if there were no inductance in the circuit.

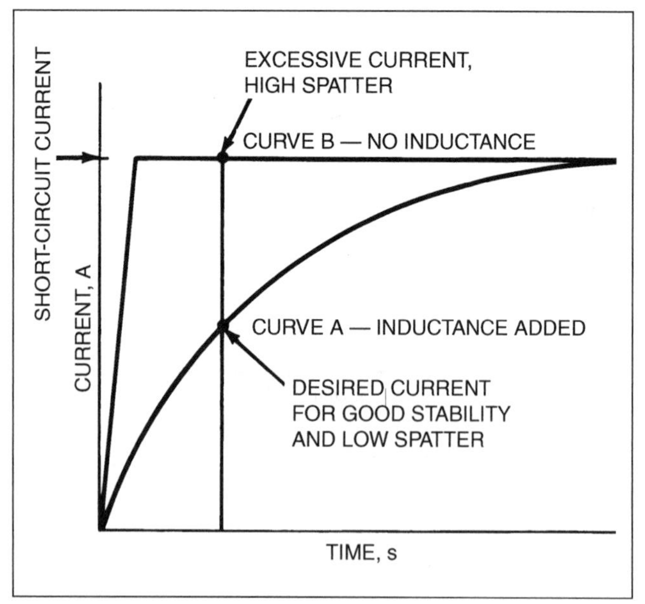

Source: Adapted from American Welding Society (AWS) Committee on Arc Welding and Cutting, 1994, *Recommended Practices for Gas Metal Arc Welding*, ANSI/AWS C5.6-94R, Miami: American Welding Society, Figure 16.

Figure 4.26—Change in Rate of Current Rise Due to Added Inductance

The maximum amount of pinch effect is determined by the final short-circuit current level. The instantaneous pinch effect is controlled by the instantaneous current, and therefore the shape of the current-time curve is significant. The inductance in the circuit controls the rate of current rise. Without inductance, the pinch effect is applied rapidly, and the molten drop is violently squeezed off the electrode, causing excessive spatter. Higher inductance results in a decrease in the short-circuits per second and an increase in the arc-on time. Increased arc-on time makes the weld pool more fluid and results in a flatter, smoother weld bead. The higher inductance, however, can adversely affect arc initiation.

In the spray transfer mode, the addition of some inductance to the power source produces a softer arc start without affecting the steady-state welding conditions. Power source adjustments required for minimum spatter conditions vary with the electrode material and diameter. As a rule, higher short-circuit currents and higher inductance are needed for larger-diameter electrodes. Power sources are available with fixed, incremental, or continuously adjustable inductance levels.

Pulsed Power Sources

With the advent of electronics, it became possible to build power sources with current outputs that could be varied at a very rapid rate, or "pulsed." When these were first introduced, they essentially consisted of two power sources in one cabinet. One power source was switched on all the time, supplying the background current and voltage. The second power source would switch on at either 60 or 120 times per second to supply the peak current and voltage (see Figure 4.8). For limited applications, these units worked well, but they lacked versatility and were quite expensive.

As control systems became more sophisticated, the pulse machines reverted to one power source that had the function of switching the welding output from the background settings to the peak settings very quickly. These units gave the welding operator the ability to regulate the background settings, the peak settings, and the number of cycles per second (frequency) independently. This capability increased the potential applications for pulsed gas metal arc welding, but the power sources were difficult to set because of the increased number of controls.

To overcome this limitation, the next generation of power sources featured synergic, or one-knob, control systems. When the operator changed the wire feed speed rate, the peak output, background output and frequency were all adjusted automatically. This development simplified the setting of machines, which led to their increased acceptance in the marketplace.

The power sources that are currently available have the capacity to operate as a synergic power source when needed, but also permit the operator to adjust each parameter separately. This feature allows the welding arc to be customized for optimum performance for each application.

Shielding Gas Regulators

A system is required to provide a constant shielding gas flow rate at atmospheric pressure during welding. Gas regulators reduce the source gas pressure to a constant working pressure regardless of variations at the source. Regulators may be single or dual stage and may have a built-in flow meter. Dual-stage regulators deliver gas at a more consistent pressure than single-stage regulators when the source pressure varies.

Shielding Gas Supply

The shielding gas can be supplied from a high-pressure cylinder, a liquid-filled cylinder, or a bulk-liquid system. Premixed gases are available in single cylinders. Mixing devices are used to obtain the correct proportions when two or more gas or liquid sources are used. The size and type of the gas storage source is determined by the user, based on the volume of gas required.

Electrode Supply

The gas metal arc welding process uses a continuously fed electrode (wire) that is consumed at relatively high speeds. The electrode source must therefore provide a large volume of wire that can readily be fed to the gun to provide maximum process efficiency. This source usually takes the form of a spool or coil that holds approximately 4.5 kilograms (kg) to 27 kg (10 pounds [lb] to 60 lb) of wire. The spool is wound to allow free feeding without kinks or tangles. Larger spools of up to 114 kg (250 lb) are also available, and wire can be provided in drums or reels of 340 kg to 450 kg (750 lb to 1000 lb). These larger packages are good for use in high-volume applications because they reduce the labor associated with changing to new packages as material is consumed. For spool-on-gun equipment, small spools (0.45 kg to 0.9 kg [1 lb to 2 lb]) are used. The applicable American Welding Society or military electrode specification defines standard packaging requirements. Normally, special requirements are agreed upon by the user and the supplier.

The electrode source can be located close to the wire feeder, or it can be positioned some distance away with the wire fed through special dispensing equipment. The electrode source should normally be as close as possible to the gun to minimize feeding problems, yet far enough away to provide the welder flexibility and accessibility.

AUTOMATED GMAW INSTALLATIONS

The gas metal arc welding process is easily mechanized. The major components in a mechanized, automatic or robotic installation are nearly identical to those of the semiautomatic installation described above. A motorized carriage device is often added to carry the welding head (gun and wire feeder). This arrangement is depicted in Figure 4.27, which illustrates the typical components in an automatic or mechanized installation for welding straight seams. With another type of installation the workpiece can be mounted on a turntable or positioner and moved under a stationary welding head. The welding head can be equipped with an oscillator that provides arc movement transverse to the welding direction and a seam tracker, which positions the electrode in the weld joint. The degree of motion and process control for a given system will dictate whether that system is classified as mechanized, automatic, robotic or adaptive control.

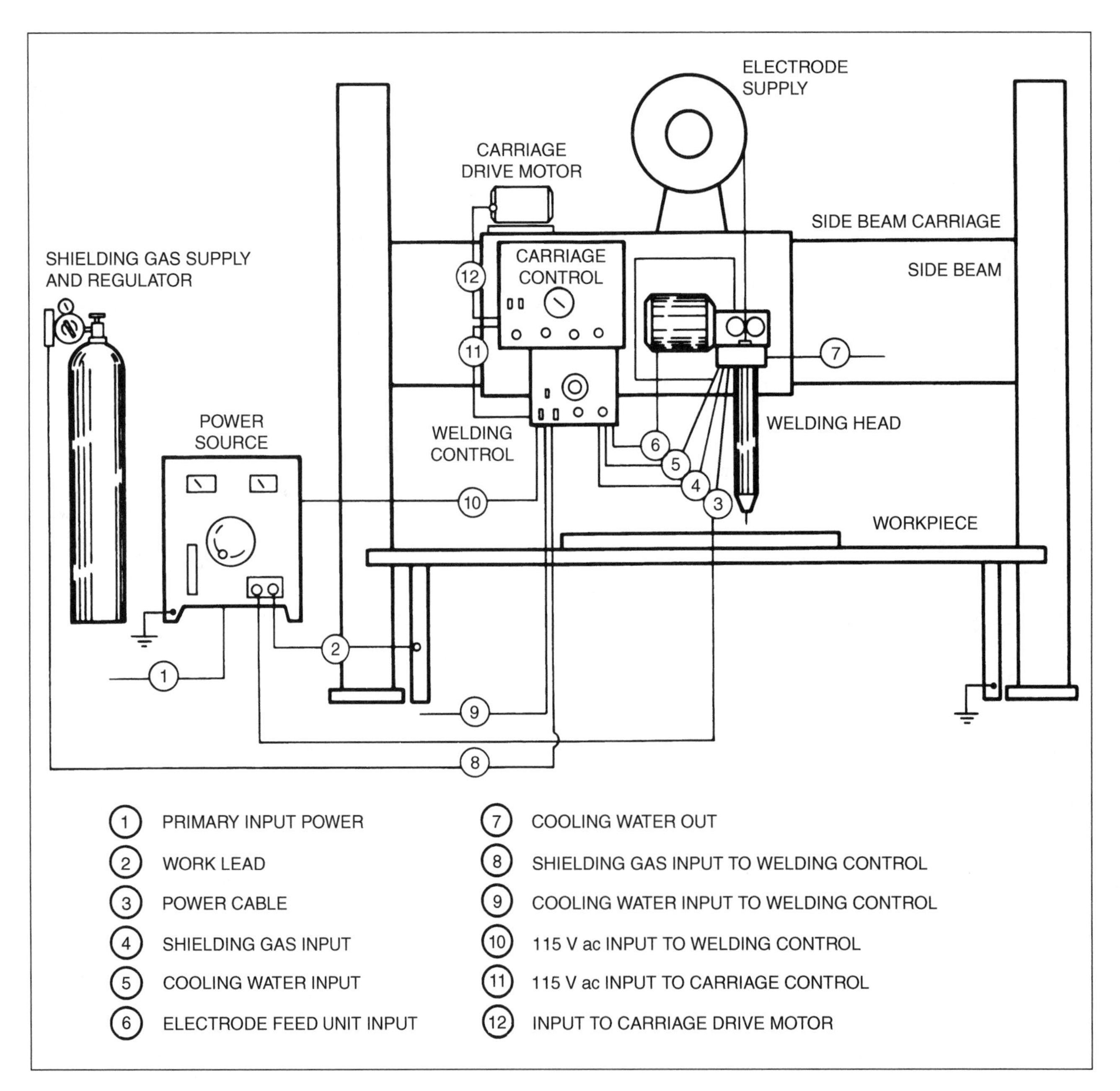

Figure 4.27—Mechanized Gas Metal Arc Welding Installation

EQUIPMENT SELECTION

When selecting equipment, the buyer should consider application requirements, the range of power output, static and dynamic electrical characteristics, and expected range of wire feed speeds. If a major part of production involves small-diameter aluminum wire, for example, the fabricator should consider a push-pull type of wire feeder. If out-of-position welding is contemplated, the user should look into pulsed-power welding machines. For the welding of light-gauge stainless steel, a power source with adjustable slope and inductance may be considered.

When new equipment is to be purchased, some consideration should be given to the versatility of the equipment and to standardization. The selection of equipment for single-purpose or high-volume production can usually be based on the requirements of that particular application only. However, if a multitude of jobs are to be performed, many of which may be unknown at the time of selection (as is the case in a job shop operation, for example), versatility may be very important.

The equipment presently in use at the facility should also be considered. Standardizing certain components and complementing existing equipment minimizes parts inventory requirements and provides for maximum efficiency of the overall operation.

Table 4.5
Specifications for Various GMAW Electrodes

Base Material	AWS Specification*
Carbon steel	*Specification for Carbon Steel Electrodes and Rods for Gas Shielded Arc Welding*, AWS A5.18/A5.18M
Low-alloy steel	*Specification for Low-Alloy Steel Electrodes and Rods for Gas Shielded Arc Welding*, ANSI/AWS A5.28
Aluminum alloys	*Specification for Bare Aluminum and Aluminum Alloy Welding Electrodes and Rods*, ANSI/AWS A5.10/A5.10M
Copper alloys	*Specification for Copper and Copper Alloy Bare Welding Rods and Electrodes*, AWS A5.7
Magnesium	*Specification for Magnesium Alloy Welding Electrodes and Rods*, ANSI/AWS A5.19
Nickel alloys	*Specification for Nickel and Nickel-Alloy Bare Welding Electrodes and Rods*, ANSI/AWS A5.14/A5.14M
300-Series stainless steel	*Specification for Bare Stainless Steel Welding Electrodes and Rods*, ANSI/AWS A5.9
400-Series stainless steel	*Specification for Bare Stainless Steel Welding Electrodes and Rods*, ANSI/AWS A5.9
Titanium	*Specification for Titanium and Titanium Alloy Welding Electrodes and Rods*, ANSI/AWS A5.16

*The facts of publication for the specifications listed here are included in the Bibliography.

MATERIALS AND CONSUMABLES

In addition to equipment components—such as contact tips and conduit liners, which wear out and have to be replaced—the consumables in gas metal arc welding consist of electrodes and shielding gases. The chemical composition of the electrode, the base metal, and the shielding gas determine the chemical composition of the weld metal. This composition largely determines the metallurgical and mechanical properties of the weldment. The following are factors that influence the selection of the shielding gas and the welding electrode:

1. Base metal type,
2. Required weld metal mechanical properties,
3. Base metal condition and cleanliness,
4. Type of service or applicable specification requirement,
5. Welding position, and
6. Intended mode of metal transfer.

ELECTRODES

The electrodes (filler metals) for gas metal arc welding are specified by various American Welding Society filler metal specifications. Other standards-writing societies also publish filler metal specifications for specific applications. An example is the Aerospace Materials Specifications written by the Society of Automotive Engineers (SAE). The AWS specifications, designated as A5.XX standards, are presented in Table 4.5.[1] They define requirements for sizes and tolerances, packaging, chemical composition, and in some cases, mechanical properties. The American Welding Society also publishes *Filler Metal Comparison Charts*, which list trade names for each of the filler metal classifications.[2]

1. At the time of the preparation of this chapter, the referenced codes and other standards were valid. If a code or other standard is cited without a date of publication, it is understood that the latest edition of the document referred to applies. If a code or other standard is cited with the date of publication, the citation refers to that edition only, and it is understood that any future revisions or amendments to the code or standard are not included; however, as codes and standards undergo frequent revision, the reader is encouraged to consult the most recent edition.
2. American Welding Society (AWS) Committee on Filler Metals and Allied Materials, *Filler Metal Comparison Charts*, FMC:2000, Miami: American Welding Society.

Composition

For most welding applications, the composition of the electrode (filler metal) is generally similar to that of the base metal. The filler metal composition may be altered slightly to compensate for losses that occur in the welding arc or to provide for deoxidation of the weld pool. In some cases, this involves very little modification from the base metal composition. In certain applications, however, obtaining satisfactory welding characteristics and weld metal properties requires an electrode with a different chemical composition from that of the base metal. For example, the most satisfactory electrode for use in the gas metal arc welding of manganese bronze, a copper-zinc alloy, is either an aluminum bronze or a copper-manganese-nickel-aluminum alloy electrode.

The electrodes that are most suitable for welding high-strength aluminum alloys are often different in composition from the base metals on which they are to be used because aluminum alloy compositions such as 6061 are unsuitable as weld filler metals. Accordingly, electrode alloys are designed to produce the desired weld metal properties and have acceptable operating characteristics.

In other applications, the gas metal arc welding process is used for surfacing operations in which an overlaid weld deposit may provide desirable wear or corrosion resistance or other properties. Overlays are normally applied to carbon or manganese steels and must be carefully engineered and evaluated to assure satisfactory results. During surfacing, the weld metal dilution with the base metal becomes an important consideration; it is a function of arc characteristics and technique. With gas metal arc overlaying, dilution rates from 10% to 50% can be expected depending on the transfer mode, current level, and speed. Multiple layers are normally required, therefore, to obtain suitable deposit chemistry at the surface. Most weld metal overlays are deposited automatically to precisely control dilution, bead width, bead thickness, and overlaps.

Whatever other modifications are made in the composition of electrodes, deoxidizers or other scavenging elements are generally added. These elements are added to minimize porosity in the weld or to assure satisfactory weld metal mechanical properties. The addition of appropriate deoxidizers in the right quantity is essential to the production of sound welds. Deoxidizers most commonly used in steel electrodes are manganese, silicon, and aluminum. Titanium and silicon are the principal deoxidizers used in nickel alloy electrodes. Copper alloy electrodes may be deoxidized with titanium, silicon, or phosphorus.

Electrode Size and Feed Rates

The electrodes used for gas metal arc welding are small in diameter compared to those used for submerged arc or flux cored arc welding. Wire diameters of 0.9 mm to 1.6 mm (0.035 in. to 0.062 in.) are common. However, electrode diameters as small as 0.5 mm (0.020 in.) and as large as 3.2 mm (1/8 in.) may be used. Because the electrode sizes are small and the currents comparatively high, GMAW wire feed rates are high. The rates range from approximately 42 mm/s to 425 mm/s (100 in./min to 1000 in./min) for most metals except magnesium, for which rates up to 590 mm/s (1400 in./min.) may be required. For such wire speeds, electrodes are provided as long, continuous strands of suitably tempered wire of a uniform diameter that can be fed smoothly and continuously through the welding equipment.

Packaging and Handling

The electrode (wire) is normally wound on conveniently sized spools or coils. The uniformity of winding and freedom from kinks or bends are important considerations in the proper feeding of the electrode. Other important characteristics resulting from the winding operation are the *cast* and *helix* of the electrode. The cast refers to the diameter of one loop of wire made when enough wire is cut from the spool or coil to form a loop and is laid unrestrained on a flat surface, as shown in Figure 4.28. The helix is a measure of the amount of rise of the end of the wire above the flat surface. The larger the cast, the more uniform the wire feeding will be. This is a result of the reduction of frictional force as the wire exits the contact tube. A smaller cast can cause the tip of the electrode to wander. A larger helix can cause the tip of the electrode to spiral or flip suddenly as it exits the contact tip. Either of these conditions can lead to erratic weld bead contour and inconsistent penetration, especially in groove welds and welds made with automatic equipment. Specifications for steel wire generally call for an acceptable range for cast and a maximum for the helix. For aluminum wire,

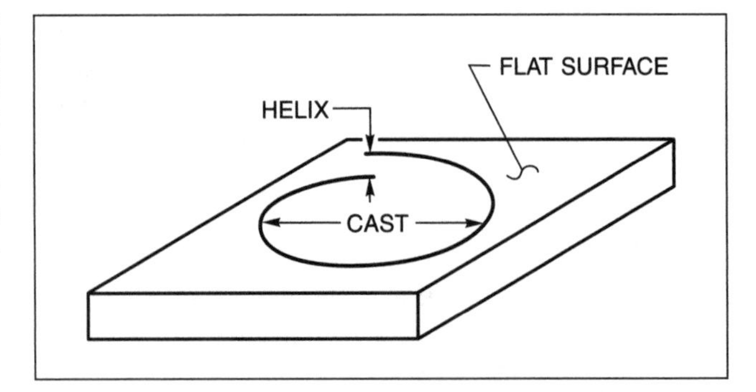

Figure 4.28—Cast and Helix in a Continuous Electrode

more general requirements are specified, such as "filler metal will feed in an uninterrupted manner in automatic or semiautomatic equipment."

Electrodes have high surface-to-volume ratios because of their relatively small size. Any drawing compounds or lubricants inadvertently worked into the surface of the electrode may adversely affect the weld metal properties. These foreign materials may result in weld metal porosity in aluminum and steel alloys, and weld metal or heat-affected zone cracking in high-strength steels. Consequently, the electrodes should be manufactured with a high-quality surface and handled carefully to preclude the collection of contaminants in seams or laps.

Electrode (Filler Metal) Selection

In the engineering of weldments, the objective is to select filler metals that produce a weld deposit with the following two basic characteristics:

1. A deposit that either closely matches the mechanical and physical properties of the base metal or provides some enhancement to the base material, such as corrosion or wear resistance; and
2. A sound weld deposit that is free from unacceptable discontinuities.

In the first case, the weld deposit—even one with a composition nearly identical to that of the base metal—will possess unique metallurgical characteristics. This is dependent on factors such as the energy input and weld bead configuration. The second characteristic is generally achieved through the use of a formulated filler metal electrode—for example, one containing deoxidizers that produce a relatively discontinuity-free deposit.

Compatibility. The electrode must meet certain demands of the process relative to arc stability, metal transfer behavior, and solidification characteristics. It must also provide a weld deposit that is compatible with one or more of the following base metal characteristics:

1. Chemistry,
2. Strength,
3. Ductility,
4. Toughness, and
5. Other properties dictated by specific service conditions or environments.

Consideration should be given to other properties such as corrosion, heat-treatment response, wear resistance, and color match. All such considerations, however, are secondary to the metallurgical compatibility of the base metal and the filler metal.

The American Welding Society has established specifications for the filler metals in common usage. Table 4.6 provides a basic guide to the selection of the appropriate filler metal types for the listed base metals, along with each applicable AWS filler metal specification.

Electrode Type. Both solid and tubular wire electrodes are used with gas metal arc welding. The tubular wires have a powdered metallic core that includes small amounts of arc stabilizing compounds and the appropriate alloying elements. These wires have good arc stability and deposition efficiencies similar to those offered by solid wire. In many cases, their deposition rates exceed those of solid wire of the same size. This tubular approach permits the manufacture of high-efficiency metallic electrodes in compositions that would be difficult and costly to manufacture as a solid wire. In addition, the production of smaller quantities of material for special applications may be advantageous.

SHIELDING GASES

The primary function of the shielding gas is to exclude the atmosphere from contact with the molten weld metal. This is necessary because most metals, when heated to their melting point in air, exhibit a strong tendency to form oxides and, to a lesser extent, nitrides. Oxygen also reacts with carbon in molten steel to form carbon monoxide and carbon dioxide. These varied reaction products may result in weld discontinuities such as slag inclusions, porosity, and weld metal embrittlement. Reaction products are easily formed in the atmosphere unless precautions are taken to exclude nitrogen and oxygen.

In addition to providing a protective environment, the shielding gas and flow rate also have a pronounced effect on the following:

1. Arc characteristics,
2. Mode of metal transfer,
3. Penetration and weld bead profile,
4. Speed of welding,
5. Undercutting tendency,
6. Cleaning action, and
7. Weld metal mechanical properties.

Generally, the manufacturer of the electrodes chosen for the welding application is a good source of information on selecting the optimum shielding gas for use with their electrodes.

The principal gases used in the spray arc mode of gas metal arc welding are shown in Table 4.7. Most of these are mixtures of inert gases that may also contain small quantities of oxygen or carbon dioxide. The use of nitrogen in welding copper is an exception. Table 4.8 lists gases used for short-circuit gas metal arc welding.

Table 4.6
Recommended Electrodes for Gas Metal Arc Welding

Base Material			
Type	Classification	Electrode Classification	AWS Electrode Specification
Aluminum and aluminum alloys	1100	ER4043	*Specification for Bare Aluminum and Aluminum Alloy Welding Electrodes and Rods*, ANSI/AWS A5.10
	3003, 3004	ER5356	
	5052, 5454	ER5554, ER5556, or ER5183	
	5083, 5086, 5456	ER5556 or ER5356	
	6061, 6063	ER4043 or ER5356	
Magnesium and magnesium alloys	AZ10A	ERAZ61A, ERAZ92A	*Specification for Magnesium Alloy Welding Electrodes and Rods*, ANSI/AWS A5.19
	AZ31B, AZ61A, AZ80A	ERAZ61A, ERAZ92A	
	ZE10A	ERAZ61A, ERAZ92A	
	ZK21A	ERAZ92A	
	AZ63A, AZ81A, AZ91C	EREZ33A	
	AZ92A, AM100A	EREZ33A	
	HK31A, HM21A, HM31A	EREZ33A	
	LA141A	EREZ33A	
Copper and copper alloys	Commercially pure	ERCu	*Specification for Copper and Copper Alloy Bare Welding Rods and Electrodes*, ANSI/AWS A5.7
	Brass	ERCuSi-A, ERCuSn-A	
	Cu-Ni alloys	ERCuNi	
	Manganese bronze	ERCuAl-A2	
	Aluminum bronze	ERCuAl-A2	
	Bronze	ERCuSn-A	
Nickel and nickel alloys	Commercially pure	ERNi	*Specification for Nickel and Nickel Alloy Bare Welding Electrodes and Rods*, ANSI/AWS A5.14/A5.14M
	Ni-Cu alloys	ERNiCu-7	
	Ni-Cr-Fe alloys	ERNiCrFe-5	
Titanium and titanium alloys	Commercially pure	ERTi-1, -2, -3, -4	*Specification for Titanium and Titanium Alloy Welding Electrodes and Rods*, ANSI/AWS A5.16
	Ti-6 AL-4V	ERTi-6Al-4V	
	Ti-0.15Pd	ERTi-0.2Pd	
	Ti-5Al-2 5Sn	ERTi-5Al-2.5Sn	
	Ti-13V-11Cr-3AL	ERTi-13V-11Cr-3AL	
Austenitic stainless steels	Type 201	ER308	*Specification for Bare Stainless Steel Welding Electrodes and Rods*, ANSI/AWS A5.9
	Types 301, 302, 304, and 308	ER308	
	Type 304L	ER308L	
	Type 309	ER 309	
	Type 310	ER310	
	Type 316	ER316	
	Type 321	ER321	
	Type 347	ER347	
Carbon steels	Hot- and cold-rolled plain carbon steels	ER70S-3	*Specification for Carbon Steel filler Metals for Gas Shielded Arc Welding*, ANSI/AWS A5.18
		ER70S-2, ER70S-4	
		ER70S-5C, ER70S-6C	
		ER70S-7	
		E70C-3X, E70C-6X	

Source: Adapted from American Welding Society (AWS) Committee on Arc Welding and Cutting, 1994, *Recommended Practices for Gas Metal Arc Welding*, ANSI/AWS C5.6-94R, Miami: American Welding Society, Table 5.

Table 4.7
GMAW Shielding Gases for Spray Transfer

Metal	Shielding Gas	Characteristics
Aluminum	100% argon	Best metal transfer and arc stability; least spatter; good cleaning action.
	35% argon–65% helium	Higher heat input than 100% argon; improved fusion characteristics on thicker material; minimizes porosity.
	25% argon–75% helium	Highest heat input; minimizes porosity; least cleaning action.
Magnesium	100% argon	Excellent cleaning action; stable arc.
	Argon +20%–70% helium	Improved wetting; less chance of porosity.
Carbon steel	1%–5% oxygen, balance argon	Improves arc stability; produces a more fluid and controllable weld pool; good fusion and bead contour; minimizes undercutting; permits higher speeds than pure argon.
	5%–15% carbon dioxide, balance argon	High-speed mechanized welding; low-cost manual welding.
Low-alloy steel	98% argon–2% oxygen	Minimizes undercutting; provides good toughness.
Stainless steel	99% argon–1% oxygen	Improves arc stability; produces a more fluid and controllable weld pool, good fusion and bead contour; minimizes undercutting on heavier stainless steels.
	98% argon–2% oxygen	Provides better arc stability, coalescence, and welding speed than 1% oxygen mixture for thinner stainless steel materials.
Nickel, copper, and their alloys	100% argon	Provides good wetting; decreases fluidity of weld metal.
	Argon-helium	Higher heat inputs of 50% and 75% helium mixtures offset high heat dissipation of heavier gauges.
Titanium	100% argon	Good arc stability; minimum weld contamination; inert gas backing is required to prevent air contamination on back of weld area.

Source: Adapted from American Welding Society (AWS) Committee on Arc Welding and Cutting, 1994, *Recommended Practices for Gas Metal Arc Welding*, ANSI/AWS C5.6-94R, Miami: American Welding Society, Table 3.

Table 4.8
GMAW Shielding Gases for Short-Circuiting Transfer

Metal	Shielding Gas	Characteristics
Carbon steel	75% argon + 25% carbon dioxide	High welding speeds with minimum melt-through; minimum spatter; clean weld appearance; good pool control in vertical and overhead positions.
	100% carbon dioxide	Deeper penetration; faster welding speeds; high spatter levels.
Stainless steel	90% helium + 7.5% argon + 2.5% carbon dioxide	No effect on corrosion resistance; small heat-affected zone; minimizes undercut.
Low-alloy steel	60% to 70% helium + 25% to 35% argon + 4.5% carbon dioxide	Minimum reactivity; excellent toughness; excellent arc stability, wetting characteristics, and bead contour; little spatter.
	75% argon +25% carbon dioxide	Fair toughness; excellent arc stability, wetting characteristics, and bead contour; little spatter.
Aluminum, copper magnesium, nickel, and their alloys	Argon and argon + helium	Argon satisfactory on sheet metal; argon-helium preferred for thicker base material.

Source: American Welding Society (AWS) Committee on Arc Welding and Cutting, 1994, *Recommended Practices for Gas Metal Arc Welding*, ANSI/AWS C5.6-94R, Miami: American Welding Society, Table 4.

Inert Shielding Gases: Argon and Helium

Argon and helium are inert gases. These gases are inert due to their electron structure and they generally do not react with the metals being joined. Argon and helium or mixtures of the two are used to weld nonferrous metals and stainless, carbon, and low-alloy steels. The physical differences between argon and helium are density, thermal conductivity, and ionization potential, all of which influence arc characteristics.

Argon is approximately 1.4 times denser than air, while the density of helium is approximately 0.14 times that of air. The heavier argon is most effective at shielding the arc and blanketing the weld area in the flat position. Helium requires approximately two to three times higher flow rates than argon to provide equal protection.

Helium has a higher thermal conductivity than argon and produces an arc plasma with more uniformly distributed arc energy. The argon arc plasma, on the other hand, is characterized by a high-energy inner core and an outer zone of less energy. This difference strongly affects the weld bead profile. A welding arc shielded by helium produces a broad, parabolic weld bead. An arc shielded by argon produces a bead profile characterized by narrow, deep penetration, and improved weld metal/ base metal wetting at the weld toes. Typical bead profiles for argon, helium, argon-helium mixtures, and carbon dioxide are illustrated in Figure 4.29.

Helium has a higher ionization potential than argon, and thus produces a higher arc voltage when other variables are held constant. As a result, helium can present problems during arc initiation. Arcs shielded only by helium do not exhibit true axial spray transfer at any current level. The result is that helium-shielded arcs produce more spatter and rougher bead surfaces than argon-shielded arcs. Argon shielding (including mixtures with as low as 80% argon) produces axial spray transfer when the current is above the transition current.

Mixtures of Argon and Helium. Pure argon shielding is used in many applications to weld nonferrous materials. The use of pure helium is generally restricted to specialized applications because an arc in helium has limited arc stability. However, the desirable weld profile characteristics (i.e., a broad, parabolic weld) obtained with the helium arc are quite often the objective in using an argon-helium shielding gas mixture. The result, as Figure 4.29 shows, is an improved weld bead profile in addition to the desirable axial spray metal transfer characteristic of argon.

In short-circuiting transfer, argon-helium mixtures of from 60% to 90% helium are used to obtain higher heat input into the base metal for better fusion characteristics. For some metals, such as the stainless and low-alloy steels, helium additions are chosen instead of carbon dioxide additions because the latter may adversely affect the mechanical properties of the deposit.

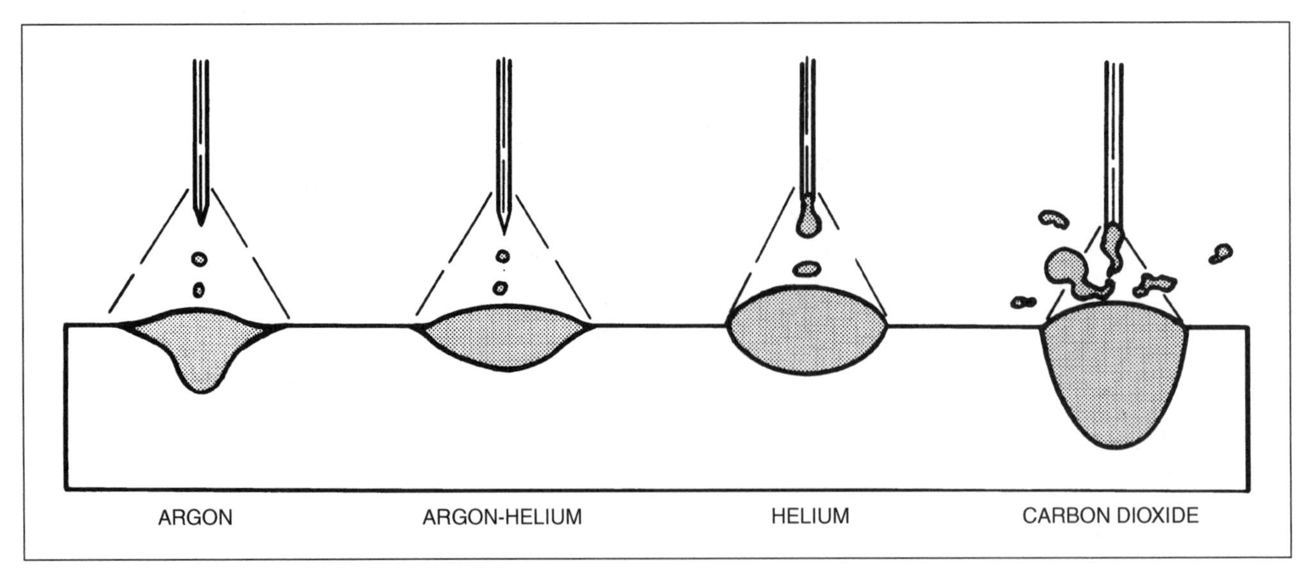

Source: Adapted from American Welding Society (AWS) Committee on Arc Welding and Cutting, 1994, *Recommended Practices for Gas Metal Arc Welding,* ANSI/AWS C5.6-94R, Miami: American Welding Society, Figure 17.

Figure 4.29—Bead Contour and Penetration Patterns for Various Shielding Gases

Mixtures of argon and 50% to 75% helium increase the arc voltage (for the same arc length) over that achieved when using pure argon. These mixtures are used for welding aluminum, magnesium, and copper because the higher heat input (from the higher voltage) reduces the effect of the high thermal conductivity of these base metals.

Oxygen and Carbon Dioxide Additions to Argon and Helium. Pure argon and, to a lesser extent, pure helium produce excellent results in welding nonferrous metals. However, pure argon shielding on ferrous alloys results in an erratic arc and a tendency for undercut to occur. The additions of 1% to 5% oxygen or 3% to 25% carbon dioxide to argon produces a noticeable improvement in arc stability and reduces the tendency for undercut. This is due to the elimination of "arc wander" on the tip of the electrode.

The optimum amount of oxygen or carbon dioxide to be added to the inert gas is a function of the workpiece surface condition (e.g., the presence or absence of mill scale or oxides), the joint geometry, the welding position or technique, and the base metal composition. Generally, 2% oxygen or 8% to 10% carbon dioxide is considered a good compromise to cover a broad range of these variables.

Carbon dioxide additions to argon may also enhance the weld bead configuration by producing a more readily defined pear-shaped profile, as illustrated in Figure 4.30. Adding between 1% and 9% oxygen to the gas improves the fluidity of the weld pool, penetration, and the arc stability. Oxygen also lowers the spray transition current at which a spray transfer is achieved. The tendency to undercut is reduced, but greater oxidation of the weld metal occurs, with a noticeable loss of alloying elements such as silicon and manganese.

Argon-carbon dioxide mixtures (up to 25% CO_2) are used on carbon and low-alloy steels, and to a lesser extent on stainless steels. The addition of carbon dioxide may produce adverse effects such as an increase in spray transition current, increased spatter, deeper penetration, and decreased arc stability. Argon-carbon dioxide mixtures are primarily used in short-circuiting transfer applications, but are also usable in spray transfer and when pulsed currents are employed.

A mixture of argon with 5% carbon dioxide has been used extensively for pulsed-current welding with solid carbon steel wires. Mixtures of argon, helium, and carbon dioxide are favored for pulsed-current welding with solid stainless steel wires.

Argon-Oxygen-Carbon Dioxide Shielding Gas Mixtures

Gas mixtures of argon with up to 20% carbon dioxide and 3% to 5% oxygen are versatile. They provide adequate shielding and desirable arc characteristics for both the spray and short-circuiting modes of gas metal arc welding. Mixtures with 10% to 20% carbon dioxide are not in common use in the United States but are popular in Europe.

Argon-Helium-Carbon Dioxide Shielding Gas Mixtures

Mixtures of argon, helium, and carbon dioxide, commonly referred to as *tri-mix,* are used with the

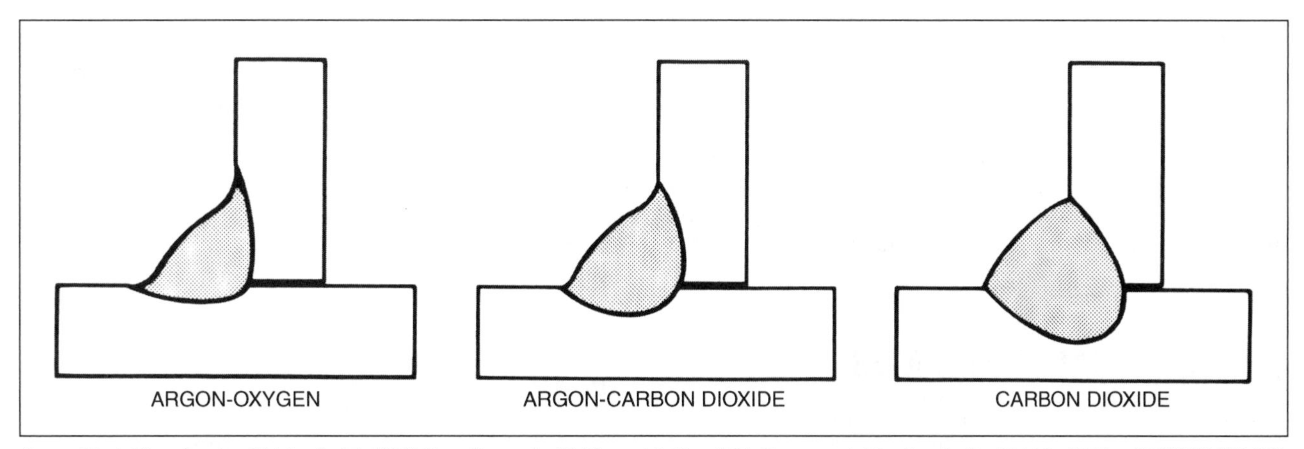

Source: Adapted from American Welding Society (AWS) Committee on Arc Welding and Cutting, 1994, *Recommended Practices for Gas Metal Arc Welding,* ANSI/AWS C5.6-94R, Miami: American Welding Society, Figure 18.

Figure 4.30—Relative Effect of Oxygen versus Carbon Dioxide Additions to the Argon Shield

short-circuiting and pulsed-current welding of carbon, low-alloy, and stainless steels. Mixtures in which argon is the primary constituent are used for pulsed-current welding, and those in which helium is the primary constituent are used for short-circuit gas metal arc welding.

Argon-Helium-Carbon Dioxide-Oxygen Shielding Gas Mixtures

The mixture of argon, helium, carbon dioxide, and oxygen, commonly referred to as *quad-mix*, is popular for high-deposition gas metal arc welding using a very high current density. This mixture yields good mechanical properties and operability over a wide range of deposition rates. Its major application is for the welding of low-alloy, high-tensile strength base materials, but it has been used on mild steel for high-production welding. Welding economics are an important consideration in using this gas to weld mild steel.

Reactive Shielding Gas: Carbon Dioxide

Carbon dioxide is a reactive gas widely used in its pure form for the gas metal arc welding of carbon and low-alloy steels. It is the only reactive gas suitable for use alone as a shield in the gas metal arc welding process. Higher welding speed, greater joint penetration, and lower cost are general characteristics that have promoted the use of carbon dioxide shielding gas.

With a carbon dioxide shield, the metal transfer mode is either short-circuiting or globular. Axial spray transfer requires an argon shield; therefore, this mode cannot be achieved with a carbon-dioxide shield. With globular transfer, the arc is quite harsh and produces a higher level of spatter. To minimize spatter, the welding conditions can be set with high current and low voltage to provide a very short arc, called a *buried arc* (i.e., the tip of the electrode is actually below the surface of the workpiece).

In overall comparison to the argon-rich shielded arc, the carbon dioxide-shielded arc produces a weld bead of excellent penetration with a rougher surface profile and much less wetting action at the sides of the weld bead due to the buried arc. Very sound weld deposits are achieved, but mechanical properties may be adversely affected due to the oxidizing nature of the arc.

SHIELDING GAS FLOW RATES

The flow rate of the shielding gas also has an influence over the quality of the resulting weld deposit. If the flow rate is too low the weld may not be properly protected from atmospheric contamination. If the flow rate is too high, turbulence may be created as the gas exits the gun nozzle, causing oxygen and nitrogen to be drawn into the weld zone. In either case, the result may be welds contaminated with entrapped oxides, nitrides, or porosity.

The selection of the proper flow rate is dependent on the gas being used and the size of the gun nozzle. When a light inert gas such as helium is used, a higher flow rate will be required to achieve the same shield integrity as that obtained with argon, a heavier gas. As the gun nozzle size is increased the flow rate requirement will generally increase proportionally.

The selection of the proper flow rate is subject to some trial and error for the particular application. Factors such as material, joint design, welding position, travel speed and type of electrode must be considered. Typical flow rates for various welding modes and materials are shown in Table 4.10 and Tables 4.12 through 4.16 in the next section, "Process Variables."

PROCESS VARIABLES

Numerous variables can affect weld penetration, bead geometry, and overall weld quality. These include the following, which are discussed in more detail below:

1. Welding amperage (electrode feed speed),
2. Polarity (electrode positive or electrode negative),
3. Arc voltage (arc length),
4. Travel speed,
5. Electrode extension (beyond the contact tip),
6. Electrode orientation (work angle, travel angle),
7. Weld joint position,
8. Electrode diameter, and
9. Shielding gas composition and flow rate.

Knowledge and control of these variables is essential to the consistent production of welds of satisfactory quality. It should be noted that these variables are not completely independent; changing one generally requires changing one or more of the others to produce the desired results. Considerable skill and experience are needed to select the optimum settings for each application. The optimum values are affected by (1) the type of base metal, (2) the electrode composition, (3) welding position, and (4) quality requirements. Thus, no single set of parameters gives optimum results in every case.

WELDING AMPERAGE

When all other variables are held constant, the welding amperage varies directly with the electrode feed speed or the electrode melting rate. As the electrode feed speed is varied, the welding amperage varies in a like manner if a constant-voltage power source is used. The magnitude depends on the polarity, electrode composition, and other factors. This relationship of welding current to wire feed speed for carbon steel electrodes is shown in Figure 4.31.

At the low-current levels for each electrode size, the curve is nearly linear. However, at higher welding currents, particularly with small diameter electrodes, the curves become nonlinear, progressively increasing at a higher rate as welding amperage increases. This is attributed to the resistance heating of the electrode extension beyond the contact tip. The curves can be approximately represented by the following equation:

$$EFS = aI + bLI^2 \quad (4.1)$$

where

EFS = Electrode feed speed, mm/s (in./min);
a = Constant of proportionality for anode or cathode heating, mm/(s · amperes) (in./[min · amperes]);
b = Constant of proportionality for electrical resistance heating, sec^{-1} $amperes^{-2}$ (min^{-1} $amperes^{-2}$);
L = Electrode extension, mm (in.); and
I = Welding current, amperes.

When the diameter of the electrode is increased while maintaining the same electrode feed speed, a higher welding current is required, as shown in Figures 4.31, 4.32, 4.33, and 4.34. This relationship between the electrode feed speed and the welding current is affected by the electrode chemical composition. This effect can be seen by comparing Figures 4.31, 4.32, 4.33, and 4.34, which are for carbon steel, aluminum, stainless steel, and copper electrodes, respectively. The different positions and slopes of the curves are due to differences in the melting temperatures and electrical resistivities of the metals. Electrode extension also affects these relationships because of the resistance heating of the electrode.

When all other variables are held constant, an increase in welding current (wire feed speed) results in the following:

1. An increase in the depth and width of the weld penetration,
2. An increase in the deposition rate (amount of weld metal deposited per unit of time), and
3. An increase in the size of the weld bead.

Arc force and deposition rate are exponentially dependent on amperage. As a result, operation above

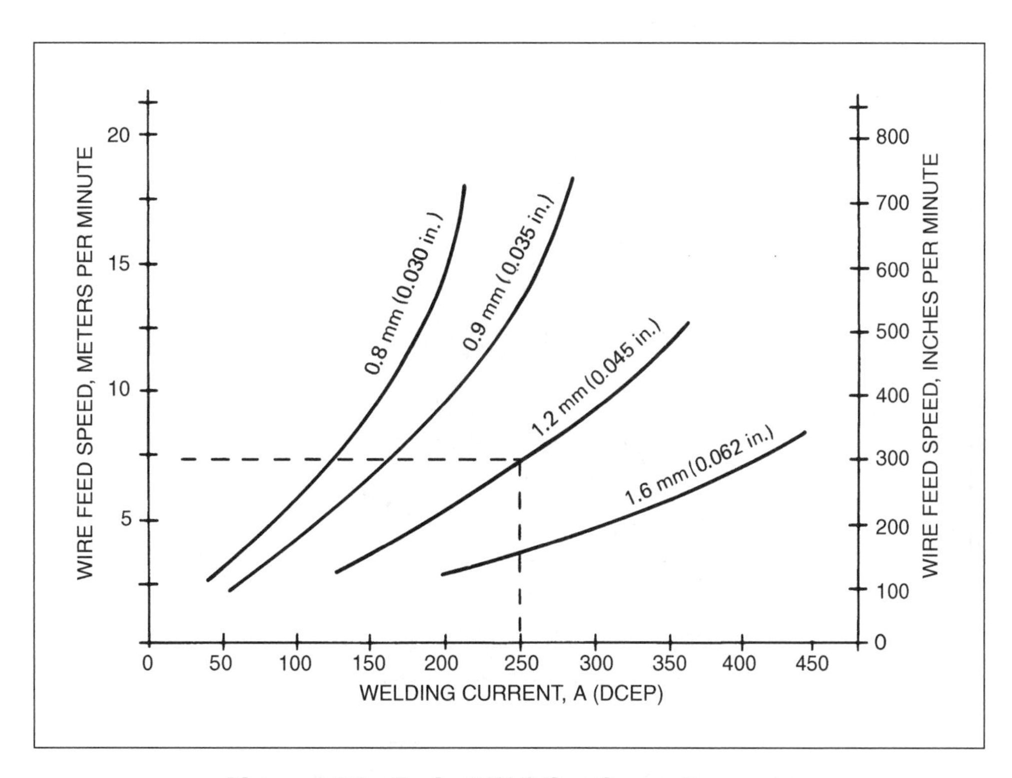

Figure 4.31—Typical Welding Currents versus Wire Feed Speeds for Carbon Steel Electrodes

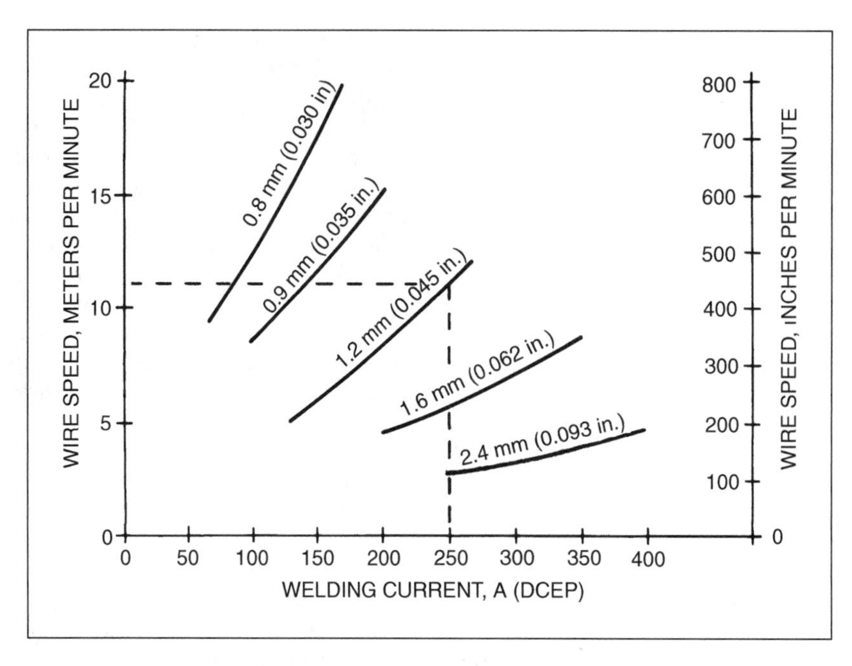

Figure 4.32—Welding Currents versus Wire Feed Speed for ER4043 Aluminum Electrodes

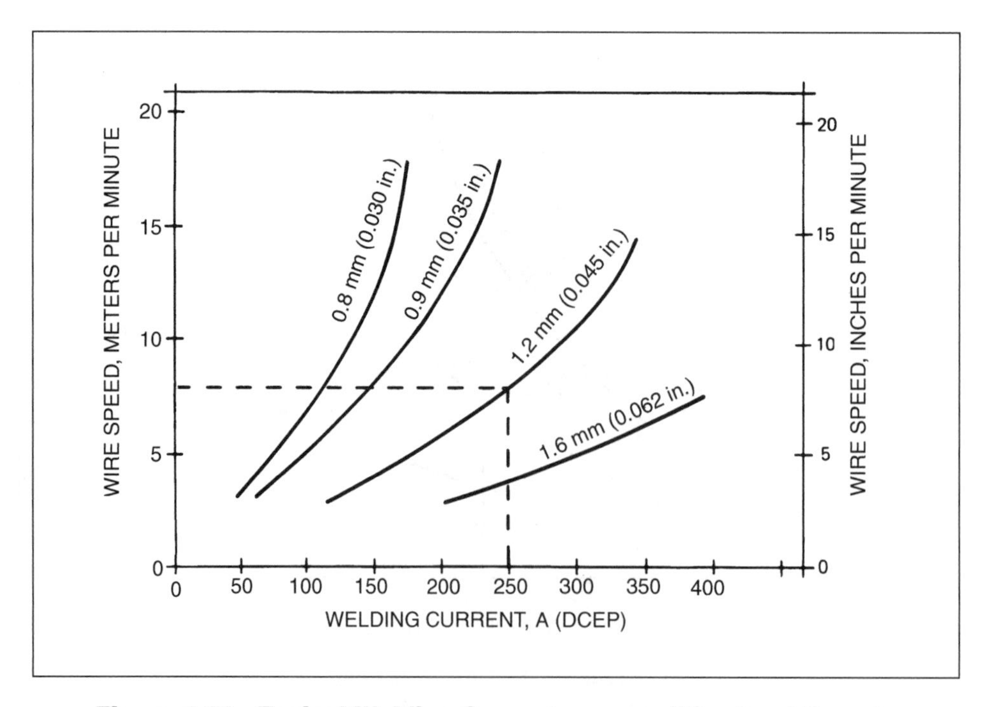

Figure 4.33—Typical Welding Currents versus Wire Feed Speeds for 300-Series Stainless Steel Electrodes

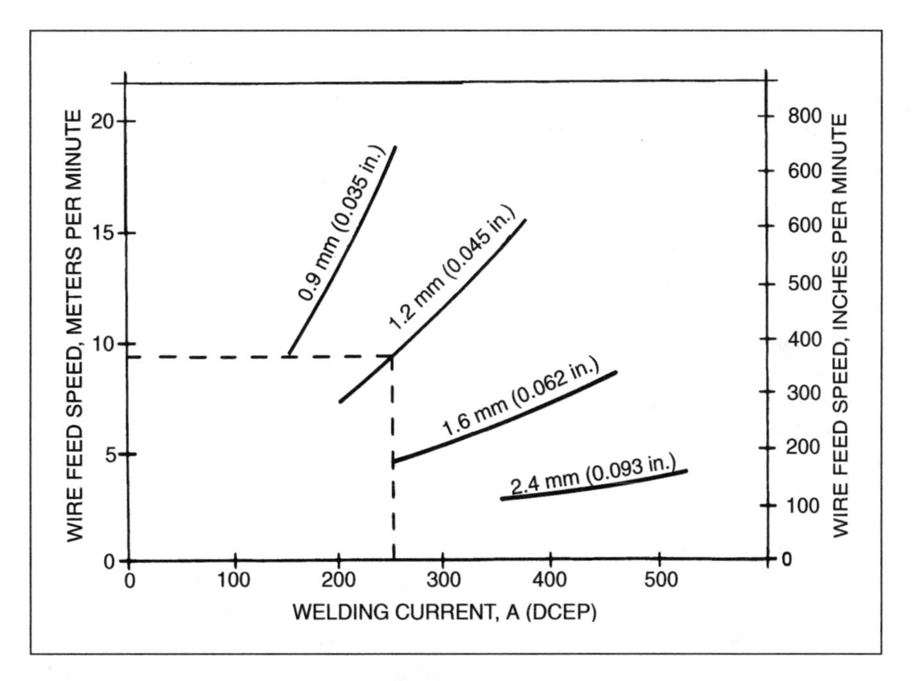

Figure 4.34—Welding Currents versus Wire Feed Speed for ECu Copper Electrodes

the spray transition current often makes the process unsuitable for use in the vertical and overhead positions. However, with power sources that pulse the welding current, both the arc force and deposition rates can be reduced, allowing welds to be made in all positions and in thin sections.

Polarity

The term *polarity* is used to describe the electrical connection of the welding gun with relation to the terminals of a direct-current power source. When the power lead of the gun is connected to the positive terminal, the polarity is designated as *direct current electrode positive* (DCEP). When the gun is connected to the negative terminal, the polarity is designated as *direct current electrode negative* (DCEN). The vast majority of gas metal arc welding applications, whether the mode is short-circuiting, globular or spray transfer, use DCEP. This condition yields a stable arc, smooth metal transfer, relatively low spatter, good weld bead characteristics, and the greatest depth of penetration for a wide range of welding amperage.

Direct current electrode negative (DCEN) is seldom used because axial spray transfer is not possible without modifications that have had little commercial acceptance. DCEN has a distinct advantage of high electrode melting rates, but this generally cannot be exploited because the transfer is globular. With steels, the transfer can be improved by adding a minimum of 5% oxygen to the argon shield (requiring special alloys to compensate for oxidation losses) or by treating the wire to make it thermionic (adding to the cost of the filler metal). In both cases, the deposition rates drop, eliminating the only real advantage of changing polarity. However, because of the high deposition rate and reduced penetration, DCEN has found some use in surfacing applications.

Historically, attempts to use alternating current (ac) with the gas metal arc welding process have generally been unsuccessful. The cyclic waveform creates arc instability due to the tendency of the arc to extinguish as the current passes through the zero point. Special wire surface treatments have been developed to overcome this problem, but the expense of applying them has made this approach uneconomical.[3] Recently (in the early 2000s) at least one Japanese equipment manufacturer has developed an inverter-based power source that is capable of producing a waveform that is quite usable and offers a number of advantages, especially for sheet metal applications.

3. As of the writing of this chapter, several special-purpose alternating-current power sources have been developed and made commercially available.

ARC VOLTAGE (ARC LENGTH)

The terms *arc voltage* and *arc length* are often used interchangeably. It should be noted, however, that these terms have different connotations even though they are directly related. With gas metal arc welding, arc length is a critical variable that must be carefully controlled. For example, in the spray-arc mode with argon shielding, an arc that is too short experiences momentary short-circuits. These short-circuits cause pressure fluctuations that pump air into the arc stream, producing porosity or embrittlement due to absorbed nitrogen. If the arc is too long, it tends to wander, affecting both the penetration and surface bead profiles. A long arc can also disrupt the gas shield. In the case of a buried arc with a carbon-dioxide shield, a long arc results in an unburied condition and produces excessive spatter and porosity. If the arc is too short, the electrode tip short-circuits the weld pool, causing instability.

Arc voltage depends on the arc length as well as many other variables, such as the electrode composition and dimensions, the shielding gas, the welding technique, electrode extension, and even the length of the welding cable, since arc voltage often is measured at the power source. As shown in Figure 4.35, arc voltage is an approximate means of stating the physical arc length in electrical terms even though the measured arc voltage also includes the voltage drop in the electrode extension beyond the contact tip.

With all variables held constant, arc voltage is directly related to arc length. Even though the arc length is the variable of interest and the variable that should be controlled, the voltage is more easily monitored. Because of this factor, as well as the requirement that the arc voltage be specified in the welding procedure, the term *arc voltage* is commonly used.

Arc voltage settings vary depending on the material, shielding gas, and transfer mode. Typical values are shown in Table 4.9. Trial runs are necessary to adjust the arc voltage to produce the most favorable arc characteristics and weld bead appearance. Trials are essential because the optimum arc voltage is dependent upon a variety of factors, including metal thickness, the type of joint, welding position, electrode size, shielding gas

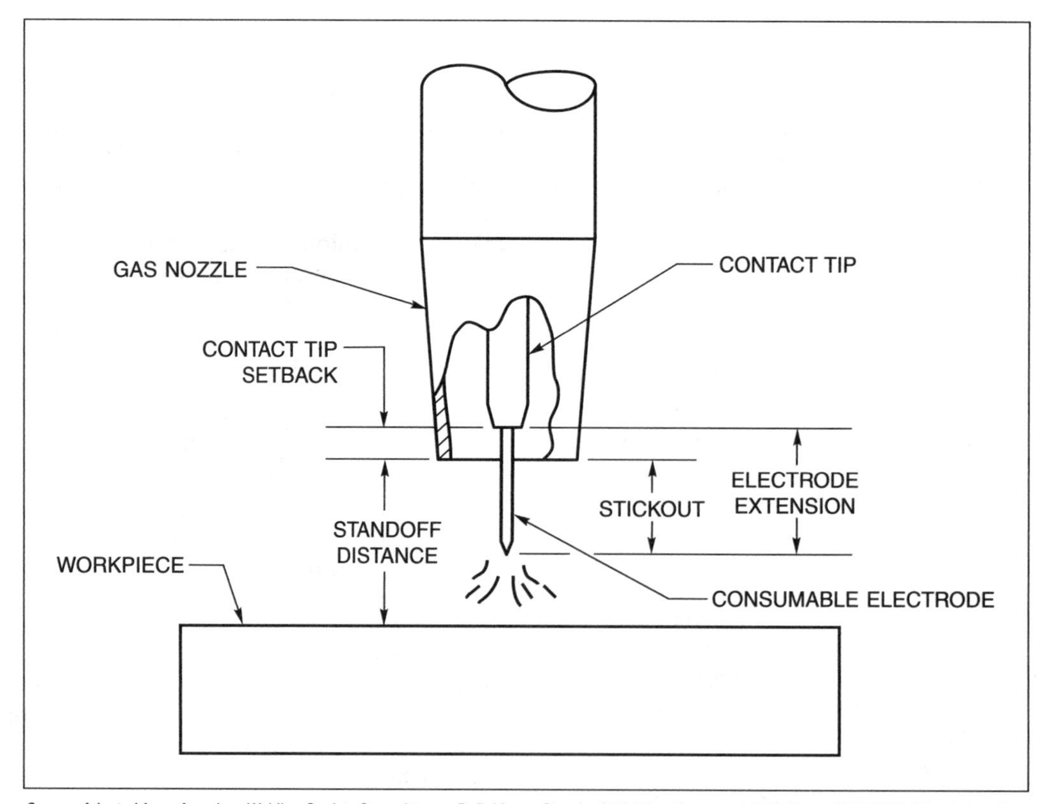

Source: Adapted from American Welding Society Committee on Definitions, *Standard Welding Terms and Definitions*, A3.0:2001, Miami: American Welding Society, Figure 38(A).

Figure 4.35—Arc Welding Gun Nomenclature

Table 4.9
Typical Arc Voltages for the Gas Metal Arc Welding of Various Metals*

	Globular and Spray† Globular Transfer					Short-Circuiting Transfer			
	1.6 mm (1/16 in.) Diameter Electrode					0.9 mm (.035 in.) Diameter Electrode			
Metal	Argon	Helium	25% Argon–75% Helium	Argon–Oxygen (1%–5% Oxygen)	Carbon Dioxide	Argon	Argon–Oxygen (1%–5% Oxygen)	75% Argon–25% Carbon Dioxide	Carbon Dioxide
Aluminum	25	30	29	—	—	19	—	—	—
Magnesium	26	—	28	—	—	16	—	—	—
Carbon steel	—	—	—	28	30	17	18	19	20
Low-alloy steel	—	—	—	28	30	17	18	19	20
Stainless steel	24	—	—	26	—	18	19	21	—
Nickel	26	30	28	—	—	22	—	—	—
Nickel-copper alloy	26	30	28	—	—	22	—	—	—
Nickel-chromium-iron alloy	26	30	28	—	—	22	—	—	—
Copper	30	36	33	—	—	24	22	—	—
Copper-nickel alloy	28	32	30	—	—	23	—	—	—
Silicon bronze	28	32	30	28	—	23	—	—	—
Aluminum bronze	28	32	30	—	23	—	—	—	
Phosphor bronze	28	32	30	23	—	23	—	—	—

*Plus or minus approximately 10%. The lower voltages are normally used on thin-gauge material and at low amperage; the higher voltages are used on thick sections of material at high amperage.
† For pulsed-current spray welding, the arc voltage would be from 18 V to 28 V depending on the amperage range used.

composition, and the type of weld. From any specific value of arc voltage, a voltage increase tends to flatten the weld bead and increase the width of the fusion zone. Excessively high voltage may cause porosity, spatter, and undercut. A reduction in voltage results in a narrower weld bead with a higher crown and deeper penetration. Excessively low voltage may cause the "stubbing" of the electrode, a condition in which the electrode dips into the weld pool and then solidifies into place due to the lack of a short-circuit current.

TRAVEL SPEED

The term *travel speed* is defined as the linear rate at which the arc is moved along the weld joint. With all other conditions held constant, weld penetration is maximum at an intermediate travel speed. When the travel speed is decreased, the deposition of filler metal per unit length increases. At very slow speeds, the welding arc impinges on the weld pool rather than on the base metal, thereby reducing the effective penetration. A wide weld bead is also a result.

As the travel speed is increased, the thermal energy per unit length of weld transmitted to the base metal from the arc is at first increased because the arc acts more directly on the base metal. With further increases in travel speed, less thermal energy per unit length of weld is imparted to the base metal. Therefore, the melting rate of the base metal first increases and then decreases with increasing travel speed. As the travel speed is further increased, undercutting tends to occur along the toe of the weld bead because insufficient filler metal has been deposited to fill the path melted by the arc.

ELECTRODE EXTENSION

The electrode extension is the distance between the end of the contact tip and the end of the electrode, as shown in Figure 4.35. An increase in the electrode extension results in an increase in its electrical resistance. In turn, resistance heating causes the temperature of the electrode to rise and results in a small increase in the electrode melting rate. Overall, the increased electrical resistance produces a greater voltage drop from the contact tip to the workpiece. This is sensed by the power source, which compensates by decreasing the current. The electrode melting rate is immediately reduced, permitting the electrode to shorten the physical arc length. Thus, unless the voltage is increased at the welding machine, the filler metal is deposited as a narrow, high-crowned weld bead.

The desirable electrode extension typically ranges from 6 mm to 13 mm (1/4 in. to 1/2 in.) for short-circuiting transfer and from 13 mm to 25 mm (1/2 in. to 1 in.) for globular and spray transfer.

ELECTRODE ORIENTATION

As with all arc welding processes, the orientation of the welding electrode in relation to the weld joint affects the weld bead shape and penetration. Electrode orientation affects bead shape and penetration to a greater extent than arc voltage or travel speed. The orientation of the electrode is described in two ways—by the relationship of the electrode axis to the direction of travel (the travel angle) and by the angle between the electrode axis and the adjacent workpiece surface (the work angle). The electrode orientation and its effect on the width and penetration of the weld are illustrated in Figure 4.36(A), (B), and (C). When the electrode points in the direction of travel, the technique is referred to as *forehand welding* with a *push angle* (A). When the electrode points opposite from the direction of travel, the technique is termed *backhand welding* with a *drag angle* (C).

When the electrode is changed from the perpendicular (B) to a push-angle (forehand) technique with all other conditions unchanged, the penetration decreases and the weld bead becomes wider and flatter (A).

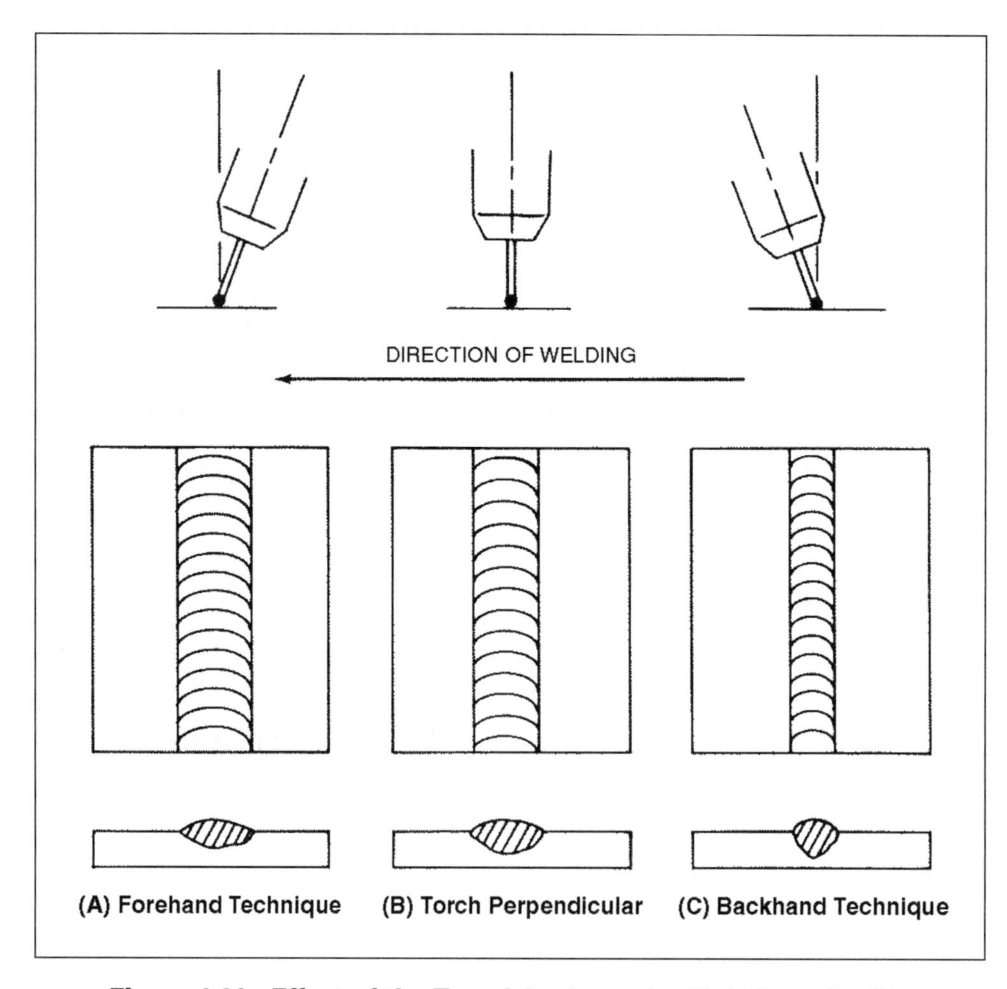

Figure 4.36—Effect of the Travel Angle on the Weld Bead Profile

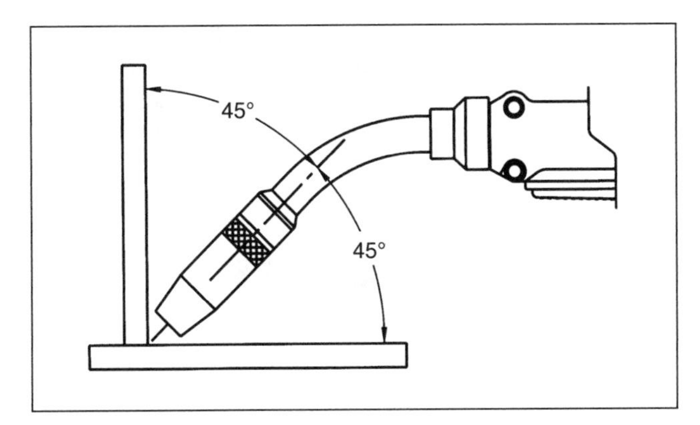

Figure 4.37—Normal Work Angle for Fillet Welds

Maximum penetration is obtained in the flat position with the backhand technique at a drag angle of about 25° from perpendicular. The drag-angle technique produces a more convex and narrower bead (C), a more stable arc, and less spatter on the workpiece. For all positions, the electrode travel angle normally used is a drag angle in the range of 5° to 15° for good control and shielding of the weld pool.

For some materials, such as aluminum, a push-angle (forehand technique) is preferred. This technique provides a cleaning action ahead of the molten weld metal, which promotes wetting and reduces base metal oxidation. A push-angle technique can also be used on thin-gauge material to minimize burn-through.

When producing fillet welds in the horizontal position, the electrode should be positioned at a work angle of approximately 45° (or slightly greater for larger welds) to the vertical member, as illustrated in Figure 4.37.

WELD JOINT POSITION

Most spray-mode gas metal arc welding is done in the flat or horizontal positions. At the lower energy levels achieved with pulsed power sources or the short-circuit mode of transfer, the process can be used in all positions. Fillet welds made in the flat position with spray transfer are usually more uniform, less likely to have unequal legs and convex profiles, and less susceptible to undercutting than similar fillet welds made in the horizontal position.

To minimize the effect of gravity on the weld metal in the vertical and overhead positions of welding, smaller weld beads are deposited with a lower energy input. This is accomplished using small-diameter electrodes with either short-circuiting metal transfer or spray transfer with pulsed direct current. Electrode diameters of 1.1 mm (0.045 in.) and smaller are best suited for this out-of-position welding, where the low heat input allows the molten pool to freeze quickly. Downward welding progression is usually effective on sheet metal in the vertical position.

When welding is done in the flat position, the inclination of the weld axis in relation to the horizontal plane influences the weld bead shape, penetration, and travel speed. In flat-position circumferential welding, the workpiece rotates under the welding gun and an inclination is obtained by moving the welding gun in either direction from top dead center.

Excessive weld reinforcement can be decreased by positioning linear joints with the weld axis at 15° to the horizontal and welding downhill. The same welding conditions would produce excessive reinforcement when the workpiece is in the flat position. Speeds can also usually be increased when using downhill travel. At the same time, penetration is lower, which is beneficial when welding sheet metal.

The effect of inclination on the shape of the weld bead is illustrated schematically in Figure 4.38. Downhill welding affects the weld contour and penetration, as shown in Figure 4.38(A). The weld pool tends to flow toward the electrode and preheats the base metal, particularly at the surface, producing an irregularly shaped fusion zone. As the angle of inclination increases, the middle surface of the weld is depressed, penetration decreases, and the width of the weld increases. Although downhill welding may have its benefits, this technique is not recommended for aluminum due to the loss of cleaning action and of adequate shielding.

Uphill welding affects the fusion zone contour and the weld surface, as illustrated in Figure 4.38(B). The force of gravity causes the weld pool to flow backward and lag behind the electrode. The edges of the weld lose metal because the molten metal flows to the center. As the angle of inclination increases, face reinforcement and penetration increase, and the width of the weld decreases. The effects are exactly the opposite of those produced by downhill welding. When higher welding currents are used, the maximum usable angle decreases.

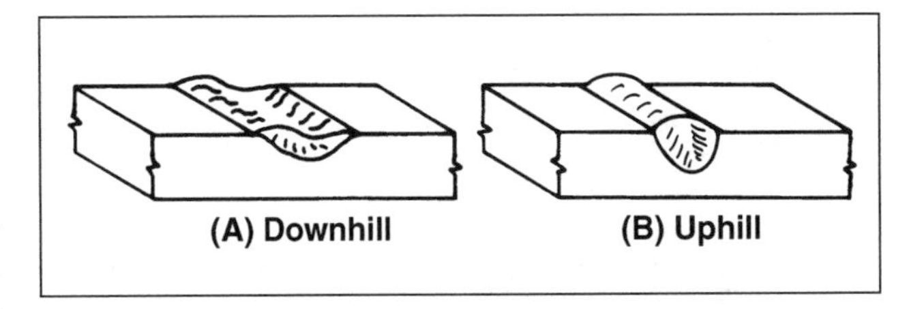

Figure 4.38—Effect of Inclination on Weld Bead Shape

ELECTRODE SIZE

The electrode size (diameter) influences the weld bead configuration. A larger electrode requires higher minimum current than a smaller electrode for the same metal transfer characteristics. Higher currents, in turn, produce additional electrode melting and larger, more fluid weld deposits. As a consequence, vertical and overhead welding are usually performed with smaller-diameter electrodes and lower currents even though higher currents result in higher deposition rates and greater penetration.

SHIELDING GAS COMPOSITION AND FLOW RATES

A variety of combinations of inert and reactive shielding gases (argon, helium, carbon dioxide, and oxygen) can be used and the selection has a significant influence on arc characteristics and weld properties. For information on the various mixtures of shielding gases, applications, and flow rates, refer to the previous section, "Shielding Gases."

SELECTION OF PARAMETERS

The selection of the process parameters—amperage, voltage, travel speed, shielding gas and flow rate, electrode extension, and so on—requires some experimentation to determine an acceptable set of conditions. This task is complicated by the fact that there is an interdependence of several of these variables. For example, a change in amperage (wire feed speed) may require an adjustment in the arc voltage in order to maintain the desired arc length. Typical ranges of variables have been established and are listed in Tables 4.10 through 4.16 for various base metals.

Table 4.10
Typical Conditions for the Gas Metal Arc Welding of Carbon and Low-Alloy Steels in the Flat Position (Short-Circuiting Transfer)

Material Thickness		Type of Weld	Electrode Diameter		Amperage	Voltage*	Electrode Feed Speed		Shielding Gas†	Gas Flow	
mm	in.		mm	in.	A	V	mm/s	in./min.		L/min.	ft³/h
1.6	0.062	Butt‡	0.9	0.035	95	18	64	150	Argon 75%, carbon dioxide 25%	12	25
3.2	0.125	Butt‡	0.9	0.035	140	20	106	250	Argon 75%, carbon dioxide 25%	12	25
4.7	0.187	Butt‡	0.9	0.035	150	20	112	265	Argon 75%, carbon dioxide 25%	12	25
6.4	0.250	Butt‡	0.9	0.035	150	21	112	265	Argon 75%, carbon dioxide 25%	12	25
6.4	0.250	Butt§	1.1	0.045	200	22	106	250	Argon 75%, carbon dioxide 25%	12	25

*Direct current electrode positive.
† Welding-grade carbon dioxide may also be used.
‡ Root opening of 0.8 mm (0.03 in.).
§ Root opening of 1.6 mm (0.062 in.).

Table 4.11
Typical Conditions for the Gas Metal Arc Welding of Carbon and Low-Alloy Steels (Spray Transfer)*

Material Thickness			Electrode Diameter		Amperage	Voltage**	Electrode Feed Speed		Travel Speed	
mm	in.	Type of Weld	mm	in.	A	V	mm/s	in./min.	mm/s	in./min.
3.2	0.125	Butt	0.89	0.035	190	26	148	350	8–11	20–25
6.4	0.25	Butt	1.1	0.045	320	29	169	400	7–9	17–22
9.5	0.375	Butt	1.1	0.045	300	29	154	365	5–7	11–16
9.5	0.375	Fillet	1.6	0.063	300	26	87	205	4–6	10–15
12.7	0.500	Butt	1.6	0.063	320	26	82	195	7–9	17–22
19.1	0.750	Fillet	1.6	0.063	360	27	99	235	4–6	10–15

* Shielding gas is composed of 98% argon and 2% oxygen at 19–24 L/min. (40 to 50 ft^3/h). The welding position is flat for groove welds and horizontal for fillet welds.

** Direct current electrode positive.

Table 4.12
Typical Conditions for the Gas Metal Arc Welding of Aluminum in the Flat Position (Spray Transfer Mode)

Material Thickness		Type of Weld	Electrode Diameter		Amperage	Voltage*	Electrode Feed Speed		Shielding Gas	Gas Flow	
mm	in.		mm	in.	A	V	mm/s	in./min.		L/min.	ft^3/h
3.2	0.125	Butt	0.8	0.030	125	20	186	440	Argon	14	30
4.8	0.187	Butt	1.1	0.045	160	23	116	275	Argon	16	35
6.4	0.250	Butt	1.1	0.045	205	24	142	335	Argon	16	35
9.5	0.375	Butt	1.6	0.063	240	26	91	215	Argon	19	40
19	0.750	Butt	1.6	0.063	240	28	91	215	75% Helium–25% Argon	42	90
76	3.0	Butt	2.4	0.093	400	33	76	180	75% Helium–25% Argon	70	150

*Direct current electrode positive.

Table 4.13
Typical Conditions for the Gas Metal Arc Welding of Austenitic Stainless Steel Using a Spray Arc in the Flat Position

Material Thickness			Electrode Feed Speed		Amperage	Voltage*	Wire Feed Speed		Shielding Gas	Gas Flow	
mm	in.	Type of Weld	mm	in.	A	V	mm/s	in./min.		L/min.	ft^3/h
3.2	0.125	Butt—with backing	1.6	0.062	225	24	55	130	98% Argon–2% Oxygen	14	30
6.4	0.250	V-groove butt—60°	1.6	0.062	275	26	74	175	98% Argon–2% Oxygen	16	35
9.5	0.375	V-groove butt—60°	1.6	0.062	300	28	102	240	98% Argon–2% Oxygen	16	35

*Direct current electrode positive.

Table 4.14
Typical Conditions for the Gas Metal Arc Welding of Austenitic Stainless Steel Using a Short-Circuiting Arc

Material Thickness		Type of Weld	Electrode Diameter		Amperage	Voltage*	Electrode Feed Speed		Shielding Gas	Gas Flow	
mm	in.		mm	in.	A	V	mm/s	in./min.		L/min.	ft³/h
1.6	0.062	Butt	0.8	0.030	85	21	78	185	90% Helium–7.5% Argon–2.5% Carbon Dioxide	14	30
2.4	0.093	Butt	0.8	0.030	105	23	97	230	90% Helium–7.5% Argon–2.5% Carbon Dioxide	14	30
3.2	0.125	Butt	0.8	0.030	125	24	118	280	90% Helium–7.5% Argon–2.5% Carbon Dioxide	14	30

*Direct current electrode positive.

Table 4.15
Typical Conditions for the Gas Metal Arc Welding of Copper Alloys in the Flat Position

Material Thickness			Electrode Diameter		Amperage	Voltage*	Electrode Feed Speed		Shielding Gas	Gas Flow	
mm	in.	Type of Weld	mm	in.	A	V	mm/s	in./min.		L/min.	ft³/h
3.2	0.125	Butt	0.89	0.035	175	23	430	182	Argon	12	25
4.8	0.187	Butt	1.1	0.045	210	25	240	101	Argon	14	30
6.40	0.250	Butt, spaced	1.6	0.062	365	26	240	101	Argon	16	35

*Direct current electrode positive.

Table 4.16
Typical Variable Settings for the Gas Metal Arc Welding of Magnesium

Material Thickness			Electrode Diameter		Amperage	Voltage*	Electrode Feed Speed		Argon Flow	
mm	in.	Type of Weld	mm	in.	A	V	mm/s	in./min.	L/min.	ft³/h
1.6	0.062	Square groove or fillet	1.6	0.062	70	16	68	160	24	50
2.3	0.090	Square groove or fillet	1.6	0.062	105	17	104	245	24	50
3.2	0.125	Square groove or fillet	1.6	0.062	125	18	123	290	24	50
6.4	0.250	Square groove or fillet	1.6	0.062	265	25	254	600	28	60
9.5	0.375	Square groove or fillet	2.4	0.094	335	26	157	370	28	60

*Direct current electrode positive.

WELD JOINT DESIGNS

Gas metal arc welding can be used to weld a wide variety of metals and configurations. Typical weld joint geometries and dimensions for the gas metal arc welding process, as used in the welding of steel, are shown Figure 4.39. The dimensions shown normally produce complete joint penetration and acceptable face reinforcement or root reinforcement when suitable welding procedures are used. Similar joint configurations may be used on other metals, although the more thermally conductive types (e.g., aluminum and copper) should have larger groove angles to minimize problems with incomplete fusion. The deep penetration characteristics of spray transfer gas metal arc welding may permit the use of smaller included angles. This reduces the amount of filler metal required and the labor hours needed to fabricate weldments.

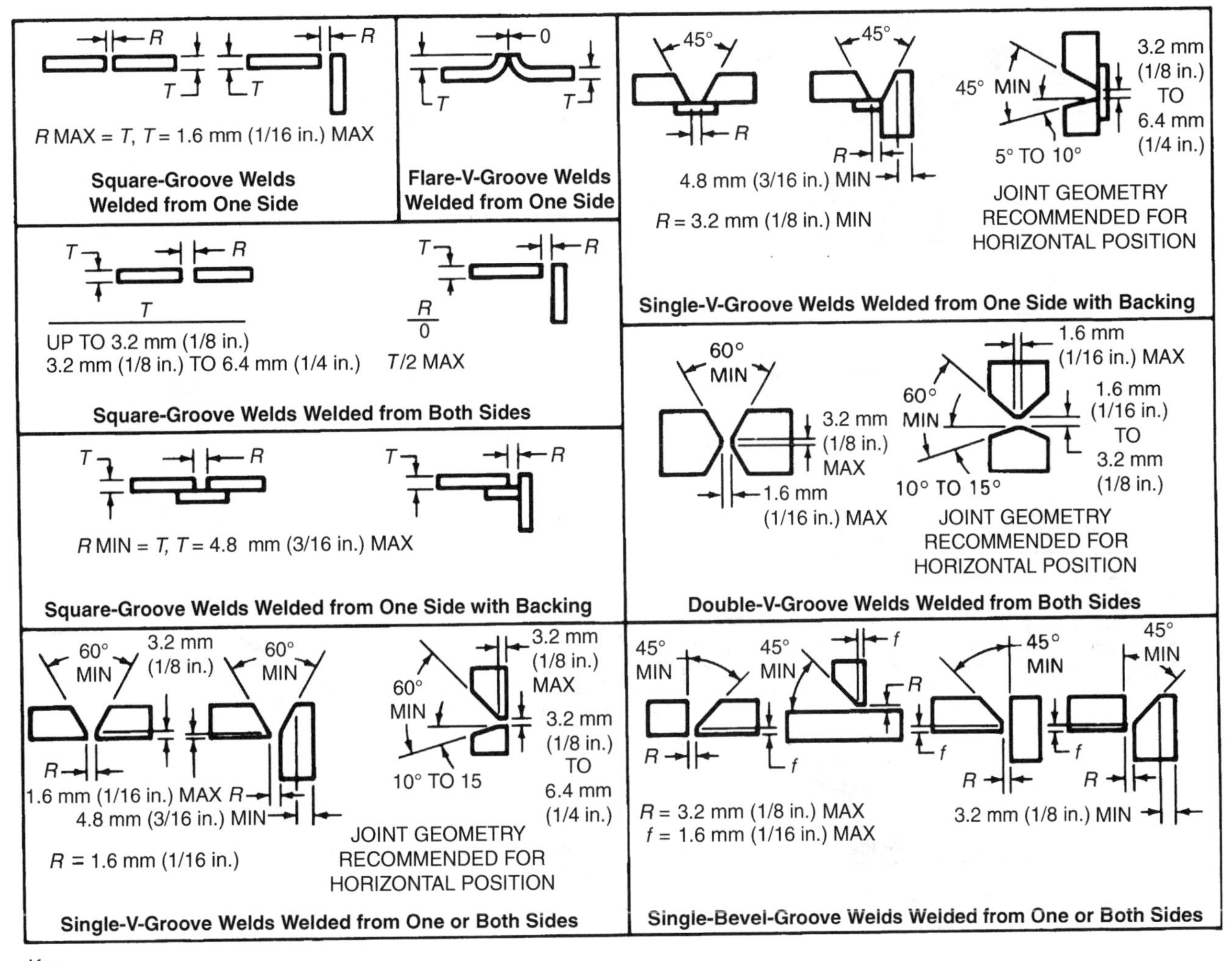

Key:
α = Groove angle, radians (degrees)
f = Root face, mm (in.)
r = Bevel radius, mm (in.)
R = Root opening, mm (in.)
T = Thickness

Source: Adapted from American Welding Society (AWS) Committee on Arc Welding and Cutting, 1994, *Recommended Practices for Gas Metal Arc Welding*, ANSI/AWS C5.6-94R, Miami: American Welding Society, Figure 34.

Figure 4.39—Typical Weld Joint Geometries and Dimensions for the Gas Metal Arc Welding Process as Used for Steel

INSPECTION AND WELD QUALITY

Weld quality control procedures for gas metal arc welding (GMAW) are quite similar to those used for other processes. Depending on the applicable specifications, inspection procedures should provide for the following: (1) determining the adequacy of welder and welding operator performance qualification tests, (2) qualification and implementation of a satisfactory welding procedure, and (3) a complete examination of the final welded product.

Thorough, in-process inspection of welding operations (e.g., adherence to established and qualified procedures) is one of the most effective inspection tools. This is because final weld inspection on the assembled product is often limited to nondestructive examination methods (e.g., visual, liquid penetrant, and magnetic

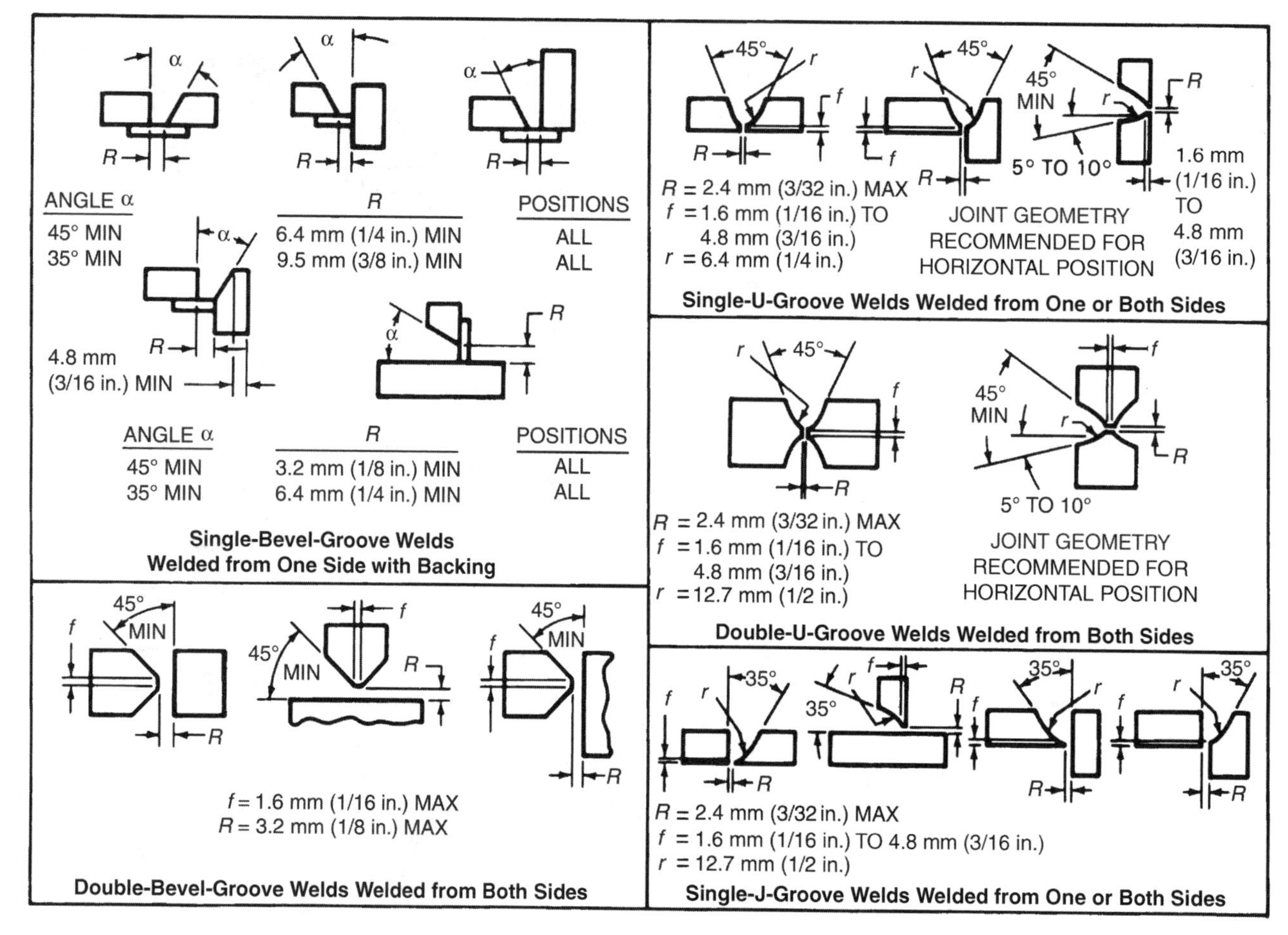

Key:
α = Groove angle, radians (degrees)
f = Root face, mm (in.)
r = Bevel radius, mm (in.)
R = Root opening, mm (in.)
T = Thickness

Source: Adapted from American Welding Society (AWS) Committee on Arc Welding and Cutting, 1994, *Recommended Practices for Gas Metal Arc Welding*, ANSI/AWS C5.6-94R, Miami: American Welding Society, Figure 34.

Figure 4.39 (Continued)—Typical Weld Joint Geometries and Dimensions for the Gas Metal Arc Welding Process as Used for Steel

particle), all of which are restricted to surface evaluation. If the design of the weld joints is appropriate for radiographic and ultrasonic inspection, these techniques can be used. Destructive testing, including tensile, shear, fatigue, impact, bend, fracture, peel, cross-section, or hardness tests, is usually confined to engineering development, welding procedure qualification, and welder and welding operator performance qualification tests.

POTENTIAL QUALITY CONCERNS

The gas metal arc welding process is capable of depositing very high quality welds under the right conditions. However, as with any process, the potential for discontinuities such as cracks, incomplete fusion, and porosity in the final weld is always present. Discontinuities that are of particular concern when using the gas metal arc welding process, as well as possible causes and corrective actions, are discussed in this section.

Hydrogen Cracking

An awareness of the potential problems related to hydrogen contamination during welding is important, even though it is less likely to occur with gas metal arc welding, since no hygroscopic flux or coating is used. However, other hydrogen sources must be considered. For example, shielding gas must be sufficiently low in moisture content. The moisture content in shielding gas is typically specified in terms of a dew point temperature. Lower values of dew point indicate lower moisture content. A common value, e.g., as specified in AWS D1.1)[4] is –40°C (–40°F). This is usually well controlled by the gas supplier, but may need to be checked. In steels, especially high-strength alloys, hydrogen absorption by the molten weld metal can result in heat-affected-zone cracking and weld metal cracking. Referred to as *delayed cracking* or *underbead cracking*, it may not manifest itself until hours, or even days, after the weld has cooled.

Oil, grease, and drawing compounds on the electrode or the base metal are potential sources for hydrogen pickup in the weld metal. Electrode manufacturers are aware of the need for cleanliness and normally take special care to provide a clean electrode. However, contaminants may be introduced during handling in the user's facility. Users who are aware of such possibilities take steps to avoid serious problems, particularly in welding hardenable steels. The same awareness is necessary in the welding of aluminum, except that the potential problem is porosity caused by the relatively low solubility of hydrogen in solidified aluminum, rather than hydrogen embrittlement.

Oxygen and Nitrogen Contamination

Oxygen and nitrogen contamination is potentially a greater cause of concern than hydrogen contamination in the gas metal arc welding process. If the shielding gas is not completely inert or adequately protective, oxygen and nitrogen may be readily absorbed from the atmosphere. The oxides and nitrides that are formed can reduce weld metal notch toughness. Because of this, weld metal deposited by gas metal arc welding (GMAW) is not as tough as weld metal deposited by gas tungsten arc welding (GTAW). It should be noted, however, that oxygen in percentages of up to 5% and more can be added to the shielding gas without adversely affecting weld quality when used on ferrous materials.

4. American Welding Society (AWS) Committee on Structural Steel, D1.1 and D1.1M, *Structural Welding Code—Steel*, AWS D1.1/D1.1M:2002, Miami: American Welding Society.

Cleanliness

Base metal cleanliness when using gas metal arc welding is more critical than with shielded metal arc welding (SMAW), flux cored arc welding (FCAW) or submerged arc welding (SAW). The fluxing compounds present in these processes scavenge and cleanse the molten weld deposit of oxides and gas-forming compounds. Such fluxing agents are not present in gas metal arc welding. This places a premium on doing a thorough job of preweld and interpass cleaning. This is particularly true for aluminum, where elaborate procedures for chemical cleaning or mechanical removal of metallic oxides, or both, are applied.

WELD DISCONTINUITIES

Some of the more common weld discontinuities that may occur with the gas metal arc welding process are described below.

Excessive Melt-Through

Melt-through is the visible root reinforcement produced in a joint that is welded from one side. This can be a concern in gas metal arc welding because of the high current density and the penetrating nature of the arc. Although this root reinforcement may be acceptable in many cases, an excessive amount may lead to a poor surface contour and cracking. Figure 4.40 illustrates excessive melt-through. The possible causes of this discontinuity and the corrective actions that can be taken are summarized in Table 4.17.

Heat-Affected Zone Cracking

Cracking in the heat-affected zone of the base metal is usually associated with hardenable steels. It is no more likely to occur with gas metal arc welding than with other processes. The possible causes and corrective actions are summarized in Table 4.18.

Incomplete Fusion

The reduced heat input common to the short-circuiting mode of gas metal arc welding results in minimum melting of the base metal. This is desirable on thin-gauge materials and for out-of-position welding. However, an improper welding technique may result in incomplete fusion, especially in root areas or along groove faces. This may also occur when using the spray mode of transfer if the arc force is directed at the weld pool instead of at the base material.

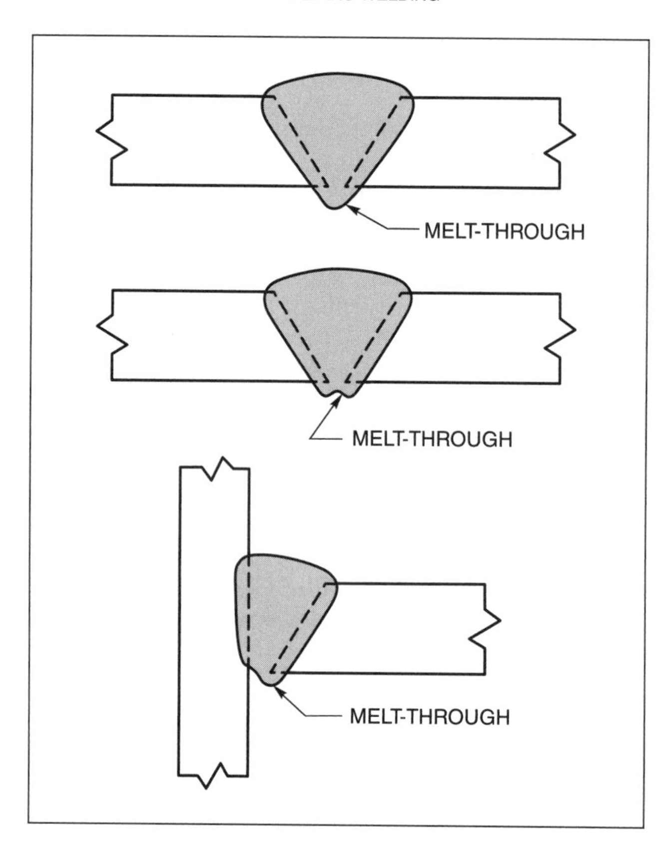

Figure 4.40—Excessive Melt-Through in a GMAW Weld

Table 4.17
Possible Causes and Remedies for Excessive Melt-Through

Possible Causes	Corrective Actions
Excessive heat input	Reduce the wire feed speed (welding current) and the voltage. Increase the travel speed.
Improper joint configuration	Reduce the root opening. Increase the root face dimension.

Table 4.18
Possible Causes and Remedies for Heat-Affected Zone Cracking

Possible Causes	Corrective Actions
Hardening in the heat-affected zone	Preheat to retard the cooling rate.
Residual stresses too high	Use stress-relief heat treatment.
Hydrogen embrittlement	Use a clean electrode and dry shielding gas. Remove contaminants from the base metal. Hold the weld at elevated temperatures for several hours before cooling (the temperature and time required to diffuse hydrogen are dependent on the type of base metal).

A cross section of a weld exhibiting incomplete fusion is shown in Figure 4.41. The possible causes of this discontinuity and corrective actions that can be taken are shown in Table 4.19.

Incomplete Joint Penetration

Incomplete joint penetration is a condition in the root area of a groove weld in which weld metal does not extend through the joint thickness, as illustrated in

Photograph courtesy of Northrop Grumman Marine Systems

Figure 4.41—Incomplete Fusion in a Weld Produced with Gas Metal Arc Welding

Table 4.19
Possible Causes and Remedies for Incomplete Fusion

Possible Causes	Corrective Actions
Weld zone surfaces not free of film or excessive oxides.	Clean all groove faces and weld zone surfaces of any mill scale impurities prior to welding.
Insufficient heat input	Increase the wire feed speed. Reduce electrode extension.
Too large a weld pool	Minimize excessive weaving to produce a more controllable weld pool. Increase the travel speed.
Improper weld technique	When using a weaving technique, dwell momentarily on the faces of the groove. Provide improved access at root of joints. Keep electrode directed at the leading edge of the pool.
Improper joint design (see Figure 4.38)	Use a groove angle large enough to allow access to the bottom and groove faces while maintaining proper electrode extension. Use a "J" or "U" groove.

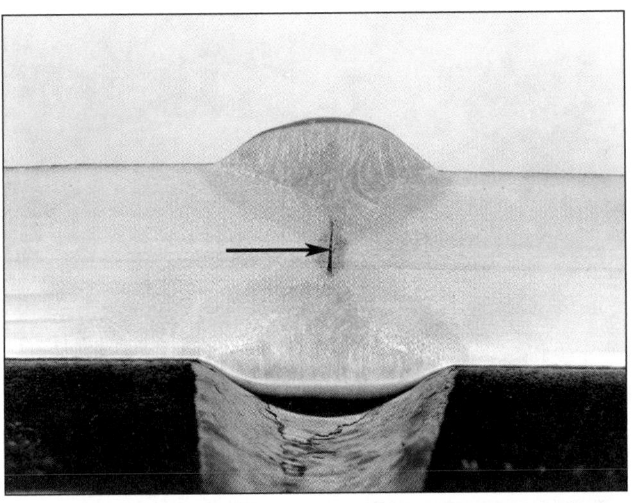

Photograph courtesy of Northrop Grumman Marine Systems

Figure 4.42—Incomplete Joint Penetration

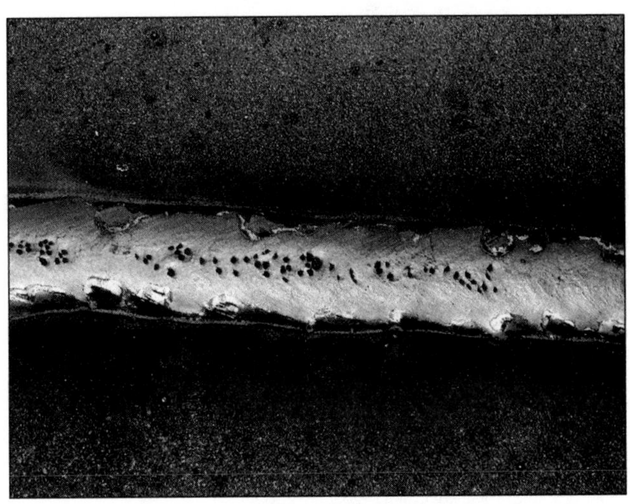

Photograph courtesy of Northrop Grumman Marine Systems

Figure 4.43—Weld Surface Porosity

Table 4.20
Possible Causes and Remedies for Incomplete Joint Penetration

Possible Causes	Corrective Actions
Improper joint preparation	Joint design must provide proper access to the bottom of the groove while maintaining proper electrode extension. Reduce excessively large root face. Increase the root opening in butt joints, and increase depth of backgouge.
Improper weld technique	Maintain 5°–15° drag angle to achieve maximum penetration. Keep arc on leading edge of the pool.
Inadequate welding current	Increase the wire feed speed (welding current).
Excessive travel speed	Reduce travel speed.
Travel speed too slow	Increase travel speed.

Figure 4.42. This condition can be a problem particularly with the low-energy forms of gas metal arc welding such as short-circuiting transfer. The possible causes of incomplete joint penetration and corrective actions that can be taken are summarized in Table 4.20.

Porosity

Porosity is a cavity-type discontinuity formed by gas entrapment during the solidification of a weld deposit. These cavities vary in size and may contain oxygen, nitrogen, or hydrogen, or even the shielding gas that is being used. Porosity on the surface of a weld is shown in Figure 4.43. The existence of internal porosity is exhibited in Figure 4.44. Because the gas metal arc welding process is sensitive to contamination and relies on the proper application of a gas shield, it can be subject to porosity to a greater degree than other processes that utilize fluxing agents. The possible causes of porosity and corrective actions that can be taken are outlined in Table 4.21.

Undercut

As shown in Figure 4.45, undercut is a groove melted into the base metal adjacent to the weld toe or root and left unfilled by weld metal. Because of the inherently high fluidity of the weld deposit and the ability of the process to weld at high travel speeds, gas metal arc welds can be susceptible to undercut when improper techniques are used. The possible causes of undercut and corrective actions that can be taken are presented in Table 4.22.

Weld Metal Cracks

A weld metal crack is one of the most serious discontinuities that may be present in a weld. Cracks often result from the metallurgical conditions that exist, and

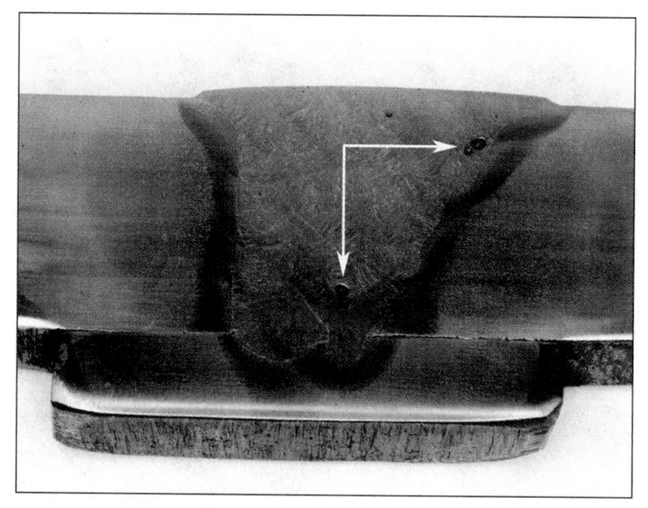

Photograph courtesy of Northrop Grumman Marine Systems

Figure 4.44—Internal Porosity

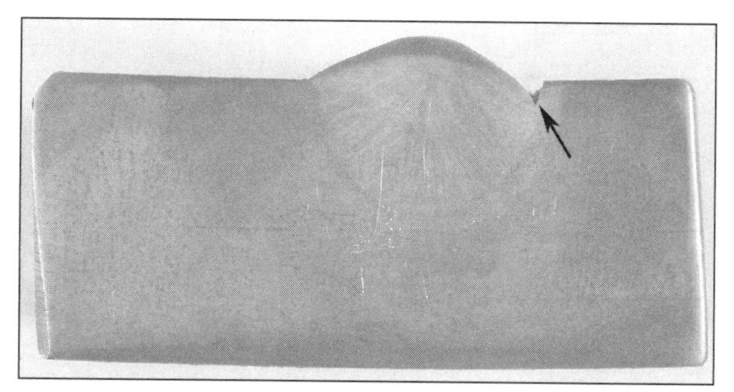

Figure 4.45—Undercut at the Toe of a GMAW Weld

Table 4.21
Possible Causes and Remedies for Porosity

Possible Causes	Corrective Actions
Inadequate shielding gas coverage	Optimize the gas flow. Increase gas flow to displace all air from the weld zone. Decrease excessive gas flow to avoid turbulence and the entrapment of air in the weld zone. Eliminate any leaks in the gas line. Eliminate drafts (from fans, open doors, etc.) blowing into the welding arc. Eliminate frozen (clogged) regulators in carbon dioxide welding by using heaters. Reduce travel speed. Reduce nozzle-to-workpiece distance. Hold gun at end of weld until molten metal solidifies.
Gas contamination	Use welding grade shielding gas.
Electrode contamination	Use only clean and dry electrode.
Workpiece contamination	Remove all grease, oil, moisture, rust, paint, and dirt from workpiece surface before welding. Use more highly deoxidizing electrode.
Arc voltage too high	Reduce voltage.
Excess contact tip-to-work distance	Reduce electrode extension.

Table 4.22
Possible Causes and Remedies for Undercut

Possible Causes	Corrective Actions
Travel speed too high	Use slower travel speed.
Welding voltage too high	Reduce the voltage.
Excessive welding current	Reduce wire feed speed.
Insufficient dwell	Increase dwell at edge of weld pool.
Gun angle	Change gun angle so arc force can aid in metal placement.

they are very dependent on the material being welded. However, the gas metal arc welding process can be responsible for or contribute to the incidence of cracking.

Typical weld metal cracks are shown in Figures 4.46 and 4.47. The possible causes of weld metal cracks and corrective actions are outlined in Table 4.23.

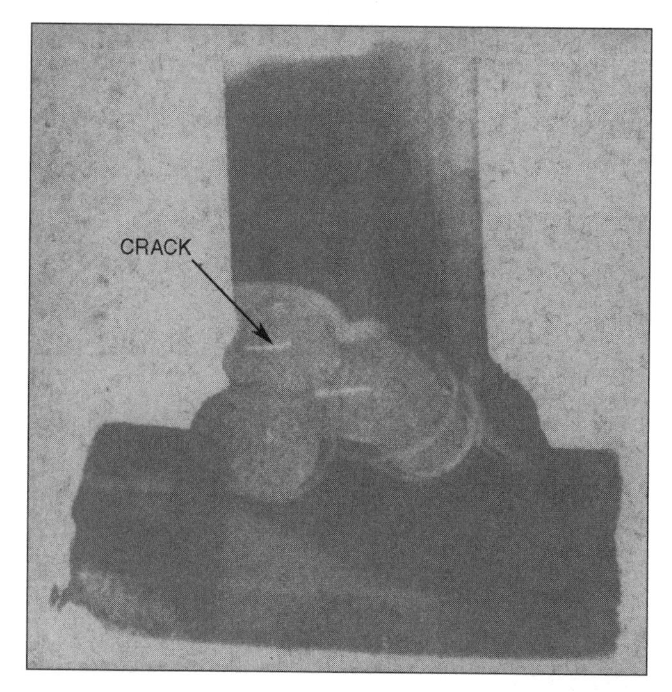

Figure 4.46—Cracks in a Weld as a Result of Excessive Depth-to-Width Ratio

Photograph courtesy of Northrop Grumman Marine Systems

Figure 4.47—Crater Crack in Weld Metal at the End of a Gas Metal Arc Weld

TROUBLESHOOTING

Troubleshooting of any process requires a thorough knowledge of the equipment and the function of the various components, the materials involved, and the process itself. It is a more complicated task with gas metal arc welding than with manual processes such as shielded metal arc welding and gas tungsten arc welding because of the complexity of the equipment, the number of variables, and the interrelationship of these variables.

The difficulties that arise can be categorized as electrical, mechanical, and process-related. Tables 4.24 through 4.26 address these categories, listing the problems that are likely to be encountered, their possible causes, and remedies. It should be noted that the difficulties referred to in these tables occur during the welding operation or prevent the making of the weld as opposed to those that are discovered while inspecting the final product. The latter type is covered in the section titled "Inspection and Weld Quality" in this chapter.

Table 4.23
Possible Causes and Remedies for Weld Metal Cracking

Possible Causes	Corrective Actions
Improper joint design	Maintain proper groove dimensions to allow deposition of adequate filler metal or weld cross section to overcome restraint conditions.
Too high a weld depth-to-width ratio (see Figure 4.45)	Increase the arc voltage or decrease the current, or both, to widen the weld bead or decrease the penetration.
Small or concave weld bead (particularly fillet and root beads)	Decrease the travel speed to increase cross section of deposit.
Heat input too high, causing excessive shrinkage and distortion	Reduce the current or voltage, or both; increase the travel speed.
Hot-shortness	For carbon steel, use an electrode with a higher manganese content (use a shorter arc length to minimize the loss of manganese across the arc); adjust the groove angle to allow an adequate percentage of filler metal addition; adjust the pass sequence to reduce the restraint on the weld during cooling; change to another filler metal that provides the desired characteristics.
High restraint of the joint members	Use preheat to reduce the magnitude of residual stresses Adjust the welding sequence to reduce the restraint conditions.

Table 4.24
Troubleshooting Electrical Problems Encountered in Gas Metal Arc Welding

Problem	Possible Cause	Remedy
Difficult arc starting	Wrong polarity	Check polarity; reverse the leads if necessary.
	Poor workpiece lead connection	Secure the workpiece lead connection.
Irregular wire feed and burn-back	Power circuit fluctuations	Check the line voltage.
	Polarity wrong	Check the polarity; reverse the leads if necessary.
Welding cables overheating	Cables are too small or too long	Check the current-carrying requirements and replace or shorten if necessary.
	Cable connections loose	Tighten.
No wire feed speed control circuit	Broken or loose wires in control	Check and repair if necessary.
	Defective PC board in governor	Replace the PC board.
Unstable arc	Cable connections are loose	Tighten the connections.
Electrode won't feed	Control circuit fuse blown	Replace the fuse.
	Fuse blown in power source	Replace the fuse.
	Defective gun trigger switch or broken wire leads	Check the connections; replace the switch.
	Drive motor burned out	Check and replace.
Wire feeds but no gas flows	Failure of gas valve solenoid	Replace.
	Loose or broken wires to gas valve solenoid	Check and repair if necessary.
Electrode wire feeds but is not energized (no arc)	Poor workpiece connection	Tighten if loose; clean workpiece of paint, rust, or other contaminant.
	Loose cable connections	Tighten.
	Primary contactor coil or points defective	Repair or replace.
	Contactor control leads broken	Repair or replace.
Porosity in weld	Loose or broken wires to gas solenoid valve	Repair or replace.

Table 4.25
Troubleshooting Mechanical Problems Encountered in Gas Metal Arc Welding

Problem	Possible Cause	Remedy
Irregular wire feed and burnback	Insufficient drive roll pressure	Adjust.
	Contact tube plugged or worn	Clean or replace.
	Kinked electrode wire	Cut out; replace the spool.
	Coiled gun cable	Straighten the cables; hang the wire feeder.
	Conduit liner dirty or worn	Clean or replace.
	Conduit too long	Shorten, or use the push-pull drive system.
Electrode wire wraps around drive roll ("birdnesting")	Excessive feed roll pressure	Adjust.
	Incorrect conduit liner or contact tip	Match the liner and contact tip to the electrode size.
	Misaligned drive rolls or wire guides	Check and align properly.
	Restriction in gun or gun cable	Remove the restriction.
Heavily oxidized weld deposit	Air/water leaks in gun and cables	Check for leaks and repair or replace as necessary.
	Restricted shield gas flow; defective gas solenoid valve	Check and clean the nozzle and repair or replace.
Electrode wire stops feeding while welding	Excess or insufficient drive roll pressure	Adjust.
	Wire drive rolls misaligned or worn	Realign or replace, or both.
	Liner or contact tip plugged	Clean or replace.

Table 4.25 (Continued)
Troubleshooting Mechanical Problems Encountered in Gas Metal Arc Welding

Problem	Possible Cause	Remedy
Wire feeds but no gas flows	Gas cylinder is empty	Replace and purge the lines before welding.
	Gas cylinder valve closed	Open cylinder valve.
	Flow meter not adjusted	Adjust to render the flow specified in the procedure.
	Restriction in gas line or nozzle	Check and clean.
Porosity in the weld bead	Failed gas valve solenoid	Repair or replace.
	Gas cylinder valve closed	Turn valve on.
	Insufficient shielding gas flow	Check for restrictions in the gas line or nozzle and correct.
	Leaks in gas supply lines (including the gun)	Check for leaks (especially at the connections) and correct.
Wire feed motor operates but wire does not feed	Insufficient drive roll pressure	Adjust.
	Incorrect wire feed rolls	Match the feed rolls to the wire size and type.
	Excessive pressure on wire spool brake	Decrease the brake pressure.
	Restriction in the conduit liner or gun	Check the liner and contact tip; clean or replace, or both.
	Incorrect liner or contact tip	Check and replace with correct size.
Welding gun overheats	Pinched or clogged coolant line	Check and correct.
	Low coolant level in pump reservoir	Check and add coolant as necessary.
	Water pump not functioning correctly	Check and repair or replace.

Table 4.26
Troubleshooting Process-Related Problems Encountered in Gas Metal Arc Welding

Problem	Possible Cause	Remedy
Unstable arc	Weld joint area dirty	Clean to remove scale, rust, etc.
Heavily oxidized weld deposit	Improper gun angle	Use approximately 15° push angle or trail angle.
	Excessive nozzle-to-workpiece distance	Reduce distance; distance should be approximately 12.7 mm to 19.1 mm (1/2 in. to 3/4 in.).
	Air drafts	Protect the weld area from drafts.
	Contact tube not centered in the gas nozzle distance	Center the contact tip.
Porosity in the weld bead	Dirty base material	Clean to remove scale, rust, etc.
	Excessive wire feed speed	Reduce.
	Moisture in the shielding gas	Replace the gas cylinder.
	Contaminated electrode	Keep the wire protected while using. Clean wire before it enters feeder.
	Gas flow rate too high or too low	Adjust.
Electrode wire stubs into the workpiece	Excessive wire feed speed	Reduce the speed.
	Arc voltage too low	Increase the voltage.
	Excessive slope set on power source (for short-circuiting transfer)	Reset to reduce slope.
Excessive spatter	Excessive arc voltage	Reduce the voltage.
	Insufficient slope set on power source (for short-circuiting transfer)	Increase slope setting.
	Contact tip recessed too far in nozzle	Adjust or replace with a longer contact tip.
	Excessive gas flow rate	Reduce flow.

ECONOMICS

Several economic advantages can be attributed to gas metal arc welding: the process is flexible, efficient, and capable of depositing high-quality welds on a variety of materials and configurations. Gas metal arc welding is characterized by high deposition rates and a high operator factor, i.e., the ratio of arc time or actual weld deposition time to the total work time required of the welder or operator.[5]

Although the cost of gas metal arc welding equipment is significantly higher than the cost of the equipment used in manual welding processes, e.g., shielded metal arc welding, this expenditure can normally be justified through savings in the cost of labor. The process can be easily automated to achieve even higher efficiencies. In addition, the cost of training personnel to use the process is typically lower than the cost of training for the manual processes.

Users should perform a cost analysis of their particular situation to determine whether the gas metal arc welding process is beneficial in comparison to competing processes. Table 4.27 lists typical electrode (wire) weights, in inch-pound units, for various electrode wire compositions. Table 4.28 lists the factors that yield the total weld costs and the formulas for calculating the cost of each item.

Table 4.29 shows how the formulas are used to calculate weld costs based on a given set of welding conditions. It should be noted that the calculations in Table 4.28 and 4.29 are for the conventional application of the process and do not cover the cost of procedures that are often necessary in the completion of a weldment, including material preparation, preheating, postweld heat treatment, and so forth. The formulas in Table 4.28 and the calculations in Table 4.29 are in inch-pound units; however, the formulas can be readily adapted for SI calculations.

5. For additional information on calculating welding costs, see Chapter 12, "Economics of Welding" in American Welding Society (AWS) Welding Handbook, *Welding Science and Technology*, Volume 1 of Welding Handbook, 9th ed. AWS Welding Handbook Committee, Jenney, C. L., and A. O'Brien, eds, 2001, Miami: American Welding Society.

Table 4.27
Inches of Electrode (Wire) per Pound of Material

Wire Diameter			Material								
Decimal		Fraction									
mm	(in.)	(in.)	Mag.	Alum.	Aluminum-Bronze (10%)	Mild Steel	Stainless Steel 300 Series	Si. Bronze	Copper-Nickel	Nickel	De-ox. Copper
0.51	0.020		50500	32400	11600	11100	10950	10300	9950	9900	9800
0.76	0.030		22400	14420	5150	4960	4880	4600	4430	4400	4360
0.89	0.035		16500	10600	3780	3650	3590	3380	3260	3240	3200
1.02	0.040		12600	8120	2900	2790	2750	2580	2490	2480	2450
1.14	0.045		9990	6410	2290	2210	2170	2040	1970	1960	1940
1.57	0.062	1/16	5270	3382	1220	1160	1140	1070	1040	1030	1020
2.38	0.094	3/32	2350	1510	538	519	510	480	462	460	455

Table 4.28
Weld Cost Formulas Used to Determine the Cost of a Pound of Completed Weld (Not Including Such Items as Material Preparation, Preheating, and Postweld Cleanup)

$$\text{Total cost} = \frac{\text{Labor} + \text{Overhead} + \text{Electrode} + \text{Shielding gas} + \text{Power}}{\text{Operator factor}} = \text{\$/lb of weld deposited} \quad (1)$$

$$\text{Labor} = \frac{\text{Welder rate, \$/h}}{\text{Deposition rate lb/h}} = \text{\$/lb} \quad (2)$$

$$\text{Overhead} = \frac{\text{Overhead rate, \$/h}}{\text{Deposition rate, lb/h}} = \text{\$/lb} \quad (3)$$

$$\text{Electrode} = \frac{\text{\$/lb}}{\text{Deposition efficiency}} = \text{\$/lb} \quad (4)$$

$$\text{Shielding gas} = \frac{\text{Gas flow rate, ft}^3\text{/h}}{\text{Depositon rate, lb/h}} \times \frac{\$}{\text{ft}^3} = \text{\$/lb} \quad (5)$$

$$\text{Power} = \frac{\text{\$/kilowatt (kW) h}}{\text{Depositon rate, lb/h}} \times \frac{A \times V}{1000 \times \text{Power source efficiency}} = \text{\$/lb} \quad (6)$$

where

$$\text{Deposition rate, lb/h} = \text{Deposition efficiency} \times \text{Electrode (wire) melting rate, lb/h}$$

$$\text{Electrode (wire) melting rate, lb/h} = \frac{\text{Electrode (wire) feed speed, in./min} \times 60}{\text{Electrode (wire) weight, in./lb}}$$

$$\text{Deposition efficiency} = 0.9 - 0.95 \text{ (for GMAW)}$$

Electrode (wire) feed speed, in./min. = Measure or use approximation from Tables 4.30 through 4.33

$$\text{Power source efficiency} = \frac{\text{Output power } (V \times A)}{\text{Input power } (V \times A)}$$

$$\text{Operator factor} = \frac{\text{Arc time, h}}{\text{Welder's total labor time, h}}$$

Electrode (wire) weight, in./lb = Input from Table 4.28

Table 4.29
Welding Conditions and Cost Calculations for a Sample Weld (Inch-Pound Units)

Welding Conditions

Material	**Mild steel**
Wire size	0.45 in.
Wire feed speed	365 in./min.
Amperage	300 A
Voltage	29 V
Electrode (wire) cost	\$0.75/lb
Gas	Argon, 2% oxygen
Gas cost	\$0.05/ft^3
Gas flow rate	40 ft^3/hr
Deposition efficiency	0.9%
Welder labor rate	\$16.00/hr
Overhead rate	\$24.00/hr
Power cost	\$0.09/kwh
Power source efficiency	0.5 (50%)
Operator factor	0.6 (60%)

Cost Calculations

$$\text{Total cost} = \frac{\text{Labor} + \text{Overhead} + \text{Electrode} + \text{Shielding gas} + \text{Power}}{\text{Operator factor}} = \text{\$/lb of weld deposited}$$

$$\text{Labor} = \frac{\$16/\text{h}}{\$\ 8.92\ \text{lb/h}} = \$1.79/\text{lb}$$

$$\text{Overhead} = \frac{\$24/\text{h}}{\$\ 8.92\ \text{lb/h}} = \$2.69/\text{lb}$$

$$\text{Electrode} = \frac{\$.75/\text{lb}}{.90} = \$.83/\text{lb}$$

$$\text{Shielding gas} = \frac{40\ \text{ft}^3/\text{h}}{\$\ 8.92\ \text{lb/h}} \times \$.05/\text{ft}^3 = \$.22/\text{lb}$$

$$\text{Power} = \frac{\$.09/\text{kW h}}{\$\ 8.92\ \text{lb/h}} \times \frac{33\ \text{A} \times\ 29\ \text{V}}{1000 \times .5} = \$.17/\text{lb}$$

$$\text{Cost} = \frac{\$1.79/\text{lb} + \$2.69/\text{lb} + \$.83/\text{lb} + \$.22/\text{lb} + \$.17/\text{lb}}{.6} = \frac{\$5.70/\text{lb}}{.6}$$

SAFE PRACTICES

The general subject of safety in welding, cutting, and allied processes is addressed in the American National Standards Institute document *Safety in Welding, Cutting, and Allied Processes,* ANSI Z49.1,[6] and *Fire Prevention in the Use of Welding and Cutting Processes,* ANSI Z49.2,[7] as well as in Chapter 17 in *Welding Science and Technology,* Volume 1 of the *Welding Handbook,* 9th edition.[8] The safe handling of compressed gases is covered in *Safe Handling of Gas in Containers,* CGA P-1.[9] Personnel should be familiar with the safe practices outlined in these documents and those specified in the pertinent equipment operating manuals and material safety data sheets (MSDS) for consumables. A list of codes, standards, and other documents applicable to safe practices in the welding industry is presented in Appendix B. The publishers and their contact information are included.

Other potential hazards in gas metal arc welding—including the handling of cylinders and regulators, fumes and gases, radiant energy, noise, and electric shock—warrant consideration and are discussed in this section.

SAFE HANDLING OF GAS CYLINDERS AND REGULATORS

Compressed gas cylinders should be carefully handled and adequately secured when stored or in use. Knocks, falls, or rough handling may damage cylinders, valves, and fuse plugs, and cause leakage or an accident. Valve-protecting caps should be kept in place (hand tight) until the cylinder is connected to the equipment being used, as per the standard *Safe Handling of Gas in Containers,* CGA P-1.[10]

The following precautions should be observed when setting up and using cylinders containing shielding gas:

1. The cylinder should be properly secured;
2. Before connecting a regulator to the cylinder valve, the valve should momentarily be opened slightly and then closed immediately to clear the valve of dust or dirt that otherwise might enter the regulator. This procedure is often referred to as *cracking*. The valve operator should stand to one side of the regulator gauges, never in front of them;
3. After the regulator is attached, the pressure adjusting screw should be released by turning it counter-clockwise. The cylinder valve should then be opened slowly to prevent a rapid surge of high-pressure gas into the regulator. The adjusting screw should then be turned clockwise until the proper pressure is obtained; and
4. The source of the gas supply (i.e., the cylinder valve) should be shut off if it is to be left unattended, and the adjusting screw should be turned counterclockwise to the full open position.

6. American National Standards Institute (ANSI) Accredited Standards Committee Z49, *Safety in Welding, Cutting, and Allied Processes,* ANSI Z49.1, Miami: American Welding Society.
7. American National Standards Institute, *Fire Prevention in the Use of Welding and Cutting Processes,* ANSI Z49.2, New York, American Standards Institute.
8. American Welding Society (AWS) Welding Handbook Committee, *Welding Science and Technology,* Volume 1 of *Welding Handbook,* 9th ed., Jenney, C. L., and A. O'Brien, eds., 2001, Miami: American Welding Society.
9. Compressed Gas Association (CGA), *Safe Handling of Gas in Containers,* CGA P-1, Arlington, Virginia: Compressed Gas Association.
10. See Reference 9.

GASES

The major toxic gases associated with gas metal arc welding are ozone, nitrogen dioxide, and carbon monoxide. Phosgene gas could also be present as a result of thermal or ultraviolet decomposition of chlorinated hydrocarbon cleaning agents located in the vicinity of welding operations. Two such solvents are trichlorethylene and perchlorethylene. Degreasing or other cleaning operations involving chlorinated hydrocarbons should be located so that vapors from these operations cannot be reached by radiation from the welding arc or breathed by the welder.

Ozone

The ultraviolet light emitted by the GMAW arc acts on the oxygen in the surrounding atmosphere to produce ozone, the amount of which depends upon the intensity and the wave length of the ultraviolet energy, the humidity, the amount of screening afforded by any welding fumes, and other factors. With the use of argon as the shielding gas and when welding highly reflective metals, the ozone concentration generally rises with an increase in welding current. If the ozone cannot be reduced to a safe level by means of ventilation or process variations, it may be necessary to supply fresh air to the welder with either an air-supplied respirator or other means.

Nitrogen Dioxide

High concentrations of nitrogen dioxide are usually found only within 152 mm (6 in.) of the arc. With natural ventilation, these concentrations are quickly reduced to safe levels in the welder's breathing zone provided that the welder's head is kept out of the plume of fumes and thus out of the plume of welding-generated gases. Nitrogen dioxide is not considered a hazard in gas metal arc welding.

Carbon Monoxide

Carbon monoxide is formed when the carbon dioxide shielding used with the gas metal arc welding process is dissociated by the heat of the arc. Only a small amount of carbon monoxide is created by the welding process, although relatively high concentrations are formed temporarily in the plume of fumes. However, the hot carbon monoxide oxidizes to carbon dioxide so that the concentrations of carbon monoxide become insignificant at distances of more than 76 mm to 102 mm (3 in. to 4 in.) from the welding plume.

Under normal welding conditions, this source presents no hazards. Ventilation adequate to deflect the plume or remove the fumes and gases must be provided when the welder must work with his or her head over the welding arc, with natural ventilation moving the plume of fumes toward the breathing zone, or in areas where welding is performed in a confined space. Because shielding gas can displace air, special care must be taken to ensure that breathing air is safe when welding in a confined space. Further information on this topic is provided in *Safety in Welding, Cutting, and Allied Processes,* ANSI Z49.1.[11]

METAL FUMES

The welding fumes generated by gas metal arc welding can be controlled by general ventilation, local exhaust ventilation, or by respiratory protective equipment, as described in ANSI Z49.1.[12] The method of ventilation required to maintain the level of toxic substances in the welder's breathing zone at acceptable concentrations is directly dependent on a number of factors. Among these are the material being welded, the size of the work area, and the degree of confinement or obstruction to normal air movement in the area where the welding is being done. Each operation should be evaluated on an individual basis to determine the requirements of maintaining a safe breathing zone. Material safety data sheets are available from the manufacturers of metals and materials and should be consulted.

Acceptable exposure levels to the toxic substances associated with welding, designated as time-weighted average threshold limit values (TLV®) and permissible exposure limits (PEL), have been established by the American Conference of Governmental Industrial Hygienists (ACGIH)[13] and by the Occupational Safety and Health Administration (OSHA), respectively. Compliance with these acceptable levels of exposure can be determined by sampling the atmosphere under the welder's helmet or in the immediate vicinity of the welder's breathing zone. Sampling should be in accordance with the most recent edition of *Method for Sampling Airborne Particulates Generated by Welding and Allied Processes*, ANSI/AWS F1.1.[14]

RADIANT ENERGY

The total radiant energy produced by the gas metal arc welding process can be higher than that produced by the shielded metal arc welding process, because of its higher arc energy, significantly lower welding fume and the more exposed arc. Generally, the highest ultraviolet radiant energy intensities are produced when using an argon shielding gas and when welding on aluminum.

The suggested filter glass shades for gas metal arc welding, as presented in ANSI Z49.1,[15] and AWS F2.2, *Lens Shade Selector*[16] are shown in Table 4.30. To select the best shade for an application, a very dark shade should be the first selection. If it is difficult to see the operation properly, successively lighter shades should be selected until the operation is sufficiently visible for good control. Shades lower than the lowest recommended shade number should not be selected.

The intensity of the ultraviolet radiation can cause the rapid disintegration of cotton clothing. Thus, to reduce the reflection that could cause ultraviolet burns to the face and neck underneath the helmet while performing gas metal arc welding, dark leather or wool clothing is recommended.

---

11. See Reference 6.
12. See Reference 6.
13. American Conference of Governmental Industrial Hygienists (ACGIH). 2002. *2002 TLVs® and BEIs®: Threshold limit values for chemical substances and physical agents in the workroom environment.* Cincinnati: American Conference of Governmental Industrial Hygienists. (Published annually; the most recent edition should be consulted).
14. American Welding Society (AWS) Committee on Fumes and Gases, *Method for Sampling Airborne Particulates Generated by Welding and Allied Processes,* ANSI/AWS F1.1, Miami: American Welding Society.
15. American National Standards Institute (ANSI) Accredited Standards Committee Z49, 1999, *Safety in Welding, Cutting, and Allied Processes,* ANSI Z49.1:1999, Miami: American Welding Society, Table 1, p 7.
16. American Welding Society (AWS) Committee on Safety and Health, 2001, *Lens Shade Selector,* AWS F2.2:2001, Miami: American Welding Society.

Table 4.30
Suggested Filter Glass Shades for GMAW

Welding Current, A	Lowest Shade Number	Comfort Shade Number
Under 60	7	9
60–160	10	11
160–250	10	12
250–500	10	14

Source: Adapted from American National Standards Institute (ANSI) Accredited Standards Committee Z49, 1999, *Safety in Welding, Cutting, and Allied Processes*, ANSI Z49.1:1999, Miami: American Welding Society, Table 1.

NOISE—HEARING AND EAR PROTECTION

Personnel must be protected against exposure to noise generated in welding and cutting processes in accordance with Paragraph 1910.95 of the Occupational Safety and Health Administration's (OSHA) *Occupational Noise Exposure*.[17]

Ear protection is also necessary to prevent spatter from entering the ear and to shield the ears from radiant energy from the arc, as recommended in *Safety in Welding, Cutting, and Allied Processes*, ANSI Z49.1.[18]

ELECTRIC SHOCK

Welders and welding operators should be trained in safe practices and must be cognizant of the potential hazards related to working with electricity. In addition to using precautionary procedures, they should also read and implement the manufacturer's safety recommendations for the specific equipment they are using. Safe operation of gas metal arc welding includes but is not limited to the precautions noted in this section.[19]

The welder or welding operator should be aware that the electrode and the ground circuits are energized ("live" or "hot") when the welding machine is turned on and should not be touched with bare skin or wet clothing. Welders and welding operators should assure that they are insulated from the workpieces and ground. The insulation should cover the full area of physical contact with the workpiece and ground circuits. They should wear sturdy clothing and leather gloves in good condition (no holes).

In semiautomatic or automatic operations, the electrode, electrode reel, welding head, nozzle or semiautomatic welding gun are also electrically charged and cannot be touched with bare skin.

The following conditions increase the hazard potential for arc welding:

1. Welding in damp locations,
2. Wearing wet clothing during welding,
3. Standing on wet surfaces or metal structures such as floors, grates, or scaffolds while welding, and
4. Welding in cramped positions, while sitting, kneeling or lying down.

When working above floor level, a safety belt should be used to protect the welder from a fall in case of an accidental electrical shock.

A welding machine with dc output should be used in confined spaces or in high-risk situations in which contact with the energized workpiece or ground may be unavoidable or could be accidental. A semiautomatic dc constant voltage (wire) welding machine or a dc manual (stick) welding machine should be used. Welding machines with ac output can be used if the machine is equipped with a voltage reducer and remote output control.

Electrical equipment must be correctly installed, used, and maintained according to the manufacturer's instruction or safety manual. The workpiece cable must make a good electrical connection with the workpiece. The connection should be as close as possible to the area being welded. The workpiece connection should be grounded to an earth ground. The electrode holder, work clamp, welding cable and welding machine should be inspected periodically and maintained in good, safe operating condition.

Welders or welding operators must never simultaneously touch electrically hot parts of the electrode holders when the holders are connected to two welding machines because the voltage between the two can be the total of the open circuit voltage of both machines. Damaged insulation should be replaced. The electrode should never be dipped in water for cooling.

17. Occupational Safety and Health Administration (OSHA), 1999, *Welding, Cutting, and Brazing* (available on line at http://www.osha.gov/oshainfo/priorities/welding).
18. American National Standards Institute (ANSI) Accredited Standards Committee Z49, 1999, *Safety in Welding, Cutting, and Allied Processes*, ANSI Z49.1: 1999, Miami: American Welding Society. 8-10.
19. American Welding Society (AWS) Safety and Health Fact Sheet No. 5, Electrical Hazards, 1997, Miami: American Welding Society.

CONCLUSION

Gas metal arc welding is one of the most versatile and efficient joining processes. With the proper selection of equipment, electrical characteristics, and shielding gas, it can be used on a wide variety of materials and material thicknesses, applied to work requiring out-of-position welding, and used for special applications such as spot welding. Because it employs a continuously fed solid or alloy-cored electrode at high current densities, it will deposit welds at a significantly higher rate than the shielded metal arc welding process and will generally exhibit a high percentage of arc-on time. Little post-weld cleanup is required. The process can be easily and economically automated and has become the process of choice for robotic installations.

However, the process is more complicated than the manual processes, such as shielded metal arc welding or gas tungsten arc welding, and the equipment is more expensive. Gas metal arc welding requires that those involved in its application have a good understanding of the various equipment and consumable material components and their functions, and how selection of the various parameters affects arc characteristics and usability. After the initial setup, the ability of the welder or welding operator to recognize when the process is performing correctly and what to do if it is not will be the keys to a successful application.

BIBLIOGRAPHY[20]

American Conference of Governmental Industrial Hygienists (ACGIH). 1999. *1999 TLVs® and BEIs®: Threshold limit values for chemical substances and physical agents in the workroom environment.* Cincinnati: American Conference of Governmental Industrial Hygienists. (Editions of this publication are also available in Greek, Italian, and Spanish).

American National Standards Institute (ANSI) Accredited Standards Committee Z49. 1999. *Safety in welding, cutting, and allied processes.* ANSI Z49.1: 1999. Miami: American Welding Society.

American Welding Society (AWS) Committee on Filler Metals and Allied Materials. 2000. *Filler Metal Comparison Charts.* Miami: American Welding Society.

American Welding Society (AWS) Committee on Arc Welding and Cutting. 1994. *Recommended practices for gas metal arc welding.* ANSI/AWS C5.6-94R. Miami: American Welding Society.

American Welding Society (AWS). Committee on Fumes and Gases. 1999. *Methods for sampling airborne particulates generated by welding and allied processes.* ANSI/AWS F1.1-99. Miami: American Welding Society.

American Welding Society (AWS) Committee on Filler Metals and Allied Materials. 2001. *Specification for carbon steel electrodes and rods for gas shielded arc welding.* ANSI/AWS A5.18-01. Miami: American Welding Society.

American Welding Society (AWS) Committee on Filler Metals and Allied Materials. 1996. *Specification for low-alloy steel electrodes and rods for gas shielded arc welding.* ANSI/AWS A5.28-96. Miami: American Welding Society.

American Welding Society (AWS) Committee on Filler Metals and Allied Materials. 1999. *Specification for bare aluminum and aluminum alloy welding electrodes and rods.* AWS A5.10/A5.10M:1999. Miami: American Welding Society.

American Welding Society (AWS) Committee on Filler Metals and Allied Materials. 2000. *Specification for copper and copper alloy bare welding rods and electrodes.* ANSI/AWS A5.7-84. Miami: American Welding Society.

American Welding Society (AWS) Committee on Filler Metals and Allied Materials. 1992. *Specification for magnesium alloy welding electrodes and rods.* ANSI/AWS A5.19-92. Miami: American Welding Society.

American Welding Society (AWS) Committee on Filler Metals and Allied Materials. 1997. *Specification for nickel and nickel-alloy bare welding electrodes and rods.* ANSI/AWS A5.14/A5.14M-97. Miami: American Welding Society.

American Welding Society (AWS) Committee on Filler Metals and Allied Materials. 1993. *Specification for bare stainless steel welding electrodes and rods.* ANSI/AWS A5.9-93. Miami: American Welding Society.

American Welding Society (AWS) Committee on Filler Metals and Allied Materials. 1990. *Specification for titanium and titanium alloy welding electrodes and rods.* ANSI/AWS A5.16-90. Miami: American Welding Society.

American Welding Society (AWS) Welding Handbook Committee. Jenney, C. L., and A. O'Brien, eds. 2001.*Welding Science and Technology.* Vol. 1 of *Welding handbook.* 9th ed. Miami: American Welding Society.

Compressed Gas Association (CGA). 1999. *Safe handling of gas in containers.* CGA P-1:1999. Arlington, Virginia: Compressed Gas Association.

Occupational Safety and Health Administration (OSHA). 1999. *Occupational noise exposure.* In *Code of Federal Regulations (CFR),* Title 29 CFR 1910.95. Washington D.C.: Superintendent of Documents, U.S. Government Printing Office.

20. The dates of publication given for the codes and other standards listed here were current at the time the chapter was prepared. The reader is advised to consult the latest edition.

SUPPLEMENTARY READING LIST

Adam, G., and T. A. Siewert. 1990. Sensing of GMAW droplet transfer modes using an ER100S-1 electrode. *Welding Journal* 69(3): 103-s–108-s.

Aldenhoff, B. J., J. B. Stearns, and P. W. Ramsey. 1974. Constant potential power sources for multiple operation gas metal arc welding. *Welding Journal* 53(7): 425–429.

Althouse, A. D., C. H. Turnquist, W. A. Bowditch, and K. E. Bowditch. 1984. *Modern welding*. South Holland, Illinois: The Goodheart-Willcox Company.

Altshuller, B. 1998. A guide to GMA welding of aluminum. *Welding Journal* 77(6): 49.

Baujet, V., and C. Charles. 1990. Submarine hull construction using narrow-groove GMAW. *Welding Journal* 69(8): 31–36.

Bosworth, M. R. 1991. Effective heat input in pulsed current gas metal arc welding with solid wire electrodes. *Welding Journal* 70(5): 111-s–117-s.

Bruss, R. A. 1996. Designing GMA welding guns with the welder's comfort in mind. *Welding Journal* 75(10): 31–33.

Butler, C. A., R. P. Meister, and M. D. Randall. 1969. Narrow gap welding—a process for all positions. *Welding Journal* 48(2): 102–108.

Cary, H. B. 1994. *Modern welding technology.* 3rd ed. Englewood Cliffs, New Jersey: Regents Prentice-Hall.

Castner, H. R. and R. Singh. 1997. Pulsed vs. steady current GMAW: Which is louder? *Welding Journal* 76(11): 39–51.

Castner, H. R. 1995. Gas metal arc welding fume generation using pulsed current. *Welding Journal* 74(2): 59-s–68-s.

Chandiramani, D. 1994. Hydrogen reduced in wet underwater GMA welds. *Welding Journal* 73(3): 45–49.

Choi, S. K., C. D. Yoo, and Y. S. Kim. 1998. Dynamic simulation of metal transfer in GMAW, Part 1: Globular and spray transfer modes. *Welding Journal* 77(1): 38-s–44-s.

Choi, S. K., C. D. Yoo, and Y. S. Kim. 1998. Dynamic simulation of metal transfer in GMAW, Part 2: Short-circuit transfer mode. *Welding Journal* 77(1): 45-s–51-s.

DeSaw, F. A., and J. E. Rodgers. 1981. Automated welding in restricted areas using a flexible probe gas metal arc welding torch. *Welding Journal* 60(5): 17–22.

Dillenbeck, V. R., and L. Castagno. 1987. The effects of various shielding gases and associated mixtures in GMA welding of mild steel. *Welding Journal* 66(9): 45–49.

DiPietro, D., and J. Young. 1996. Pulsed GMAW helps John Deere meet fume requirements. *Welding Journal* 75(10): 57–58.

Dorling, D. V., A. Loyer, S. N. Russell, and T. S. Thompson. 1992. Gas metal arc welding used on mainline 80 ksi pipeline in Canada. *Welding Journal* 71(5): 55–61.

Eickhoff, S. T., and T. W. Eagar. 1990. Characterization of spatter in low-current GMAW of titanium alloy plate. *Welding Journal* 69(10): 382-s–388-s.

Fang, C. K., E. Kannatey-Asibu, Jr., and J. R. Barber. 1995. Acoustic emission investigation of cold cracking in gas metal arc welding of AISI 4340 steel. *Welding Journal* 74(6): 177-s–184-s.

Farson, D., C. Conrardy, J. Talkington, K. Baker, T. Kershbaumer and F. Edwards. 1998. Arc initiation in gas metal arc welding. *Welding Journal* 77(8): 315-s–321-s.

Feree, S. E. 1995. New generation of cored wires creates less fume and spatter. *Welding Journal* 74(12): 45–49.

French, I. E., and M. R. Bosworth. 1995. A comparison of pulsed and conventional welding with basic flux cored and metal cored welding wires. *Welding Journal* 74(6): 197-s–205-s.

Hilton, D. E., and J. Norrish. 1988. Shielding gases for arc welding. *Welding and Metal Fabrication* 189–196.

Hussain, H. M., P. K. Ghosh, P. C. Gupta, and N. B. Potluri. 1996. Properties of pulsed current multipass GMA welded Al-Zn-Mg alloy. *Welding Journal* 75(7): 209-s–215-s.

Irving, B. 1997. GMA welding goes to work on the Z3 Roadster's exhaust system. *Welding Journal* 76(2): 29–33.

Irving, B. 1993. Laser beam and GMA welding lines go on-stream at Arvin Industries, *Welding Journal* 72(11): 47–50.

Irving, B. 1992. Inverter power sources cut fume emissions in GMAW. *Welding Journal* 71(2): 53–57.

Johnson, J. A, N. M. Carlson, H. B. Smartt, and D. E. Clark. 1991. Process control of GMAW: Sensing of metal transfer mode. *Welding Journal* 70(4): 91-s–99-s.

Jonsson, P. G., A. B. Murphy, and J. Szekely. 1995. The influence of oxygen additions on argon-shielded gas metal arc welding processes. *Welding Journal* 74(2): 48-s–58-s.

Jonsson, P. G., J. Szekely, R. B. Madigan, and T. P. Quinn. 1995. Power characteristics in GMAW: Experimental and numerical investigation. *Welding Journal* 74(3): 93-s–102-s.

Kim, J. W., and S. J. Na. 1995. A study on the effect of contact tube-to-workpiece distance on weld bead shape in gas metal arc welding. *Welding Journal* 74(5): 141-s–152-s.

Kim, J. W., and S. J. Na. 1993. A self-organizing fuzzy control approach to arc sensor for weld joint tracking in gas metal arc welding of butt joints. *Welding Journal* 72(2): 60-s–66-s.

Kim, J. W., and S. J. Na. 1991. A study on an arc sensor for gas metal arc welding of horizontal fillets. *Welding Journal* 70(8): 216-s–221-s.

Kim, Y. S., and T. W. Eagar. 1993. Metal transfer in pulsed current GMAW. *Welding Journal* 72(7): 279-s–287-s.

Kim, Y. S., and T. W. Eagar. 1993. Analysis of metal transfer in gas metal arc welding. *Welding Journal* 72(6): 269-s–278-s.

Kim, Y. S., D. M. McEligot, and T. W. Eagar. 1991. Analysis of electrode heat transfer in gas metal arc welding. *Welding Journal* 70(1): 20-s–31-s.

Kimura, S., I. Ichihara and Y. Nagai. 1979. Narrow-gap gas metal arc welding process in flat position. *Welding Journal* 58(7): 44–52.

Kiyohara, M., T. Okada, Y. Wakino and H. Yamamoto. 1977. On the stabilization of GMA welding of aluminum. *Welding Journal* 56(3): 20–28.

Kjeld. F. 1990. Gas metal arc welding for the collision repair industry. *Welding Journal* 69(4): 39.

Kluken, A. O., and B. Bjorneklett. 1997. A study of mechanical properties for aluminum GMA weldments. *Welding Journal* 76(2): 39–44.

Lesnewich, A. 1991. Technical commentary: Observations regarding electrical current flow in the gas metal arc. *Welding Journal* 70(7): 171-s–172-s.

Lesnewich, A. *MIG welding with pulsed power.* 1972. Welding Research Council Bulletin 170. New York; Welding Research Council.

Lesnewich, A. 1958. Control of melting rate and metal transfer in gas-shielded metal-arc welding. *Welding Journal* 37(8): 343–353.

Lincoln Electric Company, The. 1994. *The procedure handbook of arc welding.* 13th ed. Cleveland, Ohio: The Lincoln Electric Company.

Lincoln Electric Company, The. 1992. Switch to metal cored electrodes helps crane fabricator gain productivity. *Welding Journal* 71(6): 75–77.

Liu, S., and T. A. Siewart. 1989. Metal transfer in gas metal arc welding: Droplet rate. *Welding Journal* 68(2): 52-s–58-s.

Lu, M. J., and S. Kou. 1989. Power inputs in gas metal arc welding of aluminum. Part 2. *Welding Journal* 68(11): 452-s–456-s.

Lu, M. J., and S. Kou. 1989. Power inputs in gas metal arc welding of aluminum. Part 1. *Welding Journal* 68(9): 382-s–388-s.

Lyttle, K. A. 1983. GMAW—A versatile process on the move. *Welding Journal* 62(3): 15–23.

Lyttle, K. A. 1982. Reliable GMAW means understanding wire quality, equipment and process variables. *Welding Journal* 61(3): 43–48.

Malin, V. Y. 1983. The state-of-the-art of narrow gap welding, Part II. *Welding Journal* 62(6): 37–46.

Malin, V. Y. 1983. The state-of-the-art of narrow gap welding, Part I. *Welding Journal* 62(4): 22–30.

Manz, A. F. 1990. The dawn of gas metal arc welding. *Welding Journal* 69(1): 67–68.

Manz, A. F. 1973. *The welding power handbook.* Miami: American Welding Society.

Manz, A. F. 1969. Inductance vs. slope for control for gas metal arc power. *Welding Journal* 48(9): 707–712.

Mitchie, K., S. Blackman, and T. E. B. Ogunbiyi. 1999. Twin-wire GMAW: process characteristics and applications. *Welding Journal* 78(5): 31–34.

Miyazaki, H., H. Miyauchi, Y. Sugiyama, and T. Shinoda. 1994. Puckering phenomenon and its prevention in GMA welding of aluminum alloys. *Welding Journal* 73(12): 277-s–284s.

Modensi, P. J., and J. H. Nixon. 1994. Arc instability phenomena in GMA welding. *Welding Journal* 73(9): 219-s–224-s.

Morris, R. W. 1968. Application of multiple electrode gas metal arc welding to structural steel fabrication. *Welding Journal* 47(5): 379–385.

Nadeau, F. 1990. Computerized system automates GMA pipe welding. *Welding Journal* 69(6): 53–59.

Occupational Safety and Health Administration (OSHA). 1999. *Occupational safety and health standards for general industry.* In *Code of Federal Regulations (CFR),* Title 29 CFR 1910, Subpart Q. Washington D.C.: Superintendent of Documents, U.S. Government Printing Office.

Ohring, S. and H. J. Lugt. 1999. Numerical simulation of a time-dependent 3-D GMA weld pool due to a moving arc. *Welding Journal* 78(12): 416-s–424-s.

Pan, J. L., R. H. Zhang, Z. M. Ou, Z. Q. Wu and Q Chen. 1989. Adaptive control GMA welding—A new technique for quality control." *Welding Journal* 68(3): 73–76.

Pierre, E. R. 1985. *Welding processes and power sources.* 3rd ed. Minneapolis: Burgess Publishing Company.

Quinn, T. P., C. Smith, C. N. McGowan, E. Blachowiak, and R. B. Madigan. 1999. Arc sensing for defects in constant-voltage gas metal arc welding. *Welding Journal* 78(9): 322-s–328-s.

Quinn, T. P., R. B. Madigan, and T. A. Siewert. 1994. An electrode extension model for gas metal arc welding. *Welding Journal* 73(10) 241-s–248-s.

Quinn, T. P., R. B. Madigan, M. A. Mornis, and T. A. Siewert. 1995. Contact tube wear detection in gas metal arc welding. *Welding Journal* 74(4): 141-s–121-s.

Rajasekaran, S. 1999. Weld bead characteristics in pulsed GMA welding of Al-Mg alloys, *Welding Journal* 78(12): 416-s–424-s.

Rajasekaran, S., S. D. Kulkarn, U. D. Mallya, and R. C. Chaturvedi. 1998. Droplet detachment and plate fusion characteristics in pulsed current gas metal arc welding. *Welding Journal* 77(6): 254-s–268-s.

Reilly. R. 1990. Real-time weld quality monitor controls GMA welding. *Welding Journal* 69(3): 36.

Rhee, S., and E. Kannatey-Asibu, Jr. 1992. Observation of metal transfer during gas metal arc welding. *Welding Journal* 71(10): 381-s–386-s.

Richardson, I. M., P. W. Bucknall, and I. Stares. 1994. The influence of power source dynamics on wire melting rate in pulsed GMA welding. *Welding Journal* 73(2): 32-s–37-s.

Sadler, H. 1999. A look at the fundamentals of gas metal arc welding. *Welding Journal* 78(5): 45–47.

Sampath, K., R. S. Green, D.A. Civis, B. E. Williams, and P. J. Konkol. 1995. Metallurgical model speeds development of GMA welding wire for HSLA steel. *Welding Journal* 74(12): 69–76.

Shackleton, D. N., and W. Lucas. 1974. Shielding gas mixtures for high quality mechanized GMA welding of Q and T steels. *Welding Journal* 53(12): 537-s–547-s.

Smartt, H. B., and C. J. Einerson. 1993. A model for mass input control in gas metal arc welding. *Welding Journal* 72(5): 217-s–229-s.

Stanzel, K. 1999. Nothing complicated—Just basic GMA welding. *Welding Journal* 78(5): 36–38.

Sullivan, D. 1998. The gas metal arc welding process celebrates a 50th anniversary. *Welding Journal* 77(9): 53–54.

Tekriwal, P., and J. Mazumder. 1988. Finite element analysis of three-dimensional transient heat transfer in GMA welding. *Welding Journal* 67(7): 150-s–156-s.

Tsao, K. C., and C. S. Wir. 1988. Fluid flow and heat transfer in GMA weld pools. *Welding Journal* 67(3): 70-s–75-s.

Union Carbide Corporation. 1984. *MIG Welding handbook*. Danbury, Connecticut: Union Carbide Corporation, Linde Division.

Villafuerte, J. 1999. Understanding contact tip longevity for gas metal arc welding. *Welding Journal* 78(12): 29–35.

Waszink, J. H., and G. J. P. M. Van Den Heurel. 1982. Heat generation and heat flow in the filler metal in GMA welding. *Welding Journal* 61(8): 269-s–282-s.

Zhu, P., M. Rados, and S. W. Simpson. 1997. Theoretical predictions of the start-up phase in GMA welding. *Welding Journal*, 76(7): 269-s–274-s.

CHAPTER 5

FLUX CORED ARC WELDING

Prepared by the Welding Handbook Chapter Committee on Flux Cored Arc Welding:

D. B. Arthur, Chair
J. W. Harris Company

B. A. Morrett
ITW/Hobart Brothers Company

J. E. Beckham
ThermoKing-Ingersoll Rand

D. Sprenkel
Consultant

Welding Handbook Committee Member:

C. E. Pepper
ENGlobal Engineering, Inc.

Contents

CHAPTER 5

FLUX CORED ARC WELDING

INTRODUCTION

Flux cored arc welding (FCAW) is a welding process that uses an arc between a continuous filler metal electrode and the weld pool. The process is used with shielding from a flux contained within the tubular electrode, with or without additional shielding from an externally supplied gas, and without the application of pressure.[1,2]

The remarkable operating characteristics and weld properties that distinguish FCAW from other arc welding processes are attributable to the continuously fed flux cored electrode. The tubular electrode is a filler metal composite consisting of a metal sheath and a core of various powdered materials manufactured in the form of wire. During welding, an extensive protective slag cover is produced on the face of the weld bead.

Flux cored arc welding offers two major variations, self-shielded (FCAW-S) and gas-shielded (FCAW-G), which add great flexibility to the process. These variations differ in the method of shielding the arc and weld pool from atmospheric contamination (oxygen and nitrogen).

Flux cored arc welding is an efficient welding process readily adaptable to semiautomatic or automatic welding operations and capable of producing high-quality weld metal at a high deposition rate. Many industries rely on flux cored arc welding to produce high-integrity welds. Users of the process include manufacturers or builders of pressure vessels, submarines, aircraft carriers, earth-moving equipment, and buildings and other structures.

This chapter covers the fundamental operating principles of the flux cored arc welding process and describes the necessary equipment and materials. Significant information is included for a variety of flux cored electrodes used in the major applications of FCAW. Welding procedures, process control, and weld quality are discussed. The chapter ends with comments on the economics of the process and important information on safe practices.

FUNDAMENTALS

The history of gas-shielded arc welding provides the background for the technology and evolution of flux cored arc welding. Gas-shielded metal arc welding processes have been in use since the early 1920s, when it was demonstrated that a significant improvement of weld metal properties could be produced if the arc and molten weld metal were protected from atmospheric contamination. However, the development of covered electrodes in the late 1920s diminished interest in gas-shielding methods. Interest was renewed in the early 1940s with the introduction and commercial acceptance of gas tungsten arc welding (GTAW) and, later in the same decade, gas metal arc welding (GMAW). Argon and helium were the two primary shielding gases used at that time.

Research conducted on manual welds made with covered electrodes focused on the analysis of the gas produced in the disintegration of electrode coverings. Results confirmed that carbon dioxide (CO_2) was the predominant gas given off by electrode coverings. This

1. American Welding Society (AWS) Committee on Definitions and Symbols, 2001, *Standard Welding Terms and Definitions*, A3.0:2001, Miami: American Welding Society.
2. At the time of preparation of this chapter, the referenced codes and other standards were valid. If a code or other standard is cited without a date of publication, it is understood that the latest edition of the document referred to applies. If a code or other standard is cited with the date of publication, the citation refers to that edition only, and it is understood that any future revisions or amendments to the code or standard are not included. As codes and standards undergo frequent revision, the reader is advised to consult the most recent edition.

discovery quickly led to the use of CO_2 as a shielding gas for welds made on carbon steels with GMAW. Although early experiments with CO_2 were unsuccessful, techniques were eventually developed which permitted its use. Gas metal arc welding using CO_2 became available in the mid-1950s.

In concurrent research, CO_2 shielding was combined with a flux-containing tubular electrode, which overcame many of the problems previously encountered. Operating characteristics were improved by the addition of the core materials, and weld quality was improved by eliminating atmospheric contamination. These experiments resulted in the development of flux cored arc welding. This new process was introduced at the American Welding Society (AWS) Exposition in Buffalo, New York, in 1954. By 1957 the electrodes and equipment were refined, and the process was introduced commercially in essentially its present form.

During the 1990s significant improvements were made in both gas-shielded and self-shielded electrode arc stability that resulted in much less spatter than the earlier electrodes produced. The impact resistance of FCAW electrodes was also significantly improved. The development and production of alloy electrodes and small-diameter electrodes, down to 0.8 millimeters (mm) (0.030 inches [in.]), were other advances.

Improvements continue to be made to the FCAW process. Modern power sources and electrode (wire) feeders are greatly simplified and more dependable than their predecessors. Welding guns are lightweight and rugged, and electrodes undergo continuous improvement.

PROCESS VARIATIONS

The two major variations of the FCAW process, the self-shielded and the gas-shielded versions, are shown in Figure 5.1. Both illustrations in Figure 5.1 emphasize the melting and deposition of filler metal and flux and show the formation of a slag covering the weld metal. Cross sections of examples of FCAW electrodes also are shown in Figure 5.1.

In the gas-shielded method, the shielding gas (CO_2 or a mixture of argon and CO_2) protects the molten metal from the oxygen and nitrogen present in air by forming an envelope of gas around the arc and over the weld pool. Little need exists for denitrification of the weld metal because air is mostly excluded, along with the nitrogen it contains. However, some oxygen may be generated from the dissociation of CO_2, which forms carbon monoxide and oxygen. The compositions of the electrodes are formulated to provide deoxidizers that combine with small amounts of oxygen in the gas shield.

Self-shielded flux cored arc welding is often the process of choice for field welding because it can tolerate stronger air currents than the gas-shielded variation. The main reason for this distinction is that some shielding is provided by the high-temperature decomposition of some of the electrode core ingredients. The vaporization of these ingredients displaces air from the area immediately surrounding the arc. In addition, the wire contains a large proportion of scavengers (deoxidizers and denitrifiers) that combine with undesirable elements that might contaminate the weld pool. A slag cover protects the metal from the air surrounding the weld.

APPLICATIONS

Both self-shielded and gas-shielded flux cored arc welding can be used in most welding applications. However, the specific characteristics of each method make each suitable for different operating conditions. The process is used to weld carbon- and low-alloy steels, stainless steels, cast irons, and nickel and cobalt alloys. It is also used for the arc spot welding of lap joints in sheet and plate, as well as for cladding and hardfacing.

Flux cored arc welding is widely used in fabrication shops, for maintenance applications, and in field erection work. An example of field erection work is shown in Figure 5.2, in which both self-shielded and gas-shielded FCAW are used in the fabrication of an offshore oil drilling structure.

Flux cored arc welding can be used to produce weldments that conform to the *ASME Boiler and Pressure Vessel Code*,[3] the rules of the American Bureau of Shipping,[4] and *Structural Welding Code—Steel*, AWS D1.1.[5] The process is given prequalified status in AWS D1.1. Stainless steel, self-shielded, and gas-shielded flux cored electrodes are used in general fabrication, surfacing, joining dissimilar metals, and maintenance and repair.

Figure 5.3, which shows the fabrication of a suction filter used in the pulp and paper industry, illustrates the versatility of the FCAW process. The base material in this application was ST-360-C; E308LT1 electrodes were used.

The self-shielded method can often be used for applications that are normally welded with the shielded metal arc welding (SMAW) process. Gas-shielded FCAW can also be used for some applications that are welded by the GMAW process. The selection of self-shielded or gas-shielded FCAW depends on the type of electrodes available, the type of welding equipment available, the environment in which the welding is to be

3. American Society of Mechanical Engineers (ASME) Boiler and Pressure Vessel Code Committee, 1998, *Boiler and Pressure Vessel Code*. New York: American Society of Mechanical Engineers.
4. American Bureau of Shipping (ABS) Group, ABS Plaza, 166855 Northchase Drive, Houston, TX 77060-6008.
5. American Welding Society (AWS) Committee on Structural Welding, 2002, *Structural Welding Code—Steel*, AWS D1.1/D1.1M:2002, Miami: American Welding Society.

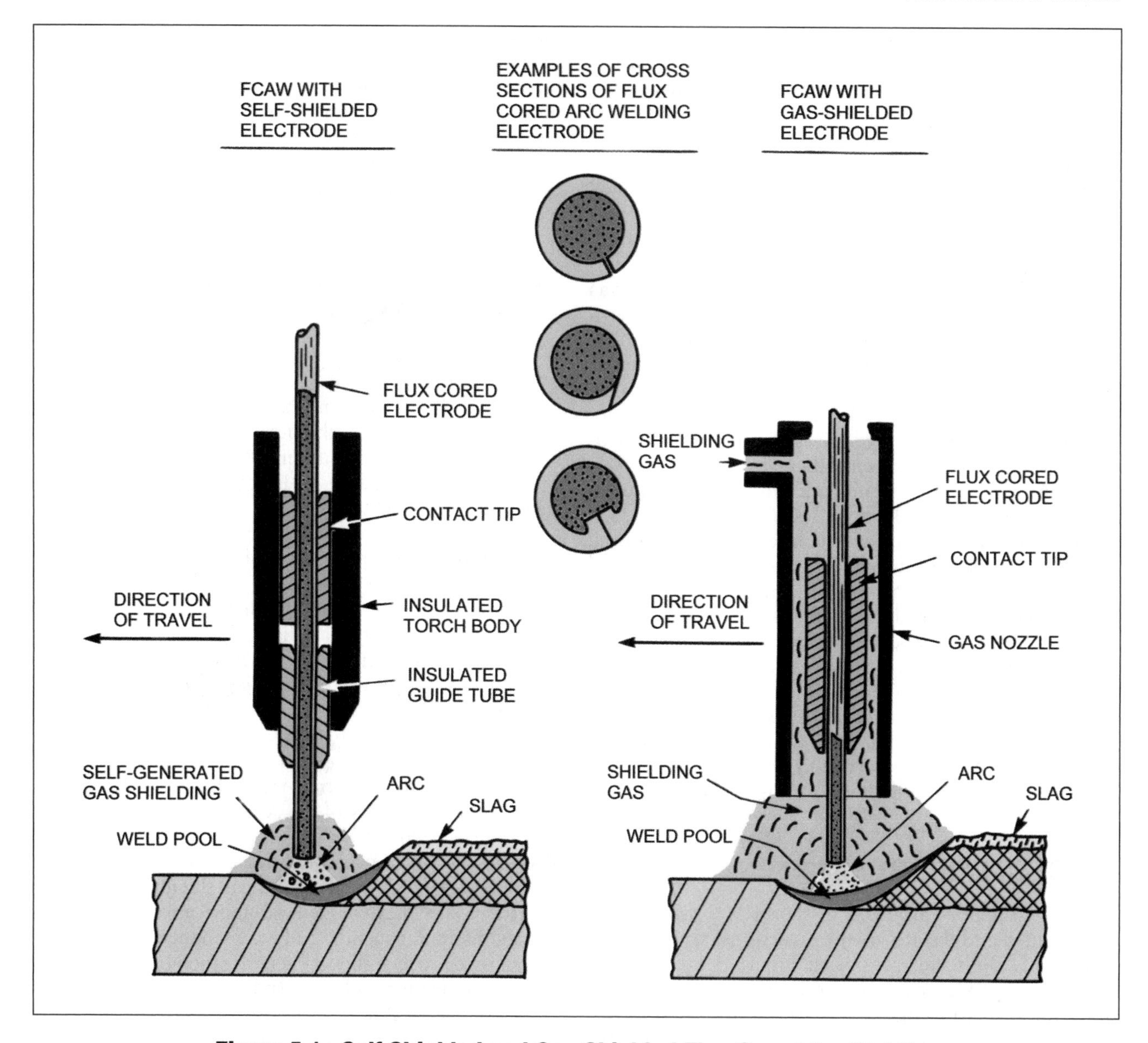

Figure 5.1—Self-Shielded and Gas-Shielded Flux Cored Arc Welding

done, the mechanical property requirements of the welded joints, and the joint design and fitup. The advantages and disadvantages of FCAW should be compared to those of other processes when it is evaluated for a specific application.

ADVANTAGES

When compared to SMAW, higher productivity is the chief advantage of flux cored arc welding for many applications. This generally translates into lower overall costs per pound of metal deposited in joints that permit continuous welding and easy FCAW gun and equipment accessibility. The advantages are higher deposition rates, higher operating factors, and higher deposition efficiency (no stub loss).

In addition to the advantages of FCAW over the manual SMAW process, FCAW also provides certain advantages over submerged arc welding (SAW) and GMAW. In many applications, FCAW produces high-quality weld metal at lower cost with less effort on the part of the welder than SMAW. Flux cored arc welding

Figure 5.2—Offshore Drilling Structure Fabricated with Self-Shielded and Gas-Shielded Flux Cored Arc Welding

is more forgiving of minor disparities in procedures and differences in welder skill than GMAW, and it is more flexible and adaptable than SAW. Among the benefits offered by FCAW are the following:

1. High-quality weld metal deposit,
2. Excellent weld appearance,
3. Welds many steels in a wide thickness range,
4. High operating factor and easily mechanized,
5. High deposition rate (up to four times greater than SMAW) and high current density,
6. Relatively high electrode deposit efficiency,
7. Allows economical engineering of joint designs,
8. Visible arc contributes to easy use,
9. Requires less precleaning than GMAW,
10. Often results in less distortion compared to SMAW,
11. Exceptionally good fusion when used with shielding gas compared to GMAW-S,
12. High tolerance for contaminants that may cause weld cracking,
13. Resistance to underbead cracking,

Photograph courtesy of Bohler Thyssen Welding USA, Inc.

Figure 5.3—Flux Cored Arc Welding of a Suction Filter Used in the Pulp and Paper Industry

14. Self-shielding characteristic of electrodes eliminates the need for flux handling and gas apparatus, and
15. Self-shielding tolerates windy conditions in outdoor applications. (See No. 6 relative to gas shields under "Limitations" in the next section).

An example of the good sidewall fusion, deep penetration and smooth weld profile that can be obtained with gas-shielded FCAW is shown in Figure 5.4

LIMITATIONS

Compared to the SMAW process, the major limitations of FCAW are the higher cost of the equipment, the relative complexity of setup and control of the equipment, and the restriction on operating distance from the electrode wire feeder. Flux cored arc welding may generate large volumes of welding fumes and requires suitable exhaust equipment, except in field work. Compared to the slag-free GMAW process, the need for removing slag between passes is an added labor cost when using FCAW. This is especially true in making root pass welds.However, in most cases, slag is easily removed and cleanup time is minimized, as shown in Figure 5.5. The limitations of FCAW are summarized as follows:

1. FCAW is limited to welding ferrous metals and nickel-base alloys;
2. The process produces a slag covering that must be removed;
3. FCAW electrode wire is more expensive on a weight basis than solid electrode wires, except for some high-alloy steels;
4. The equipment is more expensive and complex than that required for SMAW, however, increased productivity usually compensates for this;

Photograph courtesy of Bohler Thyssen Welding USA, Inc.

Figure 5.4—Flux Cored Arc Weld Profile

Photograph courtesy of Bohler Thyssen Welding USA, Inc.

Figure 5.5—Self-Peeling Slag Reveals a Clean Flux Cored Arc Weld

5. The wire feeder and power source must be fairly close to the point of welding;
6. For gas-shielded FCAW, the external shield may be adversely affected by breezes and drafts;
7. Equipment is more complex than that used for SMAW, so more maintenance is required; and
8. More smoke and fumes are generated by FCAW than by GMAW and SAW.

It should be noted that self-shielded FCAW is not adversely affected by windy conditions, except in very high winds, because the shield is generated at the end of the electrode exactly where it is required.

EQUIPMENT

The basic equipment setup for flux cored arc welding is shown in Figure 5.6. Equipment consists of a power source, electrode feed and current controls, a shielding gas source, a wire electrode feeding system, a welding gun, and the associated cables and gas hoses. In addition, appropriate fume extraction equipment may be needed. Proper ventilation or some means of fume removal is necessary for FCAW.

SEMIAUTOMATIC EQUIPMENT

Control equipment for semiautomatic self-shielded and gas-shielded flux cored arc welding is similar. The major difference between the shielding variations is the provision for supplying and metering gas to the arc of the electrode in the gas-shielded method.

Power Source

The recommended power source is the direct current (dc) constant-voltage type, similar to power sources used for GMAW. The power source should be capable of operating at the maximum current required for the specific application. Most semiautomatic applications use less than 500 amperes (A). The voltage control should be capable of adjustments in increments of one volt or less. Constant-current dc power sources of adequate capacity with appropriate controls and wire feeders are sometimes used, but applications are rare.

Electrode Feed Control

The purpose of the electrode (wire) feed control is to supply the continuous electrode to the welding arc at a constant preset rate. The rate at which the electrode is fed into the arc determines the welding amperage supplied by a constant-voltage power source. If the

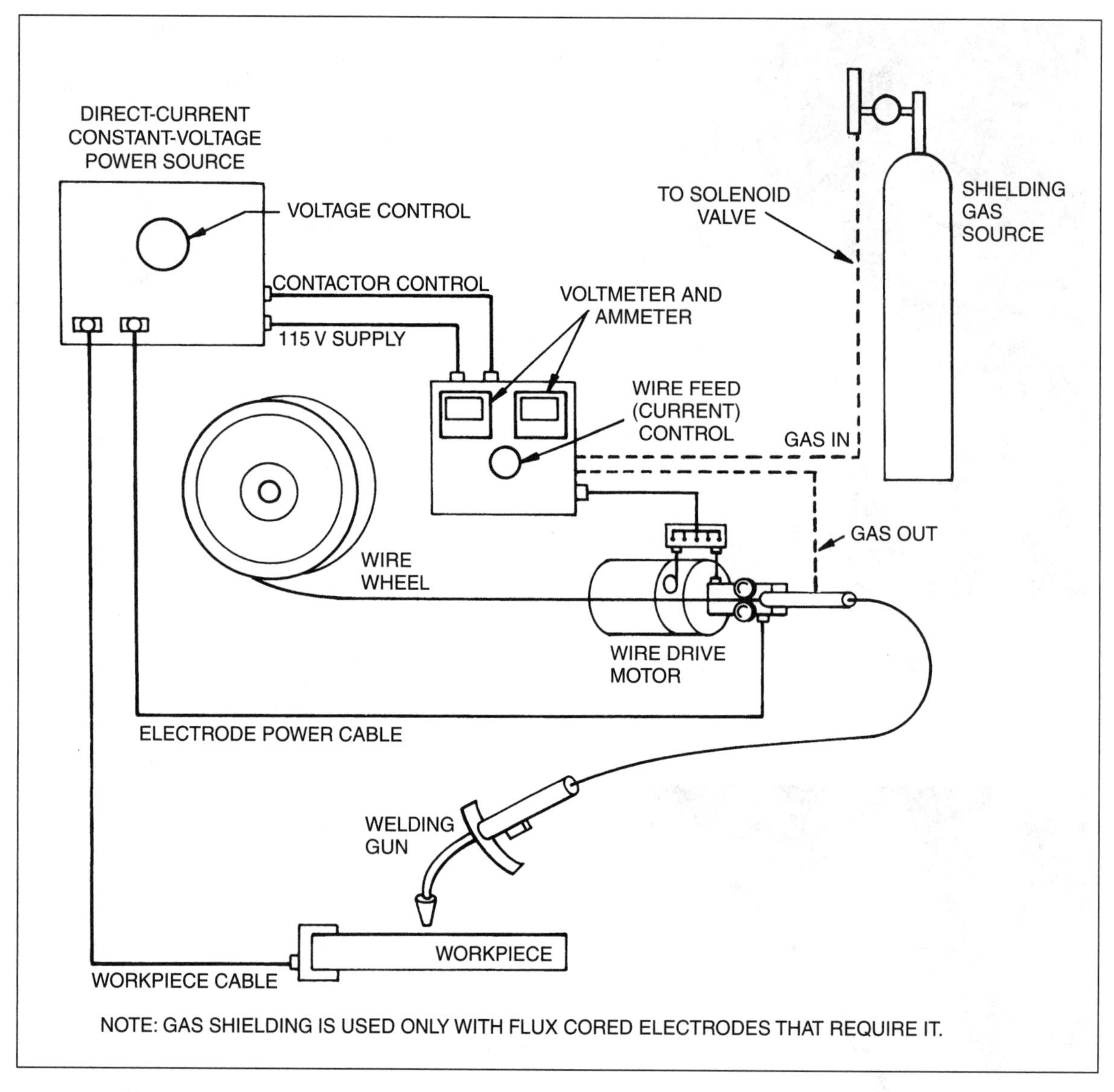

Figure 5.6—Typical Equipment for Semiautomatic Flux Cored Arc Welding

electrode feed rate is changed, the welding machine automatically adjusts to maintain the preset arc voltage. The electrode feed rate can be controlled by mechanical or electronic means.

Semiautomatic flux cored arc welding requires the use of drive rolls that will not flatten or otherwise distort the tubular electrode. Various grooved and knurled feed roll surfaces are used to advance the electrode. Some wire feeders have a single pair of drive rolls; others have two pairs of rolls with at least one roll of each pair being driven. When all rolls are driven, the wire can be advanced with less pressure on the rolls.

Welding Guns

Typical guns for semiautomatic welding are shown in Figure 5.7 and Figure 5.8. They are designed for handling comfort, easy manipulation, and durability. The guns provide internal contact with the electrode to conduct the welding current. The welding current and electrode feed are actuated by a switch mounted on the gun.

Welding guns may be either gas-cooled or water-cooled. Gas-cooled (including air-cooled) guns are favored because a water delivery system is not required; however, water-cooled guns are more compact, lighter

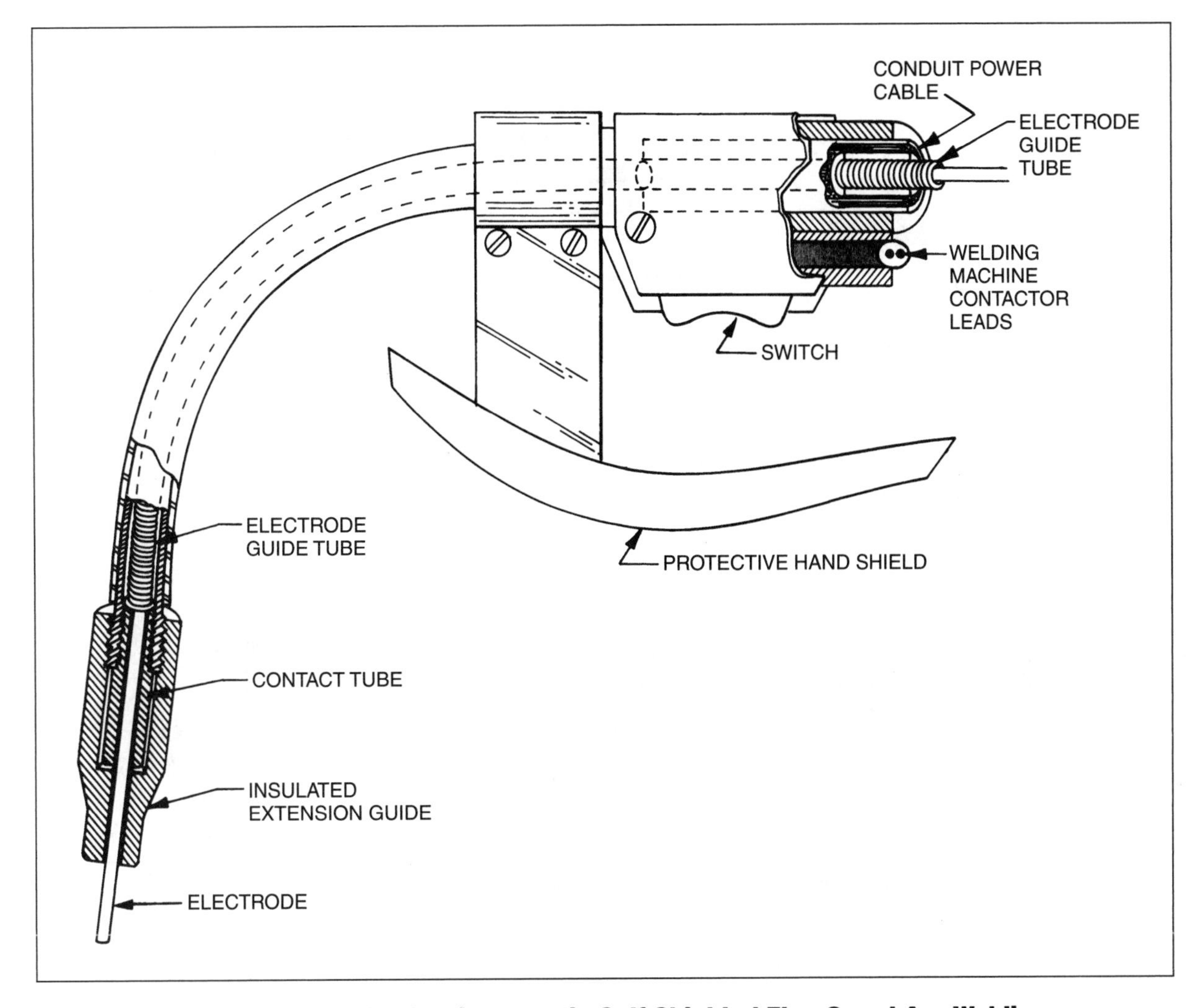

Figure 5.7—Gun for Semiautomatic Self-Shielded Flux Cored Arc Welding

in weight, and require less maintenance than gas-cooled guns. Water-cooled guns generally have higher current ratings (up to 700 A, continuous duty). Current ratings for gas-cooled guns are based on using CO_2. If argon-based gas is used, the gun current rating should be decreased 30%. Guns have either straight or curved nozzles. The curved nozzle may vary from 10° to 90°. In some applications, the curved nozzle enhances flexibility and ease of electrode manipulation.

Some self-shielded flux cored electrodes require a specific minimum electrode extension to develop proper shielding. Welding guns for these electrodes generally have guide tubes with an insulated extension guide to support the electrode and assure a minimum electrode extension. Details of a self-shielded electrode nozzle showing the insulated guide tube are illustrated in Figure 5.9.

AUTOMATIC EQUIPMENT

Figure 5.10 shows the equipment layout for an automatic flux cored arc welding installation. A direct-current power source with constant-voltage designed for 100% duty cycle is recommended for automatic operation. The size of the power source is determined by the current required for the work to be performed. Because large electrodes, high electrode feed rates, and long welding times may be required, electrode feeders for automatic operation necessarily have higher-capacity drive motors and heavier-duty components than similar equipment for semiautomatic operation.

Two typical nozzle assemblies for automatic gas-shielded flux cored arc welding are shown in Figure 5.11. Nozzle assemblies are designed for side shielding or for concentric shielding of the electrode. Side shielding

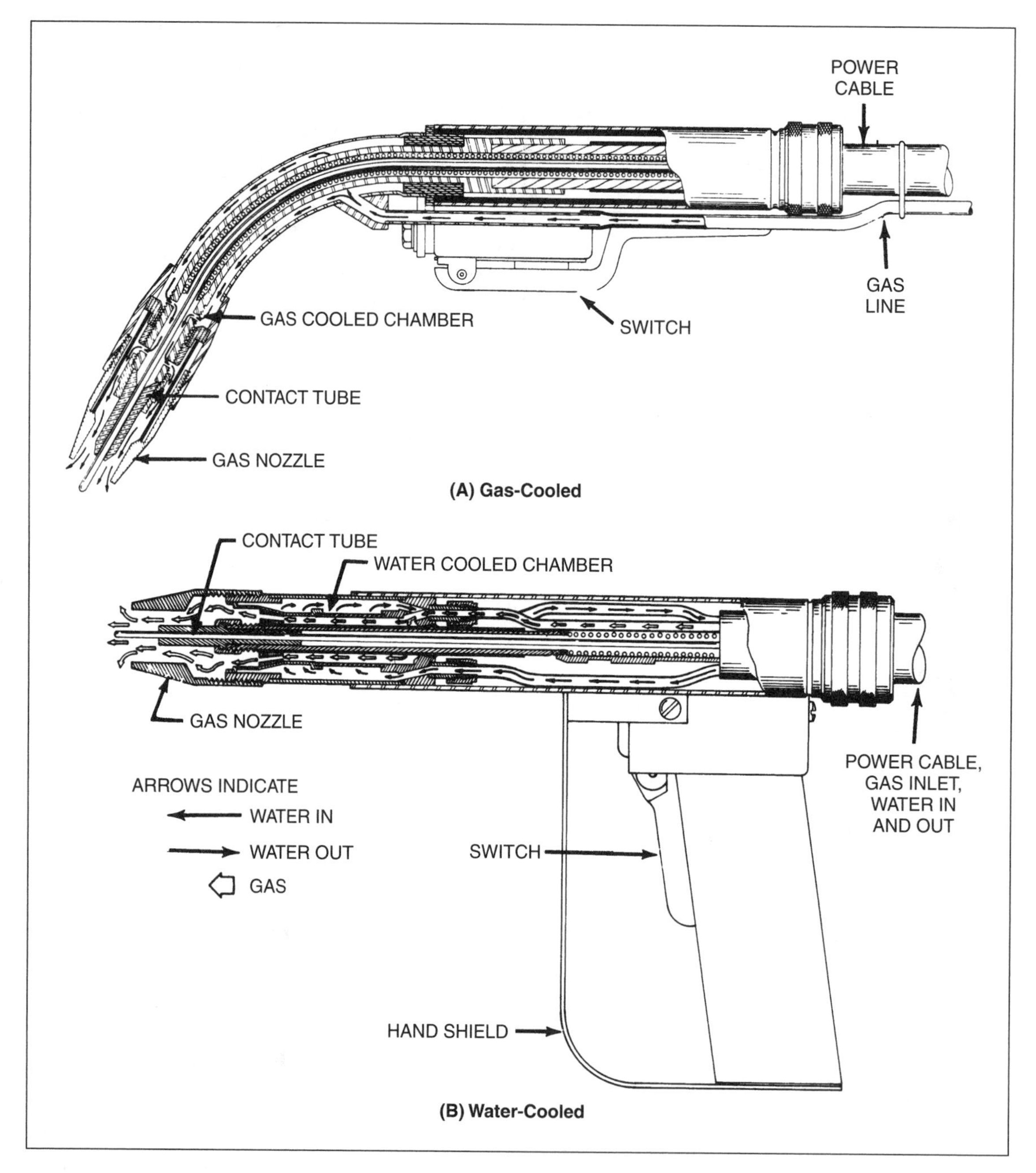

Figure 5.8—Typical Guns for Gas-Shielded Flux Cored Arc Welding

permits welding in deep, narrow grooves and minimizes spatter buildup in the nozzle. Nozzle assemblies are air-cooled or water-cooled. In general, air-cooled nozzle assemblies are preferred for operation with welding currnts up to 600 A. Water-cooled nozzle assemblies are recommended for currents above 600 A. For higher deposition rates with gas-shielded electrodes, tandem welding guns can be used, as shown in Figure 5.12.

For large-scale surfacing applications, automatic multiple-electrode oscillating equipment can be used to increase productivity. The equipment for these installations may include a track-mounted manipulator supporting a multiple-electrode oscillating welding head with individual electrode feeders and a track-mounted, power-driven turning roll, in addition to the power source, electronic controls, and an electrode supply sys-

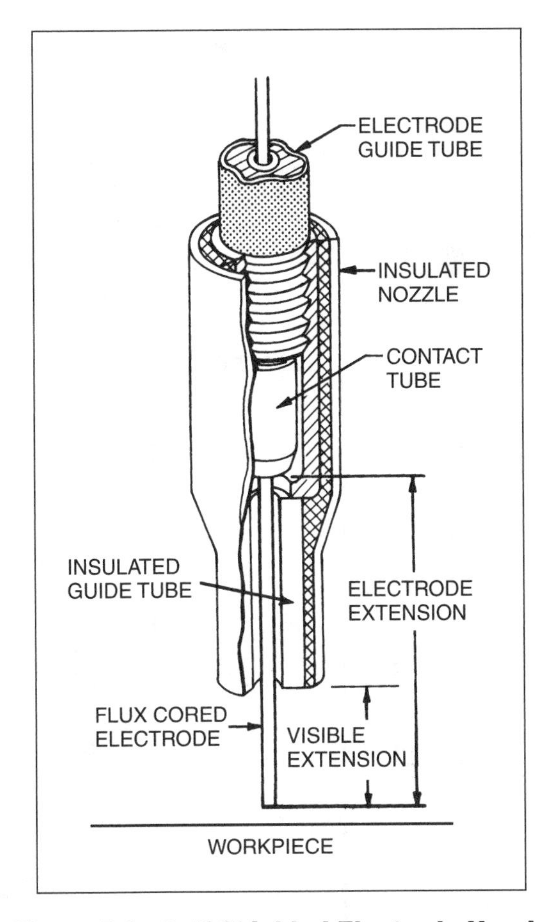

Figure 5.9—Self-Shielded Electrode Nozzle

tem. Figure 5.13 illustrates the operating details of a six-electrode oscillating system for the self-shielded surfacing of a vessel shell with stainless steel.

MATERIALS

Like GMAW electrodes, gas-shielded FCAW electrodes require gas shielding in addition to the shielding provided by the internal flux. Gas shielding equipment includes a gas source, a pressure regulator, a flow metering device, and the necessary hoses and connectors. Shielding gases are dispensed from cylinders, cylinder manifolds, or bulk tanks from which gases are piped to individual welding stations. Regulators and flow meters are used to control pressure and flow rates. Because regulators can freeze during rapid withdrawal of CO_2 from storage tanks, heaters are available to prevent that complication. Welding-grade gas purity is required because small amounts of moisture can result in porosity or hydrogen absorption in the weld metal. The dew point of shielding gases should be below –40°F (–40°C).

The fumes generated during flux cored arc welding can be hazardous. To assure adequate ventilation, portable fume extraction systems and welding guns with integrated fume extractors are available. A welding gun fume extractor usually consists of an exhaust nozzle that encircles the gun nozzle. It can be adapted to gas-shielded and self-shielded guns. The nozzle is ducted to a filter canister and an exhaust pump. The aperture of the fume-extracting nozzle is located at a sufficient distance behind the top of the gun nozzle to draw in the fumes rising from the arc without disturbing the shielding gas flow. The chief advantage of this fume extraction system is that it remains close to the fume source wherever the welding gun is used. In contrast, a portable fume exhaust system generally cannot be positioned as closely to the fume source and requires repositioning the exhaust hood for each significant change in welding location.

FUME EXTRACTION

One disadvantage of the welding gun fume extractor system is that the added weight and bulk make semiautomatic welding more cumbersome for the welder. If not properly installed and maintained, fume extractors may cause welding problems by disturbing the gas shielding. In a well-ventilated welding area, a fume-extractor and welding gun combination may not be necessary. Additional information on proper ventilation is presented in the "Safe Practices" section of this chapter.

MATERIALS

The base metals commonly welded with flux cored arc welding, the shielding gases used, and electrodes appropriate for various applications are described in this section.

BASE METALS

Most of the commonly used types of ferrous plate, pipe, and castings and many nickel alloys can be welded using the FCAW process. The categories of base metals generally welded with FCAW are mild steels, high-strength steels, chrome-molybdenum steels, stainless steels, abrasive-resistant steels, cast steels, and nickel alloys.

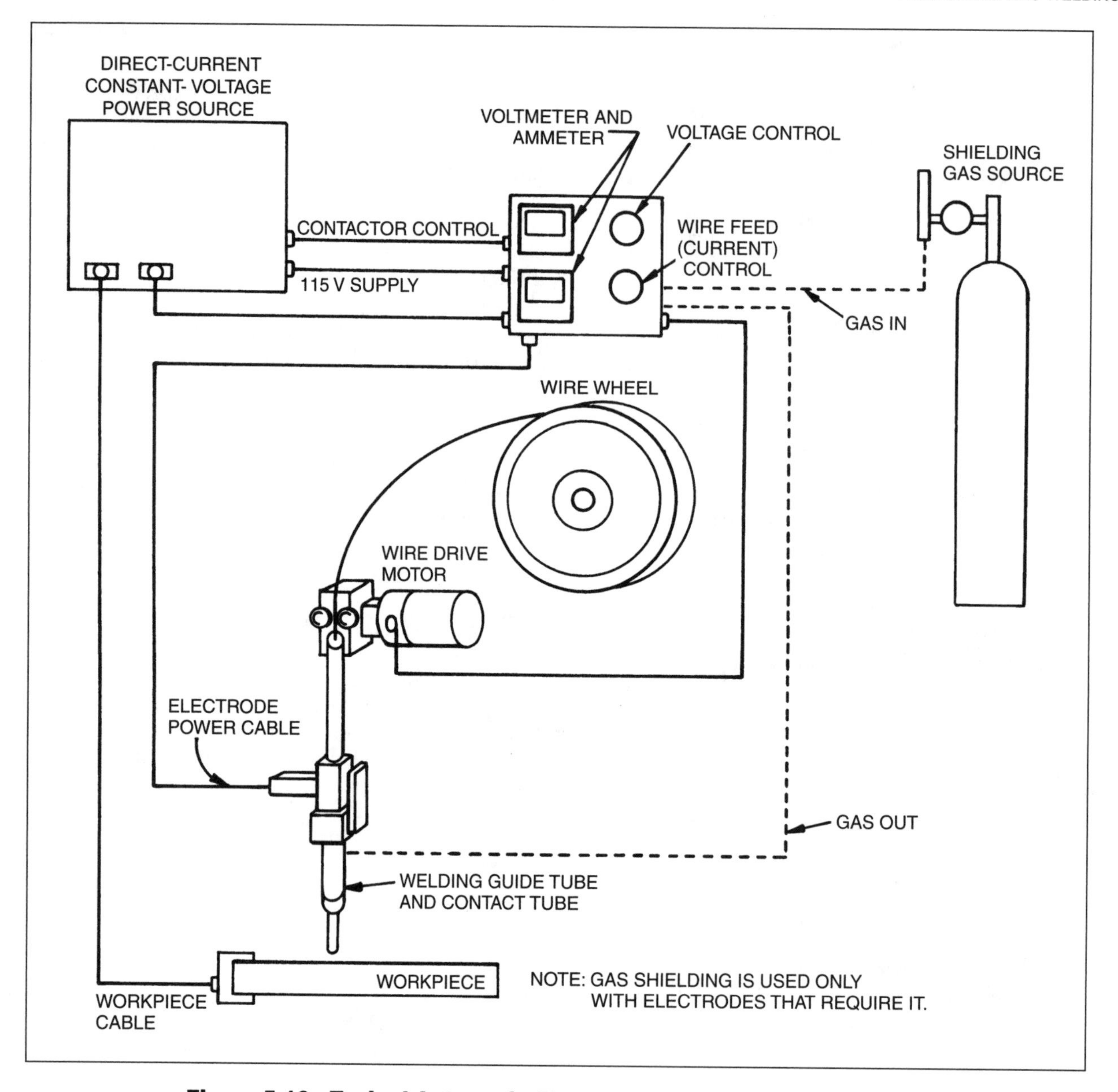

Figure 5.10—Typical Automatic Flux Cored Arc Welding Equipment

Mild Steels

Structural and pressure-vessel grades of mild steel, such as A36, A515, and A516 are the steels most often welded with FCAW. Pipe and castings of similar composition also are welded using this process. These steels are relatively easy to weld with FCAW using minimal precautions except under extreme environmental conditions. Potential moisture pick-up must be considered in very humid environments. (See "Protection from Moisture" in the "Electrodes" section of this chapter. When the base metal is very cold or welding is being done on thick sections, preheating of the base metal may be necessary.

High-Strength Steels

This category includes high-strength low-alloy (HSLA) steels, such as ASTM A441, A572, and A588. The high yield strength, quench-and-tempered (Q&T) steels ASTM A514 and A517 are also welded with the FCAW process. Welding these classes of steel is increas-

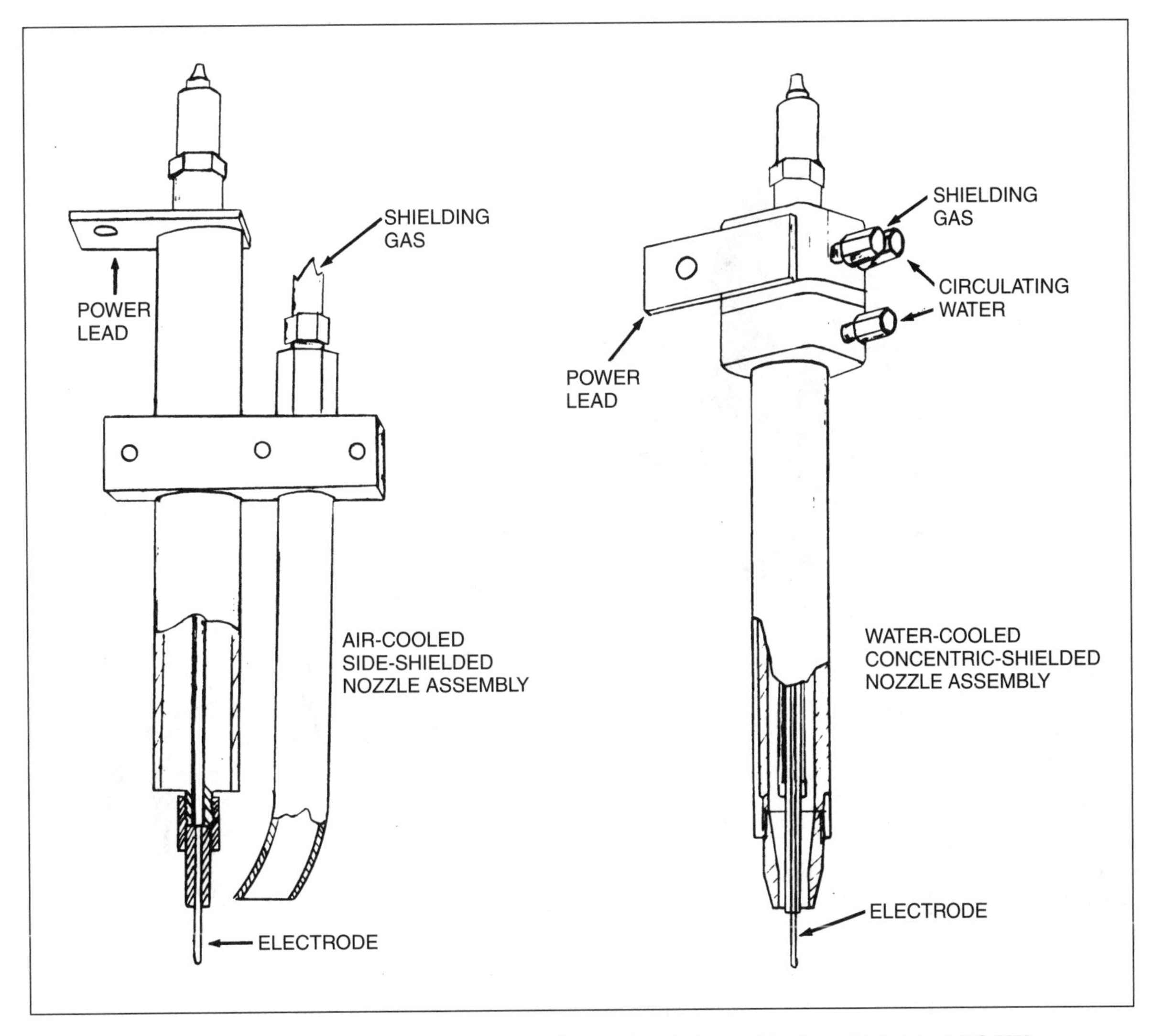

Figure 5.11—Typical Nozzle Assemblies for Automatic Gas-Shielded FCAW

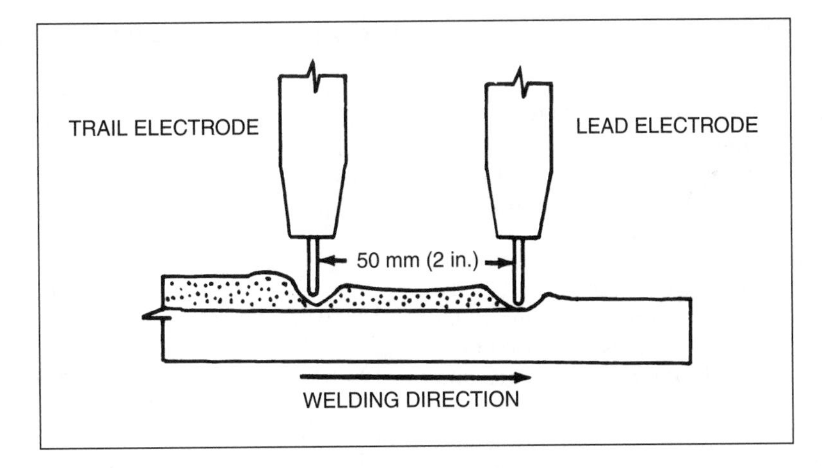

Figure 5.12—Automatic Tandem Arc Welding with Two Gas-Shielded Flux Cored Electrodes

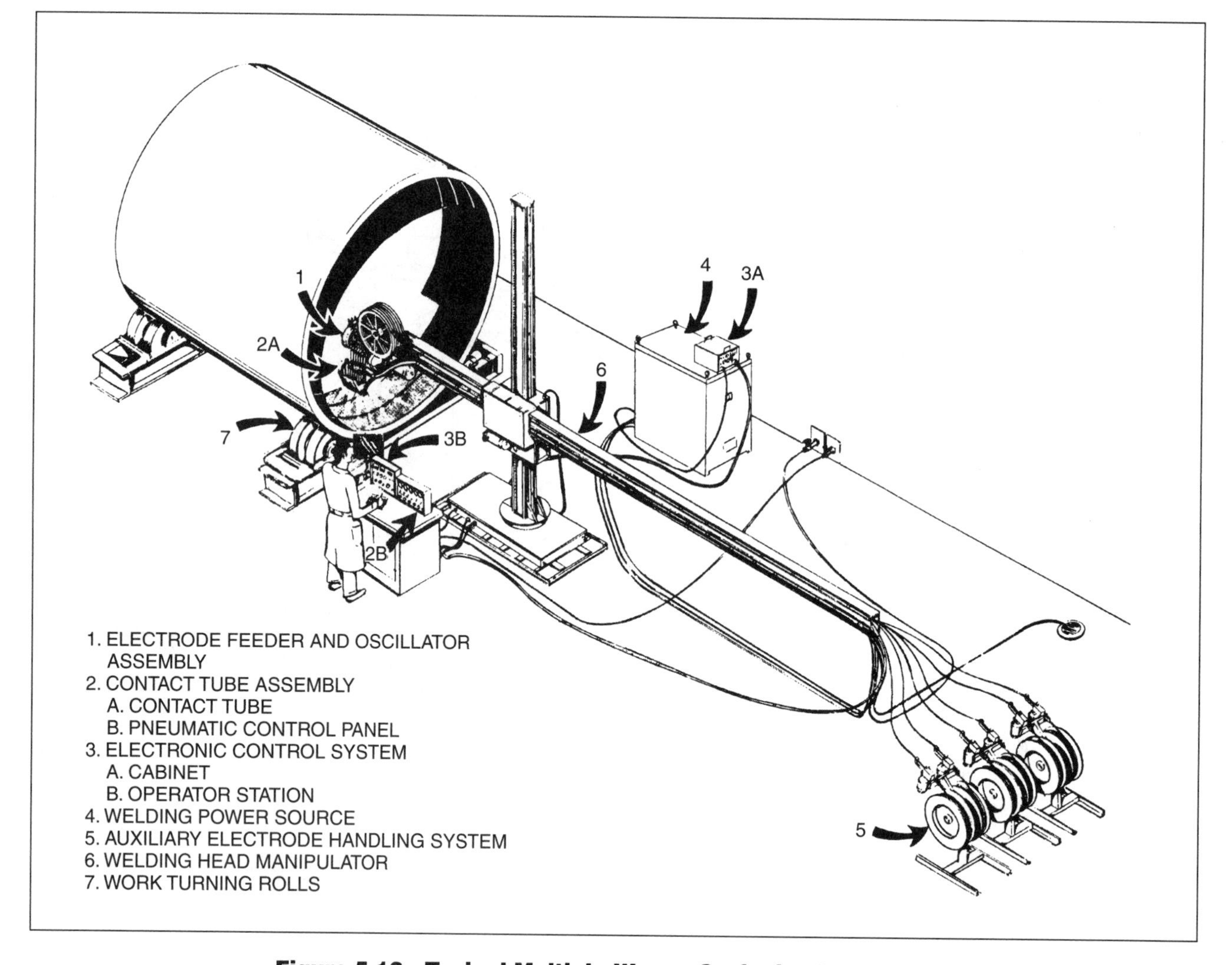

Figure 5.13—Typical Multiple-Weave Surfacing Installation

ing as manufacturers produce steels with increasingly higher strength-to-weight ratios.

Precautions recommended by the base metal and filler metal manufacturers must be followed when welding high-strength steels. The rapid cooling rates associated with welding alter the metallurgical structure and properties of the heat-affected zone (HAZ) of the base metal in the weld joint. Generally, as alloying content increases (especially carbon), there is increased need for precautions such as preheat and postweld heat treatment. The change in properties in the HAZ must be anticipated during weldment design. Further information on these precautions is presented in the AWS *Welding Handbook*, *Materials and Applications—Part 2*, Volume 4, 8th edition.[6]

6. American Welding Society (AWS) Welding Handbook Committee, W. R. Oates and A. M. Saitta, eds., 1998, *Materials and Applications—Part 2*, Vol. 4 of *Welding Handbook*, 8th ed. Miami: American Welding Society.

Chromium-Molybdenum Steels

Chrome-molybdenum (Cr-Mo) steels such as 1-1/4% Cr–1/2% Mo and 2-1/4% Cr–1% Mo and 9% Cr–1% Mo (Grade 91) are welded with the FCAW process. As with high-strength steels, precautions must be taken to allow for hardenability.

Stainless Steels

Most corrosion-resistant stainless steels such as AISI types 304, 309, 316, 409, 410, and 17-4 PH are weldable with the FCAW process. Stainless steel castings are also welded using flux cored electrodes. Discussion of the metallurgy and weldability of these steels is beyond the scope of this chapter. More detailed information is presented in the AWS *Welding Handbook*, *Materials and Applications—Part 2*, Volume 4, 8th edition.[7]

Abrasion-Resistant Steels

Abrasion-resistant steels are often welded with the FCAW process. These steels have very high hardness and high tensile strength. The welding consumables (electrodes and fluxes) used to weld abrasion-resisting steels usually have neither the structural strength nor the abrasion resistance of the base metal. The lack of strength is generally not a great concern because these steels are not normally used for structural applications. The intended function of the weld metal is mainly to hold the plates in position rather than to provide structural strength. If the weld requires abrasion resistance equal to the base plate, a hardsurfacing electrode should be used on the surface of the weld after the plates are welded in position.

Nickel Alloys

While nickel alloys are welded using the FCAW process, their metallurgy and weldability are beyond the scope of this chapter. More detailed information is presented in the AWS *Welding Handbook*, *Materials and Applications—Part 1*, Volume 3, 8th edition.[8]

SHIELDING GASES

Carbon dioxide (CO_2) and mixtures of argon and CO_2 are the preferred shielding gases for flux cored arc welding.

Carbon Dioxide

Carbon dioxide is widely used as a shielding gas for flux cored arc welding. This gas usually provides a globular metal transfer, although some flux formulations produce a spray-like metal transfer in CO_2. It promotes deep weld penetration and is lower in cost than mixed gases.

Carbon dioxide is relatively inactive at room temperature. When heated to high temperature by the welding arc, CO_2 dissociates to form carbon monoxide (CO) and oxygen (O_2), as indicated by the following chemical equation:

$$2\ CO_2 \rightarrow 2\ CO + O_2 \tag{5.1}$$

Thus, the arc atmosphere contains a considerable amount of oxygen that reacts with elements in the molten metal. The oxidizing tendency of CO_2 shielding gas is recognized in the formulation of flux cored electrodes. Deoxidizing materials are added to the core of the electrode to compensate for the oxidizing effect of the CO_2.

In addition, molten iron reacts with CO_2 and produces iron oxide and carbon monoxide in a reversible reaction:

$$Fe + CO_2 \rightleftarrows FeO + CO \tag{5.2}$$

At red heat temperatures, some of the carbon monoxide dissociates to carbon and oxygen, as follows:

$$2CO \rightleftarrows 2C + O_2 \tag{5.3}$$

The effect of CO_2 shielding on the carbon content of mild and low-alloy steel weld metal is unique. Depending on the original carbon content of the base metal and the electrode, the CO_2 atmosphere can behave either as a carburizing or decarburizing medium. Whether the carbon content of the weld metal will be increased or decreased depends on the carbon present in the electrode and the base metal. If the carbon content of the weld metal is below approximately 0.05%, the weld pool will tend to pick up carbon from the CO_2 shielding atmosphere. Conversely, if the carbon content of the weld metal is greater than approximately 0.10% the weld pool may lose carbon. The loss of carbon is attributed to the formation of carbon monoxide caused by the oxidizing characteristics of CO_2 when used as shielding gas at high temperatures.

When this reaction occurs, the carbon monoxide can be trapped in the weld metal and will create porosity. This tendency can be minimized by using an electrode that provides an adequate level of deoxidizing elements in the core. Oxygen reacts with the deoxidizing elements rather than the carbon in the steel. That reaction results in the formation of solid oxide compounds that float to the surface of the weld pool where they form part of the slag covering.

Gas Mixtures

Gas mixtures used in flux cored arc welding may combine the separate advantages of two or more gases, including carbon dioxide, oxygen and argon. The higher the percentage of inert gas in mixtures with CO_2 or oxygen, the higher the transfer efficiencies of the deoxidizers contained in the core will be. Argon is capable of protecting the weld pool at all welding temperatures. The presence of argon in sufficient quantities in a

---

7. See Reference 6.
8. American Welding Society (AWS) Welding Handbook Committee, W. R. Oates, ed., 1996, *Materials and Applications—Part 1*, Vol. 3 of *Welding Handbook*, 8th ed. Miami: American Welding Society.

shielding gas mixture results in less oxidation than occurs with 100% CO_2 shielding.

The mixture commonly used in gas-shielded FCAW is 75% argon and 25% carbon dioxide. When welding with this mixture, a spray-transfer arc is achieved. The 75% argon/25% CO_2 mixture provides better arc characteristics than 100% CO_2, resulting in greater operator appeal.

Weld metal deposited with this mixture generally has higher tensile strength and yield strength than weld metal deposited with 100% CO_2 shielding because of high transfer efficiencies. Manganese and silicon are transferred into the weld pool and remain as alloying elements instead of combining with oxygen. When shielding gas mixtures with high percentages of inert gases are used with electrodes designed for CO_2, the shielding gas mixture may cause an excessive buildup of manganese, silicon, and other deoxidizing elements in the weld metal. The resulting higher alloy content of the weld metal changes the mechanical properties. Therefore, the electrode manufacturer should be consulted to ascertain the mechanical properties of weld metal obtained with specific shielding gas mixtures. If data are not available, tests should be made to determine the mechanical properties for the particular application.

ELECTRODES

As previously described, the flux cored electrode is a composite tubular filler-metal electrode consisting of a metal sheath and a core of various powdered materials. Both the sheath and core contain ingredients that contribute to the highly desirable operating characteristics and weld properties of the process. The use of this electrode differentiates the FCAW process from other arc welding processes.[9]

Flux cored arc welding owes much of its versatility to the wide variety of ingredients that can be included in the core of the tubular electrode. For ferrous alloys, the electrode usually consists of a low-carbon steel or an alloy-steel sheath surrounding a core of fluxing and alloying materials. The composition of the flux core varies according to the electrode classification and the particular manufacturer of the electrode.

Most flux cored electrodes are made by passing a steel strip through rolls that form it into a U-shaped cross section. The formed strip is filled with a measured amount of granular core material (alloys and flux); then a closing roll rounds the filled shape and closes it. The round tube is pulled through drawing dies or rolls that reduce the diameter and compress the core. The electrode is drawn to final size, and then wound (as wire) on spools or in coils. Other manufacturing methods are also used.

Manufacturers generally consider the precise composition of their cored electrodes to be proprietary information. By proper development and selection of the core ingredients (in combination with the composition of the sheath), manufacturers have achieved the following:

1. Electrodes with welding characteristics ranging from high deposition rates in the flat position to proper fusion and bead shape in the overhead position,
2. Electrodes for various gas shielding mixtures and for self shielding, and
3. Variations in the alloy content of the weld metal from mild steel for certain electrodes to high-alloy stainless steel for others.

The primary functions of the flux core ingredients are to accomplish the following:

1. Provide the mechanical, metallurgical, and corrosion-resistant properties of the weld metal by adjusting the chemical composition;
2. Promote weld metal soundness by shielding the molten metal from oxygen and nitrogen in the air or, in the case of self-shielded FCAW, to react with nitrogen or oxygen, or both, in the air and render it harmless;
3. Scavenge impurities from the molten metal through the use of fluxing reactions;
4. Produce a slag cover to protect the solidifying weld metal from the air.
5. Control the shape and appearance of the bead in the different welding positions for which the electrode is suited; and
6. Stabilize the arc by providing a smooth electrical path to reduce spatter and facilitate the deposition of uniformly smooth, properly sized beads.

Table 5.1 lists most of the elements commonly found in the flux core, the form in which they are integrated, and the purposes for which they are used.

In mild steel and low-alloy steel electrodes, a proper balance of deoxidizers and denitrifiers (in the case of self-shielded electrodes) must be maintained to provide a sound weld deposit with adequate ductility and toughness. Deoxidizers, such as silicon and manganese, combine with oxygen and form stable oxides. This helps control the loss of alloying elements through oxidation and the formation of carbon monoxide, which otherwise could cause porosity. The denitrifiers, such as

9. The electrogas welding (EGW) process uses a flux cored electrode to make single-pass welds in the vertical position. See Chapter 8, "Electrogas Welding." Flux cored electrodes are also used in gas tungsten arc welding (Chapter 3) for some applications. It should be noted that metal-cored electrodes do not match the definition of flux cored electrodes and are not discussed in this chapter. Metal-cored electrodes are described in Chapter 4, "Gas Metal Arc Welding."

Table 5.1
Common Core Elements in Flux Cored Electrodes

Element	Usually Present As	Purpose in Weld
Aluminum	Metal powder	Deoxidize and denitrify
Boron	Ferroboron	Grain refinement
Calcium	Minerals such as fluorspar (CaF_2) and limestone ($CaCO_3$)	Provide shielding and form slag
Carbon	Element in ferroalloys such as ferromanganese	Increase hardness and strength
Chromium	Ferroalloy or metal powder	Alloying to improve creep resistance, hardness, strength, and corrosion resistance
Iron	Ferroalloys and iron powder, sheath	Alloy matrix in iron-base deposits, alloy in nickel-base and other nonferrous deposits
Manganese	Ferroalloy such as ferromanganese or as metal powder	Deoxidize; prevent hot shortness by combining with sulfur to form manganese sulfide; increase hardness and strength; form slag
Molybdenum	Ferroalloy	Alloying to increase hardness and strength; in austenitic stainless steels to increase resistance to pitting-type corrosion
Nickel	Metal powder	Alloying to improve hardness, strength, toughness and corrosion resistance
Potassium	Minerals such as potassium-bearing feldspars and silicates in frits	Stabilize the arc and form slag
Silicon	Ferroalloy such as ferrosilicon, or silicomanganese; mineral silicates such as feldspar	Deoxidize and form slag
Sodium	Minerals such as sodium-bearing feldspars and silicates in frits	Stabilize the arc and form slag
Vanadium	Oxide or metal powder	Increase strength
Titanium	Ferroalloy such as ferrotitanium; in mineral, rutile (titanium dioxide)	Deoxidize and denitrify; form slag; stabilize carbon in some stainless steels
Zirconium	Oxide or metal powder	Deoxidize and denitrify; form slag

aluminum, combine with nitrogen and tie it up as stable nitrides. This prevents nitrogen porosity and the formation of other nitrides that might be harmful.

Electrode Classifications

The American Welding Society has developed a system of electrode classifications for the various welding processes and the most commonly used metals and materials. The Society maintains the classifications with current information and publishes specifications for the various electrode classes. The descriptions of the various electrodes in this section are contributed by the American Welding Society Committee on Filler Metals and Allied Materials.

Mild Steel Electrodes

Most mild steel FCAW electrodes are classified according to the requirements of the latest edition of ANSI/AWS A5.20, *Specification for Carbon Steel Electrodes for Flux Cored Arc Welding.*[10] The identification system follows the general pattern for electrode classification and is illustrated in Figure 5.14.

The classification system can be explained by considering a typical designation, E70T-1. The prefix "E" indicates an electrode, as in other electrode classification systems. The first number refers to the minimum as-welded tensile strength in 10,000 pounds per square inch (psi) units. In this example, the number "7" indicates that the electrode has a minimum tensile strength of 70,000 psi. The second number indicates the welding positions for which the electrode is designed. Here, the "0" means that the electrode is designed for flat groove and fillet welds, and horizontal groove and fillet welds.

10. American Welding Society (AWS) Committee on Filler Metals and Allied Materials, *Specification for Carbon Steel Electrodes for Flux Cored Arc Welding*, ANSI/AWS A5.20, Miami: American Welding Society.

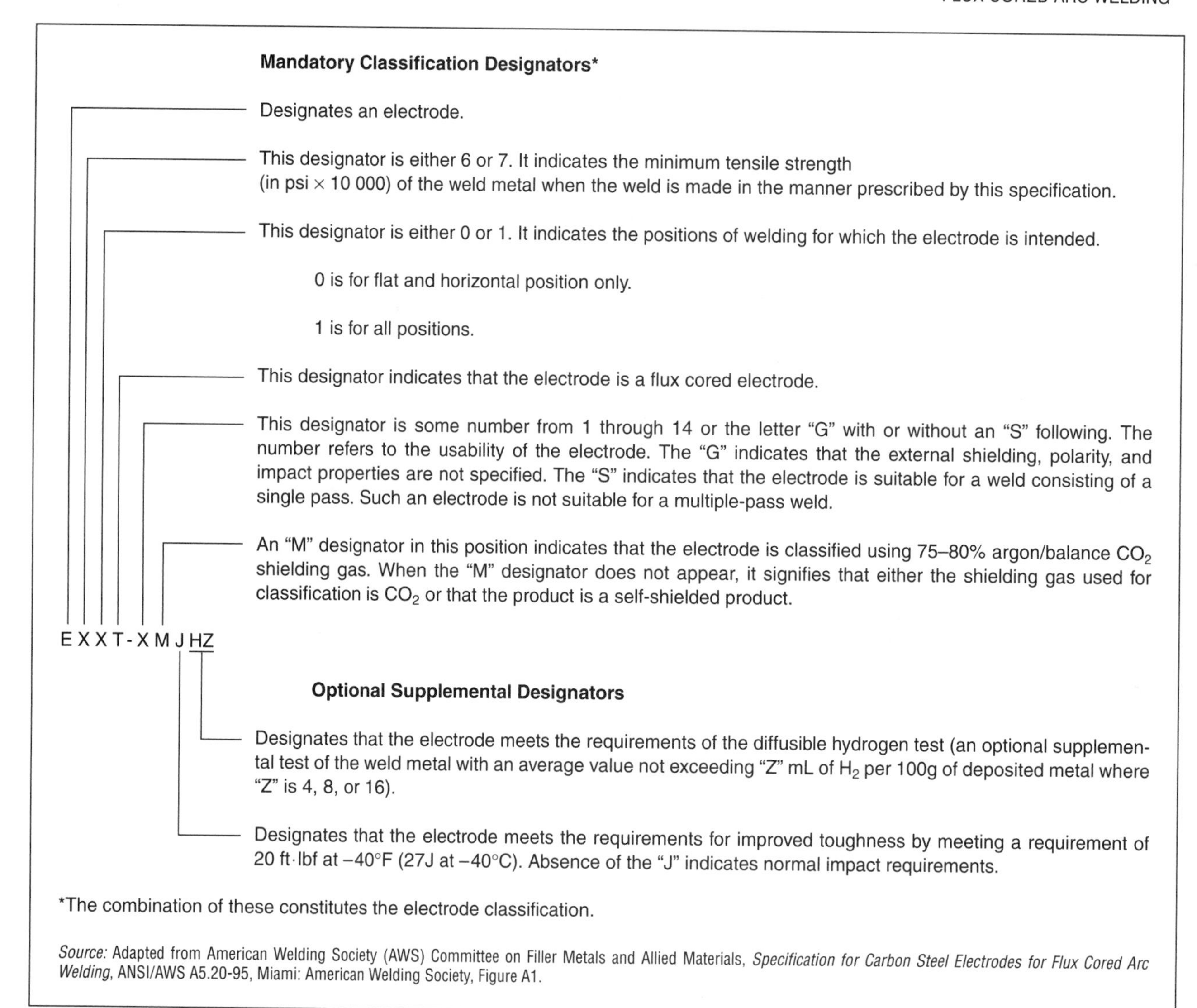

Figure 5.14—Classification System for Mild Steel FCAW Electrodes

However, some classifications may be suitable for vertical or overhead positions, or both. In those cases, a "1" would be used instead of the "0" to indicate all-position capability.

The letter "T" indicates that the electrode is of a tubular construction (a flux cored electrode). The suffix number "1" (in this example) indicates a general grouping of electrodes that contain similar flux or core components and have similar usability characteristics. The group designators described in Figure 5.14 are expanded in Table 5.2, which lists usability characteristics of mild steel FCAW electrodes. Other possible designators can be included with this suffix, such as an "M," indicating that the electrode was tested in mixed gas, "J" to designate enhanced impact properties, or an "H" followed by a number signifying diffusible hydrogen testing results.

As shown in Figure 5.14, mild steel FCAW electrodes are classified on the basis of whether they are self-shielded or whether they are intended to be used with CO_2 or mixed gas, which is usually considered to be 75% to 80% argon and the balance CO_2. The classification also specifies the type of current, usability out of position, the chemical composition, and the as-welded mechanical properties of deposited weld metal. Some classifications are listed as being suitable for multiple- and single-pass welding, while others are stated to be suitable for single-pass welding only. Electrodes for

Table 5.2
Shielding and Polarity Requirements for Mild Steel FCAW Electrodes

AWS Classification	Recommended Weld Passes	External Shielding Medium	Current and Polarity
EXXT-1	Multiple	CO_2	DCEP[c]
EXXT-1M	Multiple	Mixed gas[a]	DCEP
EXXT-2	Single	CO_2	DCEP
EXXT-2M	Single	Mixed gas[a]	DCEP
EXXT-3	Single	None	DCEP
EXXT-4	Multiple	None	DCEP
EXXT-5	Multiple	CO_2	DCEP
EXXT-5M	Multiple	Mixed gas[a]	DCEP
EXXT-6	Multiple	None	DCEP
EXXT-7	Multiple	None	DCEN[d]
EXXT-8	Multiple	None	DCEN
EXXT-9	Multiple	CO_2	DCEP
EXXT-9M	Multiple	Mixed gas[a]	DCEP
EXXT-10	Single	None	DCEN
EXXT-11	Multiple	None	DCEN
EXXT-12	Multiple	CO_2	DCEP
EXXT-12M	Multiple	Mixed gas[a]	DCEP
EXXT-13	Single	None	DCEN
EXXT-14	Single	None	DCEN
EXXT-G	Multiple	Note b	Note b
EXXT-GS	Single	Note b	Note b

Notes:
a. Mixed gas normally refers to 75% to 80% argon/balance CO_2
b. As agreed upon by supplier and user
c. Direct current electrode positive.
d. Direct current electrode negative.

single-pass welding have more deoxidizing elements such as manganese and silicon and can be used to weld over mill scale or rust without resulting in porosity. When these electrodes are used for more than a single pass, these deoxidizers will increase the effective alloy content of the weld metal, excessively increasing hardness and reducing ductility. These same effects will also be observed when an electrode classified with CO_2 shielding gas is used with a less reactive gas (argon or combinations containing argon, for example). Electrodes are designed to produce weld metal having specified chemical composition and mechanical properties when the welding and testing are performed according to the specification requirements.

Electrodes are produced in standard diameter sizes ranging from 0.8 to 4.0 mm (0.030 to 5/32 in.). Special sizes may also be available. Weld properties may vary appreciably, depending on a number of conditions, including electrode size, welding amperage, plate thickness, joint geometry, preheat and interpass temperatures, surface conditions, base metal composition and admixture with the deposited metal, and shielding gas (if required). Many electrodes are designed primarily for welding in the flat and horizontal positions. They may also be suitable for use in other positions, depending on electrode diameter, choice of heat input, and the level of operator skill. Selected electrodes with diameters below 2.4 mm (3/32 in.) may be used for out-of-position welding at welding currents on the low side of the manufacturer's recommended range.

The classifications, descriptions, and intended uses of mild steel electrodes as designated in ANSI/AWS A5.20 are described below.

EXXT-1 and EXXT-1M. Electrodes of the EXXT-1 group are classified with CO_2 shielding gas. However, other gas mixtures, such as argon and CO_2, may be used to improve the arc characteristics, especially for out-of-position work, when recommended by the manufacturer. Increasing the amount of argon in the argon/CO_2 mixture will increase the manganese and silicon contents in the weld metal. The increase in manganese and silicon will increase the yield strength and tensile strength and may affect impact properties.

Electrodes of the EXXT-1M group are classified with 75% to 80% argon/balance CO_2 shielding gas. The use of these electrodes with argon/CO_2 shielding gas mixtures with reduced amounts of argon, or with CO_2 shielding gas alone, may result in some deterioration of arc characteristics and out-of-position welding characteristics. In addition, a reduction of the manganese and silicon contents in the weld will reduce yield and tensile strengths and may affect impact properties.

Both the EXXT-1 and EXXT-1M electrodes are designed for single- and multiple-pass welding using direct current electrode positive (DCEP) polarity. The larger diameters (usually 2.0 mm [5/64 in.] and larger) are used for welding in the flat position and for welding fillet welds in the horizontal position (EX0T-1 and EX0T-1M). The smaller diameters (usually 1.6 mm [1/16 in.] and smaller) are generally used for welding in all positions (EX1T-1 and EX1T-1M). The EXXT-1 and EXXT-1M electrodes are characterized by a spray transfer, low spatter loss, flat-to-slightly convex bead contour, and a moderate volume of slag that completely covers the weld bead. Most electrodes of this classification have a rutile-base slag and produce high deposition rates.

EXXT-2 and EXXT-2M. Electrodes of these classifications are essentially EXXT-1 and EXXT-1M with higher percentages of manganese or silicon, or both, and are designed primarily for single-pass welding in the flat position and for welding fillet welds in the horizontal position. The higher levels of deoxidizers in these classifications allow the single-pass welding of heavily oxidized or rimmed steel.

Weld metal composition requirements are not specified for single-pass electrodes, since checking the composition of the undiluted weld metal will not provide an indication of the composition of a single-pass weld. These electrodes provide good mechanical properties in single-pass welds.

Should the user choose to make multiple-pass welds using EXXT-2 and EXXT-2M electrodes, it should be noted that both the manganese content and the tensile strength of the weld metal made with this filler metal will be high. These electrodes can be used for welding base metals that have heavy mill scale, rust, or other foreign matter that cannot be tolerated by some electrodes of the EXXT-1 and EXXT-1M classifications. The arc transfer, welding characteristics and deposition rates of these electrodes are similar to those of the EXXT-1 and EXXT-1M classifications.

EXXT-3. Electrodes of this classification are self-shielded, used with DCEP, and produce a spray transfer. The slag system is designed to make very high welding speeds possible. The electrodes are used for single-pass welds in the flat, horizontal, and vertical positions (up to a 20° incline, downward progression) on sheet metal. Since these electrodes are sensitive to the effects of base metal quenching, they are not generally recommended for T-joints or lap joints in material thicker than 4.8 mm (3/16 in.) and butt, edge, or corner joints in materials thicker than 6.4 mm (1/4 in.). The electrode manufacturer should be consulted for specific recommendations.

EXXT-4. Electrodes of this classification are self-shielded, operate on DCEP, and have a globular transfer. The slag system is designed to make very high deposition rates possible and to produce a weld that is very low in sulfur, which makes the weld highly resistant to hot cracking. These electrodes are designed for low penetration beyond the root of the weld, making them suitable for use on joints that are poorly fitted and for single- and multiple-pass welding.

EXXT-5 and EXXT-5M. Electrodes of the EXXT-5 classification are designed to be used with CO_2 shielding gas; however, as with the EXXT-1 classification, argon-CO_2 mixtures may be used to reduce spatter in accordance with the manufacturer's recommendations. Electrodes of the EXXT-5M classification are designed for use with 75% to 80% argon-balance CO_2 shielding gas. Electrodes of the EX0T-5 and EX0T-5M classifications are used primarily for single- and multiple-pass welds in the flat position and for welding fillet welds in the horizontal position. These electrodes are characterized by a globular transfer, slightly convex bead contour, and a thin slag that may not completely cover the weld bead. These electrodes have a lime-fluoride base slag. Weld deposits produced by these electrodes typically have impact properties and resistance to hot and cold cracking that are superior to those obtained with rutile-base slags. The EX1T-5 and EX1T-5M electrodes, using direct current electrode negative (DCEN), can be used for welding in all positions. However, these electrodes have less operator appeal than electrodes with rutile-base slags.

EXXT-6. Electrodes of this classification are self-shielded, operate on DCEP, and have a spray transfer. The slag system is designed to give good low-temperature impact properties, good penetration into the root of the weld, and excellent slag removal, even in a deep groove. These electrodes are used for single-

and multiple-pass welding in the flat and horizontal positions.

EXXT-7. Electrodes of this classification are self-shielded, operate on DCEN, and have a transfer range from small droplet transfer to a spray transfer. The slag system is designed to allow the large droplets to be used for high deposition rates in the horizontal and flat positions, and to allow the smaller spray particles to be used for all welding positions. The electrodes are used for single- and multiple-pass welding and produce very low-sulfur weld metal, which is highly resistant to cracking.

EXXT-8. Electrodes of this classification are self-shielding, operate on DCEN, and have a small droplet or spray-type transfer. The electrodes are suitable for all welding positions, and the weld metal has very good low-temperature notch toughness and crack resistance. The electrodes are used for single- and multiple-pass welds.

EXXT-9 and EXXT-9M. Electrodes of the EXXT-9 group are classified with CO_2 shielding gas. However, gas mixtures of argon and CO_2 are sometimes used to improve usability, especially for out-of-position applications when recommended by the manufacturer. Increasing the amount of argon in the argon/CO_2 mixture will affect the weld metal analysis and mechanical properties of weld metal deposited by these electrodes, just as it will for weld metal deposited by EXXT-1 and EXXT-1M electrodes.

Electrodes of the EXXT-9M group are classified with a 75% to 80% argon/balance CO_2 shielding gas. The use of these electrodes with argon/CO_2 shielding gas mixtures with reduced amounts of argon, or with 100% CO_2 shielding gas, may result in some deterioration of arc characteristics and out-of-position welding characteristics. In addition, a reduction of the manganese and silicon contents in the weld will have some effect on the properties of weld metal from these electrodes, just as it will on properties of weld metal deposited by EXXT-1M electrodes.

Both the EXXT-9 and EXXT-9M electrodes are designed for single- and multiple-pass welding. The larger diameters (usually 2.0 mm [5/64 in.] and larger) are used for welding in the flat position and for welding fillet welds in the horizontal position. The smaller diameters (usually 1.6 mm [1/16 in.] and smaller) are often used for welding in all positions.

The arc transfer, welding characteristics, and deposition rates of the EXXT-9 and EXXT-9M electrodes are similar to those of the EXXT-1 and EXXT-1M classifications. EXXT-9 and EXXT-9M electrodes are essentially EXXT-1 and EXXT-1M electrodes that deposit weld metal with improved impact properties.

Some electrodes in this classification require that joints be relatively clean and free of oil, excessive oxide, and scale in order to obtain welds of radiographic quality.

EXXT-10. Electrodes of this classification are self-shielded, operate on direct current electrode negative (DCEN), and have a small droplet transfer. The electrodes are used for single-pass welds at high travel speeds on material of any thickness in the flat, horizontal, and vertical (up to 20° incline) positions.

EXXT-11. Electrodes of this classification are self-shielded, operate on DCEN, and have a smooth, spray-type transfer. They are general-purpose electrodes for single- and multiple-pass welding in all positions. These electrodes are generally not recommended for welding on thicknesses greater than 19 mm (3/4 in.) unless preheat and interpass temperature control is maintained. The electrode manufacturer should be consulted for specific recommendations.

EXXT-12 and EXXT-12M. Electrodes of these classifications are essentially EXXT-1 and EXXT-1M electrodes that have been modified to improve impact toughness and to meet the lower manganese requirements of the A-1 Analysis Group in the ASME *Boiler and Pressure Vessel Code*, Section IX.[11] Therefore, they have an accompanying decrease in tensile strength and hardness. Since welding procedures influence all weld metal properties, users should check hardness on any application in which a specific hardness level is a requirement.

The arc transfer, welding characteristics, and deposition rates of the EXXT-12 and EXXT-12M electrodes are similar to those of the EXXT-1 and EXXT-1M classifications.

EXXT-13. Electrodes of this classification are self-shielded, operate on DCEN, and are usually welded with a short-arc transfer. The slag system is designed so that these electrodes can be used in all positions for the root pass on circumferential welds on pipe. The electrodes can be used on all pipe wall thicknesses, but are recommended for the first pass only. They generally are not recommended for multiple-pass welding.

EXXT-14. Electrodes of this classification are self-shielded, operate on DCEN, and have a smooth spray-type transfer. The slag system is designed with characteristics so that these electrodes can be used to weld in all positions and also to make welds at high speed. They are used to make welds on sheet metal up to

11. American Society for Mechanical Engineers (ASME) Boiler and Pressure Vessel Code Committee, 1998, *Welding and Brazing Qualifications*, Section IX of *Boiler and Pressure Vessel Code*, New York: American Society of Mechanical Engineers.

4.8 mm (3/16 in.) thick and are often specifically designed for galvanized, aluminized, or other coated steels. Since these welding electrodes are sensitive to the effects of base metal quenching, they are not generally recommended for T-joints or lap joints in materials thicker than 4.8 mm (3/16 in.) and butt, edge, or corner joints in materials thicker than 6.4 mm (1/4 in.). The electrode manufacturer should be consulted for specific recommendations.

EXXT-G. This classification is for multiple-pass electrodes that are not covered by any presently defined classification. Except for chemical requirements to assure a carbon-steel deposit and the tensile strength, which is specified, the requirements for this classification are not specified. The requirements are agreed upon by the purchaser and the supplier.

EXXT-GS. This classification is for single-pass electrodes that are not covered by any presently defined classification. Except for the tensile strength, which is specified, the requirements for this classification are not specified. The requirements are agreed upon by the purchaser and supplier.

Low-Alloy Steel Electrodes

Flux cored electrodes are commercially available for welding low-alloy steels. They are classified in the latest edition of ANSI/AWS A5.29, *Specification for Low Alloy Steel Electrodes for Flux Cored Arc Welding.*[12] These electrodes are generally used to weld low-alloy steels of similar chemical composition. Some electrode classifications are designed for welding in all positions while others are limited to flat and horizontal positions. The American Welding Society designates and describes the various classifications as shown in Figure 5.15, which lists the components of these designations.

The specification ANSI/AWS A5.29 contains many different classifications of flux cored electrodes. The suffix in each classification (1, 4, 5, 6, 7, 8, 11, or G) indicates a general grouping of electrodes that contain similar flux or core components and that have similar usability characteristics, except for the "G" classification in which usability characteristics may differ between similarly classified electrodes.

The steels commonly welded with low-alloy electrodes are usually used for specific purposes. The welding of these steels requires an understanding of their properties and heat treatment beyond that which can be covered in this publication. Users not familiar with the characteristics of low-alloy steels are referred to the AWS Welding Handbook, *Materials and Applications—Part 2,* Volume 4, 8th edition[13] and other publications on low-alloy steels.

EXXT1-X and EXXT1-XM. Electrodes of the EXXT1-X group are classified with CO_2 shielding gas. However, other gas mixtures (such as argon/CO_2) may be used to improve usability, especially for out-of-position applications, when recommended by the manufacturer. Increasing the amount of argon in the argon/CO_2 mixture will increase the manganese and silicon contents, along with certain other elements, such as chromium, in the weld metal. The increase in manganese, silicon, or other alloying elements will increase the yield and tensile strengths and may affect impact properties.

Electrodes of the EXXT1-XM group are classified with 75% to 80% argon/balance CO_2 shielding gas. The use of these electrodes with argon/CO_2 shielding gas mixtures having reduced amounts of argon or with CO_2 shielding gas may result in some deterioration of arc characteristics and out-of-position welding characteristics. In addition, a reduction of the manganese, silicon, and certain other alloy contents in the weld metal will reduce yield and tensile strengths and may affect impact properties.

Both the EX1T1-X and EX1T1-XM electrodes are designed for single-pass and multiple-pass welding using DCEP polarity. The larger diameters (usually 2.0 mm [5/64 in.] and larger) are used for welding in the flat position and for welding fillet welds in the horizontal position (EX0T1-X and EX0T1-XM). The smaller diameters (usually 1.6 mm [1/16 in.] and smaller) are used for welding in all positions (EX1T1-X and EX1T1-XM). The EXTT1-XM electrodes are characterized by a spray transfer, low spatter loss, flat-to-slightly convex bead contour, and a moderate volume of slag that completely covers the weld bead. Electrodes of this classification have a rutile slag and produce high deposition rates.

EX0T4-X. Electrodes of this classification are self-shielded, operate on DCEP, and have a globular transfer. The slag system is designed to make very high deposition rates possible and to produce a weld that is very low in sulfur, which makes the weld highly resistant to hot cracking. These electrodes are designed for low penetration beyond the root of the weld, making them useful for welding poorly fitted joints and for single-pass and multiple-pass welding.

EXXT5-X and EXXT5-XM. Electrodes of the EXXT5-X classifications are designed to be used with CO_2 shielding gas; however, as with the EXXT1-X classifications,

12. American Welding Society (AWS) Committee on Filler Metals and Allied Materials, 1998, *Specification for Low-Alloy Steel Electrodes for Flux Cored Arc Welding,* ANSI/AWS A5.29:98, Miami: American Welding Society.

13. See Reference 8.

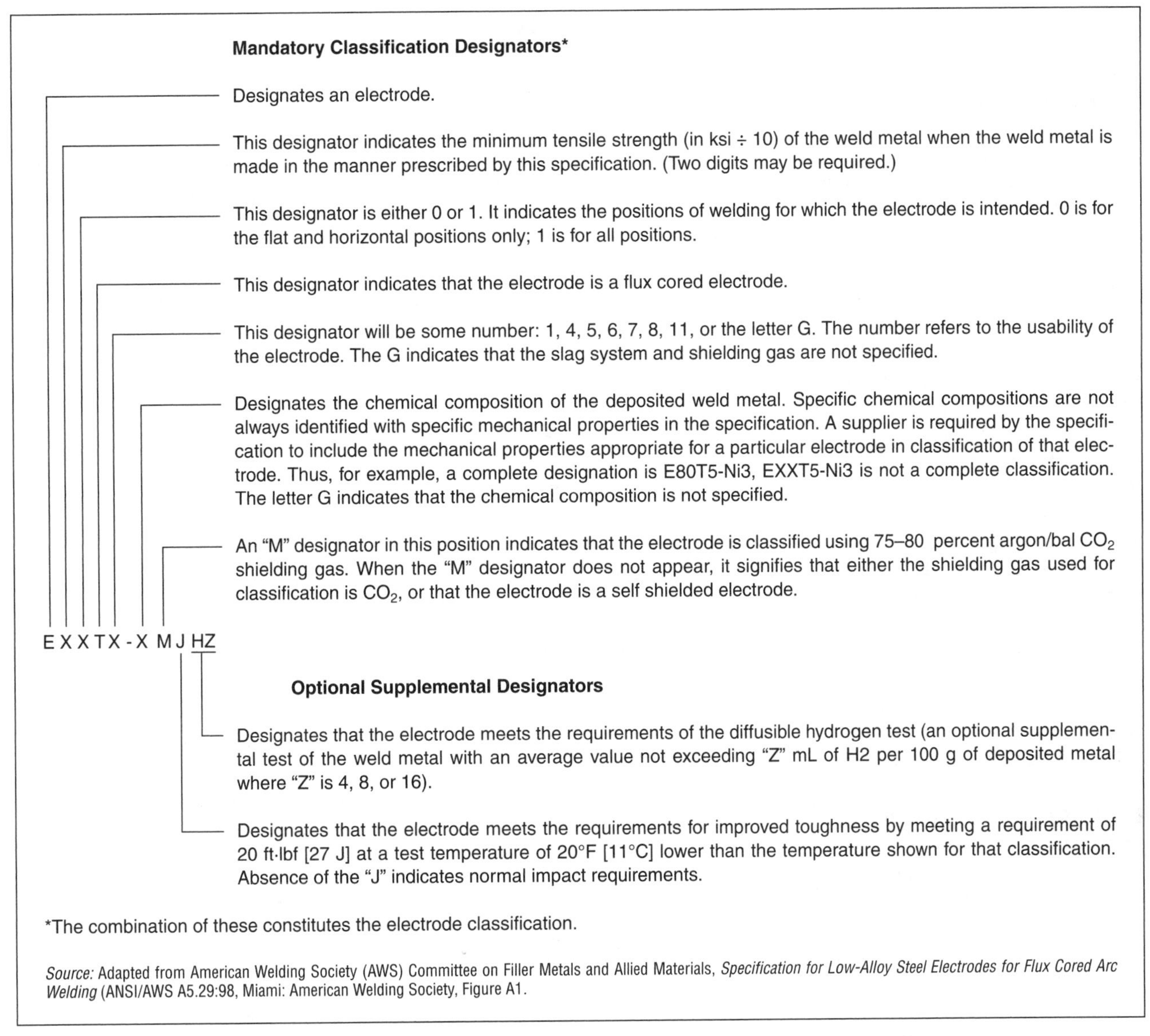

Figure 5.15—Classification System for Low-Alloy Steel FCAW Electrodes

argon/CO_2 mixtures may be used to reduce spatter when recommended by the manufacturer. Increasing the amount of argon in the argon/CO_2 mixture will increase the manganese and silicon, along with certain other alloy contents, which will increase the yield and tensile strengths and may affect impact properties.

Electrodes of the EXXT5-XM classification are designed for use with 75% to 80% argon/balance CO_2 shielding. The use of these electrodes with gas mixtures with reduced amounts of argon or with CO_2 shielding gas will result in some deterioration in arc characteristics, an increase in spatter, and a reduction of manganese, silicon, and certain other alloy elements in the weld metal. This reduction in manganese, silicon, or other alloy elements will decrease the yield and tensile strengths and may affect impact properties.

EX0T5-X and EX0T5-XM. Electrodes of the EX0T5-X and EX0T5-XM classifications are used primarily for single-pass and multiple-pass welds in the flat position and for fillet welds in the horizontal position using DCEP or DCEN, depending on the manufacturer's recommendation. These electrodes are characterized by a globular transfer, slightly convex bead contour, and a thin slag that may not completely cover the weld bead. These electrodes have a lime-fluoride-base slag. Weld deposits produced by these electrodes typically have impact properties and hot-crack and cold-crack resistance that are superior to those obtained with rutile-based slags. The EX1T5-X and EX1T5-XM electrodes, using DCEN, can be used for welding in all positions. However, these electrodes have less operator appeal than electrodes with rutile-based slags.

EXXT6-X. Electrodes of this classification are self-shielded, operate on DCEP, and have a spray transfer. The slag system is designed to produce good low-temperature impact properties, good penetration into the root of the weld, and excellent slag removal, even in a deep groove. These electrodes are used for single-pass and multiple-pass welding in the flat and horizontal positions.

EXXT7-X. Electrodes of this classification are self-shielded, operate on DCEN and have a small droplet-to-spray transfer. The slag system is designed to allow the larger sizes to be used for high deposition rates in the horizontal and flat positions and to allow the smaller sizes to be used for all welding positions. These electrodes are used for single-pass and multiple-pass welding and produce very low-sulfur weld metal, which is highly resistant to hot cracking.

EXXT8-X. Electrodes of this classification are self-shielded, operate on DCEN, and have a small droplet or spray-type transfer. These electrodes are suitable for all welding positions, and the weld metal has very good low-temperature notch toughness and crack resistance. These electrodes are used for single-pass and multiple-pass welds.

EXXT11-X. Electrodes of this classification are self-shielded, operate on DCEN, and have a smooth spray-type transfer. The electrodes are intended for single-pass and multiple-pass welding in all positions. The manufacturer should be consulted concerning any plate thickness limitations.

EXXTX-G, EXXTG-X, and EXXTG-G. These classifications are for multiple-pass electrodes that are not covered by any presently defined classification. The mechanical properties can be anything covered by this specification. Requirements are established by the digits chosen to complete the classification. Placement of the "G" in the classification designates that the alloy requirements or the shielding gas/slag system, or both, are not defined and are as agreed upon between supplier and purchaser.

Chemical Composition

The chemical composition of the weld metal produced is often the primary consideration for electrode selection. The suffixes, which are part of each alloy electrode classification, identify the chemical composition of the weld metal produced by the electrode. The following paragraphs give a brief description of the classifications, intended uses, and typical applications.

EXXTX-A1 (C-Mo Steel) Electrodes. These electrodes are similar to the E7XT-X carbon-steel electrodes classified in ANSI/AWS A5.20, *Specification for Carbon Steel Electrodes for Flux Cored Arc Welding*,[14] except that 1/2% molybdenum has been added. This addition increases the strength of the weld metal, especially at elevated temperatures, and provides some increase in corrosion resistance; however, it may reduce the notch toughness of the weld metal. This type of electrode is commonly used in the fabrication and erection of boilers and pressure vessels. Typical applications include the welding of C-Mo steels such as ASTM A 161, A 204 and A 302 Gr. A plate, and A 335-P1 pipe.

EXXTX-BX, EXXTX-BXL, and EXXTX-BXH (Cr-Mo Steel) Classifications. These electrodes produce weld metal that contains between 1/2% and 9% chromium, and between 1/2% and 1% molybdenum. They are designed to produce weld metal for high-temperature service and for matching the properties of the typical base metals shown in Table 5.3.

Low-carbon EXXTX-BXL classifications have been established for two of these Cr–Mo electrode classifications. While regular Cr–Mo electrodes produce weld metal with 0.05% to 0.12% carbon, the L-Grades ("L" signifies low carbon) are limited to a maximum of 0.05% carbon. The lower percentage of carbon in the weld metal will improve ductility and lower hardness; however, it will also reduce the high-temperature strength and creep resistance of the weld metal.

Several of these electrodes also have had high-carbon grades (EXXTX-BXH) established. In these cases, the electrode produces weld metal with 0.10% to 0.15% carbon, which may be required for high-temperature strength in some applications.

Since all Cr-Mo electrodes produce weld metal that will harden in still air, both preheat and postweld heat treatment (PWHT) are required for most applications.

No minimum notch toughness requirements have been established for any of the Cr-Mo electrode classifications. While it is possible to obtain Cr-Mo electrodes with minimum toughness values at ambient tempera-

14. See Referencd 10.

Table 5.3
Cr-Mo Steel Electrodes for High-Temperature Service for Pipe and Plate

AWS Classification	Matching Base Metals
EXXTX-B1	ASTM A 335-P2 pipe ASTM A 387 Gr. 2 plate
EXXTX-B2	ASTM A 335-P11 pipe ASTM A 387 Gr. 11 plate
EXXTX-B2L	Thin-wall A 335-P11 pipe or tube for use in the as-welded condition or for applications in which low hardness is a primary concern.
EXXTX-B3	ASTM A 335-P22 pipe ASTM A 387 Gr. 22 plate
EXXTX-B3L	Thin-wall ASTM A 335-P22 pipe for use in the as-welded condition or for applications in which low hardness is of primary concern.
EXXTX-B6	ASTM A 213-T5 tube ASTM A 335-P5 pipe
EXXTX-B8	ASTM A 213-T9 tube ASTM A 335-P9 pipe

tures down to 32°F (0°C), specific values and testing must be agreed upon by the supplier and the purchaser.

EXXTX-DX (Mn-Mo Steel) Electrodes. These electrodes produce weld metal that contains about 1–1/2% to 2% manganese and between 1/3% and 2/3% molybdenum. This weld metal provides higher strength and better notch toughness than the C-1/2% Mo and 1% Ni-1/2% Mo steel weld metal discussed in EXXTX-A1 and EXXTX-K. However, the weld metal from these Mn-Mo steel electrodes is quite air-hardenable and usually requires preheat and postweld heat treatment. The individual electrodes classified under this electrode group are designed to match the mechanical properties and corrosion resistance of the high[-strength, low-alloy pressure vessel steels, such as ASTM A 302 Gr. B and HSLA steels, and manganese molybdenum castings such as ASTM A 49, A 291, and A 735.

EXXTX-K(X) (Various Low-Alloy Steel) Electrodes. This group of electrodes produces weld metal of several different chemical compositions. These electrodes are primarily intended for as-welded applications. Table 2 in *Specification for Low-Alloy Steel Electrodes for Flux Cored Arc Welding*, AWS A5.29[15] provides a comparison of the toughness levels obtained for each classification.

15. See Reference 12, p.3.

EXXTX-K1 Electrodes. Electrodes of this classification produce weld metal with nominally 1% nickel and 1/2% molybdenum. These electrodes can be used for long-term stress-relieved applications or for welding low-alloy high-strength steels, particularly 1% nickel.

EXXTX-K2 Electrodes. Electrodes of this classification produce weld metal that will have a chemical composition of 1-1/2% nickel and up to 0.35% molybdenum. These electrodes are used on many high-strength applications ranging from 550 to 760 MPa (80 to 110 thousand pounds per square inch [ksi]) minimum yield strength. Typical applications include the welding of submarines, aircraft carriers, and other structural applications where excellent low-temperature toughness is required. Steels welded include HY-80, HY-100, ASTM A 710, A 514, and other similar high-strength steels.

EXXTX-K3 Electrodes. Electrodes of this type produce weld deposits with higher levels of Mn, Ni, and Mo than the EXXTX-K2 types. They are usually higher strength than the –K1 and –K2 types. Typical applications include the welding of HY-100 and A 514 steels.

EXXTX-K4 Electrodes. Electrodes of this classification deposit weld metal similar to that of the –K3 electrodes, with the addition of approximately 0.5% chromium. The additional alloy provides the higher strength required for many applications that need tensile strength in excess of 830 MPa (120 ksi), such as armor plate.

EXXTX-K5 Electrodes. Electrodes of this classification produce weld metal that is designed to match the mechanical properties of steels such as SAE 4130 and 8630 after the weldment is quenched and tempered. The classification requirements stipulate only as-welded mechanical properties; therefore, the end user is encouraged to perform qualification testing.

EXXTX-K6 Electrodes. Electrodes of this classification produce weld metal that utilizes less than 1% nickel to achieve excellent toughness in the 410 and 480 MPa (60 and 70 ksi) tensile strength ranges. Applications include structural work, offshore construction, and circumferential pipe welding.

EXXTX-K7 Electrodes. This electrode classification produces weld metal similar to the weld metal produced with EXXTX-Ni2 and EXXTX-Ni3 electrodes. The weld metal has approximately 1-1/2% manganese and 2-1/2% nickel.

EXXTX-K8 Electrodes. This classification was designed for electrodes intended for use in circumferential

Table 5.4
Shielding and Polarity Requirements for Low-Alloy Flux Cored Arc Welding Electrodes (ANSI/AWS A5.29 Classifications)

AWS Classification	Recommended Weld Passes	External Shielding Medium	Current and Polarity
EXXT1-X	Multiple	CO_2	DCEP
EXXT1-XM	Multiple	Mixed gas[a]	DCEP
EXXT4-X	Multiple	None	DCEP
EXXT5-X	Multiple	CO_2	DCEP or DCEN
EXXT5-XM	Multiple	Mixed gas[a]	DCEP or DCEN
EXXT6-X	Multiple	None	DCEP
EXXT7-X	Multiple	None	DCEN
EXXT8-X	Multiple	None	DCEN
EXXT11-X	Single	None	DCEN
EXXTX-G	Multiple	Note b	Note b
EXXTG -X	Multiple	Note b	Note b
EXXTG-G	Multiple	Note b	Note b

Notes:
a Mixed gas normally refers to 75% to 80% argon/balance CO_2
b As agreed upon by supplier and user.

girth welding of line pipe. The weld deposit contains approximately 1-1/2% manganese, 1.0% nickel, and small amounts of other alloying elements. It is especially intended for use on API 5LX80 pipe steels.

EXXTX-K9 Electrodes. This electrode produces weld metal similar to that produced by –K2 and –K3-type electrodes, but is intended to be similar to the military requirements of MIL-101TM and MIL-101TC electrodes in MIL-E-24403/2C. The electrode is designed for welding HY-80 steel.

EXXTX-NiX (Ni-Steel) Electrodes. These electrodes are designed to produce weld metal with increased strength without being air-hardenable or with increased notch toughness at temperatures as low as –73°C (–100°F). They have been specified with nickel contents, which fall into three nominal levels of 1% Ni, 2-1/4% Ni, and 3-1/4% Ni in steel.

With carbon levels of up to 0.12%, strength increases and permits some of these Ni-steel electrodes to be classified as E8XTX-NiX and E9XTX-NiX. However, some classifications may produce low-temperature notch toughness to match the base metal properties of nickel steels, such as ASTM A 203 Gr. A and ASTM A 352 Gr. LC1 and LC2. The manufacturer should be consulted for specific Charpy V-notch impact properties. Typical base metals would also include ASTM A 302, A 572, A 575, and A 734.

Many low-alloy steels require postweld heat treatment to stress-relieve the weld or temper the weld metal and heat-affected zone to achieve increased ductility. However, for many applications, nickel-steel weld metal can be used without PWHT. If PWHT is to be specified for a nickel-steel weldment, the holding temperature should not exceed the maximum temperature (listed in Table 8 in *Specification for Low-Alloy Steel Electrodes for Flux Cored Arc Welding,* AWS A5.29)[16] for the classification considered, since nickel steels can be embrittled at higher temperatures.

Electrodes of the EXXTX-NiX type are often used in structural applications where excellent toughness (Charpy V-notch or crack tip opening displacement) is required.

EXXTX-WX (Weathering Steel) Electrodes. These electrodes are designed to produce weld metal that matches the corrosion resistance and the coloring of the ASTM weathering-type structural steels. These special properties are achieved by the addition of about 1/2% copper to the weld metal. To meet strength, ductility, and notch toughness in the weld metal, some chromium and nickel additions also are made. These electrodes are used to weld typical weathering steel such as ASTM A 242 and A 588.

EXXTX-G (General Low-Alloy Steel) Electrodes. These electrode classifications may be either modifications of other discrete classifications or totally new classifications. These electrodes are described in Table 5.4. The purchaser and user should determine the

16. See Reference 12, pp. 13–14.

description and intended use of the electrode from the supplier.

Electrodes for Surfacing

Flux cored electrodes are produced for certain types of surfacing applications such as restoring usable service parts and hardfacing. Surfacing electrodes have many of the advantages of the electrodes used for welding, but there is less standardization of weld metal analysis and performance characteristics. Literature from various manufacturers should be consulted for details on flux cored surfacing electrodes. The AWS publication *Specification for Bare Electrodes and Rods for Surfacing*, AWS A5.21[17] includes classification of flux cored electrodes for surfacing.

Flux cored surfacing electrodes deposit iron-base alloys that may be ferritic, martensitic, or austenitic. They may deposit weld metal that is high in carbides. The electrodes are variously designed to produce surfaces with corrosion resistance, wear resistance, toughness, or anti-galling properties. They may be used to restore worn parts to original dimensions.

Stainless Steel Electrodes

Three major changes were made to the classification system specification for stainless steel electrodes by the American Welding Society in the latest edition of *Specification for Stainless Steel Electrodes for Flux Cored Arc Welding and Stainless Steel Flux Cored Rods for Gas Tungsten Arc Welding*, ANSI/AWS A5.22-95.[18] The changes are as follows:

1. Flux cored rods for root pass welding with the gas tungsten arc process are included.
2. The EXXT-2 classification, intended for use in 98% argon and 2% oxygen, is not included. The combination of the slag covering and this shielding gas was found to be inappropriate for flux cored arc welding.
3. A classification of gas-shielded electrodes intended for use only in the flat and horizontal positions is included.

The system for identifying the electrode and rod classifications in this specification follows the standard pattern used in other AWS filler metal specifications. The letter "E" at the beginning of each classification designation stands for electrode, and the letter "R" indicates a welding rod. The chemical composition is identified by a three-digit or four-digit number, and in some cases, additional chemical symbols and the letters "L" or "H." The numbers generally follow the pattern of the American Iron and Steel Institute (AISI) numbering system for heat- and corrosion-resisting steels; however, there are exceptions. In some classifications additional chemical symbols are used to signify modifications of basic alloy types. The letter "L" denotes a low-carbon content in the deposit. The letter "H" denotes carbon content in the upper part of the range that is specified for the corresponding standard alloy type. The letter "K" in the E316LKT0-3 classification signifies that weld metal deposited by these electrodes is designed for cryogenic applications. The following letter "T" indicates that the product is a flux cored electrode or rod. The "1" or "0" following the "T" indicates the recommended position of operation. Figure 5.16 illustrates the AWS classification system.

These electrodes are classified on the basis of the chemical composition of the deposited weld metal and the shielding medium to be employed during welding. Table 5.5 identifies the shielding designations used for classification, indicates the respective current and polarity characteristics, and specifies the welding process for which the classification is intended.

Electrodes classified EXXXTX-1 that use CO_2 shielding undergo some minor loss of oxidizable elements and some increase in carbon content. Electrodes with the EXXXT-3 classifications are used without external shielding and undergo some loss of oxidizable elements and a pickup of nitrogen, which may be significant. Low welding currents coupled with long arc lengths (high arc voltages) increase the nitrogen pickup. Nitrogen stabilizes austenite and may therefore reduce the ferrite content of the weld metal.

The requirements of the EXXXT-3 classifications are different from those of the EXXXT-1 classifications because shielding with a flux system alone is not as effective as shielding with both a flux system and a separately applied external shielding gas. The EXXXT-3 deposits, therefore, usually have a higher nitrogen content than the EXXXT-1 deposits. This means that to control the ferrite content of the weld metal, the chemical compositions of the EXXXT-3 deposits must have different Cr/Ni ratios than those of the EXXXT-1 deposits. In contrast to self-shielded mild steel or low-alloy steel electrodes, EXXXT-3 stainless steel electrodes generally do not contain strong denitriding elements such as aluminum.

Although welds made with electrodes meeting AWS specifications are commonly used in corrosion or heat-resisting applications, it is not practical to require

17. American Welding Society (AWS) Committee on Filler Metals and Allied Materials, 2001, *Specification for Bare Electrodes and Rods for Surfacing*, AWS A5.21:2001, Miami: American Welding Society.
18. American Welding Society (AWS) Committee on Filler Metals and Allied Materials, 1995, *Specification for Stainless Steel Electrodes for Flux Cored Arc Welding and Stainless Steel Flux Cored Rods for Gas Tungsten Arc Welding*, AWS A5.22-95, Miami: American Welding Society.

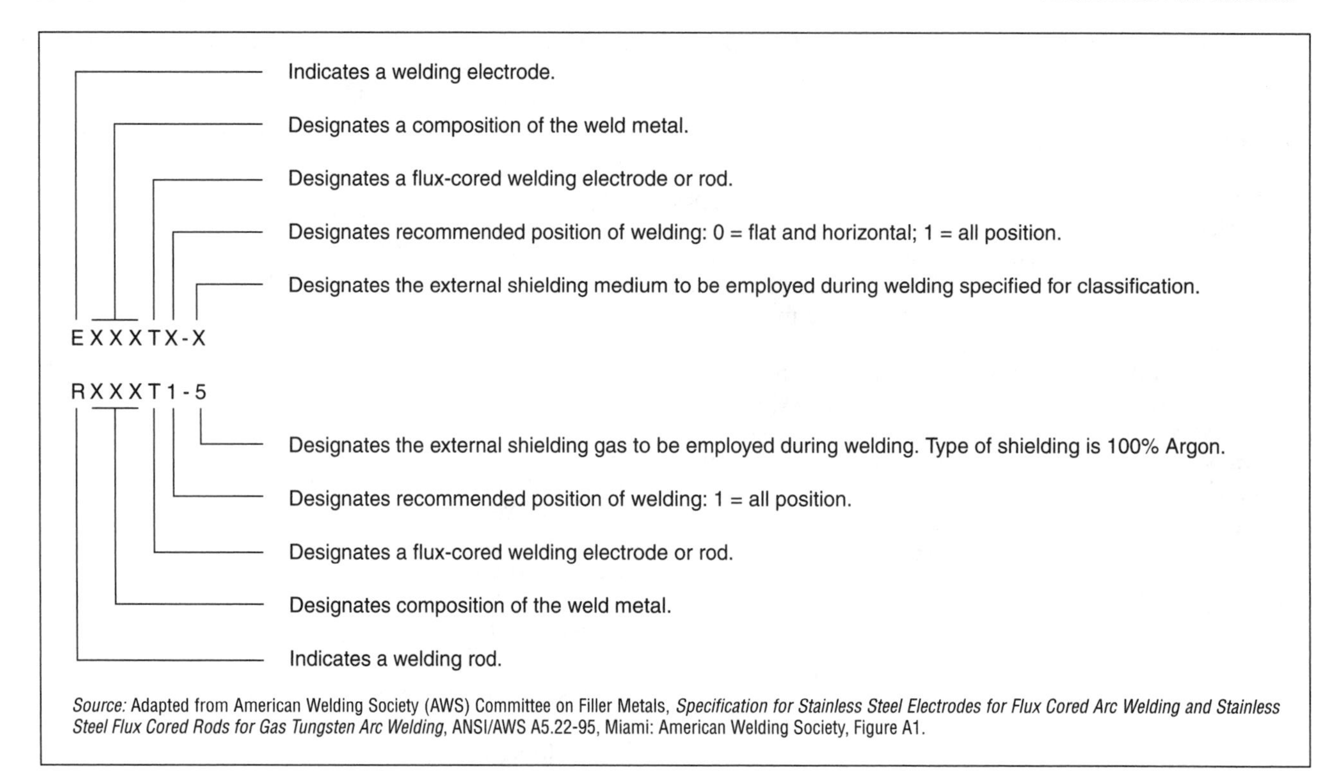

Source: Adapted from American Welding Society (AWS) Committee on Filler Metals, *Specification for Stainless Steel Electrodes for Flux Cored Arc Welding and Stainless Steel Flux Cored Rods for Gas Tungsten Arc Welding*, ANSI/AWS A5.22-95, Miami: American Welding Society, Figure A1.

Figure 5.16—Classification System for Stainless Steel FCAW Electrodes

Table 5.5
Shielding, Polarity, and Process Requirements for Stainless Steel Flux Cored Electrodes Classified to ANSI/AWS A5.22

AWS Designations (All Classifications)	External Shielding Medium	Current and Polarity	Welding Process
EXXXTX-1	CO_2	DCEP	FCAW
EXXXTX-3	None	DCEP	FCAW
EXXXTX-4	75% to 80% argon/balance CO_2	DCEP	FCAW
RXXXT1-5	100% argon	DCEN	GTAW
EXXXT-G	None specified	Not specified	FCAW
RXXXT1-G	None specified	Not specified	GTAW

electrode qualification tests for corrosion or scale resistance on welds or weld metal specimens. Special tests that are pertinent to an intended application should be established by agreement between the electrode manufacturer and the user.

Protection from Moisture

Protection from moisture pickup is essential with most flux cored electrodes. Electrodes should be handled and stored according to directions from the manufacturer. It is recommended that electrodes be returned to the original package for overnight storage.

Wire exposed to a moist environment can be reconditioned by heating in a drying oven and baking at 150° to 315°C (300°F to 600°F), as recommended by certain manufacturers. This assumes that the wire is spooled or coiled on a metal device; other materials might melt. Precise temperatures and drying times should be obtained from the manufacturer of the electrode.

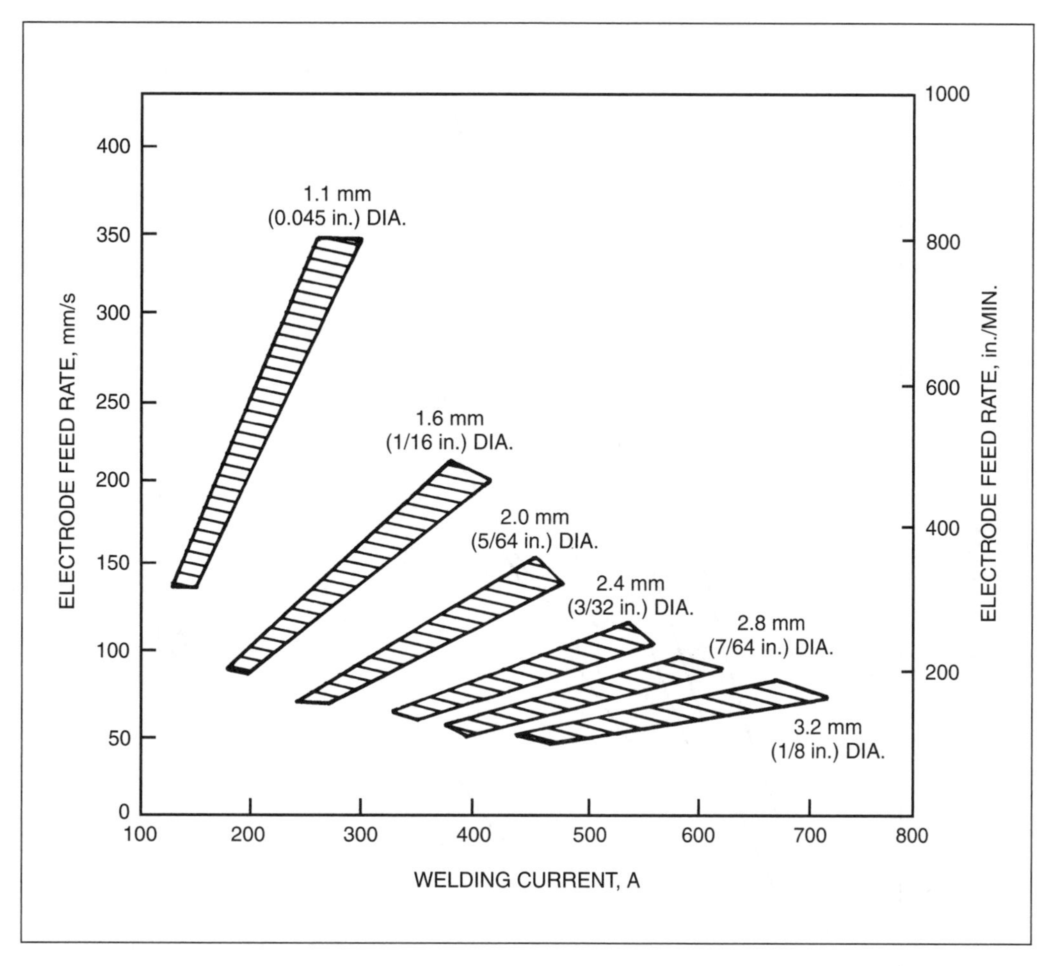

Figure 5.17—Electrode Feed Rate Versus Welding Current Range for E70T-1 Steel Electrodes with CO_2 Shielding

Moisture pickup can result in porosity or "worm tracks" in the weld bead. Moisture can also cause hydrogen cracking, especially in alloyed steels. More information on the risks of hydrogen cracking is provided in the AWS *Welding Handbook, Materials and Applications—Part* 2, Volume 4, 8th edition[19] and the Supplementary Reading List at the end of this chapter.

PROCESS CONTROL

The welding process variables that must be controlled to achieve high-quality welds are welding current (wire feed speed), polarity, electrode extension, travel speed, welding position, electrode angle, and shielding gas composition and flow. Control of these variables contributes to the deposition rate, deposition efficiency, and high-quality welds.

WELDING CURRENT

Welding current is proportional to electrode feed rate for a specific electrode diameter, composition, and electrode extension. The relationship between electrode feed rate and welding current for typical mild steel gas-shielded electrodes, self-shielded mild steel electrodes, and self-shielded stainless steel electrodes are presented in Figures 5.17, 5.18, and 5.19. A constant-voltage power source of the proper size is used to melt the electrode at a rate that maintains the preset output

19. See Reference 6.

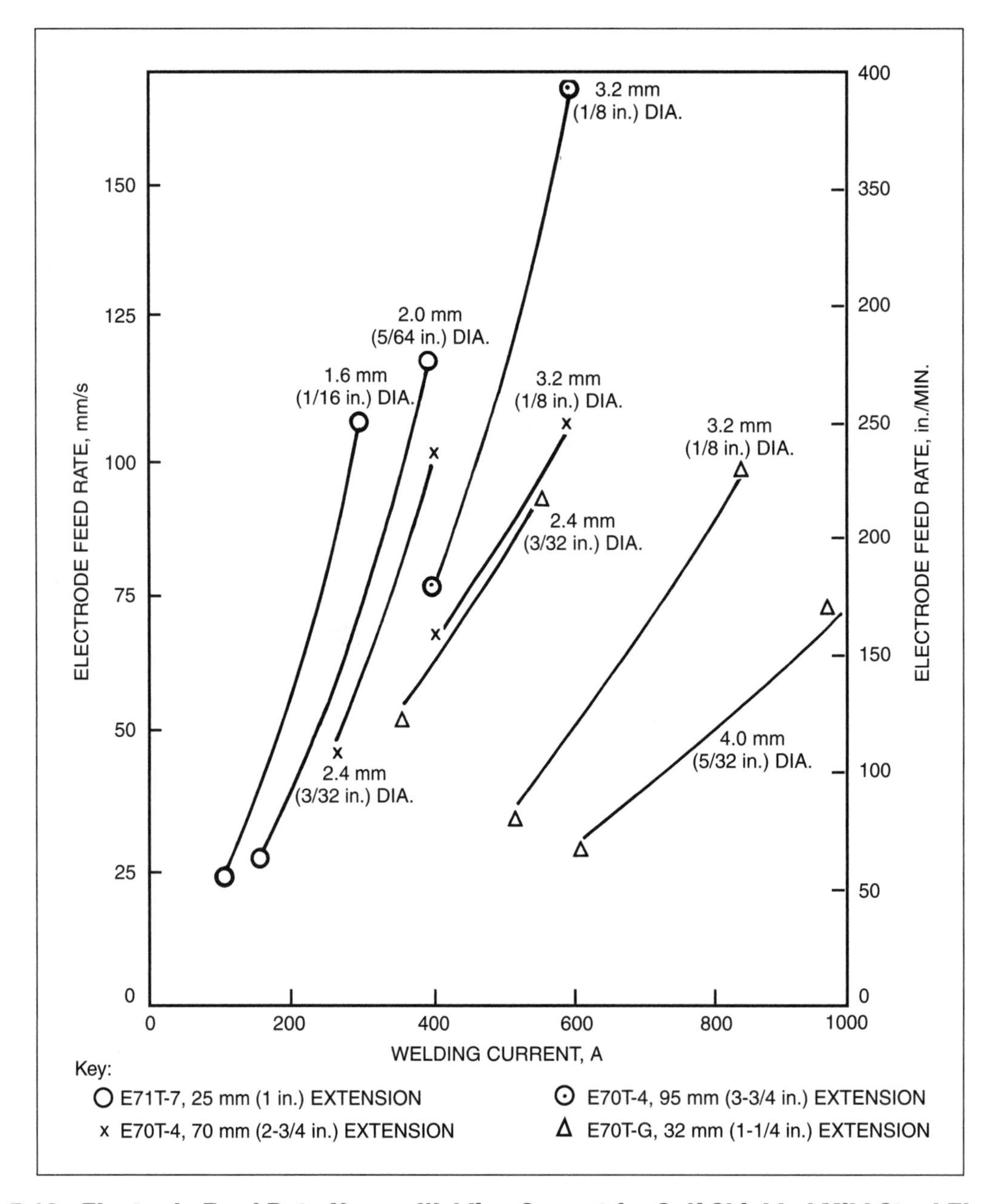

Figure 5.18—Electrode Feed Rate Versus Welding Current for Self-Shielded Mild Steel Electrodes

voltage (arc length). When other welding variables are held constant for a given diameter of electrode, changing the welding current will generally have the following effects:

1. Increasing current increases electrode deposition rate;
2. Increasing current increases penetration (see "Travel Speed" section in this chapter for interaction effects);
3. Excessive current produces convex weld beads with poor appearance;
4. Insufficient current produces large droplet transfer and excessive spatter; and
5. Insufficient current can result in pickup of excessive nitrogen and also porosity in the weld metal when welding with self-shielded flux cored electrodes.

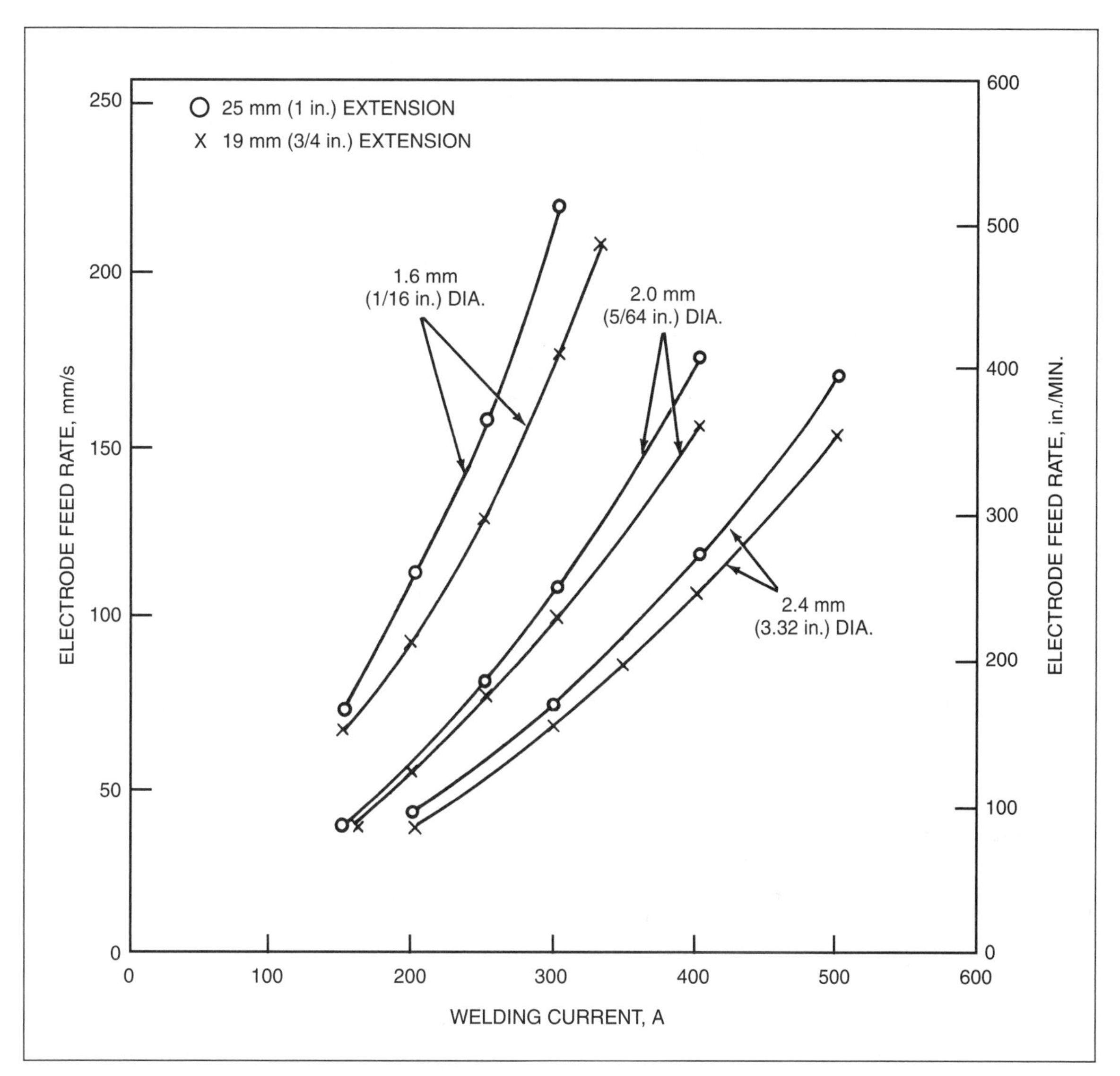

Figure 5.19—Electrode Feed Rate Versus Welding Current for Self-Shielding E308T-3

As welding current is increased or decreased by changing the electrode feed rate, the power source output voltage should be changed to maintain the optimum relationship of arc voltage to current. For a given electrode feed rate, measured welding current varies inversely with the electrode extension. As the electrode extension increases, welding current decreases; as the electrode extension decreases, welding current increases.

ARC VOLTAGE

Arc voltage and arc length are closely related. The voltage shown on the meter of the welding power source is the sum of the voltage drops throughout the welding circuit. This includes the drop through the welding cables, the electrode extension, the arc, and the workpiece. Therefore, arc voltage will be proportional to the meter reading, provided all other circuit elements (including temperatures) remain constant.

The appearance, soundness, and properties of welds made with flux cored electrodes can be affected by the arc voltage. Too high an arc voltage (too long an arc) can result in excessive spatter and wide, irregularly shaped weld beads. With self-shielded electrodes, too high an arc voltage will result in excessive nitrogen pickup. With mild steel electrodes, this may cause porosity. With stainless steel electrodes, voltage that is too high will reduce the ferrite content of the weld metal, and this in turn may result in cracking. Too low an arc voltage (too short an arc) will result in narrow, convex beads with excessive spatter and reduced penetration.

POLARITY

Polarity is one of the variables that must be considered when selecting FCAW electrodes. Some flux cored electrodes are designed to be used with DCEP (reverse polarity) and others for DCEN (straight polarity). Some of the self-shielded flux cored electrode classifications specify DCEN polarity. This polarity results in less base metal penetration. Consequently, small-diameter electrodes such as 0.8 mm (0.030 in.), 0.9 mm (0.035 in.), and 1.2 mm (0.045 in.) have proven to be quite successful for work on thin-gauge materials.

ELECTRODE EXTENSION

The unmelted electrode that extends beyond the contact tube during welding, called the *electrode extension*, is resistance heated in proportion to its length, assuming other variables remain constant. As previously explained, electrode temperature affects arc energy, electrode deposition rate, and weld penetration. Electrode temperature also can affect weld soundness and arc stability.

A characteristic of some self-shielded electrodes is a long electrode extension. Self-shielded electrode extensions of 19 to 95 mm (1/2 to 3-3/4 in.) are generally used, depending on the application. Increasing the electrode extension increases the resistance heating of the electrode. This preheats the electrode and lowers the voltage drop across the arc. At the same time, the welding current decreases, which lowers the heat available for melting the base metal. The resulting weld bead is narrow and shallow. This makes the process suitable for welding light-gauge material and for bridging gaps caused by poor fitup. If the arc length (voltage) and welding current are maintained (by higher voltage settings at the power source and higher electrode feed rates), longer electrode extension will increase the deposition rate.

The effect of electrode extension as an operating factor in FCAW introduces a new variable that must be held in balance with the shielding conditions and the related welding variables. For example, the melting and activation of the core ingredients must be consistent with that of the containment tube, as well as with arc characteristics. Other things being equal, too long an extension produces an unstable arc with excessive spatter. Too short an extension may cause excessive arc length at a particular voltage setting. With gas-shielded electrodes, an extension that is too short may result in excessive spatter buildup in the gun nozzle, which can interfere with the gas flow. Poor shielding gas coverage may cause weld metal porosity and excessive oxidation. Too short an extension may also result in inadequate preheating of the electrode, which may result in increased diffusible hydrogen contents.

Most manufacturers recommend an extension of 19 to 38 mm (3/4 to 1-1/2 in.) for gas-shielded electrodes and from approximately 19 to 95 mm (3/4 to 3-3/4 in.) for self-shielded types, depending on the application. The electrode manufacturer should be consulted for optimum settings in these ranges.

TRAVEL SPEED

Travel speed is defined as the linear rate at which the arc is moved along the weld joint. Travel speed influences weld bead penetration and contour. When other factors remain constant, weld penetration is at a maximum at an intermediate travel speed. With a normal current and voltage setting, the arc will impinge on the weld pool and reduce penetration if travel is slowed too much. As travel speed is increased from this excessively slow rate, it will increase to a maximum when the arc is impinging on the base metal immediately ahead of the weld pool and the weld bead will become narrow and rope-like. Low travel speeds at high currents can result in overheating the weld metal. This will cause a rough-appearing weld with the possibility of mechanically trapping slag or melting through the base metal.

ELECTRODE ANGLE AND WELDING POSITION

The angle at which the electrode is held during welding determines the direction in which the arc force is applied to the molten metal of the weld pool. When welding variables are properly adjusted for the application involved, the arc force can be used to oppose the effects arising from the force of gravity. In the FCAW process, the arc force is used not only to help shape the desired weld bead, but also to prevent the slag from running ahead of the arc and becoming entrapped in the weld metal.

When making groove and fillet welds in the flat position, gravity tends to cause the molten metal to run ahead of the weld. To counteract this, the electrode is held at an angle to the vertical with the electrode tip pointing away from the direction of travel. This travel angle, defined as the *drag angle*, is measured from a vertical line in the plane of the weld axis, as shown in Figure 5.20A.

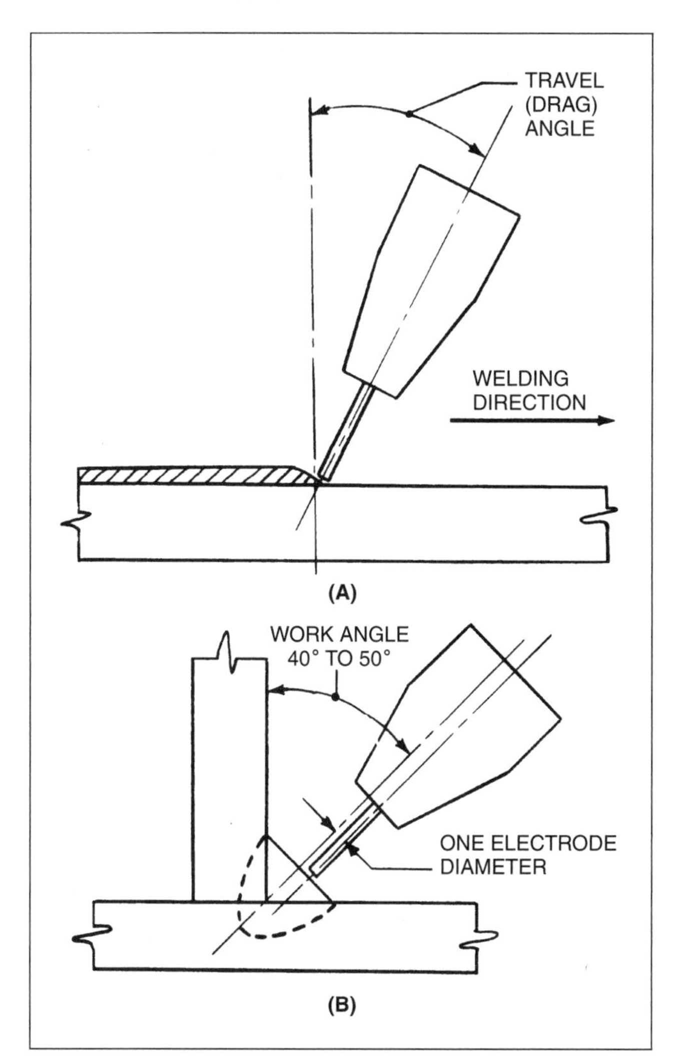

Figure 5.20—Welding Electrode Positions

The proper drag angle depends on the FCAW method used, the base metal thickness, and the position of welding. For the self-shielded process, drag angles should be about the same as those used with shielded metal arc welding electrodes. Drag angles can vary from approximately 20° to 45° for the flat and horizontal positions. Larger angles are used for thin sections to decrease penetration.

With gas-shielded FCAW, the drag angle should be small, preferably 0° to 15°, but no more than 35°. If the drag angle is too large, the effectiveness of the shielding gas will be lost.

In uphill (vertical-up) welding the arc force helps hold the molten metal in place and shape the resulting weld bead. A push angle of 5° to 35° is recommended.

When fillet welds are made in the horizontal position, the weld pool tends to flow both in the direction of travel and at right angles to it. To counteract the latter, the electrode should point at the bottom plate close to the corner of the joint, and in addition to the drag angle, a work angle of 45° to 50° from the vertical member should be used. Figure 5.20B shows the electrode offset and the work angle used for horizontal fillet welds.

SHIELDING GAS FLOW

For gas-shielded electrodes, the gas flow rate is a variable that can affect weld quality. Inadequate flow will result in poor shielding of the weld pool, causing weld porosity and oxidation. Excessive gas flow can result in turbulence and mixing with air. The effect on the weld quality will be the same as inadequate flow. Either extreme will increase weld metal impurities. Correct gas flow depends on the type and the diameter of the gun nozzle, the distance of the nozzle from the work, and air movements in the immediate region of the welding operation.

DEPOSITION RATE AND EFFICIENCY

The deposition rate in any welding process is the weight of material deposited per unit of time. Deposition rate is dependent on welding variables such as electrode diameter, electrode composition, electrode extension, and welding current. Deposition rates versus welding current for various diameters of gas-shielded and self-shielded mild steel electrodes and for self-shielded stainless steel electrodes are presented in Figures 5.21, 5.22, and 5.23, respectively.

Deposition efficiency is the ratio of the weight of the metal deposited to the weight of the electrode consumed. Deposition efficiencies of FCAW electrodes range from 80% to 90% for those used with gas shielding, and from 78% to 87% for self-shielded electrodes.

JOINT DESIGNS AND WELDING PROCEDURES

The joint designs and welding procedures appropriate for flux cored arc welding depend on whether the gas-shielded or self-shielded method is used. However, all the basic weld joint shapes can be welded by both FCAW methods. There may be some differences between the two methods of FCAW in specific groove dimensions for a specific joint. For example, the

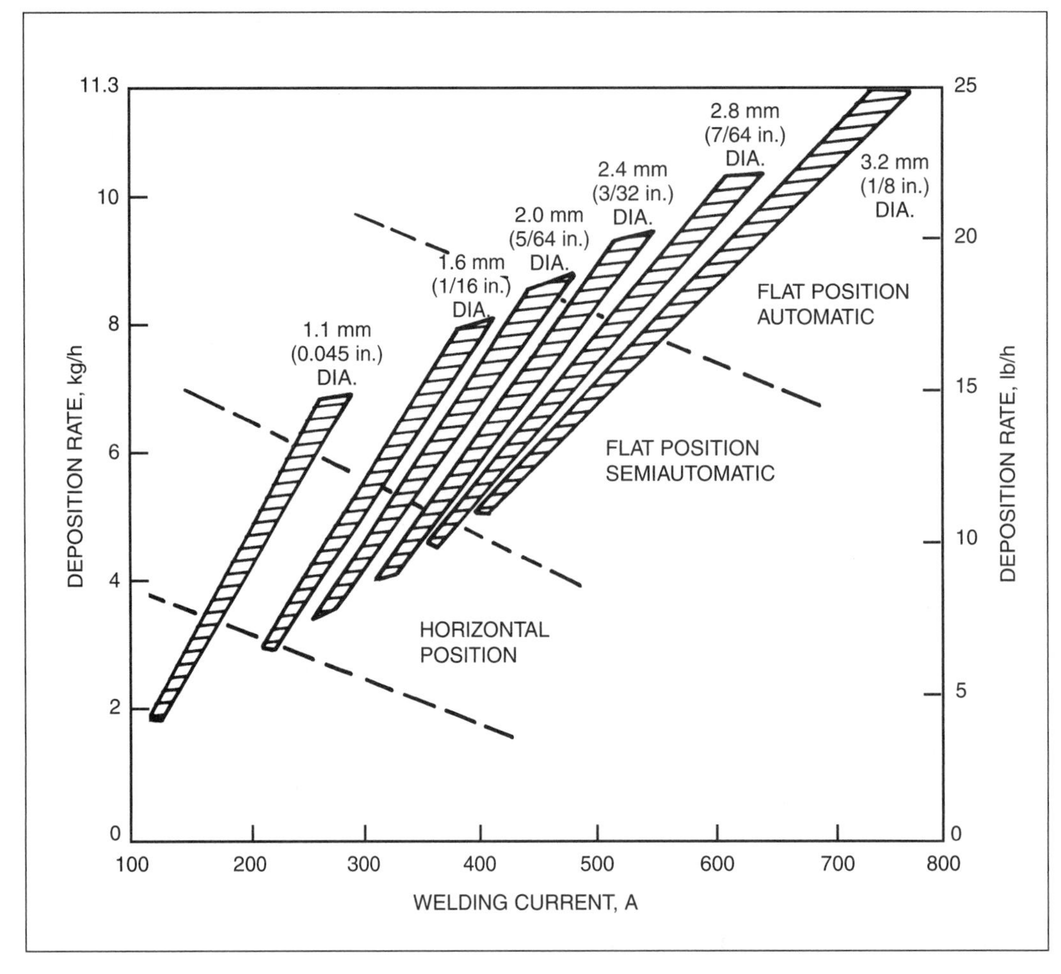

Figure 5.21—Deposition Rate Versus Welding Current for E70T-1 Mild Steel Electrodes with CO_2 Shielding

self-shielded FCAW process may allow smaller included angles, since there is no need to provide room for the gas nozzle. Also, the process has relatively shallow penetration, and might require a smaller land in the joint than gas-shielded FCAW. Because FCAW electrode formulations, intended purpose, and operating characteristics differ between classifications, the values of their welding procedure variables may also differ. (See the Supplementary Reading List for more specific information.)

The welding procedures for FCAW vary from one electrode product to another more than they vary for any other major process. This happens because the differing constituents and quantities present in the core, the thickness of the strip, and other variables can affect the current-carrying ability of the electrode (wire). General conditions can be given for different diameters of wire, but the electrode manufacturers' literature should be consulted for the best operating conditions for a specific product.

DESIGNS FOR GAS-SHIELDED MILD STEEL AND LOW-ALLOY ELECTRODES

In general, joints can be designed to take advantage of the penetration achieved by high current densities. Designs with relatively narrow grooves with small groove angles, narrow root openings, and larger root faces can be used with the gas-shielded method.

For basic butt joint designs, the following points should be considered:

1. The joint should be designed so that a constant electrode extension can be maintained when welding successive passes in the joint;

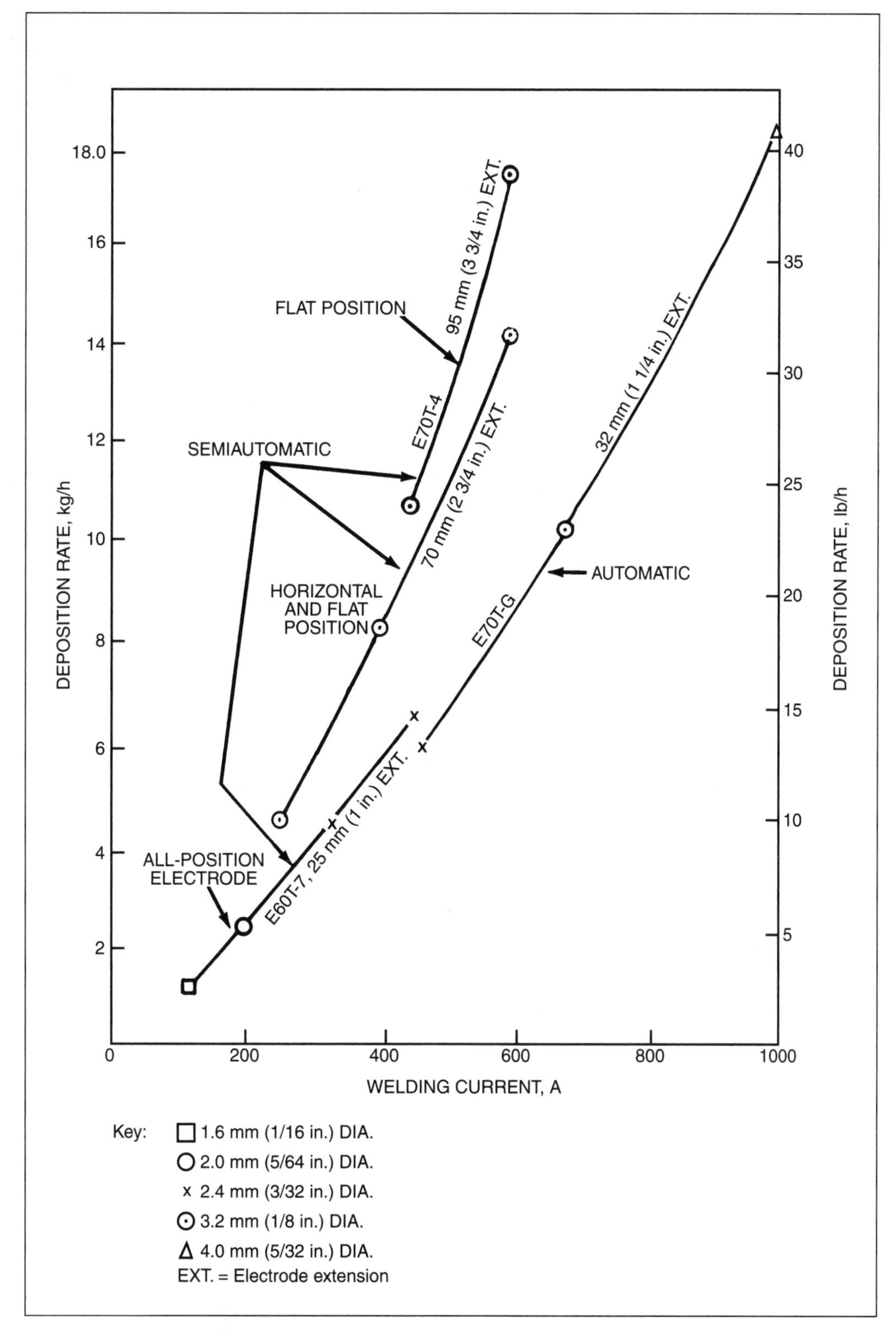

Figure 5.22—Deposition Rate Versus Welding Current for Self-Shielded Mild Steel Electrodes

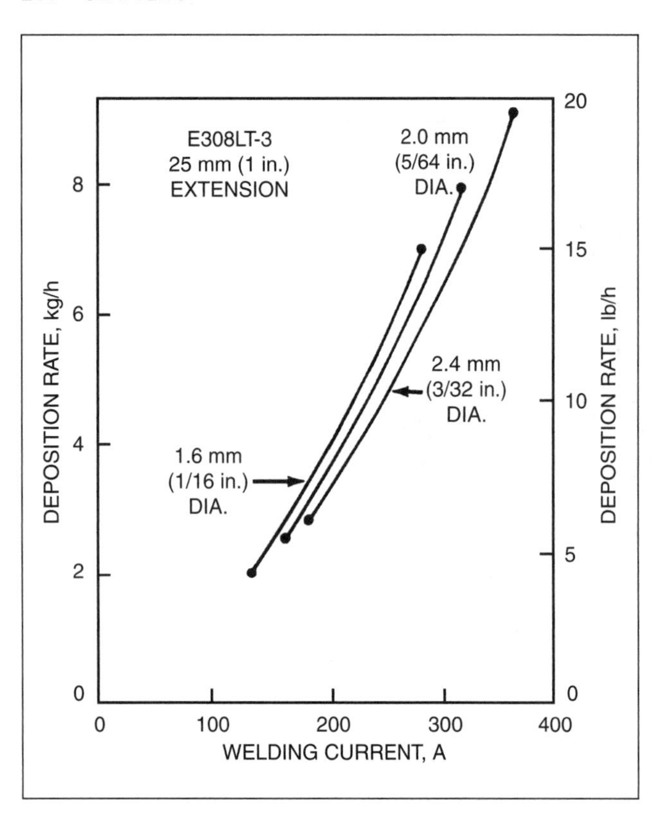

Figure 5.23—Deposition Rate versus Welding Current for Self-Shielded E308LT-3 Stainless Steel Electrodes

2. The joint should be designed so that the root is accessible, so that any necessary electrode manipulation during welding can be done easily.

Groove angles for various metal thicknesses are properly designed when they provide welding accessibility for the appropriate gas nozzle and electrode extension. Side-shielding nozzles for automatic welding permit better accessibility into narrower joints and also permit smaller groove angles than concentric nozzles allow. Sound welds can be obtained with the proper welding procedure.

Table 5.6 lists typical parameters for welding with mild steel and low-alloy steel all-position FCAW electrodes. Table 5.7 lists typical parameters for mild steel and low-alloy steel electrodes for the flat and horizontal positions. The parameters may vary from one manufacturer to another. Users should consult the manufacturers' literature to confirm parameters and consider any that may not be included in this chapter. The Supplementary Reading List at the end of this chapter should be consulted for additional information on FCAW procedures.

Adequate gas shielding is required to obtain sound welds. Flow rates required depend on nozzle size, draft conditions, and electrode extension. Welding in still air requires flow rates in the range of 14 to 19 liters per minute (L/min) (30 to 40 cubic feet per hour [ft^3/h]). When welding in moving air or when electrode extension is longer than normal, flow rates of up to 26 L/min (55 ft^3/h) may be needed. Flow rates for side-shielded

Table 5.6
Typical Parameters for Mild Steel and Low-Alloy Steel Electrodes (EX1T-1 Types) for Welding in All Positions

Diameter mm (in.)	Position	Wire Feed Speed cm/min (in./min)	Current, A	Volts, V
0.9 (0.035)	Flat	1829 (720)	250	32
	Uphill	813 (320)	150	26
	Overhead	813 (320)	150	26
1.2 (0.045)	Flat	953 (375)	250	28
	Uphill	660 (260)	200	25
	Overhead	660 (260)	200	26
1.3 (0.052)	Flat	914 (360)	300	28
	Uphill	610 (240)	225	25
	Overhead	610 (240)	225	26
1.6 (1/16)	Flat	762 (300)	350	29
	Uphill	406 (160)	225	25
	Overhead	406 (160)	225	26
2.0 (5/64)	Flat	635 (250)	400	30
2.4 (3/32)	Flat	457 (180)	450	30

Table 5.7
Typical Parameters for Mild Steel and Low-Alloy Steel Electrodes (EX70T-1 Types) for Welding in the Flat and Horizontal Positions

Diameter, mm (in.)	Welding Position	Wire Feed speed cm/min (in./min)	Current, A	Volts, V
1.6 (1/16)	Flat and horizontal	762 (300)	350	29
2.0 (5/64)	Flat and horizontal	635 (250)	400	30
2.4 (3/32)	Flat and horizontal	457 (180)	450	30
2.8 (7/64)	Flat and horizontal	381 (150)	550	30
3.2 (1/8)	Flat and horizontal	318 (125)	600	33

nozzles are generally the same or slightly higher than those for concentric nozzles. Nozzle openings must be maintained free from excessive adhering spatter. Gas nozzles may become obstructed by weld spatter or other build-up, limiting shielding gas flow. Regular cleaning of the inside of the gas nozzle and contact tip is recommended to prevent obstruction of gas flow.

If there is brisk air movement in the welding area, such as when welding outdoors, curtains should be used to screen the weld zone and to avoid loss of gas shielding.

Self-Shielded Mild and Low-Alloy Steel Electrodes

The basic joint types that are suitable for the gas-shielded FCAW and shielded metal arc welding (SMAW) processes are also suitable for self-shielded FCAW. Although the general shapes of weld grooves are similar to those used for SMAW, specific groove dimensions may differ. The differences are largely required by the higher deposition rates and shallower penetration available with self-shielded FCAW. This usually means that the groove angle needs to be more open than with gas-shielded FCAW.

Table 5.8 lists typical welding parameters for self-shielded FCAW electrodes.

Electrode extension introduces another welding procedure variable that may influence joint design. When using a long electrode extension for making flat-position groove welds without backing, welding of the first pass in the groove may be accomplished best with the SMAW process for better control of fusion and penetration. Similarly, in grooves with backing, the root opening must be sufficient to permit complete fusion by globular metal transfer.

When welding in the flat position, techniques similar to those used with covered low-hydrogen electrodes are followed. When making vertical welds on plates 19 mm (3/4 in.) and thicker, the root pass may be deposited downhill (vertical-down) for joints without backing and uphill (vertical-up) for joints with backing. With some self-shielded electrodes, root passes may be deposited in any position without backing. Subsequent passes are deposited uphill (vertical-up) using a technique similar to that used for covered low-hydrogen electrodes. It may not be advisable to use self-shielded electrodes for root passes followed by gas-shielded electrodes for fill passes because self-shielded mild steel and low-alloy steel electrodes (but not self-shielded stainless steel electrodes) contain considerable denitrifiers which may have undesirable metallurgical effects when diluted into gas-shielded deposits. Before attempting such a procedure, the electrode manufacturer should be consulted for recommendations.

Stainless Steel Electrodes

Stainless steel FCAW electrodes are available in both the gas-shielded and self-shielded variety. Typical welding parameters for gas-shielded stainless steel electrodes are listed in Table 5.9. The parameters may vary from one manufacturer to another. Users should consult the manufacturers' literature to confirm parameters and become aware of any that may not be included in this chapter. In general, joint geometry when using self-shielded stainless steel electrodes should be approximately the same as that used for SMAW. Joint geometries for the gas-shielded stainless steel electrodes are similar to gas-shielded mild steel FCAW electrodes. The Supplementary Reading List at the end of this chapter provides additional sources of information on FCAW joint designs.

EDGE PREPARATION AND FITUP TOLERANCES

Edge preparation for welding with flux cored electrodes can be accomplished by oxyfuel gas cutting,

Table 5.8
Typical Welding Parameters for Mild and Low-Alloy Steel Self-Shielded FCAW Electrodes

AWS Classification	Diameter, mm (in.)	Welding Position	Wire Feed Speed cm/min (in./min)	Volts, V	Current, A
EXXT-3	2.4 (3/32)	Flat	610 (240)	24	510
	3.0 (0.120)	Flat	508 (200)	24	700
	4.0 (5/32)	Flat	330 (130)	23	780
EXXT-4	2.0 (5/64)	Flat	508 (200)	31	275
	2.4 (3/32)	Flat	483 (190)	31	355
	3.0 (0.120)	Flat	406 (160)	30	500
EXXT-6	2.0 (5/64)	Flat	762 (300)	28	450
	2.4 (3/32)	Flat	762 (300)	28	480
EXXT-7	2.0 (5/64)	Flat	508 (200)	26	300
	2.4 (3/32)	Flat	381 (150)	24	350
	2.8 (7/64)	Flat	318 (125)	26	400
EXXT-8	2.0 (5/64)	Flat	305 (120)	21	280
	2.4 (3/32)	Flat	254 (100)	22	340
EXXT-10	2.4 (3/32)	Flat	636 (250)	26	570
EXXT-11	0.9 (0.035)	Flat	330 (130)	17	140
		Uphill	254 (100)	16	90
	1.2 (0.045)	Flat	305 (120)	17	150
		Uphill	241 (95)	16	110
	1.6 (1/16)	Flat	305 (120)	17	230
		Uphill	241 (95)	16	175
	2.0 (5/64)	Flat	305 (120)	20	290
	2.4 (3/32)	Flat	254 (100)	20	365
EXXT-14	1.2 (0.045)	All	254 (100)	16	145
	1.6 (1/16)	All	178 (70)	16	185
	2.0 (5/64)	All	152 (60)	18	210
EXXT-GS	0.8 (0.030)	Flat	572 (225)	16	125
		Uphill	432 (170)	15	100
	0.9 (0.035)	Flat	635 (250)	18	150
		Uphill	635 (250)	18	150
	1.2 (0.045)	Flat	483 (190)	17	200
		Uphill	406 (160)	17	175

Table 5.9
Typical Parameters for Gas-Shielded Stainless Steel FCAW Electrodes

AWS Classification	Diameter, mm (in.)	Welding Position	Wire Feed Speed cm/min (in./min)	Current, A	Volts, V
EXXXT0-1/-4	1.2 (0.045)	Flat & horizontal	991 (390)	190	26
EXXXT0-1/-4	1.6 (1/16)	Flat & horizontal	635 (250)	250	29
EXXXT1-1/-4	0.9 (0.035)	Flat	1016 (400)	140	24
		Uphill	699 (275)	95	23
EXXXT1-1/-4	1.2 (0.045)	Flat	546 (215)	390	29
		Uphill	406 (160)	250	26
EXXXT1-1/-4	1.6 (1/16)	Flat	660 (260)	300	29
		Uphill	521 (205)	250	26

plasma arc cutting, air carbon arc gouging, or machining, depending on the type of base metal and joint design required. For best radiographic quality, dross from cutting or gouging and any machining lubricants should be completely removed before welding. Fitup tolerances for welding will depend on the following:

1. Overall tolerance of completed assembly;
2. Level of quality required for the joint;
3. Method of welding (gas-shielded, self-shielded, automatic, or semi-automatic);
4. Thickness of the base metal welded;
5. Type and size of electrode;
6. Position of welding; and
7. Welder skill.

In general, mechanized and automatic flux cored arc welding preparations require close tolerances in joint fitup. Welds made with semiautomatic equipment can accept somewhat wider tolerances.

WELD QUALITY

The quality of welds that can be produced with the FCAW process depends on the type of electrode used, the method (gas-shielded or self-shielded), condition of the base metal, weld joint design, welding conditions, and the competency of the welder or welding operator. Particular attention must be given to each of these factors to produce sound welds with the best mechanical properties.

The impact properties of mild steel weld metal may be influenced by the welding method. Some self-shielded electrodes are highly deoxidized types that may produce weld metal with relatively low notch toughness. Other self-shielded electrodes have excellent impact properties. Gas-shielded and self-shielded electrodes are available that meet the Charpy V-notch impact requirements of specific classifications of the AWS filler metal specifications. Notch toughness requirements should be considered before selecting the FCAW method and the specific electrode for an application.

A few FCAW mild steel electrodes are designed to tolerate a certain amount of mill scale and rust on the base metals. Some deterioration of weld quality should be expected when welding dirty materials. When these electrodes are used for multiple-pass welding, cracking caused by accumulated deoxidizing agents in the weld metal may occur.

In general, sound FCAW welds can be produced in mild and low-alloy steels that will meet the requirements of several construction codes. If particular attention is given to all factors affecting weld quality, meeting code requirements can be assured.

When less stringent requirements are imposed, the advantages of higher welding speeds and currents can be realized. Minor discontinuities may be permitted in such welds if they are not objectionable from the standpoints of design and service.

Flux cored arc welds can be produced in stainless steels with qualities equivalent to those of gas metal arc welds. Welding position and arc length are significant factors when self-shielded electrodes are used. Out-of-position welding procedures should be carefully evaluated with respect to weld quality. Excessive arc length generally causes increased nitrogen pickup in the weld metal. Because nitrogen is an austenite stabilizer, absorption of excess nitrogen into the weld may prevent the formation of sufficient ferrite and thus may increase susceptibility to microfissuring.

Low-alloy steels can be welded with the gas-shielded method using T1X-1 or T5-X electrode core formulations when good low-temperature toughness is required. The combination of gas shielding and proper flux formulation generally produces sound welds with good mechanical properties and notch toughness. Self-shielded electrodes containing nickel for good strength and impact properties and with aluminum as a denitrider are also available. In general, the composition of the deposited metal should be similar to that of the base metal.

TROUBLESHOOTING

Several types of discontinuities can result from improper procedures or practices. Although many of the discontinuities are innocuous, they adversely affect the weld appearance, and therefore adversely affect the reputation of FCAW. These problems and discontinuities, along with their causes and remedies, are shown in Table 5.10.

ECONOMICS

Flux cored arc welding often provides an economic advantage because it is capable of producing high-quality welds at a lower cost than some of the other processes. FCAW electrodes offer a particular advantage in out-of-position welding because they provide the benefit of a full slag cover. Additional savings may be realized because of the high quality of welds deposited with FCAW electrodes compared to other processes. This high quality can result in fewer rejects and rework.

Table 5.10
Flux Cored Arc Welding Troubleshooting

Problem	Possible Cause	Corrective Action
Porosity	Low gas flow	Increase gas flowmeter setting. Clean spatter-clogged nozzle.
	High gas flow	Decrease to eliminate turbulence.
	Excessive wind drafts	Shield weld zone from draft or wind.
	Contaminated gas	Check gas source. Check for leak in hoses and fittings.
	Contaminated base metal	Clean weld joint faces.
	Contaminated filler wire	Remove drawing compound on wire. Clean oil from rollers. Avoid shop dirt. Rebake filler wire.
	Insufficient flux in core	Change electrode.
	Excessive voltage	Reset voltage.
	Excess electrode stickout	Reset stickout and balance current.
	Insufficient electrode stickout (self-shielded electrodes)	Reset stickout and balance current.
	Excessive travel speed	Adjust speed.
Incomplete fusion or penetration	Improper manipulation	Direct electrode to the joint root.
	Improper parameters	Increase current. Reduce travel speed. Decrease stickout. Reduce wire size. Increase travel speed (self-shielded electrodes).
	Improper joint design	Increase root opening. Reduce root face.
Cracking	Excessive joint restraint	Reduce restraint. Preheat. Use more ductile weld metal.
	Improper electrode	Check formulation and content of flux.
	Insufficient deoxidizers or inconsistent flux-fill in core	Check formulation and content of flux.
Electrode feeding	Excessive contact tip wear	Reduce drive roll pressure.
	Melted or stuck contact tip	Reduce voltage. Adjust backburn control. Replace worn liner.
	Dirty wire conduit in cable	Change conduit liner. Clean out with compressed air.

Flux cored arc welding and other processes using continuously fed electrodes produce much higher deposition rates than the electrodes used in manual SMAW. When compared to GMAW electrodes, the higher deposition rates of FCAW electrodes result in lower costs per pound of deposited weld metal.

Self-shielded FCAW electrodes offer an obvious advantage over gas-shielded FCAW and other gas-shielded processes because there is no shielding cost or cost of associated gas apparatus. However, the deposition rates when using self-shielded electrodes in positions other than the flat position are typically lower than when using gas-shielded electrodes.

Table 5.11 provides a method of calculating the cost of depositing one pound or other unit weight of weld metal and comparing the cost of weld metal produced by alternative materials and processes. In this example, two electrodes, one currently in use and another proposed, are evaluated. To use this model, costs must be calculated for labor, overhead, deposition rate, operat-

Table 5.11
Sample Worksheet Used in Cost Evaluation

Formulas for Calculating Cost per Pound Deposited Weld		(1) Proposed Method Cost Calculation E71T-1 1.6 mm (1/16 in.) @ 300 Amps		(2) Present Method Cost Calculation E7018 4.8 mm (3/16 in.) @ 250 Amps		Result (Cost Reduction) Cost Increase
(3)		(4)	(5)	(4)	(6)	(5–6)
Labor & Overhead = $\frac{\text{Labor \& Overhead Cost/Hr}}{\text{Deposition Rate} \times \text{Operating Factor}}$	=	$\frac{\$45.00}{10.2 \times 0.45} = \frac{\$45.00}{4.59}$	= \$9.80	$\frac{\$45.00}{5.4 \times 0.3} = \frac{\$45.00}{1.62}$	= \$27.78	**(\$17.98)**
(3)		(4) (9)	(5)	(4)	(6)	(5–6)
Electrode $\frac{\text{Electrode Cost/lb}}{\text{Deposition Efficiency}}$	=	$\frac{1.78}{0.87}$	= 2.05	$\frac{0.85}{0.64}$ 2 in. (50 mm) stub	= 1.33	**0.72**
(3)		(4)	(5)	(4)	(6)	(5–6)
Gas $\frac{\text{Gas Flow Rate (Cu ft/hr)} \times \text{Gas Cost/Cu ft}}{\text{Deposition Rate (lbs/hr)}}$	=	$40 \times 0.02 = \frac{0.8}{10.2}$	= 0.08	0 × 0 =	$\frac{0}{5.4} = 0$	**0.08**
			(7)		(8)	(7–8)
Sum of the Above		Total Variable Cost/lb Deposited Weld Metal	= **\$11.93**	Total Variable Cost/lb Deposited Weld Metal	= **\$29.11**	**(\$17.18) Total**

Pounds of Weld Metal and Welder Hours Required to Amortize Equipment Cost

Equipment Cost Power Supply Wire Feeder Gun & Accessories	$\frac{\text{Equipment Cost in Dollars}}{\text{Savings/lb}}$		Pounds of Weld Metal Required to Amortize Cost	÷	Deposition Factor		Welder Hours Required to Amortize Cost
					(9)		
Total \$2,900	$\frac{\$2,900}{\$17.18}$	=	168.8	÷	4.59	=	37 Hours

Notes:
1. Enter descriptions of proposed and present products at (1) and (2).
2. Calculate labor, overhead, deposition rate, operating factor, electrode cost, deposition efficiency, gas flow rate, and gas cost required for the formulas (3) and enter data into (4).
3. Calculate costs for "Proposed Method" and "Present Method" and enter data into (5) and (6).
4. Add calculated costs and enter sums into (7) and (8).
5. Subtract "Present Method" cost (6) from "Proposed Method" (5) for Labor, Overhead, Electrode, and Gas. Enter differences in last column (5–6).
6. For "Total Variable Cost" subtract "Present Method" cost (8) from "Proposed Method" cost (7). Enter the difference under "Total." (If negative number, enclose in parentheses). The negative number represents the cost reduction per pound of weld metal deposited.
7. For equipment amortization, use the formula in the bottom section of the chart. The deposition factor (9) is the deposition rate multiplied by the operator factor.
8. Multiply deposition rate times operating factor to calculate deposition factor; enter in (9).

Source: Adapted from ESAB Welding and Cutting Products, *Filler Metal Data Book*; Florence, S.C.: ESAB Welding and Cutting Products, 7.9.

ing factor, electrode cost per unit weight deposition efficiency, gas flow rate, i.e., liters per minute (cubic feet per hour), and gas cost per unit volume.

Labor and overhead rates can be supplied by the company's accountants or managers. However, if the actual labor and overhead rate is not known, a reasonable rate equivalent to local rates can be selected.

The deposition rate is the weight of weld metal that can be deposited in one hour at a specified welding current, assuming welding takes place 100% of the hour.

The operating factor is the percentage of a welder's time actually spent welding. If the operating factor is not known, a 45% operating factor for semiautomatic FCAW and a 60% to 80% operating factor for automatic FCAW can be assumed.

The electrode cost should be based on the most appropriate quantity price bracket.

Deposition efficiency (DE) is the percent ratio of electrode used to the amount of weld metal deposited, as follows:

$$DE = \frac{\text{Weight of Weld Metal}}{\text{Weight of Electrode Used}} \quad (5.4)$$

The shielding gas flow rate can be estimated at 16 L/min (35 ft^3/h) for small-diameter flux cored electrodes (l.6 mm [1/16 in.] diameter and smaller) and 18 to 21 L/min (40 to 45 ft^3/h) for large-diameter flux cored electrodes (2 mm [5/64 in.] and over). Gas price is based on the type of gas shielding required and local prices.

SAFE PRACTICES

Potential hazards in arc welding and cutting include the handling of cylinders and regulators, gases, fumes, radiant energy, noise, and electric shock. Safety concerns that may be associated with the FCAW process are briefly discussed in this section.

Safe practices are covered in ANSI Z49.1, *Safety in Welding, Cutting, and Allied Processes,*[20] and other standards such as *Fire Prevention During Welding, Cutting, and Other Hot Work,* NFPA 51B.[21] Chapter 17, "Safe Practices," in Volume 1 of the *Welding Handbook*, 9th edition, also provides comprehensive information.[22]

The Occupational Safety and Health Administration (OSHA) of the federal government publishes mandatory standards in *Occupational Safety and Health Standards for General Industry,* Subpart Q of *Code of Federal Regulations (CFR),* Title 29 CFR 1910.[23] Other sources of safety information are listed in Appendix B of this volume.

Personnel should be familiar with the safe practices discussed in these documents and in addition should consult manufacturers' operating manuals and suppliers' Material Safety Data Sheets specific to the equipment and materials they are using.

SAFE HANDLING OF GAS CYLINDERS AND REGULATORS

Compressed gas cylinders should be handled carefully and should be adequately secured when stored or in use. Knocks, falls, or rough handling may damage cylinders, valves and fuse plugs, and can cause leakage or an accident. Cylinder caps should be kept in place (hand tight) to protect the valves unless a regulator is attached to the cylinder.

The following guidelines should be observed when setting up and using cylinders of shielding gas:

1. Secure the cylinder properly.
2. Before connecting a regulator to the cylinder valve, the valve should momentarily be opened slightly and closed immediately (called *cracking*) to clear the valve of dust or dirt that otherwise might enter the regulator. The valve operator should stand to one side of the regulator gauges, never in front of them.
3. After the regulator is attached, the pressure adjusting screw should be released by turning it counter-clockwise. The cylinder valve should then be opened slowly to prevent a rapid surge of high-pressure gas into the regulator. The adjusting screw should then be turned clockwise until the proper pressure is obtained.
4. The source of the gas supply (i.e., the cylinder valve) should be shut off if it is to be left unattended, and the adjusting screw should be in the open position.

20. American National Standards Institute (ANSI) Accredited Standards Committee Z49, 1999, *Safety in Welding, Cutting, and Allied Processes,* ANSI Z49.1:1999, Miami: American Welding Society.

21. National Fire Prevention Association (NFPA), 1999, *Fire Protection During Welding, Cutting and Other Hot Work,* NFPA 51B, Quincy, Massachusetts: National Fire Protection Association.

22. American Welding Society (AWS) Welding Handbook Committee, C. L. Jenney and A. O'Brien, eds., 2001, *Welding Science and Technology,* Vol. 1 of *Welding Handbook*, Miami: American Welding Society, Chapter 17 and Bibliography, Chapter 17.

23. Occupational Safety and Health Administration (OSHA), 1999, *Occupational Safety and Health Standards for General Industry,* in *Code of Federal Regulations, (CFR)* Title 29, CFR 1910, Subpart Q, Washington, D.C.: Superintendent of Documents, U.S. Government Printing Office.

The Compressed Gas Association (CGA) provides further information in CGA Pamphlet P-1, *Safe Handling of Gas in Containers*.[24]

GASES

The major toxic gases associated with FCAW welding are ozone, nitrogen dioxide, and carbon monoxide. Phosgene gas could also be present as a result of thermal or ultraviolet decomposition of chlorinated hydrocarbon cleaning agents located in the vicinity of welding operations. Two such solvents are trichlorethylene and perchlorethylene. Degreasing or other cleaning operations involving chlorinated hydrocarbons should be located so that vapors from these operations cannot be reached by radiation from the welding arc.

Ozone

The ultraviolet light emitted by the FCAW arc acts on the oxygen in the surrounding atmosphere to produce ozone in an amount that depends on the intensity and the wave length of the ultraviolet energy, the humidity, the amount of screening afforded by any welding fumes, and other factors. The ozone concentration will generally increase with an increase in welding current; it also increases with the use of argon as the shielding gas and when welding highly reflective metals. If the ozone cannot be reduced to a safe level by ventilation or process variations, it is necessary to supply fresh air to the welder either with an air-supplied respirator or by other means. The National Institute for Occupational Safety and Health (NIOSH) publishes *Guide to Industrial Respiratory Protection,* Publication No. 87-116, to cover such ventilation issues.[25]

Nitrogen Dioxide

Some test results show that high concentrations of nitrogen dioxide are found only within 6 in. (150 mm) of the arc. With normal or natural ventilation, these concentrations are quickly reduced to safe levels in the welder's breathing zone, as long as the welder's head is kept out of the plume of fumes (and thus out of the plume of welding-generated gases). This does not apply to overhead welding, when the fumes may fall back into the welder's breathing zone. Nitrogen dioxide is not considered to be a hazard in FCAW.

24. Compressed Gas Association (CGA), 1999, *Safe Handling of Gas in Containers*, CGA P-1, Arlington, Virginia: Compressed Gas Association.

25. National Institute for Occupational Safety and Health (NIOSH), 1987, *NIOSH Guide to Industrial Respiratory Protection*, U.S. Department of Human Services (DHHS), Publication No. 87-116, http://www.cdc.gov/niosh/87-116.html.

Carbon Monoxide

Carbon dioxide shielding gas used with the FCAW process is dissociated by the heat of the arc and forms carbon monoxide. Only a small amount of carbon monoxide is created by the welding process, although relatively high concentrations are temporarily present in the plume of fumes. However, the hot carbon monoxide oxidizes to carbon dioxide so that the concentrations of carbon monoxide become insignificant at distances of more than 3 or 4 in. (75 or 100 mm) from the welding plume.

Under normal welding conditions, there should be no hazard from this source. When welders must work over the welding arc, or with natural ventilation that may move the plume of fumes toward the welder's breathing zone, or where welding is performed in a confined space, ventilation adequate to deflect the plume or remove the fumes and gases should be provided (See ANSI Z49.1, *Safety in Welding, Cutting, and Allied Processes*).[26]

METAL FUMES

The welding fumes generated by FCAW can be controlled by general ventilation, local exhaust ventilation, or by respiratory protective equipment as described in ANSI Z49.1. The method of ventilation required to keep the level of toxic substances within the welder's breathing zone below threshold concentrations is directly dependent on a number of factors. Among these are the material being welded, the size of the work area, and the degree of confinement or obstruction to normal air movement where the welding is being done. Each operation should be evaluated on an individual basis in order to determine what will be required.

Acceptable exposure levels to substances associated with welding and designated as time-weighted average threshold limit values (TLV), permissible exposure limits (PEL) and ceiling values have been established by the American Conference of Governmental Industrial Hygienists (ACGIH)[27] and by the Occupational Safety and Health Administration (OSHA).[28] Compliance with these acceptable levels of exposure can be checked by sampling the atmosphere under the welder's helmet or in the immediate vicinity of the welder's breathing zone. Sampling should be in accordance with *Method*

26. See Reference 21.

27. American Conference of Governmental Industrial Hygienists (ACGIH), *TLVs and BEIs: Threshold Limit Values for Chemical Substances and Physical Agents and Biological Exposure Indices,* Cincinnati: American Conference of Governmental Industrial Hygienists.

28. See Reference 23.

for Sampling Airborne Particulates Generated by Welding and Allied Processes, ANSI/AWS F1.1.[29]

RADIANT ENERGY

The total radiant energy produced by the FCAW process can be high because of its high arc energy. Generally, the highest ultraviolet radiant energy intensities are produced when using argon shielding gas mixtures.

The suggested filter glass shades for FCAW are listed in Appendix A, *Lens Shade Selector.*[30] To select the best shade for an application, the first selection should be a very dark shade. If it is difficult to see the operation properly, successively lighter shades should be selected until the operation is sufficiently visible for good control. However, the welder must not go below the lowest recommended number.

Dark leather or other sturdy clothing is recommended for FCAW to reduce reflection, which could cause ultraviolet burns to the face and neck under the helmet. The great intensity of the ultraviolet radiation can cause rapid disintegration of cotton clothing.

NOISE

Personnel should wear properly fitting earplugs and otherwise protect against noise generated in welding and cutting processes. Allowable noise exposure levels are regulated by OSHA in *Occupational Safety and Health Standards for General Industry,* Title 29 CFR 1910.95,"Occupational Noise Exposure."[31]

ELECTRIC SHOCK

Line voltages to power supplies and auxiliary equipment used in FCAW range from 110 to 575 volts. Welders and service personnel should exercise caution not to come in contact with these voltages. The equipment manufacturers' instructions for installation, use, and maintenance of the equipment should be followed. Additional precautionary information is published in *Safety in Welding, Cutting, and Allied Processes*, ANSI Z49.1.[32]

29. American Welding Society (AWS) Safety and Health Committee, 1999. *Method for Sampling Airborne Particulates Generated by Welding and Allied Processes,* Miami: American Welding Society.
30. American Welding Society (AWS) Safety and Health Committee, 2001. Lens Shade Selector (Chart), AWS F2.2:2001, Miami: American Welding Society.
31. See Reference 24.
32. See Reference 21.

CONCLUSION

The use of flux cored arc welding continues to increase because of the versatility of the process. Compared to other semiautomatic processes, FCAW requires less operator skill (less training) and provides excellent out-of-position usability and deposition rates. Flux cored arc welding electrodes currently available can deposit weld metal with mechanical properties equal to or better than other welding processes. Welding with FCAW electrodes results in high weld quality, especially at high travel speeds.

Two issues with FCAW in the past were inconsistent flux fill and electrode feeding. Most major manufacturers have solved these former problems; they are seldom seen when the wire is produced with modern manufacturing techniques. The use of the flux cored arc welding process is expected to continue to expand.

BIBLIOGRAPHY[33]

American Conference of Governmental Industrial Hygienists (ACGIH). *TLVs and BEIs: Threshold limit values for chemical substances and physical agents and biological exposure indices*, Cincinnati, Ohio: American Conference of Governmental Industrial Hygienists.

American National Standards Institute (ANSI) Accredited Standards Committee Z49. 1999. *Safety in welding, cutting, and allied processes,* ANSI Z49.1:1999. Miami: American Welding Society.

American Society for Mechanical Engineers (ASME) Boiler and Pressure Vessel Code Committee. 1998. *Welding and brazing qualifications,* Section IX of *Boiler and pressure vessel code.* New York: American Society of Mechanical Engineers.

American Welding Society (AWS) Committee on Definitions and Symbols. 2001. *Standard welding terms and definitions.* A3.0:2001. Miami: American Welding Society.

American Welding Society (AWS) Committee on Filler Metals and Allied Materials. 1995. *Specification for carbon steel electrodes for flux cored arc welding.* ANSI/AWS A5.20-95. Miami: American Welding Society.

33. The dates of publication given for the codes and other standards listed here were current at the time the chapter was prepared. The reader is advised to consult the latest edition.

American Welding Society (AWS) Committee on Filler Metals and Allied Materials. 1995. *Specification for stainless steel electrodes for flux cored arc welding and stainless steel flux cored rods for gas tungsten arc welding, AWS A5.22-95. Miami: American Welding Society.*

American Welding Society (AWS) Committee on Filler Metals and Allied Materials. *Specification for low alloy steel electrodes for flux cored arc welding*, ANSI/AWS A5.29:1998. Miami: American Welding Society.

American Welding Society (AWS) Committee on Filler Metals and Allied Materials. 2001. *Specification for bare electrodes and rods for surfacing*, ANSI/AWS A5.21:2001. Miami: American Welding Society.

American Welding Society (AWS) Welding Handbook Committee. W. R. Oates and A. Saitta, eds. 1998. *Materials and applications—Part 2*, Vol. 4 of *Welding Handbook*, 8th ed. Miami: American Welding Society.

American Welding Society (AWS) Committee on Structural Welding. 2002. *Structural welding code—steel*, AWS D1.1/D1.1M:2002. Miami: American Welding Society.

American Welding Society (AWS) Safety and Health Committee. 1999. *Method for sampling airborne particulates generated by welding and allied processes*, F1.1:1999. Miami: American Welding Society.

American Welding Society (AWS) Safety and Health Committee. 2001. *Lens shade selector* (chart), AWS F2.2:2001. Miami: American Welding Society.

American Welding Society (AWS) Welding Handbook Committee. 2001. C. L. Jenney and A. O'Brien, eds. *Welding science and technology*, Vol. 1 of *Welding handbook*, 9th ed. Miami: American Welding Society.

Compressed Gas Association (CGA). 1999. *Safe handling of gas in containers*, CGA P-1-1999. Arlington, Virginia: Compressed Gas Association.

National Fire Protection Association (NFPA). 1999. *Fire prevention during welding, cutting, and other hot work*. NFPA 51B. Quincy, Massachusetts: National Fire Protection Association.

National Institute for Occupational Safety and Health (NIOSH). 1997. *NIOSH respirator user notice*. In *NIOSH guide to selection and use of particulate respirators* (Certified under 42 CFR 84). U.S. Department of Health and Human Services (DHHS) Publication No. 96-101. Washington, D.C.: Superintendent of Documents, U.S. Government Printing Office.

Occupational Safety and Health Administration (OSHA). 1999. *Occupational safety and health standards for general industry*, in *Code of federal regulations, (CFR)* Title 29, CFR 1910, Subpart Q. Washington, D.C.: Superintendent of Documents, U.S. Government Printing Office.

SUPPLEMENTARY READING LIST

Amata, M. A., and S. R. Fiore. 1996. Choosing the proper self-shielded flux-cored wire. *Welding Journal* 75(6):33–39.

ASM International. 1993. *Welding, brazing, and soldering*. Volume 6 of *ASM handbook, 10th ed.* Materials Park, Ohio: ASM International.

American Welding Society (AWS) Committee on Procedure and Performance Qualification. *Standard for welding procedure and performance qualification*, B2.1:2000. Miami: American Welding Society.

American Welding Society (AWS) Committee on Procedure and Performance Qualification. 1994. *Standard welding procedure specification (WPS) for 75% A/25% CO_2 shielded flux cored arc welding of carbon steel (M-1/P-1/S-1, Groups 1 or 2), 1/8 through 1-1/2 inch thick, E70T-1 and E71T-1, as-welded or PWHT condition*, AWS B2.1-1-020-94. Miami: American Welding Society.

American Welding Society (AWS) Committee on Procedure and Performance Qualification. 1998. *Standard welding procedure specification (WPS) for self-shielded flux cored arc welding of carbon steel (M-1/P-1/S-1, Group 1 or 2) 1/8 through 1/2 inch thick E71T-11, as-welded condition*, AWS B2.1-1.027:1998. Miami: American Welding Society.

Babu, S. S., S. A. David, and M. A. Quintana. 2001. Modeling microstructure development in self-shielded flux cored arc welds. *Welding Journal* 80(4): 91-s–97-s.

Boneszewski, T. 1992. *Self-shielded arc welding*. Cambridge, England: Abington Press.

Compressed Gas Association (CGA). 1994. *Standard connections for regulator outlets, torches, and fitted hose for welding and cutting equipment*. Arlington, Virginia: Compressed Gas Association.

Compressed Gas Association (CGA). *Compressed gas cylinder valve outlet and inlet connections*, ANSI/CGA B57.1. Arlington, Virginia: Compressed Gas Association.

Edison Welding Institute. 1998. EWI Project No. 41722IRP, Report No. MR9810. *Effects of welding parameters and electrode atmospheric exposure on the diffusible hydrogen content of gas-shielded flux cored arc welds*. Columbus, Ohio: Edison Welding Institute.

Ellis, T. and G. G. Garret. 1986. Influence of process variables in flux cored arc welding of hardfacing deposits. *Surface Engineering*. 2(1): 55–66.

Ferree, S. E. 1996. Status report on small-diameter cored stainless steel wires. *Svetsaren*. 51(1-2) 27–34.

French, I. E. and M. R. Bosworth. 1995. A comparison of pulsed and conventional welding with basic flux cored and metal cored welding wires. *Welding Journal.* 74(6): 197-s–205-s.

Harwig, D. D., D. P. Longnecker, and J. H. Cruz. 1999. Effects of welding parameters and electrode atmosphere exposure on the diffusible hydrogen content of gas-shielded flux cored arc welds. *Welding Journal.* 78(9): 314-s–321-s.

Kotecki, D. J. 1978 Welding parameter effect on open-arc stainless steel weld metal ferrite. *Welding Journal.* 57(4): 109-s–117-s.

Lathabai, S. and R. D. Stout. 1985. Shielding gas and heat input effects on flux cored weld metal properties. *Welding Journal.* 64(11): 303-s–313-s.

Lincoln Electric Company, The. 1973. *The procedure handbook of arc welding*, 12th Ed. Cleveland, Ohio: The Lincoln Electric Company.

Liu, S. 1998. Arc welding consumables—covered and cored electrodes: a century of evolution. ASM Conference Paper: *Trends in Welding Research.* (6) 1998. Materials Park, Ohio: ASM International.

Quintana, M. A., J. McLane, S. S. Babu, and S. A. David. 2001. Inclusion formation in self-shielded flux cored arc welds. *Welding Journal.* 80(4):98–105.

Rodgers, K. J. and J. C. Lockhead. 1987. Self-shielded flux cored arc welding—the route to good fracture toughness. *Welding Journal*, 66(7): 49–59.

Wei, Q., Q. Hu, F. Guo, and D. Xiong. 2002. A study of weld pore sensitivity of self-shielded flux cored electrodes. *Welding Journal* 81(6): 90-s–94-s.

Yeo, R. B. G. 1993. Electrode extension often neglected when using self-shielded cored wires. *Welding Journal.* 72(1): 51–53.

Yeo, R. B. G. 1989. Cored wires for lower cost welds. *Joining and Materials* 20(2): 68-72.

CHAPTER 6

SUBMERGED ARC WELDING

Prepared by the Welding Handbook Chapter Committee on Submerged Arc Welding:

R. A. Swain, Chair
Euroweld, Limited

H. K. Zentner
Bavaria Schweisstechnik, GmbH

J. F. Hunt
BWX Technologies

J. R. Scott
Wyman Gordon Fabrications

Welding Handbook Volume 2 Committee Member:

D. R. Amos
Siemens Westinghouse Power Corporation

Photograph courtesy of BAE Systems

Contents

CHAPTER 6

SUBMERGED ARC WELDING

INTRODUCTION

Submerged arc welding (SAW) and its process variation, series submerged arc welding (SAW-S), produce the coalescence of metals by heating them with an arc or arcs between one or more bare metal electrodes and the workpieces. The arc and molten metal are submerged in a blanket of granular fusible flux on the workpiece. Pressure is not used, and filler metal is obtained from the electrode and sometimes from a supplemental source such as a welding rod, flux, or metal granules.

The distinguishing feature of submerged arc welding is that the flux covers the arc and prevents fumes, sparks, spatter, and radiation from escaping. Flux is fundamental to the process in that the stability of the arc is dependent on the flux, the mechanical and chemical properties of the final weld deposit can be controlled by the flux, and the quality of the weld can be affected by the control and handling of the flux.

Submerged arc welding is a versatile production welding and cladding process that operates on alternating current (ac) or direct current (dc) up to 2000 amperes (A) using single or multiple wires or a strip of various alloys. Both alternating-current and direct-current power sources may be used on the same weld simultaneously when using a multiple-wire technique with wires of the same or different chemistry.

Submerged arc welding is used in a wide range of industrial applications. It is typically used to weld butt joints, create fillet welds, and deposit surfacing in the flat position. With special tooling and fixturing, lap and butt joints can be welded in the horizontal position. Excellent weld quality, high deposition rates, controllable penetration (deep or shallow), and adaptability to automated operation make the process ideal for the fabrication of large weldments. It is used extensively in pressure vessel fabrication, ship and barge building, railroad car fabrication, pipe manufacturing, and the fabrication of structural members where long welds are required. Automated submerged arc welding installations manufacture mass-produced assemblies joined with repetitive short welds.

FUNDAMENTALS

In the submerged arc welding process, the end of a continuous bare wire electrode is inserted into a mound of flux that covers the area or joint to be welded. An arc is initiated and a wire feeding mechanism then begins to feed the electrode (wire) toward the joint at a controlled rate. The feeder is moved manually or automatically along the weld seam. In mechanized or automated welding, the workpiece is moved under a stationary wire feeder or the welding head moves over the stationary workpiece.

Additional flux is continuously fed either in front of or all around the electrodes and continuously distributed over the joint. The heat evolved by the electric arc progressively melts some of the flux, the end of the wire, and the adjacent edges of the base metal, creating a pool of molten metal (the weld pool) beneath a layer of liquid slag and unmelted flux. The melted bath near the arc is in a highly turbulent state because of the arc pressure. Gas bubbles are quickly swept to the surface of the weld pool. The main portion of the liquid flux-slag floats on the molten metal and completely shields the welding zone from the atmosphere.

The liquid slag may conduct some electric current between the wire and base metal, but the electric arc is in a gaseous environment and is the predominant heat source. The flux blanket on the top surface of the weld pool prevents atmospheric gases from contaminating the weld metal and dissolves impurities in the base metal and electrode, then floats them to the surface.

The flux can also add or remove certain alloying elements to or from the weld metal.

As the welding zone progresses along the seam, the weld metal and then the liquid flux cool and solidify, forming the weld bead and a protective slag shield over it. Before making another weld pass, the slag must be completely removed. The submerged arc welding process is illustrated schematically in Figure 6.1.

Factors that should be considered when determining whether submerged arc welding can or should be used for a given application include the following:

1. The chemical composition and mechanical properties required of the final weld deposit,
2. Thickness of base metal and alloy to be welded,
3. Joint accessibility,
4. Length of the joint,
5. Position in which the weld is to be made,
6. Frequency or volume of welding to be performed, and
7. The availability of capital for the submerged arc welding equipment expenditure.

ADVANTAGES AND LIMITATIONS

The main advantage of using the submerged arc welding process is high quality and productivity. The process can be implemented in three different operational modes—semiautomated, mechanized, and automated.

The main disadvantage of submerged arc welding is that it can be used only in the flat or horizontal welding positions (test positions 1G or 2G) for plate and pipe welding.

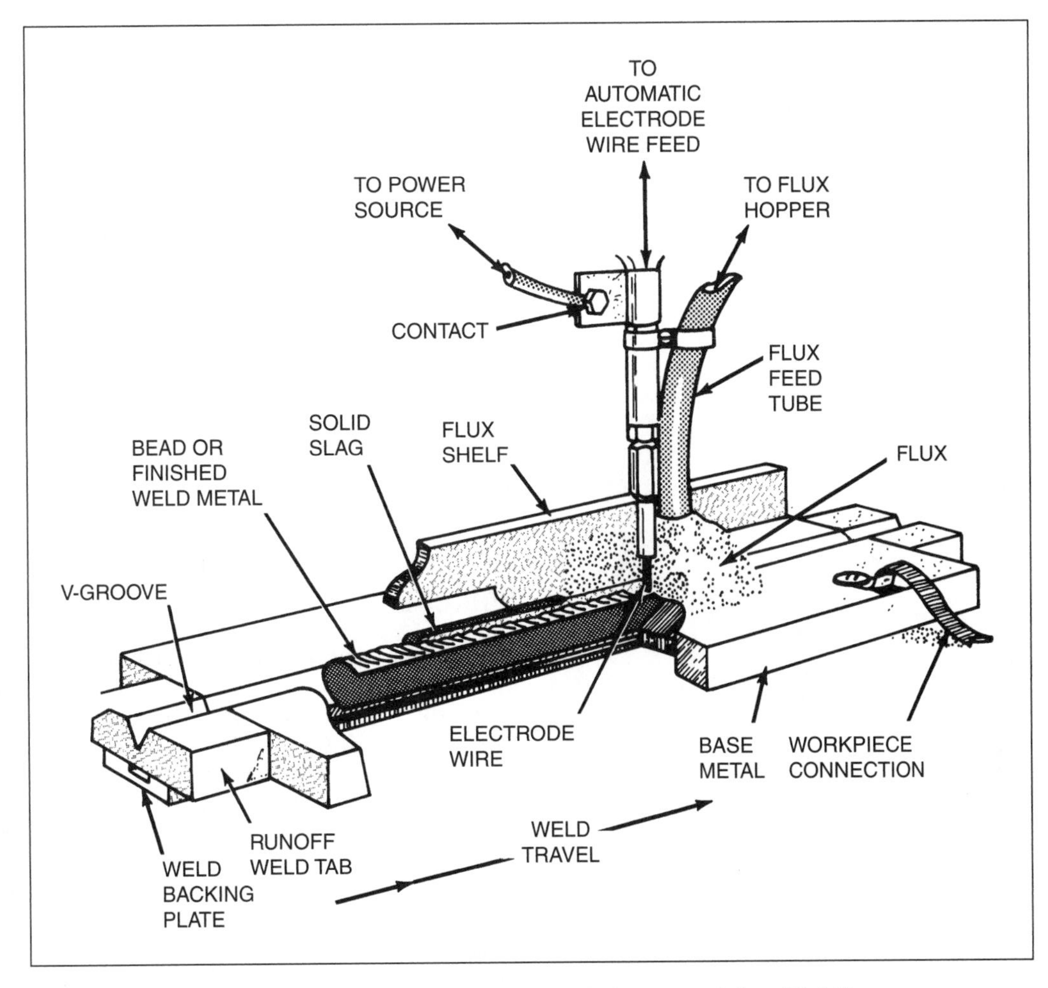

Figure 6.1—Schematic View of Submerged Arc Welding

EQUIPMENT

The equipment used for semiautomated, mechanized, and automated welding is described in this section. The basic equipment consists of a power source, an electrode delivery system, a flux distribution system, a travel arrangement, and a process control system. Optional equipment includes a flux recovery system and positioning or manipulating equipment, depending on the requirements of the application.

SEMIAUTOMATED WELDING

Semiautomated welding is performed with a hand-held welding gun that delivers both flux and the electrode (in wire form). The electrode is driven by a wire feeder. Flux may be supplied by a gravity hopper mounted on the gun or may be pressure-fed through a hose. This method uses relatively small-diameter electrodes and moderate travel speeds, and it requires manual guidance. The travel may be controlled manually or driven by a small gun-mounted driving motor, as shown in Figure 6.2. This equipment is relatively inexpensive and can easily be adapted for mechanized applications.

MECHANIZED WELDING

The equipment used in mechanized welding performs the entire welding operation. However, the process must be monitored by a welding operator, who must position the work, start and stop welding, adjust the controls, replenish the flux and set the travel speed of each weld. Figure 6.3 shows a typical mechanized submerged arc welding operation, in which a 76 mm (3 in.) thick 3.8 m (150 in.) diameter bridge pile casing is welded for a seismic retrofit project.

Figure 6.2—Hand-Held Submerged Arc Welding Gun

AUTOMATIC WELDING

Automatic welding is accomplished with equipment that performs the welding operation without requiring a welding operator to continually monitor and adjust the controls. Self-regulating equipment, although expensive, can often be justified in order to achieve high production rates. An example of an automatic submerged arc welding production system is shown in Figure 6.4. A tandem-wire submerged arc welding system is used to weld a pressure hull frame to table joints in the construction of a submarine.

POWER SOURCES

The power source for a submerged arc welding system is of major importance and should be carefully selected. Submerged arc welding is typically a high-current process with a high duty cycle, thus a power source capable of providing high amperage at 100% duty cycle is recommended. Two general types of power sources are suitable. Direct-current power sources, which may be transformer rectifiers, motor or engine generators, provide a constant-voltage (CV), constant-current (CC), or a selectable constant-voltage/constant-current output. Alternating-current power sources (generally transformer types) can provide either a constant-current output or a constant-voltage square-wave output.

Direct-Current Constant-Voltage Power Sources

Direct-current constant-voltage power sources are available in both transformer-rectifier and motor-generator models. They range in output from 400 amperes (A) to 1500 A. The lower-amperage power sources may also be used for gas metal arc and flux cored arc welding. These power sources are used for semiautomatic submerged arc welding at currents ranging from about 300 A to 600 A with electrode diameters of 1.6 millimeters (mm), 2.0 mm, and 2.4 mm (1/16 inch [in.], 5/64 in., and 3/32 in.). Automated and mechanized welding require currents ranging from 300 A to over 1000 A, with electrode diameters generally ranging from 2.0 mm to 6.4 mm (5/64 in. to 1/4 in.). Appli-

Photograph courtesy of XKT Engineering

Figure 6.3—Mechanized Submerged Arc Weld Made on a Bridge Pile Casing

cations for direct-current welding over 1000 A are limited, however, because severe arc blow—the deflection of the arc from its normal path—may occur at these high currents. A typical direct-current constant-voltage power source is shown in Figure 6.5.

A constant-voltage power source is self-regulating: the wire feed speed and wire diameter control the arc current and the power source controls the arc voltage. Once the arc length is established by the voltage adjustment, any changes in arc length caused by welding conditions are automatically compensated for by an increase or decrease in current. This, in turn, increases or decreases the strip or wire burn-off rate and the arc is returned to its original setting. Constant-voltage power sources are intended for use with constant-speed wire feeders. Because voltage or current sensing is not required to maintain a stable arc, very simple wire feed speed controls that assure constant wire feed can be used. Constant-voltage dc power sources are the most commonly used power sources for submerged arc welding and are the best choices for the welding of stainless steel, the high-speed welding of thin steel, for all cladding applications, and for use with flux cored wire. This power source can also be used for carbon arc cutting and gouging.

Direct-Current Constant-Current Power Sources

Direct-current constant-current power sources are available in both transformer-rectifier and motor-generator models with rated outputs up to 1500 A. Some constant-current dc power sources can also be used for gas tungsten arc welding (GTAW), shielded metal arc welding (SMAW), and air carbon arc cutting and gouging (CAC-A). With the exception of the high-speed welding of thin steel, constant-current dc sources can be used for the same range of applications as constant-voltage dc power sources.

Constant-current power sources are not self-regulating, so they must be used with a voltage-sensing variable wire feed speed control. This type of control adjusts the wire feed speed in response to changes in arc voltage. The voltage is monitored to maintain a constant arc length. With this system, the arc voltage is dependent on the wire feed speed and the wire diameter. The power source controls the arc current. Because voltage-sensing variable wire feed speed controls are more complex, they are also more expensive than the simple constant wire feed speed controls that can be used with constant-voltage systems.

Photograph courtesy of BAE Systems

Figure 6.4—Automatic Tandem Wire Submerged Arc Welding System Used to Weld a Submarine Hull

Combination Power Sources

Some power sources used for submerged arc welding can be switched between constant-voltage and constant-current modes. Power sources rated at up to 1500 A are available, but machines rated at 650 A or less are much more common. The value of these power sources is in their versatility, as they can also be used for shielded metal arc welding, gas metal arc welding, gas tungsten arc welding, flux cored arc welding, air carbon arc cutting, and arc stud welding.

Alternating-Current Power Sources

The power sources used for ac submerged arc welding are most commonly transformers. Power sources rated for 800 A to 1500 A at 100% duty cycle are available. If higher amperages are required, these machines can be connected in parallel.

Conventional ac power sources are the constant-current type. The output voltage of these machines approximates a sine wave, as shown in Figure 6.6(A). The output of these machines drops to zero with each

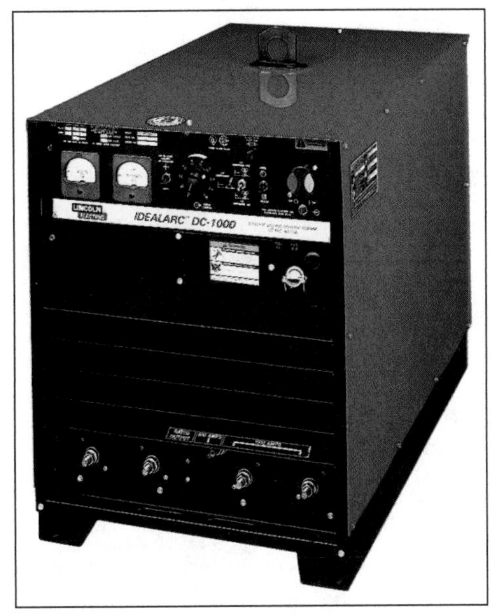

Photograph courtesy of The Lincoln Electric Company

Figure 6.5—Typical DC Constant-Voltage Power Source for Submerged Arc Welding

polarity reversal, so a high open-circuit voltage (greater than 80 volts [V]) is required to ensure reignition of the arc. Even at that high open-circuit voltage, arc reignition problems are sometimes encountered with fully basic fluxes that are not designed for alternating current. Because these power sources are the constant-current type, the speed controls must be the voltage-sensing, variable wire feed type.

With constant-voltage square-wave ac power sources, the output current and the output voltage approximate a square wave. Because polarity reversals are instantaneous with square-wave power sources, as shown in Figure 6.6(B), arc reignition problems are not as severe as those encountered with conventional ac sources. Hence, some of the fluxes that do not work with conventional ac power sources can be used with square-wave ac sources. Relatively simple constant wire feed speed controls can be used with square-wave power sources because they provide constant voltage.

The most common uses of ac power for submerged arc welding are high-current applications, multiple-wire applications, narrow groove welding, and applications in which arc blow is a problem.

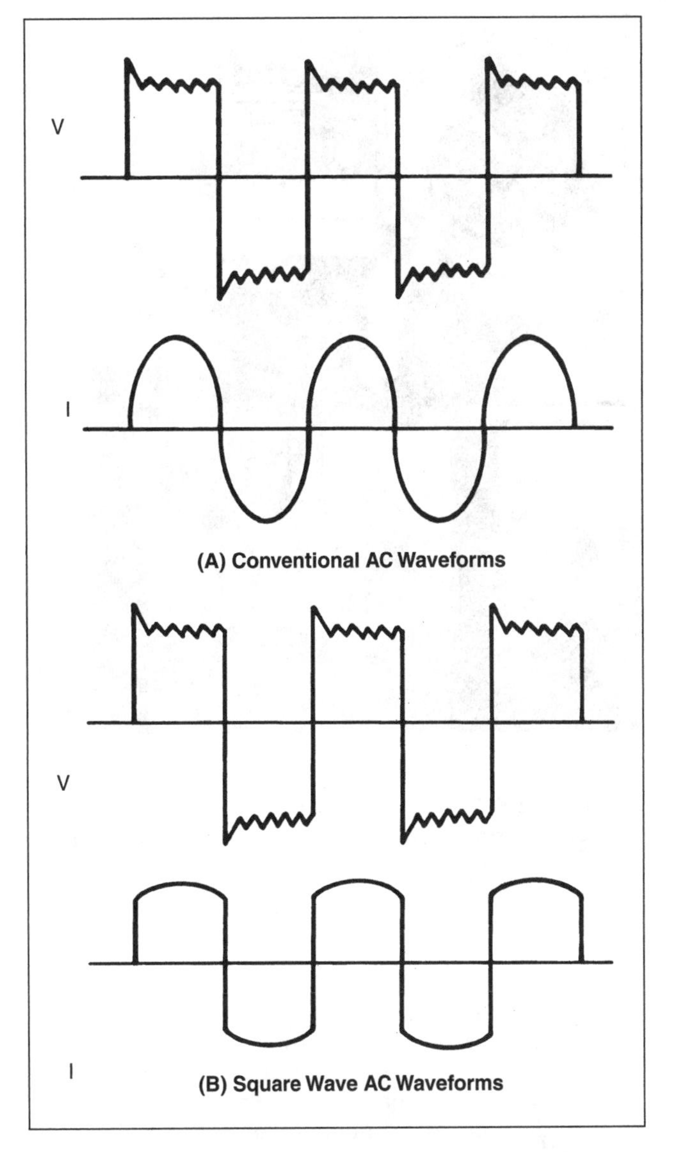

Figure 6.6—Difference Between Conventional and Square-Wave AC Waveforms

CONTROL SYSTEMS

Basic control systems for submerged arc welding consist of the following:

(1) Wire-feed speed control, which adjusts the current in constant voltage systems and controls the voltage in constant-current systems;
(2) Power source control, which adjusts the voltage in CV systems and adjusts the current in CC systems;
(3) Weld start/stop switch;
(4) Manual or automatic travel on/off switch; and
(5) Cold wire feed up/down.

The control systems used in semiautomated submerged arc welding are simple wire feed speed controls.

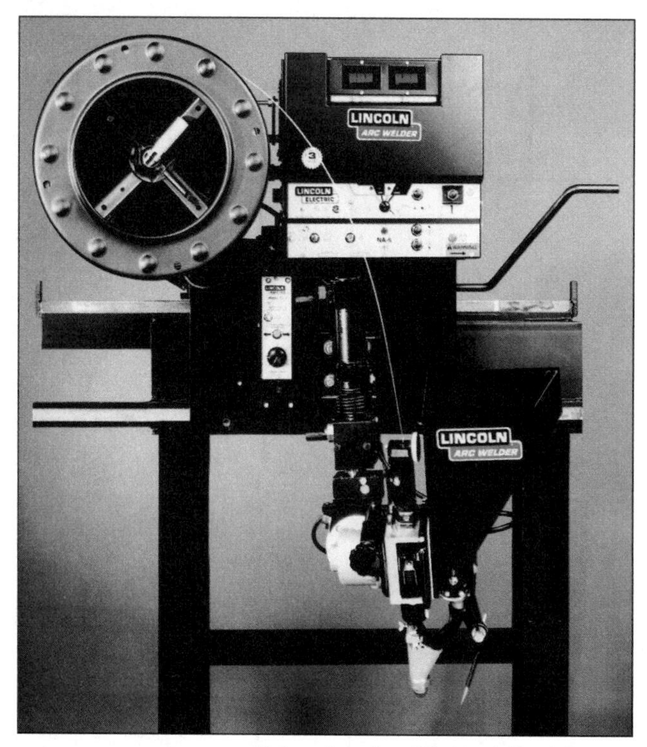

Photograph courtesy of The Lincoln Electric Company

Figure 6.7—Wire Feeder Showing Control, Wire Feed Motor, and Wire Drive Assembly

Figure 6.8—Digital Control for Two-Wire Submerged Arc Wire Feeder

The controls used with CV power sources maintain a constant wire feed speed, whereas those used with CC power sources monitor the arc voltage and adjust the wire feed speed to maintain a constant voltage. The simplest wire feeders have one-knob analog controls that maintain a constant wire feed speed. Figure 6.7 shows a typical control unit.

Digital current, voltage, and wire feed speed meters are standard features of digital control systems. The controls typically provide for wire feed speed adjustment (current control), power source adjustment (voltage control), weld start-stop, automatic and manual travel on-off, cold wire feed up-down, run-in and crater fill control, burnback, and flux feed on-off. The state-of-the-art wire feeders used in automated submerged arc welding have microprocessor-based digital controls, as shown in Figure 6.8. These controls have feedback loops interfaced with the power source and wire feed motor to maintain the welding voltage and wire speed at preset values. The great advantage of digital controls is precise control of the welding process; however, they are not compatible with all power sources.

WELDING HEADS AND GUNS

A submerged arc welding head incorporates the wire feed motor and feed-roll assembly, the gun assembly and contact tip, and accessories for mounting and positioning the head. Wire feed motors are typically heavy-duty, permanent magnet motors with an integral reducing gearbox. They are capable of feeding wire at speeds in the range of 0.5 to 14 m/min (20 in./min to 46 ft./min).

The feed roll assembly may have one drive and one idler roll, two drive rolls, or four drive rolls. Four-roll drive assemblies provide positive feeding with the least wire slippage. Feed rolls may be of the knurled V-groove or smooth U-groove type. The knurled V-groove rolls are the most common. When the wire is being pushed through a conduit, smoother feeding results if smooth V-groove rolls are used.

Gun assembly designs are numerous, but their purpose is always the same. The gun assembly guides the wire or strip through the contact tip to the weld zone and delivers welding power to the wire or strip at the contact tip.

Special equipment is needed when performing narrow groove welding with the submerged arc process or when welding with strip electrodes. Parallel wire devices incorporate special feed-roll and gun assemblies that provide positive feeding of two wires through one gun body. Strip-electrode submerged arc welding requires a special feed-roll assembly. The strip heads that feed the strip are generally water-cooled. They can accommodate several sizes of strip, typically 30 mm (1.18 in.), 45 mm (1.77 in.), 60 mm (2.36 in.), and 90 mm (3.54 in.) wide by 0.5 mm (0.020 in.) thick. The assemblies for parallel wire and strip electrode submerged arc welding are generally designed for mounting on standard welding heads with little or no modification.

The special submerged arc narrow groove welding equipment has long, narrow gun assemblies and long, narrow flux nozzles to deliver the flux and wire to the bottom of deep, narrow grooves. These systems may also have some means to bend the wire to assure good groove face fusion in the narrow groove. Simple SAW narrow-groove adapters can be mounted directly on standard welding heads. More complex integrated systems are available as complete welding head assemblies.

For semiautomatic submerged arc welding, the weld head may be a GMAW-type wire feeder that pushes the electrode through a conduit to the gun assembly. These wire feeders accept any of the previously mentioned drive roll systems and are generally capable of feeding wire up to 2.4 mm (3/32 in.) in diameter at wire feed speeds over 14 m/min (43 ft/min). The gun-conduit assembly allows for welding up to 4.6 m (14 ft) from the wire feeder. A typical semiautomated submerged arc welding system is shown in Figure 6.2.

FLUX DELIVERY SYSTEMS

Pneumatic (pressurized) flux feed is commonly used in semiautomated submerged arc welding and frequently in mechanized and automated submerged arc welding. A flux nozzle is usually mounted on the welding head to deposit the flux either slightly ahead of or concentric with the welding wire. For semiautomated operation, flux feed is provided either by a small 1.8 kilograms (kg) (4 pounds [lb]) gravity-feed flux hopper mounted on the gun or by a remote flux tank that uses compressed air to push the flux toward the weld zone. In both cases, the flux is delivered through the gun surrounding the welding wire. It may be necessary to include an auxiliary trailing or leading flux feed to eliminate arc flashing, which occurs if the flux does not feed properly. It is important to ensure that pneumatic systems have air filters and dryers to prevent moisture, oil, and other contaminants from entering the flux system.

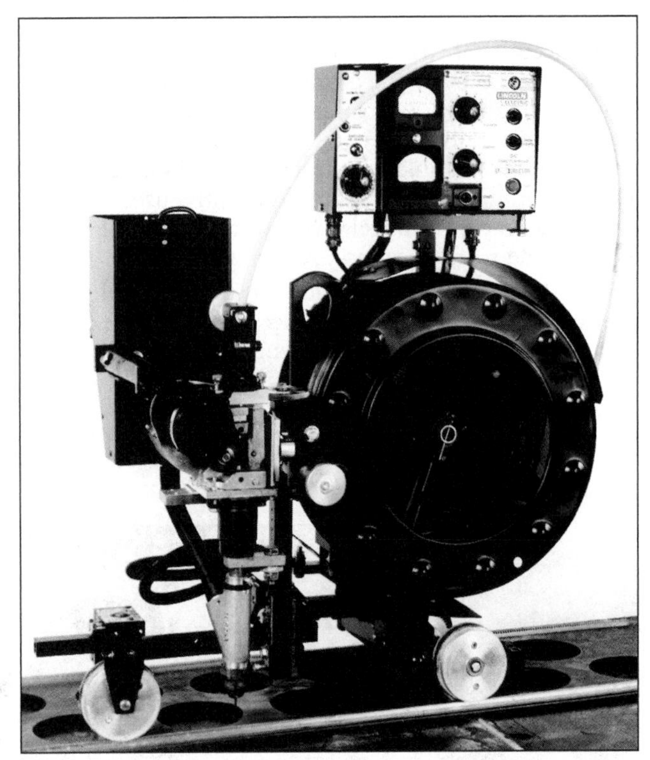

Photograph courtesy of The Lincoln Electric Company

Figure 6.9—Submerged Arc Welding Head, Control, Wire Supply and Flux Hopper Mounted on a Tractor Carriage

ACCESSORIES

Accessory equipment commonly used with submerged arc welding includes travel equipment, flux delivery and recovery equipment, baking ovens, and fixturing and positioning equipment.

Travel Equipment

Welding head travel in submerged arc welding is generally provided by a tractor carriage, a side-beam carriage, or a manipulator. Additional information about this equipment is presented in "Weldment Tooling and Positioning," Chapter 10 of *Welding Science and Technology*, Volume 1 of the *Welding Handbook*, 9th edition.[1]

A tractor carriage, as shown in Figure 6.9, provides travel along straight or slightly curved weld joints by

1. American Welding Society (AWS) Welding Handbook Committee, Jenney, C. L., and A. O'Brien, eds., 2001, *Welding Science and Technology*, Vol. 1 of *Welding Handbook*, 9th ed., Miami: American Welding Society.

riding on tracks set up along the joint or riding on the workpiece itself. Trackless units use guide wheels or some other type of mechanical joint-tracking device. The welding head, control, wire supply, and flux hopper are generally mounted on the tractor. Maximum travel speeds possible with tractors are about 2.5m/min (100 in./min.). Tractors are frequently used in field welding, where their relative portability is necessary because the workpiece cannot be moved.

Side-beam carriages provide linear travel only. They are capable of travel speeds in excess of 5 m/min (200 in./min.). Because side-beam systems are generally fixed and the workpiece must be brought to the weld station, their greatest use is for in-shop welding. The welding head, wire, flux hopper, and sometimes the controls are mounted on the carriage. Figure 6.10 shows two welding heads mounted on a single carriage for a cladding operation.

Manipulators, like side-beam carriages, are fixed, and the workpiece must be brought to the welding head. Manipulators are more versatile than side beams because they are capable of linear motion on three axes. The welding head, wire, flux hopper, and often the control unit and operator ride on the manipulator. Figures 6.11 and 6.12 show typical applications. In Figure 6.11, a positioner rotates the fixture beneath a stationary

Figure 6.10—Two Submerged Arc Welding Heads Mounted on a Side-Beam Carriage for a Cladding Operation

Figure 6.11—Positioner Rotates a Heavy-Wall Fixture Used to Test Oil Well Blowout Preventers

head, whereas in Figure 6.12, the welding head rides on a beam-mounted travel carriage.

Flux recovery units that collect the used flux for recycling during welding are frequently employed for economic reasons to maximize flux utilization and minimize manual cleanup. These systems can accomplish any combination of the following:

1. Remove unfused flux, small particles, and fused slag from behind the welding head,
2. Screen out fused slag and other oversized material,
3. Remove magnetic particles,
4. Remove fines (small dusty particles),
5. Recirculate flux back to a hopper for reuse, and
6. Heat flux in a hopper to keep it dry.

Moisture and oil must be removed from the compressed air in the flux recovery system, as these can contaminate the flux and can cause porosity in the deposited weld metal.

Rebaking Ovens

Flux must be kept dry and maintained at a consistent temperature. Flux should be dried in a properly vented

Photograph courtesy of The Lincoln Electric Company

Figure 6.12—Welding Head Rides on Beam-Mounted Travel Carriage

oven to assure that moisture is removed and consistent control of the overall flux temperature is maintained.

Positioners and Fixtures

Each operational mode of submerged arc welding requires that the workpiece be positioned in a manner that permits the flux and the weld pool to remain in place until they have solidified. Many types of fixtures and positioning equipment are available or can be fabricated to satisfy this requirement.

Because submerged arc welding is limited to nearly flat- or horizontal-position welding, positioners and related fixturing equipment are commonly used. Headtailstock units or turning rolls, or both, are used to rotate cylindrical workpieces under the welding head, as shown in Figure 6.13. Tilting-rotating positioners are used to bring the area to be welded on irregular workpieces into the flat position. As illustrated in Figure 6.14, a positioner rotates the workpiece beneath the welding head. Custom fixturing often includes positioners to aid in setting up, positioning, and holding the workpiece. Turnkey systems are available.

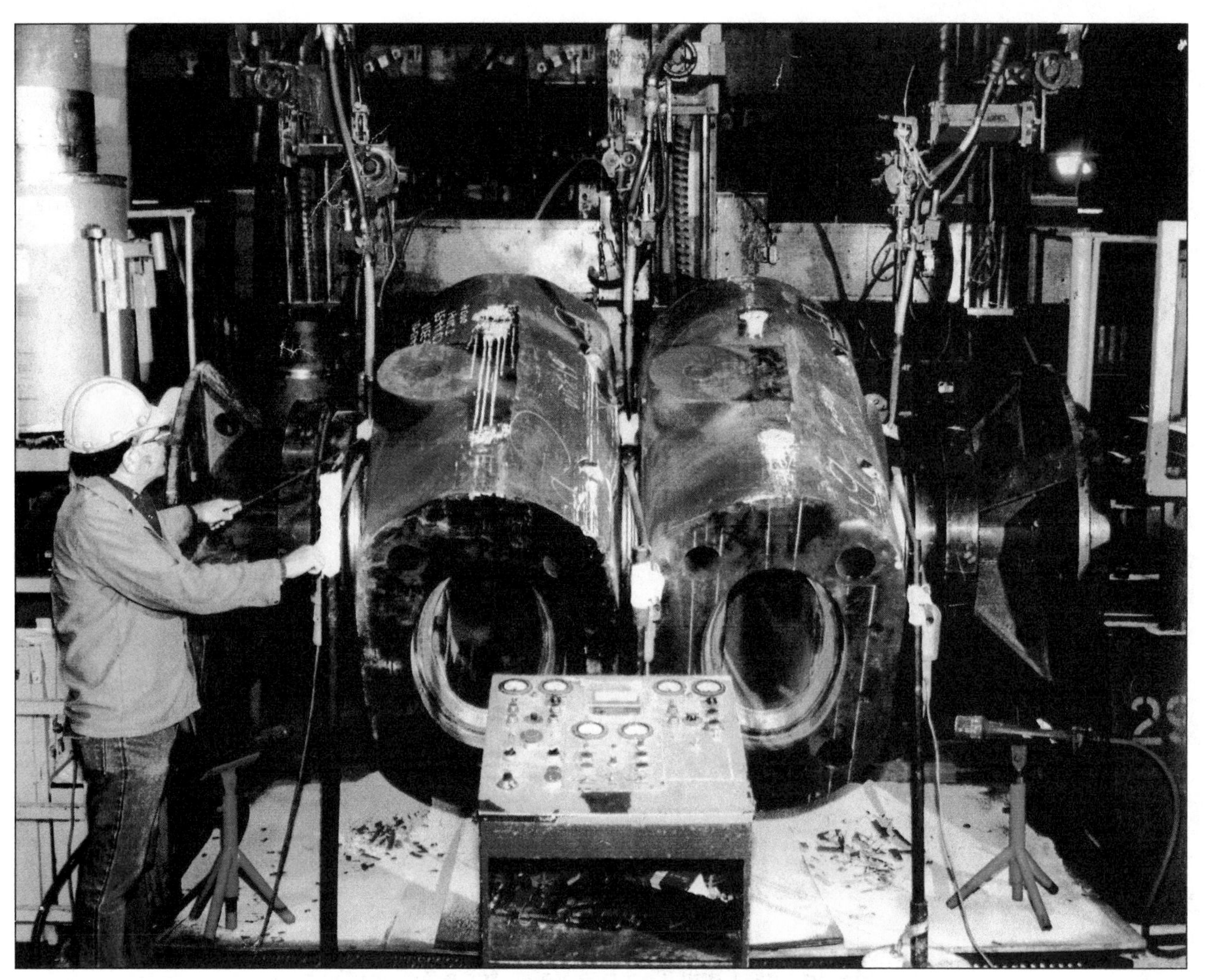

Figure 6.13—Three Circumferential Submerged Arc Welds Made Simultaneously on Heavy-Wall Forgings

The main purpose of fixturing is to hold the workpiece assembly in proper alignment during handling and welding. In order to maintain the shape of some assemblies, stiffening fixtures may be required. In addition, some type of clamping or fixturing may be required to hold the joint alignment for welding and to prevent warping and buckling from the heat of welding.

For assemblies that are inherently rigid, tack welding alone may suffice. The thicknesses of heavy sections offer considerable restraint against buckling and warping. In intermediate cases, a combination of tack welding, fixturing, and weld sequencing may be required. For joints of low restraint in light-gauge materials, clamping is needed. Clamping bars maintain alignment and remove heat to reduce or prevent warping. Tack welds are usually necessary.

Fixtures also include the tooling used to facilitate the welding operation. Weld seam trackers and travel carriages are used to guide mechanized or automatic welding heads. Turning rolls are used to rotate cylindrical workpieces during fitup and welding. Rotating turntables with angular adjustment are used to position weld joints in the most favorable position for welding. Manipulators with movable booms are used to position the welding head and sometimes the welding operator for hard-to-reach locations.

Figure 6.14—Multiple-Pass Deep-Groove Submerged Arc Weld Being Made on a Steam Chest and Throttle Valve for a Steam Turbine

MATERIALS

The submerged arc welding process is capable of welding most materials, from plain carbon steels to exotic nickel-based alloys. Most steels and alloys are readily weldable with commercially available wires and fluxes. However, some metals require special heats of electrode wire manufactured to precise chemistries and special fluxes designed to obtain specific weld joint properties.

BASE METALS

The submerged arc welding process is used to weld the following general classes of base metals, including weldable castings of each class:

1. Carbon steels,
2. Low-alloy steels,
3. Chromium-molybdenum steels,
4. Stainless steels, and
5. Nickel-based alloys.

The number of alloy compositions and castings that can be welded using this process has increased with the availability of suitable electrodes and fluxes.

ELECTRODES

Submerged arc electrodes, used in combination with the appropriate fluxes, produce weld deposits that match carbon steel, low-alloy steel, high-carbon steels, special alloy steels, stainless steels, nickel alloys, or special alloys for surfacing applications. The electrodes are supplied as bare solid wire or strip, composite metal-cored electrodes (similar to flux cored arc welding electrodes), and sintered strip. Electrode manufacturers prepare composite electrodes that duplicate complex alloys by enclosing the required alloying elements in a tube of more available composition (e.g., stainless steel or other metals).

Electrodes are packaged to ensure long shelf life when stored indoors under normal conditions. Electrodes are usually available in coils or drums ranging in weight from 11 kg to 454 kg (25 lb to 1000 lb). Large electrode packages are economical in that they increase operating efficiency and eliminate end-of-coil waste.

Steel electrodes are generally copper-coated, except those used for the welding of corrosion-resisting materials or for limited special applications. The copper coating provides a longer shelf life, decreases contact tip wear, and improves electrical conductivity.

Submerged arc welding electrodes normally vary in size from 1.6 mm to 6.4 mm (1/16 in. to 1/4 in.) in diameter. General guidelines for amperage range selection are presented in Table 6.1. It should be noted that wide amperage ranges are typical for submerged arc welding.

Table 6.1
Submerged Arc Welding Wires: Diameters versus Current Range

Wire Diameter		Current Range
mm	in.	Amperes
1.6	1/16	150–350
2.0	5/64	200–500
2.4	3/32	300–600
3.2	1/8	350–800
4.0	5/32	400–900
4.8	3/16	500–1200
5.6	7/32	600–1300
6.4	1/4	700–1600

FLUXES

Fluxes are granular, fusible mineral compounds mixed according to various formulations. The flux shields the weld pool from the atmosphere by covering the metal with molten slag (fused flux). Fluxes deoxidize and clean the weld pool, modify the chemical composition of the weld metal, and influence the shape of the weld bead and its mechanical properties.

Additional information on the topic of fluxes is provided in Annex A6 of *Specification for Carbon Steel Electrodes and Fluxes for Submerged Arc Welding,* ANSI/AWS A5.17/A5.17M-97T[2,3] and in Annex A6 of *Specification for Low-Alloy Steel Electrodes and Fluxes for Submerged Arc Welding,* ANSI/AWS A5.23/A5.23M: 1997.[4]

Classification

When classified according to the manner in which they modify the composition of the weld metal, fluxes can be categorized as neutral, active, or alloy fluxes. When classified according to their method of manufacture, the different types of fluxes are fused, agglomerated or bonded, mechanically mixed, and with restrictions, crushed slag. Additional flux categories have been created by various worldwide organizations with respect to the main components, chemical reactions, and the welding application for the specific purpose of identifying, categorizing, and classifying fluxes based on how they are manufactured.

Neutral Fluxes

Neutral flux is formulated to produce no significant modification in the chemical analysis of the weld metal as a result of a significant change in arc voltage and, thus, arc length.

The Wall Neutrality Number is a measure of flux neutrality and is applicable to multiple-layer fluxes. This number is derived by analyzing and assigning a value to the fluxes according to their manganese (Mn) or silicon (Si) content. Fluxes with a Wall Neutrality Number of 35 or lower that do not add appreciable amounts of manganese or silicon to the weld metal are considered neutral.

Active Fluxes (Custom Mixtures)

Active flux contains small amounts of manganese or silicon, or both, which is added to improve resistance to porosity and weld cracking, especially when welding oxidized base metal for high-speed single-pass welding.

2. American Welding Society (AWS) Committee on Filler Metals, 1997, *Specification for Carbon Steel Electrodes and Fluxes for Submerged Arc Welding,* ANSI/AWS A5.17/A5.17M-97, Miami: American Welding Society.
3. At the time of the preparation of this chapter, the referenced codes and other standards were valid. If a code or other standard is cited without a date of publication, it is understood that the latest edition of the document referred to applies. If a code or other standard is cited with the date of publication, the citation refers to that edition only, and it is understood that any future revisions or amendments to the code or standard are not included; however, as codes and standards undergo frequent revision, the reader is encouraged to consult the most recent edition.
4. American Welding Society (AWS) Committee on Filler Metals and Allied Materials, 1997, *Specification for Low-Alloy Steel Electrodes and Fluxes for Submerged Arc Welding,* ANSI/AWS A5.23/A5.23M: 1997, Miami: American Welding Society.

This type of flux is formulated to produce a weld metal composition that is different from the wire and dependent on the welding parameters, particularly arc voltage.

Alloy Fluxes

Alloy fluxes contain ingredients that can be used to add specific alloys to the weld metal. Alloy fluxes can provide a cost-effective method of changing the deposited alloy when using an inexpensive carbon steel or other alloyed electrode.

Fused Fluxes

A fused flux is manufactured by mixing the raw materials and fully melting them in an electric furnace. After melting, the furnace charge is poured and cooled. Cooling is accomplished by shooting the melt through a stream of water or by pouring it over large chill blocks. The result is a product with a glassy appearance that is crushed, screened to size, and packaged.

Fused fluxes offer the following advantages:

1. Good chemical homogeneity;
2. Easy removal of the fines without affecting the flux composition;
3. Less hygroscopic, which simplifies storage and handling;
4. Readily recycled through feeding and recovery systems without significant change in particle size or composition; and
5. Consistent metallurgical reactions.

The principal limitation associated with fused fluxes is the difficulty presented by the addition of de-oxidizers and ferroalloys during their manufacture without incurring segregation or extremely high losses. The high temperatures needed to melt the raw ingredients limit the range of flux compositions.

Agglomerated Fluxes and Bonded Fluxes

To manufacture an agglomerated or bonded flux, the raw materials are first measured, dry mixed, and bonded with potassium silicate or sodium silicate, or a mixture of the two. After bonding, the wet mix is formed into pellets and baked at a temperature lower than that used for fused fluxes. The pellets are then broken up, screened to size, and packaged.

The advantages of agglomerated or bonded fluxes include the easy addition of deoxidizers and alloying elements. Alloying elements are added as ferroalloys or elemental metals to produce alloys not readily available as electrodes or to adjust weld metal compositions.

Agglomerated or bonded fluxes can exhibit the following:

1. A tendency to absorb moisture in a manner similar to the coatings on some shielded metal arc electrodes;
2. Possible change in flux composition due to segregation or the removal of fine mesh particles and alloying elements used in custom-designed fluxes; and
3. Alloys transferred from the flux to the weld pool are welding-parameter dependent.

Mechanically Mixed Fluxes

To produce a mechanically mixed flux, two or more fused, agglomerated or bonded fluxes are mixed in any ratio necessary to yield the desired results. The advantage offered by mechanically mixed fluxes is that several commercial fluxes may be mixed for highly critical or proprietary welding operations. The disadvantages include the following:

1. Segregation of the combined fluxes may take place during shipment, storage, and handling;
2. Segregation may occur in the flux delivery and recovery systems during the welding operation; and
3. Inconsistency in the combined fluxes may be present from mix to mix and thus do not lend themselves to classification.

Crushed Slag

After solidification, slag resulting from submerged arc welding can be crushed and used as a welding flux, but even if virgin flux of the same type that created the slag is added, the result is a new and chemically different flux. It cannot be assumed that this resulting flux conforms to the classification of either component and it cannot be considered the same as virgin flux. It is the responsibility of the manufacturer (the crusher) to verify that the blend of crushed slag and the original brand of virgin flux is in conformance with classification requirements.

Flux Basicity

Basicity may be an additional classification of fluxes. Fluxes are similar to the coating on SMAW electrodes and are identified as chemically basic, chemically acid, or chemically neutral. The basicity or acidity of a flux is related to the ease with which the component oxides of the flux ingredients dissociate into a metallic cation and an oxygen anion. Chemically basic fluxes are normally high in manganese oxide (MgO) or calcium oxide (CaO), while chemically acid fluxes are normally high in silicon dioxide (SiO_2).

The basicity or acidity of a flux is often referred to as the ratio of CaO or MgO to SiO_2. Fluxes having ratios greater than one are considered chemically basic. Ratios near unity are chemically neutral. Ratios less than unity are chemically acidic.

Basic fluxes have become the prime fluxes for welding in critical applications in which precise control of deposit properties and chemistry is required. Most of the basic groups are formulated for specific wire deposits. They limit transfers of silicon/manganese/oxygen from the slag to the weld metal.

Fluxes independent of basicity are available to suit any material weldable by the submerged arc process when used in combination with an appropriate electrode. Further information is presented in Annex A6 of *Specification for Low-Alloy Steel Electrodes and Fluxes for Submerged Arc Welding*, ANSI/AWS A5.23/A5.23M: 1997.[5]

Storage and Handling

Plastic or paper bags for flux are designed to reduce the amount of moisture absorbed during shipping and storage when the bags of flux are completely closed. They are designed to help retard moisture penetration. However, when the bags are opened or if they become punctured, they should be kept in a closed container in order to ensure that the product maintains a low moisture content.

If the flux is exposed to humidity, it can be dried in a vented oven. Drying can often restore the flux to its original as-manufactured condition. A minimum drying temperature of 149°C (300°F) is recommended for fused submerged arc welding fluxes to assure that the moisture is reduced to its original levels. For agglomerated or bonded fluxes, a minimum of 260°C (500°F) is recommended. For agglomerated or bonded fluxes, it is important not to exceed 454°C (850°F) or the welding characteristics of the flux may be affected.

When drying the flux, it is important that the complete mass be brought to the recommended temperature. If the flux is held in large containers, this can take a very long time, as flux is an insulator. Reduction in moisture can be accomplished in as little as one hour if the flux is arranged in thin layers of, for example, 25 or 50 mm (1 or 2 in.) thick. The drying temperature and time are dependent on the amount of moisture contamination and the thickness of the flux bed being dried.

ELECTRODE-FLUX COMBINATIONS

In some cases, electrode and flux manufacturers produce electrode-flux combinations for submerged arc welding that are formulated to meet specific chemical and mechanical property requirements and weldability conditions. When composite electrodes or active fluxes are used, it is recommended that they always be purchased from the same manufacturer, whereas with solid electrodes, the fabricator can choose from among available fluxes formulated for use with a given AWS electrode classification.

When selecting electrode-flux combinations, it should be noted that both the electrode and flux, each in varying degrees, influence the chemistry of the deposited weld metal, which influences the mechanical properties. The flux, however, has the greatest influence on the overall weldability of the electrode-flux combination. Fluxes are classified on the basis of the chemical composition and mechanical properties of the deposited weld metal with some particular classification of electrode. The selection of submerged arc welding electrode-flux combinations depends on the chemical and mechanical properties required for the component being fabricated, the welding position (1G, 2G, 2F), and any surface preparation required of the material to be welded.

A number of factors must be considered when selecting submerged arc welding electrode-flux combinations. The chemical composition and mechanical properties of the base metal primarily affect the choice of the electrode, whereas fluxes should be considered as a shielding and process controlling medium. As previously explained, a neutral flux adds few or no alloying elements to the deposited weld metal, whereas an active flux adds alloying elements. Active fluxes are usually chosen for single-pass welding operations; their application for multipass use may be limited by engineering specifications because excessive alloy build-up may occur in the deposited weld metal.

The fluxes being considered must be correctly balanced in chemical composition for use with a given electrode classification. The mechanical properties required, including CVN impact properties as well as the strength and ductility of the resulting deposit, must be considered. Moreover, the factors determining the usability of a given electrode-flux combination—the wetting of groove or bevel faces without undercut or cold lap, the ability to weld over rust and scale, the ease of slag removal, and bead appearance, for example—should be evaluated.

Electrode-flux combinations are usually classified by code specifications. Data on special electrode/flux combinations for use with base metals that are not in common use can be obtained from flux manufacturers.

Electrodes and Fluxes for Use with Carbon Steels

Carbon steels are defined as steels that contain additions of carbon up to 0.29%, manganese up to 1.65%,

5. See Reference 4.

silicon up to 0.60%, and copper up to 0.60% with no specified range of other alloying elements. The standard *Specification for Carbon Steel Electrodes and Fluxes for Submerged Arc Welding,* ANSI/AWS A5.17/A5.17M-97,[6] prescribes the requirements for electrodes and fluxes for submerged arc welding of carbon steels. Solid electrodes are classified on the basis of chemical composition (as manufactured), while composite electrodes are based on deposit chemistry. Fluxes are classified on the basis of the weld metal properties obtained when used with specific electrodes, the heat treatment conditions under which these weld metal properties are obtained, and the chemical composition of the electrode. Table 6.2 explains the classification system for flux-electrode combinations.

The International Organization for Standardization (ISO/DIS 14171) publishes a counterpart of the ASME/AWS A5.17 and A5.23 documents classifying combinations of fluxes and electrodes for submerged arc welding in accordance with the European Committee for Standardization (CEN) standard EN756.[7]

Carbon steel materials are usually welded with the electrode and flux combinations classified in *Specification for Carbon Steel Electrodes and Fluxes for Submerged Arc Welding, ANSI/AWS* A5.17/A5.17M-97.[8] Typical steels that are welded with these consumables are listed as Group I and Group II classifications in *Structural Welding Code—Steel*, AWS D1.1.[9] These steels include ASTM A106 Grade B, A36, A516 Grades 55 to 70, A537 Class 1, A570 Grades 30 to 50, API 5LX Grades X42 to X52, and ABS Grades A to EH36. These steels are usually supplied in the as-rolled or the normalized condition.

Table 6.3 lists minimum mechanical properties for various electrode-flux combinations. When selecting these consumables for submerged arc welding, it is recommended that both the minimum tensile and minimum yield strengths as well as the notch toughness properties (when required) of the weld metal be matched with the base metal. The American Welding Society (AWS) publication *Filler Metal Comparison Charts*[10] shows the commercial products that meet the AWS wire-flux classifications listed in Table 6.3.

In special applications, particularly carbon steel weldments subject to long-term postweld heat treatment, the low-alloy submerged arc welding consumables covered by *Specifications for Low-Alloy Steel Electrodes and Fluxes,* ANSI/AWS A5.23/A5.23M: 1997, may be required to meet the tensile properties of the base metal.[11] This standard lists the welding electrodes and fluxes used with carbon steel base metals to meet special notch toughness requirements.

Preheat Requirements. Preheat, the act of bringing the material to a specified minimum temperature before welding begins, is used to achieve a variety of operating factors, including the following:

1. Slow the cooling rate,
2. Reduce the hardness of the weld metal and the HAZ,
3. Minimize the chance of delayed (hydrogen) cracking,
4. Reduce residual stresses,
5. Minimize distortion,
6. Assist in the diffusion of hydrogen, and
7. Achieve a desired microstructure or avoid an undesired microstructure.

It should be noted that these operating factors are not exclusive of one another. Temperature requirements for the material under consideration are usually addressed in the applicable codes or other standards. Standards are most often written in terms of minimums and therefore do not preclude the use of sound engineering judgment if a more stringent requirement is necessary. As an example, it must be recognized that certain joint designs and high-strength welding electrodes and fluxes result in high residual stress when used on carbon-steel base materials and may require higher preheats to avoid delayed (hydrogen) cracking.

Interpass Temperature Requirements. In most instances, the interpass minimum is merely a continuance of the minimum preheat. As with the preheat, sound engineering judgment may indicate a more stringent requirement than that specified in the applicable code or standard.

The maximum interpass temperature for welding carbon steels is typically limited to 260°C (500°F). Studies have shown that as preheat and maximum interpass temperatures increase, the width of the heat-affected zone (HAZ) increases, and the tensile strength and the notch toughness properties of both the deposited weld metal and the heat-affected zone decrease. Lower maximum interpass temperatures may be specified when special mechanical properties must be obtained, such as minimum notch toughness properties at low temperatures.

6. See Reference 2.
7. International Organization for Standardization (ISO), Geneva 20, Switzerland, and European Committee for Standardization, Brussels B1050, Belgium.
8. See Reference 2.
9. American Welding Society Committee on Structural Welding, *Structural Welding Code—Steel*, AWS D1.1, Miami: American Welding Society.
10. American Welding Society Committee on Filler Metals and Allied Materials. 2000, *Filler Metal Comparison Charts, FMC:2000,* Miami: American Welding Society.

11. See Reference 4.

Table 6.2
Mandatory Classification Designators for Flux-Electrode Combinations*

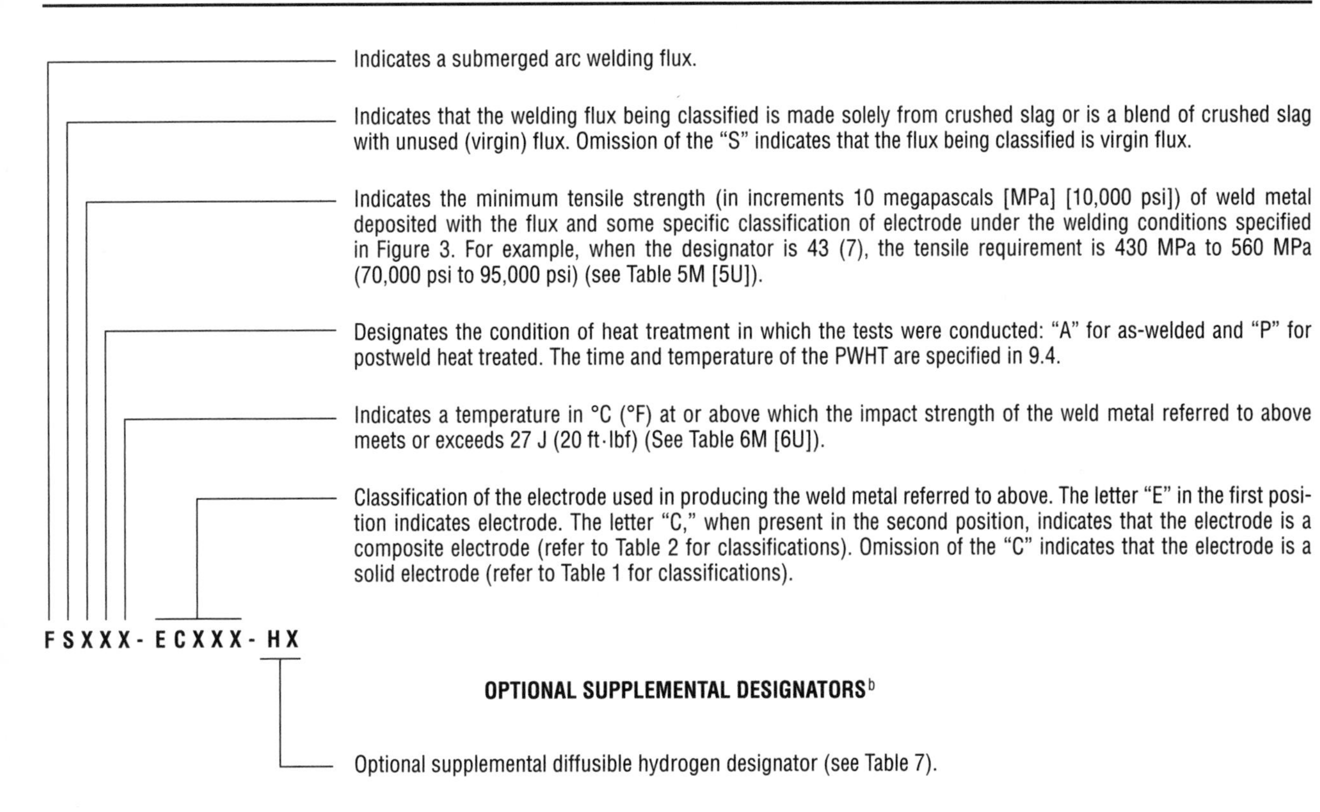

Notes:
(a) The combination of these designators constitutes the flux-electrode classification.
(b) These designators are optional and do not constitute a part of the flux-electrode classification.

Examples

F7A6-EM12K is a complete designation for a flux-electrode combination. It refers to a flux that will produce weld metal which, in the as-welded condition, will have a tensile strength of 430 MPa to 560 MPa (70,000 psi to 95,000 psi) and Charpy V-notch impact strength of at least 27 J at –20°C (20 ft·lbf at –60°F) when produced with an EM12K electrode under the conditions called for in this specification. The absence of an "S" in the second position indicates that the flux being classified is a virgin flux.

F7P4-EC1 is a complete designation for a flux-composite electrode combination when the trade name of the electrode used in the classification is indicated as well [see 17.4.1(3)]. It refers to a virgin flux that will produce weld metal with that electrode which, in the postweld heat treated condition, will have a tensile strength of 430 MPa to 560 MPa (70,000 psi to 95,000 psi) and Charpy V-notch energy of at least 27 J at –60°C (20 ft·lbf at –40°F) under the conditions called for in this specification.

*Table, figure, and paragraph numbers refer to American Welding Society (AWS) Committee on Filler Metals, 1997, *Specification for Carbon Steel Electrodes and Fluxes for Submerged Arc Welding*, ANSI/AWS A5.17/A5.17M-97, Miami: American Welding Society.

For further information, see American Welding Society (AWS) Committee on Filler Metals, 1997, *Specification for Low-Alloy Steel Electrodes and Fluxes for Submerged Arc Welding*, ANSI/AWS A5.23/A5.23M:1997, Miami: American Welding Society.

Source: Adapted from American Welding Society (AWS) Committee on Filler Metals, 1997, *Specification for Carbon Steel Electrodes and Fluxes for Submerged Arc Welding*, ANSI/AWS A5.17/A5.17M-97, Miami: American Welding Society.

Table 6.3
Minimum Mechanical Properties with the Carbon Steel Electrodes and Fluxes

AWS Classification	Welding Condition	Tensile Strength		Yield Strength		% Elongation in 50 mm (2 in.)	Charpy Impact Values		
		MPa	ksi	MPa	ksi		J	ft·lb	Test Temperature
F6A2-EL12	AW[a]	414	60	331	48	22	27	20	29°C (–20°F)
F6A6-EL12	AW	414	60	331	48	22	27	20	51°C (–60°F)
F7A2-EL12	AW	483	70	400	58	22	27	20	29°C (–20°F)
F6P4-EM12K	SR[b]	414	60	331	48	22	27	20	4°C (40°F)
F7A2-EM12K	AW	483	70	400	58	22	27	20	–7°C (20°F)
F7A6-EM12K	AW	483	70	400	58	22	27	20	16°C (60°F)
F7A2-EH14	AW	483	70	400	58	22	27	20	–7°C (20°F)

[a] AW = As-welded
[b] SR = Stress Relieved

Notes:
1. Actual mechanical properties obtained may significantly exceed the minimum values shown.
2. The type of welding flux (manufacturer) greatly influences Charpy V-notch impact properties of the weld metal.
3. Caution should be used when these weld deposits are stress relieved; they may fall below base metal strengths.
4. Test data is based on 25 mm (1 in.) thick plate (ASTM A 36 plate).
5. Stress-relieved condition (621°C [1150°F]) for 1 hour.

Source: Adapted from American Welding Society (AWS) Committee on Filler Metals and Allied Materials, 1997, *Specification for Carbon Steel Electrodes and Fluxes for Submerged Arc Welding*, ANSI/AWS A5.17/A5.17M-97, Miami: American Welding Society, Tables 5M and 5U, and 6M and 6U.

Electrodes and Fluxes for Use with Low-Alloy Steels

Low-alloy steels usually have less than 10% of any one alloying element. Low-alloy steel weld metal may be deposited using solid alloy steel electrodes, fluxes containing the alloying elements, and composite electrodes with cores containing the alloying elements. Alloy steel electrodes and composite electrodes are normally welded under a neutral flux. Alloy-bearing fluxes may also be used with carbon steel electrodes to deposit alloyed weld metal. Many electrode-flux combinations are available.

The standard *Specification for Low-Alloy Steel Electrodes and Fluxes for Submerged Arc Welding*, ANSI/AWS A5.23/A5.23M:1997[12] prescribes the requirements for solid and composite electrodes and fluxes for the welding of low-alloy steels. The fluxes are classified according to the weld metal properties obtained when used with specific electrodes. The required chemical compositions for the electrodes or deposits, or both, and other information are detailed in the latest edition of the specification.

Low-alloy steel materials are welded with electrode and flux combinations classified under ANSI/AWS A5.23. Low-alloy steels are divided into many subgroups. The chemical composition and mechanical properties of the steels to be welded determine the submerged arc welding consumables that should be selected to weld the joints. In order to meet service requirements, it may be essential that the composition of the weld metal be similar to that of the base metal in terms of alloying elements. This is particularly true for components that are to be used at temperatures higher than 345°C (650°F), where oxidation resistance and elevated tensile properties are important. Major groupings of low-alloy steels are described below.

High-Strength Low-Alloy Steels. High-strength low-alloy steels are steels with relatively low chemical additions (usually less than 1% of the chemical composition) of chromium, molybdenum, copper, nickel, niobium (columbium), and vanadium. These steels are usually supplied in the as-rolled, thermo-mechanically treated, normalized, or quenched-and-tempered condition from the manufacturer, depending on the material specification requirements. Steels that are welded with submerged arc welding electrodes and fluxes covered by ANSI/AWS A5.23[13] include ASTM specifications A 242,

12. See Reference 4.

13. See Reference 4.

A 533, A 537 Class 1 and 2, A 572 Grades 42-65, A588, and A 633 Grades A through E.

It must be noted that some of these steels can also be welded with electrodes and fluxes specified in ANSI/AWS A5.17/A5.17M-97.[14] The choice depends on the chemical and mechanical property requirements. ASTM A 242 and A 588 steels are resistant to rusting. Some applications require welding these metals with electrode-flux combinations that yield the same appearance and oxidation resistance as the base metal.

Some of the steels listed above can be produced in micro-alloyed versions that are generally available from Japan or Europe. The micro-alloy additions are amounts less than 0.1% of boron, niobium, titanium and vanadium. Many of the weldable micro-alloyed steels have tensile strengths up to approximately 552 MPa (80 ksi). Electrodes with a similar chemical composition should be selected and combined with a basic flux (hydrogen-controlled).

High Yield Strength Quenched-and-Tempered Low-Alloy Steels. High yield strength quenched-and-tempered low-alloy steels are similar to the high-strength low-alloy steels mentioned above, except that they have higher amounts of alloy additions. They may contain up to approximately 2% each of chromium, copper, nickel, niobium and vanadium. These steels are always supplied in the heat-treated condition.

Typical steels that are welded with submerged arc welding electrodes and fluxes specified in ANSI/AWS A5.23/A5.23M:1977[15] include ASTM A 514 and A 517. Many different grades of these steels with varying chemical compositions are available. The manufacturer of the steel should be consulted when welding these steels to determine which operating factors (heat input, preheat or postheat, and postweld heat treatment [PWHT]) are recommended. Selection of the appropriate electrode-flux combination depends on the thickness of the steels and the mechanical properties required, including notch toughness properties.

Carbon-Molybdenum Steels. Carbon-molybdenum steels are similar to the carbon steels except they have an addition of approximately 0.5% of molybdenum. These steels are used in pressure vessels or pipelines operating at elevated temperatures. They include ASTM A 204 Grades A, B, C, and ASTM A 182 Grade F1 forgings. Carbon-molybdenum steels are produced in the as-rolled or normalized conditions. After welding is completed, the weldments should usually be postweld heat-treated. Electrodes typically used are classifications EA1- through A4 and basic fluxes.

14. See Reference 2.
15. See Reference 4.

Chromium-Molybdenum Steels. Chromium-molybdenum steels contain varying amounts of chromium, to a nominal 9% chemical composition, and molybdenum to a nominal 1% chemical composition. These steels usually arrive from the steel manufacturer in the annealed, normalized and tempered, or quenched-and-tempered condition. These steels are also used in pressure vessels and pipelines operating at elevated temperatures. They include ASTM A 387 grades 2, 5, 7, 9, 91, 11, 12, 21, and 22 and ASTM A 182 forging grades F2, F5, F7, F9, F91, F11, F12, F21, and F22. Basic fluxes with low hydrogen potential should be used with electrodes classified as EB1 through EB9.

Other Alloy Steels. The consumables listed in *Specification for Low-Alloy Steel Electrodes and Fluxes for Submerged Arc Welding*, ANSI/AWS A5.23/A5.23M-1997, are used to weld a large number of nickel, nickel-molybdenum, nickel-chromium-molybdenum, and other steels.[16]

Fluxes and Electrodes for Use with Stainless Steels

The standard *Specification for Bare Stainless Steel Welding Electrodes and Rods*, ANSI/AWS A5.9,[17] prescribes filler metals for welding corrosion- or heat-resisting chromium and chromium-nickel steels. This specification includes steels in which chromium exceeds 4% and nickel does not exceed 50% of the composition. Solid wire electrodes are classified on the basis of the chemical composition, as manufactured, and composite electrodes on the basis of the chemical analysis of a fused sample. The American Iron and Steel Institute (AISI) numbering system is used for these alloys.

Submerged arc fluxes are available in fused agglomerated or bonded types for welding stainless alloys. Because of over-alloyed electrodes, fluxes for welding stainless steel are commonly neutral or nonalloyed. Some agglomerated or bonded fluxes contain alloys such as chromium, nickel, molybdenum, or niobium to compensate element loss across the arc. Those fluxes become depleted in compensating elements like chromium and can affect the deposited ferrite number (FN). The manufacturer's recommendation should be consulted regarding the handling and recycling of flux.

Stainless steels are capable of meeting a wide range of service needs such as corrosion resistance, strength at elevated temperatures, and toughness at cryogenic temperatures. They are selected for a broad range of

16. See Reference 4.
17. American Welding Society Committee on Filler Metals, *Specification for Bare Stainless Steel Welding Electrodes and Rods,* ANSI/AWS A5.9, Miami: American Welding Society.

applications. The stainless steels most widely used for welded industrial applications are classified as follows:

1. Martensitic,
2. Ferritic,
3. Austenitic,
4. Precipitation-hardening, and
5. Duplex (ferritic-austenitic).

The filler metals for fabricating these steels are specified in *Specification for Bare Stainless Steel Welding Electrodes and Rods*, ANSI/AWS A5.9.[18]

Not all stainless steels are readily weldable by the submerged arc process, and some require that special considerations be followed. In stainless steels and nickel base alloys, the main advantage of submerged arc welding is its high deposition rates. These high deposition rates sometimes become a disadvantage; as deposition rates increase, so does heat input, and in stainless alloys, high heat inputs may cause deleterious microstructural changes. Brief comments about each class of stainless steels and pertinent welding considerations are presented in the following sections. Metallurgical summaries are presented in *Materials and Applications—Part 2*, Volume 4 of the *Welding Handbook*, 8th edition.[19]

Martensitic Stainless Steels. The martensitic AISI 400-series stainless steels have 11.5% to 18% chromium as the major alloying element. AISI type 410, which may also be produced as a casting known as CA-15, has 11.5% to 13.5% chromium. Typical flux-wire combinations used to achieve an ER41NiMo deposit on steel mill rolls vary with the application. For joining, neutral or basic flux is used with ER410. For hardsurfacing, alloyed or customized flux is used with ER430.

The AISI 500-series (e.g., type 502 [B6] with 5 Cr-0.5 Mo, type 505 [B8] with 9 Cr-1 Mo) and type 505 (B9) with 9Cr-1Mo-V are heat-resisting steels, although not classed as stainless because their chromium content is well under the required 11% minimum. They are martensitic, nevertheless. These are described in *Specification for Stainless Steel Welding Electrodes for Shielded Metal Arc Welding*, ANSI/AWS A5.4,[20] and *Specification for Low-Alloy Steel Electrodes and Fluxes for Submerged Arc Welding*, ANSI/AWS A5.23/A5.23M: 1997.[21]

Martensitic stainless steels are not easily welded and require relevant heat control to avoid the development of brittle structure when rapidly cooled. Preheating the base metal retards the rate of cooling, thus reducing shrinkage stresses and allowing dissolved hydrogen to escape.

Ferritic Stainless Steels. The AISI 400 series also covers the ferritic stainless steels. As chromium increases beyond 18% in steels, the predominant metallurgical structure is ferrite, even at elevated temperatures, if the carbon content is low. Ferritic stainless steels are relatively nonhardening. Because they are highly magnetic, they are subject to arc blow during welding. If the heat of welding causes carbon and nitrogen to combine with chromium to form carbides and nitrides at grain boundaries, the result may be intergranular corrosion, although not to the extent experienced with austenitic stainless steels. Carbide precipitation is discussed in detail in *Materials and Applications—Part 2*, Volume 4 of the *Welding Handbook*, 8th edition.[22]

Because of embrittlement problems, the submerged arc welding of the ferritic stainless steels is considered difficult. When standard filler metals are not readily available, metal powder-composite stainless wire can be purchased on special order from several manufacturers. Austenitic filler metals 309, 310, and 312 are often used when the application can reconcile the different corrosion-resistant characteristics and the greater coefficients of linear expansion of an austenitic weld metal. When postweld annealing at 790°C (1450°F) is specified, the austenitic filler metal should be a stabilized grade or a low-carbon grade to avoid carbide precipitation.

Austenitic Stainless Steels. Austenitic stainless steels are essentially chromium-nickel alloys. They are covered by AISI 300 classifications. Corresponding AWS ER3XX filler metals or overmatching 300-series filler metals are used in over 90% of stainless steel welding applications. Such welds have good corrosion resistance and excellent strength at both low and high temperatures and are characterized by a high degree of toughness even in the as-welded condition. Being austenitic, these weldments are practically nonmagnetic.

Although the 300-series austenitic stainless steels are welded with greater ease than the 400 series, several factors peculiar to the 300-series stainless steels must be considered to ensure the production of satisfactory weldments. When compared with plain carbon steels, low-alloy and 400-series stainless steels, the austenitic stainless steels have lower melting points, higher electrical resistance, one-third the thermal conductivity, and as much as 50% greater coefficients of expansion. For

18. See Reference 17.
19. American Welding Society (AWS) Welding Handbook Committee, Oates, W. R., and A. Saitta, eds., 1998, *Materials and Applications—Part 2*, Vol. 4 of *Welding Handbook*, 8th ed., Miami: American Welding Society.
20. American Welding Society (AWS) Committee on Filler Metals and Allied Materials, *Specification for Stainless Steel Welding Electrodes for Shielded Metal Arc Welding*, ANSI/AWS A5.4, Miami: American Welding Society.
21. See Reference 4.
22. See Reference 19.

these reasons, less heat input (less current or higher travel speeds) is required for fusion, and the heat concentrates in a small zone adjacent to the weld. Greater thermal expansion may result in warping or distortion, especially in thin sections, suggesting greater need for tooling to maintain dimensional control.

The structure of austenitic stainless steel weld metals varies from fully austenitic in type 310 to dual phase austenitic-ferritic in types 308, 309, 312, and 316. Some ferrite is desired in such weldments for crack resistance. Techniques for measuring the ferrite number of stainless steel weld metals are described in *Standard Procedures for Calibrating Magnetic Instruments to Measure the Delta Ferrite Content of Austenitic and Duplex Ferritic-Austenitic Stainless Steel Weld Metal*, ANSI/AWS A4.2M/A4.2.[23] The ferrite number should exceed 3 to ensure adequate crack resistance and good low-temperature impact strength. A ferrite number under 8 may prevent the formation of sigma phase when the weldment is exposed to temperatures of approximately 650°C (1200°F).

Precipitation-Hardening Stainless Steels. The precipitation-hardening (PH) stainless steels are a group of iron-chromium-nickel alloys with additives such as copper, molybdenum, niobium, titanium, and aluminum. Precipitation hardening results when the metallurgical structure of a super-cooled solid solution changes on aging (solution-annealing). The advantage of precipitation-hardening steel is that components can be fabricated in the annealed condition and subsequently hardened (strengthened) by treatment at 480°C to 590°C (900°F to 1100°F), minimizing the problems associated with high-temperature quenching. Strength levels up to 1800 MPa (260 ksi) can be achieved (exceeding the strength of martensitic stainless steels), with corrosion resistance similar to that of type 304.

Precipitation-hardening steels are categorized within three general groups: martensitic PH, semiaustenitic PH, and austenitic PH stainless steels. No AWS specifications for precipitation-hardening stainless steels have been issued. Proprietary electrodes are available for joining this unique group of steels.

Duplex Stainless Steels. Duplex stainless steels have a microstructure that is part austenitic and part ferritic. Many austenitic stainless steel grades are dual-phase, with ferrite levels from about 3 FN (in type 308) to over 28 FN (in type 312). The duplex grades have approximately 50% ferrite. They are low in carbon and usually contain about 0.12% to 0.20% nitrogen. Weldments in duplex stainless steels that do not contain nitrogen may have wholly ferritic heat-affected zones and may contain the brittle sigma phase. Type 312 is not a duplex grade because of its high carbon content.

Duplex stainless steels combine some of the better features of austenitic stainless steels (which are vulnerable to stress-corrosion cracking in chloride environments) and ferritic stainless steels (which are brittle). Compared with lower ferrite austenitic grades, duplex austenite-ferrite filler metal grades exhibit higher strength (more than twice the yield strength), dramatically better resistance to stress corrosion cracking in chloride solutions, but lower ductility and lower toughness than the 300-series stainless steels.

Duplex stainless steel filler metals (ER2209/ER2553) have been successfully used in many applications. These filler metals are intentionally rich in austenitizers, approximately 9% nickel and more than 0.12% nitrogen. For some applications, nickel-based filler metal like E/ER NiCrMo-3 can be used.

Nickel and Nickel-Alloy Electrodes and Fluxes

Nickel and nickel-alloy electrodes in wire or strip form are available for submerged arc welding. The standard *Specification for Nickel and Nickel Alloy Bare Welding Electrodes and Rods*, ANSI/AWS A5.14/A5.14M-97,[24] covers nickel and nickel-alloy filler metals. The electrodes are classified according to their chemical compositions as manufactured. Detailed information on these electrodes is available in this specification. Fluxes for the submerged arc welding of nickel and nickel-alloy base metals are internationally classified and available. Flux manufacturers should be consulted for recommendations.

In many applications, nickel and nickel-alloy electrode-flux combinations are used for joining dissimilar metals (low-alloy steel to stainless steel) or for joining 5% to 9% nickel steels in liquefied natural gas (LNG) applications.

Nickel-alloy steels contain between 1% and 9% nickel. These steels are used in low-temperature applications (below –46°C [–50°F]) because they have good notch toughness properties at low temperatures. These steels include ASTM A203 Grades A, B, C, D, E, and F and ASTM A350 Grades LF3, LF5, and LF9.

When electrode-flux combinations are selected, the important properties that should be matched are the minimum tensile strength and the minimum notch

23. American Welding Society (AWS) Committee on Filler Metals and Allied Materials, *Standard Procedures for Calibrating Magnetic Instruments to Measure the Delta Ferrite Content of Austenitic and Duplex Ferritic-Austenitic Stainless Steel Weld Metal*, ANSI/AWS A4.2M/A4.2, Miami: American Welding Society.

24. American Welding Society (AWS) Committee on Filler Metals and Allied Materials, *Specification for Nickel and Nickel Alloy Bare Welding Electrodes and Rods*, ANSI/AWS A5.14/A5.14M-97, Miami: American Welding Society.

toughness of the base metal. It is usually difficult to produce impact values exceeding 30 joules (J) (20 ft lb-f) at –101°C (–100°F) in weldments produced with submerged arc welding.

PROCESS VARIABLES

Control of the operating variables in submerged arc welding is essential if high production rates and welds of consistent quality are to be achieved. These variables include the following:

1. Welding current,
2. Type of flux and particle distribution,
3. Welding voltage,
4. Travel speed,
5. Electrode type and size,
6. Electrode extension, and
7. Width and depth of the layer of flux.

The operator must know how the variables affect the welding action and what changes should be made to them. It is of utmost importance to realize that these variables are interrelated. An adjustment in one parameter may make it necessary to adjust one or more of the others.

WELDING CURRENT

The welding current or amperage controls the deposition rate, the depth of penetration, and the amount of base metal melted. If the current is too high at a given travel speed, the depth of fusion or penetration may be excessive. The resulting weld may melt through the metal being joined. Excessively high current also leads to the waste of electrodes in the form of excessive reinforcement or overwelding. Overwelding increases weld shrinkage and causes greater distortion.

Conversely, if the welding current is too low, inadequate penetration or incomplete fusion may result. The effect of current variation is shown in Figure 6.15.

Adjusting the welding current produces two effects. Increasing the current increases penetration and melting rate and increases the consumption of flux. Decreasing to a welding current that is too low produces an unstable arc.

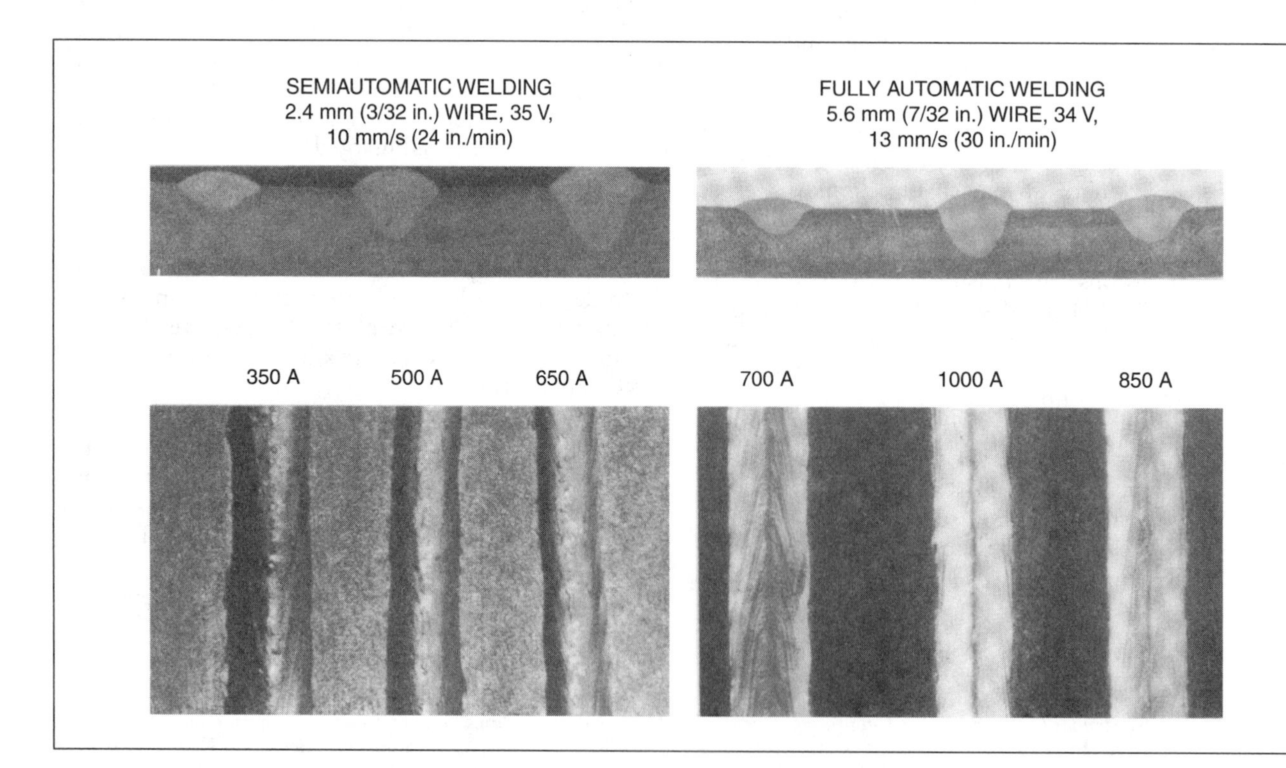

Figure 6.15—Effect of Amperage Variation on Weld Bead Shape and Penetration

WELDING VOLTAGE

Welding voltage adjustments vary the length of the arc between the electrode and the weld pool. If the overall voltage is increased, the arc length increases; if the voltage is decreased, the arc length decreases.

Voltage has little effect on the electrode deposition rate, which is determined by welding current. The voltage mainly determines the shape of the weld bead cross section and its external appearance. Figure 6.16 illustrates this effect.

Increasing the constant-current welding voltage and the travel speed may have the effect of producing a wider concave bead, increasing flux consumption, increasing porosity caused by rust or scale on steel, and increasing pickup of alloying elements from an alloy flux.

Excessively high arc voltage may produce the following:

1. A wide bead shape that is subject to cracking,
2. Difficult slag removal in groove welds,
3. A concave-shaped weld that may be subject to cracking, and
4. Increased undercut along the edges of fillet welds.

Lowering the voltage produces a forceful, stiff arc that improves penetration in a deep weld groove and also resists arc blow. An excessively low voltage produces a high, narrow bead and causes difficult slag removal along the bead edges.

TRAVEL SPEED

With any combination of welding current and voltage, the effects of changing the travel speed conform to a general pattern. If the travel speed is increased, power or heat input per unit length of weld is decreased, and less filler metal is applied per unit length of weld, resulting in less weld reinforcement. Thus, the weld bead becomes smaller, as shown in Figure 6.17.

Weld penetration is affected more by travel speed than by any variable other than current. This is true except for excessively slow speeds when the weld pool is beneath the welding electrode. Then the penetrating force of the arc is cushioned by the molten metal. Excessive speed may cause undercutting.

Within limits, travel speed can be adjusted to control weld size and penetration. In these respects, it is related to current and the type of flux. Excessively high travel

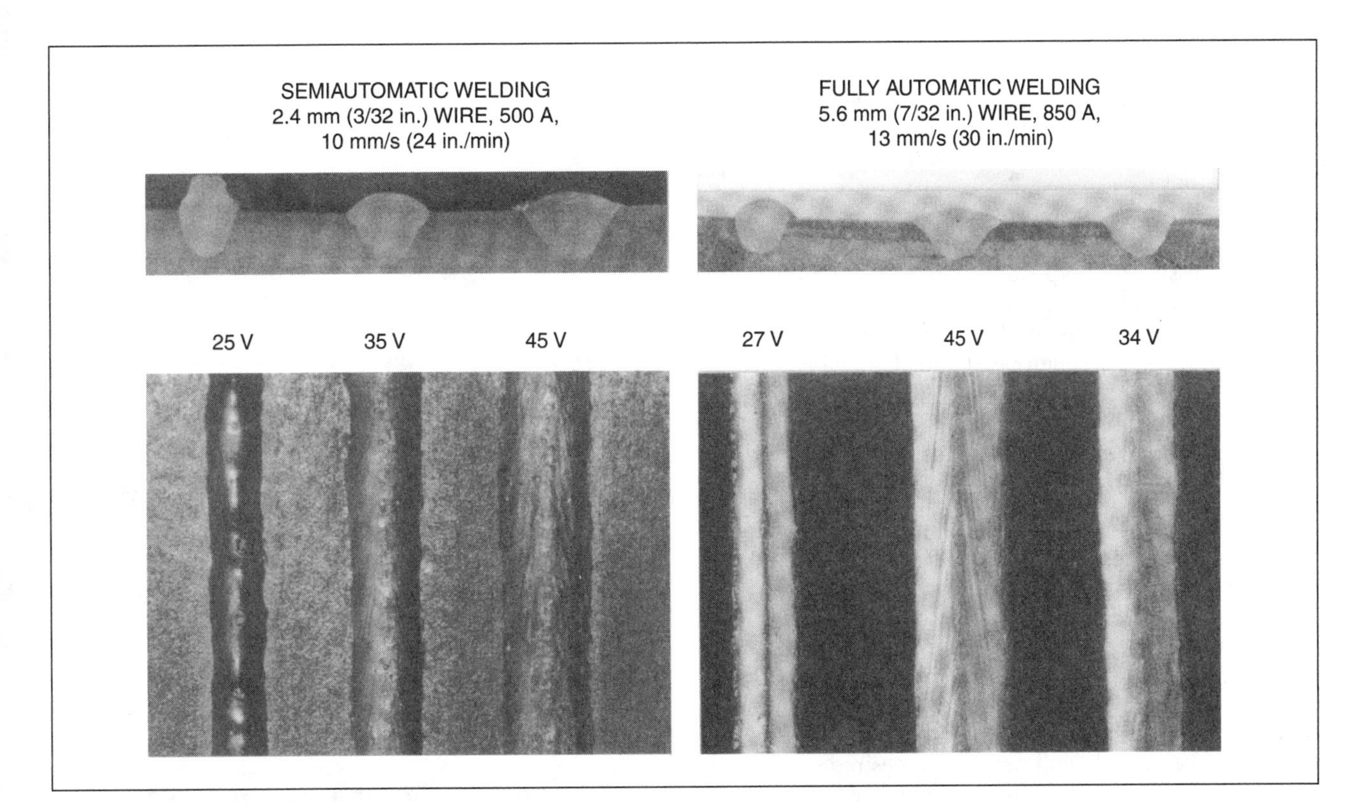

Figure 6.16—Effect of Arc Voltage Variations on Weld Bead Shape and Penetration

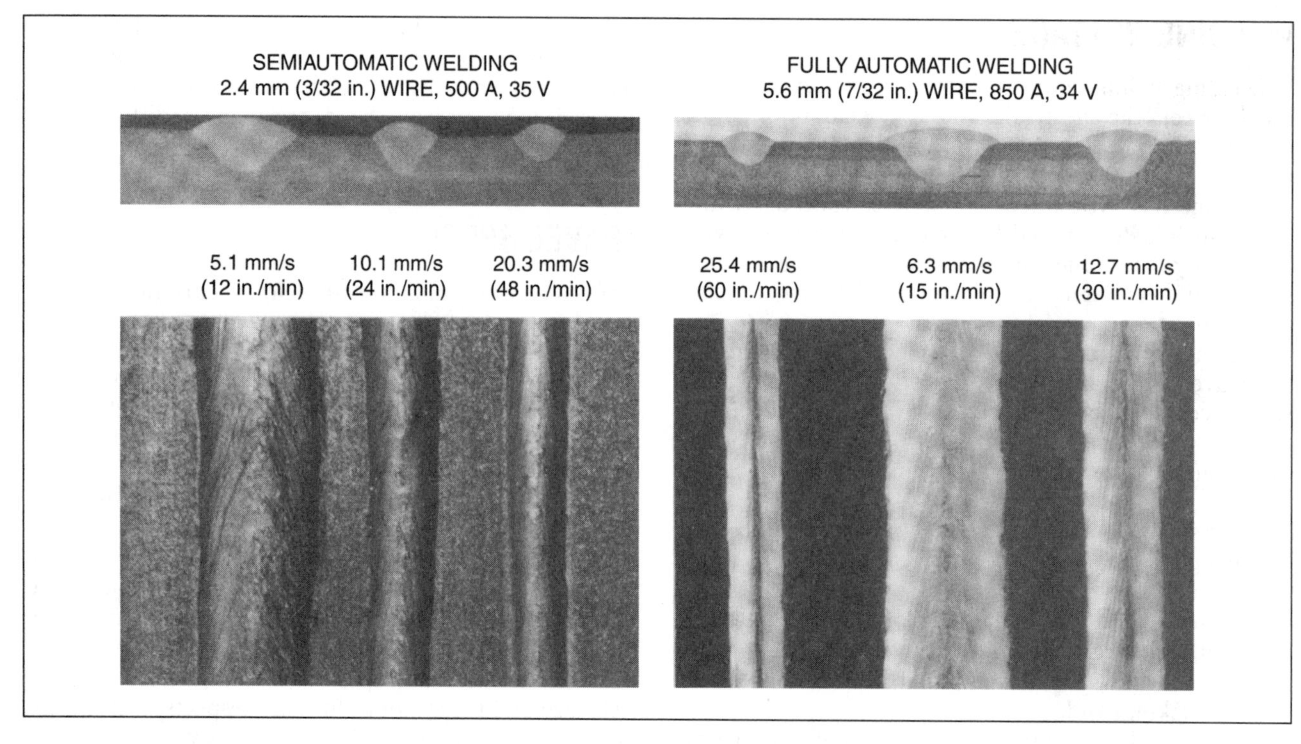

Figure 6-17—Effect of Travel Speed Variation on Weld Bead Shape and Penetration

speeds promote undercut, arc blow, porosity, and an uneven bead shape. Relatively slow travel speeds provide time for gases to escape from the weld pool, thus reducing porosity. Excessively slow speeds may produce the following:

1. A bead shape that is subject to uneven wetting;
2. Excessive arc exposure, which is uncomfortable for the welding operator; and
3. A large weld pool that flows around the arc, resulting in a rough bead and slag inclusions.

ELECTRODE SIZE

Electrode size also influences the deposition rate. At any given current, a small-diameter electrode will have a higher current density and a higher deposition rate than a larger electrode. However, a larger-diameter electrode can carry more current than a smaller electrode and produce a higher deposition rate at higher amperage. If a desired electrode feed rate is higher (or lower) than the feed motor can maintain, changing to a larger (or smaller) electrode will permit the desired deposition rate. Electrode size also affects weld bead shape and penetration, as shown in Figure 6.18.

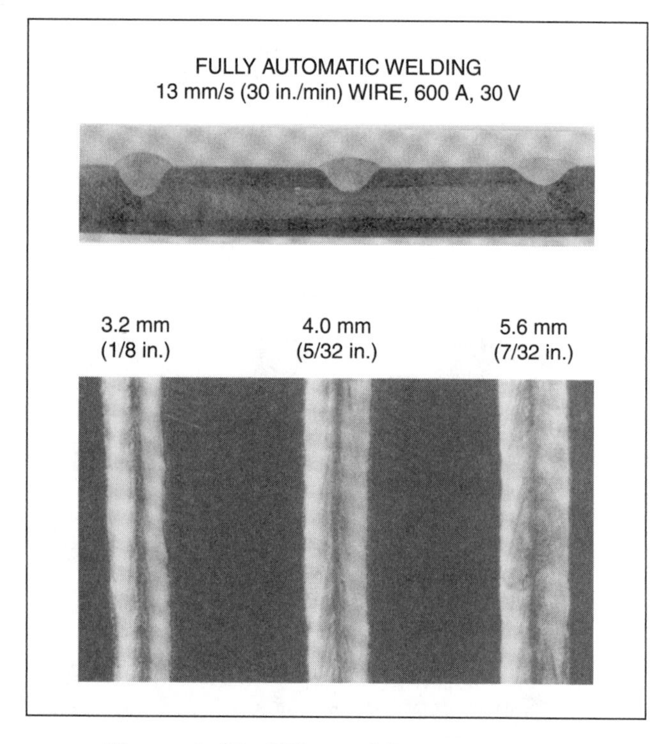

Figure 6.18—Effect of Electrode Size on Weld Bead Shape and Penetration

ELECTRODE EXTENSION

Electrode extension is the length the electrode extends beyond the contact tip. In developing a procedure, an electrode extension of approximately eight times the electrode diameter is a good starting point. As the procedure is developed, the length is modified to achieve the optimum electrode melting rate with fixed amperage. The condition of the contact tip influences the effective electrode extension. Contact tips should be replaced at predetermined intervals to ensure consistent welding conditions.

Increased electrode extension adds a resistance element in the welding circuit and consumes some of the energy previously supplied to the arc. That is, an increased voltage drop occurs in the electrode between the contact tip and the arc. With lower voltage across the arc, bead width and penetration decrease, as shown in Figure 6.16. Additionally, decreases in the arc voltage promote a convex bead. Therefore, when the electrode extension is increased to take advantage of the higher melting rate, the voltage setting on the power source voltmeter should be increased to maintain proper arc length and bead shape.

Other advantages of longer electrode extension include lower heat input, higher impact properties, narrower HAZ, decreased penetration when welding thin sections or welding over root passes, and lower dilution levels. However, care must be exercised with the decreased penetration to avoid slag inclusions.

Electrode extension is an important variable at high current densities. Resistance heating of the electrode between the contact tip and the arc increases the melting rate of the electrode. The longer the extension, the greater is the amount of heating and the higher the melting rate. This resistance heating is commonly referred to as I^2R heating, where I denotes the current expressed in amperes, and R denotes the resistance expressed in ohms.

Deposition rates can be increased from 25% to 50% by using long electrode extensions with no change in welding amperage. With single-electrode (wire or strip) automated submerged arc welding, the deposition rate may approach that of the two-wire method with two power sources.

An increase in deposition rate is accompanied by a decrease in penetration, therefore changing to a long electrode extension is not recommended when deep penetration is needed. When melt-through (visible root reinforcement in a joint welded from one side) is a problem, as may be encountered when welding thin-gauge material, increasing the electrode extension may be beneficial. However, as the electrode extension increases, it becomes more difficult to maintain the electrode tip in the correct position relative to the joint. It is also difficult to predict the response of the resistance-heated wire under the flux cover if extensions approaching the maximum values listed below are desired. Nonconducting guides that attach to the contact tip become necessary to control the wire.

Following are suggested maximum electrode extensions for solid steel electrodes for submerged arc welding:

1. 75 mm (3 in.) for 2.0 mm, 2.4 mm, and 3.2 mm (5/64 in., 3/32 in., and 1/8 in.) wire electrodes; and
2. 125 mm (5 in.) for 4.0 mm, 4.8 mm, and 5.6 mm (5/32 in., 3/16 in., and 7/32 in.) wire or 30 mm, 60 mm, and 90 mm (1-1/5 in., 2-2/5 in., and 3-1/2 in.) strip electrodes.

FLUX BURDEN

The width and depth of the layer of granular flux influence the welding action as well as the appearance and soundness of the finished weld. If the granular layer is too deep, the arc becomes confined. Thus, the gases generated during welding cannot readily escape and the surface of the weld pool becomes irregularly distorted, resulting in a weld with a rough rope-like appearance.

If the granular layer is too shallow, the arc fails to submerge entirely in the flux, producing flashing and spattering. The resulting weld may have a poor appearance, and it may contain porosity. Figure 6.19 shows the effects on weld bead surface appearance of proper and shallow depths of flux.

CORRECT DEPTH

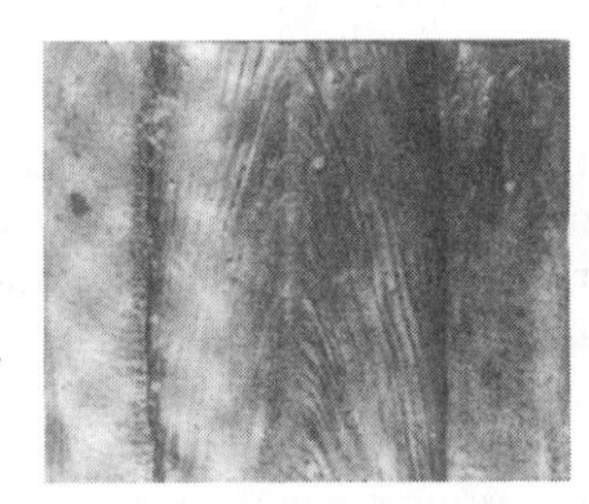

19 mm (3/4 in.) DEPTH
16 mm (5/8 in.) PLATE
RESULT: SMOOTH TOP; SOUND WELD STRUCTURE

TOO SHALLOW

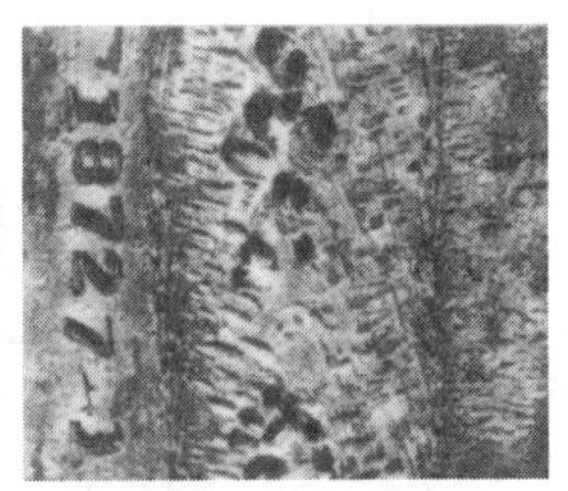

6 mm (1/4 in.) DEPTH
16 mm (5/8 in.) PLATE
RESULT: GAS POCKETS IN WELD; OPEN ARCING OCCURRED

Figure 6.19—Effect of Proper and Improper Depth of Flux on Weld Appearance

An optimum depth of flux exists for any set of welding conditions. This depth can be established by slowly increasing the flow of flux until the welding arc is submerged and flashing no longer occurs. The gases will then puff up quietly around the electrode, sometimes igniting.

During welding, the unfused granular flux can be removed a short distance behind the welding zone after the fused flux has solidified. However, it may be best not to disturb the flux until the heat from welding has been evenly distributed throughout the section thickness.

Molten slag should not be forcibly loosened while the weld metal is at a high temperature. Allowed to cool, the slag will readily detach itself. Then it can be brushed away with little effort. If needed, a small section may be forcibly removed for quick inspection of the weld surface appearance.

It is important that no foreign material be picked up when reclaiming the flux. To prevent this, a space approximately 300 mm (12 in.) wide should be cleaned on both sides of the weld joint before the flux is applied. If the recovered flux contains fused pieces, it should be filtered through a screen with openings no larger than 3.2 mm (1/8 in.) to remove the coarse particles.

Flux is thoroughly dry when packaged by the manufacturer. After exposure to high humidity, it should be dried again by baking before use, as moisture in the flux may cause porosity in the weld. The manufacturer's recommendations should be followed.

JOINT DESIGN AND EDGE PREPARATION

The types of joints appropriate for submerged arc welding are mainly butt joints, T-joints, and lap joints, although edge and corner joints can also be welded. The principles of joint design and the methods of edge preparation required for submerged arc welding are similar to those used for other arc welding processes. Typical welds, described in greater detail in the section titled "Weld Types," include fillet, square groove, single- and double-V-groove, and single- and double-U-groove welds.

Joint designs, especially for plate welding, often call for a root opening of 0.8 mm to 1.6 mm (1/32 in. to 1/16 in.) to prevent angular distortion or cracking due to shrinkage stresses, although root openings are most often dictated by the necessities of the process used in the root. A root opening that is larger than that required for proper welding may increase welding time and costs. This is true for both groove and fillet welds.

Edge preparation can be accomplished by means of any of the thermal cutting methods or by machining. The accuracy of edge preparation is important, especially for mechanized or automated welding. For example, if a joint designed with a 6.4 mm (1/4 in.) root face were produced with a root face that tapered from 7.9 mm to 3.2 mm (5/16 in. to 1/8 in.) along the length of the joint, the weld might be unacceptable because of incomplete penetration at the start and excessive melt-through at the end. In such a case, the capability of the cutting equipment, as well as the skill of the operator, should be checked and corrected if necessary.

STARTING WELD TABS AND RUNOFF WELD TABS

For longitudinal welding, it is necessary to provide a means of supporting the weld metal, flux, and molten slag to assure consistent weld shape at the start and finish of the weld and to prevent spillage. Starting weld tabs (also called *run-on tabs*) and runoff tabs are commonly used. The arc is initiated on a starting weld tab that is tack welded to the start end of the weld. It is terminated on a runoff tab at the finish end of the weld. The tabs are large enough to assure that the weld metal on the workpiece itself is properly shaped at the ends of the joint.

When the tabs are prepared, the groove in the tab should be similar to the groove in the workpiece. The groove must be wide enough to support the flux. The removal of starting tabs and runoff tabs or other temporary attachments should be performed by a method that does not adversely affect the properties of the base metal and the weld deposit.

OPERATING PROCEDURES

Operating procedures and techniques that contribute to successful submerged arc welding include joint fitup and weld backing, welding position, power source and workpiece connections, arc initiation and termination, and electrode position. Procedures for welds in plate and for circumferential welds are discussed in this section.

JOINT FITUP

Joint fitup is an important part of the assembly or subassembly procedure. Fitup can materially affect the quality, strength, and appearance of the finished weld. When welding plate, the deep penetration that is characteristic of the submerged arc process prompts the need for close control of fitup. Uniformity of joint alignment and of the root opening must also be maintained.

WELD BACKING

Submerged arc welding creates a large volume of molten weld metal that remains fluid for an appreciable period of time. This molten metal must be supported and contained until it has solidified. Several methods are commonly used to support molten weld metal when complete joint penetration is required. These include the following:

1. Backing strips,
2. Backing welds (root and cover passes),
3. Copper backing bars,
4. Flux backing, and
5. Flux/ceramic backing tapes.

In the first two methods, the backing may become a part of the completed joint. However, the removal of the backing may be specified, depending on the design requirements of the joint. The last three methods listed employ a temporary backing that is removed after the weld is completed.

In many joints, the root face is designed to be thick enough to support the first pass of the weld. This method can be used for butt welds (partial joint penetration), fillet welds, and plug or slot welds. Supplementary backing or chilling is sometimes used. It is most important that the root faces of groove welds be tightly butted at the point of maximum penetration of the weld.

Backing Strips

With this method, the weld penetrates into a backing strip and fuses with it, which temporarily or permanently becomes an integral part of the assembly. Backing strips must be compatible with the metal being welded. When the design permits, the joint should be located so that a part of the structure forms the backing. It is important that the contact surfaces are clean and close together; otherwise, porosity and leakage of molten weld metal may occur.

Backing Welds

In a joint backed by weld metal, the backing pass (root and cover pass) is usually made with another process such as flux cored arc welding, gas metal arc welding, or shielded metal arc welding. This backing pass forms a support for subsequent submerged arc weld passes made from the same or the opposite side. Manual, semiautomated, or mechanized welds are used as backing for submerged arc welds when alternate backing methods are not convenient because of inaccessibility, poor joint penetration or fitup, higher productivity in full penetration joints, or difficulty in turning the weldment.

The weld backing may remain as a part of the completed joint or may be removed by thermal gouging, chipping, grinding, or machining after the submerged arc weld has been completed. It may then be replaced by a permanent submerged arc surfacing bead.

Copper Backing Bars

With some joints, a copper backing bar, usually water-cooled, is used to support the weld pool but does not become a part of the weld. Copper is used because of its high thermal conductivity, which prevents the weld metal from fusing to the backing bar. When it is desirable to reinforce the underside of the weld, the backing bar can be grooved to the desired shape of the reinforcement. The backing bar must have enough mass to prevent it from melting beneath the arc, which would contaminate the weld with copper.

Caution must be used to prevent copper pickup in the weld, which may be caused by harsh arc initiation. Water is sometimes passed through the interior of the copper backing bar to keep it cool, particularly in high-production welding applications. Care must be exercised to prevent water condensation from forming on the copper backing bar.

Copper backing is sometimes designed to slide, so a relatively short length can be used in the vicinity of the arc and the weld pool. In other applications, the copper backing may take the form of a rotating wheel.

Flux Backing

Flux, under moderate pressure, can be used as backing material for submerged arc welds. In one method, an inflatable rubberized canvas fire hose is placed in a trough. Loose granular flux is placed on a thin piece of flexible sheet material and placed over the hose. The workpieces are positioned over the trough-flux assembly, as illustrated in Figure 6.20. The hose is inflated to a maximum of 35 kPa to 70 kPa (5 psi to 10 psi) to apply moderate flux pressure on the back side of the weld.

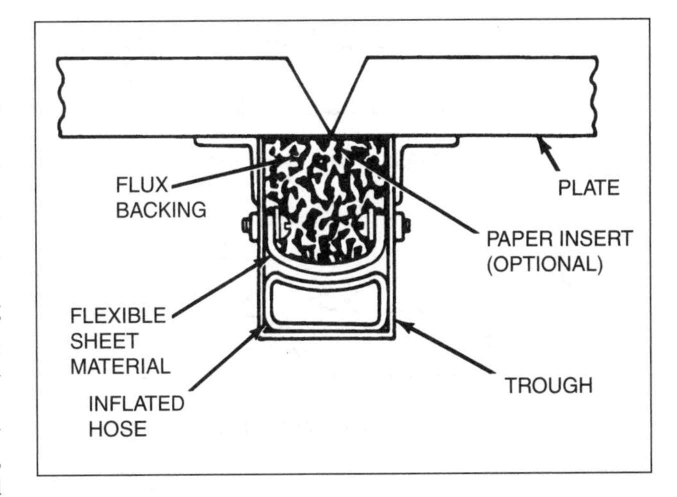

Figure 6.20—Air Pressure Method of Supporting Flux Backing for Submerged Arc Welding

A closed or open weld joint can also be taped with flux/ceramic backing tape to support the weld pool.

INCLINATION OF THE WORKPIECE

Figure 6.21 illustrates how the inclination of the workpiece during welding can affect the weld bead shape. Most submerged arc welding is done in the flat position, resulting in the bead shape shown in 6.21(A). However, it is sometimes necessary or desirable to perform the welding operation with the workpiece slightly inclined so that the weld progresses downhill or uphill. For example, in the high-speed welding of 1.3 mm (0.050 in.) steel sheet, a better weld results when the workpiece is inclined 15° to 18° and the welding is performed downhill. Inclining the workpiece in this manner results in less penetration than when the sheet is welded on a horizontal plane. To augment penetration, the angle of inclination should be decreased as plate thickness increases.

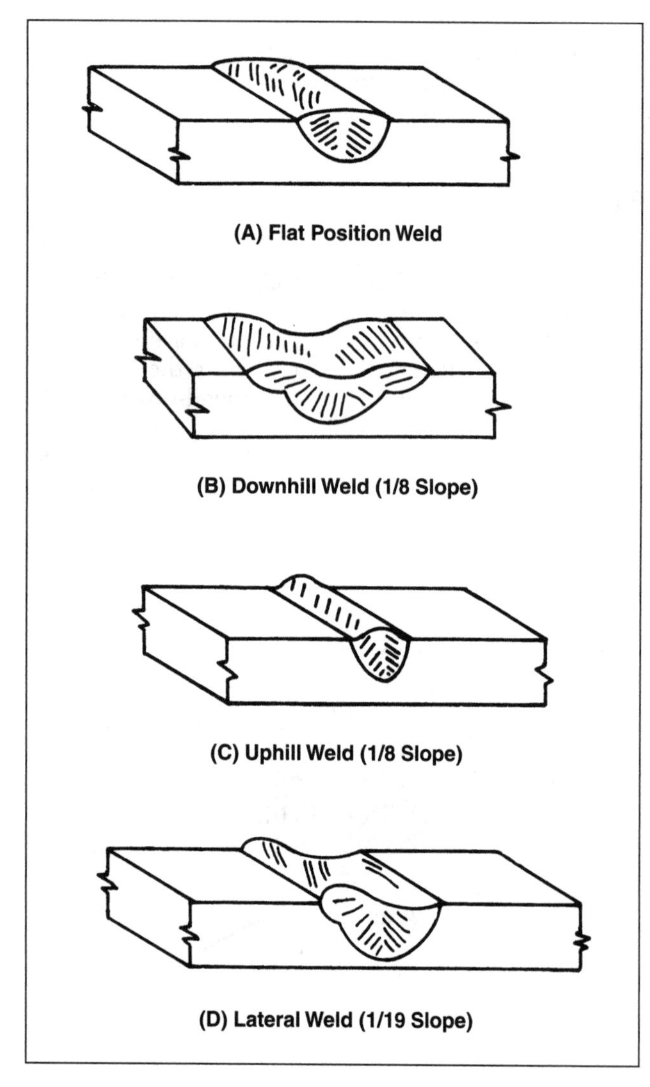

Figure 6.21—Effect of Work Inclination on Weld Bead Shape

As shown in Figure 6.21(B), downhill welding produces a wide, concave weld. The weld pool tends to flow under the arc and preheat the base metal, particularly at the surface. This produces an irregularly shaped fusion zone. As the angle of inclination increases, the middle surface of the weld is depressed, penetration decreases, and the width of the weld increases.

Uphill welding affects the fusion zone contour and the weld surface, as illustrated in 6.21(C). The force of gravity causes the weld pool to flow backward and lag behind the welding electrode. The edges of the base metal melt and flow toward the middle. As the angle of inclination increases, reinforcement and penetration increase, and the width of the weld decreases. Also, the larger the weld pool, the greater is the penetration and center buildup. These effects are exactly the opposite of those produced by downhill welding.

The limiting angle of inclination when welding uphill with currents up to 800 A is about 6°. When higher welding currents are used, the maximum workable angle decreases. Inclination greater than approximately 6° makes the weld uncontrollable. Lateral inclination of the workpiece produces the effects shown in Figure 6.21(D). The limit for a lateral slope is approximately 3°, although the permissible lateral slope varies somewhat, depending on the size of the weld pool.

POWER SOURCE AND WORKPIECE CONNECTIONS

The electric welding circuit consists of a conductor that connects the power source to the welding head and another conductor, called the *workpiece lead,* that connects the workpiece to the power source. The point at which the workpiece lead is attached to the workpiece is called the *workpiece connection.* The workpiece lead is sometimes incorrectly referred to as the *ground lead.* The workpiece may also be grounded to the building or earth, but for best results in submerged arc welding, a workpiece lead connection directly to the power source is required.

The method of attachment and the location of the workpiece lead connection are important considerations in submerged arc welding because they can affect the arc action, the quality of the weld, and the speed of welding. A poor workpiece lead location can cause or increase arc blow, resulting in porosity, incomplete penetration, and poor bead shape. Testing the location may be necessary, since it is often difficult to predict the effect of the workpiece lead location. Generally, the best direction of welding is away from the workpiece connection.

The welding current may tend to change slowly when welding long seams. This occurs because the path and electrical characteristics of the circuit change as the weld progresses. A uniform weld can frequently be obtained by attaching workpiece leads to both ends of the object being welded.

When the longitudinal seam of a light-gauge cylinder is welded in a clamping fixture with copper backing, it is usually best to connect the workpiece lead on the bottom of the cylinder at the start end. If this is not possible, the workpiece lead should be attached to the fixture at the start end.

It is undesirable to connect the workpiece lead to a copper backing bar because the welding current will enter or leave the work at the point of best electrical contact, not necessarily beneath the arc. If the current sets up a magnetic field around some length of the backing bar, arc blow may result. When the current return is directed through a sliding shoe, two or more shoes should always be used to prevent interruptions of the current.

ARC INITIATION

The method used to start the arc in a particular application depends on factors such as the time required for starting relative to the total setup and welding time, the number of pieces to be welded, and the importance of starting the weld at a particular place on the joint. Arc initiation methods are described below.

Sharp Wire Start

To implement the sharp wire start, the welding electrode, which protrudes from the contact tip, is snipped with wire cutters. This forms a sharp chisel-like configuration at the end of the wire. The electrode is then lowered until the end lightly contacts the workpiece. The flux is applied and welding commences. The chisel point melts away rapidly to start the arc.

Scratch Start

In the scratch start, the welding electrode is lowered until it is in light contact with the work, and the flux is applied. Next, the carriage is started and the welding current is immediately applied. The motion of the carriage prevents the welding wire from fusing to the workpiece.

Molten Flux Start

When a pool of molten flux is present, an arc can be started by simply inserting the electrode into the pool and applying the welding current. This method is regularly used in multiple-electrode welding. When two or more welding electrodes are separately fed into one weld pool, it is necessary to start only one electrode to establish the weld pool. The remaining electrodes will arc when they are fed into the pool.

Wire Retract Start

Although specially designed welding equipment must be used, wire retract arc starting is cost-effective when frequent starts are made and the starting location is important.

The first step involves moving the electrode downward until it lightly contacts the workpiece. When the end of the electrode is covered with flux, the welding current is turned on. The low voltage between the electrode and the workpiece signals the wire feeder to withdraw the tip of the electrode from the surface of the workpiece. An arc is initiated as this action takes place. As the arc voltage increases, the wire feed motor quickly reverses direction to feed the welding electrode toward the surface of the workpiece. The electrode feed speeds up until the electrode melting rate and arc voltage stabilize at the preset value.

If the workpiece is light-gauge metal, the electrode should make only light contact, consistent with good electrical contact. The welding head should be rigidly mounted. The end of the electrode must be clean and free of fused slag. Wire cutters can be used to snip off the tip of the electrode (preferably shaping a point) before each weld is made. The electrode size should be chosen to permit operation with high current densities, as they facilitate arc initiation.

High-Frequency Start

The high-frequency start requires special equipment but no manipulation by the operator other than closing a starting switch. It is particularly useful as a starting method for intermittent welding or welding at high production rates, when frequent starts are required.

When the welding electrode approaches within approximately 1.6 mm (1/16 in.) of the workpiece, a high-frequency, high-voltage generator in the welding circuit causes a spark to jump from the electrode to the workpiece. The spark produces an ionized path through which the welding current can flow, and the welding action begins. This method is generally used when the arc must be initiated at an exact starting point, as in strip cladding.

ARC TERMINATION

In some electrical systems, the travel and the electrode feed cease at the same time the "stop weld" button is pushed. Other systems stop the travel, but the electrode continues to feed for a controlled length of time. A third type of system reverses the direction of travel for a controlled length of time while welding continues. The latter two systems fill the weld crater.

ELECTRODE POSITION

In determining the proper position of the welding electrode, the following three factors must be considered:

1. The alignment of the welding electrode in relation to the joint;
2. The work angle (the angle of tilt in the lateral direction, e.g., the tilt in a plane perpendicular to the joint; and
3. The travel angle (the forward or backward direction in which the welding electrode points).

As the travel progresses in the forward direction, a forward-pointing electrode is one that makes an acute angle with the finished weld. A backward-pointing electrode makes an obtuse angle with the finished weld. Most submerged arc welds are fabricated with the electrode axis in a vertical position. Pointing the electrode forward or backward becomes important when multiple arcs are being used, while performing surfacing operations, and when the workpiece cannot be inclined. Pointing the electrode forward results in a weld configuration similar to that produced with downhill welding; pointing the electrode backward results in a weld similar to that produced with uphill welding. Pointing the electrode forward or backward does not affect the weld configuration as much as the uphill or downhill positioning of the workpieces.

When butt joints are welded between plates of equal thicknesses, the electrode should be aligned with the joint centerline, as shown in Figure 6.22(A). Improper alignment may cause incomplete penetration, as shown in Figure 6.22(B). When unequal thicknesses are butt welded, the electrode must be located over the thick section to melt it at the same rate as the thin section. Figure 6.22(C) shows this requirement.

When welding horizontal fillets, the centerline of the electrode should be aligned below the root of the joint and toward the horizontal piece at a distance equal to one-fourth to one-half the electrode diameter. The greater distance is used when making larger-sized fillet welds. Careless or inaccurate alignment may cause undercut in the vertical member or produce a weld with unequal legs.

When horizontal fillet welds are made, the electrode is tilted between 20° and 45° from the vertical (work angle) position. One or both of the following factors determine the exact angle:

1. Clearance for the welding gun, especially when structural sections are being welded to plate; and
2. The relative thicknesses of the members forming the joint (if the possibility of melting through one of the members exists, it is necessary to direct the electrode toward the thicker member).

Normally, horizontal fillet welding should be conducted with the welding electrode positioned perpendicular to the axis of the weld.

When fillet welding is performed in the flat position, the electrode axis is normally in a vertical position and bisects the angle between the workpieces, as illustrated in Figure 6.23(A). When making positioned fillet welds in which greater than normal penetration is desired, the electrode and the workpieces are positioned as shown in Figure 6.23(B). The electrode is positioned so that its centerline intersects the joint near its center. The electrode can be tilted to avoid undercutting.

CIRCUMFERENTIAL WELDS

Circumferential welds differ from those made in the flat position because of the tendency of the molten flux and weld metal to flow away from the arc as rotation progresses. To prevent spillage or distortion of the bead shape, welds must solidify as they pass the six o'clock or twelve o'clock positions. Figure 6.24 illustrates the bead shapes that result from various electrode positions relative to these positions. Correct displacements are shown in Figure 6.24(A).

Figure 6.24(B) illustrates insufficient arc displacement on an outside weld or excessive arc displacement on an inside weld, which produces deep penetration and a narrow, excessively convex bead shape. Undercutting may also take place. Excessive displacement on an outside weld or insufficient displacement on an inside weld, as shown in Figure 6.24(C), produces a shallow, concave bead.

As the flux is granular, it may spill off small-diameter workpieces if it is not contained. If spilling occurs, the uncovered arc produces welds of poor quality. One method of overcoming spillage is to use a nozzle assembly that pours the flux concentric to the arc, permitting it less chance to spill. Another method is to attach a wire brush or some other flexible heat-resisting material so that it rides on the workpiece ahead of the arc and contains the flux.

Regardless of the position of the electrode, if the weld pool is too large for the diameter of the workpiece, the molten weld metal will spill because it cannot

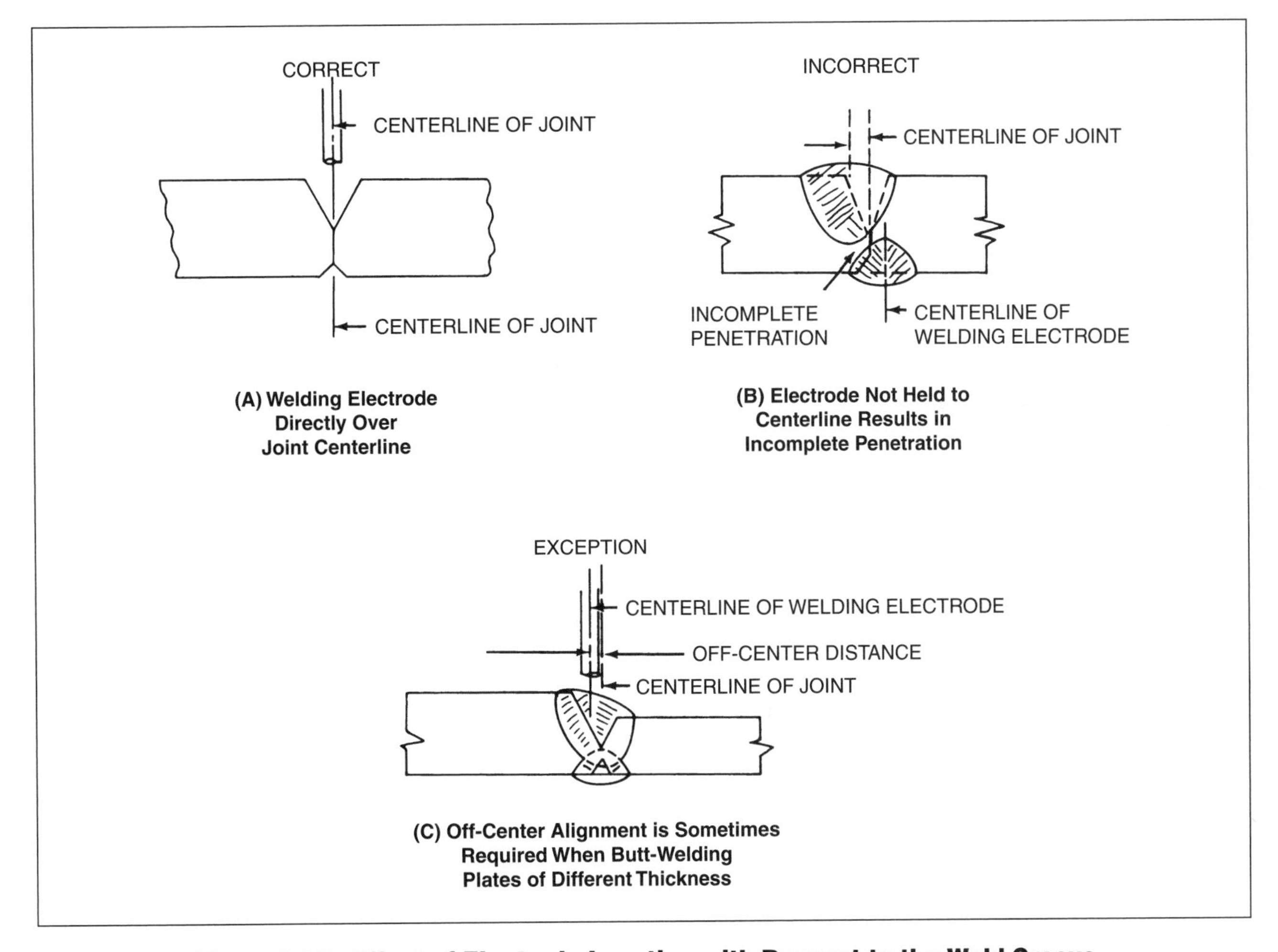

Figure 6.22—Effect of Electrode Location with Respect to the Weld Groove

freeze fast enough. Bead size, as measured by the volume of deposited metal per unit length of weld, depends on the amperage, the electrode diameter, electrode position and the travel speed employed. The use of lower amperages, smaller electrodes, and higher travel speeds will reduce the size of the bead.

SLAG REMOVAL

On multipass welds, slag removal becomes important because no subsequent passes should be made where slag is present. The factors that are particularly important in dealing with slag removal are bead size and bead shape. Smaller beads tend to cool more quickly, and slag adherence is reduced. Flat to slightly convex beads that blend evenly with the base metal make slag removal much easier than excessively concave or excessively convex or undercut beads. For this reason, a decrease in voltage improves slag removal in narrow grooves. On the first pass of two-pass welds, a concave bead that blends smoothly to the top edges of the joint is much easier to clean than a convex bead that does not blend well.

PROCESS VARIATIONS AND TECHNIQUES

Submerged arc welding lends itself to a variety of strip, wire, and flux combinations; single- and multiple-electrode arrangements; and the use of ac or dc constant-current or constant-voltage welding power sources. The process has been adapted to a wide range

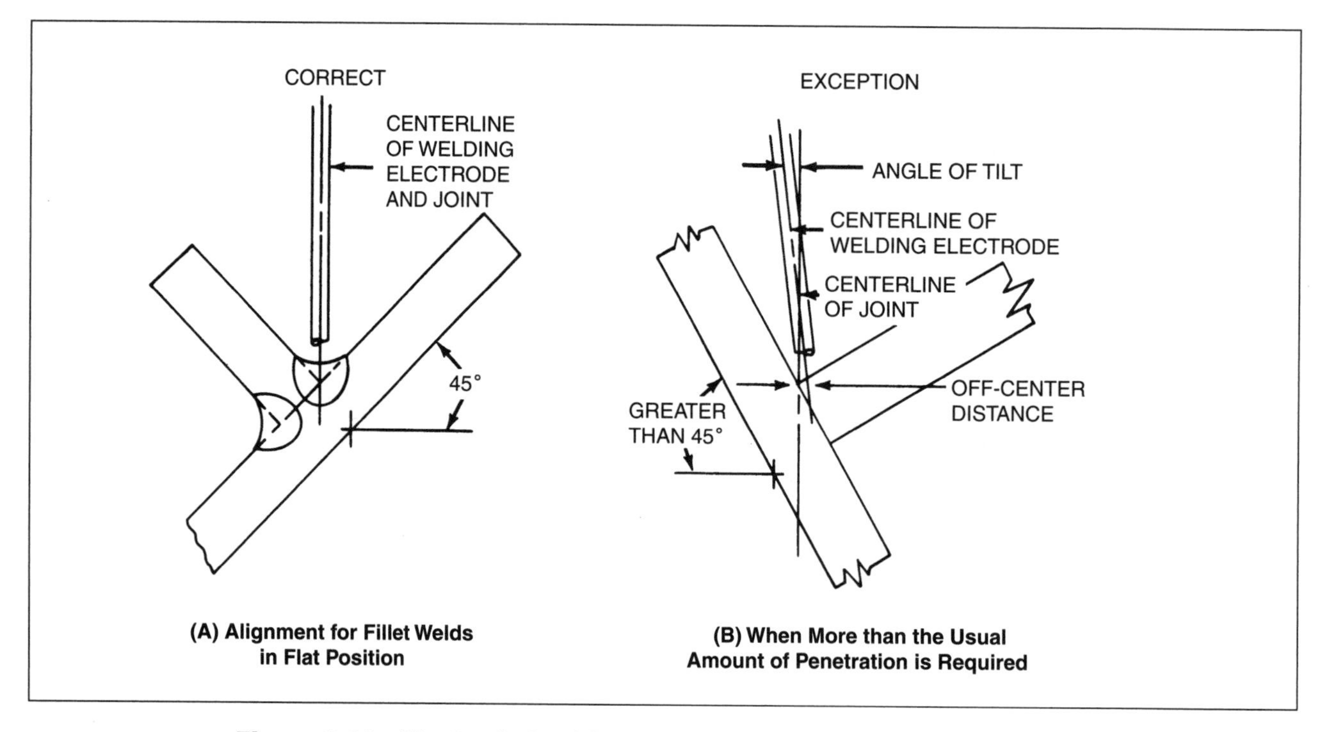

Figure 6.23—Electrode Positions for Fillet Welds in the Flat Position

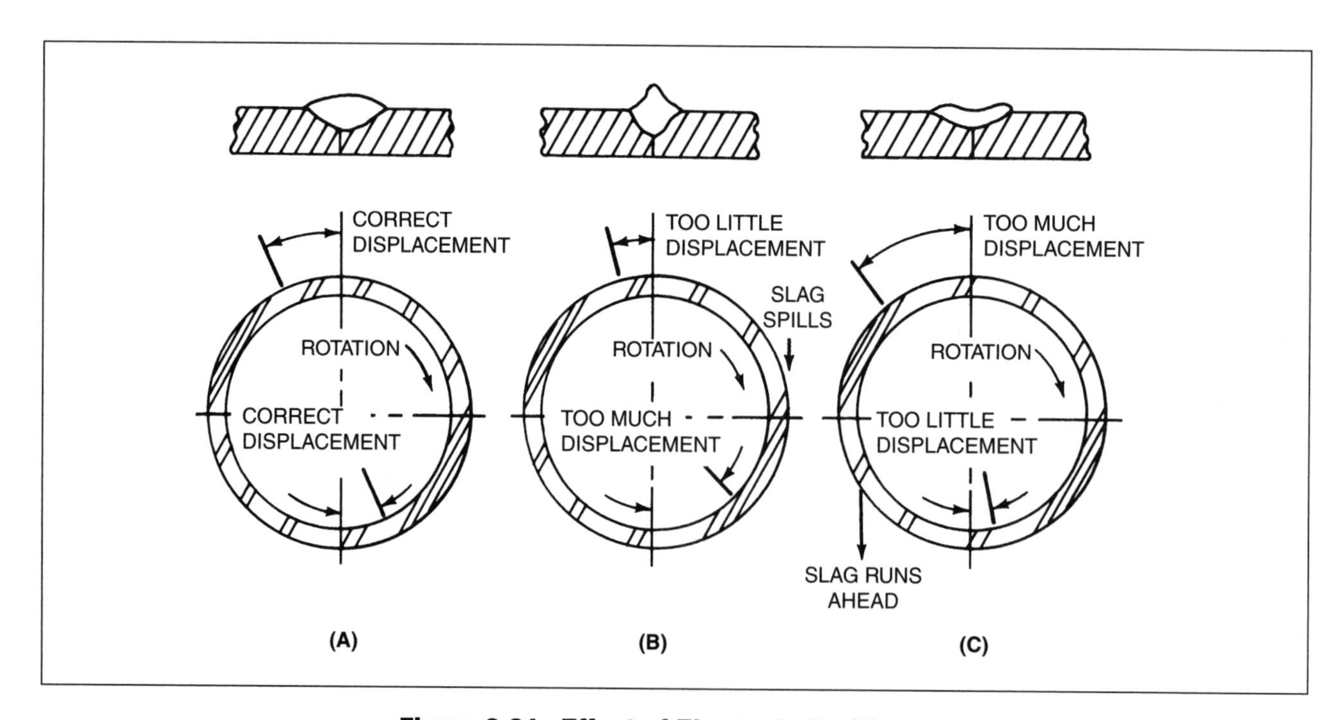

Figure 6.24—Effect of Electrode Position on Weld Bead Shape in Circumferential Welding

of materials and thicknesses. Various multiple-arc configurations are used to control the weld profile and increase the deposition rates achieved by single-arc operations. Weld deposits range from the wide beads with shallow penetration necessary for weld surfacing to narrow beads with deep penetration for welding thick joints.

The versatility of submerged arc welding is due, in part, to the use of alternating-current power supplies. The principles that favor the use of alternating current to minimize arc blow in single-arc welding are often applied in multiple-arc welding to create favorable arc deflection. The current flowing in adjacent electrodes sets up interacting magnetic fields that can either reinforce or diminish each other. These magnetic fields exist in the space between the arcs and are used to produce forces that deflect the arcs (and thus distribute the heat) in directions beneficial to the specific welding application.

Various types of power sources and related equipment are designed and manufactured especially for multiple-arc welding. These relatively sophisticated machines are intended for high production on long runs of repetitive applications.

Series submerged arc welding, a submerged arc welding process variation, and several modes and techniques are discussed in this section. These include narrow groove welding; single and multiple-wire welding; and twin-electrode, tandem-arc and triple-arc submerged arc welding. Modes include those with additions of cold wire, hot wire, and metal powder. Also discussed are submerged are techniques for cladding and overlaying.

SERIES SUBMERGED ARC WELDING

Series submerged arc welding (SAW-S) is a process variation in which the arc is established between two consumable electrodes that meet just above the surface of the workpieces, which are not part of the welding current circuit.[25] This variation involves a configuration of welding circuit connections that is different from conventional submerged arc welding and results in high deposition rates with very little dilution.

In this SAW variation, two electrodes are connected in series to two welding heads. The two electrodes may be synchronized to operate from the same wire-feed control unit or the electrodes may operate independently with separate wire feed control units. The electrode lead is connected to one welding head. The cable that is normally the workpiece lead is connected to the second welding head, directing the welding current from one electrode to the other through an arc over the weld pool. The circuit does not include a connection between the power supply and the workpiece; therefore almost all of the power is used to melt the electrode, with little power (heat) entering the workpiece. With the electrodes in the transverse position, series SAW produces the broad, shallow weld bead typically required for surfacing overlays.[26]

TECHNIQUES AND MODES

Typical submerged arc welding process configurations used in modern production welding can be used, within the limitations previously noted, for welding carbon and low-alloy steels, high-alloy steels such as stainless steel, or nickel-base alloys.

Narrow Groove Welding

Narrow groove welding, a submerged arc welding technique using a configuration that employs multipass welding with filler metal, is often used to weld material 50 mm (2 in.) thick and greater with a root opening between 13 mm and 25 mm (1/2 in. and 1 in.) wide at the bottom of the groove and a total included groove angle between 0° and 8°. This SAW configuration usually involves a single electrode powered by either direct current electrode positive (DCEP) or alternating current, depending on the type of electrode and flux being used. It is essential to use welding fluxes that have been developed for narrow groove welding because of the difficulty in removing slag. These special fluxes also reduce groove or bevel face undercut and enhance the easier removal of fused slag from the narrow groove.

Single-Electrode Welding

Single-electrode welding, in which one electrode and one power source are used, is the most common of all submerged arc welding process techniques. It is normally used with DCEP polarity, but may also be used with direct current electrode negative (DCEN) polarity when less penetration of the base metal is required, e.g., for overlay welding. Single-electrode welding may be used in semiautomated operations, in which the welder manipulates the electrode, or in mechanized welding.

A single electrode is frequently used with special welding equipment to complete horizontal (three o'clock) groove welds in large storage tanks and pressure vessels. The unit rides on the top of each ring as it is constructed and welds the circumferential joint below it. A special flux belt or other equipment is used to hold the flux in place against the shell ring. In addition, both

25. American Welding Society (AWS) Committee on Definitions and Symbols, 2001, *Standard Welding Terms and Definitions, AWS* A3.0, Miami: American Welding Society.

26. ASM International, 1993, Metals Handbook, Ninth Edition, Volume 6, *Welding, Brazing and Soldering*, Materials Park, Ohio: ASM International.

sides of the joint (inside and outside) are usually welded simultaneously to reduce fabrication time.

Multiple-Wire Welding

Multiple-wire configurations combine two or more welding wires feeding into the same weld pool. The wires may be current-carrying electrodes or cold fillers. They may be supplied from single or multiple power sources. The power sources may be direct current, alternating current, or both.

Multiple-wire welding systems not only increase weld metal deposition rates but also improve operating flexibility and provide more efficient use of available weld metal. This increased control of metal deposition can also result in welding speeds up to five times higher than those obtainable with a single wire. Figure 6.25 shows the automatic submerged arc welding of pipe using five welding heads.

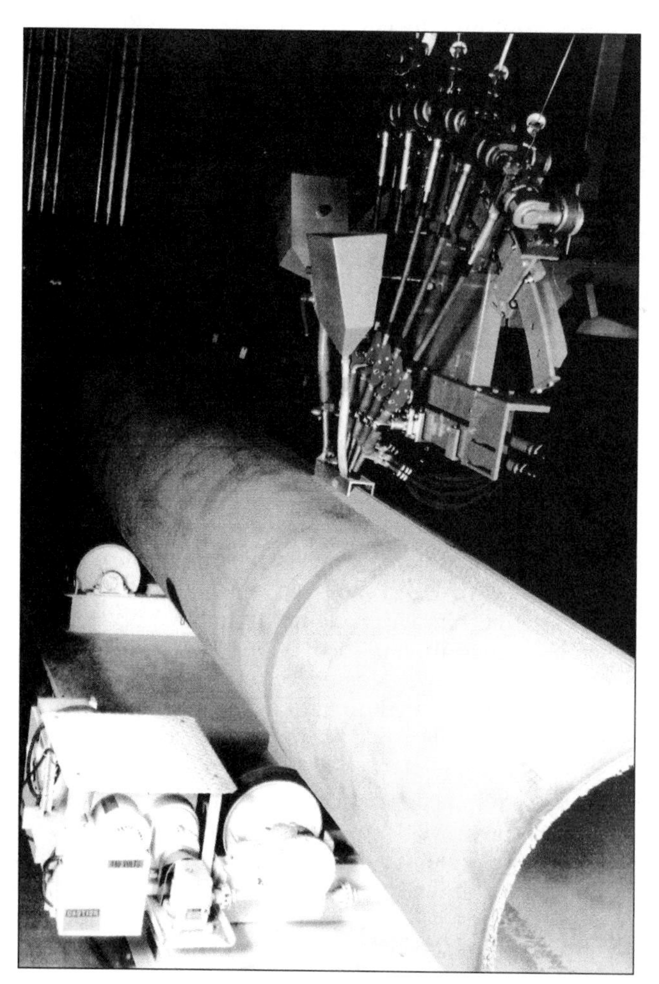

Figure 6.25—Automatic Submerged Arc Welding of Pipe Using Five Weld Heads

Twin-Electrode Submerged Arc Welding

Twin-electrode submerged arc welding utilizes two electrodes that feed into the same weld pool. The two electrodes are connected to a single power source and wire feeder and are normally used with DCEP. Because two electrodes are melted, this mode offers increased deposition rates compared to single-electrode submerged arc welding. This configuration is used in the mechanized or automated welding modes and can be used for flat groove welds, horizontal fillet welds, and for hardsurfacing in combination with solid or metal-core wire and the appropriate flux.

Tandem Arc Submerged Arc Welding

Two techniques are used for tandem arc submerged arc welding. One configuration uses DCEP for the leading electrode and an ac trailing electrode. The other uses DCEP for the lead electrode and DCEP for the trail electrode. The electrodes are separated 19 mm (3/4 in.) but are active in the same weld pool. This configuration offers higher deposition rates—up to 20 kg (44 lb) per hour when using 4.0 mm (5/32 in.) electrodes—compared to the single-electrode submerged arc welding process.

This configuration is used in mechanized or automated operations to weld thick materials (25 mm [1 in.] and greater) in the flat welding position. The DCEP-with-ac configuration is shown in Figure 6.26. It should be noted that additional alternating current trailing electrodes can be added to the configuration to improve operating flexibility and to augment deposition rates.

The second configuration uses two ac power sources electrically connected as shown in Figure 6.27. This configuration is termed a *Scott connection*, in which the interaction of the magnetic fields of the two arcs result in a forward deflection of the trailing arc. The forward deflection allows for greater welding speeds without undercutting the base metal.

Triple-Arc Submerged Arc Welding

Triple-arc submerged arc welding is a variation of the tandem-arc technique in which three electrodes are used instead of two. Two configurations of triple-arc submerged arc welding are commonly employed. In one configuration, all three electrodes are connected to ac transformers. The transformers are connected to the three-phase primary as shown in Figure 6.28. The first

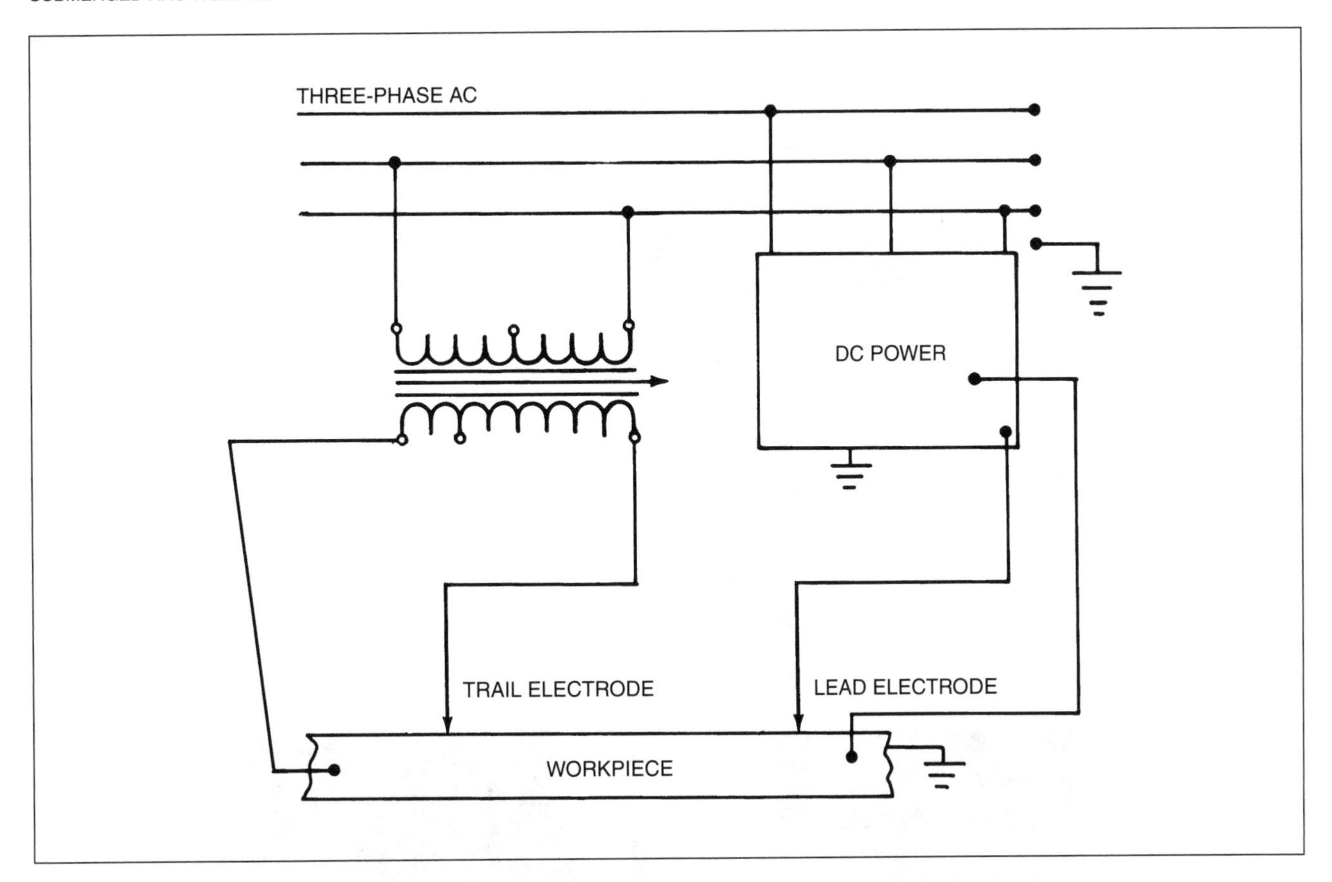

Figure 6.26—DCEP-AC Multiple-Arc Welding

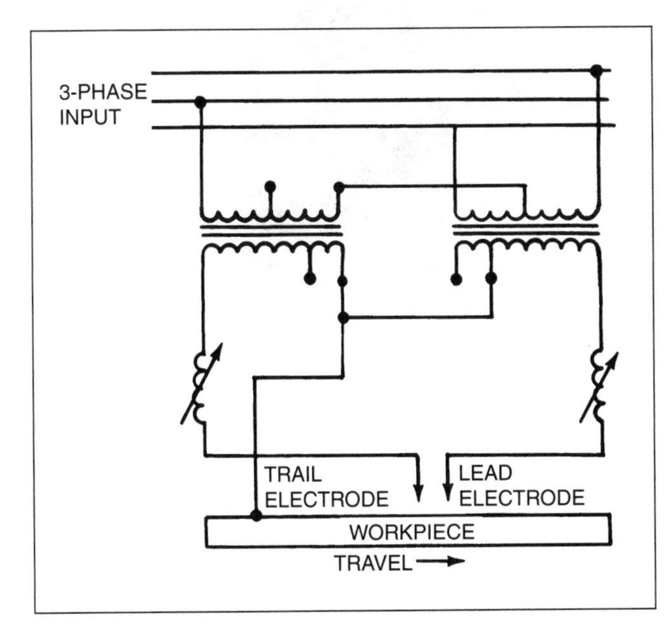

Figure 6.27—Scott Connection: AC Arc Welding Transformers

electrodes in this system are connected using a Scott connection like the one shown in Figure 6.27, and the trail electrode is in phase with the lead electrode. This connection results in a powerful forward deflection of the trail arc and promotes high travel speeds. This configuration is used in many pipe mills and is also popular in shipbuilding. It is often used in one-side welding applications.

The second configuration of triple arc submerged arc welding uses a DCEP lead arc and two ac trail arcs connected via a Scott connection, as illustrated in Figure 6.27.

The welding of line pipe is a typical application of triple-arc submerged arc welding, as shown in Figure 6.29.

Cold Wire Addition

The equipment required for cold wire additions is the same as that used for any multiple-wire application, but in this technique, one wire is not connected to a power source, hence the term *cold wire*. Cold wire

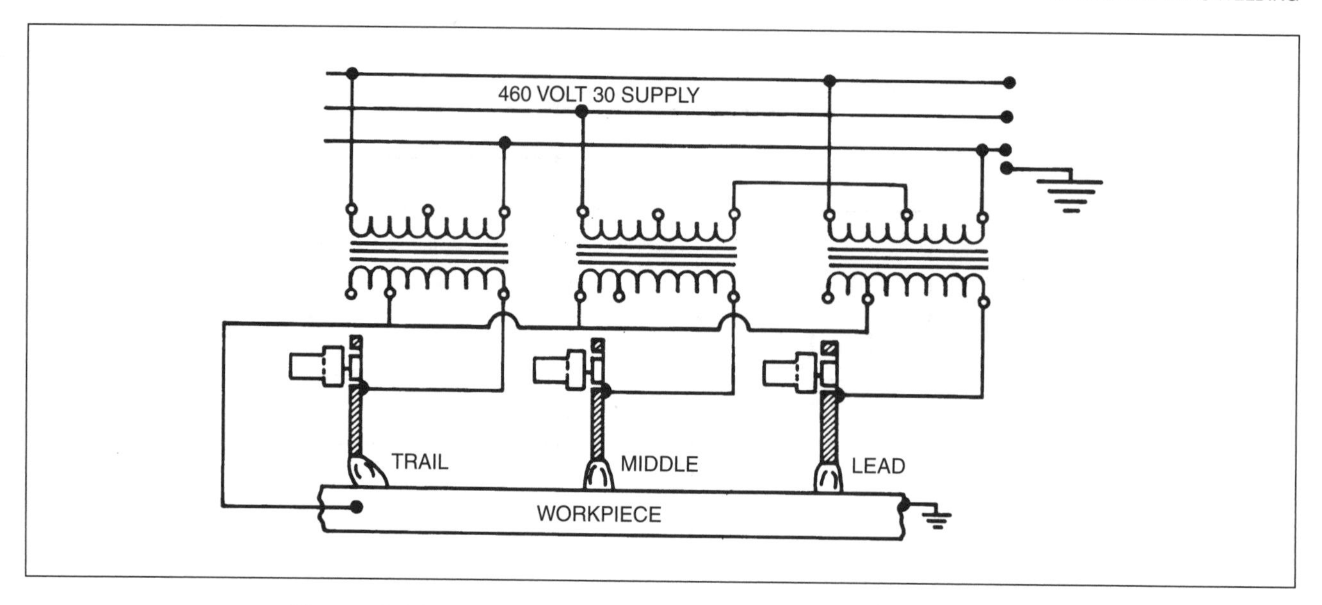

Figure 6.28—Three-Wire AC-AC-AC Submerged Arc Welding

Figure 6.29—Triple-Arc Submerged Arc Welding of Line Pipe

additions have proven feasible using both solid and flux-cored wires without deterioration in weld properties. The technique has not gained widespread industrial use because travel speeds and penetration may be diminished, even though deposition rates are increased. Increases in deposition rates to 73% are possible; rates from 35% to 40% greater are consistently achievable. The higher deposition at fixed heat input results in less penetration.

Hot Wire Addition

A hot wire addition is accomplished with an electrode wire that is resistance-heated as it is fed into the weld pool. Hot wire additions are much more efficient than cold wire or using an additional arc because the current introduced is used entirely to heat the filler wire and not to melt base material or flux. Deposition can be increased 50% to 100% without impairing the weld metal properties. The use of hot wire requires additional welding equipment, additional control of variables, considerable setup time and close operator attention.

Metal Powder Addition

Metal powders are specifically formulated to work with certain designations of electrodes. It is important to control the chemistry of the metal powder carefully in order to achieve the desired deposition chemistry. Metal powder additions may increase deposition rates up to 70%. The technique yields smooth fusion, improved bead appearance, and reduced penetration and dilution.

Metal powders can also modify the chemical composition of the final weld deposit. These powders can be added ahead of the weld pool or directly into the pool using either gravity feed or the magnetic field surrounding the wire to transport the powder. A setup for the addition of metal powder is illustrated in Figure 6.30.

Testing of metal powder additions has confirmed that the increase in deposition rate requires no additional arc energy and causes no deterioration of weld metal toughness or increased risk of cracking. Tests have also indicated that weld properties may be enhanced by the control of resulting grain structures due to the lower heat input and restoration of diluted weld metal chemistries. Troyer and Mikurak published a study of iron powder addition as joint fill showing that a 67% increase in deposition rate, usually in a

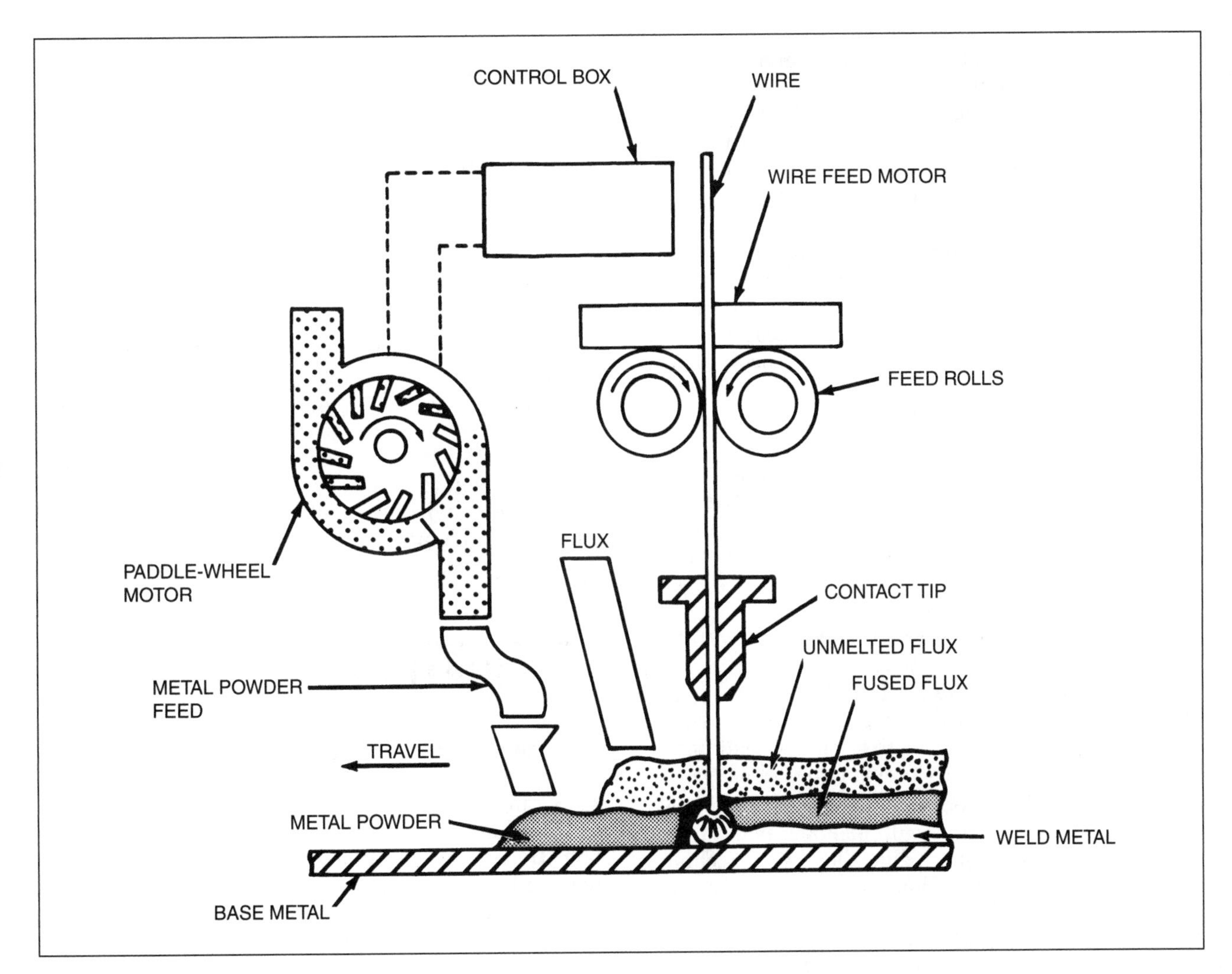

Figure 6.30—Typical Setup for Metal Powder Addition

single pass, a joint fill of 60% powder and 40% wire was achieved, with mechanical properties exceeding the requirements of AWS, ABS, and ASME qualification tests.[27]

WELD SURFACING AND CLADDING

The terms *surfacing* and *cladding,* as used with submerged arc welding, refer to the application by welding of a single or multiple-layer overlay to a surface to obtain the desired properties or dimensions, as opposed to making a joint on a deposited surface to obtain desired properties.

The submerged arc welding process is often used to apply a surface of stainless steel or nickel-based alloys to carbon steel as an economical way to add a corrosion-resistant layer to a steel workpiece. To create an overlay of a specified composition, the filler metal must be enriched sufficiently to compensate for dilution. For any given filler metal composition, changes in weld procedure can cause variations in dilution and undesirable surfacing compositions. To ensure consistently satisfactory results, the welding procedure and the type of flux used must be carefully controlled. A method for calculating weld dilution and the approximate weld metal content of any element is presented in Figure 6.31.

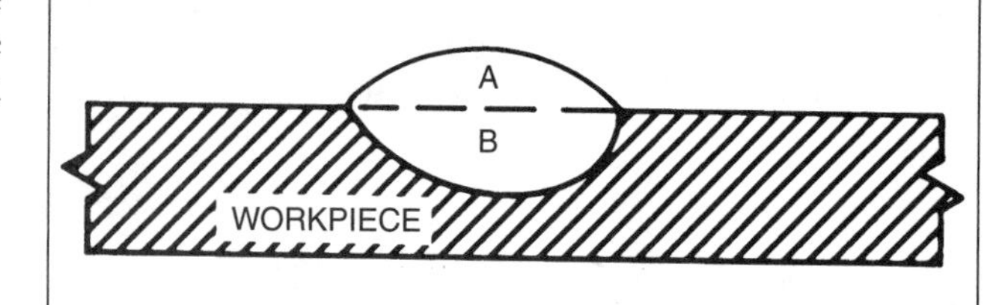

$$Z_{WM} = \frac{Z_{BM} \times B}{A + B} + \frac{Z_{FM} \times A}{A + B}$$

where

A = Filler metal
B = Base metal
Z = Weight, %
Z_{WM} = Weld metal content of element Z
Z_{BM} = Base metal content
Z_{FM} = Filler metal content

In a weld bead containing both filler metal A and base metal B, the percent of base metal B in the total weld $(A + B)$ represents the percentage by which the filler metal has been diluted:

$$\% \text{ Dilution} = \frac{B}{A + B} \times 100$$

Figure 6.31—Calculation of Weld Dilution

Welding variables and procedure deviations may result in variation in dilution, with consequent variation in composition. Procedure qualification records (PQRs) should include a chemical analysis of the weld deposit. The welding procedure specification (WPS) should contain the acceptable limits of chemical composition of the deposit.

When welding clad steels, the dilution effects discussed above are equally important. Stainless clad carbon steel or low-alloy steel plates are sometimes welded with stainless filler metal throughout the plate thickness. Carbon- or low-alloy steel filler metal is used on the unclad side, followed by the removal of a portion of the cladding and the completion of the joint with stainless or nickel-based filler metal.

Fabricators should consult the manufacturer of the clad steel for recommendations regarding detailed welding procedures and postweld heat treatments. Joining clad steel to unclad steel sections normally requires making the butt weld and restoring the clad section with a technique similar to joining two clad plates.

A typical welding procedure for stainless clad steel is shown in Figure 6.32. It should be noted that alternate Step No. 6 shows a center "annealing" bead technique for use when 400-series stainless cladding is involved.

27. Troyer, W. and J. Mikurak. 1974. *High Deposition Submerged Arc Welding Rates with Iron Powder Joint Fill,* Welding Journal, 53(8): 494-s–504-s.

Careful consideration of filler metals is required. The unclad side requires a low-carbon or low-alloy filler metal of appropriate composition (XX18 for manual shielded electrodes). For the clad side, type 309L can be used for 405, 410, 430, 304, or 304L, and type 309 Nb can be used for 321 or 347. Type 310 can be used for 310, and 309 Mo can be used for 316. Some applications may require an essentially matching composition in the case of type 400 clad surfaces. The user should consider the manufacturer's recommendations in choosing filler metals.

APPLICATIONS

Submerged arc welding is widely used to weld carbon steels, low-alloy structural steels, and stainless steels ranging from 1.5 mm (0.06 in.) sheet to thick, heavy weldments. The process is also used to join some high-strength structural steels, high-carbon steels, and nickel alloys. Better joint properties may be obtained with these metals when the heat input can be limited. This may suggest selecting a welding process that provides a lower heat input to the base metal, such as gas metal arc welding (GMAW), although heat input can be

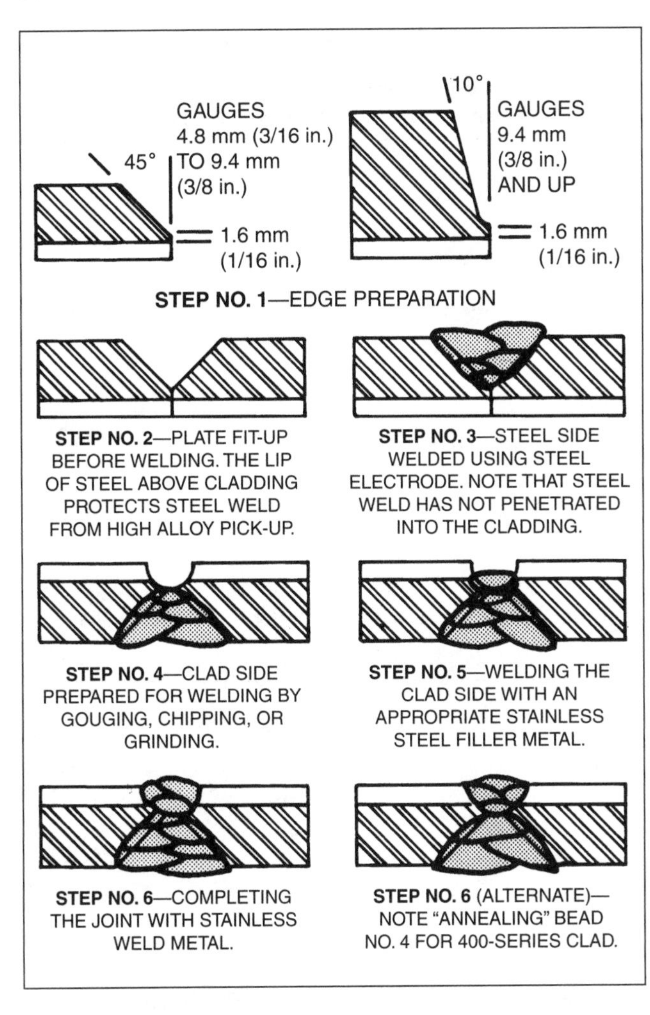

Figure 6.32—Typical Welding Procedure for Stainless Clad Steel (The Steel Side is Welded First)

limited when using submerged arc welding. Avoiding the introduction of humidity is another requisite of good joint properties that may be better achieved with GMAW.

SPECIAL SERVICE CONDITIONS

Some low-alloy steel components that are welded with the submerged arc process are destined for use in special service conditions in which weld metal, heat-affected zone, and plate hardness must not exceed maximum engineering specification hardness requirements, similar to those previously discussed for carbon steel components. These special service conditions are typically required for oil industry components that are exposed to wet hydrogen sulfide gas.

Table 6.4
Maximum Hardness Requirements of C-Mo and Cr-Mo Steels When Exposed to Wet Hydrogen Sulfide Gas

	Typical Maximum Hardness, BHN*
Carbon-1/2% molybdenum steels	225
1-1/4% chromium—1/2% molybdenum steels	225
2-1/4% chromium—1% molybdenum steels	225
5% chromium—1/2% molybdenum steels	235
9% chromium—1% molybdenum steels	241

*BHN = Brinell Hardness Number

It has been found that if hardness is kept below a prescribed level, depending on the type of material and the service condition, cracking due to exposure to hydrogen does not generally occur. Typical maximum hardness requirements for carbon-molybdenum steels and chromium-molybdenum steels as established by the petroleum industry through the American Petroleum Institute (API)[28] and NACE International (National Association of Corrosion Engineers)[29] are shown in Table 6.4.

Hardness may be reduced in several ways:

1. Selecting consumables that produce low weld metal hardness,
2. Raising the preheat temperature or welding heat input to produce weld metals and heat-affected zones with softer microstructures, and
3. Implementing postweld heat treatment.

WELD TYPES

Submerged arc welding is used to make groove, fillet, plug, and surfacing welds. Groove welds are usually made in the flat position; fillet welds are typically made in the flat and horizontal positions. These positions are used because the weld pool and the flux are most easily contained in these positions. However, simple techniques are available to produce groove welds in the horizontal welding position. Good submerged arc welds can be made downhill at angles up to 15° from the horizontal. Surfacing and plug welding are commonly performed in the flat position.

28. American Petroleum Institute, 1220 L Street N.W., Washington, DC 20005.
29. NACE International, P.O. Box 218340, Houston, TX 77218-8340.

Table 6.5
Typical Welding Conditions for Single-Electrode Mechanized Submerged Arc Welding of Steel Plate Using One Pass (Square-Groove Weld)

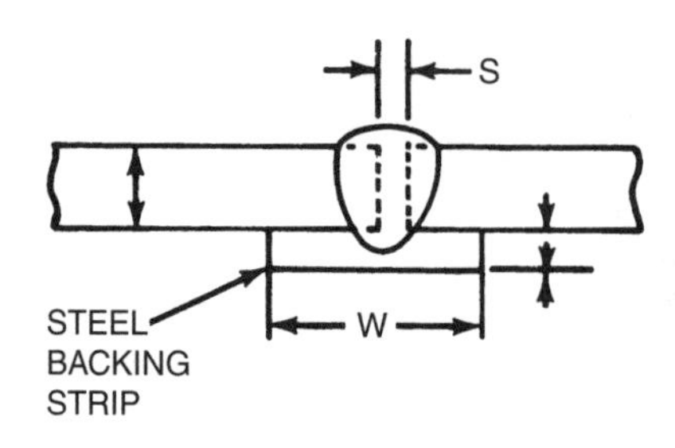

S = Root opening, mm (in.)
W = Width of backing strip, mm (in.)
T = Base metal thickness, mm (in.)

Plate Thickness, T		Root Opening, S		Current	DCEP Voltage	Travel Speed		Electrode Diameter		Electrode Consumption	
mm	in.	mm	in.	A	V	cm/s	in./min	mm	in.	kg/m	lb/ft
3.6	10 ga.	1.6	1/16	650	28	122	48	3.2	1/8	0.104	0.070
4.8	3/16	1.6	1/16	850	32	91	36	4.8	3/16	0.194	0.13
6.4	1/4	3.2	1/8	900	33	66	26	4.8	3/16	0.248	0.20
9.5	3/8	3.2	1/8	950	33	61	24	5.6	7/32	0.357	0.24
12.7	1/2	4.8	3/16	1100	03	53	21	5.6	7/32	0.685	0.46

Groove Welds

Groove welds are commonly made in butt joints ranging from 1.2 mm (0.05 in.) sheet metal to thick plate. The greater penetration capability inherent in submerged arc welding permits square-groove joints in base metal thicknesses up to 13 mm (1/2 in.) to be completely welded from one side, provided some form of backing is used to support the molten metal. Table 6.5 presents typical welding conditions for mechanized submerged arc welding using a single electrode to produce a square-groove weld in steel plate in one pass. Single-pass welds up to 7.9 mm (5/16 in.) thickness can be made. Table 6.6 presents typical welding conditions for two passes, starting with a backing pass, using a single electrode to weld steel plate with a square-groove weld. Two-pass welds can be made in steel up to 15.9 mm (5/8 in.) with a square-groove butt joint, no root opening, and a weld backing.

With multipass welding using single or multiple electrodes, any plate thickness can be welded. Welds in thick material, when deposited from both sides, can use V- or U-grooves on one or both sides of the plate.

Groove welding of butt joints in the horizontal position is known as *three o'clock welding*. This type of weld may be made from both sides of the joint simultaneously, if desired. In most cases, the electrodes are positioned at a 10° to 30° angle above the horizontal position. To support the flux and molten metal, some form of sliding support or a proprietary moving belt is used.

Fillet Welds

Using a single electrode, fillet welds with throat sizes up to 9.5 mm (3/8 in.) can be made in the horizontal position with one pass. Larger single-pass, horizontal-position fillet welds may be made with multiple electrodes. Because of the effects of gravity, welds larger than 7.9 mm (5/16 in.) are usually made in the flat position or by means of multiple passes in the horizontal position.

Fillet welds made by the submerged arc welding process can have greater penetration than those made by shielded metal arc welding, thereby exhibiting higher shear strength for the same size weld. This allows a reduction in throat size of the bead, which results in less distortion compared to fillet welds made with the shielded metal arc welding process.

Table 6.6
Typical Welding Conditions for Single-Electrode, Two-Pass Submerged Arc Welding of Steel Plate (Square Groove)

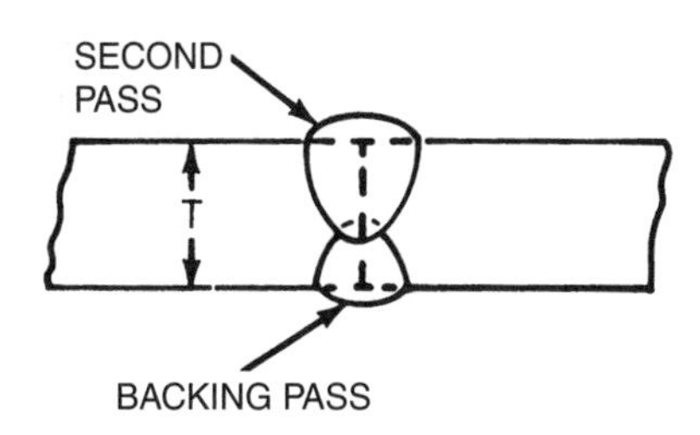

T = Base metal thickness, mm (in.)

		Second Pass						Backing Pass							
Electrode Thickness, T		DCEP Current	DCEP Voltage	Travel Speed		Electrode Diameter		DCEP Current	Voltage	Travel Speed		Electrode Diameter		Electrode Consumption	
in.	mm	A	V	in./min	cm/s	in.	mm	A	V	in./min	cm/s	in.	mm	lb/ft	kg/m
Semiautomatic Welding															
10 gauge	3.6	325	27	50	127	1/16	1.6	250	25	50	127	1/16	1.60	0.070	0.104
3/16	4.8	350	32	46	117	1/16	1.6	300	29	46	117	1/16	1.60	0.088	0.131
1/4	6.4	375	33	42	107	1/16	1.6	325	34	42	107	1/16	1.60	0.106	0.158
3/8	9.5	475	35	28	71	5/64	2.0	425	33	28	71	5/64	2.00	0.18	0.268
1/2	12.7	500	36	21	53	5/64	2.0	475	34	21	53	5/64	2.00	0.28	0.417
5/8	15.9	500	37	16	40	5/64	2.0	500	35	16	40	5/64	2.00	0.43	0.640
Mechanized Welding															
1/4	6.4	575	32	48	122	5/32	4.0	475	29	48	122	5/32	4.00	0.11	0.164
3/8	9.5	850	35	32	81	5/32	4.0	500	33	32	81	5/32	4.00	0.23	0.343
1/2	12.7	950	36	27	68	3/16	4.8	700	35	27	68	3/16	4.80	0.34	0.506
5/8	15.9	950	36	22	56	3/16	4.8	900	36	22	56	3/16	4.80	0.50	0.745

Plug Welds

Submerged arc welding is used to make high-quality plug welds. The electrode is positioned in the center of the hole and remains in this position until the weld is complete. The time required is dependent on welding amperage and hole size. Because of the deep penetration obtained with this process, it is essential to have adequate thickness in the weld backing.

Surfacing Welds

Single- and multiple-electrode and strip submerged arc welding methods are used to provide a base metal with special surface properties. The purpose may be to repair or reclaim worn equipment that is otherwise serviceable or to impart desired properties to surfaces of original equipment. The high deposition rates achieved by submerged arc welding are well suited to large-area surfacing applications.

WELD QUALITY

Submerged arc welding can produce high-quality welds with fewer weld defects than other processes because of the excellent protection of the weld metal afforded by the blanket of molten slag. However, as in other processes with many combinations of variables, the problems of porosity, slag inclusions, incomplete fusion, and cracking also occur in submerged arc welding. The causes of these problems and remedies are discussed in this section.

POROSITY

Weld metal deposited by the submerged arc process is usually clean and free of harmful porosity, but when porosity does occur, it may be found on the weld bead surface or beneath a sound surface. The possible causes of porosity include the following:

1. Contaminants in the joint such as paint residue, hydrocarbons from oil-based products, or manufacturing coatings;
2. Electrode contamination, such as rust or oils from improper storage;
3. Insufficient flux coverage;
4. Contaminants in the flux, especially in recycled fluxes that may be reused without proper screening;
5. Entrapped slag at the bottom of the joint;
6. Segregation of constituents in the weld metal;
7. Wrong welding parameters, such as high voltage and excessive travel speed, which cause fast solidification and prevents gases from escaping;
8. Slag residue that has oxidized from tack welds made with covered electrodes (residue may also form gases, inhibit fusion, and create voids);
9. Moisture in the flux, which can create both hydrogen and oxygen gas pockets;
10. High flux burden; and
11. Arc blow.

As with other welding processes, the base metal and electrode must be clean and dry in submerged arc welding. High travel speeds and associated fast weld metal solidification do not provide time for gas to escape from the molten weld metal. The travel speed can be reduced, but other solutions should be investigated first to avoid higher welding costs. Porosity from covered electrode tack welds can be avoided by using electrodes that leave no porosity-causing residue. Recommended tack weld electrodes are E6010, E6011, E7015, E7016, and E7018.

INCLUSIONS

An inclusion in a weld is defined as the entrapment of solid foreign material, such as slag, flux, or oxide. As with all flux-shielded processes, submerged arc welding, if not properly applied, is not immune to slag inclusions. Inclusions can be found primarily in the root overlap or between previous passes, and generally near bevel faces or groove faces. The risk of inclusions is greater under the following circumstances:

1. In downhill orientations, which may allow the molten flux to race ahead of the weld pool and then roll under to become entrapped during solidification;
2. In grooves along the edges of previously deposited beads, especially if they are convex;
3. If there is undercut along the edges of the weld bead;
4. When arc destabilization inhibits the slag from rising to the top of the solidifying weld pool;
5. When a flux too high in viscosity is used, impeding the solidification of the weld pool;
6. When a clean prior bead surface is not maintained;
7. In the case of improper welding head displacement, particularly in the downhill orientation;
8. In the case of reduced penetration due to reduced heat input, supplemental wires or powders, increased electrode extension, excessive travel speed, or reduced arc density; and
9. In the case of excessive travel speed, which causes faster cooling and may inhibit the slag from rising to the top of the weld pool.

INCOMPLETE FUSION

As with porosity and inclusions, incomplete fusion can occur at any depth of the weld. It can occur between either the present or previous beads or the bead and the bevel or groove faces. Specific conditions may foster incomplete fusion. Most of these are conditions that inhibit the melting and the fusible contact between surfaces, including the following:

1. Incorrect joint preparation or procedure,
2. Presence of slag or oxide residue,
3. Arc instability,
4. High travel speeds,
5. Insufficient heat input, and
6. Improper displacement, angle, or position of the welding head.

CRACKING

Weldment cracking is generally considered the most serious of discontinuities because of the potential for catastrophic service failure and the often-experienced difficulty in radiographic inspection. Cracking can be categorized according to several criteria, including location (weld metal, heat-affected-zone, or base metal cracking [hot or cold cracking]), but it is generally assumed to have two basic causes—an imposed strain and an inability to accommodate that strain.[30]

Hot and Cold Cracking

The types of hot cracking include solidification, liquation, and stress-relief cracking. This discontinuity is usually associated with the weld metal. Several conditions associated with strains and the ability to accommodate them, predominantly of cross-sectional geometry and chemistry, are listed below:

1. Excess depth-to-width ratio (high-penetration processes with single-pass procedures such as submerged arc welding are especially susceptible;
2. Excessive concavity (inability to accommodate strains) especially with fillet welds;
3. Inadequate fill at weld bead stopping points (decreases ability to accommodate strain), which can cause crater cracks; and
4. Chemical contamination, especially from sulfur or hydrocarbon-bearing compounds.

Excessive heat input may decrease the cooling rate and increase the time for metallurgical reactions in a crack-susceptible temperature range.

Like hot cracking, cold cracking manifests itself in a number of ways. These include hydrogen-induced cracking or stress-corrosion cracking. Most cold cracking is associated with the heat–affected zone, base metal, or weld metal. Causes of cracking and remedial actions are discussed in Chapter 13 of the Welding Handbook, *Welding Science and Technology,* Volume 1.[31]

ECONOMICS

Cost savings can be achieved with the efficient use and management of the submerged arc welding process. This process is an economical asset when used to surface carbon steel with stainless steel or nickel-base alloys or to apply a corrosion-resistant layer to a steel workpiece. Other techniques associated with economical welding are the following:

1. Alloy fluxes containing alloying additions can be used to change the deposited alloy when using an inexpensive carbon steel or other alloyed electrode;
2. Wire retract arc starting can be used to economic advantage when frequent starts are made and starting location is important;
3. Root openings of the correct size, as opposed to those larger than required for proper groove and fillet welds, may maintain normal welding time and costs;
4. High travel speeds and the associated fast weld metal solidification may produce unacceptable welds, and although the travel speed can be reduced, other solutions should be investigated first to avoid higher welding costs;
5. Flux recovery units, which collect the unused flux for recycling, can reduce expense, maximize flux utilization and minimize manual cleanup; and
6. The use of large electrode packages can increase operating efficiency and eliminate end-of-coil waste.

SAFE PRACTICES

The potential hazards associated with welding processes are somewhat lessened in submerged arc welding because the blanket of flux covering the arc prevents most of the fumes, sparks, spatter, and radiation from escaping. This provides a cleaner and safer atmosphere for the welder and welding operators; however, general welding safety precautions must be carefully observed. For detailed safety information, the reader should follow instructions in manuals provided by equipment manufacturers for the safe installation and operation of equipment. Instructions provided in Materials Safety Data Sheets (MSDS), which can be obtained from manufacturers of welding materials and consumables, should be followed. The standard *Safety in Welding, Cutting, and Allied Processes,* ANSI Z49.1[32] should be consulted. Mandatory federal safety regulations as established by the U.S. Department of Labor's Occupational Safety and Health Administration in the *Code of Federal Regulations (CFR)*, Title 29, Part 1910, Subpart Q should be observed.[33]

30. For additional information on cracking, see "Weld Quality," Chapter 13 in American Welding Society (AWS) Welding Handbook Committee, Jenney, C. L., and A. O'Brien, eds., 2001, *Welding Science and Technology,* Vol. 1 of *Welding Handbook*, 9th ed., Miami: American Welding Society.

31. American Welding Society (AWS) Welding Handbook Committee, Jenney, C. L. and A. O'Brien, eds., 2001, *Welding Science and Technology,* Volume 1 of Welding Handbook, 9th ed., Miami: American Welding Society.

32. American National Standards Institute (ANSI) Accredited Standards Committee Z49, *Safety in welding, cutting, and allied processes,* ANSI Z49.1:1999, Miami: American Welding Society.

33. Occupational Safety and Health Administration (OSHA), 1999, *Occupational Safety and Health Standards for General Industry,* in *Code of Federal Regulations (CFR),* Title 29 CFR 1910, Subpart Q. Washington, D.C.: Superintendent of Documents, U.S. Government Printing Office.

Health and safety codes and standards and the publishers of these documents are listed in Appendix B.

PERSONAL PROTECTIVE EQUIPMENT

Welders are exposed to molten metal, sparks, slag and hot surfaces. Welders and welding operators should wear fire-resistant protective clothing, as described in ANSI/AWS A5.17/A5.17M-97.[34] Appropriate eye protection should be used, as well as a helmet or a face shield to protect the face, neck, and ears from weld spatter, arc glare, and flying slag particles. Ear plugs or other equipment should be used to protect against excessive noise.

ELECTRICAL SAFETY

The equipment manufacturer's instructions and recommended safe practices should be followed. To reduce the risk of electric shock, welding power sources and accessory equipment such as wire feeders should be properly grounded. A separate connection is required to ground the workpiece. The workpiece lead completes the welding circuit but is not a ground lead, nor should the workpiece be mistaken for a ground connection. Welding cables should be kept in good condition. Open-circuit voltage should be avoided to decrease the potential hazard present when several welders are working with arcs of different polarities, or when several alternating current machines are in use. The standards *Safety in Welding, Cutting and Allied Processes*, ANSI Z49.1,[35] and National Electrical Code, NFPA No. 70,[36] must be carefully followed.

FUMES AND GASES

Certain elements, when vaporized, can be potentially hazardous. Alloy steels, stainless steels, and nickel alloys contain such elements as chromium, cobalt, manganese, nickel, and vanadium. Filler metals and fluxes generate fumes and gases during the welding process. Material safety data sheets should be obtained from the manufacturers to determine the content of the potentially hazardous elements and the threshold limit values. Exposure limits are included in the OSHA *Code of Federal Regulations*, Title 29, Part 1910, Subpart Q.[37]

The submerged arc welding process greatly limits exposure of operators to air contaminants because few welding fumes escape from the flux overburden. Adequate ventilation will generally keep the welding area clear of hazards from fumes and gases. The type of fan, exhaust, or other air movement system will be dependent on the work area to be cleared. The various manufacturers of ventilation equipment should be consulted for particular applications.

CONCLUSION

Submerged arc welding can bring valuable benefits to the welder or fabricator. The many configurations of submerged arc welding make this process an extremely versatile and adaptable medium for countless welding and surfacing applications. Submerged arc welding is capable of producing high-quality welds in manual, semiautomated and automated modes in many metals and alloys, enhanced with selections from a comprehensive array of wire electrodes and fluxes.

The high level of weld quality that can be achieved is rewarding. However, the user must be aware that the overall quality of the welded joint or overlay will be greatly affected, individually and collectively, by the flux-wire technique used; the electrode type, size, and extension; the type of flux and particle distribution; welding equipment; and variables such as current, voltage, travel speed and others. Each must be given due consideration. The information in this chapter, the referenced resources, and the supplementary reading list will help in coordinating all the parameters necessary for achieving optimum weld quality.

BIBLIOGRAPHY[38]

American National Standards Institute (ANSI) Accredited Standards Committee Z49. 1999. *Safety in welding, cutting, and allied processes.* ANSI Z49.1:1999. Miami: American Welding Society.

American Welding Society (AWS) Committee on Structural Welding. 2000. *Structural welding code—Steel.* AWS D1.1:2000. Miami: American Welding Society.

34. American Welding Society (AWS) Committee on Filler Metals and Allied Materials, 1997, *Specification for Carbon Steel Electrodes and Fluxes for Submerged Arc Welding*, ANSI/AWS A5.17/A5.17M-97, Miami: American Welding Society, p 25.
35. See Reference 32.
36. National Fire Protection Association, *National Electrical Code*, NFPA 70, Quincy, Massachusetts: National Fire Protection Association.
37. Occupational Safety and Health Administration (OSHA), 1999, *Occupational Safety and Health Standards for General Industry*, in *Code of Federal Regulations (CFR)*, Title 29 CFR 1910, Subpart Q, Washington, D.C.: Superintendent of Documents, U.S. Government Printing Office.
38. The dates of publication given for the codes and other standards listed here were current at the time this chapter was prepared. The reader is advised to consult the latest edition.

American Welding Society (AWS) Committee on Filler Metals and Allied Materials. 1997. *Specification for carbon steel electrodes and fluxes for submerged arc welding. ANSI/AWS* A5.17/A5.17M-97. Miami: American Welding Society.

American Welding Society (AWS) Committee on Filler Metals and Allied Materials. 1997. *Specification for low-alloy steel electrodes and fluxes for submerged arc welding.* ANSI/AWS A5.23/A5.23M. Miami: American Welding Society.

American Welding Society (AWS) Committee on Filler Metals and Allied Materials. 1997. *Standard procedures for calibrating magnetic instruments to measure the delta ferrite content of austenitic and duplex ferritic-austenitic stainless steel weld metal.* ANSI/AWS A4.2M/A4.2:1997. Miami: American Welding Society.

American Welding Society (AWS) Committee on Filler Metals and Allied Materials. 1997. *Specification for nickel and nickel alloy bare welding electrodes and rods*, ANSI/AWS A5.14/A5.14M-97. Miami: American Welding Society.

American Welding Society (AWS) Committee on Filler Metals and Allied Materials. 1993. *Specification for bare stainless steel welding electrodes and rods*, ANSI/AWS A5.9-93. Miami: American Welding Society.

American Welding Society (AWS) Committee on Filler Metals and Allied Materials. 1992 (R 2000). *Specification for stainless steel welding electrodes for shielded metal arc welding.* AWS A5.4-92(R 2000). Miami: American Welding Society.

American Welding Society (AWS) Committee on Filler Metals and Allied Materials. 2000. *Filler metal comparison charts.* Miami: American Welding Society.

American Welding Society (AWS) Welding Handbook Committee, Jenney, C. L., and A. O'Brien, eds. 2001. *Welding science and technology.* Vol. 1 of *Welding handbook*. 9th ed. Miami: American Welding Society.

American Welding Society (AWS) Welding Handbook Committee. Oates, W. R., and A. Saitta, eds. 1998. *Materials and applications—Part 2*. Vol. 4 of *Welding handbook*. 8th ed. Miami: American Welding Society.

National Fire Protection Association. *National electrical code*, NFPA 70. Quincy, Massachusetts: National Fire Protection Association.

Troyer, W. and J. Mikurak. 1974, High deposition submerged arc welding rates with iron powder joint fill. *Welding Journal*, 52(8): 494-s—504-s.

Occupational Safety and Health Administration (OSHA). 1999. *Occupational safety and health standards for general industry,* in *Code of Federal Regulations (CFR),* Title 29, CFR 1910, Subpart Q. Washington, D.C.: Superintendent of Documents, U.S. Government Printing Office.

SUPPLEMENTARY READING LIST

Allen, D. J., B. Chew, and P. Harris. 1981. The formation of chevron cracks in submerged arc weld metal. *Welding Journal* 61(7): 212–s-221-s.

ASM International. 1983. Vol. 6 of *Metals handbook, welding and brazing*. 9th ed. Materials Park, Ohio: American Society for Metals.

Bailey, N., and S. B. Jones. 1978. The solidification mechanics of ferritic steel during submerged arc welding. *Welding Journal* 57(8): 217-s–231-s.

Butler, C. A., and C. E. Jackson. 1967. Submerged arc welding characteristics of the CaO-TiO_2-SiO_2 system. *Welding Journal* 46(10): 448-s–456-s.

Chandel, R. S. 1987. Mathematical modeling of melting rates for submerged arc welding. *Welding Journal* 66(5): 135-s–139-s.

Dallam, C. B., S. Liu, and D. L. Olson. 1985. Flux composition dependence of microstructure and toughness of submerged arc HSLA weldments. *Welding Journal* 64(5): 1405-s–1515-s.

Dittrich, S., G. Grote, K. Wilhelmsberger, and A. Heuser. 1994. Hydrogen in heavy wall SAW joints of 2-1/4 Cr-1 Mo steel after different dehydrogenation heat treatments. Vienna, Austria: Second International Conference on Interaction of Steels with Hydrogen in Petroleum Industry Pressure Vessel and Pipeline Service.

Eager, T. W. 1978. Sources of weld metal oxygen contamination during submerged arc welding. *Welding Journal* 57(3): 76-s–80-s.

Ebert, H. W., and F. J. Windsor. 1980. Carbon steel submerged arc welds—Tensile strength vs. corrosion resistance. *Welding Journal* 59(7): 193-s–198-s.

Fleck, N. A., O. Grong, G. R. Edwards and D. K. Matlock. 1986. The role of filler metal wire and flux composition in submerged arc weld metal transformation kinetics. *Welding Journal* 65(5): 113-s–121-s.

Gowrisankar, I., A. K. Bhaduri, V. Seetharaman, D. D. N. Verma and R. G. Achar. 1987. Effect of the number of passes on the structure and properties of submerged arc welds of AISI type 316L stainless steel. *Welding Journal* 66(5): 147-s–151-s.

Hantsch, H., K. Million and H. Zimmerman. 1982. Submerged arc narrow-gap welding of thick walled components. *Welding Journal* 61(7): 27–34.

Hinkel, J. E., and F. W. Forsthoefel. 1976. High current density submerged arc welding with twin electrodes. *Welding Journal* 55(3): 175–180.

Indacochea, J. E., 1989. Submerged arc welding: Evidence for electrochemical effects on the weld pool. *Welding Journal* 68(3): 77-s–81-s.

Jackson, C. E. 1973. *Fluxes and slags in welding*. Welding Research Council Bulletin 190. New York: Welding Research Council.

Keith, R. H. 1975. Weld backing comes of age. *Welding Journal* 54(6): 422-s–430-s.

Konkol, P. J., and G. F. Koons. 1978. Optimization of parameters for two-wire ac-ac submerged arc welding. *Welding Journal* 57(12): 367-s–374-s.

Kubli, R. A., and W. B. Sharav. 1961. Advancements in submerged arc welding of high impact steels. *Welding Journal* 40(11): 497-s–502-s.

Lau, T., G. C. Weatherly, and A. McLean. 1986. Gas/metal/slag reactions in submerged arc welding using $CaO-Al_2O_3$ based fluxes. *Welding Journal* 65(2): 31-s–38-s.

Lau, T., G. C. Weatherly, and A. McLean. 1985. The sources of oxygen and nitrogen contamination in submerged arc welding using $CaO-Al_2O_3$ based fluxes. *Welding Journal* 64(12): 343-s–348-s.

Lewis, W. J., G. E. Faulkner, and P. J. Rieppel. 1961. Flux and filler wire developments for submerged arc welding HY-80 steel. *Welding Journal* 40(8): 337-s–345-s.

The Lincoln Electric Company. 1973. *The procedure handbook of arc welding*. 13th ed. Cleveland, Ohio: Lincoln Electric Company.

Majetich, J. C. 1985. Optimization of conventional SAW for severe abrasion—Wear hardfacing application. *Welding Journal* 64(11): 3145-s–3215-s.

Mallya, V. D., and H. S. Srinivas. 1989. Bead characteristic in submerged arc strip cladding. *Welding Journal* 68(12): 30.

McKeighan, J. S. 1955. Automatic hard facing with mild steel electrodes and agglomerated alloy fluxes. *Welding Journal* 34(4): 301s–308-s.

Newell, W. F. Jr., and R. A. Swain. 1996. Hardfacing and surfacing using the SAW/ESW strip cladding process. *Welding Journal* 75(2): 55–57.

North, T. H, H. B. Bell. A. Nowicki and I. Craig. 1978. Slag/metal interaction, oxygen, and toughness in submerged arc welding. *Welding Journal* 57(3): 63-s–75-s.

Patchett, B. M. 1974. Some influences of slag composition on heat transfer and arc stability. *Welding Journal* 53(5): 203-s–210-s.

Polar, A., J. E. Indacochea, and M. Blander. 1990. Electrochemically generated oxygen contamination in submerged arc welding. *Welding Journal* 69(2): 68-s–74-s.

Renwick, B. G., and B. M. Patchett. 1976. Operating characteristics of the submerged arc process. *Welding Journal* 55(3): 69-s–76-s.

Smith, N. J., J. T. McGrath, J. A. Gianetto and R. F. Orr. 1989. Microstructure/mechanical property relationships of submerged arc welds in HSLA 80 steel. *Welding Journal* 68(3): 112-s–120-s.

Troyer, W., and J. Mikurak. 1974. High deposition submerged arc welding with iron powder joint-fill. *Welding Journal* 53(8): 494-s–504-s.

Union Carbide Corporation. 1974. *Submerged arc welding handbook*. New York: Union Carbide Corporation, Linde Division.

Uttrachi, G. D., and J. E. Messina. 1968. Three-wire submerged arc welding of line pipe. *Welding Journal* 47(6): 475-s–481-s.

Wittstock, G. G. 1976. Selecting submerged arc fluxes for carbon and low-alloy steels. *Welding Journal* 55(9): 733-s–741-s.

Wilson, R. A. 1956. A selection guide for methods of submerged arc welding. *Welding Journal* 35(6): 549-s–555-s.

CHAPTER 7

PLASMA ARC WELDING

Photograph courtesy of Thermal Arc

Prepared by the Welding Handbook Chapter Committee on Plasma Arc Welding:

W. L. Roth, Chair
Procter & Gamble

G. L. Reid
Thermal Arc

R. P. Webber
Stellite Coatings

S. D. Kiser
Special Metals Company

Welding Handbook Volume 2 Committee Member:

B. R. Somers
Lucius Pitkin, Incorporated

Contents

CHAPTER 7

PLASMA ARC WELDING

INTRODUCTION

Plasma arc welding (PAW) is an arc welding process in which coalescence of metals is produced by a constricted arc between a non-consumable electrode and the workpiece (transferred arc mode) or between the electrode and the constricting nozzle (nontransferred arc mode). Pressure is not applied, and filler metal may or may not be added. The arc is concentrated in a column of ionized plasma issuing from the torch. While the orifice gas (also known as plasma gas) provides some shielding, it is normally supplemented by a separate source of shielding gas. Shielding gas may be a single inert gas or a mixture of inert gases.

Plasma arc welding is essentially an extension of gas tungsten arc welding (GTAW),[1] but utilizes a slightly different mechanism to deliver the heat for welding. Like GTAW, plasma arc welding uses a non-consumable electrode; however, the plasma arc welding torch differs from the GTAW torch in that it has a nozzle that creates a gas chamber surrounding the electrode. The arc heats the orifice gas that is fed into the chamber to a temperature at which it becomes ionized and conducts electricity. This ionized gas is defined as *plasma*. Plasma issues from the nozzle orifice at a temperature of about 16 700°C (30,000°F), creating a narrow, constricted arc pattern that provides excellent directional control and produces a very favorable depth-to-width weld profile.

Plasma arc welding can be used to join most metals in all positions. While it has many similarities to gas tungsten arc welding, it has additional features that make it the process of choice for many industries. The automotive industry uses PAW for air-bag assemblies, exhaust systems, and torque converters. The aerospace community uses PAW for many airframe components, fuel-storage vessels, guidance systems and repair of gas turbine components. In the medical industry, catheters, batteries, surgical tools and other medical instruments are routinely fabricated by PAW. The electrical industry also uses PAW to great advantage to fabricate filaments and connections for light bulbs, laminations and shading coils for motor-generator windings, relay coils, and electrical contactors.

Because of the numerous similarities between plasma arc welding and gas tungsten arc welding, an understanding of the GTAW process is helpful in appreciating the many comparisons of the two processes made in this chapter. Information on the GTAW process can be found in Chapter 3 of this volume. More detailed information is provided in this chapter on the areas in which plasma arc welding differs from gas tungsten arc welding.

Also discussed in this chapter are plasma arc welding equipment and materials, process variations, applications, procedures, techniques, and weld quality, followed by brief comments on comparative costs. The final section of this chapter provides information and references for safe practices in plasma arc welding.

HISTORY OF THE PLASMA ARC

One of the earliest plasma arc systems was a gas vortex stabilized device introduced by Schonherr in 1909.[2] In this unit, gas was blown tangentially into a tube through which an arc was struck. The centrifugal force of the gas stabilized the arc along the axis of the tube by creating a low-pressure axial core. Arcs up to several meters in length were produced, and the system proved useful for arc studies.

Gerdien and Lotz built a water vortex arc-stabilizing device in 1922. In this device, water injected tangentially into the center of a tube was swirled around the inner surface and ejected at the ends. When an arc struck between carbon electrodes was passed through the tube, the water concentrated the arc along its axis, producing higher current densities and temperatures than were otherwise available. Although the concept was promising, the Gerdien and Lotz invention had no practical metalworking applications because of the

1. See Chapter 3, "Gas Tungsten Arc Welding," in this volume.

2. Springer-Verlag, 1956, *Encyclopedia of Physics*, Volume XXII, Berlin: Springer-Verlag, p. 300.

rapid consumption of the carbon electrodes and the presence of water vapor in the plasma jets.

While working on the arc melting of refractory metals in 1953, Gage[3] observed the similarity in appearance between a long electric arc and an ordinary gas flame. Efforts to control the heat intensity and velocity of the arc led to the development of the modern plasma arc torch.

The first practical plasma arc metalworking tool was a cutting torch introduced in 1955. This device was similar to a gas tungsten arc welding torch in that it used a tungsten electrode. However, the electrode was recessed in the torch; it used a separate plasma gas; and the arc was constricted as it passed through an orifice in the torch nozzle. The usual circuitry for gas tungsten arc welding was supplemented in the plasma cutting equipment with a pilot arc circuit for arc initiation.

Commercial equipment for plasma arc surfacing emerged in 1961, and plasma arc welding was introduced in 1963. In the early 1980s, the introduction of variable polarity power sources made it possible to use the plasma arc for the welding of heavy sections in one pass with remarkable quality.

FUNDAMENTALS

In plasma arc welding, the orifice gas, also called *plasma gas*, is directed through the torch and surrounds the electrode. The gas becomes ionized in the arc to form the plasma and issues from the orifice in the torch nozzle as a plasma jet stream. This plasma stream becomes the conductor for the transferred arc. For most operations, shielding gas is provided through an outer gas nozzle in a manner similar to gas tungsten arc welding. The purpose of the shielding gas is to blanket the area of arc plasma impingement on the workpiece to avoid contamination of the weld pool.[4]

The constricting nozzle through which the arc plasma passes has two main dimensions: the orifice diameter and the constricting orifice length. The orifice shapes the plasma arc column; it may be cylindrical or may have a converging or diverging taper. The distance that the electrode is recessed from the face of the constricting nozzle is called the *electrode setback*. The dimension from the outer face of the constricting nozzle to the workpiece is known as the *torch standoff distance*. Figure 7.1 shows a cross section of a typical plasma arc torch.

3. Gage, R. M., U.S. Patent No. 2,806,124.
4. American Welding Society (AWS) Committee on Plasma Arc Welding, 1973, *Recommended Practices for Plasma Arc Welding*, C5.1-73, Miami: American Welding Society.

The *plenum* or *plenum chamber* is the space between the inside wall of the constricting nozzle and the electrode. The orifice gas is directed into this chamber and then through the orifice toward the workpiece. The plenum often has small vanes inside which cause the plasma gas to swirl. The effect of this swirling motion is similar to that of a miniature tornado; it creates a tight, focused column of plasma.

A comparison of the gas tungsten arc welding and plasma arc welding processes is illustrated in Figure 7.2. The electrode in the gas tungsten arc welding torch extends beyond the end of the shielding gas nozzle. The gas tungsten arc is not constricted; thus, it assumes an approximately conical shape, producing a relatively wide heat pattern on the workpiece. For a given welding current and electrode tip preparation angle, the area of impingement of the cone-shaped arc on the workpiece varies with the electrode-to-workpiece distance. Thus, a small change in arc length produces a relatively large change in heat input per unit area.

By contrast, the electrode in the plasma arc torch is normally recessed within the constricting nozzle. The arc is collimated and focused by the constricting nozzle on a relatively small area of the workpiece. Because the shape of the plasma stream is essentially cylindrical, very little change occurs in the area of contact on the workpiece as torch standoff varies. Thus, the plasma arc welding process is significantly less sensitive to variations in torch-to-workpiece distance than the gas tungsten arc welding process.

Since the electrode of the plasma arc torch is normally recessed inside the arc-constricting nozzle, it is not possible for the electrode to touch the workpiece. This feature greatly reduces the possibility of contaminating the weld with electrode metal and damaging the electrode. Tungsten electrode contamination is a significant source of downtime in gas tungsten arc welding, thus emphasizing a major advantage of the plasma arc welding process.

The arc heats the orifice gas as it passes through the plenum of the plasma arc torch. This heating expands the gas, and it exits through the constricting orifice at higher velocity. Since a gas jet of very high flow rate would cause turbulence in the weld pool, orifice gas flow rates are generally held to within 0.1 liter per minute (L/min) to 5 L/min (0.2 cubic foot per hour [ft^3/h] to 11 ft^3/h) for most plasma arc welding applications. The orifice gas alone is generally not adequate to shield the weld pool from atmospheric contamination. Therefore a separate shielding gas must be provided through an outer gas nozzle. The shielding gas may or may not be the same composition as the orifice gas. Typical shielding gas flow rates are in the range of 5 L/min to 30 L/min (11 ft^3/h to 64 ft^3/h).

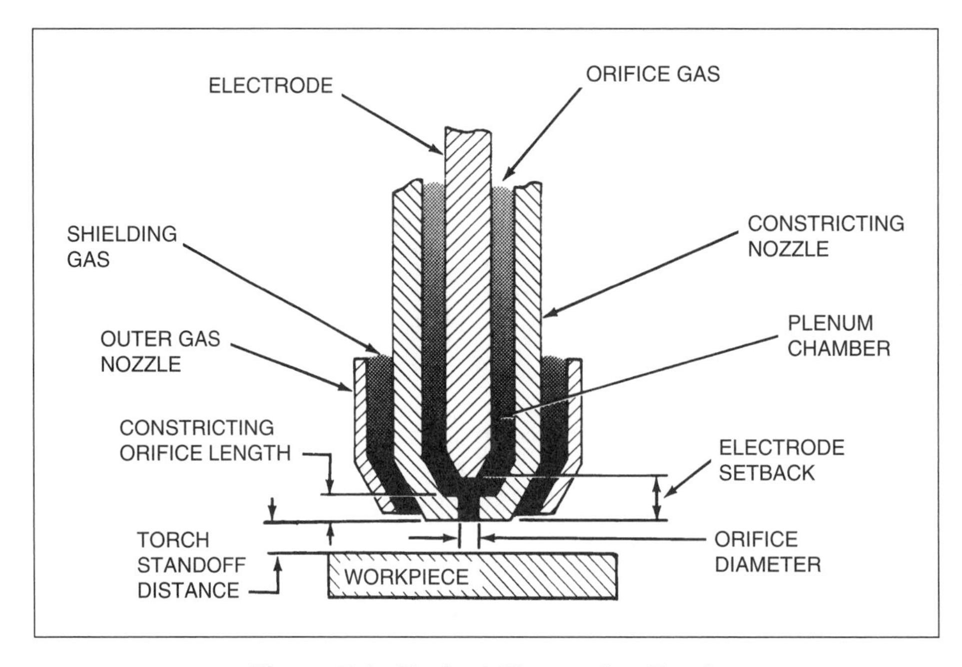

Figure 7.1—Typical Plasma Arc Torch

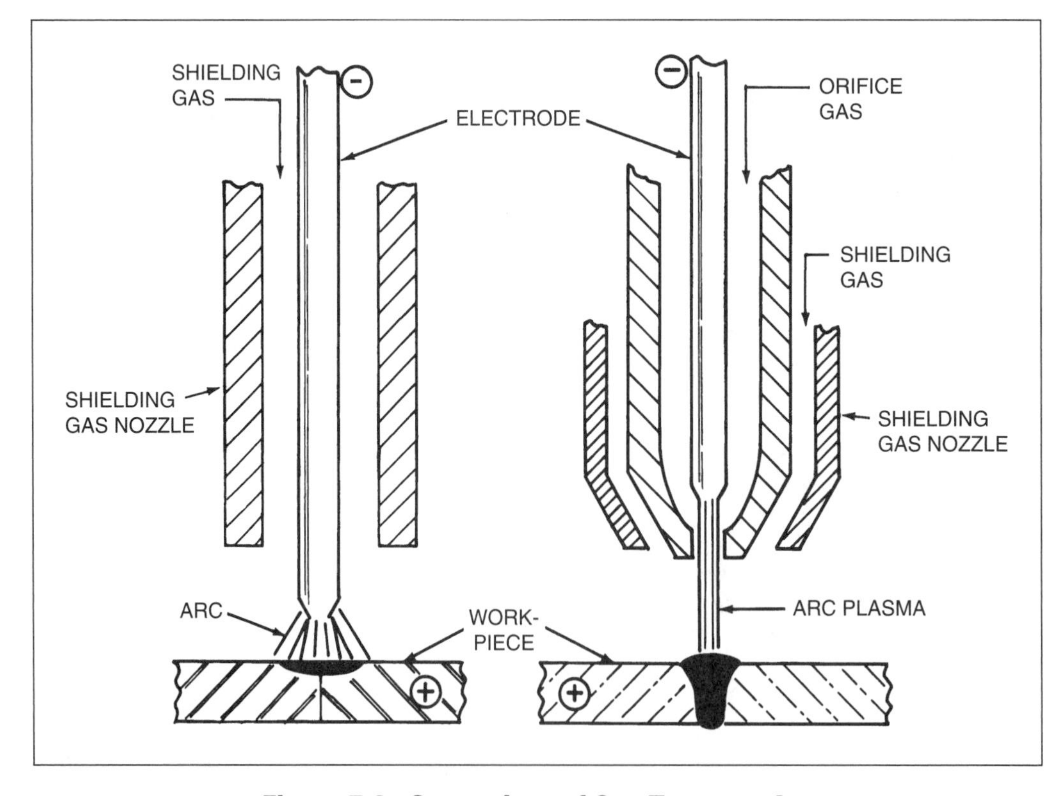

Figure 7.2—Comparison of Gas Tungsten Arc and Plasma Arc Welding Processes

ADVANTAGES OF ARC CONSTRICTION

Several improvements in performance over an open-arc operation, such as gas tungsten arc welding, are obtained when the arc is constricted as it passes through a small orifice. The most noticeable improvement is the directional stability of the plasma jet stream. A conventional gas tungsten arc can be easily deflected by low-strength magnetic fields, whereas a plasma jet is comparatively stiff and tends to be less affected by magnetic fields.

High current densities and high energy concentration are produced by arc constriction. The higher current densities result in higher temperatures in the plasma arc. The higher temperatures and electrical changes brought about by the constricted arc are compared in Figure 7.3. The left half of this figure represents a normal nonconstricted arc operating at 200 amperes (A), direct current electrode negative (DCEN), in argon at a flow rate of 19 L/min (40 ft^3/h). The right side illustrates an arc with the same current and gas flow that is constricted by its passage through an orifice 4.8 millimeters (mm) (3/16 inch [in.]) in diameter. Under these conditions, the constricted arc shows a 100% increase in arc power and a 30% increase in temperature over the open arc.

The increased temperature of the constricted arc is not the major advantage, however, since the temperature in the gas tungsten arc far exceeds the melting points of the metals generally welded by either arc welding process. The main advantages of plasma arc welding are consistent arc starting without the use of high-frequency initiation or touch (scratch) starting, the directional stability and focusing effect brought about by arc constriction, and the relative insensitivity of the constricted arc to variations in torch standoff distance.

The plasma arc efficiently uses the supplied arc energy. The degree of arc collimation, arc force, energy density on the workpiece, and other characteristics are primarily functions of the following welding variables:

1. Plasma current;
2. Orifice diameter, shape, and length;
3. Tungsten-to-orifice relationship (electrode setback);
4. Electrode tip geometry, such as bevel angle and truncation of the tip;
5. Type of orifice gas;
6. Flow rate of orifice gas; and
7. Type of shielding gas.

The fundamental differences among the plasma arc metalworking processes arise from the relationships among these seven factors. These factors can be adjusted independently to provide very high or very low thermal energies. For example, the high energy concentration and high jet velocity necessary for plasma arc cutting require a high arc current, a small-diameter orifice, high orifice gas flow rate and an orifice gas that has high thermal conductivity. Conversely, most welding applications require a low-velocity plasma jet to prevent the expulsion of weld metal from the workpiece. This calls for larger orifices, considerably lower gas flow rates, and lower arc currents.

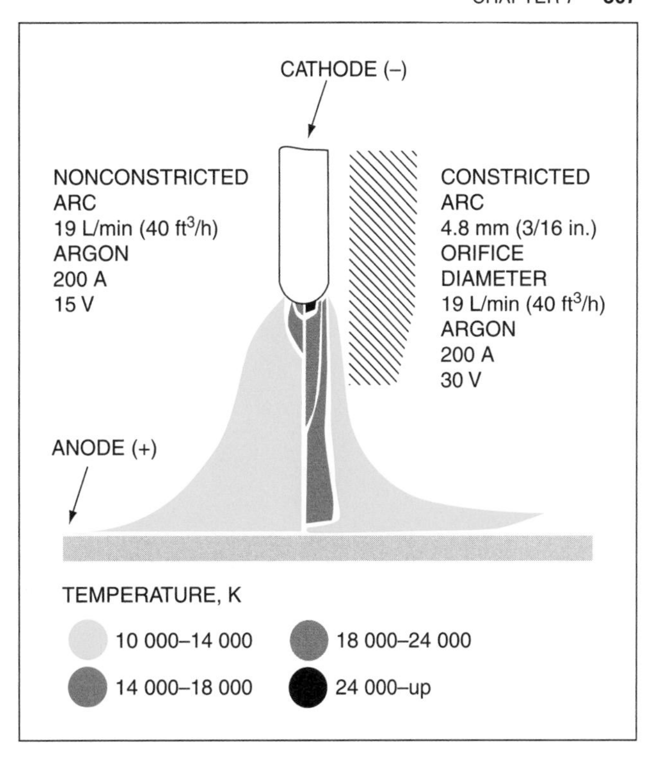

Figure 7.3—Effect of Arc Constriction on Temperature

The constricted arc is much more effective than an open arc for heating the gas to be used for a particular operation. When the gas passes directly through a constricted arc, it is exposed to higher energy concentrations than when it passes around a conventional gas tungsten arc, as shown in Figure 7.3.

ARC MODES

Two arc modes are used in plasma arc welding: transferred arc and nontransferred arc. With a transferred arc, the electric current travels directly from the electrode to the workpiece. The workpiece is part of the electrical circuit (but the nozzle is not), and heat is obtained from the anode spot on the workpiece and from the plasma jet.

In the nontransferred arc mode, the arc is established and maintained between the electrode and the constricting orifice. The arc plasma is forced through the orifice by the plasma gas. The workpiece is not in the arc circuit. Useful heat is obtained only from the plasma jet. Nontransferred arcs are useful for cutting and joining nonconductive workpieces or for applications in which a relatively low energy concentration is desirable. Figure 7.4 illustrates the two arc modes.

Transferred arcs have the advantage of greater energy transfer to the workpiece for a given current, voltage, and travel speed. This results from the additional energy transferred into the workpiece as an element of the electrical circuit. Nearly all industrial welding is performed with a transferred arc.

In the transferred arc mode, if there is insufficient orifice gas flow or excessive arc current for a given nozzle geometry, or if the nozzle touches the workpiece, the nozzle may be damaged by a phenomenon known as *double arcing*. In double arcing, the metallic torch nozzle forms part of the current path from the electrode back to the power source. In essence, two arcs are formed, as shown in Figure 7.5. The first arc is from the electrode to the constricting nozzle, and the second is from the constricting nozzle to the workpiece. Heat generated at the cathode and anode spots, formed where the two arcs attach to the constricting nozzle, invariably causes damage to the constricting nozzle. To correct double arcing, the transferred arc must be broken and restarted. To prevent double arcing, the welding current must be reduced or the constricting nozzle must be replaced by a nozzle with a larger orifice.

TYPES OF WELDING CURRENT

Various current polarities and waveforms are used for plasma arc welding. Direct current electrode negative power is most commonly used for PAW applications, generally with a thoriated tungsten electrode and a transferred arc. Other types of tungsten electrodes that may also be used in limited applications include ceriated, lanthanated, zirconiated, and pure tungsten

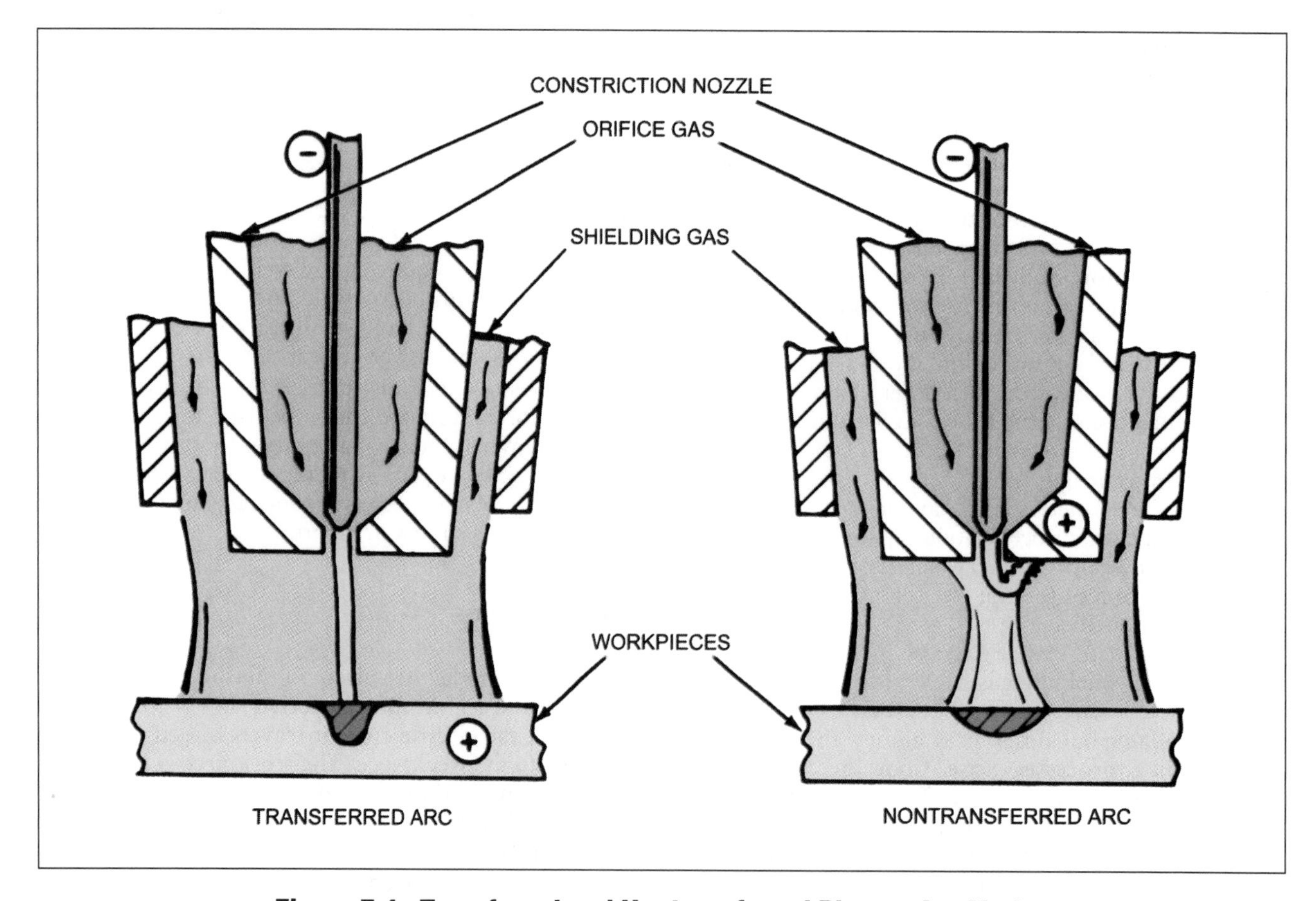

Figure 7.4—Transferred and Nontransferred Plasma Arc Modes

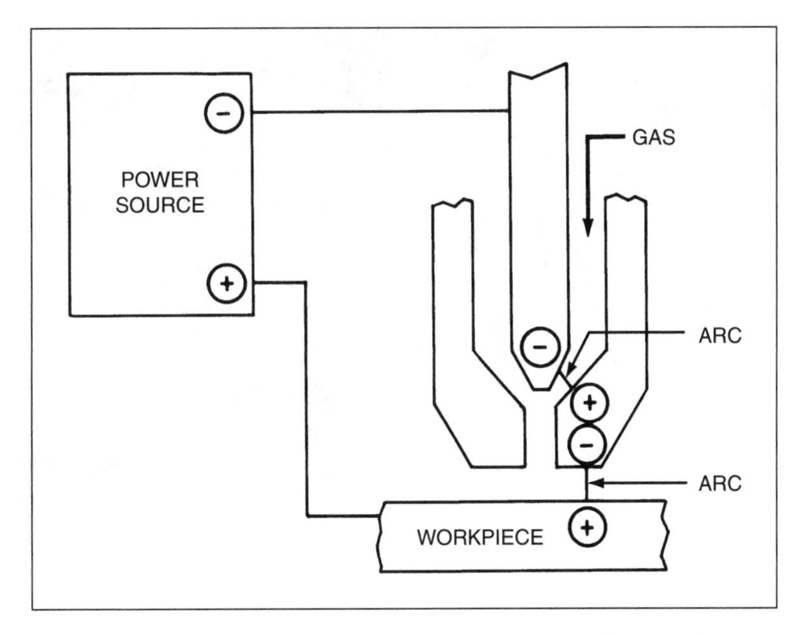

Figure 7.5—Schematic Illustration of Double Arcing

electrodes. Ceriated and lanthanated electrodes are occasionally used for DCEN applications because of potential health concerns associated with thoriated electrodes. Pure or zirconiated tungsten electrodes are used in very few PAW applications because these electrodes have limited ability to maintain a ground tip shape during operation. The current for DCEN plasma arc welding ranges from approximately 0.1 A to 500 A. Pulsed welding current is often used to reduce excessive heat input and improve the weldability of thin alloys. Steels (including stainless), nickel alloys, and titanium are commonly welded with PAW using DCEN. Aluminum and magnesium alloys can be welded with DCEN, however extreme cleaning measures must be taken due to the tenacious surface oxides formed on these alloys.

Sine-wave alternating current (ac) with continuous high-frequency stabilization can be used to weld light-gauge aluminum and magnesium alloys. The current is generally limited to a range between 10 A and 120 A, depending on the electrode diameter and composition. Higher amperages generally cannot be used because of excessive electrode deterioration during the electrode-positive half-cycle of the current. Unlike the GTAW process, "balling" of the tungsten electrode is not possible with PAW because the electrode must maintain the required shape inside the constricting nozzle. (Balling of a tungsten electrode refers to the GTAW technique of heating the electrode tip white-hot by applying a large DCEP current. The current is applied until the electrode forms a round molten ball approximately 1 to 1.5 times the electrode diameter. The current is then rapidly removed to preserve the hemispherical shape. This end shape improves alternating-current arc properties.)

The primary reason for using alternating current when welding magnesium and aluminum alloys is the cleansing effect of the DCEP half-cycle. During the electrode-positive half-cycle of alternating current, positive ions are released from the electrode, and they bombard the oxides on the surface of the workpiece. This bombardment, termed *cathodic etching*, removes the surface oxides and exposes clean aluminum alloy to be welded. Square-wave ac has largely replaced sine-wave ac for welding aluminum and magnesium alloys.

Square-wave ac, with unbalanced electrode positive and negative current half-cycles (also called *variable polarity*), plasma arc welding is highly efficient for joining magnesium and aluminum alloys. High-frequency stabilization is not required due to the rapid switching of arc polarity. Variable polarity minimizes the DCEP arc time to reduce electrode heating and achieve exactly the desired level of cathodic etching for oxide removal, while spending the majority of the arc time in the DCEN mode for maximum heat transfer.

Direct current electrode positive (DCEP) is used to some extent for welding aluminum. Excessive electrode heating is the primary limitation to the use of electrode-positive polarity. The maximum current is usually less than 100 A.

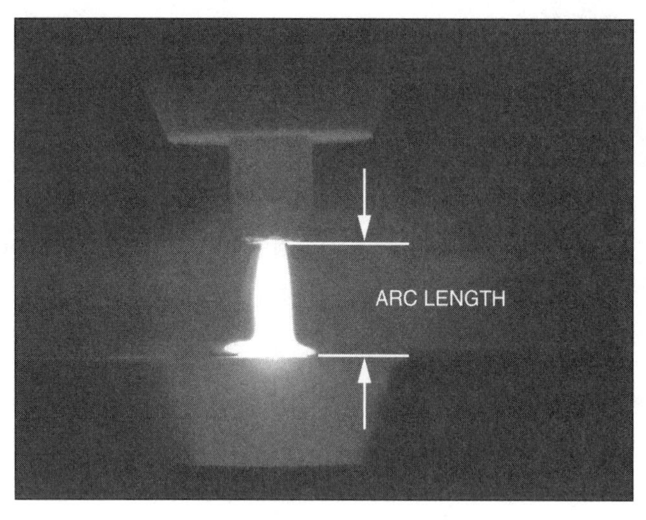

Plasma Arc, 6.3 mm (1/4 in.)

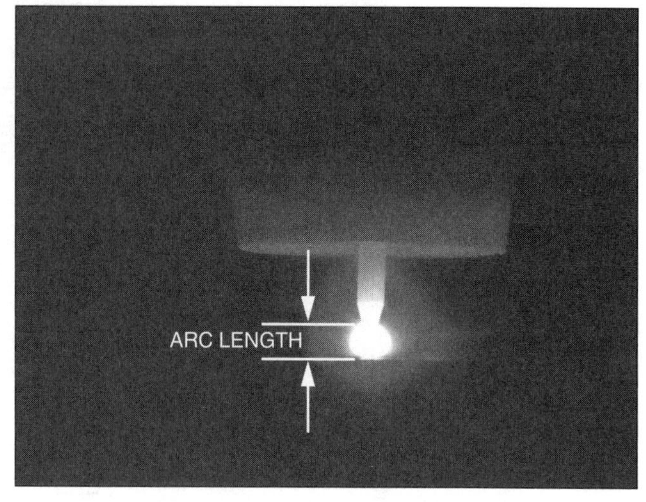

Gas Tungsten Arc, 1.6 mm (1/16 in.)

Photographs courtesy of Edison Welding Institute

Figure 7.6—Comparison of Typical Plasma Arc Length and Gas Tungsten Arc Length Used for Welding Very Thin Metal Sections at 15 A

ARC LENGTH

The columnar nature of the constricted arc makes the plasma arc process less sensitive to variations in arc length than the gas tungsten arc process with its nonconstricted conical arc shape. In GTAW, the area of heat input to the workpiece varies inversely as the square of the arc length. In other words, a small change in arc length causes a relatively large change in the area in which the heat is applied, much like a flashlight illuminates a larger circle as the light is moved away from the target. However, with the essentially cylindrical plasma jet in PAW, as the arc length is varied within normal limits, the area of heat input to the workpiece and the intensity of the arc are virtually constant, similar to a laser beam projected on a target.

The collimated plasma jet permits the use of a much longer torch standoff distance than is possible with the gas tungsten arc welding process. This reduces the level of operator skill required to manipulate the torch for some types of welded joints. Typical arc lengths used to weld thin-gauge material at approximately 15 A are shown in Figure 7.6. The plasma arc is approximately 6.3 mm (1/4 in.) long compared to the 1.6 mm (1/16 in.) gas tungsten arc. Because the torch standoff distance is proportional to arc voltage, PAW requires higher arc voltages than GTAW.

EQUIPMENT

The basic equipment used in plasma arc welding is shown in Figure 7.7. Plasma arc welding can be performed in manual, mechanized, or robotic operations. A complete system for manual plasma arc welding consists of a torch, control console, power source, orifice and shielding gas supplies, a source of torch coolant, and accessories such as an on-off switch, gas flow timers, and remote current control. Unlike most gas tungsten arc welding systems, the plasma arc process requires a liquid cooling system. Equipment is commercially available for manual operation in the current range of 0.1 A to 220 A, DCEN. While power sources are available for higher currents, manual torches are not commercially available for above 220 A. Figure 7.8 shows manual plasma arc welding being performed with the use of a rotary positioner.

Mechanized or robotic equipment can help to maximize the high welding speeds and deep penetration advantages associated with high-current plasma arc welding. Both systems, mechanized and robotic, require a power source, control unit, mechanized welding torch, coolant source, high-frequency power generator, and supplies of shielding gases. Accessory units such as an arc voltage control and a filler-wire feeding

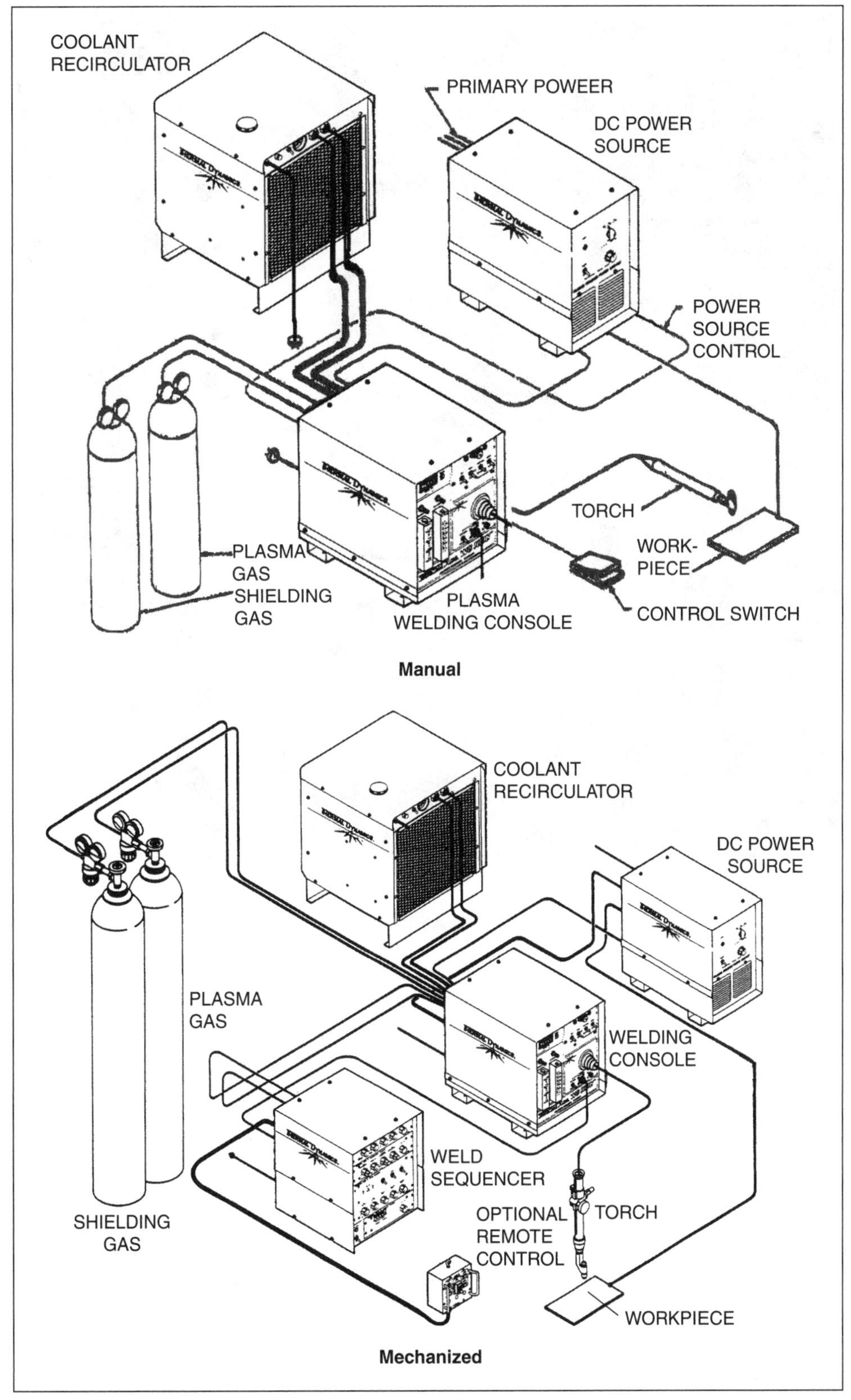

Source: Thermal Arc.

Figure 7.7—Typical Equipment Setup for Manual and Mechanized Plasma Arc Welding

Photo courtesy of Thermal Arc

Figure 7.8—Manual Plasma Arc Welding

system can be used as required. Mechanized welding torches are available for welding with DCEN currents up to 500 A. Mechanized systems use a torch holder mounted to a motion device (i.e., a travel carriage) with the workpiece either stationary or moved with synchronized or coordinated motion. Robotic systems typically operate with multiple-axis coordinated motion.

One additional differentiating characteristic of mechanized plasma arc welding is that operator intervention is required to initiate the transferred arc, whereas a robotic system automatically initiates the welding process when the control system determines that all necessary components are ready and all safety interlock requirements are met.

ARC INITIATION

The plasma arc cannot be started with all the normal techniques used with gas tungsten arc welding. Since the electrode is recessed or flush in the constricting nozzle, it cannot be touch-started against the workpiece. Therefore, to initiate a transferred plasma arc, it is necessary to ignite a nontransferred, low-current pilot arc between the electrode and the constricting nozzle to ionize the gas and facilitate the start of the welding arc. Pilot arc power is normally provided by a separate power source within the control console, but it can be supplied by the welding power source itself. The pilot arc is generally initiated by using high-frequency ac or by a high-voltage direct current (dc) pulse superimposed on the pilot-arc circuit. The high frequency helps to initiate the pilot arc by ionizing the orifice gas so that it will conduct the pilot-arc current.

The basic circuitry for a plasma arc welding system with high-frequency pilot-arc initiation is shown in Figure 7.9. The constricting nozzle is connected to the positive terminal of the power source through a current-limiting resistor. A low-current pilot arc is initiated between the electrode and the nozzle by the high-frequency generator. The electrical circuit is completed through the resistor. The ionized gas from the pilot arc forms a low-resistance path between the electrode and workpiece and thus becomes the preferred path of conductance. When the power source is energized, the main arc is initiated between the electrode and the workpiece. The pilot arc is used only to assist in starting the transferred arc. After the transferred arc starts, the pilot arc may be extinguished, but is often left on to assist in stabilizing the transferred arc and prevent contamination of the tungsten electrode when the transferred arc is terminated.

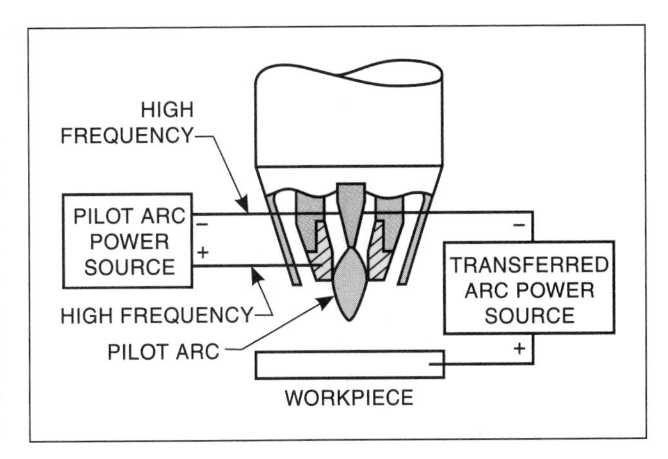

Figure 7.9—Plasma Arc Welding System with High-Frequency Pilot Arc Initiation

POWER SOURCES

The power sources used for plasma arc welding are similar to those for gas tungsten arc welding. Both processes use constant-current power sources and a high-frequency source for arc starting. The current output of the power source can be either pulsed or nonpulsed.

Inverter units are generally preferred over rectifiers or motor-generator units because inverters provide greater consistency in electrical output characteristics and better waveform control. However, some applications benefit from the "overshooting" spike of the rectifiers in variable polarity applications.

Pulsed-Current Power Sources

The use of pulsed current is essential for some plasma arc welding applications. Pulsed-current power sources similar to those used with gas tungsten arc welding are used for plasma arc welding. In the pulsing mode, the current is consistently fluctuated between a higher and a lower amperage setting, which allows the weld pool to partially solidify at the lower level. The pulsing mode improves weld pool control for out-of-position welding and helps alleviate problems of distortion by reducing the total heat input along the weld joint. A pulsed-current power source is a conventional drooping volt-ampere characteristic power source that has the capability of pulsing to a high level, referred to as a high-pulse current. Pulsed-current power sources used for plasma arc welding normally have variable pulse time and current levels. Figure 7.10 is a schematic representation of a typical pulsed-current power source output.

Transistorized, inverter, and silicon-controlled rectifier (SCR) power sources are available with built-in pulsed-current capabilities. For conventional nonpulsed current power sources, add-on modules are available that provide pulsed current within a limited range of pulse frequencies. Pulsed-current power sources may also include upslope, taper-current, and downslope controls.

Nonpulsed Current Power Sources

Conventional power sources with a drooping volt-ampere characteristic are used for DCEN plasma arc welding. They are typically the same power sources that are used for gas tungsten arc welding. These constant-current dc power sources are available with varying amperage capacities, ranging from 0.1 ampere to several hundred amperes, and with various duty cycles.

Typical power sources with an open-circuit voltage in the range of 65 volts (V) to 80 V are satisfactory for plasma arc welding with argon or with argon-hydrogen shielding gas mixtures containing up to 7% hydrogen. However, if helium or an argon-hydrogen gas mixture containing more than 7% hydrogen is used, additional open-circuit voltage may be required for reliable arc ignition. The additional voltage can be obtained by connecting two power sources in series. If erratic arc ignition is experienced, another approach is to initiate the arc in pure argon and then switch over to the desired argon-hydrogen mixture or to helium for the welding operation. Constant-current power sources are available with several options such as a programmed upslope of current, a programmed taper or decay of weld

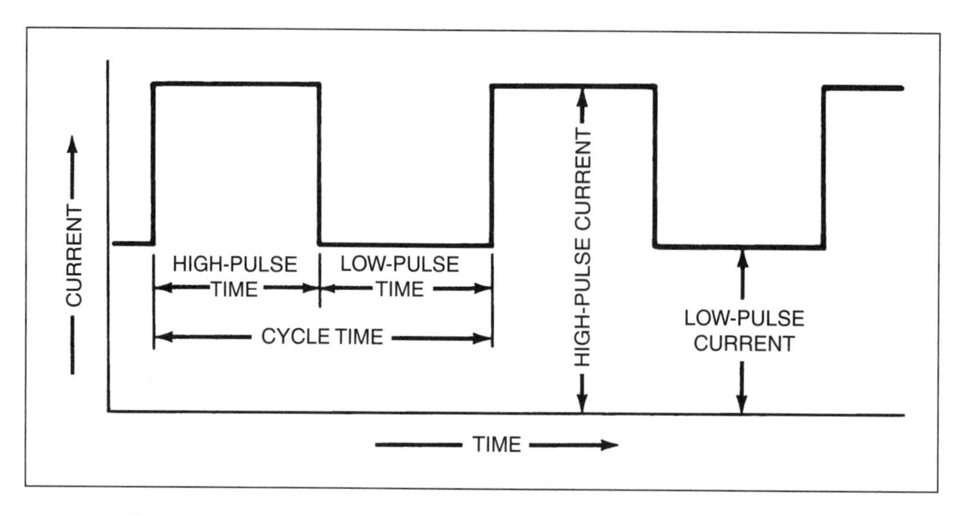

Figure 7.10—Typical Pulsed-Current Power Source Output

current, and a programmed downslope of weld current. These special features of the power source are used for various applications, primarily in mechanized and robotic welding.

PLASMA CONTROL CONSOLE

The main plasma arc welding control systems are contained in a plasma console. The console is normally integrated with the primary power source, but it can also be a stand-alone unit. The stand-alone unit allows a conventional constant-current or pulsed-power source to be used for PAW. A typical plasma console includes controls for setting plasma gas flow, shielding gas flow, and the pilot arc current. In addition, a stand-alone console usually provides a junction box for gas and water hoses, a high-frequency circuit for starting the pilot arc, and a small power source to supply current to the pilot arc. Other features that might be provided on either integrated or stand-alone units are high-low flow-rate options, with upslope and downslope of the plasma gas, so that the plasma gas flow rate can be easily switched between the melt-in mode and the keyhole mode. The upslope and downslope feature is essential to proper opening and closing of the plasma keyhole. Both keyhole and melt-in modes of PAW are further described in this chapter under "Process Variations." Some consoles may include an arc pressure gauge that measures the plasma gas backpressure at the orifice. Most units also include equipment-protection interlocks such as sensors for low coolant flow and low gas pressure. A signal from one of these protective devices would normally extinguish both the pilot and transferred arcs to prevent torch damage.

Some plasma consoles have the built-in capability for programmed upslope and downslope of the plasma gas to start and close a keyhole. Many systems also use an integrated water circulator. Figure 7.11 shows a typical plasma control console.

WELDING TORCHES

The complex functions of the plasma arc welding torch are to direct the current to the fixed electrode, position the electrode, and to direct the flow of shielding gas, orifice gas, and liquid coolant. To accommodate these functions, plasma arc welding torches are constructed with a series of passages necessary to supply the torch with plasma gas, shielding gas, and coolant. In most instances, two dual-function cables provide both the electrical energy and circulating coolant. Both cables supply a current path for the pilot arc, while only one supplies the transferred arc current. Two additional hoses provide the orifice and shielding gases. A coolant is necessary to dissipate the heat generated in the constricting nozzle by both the pilot arc and transferred arc and the heat generated in the current-carrying cables.

The electrode holder assembly in a plasma arc welding torch may be made from a variety of copper alloys. Most are designed to mechanically center the electrode automatically within the central section of the torch nozzle. However, some torches require the electrode to be centered manually. The latter type relies on the initiation of the high-frequency source to create a low-power arc between the constricting nozzle and the electrode. The operator then adjusts the electrode position until the high-frequency arc is evenly distributed around the electrode, thus making the electrode electrically centered. While it is more time consuming, electrically centering the tungsten electrode can extend the life of both the electrode and the constricting nozzle. Any misalignment of the electrode can cause the arc to concentrate on one portion of the torch. This can result in rapid deterioration or melting of the copper nozzle near the orifice, possible weld contamination, and undercutting.

Shielding gas is necessary because the low flow rate of the orifice gas supplied to the torch does not provide sufficient gas volume to protect the weld pool from atmospheric contamination. In addition, turbulence caused by the high velocity of the plasma stream during keyhole welding further reduces the effectiveness of the plasma gas coverage. The shielding gas is supplied through the shielding gas nozzle that surrounds the constricting nozzle of the torch. In some applications, additional trailing gas shields are required for further protection of the workpiece.

Manual Torches

A cross-sectional view of a typical torch design for manual plasma arc welding is shown in Figure 7.12. The torch is generally lightweight and has a handle; a

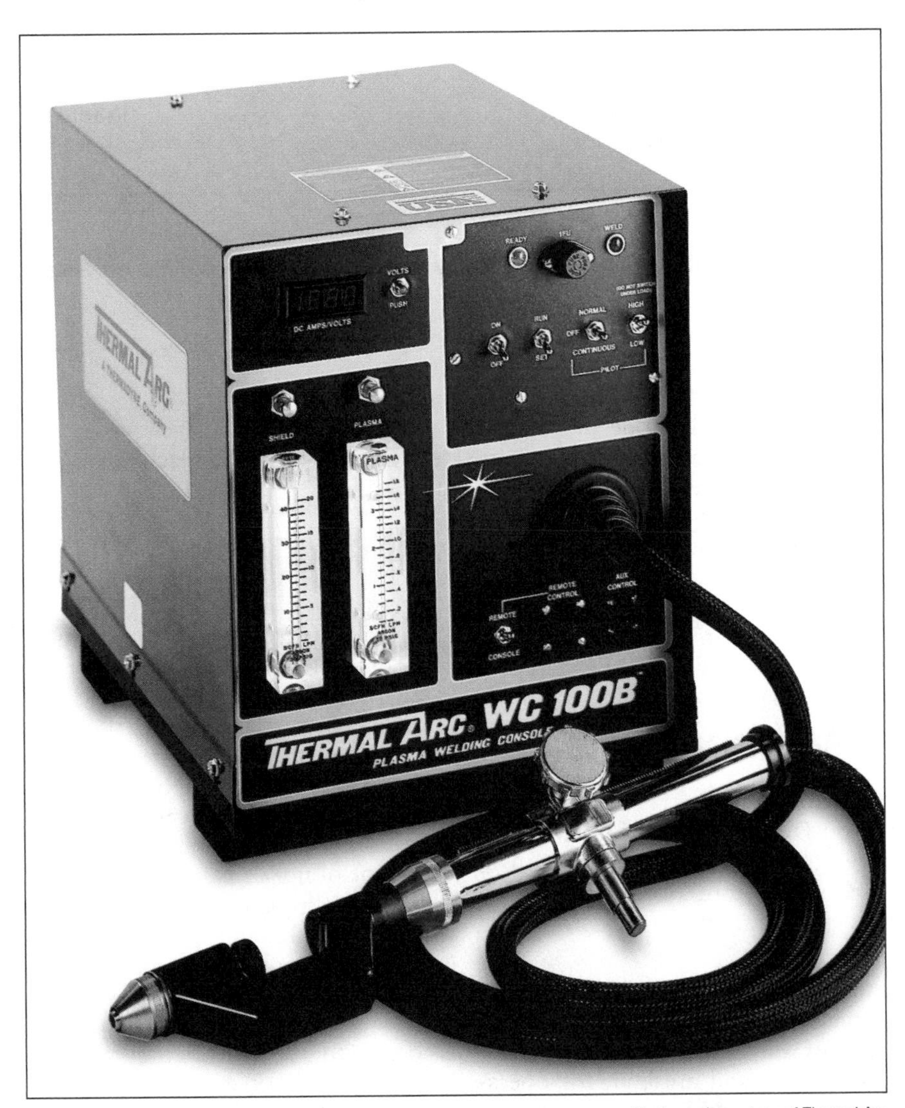

Photograph courtesy of Thermal Arc

Figure 7.11—Typical Plasma Arc Welding Control Console (Shown with Mechanized Torch)

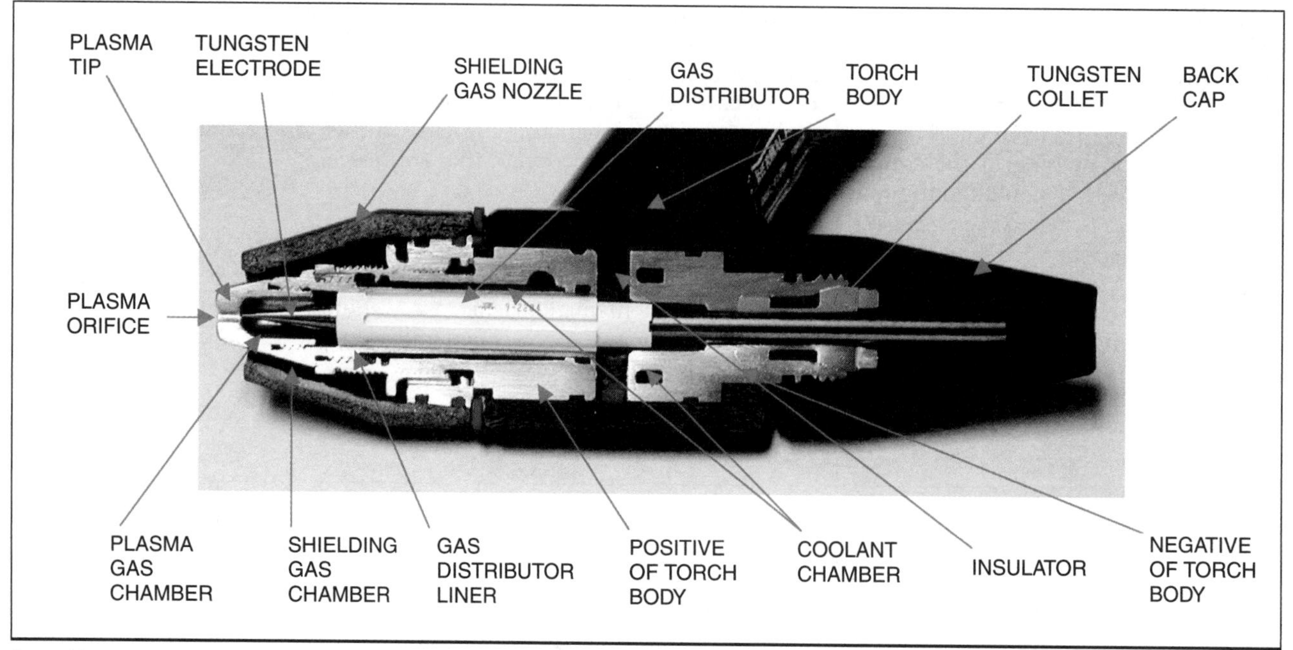

Source: Thermal Arc

Figure 7.12 — Cross-Sectional View of a Typical Manual Torch Head

device for securing the tungsten electrode in position and conducting current to it; separate passages for orifice gas, shielding gas, and coolant; and a shielding gas nozzle (usually made of ceramic material). Manual plasma arc welding torches are commercially available with head angles of 70° and 90°. Manual torches are available for operation with DCEN up to 220 A, but they may be used for DCEP or square-wave ac at reduced current ranges.

Mechanized and Automated Torches

Mechanized plasma arc welding torches are available for operation with either DCEN, DCEP, or square-wave ac with current ratings generally ranging from 10 A to 500 A. DCEN is used with a thoriated tungsten electrode for most welding applications, often with pulsed welding current.

The torches for automated plasma arc welding are similar to manual torches. However, most are designed with straight (in-line) or offset configurations. They usually have larger diameters and are built for operation at significantly higher currents than manual torches. These differing configurations are more compatible with the standard commercial torch-holding devices used with mechanized and automated welding equipment. Figure 7.13 shows a cross-sectional view of a typical mechanized torch head.

ARC CONSTRICTING NOZZLES

A wide variety of constricting nozzles are designed for use in plasma arc welding, including single-orifice and multiple-orifice nozzles, with holes arranged in circles, rows, and other geometric patterns. Each design brings unique advantages to the PAW process.

Single-orifice nozzles are the most widely used and are usually the least expensive. They have only one opening located at the center of the nozzle. The arc and all of the plasma gas pass through the single orifice with this type of nozzle.

The multiple-port (also called *multiport*) nozzle allows the arc and some of the orifice gas to pass through the larger center orifice while the remainder of the gas moves through the smaller auxiliary orifices. The most widely used multiple-port nozzle designs are those with the constricting orifice bracketed by two smaller auxiliary gas orifices with the centers of all three orifices on the same plane. Single- and multiple-port nozzles are illustrated in Figure 7.14.

Multiple-port nozzles can be advantageous for several types of weld joints. When the multiple-port nozzle

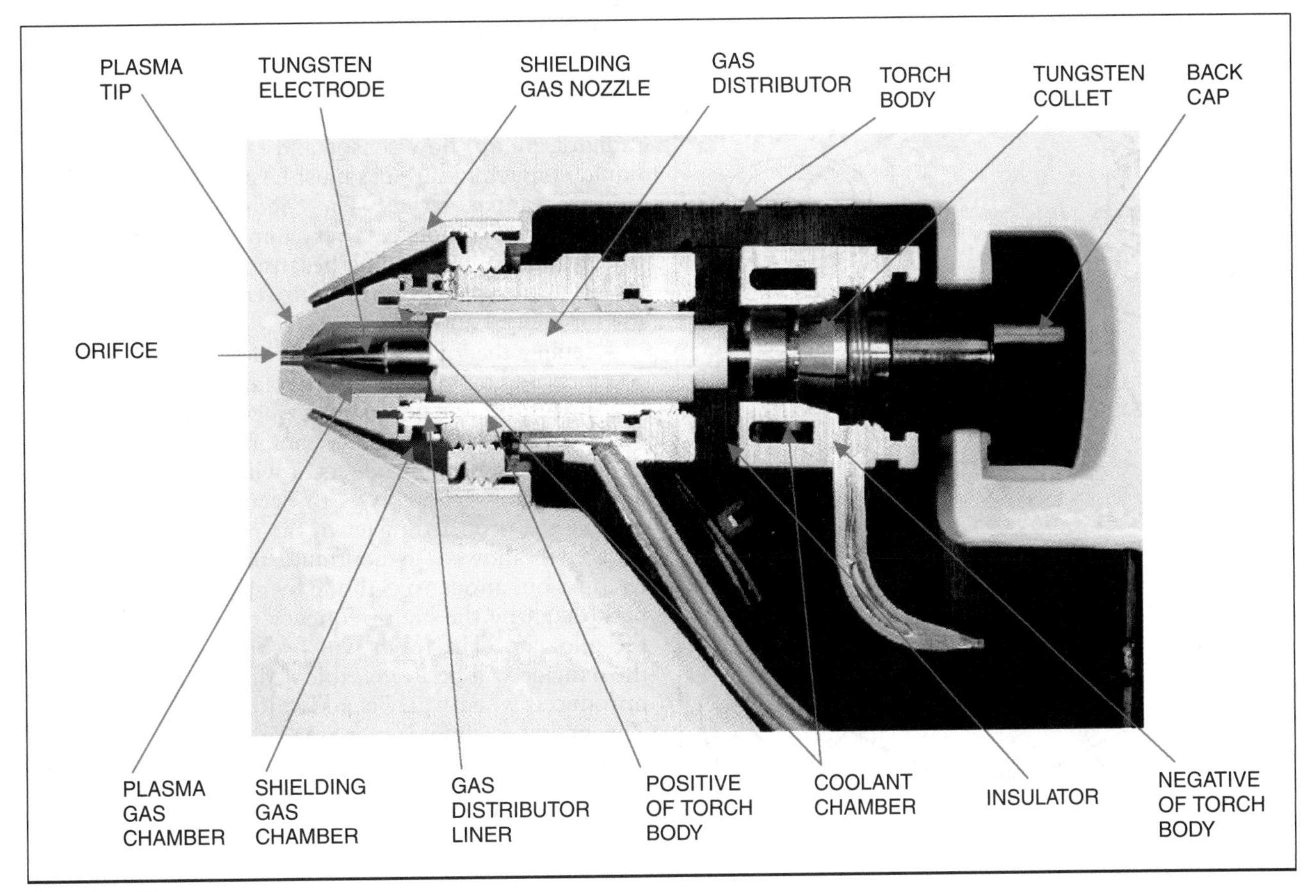

Source: Thermal Arc

Figure 7.13 — Cross-Sectional View of a Typical Mechanized Torch Head

is aligned to place the common centerline of all three ports perpendicular to the weld groove, the plasma jet is concentrated on the joint by the adjacent gas streams. The results are a narrowed weld bead and higher welding speeds.

Each orifice size and orifice gas flow rate has a maximum current rating. For example, an orifice with 2.1 mm (0.083 in.) diameter might be rated for 75 A with an argon flow rate of 0.9 L/min (1.9 ft^3/hr). If the flow rate of the orifice gas dropped below 0.9 L/min (1.9 ft^3/hr), the maximum current rating of the orifice would also decrease.

During normal operation, the arc column within the torch nozzle is surrounded by a layer of nonionized gas. This layer of relatively cool nonconductive gas at the nozzle wall provides thermal and electrical insulation that protects the inside surface of the nozzle. The most commonly used nozzle materials are alloys of copper. A properly cooled copper nozzle can be used to constrict a plasma arc with temperature in excess of 16 600°C (30,000°F). If the protective layer of gas is disturbed, which happens when there is insufficient orifice gas flow or if there is excessive arc current for a particular nozzle geometry, double arcing may occur. Double arcing may result in damage to the nozzle, as described in the previous section, "Arc Modes."

GAS SUPPLY EQUIPMENT

Orifice gas (plasma gas), shielding gas, and trailing gas supplies can be provided by any of the methods used for GTAW. These include locally filled high-pressure gas cylinders, liquid cylinders, or manifold systems. Manifold systems can be supplied by high-pressure gas cylinders, liquid cylinders or fixed cryogenic systems (i.e., bulk tanks). Care must be exercised with all gas supply systems to ensure that contaminants

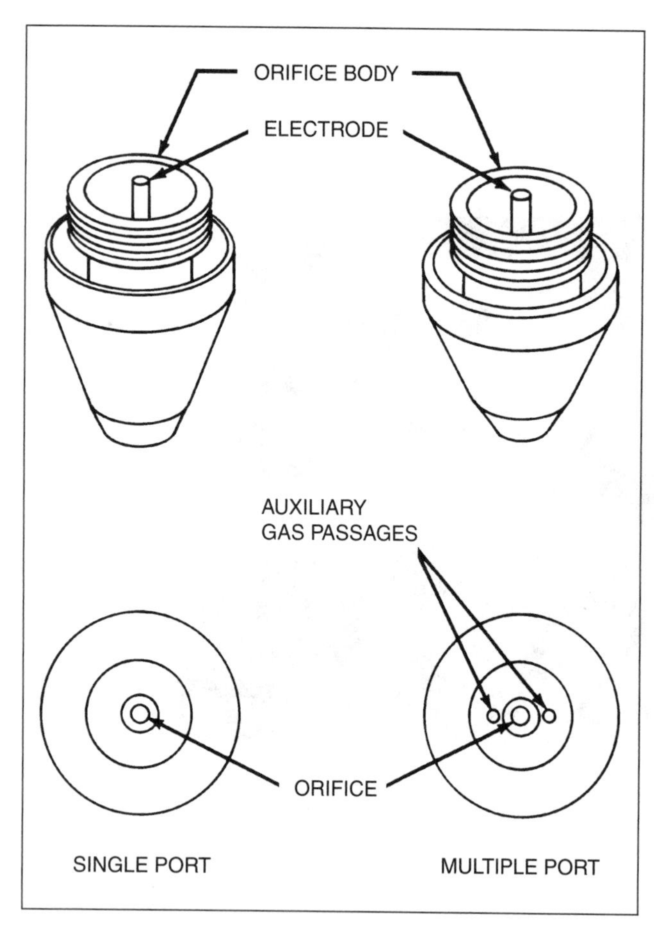

Figure 7.14—Top View (A) and End View (B) of Single- and Multiple-Port Constricting Nozzles

are not introduced through the distribution system. Gas line hoses, pipes, and tubes should be checked regularly for mechanical integrity. Natural rubber hoses, such as oxyfuel gas hoses, should not be used for gas supply lines because of the inherent characteristic of rubber to aspirate atmospheric air into the gas supply. When flexible lines are needed, hoses made of synthetic material should be used. Manifold system installations should not be constructed of carbon steel or other corrosion-prone materials. Moisture and other gas contaminants can provide an environment that allows these materials to corrode and can introduce corrosion products into the shielding gas. This can result in a variety of weld defects. To avoid this type of contamination, manifold systems should be constructed from corrosion-resistant materials such as stainless steel or copper.

COOLING SYSTEM

A liquid cooling system is required for plasma arc welding. It should consist of a coolant reservoir, radiator, pump, flow sensor and control switches. The liquid-contacting surfaces must be constructed of corrosion-resistant materials. The condition of the coolant system and the coolant is very important to the operation of plasma arc welding because the process depends on an air-to-water exchange to remove the heat from the torch head and cables.

Keeping the radiator free of debris and corrosion products and maintaining the coolant in good condition are imperative for continuous operation of the equipment. If the coolant becomes contaminated with foreign objects or corrosion products, it will conduct electricity. This will allow electrolysis to occur in the torch, preventing the proper transfer of the pilot arc to the torch nozzle. If allowed to continue, the torch will short-circuit from anode to cathode by an incorrect electrical path taken by the high-frequency current used to start the pilot arc. The torch will become inoperative, and the damage will be irreparable. Contaminants are often introduced when water is added to the coolant system. The operator should exercise caution to prevent the introduction of foreign objects when adding coolant. Only deionized water should be used in PAW cooling systems.

ACCESSORIES

The accessories available for use with plasma arc welding enhance productivity, improve quality, and expand the capabilities of the process. Accessories include wire feeders, arc voltage control systems, torch oscillators, and positioning equipment. These are the same accessories as those used with gas tungsten arc welding systems.

WIRE FEEDERS

As with the gas tungsten arc welding process, conventional filler wire-feed systems (cold wire) can be used with the plasma arc welding process. The filler metal is added to the leading edge of the weld pool or the keyhole. Most systems use a predetermined feed rate, although more sophisticated controllers like those used in robotic systems can manipulate the wire-feed speed to provide a consistent weld profile. A wire-feed system can alleviate occurrences of undercut or underfill when welding thick materials and can also improve the uniformity of the weld bead by providing a steady addition of filler metal.

Hot-wire feed systems can also be used and should feed the wire into the trailing edge of the weld pool. As with cold wire, the initiation and termination of wire feed can be controlled and programmed with automatic welding equipment.

A popular technique when using pulsed-current welding is to place the filler material into the weld joint in synchronism with the pulsing of the plasma arc current. When automated or mechanized, this process is colloquially known as *dabber welding* because of the back-and-forth movement of the filler wire as it is "dabbed" into the weld pool. This technique is widely used by the aerospace industry to repair thin sections, for example, in building up the edges of labyrinth seals on gas turbine engines.

Arc Voltage Controllers

Arc voltage control (AVC) systems measure transferred arc voltage in mechanized or robotic operations, thus maintaining a programmed value by regulating the torch standoff distance via torch movement. Since the plasma arc welding process is relatively insensitive to arc length variations, AVC equipment is not necessary for many applications. However, arc voltage control can be beneficial when joining uneven or contoured joint geometries using the melt-in mode or when keyhole welding is performed. The control must be deactivated or "locked out" when the current or plasma gas flow rate is sloped during weld starts or crater filling, because changing these variables also causes a change in the arc voltage. AVC systems are used only with robotic or mechanized operations.

Torch Oscillators

Torch oscillators are often used in robotic or mechanized plasma arc welding when a weave bead is desired. Some commercially available systems incorporate the normal robotic motion axis; others can be purchased as stand-alone units. The speed, width, and dwell of oscillation travel are usually adjustable to accommodate the user's specific weld dimensions. The speed is the rate at which the oscillator moves the torch from one side of the joint to the other. The width of oscillation, called the *amplitude*, is the distance between the two stop points of torch travel across the joint. Dwell is the length of time the oscillator stops at one side of the joint before moving back toward the opposite side. Oscillators are most commonly used on PAW equipment for surfacing or hardfacing applications, for example, with the plasma transferred arc method described in the section "Plasma Arc Surfacing."

Positioning Equipment

Positioning equipment for plasma arc welding is similar to that used for gas tungsten arc welding. Depending on the application, the workpiece can be manipulated or the torch motion can be controlled, or both. Workpiece manipulation generally involves a rotary positioner with the capability of tilt control. Moving the torch while the workpiece remains stationary requires a carriage on tracks, a side-beam carriage, or a boom-and-mast manipulator for following linear joints. Combining the movement of the torch and workpiece as a unit normally requires the use of computer programming to coordinate and synchronize the operations (called *coordinated motion control).* Robotic systems are also used to control these complex movements.

Fitup and fixturing equipment for plasma arc welding is the same as that used for gas tungsten arc welding. Joint edges for butt welds should be in intimate contact and should be sufficiently restrained during welding to prevent movement and reduce distortion. Burrs, notches, and gaps along the weld joint should be eliminated (unless a gap is required by the welding procedure specification), as they can cause meltback and separation of the weld joint, underfill (excess concavity), and excessive root reinforcement defects.

Fitup, fixturing, and cleanliness become critical when welding very thin materials such as 0.1 mm (0.004 in.) stainless steel. Joint fitup must be precise, and consideration should be given to the use and placement of chill bars to remove heat from the joint. The graph in Figure 7.15 provides general guidelines of fixturing and clamping requirements when welding thin sections of Type 304 stainless steel.

MATERIALS

Plasma arc welding can be used to join one of the widest ranges of materials that are weldable by an arc welding process. When planning for plasma arc welding, the properties of the base metals, filler wires or rods, and the consumable inserts must be evaluated to achieve the desired outcome.

BASE METALS

Metals welded by the plasma arc welding process include carbon steels, stainless steels, other ferrous alloys, aluminum, magnesium, copper and copper alloys (copper-nickel, bronze, and brass), and nickel and nickel alloys. Reactive metals such as titanium, zirconium, tantalum and others can also be welded with PAW.

Most material thicknesses from 0.3 to 6.4 mm (.012 to 1/4 in.) can be welded in one pass with a standard DCEN transferred arc. All metals except aluminum and

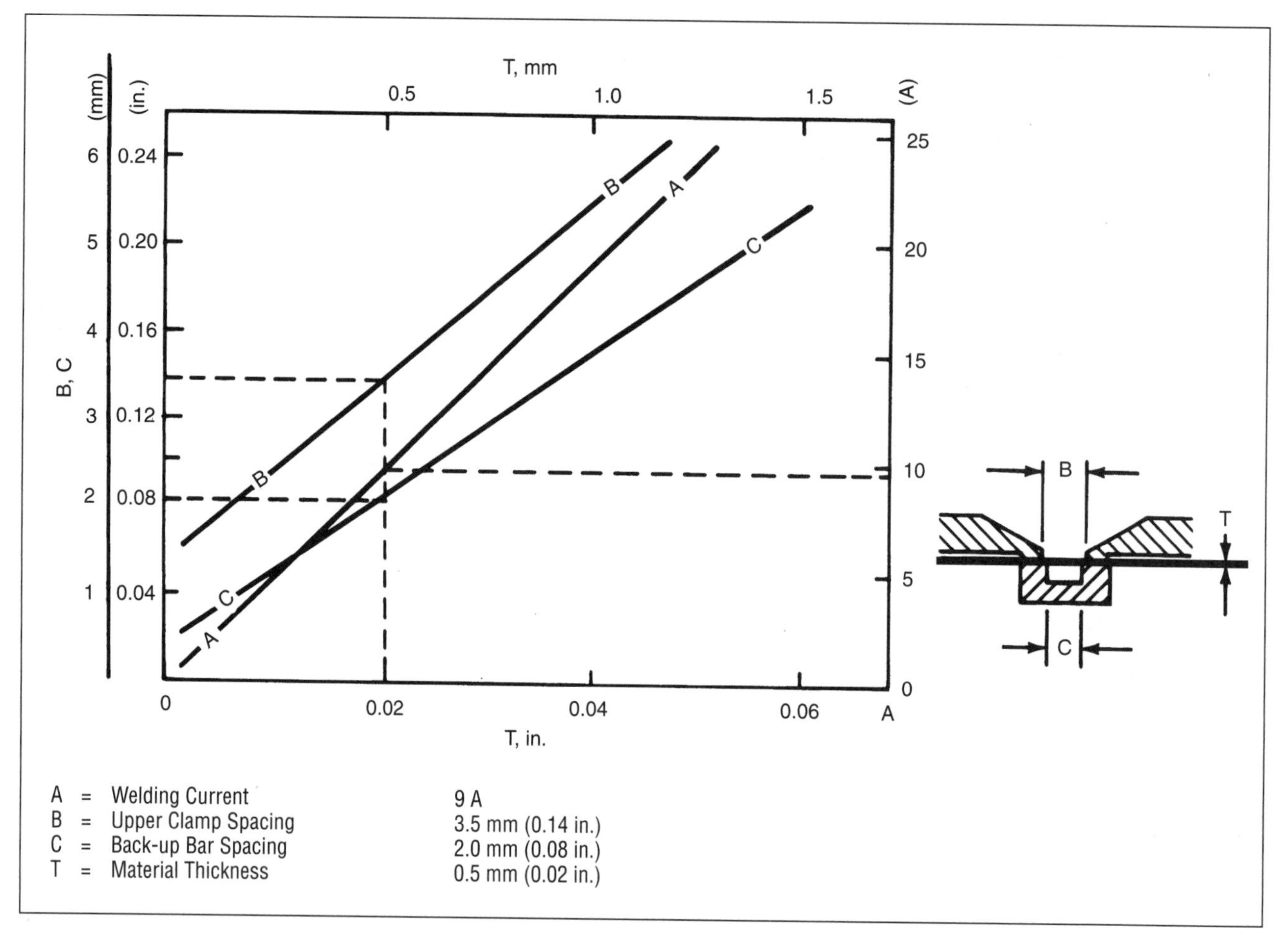

Figure 7.15—Guidelines for Low-Current Groove Welding of Stainless Steel with Cross Section of Butt Joint Geometry

magnesium and alloys of these metals are normally welded with DCEN. When welding aluminum and magnesium, alternating current is effectively used to remove the tenacious refractory oxides typical of these metals. Either conventional or variable polarity (square-wave) alternating current will provide this oxide removal. However, conventional alternating current will significantly reduce the current capacity of the electrode. Variable polarity alternating current minimizes the duration of the electrode positive portion of the current cycle. High quality single-pass welds can be made in aluminum alloys up to 19 mm (3/4 in.) thick with variable polarity plasma arc welding operating in the keyhole mode. Variable polarity plasma arc (VPPA) welding is discussed in the section "Process Variations."

The metallurgical effects of the heat input from plasma arc welding and gas tungsten arc welding are similar except that the smaller-diameter, more focused plasma arc will usually melt a smaller, deeper weld pool in the base metal, thus achieving narrower and deeper penetration. The plasma arc also creates a smaller heat-affected zone (HAZ) than a gas tungsten arc weld of the same type. Preheat, postheat, and gas shielding procedures are similar for both processes. Each base metal has unique requirements for maximizing weld quality.

ELECTRODES

The electrodes used in plasma arc welding are the same as those used in gas tungsten arc welding. Electrodes are produced in accordance with *Specification for Tungsten and Tungsten Alloy Electrodes for Arc*

Table 7.1
AWS Specifications for Filler Metals Used for Plasma Arc Welding

AWS Specification	Filler Metals
A5.7	*Specification for Copper and Copper Alloy Bare Welding Rods and Electrodes*
A5.9	*Specification for Bare Stainless Steel Welding Electrodes and Rods*
A5.10	*Specification for Bare Aluminum and Aluminum Alloy Welding Electrodes and Rods*
A5.13	*Specification for Solid Surfacing Welding Rods and Electrodes*
A5.14	*Specification for Nickel and Nickel Alloy Bare Welding Electrodes and Rods*
A5.15	*Specification for Welding Electrodes and Rods for Cast Iron*
A5.16	*Specification for Titanium and Titanium Alloy Welding Electrodes and Rods*
A5.18	*Specification for Carbon Steel Filler Metals for Gas Shielded Arc Welding*
A5.19	*Specification for Magnesium Alloy Welding Electrodes and Rods*
A5.21	*Specification for Composite Surfacing Welding Rods and Electrodes*
A5.22	*Specification for Stainless Steel Electrodes for Flux Cored Arc Welding and Stainless Steel Flux Cored Welding Rods for Gas Tungsten Arc Welding*
A5.24	*Specification for Zirconium and Zirconium Alloy Welding Electrodes and Rods*
A5.28	*Specification for Low Alloy Steel Electrodes and Rods for Gas Shielded Metal Arc Welding*
A5.30	*Specification for Consumable Inserts*

Welding and Cutting, A5.12/A5.12M.[5,6] Tungsten electrodes with small additions of thorium, lanthanum, or cerium can be used for DCEN welding, with the thoriated variety being the most common. Pure tungsten and zirconiated electrodes are rarely used for plasma arc welding because the electrode tip geometry cannot be maintained.

The work end of the electrode is normally ground to a cone shape with an included angle of 20 to 60 degrees, as specified by the torch manufacturer. It is essential to have a smooth concentric shape for the cone in order to provide a uniform current-carrying surface on the electrode. Thus, precision machine grinding of electrodes is strongly recommended. Improved electrical characteristics are achieved when the grinding direction is parallel to the long axis of the electrode or when the ground surface has a finish of 0.813 microns (32 microinches) or smoother. For square-wave ac or VPPA welding, the electrode is usually prepared with a flat or blunt area at the end of the cone. The appropriate shape helps prevent electrode overheating, improves arc stabilization, and provides greater current-carrying capability.

FILLER MATERIALS

The consumable materials used for plasma arc welding are identical to those used for gas tungsten arc welding. They include filler wires, rods, and consumable inserts.

Filler Wires and Rods

Filler metals are added in rod form for manual welding or wire form for mechanized and robotic welding. Table 7.1 lists the American Welding Society (AWS) specifications for appropriate filler metals developed by various subcommittees of the AWS Committee on Filler Metals and Allied Materials.

Consumable Inserts

Consumable inserts are specially formed or machined sections of filler metal. They are placed into the weld joint prior to arc initiation, subsequently fused

5. American Welding Society (AWS) Committee on Filler Metals and Allied Materials, 1998, *Specification for Tungsten and Tungsten Alloy Electrodes for Arc Welding and Cutting*, AWS A5.12/A5.12M: 1998. Miami: American Welding Society.
6. At the time of the preparation of this chapter, the referenced codes and other standards were valid. If a code or other standard is cited without a date of publication, it is understood that the latest edition of the document referred to applies. If a code or other standard is cited with the date of publication, the citation refers to that edition only, and it is understood that any future revisions or amendments to the code or standard are not included; however, as codes and standards undergo frequent revision, the reader is advised to consult the most recent edition.

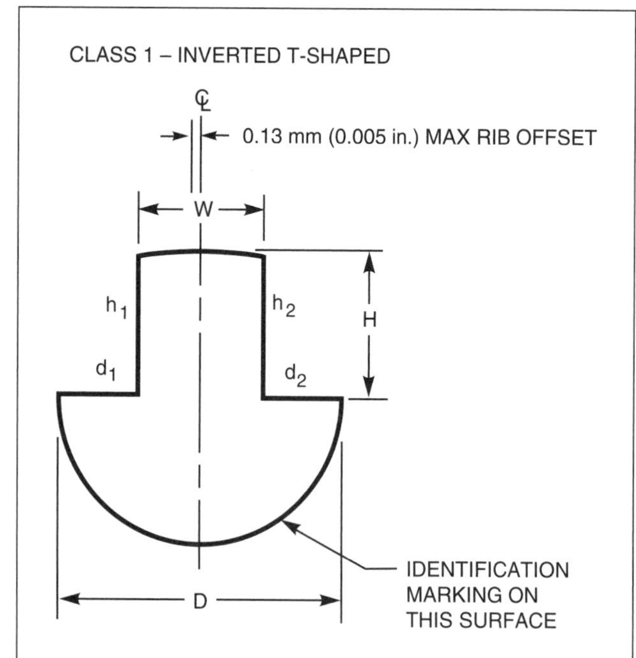

Source: Adapted from American Welding Society (AWS) Committee on Filler Metals and Allied Materials, 1997, *Specification for Consumable Inserts*, A5.30-97, Miami: American Welding Society.

Figure 7.16—Class 1 Consumable Insert for Heavy-Wall Pipe Welding

into the base metal, and consumed by the weld pool similarly to filler wire or rod. Consumable inserts differ from backing bars or plates in that consumable inserts are placed in the root of the weld joint; there they are consumed by the weld pool and become an integral part of the weld. Consumable inserts are most widely used for circumferential butt joints in heavy-wall pipe. Figure 7.16 shows a cross section of a Class 1 insert commonly used for heavy-wall pipe welding.

GASES

The choice of gases to be used for plasma arc welding depends on welding position, joint configuration, and the metal to be welded. For many plasma arc welding applications, the shielding gas is often the same as the orifice gas. However, some advantages can be observed when a different gas is used for certain applications. Typical gases used to weld various metals with the melt-in mode are shown in Table 7.2.

When the gas flow and current must be varied during the weld, such as at the start and end of a keyhole weld, a programmable electronic gas control system is used. When welding in the keyhole mode, the sloping of the orifice gas flow rate and current are essential to the proper opening and closing of the keyhole.

Orifice Gas

Purity of the orifice gas is critical to electrode life and should be at least 99.99% pure. When cleanliness is critical for the materials being joined, welding operations may be improved by using an ultra-high-purity gas (99.999%). The orifice gas must be inert with respect to the tungsten electrode in order to avoid rapid deterioration of the electrode.

Shielding Gas

The shielding gases are generally inert. However, an active gas can be used for shielding if it does not adversely affect weld properties. The gas provided through the shielding gas nozzle can be argon, an argon-hydrogen mixture, or an argon-helium mixture, depending on the welding application. Shielding gas flow rates are usually in the range of 5 L/min to 15 L/min (11 ft^3/h to 32 ft^3/h) for low-current applications. For high-current welding, flow rates of 15 L/min to 32 L/min (32 ft^3/h to 64 ft^3/h) are used.

Argon. Argon is the preferred orifice gas for most plasma arc welding because the low ionization potential of argon assures a dependable pilot arc and reliable arc starting. Since the pilot arc is used only to maintain ionization in the plenum chamber, pilot arc current is not critical; it can remain fixed for a wide variety of operating conditions. The recommended orifice gas flow rates are normally 0.25 to 5 L/min (0.5 to 11 ft^3/h). The pilot arc current can range from 5 A to 30 A. Both gas flow rate and pilot arc current depend on the size of the orifice in the constricting nozzle. Argon is also the most economical of the standard welding gases or gas mixtures.

Argon-Hydrogen Mixtures. Argon-hydrogen mixtures can be used as the orifice gas and shielding gas for making welds in stainless steel, nickel-base alloys, and copper-nickel alloys. Permissible hydrogen percentages vary from 5%, used on 6.4 mm (1/4 in.) stainless steel, to 10% used in some tube mills to obtain the highest welding speeds on 3.8 mm (5/32 in.) wall thickness and thinner-wall stainless tubing. In general, the thinner the workpiece the higher the permissible percentage of hydrogen in the gas mixture is, up to 10% maximum. However, when argon-hydrogen mixtures are used as an orifice gas, the rating of the orifice diameter for a

Table 7.2
Gas Selection Guide for Plasma Arc Welding (Melt-in Mode)*

Metal		Thickness mm	Thickness in.	Orifice (Plasma) Gas	Shielding Gas	Backing or Trailing Gas
Carbon steel (aluminum killed)	Under	3.2	1/8	Ar	Ar	Ar
	Over	3.2	1/8	Ar	Ar75% He-25% Ar	Ar
Low-alloy steel	Under	3.2	1/8	Ar	Ar	Ar
	Over	3.2	1/8	Ar	Ar75% He-25%Ar	Ar
Stainless steel	Under	3.2	1/8	Ar, 98% Ar-2% H_2**	Ar, 95% Ar-5% H_2	Ar
	Over	3.2	1/8	Ar, 98% Ar-2% H_2**	Ar, 95% Ar-5% H_2	Ar
Aluminum	Under	2.8	3/32	Ar	Ar, He, 75% He-25% Ar	N/A***
	Over	2.8	3/32	Ar	He, 75% He-25% Ar	N/A
Copper	Under	2.8	3/32	Ar	Ar-He, 75% He-25% Ar	Ar
	Over	2.8	3/32	Ar	He, 75% He-25% Ar	Ar
Nickel alloys	Under	3.2	1/8	Ar, 98% Ar-2% H_2**	Ar, 95% Ar-5% H_2	Ar
	Over	3.2	1/8	Ar, 98% Ar-2% H_2**	Ar, 75% He-25% Ar95% Ar, 5% H_2	Ar
Reactive metals (e.g., titanium, tantalum)	Under	6.4	1/4	Ar	Ar, 75% He-25% Ar	Ar
	Over	6.4	1/4	Ar	Ar, He, 75% He-25% Ar	Ar

* Gas types and percentages are for general guidance. Moderate adjustments may yield improved results when the Welding Procedure Specification is properly qualified.
** Hydrogen in the plasma gas may have a detrimental effect on arc starting performance. Amounts greater than 5% in plasma gas can lead to accelerated erosion of the constricting nozzle.
*** Some highly reactive aluminum alloys may require a backing or trailing gas.

given welding current is usually reduced because of the higher arc temperature. Operation at the higher, argon-only ratings with argon-hydrogen gas mixtures for the orifice gas will reduce the service life of the orifice.

Additions of hydrogen to argon shielding gas produce a hotter arc and more efficient heat transfer to the workpiece than pure argon. Argon-hydrogen gas mixtures facilitate higher welding speeds with a given arc current. However, some alloys, such as aluminum, cannot be welded with hydrogen mixed into the gas. The amount of hydrogen that can be used in the mixture is limited because excessive hydrogen additions tend to cause porosity or cracking in the weld bead or heat-affected zone, or both. When welding with the plasma keyhole mode, a given metal thickness can be welded with higher percentages of hydrogen than are possible with the melt-in mode of PAW or the GTAW process. The ability to use these higher percentages of hydrogen may be associated with the keyhole effect, in which contaminants are removed in the plasma gas exiting the root of the weld joint. It may also be associated with the different solidification pattern the keyhole technique produces.

Argon-Helium. Helium additions to argon shielding gas produce a hotter arc for a given arc current than pure argon. However, the mixture must contain at least 50% helium before a significant change in heat can be detected; performance with mixtures containing over 75% helium is about the same as with pure helium. Argon-helium mixtures containing between 50% and 100% helium are generally used for making keyhole welds in heavy titanium and aluminum sections and for fill passes on all metals when the additional heat is desirable.

Carbon Dioxide. Since the shielding gas does not come in contact with the tungsten electrode, reactive gases such as carbon dioxide (CO_2) can sometimes be used. One application of this is in the plasma arc welding of transformer lamination stacks, often using 75% Ar-25% CO_2 as the shielding gas and flow rates in the range of 10 L/min to 15 L/min (21 ft^3/h to 32 ft^3/h). However, most plasma arc welding applications do not use CO_2.

Back Purge Gas and Trailing Shield Gas

When welding reactive metals such as titanium, zirconium, or tantalum, it is essential to shield the hot metals from atmospheric contamination until they have

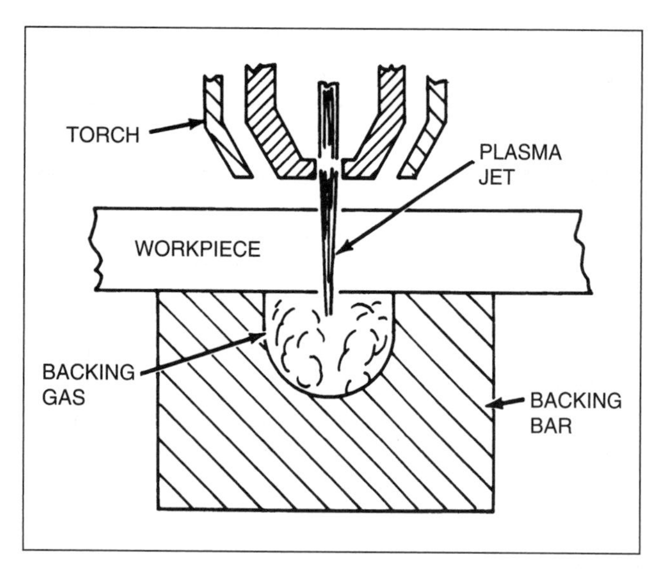

Figure 7.17—Typical Backing Bar Used for Back-Purging During Keyhole Plasma Arc Welding

cooled below the temperature at which they react with oxygen. The same is true for other corrosion-resistant metals such as stainless steel and nickel alloys. Auxiliary shielding provided by backup and trailing shields may be necessary. A trailing shield device can be attached to the rear of a plasma torch or on the motion device behind the torch. The trailing shield provides increased gas coverage while the workpiece is cooling, preventing oxidation or other metal reactions with the atmosphere.

The plasma arc welding of highly reactive materials can be performed in a welding chamber (colloquially, a *glove box*) similar to gas tungsten arc welding. When welding metals such as stainless steel with the root of the weld exposed to the atmosphere, a back purge of a gas is normally used. A back purge can also be applied through a specially designed backing bar. Figure 7.17 shows a cross section of a backing bar used for plasma keyhole welding.

Weld Backing

Weld backing is a material or device placed at the back side of a joint near the root to position the workpieces, absorb heat from the weld, and shield the area of the weld root. Weld backing also supports the molten weld metal and allows greater ease in making full-penetration groove welds on butt joints.

Chill clamping and gas backup can also be used to help reduce the heat-affected zone of the weld. Chill clamping is a general term for the placement (normally near the weld) of thermally conductive metal pieces, such as copper or aluminum, in intimate contact with the base metal to accelerate heat removal. The type of material being welded largely determines if gas backup is required. Metals such as titanium, zirconium, nickel, stainless steels, and tantalum need inert gas backup. Using a back purge gas can also help cool and protect items such as embedded electronic components that might be inside a workpiece during welding. Some weld joints may not allow gas backup because of design constraints.

APPLICATION METHODS

Plasma arc welding is performed in the same manner as gas tungsten arc welding, with some variations and additional techniques. The basic techniques are the same for manual, mechanized, and robotic modes. The greatest difference in possible application methods is derived from the options provided by the melt-in mode or the keyhole mode. Whereas the melt-in mode can be performed manually, mechanized or robotically, the keyhole mode cannot practically be done manually. For other applications, however, manual plasma arc welding can be superior to gas tungsten arc welding because torch standoff distance is not as critical.

MANUAL PLASMA ARC WELDING

Manual plasma arc welding is usually best when performed in the low-current range from 0.1 A to 50.0 A. Manual welders must recognize that the plasma arc is very directional because of the columnar nature of the plasma stream, which makes it more difficult to keep the arc centered on the weld joint. The arc does not bend to follow the centerline of the joint. Side-to-side angular variations of the torch must be extremely limited. Fortunately, the plasma arc is forgiving of variations in torch standoff distance; thus, such variations are not as critical as in gas tungsten arc welding.

MECHANIZED PLASMA ARC WELDING

Mechanized welding, in general, involves preset control of welding parameters, but it requires operator action to initiate the transferred arc. One benefit of using plasma arc welding in preference to gas tungsten arc welding with any type of automation is the elimination of the high voltage needed by gas tungsten arc welding for transferred arc initiation. Because high frequency is used only to start the pilot arc, there is no high-frequency burst when the arc is transferred for

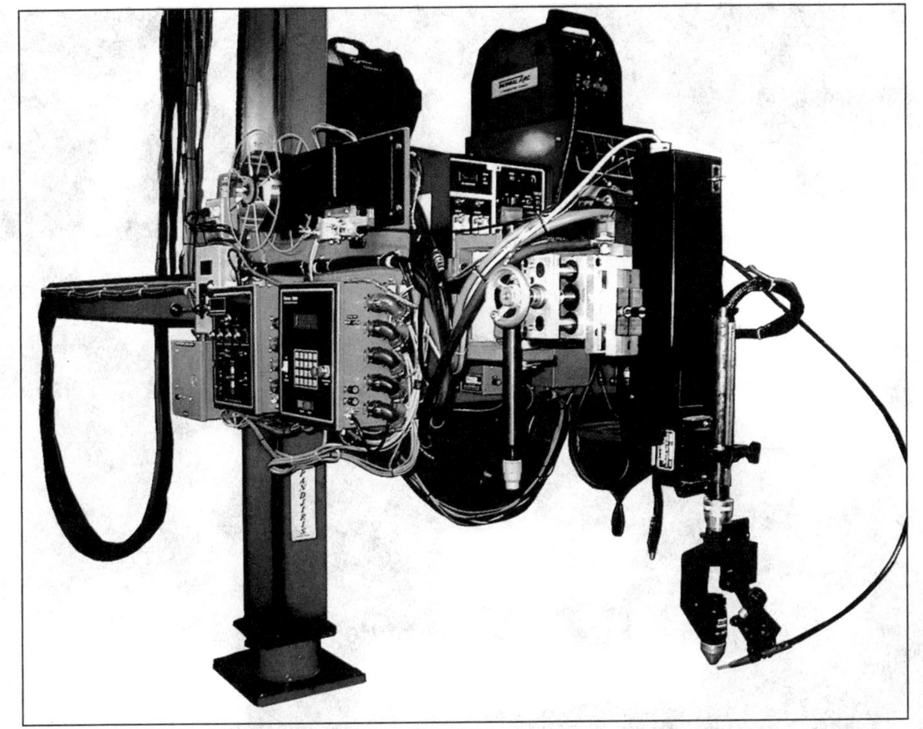

Photograph courtesy of Pandjiris Inc.

Figure 7.18—Mechanized Plasma Arc Welding with Wire Feeder and Column-and-Boom Manipulator

welding. This allows the plasma arc to be used more easily around sensitive electronic equipment such as robots, microprocessor controls, electronic testing equipment, and programmable controls. Figure 7.18 shows a mechanized plasma arc welding system with wire feeder. This system has a column-and-boom manipulator that provides the torch motion for straight-line welding. A rotary positioner, turning rolls, or other workpiece motion device is often used in conjunction with a manipulator to perform circumferential and other nonlinear welds.

AUTOMATED PLASMA ARC WELDING

Both weld quality and production rates can be greatly improved for many plasma arc welding applications by implementing automated systems. Automated systems can be either mechanized or robotic. These systems provide the most precise control methods for the numerous parameters required for plasma arc welding. This is especially true for the keyhole welding mode, when precise control of the plasma gas flow rate, current, wire feed, and travel speed must be coordinated to properly open and close the keyhole.

ROBOTIC PLASMA ARC WELDING

Robotic systems often include various feedback mechanisms that assist the control system in adjusting to welding parameters, without operator action, to accommodate unforeseen changes in the weld joint geometry or location. These changes include variables such as manufacturing tolerances beyond specifications, poor fitup, or excessive distortion. Some robotic systems utilize real time digital images of the weld pool to make changes to welding parameters (e.g., current, travel speed, and wire feed) to prevent a poor-quality weld. Figure 7.19 shows a robotic application of plasma arc welding using an articulate arm robot in a work cell.

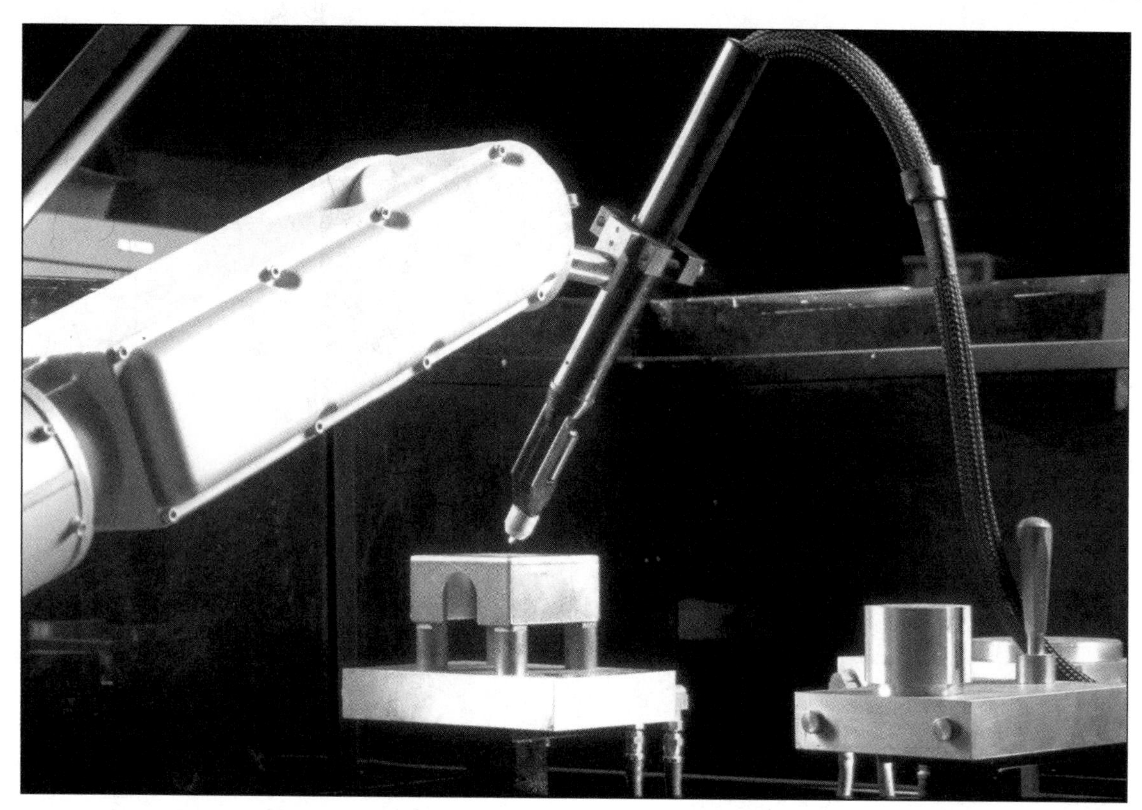

Photograph courtesy of Thermal Arc

Figure 7.19—Robotic Plasma Arc Welding

PROCESS VARIATIONS

Plasma arc welding can be performed in all positions with limitations similar to those of gas tungsten arc welding. Through its several variations—high-current and low-current melt-in modes, the keyhole mode, hot-wire welding, variable polarity plasma arc welding, and surfacing—the process offers fabrication flexibility and economy while maintaining weld joint quality and reliability. All metals weldable with the gas tungsten arc welding process can be satisfactorily welded with the plasma arc process; therefore, few exceptions are required in the establishment of weldment acceptance specifications.

MELT-IN WELDING

Mechanized melt-in plasma arc welding is very popular, especially for welding the small, intricate components of medical equipment, lighting instruments, batteries, wires, and bellows. It is also commonly used for applications in which the greater energy concentration improves overall weld quality. The plasma arc melt-in mode is often the preferred method for welding thin components when the initial energy burst of arc initiation in gas tungsten arc welding would tend to melt through or vaporize these thin sections. In many applications, microprocessor controls are used to regulate, via feedback loops, parameters such as initial current, upslope, pulsation, downslope, and final current.

High-Current Mode

Welding procedure specifications for high-current welding in the range of 50 A to 500 A often use the melt-in mode, which produces a weld similar to that obtained with conventional gas tungsten arc welding. The melt-in mode is generally preferred over the gas tungsten arc process in mechanized applications for consistent control of weld quality. Again, due to arc stability and stiffness, arc penetration into the weld joint is more controlled and welding time is reduced.

The high-current melt-in mode of the plasma arc welding process is widely used for welding and cladding operations in pipe and tube mills, as well as in the aerospace and nuclear industries. The welding of cover passes on keyhole welds is also an important application.

Low-Current (Microplasma) Mode

The low-current melt-in mode of plasma arc welding is typically called *microplasma*. While there is no firm differentiation of high or low current values, users typically consider microplasma as operating in the range of 0.1 A to 20 A, using precise current control. Modern inverter power sources provide a very stable and controllable arc at low currents for welding thin materials. A pilot arc provides a reliable means of transferred arc initiation at these low current levels. The columnar arc produces uniform bead contours in edge joints with manual and automated welding.

Applications include the welding of turbine blades, gas turbine rotor seal teeth edges, bellows, pacemakers, diaphragms, electronic components, and many other fabrications using thin material. Plasma arc welding is often the economical choice over laser welding for these applications because initial equipment costs and operating costs are lower. The low-current mode is also a superior substitute for many applications that previously required brazing.

Advantages of the Melt-in Mode

The low-current and high-current melt-in modes have several advantages over gas tungsten arc welding resulting from the greater energy concentration. The melt-in modes reduce welding time in many applications. Lower current is needed to produce a given weld, which results in lower heat input, less shrinkage, and reduced distortion. Penetration is less affected by changes in torch standoff distance, and arc starting is more reliable at lower currents.

Other advantages are the following:

1. Arc stability is improved at very low currents;
2. Narrower beads (higher depth-to-width ratio) for a given penetration result in less distortion;
3. The arc column has greater directional stability;
4. The need for fixturing is reduced for some applications;
5. The addition of filler metal (if required) is easier than with GTAW because of the increased torch standoff distance and the recessed electrode, which cannot readily make contact with the filler metal or weld pool;
6. The increased torch standoff distance results in less downtime to regrind the point on the tungsten electrode and also reduces the chance for tungsten contamination of the weld; and
7. Out-of-position welding is much easier because reasonable variations in torch standoff distance have little effect on bead width or heat concentration at the workpiece.

Limitations of the Melt-in Mode

Several limitations are associated with high-current and low-current melt-in plasma arc welding. The narrow, constricted arc allows little tolerance for joint misalignment. Because the torches are larger for manual plasma arc welding, they are sometimes slightly more difficult to manipulate and more expensive than a comparable gas tungsten arc welding torch. For consistent weld quality, the constricting nozzle must be well maintained and regularly inspected for signs of deterioration.

KEYHOLE WELDING

In the plasma arc keyhole welding mode, the orifice gas flow rate is gradually increased after the weld pool is established to displace the molten metal and form a hole (the keyhole). The hole fully penetrates the base metal. As the plasma arc torch moves along the weld joint, metal flows from the leading edge of the keyhole, around the plasma stream, and to the rear where the weld pool progressively solidifies. Appropriate combinations of plasma gas flow rate, arc current, and weld travel speed must be coordinated to produce the keyhole. Figure 7.20 is a sketch of a typical keyhole weld.

The principal advantages of keyhole welding are that it makes very high quality, complete-penetration welds on most thicknesses in a single pass, and normally does not require the beveling of the plate edges before welding.

Plasma arc keyhole welding is generally performed in the flat position on material thicknesses ranging from 1.6 mm to 9.5 mm (1/16 in. to 3/8 in.). However, with appropriate welding conditions, keyhole welding can be done in any position and on some metal thicknesses up to 19 mm (3/4 in.). The open keyhole provides an escape path through its liquid edges for impurities to flow to the surface and gases to be expelled before solidification of the weld pool. The maximum volume of the weld pool and the resulting underbead root surface profile are largely determined by the force balance between the welding position, surface tension of the molten weld metal, the plasma arc current, and the velocity of the ionized gas exiting the orifice.

The high-current keyhole welding technique operates just below conditions that would actually cut and expel metal rather than weld. For cutting, a slightly higher orifice gas velocity blows the molten metal away. In

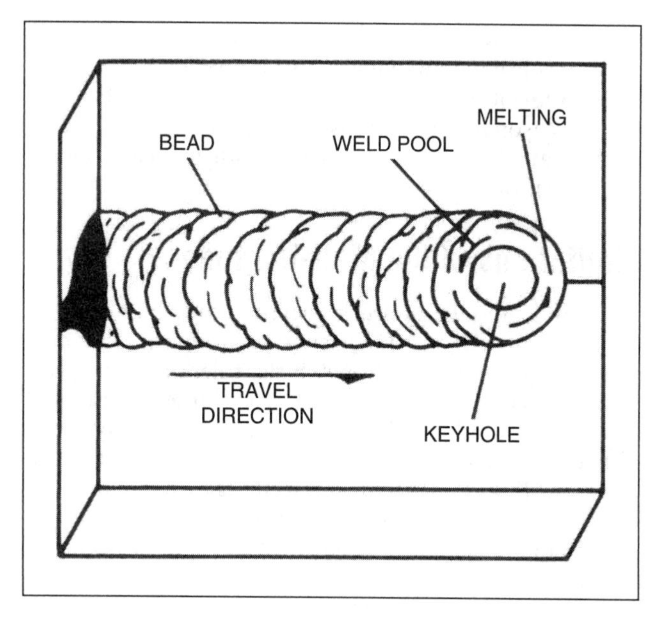

Figure 7.20—Schematic Representation of the Keyhole in Plasma Arc Welding

welding, a lower gas velocity allows surface tension to hold the molten metal in the joint. Consequently, orifice gas flow rates for keyhole welding are critical and must be closely controlled. Orifice gas flow of less than 0.62 L/min (1.3 ft^3/h) is normally recommended, depending on the size of the orifice in the constricting nozzle. Table 7.3 lists typical orifice gases and shielding gases for keyhole welding.

Plasma arc keyhole welding is best when performed in an automated mode because of the requirement for the accurate control of travel speed, plasma gas flow, and wire feed speed. The continuing development of more accurate and reliable mass-flow controllers continues to provide improvements to the precise control of the plasma gas during welding that is critical to successful keyhole welding.

Advantages and Limitations of Plasma Arc Keyhole Welding

Plasma arc keyhole welding has several important advantages that result in improved operations com-

Table 7.3
Gas Selection Guide for Plasma Arc Welding (Keyhole Mode)[a]

Metal		Thickness mm	Thickness in.	Orifice (Plasma) Gas	Shielding Gas	Backing or Trailing Gas
Carbon steel	Under	3.2	1/8	Ar[b]	Ar	Ar
(aluminum killed)	Over	3.2	1/8	Ar	Ar, 75% He-25% Ar	Ar
Low-alloy steel	Under	3.2	1/8	Ar[b]	Ar,	Ar
	Over	3.2	1/8	Ar	Ar, 75% He-25% Ar	Ar
Stainless steel	Under	3.2	1/8	Ar, 98% Ar-2% H_2[c]	Ar, 95% Ar-5% H_2	Ar
	Over	3.2	1/8	Ar, 98% Ar-2% H_2[c]	Ar, 95% Ar-5% H_2	Ar
Aluminum	Under	2.8	3/32	Ar[b]	Ar, He, 75% He-25% Ar	Ar
	Over	2.8	3/32	Ar	He, 75% He-25% Ar	Ar
Copper	Under	2.8	3/32	Ar[b]	Ar, He, 75% He-25% Ar	Ar
	Over	2.8	3/32	Not Recommended[d]	Not Recommended[d]	Not Recommended[d]
Nickel alloys	Under	3.2	1/8	Ar, 98% Ar-2% H_2[c]	Ar, 95% Ar-5% H_2	Ar
	Over	3.2	1/8	Ar, 98% Ar-2% H_2[c]	Ar, 75% He-25% Ar, 95% Ar-5% H_2	Ar
Reactive metals	Under	6.4	1/4	Ar	Ar, He, 75% He-25% Ar	Ar
(e.g., titanium and tantalum)	Over	6.4	1/4	Ar	Ar, He, 75% He-25% Ar	Ar

a. Gas types and percentages are for general guidance. Moderate adjustments may yield improved results when the Welding Procedure Specification is properly qualified.
b. Not recommended for material less than 1.6 mm (1/16 in.) thick.
c. Hydrogen in plasma gas may have a detrimental effect on arc starting performance. Amounts greater than 5% in plasma gas can lead to accelerated constricting nozzle erosion.
d. The underbead will not form correctly. The technique can be used for copper-zinc alloys only.

pared to gas tungsten arc welding. The advantages and limitations are summarized in the next sections.

Advantages

1. The plasma gas flushing through the open keyhole helps remove gases that would, under other circumstances, be trapped as porosity in molten metal;
2. The symmetrical fusion zone of the keyhole weld reduces the tendency for transverse distortion;
3. The greater joint penetration permits a reduction in the number of passes required for a given joint, thus further reducing distortion;
4. Many welds can be completed in a single pass, reducing weld time; and
5. Square butt joints are generally used, thus reducing joint preparation and machining costs.

Limitations. Plasma welding keyhole welding procedures involve more process variables and tolerate less deviation in them. Except when welding aluminum alloys, most keyhole plasma arc welding is restricted to the flat position. Also, the plasma torch must be well maintained for consistent operation, which may add to costs.

Electrode-to-Orifice Relationship

The electrode setback (the distance the tungsten electrode is recessed into the torch) is normally specified by the torch manufacturer. A gauge is usually provided to adjust the setback to the recommended position, however, some operational benefits can be achieved when the electrode tip is positioned closer to the orifice, thus reducing the electrode setback. This produces an arc phenomenon commonly called a *soft plasma* arc. A soft plasma arc is most often used for applications such as transformer laminations, outside corner welds, and any other areas where an arc that performs like a gas tungsten arc is beneficial. The advantage of the soft plasma arc over GTAW is reliable arc starting. When the electrode is essentially flush with the orifice, an arc similar to a gas tungsten arc can be achieved, but the need for high-frequency arc initiation with each weld can be eliminated. This is often used when GTAW is the desired process, but arc initiation or the use of high frequency is a concern. Caution should be exercised when operating with a reduced setback, as the operational life of the orifice may be shorter than that experienced in the standard plasma arc welding operation. Electrode setback should not be changed beyond the manufacturers' recommendations without consulting the manufacturer.

VARIABLE POLARITY PLASMA ARC KEYHOLE WELDING OF ALUMINUM

Keyhole plasma arc welding of aluminum can be accomplished by using square-wave alternating current with a variable polarity plasma arc (VPPA). The variable polarity waveform is shown in Figure 7.21. This type of waveform, in which the duration and magnitude of the DCEN and DCEP current excursions can be controlled separately, is obtainable using solid-state technology. Most modern VPPA systems use inverter power sources, although many fabricators continue to use silicon-controlled rectifier (SCR) power sources. In some applications, fewer problems with arc rectification are

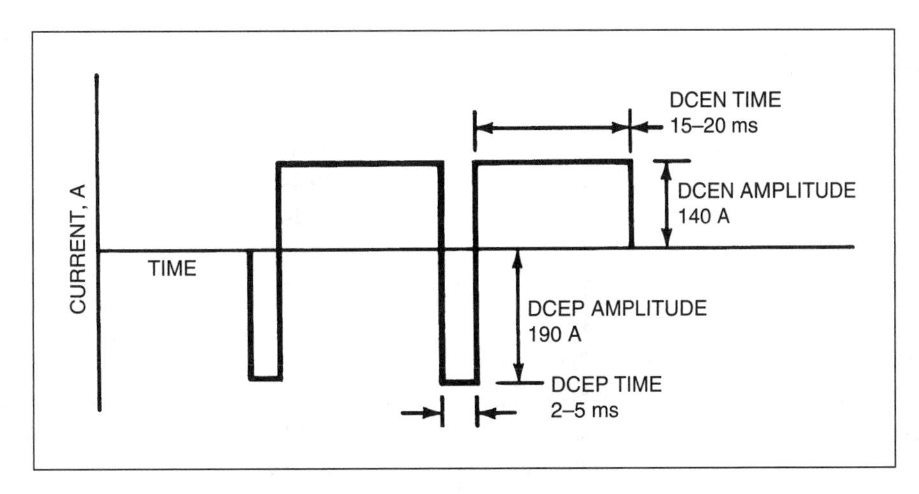

Figure 7.21—Typical Variable Polarity Current Waveform

experienced with the SCR power sources than the inverters. This may be linked to the slight current spike or "overshoot" an SCR goes through when switching polarities. A description of power sources using inverters and those using silicon-controlled rectifiers is provided in Chapter 1 of this volume.

The cleanliness of the aluminum workpiece surface is of the utmost importance in avoiding porosity in the weld. Usual cleaning procedures involve mild alkaline solutions or vapor degreasing. Welding should occur soon after cleaning. With variable polarity welding, oxide removal before welding is not required for most aluminum alloys. However, the 5000-series aluminum alloys typically have extremely tenacious surface oxides that may require removal by scraping, grinding, or machining prior to welding.

Continuous high frequency is needed in alternating-current GTAW to re-ignite the arc, which tends to extinguish as voltage and current go through zero during the polarity switch. Since a direct-current pilot arc is maintained during VPPA welding, continuous high frequency is not required.

The most important variable in the keyhole plasma arc welding of aluminum is the duration of DCEN and DCEP. From empirical testing, best results have normally been obtained with DCEN current flowing from 15 milliseconds (ms) to 20 ms and DCEP current flowing from 2 ms to 5 ms. These welding conditions are shown in Figure 7.21. With DCEP durations shorter than 2 ms, sufficient cleaning is not usually achieved and the weld tends to be porous. When the DCEP time exceeds 6 ms, tungsten deterioration and double-arcing tendencies become more apparent.

It should be noted that the DCEP current amplitude shown in Figure 7.21 is greater than the DCEN current amplitude. The added DCEP current provides an additional spike of cleaning action to break up surface oxides on the workpiece, yet produces minimal heat input on the electrode and torch orifice. Appropriate cleaning of the weld face and root face can be accomplished by increasing the DCEP current an additional 30 A to 100 A. This additional DCEP current also reduces the DCEP time required for proper base metal cleaning.

Typical keyhole VPPA welding conditions for 6.4 mm (1/4 in.) thick aluminum in the flat, horizontal, and overhead positions are shown in Table 7.4. This process has been used to make single-pass, full-penetration welds on space shuttle fuel tanks, hulls for hydrofoil boats, and aluminum tanks.

HOT-WIRE WELDING

Hot-wire welding, a variation in which filler-metal wire is resistance heated by the current flowing through the wire, can be implemented with the plasma arc process. Hot-wire welding is beneficial because it improves

Table 7.4
Variable Polarity Plasma Arc Welding Conditions for 6.35 mm (0.250 in.) Aluminum in the Flat, Horizontal, and Overhead Positions

	Position		
Welding Parameter	Flat	Horizontal	Overhead
Material thickness, mm (in.)	6.4 (1/4)	6.4 (1/4)	6.4 (1/4)
Type of aluminum	2219	3003	1100
Filler metal diameter, mm (in.)	1.6 (1/16)	1.6 (1/16)	1.6 (1/16)
Filler metal grade	2319	4043	4043
DCEN Welding current, A	140	140	170
DCEN Welding time, ms	19	19	19
Additional DCEP current, A	50	60	80
DCEP Current time, ms	3	4	4
Plasma gas, L/min (cfh) (start)	Ar, 0.9 (2)	Ar, 1.2 (2.5)	Ar, 1.2 (2.5)
Plasma gas, L/min (cfh) (run)	Ar, 2.4 (5)	Ar, 2.1 (4.5)	Ar, 2.4 (5)
Shielding gas flow, L/min (cfh)	Ar, 14 (30)	Ar, 19 (40)	Ar, 21 (45)
Electrode size, mm (in.)	3.2 (1/8)	3.2 (1/8)	3.2 (1/8)
Travel speed, mm/s (in./min)	3.4 (8)	3.4 (8)	3.2 (7.5)

travel speed and can minimize dilution. Hot-wire plasma arc welding is normally used in the high-current melt-in mode.

SURFACING

The plasma arc process can also be used for surfacing operations. Surfacing is the application by welding, brazing, or thermal spraying of a layer or layers of material to a surface to obtain desired properties or dimensions. Plasma arc spraying can be performed either with a transferred arc or nontransferred arc.

PLASMA SPRAYING

Plasma spraying (PSP) is a thermal spraying process in which a nontransferred arc is used to create an arc plasma that melts and propels the surfacing material to the substrate. A variation of this process is vacuum plasma spraying (VPSP), in which the plasma spraying gun is confined to an enclosure that is partially evacuated. Plasma spraying is not a true welding process but is similar to brazing or braze welding because adhesion between the surfacing metal to the workpiece is achieved through intermetallic bonding. The surfacing material may be in wire or powder form.[7]

PLASMA TRANSFERRED ARC SURFACING

The plasma arc welding process is used in a surfacing technique commonly called *plasma transferred arc (PTA)* surfacing. The term *PTA* is not technically specific to a surfacing operation, but rather defines the mode of the plasma arc. When surfacing material is applied in powder form, users in the industry often refer to the process as *PTA*. Using a plasma transferred arc, the surfacing material is continuously introduced by a mechanized powder feeder. It can also be added in wire or rod form, but this is not common. Figure 7.22 shows an example of a surfacing operation being performed on flat plate using a transferred plasma arc and a surfacing alloy in powdered form.

Photograph courtesy of Stellite Coatings

Figure 7.22—A Plasma Transferred Arc Surfacing Operation Applying a Powdered Metal Alloy

Powdered metal has several advantages over solid wire. Many alloys are used for surfacing to achieve unique mechanical or chemical properties, such as resistance to wear or corrosion. Unfortunately, many of these alloys cannot be drawn into a wire or rod because of these same unique mechanical properties. It is also much easier and more economical to obtain specialized formulas in powder form. However, the greatest benefit from the use of powdered surfacing material is that it reduces dilution of the weld deposit.

In a surfacing operation using wire or rod, a relatively large weld pool must be formed from the base metal. The surfacing wire or rod is then inserted into the weld pool as a solid, where it melts and becomes liquid. When powdered metal is used, the material is injected into the plasma stream after it leaves the constricting nozzle and before it strikes the weld pool. Compared to wire or rod, the same amount of material in powder form covers a much greater surface area. These two factors, greater surface coverage and injection into the plasma stream, combine so that nearly all the powder is heated to a molten state when it strikes the weld pool. This greatly reduces the size of the weld

7. American Welding Society (AWS) Committee on Definitions, 2001, *Standard Welding Terms and Definition sand Symbols*, A3.0:2001, Miami: American Welding Society.

pool needed and thus results in a significant reduction in weld deposit dilution compared to other surfacing processes. Dilution rates as low as 5% can be obtained in some alloys using powdered metal applied by the plasma transferred arc technique, compared to 15% to 50% with other processes.

WELDING PROCEDURES

A welding procedure specification (WPS) is necessary for a reliable and repeatable welding operation. The WPS is a document that provides the required welding variables and their allowable ranges for a specific application.[8] The usefulness and effectiveness of a WPS are so critical that one is required by nearly every welding fabrication code. Some typical welding parameters for a variety of materials are shown in Tables 7.5 through 7.8. While these values are good approximations for WPS development, they are not to be used without the required WPS testing.

WELD QUALITY

Plasma arc weld discontinuities include both surface and subsurface types. Most are identical to those encountered in the gas tungsten arc welding process. This section will discuss only the discontinuities unique to plasma arc welding: porosity in keyhole closure welds and subsurface contamination.

POROSITY

Keyhole closure porosity can be a significant problem in plasma arc welding. As the name suggests, it occurs only when welding in the keyhole mode. Keyhole porosity can normally be discerned from typical weld porosity by its shape. Because of the forces present during its formation, keyhole porosity tends to be oblong in shape, whereas typical out-gassing porosity is more spherical in shape.

The source of the problem is the same as for other forms of porosity: gases trapped in solidifying weld metal. During the closing of the keyhole, both current rate and plasma gas flow rate are gradually reduced, or downsloped. Porosity is formed when the arc current and the plasma gas flow rate are reduced at the same time and too abruptly. As the keyhole shrinks, the speed of the plasma gas going through the keyhole may increase, despite the reduced flow rate, and cause a pressure drop on the torch side of the keyhole. The drop in pressure increases the risk of air being drawn through the argon shielding and into the keyhole, which would contaminate the weld pool. Additionally, at the exact point where the keyhole closes on the root side, sufficient gas pressure is still present on the face side to create a dimple in the weld pool. The dimple can force some of the liquid weld metal to be expelled from the weld pool and can leave a void that may be left unfilled.

During keyhole closure, an increase in the filler wire feed speed can help reduce the likelihood of keyhole closure porosity. This is a function of the deoxidizing agents that are present in most welding filler metals. In linear welding applications, the use of runoff weld tabs can improve weld quality. A runoff tab is a relatively short piece of the same metal as the workpiece, abutted and centered at the end of the weld joint. During welding, the arc and keyhole are terminated on the runoff tab, leaving any possible defects from keyhole closure outside of the final weld joint when the runoff tab is removed.

SUBSURFACE CONTAMINATION

Subsurface contamination of the plasma arc weld can result when copper from a molten torch nozzle is expelled into the weld. This condition usually arises when the torch nozzle gets too close to the weld or excessive current is used that overheats the constricting nozzle, causing copper to melt into the weld pool. The resulting contamination, which may be detrimental, may not be detectable by conventional nondestructive examination (NDE) procedures in some metals. The best way to avoid copper contamination is to use proper welding procedure specifications and provide operator training to help develop good torch manipulation techniques.

ECONOMICS

Plasma arc welding has significant economic advantages over gas tungsten arc welding. Higher welding speeds not only increase productivity, but they also can reduce workpiece distortion through lower overall heat input. The deeper penetration typical of plasma arc welds will often produce high-quality welds with less weld joint preparation time and with fewer weld passes. These reduced time requirements not only lower labor and fabrication costs but also help to further reduce distortion.

8. See Reference 7.

Table 7.5
Typical Plasma Arc Welding Parameters for Butt Joints in Stainless Steel

Thickness		Travel Speed		Current (DCEN),	Arc Voltage,	Nozzle	Gas Flow[b]				
							Orifice[c] Gas		Shielding Gas[c]		
mm	in.	mm/s	in./min	A	V	Type[a]	L/min	ft³/h	L/min	ft³/h	Remarks[d]
2.4	0.092	10	24	115	30	111M	3	6	17	36	Keyhole, square-groove weld
3.2	0.125	13	30	145	32	111M	5	10	17	36	Keyhole, square-groove weld
4.8	0.187	7	16	165	36	136M	6	13	21	44	Keyhole, square-groove weld
6.4	0.250	6	14	240	38	136M	8	17	24	50	Keyhole, square-groove weld

a. Number designates orifice diameter (in thousandths of an inch). Diameter is stated only in U.S. customary units to reflect the manufacturers' use of inch-pound units of measure. "M" designates multiple-port design.
b. Gas underbead shielding is required for all welds.
c. Gas used is 95% Ar-5% H_2.
d. Torch standoff distance is 4.8 mm (3/16 in.)

Table 7.6
Typical Plasma Arc Welding Parameters for Butt Joints in Carbon and Low-Alloy Steels

Metal	Thickness, mm (in.)	Travel Speed, mm/s (in./min)	Current (DCEN), A	Arc Voltage, V	Nozzle Type[a]	Gas Flow[b] Orifice Gas,[c] L/min (ft³/h)	Gas Flow[b] Shielding Gas,[c] L/min (ft³/h)	Remarks[d]
Mild steel	3.2 (0.125)	5 (12)	185	28	111M	6 (13)	28 (59)	Keyhole, square-groove weld
4130 steel	4.3 (0.170)	4 (9)	200	29	136M	6 (13)	28 (59)	Keyhole, square-groove weld, 1.2 mm (3/64 in.) diameter filler wire added at 30 in./min (13 mm/s)
D6AC steel	6.4 (0.250)	6 (14)	275	33	136M	7 (15)	28 (59)	Keyhole, square-groove weld, 600°F (315°C) preheat

a. Number designates orifice diameter (in thousandths of an inch). Diameter is stated only in U.S. customary units to reflect the manufacturers' use of inch-pound units of measure. "M" designates multiple-port design.
b. Gas underbead shielding is required for all welds.
c. Gas used is argon.
d. Torch standoff distance is 1.2 mm (3/64 in.) for all welds.

Table 7.7
Typical Plasma Arc Welding Parameters for Butt Joints in Titanium

Thickness, mm (in.)	Travel Speed, (mm/s) in./min	Current (DCEN), A	Arc Voltage, V	Nozzle Type[a]	Gas Flow[b] Orifice Gas, L/min (ft³/h)	Gas Flow[b] Shielding Gas, L/min (ft³/h)	Remarks[c]
3.2 (0.125)	8.5 (20)	185	21	111M	4 (8[d])	28 (59[d])	Keyhole, square-groove weld
4.8 (0.189)	5.5 (13)	175	25	136M	9 (19[d])	28 (59[d])	Keyhole, square-groove weld
9.9 (0.390)	4.2 (10)	225	38	136M	15 (32[e])	28 (59[e])	Keyhole, square-groove weld
12.7 (0.500)	4.2 (10)	270	36	136M	13 (27[f])	28 (59[f])	Keyhole, square-groove weld

a. Number designates orifice diameter (in thousandths of an inch). Diameter is stated only in U.S. customary units to reflect the manufacturers' use of inch-pound units of measure. "M" designates multiple-port design.
b. Gas underbead shielding is required for all welds.
c. Torch standoff distance is 4.8 mm (3/16 in.)
d. Gas used is argon.
e. Gas used is 75% He-25% Ar.
f. Gas used is 50% He-50% Ar.

Table 7.8
Typical Plasma Arc Welding Parameters for Welding Stainless Steels-Low Amperage

Thickness, mm (in.)	Type of Weld	Travel Speed, mm/s (in./min)	Current (DCEN), A	Orifice Diameter, mm (in.)	Orifice Gas Flow, L/min (ft³/h)[a, b, c]	Torch Standoff Distance, mm (in.)	Electrode Diameter, mm (in.)	Remarks
0.76 (0.030)	Square-groove weld, butt joint	2 (5)	11	0.76 (0.030)	0.3 (0.6)	6.4 (1/4)	1.0 (0.040)	Mechanized
1.5 (0.060)	Square-groove weld, butt joint	2 (5)	28	1.2 (0.047)	0.4 (0.8)	6.4 (1/4)	1.5 (0.060)	Mechanized
0.76 (0.030)	Fillet weld, T-joint	—	8	0.76 (0.030)	0.3 (0.6)	6.4 (1/4)	1.0 (0.040)	Manual, filler metal[d]
1.5 (0.060)	Fillet weld, T-joint	—	22	0.047 (1.2)	0.4 (0.8)	6.4 (1/4)	1.5 (0.060)	Manual, filler metal[d]
0.76 (0.030)	Fillet weld, lap joint	—	9	0.76 (0.030)	0.3 (0.6)	9.5 (3/8)	1.0 (0.040)	Manual, filler metal[d]
1.5 (0.060)	Fillet weld, lap joint	—	22	(0.047) 1.2	0.4 (0.8)	9.5 (3/8)	1.5 (0.060)	Manual, filler metal[e]

a. Orifice gas is argon.
b. Shielding gas is 95% Ar-5% H at 10 L/min (20 ft³/h).
c. Gas underbead shielding is argon at 5 L/min (10 ft³/h).
d. Filler wire is 1.1 mm (0.045 in.) diameter 310 stainless steel.
e. Filler wire is 1.4 mm (0.055 in.) diameter 310 stainless steel.

Higher initial equipment cost and a slightly higher inert gas consumption rate due to the orifice gas requirement is normal for plasma arc welding. However, users often find that the capital expenditures and operating costs for plasma arc welding are rapidly offset by the high quality of welds and the savings in labor time for many applications.

SAFE PRACTICES

The potential hazards involved with plasma arc welding are similar to those of other processes and are especially similar to those encounterd in gas tungsten arc welding. Among these safety concerns are high-pressure cylinders, gas, metal fumes, thermal burns, radiant energy, electrical shock, and equipment handling. Please refer to the "Safe Practices" section of Chapter 3, "Gas Tungsten Arc Welding," in this volume. Related safety codes, standards, specifications, pamphlets, and books and their publishers are listed in Appendix B. Information specific to plasma arc welding is summarized in this section.

Both the pilot arcs and transferred arcs used in plasma arc welding emit bright visible light and ultraviolet light. The level of light depends on the amount of current in the arc. Therefore, precautions must be taken to protect the eyes and body from the effects of these light spectrums. Too often, welders and welding operators may fail to take precautions around the pilot arc because of the low current and less brilliant light. However, the radiant energy of the pilot arc can cause eye damage and reddening of the skin from ultraviolet light exposure.

A standard welding helmet with the appropriate shade of filter lens for the current being used is required for eye and face protection. Table 7.9 provides a guide for lens shade selection for various current ranges from ANSI Z49.1.[9] It is normally suggested that the welder or welding operator choose the darkest lens available when selecting the correct lens shade, and then change to lighter shades until the weld zone is sufficiently visi-

Table 7.9
Plasma Arc Welding
Shaded Lens Selection Guide

Arc Current (A)	Minimum Shade	Suggested Shade
Less than 20	6	6–8
20–100	8	10
100–400	10	12
400–800	11	14

ble. However, the welder or operator should not change to a lens that is lighter than the minimum shade listed in Table 7.9.

When a pilot arc is operated continuously, normal precautions should be used for protection against arc flash and heat burns. Suitable clothing must be worn to protect exposed skin from arc radiation. Welding power should be turned off before electrodes are adjusted or replaced. Adequate eye protection should be used when observing the high-frequency discharge that occurs when centering the electrode in some torches.

Accessory equipment such as wire feeders, arc voltage controls, and oscillators should be properly grounded. If they are not grounded, insulation breakdown might cause these units to become electrically "hot" with respect to ground.

Adequate ventilation should always be assured. This is of particular concern when welding metals with high copper, lead, zinc, chromium, or beryllium contents.

Detailed safety information is provided in the manufacturers' instructions for the safe operation and maintenance of the equipment and in the suppliers' material safety data sheet (MSDS) for the safe handling, use, and storage of materials and gases used in the process. The standard *Safety in Welding, Cutting, and Allied Processes*, ANSI Z49.1, should be consulted.[10] Mandatory federal safety regulations established by the Occupational Safety and Health Administration (OSHA) of the U.S. Labor Department are provided in the latest edition of *Occupational Safety and Health Standards for General Industry, Code of Federal Regulations*, Title 29 Part 1910.[11]

CONCLUSION

As an extension of gas tungsten arc welding, plasma arc welding produces the same high-quality, highly reliable welds, but adds several important improvements over GTAW. These advancements include the columnar arc, more reliable arc starting, deeper weld penetration, and additional modes available for welding.

Equipment for plasma arc welding has more required components and thus is somewhat more expensive than gas tungsten arc equipment. It has additional inert gas and liquid cooling requirements. However, plasma arc welding can be more economical than gas tungsten arc welding in many applications because of the more focused arc, the more reliable transferred arc initiation, and the versatility of the process.

BIBLIOGRAPHY[12]

American National Standards Institute (ANSI) Accredited Standards Committee. 1999. *Safety in Welding, Cutting, and Allied Processes*, Z49.1: 1999. Miami: American Welding Society.

American Welding Society (AWS) Committee on Filler Metals and Allied Materials. 1997. *Specification for Consumable Inserts*, A5.30-97. Miami: American Welding Society.

American Welding Society (AWS) Committee on Filler Metals and Allied Materials. 1998. *Specification for Tungsten and Tungsten Alloy Electrodes for Arc Welding and Cutting*, AWS A5.12/A5.12M: 1998. Miami: American Welding Society.

American Welding Society (AWS) Committee on Plasma Arc Welding. 1973. *Recommended Practices for Plasma Arc Welding*, C5.1-73. Miami: American Welding Society.

Occupational Safety and Health Administration (OSHA). 1999. *Occupational Safety and Health Standards for General Industry*, in *Code of Federal Regulations (CFR)*, Title 29, CFR 1910. Washington D.C.: Superintendent of Documents, U.S. Government Printing Office.

Springer-Verlag. 1956. *Encyclopedia of Physics*, Volume XXII. Berlin: Springer-Verlag.

SUPPLEMENTARY READING LIST

Alexandrov, O. A., O. I. Steklov, and A. V. Alexeev. 1993 Use of plasma arc welding process to combat hydrogen metallic disbonding of austenitic stainless steel claddings. *Welding Journal.* 72(11): 506-s to 516-s.

Boucher, C., F. Messager, F. Gaillard, and J. L. Heuze. 1992. Plasma arc welding of TA6V titanium alloy Titanium. '92 Science and Technology. *Proceedings, Symposium at 7th World Titanium Conference,* San

9. American National Standards Institute (ANSI) Accredited Standards Committee, 1999, *Safety in Welding, Cutting, and Allied Processes*, ANSI Z49.1, Miami: American Welding Society.
10. See Reference 9.
11. *Occupational Safety and Health Administration (OSHA), 1999, Occupational Safety and Health Standards for General Industry, in Code of Federal Regulations (CFR)*, Title 29, CFR 1910, Washington D.C.: Superintendent of Documents, U.S. Government Printing Office. Online bookstore can be reached at http://bookstore.gpo.gov.

12. The dates of publication given for the codes and other standards listed here were current at the time this chapter was prepared. The reader is advised to consult the latest edition.

Diego, June 29–July 2,1992. ISBN 0-87339-222-1. F. H. Froes and I. L. Caplan, eds. Warrendale, Pa. The Minerals, Metals and Materials Society. Vol. 2: 1461–1468.

Chruszez, M., H. Schmalenstroth, T. Kasting, M. Kuhnel, and J. Marksmann. 2000. Application and optimization of the plasma-arc welding process with a powder filler metal in the case of the manufacture of joints on X6CrNiMoTi17122. *Schweissen und Schneiden* 52(1): E17-E20.

Craig, E. 1988. The plasma arc process: A review. *Welding Journal* 67(2):19–25.

Cullison, A. T., ed. 1998. Gas tungsten TIG and plasma arc welding. Welding workbook datasheets 220a, 220b. *Welding Journal* 77(9): 61-62.

Deng, X. H. 1999. *Analysis of the mechanical characteristics of aluminium-lithium alloy 2195 variable polarity plasma arc (VPPA) weldments.* Thesis (Ph. D). Wichita State University, Wichita, Kans. 123 pp.

Dilthey, U., and L. Kabatnik. 2001. Plasma arc welding of aluminium alloys with reverse polarity DC electrode positive in high power ranges. *Schweissen und Schneiden* 53(3): 156, 160–161 (English translation of text and captions: E59-E62).

Dowden, J., and P. Kapadia. 1994. Plasma arc welding: a mathematical model of the arc. *Journal of Physics D: Applied Physics. (*275): 902–910.

Evans, D. M., D. Huang, J. C. McClure, and A. C. Nunes. 1998. Arc efficiency of plasma arc welding. *Welding Journal.* 77(2): 53-s to 58-s.

Fan, H. G., and R. Kovacevic. 1999. Keyhole formation and collapse in plasma arc. *Journal of Physics D: Applied Physics.* 32.22.21(11): 2902–2909.

Fuehrsbach, P. W. 1998. Cathodic cleaning and heat input in variable polarity plasma arc welding of aluminum. *Welding Journal.* 77(2): 76-s to 85-s.

Hou, R., D. M. Evans, J. McCure, A. C. Nunes, and G. Garcia. 1996. Shielding gas and heat transfer efficiency in plasma arc welding. *Welding Journal.* 75(10): 305-s to 310-s.

Hung, R. J., C. Lee, and J. W. Liu. 1990. Characteristics and performance of the variable polarity plasma arc welding process used in the space shuttle external tank. Report NASA-CR-184226 (N92-13431/1; NAS 1.26:184226). Washington, D.C. National Aeronautics and Space Administration; July 1990. 73 pp.

Jernstrom, P., and J. Martikainen. 1998. Influence of base parent material on the selection of weld gases in plasma arc keyhole welding. *Proceedings, Third European Conference on Joining Technology*, Bern, 30. (3 and 4). March 30–April 1, 1998. Basel, Switzerland: Swiss Welding Association (SVS). Part 2: 643–648.

Keanini, R. G., and B. Rubinsky. 1990. Plasma arc welding under normal and zero gravity. *Welding Journal* 69(6): 41–50.

Kippes, W., T. Petitjean, and H. Frauenfelder. 1999. The plasma arc as a tool for fine joining. DVS Berichte, No. 204. *Welding and Cutting '99. Proceedings, Welding Conference,* Weimer, 15–17: 133–136.

Langford, G. J. 1968. Plasma arc welding of structural titanium joints. *Welding Journal.* 47(2): 102–113.

Liu, Z. H., Q. L. Wang, and Z. D. Jia. 2000. Process control based on double-side image sensing of keyhole puddle for the VPPA variable polarity plasma arc welding of aluminum alloys. *China Welding* 9.2(11): 143–151.

Martikainen, J. K. 1994. Plasma arc keyhole welding of high strength structural steels. *International Journal for the Joining of Materials.* 6.3(9): 93–99.

Martikainen, J. K., and T. J. I. Mìoisio. 1993. Investigation of the effect of welding parameters on weld quality of plasma arc keyhole welding of structural steels. *Welding Journal.* 72(7): 329-s to 340-s.

McClintock, A. 1996. Plasma arc welding revisited—new research confirms this process as particularly cost effective for high quality stainless steel fabrication. *Australasian Welding Journal.* 41(2). Fourth Quarter: 15–28.

McCutcheon, K. D., S. S. Gordon, and P. A. Thompson. VPPA variable polarity plasma arc welding weld model evaluation Report NASA-CR-184460 (N93-12919/5; NAS 1.26:184460). Washington, D.C. National Aeronautics and Space Administration. July 31, 1992. 151 pp.

Metcalfe, J. C., and M. B. C. Quigley. 1975. Keyhole stability in plasma arc welding. *Welding Journal.* 54(11): 401-s to 404-s.

Micheli, J., and C. Pilcher. 2000. Advanced variable-polarity plasma arc welding. *Fabricator* 30(11): 46-48.

Walduck, R. P. 1994. Development of a robotic plasma arc spot welding technique for Jaguar cars. *Welding and Metal Fabrication.* 62(2): 51– 54.

White, R. A., R. Fusaro, M. G. Jones, R. R. Milian-Rodriguez, and H. D. Solomon. 1997. Underwater cladding with laser beam and plasma arc welding. *Welding Journal.* 76(1): 57–61.

Zheng, B., H. J. Wang, Q. L. Wang, and R. Kovacevic. 2000. Control for weld penetration in variable polarity plasma arc (VPPA) welding of aluminum alloys using the front weld pool image signal. *Welding Journal.* 79(12): 363-s to 371-s.

Zheng, B., Q. L. Wang, and X. LI. 1995. A new type of pilot arc power source used for AC plasma arc welding of aluminum alloys. *China Welding* 4.2(11): 89–97.

Zhang, S. B., and Y. M. Zhang. 2001. Efflux plasma charge-based sensing and control of joint penetration during keyhole plasma arc welding. *Welding Journal.* 80(7): 157-s to 162-s.

CHAPTER 8

ELECTROGAS WELDING

Photograph courtesy of CB & I

Prepared by the Welding Handbook Chapter Committee on Electrogas Welding:

D. A. Fink, Chair
The Lincoln Electric Company

J. H. Devletian
Portland State University

J. R. Hannahs
Edison Community College

D. K. Hartman
The Lincoln Electric Company

R. H. Jeurs
Naval Surface Warfare Center

W. H. Kavicky
Trans Bay Steel Corporation

D. Y. Ku
American Bureau of Shipping

J. S. Lee
Chicago Bridge and Iron Company

K. Y. Lee
The Lincoln Electric Company

R. J. Sowko
The Lincoln Electric Company

R. D. Thomas, Jr.
R. D. Thomas & Company

Welding Handbook Volume 2 Committee Member:

D. B. Holliday
Northrop Grumman Marine Systems

Contents

CHAPTER 8

ELECTROGAS WELDING

INTRODUCTION

The first welding process available for thick-plate, single-pass vertical applications was electroslag welding (ESW). With its demonstrated success at welding thick sections in the vertical position, demand immediately arose for equipment that would adapt the process for use with thinner sections. Most vertical joints in thin plate were being welded with the manual shielded metal arc (SMAW) process or semiautomatic gas metal arc welding (GMAW). In 1961, laboratory studies involving an electroslag welding machine that had been adapted to feed auxiliary gas shielding around a flux cored electrode demonstrated that plate as thin as 13 millimeters (mm) (1/2 inch [in.]) could be satisfactorily welded in the vertical position in a single pass. As a continuation of these studies, a process was developed and introduced as *electrogas welding* (EGW).

The mechanical aspects of electrogas welding are strikingly similar to those of the electroslag process from which it was developed. Two types of electrodes are commonly used with the electrogas welding process: a solid, continuous electrode (similar to those used in gas metal arc welding) or a flux cored electrode (similar to those used in flux cored arc welding [FCAW]). Welding with either electrode variation requires backing shoes (dams) to confine the molten weld metal, which permits welding in the vertical position. Gas shielding, when required, is introduced through inlet ports in the dams or a gas nozzle around the electrode, or both. When a self-shielded flux cored arc welding electrode is used, no gas is added.

Two basic process variations, the moving shoe and the consumable guide methods, are commonly used. In the moving shoe method, the electrode is used with a track-guided nozzle and at least one moving backing shoe that slides along the joint during welding. In the consumable guide method, the electrode is used with a stationary consumable guide, generally with fixed or stationary backing shoes.

Applications for the use of electrogas welding grew steadily in the United States. The process was initially popular because it used simple base metal edge preparations and yielded high deposition rates and efficiencies. Electrogas welds normally exhibit excellent weld metal soundness. The process also offered economic benefits and quality enhancements. However, the use of electrogas welding diminished during the 1990s because of perceived and unrelated quality problems concerning weld toughness and cracking associated with the *electroslag* process.

This chapter presents a description of electrogas welding, including a discussion of the fundamentals of the process, equipment, materials, variables, weld quality, economics, and safe practices.[1] Applications involving the welding of plain carbon steels, structural steels, and pressure vessel steels are discussed in detail.

FUNDAMENTALS

Electrogas welding is an arc welding process that uses an arc between a continuous filler metal electrode and the weld pool, employing approximately vertical welding progression with a metallic or nonmetallic backing to confine the molten weld metal. The backing is a dam-like device that is placed against the back side of the weld joint or on both sides of the weld joint to support and retain the molten weld metal. Depending on the type of backing used, it may be fused to the weld joint and become part of the weld, or remain unfused and removed after welding.[2, 3] The process is used with

1. This chapter is adapted from American Welding Society (AWS) C5 Committee on Arc Welding and Cutting, 2000, *Recommended Practices for Electrogas Welding,* AWS C5.7:2000, Miami: American Welding Society.
2. American Welding Society (AWS) C5 Committee on Arc Welding and Cutting, 2000, *Recommended Practices for Electrogas Welding,* AWS C5.7:2000, Miami: American Welding Society, p. 3.
3. At the time of the preparation of this chapter, the referenced codes and other standards were valid. If a code or other standard is cited without a date of publication, it is understood that the latest edition of the document referred to applies. If a code or other standard is cited with the date of publication, the citation refers to that edition only, and it is understood that any future revisions or amendments to the code or standard are not included; however, as codes and standards undergo frequent revision, the reader is advised to consult the most recent edition.

or without an externally supplied shielding gas and without the application of pressure. The electrode for EGW is usually in the form of solid or cored wire, often with a consumable guide tube.

Typically, a square-groove or single-V groove joint is specified and positioned with the axis or length of the weld vertical. No repositioning of the joint occurs once welding has started; welding continues to completion, so that the weld is made in one pass.

The consumable electrode, either solid or flux cored, is fed downward into the joint root opening, a cavity formed by the base metals to be welded and the backing shoes. A starting weld tab (sump) is required to seal the bottom of the weld joint to allow the process to stabilize and to support the molten weld metal until it reaches the workpiece. An arc is initiated between the electrode, the consumable guide tube (when one is used), and the starting weld tab. The heat generated by the arc melts the continuously fed electrode and the groove faces. Melted filler metal and base metal collect in a pool beneath the arc and solidify to form the weld. As the weld metal fills the joint, the weld pool rises, progressing in the uphill direction.

Thicknesses of 13 mm (1/2 in.) to 38 mm (1-1/2 in.) are typically welded with electrogas welding. When thicker sections are welded, the electrode may be oscillated horizontally through the joint for uniform distribution of the heat and the weld metal. If moving shoes are used, one or both shoes may move upward as the cavity fills. Although the weld travel is vertical, the weld metal is actually deposited in the flat position at the bottom of the cavity.

Electrogas welding is a mechanized welding process. The nature of the melting and solidification during welding results in a high-quality weld deposit. Little or no angular distortion of the base metal occurs with single-pass welds. For these reasons, major applications of electrogas welding have historically been in tank fabrication and in shipbuilding, as illustrated in Figure 8.1.

ADVANTAGES AND LIMITATIONS

Several of the advantages associated with electrogas welding, such as high deposition rates and operating factors, have resulted in considerable cost savings, particularly when welding thicker metals. Savings have been achieved in applications in which components can be joined in the vertical position with a continuous vertical weld. For thicker materials, electrogas welding is often less expensive than the more conventional joining methods, such as submerged arc welding (SAW) and flux cored arc welding (FCAW). Even in some applications involving thinner base materials, electrogas welding may result in cost savings because of its efficiency and the simple requirements for joint preparation.

The electrogas welding process has several limitations, including the following:

1. The training of operators is time consuming and is critical to the successful use of the process;
2. The initial cost of equipment is high and setup time can be lengthy;
3. The high heat input may cause lower toughness in the weld and heat-affected zone (HAZ), as measured by Charpy tests; and
4. If the weld is not completed in one continuous pass, the resulting restarts usually require that a repair be made. The reworking of problem welds is difficult.

Electrogas welding is not generally used for applications involving aluminum alloys and stainless steel, although a few successful examples have been reported.

ELECTRODE VARIATIONS

In electrogas welding, the electrode, which may be either a solid or flux cored type, is a filler metal component. It completes the welding circuit through which current is conducted from the electrode guide (discussed in the next section) to the weld pool.[4]

Solid Electrode

A schematic view of a typical moving-shoe electrogas welding installation using a solid electrode is shown in Figure 8.2. The electrode is fed through a welding gun, referred to as a *nonconsumable guide*. The electrode may be oscillated horizontally to weld thicker materials. Gas shielding—typically carbon dioxide (CO_2) or an argon-carbon dioxide ($Ar-CO_2$) mixture, class SG-C or class SG-AC-X, respectively, as described in *Specification for Welding Shielding Gas,* AWS A5.32/A5.32M—is provided to the weld cavity through gas ports in the moving backing shoe, gas boxes, or nozzles.[5]

Water-cooled copper backing shoes are normally used on each side of the joint to retain the molten weld metal. The shoes, which are usually attached to the welding machine, automatically move vertically upward with the welding gun. The vertical movement of the welding machine must be consistent with the deposition rate. This movement is controlled by the welding operator.

The electrode diameters most commonly used are 1.6 mm, 2.0 mm, 2.4 mm, and 3.2 mm (1/16 in., 5/64 in., 3/32 in., and 1/8 in.). Electrogas welding with solid

4. See Reference 2.
5. American Welding Society (AWS) Committee on Filler Metals and Allied Materials, 1997, *Specification for Welding Shielding Gas,* AWS A5.32/A5.32M, Miami: American Welding Society.

Figure 8.1—Electrogas Welding of a Ship Hull Using Two Machines Controlled by One Operator

electrodes can be used to weld base metals ranging in thickness from approximately 10 mm (3/8 in.) to 100 mm (4 in.). The base metal thicknesses most commonly welded are between 13 mm (1/2 in.) and 76 mm (3 in.).

Flux Cored Electrode

The principles of operation when using flux cored electrodes are nearly identical to those used with the solid electrode variation, except that, typically, no separate gas shielding is needed, depending on the electrode type. The flux cored electrode creates a thin layer of slag between the weld metal and copper shoes that provides a smooth weld surface. If excessive slag is generated by the welding electrode, it may be necessary to drain some of the slag periodically throughout the welding process. The slag may be drained over the top of a non-stationary shoe or through drainage slots on a stationary dam or shoe. Figure 8.3 is a schematic illustration of electrogas welding using a self shielding

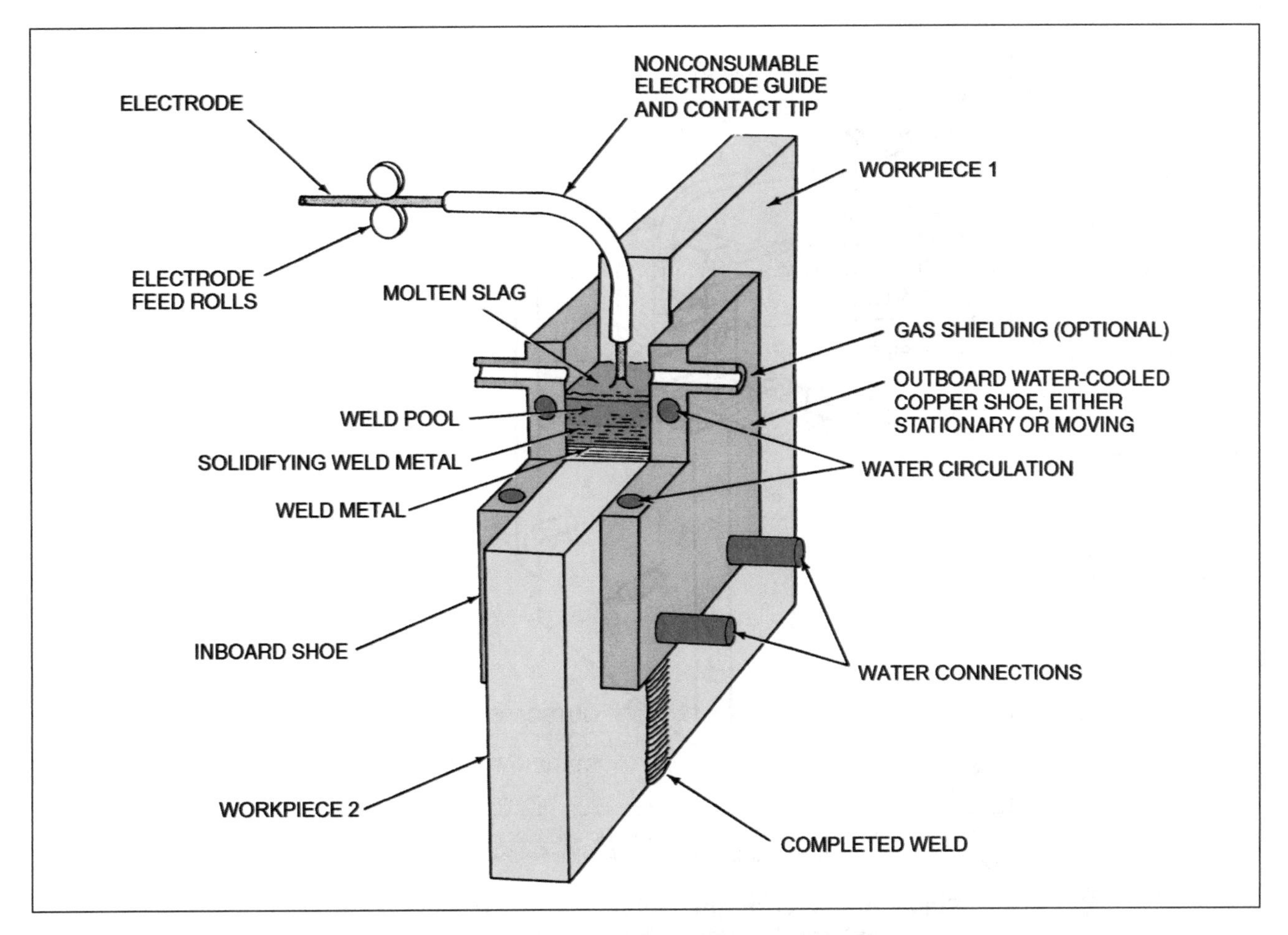

Figure 8.2—Electrogas Welding with Moving Shoes

flux-cored electrode. Figure 8.4 shows an electrogas weld setup in progress, using two water-cooled copper shoes.

Electrogas welding with a flux-cored electrode can be performed with an external gas shield or without a shielding gas, depending on the type of electrode. Self-shielded electrodes operate at higher current levels and deposition rates than the gas-shielded types.

The diameters of flux cored electrodes commonly used vary from 1.6 mm to 3.2 mm (1/16 in. to 1/8 in.). The wire (electrode) feeder must be capable of smooth, continuous feeding of small-diameter wires at high speeds and larger-diameter wires at slower speeds.

CONSUMABLE GUIDE PROCESS

Electrogas welding with a consumable guide is similar to consumable guide electroslag welding. This variation of electrogas welding is primarily used for short weldments, typically 1 meter (m) to 1.5 m [3 feet (ft) to 4 ft] in shipbuilding and in column and beam fabrication. Consumable guide electrogas welding uses relatively simple equipment, as shown in Figure 8.5 and described in the section titled "Equipment."

The principal difference between conventional (moving shoe) electrogas welding and consumable guide electrogas welding is that in the consumable guide mode none of the equipment moves vertically. Instead, the electrode is fed through a stationary hollow consumable guide tube that extends to about 25 mm (1 in.) from the bottom of the joint. As the weld progresses vertically, the electrode melts back to the guide tube while the arc is maintained. Initially, the wire electrode extends about 25 mm (1 in.) beyond the end of the guide tube. A steady-state relationship then develops between the melting of the end of the guide tube and the electrode wire, as shown in Figure 8.6. The relationship indicated in Figure 8.6 remains as the weld is completed.

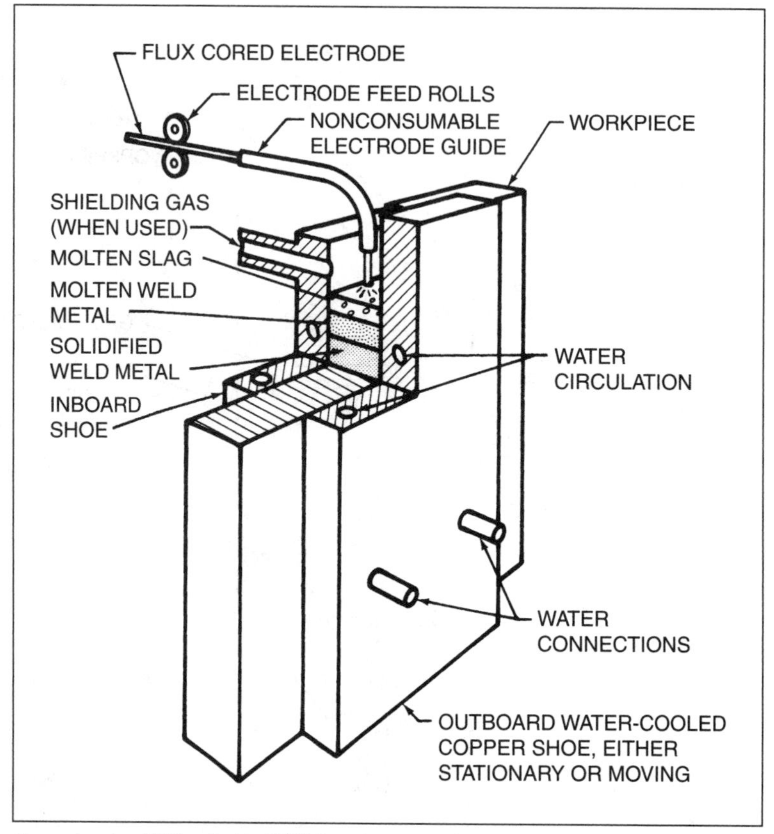

Source: American Welding Society (AWS) Committee on Arc Welding and Cutting, 2000, *Recommended Practices for Electrogas Welding,* AWS C5.7:2000, Miami: American Welding Society, Figure 4.

Figure 8.3—Electrogas Welding with a Moving Shoe and Self-Shielded Flux Cored Electrode

The consumable guide provides approximately 5% to 30% of the deposited metal; the balance is supplied from the wire electrode. The consumable guide is selected so that its chemical composition is compatible with the base material. Guide tubes are usually 13 mm to 16 mm (1/2 in. to 5/8 in.) in outer diameter (O.D.) and 3 mm to 5 mm (1/8 in. to 3/16 in.) in inner diameter (I.D.). However, smaller-sized tubes (6 mm to 9 mm [1/4 in. to 3/8 in.]) are sometimes used on thinner base material. Guide tubes are available in various lengths, usually up to 3 m (10 ft), and for some applications may be welded together to accommodate longer joints.

The arc is initiated in the same general fashion as in the moving-shoe process. A collar or clamp device, usually made of copper or brass, holds the stationary guide tube in position.

Oscillation may be used for thicker plate; however, oscillation with a consumable guide tube over 1 m (3 ft) long becomes increasingly more difficult as weld length increases. This is because the bottom end of the tube tends to whip and may distort and lose its rigidity due to heating. Thus, the preferred method of welding plate is to use one or more additional electrodes, thereby eliminating the oscillation. When one or more electrodes and tubes are added, the polarity of the added electrode may need to be alternated, i.e., one direct current electrode positive (DCEP) and the next direct current electrode negative (DCEN), to reduce the buildup of excessive heat and allow for better arc action. Circular insulators may be spaced at intervals of approximately 150 mm to 450 mm (6 in. to 18 in.) along the length of the tube to prevent shorting of the tube to the wall of the groove.

In consumable guide electrogas welding, the backing shoes are independent of the welding gun, as indicated in Figure 8.6. The shoes can be held in place with wedges placed between the shoes and strongbacks (devices to maintain alignment) welded to the workpiece. On short welds, the shoes may be the same length

Photograph courtesy of CB & I

Figure 8.4—An Electrogas Weld Setup in Progress Using Two Water-Cooled Copper Shoes

as the joint. Several sets of shoes may be stacked for longer joints.

One complication in consumable guide electrogas welding when using a flux cored electrode is the buildup of slag on the top of the weld pool. Excessive slag buildup can lead to incomplete penetration or incomplete fusion. Slag buildup can be controlled by providing intermittent vertical slots in the backing shoes or by shimming and gapping the backing shoes intermittently along the length of the weld. Both methods permit the excess slag to escape from the weld groove.

Consumable guide tubes can sometimes be used for complex shaped joints. For such applications, the guides may be preformed to suit the required shapes.

EQUIPMENT

The basic equipment for electrogas welding consists of a direct-current power source, a wire feeder, shoes for retaining molten metal, an electrode guide, a mechanism to oscillate the electrode guide, and equipment needed for supplying shielding gas, when used. As shown in Figure 8.7, the essential components (with the exception of the power source) in a typical moving-shoe electrogas welding system are incorporated in an assembly that moves vertically as welding progresses. The equipment used for the consumable guide mode is

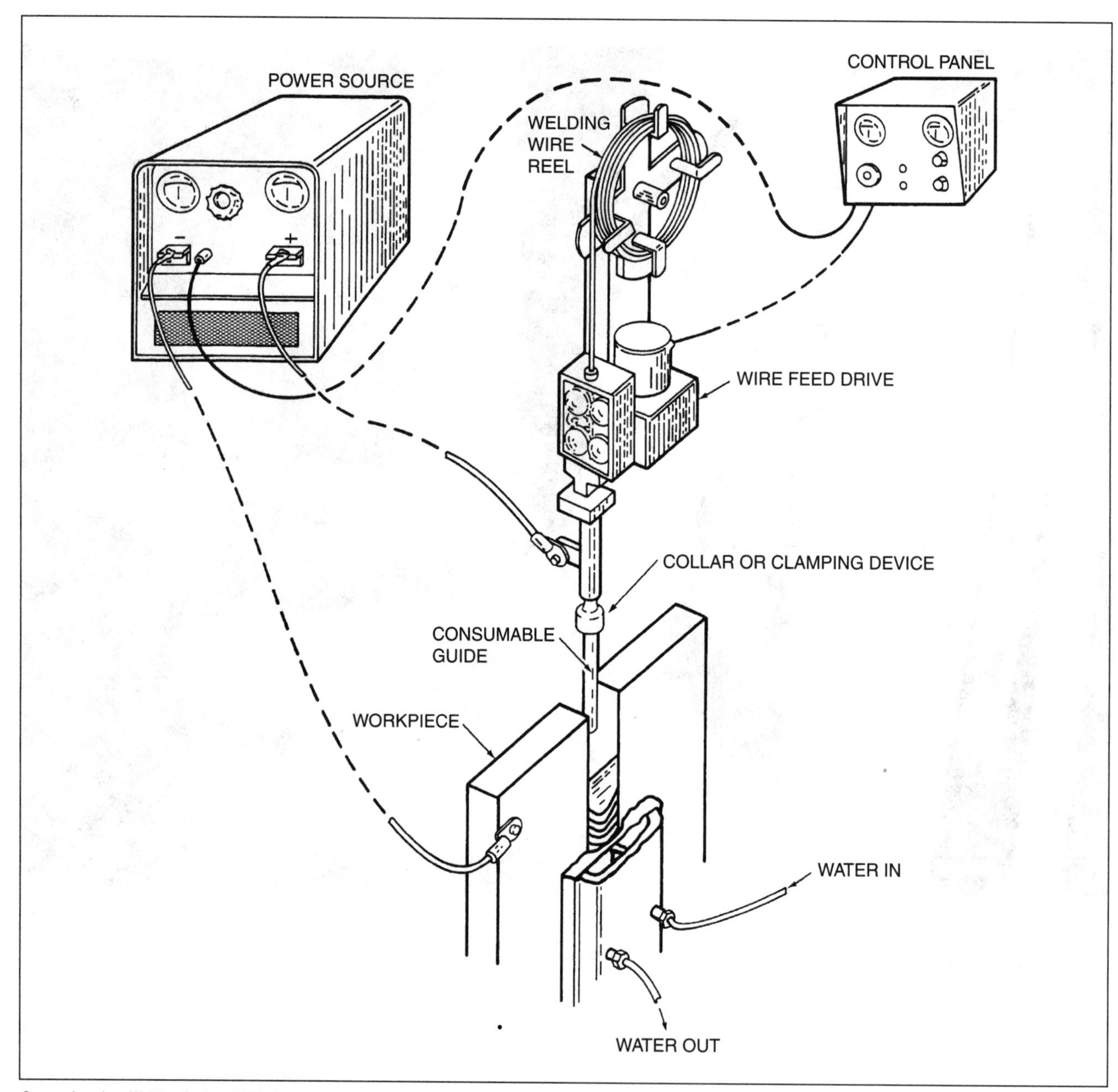

Source: American Welding Society (AWS) Committee on Arc Welding and Cutting, 2000, *Recommended Practices for Electrogas Welding,* AWS C5.7:2000, Miami: American Welding Society, Figure 5.

Figure 8.5—Consumable Guide Electrogas Welding Equipment

essentially the same, except that it remains fixed. The manufacturer should be consulted for technical assistance in coordinating the equipment with the EGW application.

Figure 8.8 shows the equipment installation required for the long vertical welds typically made with electrogas welding. Figure 8.9 shows a portable elevator cage containing a welding unit.

POWER SOURCE

Direct current electrode positive (reverse polarity) is normally used for electrogas welding, with the power source being either the constant-voltage or, in some circumstances, the constant-current type. The constant-voltage type is generally preferred. The power source should be capable of delivering the required current

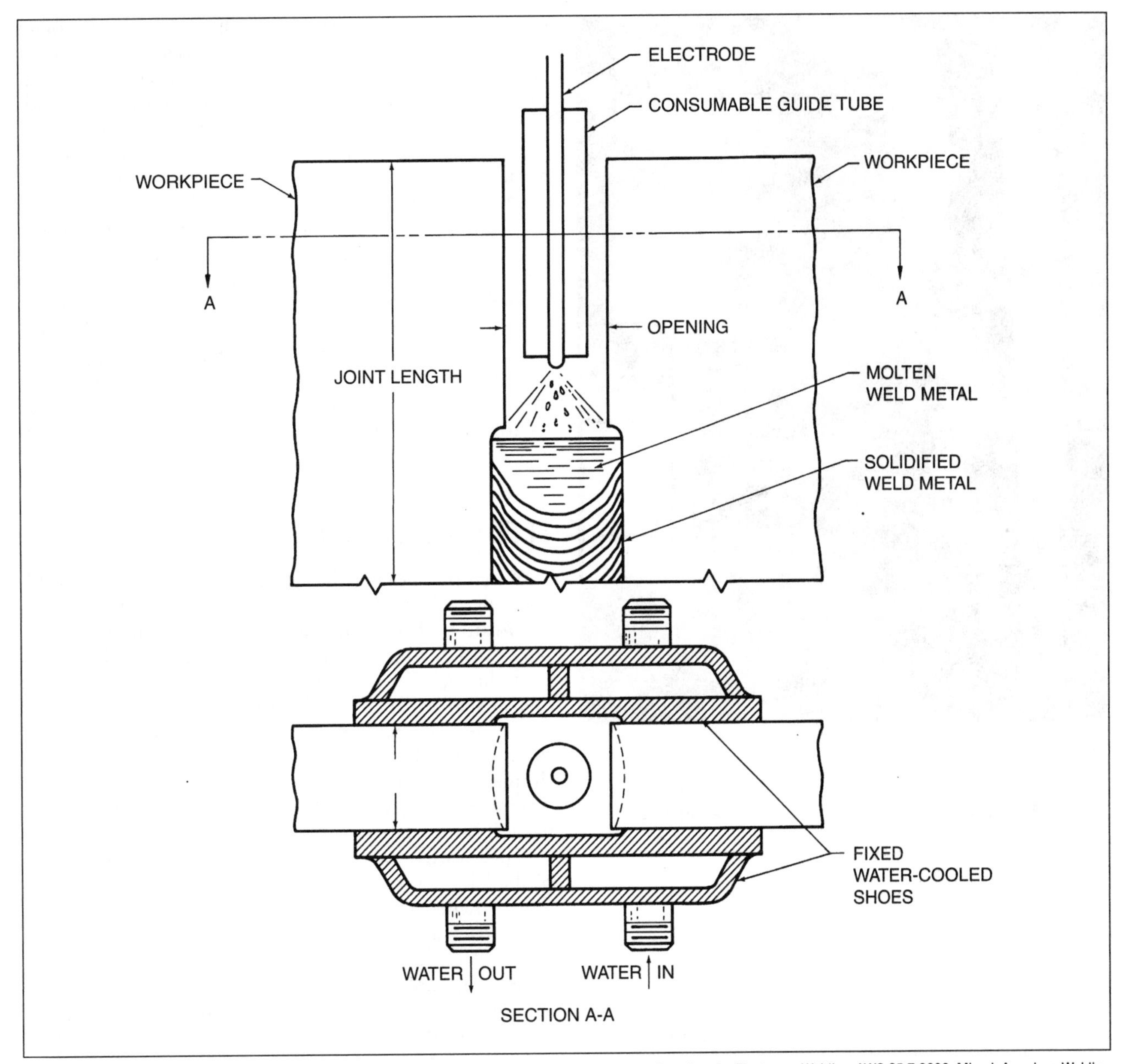

Source: American Welding Society (AWS) Committee on Arc Welding and Cutting, 2000, *Recommended Practices for Electrogas Welding,* AWS C5.7:2000, Miami: American Welding Society, Figure 6.

Figure 8.6—Schematic Illustration of the Consumable Guide Electrogas Welding Process

without interruption during the welding of a joint of considerable length that may require hours of continuous operation.

The power sources used for electrogas welding usually have capacities of 750 amperes (A) to 1000 A at 30 volts (V) to 55 V and 100% duty cycle. Direct current (dc) is usually supplied by transformer-rectifier power sources, although motor-driven and engine-driven generators may be used, especially on construction sites.

ELECTRODE FEEDER

The wire feeder for the electrode is usually the push type, such as those used in automated gas metal arc welding or flux cored arc welding. The wire feeder can

Photograph Courtesy of Metro Machine Corporation

Figure 8.7—Electrogas Welding Equipment with a Moving Shoe Using Self-Shielded Flux Cored Electrode

be mounted as an integral part of the vertical-moving welding machine. Wire feed speeds may range up to 230 millimeters per second (mm/s) (550 inches per minute [in./min]) with a 3.2 mm (1/8 in.) electrode.

The wire feed system may include a wire straightener to minimize the cast and helix set in the electrode, thus minimizing electrode wander at the joint, especially with the moving-shoe process. Since the electrode extension in moving-shoe electrogas welding is relatively long, typically up to 75 mm (3 in.) or more, a straight length of electrode projecting from the guide is necessary to assure accurate wire placement and arc location in the joint.

ELECTRODE GUIDE

The electrode guide is a conductive supporting component through which the solid or cored electrode passes. The guide is usually tubular in shape, although other shapes may be used.[6] In many respects, electrode guides are similar to the welding guns used for semiautomatic gas metal arc welding or flux cored arc welding. The guide may have a shielding gas outlet to deliver gas around the protruding electrode (usually supplementary to the shielding gas supplied by ports in the shoes), and it may or may not be cooled with water.

The electrode guide (or at least part of it, in the case of a curved guide) rides inside the root opening directly above the weld pool. Therefore, it must be narrow enough to fit in the root opening with clearance to permit horizontal oscillation, if used, between the two shoes. For this reason, the width of an electrode guide is usually limited to 10 mm (3/8 in.) or smaller to accommodate the 18 mm (11/16 in.) minimum root openings customarily used for the moving-shoe mode of electrogas welding.

ELECTRODE GUIDE OSCILLATORS

When the moving-shoe mode of the electrogas welding process is used with base metals 30 mm to 100 mm (1-1/4 in. to 4 in.) thick, it is necessary to move the arc back and forth, oscillating it between the shoes and over the weld pool to obtain uniform metal deposition and ensure fusion to both groove faces. Oscillation of the arc is not usually needed for joints in base metal less than 30 mm (1-1/4 in.) thick but is sometimes used with thinner base metals to minimize penetration into the base metal and to improve weld properties by modifying the solidification pattern.

This horizontal back-and-forth motion is accomplished by a system that oscillates the electrode guide and provides adjustable dwell times at both endpoints of the oscillation. The term *dwell* refers to the duration of the electrode's pause at the extreme transverse positions.[7] The arc is normally oscillated to about 1/4 in. (6 mm) from the backing shoe on each side of the joint. The dwell time is then adjusted to produce good fusion at both groove faces.

BACKING SHOES

The backing shoes used in electrogas welding, also called *dams,* are typically made of copper and do not fuse during welding.[8] The backing shoes are pressed against the workpieces on each side of the root opening to retain (dam) the molten weld metal in the groove. One or both shoes may move upward as welding progresses. In some weldments, a steel backing bar,

6. See Reference 2.
7. See Reference 2.
8. See Reference 2.

Photograph Courtesy of CB&I

Figure 8.8–Electrogas Welding Equipment Installation Used in Storage Tank Construction

which fuses to the weld, may replace the outboard shoe, which is the backing shoe located on the side of the plate farthest from the main portion of the welding machine.[9] Ceramic backings that do not fuse have also been used.

Copper backing shoes are usually water-cooled; steel backing bars are not water-cooled. The use of ceramic backings or shoes that are not water-cooled may affect mechanical properties because of the slower cooling rates. The copper shoes are usually grooved to obtain the desired weld face reinforcement. When using gas-shielded electrodes, sliding shoes are available with gas ports through which shielding gas is supplied directly

9. See Reference 2.

Figure 8.9—A Portable Electrogas Welding Unit Housed in a Special Elevator Cage

into the cavity formed by the shoes and the weld groove. When gas ports are not used in the shoes, a gas box arrangement can be mounted on the shoes to surround the electrode and the welding arc with shielding gas. Gas ports in the shoes and gas boxes are not required with self-shielded flux cored electrodes.

CONTROLS

With the exception of the vertical travel control, moving-shoe electrogas welding controls are primarily adaptations of the devices used in gas metal arc welding and flux cored arc welding. Electrical, optical, or manual vertical travel controls maintain a fixed electrode extension and keep the top of the movable shoe a specific distance above the weld pool.

On equipment using a self-shielded flux cored electrode, starting procedures must be carefully controlled to minimize porosity in the start area of the starting weld tab. Welds may be started at low wire-feed speeds and high voltages. These lower start settings must be maintained until the arc stabilizes and the starting area is sufficiently heated. After a preset time, the equipment automatically changes the feed speed and voltage to the operating welding speed and voltage. Automatic rather than manual changing is recommended as it reduces the probability of starting porosity.

MATERIALS

Many different types of materials can be welded with electrogas welding, including carbon steels, structural steels, and pressure vessel steels. Various classes of electrodes and types of shielding gases are required to accommodate this wide range of applications. Consumable guides and insulators may also be required.

ELECTRODES

Both solid and flux cored electrodes can be used with the electrogas welding process. The solid electrodes used in electrogas welding are generally identical to those used for gas metal arc welding. The electrogas electrodes contain a lower percentage of slagging compounds than the typical electrodes used in the flux cored arc welding process and may be either gas-shielded or self shielded. These special electrodes allow a thin layer of slag to form between the shoes and the weld to provide a smooth weld surface.

Solid electrodes are usually supplied in sizes from 0.8 mm to 3.2 mm (1/32 in. to 1/8 in.). Flux cored electrodes are typically supplied in sizes from 1.6 mm to 3.2 mm (1/16 in. to 1/8 in.). Only those flux cored electrodes specifically designated and classified for electrogas welding should be used.

Solid and flux cored electrodes are available in various chemical compositions designed to introduce the necessary alloying elements to achieve strength, impact properties, or appropriate combinations of these and other properties in the deposited weld metal. For the welding of steel, these additional elements include manganese, silicon, and nickel. The required properties are normally obtained in the as-welded condition.

The specification approved by the American National Standard Institute (ANSI), *Specification for Carbon and Low-Alloy Steel Electrodes for Electrogas Welding,* ANSI/AWS A5.26/A5.26M,[10] prescribes the

10. American Welding Society (AWS) Committee on Filler Metals and Allied Materials, *Specification for Carbon and Low-Alloy Steel Electrodes for Electrogas Welding,* ANSI/AWS A5.26/A5.26M, Miami: American Welding Society.

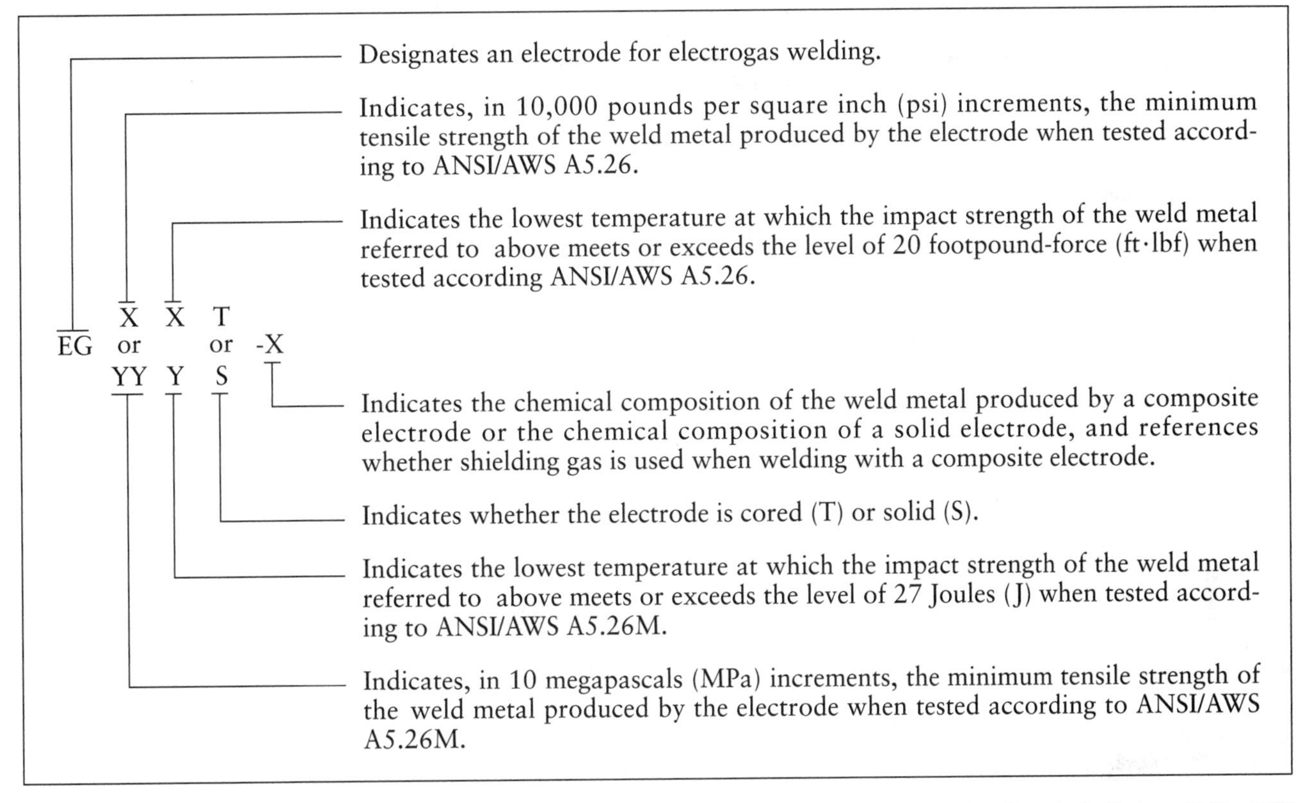

Source: Adapted from American Welding Society (AWS) Committee on Filler Metals, 1997. *Specification for Carbon and Low-Alloy Steel Electrodes for Electrogas Welding,* ANSI/AWS A5.26/A5.26M-97, Miami: American Welding Society, Annex A, Figure 1.

Figure 8.10—Electrogas Welding Electrode Classification System

requirements for the solid and flux cored electrodes used in electrogas welding. This specification classifies solid electrodes based on as-manufactured chemical composition and weld metal mechanical properties in the as-welded condition. It classifies flux cored electrodes on the basis of whether an external shielding gas is required, the chemical composition of the weld metal, and the mechanical properties of the as-deposited weld metal. The general classification system of electrodes used in electrogas welding is illustrated in Figure 8.10.

The classification requirements for the chemical composition of as-manufactured solid wires and as-deposited flux cored electrogas weld metals are shown in Tables 8.1 and 8.2, respectively. The required as-deposited weld metal mechanical properties are shown in Tables 8.3 and 8.5 for International System of Units (SI) and Tables 8.4 and 8.6 for U.S. customary (inch-pound) units. It should be noted that three classifications are used for strength in both the SI and U.S. customary specifications and that different base metals are employed in the standard test weldment for each classification.

Solid and flux cored electrodes can be categorized in any of the listed mechanical properties classifications, thus specifying tensile strength, ductility and notch toughness. The use of specific base metals for electrode classification tests takes into account the high dilution rate characteristic of single-pass electrogas welding. For each level of strength, the electrode classifications define three levels of minimum toughness as determined by the Charpy V-notch impact test.

SHIELDING

Self-shielded flux cored electrodes for electrogas welding contain core materials that form a slag and shield the molten weld metal from oxygen and nitrogen. Other flux cored electrodes require an additional external shielding gas (normally carbon dioxide class SG-C

Table 8.1
Chemical Composition Requirements for Solid Electrode

AWS Classification[c]	UNS Number[d]	Weight percent[a,b]											
		C	Mn	S	P	Si	Ni	Mo	Cu[e]	Ti	Zr	Al	Other Elements, Total
EGXXS-1	K01313	0.07–0.19	0.90–1.40	0.035	0.025	0.30–0.50	—	—	0.35	—	—	—	0.50
EGXXS-2	K10726	0.07	0.90–1.40	0.035	0.025	0.40–0.70	—	—	0.35	0.05–0.15	0.02–0.12	0.05–0.15	0.50
EGXXS-3	K11022	0.06–0.15	0.90–1.40	0.035	0.025	0.45–0.75	—	—	0.35	—	—	—	0.50
EGXXS-5	K11357	0.07–0.19	0.90–1.40	0.035	0.025	0.30–0.60	—	—	0.35	—	—	0.50–0.90	0.50
EGXXS-6	K11140	0.06–0.15	1.40–1.85	0.035	0.025	0.80–1.15	—	—	0.35	—	—	—	0.50
EGXXS-D2	K10945	0.07–0.12	1.60–2.10	0.035	0.025	0.50–0.80	0.15	0.40–0.60	0.35	—	—	—	0.50
EGXXS-G	—	Not specified[f]											

Notes:
a. The filler metal must be analyzed for the specific elements for which values are shown in this table. If the presence of other elements is indicated in the course of this work, the amount of those elements must be determined to ensure that their total (excluding iron) does not exceed the limit specified for "Other Elements, Total" in the last column of the table.
b. Single values are maximums.
c. The letters "XX" as used in the AWS classification column of this table refer respectively to the designator for the tensile strength of the weld metals (see Tables 8.3 and 8.4) and the designator for impact strength (see Tables 8.5 and 8.6).
d. Society of Automotive Engineers (SAE)/American Society for Testing and Materials (ASTM) Unified Numbering System for Metals and Alloys.
e. The copper limit includes copper that may be applied as a coating on the electrode.
f. The composition must be reported; the requirements are those agreed to by the purchaser and the supplier.

Source: Adapted from American Welding Society (AWS) Committee on Filler Metals and Allied Materials, 1997, *Specification for Carbon and Low-Alloy Steel Electrodes for Electrogas Welding*, ANSI/AWS A5.26/A5.26M-97, Miami: American Welding Society, Table 1.

per *Specification for Welding Shielding Gases*, AWS A5.32/A5.32M,[11] although mixtures of argon and carbon dioxide are also used).

When shielding gas is used, recommended gas flow rates range from 14 liters per minute (L/min) to 66 L/min (30 cubic feet per hour [ft^3/h] to 140 ft^3/h). The recommended gas flow rates depend on the design of the equipment, so the manufacturer's recommendations should be followed. Mixtures of argon and carbon dioxide (AWS class SG-AC-X) are normally used for the electrogas welding of steel with solid electrodes and may be used with flux cored electrodes.

PROCESS VARIABLES

Process variables for electrogas welding include arc voltage, current, polarity, electrode feed speed, electrode extension, electrode oscillation and dwell, and root opening. The proper control of these variables influences the operation and economics of the process and the resulting weld quality.

The effects of each electrogas welding variable must be fully understood, since they differ significantly from gas metal arc and flux cored arc welding. As an example, joint penetration or root penetration and depth of fusion in conventional arc welding processes (gas metal arc welding and flux cored arc welding) are in line with the axis of the electrode. Penetration and depth of fusion are generally increased by increasing the welding current and decreasing the voltage. However, in electrogas welding, the joint faces of the base metal are parallel to the axis of the electrode. Increasing the welding current or decreasing the voltage in electrogas welding results in a deeper weld pool and a narrower depth of fusion (bevel-face or groove-face penetration), and thus less penetration into the base metal.

The process variables detailed in this section apply primarily to the moving-shoe mode of the electrogas

11. See Reference 5.

Table 8.2
Chemical Composition Requirements for Weld Metal from Composite Flux Cored and Metal Cored Electrode

AWS Classification[a]				Weight Percent[b, c]										
A5.26M	A5.26	UNS Number[d]	Shielding Gas	C	Mn	P	S	Si	Ni	Cr	Mo	Cu	V	Other Elements, Total
EG43XT-1	EG6XT-1	W06301	None	See Note e	1.7	0.03	0.03	0.50	0.30	0.20	0.35	0.35	0.08	0.50
EG48XT-1	EG7XT-1	W07301	None	See Note e	1.7	0.03	0.03	0.50	0.30	0.20	0.35	0.35	0.08	0.50
EG55XT-1	EG8XT-1	—	None	See Note e	1.8	0.03	0.03	0.90	0.30	0.20	0.25–0.65	0.35	0.08	0.50
EG43XT-2	EG6XT-2	W06302	CO_2	See Note e	2.0	0.03	0.03	0.90	0.30	0.20	0.35	0.35	0.08	0.50
EG48XT-2	EG7XT-2	W07302	CO_2	See Note e	2.0	0.03	0.03	0.90	0.30	0.20	0.35	0.35	0.08	0.50
EGXXXT-Ni1	EGXXT-Ni1	W21033	CO_2	0.10	1.0–1.8	0.03	0.03	0.50	0.70–1.10	—	0.30	0.35	—	0.50
EGXXXT-NM1	EGXXT-NM1	W22334	Ar/CO_2 or CO_2	0.12	1.0–2.0	0.02	0.03	0.15–0.50	1.5–2.0	0.20	0.40–0.65	0.35	0.05	0.50
EGXXXT-NM2	EGXXT-NM2	W22333	CO_2	0.12	1.1–2.1	0.03	0.03	0.20–0.60	1.1–2.0	0.20	0.10–0.35	0.35	0.05	0.50
EGXXXT-W	EGXXT-W	W20131	CO_2	0.12	0.5–1.3	0.03	0.03	0.30–0.80	0.40–0.80	0.45–0.70	—	0.30–0.75	—	0.50
EGXXXT-G	EGXXT-G	—						Not specified[f]						

Notes:

a. The letters "XX" or "XXX" as used in the AWS classification column of this table refer respectively to the designator(s) for the tensile strength of the weld metal (see Tables 8.3 and 8.4 in this chapter) and the designator for impact strength (see Tables 8.5 and 8.6 in this chapter). The single letter "X" as used in the AWS Classification column refers to the designator for impact strength (see Tables 8.5 and 8.6 in this chapter).

b. The weld metal must be analyzed for the specific elements for which values are shown in this table. If the presence of other elements is indicated in the course of this work, the amount of those elements must be determined to ensure that their total (excluding iron) does not exceed the limit specified for "Other Elements, Total" in the last column of the table.

c. Single values are maximums.

d. Society of Automotive Engineers (SAE)/American Society for Testing and Materials (ASTM) Unified Numbering System for Metals and Alloys.

e. The composition range of carbon is not specified for these classifications, but the amount must be determined and reported.

f. The composition must be reported; the requirements are those agreed to by the purchaser and the supplier.

Source: Adapted from American Welding Society (AWS) Committee on Filler Metals and Allied Materials, 1997, *Specification for Carbon and Low-Alloy Steel Electrodes for Electrogas Welding*, ANSI/AWS A5.26/A5.26M-97, Miami: American Welding Society, Table 4.

welding process and may differ for the consumable-tube mode.

ARC VOLTAGE

Welding voltage is one of the major variables affecting weld width and base metal melting in electrogas welding. Welding voltages of 35 V to 45 V are normally used. Increasing the voltage increases the bevel-face or groove-face penetration and width of the weld, as illustrated in Figure 8.11. For thicker base metals or higher deposition rates, the voltage should be increased. However, excessively high voltage may cause the electrode to arc to the bevel faces or groove faces of the joint above the weld pool, causing unstable operation. Excessively high voltages used with self-shielded electrodes will result in porosity. Thus, it is important to stay within the manufacturer's recommended voltage range or the range that results in the desired mechanical properties.

Table 8.3
AWS A5.26M Tension Test Requirements (As-Welded)

AWS A5.26M Classification[a]	Tensile Strength, (MPa)	Yield Strength, minimum (MPa)[b]	Elongation, minimum[b] (%)
EG43ZX-X EG432X-X EG433X-X	430–550	250	24
EG48ZX-X EG482X-X EG483X-X	480–650	350	22
EG55ZX-X EG552X-X EG553X-X	550–700	410	20

Notes:
a. The letters "X-X" as they are used in the "AWS Classification A5.26M" column in this table refer respectively to "S" or "T" (whether the electrode is solid or composite), which replaces the first "X," and "1, 2, 3, 5, 6, D2, Ni1, NM2, W, or G" (designation for chemical composition and shielding gas requirements for composite electrodes only), which replaces the second "X."
b. Yield strength at 0.2% offset and elongation in 50 mm (2 in.) gauge length.

Source: Adapted from American Welding Society (AWS) Committee on Filler Metals and Allied Materials, 1997, *Specification for Carbon and Low-Alloy Steel Electrodes for Electrogas Welding,* ANSI/AWS A5.26/A5.26M-97, Miami: American Welding Society, Table 2M.

Table 8.4
AWS A5.26 Tension Test Requirements (As-Welded)

AWS A5.26 Classification[a]	Tensile Strength (psi)	Yield Strength, minimum (psi)	Elongation, minimum[b] (%)
EG6ZX-X EG60X-X EG62X-X	60,000–80,000	36,000	24
EG7ZX-X EG70X-X EG72X-X	70,000–95,000	50,000	22
EG8ZX-X EG80X-X EG82X-X	80,000–100,000	60,000	20

Notes:
a. The letters "X-X" as they are used in the "AWS Classification A5.26" column in this table refer respectively to "S" or "T" (whether the electrode is solid or composite), which replaces the first "X" and "1, 2, 3, 5, 6, D2, Ni1, NM2, W, or G" (designation for chemical composition and shielding gas requirements for composite electrodes only), which replace the second "X."
b. Yield strength at 0.2% offset and elongation in 50 mm (2 in.) gauge length.

Source: Adapted from American Welding Society (AWS) Committee on Filler Metals and Allied Materials, 1997, *Specification for Carbon and Low-Alloy Steel Electrodes for Electrogas Welding,* ANSI/AWS A5.26/A5.26M-97, Miami: American Welding Society, Table 2.

Table 8.5
AWS A5.26M Impact Test Requirements (As-Welded)[a]

AWS A5.26M Classification[a]	Average Impact Strength, min.[b] (J)
EG43ZX-X EG48ZX-X EG55ZX-X	Not specified
EG432X-X EG482X-X EG552X-X	27 @ –20°C
EG433X-X EG483X-X EG553X-X	27 @ –30°C

Notes:
a. An electrode combination that meets the impact requirements at a given temperature also meets the requirement at all higher temperatures in this table. Accordingly, EGXX3X-X can also be classified as EGXX2X-X and EGXXZX-X, and EGXX2X-X can be classified as EGXXZX-X.
b. Both the highest and the lowest of the five test values obtained must be disregarded in computing the impact strength. Two of the remaining three values must equal or exceed 27 J, and one of the remaining values may be lower than 27 J, but not lower than 20 J. The average of the three must not be less than the 27 J specified.

Source: Adapted from American Welding Society (AWS) Committee on Filler Metals and Allied Materials, 1997, *Specification for Carbon and Low-Alloy Steel Electrodes for Electrogas Welding,* ANSI/AWS A5.26/A5.26M-97, Miami: American Welding Society, Table 3M.

Table 8.6
AWS A5.26 Impact Test Requirements (As-Welded)[a]

AWS A5.26 Classification[a]	Average Impact Strength, min.[b] (ft·lb)
EG6ZX-X EG7ZX-X EG8ZX-X	Not specified
EG60X-X EG70X-X EG80X-X	20 @ 0°F
EG62X-X EG72X-X EG82X-X	20 @ –20°F

Notes:
a. An electrode combination that meets the impact requirements at a given temperature in this table also meets the temperature requirements of all higher temperatures in this table. Accordingly, EGX2X-X can also be classified as EGX0X-X and EGXZX-X, and EGX0X-X can be classified as EGXZX-X.
b. Both the highest and the lowest of the five test values obtained must be disregarded in computing the impact strength. Two of the remaining three values must equal or exceed 20 ft·lbf, and one of the remaining values may be lower than 20 ft·lbf, but not lower than 15 ft·lbf. The average of the three must not be less than the 20 ft·lbf specified.

Source: Adapted from American Welding Society (AWS) Committee on Filler Metals and Allied Materials, 1997, *Specification for Carbon and Low-Alloy Steel Electrodes for Electrogas Welding,* ANSI/AWS A5.26/A5.26M-97, Miami: American Welding Society, Table 3.

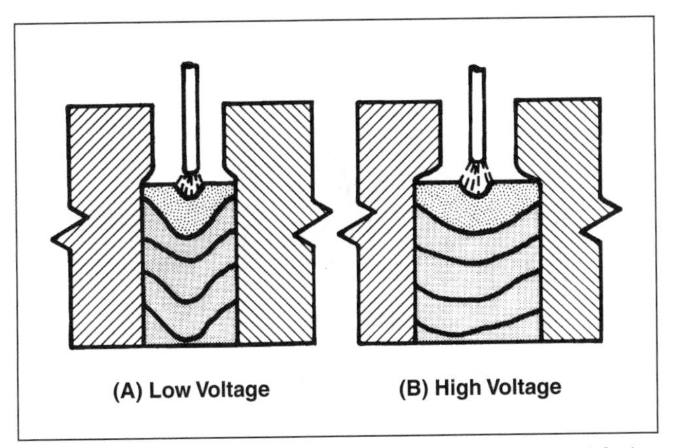

Source: American Welding Society (AWS) Committee on Arc Welding and Cutting, 2000, *Recommended Practices for Electrogas Welding,* AWS C5.7:2000, Miami: American Welding Society, Figure 18.

Figure 8.11—Effect of the Welding Voltage on the Shape of the Weld Pool

WELDING CURRENT AND ELECTRODE FEED SPEED

Welding current and electrode feed speed are proportional for a given electrode type, diameter, and extension. Thus, increasing the electrode feed speed increases the deposition rate, welding current, and vertical travel speed (fill rate).

For a given set of conditions in the moving-shoe mode of electrogas welding, increasing the current decreases the bevel-face or groove-face penetration and the width of the weld, as shown in Figure 8.12. Low current levels result in slow vertical travel speeds and wide welds. Excessive welding current may cause a severe reduction in weld width and bevel-face or groove-face penetration. Excessive welding current also causes a low form factor (see next section), which contributes to susceptibility to weld cracking at the centerline. Welding currents of 300 A to 400 A for 1.6 mm (1/16 in.), 400 A to 800 A for 2.4 mm (3/32 in.), and 500 A to 1000 A for 3.2 mm (1/8 in.) diameter electrodes are commonly used. The power source and welding cables must be sufficiently rated for the high currents and extended arc times typical of this welding process.

FORM FACTOR

The resistance of the weld to centerline cracking is greatly influenced by the manner in which it solidifies. As heat is removed from the molten weld metal by the base metal and backing shoes, solidification begins at these cooler areas and progresses toward the center of the weld. Since filler metal is continuously added, a progressive solidification takes place from the bottom of the joint, and molten weld metal is always present above the solidifying weld metal. The resulting solidification pattern is shown by the sketch of a vertical cross section of an electrogas weld in Figure 8.13.

The pattern of weld pool solidification can be expressed by the term *form factor.* The form factor is the

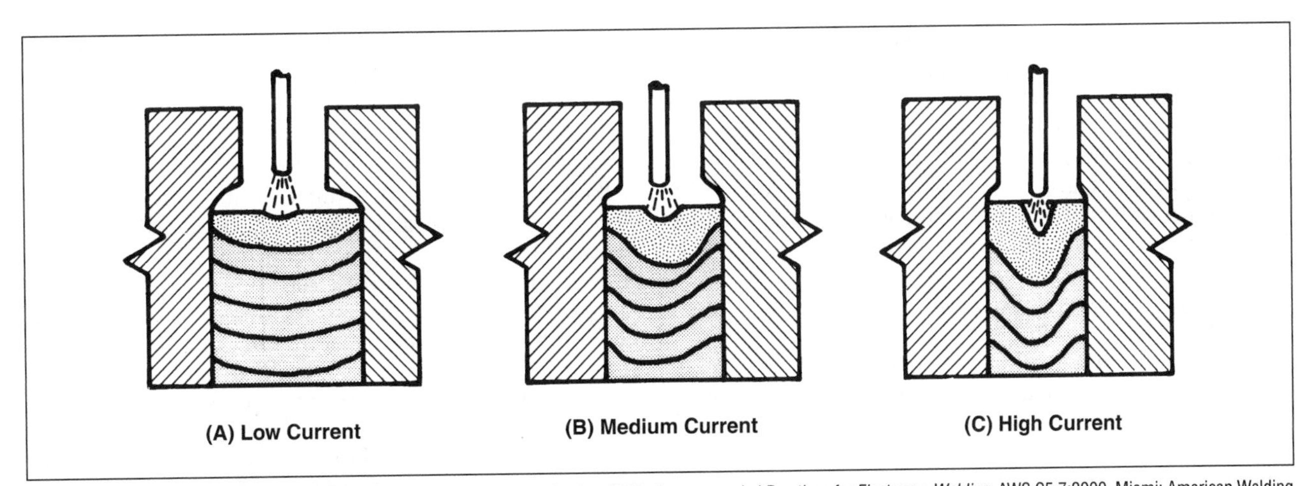

Source: American Welding Society (AWS) Committee on Arc Welding and Cutting, 2000, *Recommended Practices for Electrogas Welding,* AWS C5.7:2000, Miami: American Welding Society, Figure 19.

Figure 8.12—Effect of the Welding Current on the Shape of the Weld Pool

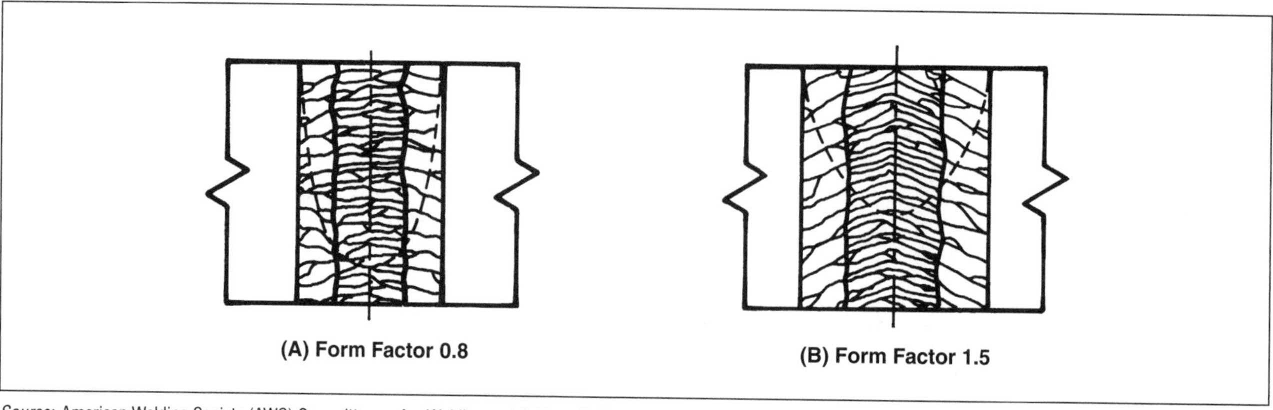

Source: American Welding Society (AWS) Committee on Arc Welding and Cutting, 2000, *Recommended Practices for Electrogas Welding,* AWS C5.7:2000, Miami: American Welding Society, Figure 20.

Figure 8.13—Sketches of the Weld Macrostructure of a Vertical Section of an Electrogas Weld

ratio of the maximum weld pool width (root opening plus bevel face or groove face penetration) to its maximum depth. The form factor is measured on a vertical section of the electrogas weld, usually taken at the mid-thickness of the base metal. Welds having high (approximately 1.5) form factors (wide width and shallow depth) have maximum resistance to centerline cracking. Welds having low (approximately 0.5) form factors (narrow width and deep weld pool) tend to have low resistance to centerline cracking. Thus, the form factor is an empirical number that provides a relative measure of resistance to centerline cracking and describes the shape of the weld metal solidification pattern. However, form factor alone does not determine cracking propensity. The chemical composition of the filler metal and the base metal (especially carbon content) and joint restraint also contribute to centerline crack susceptibility.

The shape of the weld pool and its resultant form factor are controlled by the welding variables. In general, increasing root opening or the voltage increases the form factor, while increasing the current or decreasing the root opening decreases the form factor. Since the welding variables are usually specified in a qualified welding procedure, the form factor is rarely required to be measured or recorded.

ELECTRODE EXTENSION

In externally gas-shielded electrogas welding with a moving shoe, an electrode extension of about 40 mm (1-1/2 in.) is recommended for solid and flux cored electrodes. For most self-shielded flux cored electrogas welding, an extension of 50 mm to 75 mm (2 in. to 3 in.) is recommended.

The power source modifies the effect of electrode extension. When using a recommended constant-voltage power source, increasing electrode extension by jogging up (slightly raising) the electrode guide will reduce the current. The width of the resulting weld will be reduced, as indicated in Figure 8.14. Increasing electrode extension by increasing the electrode feed speed increases the deposition rate and decreases the bevel face or groove face penetration and weld width.

For a constant-current (or drooping) power source, increasing the electrode extension by jogging up the welding head causes an increase in the voltage, resulting in a wider weld. An increase in the electrode extension by increasing the wire feed speed will decrease the arc voltage and result in a higher deposition rate (and fill rate) with a narrower weld width.

ELECTRODE OSCILLATION

The necessity for electrode oscillation depends on the welding conditions, including the current, voltage, and electrode diameter. Therefore, the following comments are general, and users should develop procedures that best suit their individual operations.

Base metals up to 19 mm (3/4 in.) thick are commonly welded with a stationary electrode. Base metals between 19 mm and 32 mm (3/4 in. and 1-1/4 in.) can be welded with stationary electrodes, but oscillation is commonly employed for greater thicknesses. Base metals with a thickness between 32 mm and 40 mm (1-1/4 in. and 1-1/2 in.) have been welded with a stationary electrode, but high voltages and special techniques are required to prevent incomplete fusion.

Electrode oscillation should be controlled so that the arc stops at a distance of approximately 10 mm (3/8 in.) from each shoe or dam. Oscillation speed is commonly 13 mm/s to 16 mm/s (20 in./min to 50 in./min). A period of dwell time at each end of the oscillation travel

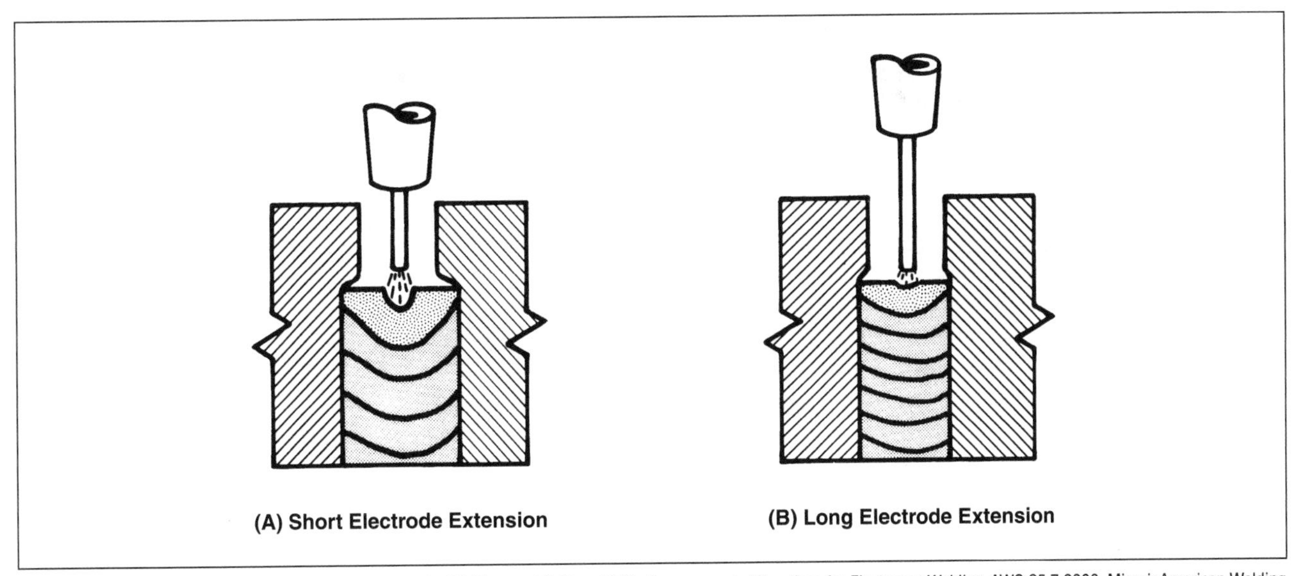

Source: American Welding Society (AWS) Committee on Arc Welding and Cutting, 2000, *Recommended Practices for Electrogas Welding,* AWS C5.7:2000, Miami: American Welding Society, Figure 21.

Figure 8.14—Effect of Electrode Extension (with a Constant-Voltage Power Source) on the Shape of the Weld Pool

is often needed to assure complete fusion across the weld faces. The dwell time may vary from 1/2 second to 3 seconds, depending on joint thickness.

Oscillation with a consumable guide tube over 1 m (3 ft) long becomes difficult because the bottom end of the tube tends to whip, distort, and lose its rigidity due to heating. For these welds, greater reliability has been obtained using multiple guide tubes and electrodes.

ROOT OPENING

A minimum root opening (the separation between the workpieces) is needed to give sufficient clearance for the electrode guide, which travels inside the joint. Increasing the root opening increases the weld width and form factor. Excessive root openings increase the welding time and consume extra filler metals and gases, which increase welding costs. Excessive root openings may also cause incomplete fusion if the voltage is not increased. Root openings for the square-groove weld design are generally 17 mm to 32 mm (11/16 in. to 1-1/4 in.). Root openings for the single-V-groove weld design are generally 17 mm to 32 mm (11/16 in. to 1-1/4 in.) at the face and 4 mm to 10 mm (5/32 in. to 3/8 in.) at the root.

WORKPIECE LEAD

The workpiece lead, which carries the welding current, is usually connected to the starting weld tab, but it can be split and attached to each side of the joint at the bottom. Although this workpiece lead connection location has proved satisfactory in most applications, in some situations, severe arc blow can cause excessive starting porosity and incomplete fusion on one or both sides of the joint, or both. The optimum location for attaching the workpiece lead must be determined by the user to suit the specific conditions of each application. For example, it is often attached at the top of the joint when welding with self-shielded flux cored electrogas electrodes.

SETTINGS FOR ELECTROGAS WELDS

Commonly used settings for electrogas welds are shown in Tables 8.7, 8.8, and 8.9. Users should optimize these settings to meet their specific requirements. In addition to the notes below the tables, the following definitions pertain to these tables:

1. The drag angle is the angle formed between the electrode and the vertical plane of the weld joint;
2. The cycle period is the total time for the electrode to complete one transverse oscillation from one extreme to the other and back, including dwell time at both extremes; and
3. The oscillation distance is the horizontal distance in one direction over which the oscillating electrode traverses.

Table 8.7
Typical Settings for Electrogas Welds with AWS Class EG72T-1 Electrode with Moving Shoe(s)

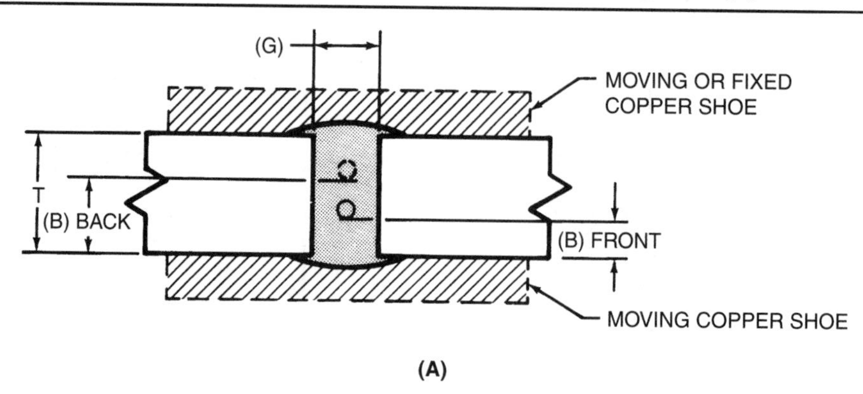

(A)

Key:
G = Root opening, mm (in.)
B = Electrode location, mm (in.)
T = Thickness of the base metal, mm (in.)

	Base Metal Thickness, mm (in.)			
	19 (3/4)	25 (1)	32 (1-1/4)	38 (1-1/2)
Current, A (approx.) Voltage, V (DCEP)	◄ 645–685 40–42 ►			
Wire feed speed, mm/s (in./min)	◄ 148 (350) ►			
Travel speed, (approx.) mm/s (in./min)	1.7 (4.1)	1.4 (3.4)	1.1 (2.6)	0.7 (1.7)
Electrode class Electrode size, mm (in.) Electrode extension, mm (in.)	◄ EG72T-1 3.0 (0.120) 70 ± 6 (2-3/4 ± 1/4) ►			
Electrode location[a] [(B) Front], mm (in.) [(B) Back], mm (in.)	6.4 (1/4)	8.0 (5/16)	10 (3/8)	6.4 (1/4) 25 (1)
Electrode drag angle[b]	◄ 8–10° ►			
Oscillation distance,[c] mm (in.) Cycle period,[d] s Dwell, s Traverse time, s	No oscillation	No oscillation	No oscillation	19 (3/4) 5 2 front, 2 back 0.5
Root opening (G), mm (in.)	◄ 19 + 3, – 0 (3/4 + 1/8, – 0) ►			
Start sequence "on,"[e] s	2	5	8	10

Notes:
a. The electrode location is measured from the open face of the joint to the near side of the electrode tip after establishing 70 mm ± 6 mm (2-3/4 in. ± 1/4 in.) electrode extension. The location should be set carefully and monitored during welding.
b. As defined in AWS C5.7:2000, the drag angle is the angle formed between the electrode and the vertical plane of the weld joint. See Item 11d in Table 8.13 in this chapter as well as in Table 13 in American Welding Society (AWS) Committee on Arc Welding and Cutting, 2000, *Recommended Practices for Electrogas Welding*, AWS C5.7:2000, Miami: American Welding Society.
c. As defined in AWS C5.7:2000, the oscillation distance is the horizontal distance in one direction over which the oscillating electrode traverses (p. 3).
d. As defined in AWS C5.7:2000, the cycle period is the total time for the electrode to complete one transverse oscillation from one extreme to the other and back, including dwell time at both extremes (p. 3).
e. The starting circuit should deliver 64 mm/s (150 in./min) wire feed speed at 29 arc volts. These conditions should be maintained for the number of seconds shown in the table.

Source: Adapted from American Welding Society (AWS) Committee on Arc Welding and Cutting, 2000, *Recommended Practices for Electrogas Welding*, AWS C5.7:2000, Miami: American Welding Society, Table 7, p. 31.

Table 8.7 (Continued)
Typical Settings for Electrogas Welds with AWS Class EG72T-1 Electrode with Moving Shoe(s)

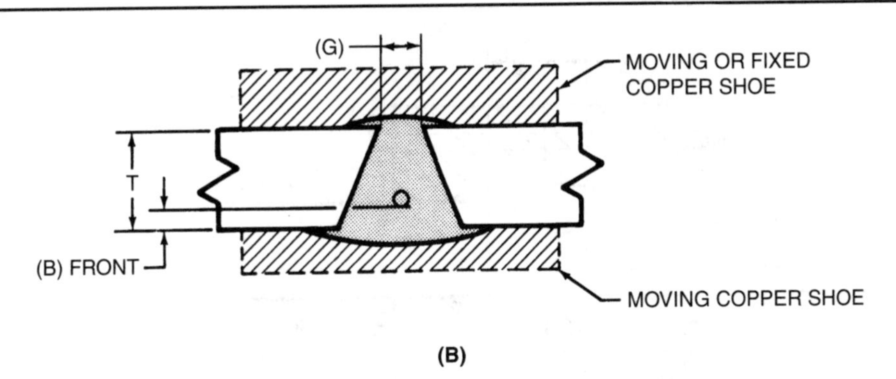

(B)

Key:
G = Root opening, mm (in.)
B = Electrode location, mm (in.)
T = Thickness of the base metal, mm (in.)

	Base Metal Thickness, mm (in.)			
	13 (1/2)	16 (5/8)	19 (3/4)	25 (1)
Current, A (approx.) Voltage, V (DCEP)	450–500 35–37	475–525 36–38	525–575 37–39	525–575 37–39
Wire feed speed, mm/s (in./min)	127 (300)	144 (340)	161 (380)	161 (380)
Travel speed, (approx.) mm/s (in./min)	1.7 (4.1)	1.4 (3.4)	1.1 (2.7)	0.7 (1.7)
Electrode class Electrode size, mm (in.) Electrode extension, mm (in.)	← EG72T-1 2.4 (3/32) 51 ± 3 (2 ± 1/8) →			
Electrode location[a] [(B) Front], mm (in.) Electrode drag angle[b]	3.2 (1/8) 8°	4.8 (3/16) 8°	6.4 (1/4) 8°	8.0 (5/16) 8°
Root opening G (mm) +3.2 – 0 (in.) +1/8 – 0	13 (1/2)	16 (5/8)	19 (3/4)	19 (3/4)
Start sequence "on 3,"[c] s	2	3	4	6

Notes:

a. The electrode location is measured from the open face of the joint to the near side of the electrode tip after establishing 51 mm (2 in.) electrode extension. The location should be set carefully and monitored during welding. The guide tip should be 51 mm (2 in.) above the top of the shoe.

b. See Table 13 (11d) in American Welding Society (AWS) Committee on Arc Welding and Cutting, 2000, Recommended Practices for Electrogas Welding, AWS C5.7:2000, Miami: American Welding Society.

c. The starting circuit should deliver 64 mm/s (150 in./min) wire feed speed at 29 arc volts. These conditions should be maintained for the number of seconds shown in the table.

Source: Adapted from American Welding Society (AWS) Committee on Arc Welding and Cutting, 2000, *Recommended Practices for Electrogas Welding,* AWS C5.7:2000, Miami: American Welding Society, Table 7 (Continued), p. 32.

Table 8.8
Commonly Used Settings for Electrogas Welds with AWS Class EG72T-1 Electrode with Consumable Guide Tube

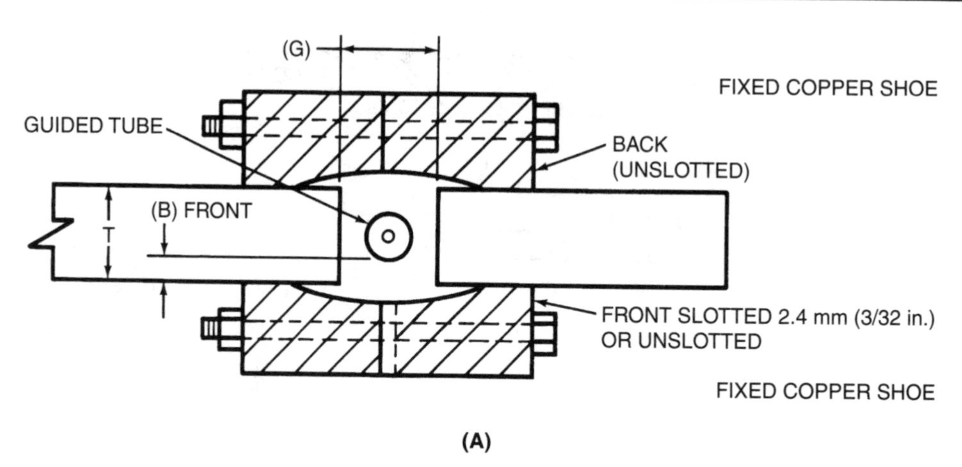

(A)

Key:
G = Root opening, mm (in.)
B = Guide tube location, mm (in.)
T = Thickness of the base metal, mm (in.)

	Base Metal Thickness, mm (in.)			
	19 (3/4)	**25 (1)**	**32 (1-1/4)**	**38 (1-1/2)**
Current, A (approx.) Voltage, V (DCEP)	650–750 35–38	675–775 38–40	700–800 40–42	725–825 42–44
Wire feed speed, mm/s (in./min)	85 (200)	97 (230)	112 (265)	127 (300)
Travel speed, (approx.) mm/s (in./min)	1.3 (3)	1.1 (2.5)	0.85 (2)	0.85 (2)
Electrode class Electrode size, mm (in.)	← EG72T-1 3.0 (0.120) →			
Guide tube	13 mm (1/2 in.) O.D. × 4.0 mm (5/32 in.) I.D. C1008/C1020}			
Guide tube location[a] [(B) Front], mm (in.)	3.2 (1/8)	6.4 (1/4)	10 (3/8)	13 (1/2)
Oscillation distance, mm (in.) Cycle period, s	← No oscillation →			
Root opening (G), mm (in.)		22 + 3.2, − 0	(7/8 + 1/8, − 0)	
Start time "on,"[b] s	3	3	3	3
Crater time "on,"[c] s	6	10	15	20

Notes:
a. The location of the guide tube is measured from the open face of the joint to the near side of the guide tube. The location should be set carefully, and the guide tube should be checked at the top and the bottom of the joint before the shoes are attached.
b. The starting circuit should deliver 42 mm/s (100in/min) wire feed speed at 35 arc volts. The conditions should be maintained for the number of seconds shown in the table.
c. The crater circuit should deliver 42 mm/s (100 in./min) wire feed speed at 35 arc volts. The conditions should be maintained for the number of seconds shown in the table.

Source: Adapted from American Welding Society (AWS) Committee on Arc Welding and Cutting, 2000, *Recommended Practices for Electrogas Welding*, AWS C5.7:2000, Miami: American Welding Society, Table 8, p. 33.

Table 8.8 (Continued)
Commonly Used Settings for Electrogas Welds with AWS Class EG72T-1 Electrode with Consumable Guide Tube

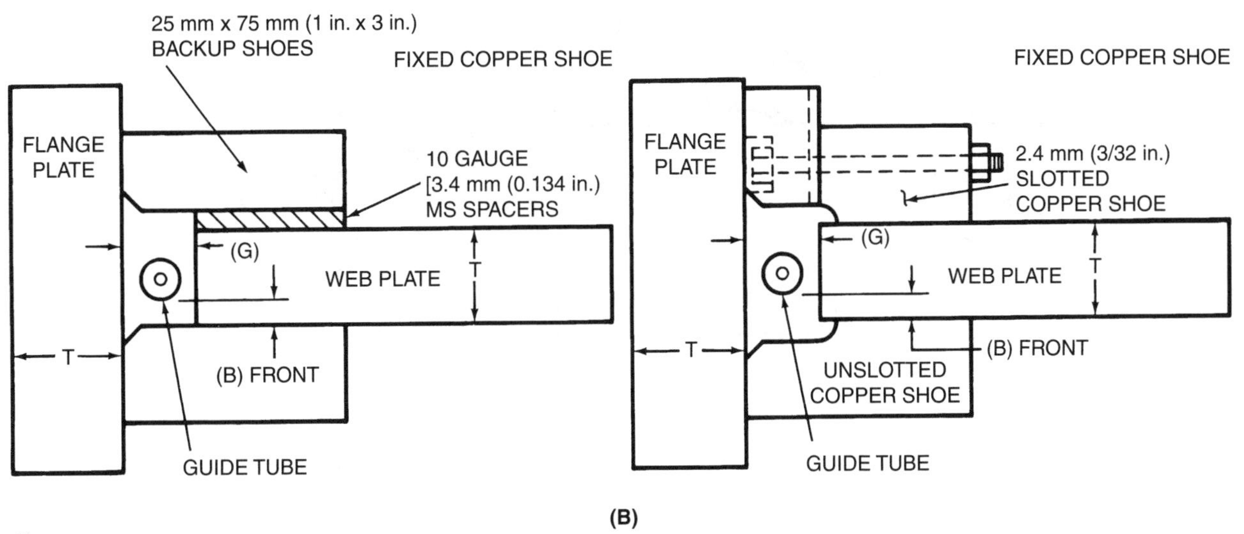

(B)

Key:
MS = Mild steel
G = Root opening, mm (in.)
B = Guide tube location, mm (in.)
T = Thickness of the base metal, mm (in.)

	Base Metal Thickness, mm (in.)			
	19 (3/4)	**25 (1)**	**32 (1-1/4)**	**38 (1-1/2)**
Current, A (approx.) Voltage, V (DCEP)	650–750 35–38	675–775 38–40	700–800 40–42	725–825 42–44
Wire feed speed, mm/s (in./min)	85 (200)	97 (230)	112 (265)	127 (300)
Travel speed, (approx.), mm/s (in./min)	0.8 (2)	0.8 (2)	0.8 (2)	0.8 (2)
Electrode class Electrode size, mm (in.)	← EG72T-1 3.0 (0.120) →			
Guide tube	13 mm (1/2 in.) O.D. × 4.0 mm (5/32 in.) I.D. C1008/C1020}			
Guide tube location[a] [(B) Front], mm (in.)	3.2 (1/8)	6.4 (1/4)	10 (3/8)	13 (1/2)
Oscillation distance, mm (in.) Cycle period, s	← No oscillation →			
Root opening (G), mm (in.)		22 + 3.2 – 0	(7/8 + 1/8 – 0)	
Start time "on,"[b] s	3	3	3	3
Crater time "on,"[c] s	6	10	15	20

Notes:
a. The location of the guide tube is measured from the face of the joint to the near side of the guide tube. The location should be set carefully, and the guide tube should be checked at the top and the bottom of the joint before the shoes are attached.
b. The starting circuit should deliver 42 mm/s (100 in./min) wire feed speed at 35 arc volts. The conditions should be maintained for the number of seconds shown in the table.
c. The crater circuit should deliver 42 mm/s (100 in./min) wire feed speed at 35 arc volts. The conditions should be maintained for the number of seconds shown in the table.

Source: Adapted from American Welding Society (AWS) Committee on Arc Welding and Cutting, 2000, *Recommended Practices for Electrogas Welding*, AWS C5.7:2000, Miami: American Welding Society, Table 8 (Continued), p. 34.

Table 8.8 (Continued)
Commonly Used Settings for Electrogas Welds with AWS Class EG72T-1 Electrode with Consumable Guide Tube

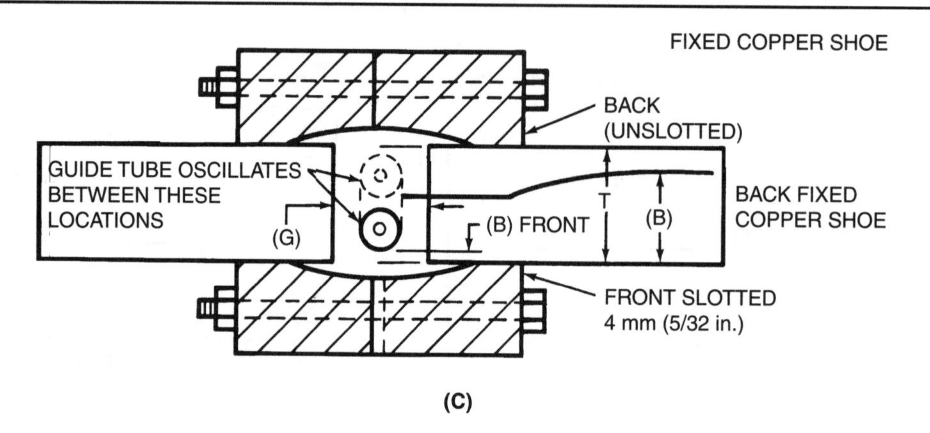

(C)

Key:
G = Root opening, mm (in.)
B = Guide tube location, mm (in.)
T = Thickness of the base metal, mm (in.)

	Base Metal Thickness, mm (in.)		
	51 (2)	76 (3)	102 (4)
Current, A (approx.) Voltage, V (DCEP)	775–875 42–44	800–900 43–45	800–900 45–47
Wire feed speed, mm/s (in./min)	148 (350)	169 (400)	169 (400)
Travel speed, (approx.) mm/s (in./min)	0.59 (1.4)	0.46 (1.2)	0.34 (0.8)
Electrode class Electrode size, mm (in.)	←	EG72T-1 3.0 (0.120)	→
Guide tube	←	13 mm (1/2 in.) O.D. × 4.0 mm (5/32 in.) I.D. C1008/C1020	→
Guide tube location[a] [(B) Front], mm (in.) [(B) Back], mm (in.)	 6.4 (1/4) 32 (1-1/4)	 6.4 (1/4) 57 (2-1/4)	 6.4 (1/4) 83 (3-1/4)
Oscillation distance, mm (in.) Cycle period, s Dwell, s Traverse time, s	25 (1) 8.0 2–3 1.5	51 (2) 12 2–3 3.5	76 (3) 18 3–4 5.5
Root opening (G), mm (in.)	←	25 + 3.2 – 0 (1 + 1/8 – 0)	→
Start time "on,"[b] s	3	3	3
Crater time "on,"[c] s	20	30	40

Notes:
a. On 51 mm (2 in.) through 100 mm (4 in.) plate, the guide tube oscillates between the locations shown and along the centerline of the root opening. The locations are 6.4 mm (1/4 in.) in from the surface of the plate to the edge of the guide tube. The oscillation should set before putting on the copper shoes so that the guide locations can be checked at the top and the bottom of the joint. The locations can be set by adjusting the stops on the oscillator.
b. The starting circuit should deliver 42 mm/s (100 in./min wire feed speed at 35 arc volts. The conditions should be maintained for the number of seconds shown in the table.
c. The crater circuit should deliver 42 mm/s (100 in./min) wire feed speed at 35 arc volts. The conditions should be maintained for the number of seconds shown in the table.

Source: Adapted from American Welding Society (AWS) Committee on Arc Welding and Cutting, 2000, *Recommended Practices for Electrogas Welding*, AWS C5.7:2000, Miami: American Welding Society, Table 8 (Continued), p. 35.

Table 8.8 (Continued)
Commonly Used Settings for Electrogas Welds with AWS Class EG72T-1 Electrode with Consumable Guide Tube

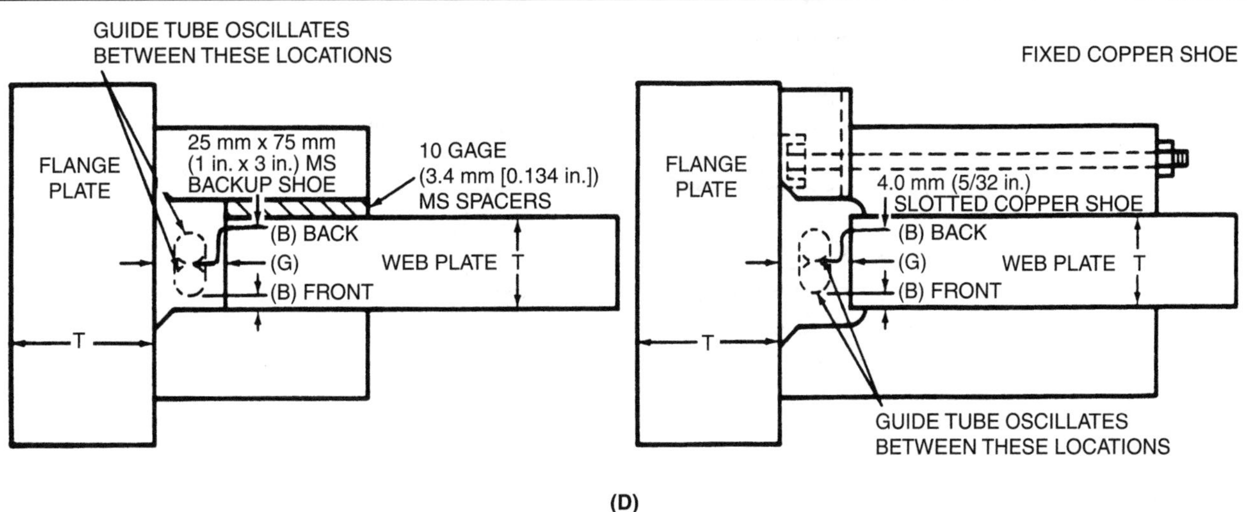

(D)

Key:
MS = Mild steel
G = Root opening, mm (in.)
B = Guide tube location, mm (in.)
T = Thickness of the base metal, mm (in.)

	Base Metal Thickness, mm (in.)		
	51 (2)	76 (3)	102 (4)
Current, A (approx.)	775–875	800–900	800–900
Voltage, V (DCEP)	42–44	43–45	45–47
Wire feed speed, mm/s (in./min)	148 (350	169 (400)	169 (400
Travel speed, (approx.) in./min (mm/s)	0.59 (1.4)	0.46 (1.1)	0.34 (0.8)
Electrode class	←	EG72T-1	→
Electrode size, mm (in.)	←	3.0 (0.120)	→
Guide tube	←	13 mm (1/2 in.) O.D. × 4.0 mm (5/32 in.) I.D. C1008/C1020	→
Guide tube location[a]			
[(B) front], mm (in.)	6.4 (1/4)	6.4 (1/4)	6.4 (1/4)
[(B) back], mm (in.)	32 (1-1/4)	57 (2-1/4)	83 (3-1/4)
Oscillation distance, mm (in.)	25 (1)	51 (2)	76 (3)
Cycle period, s	8	12	18
Dwell, s	2–3	2–3	3–4
Traverse time, s	1.5	3.5	5.5
Root opening (G), mm (in.)	←	25 + 3 – 0 (1 + 1/8 – 0)	→
Start time "on,"[b] s	3	3	3
Crater time "on,"[c] s	20	30	40

Notes:

a. On 51 mm (2 in.) through 100 mm (4 in.) plate, the guide tube oscillates between the locations shown and along the centerline of the root opening. The locations are 6.4 mm (1/4 in.) in from the surface of the plate to the edge of the guide tube. The oscillation should set before putting on the copper shoes so that the guide locations can be checked at the top and the bottom of the joint. The locations can be set by adjusting the stops on the oscillator.

b. The starting circuit should deliver 42 mm/s (100 in./min wire feed speed at 35 arc volts. The conditions should be maintained for the number of seconds shown in the table.

c. The crater circuit should deliver 42 mm/s (100 in./min) wire feed speed at 35 arc volts. The conditions should be maintained for the number of seconds shown in the table.

Source: Adapted from American Welding Society (AWS) Committee on Arc Welding and Cutting, 2000, *Recommended Practices for Electrogas Welding*, AWS C5.7:2000, Miami: American Welding Society, Table 8 (Continued), p. 36.

Table 8.9
Commonly Used Settings for Electrogas Welds with AWS Class EG72T-NM1 Electrode with Moving Shoe(s)

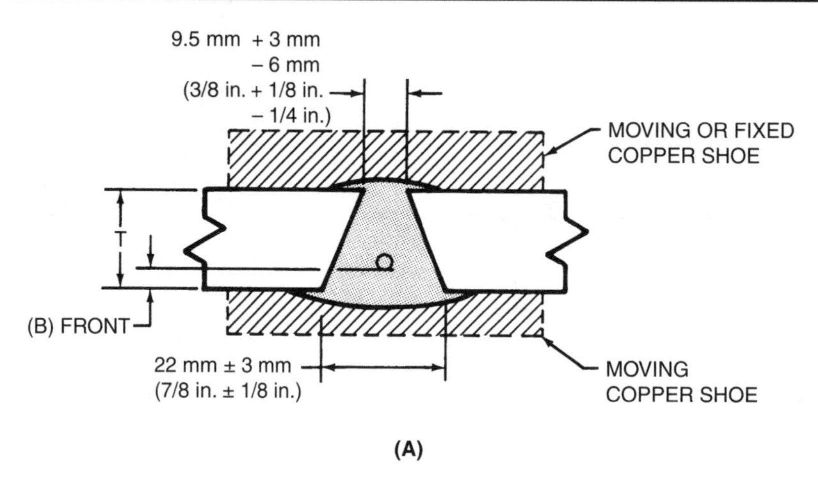

(A)

Key:
G = Root opening, mm (in.)
B = Electrode location, mm (in.)
T = Thickness of the base metal, mm (in.)

	Base Metal Thickness, mm (in.)		
	19 (3/4)	25 (1)	32 (1-1/4)
Current, amperes	650–700	700–750	700–750
Voltage, V (DCEP)	42–44	44–46	46–48
Travel speed, mm/s (in./min)	2.8–3.8 (6.7–9.0)	1.8–2.4 (4.3–5.8)	1.5–2.0 (3.6–4.8)
Electrode class	←	EG72T-NM1	→
Electrode size, mm (in.).	←	3.2 (1/8)	→
Electrode extension, mm (in.)	←	51 ± 3 (2 ± 1/8)	→
Electrode location[a]			
[(B) front], mm (in.)	8 (5/16)	11 (7/16)	16 (5/8)
[(B) back], mm (in.)	—	—	—
Electrode drag angle[b]	0°	0°	0°
Oscillation distance, mm (in.)	←	No oscillation	→
Cycle period, s			
Dwell, s			
Traverse time, s			
Gas – CO_2, L/min (CFH)	18.9–28.3 (40–60)	18.9–28.3 (40–60)	18.9–28.3 (40–60)

Notes:
a. The location of the electrode is measured from the open face of the joint to the near side of the electrode tip after establishing 51 mm (2 in.) electrode extension. The location should be set carefully and monitored during welding. The guide tip should be 25 mm (1 in.) above the top of the shoe. The top of the weld pool will then be 25 mm (1 in.) below the top of the shoe.
b. See Table 13 (11d) in American Welding Society (AWS) Committee on Arc Welding and Cutting, 2000, *Recommended Practices for Electrogas Welding*, AWS C5.7:2000, Miami: American Welding Society.

Source: Adapted from American Welding Society (AWS) Committee on Arc Welding and Cutting, 2000, *Recommended Practices for Electrogas Welding*, AWS C5.7:2000, Miami: American Welding Society, Table 9, p. 37.

Table 8.9 (Continued)
Commonly Used Settings for Electrogas Welds with AWS Class EG72T-NM1 Electrode with Moving Shoe(s)

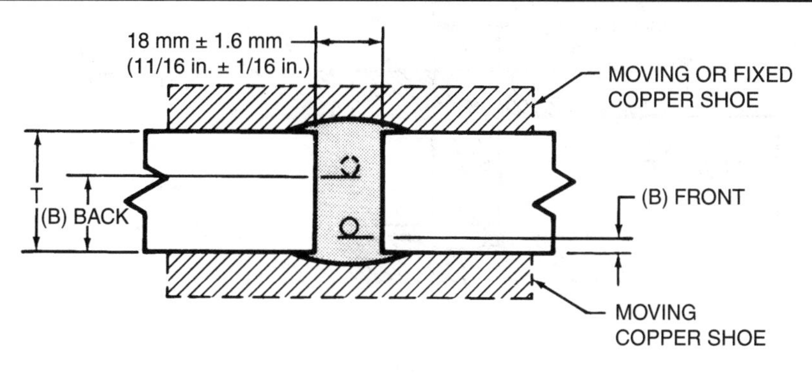

(B)

Key:
G = Root opening, mm (in.)
B = Electrode location, mm (in.)
T = Thickness of the base metal, mm (in.)

<table>
<tr><th></th><th colspan="4">Base Metal Thickness, mm (in.)</th></tr>
<tr><th></th><th>13 (1/2)</th><th>19 (3/4)</th><th>25 (1)</th><th>51 (2)</th></tr>
<tr><td>Current, amperes</td><td colspan="4">700–750</td></tr>
<tr><td>Voltage, V (DCEP)</td><td>42–44</td><td>46–48</td><td>40–41</td><td>42–43</td></tr>
<tr><td>Travel speed, mm/s (in./min)</td><td>1.4–1.9 (3.4–4.6)</td><td>1.1–1.5 (2.6–3.5)</td><td>0.8–1.0 (1.9–2.5)</td><td>0.6–0.8 (1.4–1.9)</td></tr>
<tr><td>Electrode class</td><td colspan="4">EG72T-NM1</td></tr>
<tr><td>Electrode size, mm (in.)</td><td colspan="4">3.2 (1/8)</td></tr>
<tr><td>Electrode extension, mm (in.)</td><td colspan="4">51 ± 3 (2 ± 1/8)</td></tr>
<tr><td>Electrode location[a]</td><td></td><td></td><td></td><td></td></tr>
<tr><td>[(B) front], in., (mm)</td><td>4.8 (3/16)</td><td>7.9 (5/16)</td><td>3.2 (1/8)</td><td>3.2 (1/8)</td></tr>
<tr><td>[(B) back], in., (mm)</td><td>—</td><td>—</td><td>19 (3/4)</td><td>44 (1-3/4)</td></tr>
<tr><td>Electrode drag angle[b]</td><td>0°</td><td>0°</td><td>0°</td><td>0°</td></tr>
<tr><td>Oscillation distance, mm (in.)</td><td rowspan="4">No oscillation</td><td rowspan="4">No oscillation</td><td>16 (5/8)</td><td>41 (1-5/8)</td></tr>
<tr><td>Cycle period, s</td><td>5</td><td>5</td></tr>
<tr><td>Dwell, s</td><td>2.0 front</td><td>2.0 back</td></tr>
<tr><td>Traverse time, s</td><td>1</td><td>1</td></tr>
<tr><td>Gas – CO_2, L/min (CFH)</td><td>18.9–28.3 (40–60)</td><td>18.9–28.3 (40–60)</td><td>18.9–28.3 (40–60)</td><td>18.9–28.3 (40–60)</td></tr>
</table>

Notes:
a. The location of the electrode is measured from the open face of the joint to the near side of the electrode tip after establishing 51 mm (2 in.) electrode extension. The location should be set carefully and monitored during welding. The guide tip should be 25 mm (1 in.) above the top of the shoe. The top of the weld pool will then be 25 mm (1 in.) below the top of the shoe.
b. See Table 13 (11d) in American Welding Society (AWS) Committee on Arc Welding and Cutting, 2000, *Recommended Practices for Electrogas Welding*, AWS C5.7:2000, Miami: American Welding Society.

Source: Adapted from American Welding Society (AWS) Committee on Arc Welding and Cutting, 2000, *Recommended Practices for Electrogas Welding*, AWS C5.7:2000, Miami: American Welding Society, Table 9 (Continued), p. 38.

Table 8.9 (Continued)

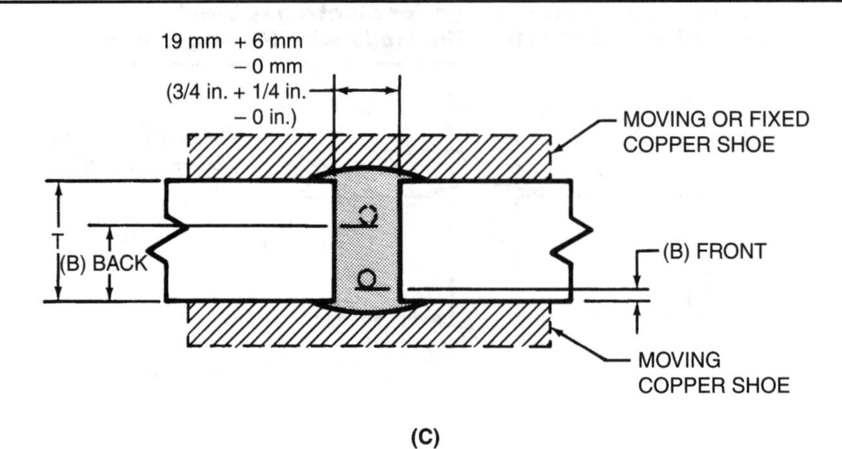

(C)

Key:
G = Root opening, mm (in.)
B = Electrode location, mm (in.)
T = Thickness of the base metal, mm (in.)

	Base Metal Thickness, mm (in.)		
	25 (1)	**32 (1-1/4)**	**38 (1-1/2)**
Current, A	305–335	335–375	355–395
Voltage, V (DCEP)	35–39	36–40	37–41
Wire feed speed, mm/s (in./min)	138 (325)	148 (350)	158 (375)
Travel speed, mm/s (in./min)	0.57–0.63 (1.35–1.50)	0.42–0.46 (1.0–1.1)	0.34–0.38 (0.80–0.90)
Electrode class		EG72T-NM1	
Electrode size, mm (in.)	←	1.6 (1/16)	→
Electrode extension, mm (in.)		38 ± 3 (1-1/2 ± 1/8)	
Electrode location[a]			
[(B) front] in., (mm)	11.9 (15/32)	8.7 (11/32)	8.7 (11/32)
[(B) back] in., (mm)	—	15 (19/32)	28 (1-3/32)
Electrode drag angle[b]	0°	0°	0°
Oscillation distance, mm (in.)		6.4 (1/4)	13 (1/2)
Cycle period, s	No	2.02	2.06
Dwell, s	oscillation	1 front, 1 back	1 front, 1 back
Traverse time, s		0.01	0.03
Gas – Ar80/CO_2, L/min (CFH)	18.9–28.3 (40–60)	18.9–28.3 (40–60)	18.9–28.3 (40–60)

Notes:
a. The location of the electrode is measured from the open face of the joint to the near side of the electrode tip after establishing 51 mm (2 in.) electrode extension. The location should be set carefully and monitored during welding. The guide tip should be 25 mm (1 in.) above the top of the shoe. The top of the weld pool will then be 25 mm (1 in.) below the top of the shoe.
b. See Table 13 (11d) in American Welding Society (AWS) Committee on Arc Welding and Cutting, 2000, *Recommended Practices for Electrogas Welding*, AWS C5.7:2000, Miami: American Welding Society.

Source: Adapted from American Welding Society (AWS) Committee on Arc Welding and Cutting, 2000, *Recommended Practices for Electrogas Welding*, AWS C5.7:2000, Miami: American Welding Society, Table 9 (Continued), p. 39.

WELD JOINT PREPARATION

The weldment must be set up appropriately in order to successfully apply the electrogas welding process. Particular attention should be given to proper fitup and assembly.

Misalignment

In electrogas welding operations, the misalignment of the surface of the base metal should be less than 3.2 mm (1/8 in.) on both front and back surfaces. Grinding may be required to remain within this tolerance. Greater misalignment creates a variety of problems, including leakage of slag and weld metal, irregular bead shape, poor tie-in, overlap, undercut, and porosity due to loss of shielding. In addition, excessive misalignment can result in melt-through when steel backing is used.

When plates of different thicknesses are to be welded, the thicker plate should be tapered, and the backing shoe should be tapered to match it. For structural welding applications, a 1 to 2-1/2 taper is satisfactory in most cases.

Strongbacks. Suitable holding clamps, brackets, or strongbacks (devices attached to the workpieces to maintain alignment) are required to maintain the joint dimensions and the workpiece alignment during welding. The U-shaped strongbacks shown in Figure 8.15 allow clearance for the moving shoe or other backing. The strongbacks should be rigid enough to prevent excessive shrinkage of the root opening during welding. The strongbacks should hold the workpieces in alignment but should not be so rigid so as to cause excessive restraint. Smaller strongbacks can generally be used for the fixed back-up applications than are needed for the clearance of a moving shoe.

Starting Weld Tabs. The starting weld tab (or sump) is a U-shaped piece of additional material with a groove similar to the weld joint. It is attached as an extension below the beginning of the joint. The tab permits initiating the arc below the actual joint so any starting discontinuities are located off the workpiece proper and can easily be removed and discarded. The starting weld tab should be long enough, typically 38 mm (1 1/2 in.) or more, to allow the process to switch from starting to welding procedures and for the arc and travel to stabilize before the weld pool rises into the joint being welded. The tab is cut off after welding.

The thickness of the starting weld tab and the groove shape should be the same as those of the workpieces being welded. The starting weld tab is attached to the workpieces as shown in Figure 8.16. The top of the tab that fits against the bottom of the workpieces should be

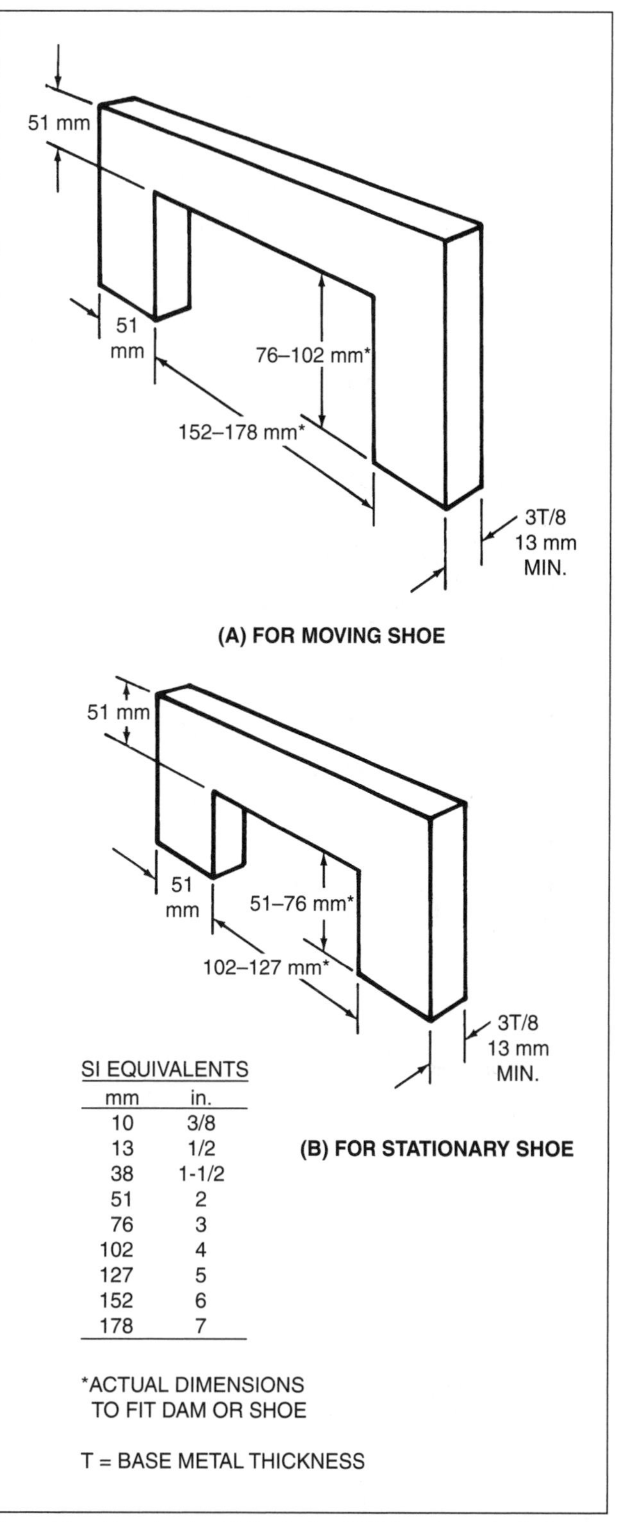

SI EQUIVALENTS

mm	in.
10	3/8
13	1/2
38	1-1/2
51	2
76	3
102	4
127	5
152	6
178	7

Source: Adapted from American Welding Society (AWS) Committee on Arc Welding and Cutting, 2000, *Recommended Practices for Electrogas Welding,* AWS C5.7:2000, Miami: American Welding Society, Figure 9.

Figure 8.15—Typical Dimensions for Electrogas Welding Strongbacks

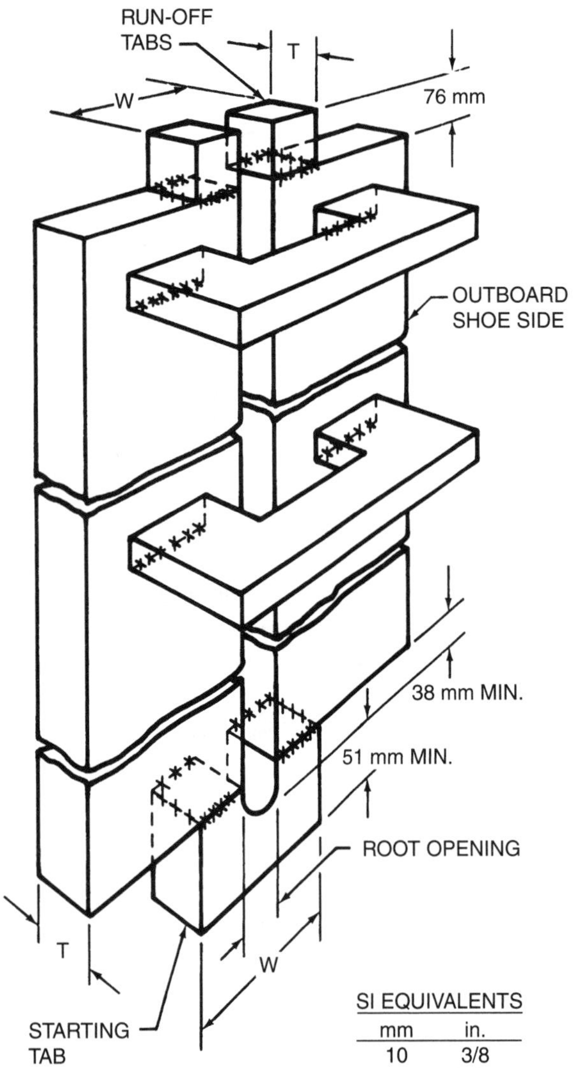

RUN-OFF TABS
T
W
76 mm
OUTBOARD SHOE SIDE
38 mm MIN.
51 mm MIN.
ROOT OPENING
T
W
STARTING TAB

SI EQUIVALENTS

mm	in.
10	3/8
13	1/2
38	1-1/2
51	2
76	3
102	4
127	5
152	6
178	7

Key
T = Thickness
W = Width of moving shoe plus 50 mm (2 in.)
XXX = Welds

Source: American Welding Society (AWS) Committee on Arc Welding and Cutting, 2000, *Recommended Practices for Electrogas Welding,* AWS C5.7:2000, Miami: American Welding Society, Figure 10.

Figure 8.16—Typical Arrangement of Starting Tabs, Runoff Tabs and Strongbacks for Electrogas Welding Joint Assemblies

clean and dry, as part of these surfaces will be consumed by the weld.

The starting tab should be seal-welded at the joint but should not be sealed all around. An opening should be left so that gases expanding between the starting weld tab and the workpiece are driven away from the molten weld metal. Insufficient seal-weld strength or thickness of material beneath the groove in the starting tab can result in weld metal leakage, which would delay the start of upward travel and may result in a defect.

Runoff Weld Tabs. Runoff weld tabs are required because the rapid cooling of the large weld pool at the end of the weld creates a shrinkage crater with a tendency to trap slag or gas. Typical runoff tabs are shown schematically in Figure 8.16. Stationary backing bars must reach to the top of the runoff area, finishing 25 mm to 75 mm (1 in. to 3 in.) above the top of the workpiece. The runoff tabs should be removed after the weld is completed.

APPLICATIONS

Virtually any vertical weld joint can be successfully welded using electrogas welding. This process overcomes the inherent difficulties presented by multiple-pass vertical welds and in addition, has only practical upper limits on the length of a weld that can be made in a single pass.

WELDABLE METALS

The metals most commonly joined by electrogas welding are low-carbon, structural, and pressure vessel steels. Examples of these steel grades are shown in Table 8.10. In addition to the steels shown in this table, electrogas welding has been used to join aluminum and some grades of stainless steels under very limited conditions.

Table 8.10
Examples of Steel Grades Commonly Joined with Electrogas Welding

Application	Grades
Low-carbon steels	AISI 1018, 1020
Structural steels	ASTM A 36, A 131, A 242, A 283, A 572, A 573, A 588
Pressure vessel steels	ASTM A 285, A 515, A 516, A 537

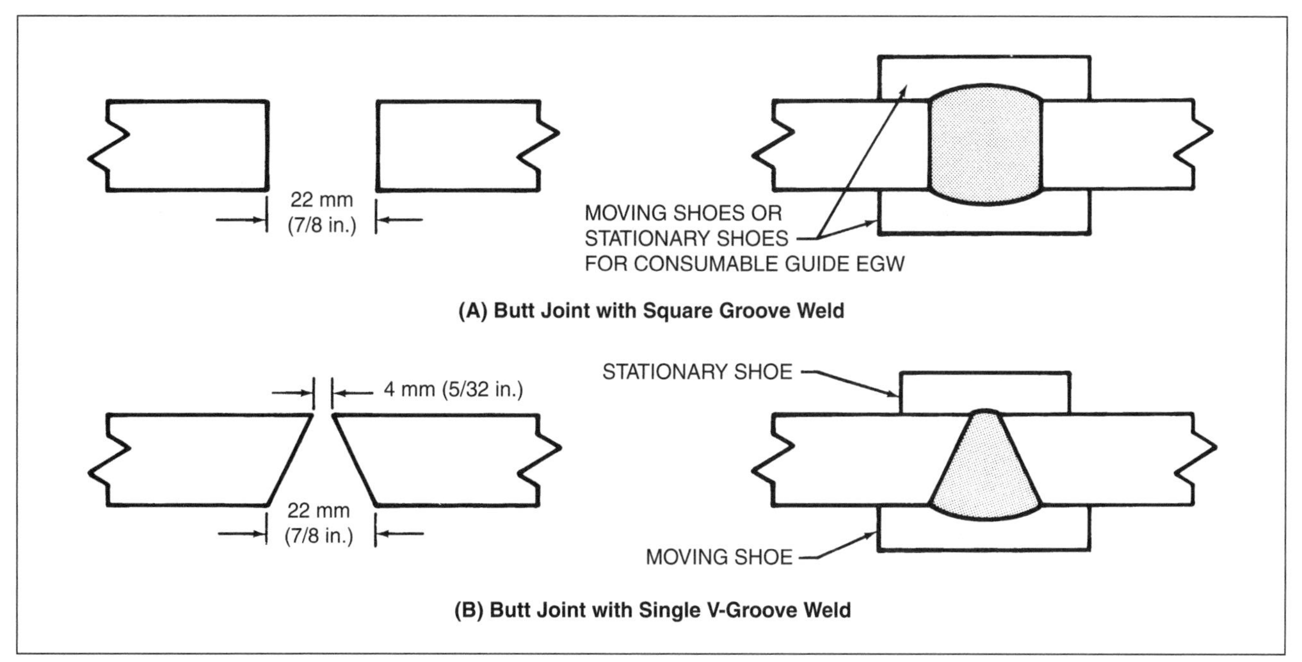

Source: Adapted from American Welding Society (AWS) Committee on Arc Welding and Cutting, 2000, *Recommended Practices for Electrogas Welding,* AWS C5.7:2000, Miami: American Welding Society, Figure 7.

Figure 8.17—Typical Joint Designs for Electrogas Welding

The electrogas welding of quenched and tempered or normalized steels may not be permitted by some codes. For example, *Structural Welding Code—Steel,* AWS D1.1/D1.1M:2002,[12] does not permit the electrogas welding of quenched-and-tempered steels, and *Bridge Welding Code,* ANSI/AASHTO/AWS D1.5,[13] contains restrictions on the use of the electrogas process on tension members and those subject to reversal of tension. The high heat input associated with electrogas welding may reduce the strength of some quenched and tempered steels in the heat-affected zone adjacent to the weld.

Thicknesses Welded

The electrogas welding process can be used for thicknesses from 13 mm (1/2 in.) to 38 mm (1-1/2 in.). Thicker sections are welded with electrode oscillation or with the use of multiple consumable guides.

12. American Welding Society (AWS) Committee on Structural Welding, 2002, *Structural Welding Code—Steel,* AWS D1.1/D1.1M:2002, Miami: American Welding Society.
13. American Welding Society (AWS) Committee on Structural Welding, 2002, *Bridge Welding Code,* ANSI/AASHTO/AWS D1.1/D1.5M:2002, Miami: American Welding Society.

JOINT DESIGN

Square butt joints with approximately 22 mm (7/8 in.) spacing are most commonly used in electrogas welding. This joint normally uses one moving shoe to increase weld travel speeds, as shown in Figure 8.17(A). Single-V-groove welds are also used with one moving and one stationary shoe, as depicted in Figure 8.17(B). The root opening of single-V-groove welds is typically about 4 mm (5/32 in.), and the face opening is normally 22 mm (7/8 in.). The moving shoe molds the metal at the weld face, and the stationary shoe molds the root face.

Examples of other joint configurations that have been used are shown in Figure 8.18. Variations of these joint configurations are also acceptable; however, the welding engineer should conduct experiments to determine the range of joint dimensions that is best suited to the application.

Applications of electrogas welding include the fabrication of storage tanks, ship hulls, structural members and pressure vessels. Electrogas welding should be considered for any joint to be welded in the vertical position in materials ranging in thickness from 10 mm to 100 mm (3/8 in. to 4 in.). Some typical electrogas welding applications are illustrated in Figures 8.19 through 8.23.

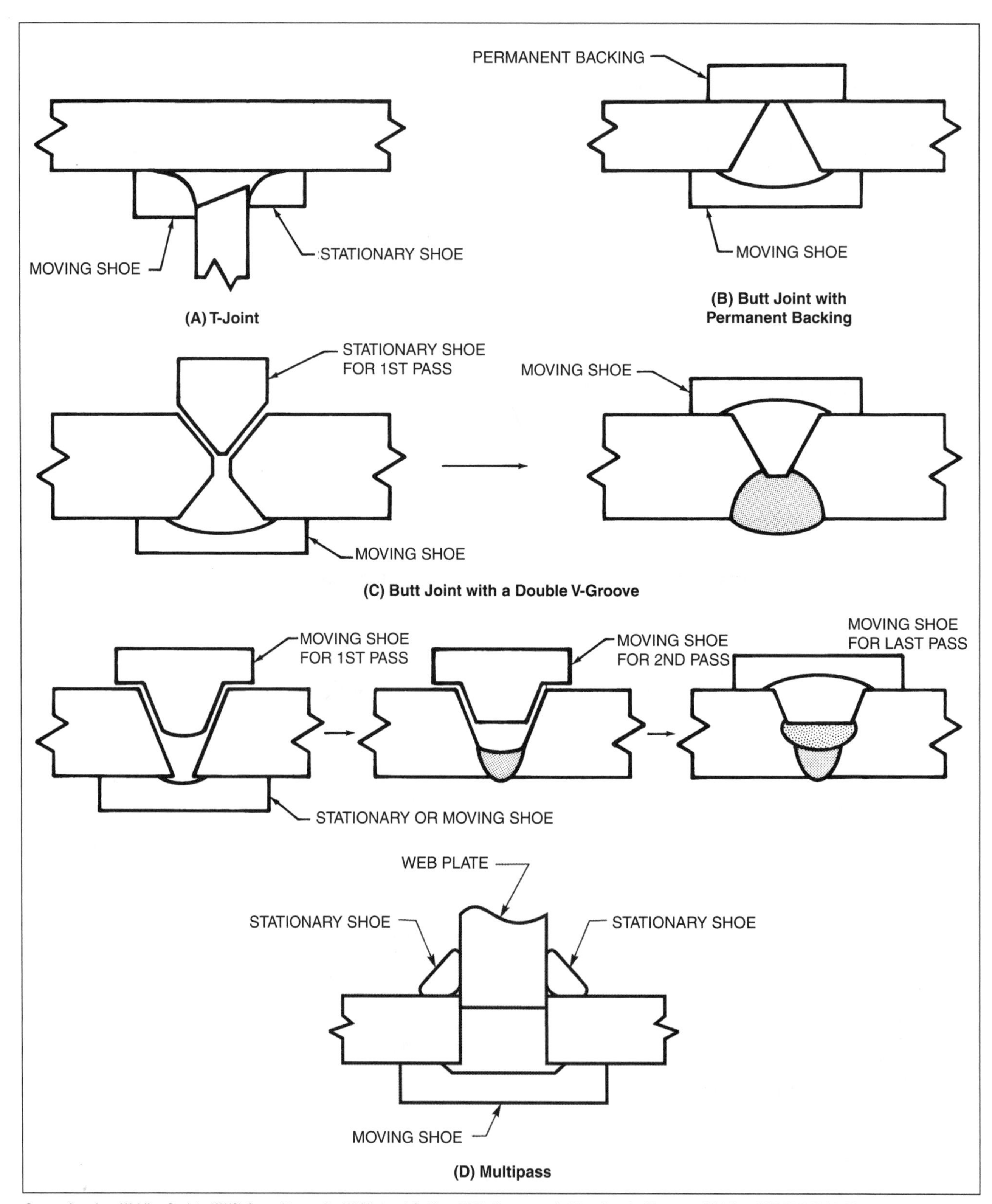

Source: American Welding Society (AWS) Committee on Arc Welding and Cutting, 2000, *Recommended Practices for Electrogas Welding,* AWS C5.7:2000, Miami: American Welding Society, Figure 8.

Figure 8.18—Alternate Joint Designs for Electrogas Welding

Figure 8.19—Electrogas Welding of a Large-Volume Storage Tank

INSPECTION AND WELD QUALITY

As with any welding or manufacturing process, it is important to inspect the work to ensure weld quality and proper workmanship. Conventional nondestructive examination (NDE) methods can be used to inspect electrogas welded joints.

METHODS OF NONDESTRUCTIVE INSPECTION

Nondestructive examination techniques normally used with electrogas welding include the following:

1. Visual examination,
2. Radiography (isotope or X-ray),
3. Magnetic particle examination,
4. Liquid penetrant examination, and

NOTE: Panels shield the electrogas unit from air currents that would disperse the shielding gas.

Figure 8.20—A Self-Contained Electrogas Welding Unit Used in Shipbuilding

5. Ultrasonic examination (large grain size may require special techniques).

All nondestructive examination activities should be performed in accordance with qualified procedures by qualified technicians. Some training in the interpretation of weld discontinuities found in electrogas welding is necessary in addition to the usual training of nondestructive examination personnel. Typical discontinuities found in electrogas welding are discussed in this chapter in the section "Possible Electrogas Weld Discontinuities."

The inspection method used depends not only on the requirements of the governing code but also on the contract specifications of the owner or purchaser.

Ultrasonic examination is the most effective means of examination to find internal discontinuities such as porosity, inclusions, cracking, and, in rare instances,

Photograph Courtesy of Metro Machine Corporation

Figure 8.21—15 m (50 ft) Electrogas Welded Ship Subassembly in Fixture

incomplete internal fusion. Magnetic particle examination can also be used to reveal cracks and incomplete fusion, but this method is limited to detecting discontinuities on the surface and the immediate subsurface. In cases where the material is too thick to inspect with radiography, ultrasonic examination is used. The large grains normally found in an electrogas weld may interfere with ultrasonic examination, resulting in false indications at the grain boundaries when no defects exist. To ensure reliable inspection, properly trained personnel experienced in examining electrogas welds are required.

Acceptance

Welding acceptance criteria are usually established by the customer or a regulatory agency, or both. Electrogas welding applications are commonly governed by

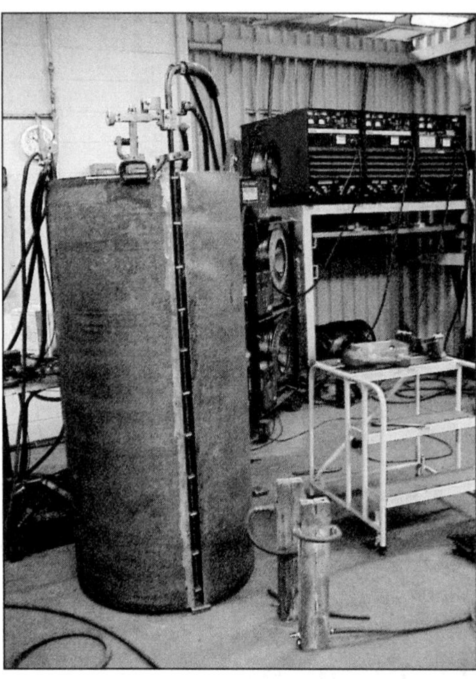

Photograph Courtesy of Vermeer Manufacturing

Figure 8.22—Drum Welded with Self-Shielded Electrogas Welding

Photograph Courtesy of Pasadena Tank Corporation

Figure 8.23—Self-Shielded Moving Shoe Electrogas Welding Equipment on a Field Storage Tank

various code requirements. Occasionally, customer specifications may add to the requirements of the regulatory agency. At the time this chapter was prepared (2000-2003), the following standards permitted the use of the electrogas welding process (sometimes with additional restrictions):

Structural Welding Code—Steel, AWS D1.1;[14]

Bridge Welding Code, AWS D1.5;[15]

Boiler and Pressure Vessel Code, American Society of Mechanical Engineers' (ASME);[16]

Design and Construction of Large, Welded, Low-Pressure Storage Tanks, API 620, and *Welded Steel Tanks for Oil Storage*, API 650;[17]

Welded Steel Construction (Metal Arc Welding), CSA W59M/W59;[18]

ABS *Rules for Building and Classing Steel Vessels*;[19] and

Lloyd's Register of Shipping Rules and Regulations for Classification of Ships.[20]

Quality Control

A written procedure that identifies the essential variables to be used in electrogas welding should be prepared. This procedure should be reviewed to determine conformance to applicable code requirements.

Only approved welding procedures should be issued to the welding supervisor or welding operator. Sample electrogas welding procedure specification (WPS) forms are shown in Tables 8.11 and Table 8.12. Any additional instructions considered necessary should be made a part of the procedure specification. Proper quality control procedures should be used to assure that the correct filler metal is being used and that the qualified welding procedure is being followed. A checklist for electrogas welding with moving shoe(s) is provided in Table 8.13.

Electrogas weldments of excellent quality can be fabricated by maintaining the equipment properly and adhering strictly to the weld procedure. Operators should inspect the equipment before each weld is started to verify that the contact tip is not worn and the welding electrode feed cable is free and straight.

14. See Reference 12.
15. See Reference 13.
16. American Society of Mechanical Engineers (ASME), *Boiler and Pressure Vessel Code,* New York: American Society of Mechanical Engineers.
17. American Petroleum Institute (API), Washington D.C.: American Petroleum Institute
18. Canadian Standards Association (CSA), *Welded Steel Construction (Metal Arc Welding),* CSA W59M/W59, Toronto: Canadian Standards Association.
19. American Bureau of Shipping (ABS), *Rules for Building and Classing Steel Vessels,* New York: American Bureau of Shipping.
20. Lloyd's Register of Shipping, *Lloyd's Register of Shipping Rules and Regulations for Classification of Ships,* London, U.K.: Lloyd's Register of Shipping.

Table 8.11
Typical EGW Procedure Specification with Moving Shoe(s)[a]

Welding Process: Welding shall be done by the electrogas welding (EGW) process, using one electrode wire. (If two [2] electrodes are used, the electrode spacing should be specified.)

Base Metal: The base metal shall conform to the specification for __________.

Base Metal Thickness: This procedure will cover the welding of the base metal from __________ mm (__________ in.) thick to __________ mm (__________ in) thick.

Filler Metal: The electrode shall conform to AWS Specification A5.26/A5.26M for classification EG __________. The diameter of the electrode shall be __________ mm (__________ in.).

Shielding Gas: Electrode is used with/without externally supplied shield gas.

Shielding gas, if used: Composition ______________________________
Flow Rate ______________________________
AWS A5.32/A5.32M class ______________________________

Position: Welding shall be done in the vertical position.

Shoes: The welding shall be done using a water-cooled copper inboard shoe. The outboard shoe shall be ______________________________.

Preheating: None required by this procedure specification. However, no welding shall be performed when the temperature of the base metal at the point of welding is below 0°C (32°F).

Postweld Heat Treatment: None required by this procedure specification.

Base Metal Preparation: The edges or surfaces of the parts to be joined shall be prepared by __________ and shall be cleaned of oil, grease, moisture, scale, rust, or other foreign material. The surfaces upon which the copper shoes must slide shall be flat and smooth.

Welding Current: The welding current shall be direct current, electrode positive.

Welding Technique: The welding shall be done in a single pass, uphill. Starting and runoff tabs shall be used.

An electrode extension of __________ mm (__________ in.) shall be maintained during welding. The welding power supply shall have a __________ characteristic.

Welding Conditions: All welding shall be performed using the conditions given below:
Plate thickness, mm (in.) ______________________________
Root opening, mm (in.) ______________________________
Welding current,[b] A ______________________________
Electrode wire feed speed,[c] mm/s (in./min.) ______________
Welding voltage, V ______________________________
Drag angle, degrees ______________________________
Electrode distance from front face, mm (in.) ______________

Oscillation:
Distance, mm (in.) ______________________________
Period (s) ______________________________
Dwell (s) ______________________________
Traverse time (s) ______________________________

Travel Speed: Speed of vertical travel is a function of deposition rate and need not be specified.

Procedure Qualification: Procedure Qualification tests have been made in accordance with ______________________________.

Joint Design: The joint design shall be as detailed below:

Notes:
a. A typical EGW Procedure Specification should include but not be limited to the items included herein.
b. On some equipment, electrode wire feed speed (WFS) is the controlling variable, and welding current need not be specified.
c. On some equipment, electrode feed speed may not be proportional to welding current and may need to be specified.

Source: Adapted from American Welding Society (AWS) Committee on Arc Welding and Cutting, 2000, *Recommended Practices for Electrogas Welding,* AWS C5.7:2000, Miami: American Welding Society, Table 11.

Table 8.12
Typical EGW Procedure Specification Using Consumable Guide Tubes and Fixed Copper Shoe(s)

Welding Process: Welding shall be done by the electrogas welding (EGW) process, using a consumable guide tube.

Base Metal: The base metal shall conform to the specification for ____ .

Base Metal Thickness: This procedure will cover the welding of the base metal from __________ mm (__________ in.) thick to __________ mm (__________ in.) thick.

Filler Metal: The electrode shall conform to AWS Specification A5.26/A5.26M for classification EG __________. The diameter of the electrode shall be __________ mm (__________ in.).

Shielding Gas: Electrode is used with/without externally supplied shield gas.

Shielding gas, if used: Composition______________________
Flow Rate______________________
AWS A5.32/A5.32M class ______________________

Consumable Guide Tube: The guide tube shall be made of __________ steel and have an O.D. of __________ mm (__________ in.) and an I.D. of __________ mm (__________ in.).

Shoes: The fixed copper shoes shall be water cooled/air cooled and shall be one piece/stacked.

Preheating: None required by this procedure specification. However, no welding shall be performed when the temperature of the base metal at the point of welding is below 0°C (32°F). Start area of weldment may be preheated to __________°C (__________°F) to improve fusion at start of weld, especially on heavier plates.

Postweld Heat Treatment: None required by this procedure specification.

Base Metal Preparation: The edges or surfaces of the parts to be joined shall be prepared by __________ and shall be cleaned of oil, grease, moisture, scale, rust, or other foreign material. The surfaces against which the copper shoes must fit shall be flat and smooth.

Welding Current: The welding current shall be direct current, electrode positive.

Welding Technique: The welding shall be done in a single pass, uphill. Starting and runoff tabs shall be used. The welding power supply shall have a __________ characteristic.

Welding Conditions: All welding shall be performed using the conditions given below:
Plate thickness, mm (in.) ______________________
Root opening, mm (in.) ______________________
Electrode wire feed speed,[b] mm/s (in./min. ______________________
Welding current,[c] A ______________________
Welding voltage, V) ______________________

Oscillation:
Distance, mm (in.) ______________________
Period (s) ______________________
Dwell (s) ______________________
Traverse time (s)______________________

Travel Speed: Speed of vertical travel is a function of deposition rate and need not be specified.

Procedure Qualification: Procedure Qualification tests have been made in accordance with ________________________________ .

Joint Design: The joint design shall be as detailed below:

Notes:
a. A typical EGW Procedure Specification should include but not be limited to the items included herein.
b. On some equipment, electrode feed speed may not be proportional to welding current and may need to be specified.
c. On most equipment, electrode wire feed speed (WFS) is the controlling variable, and welding current need not be specified.

Source: Adapted from American Welding Society (AWS) Committee on Arc Welding and Cutting, 2000, *Recommended Practices for Electrogas Welding,* AWS C5.7:2000, Miami: American Welding Society, Table 12.

Table 8.13
Typical Checklist for Electrogas Welding with Moving Shoe(s)

Electrogas welding is designed to weld the entire joint continuously, usually in a single pass. Before starting to weld, it is vital that the operator check all elements of the operation.

CHECK EACH ITEM AS JOB PROGRESSES

1. Check workpiece preparation. Make sure that the weld edge and adjacent surface have been prepared properly along entire length.
2. Fit starting tab at bottom of joint and run-off tabs at top. Minimum distance from starting surface in the tab to bottom edge of joint shall be 38 mm (1-1/2 in.).
3. Fit seam per procedure. Maintain required tolerance.
4. Attach and tighten electrode and workpiece leads.
5. Make sure the machine is properly fitted to the joint.
6. Check vertical clearance and obstruction. Make sure there are no obstructions that would prevent or stop the rise of the machine and that equipment leads are of adequate length.
7. Check quantity of electrode wire and gas (if used). Make sure there is enough to finish the weld.
8. Check water and gas connections for leaks and adequate flow.
9. Check the face of the shoe(s) for worn or irregular areas that may interfere with its vertical during welding.
10. Check wire guide contact tip. Do not start with a worn contact tip; it could cause a reduction in arc voltage, excessive wire wander, a shut-down or, at best, poor welding conditions.
11. Check wire guide and electrode alignment. Line up the head and electrode control. Specific alignment settings and tolerances are listed in each welding procedure.

11a. Electrode angle to joint is always 90°.

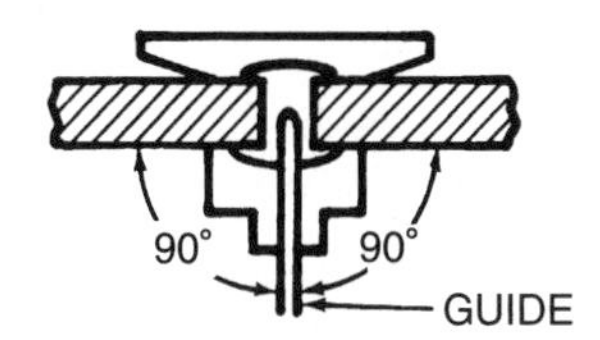

11b. Adjust the wire straightener so that the electrode is straight or has a maximum underbend of 10 mm (3/8 in.).

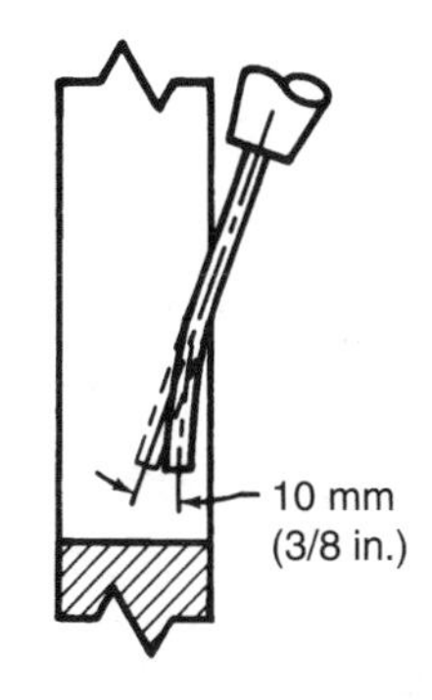

11c. Adjust electrode extension. Do *not* change the electrode extension when making other adjustments.

11d. Adjust drag angle

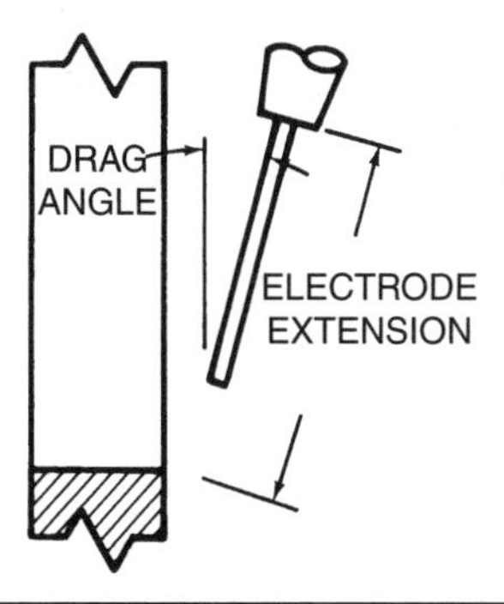

Table 8.13 (Continued)
Typical Checklist for Electrogas Welding with Moving Shoe(s)

11e. Set the electrode location by measuring from the front face of the plate after the electrode is in the seam. Note: the electrode location changes as the guide tip wears. Replace the guide tip when it becomes worn or the end is fused or deformed.

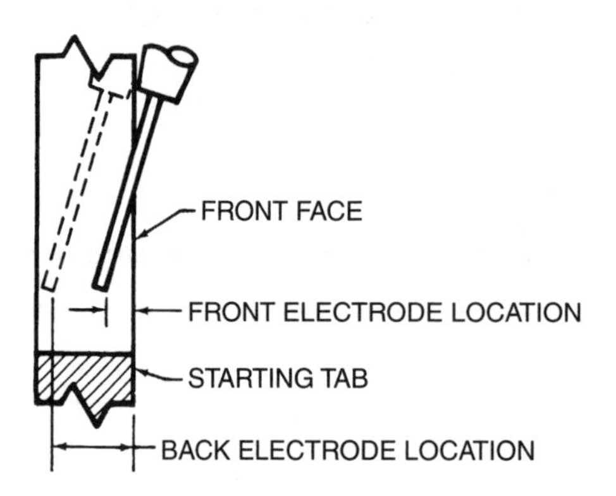

11f. Set sufficient spacing between the guide tip and top of the movable shoe. When oscillation of the electrode in the joint is required, set the front location (for the electrode), oscillation distance, oscillation time, and dwell time(s) as specified.

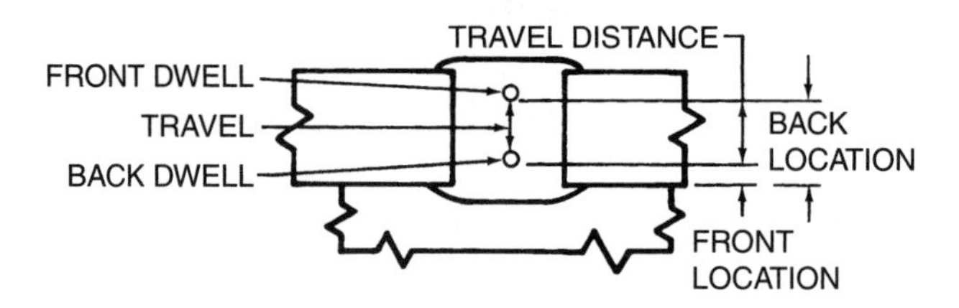

11g. Set 13 mm (1/2 in.) spacing between the end of the electrode and the starting weld tab.

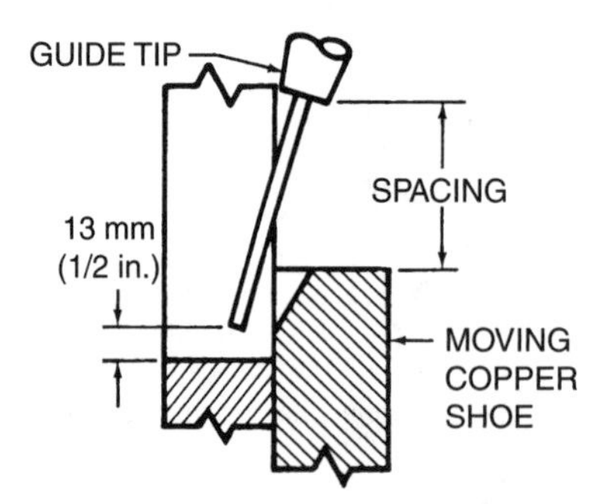

12. Check moving shoe(s). Clamp with the correct operating pressure.
13. Apply glass tape or putty to any opening between the shoe face and work caused by the weld reinforcement groove in the shoe. Excessive weld leakage may occur during the start if this is not done.
14. Set water and gas flow rates.
15. Have a few tools (e.g., pliers, screwdriver, and open-end wrench) handy.
16. Check operation of fume exhaust equipment, if used.
17. Have welding shield ready.
18. Set start voltage and wire speed speed (amperage).
19. Energize welding current and vertical travel mechanism.
20. Recheck and adjust welding conditions.
21. Turn on oscillator, if used.

Source: Adapted from American Welding Society (AWS) Committee on Arc Welding and Cutting, 2000, *Recommended Practices for Electrogas Welding,* AWS C5.7:2000, Miami: American Welding Society, Table 132.

Operators should also check the workpiece to verify that the joint is correctly aligned and that the workpiece allows free passage of the electrode guide to the top of the joint. Finally, prior to starting the weld, the electrode should be set and oscillated within the joint to assure that it is free to oscillate and that the oscillation speed and dwell are correctly adjusted. When the weld is stabilized, the arc current and voltage should be determined to be within the limits prescribed by the welding procedure specification.

POSSIBLE ELECTROGAS WELD DISCONTINUITIES

The discontinuities that could be encountered when using electrogas welding are porosity at the start and termination of a weld, porosity in the weld metal, centerline weld cracking, incomplete fusion, overlap, underfill, and melt-through in the starting weld tab. These discontinuities are shown schematically in Figure 8.24.

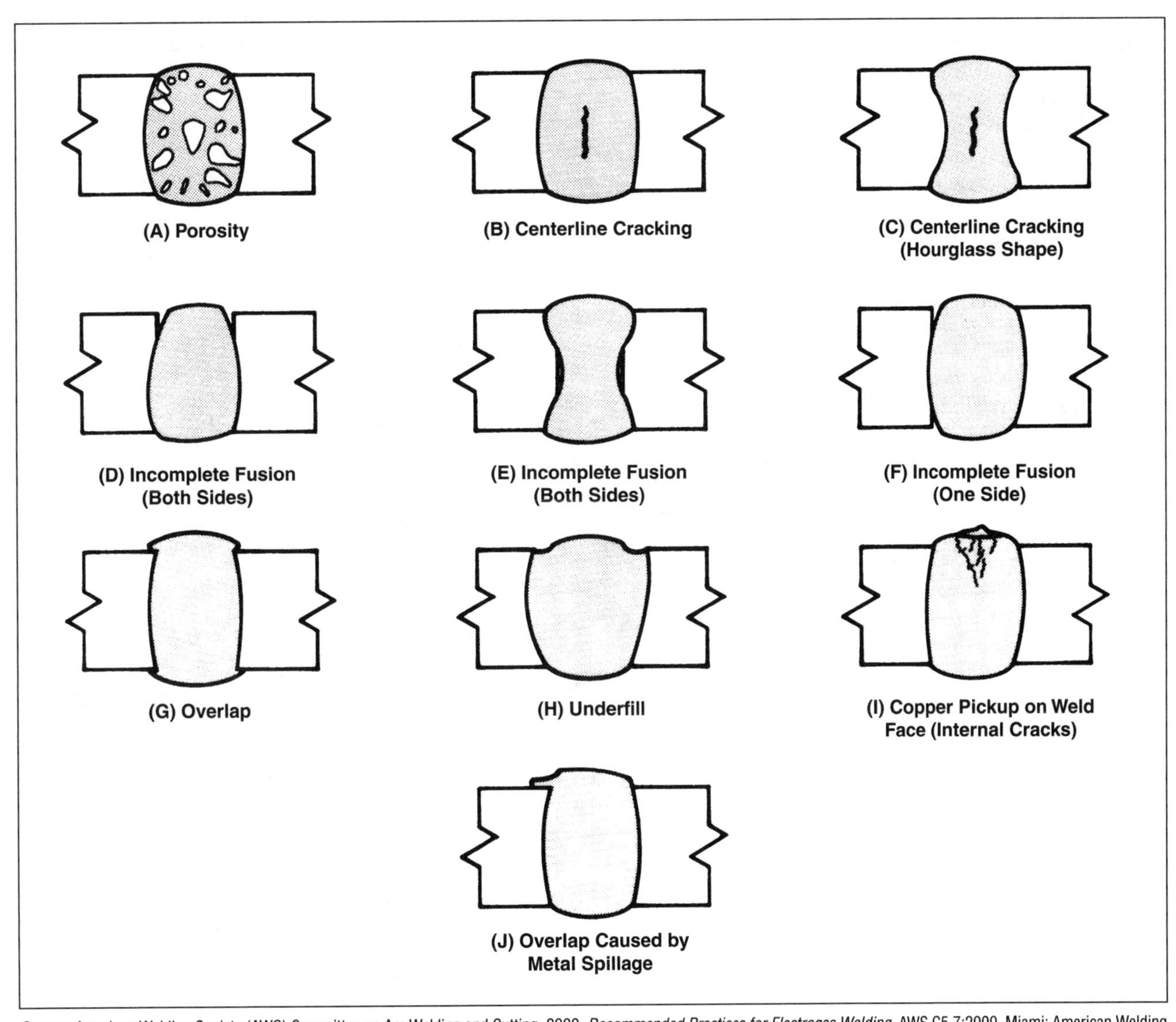

Source: American Welding Society (AWS) Committee on Arc Welding and Cutting, 2000, *Recommended Practices for Electrogas Welding,* AWS C5.7:2000, Miami: American Welding Society, Figure 23.

Figure 8.24—Weld Discontinuities in Electrogas Welds Caused by Improper Technique or Defective Equipment, or Both

Starting Porosity

Starting porosity consists of fine holes or pores within the weld metal near the start of the weld.[21] Several factors contribute to starting porosity in electrogas welds, including the following:

1. Low electrode feed speed, high voltage, short electrode extension, and insufficient time in the starting weld tab;
2. Contaminants in the starting weld tab or between the starting weld tab and the base metals;
3. Condensation on the shoes or between the shoes and the base metals;
4. Fast cooling rates caused by thick plates, large copper starting weld tabs, and low ambient temperatures;
5. Shallow starting weld tab;
6. Arc located too close to shoes;
7. Water leakage in the starting weld tab;
8. Poor fit of the shoe or starting weld tab;
9. Insufficient gas coverage or contaminated gas;
10. Erratic welding arc.
11. Moisture added to weld pool from sealants used to prevent weld metal tap-outs.

Porosity in the Weld Metal

Weld metal porosity, illustrated in Figure 8.24(A), is generally caused by expanding gases that are insoluble in the freezing metal, but gas-like voids can also be caused by mechanical problems. Some of the causes of this discontinuity in electrogas welds are the following:

1. Starting porosity that extends into the production weld;
2. Excessive voltage;
3. Low electrode feed speed;
4. Electrode extension too short;
5. A "cold" weld (low electrode feed speed and voltage);
6. Contaminants in weld area;
7. Loose fit of the shoe, allowing air to bleed into weld;
8. Insufficient gas coverage or contaminated gas;
9. Water leakage from the shoe;
10. Erratic welding arc; and
11. Moisture added to weld pool from sealants used to prevent weld metal tap-outs.

21. American Welding Society (AWS) Committee on Definitions, 2001, *Standard Welding Terms and Definitions,* A3.0:2001, Miami: American Welding Society.

Porosity at Weld Termination

Porosity at the end of the weld can be caused by the same factors as porosity within the weld. The following factors can cause porosity:

1. Short runoff tabs or retaining dams,
2. Slag leakage due to improperly attached runoff tabs, and
3. Arc blow caused by improper location of the workpiece lead.

Centerline Weld Cracking

Centerline cracking, illustrated in Figures 8.24(B) and 8.24(C), may be related to procedure variables, thermal conditions, and weld restraint. Some steel grades, particularly those with a high carbon equivalent, high sulfur, or high phosphorus are more susceptible to centerline cracking than others. For the more susceptible steels, each of the other factors becomes more critical.

As previously discussed, the procedure variables that contribute to a low form factor contribute to centerline cracking. Among these conditions are the following:

1. Excessive wire feed speed (excessive current),
2. Low arc voltage,
3. Excessively narrow root opening, and
4. Long dwell time when oscillating.

Rapid cooling rates also contribute to centerline cracking. The factors that prompt rapid cooling rates include large shoes with excessive water flow and a lack of preheat for thick plates at low ambient temperatures. Excessive restraint is a major contributing factor in all occurrences of cracking. Excessive base metal admixture may also contribute to centerline weld cracking.

Incomplete Fusion to Both Fusion Faces

Incomplete fusion is a weld discontinuity in which fusion did not occur between the weld metal and both fusion faces.[22] Incomplete fusion at both faces, illustrated in Figures 8.24(D) and 8.24(E), is caused by unsatisfactory thermal conditions that prevent the bevel faces or groove faces from melting. Poor heat distribution as well as insufficient heat can result in this discontinuity. It can also occur due to slag trapped between the bevel face or groove face and the molten filler material. The following are among the contributing factors:

1. Cold weld (low voltage or slow electrode feed speed and low voltage),

22. See Reference 21.

2. Excessive electrode feed speed (fast fill rate),
3. Narrow root opening (fast fill rate),
4. Rapid oscillation speed,
5. Excessive slag on top of weld pool, and
6. Excessive root opening.

An excessive slag burden (i.e., the volume of slag on the weld pool) can be avoided by improving the shoe design. By machining a deeper and wider groove in the shoe, more slag is permitted to coat the weld reinforcement. This reduces the burden. If the weld pool is maintained higher in the shoe, more slag escapes as spatter. This measure also reduces the slag burden.

Incomplete Fusion to One Face

Incomplete fusion to one bevel face or groove face, a weld discontinuity in which fusion did not occur between weld metal and one fusion face, is illustrated in Figure 8.24(F). It is caused by asymmetric thermal conditions, which can result from the following:

1. Arc located off-center,
2. Electrode angled toward one bevel face or groove face,
3. Arc blow caused by the improper location of the workpiece lead connection, and
4. Excessive root opening.

Overlap

Overlap is the protrusion of weld metal beyond the weld toe or root.[23] Overlap is caused by weld metal flowing out of the joint without melting the base metal. The condition is illustrated in Figures 8.24(G) and 8.24(J).

Overlap on the front face can result when the arc is located too far back; when the wire straightener, drag angle, or guide location have the improper settings; or when the contact tip is worn. In addition, the bevel angle could be too large, or the weld could be cold (low voltage or low electrode feed and voltage). Overlap on the back face occurs when the arc is located too far forward; the wire straightener, drag angle, or the guide location are improperly set; the contact tip is worn; or the weld is cold (low voltage or low electrode feed and voltage). Overlap on both faces may result when the weld is cold (low voltage or low electrode feed and voltage); when the groove in the copper shoe(s) is too wide; or from excessive quenching from the shoe(s); improper shoe design; excessive water flow; excessive travel speed; narrow root opening; arc blow; or incorrect oscillation cycle.

23. See Reference 21

The condition shown in Figure 8.24(J) is often caused by a poor fit of the base metal. It can result if the sliding shoe is lifted from contact with the base metal by foreign material such as spatter. This allows weld metal to leak and freeze against the plate.

Underfill

Underfill is a groove weld condition in which the weld face or root surface is below the adjacent surface of the base metal.[24] Underfill may be acceptable under certain conditions; nevertheless, it represents marginal quality work at best. It is easily prevented. Underfill, illustrated in Figure 8.24(H), can be caused by excessive melting of base metal beyond the shoe or too narrow a groove in the shoe.

Melt-Through in Starting Weld Tab

Melt-through is visible root reinforcement in a joint welded from one side.[25] It is weld metal that appears outside of the weld joint. Because melt-through in the starting weld tab occurs outside the production weld, it is not a weld discontinuity. It prevents satisfactory welding of the workpiece, however, due to the loss of the weld pool. Melt-through is easily prevented by using an adequate thickness of material on the bottom of the starting weld tab or by attaching a back-up plate to the starting weld tab. Poorly fitted backing shoes can result in the same problem.

Hot Cracking

Hot cracking refers to cracking formed at temperatures reached as solidification nears completion.[26] Hot cracking on the weld face can be caused by the partial dissolution of the copper backing shoes. Hot cracks are generally located at or near the surface. This type of cracking is shown in Figure 8.24(I). The presence of dissolved copper in the weld metal can be caused by arcing on shoe(s) or the melting of shoes due to poor shoe cooling. Hot cracking can also be caused by poor quality of the base metal; some steels are more susceptible to hot cracking than others.

REWORK

In general, defective joints requiring repairs can be minimized by implementing a good preventive maintenance program for the equipment, by employing

24. See Reference 21.
25. See Reference 21.
26. See Reference 21.

well-trained and qualified welding operators, and by using a sound welding procedure.

A discontinuity such as underfill can often be repaired by building up with the shielded metal arc welding process without gouging or grinding. Discontinuities such as incomplete fusion (at the joint surface), overlap, copper pickup on the weld face, and metal spillage can be repaired by superficial gouging or grinding to sound metal and rewelding with the shielded metal arc welding process. Discontinuities such as porosity, cracking, and incomplete fusion (internal), which are usually detected by radiographic or ultrasonic testing, can be repaired by gouging deeply to sound metal and rewelding with shielded metal arc welding, flux cored arc welding, or electrogas welding.

Economic considerations should dictate whether electrogas welding or some other process is used for repair. Repairs are generally made with the shielded metal arc welding process using electrodes appropriate for the base metal and in accordance with a qualified welding procedure.

The restarting of electrogas welds should be avoided if possible. However, when it becomes necessary to restart a weld, any starting defect that may result can be confined to the near surface area by employing the technique illustrated in Figure 8.25. This will produce better fusion and quality and allow easier inspection and reworking of the start if necessary. As shown in this figure, the starting cavity is sloped by air carbon arc gouging. The start area is preheated to a minimum of 135°C (300°F), and the arc initiated near the front shoe.

As the sloped crater fills, the arc is moved toward the center of the groove until the normal running position is achieved. For welds in thick plates requiring oscillation, the oscillation distance is expanded following the same principle. This technique should produce a shallow starting discontinuity on the near side. This discontinuity can be easily removed, and the cavity is filled with sound metal deposited using conventional arc welding processes.

METALLURGICAL CONSIDERATIONS

Electrogas welds produce high-quality welds when proper welding techniques are used; however, due to the nature of the process, i.e., high deposition rate and heat input, metallurgical concerns and welding techniques differ significantly from many of the other arc welding processes.

Electrogas Welding Thermal Cycle

The thermal cycle of electrogas welding is prolonged because of the relatively slow travel speeds of 0.6 mm/s to 3 mm/s (1-1/2 in./min. to 8 in./min.). Therefore, the

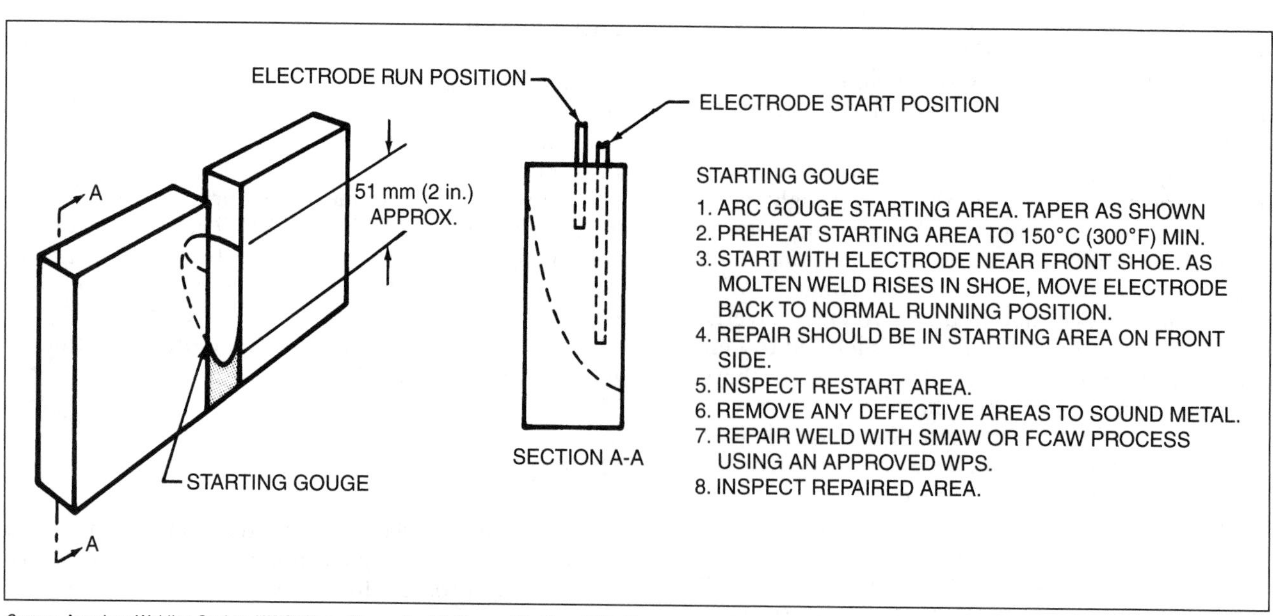

Source: American Welding Society (AWS) Committee on Arc Welding and Cutting, 2000, *Recommended Practices for Electrogas Welding,* AWS C5.7:2000, Miami: American Welding Society, Figure 22.

Figure 8.25—Electrogas Welding Restart Procedure

electrogas weld metal structure contains large grains with a marked tendency for coarse columnar growth. The heat-affected zone is wider than in conventional arc welding and contains a wider grain-coarsened region. However, the heat-affected zone and the grain-coarsened regions of electrogas welds are narrower than those of electroslag welds of comparable size. The prolonged thermal cycle results in a relatively slow weld cooling rate. The slow cooling rates produce heat-affected zones without the undesirable hard structures that often occur in conventional arc welds in carbon and low-alloy steels. These heat-affected zones may also exhibit lowered toughness values.

For purposes of comparison, the hardness profiles of an electrogas and a shielded metal arc weld are shown in Figures 8.26 and 8.27. The hardness profile of the electrogas weld is relatively uniform, whereas the shielded metal arc weld exhibits a significant increase in hardness in the heat-affected zone near the weld interface. The high heat-affected-zone hardness results from

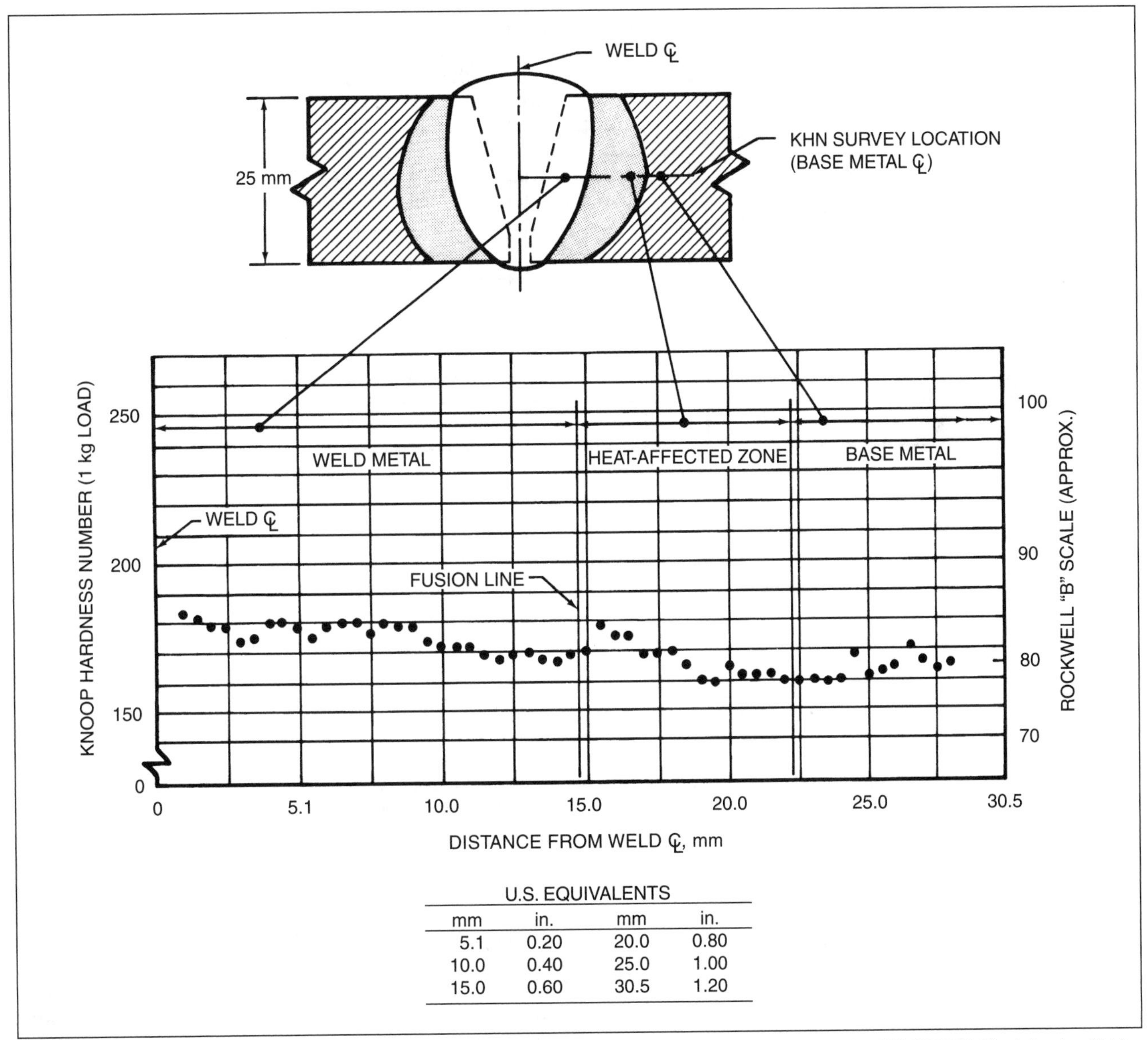

U.S. EQUIVALENTS			
mm	in.	mm	in.
5.1	0.20	20.0	0.80
10.0	0.40	25.0	1.00
15.0	0.60	30.5	1.20

Source: American Welding Society (AWS) Committee on Arc Welding and Cutting, 2000, *Recommended Practices for Electrogas Welding,* AWS C5.7:2000, Miami: American Welding Society, Figure 16.

Figure 8.26—Knoop Microhardness Survey across an Electrogas Weld in A283 Grade C Base Metal

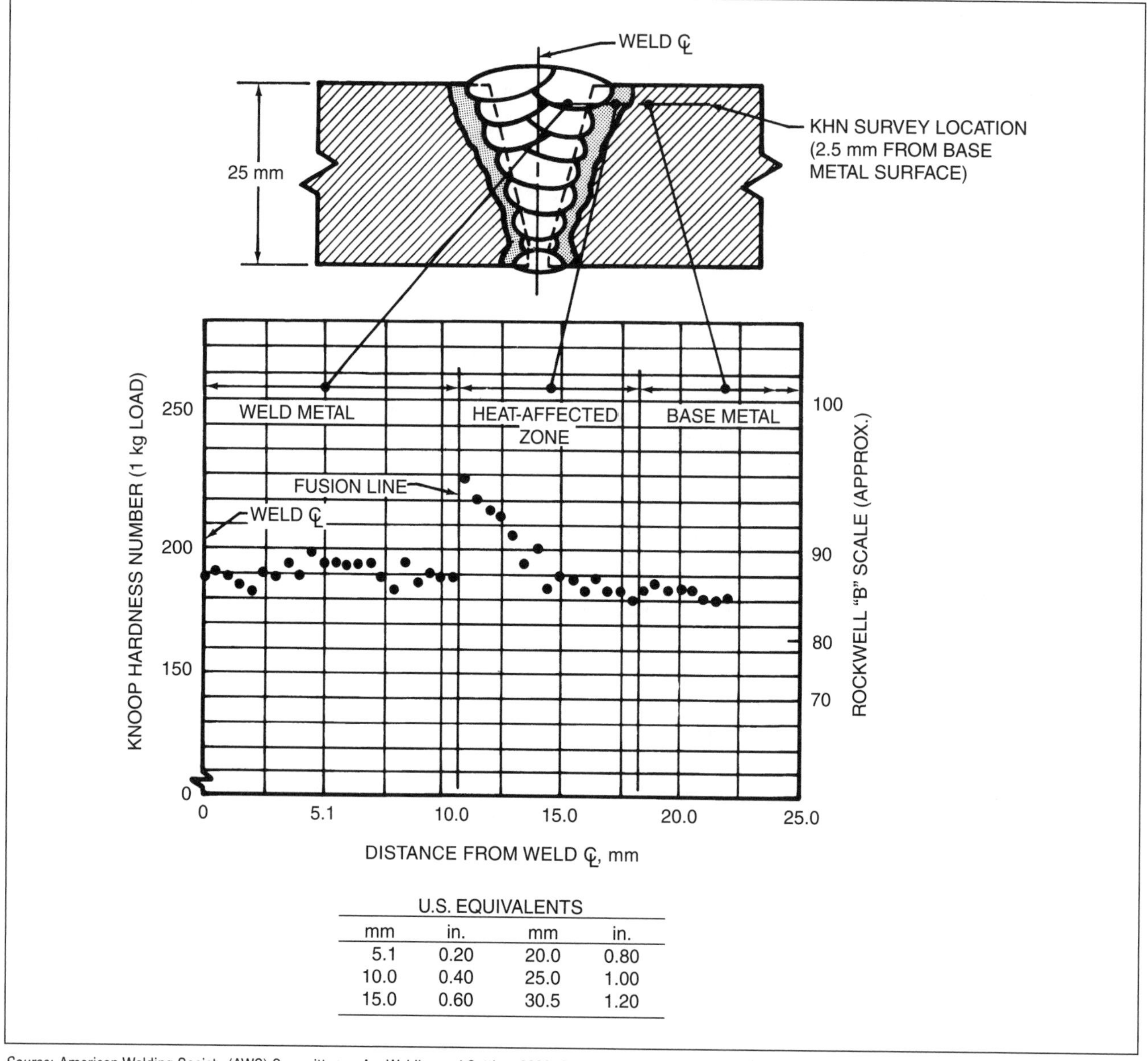

U.S. EQUIVALENTS

mm	in.	mm	in.
5.1	0.20	20.0	0.80
10.0	0.40	25.0	1.00
15.0	0.60	30.5	1.20

Source: American Welding Society (AWS) Committee on Arc Welding and Cutting, 2000, *Recommended Practices for Electrogas Welding,* AWS C5.7:2000, Miami: American Welding Society, Figure 17.

Figure 8.27—Knoop Microhardness Survey across a Shielded Metal Arc Weld (Uphill) in A283 Grade C Base Metal

the metallurgical reactions that occur when the shielded metal arc weld cools at a rapid rate from the welding temperature.

In some cases, this hardness can approach the maximum quench hardness of the material. The heat-affected-zone pattern shown in Figure 8.27 is typical for commonly used multipass welding processes such as shielded metal arc welding, submerged arc welding, flux cored arc welding, and gas metal arc welding.

In addition to slowing the cooling rate, the protracted thermal cycle of electrogas welding allows more time at high temperatures where grain growth occurs. Thus, the weld metal and the heat-affected zone of electrogas welds exhibit larger grains and larger coarse-grain regions.

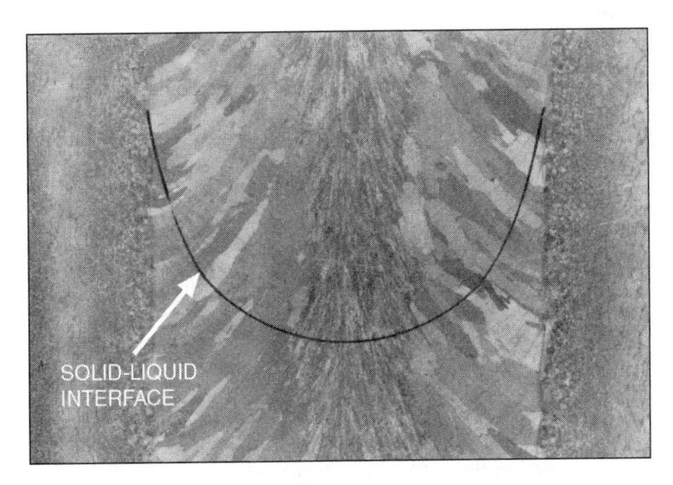

Figure 8.28—Typical Grain Growth Pattern of a Vertical Section of an Electrogas Weld

Weld Macrostructure

The dominant feature of weld macrostructure in electrogas welds is the large columnar grains that result from the slow weld solidification.[27] The contour of the solid-liquid interface that was generated during welding can be estimated from a vertical cross section of a completed weld. Columnar grain growth occurs perpendicular to the solid-liquid interface, and an etched vertical macrosection of an electrogas weld would be identical to that of an electroslag weld cross section shown in Figure 8.28. Thus, during welding, the approximate boundary between the liquid and solid metal is essentially perpendicular to the columnar grains, as shown in Figure 8.28.

The form factor—the ratio of the width of the weld pool to its depth—is most often used to describe the shape of the weld pool in electrogas welding. A high form factor is generally desirable because the freezing progression is vertical, and any impurities, segregates, and low-melting-point constituents remain in the weld pool, float upward, and freeze harmlessly in the runoff tab above the production weld metal.

In contrast, welds with low form factors may trap the low-melting-point constituents and impurities along the centerline of the weld and result in a plane of weakness. This condition increases the tendency for weld cracking, which usually occurs at high temperatures in conjunction with or immediately after solidification.

27. Weld solidification is discussed in detail in Chapter 4, "Welding Metallurgy," in American Welding Society (AWS) Welding Handbook Committee, 2001, Jenney, C. L., and A. O'Brien, eds., *Welding Science and Technology*, Vol. 1 of *Welding Handbook*, 9th ed., Miami: American Welding Society.

The weld structure resulting from electrogas welding consists of large columnar grains. The heat-affected zone is wide and has considerable grain coarsening near the fusion line. Welds with large columnar grains generally exhibit lower notch toughness and a higher ductile-to-brittle transition temperature than welds with fine equiaxed or dendritic grains. However, properly made electrogas welds in the as-welded condition typically meet the minimum impact properties specified for hot-rolled carbon and low-alloy steels. Also, the impact properties of the weld metal and heat-affected zone are often adequate for normalized steel grades. Postweld heat treatment may enhance the toughness to allow the use of electrogas welding for applications with lower service temperatures. Electrogas welds typically have finer grains and narrower heat-affected zones than electroslag welds.

Preheat

Preheating the base metal is usually unnecessary when low- and medium- carbon steels are welded with the electrogas process. Most of the heat generated in electrogas welding is conducted into the workpieces. The conducted heat serves to preheat the base metal due to the relatively slow advance of the welding arc.

Although the heat generated during welding provides adequate preheat for most applications, preheat requirements should be determined for the specific conditions of each application. Preheat may increase resistance to cracking under the following conditions:

1. Welding is performed on high-strength steels, high-carbon, or hardenable high-alloy steels;
2. Base metal thicknesses are greater than 75 mm to 100 mm (3 in. to 4 in.);
3. The restraint of the base metal is high; and
4. The temperature of the base metal is below 0°C (32°F).

Preheating the start area of the weld also improves edge wetting. It should be noted, however, that excessive preheat slows the weld cooling rate, increases joint penetration, and may contribute to weld metal leakage and melt-through.

Sometimes, because of an equipment malfunction, electrode breakage, weld metal leakage, or an unplanned interruption, welding may be terminated prior to completion, and the weld must be restarted along the joint. In this case, an insert or wedge must be jammed into the open joint immediately after ceasing the welding. The insert prevents excessive shrinkage of the joint root opening, which squeezes the head assembly in the weld groove as the workpieces contract after the removal of the welding heat.

Preheating the restart area reduces the shrinkage forces. This increases the joint clearance, allowing the welding head assembly to track up the joint. After the arc has been restarted, the weld groove root opening should expand enough to remove the wedge. It is important to note that discontinuities commonly occur as a result of the restart. The weld should be carefully examined in these regions.

Postweld Heat Treatment

Most applications of electrogas welding, particularly the welding of field-erected structures, require no postweld heat treatments. Stress-relief heat treatment of electrogas weldments typically results in a slight drop in the yield and tensile strengths and a slight improvement in weld and heat-affected-zone notch toughness. In weldments requiring optimum notch toughness, a postweld heat treatment such as normalizing may be required. Quenched-and-tempered steels may require postweld heat treatment to restore the original properties of the base metal.

Residual Stresses and Distortion

Since solidification of an electrogas weld begins at the shoes and dams, the outer surfaces of the weld are under compressive residual stresses, and the center of the weld is in tension. This unique residual stress pattern is just the reverse of the residual stresses that are likely to occur with multipass arc welding processes.

Angular distortion in the horizontal plane is virtually nonexistent in single-pass electrogas welds. This is due to the symmetry that most electrogas joint designs have about the mid-thickness of the base material, resulting in uniform shrinkage in the thickness direction. As welding progresses up the joint, the workpieces are drawn together by the weld shrinkage. Therefore, the fitup of the root opening at the top of the joint should be approximately 2.4 mm to 6.4 mm (3/32 in. to 1/4 in.) more than at the bottom to allow for this shrinkage. The factors influencing the shrinkage allowance include material type, joint thickness, joint length, and the degree of restraint of the workpieces being joined. Angular distortion is more prevalent in single-pass V-groove welds.

Angular distortion can occur during multiple-pass electrogas welding. Presetting the workpieces may compensate for it. Experience will suggest the proper amount of preset for each application.

MECHANICAL PROPERTIES

As electrogas welding is a high-dilution process, melted base metal may contribute up to approximately 35% of the total weld metal. Thus, the composition and mechanical properties of the weld may vary substantially with different types of steel and even with different heats of the same grade of steel.

Mechanical properties also vary with welding conditions such as current and voltage, joint design, type of backup, and cooling rate. Electrogas electrodes are designed to accommodate as much variation as practical, and consistent mechanical properties can be achieved when the welding conditions are controlled within the normal limitation of variables found in fabrication standards.

The publication *Standard for Welding Procedure and Performance Qualification*, AWS B2.1,[28] describes the welding procedure qualification variables and the limits that should be controlled for electrogas and other welding processes. Similar procedure qualification requirements directed at specific applications are described in some codes.

The standard *Structural Welding Code—Steel,* AWS D1.1[29] includes qualification requirements for structural welding. The results of welding procedure qualification tests are intended to be representative of mechanical properties to be expected in production welds. Typical weld metal mechanical properties for self-shielded flux cored electrogas welds in several grades of carbon steel are listed in Tables 8.14. Table 8.15 shows similar results for gas-shielded welds.

One problem encountered in electrogas welding is low notch toughness properties of welds and heat-affected zones in the as-welded condition. By using suitably alloyed electrodes and controlled welding conditions, the notch toughness properties of the weld metal can be equal to or better than those of the base metal for most structural and pressure vessel steels.

In most cases, these notch toughness properties can be achieved in the as-welded condition. For many structural and pressure vessel steel grades, the heat-affected zone notch toughness is more dependent on the notch toughness of the base metal than on the welding conditions. The base metal, heat-affected zone, and weld metal toughness of an A283 steel electrogas weldment are shown in Table 8.16.

The electrogas process should be thoroughly tested for applications that require low-temperature notch toughness or that involve a reversal of stress. The electrogas process may degrade the heat-affected zone toughness to the extent that use of the process should only be considered in connection with a full postweld heat treatment.

---

28. American Welding Society (AWS) Committee on Procedure and Performance Qualification, *Standard for Welding Procedure and Performance Qualification*, AWS B2.1, Miami: American Welding Society.
29. See Reference 12.

Table 8.14
Typical As-Welded Mechanical Properties of Structural Steels Welded with Self-Shielded Flux Cored Electrodes[a]

Materials (ASTM)	Thickness		Electrode Class (A5.26)	Gas[b]	Yield Strength		Ultimate Tensile Strength		Elongation in 51 mm (2 in.)	Test Temperature		Charpy V-Notch Impact Strength WMCL[c]	
	mm	in.			MPa	ksi	MPa	ksi	Percent	°C	°F	J	ft·lbf
A 36	13	1/2	EG72T-1	—	506	73.3	618	89.6	30	−18	0	84	62
										−29	−20	75	55
A 36	25	1	EG72T-1	—	436	63.2	559	72.1	27	−18	0	62	46
										−29	−20	46	34
A 36	25	1	EG72T-1	—	443	64.2	566	82.1	26	−18	0	46	34
										−29	−20	35	26
A 36	50	2	EG72T-1	—	402	58.3	555	80.5	28	−18	0	19	14
										−29	−20	15	11
A 131 DH36	13	1/2	EG72T-1	—	527	76.4	660	95.7	28	−18	0	57	42
										−29	−20	52	38
A 131 DH36	19	3/4	EG72T-1	—	551	79.9	639	92.6	25	−18	0	56	41
										−29	−20	47	35
A 131 DH36	25	1	EG72T-1	—	481	69.7	613	88.9	27	−18	0	37	27
										−29	−20	33	24
A 537	25	1	EG82T-G	—	454	65.9	607	88.0	28	−18	0	61	45
										−29	−20	51	38
A 572 Grade 50, Type II	19	3/4	EG72T-1	—	549	79.6	646	93.6	27	−18	0	57	42
										−29	−20	52	38
A 572 Grade 50, Type II	19	3/4	EG82T-G	—	521	75.5	638	92.5	28	−18	0	55	41
										−29	−20	44	33
A 572 Grade 50, Type I	25	1	EG72T-1	—	453	65.7	606	87.9	28	−18	0	38	28
										−29	−20	31	23
A 572 Grade 50, Type II	25	1	EG72T-1	—	508	73.6	632	91.7	27	−18	0	38	28
										−29	−20	31	23
A 572 Grade 50, Type I	25	1	EG82T-G	—	507	73.5	642	93.1	27	−18	0	34	25
										−29	−20	24	18
A 572 Grade 50, Type II	25	1	EG82T-G	—	526	76.3	653	94.7	27	−18	0	78	58
										−29	−20	60	44
A 572 Grade 60	13	1/2	EG82T-G	—	544	78.9	674	97.8	30	−18	0	91	67
										−29	−20	74	55
A 572 Grade 60	25	1	EG82T-G	—	542	78.6	654	93.5	27	−18	0	61	45
										−29	−20	51	38
A 572 Grade 60	32	1-1/4	EG82T-G	—	508	73.6	679	98.4	27	−18	0	61	45
										−29	−20	53	39
JIS SM570	25	1	EG82T-G	—	506	73.3	619	89.7	27	−5	23	95	70

Notes:
a. Some of the test results shown are from production test joints and are not necessarily in accordance with AWS A5.26.
b. No gas used.
c. The test specimen notch is located at the weld metal centerline (WMCL).

Source: Adapted from American Welding Society (AWS) Committee on Arc Welding and Cutting, 2000, *Recommended Practices for Electrogas Welding,* AWS C5.7:2000, Miami: American Welding Society, Table 4

Table 8.15
Typical As-Welded Mechanical Properties of Steels Welded with Gas Shielded Flux Cored Electrodes

Materials (ASTM)	Thickness		Electrode Class (A5.26)	Gas[a]	Yield Strength		Ultimate Tensile Strength		Elongation in 51 mm (2 in.) Percent	Notch Toughness Charpy V-Notch Impact Strength[b]					
										Test Temperature		WMCL		BM-HAZ	
	mm	in.			MPa	ksi	MPa	ksi		°C	°F	J	ft·lbf	J	ft·lbf
A 36	25	1	EG72T-NM1	CO_2	479	69.5	633	91.8	26	–30	–22	35	26	—	—
A 36	76	3	EG72T-NM1	CO_2	462	67.0	614	89.0	26	–30	–22	31	23	—	—
A 36	76	3	EG72T-NM1	$Ar80/CO_2$	—	—	601	87.2	23	–18	0	49	36	—	—
A 131C	38	1-1/2	EG72T-Ni1	CO_2	—	—	490	71.1	30	–34	–30	45	33	30	22
A 203	41	1-5/8	EG72T-NM1	$Ar80/CO_2$	365	53.0	493	71.5	32	–40	–40	28	21	57–111	42–82
A 516	38	1-1/2	EG72T-NM1	$Ar80/CO_2$	538	78.0	610	89.5	29	–29	–20	41	30	33–91	28–67
A 537 Grade 1	25	1	EG72T-NM1	CO_2	430	62.3	572	83.0	29	–29	–20	34	25	39	29
A 537 Grade 1	29	1-1/8	EG72T-NM1	$Ar80/CO_2$	510	74.0	690	100.0	26	–30	–22	46	34	—	—
A 572 Grade 50	25	1	EG72T-NM1	$Ar80/CO_2$	—	—	—	—	—	–10	14	61	45	34	25
A 588	76	3	EG72T-NM1	CO_2	—	—	658	95.4	23	–18	0	56	41	—	—

Notes:
a. Test specimen notch is located at the weld metal centerline (WMCL) and the heat-affected zone of the base metal (BM-HAZ).
b. $Ar80/CO_2$ is 80% argon and 20% carbon dioxide. $Ar80/CO_2$ is AWS A5.32 classification SG-AC-20, and CO_2 is AWS A5.32 classification SG-C. (see Reference 5).

Source: Adapted from American Welding Society (AWS) Committee on Arc Welding and Cutting, 2000, *Recommended Practices for Electrogas Welding,* AWS C5.7:2000, Miami: American Welding Society, Table 5.

Table 8.16
As-Welded Notch Toughness Properties of an Electrogas Weld in 25 mm (1 in.) Thick A283 Steel

	Charpy V-Notch* J (ft·lbf)		
	21°C (70°F)	-7°C (20°F)	-18°C (0°F)
Unaffected base plate	34, 24, 12 (25, 18, 9)	7, 7, 8 (5, 5, 6)	5, 4, 7 (4, 3, 5)
Heat-affected zone	47, 38, 16 (35, 28, 12)	7, 7, 7 (5, 5, 5)	8, 8, 7 (6, 6, 5)
Center of weld	47, 43, 24 (35, 32, 18)	30, 31, 15 (22, 23, 11)	16, 14, 11 (12, 10, 8)

*Notch perpendicular to the plate surface.

Source: Adapted from American Welding Society (AWS) Committee on Arc Welding and Cutting, 2000, *Recommended Practices for Electrogas Welding,* AWS C5.7:2000, Miami: American Welding Society, Table 6.

TRAINING AND QUALIFICATION

Properly trained operators are vital to the successful application of the electrogas welding process. Electrogas welding is generally a single-pass welding process, but with special techniques, multiple-pass welds can be made. The operator should be trained to set up and properly operate the equipment.

Electrogas welding is unlike manual and semiautomatic processes in which welding can be temporarily stopped or delayed and easily restarted. Electrogas welding is designed to be continuous from beginning to end. If the weld is not sound, the discontinuities are often continuous. Therefore, the entire weld may have to be removed and the joint rewelded. Depending on the type of application and setup, at least some portion of welding must be completed before the welding operator can visually inspect, evaluate, and take any necessary corrective action. In some applications, the entire weld is completed before visual inspection can be performed. For these reasons, electrogas operators should understand the process and follow the procedure in every detail. Operators should be patient, conscientious, alert, and sufficiently experienced to recognize when unsatisfactory welds are being deposited. Operators should follow the recommendations of the equipment manufacturer.

It is important to the success of electrogas welding to maintain the equipment properly, use trained and qualified operators, and follow welding procedures that meet the requirements of the relevant code or specifications. An operator checklist, similar to that shown in Table 8.13, should be developed for each new application.

ECONOMICS

The use of the electrogas welding process for appropriate applications can result in significant cost savings over other welding processes. This is particularly true with respect to thick or long weldments in the vertical position.

Selecting the electrogas welding process for applications with a continuous vertical weld is a sound economical decision. For some applications involving thinner base materials, electrogas welding can still result in cost savings because of its operating efficiency and the minimal joint preparation necessary.

Although EGW equipment, setup, and operator training is more expensive than that required for the more conventional processes, the lengthy vertical welds such as those required in shipbuilding or the construction of storage tanks make EGW the less expensive option. Figures 8.19, 8.20, and 8.21 illustrate the magnitude of some of the applications for which the process offers an economical advantage.

As in all processes, deviations in weld quality add to the overall cost; therefore welding variables (arc voltage, current, polarity, electrode feed speed, electrode extension, oscillation and dwell of the electrode, and root opening) must be carefully selected and monitored to assure high-quality welds. In addition to impacting weld quality, excessive root openings increase weld time and require more filler metal and shielding gases, thus increasing welding costs.

Additional information on cost-effective welding is presented in Chapter 12, "The Economics of Welding and Cutting, in Vol. 1 of the Welding Handbook, 9th ed., *Welding Science and Technology.*"

SAFE PRACTICES

Electrogas welding is as safe as other welding processes, provided that the proper procedures are followed and appropriate precautions are taken. When procedures are followed and precautions are observed, electrogas welding can be performed safely with minimal health risks.

As a general practice, the manufacturer's product literature should be consulted for specific instructions for safe operation of electrogas welding equipment. Safety instructions for all welding and cutting can be found in *Safety in Welding, Cutting, and Allied Processes,* ANSI Z49.1[30] Mandatory federal safety regulations are estab-

30. American National Standards Institute (ANSI) Accredited Standards Committee Z49, *Safety in Welding, Cutting, and Allied Processes,* ANSI Z49.1, Miami: American Welding Society.

lished by the U.S. Department of Labor's Occupational Safety and Health Administration (OSHA) in the *Code of Federal Regulations* (CFR), Title 29, Part 1910.[31]

Appendix B of this volume, "Safety and Health Codes and Other Standards" lists health and safety standards, codes, specifications, and other publications. The publishers, the letter designations of the organizations, and the facts of publication are also listed.

RADIANT ENERGY

Electrogas welding produces radiant energy (radiation) that may be harmful to health. Thus, welders and welding operators must acquaint themselves with the effects of radiant energy. The radiant energy generated by electrogas welding is non-ionizing and similar in wavelength and nature to ultraviolet, visible or infrared light. If excessive exposure occurs, radiation can produce a variety of effects, including skin burns and eye damage, depending on the wavelength and intensity of the radiant energy.

The intensity and wavelengths of non-ionizing radiant energy depend on many factors. These include the welding parameters, the composition of the electrode and base metal, the fluxes, and any coating or plating on the base material. The total radiant energy produced by electrogas welding can be higher than that generated by the shielded metal arc welding process because of the significantly lower amount of welding fumes and the more exposed arc. Generally, the highest ultraviolet radiant energy intensities are produced when an argon shielding gas is used and when welding aluminum.

Numerous measures can be taken to protect against the possible hazardous effects caused by non-ionizing radiant energy from welding. Welders should refrain from looking at the welding arc except through filter plates that meet the requirements of *Practice for Occupational and Educational Eye and Face Protection,* ANSI Z87.1.[32] The choice of filter shade may be made on the basis of visual acuity and may therefore vary from one individual to another, particularly under different current densities, workpiece materials, and electrode types. As a rule of thumb, the welder should start with a shade that is too dark to permit visibility of the weld zone. A lighter shade that provides sufficient visibility without exceeding the minimum suggested filter number should then be selected. Minimum values apply where the actual arc is hidden by the workpiece, the guide, or the shoes. Table 8.17 suggests filter glass shades for use during electrogas welding operations.[33]

Table 8.17
Recommended Shade of Filter Glass for the Electrogas Welding of Several Metals

Electrogas Welding Application	Arc Current	Minimum Protective Shade No.	Suggested Shade No. (Comfort)
Ferrous material	160–250	10	12
	250 and above	10	14
Nonferrous material	160–250	10	12
	250 and above	11	14

Source: Adapted from American Welding Society (AWS) Committee on Arc Welding and Cutting, 2000, *Recommended Practices for Electrogas Welding,* AWS C5.7:2000, Miami: American Welding Society, Page 49.

Another safety measure involves protecting exposed skin with adequate gloves and clothing as specified in *Safety in Welding, Cutting, and Allied Processes,* ANSI Z49.1.[34] Welders and welding operators must also be aware of the reflections from welding arcs and protect all persons from intense reflections. The exposure of passersby to welding operations can be minimized by the use of screens or curtains or by maintaining welding operations at an adequate distance from aisles and walkways.

BURN PROTECTION

Molten metal, sparks, slag, and hot work surfaces are generated during the electrogas welding process. These can cause burns if precautionary measures are not taken.

To protect against burns, welders must wear protective clothing made of fire-retardant material. The use of leather or wool clothing that is dark in color is recommended to reduce reflections that could cause ultraviolet burns to the face and neck underneath the helmet. The great intensity of ultraviolet radiation may cause rapid disintegration of cotton clothing unless protective chemicals are used to treat the cotton.

Pants with cuffs or open pockets or other places on clothing that can catch and retain molten metal or sparks should not be worn. High-top shoes or leather leggings and fire-retardant gloves should be worn. Pant legs should be worn over the outside of the high-top boots. Helmets or hand shields and a head covering

31. Occupational Safety and Health Administration (OSHA), 1999, *Occupational Safety and Health Standards for General Industry,* in *Code of Federal Regulations (CFR),* Title 29 CFR 1910, Subpart Q, Washington D.C.: Superintendent of Documents, U.S. Government Printing Office.

32. American National Standards Institute (ANSI), *Practice for Occupational and Educational Eye and Face Protection,* ANSI Z87.1, Des Plaines, Illinois: American Society of Safety Engineers (ASSE).

33. For further information, see Reference 31.

34. See Reference 31.

that provide protection for the head, face, neck, and ears must be used. In addition, appropriate eye protection should be used.

Clothing must be kept free of grease and oil, and combustible materials should not be carried in pockets. If a combustible substance has been spilled on clothing, the operator should change to clean fire-retardant clothing before working with open arcs. When required, aprons, cape sleeves, leggings, shoulder covers, and bibs designed for welding service must be used.

All personnel in the work area should be protected from burn hazards by means of noncombustible screens or other appropriate protection, as described in the previous paragraph. Before leaving a work area, welders should mark hot workpieces to alert other persons of this hazard, as touching hot equipment such as electrode guides, tips, and shoes can cause burns. Insulated gloves should be worn when handling these items unless an adequate cooling period has been allowed before touching.

ELECTRICAL HAZARDS

Electrical shock can kill; therefore precautions must be implemented at all times. Welders and welding operators must not touch live electrical parts. Faulty installation, improper grounding, and incorrect operation and maintenance of electrical equipment are all sources of potential hazards. The manufacturer's instructions and recommended safe practices must be carefully followed.

All electrical equipment and the workpiece must be grounded. The workpiece lead is not a ground lead; it is used only to complete the welding circuit. A separate connection is required to ground the workpiece. Accessory equipment such as wire feeders, travel mechanisms, and oscillators must also be grounded. If they are not grounded, insulation breakdown might cause the units to become electrically "hot" with respect to ground.

The correct cable size must be used because sustained overloading can cause cable failure and result in possible electric shock or a fire hazard. All electrical connections must be tight, clean, and dry. Poor connections can overheat and even melt. Furthermore, they can produce hazardous arcs and sparks. Water, grease, or dirt must not be allowed to accumulate on plugs, sockets, or electrical units. Moisture can conduct electricity. To prevent shock, the work area, equipment, and clothing must be kept dry at all times. Dry gloves and rubber-soled shoes should be used, or the operator should stand on a dry board or insulated platform.

No attempt should be made to repair or disconnect electrical equipment under load. Disconnecting under load produces arcing of the contacts and may cause burns or shocks, or both. The welding power contactor should be turned off except when welding is in progress or at service check for open-circuit voltage (when the electrode cannot feed). When the welder or operator must leave the work area or move the machine, the main power switch must be turned off before opening the control cabinet, and the shielding gas supply must be shut off at the supply source.

Cables and connectors must be maintained in good condition. Improper or worn electrical connections may permit conditions that could cause electrical shocks or short-circuits. Worn, damaged, or bare cables must never be used. Direct contact across open-circuit voltage and line voltage should be avoided.

When several welders are working with arcs of different polarities or when a number of alternating-current machines are being used, the open-circuit voltages can be additive. The added voltages increase the severity of the shock hazard.

In case of electric shock, the power should be turned off. If the rescuer must resort to pulling the victim from the live contact, nonconducting materials should be used. If the victim is not breathing, cardiopulmonary resuscitation (CPR) should be administered until breathing has been restored or until medical help has arrived. Electrical burns should be treated like thermal burns, that is, clean, cold (iced) compresses should be applied. Covering it with a clean, dry dressing should prevent contamination of the burn. A physician should be called immediately.

FUMES AND GASES

Electrogas welding produces fumes and gases that may be hazardous to health. Fumes are solid particles that originate from welding consumables, the base metal, and any coatings present on the base metal. Gases are produced during the welding process or by the effects of process radiation on the surrounding environment. Operators must acquaint themselves with the effects of these fumes and gases. The amount and compositions of these fumes and gases depend on the filler metal and base materials, the current level, the arc length, and other factors.

The possible effects of overexposure range from irritation of the eyes, skin, and respiratory system to more severe complications. The effects may become apparent immediately or at some later time. Fumes can cause symptoms such as nausea, headache, dizziness, and metal fume fever. The possibility of more serious effects exists when especially toxic materials are involved. In confined spaces, the gases may displace the breathing air and cause asphyxiation.

The operator's head must be kept out of the fume plume. Ventilation or exhaust at the arc, or both, must be provided to keep fumes and gases from the breathing zone and the general area. Adequate ventilation should be furnished, especially when flux cored electrodes are being used. If chlorinated solvents have been used to degrease or clean the workpiece, a check should be made to ensure that the solvents have been removed before welding. In some cases, natural ventilation may be questionable. Air sampling should be performed to determine whether corrective measures should be taken. The manufacturers' Material Safety Data Sheets (MSDS) should be consulted for further information.

NOISE AND HEARING PROTECTION

Exposure to excessive noise can cause a loss of hearing. The loss of hearing can be either full or partial and temporary or permanent. In electrogas welding, noise may result from the welding process, the power source, or other equipment. Excessive noise most often results from other operations (e.g., grinding and chipping) that are taking place in the vicinity of the electrogas equipment.

Excessive noise adversely affects hearing capability. This adverse effect may be a temporary threshold shift from which the ears recover if the noise source is eliminated. However, if a person is exposed to this same noise level for a prolonged period, the loss of hearing may become permanent. The amount of time required to develop permanent hearing loss depends on factors such as individual susceptibility, noise level, and the duration of the exposure.

A direct protective measure against excessive noise is to reduce the intensity of the source. Another method is to shield the source, but this has its limitations. The acoustical characteristics of a room also affect the level of noise. When engineering controls fail to reduce the noise, personal protective devices such as earmuffs or earplugs should be employed. Generally, these devices are only accepted when engineering controls are not fully effective.

The permissible noise exposure limits can be found in CRF, Title 29, Chapter XVII, Part 1910.[35] Additional information is presented in the American Conference of Governmental Industrial Hygienists' (ACGIH) *TLVs® and BEIs®: Threshold Limit Values for Chemical Substance and Physical Agents* in the *Workroom Environment*.[36] The American Welding Society publication *Arc Welding and Cutting Noise*[37] provides additional information.

35. See Reference 32.

36. American Conference of Governmental Industrial Hygienists (ACGIH), *1999 TLVs® and BEIs®: Threshold Limit Values for Chemical Substances and Physical Agents in the Workroom Environment,* Cincinnati: American Conference of Governmental Industrial Hygienists.

37. American Welding Society (AWS) Committee on Safety and Health, 1979, *Arc Welding and Cutting Noise,* Miami: American Welding Society.

CONCLUSION

It has been demonstrated that electrogas welding has the potential to produce high-quality welds while offering large cost savings when the unique advantages of the process can be utilized. The process has been proven to be particularly appropriate for applications in shipbuilding, the erection of storage tanks, and fabrications of large plates for girder flange and web sections for bridge applications (especially compression members). Benefits have also been realized in heavy-section welding on machine bases and similar applications.

The use of electrogas welding has proven successful for many applications, including those discussed in this chapter. Successful users of electrogas welding select appropriate equipment and follow the recommendations of the equipment manufacturer for its use and maintenance. They employ properly trained and qualified operators and follow welding procedures that meet the requirements of the relevant code or specifications.

BIBLIOGRAPHY[38]

American Conference of Governmental Industrial Hygienists (ACGIH). 1999. *1999 TLVs® and BEIs®: Threshold limit values for chemical substances and physical agents in the workroom environment.* Cincinnati: American Conference of Governmental Industrial Hygienists. (Editions of this publication are also available in Greek, Italian, and Spanish).

American Society of Mechanical Engineers (ASME). 2001. *Boiler and pressure vessel code.* BPV-2001. New York: American Society of Mechanical Engineers.

American National Standards Institute (ANSI) Accredited Standards Committee Z49. 1999. *Safety in welding, cutting, and allied processes.* ANSI Z49.1:1999. Miami: American Welding Society.

American National Standards Institute (ANSI). 1989. *Practice for occupational and educational eye and face protection.* ANSI Z87.1-1989. Des Plaines, Illinois: American Society of Safety Engineers (ASSE).

American Welding Society (AWS) Committee on Arc Welding and Cutting. 2000. *Recommended practices for electrogas welding.* AWS C5.7:2000. Miami: American Welding Society.

American Welding Society (AWS) Committee on Definitions. 2001. *Standard welding terms and definitions.* AWS C5.7:2000. Miami: American Welding Society.

38. The dates of publication given for the codes and other standards listed here were current at the time this chapter was prepared. The reader is advised to consult the latest edition.

American Welding Society (AWS) Committee on Filler Metals. 1997. *Specification for carbon and low-alloy steel electrodes for electrogas welding.* ANSI/AWS A5.26/A5.26M-97. Miami: American Welding Society.

American Welding Society Committee on Filler Metals. 1997. *Specification for welding shielding gases.* AWS A5.32/A5.32M. Miami: American Welding Society.

American Welding Society (AWS) Committee on Safety and Health. 1979. *Arc welding and cutting noise.* Miami: American Welding Society.

American Welding Society (AWS) Committee on Structural Welding. 2002. *Structural welding code—Steel.* AWS D1.1/D1.1M:2002. Miami: American Welding Society.

American Welding Society, (AWS) Committee on Structural Welding. 2002. *Bridge welding code,* ANSI/AASHTO/AWS D1.5.D1.5M:2002. Miami: American Welding Society.

American Welding Society (AWS) Committee on Safety and Health. 1999. *Safety and health fact sheets.* Miami: American Welding Society.

American Welding Society (AWS) Welding Handbook Committee. 2001. Jenney, C. L., and A. O'Brien, eds. *Welding science and technology.* Vol. 1 of *Welding handbook*, 9th ed. Miami: American Welding Society.

Canadian Standards Association International (CSA). 1989. *Welded steel construction (metal arc welding).* CSA W59M/W59-1989. Toronto: Canadian Standards Association.

SUPPLEMENTARY READING LIST

ASM International Handbook Committee. 1993. *Welding, brazing, and soldering.* Vol. 6 of *Metals Handbook.* Materials Park, Ohio: American Society for Metals.

American Welding Society (AWS) Committee on Filler Metals. 1993. *Filler metal procurement guidelines.* ANSI/AWS A5.01-1993. Miami: American Welding Society.

Arnold, P. C., and D. C. Bertossa. 1966. Multiple-pass automatic vertical welding. *Welding Journal* 45(8): 651–660.

Campbell, H. C. 1970. *Electroslag, electrogas, and related welding processes.* Welding Research Council Bulletin No. 154. New York: Welding Research Council.

Franz, R. J., and W. H. Wooding. 1963. Automatic vertical welding and its industrial applications. *Welding Journal* 42(6): 489–494.

Irving, R. R. 1972. Vertical welding goes into orbit. *Iron Age* 50(10): 50.

Normando, N. J., D. V. Wilcox, and R. F. Ashton. 1973. Electrogas vertical welding of aluminum. *Welding Journal* 52(7): 440–448.

Schwartz, N. B. 1970. New way to look at welded joints. *Iron Age* 20 (8): 54–55.

CHAPTER 9

ARC STUD WELDING

Photograph courtesy of Nelson Stud Welding

Prepared by the Welding Handbook Chapter Committee on Stud Welding:

H. A. Chambers, Chair
Consultant, Nelson Stud Welding

Clark Champney
Nelson Stud Welding

B. C. Hobson
Image Industries, Incorporated

C. C. Pease
C. P. Metallurgical Company

Welding Handbook Volume 2 Committee Member:

D. B. Holliday
Northrop Grumman Marine Systems

Contents

CHAPTER 9

ARC STUD WELDING

INTRODUCTION

Stud welding is a general term for joining a metal stud or similar component to a workpiece. The stud welding process is used without a filler metal, and with or without external gas or flux shielding. Partial shielding in the form of a ceramic ferrule or a ferrule of another material surrounding the stud is sometimes used. Pressure is applied after the workpieces are sufficiently heated.[1, 2] Arc stud welding (SW) is the specific process discussed in this chapter.

Arc stud welding and a variation of the process, capacitor discharge stud welding, are reliable, widely used joining processes important to manufacturers in the automotive, railroad, shipbuilding, heavy equipment, construction, and military defense industries. Studs can be welded to a vast range of materials in applications requiring the joining, fastening, mounting, securing, or anchoring of equipment, structures, or components.

Stud welding can be accomplished with a number of welding processes, including arc, resistance, friction, and percussion. These processes are used with conventionally designed equipment with special tooling for stud welding.[3] However, the equipment and techniques most suited to arc stud welding are derived from the arc welding processes.

The fundamentals of arc stud welding and capacitor discharge stud welding, including its three modes of operation, are described in this chapter. Information on stud welding equipment, applications, materials, techniques, and quality control is presented, along with guidance on process selection, economics, and safe practices. The updated Supplementary Reading List at the end of the chapter is a guide to more detailed technical information.

It should be noted that, consistent with stud welding industry standards, measurements of stud diameters and thread sizes are stated as fractions, while measurements of metal thicknesses are stated as decimals in this chapter.

FUNDAMENTALS

Studs are attachments in application-oriented configurations that can be welded to an assembly or structure to serve as anchoring, spacing, or fastening devices. The stud types include many refinements of design, such as threaded fasteners, plain or slotted pins, internally threaded fasteners, flat fasteners with a rectangular cross section, headed pins with various upsets, and headed anchors and shear connectors used in steel-to-concrete construction. It is this broad variety of stud designs and the ability to weld them that makes stud welding a valuable tool for numerous applications. Most stud styles can be rapidly applied with portable equipment. A stud welding gun and the basic operating components of the stud welding process are illustrated in Figure 9.1. Studs in a variety of designs are displayed in Figure 9.2.

Stud welding guns hold the studs and move them in proper sequence during welding. Two basic types of power sources are used to create the arc for welding studs. One type uses direct current (dc) power sources and the other type uses capacitor storage banks to sup-

1. American Welding Society (AWS) Committee on Definitions and Symbols, 2001, *Standard Welding Terms and Definitions*, AWS A3.0:2001, Miami: American Welding Society.

2. At the time of preparation of this chapter, the referenced codes and other standards were valid. If a code or other standard is cited without a date of publication, it is understood that the latest edition of the document referred to applies. If a code or other standard is cited with the date of publication, the citation refers to that edition only, and it is understood that any future revisions or amendments to the code or standard are not included. As codes and standards undergo frequent revision, the reader is advised to consult the most recent edition.

3. See Chapter 17, "Spot, Seam, and Projection Welding," Chapter 18, "Flash, Upset, and Percussion Welding," and Chapter 23, "Friction Welding" in American Welding Society (AWS) Welding Handbook Committee, O'Brien, R. L. ed., 1991, *Welding Processes* Volume 2 of *Welding Handbook*, 8th ed., Miami: American Welding Society.

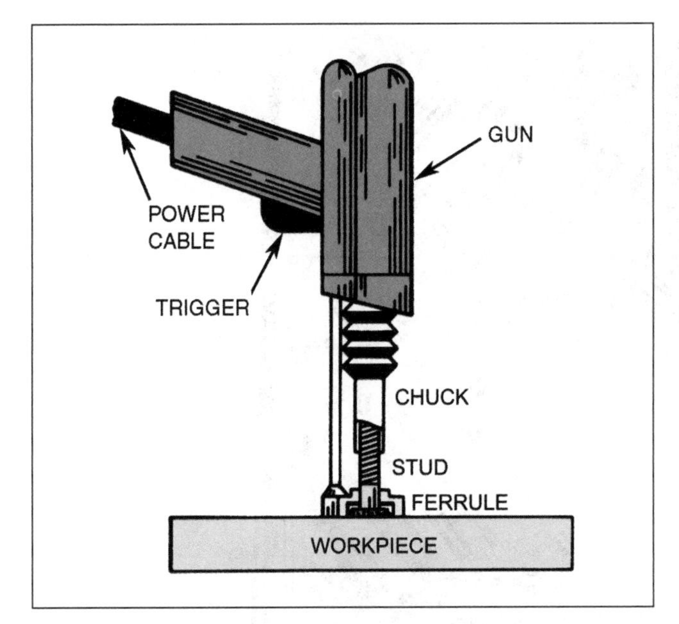

Figure 9.1— Basic Stud Welding Gun and Operating Components

ply the arc power, thus the stud welding processes are commonly known as *arc stud welding* and *capacitor discharge stud welding.*

In arc stud welding, the base (end) of the stud and the workpiece are heated by the arc drawn between the two. When the stud and the workpiece are sufficiently heated, the stud begins to melt and a weld pool is formed on the workpiece. These are brought together under low pressure by plunging the stud into the weld pool. The process is used with or without shielding gas or flux.[4] In some applications, partial shielding is provided by a ring (called a *ferrule*) made of ceramic or other material surrounding the stud.

Arc stud welding, the more widely used of the two major stud welding modes of operation, is somewhat similar to manual shielded metal arc welding. The heat necessary for the welding of studs is developed by a dc arc between the electrode (the stud) and the workpiece (the base plate) to which the stud is to be welded. The welding current is usually supplied by a dc transformer-rectifier power source much like those used for shielded metal arc welding. Stud welding systems powered by a dc motor-generator were used in the past, but for the most part, these systems are obsolete.

Welding time and the plunging of the stud into the weld pool to complete the weld are controlled automatically. This control can be a separate unit, but it is usually integrated in the power source. The stud is held in a stud welding gun and positioned by the operator, who then actuates the unit by pressing a switch. The weld is completed quickly, usually in less than one second. This process generally uses a ferrule, which surrounds the stud to contain the molten metal and shield the arc. A ferrule is not used with some special welding techniques, or with some nonferrous metals.

Capacitor discharge stud welding derives heat from an arc produced by the rapid discharge of electrical energy stored in a bank of capacitors. During or immediately following the electrical discharge, pressure is applied to the stud, plunging the base of the stud into the weld pool (the molten metal) on the workpiece.

The arc can be established by any of the following methods:

1. By the rapid resistance heating and vaporization of a projection on the stud base,
2. Drawing an arc as the stud is lifted away from the workpiece, or
3. Drawing an arc across an air gap between the stud and the workpiece.

In the first method, arc times are about three to six milliseconds; in the second method, they range from six to fifteen milliseconds. The capacitor discharge mode does not require a shielding ferrule because of the short duration of the arc and the small amount of molten metal expelled from the joint. It is suited for applications requiring small- to-medium-sized studs.

APPLICATIONS

Stud welding is widely recognized by the metalworking industries as an essential tool for many challenging joining operations. The process is used extensively in the fields of automobile manufacturing, boiler construction, building and bridge construction, farm and industrial equipment manufacturing, railroad and rail car construction, and shipbuilding. Military applications include missile containers, armored vehicles, and tanks. Studs can be used as hold-downs, standoffs, heat transfer members, insulation supports, and other fastening applications.

Typical applications are attaching wood floors to steel decks or frameworks; fastening linings or insulation in tanks, boxcars, and other containers; fastening inspection covers; mounting machine accessories; securing tubing and wire harnesses; and attaching shear connectors and concrete anchors to structures. Figure 9.3 shows a stud welding application in the construction of a bridge pier foundation. Headed shear connector studs were welded to vertical-driven sheet piling and long

4. See Reference 1.

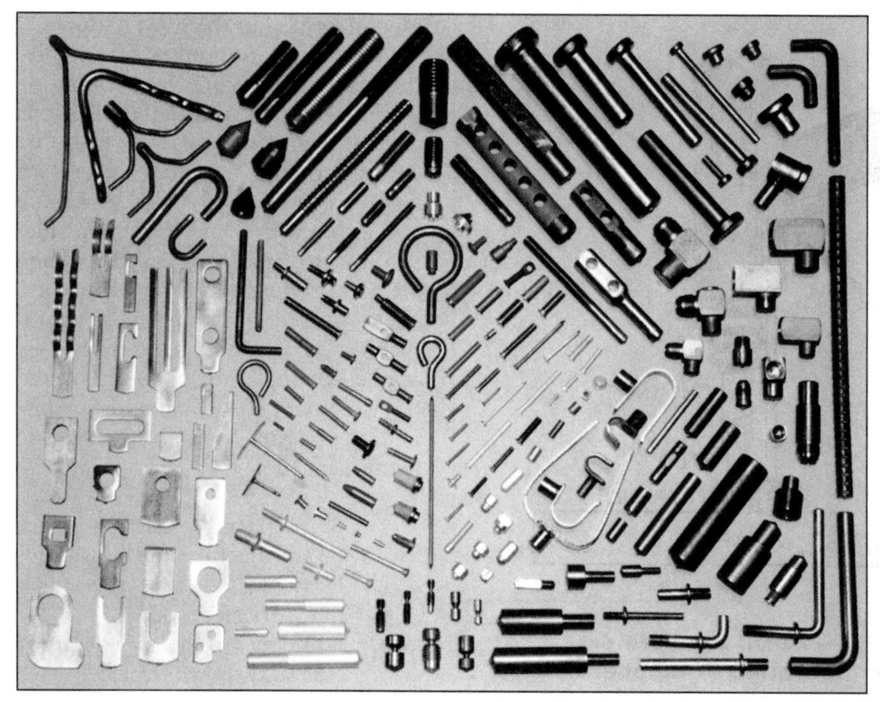

Photograph courtesy of Nelson Stud Welding

Figure 9.2—Variety of Stud Sizes and Designs

deformed bar anchor studs were welded to horizontal shelf angles prior to pouring the concrete.

Figure 9.4 shows steel beams to which studs have been welded prior to delivery to the bridge construction site.

ADVANTAGES AND LIMITATIONS

Stud welding provides versatile options for joining and fastening in many ordinary and specialized applications. The variety of studs available adds great flexibility of design.

Advantages

Because arc stud welding cycle times are very short, there is little heat input to the base metal compared to conventional arc welding. Consequently, the weld metal zone and heat-affected zone remain very narrow, and distortion of the base metal at stud locations is minimal. However, the local heat input may be harmful when studs are welded to medium- and high-carbon steels. The unheated portion of the stud and base metal cools the weld and heat-affected zones very rapidly, causing these areas to harden. The resulting lack of weld joint ductility may be detrimental under certain types of loading, such as cyclic loading. Conversely, when stud welding heat-treatable aluminum alloys, a short weld cycle minimizes overaging and softening of the adjacent base metal.

Studs can be welded at the appropriate time during construction or fabrication without access to the back side of the base member. Drilling, tapping, or riveting for installation is not required.

When using the arc stud welding process, designers need not specify thicker materials than required for the product or provide heavy bosses and flanges to accommodate the tap depths associated with threaded fasteners. Because stud welded designs can be lighter in weight, not only can material be saved, but the amount of welding and machining needed to join the components can also be reduced.

Small studs can be welded to thin sections with the capacitor discharge method. Studs have been welded to sheet as thin as 0.75 millimeter (mm) (0.03 inch [in.]) without melt-through. Studs have been joined to certain materials (e.g., stainless steel) as thin as 0.25 mm (0.01 in.). Because the depth of melting is very shallow, capacitor discharge welds can be made without damage to a prefinished opposite side. No subsequent cleaning or finishing is required.

Photograph courtesy of Nelson Stud Welding

Figure 9.3—Shear Connectors and Bar Studs Used in Bridge Piling Construction (Piling on Completed Foundation in Background)

The relatively low energy input of capacitor discharge power also permits the welding many dissimilar metals and alloys. This includes joining brass to steel, copper to steel, brass to copper, aluminum to die-cast zinc, and similar combinations.

Limitations

A stud weld does not extend through the workpiece. If a stud is required on both sides of a member, a second stud must be welded to the opposite side. Stud shape and size are limited because the stud design must permit chucking in the gun for welding. The stud base diameter is limited for thin base metals. When studs are applied by arc stud welding, a disposable ceramic ferrule around the base is usually required. In some applications it is also necessary to provide flux in the stud base or a protective gas shield to obtain a sound weld.

Most studs applied by capacitor discharge power require a close-tolerance projection on the weld base

Photograph courtesy of Nelson Stud Welding

Figure 9.4—Studs Applied to Beams Used in Bridge Construction

to initiate the arc. Stud diameters that can be attached by this method generally range from 3.2 mm to 9.5 mm (1/8 in. to 3/8 in.). Above this size, conventional arc stud welding is more economical.

For arc stud welding, 230 volt (V) or 460 V alternating-current (ac) power is required to operate the direct-current (dc) welding power source. For most capacitor discharge welding, a single-phase 110 V main supply is sufficient, but high-production units require three-phase ac 230 V or 460 V for operation. A welding power source located conveniently close to the work area is required for stud welding. Electrical resistance in weld cables and ground cables varies and increases with high welding rates. However, with adequate cable sizes, electric arc welding can typically be performed with power sources located up to 138 meters (m) (450 feet [ft]) from the welding site, while the location of capacitor discharge units usually must be placed within 3 m to 5 m (10 to 20 ft) from the site.

EQUIPMENT AND TECHNOLOGY

The arc stud welding process involves many of the same basic principles as other arc welding processes. The application of stud welding (a typical example is attaching a stud to metal plate) consists of two steps: welding heat is developed when the stud (one workpiece) is lifted off the base metal (the other workpiece) and an arc is established between the two; then the two workpieces are brought into contact when the proper temperature is reached.

Equipment for basic arc stud welding originally consisted of the stud gun, a control unit (timing device), studs and ferrules, and a separate source of dc welding current, as shown in Figure 9.5. Rotating generators and some rectifier power sources, although approaching obsolescence in the 2000s, are still in use. Continuing improvements made by manufacturers of equipment have resulted in stud welding units in which the power source, current controls and the gun timing device are integrated. The most commonly used power sources with these integrated stud welding control systems are transformer-rectifier types, as shown in Figure 9.6.

An inverter power source is not often used in stud welding because the process requires high current: 600–3000 amperes (A) is typical. These high currents require high-current-compatible components. The power components required in the inverter for switching these high welding currents make inverter power sources more expensive than others, thus limiting their use to small-diameter studs.

Modern integrated transformer-rectifier systems incorporate many features, such as current compensation, troubleshooting diagnostics, and time and current displays that assure accurate weld settings and a shutdown mode if current regulation settings are exceeded. Low-carbon steel and stainless steel studs with diameters from 2.4 mm to 32 mm (3/32 in. to 1-1/4 in.) can be welded with these systems.

Regardless of the type of power source used, the mechanics of the process are the same, as illustrated in Figure 9.7. The stud is loaded into the chuck, the ferrule (also known as an *arc shield*) is placed in position over

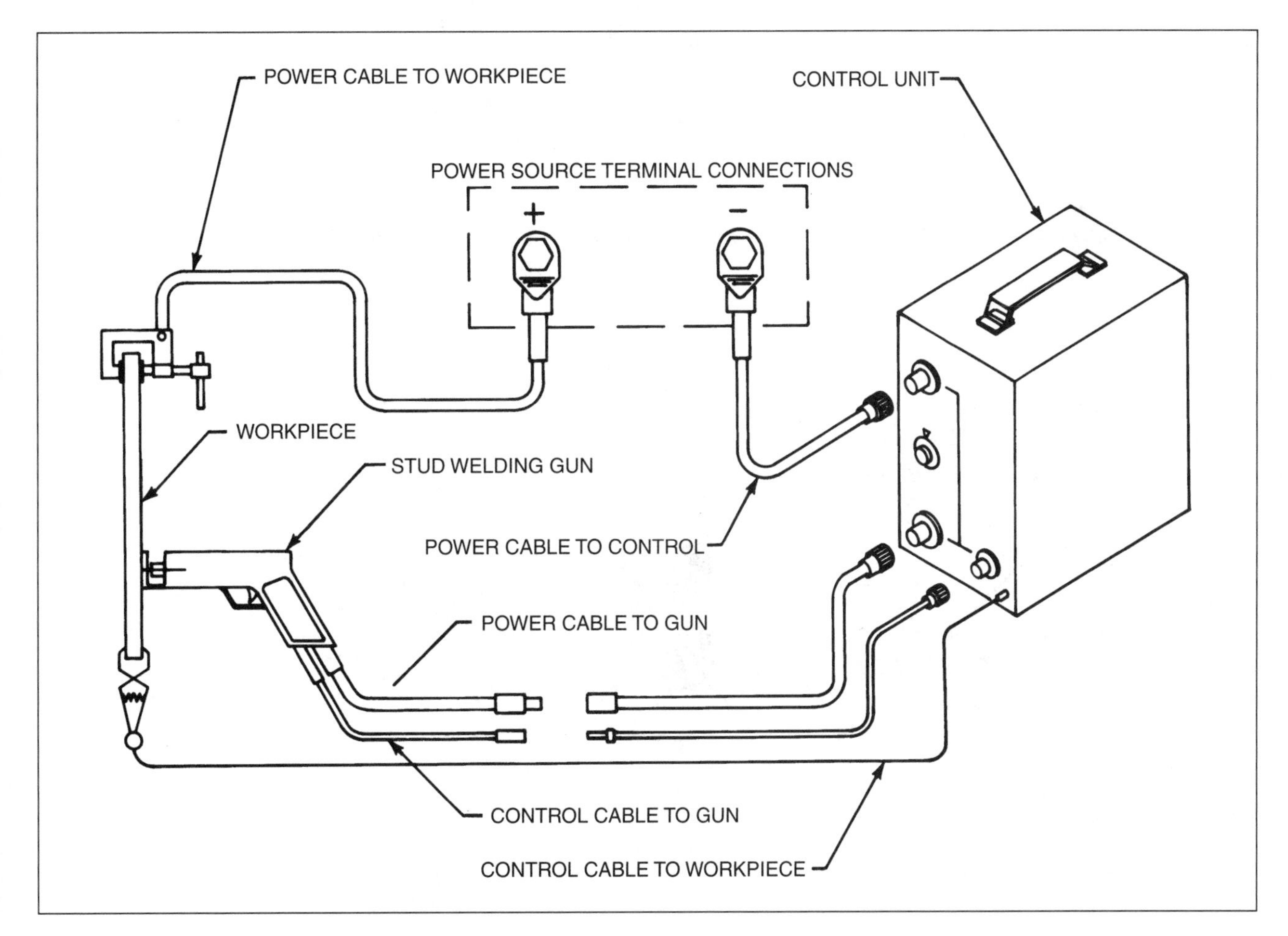

Figure 9.5—Stud Welding Power Source and Control Unit

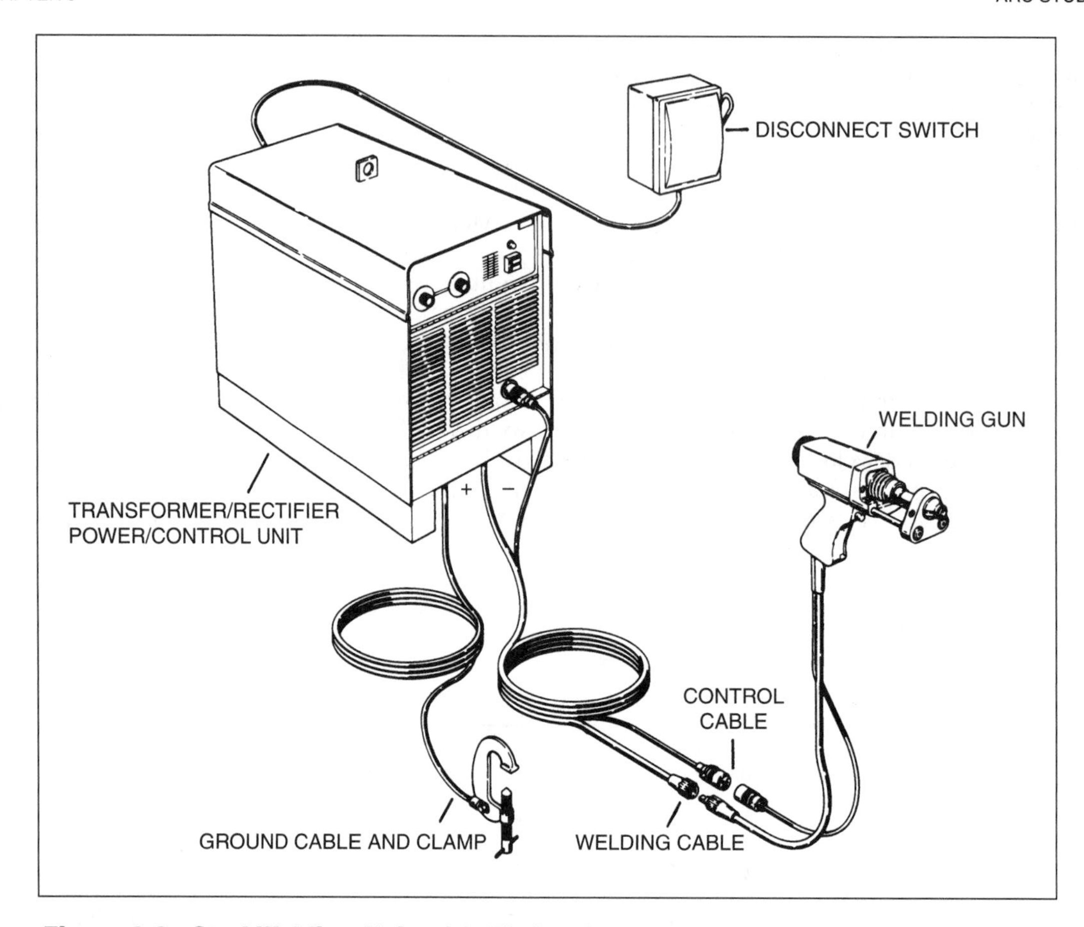

Figure 9.6—Stud Welding Unit with Timing Control Integrated into Power Source

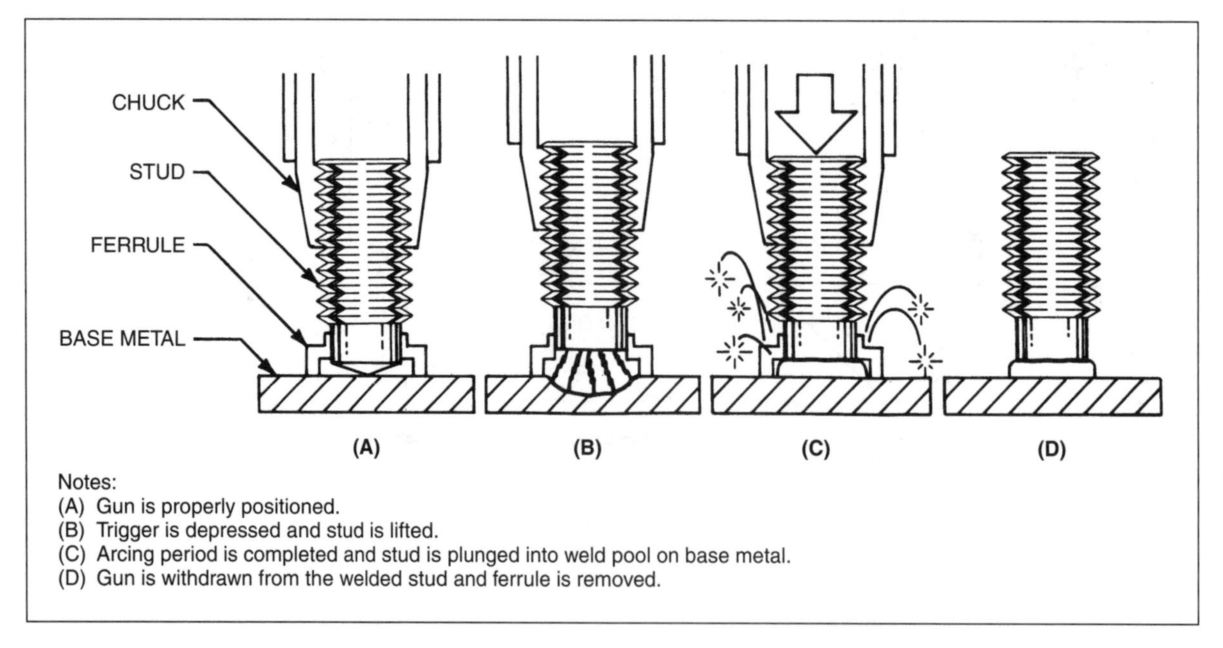

Figure 9.7—Steps in Producing an Arc Stud Weld

the end of the stud, and the gun is properly positioned for welding (A). The trigger is then depressed, starting the automatic welding cycle. A solenoid coil within the body of the gun is energized. This lifts the stud off the workpiece and at the same time creates an arc (B). The end of the stud and the target area of the workpiece are melted by the arc. When the preset arc period is completed, the welding current is automatically shut off and the solenoid is deenergized by the control unit. The mainspring of the gun plunges the stud into the weld pool on the workpiece to complete the weld (C). The gun is then lifted from the stud, and the ferrule is broken off (D).

The time required to complete a weld varies with the cross-sectional area of the stud. For example, for a 3.2 mm (1/8 in.) stud diameter the typical weld time is about 0.13 seconds; for a 22 mm (7/8 in.) stud diameter the time is approximately 0.92 seconds. An average rate is approximately 6 studs per minute, although a rate of 15 studs per minute can be achieved for some applications. Additional welding machine settings are shown in Table 9.1.

EQUIPMENT

The equipment for stud welding consists of a stud welding gun, an integrated power source control unit (or a power source and a separate control unit) to control the magnitude and time of the current flow, and appropriate connecting cables and accessories, as illustrated in Figures 9.5 and 9.6.

The portability and ease of operation of the equipment involved in stud welding compares closely with that of shielded metal arc welding (SMAW).

Stud Welding Guns

The two types of stud welding guns are portable hand-held guns and fixed guns used in production, The principles of operation are the same for both types. A schematic of a portable hand-held arc stud gun is shown in Figure 9.8(A), with an application of the gun shown in 9.8(B). A fixed gun production machine is shown in Figure 9.9.

Table 9.1
Arc Stud Welding Setups for Mild and Stainless Steel Studs and Base Materials

Stud Diameter	Stud Area	Welding Position: Flat				Welding Position: Overhead				Welding Position: Vertical			
mm (in.)	mm^2 ($in.^2$)	Amp	Sec.	Lift mm (in.)	Plunge mm (in.)	Amp	Sec.	Lift mm (in.)	Plunge mm (in.)	Amp	Sec.	Lift mm (in.)	Plunge mm (in.)
4.8 (3/16)	18.1 (0.028)	300	0.15	1.57 (0.062)	2.36 (0.093)	300	0.15	1.57 (0.062)	3.18 (0.125)	300	0.15	1.57 (0.062)	3.18 (0.125)
6.4 (1/4)	32.2 (0.049)	450	0.20	1.57 (0.062)	2.36 (0.093)	450	0.17	1.57 (0.062)	3.18 (0.125)	450	0.17	1.57 (0.062)	3.18 (0.125)
7.9 (5/16)	49.0 (0.077)	550	0.25	1.57 (0.062)	3.18 (0.125)	500	0.25	1.57 (0.062)	3.18 (0.125)	500	0.25	1.57 (0.062)	3.18 (0.125)
9.5 (3/8)	70.9 (0.111)	650	0.35	1.57 (0.062)	3.18 (0.125)	550	0.35	1.57 (0.062)	3.18 (0.125)	600	0.33	1.57 (0.062)	3.18 (0.125)
11.1 (7/16)	96.8 (0.150)	700	0.45	1.57 (0.062)	3.18 (0.125)	675	0.42	1.57 (0.062)	3.18 (0.125)	750	0.33	1.57 (0.062)	3.18 (0.125)
12.7 (1/2)	126.7 (0.196)	850	0.55	1.57 (0.062)	3.18 (0.125)	800	0.55	1.57 (0.062)	3.18 (0.125)	875	0.47	1.57 (0.062)	3.18 (0.125)
15.9 (5/8)	198.6 (0.307)	1200	0.70	1.57 (0.062)	4.75 (0.187)	1200	0.67	2.36 (0.093)	4.75 (0.187)	1275	0.60	1.57 (0.062)	4.75 (0.187)
19.1 (3/4)	286.5 (0.442)	1500	0.90	2.36 (0.093)	4.75 (0.187)	1500	0.48	2.36 (0.093)	4.75 (0.187)	1700	0.73	2.36 (0.093)	4.75 (0.187)
22.2 (7/8)	387.1 (0.601)	1750	1.10	3.18 (0.125)	6.35 (0.250)	1700	1.00	3.18 (0.125)	6.35 (0.250)	*	*	*	*
25.4 (1)	506.7 (0.785)	2000	1.40	3.18 (0.125)	6.35 (0.250)	2050	1.40	3.18 (0.125)	6.35 (0.250)	*	*	*	*

*Not recommended.

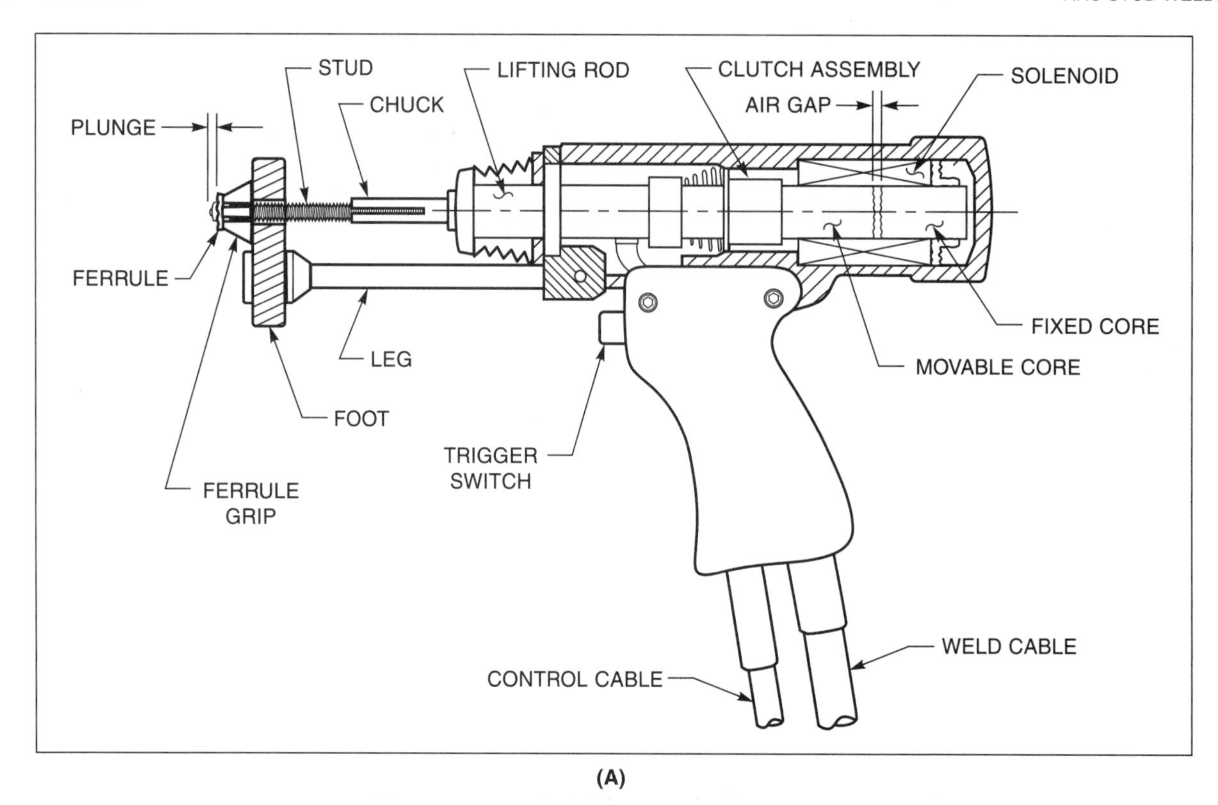

(A)

(B)

Figure 9.8—(A) Schematic of a Portable Stud Welding Gun and (B) the Gun in Operation

Photograph courtesy of Nelson Stud Welding

Figure 9.9—Stationary Arc Stud Welding Machine (Production Type)

The portable stud welding gun resembles a pistol. It is made of a tough plastic material and weighs between 2 kilograms (kg) and 4 kg (4.5 pounds [lb] and 9 lb), depending on the type and size of the gun. A small gun is used for studs from 3.2 mm to 16 mm (1/8 in. to 5/8 in.) in diameter; a larger heavy-duty gun is used for larger studs through 32 mm (1-1/4 in.) in diameter. The larger gun can be used to weld the entire range of stud sizes and designs; however, in applications where only small-diameter studs are used, it may be advantageous to use a smaller, lighter-weight gun.

A gun consists of the body, a lifting mechanism, a chuck or stud holder, an adjustable foot for the ferrule holder, and the connecting weld and control cables. The portable gun body is usually made of strong, high-impact plastic. The stud lifting mechanism consists of a solenoid, a clutch, and a mainspring. The mechanism is actuated by the solenoid to provide positive control of the lift. The lift is designed to be consistent over a range of 0.8 mm to 3.2 mm (0.03 in. to 0.125 in.), and will be constant regardless of the length of stud protrusion (within the limits of the gun). An added feature of some guns is a cushioning arrangement to control the plunging action of the stud to complete the weld. Controlled plunge eliminates the excessive spatter normally associated with the welding of large-diameter studs without plunge control.

The fixed gun used in production is mounted on an automatic positioning device, which is usually air-operated and electrically controlled. The workpiece is positioned under the gun with suitable locating fixtures. Tolerances of 0.13 mm (0.005 in.) on location and 0.254 mm (0.010 in.) in height may be obtained when a production gun is used. A production unit may contain a number of guns, depending on the nature of the job and the production rate required.

Control Unit

Fundamentally, the control unit consists of a weld timing device with associated electrical controls and a contactor suitable for conducting and interrupting the welding current from a power source. The control unit may be separate from the power source (a transformer-rectifier, motor-generator, or battery-bank) as shown in Figure 9.5. The settings of the adjustable weld timing are graduated in either cycles or seconds. Once set, the control unit maintains the proper time interval for the size of stud being welded. The time interval may be adjusted from 0.05 second to 2 seconds, depending on the diameter of the stud.

The stud welding control connects to the stud gun and the terminal of the dc welding power source, shown in Figure 9.6. A third cable is connected from the workpiece to the dc power source to serve as the ground. As with stud welding guns, control units are available in various sizes. For welding studs up to 16 mm (5/8 in.) in diameter, a small control unit can be used. A large control unit must be used for large-diameter studs.

Power Control Units

High-amperage transformer-rectifier power sources have been developed specifically for stud welding. Some are power sources only and require a separate stud welding control unit. Others include the stud welding gun control and timing circuits as an integral part of the power source. The latter are generally referred to as *power control* units. Power control units use silicon-controlled rectifiers (SCR) for initiating and interrupting the weld current. Solid-state components are used for the gun control and timing circuitry.

Power control units are available for operation with either three-phase or single-phase primary power input. Three-phase units are preferred for welding large-diameter studs because they provide a balanced load on the incoming power line and produce smoother output current. Single-phase units are low-cost, portable types for welding studs 12.7 mm (1/2 in.) diameter and smaller.

Regulated-current power control units use high-power solid-state components. Controls are designed to include current feedback circuitry that monitors and regulates the amperage output to the desired current level, regardless of changes in primary voltage or secondary circuit resistance due to changes in cable length, heat buildup or a poor ground connection. This type of unit is recommended when precise control of the weld current and time is necessary. Figure 9.6 shows a typical equipment setup for the arc stud welding of steel with an integrated power-control unit.

Power Sources

Direct-current (dc) power is used for arc stud welding; alternating current (ac) is not suitable. The three basic types of dc power sources that can be used are a transformer-rectifier, a motor-generator (motor or engine driven), and a battery.

The following characteristics are desirable in a stud welding power source:

1. High open-circuit voltage in the range of 70 V to 100 V,
2. Constant-current output, i.e., a drooping output volt-ampere characteristic,
3. A rapid output current rise to the set value, and
4. High-current output for a relatively short time.

The current requirements are higher and the duty cycle is much lower for stud welding than for many other arc welding processes, such as shielded metal arc welding (SMAW). Many standard dc arc welding power sources are available that meet these requirements and are entirely satisfactory for stud welding. However, dc welding power sources with a constant-voltage output, i.e., a flat volt-ampere characteristic (those normally used for gas metal arc welding) are not suitable for stud welding. Weld current control can be difficult with this type of power source, and it is not possible to obtain the proper weld current range for the application.

The current requirements are higher for stud welding than those used for SMAW. Because of their lower maximum current output, most standard dc power sources are generally limited to welding only 13 mm (1/2 in.) diameter and smaller studs. For large studs, two standard dc power sources wired in parallel or a single unit designed specifically for arc stud welding must be used.

When the applications require high welding currents (sometimes over 2000 A) and short weld times, power sources especially designed for arc stud welding are recommended. These special power sources yield higher efficiency not only from the standpoint of weld current output relative to their size and weight, but also from an economic perspective. They cost less than two or more standard arc welding machines.

Duty Cycle. The basis for rating special stud welding power sources is different from that of conventional arc welding machines. Because stud welding requires a high current for a relatively short time, the current output requirements of a stud welding power source are higher, but the duty cycle is much lower than the requirements for other types of arc welding. In addition, the load voltage is normally higher for stud welding. The cable voltage drop is greater with stud welding than other arc welding processes because of the higher current requirements.

The duration of a stud weld cycle is generally less than one second. Therefore, load ratings are made on the basis of one second. The rated output of a machine is its average current output at 50 V for a period of one second. Thus, a rating of 1000 A at 50 V means that during a period of one second the current output will average 1000 A, and the terminal voltage will average 50 V.

Oscillograph traces show that the current output of an unregulated stud welding power source is higher at the start of welding than at the end. It is therefore necessary to use the average current for rating purposes.

The duty cycle for arc stud welding machines is based on the following formula:

$$\% \text{ Duty cycle} = 1.7 \times \text{no. of 1-second loads per minute} \quad (9.1)$$

where the 1-second load is the rated output.

Thus, if a machine can be operated six times per minute at its rated output without causing the components to exceed their maximum allowable temperatures, then the machine would have a 10% duty cycle rating.

Welding Cable

Welding cable size and length, including both gun and ground cables, are very important in stud welding. The current available with an unregulated power source for welding at a given machine setting may vary as much as 50%, depending on the size and length of welding cables used.

In addition, power sources without current regulation or power control units may be severely hampered by the use of either very small-diameter cables or very long cables. This factor is often overlooked when the problem of inadequate welding power arises. When considering cable length, the total diameter and length of all cables in the welding circuit must be taken into account. For any given length and diameter of cable, the welding current can be increased approximately 10% by using a cable of the next larger diameter or using two parallel cables of the same length and diameter instead of one.

Other major factors to be considered when evaluating stud welding power sources and power control units are the incoming power and the cable size and length (primary power and welding cables). Both types of power sources—motor-generator and rectifier—normally operate on 230 V or 460 V ac three-phase power. Because of the high currents required for stud welding, line voltage regulation sometimes becomes a problem. Satisfactory operation of either type of equipment can be assured only if the primary power voltage does not drop below the prescribed limits while a weld is being made.

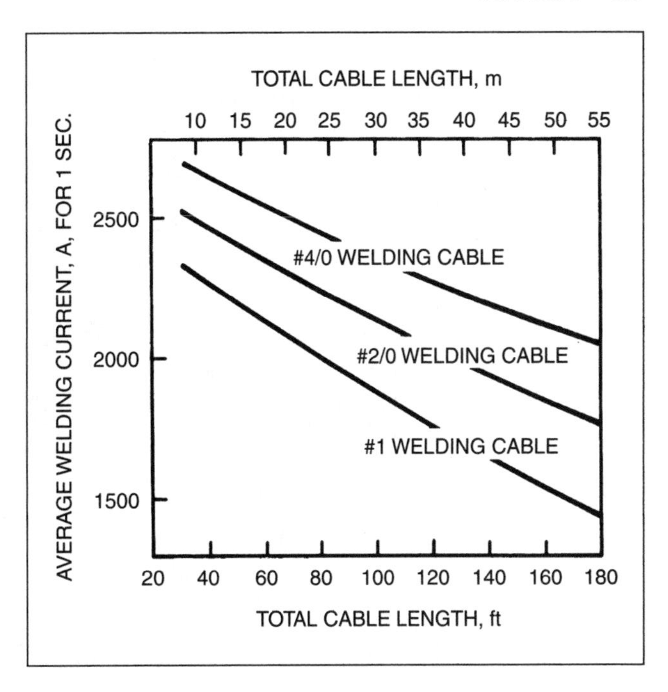

Figure 9.10—Effect of Cable Size and Length on Available Welding Current from a 2000-A Unregulated Power Source

Figure 9.10 illustrates the effect of cable size and length on the available welding current from a 2000 A unregulated power source. The tests made to determine these curves were run with a 2000 A motor-generator power source at maximum setting. Only the cable length and cable size were changed. In this case, the maximum welding current was 2360 A with 9 meters (m) (30 feet [ft]) of American Wire Gauge (AWG) #1 cable. When this same size cable was lengthened to 55 m (180 ft), the available current decreased 38% to 1450 A. Conversely, when 55 m (180 ft) of AWG #4/0 cable was used, the current was 2050 A, a decrease of only 13%. A regulated machine will maintain a constant output as cable length is added. They can only do so until the maximum output of the unit is reached. A larger cable size will extend this limit. When the distance from the power source to the welding gun increases significantly, a larger-diameter welding cable should be used.

Automatic Feed Systems

Stud welding systems with automatic stud feeding are available for both hand-held portable and fixed welding guns. As shown in Figure 9.11, studs are placed in the hopper; they progress to the stud feeder, where

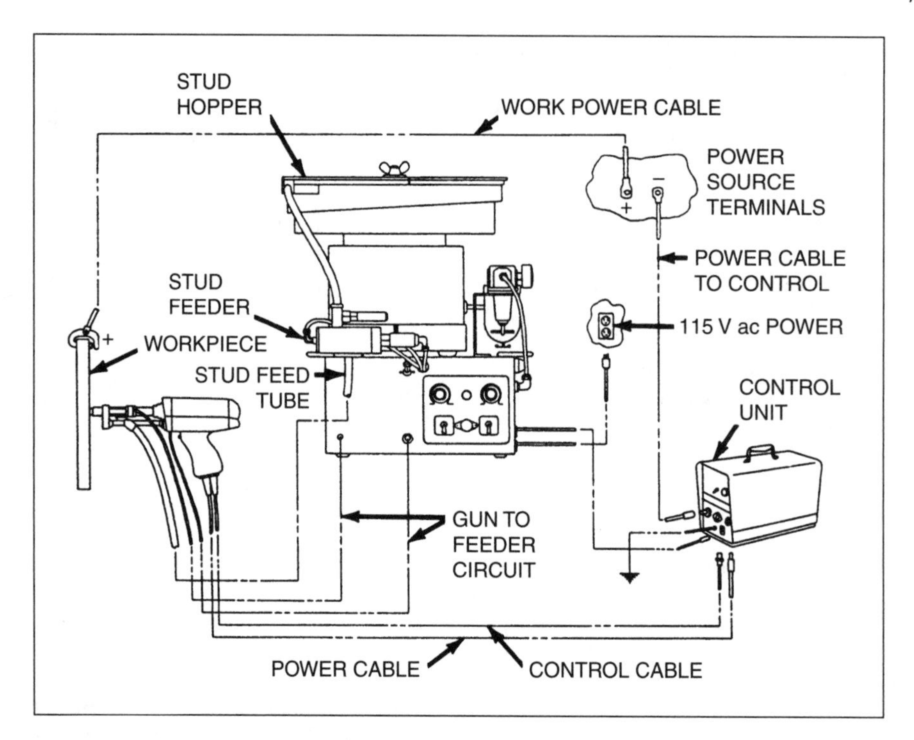

Figure 9.11—Conventional Portable Arc Stud Welding Equipment with an Automatic Stud Feed System

they are automatically oriented for transfer to the gun (usually through a flexible feed tube), and then they are loaded into the welding gun chuck. Generally, a ferrule is hand-loaded for each weld with portable guns used in automatic feed production systems. However, automatic ferrule feed is available with fixed-gun systems used in production. Figure 9.10 illustrates a portable hand-held system with automatic stud feed. Automated portable equipment with solid-state controls for use without ferrules is available for welding 9.5 mm (3/8 in.) diameter studs and smaller.

DESIGNING FOR ARC STUD WELDING

When a design calls for drilling and tapping to install fasteners or supports, arc stud welding should be considered as an alternate means for attaching them. Arc stud welding requires less thickness of the base material, or workpiece, to obtain full strength compared to the requirements of threaded fasteners. The use of arc welded studs may allow a reduction of the thickness of bosses at attachment points or may eliminate them. Cover plate flanges can be thinner than those required for threaded fasteners. Thus, a significant weight saving can be gained when the process is used. Fasteners can be stud welded with smaller edge distances than those required for through-hole threaded fasteners. However, loading and deflection requirements must be considered when stud locations are close to the edges of the flange.

The weld base diameters of steel and stainless steel studs range from 2.7 mm to 32 mm (7/64 to 1 1/4 in.). For aluminum, the range is 3.2 mm to 13 mm (1/8 in. to 1/2 in.). For design purposes, the smallest cross-sectional area of the stud should be used for load determination, and adequate safety allowances should be considered.

To develop the full strength of a stud-welded fastener, the thickness of the plate, or workpiece, should be a minimum of approximately 1/3 of the weld base diameter. Studs can be welded to thinner material if full strength is not required. As shown in Table 9.2, a minimum plate thickness is required for each stud size to permit arc stud welding without melt-through or excessive distortion. For steel, a minimum ratio of 1:5 plate thickness-to-stud base diameter is the general rule to prevent melt-through of the base plate material. For aluminum, the ratio is a minimum of 1:2.

Table 9.2
Recommended Minimum Thicknesses of Steel and Aluminum Plate for Arc Stud Welding

Stud Base Diameter		Steel		Aluminum			
		Without Backing		Without Backing		With Backing*	
mm	in.	mm	in.	mm	in.	mm	in.
4.8	3/16	1.0	0.04	3.3	0.13	3.3	0.13
6.4	1/4	1.3	0.05	3.3	0.13	3.3	0.13
7.9	5/16	1.5	0.06	4.8	0.19	3.3	0.13
9.5	3/8	2.0	0.08	4.8	0.19	4.8	0.19
11.1	7/16	2.3	0.09	6.4	0.25	4.8	0.19
12.7	1/2	3.0	0.12	6.4	0.25	6.4	0.25
15.9	5/8	3.8	0.15	—	—	—	—
19.1	3/4	4.8	0.19	—	—	—	—
22.2	7/8	6.4	0.25	—	—	—	—
25.4	1	9.5	0.38	—	—	—	—

*A metal backing is used to prevent melt-through of the high thermal conductivity aluminum plate by dissipating the heat in the weld.

BASE MATERIALS

Carbon steels, stainless steels, and many nonferrous metals—aluminum, magnesium, and nickel alloys—are commonly used in stud welding applications. Various alloys of brass, bronze, copper-nickel, chromium, and many others can be used in special applications for which the properties of these metals are needed.

Low-Carbon Steel

Low-carbon (mild) steels can be arc stud welded with no major metallurgical problems. The upper carbon limit for steel to be arc stud welded without preheat is usually 0.30%. If workpiece sections are comparatively thin relative to the stud diameters being welded (thinner than the diameters listed in Table 9.2), the carbon limit may be somewhat higher because of the decreased cooling effect of the workpiece. The most important factor relating to workpiece section thickness is that the material must be heavy enough to permit the welding of studs without melt-through.

Medium- and High-Carbon Steel

If medium- and high-carbon steels are to be stud welded, it is imperative that preheat be used to prevent cracking in the heat-affected zone. In some instances, a combination of preheating and postheating may be necessary to obtain satisfactory results. In cases where the welded assemblies are to be heat-treated after welding, the preheating or postheating operation can be eliminated if the assemblies are handled in a manner that prevents damage to the studs.

Preheat or postheat, or both, may be applied to small workpieces with a limited number of studs using a gas torch at each stud weld location. The temperature at each weld location is checked using temperature-indicating crayons. Large workpieces may require the use of an annealing furnace or an electrically wired insulating blanket. Furnaces and insulating blankets are usually equipped with temperature-sensing and reporting devices.

Low-Alloy Steel

Generally, the high-strength low-alloy steels are satisfactorily stud welded when the carbon content is 0.15% or lower. If carbon content exceeds 0.15%, it may be necessary to use a low preheat temperature to obtain desired toughness in the weld area.

When the hardness of the heat-affected zone does not exceed 30 on the Rockwell C hardness test, studs can be expected to perform well under almost any type of severe service. Although good results have been obtained when hardness ranges up to 35 Rockwell C, it is best to avoid extremely high working stresses and fatigue loading. For special cases in which microstructure properties are important, the weld should be evaluated and qualified for the specific application. Since alloy steels vary in toughness and ductility at high hardness levels, weld hardness should not be used as the sole criterion for weld evaluation.

The stud weld quality may be evaluated by performing a Stud Application Qualification Requirement test per *Structural Welding Code—Steel*, AWS D1.1 Section 7.6.[5] Additional testing such as cyclic load stress versus loading cycles may be indicated if repetitive impulse loading rather than static loading is anticipated in service.

Heat-Treated Structural Steel

Many structural steels used in shipbuilding and in other construction are heat-treated at the mill during processing. Heat-treated steels require that attention be given to the metallurgical characteristics of the heat-affected zone. Some of these steels are hardenable to the extent that the heat-affected zones will become martensitic. When this occurs, the structure will be quite sensitive to underbead cracking, and it will have insufficient ductility to carry impact loads. Therefore, for maximum toughness in these steels, a preheat of 370°C (700°F) is recommended. These steels can be arc stud welded, but consideration of the application and the service requirements of the stud will influence the welding procedures to be followed.

Stainless Steels

Most classes of stainless steel can be arc stud welded. The exceptions are the free-machining grades. However, only the austenitic stainless steels (3XX grades) are recommended for general application. The other types are subject to air hardening, and they tend to be brittle in the weld area unless heat-treated after welding. The weldable stainless steel grades include AISI Types 304, 305, 308, 309, 310, 316, 321, and 347. Types 302 HQ, 304, 305, and 316 are most commonly used for stud welding applications. The low-carbon grades of these stainless steels are also weldable.

Stainless steel studs can be welded to stainless steel or to mild steel as the application may require. The welding setup used is the same as that recommended for low-carbon steel except that an increase of approximately 10% in power is required. When stainless steel studs are welded to mild steel, best weld results are achieved when the carbon content of the base metal does not exceed 0.20%. When welding stainless steel studs to mild steel with 0.20% to 0.28% carbon or to low-carbon hardenable steels, types 308, 309, or 310 studs are recommended for better results. Because of the composition of the weld metal when chromium-nickel alloy studs are welded to mild steel, the weld zone may be quite hard. The hardness will depend on the carbon content in the base metal and whether the solidified molten metal is predominantly austenitic. It is possible to overcome this by using studs with a high alloy content such as Type 309 or 310. When welding stainless steel studs to mild steel it is also recommended that a fully annealed stainless steel stud or a stud produced from annealed-in-process material be used. The finished stud should not exceed 90 Rockwell B in order to prevent the possible occurrence of intergranular stress corrosion resulting from carbide precipitation in the weld zone, as is the case with non-annealed studs. Intergranular stress corrosion produces brittle weld failures under repetitive loading conditions.

Aluminum

The basic approach to aluminum stud welding is similar to that used for mild steel stud welding. The power sources, stud welding equipment, and controls are the same. The stud welding gun is modified slightly by the addition of a dampening device to control the plunging rate of the stud at the completion of the weld time. In addition, the foot ferrule holder is modified by adding a special gas adapter. This is used to contain the high-purity inert shielding gas during the weld cycle. Argon is generally used, but helium may be useful with large studs to take advantage of the higher arc energy it creates.

Direct current electrode positive (DCEP) is used on aluminum with the stud (the electrode) positive and the workpiece negative. An aluminum stud differs from a steel stud in that no flux is used on the weld end of the stud. The weld end has a cylindrical or cone-shaped projection that helps to initiate the long arc used for aluminum stud welding. The projection dimensions on the welding end are proportionally designed for each size stud to provide the best results.

Weld base diameters of studs for aluminum range from 6.4 mm to 13 mm (1/4 in. to 1/2 in.). The sizes and shapes are similar to steel studs.

Aluminum studs are commonly made of aluminum-magnesium alloys, including 5086 and 5356, that have a typical tensile strength of 275 MPa (40 ksi). These alloys have high strength and good ductility. They are metallurgically compatible with the majority of aluminum alloys used in industry.

In general, all aluminum plate alloys of the 1100, 3000, and 5000 series are considered excellent for stud welding; alloys of the 4000 and 6000 series are considered fair; and the 2000 and 7000 series are considered poor. The minimum aluminum plate thickness to which aluminum studs may be welded, with and without backup, are listed in Table 9.2. Typical arc stud welding conditions for aluminum studs are presented in Table 9.3. A cross section of an aluminum alloy stud weld is shown in Figure 9.12.

5. American Welding Society (AWS) Committee on Structural Welding, *Structural Welding Code—Steel,* AWS D1.1/D1.1M:2002, Miami: American Welding Society, Section 7.6.

Table 9.3
Typical Conditions for Arc Stud Welding of Aluminum Alloys

Stud Base Diameter, mm (in.)		Weld Time, seconds	Welding Current, Amperes[a]	Shielding Gas Flow, liter/min (ft³/h)[b]	
6.4	1/4	0.33	250	7.1	15
7.9	5/16	0.50	325	7.1	15
9.5	3/8	0.67	400	9.4	20
11.1	7/16	0.83	430	9.4	20
12.7	1/2	0.92	475	9.4	20

a. The currents shown are actual welding currents; they do not necessarily correspond to dial settings on the power source.
b. The shielding gas is composed of 99.95% pure argon.

Figure 9.12—Cross Section of a 9.5 mm (3/8 in.) Diameter Type 5356 Aluminum Alloy Stud Welded to a 6.4 mm (1/4 in.) Type 5053 Aluminum Alloy Plate

Magnesium

The gas shielded arc stud welding process used for aluminum also produces high-strength welds in magnesium alloys. A ceramic ferrule is not needed. Helium shielding gas and direct current electrode positive should be used. A gun with plunge dampening will avoid spattering and base metal undercutting.

The following breaking loads can be obtained with AZ3 1B alloy studs welded to 6.4 mm (1/4 in.) thick AZ31B or ZE10A base metal: up to 6.7 kilonewtons (kN) (1500 lb) for stud diameters of 6.4 mm (1/4 in.) and up to 20 kN (4500 lb.) for studs 13 mm (1/2 in.) in diameter.

Minimum base metal thicknesses and the appropriate stud diameters that may be attached without melt-through or significant loss in strength are as follows: 6.4 1/4 (in.) base metal will accommodate a stud diameter of 3.2 mm (1/8 in.), and 13 mm and (1/2 in.) base metal will accommodate a stud diameter of 6.4 mm (1/4 in.). If strength is not a consideration, 13 mm (1/2 in.) studs can be welded to 4.8 mm (3/16 in.) plate without melt-through.

Other Materials

On a moderate scale, arc stud welding is applied in industry on various brass, bronze, nickel-copper, and nickel-chromium-iron alloys. The applications are usually very specialized and require careful evaluation to assure suitability of design.

Direct current electrode positive produces best results for the stud welding of nickel, nickel-copper, nickel-chromium-iron, and nickel-chromium-molybdenum alloys. Nickel, nickel-copper, and nickel-chromium-iron alloy stud welds tend to contain porosity and crevices. The mechanical strengths should be confirmed by weld testing to determine if they are acceptable for the intended application. The weld itself should not be exposed to corrosive media.

STUDS

Studs are classified in standards published by the American Society for Testing and Materials (ASTM), the American Welding Society (AWS), the Society of

Table 9.4
Minimum Mechanical Property Requirements for Carbon Steel Studs

	Stud Type A AWS D1.1	Stud Type B AWS D1.1	Stud Type C AWS D1.1
Tensile strength (UTS)	420 MPa (61 ksi)	450 MPa (65 ksi)	552 MPa (80 ksi)
Yield strength (0.2% offset)	340 MPa (49 ksi)	350 MPa (51 ksi)	Not specified
Yield strength (0.5% offset)	Not specified	Not specified	485 MPa (70 ksi)
Elongation (% in 50.8 mm [2 in.])	17%	20%	Not specified
Elongation (% in 5 × diameter)	14%	15%	Not specified
Area reduction	50%	50%	Not specified

Automotive Engineers (SAE), and the International Organization for Standardization (ISO).

Typical low-carbon steel studs have a maximum of 0.23% carbon, 0.90% manganese, 0.040% phosphorus, and 0.050% sulfur. This conforms to *Standard Specifications for Steel Bars, Carbon, Cold Finished, Standard Quality, Grades 1010 through 1020*, ASTM A108-99.[6]

The physical properties of carbon steel studs as required by *Structural Welding Code—Steel,* AWS D1.1/D1.1M[7] are listed in Table 9.4 for three specified types of stud classifications: A, B, and C. The types are identified as follows:

1. Type A studs are general-purpose studs, of any type or size, used for purposes other than shear transfer in composite beam design and construction.
2. Type B studs are headed, bent, or of another configuration, 13 mm (1/2 in.), 16 mm (5/8 in.), 20 mm (3/4 in.), 22 mm (7/8 in.), and 25 mm (1 in.) diameter or larger, used as an essential component in the design of composite beam design and construction.
3. Type C studs are cold worked deformed steel bars manufactured in conformance with *Standard Specifications for Steel Wire, Deformed, for Concrete Reinforcement,* ASTM A-496–90a.[8]

   The nominal diameter is equivalent to the diameter of a plain wire of the same weight per foot as the deformed wire.

The physical properties of Types A and B studs also meet the requirements of the International Organization for Standardization (ISO) standard, *Welding Studs and Ceramic Ferrules for Arc Stud Welding,* ISO 13918.[9]

The specified minimum tensile strength for stainless steel studs that are made from stainless materials according to the American Society for Testing and Materials is listed in *Structural Welding Code–Stainless Steel*, AWS D1.6, and are shown in Table 9.4.[10]

Studs are required to meet the standard, *Stainless and Heat-Resisting Steel for Cold Heading and Cold Forging–Bar and Wire,* ASTM A493-95, (Reapproved 2000),[11] or *Stainless and Heat-Resisting Steel Bars and Shapes*, ASTM 276-2000a, chemical specifications for Grades XM-7, 304, 305, 309, 310 or 316 or the low-carbon versions of those alloys. The mechanical property requirements of this standard for stainless steel studs are listed in Table 9.5.[12]

High-strength studs that meet the Society of Automotive Engineers' (SAE) steel fastener Grade 5 mini-

6. American Society for Testing and Materials (ASTM) Subcommittee A108, 1999, *Standard Specifications for Steel Bars, Carbon, Cold Finished, Standard Quality, Grades 1010 through 1020*, A108-99, West Conshohocken, Pennsylvania: American Society for Testing and Materials.

7. American Welding Society (AWS) Committee on Structural Welding, 2002, *Structural Welding Code—Steel,* AWS D1.1/D1.1M:2002, Miami: American Welding Society.

8. American Society for Testing and Materials (ASTM) Subcommittee A496-90a, 1998, *Standard Specifications for Steel Wire, Deformed, for Concrete Reinforcement,* A496-90a, West Conshohocken, Pennsylvania: American Society for Testing and Materials.

9. International Organization for Standardization (ISO), Technical Committee TC-44, Subcommittee SC-10, 1998 *Welding Studs and Ceramic Ferrules for Arc Stud Welding*, ISO 13918, Geneva: International Organization for Standardization.

10. American Welding Society (AWS) Committee on Structural Welding, 1999, *Structural Welding Code—Stainless Steel*, AWS D1.6-99, Miami: American Welding Society.

11. American Society for Testing and Materials, (ASTM), 2000, *Stainless and Heat-Resisting Steel for Cold Heading and Cold Forging–Bar and Wire*, A-493-95 (Reapproved 2000), West Conshohocken, Pennsylvania: American Society for Testing and Materials.

12. American Society for Testing and Materials, (ASTM), 2000, *Stainless and Heat-Resisting Steel Bars and Shapes*, ASTM A-276-2000a, West Conshohocken, Pennsylvania: American Society for Testing and Materials.

Table 9.5
Minimum Mechanical Property Requirements for Stainless Steel Studs

Tensile strength (UTS)	490 MPa (70 ksi)
Yield strength (2% offset)	245 MPa (70 kpsi)
Elongation (% in 50.8 mm [2 in.])	40%
Area reduction (%)	Not specified

mum yield strength of 635 MPa (92 ksi) and minimum tensile strength of 825 MPa (120 ksi) are also available, as described in *Mechanical and Material Requirements for Externally Threaded Fasteners,* SAE J429.[13] These studs are made of carbon steels that are heat treated or cold worked to meet the strength requirement. Procedures for welding and for base material selection and preparation must be stringently followed.

Low-carbon steel and stainless steel studs require a quantity of welding flux within the stud or permanently affixed to the end of the stud. The main purposes of the flux are to deoxidize the weld metal and to stabilize the weld arc. Figure 9.13 shows various methods for securing the flux to the base of the stud.

Aluminum studs do not use flux on the weld end. The studs usually have a small projection on the weld end to aid arc initiation. Gas shielding with argon or helium is required to prevent oxidation of the weld metal and to stabilize the arc.

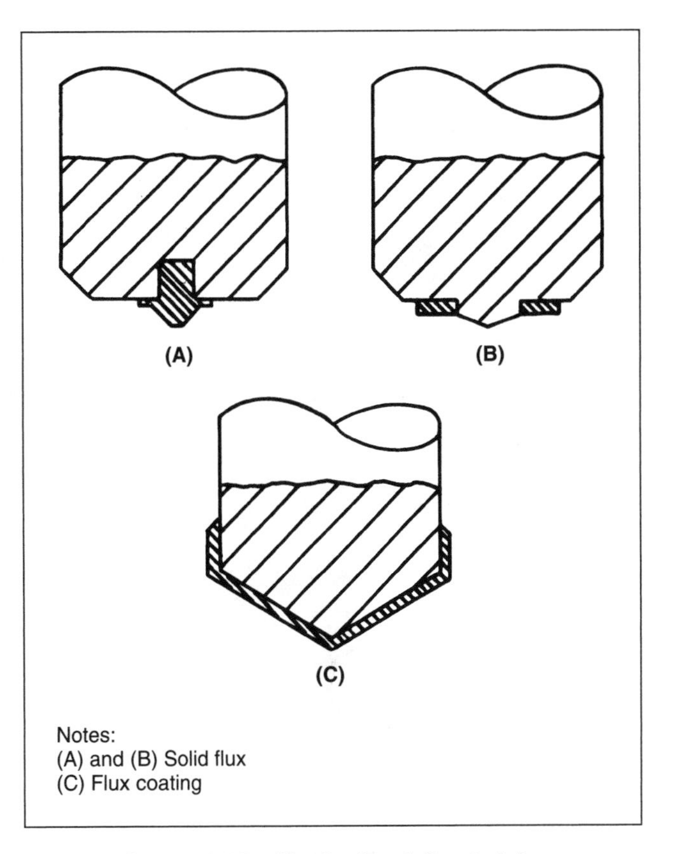

Figure 9.13—Methods of Containing Flux on the End of a Welding Stud

Stud Designs. Most stud bases are round. However, applications may require a square or rectangular stud base. To obtain satisfactory weld results with rectangular studs, the width-to-thickness ratio at the weld base should not exceed five to one. Figure 9.14 displays a variety of sizes, shapes, and types of stud weld fasteners. These include conventional straight threaded studs, and in addition, eyebolts, J-bolts, headed anchors, and punched, slotted, grooved, bent, and pointed studs.

Stud designs are limited in that welds can be made on only one end of a stud; the stud must be shaped so that a ceramic ferrule (arc shield) that fits the weld base can be produced; the cross section of the stud weld base must be within the range that can be stud welded with available equipment; and the stud size, shape, and length must permit chucking or holding the stud for welding. A number of standard stud designs are produced commercially. Stud manufacturers can provide information on both standard and special designs for various applications.

An important consideration in designing or selecting a stud is to recognize that some of its length will be lost during welding as the stud and the base metal melt. Some of the molten metal is expelled from the joint. The stud length reductions shown in Table 9.6 are typical, but they may vary to some degree depending on the materials, geometries, and weld settings involved.

Ferrules

Ferrules are usually made of a ceramic material. They are designed to be used only once and if necessary, can be easily removed by breaking them. Ferrule size is minimized for economy, and the dimensions are optimized for the application. A standard ferrule is generally cylindrical in shape and flat across the bottom for welding to flat surfaces. The base of the ferrule is serrated to vent gases expelled from the weld area. Its internal shape is designed to form the expelled molten metal into a cylindrical flash around the base of the

13. Society of Automotive Engineers (SAE), Iron and Steel Technical Committee, 2001, *Steel Fasteners*, Vol. 1 of Iron and Steel Handbook, Warren, Pennsylvania: Society of Automotive Engineers.

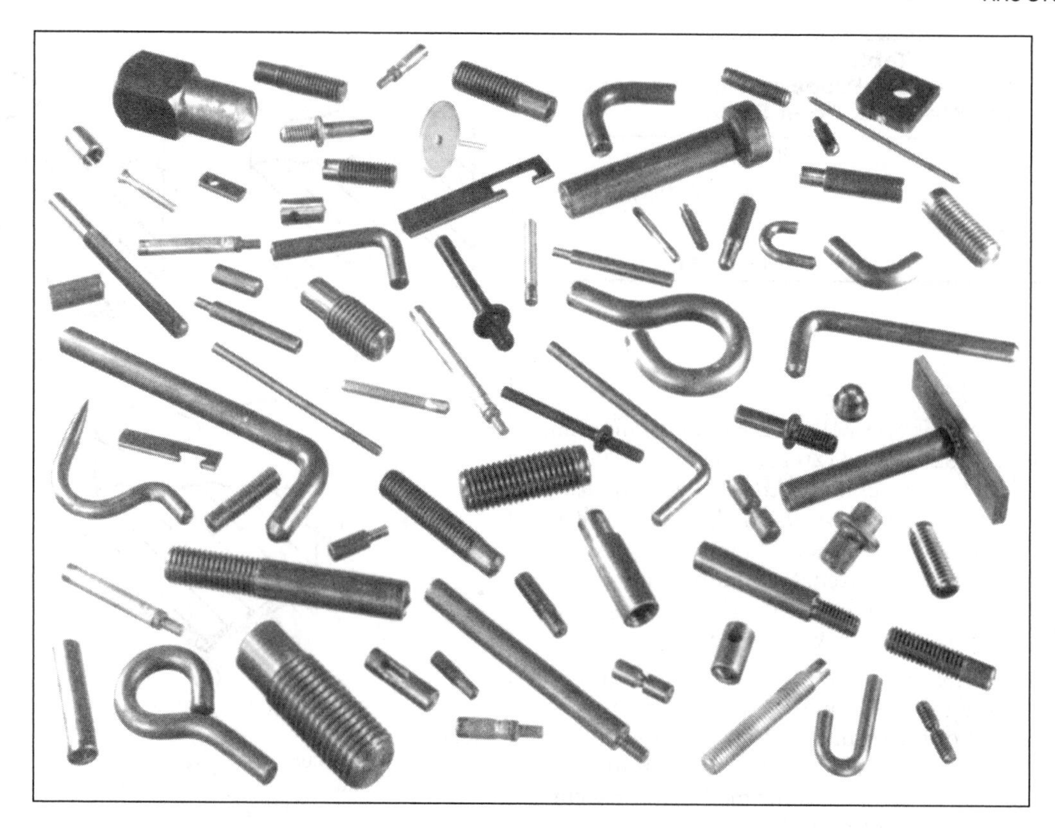

Figure 9.14—Round, Square or Rectangular Cross Section Studs and Fastening Devices Commonly Used for Arc Stud Welding

Table 9.6
Typical Length Reductions of Studs in Arc Stud Welding

Stud Diameters		Length Reductions	
mm	in.	mm	in.
5–12.7	3/16–1/2	2.4–3	3/32–1/8
16–22	5/8–7/8	4–5	5/32–3/16
25 and over	1 and over	5–6	3/16–1/4

stud. Special ferrule designs are used for special applications such as welding at angles to the workpiece, welding to the edges of steel plates, welding to the base plate in the vertical position, welding to contoured surfaces, and welding to round or square tubing. Ferrules for such applications are designed so that the bottom faces match the required surface contours of the base material.

Ferrules are required for most arc stud welding applications. A ferrule is placed over the stud at the weld end where it is held in position by a grip, or holder, on the stud welding gun. The ferrule performs the following important functions during welding:

1. Concentrates the heat of the arc in the weld area;
2. Restricts the flow of air into the area, reducing oxidation of the molten weld metal;
3. Confines the molten metal to the weld area;
4. Prevents the charring of adjacent non-metallic materials.

The ferrule also shields the operator from the welding arc. However, safety glasses with No. 3 filter lenses are recommended for eye protection. (See Appendix A for additional lens selection information.)

One of the important functions of the ferrule is to control weld "flash." Flash is material that is melted during welding and solidifies around the stud base. The term *flash* is used to distinguish this weld metal from conventional fillet weld metal, because flash is formed in a different manner. When properly formed and contained, the flash indicates complete fusion over the full cross section of the stud base. It also suggests that the weld is free of contaminants and porosity. The stud weld flash may not be fused along its vertical or horizontal legs, but this incomplete fusion is not considered detrimental to the stud weld joint quality because the weld flash metal is in excess of the area required for full strength.

When a ferrule is used, the dimensions of the flash are closely controlled by the design of the ferrule. Since the diameter of the flash is generally larger than the diameter of the stud, some consideration is required in the design of the components to be joined. Counterbore and countersink dimensions are commonly used to provide clearance for the flash of round studs, as shown in Table 9.7. The size and shape of the flash varies with stud material and ferrule clearance. Therefore, test welds should be made and checked. Three other methods of accommodating flash, (A) the use of oversized clearance holes, (B) gaskets, and (C) hold-down clips, are shown in Figure 9.15.

Stud Locating Techniques

The method of positioning and locating studs depends on the intended use of the studs and the accuracy of location required. For applications in which extreme accuracy is required, special locating fixtures and fixed (production-type) stud welding equipment is recommended. The extent of tooling necessary is a function of the required production rate as well as total unit production.

Several methods and procedures are used for positioning studs with a portable stud welding gun. The simplest and most common procedures are to lay out the work or employ a template to mark locations with a center punch. A stud is then located by placing the point of the stud in the punch mark. Cover plates that have been punched or drilled can be used as templates. Although operator skill is always a factor in accuracy, location tolerances of ±1.2 mm (± 3/64 [0.050] in.) can be maintained. When a number of pieces are to be stud welded, the common practice is to weld directly through the holes in a template without preliminary

Table 9.7
Weld Flash Clearances for Arc Stud Welds

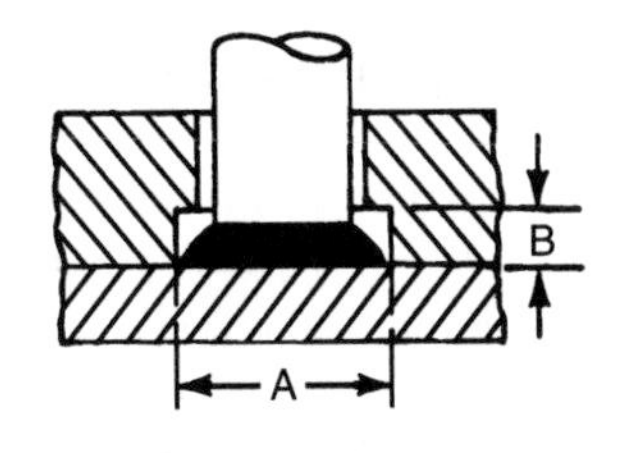

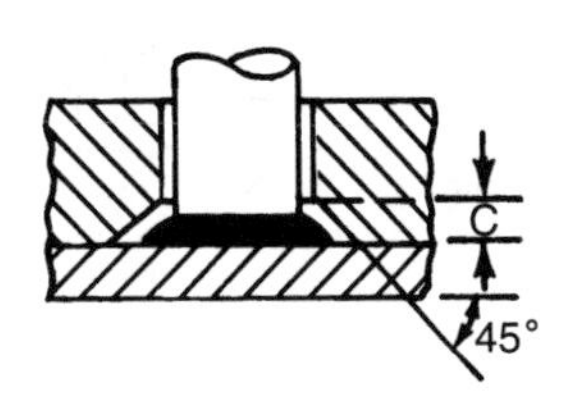

Key:
A = Nominal diameter of counterbore or countersink to accommodate flash for stud diameters.
B = Nominal depth of counterbore to accommodate stud flash height.
C = Same as B, except for countersunk hole instead of counterbored hole.

Stud Base Diameter		Counterbore				45° Countersink	
		A		B		C	
mm	in.	mm	in.	mm	in.	mm	in.
6.4	1/4	11.1	0.437	3.2	0.125	3.2	0.125
7.9	5/16	12.7	0.500	3.2	0.125	3.2	0.125
9.5	3/8	15.1	0.593	3.2	0.125	3.2	0.125
11.1	7/16	16.7	0.656	4.7	0.187	3.2	0.125
12.7	1/2	19.1	0.750	4.7	0.187	4.7	0.187
15.9	5/8	22.2	0.875	5.5	0.218	4.7	0.187
19.1	3/4	28.6	1.125	7.9	0.312	4.7	0.187

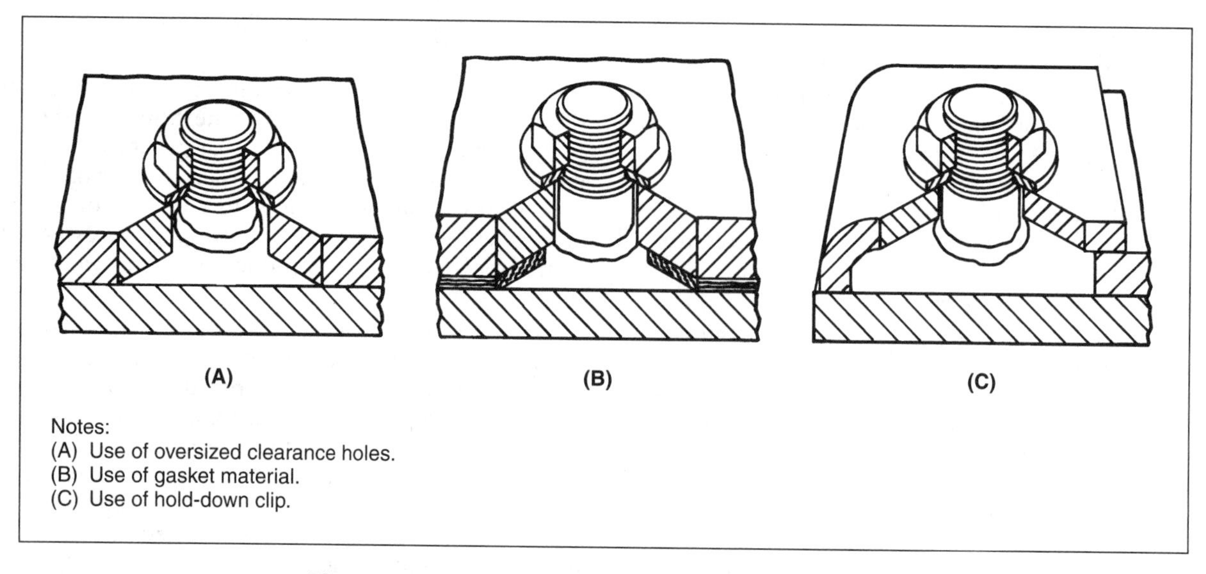

Figure 9.15—Methods of Accommodating Flash

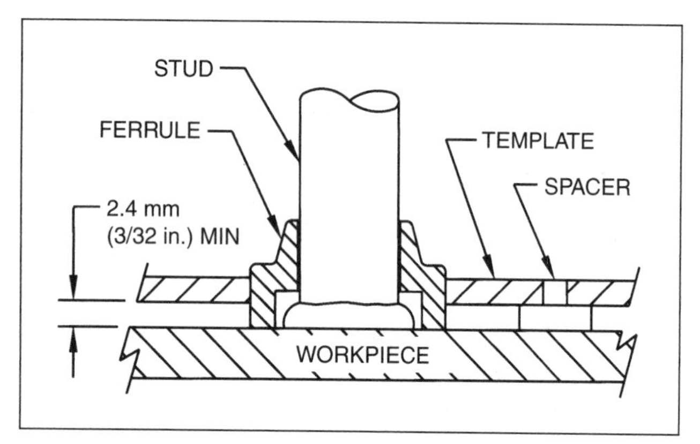

Figure 9.16—Simple Template Used to Locate Studs Within ±0.8 mm (±1/32 in.)

marking, as shown in Figure 9.16. A simple template positions the stud by locating the ferrule. Because of manufacturing tolerances on ferrules, the tolerance on stud location with this method is usually ±0.8 mm (±0.030 [1/32] in.).

When more accurate stud location and alignment are required, a tube template is used. The stud is centered indirectly by inserting a tube adapter on the gun in a locating bushing in the template. Figure 9.17 illustrates this type of template. The template has a hardened and ground bushing with a closely machined tube adapter. Because standard ferrule grips are used with this adapter, standardization of templates is possible. It is only necessary to change ferrule grips to weld studs of different diameters. With a tube template, a tolerance of ±0.4 mm (±0.015 [1/64 in.]) can be held on stud location. This method also maintains perpendicular alignment of the stud.

Welding Current and Time Relationships

The current and time required for arc stud welds are dependent on the cross-sectional area of the stud.

The same total energy input can be obtained by coordinating a range of current and time settings. It is possible, within certain limitations, to compensate for low or high welding current by changing the weld time in the opposite direction. There is a broad range of combinations for welding each stud size. Under some conditions, such as welding studs to a vertical member or to thin-gauge material, the allowable range is much smaller.

A helpful initial approach to adjusting the time/current relationship is Joule's law, as follows:

$$H = i^2Rt \tag{9.2}$$

where:

H = Heat input in joules

i = Current in amperes

R = Resistance in ohms

t = Time in seconds

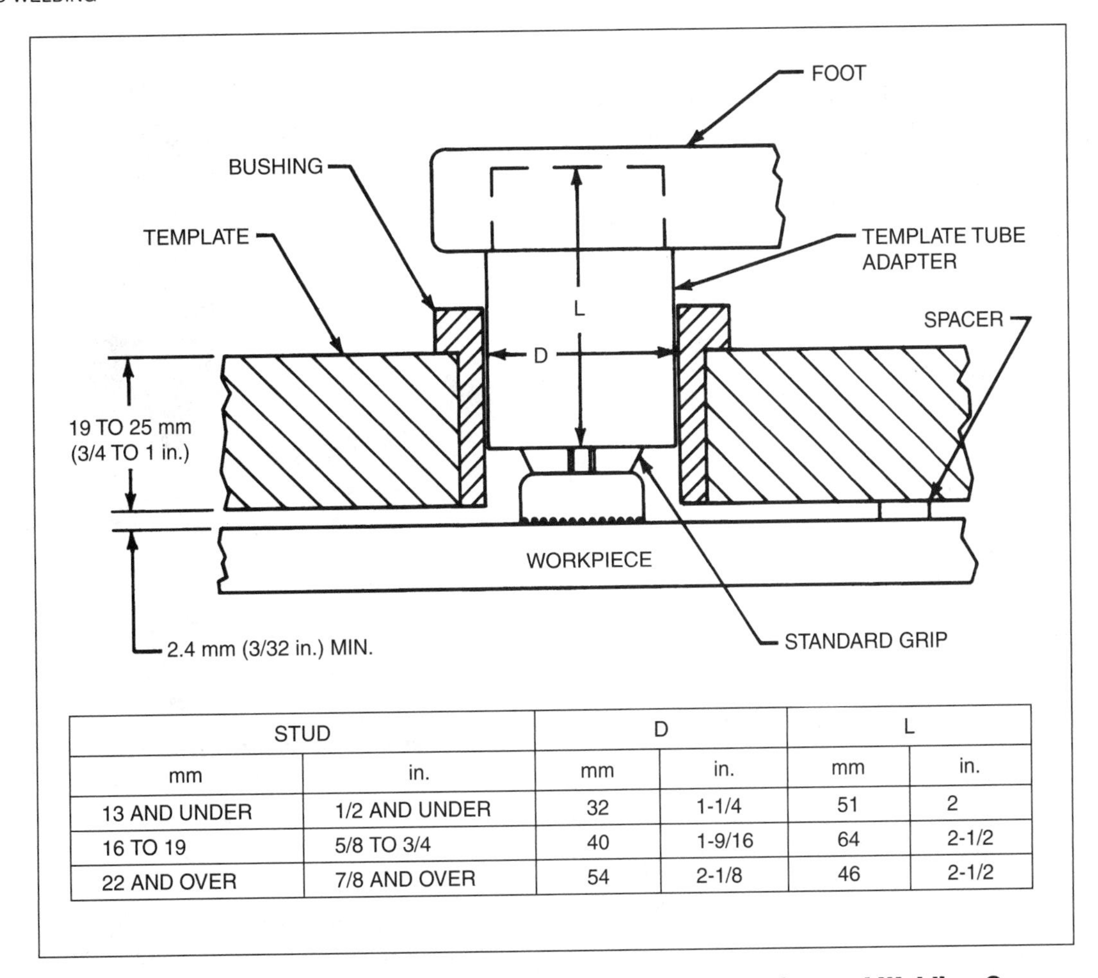

STUD		D		L	
mm	in.	mm	in.	mm	in.
13 AND UNDER	1/2 AND UNDER	32	1-1/4	51	2
16 TO 19	5/8 TO 3/4	40	1-9/16	64	2-1/2
22 AND OVER	7/8 AND OVER	54	2-1/8	46	2-1/2

Figure 9.17—Template with Hardened and Ground Bushing and Welding Gun Adapter Used to Locate Studs Within ±0.4 mm (±0.015 [±1/64] in.)

In the examples below, the same figure of 1 ohm is used because there would be no change in cables and, therefore, in resistance.

In one case, a 12.7 mm (1/2 in.) stud may have a nominal weld setting of 800 A and 0.55 seconds, so the heat input would be:

$$H = 8002 \times 1 \times 0.55 = 352{,}000 \text{ joules (333. BTU)}$$

If a shorter time, higher current setting is indicated for welding the stud to a workpiece in the vertical position, an initial setting may be calculated as:

$$
\begin{aligned}
H &= 880^2 \times 1 \times tH = 880^2 \times 1 \times t \\
& 352{,}000 = 774{,}000 \times t \\
& 352{,}000/774{,}400 = t \\
& t = 0.45 \text{ seconds}
\end{aligned}
$$

Although energy input is a major factor in obtaining satisfactory welds, it is not the only one involved. Other factors, such as arc blow, plate surface conditions (rust, scale, moisture, plating, or paint), ground clamp position, and operator technique can cause unsatisfactory welds even though the correct weld energy input was used.

Stud weld bases are qualified for use in a 60-stud weld test specified in *Structural Welding Code—Steel*, AWS D1.1/D1.1M, IX, *Manufacturers' Weld Base Qualification Requirements*,[14] where the optimum setting and amperage are determined, and then the current is varied ±10% while the optimum time is retained. Thirty of the studs are welded at one setting and the other 30 at the other setting, and then they are tested to

14. See Reference 4.

destruction by bending and tensile testing. All 60 must fail in the stud shank, or in the workpiece material, not in the weld. Varying the time and amperage too much may cause weld problems such as those illustrated in Figure 9.30(E) and 9.30(F). Good welding practice accompanied by appropriate physical and visual weld inspection should always be followed in establishing the final, optimal weld settings.

Metallurgical Considerations

The metallurgical structures encountered in arc stud welds are generally the same as those found in other arc welds for which the heat of an electric arc is used to melt both a portion of the base metal and the electrode in the course of welding. (In arc stud welding, the stud is the electrode). Acceptable mechanical properties are obtained when the stud and base material are metallurgically compatible. Properly executed stud welds are usually characterized by the absence of inclusions, porosity, cracks, and other defects.

A macrosection of a typical arc stud weld in steel, shown in Figure 9.18, illustrates how metal is pushed to the perimeter of the stud to form flash. The amount of weld metal (cast structure) in the joint is minimal. Because of the short welding cycle, the heat-affected zones common to arc welding are present, but they are small. Chapter 4, "Welding Metallurgy," *Welding Handbook*, Volume 1, 9th Edition, contains helpful information on welding metallurgy that is applicable to the arc stud welding of various materials.[15]

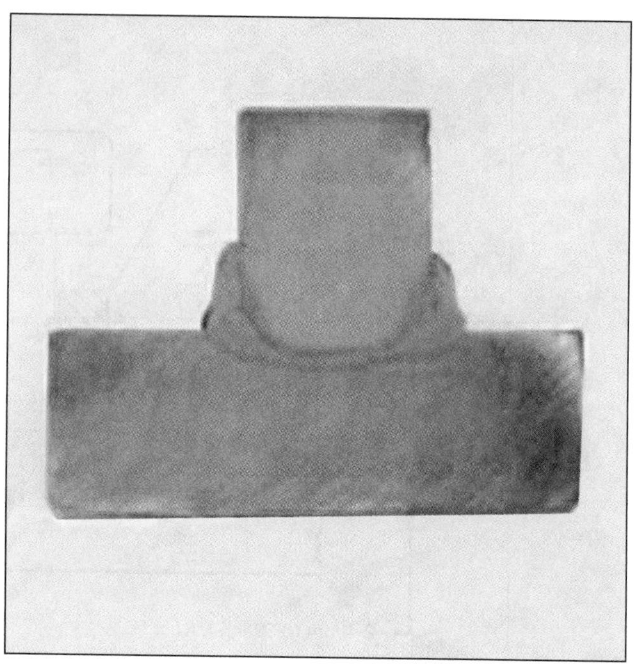

Photograph courtesy of Image Industries, Incorporated

Figure 9.18—Macrosection of a Typical Arc Stud Weld (Steel)

SPECIAL PROCESS TECHNIQUES

Special process techniques employ the basic arc stud welding process, but each is limited to very specific types of applications. One technique is referred to as *gas arc stud welding*; another is called *short cycle stud welding*.

GAS ARC STUD WELDING

The gas arc stud welding method uses the same basic principles as standard arc stud welding. The primary differences are that a shielding gas is used in place of the flux and ceramic ferrules to protect the weld arc, and shorter weld times and higher amperages are used. The gas arc presents no physical barrier like the ceramic ferrule to restrain the arc or contain the weld metal. Therefore, the gas flow rate and method of introducing the gas into the weld area are important to the weld quality. The gas flow is started before the weld cycle starts and is kept on until after the lift-plunge motion of the stud. The gas will protect the arc and molten metal until it has begun to cool and solidify.

The gas arc method is used to weld aluminum, as described previously, and can be used for mild steel and stainless steel studs up to 9.5 mm (3/8 in.) in diameter Shielding gas selection will vary with the material. Gas arc stud welding is much less tolerant of base material surface contamination than arc stud welding where ceramic ferrules are used. For this reason, this process is usually limited to mounted stud guns in production applications where consistent conditions and settings can be maintained.

This process has been used extensively in the cookware industry to weld internally tapped studs to aluminum pans to attach the handles and legs.

SHORT-CYCLE STUD WELDING

Short-cycle stud welding is based on the same principles as standard arc stud welding. The two main differences are that no ceramic ferrule is used to shield

15. American Welding Society (AWS) Welding Handbook Committee, 9th ed., 2001C. Jenney, C. L., and A. O'Brien, eds., 2001, *Welding Science and Technology*, Vol. 1 of *Welding Handbook*, Miami: American Welding Society.

and protect the weld arc or to shape and contain the molten weld metal, and shorter times and higher amperages are used. Shorter weld time reduces the amount of molten metal and the degree of oxidation. The studs used for short-cycle welding do not need to have the flux load that is used on the studs for the standard arc stud welding process. The studs used for the short-cycle method typically have a blunt, nearly flat weld end. The weld end may have a flange that is slightly larger in diameter that the stud diameter. The flange provides a weld area that is larger than the stud shank area. This ensures that the weld will have adequate strength. These flanges also facilitate the automatic feeding of the studs in high-volume applications.

The weld settings used for the short cycle method are much higher in amperage and significantly shorter in time than for the standard arc stud welding of a stud with the same diameter. Because of the reduced time, short-cycle welding has less tolerance for base material surface irregularities or contamination. Due to the high amperage and the absence of a ceramic ferrule to physically restrain the weld arc and molten metal, proper fixturing and grounding techniques are needed with the short-cycle method to achieve good weld results.

Rectifiers with solid state current and time controls are used for short-cycle stud welding to obtain the precise weld settings needed to weld studs to thin base materials. The short-cycle method is often used in conjunction with automatic stud feeding systems for high-volume automotive and industrial applications. The stud weld guns may be mounted on stationary slide assemblies or on robots. The short-cycle stud welding method can be used in low-volume applications using manually fed hand-held guns.

Time settings for short-cycle and short-cycle gas arc stud welding are much lower than those for standard electric arc stud welding. Typical settings are shown in Table 9.8. As a result of the lower time setting, the short-cycle methods may produce porosity in the finished weld. Figure 9.19 shows porosity in a short-cycle weld made without a ferrule. Welds with less porosity are produced by using shielding gas with the short-cycle technique, as shown in Figure 9.20.

Because of the shorter weld time, these welds are subject to brittleness and may not develop full strength or be suitable for repetitive loading applications. It is necessary to test these welds for suitability for the specific application being considered.

CAPACITOR DISCHARGE STUD WELDING

Capacitor discharge stud welding is an arc stud welding process variation in which direct current (dc) arc power is produced by a rapid discharge of stored electrical energy to the stud gun. Pressure is applied during or immediately following the electrical discharge by an internal spring or drop-weight system. The power source for this process is an electrostatic storage system in which the weld energy is stored in capacitors. No ferrule or flux is required.

FUNDAMENTALS

The three methods of capacitor discharge stud welding are initial contact, initial gap, and drawn arc. They differ primarily in the manner of arc initiation. Studs used with the initial contact and initial gap methods are specially designed with a small projection (tip) on the weld end of the stud to establish the weld arc. The drawn arc method utilizes a pilot arc created as the stud is lifted off the workpiece by the stud gun.

Table 9.8
Short-Cycle Welding Settings for Mild and Stainless Steel Studs and Base Materials

Diameter	Area	All Welding Positions			
mm (in.)	mm² (in.²)	Current, A	Time, s	Lift mm (in.)	Plunge mm (in.)
4 (0.156)	12.5 (0.019)	400	0.060	1.02 (0.040)	1.57 (0.062)
5 (0.198)	19.6 (0.031)	500	0.070	1.27 (0.050)	1.57 (0.062)
6 (0.237)	28.30 (0.044)	600	0.080	1.27 (0.0050)	1.57 (0.062)
7 (0.276)	38.5 (0.060)	650	0.090	1.57 (0.062)	1.57 (0.062)
8 (0.315)	50.30 (0.078)	700	0.100	1.57 (0.062)	2.03 (0.080)
9 (0.354)	63.6 (0.098)	800	0.120	1.57 (0.062)	2.54 (0.100)
10 (0.394)	78.5 (0.122)	900	0.150	1.78 (0.070)	2.54 (0.100)

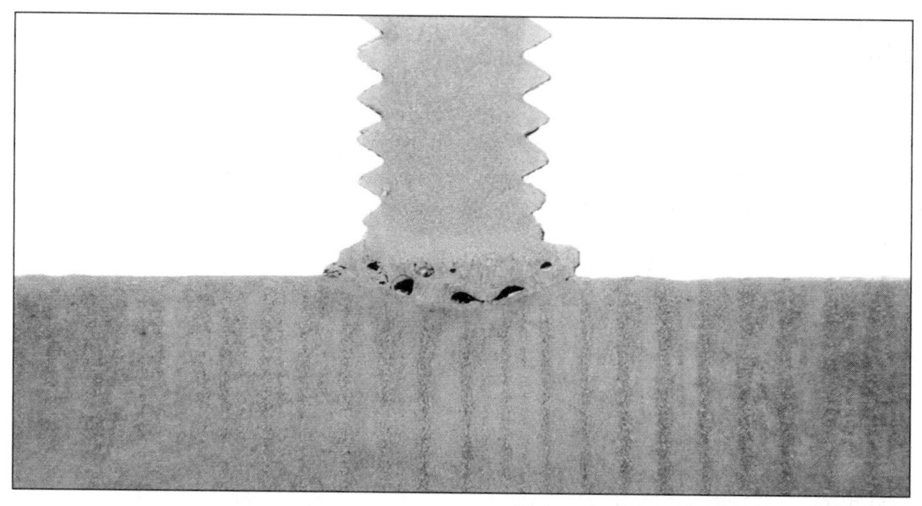

Photograph courtesy of Image Industries, Incorporated

Figure 9.19–Porosity in a Short-Cycle Weld Made without Ferrule or Gas

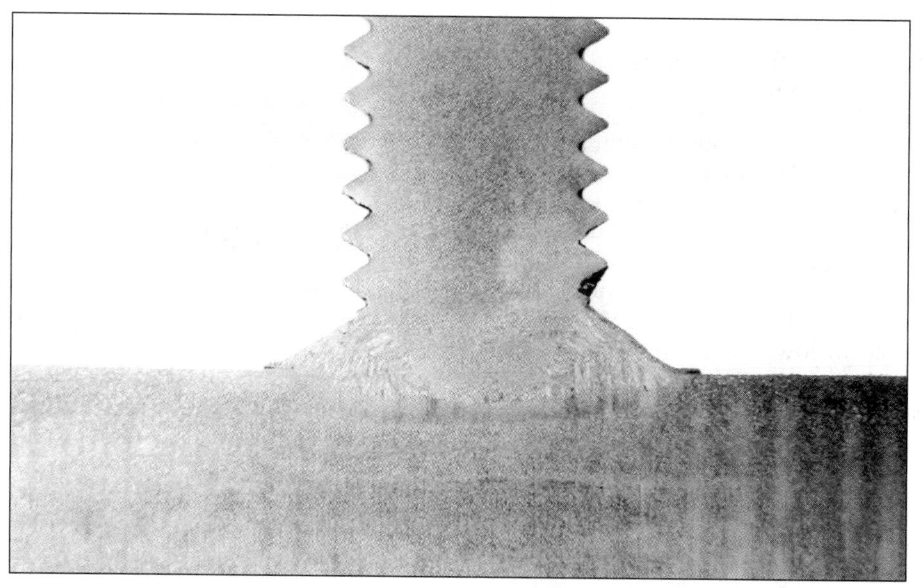

Photograph courtesy of Image Industries, Incorporated

Figure 9.20—Short-Cycle Arc Stud Weld Made with Shielding Gas

Initial Contact Method

In initial contact stud welding, a stud with a small projection on the weld end of the stud is placed against the workpiece, as shown in Figure 9.21(A). The stored energy is then discharged through the projection on the base of the stud. The small projection presents a high resistance to the stored energy, and it rapidly disintegrates from the high current density, as shown in (B). This creates an arc that melts the surfaces to be joined. During arcing (C), the pieces to be joined are brought together by the action of a spring, weight, or a pressurized air cylinder. When the two surfaces come in contact, fusion takes place (D), and the weld is complete (E).

Initial Gap Method

The sequence of events in initial gap stud welding is shown in Figure 9.22. At the start of the weld, the stud is positioned off the workpiece, leaving an arc-length

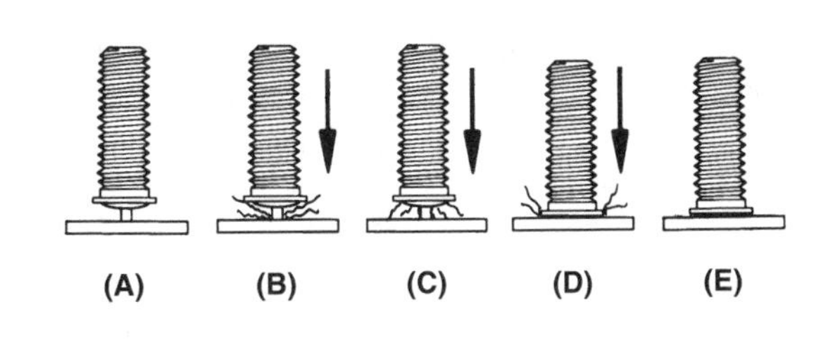

Notes:
(A) Stud is positioned.
(B) Resistance to stored energy produces an arc.
(C) Stud end and workpiece melt.
(D) Stud fuses on contact.
(E) Weld is complete.

Figure 9.21—Steps in Initial Contact Capacitor Discharge Stud Welding

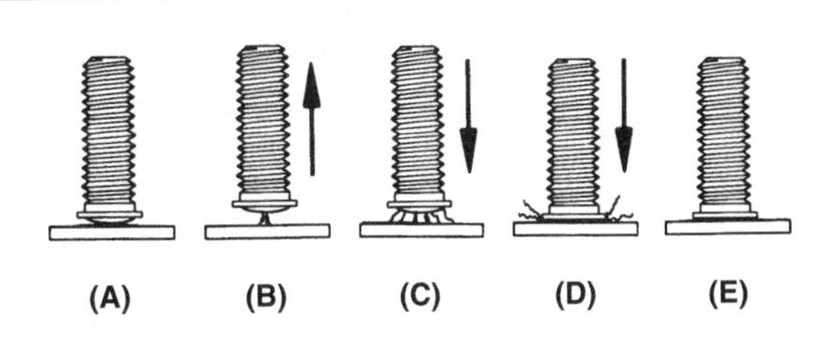

Notes:
(A) The stud is positioned.
(B) Energized welding circuit lifts the stud and initiates a low-amperage pilot arc.
(C) The deenergized circuit drops the stud; capacitors discharge, melting stud end and weld site.
(D) Stud is plunged into the workpiece.
(E) The weld is complete.

Figure 9.22—Steps in Drawn Arc Capacitor Discharge Stud Welding

space between it and the workpiece (A). The stud is released and continuously moves toward the workpiece under gravity, air pressure, or spring loading (B). At the same time, the voltage in the capacitors is applied between the stud and the workpiece. When the stud contacts the workpiece, high current flashes off the tip and initiates an arc (C). The arc melts the surfaces of the stud and workpiece (D), as the stud continues to move forward. Finally, the stud plunges into the workpiece (E), and the weld is complete (F).

With the correct selection of electrical characteristics of the weld circuit and the size of the projection, it is possible to produce a high-current arc of such short duration (about 0.001 second) that its effect on the stud and workpiece is very superficial. A surface layer only a few hundredths of a millimeter (thousandths of an inch) on each surface reaches the molten state.

Drawn Arc Method

In the drawn arc method, arc initiation is accomplished in a manner similar to that of arc stud welding. The stud does not require a projection on the weld face. An electronic control is used to sequence the operation. The welding gun is similar to the conventional arc stud welding gun and is used to lift and plunge the stud. Weld time is controlled by an electronic circuit in the unit.

The operating sequence is shown in Figure 9.22. The stud is positioned against the workpiece and the trigger switch on the stud welding gun is actuated, energizing the welding circuit and a solenoid coil in the gun body (A). The energized coil lifts the stud from the workpiece drawing a low-amperage pilot arc between them (B). When the lifting coil is deenergized, the stud starts to return to the workpiece. The welding capacitors are then discharged across the arc. The high amperage from the capacitors melts the end of the stud and the workpiece surface (C). The spring pressure of the welding gun plunges the stud into the molten metal (D) and the weld is complete (E).

DESIGNING FOR CAPACITOR DISCHARGE STUD WELDING

The capability of the capacitor discharge method to weld studs to thin sections provides important options to designers. Material as thin as 0.75 mm (0.030 in.) can be welded without melt-through. Studs have been successfully welded to some materials (stainless steel, for example) as thin as 0.25 mm (0.010 in.).

Another design advantage of this stud welding process is its capability of welding studs to dissimilar metals. The penetration into the workpiece from the arc is so shallow that there is very little mixing of the stud metal and workpiece metal. Steel welded to stainless steel, brass to steel, copper to steel, brass to copper, and aluminum to die-cast zinc are a few of the combinations that can be stud welded. Many other unusual metal combinations not normally considered weldable by fusion processes are possible with capacitor discharge stud welding.

An additional benefit of this process is the elimination of postweld cleaning or finishing operations on the face surface (the side of the base metal opposite to the stud attachment). Accordingly, the process can be used on workpieces with opposite side surfaces that have already been painted, plated, polished, or coated with ceramic or plastic.

Stud Materials

The materials that are commonly welded with the capacitor discharge method are low-carbon steel, stainless steel, aluminum, and brass. Low-carbon steel and stainless steel studs are usually the same compositions as those used for arc stud welding. For aluminum, 1xxx and 5xxx series alloys are generally used. Copper alloy studs most often used are brass compositions No. 260 and No. 268.

Stud Designs

Stud designs for capacitor discharge stud welding range from standard round shapes to complex forms for special applications. The weld base of the fastener is usually round. The shank can be almost any shape or configuration. These include threaded, plain, round, square, rectangular, tapered, grooved, and bent configurations or flat stampings. The size range is 1.6 mm to 12.7 mm (1/16 in. to 1/2 in.) diameter, with the great majority of attachments falling in the range of 3.2 mm to 9.5 mm (1/8 in. to 3/8 in.) diameter.

Initial contact and initial gap capacitor discharge studs require a tip or projection on the weld end. The size and shape of this tip is important because it is one of the variables involved in the achievement of good-quality welds. The standard tip is cylindrical in shape. For special applications, a stud with a conically shaped tip may be used. The detail of the weld base design of the stud is determined by the stud material, the base diameter, and sometimes by the particular application. The weld base is tapered slightly to facilitate the expulsion of the expanding gases that develop during the welding cycle. Usually, the stud weld base diameter is larger than that of the stud. This provides a weld area larger than the stud cross section so the weld joint strength is equal to or higher than that of the stud.

Drawn arc capacitor discharge studs are designed without a tip or projection on the weld end. However, the weld end is tapered or slightly spherical so that the arc will initiate at the center of the base. As with the other capacitor discharge methods, these studs are generally designed with a large base in the form of a flange.

Stud melt-off (reduction in length due to melting) is almost negligible when compared to the arc stud welding method. Stud melt-off is generally in the range of 0.2 mm to 0.4 mm (0.008 in. to 0.015 in.).

EQUIPMENT

Capacitor discharge stud welding requires a combination power control unit, a stud gun, and associated interconnecting cables. Both portable and stationary production units are available.

Solid-state control circuitry provides signals for automatic sequencing of several events during the welding cycle. The events include one or more of the following:

1. Energizing the gun solenoid or air cylinder for initial gap and drawn arc methods,
2. Initiating the pilot arc in the drawn arc method,
3. Discharging the welding current from the capacitor bank at the proper time in the welding sequence,
4. Deenergizing the solenoid or air cylinder of the gun, and
5. Controlling the changing voltage of the capacitor bank.

Portable Units

The hand-held stud gun is usually made of high-impact high-strength plastic. The gun positions and holds the stud for welding. A trigger initiates the welding cycle through a control cable from the power source control unit. The chuck that holds the stud can be changed to accommodate various diameters and shapes of studs.

The power source control unit provides the welding current and contains the necessary circuitry for charging the capacitors. Variable discharge currents are obtained by varying the charge voltage on the capacitors. The welding machine automatically controls the charging and discharging currents. The units generally operate on 115 V, 60-Hz power.

Typical portable capacitor discharge equipment is illustrated in Figure 9.23. The stored energy of such a unit would be in the neighborhood of 70,000 microfarads (mFd) charged to 170 V, and it would be capable of welding studs with a diameter of 6.4 mm (1/4 in.) at a rate of eight to ten per minute.

Stationary Production Equipment

Stationary production equipment consists of an air-actuated, electrically actuated, or gravity-drop stud gun (or guns) mounted above a work surface. The electrical controls for the air systems and for charging the capacitors are usually located under the worktable.

The power-control units are generally designed for a specific application because automatic sequencing of clamping, indexing, and unloading devices may be incorporated. The capacitance of production units ranges from about 20,000 mfd to 200,000 mfd. The capacitor charge voltage does not exceed 200 V, and it is isolated from the stud chuck until welding is initiated. Power input is 230 V or 460 V, single- or three-phase.

High production rates can be obtained with this equipment. Depending on the amount of automation in

Figure 9.23—Portable Capacitor Discharge Power Source and Stud Gun

the fixturing and in the feeding of studs and parts to be welded, up to 45 welds per minute can be made with a single gun.

Automatic Stud Feeding Systems

Capacitor discharge stud welding is well suited for high-speed automatic stud feeder applications because ceramic ferrules are not required. Portable drawn arc capacitor discharge equipment with automatic stud feeding is available for stud diameters ranging from 3.2 mm through 6.4 mm (1/8 in. through 1/4 in.). Using this system, weld rates of approximately 42 studs per minute can be achieved. A capacitor discharge unit is shown in Figure 9.24.

Stud Location. The method of locating studs depends on several factors. These include the accuracy and consistency of positioning required, the type of welding equipment to be used (portable or fixed), the required rate of production, and to some extent, the geometry or shape of the workpiece. In general, the stationary welding unit used in production affords greater precision in stud location than the portable hand-held unit.

Accuracy of location with a portable gun usually depends on the care used in laying out the location(s) on the workpiece. However, with the application of various types of spacers, bushings, and templates, the accuracy range can be within a tolerance of ±0.5 mm (±0.020 in.).

Standard production units are designed to provide tolerance limits of ±0.12 mm (±0.005 in.). Precision location requires not only accurate and well maintained welding equipment and tooling, but also exceptionally precise, high-quality studs.

Energy Requirements. In capacitor discharge stud welding, arc power is obtained by discharging a capacitor bank through the stud to the workpiece. Arc times are significantly shorter and welding currents are much higher than those used for arc stud welding. The very short weld time accounts for the shallow weld penetration into the workpiece and the small change in stud length caused by melt-off.

Depending on the stud size and the type of equipment used, the peak welding current can vary from about 600 A to 20,000 A. The total time to make a weld depends on the welding method used. For the drawn arc method, weld time is in the range of 4 milliseconds to 6 milliseconds. Figure 9.25 illustrates typical current-time relationships for the three capacitor discharge stud welding methods. It should be noted that the arc current required for the initial contact or initial gap method is much higher than for the drawn arc method.

MATERIALS WELDED

In general, the same metal combinations that can be joined by the arc stud welding method can also be joined by the capacitor discharge method. These include carbon steel, stainless steel, and aluminum alloys.

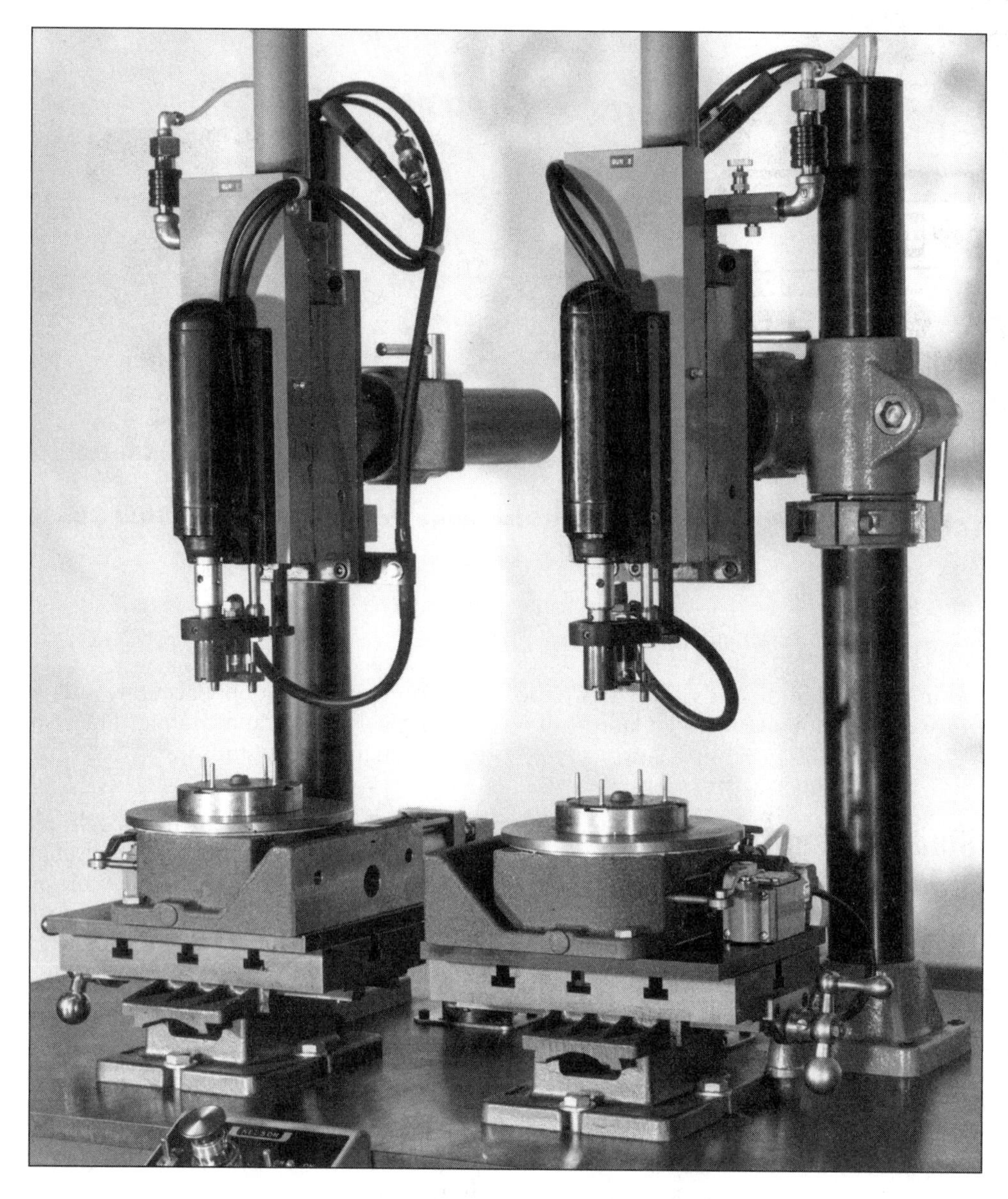

Figure 9.24—Capacitor Discharge Stud Welding Machine with Automatic Stud Feed System

In addition, some dissimilar metal combinations that present metallurgical problems with arc stud welding can be successfully welded with the capacitor discharge stud welding method because of the small volume of metal melted in the very short weld time. The small volume and its expulsion when the stud plunges into the plate result in a very thin layer of weld metal in the joint. If the weld metal is sound and strong, the stud will carry its designed load. Weld metal ductility is not a significant factor.

Weldable stud and base metal combinations of commonly used alloys are listed in Table 9.9. The possible combinations are not limited to these materials. The relative electrical conductivities or melting temperatures of the materials are not of great significance unless there are great differences between them.

Typical macrostructures of capacitor discharge stud welds of dissimilar metals are shown in Figure 9.26. Figure 9.26(A) shows a 5 mm (3/16 in.) steel stud welded to 0.6 mm (0.024 in.) mild steel sheet. In Figure

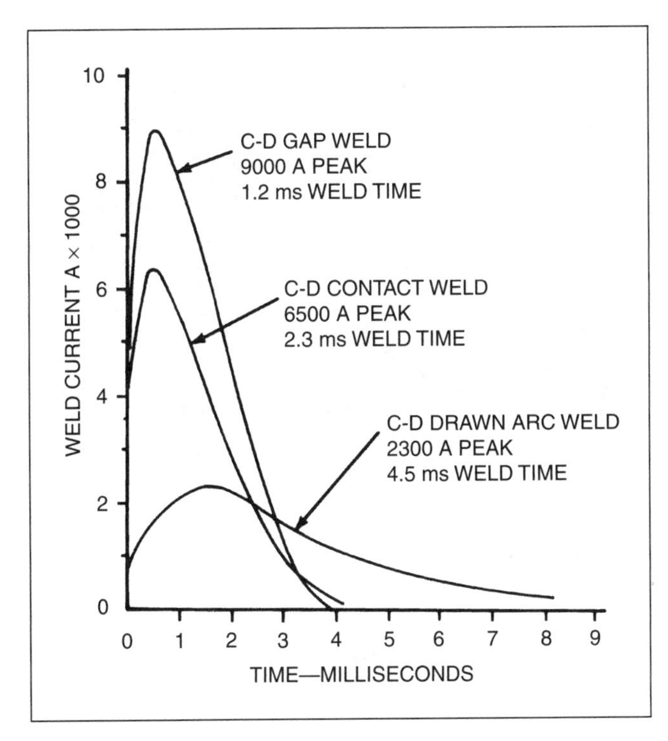

Figure 9.25—Typical Current versus Time Curves for the Three Capacitor Discharge (CD) Stud Welding Methods

9.26(B), a very narrow fusion zone between the 6.4 mm (1/4 in.) diameter brass stud and the 1.6 mm (1/16 in.) mild steel sheet is visible.

Because of the very short weld times, flux or shielding is not normally required to prevent weld metal contamination from air except when stud welding aluminum and some other metals with the drawn arc method. With these metals the arc time is long enough for harmful oxidation to occur, and argon shielding should be used. The stud gun should then be equipped with a gas adapter foot. Welding grade argon (+99.9% pure) should be used at the flow rate recommended by the equipment manufacturer.

Applications

Capacitor discharge stud welding is widely used in many industries, including aircraft and aerospace, appliances, building construction, maritime construction, metal furniture, stainless steel equipment, and transportation. Capacitor discharge welded studs are also used as fasteners or supports where appropriate. An assortment of stud designs are shown in Figure 9.27.

Table 9.9
Typical Combinations of Base Metal and Stud Metal for Capacitor Discharge Stud Welding

Base Metal	Stud Metal
Low-carbon steels, AISI 1006 to 1022	Low-carbon steels, AISI 1006 to 1010, stainless steel, series 300,* copper alloy 260 and 268 (brass)
Stainless steels, series 300* and 400	Low-carbon steels, AISI 1006 to 1010, stainless steel, series 300*
Aluminum alloys, 1100, 3000 series, 5000 series, 6061 and 6063	Aluminum alloys 1100, 5086, and 6063
ETP copper, lead free brass, and rolled copper	Low-carbon steels, AISI 1006 to 1010, stainless steel, series 300,* copper alloys 260 and 268 (brass)
Zinc alloys (die cast)	Aluminum alloys 1100 and 5086

*Except for the free-machining Type 303 stainless steel.

SELECTION OF AN ARC STUD WELDING METHOD

The capabilities of the conventional arc stud welding process and the capacitor discharge stud welding method may overlap in some types of applications, but the choice between the two methods is generally well defined. Table 9.10 presents information on stud welding process selection. If capacitor discharge stud welding has been selected, another determination, usually more difficult, must be made as to which of its modes should be used, i.e., initial contact, initial gap, or drawn arc. The basic criteria for this selection are fastener size, base metal thickness, and base metal composition. It is almost always possible to select the best method using these criteria.

FASTENER SIZE

The capacitor discharge stud welding method is limited to 8 mm (5/16 in.) diameter with hand-held guns and to 9.5 mm (3/8 in.) diameter studs with fixed equipment. The arc stud welding process must be used for applications requiring a portable system and stud diameters over 8 mm (5/16 in.). Applications suitable for the capacitor discharge method with stud diameters in the range of 8.0 mm to 9.5 mm (5/16 in. to 3/8 in.) generally involve thin base materials where avoiding reverse side marking is the foremost requirement.

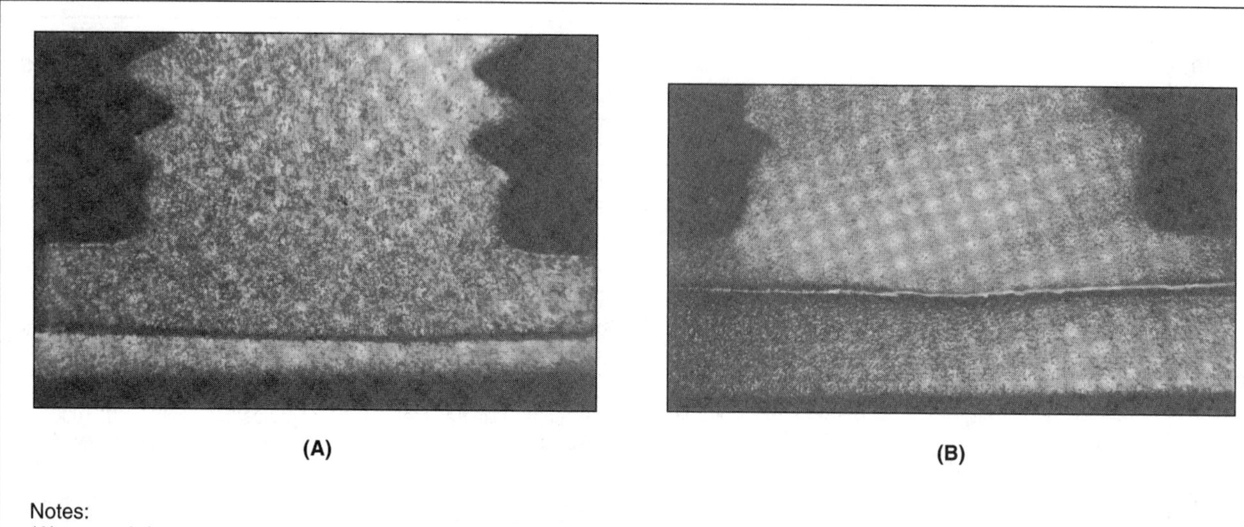

Figure 9.26—Macrostructures of Capacitor Discharge Stud Welds

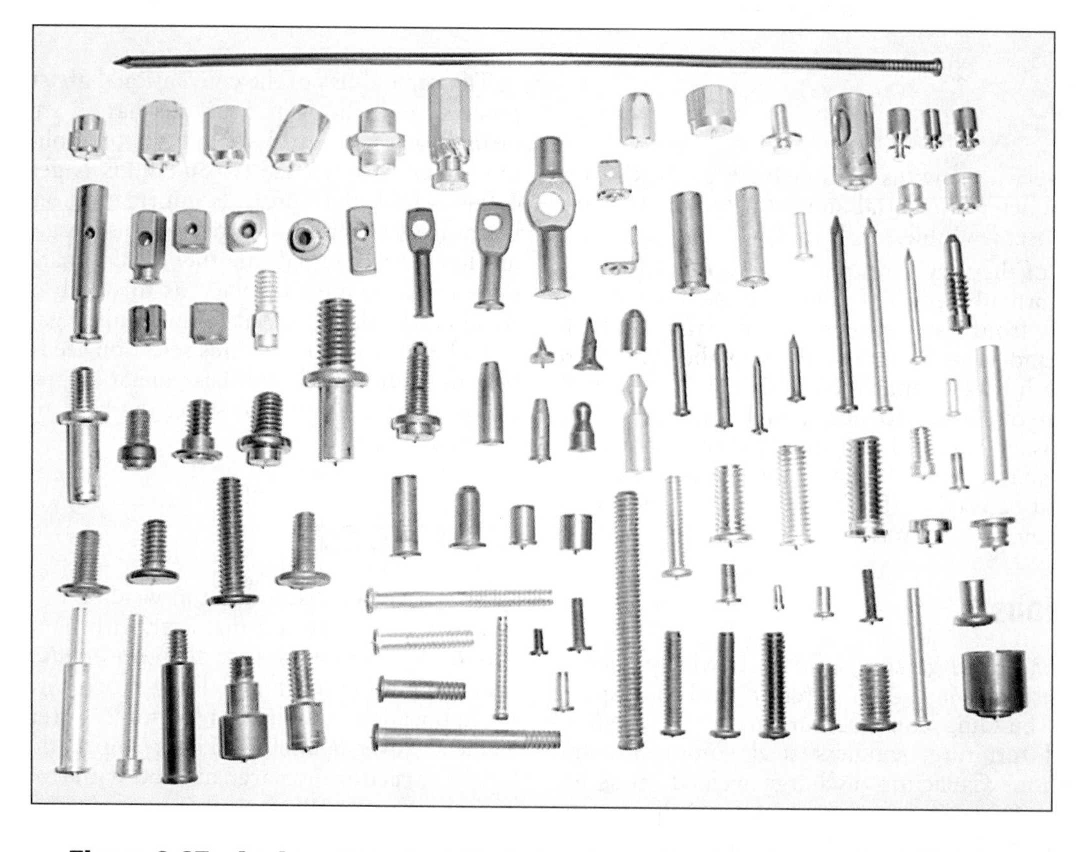

Figure 9.27—An Assortment of Studs for Capacitor Discharge Stud Welding

Table 9.10
Stud Welding Process Selection Chart

	Conventional	Capacitor Discharge Stud Welding	
Factors to be Considered	**Arc Stud Welding**	**Initial Gap and Initial Contact**	**Drawn Arc**
Stud shape			
Round	A	A	A
Square	A	A	A
Rectangular	A	A	A
Irregular	A	A	A
Stud diameter or area			
1.6 mm to 3.2 mm (1/16 in. to 1/8 in.) in diameter	D	A	A
3.2 mm to 6.4 mm (1/8 in. to 1/4 in.) in diameter	C	A	A
6.4 mm to 12.7 mm (1/4 in. to 1/2 in.) in diameter	A	B	B
12.7 mm to 25.4 mm (1/2 in. to 1 in.) in diameter	A	D	D
Up to 32.3 mm2 (0.05 in.^{2}) in area	C	A	A
Over 32.3 mm2 (0.05 in.^{2}) in area	A	D	D
Stud metal			
Carbon steel	A	A	A
Stainless steel	A	A	A
Alloy steel	B	C	C
Aluminum	B	A	B
Brass	C	A	B
Base metal			
Carbon steel	A	A	A
Stainless steel	A	A	A
Alloy steel	B	A	C
Aluminum	B	A	B
Brass	C	A	B
Base metal thickness			
Under 0.4 mm (0.015 in.)	D	A	B
0.4 mm to 1.6 mm (0.015 in. to 0.062 in.)	C	A	A
1.6 mm to 3.2 mm (0.062 in. to 0.125 in.)	B	A	A
Over 3.2 mm (0.125 in.)	A	A	A
Strength criteria			
Heat effect on exposed surfaces	B	A	A
Weld clearance	B	A	A
Strength of stud governs	A	A	A
Strength of base metal governs	A	A	A

Key:
A = Applicable without special procedures or special equipment.
B = Applicable with special techniques or on specific applications which justify preliminary trials or testing to develop welding procedure and technique.
C = Limited application.
D = Not recommended.

BASE METAL THICKNESS

For the arc stud welding process, the base metal thickness should be at least 1/3 of the weld base diameter of the stud to assure maximum weld strength. When strength is not the foremost requirement, the minimum base metal thickness can be 1/5 of the weld base diameter.

Capacitor discharge stud welding should be used for thin base metal thicknesses under 1.6 mm (1/16 in.). This method is capable of welding studs to base metal as thin as 0.5 mm (0.020 in.) without melt-through. On such thin material, the sheet will tear if the stud is loaded excessively. Reverse side marking is the principal effect of stud welding on thin materials.

BASE METAL COMPOSITION

Mild steel, austenitic stainless steel, and various aluminum alloys can be welded with either conventional arc stud welding or the capacitor discharge method. The capacitor discharge method is the best choice for copper, brass, and galvanized steel sheet metal.

CAPACITOR DISCHARGE METHOD

Using the above criteria, if the capacitor discharge method is chosen as the best one for the application, then the modes within this process, the initial contact method, the initial gap method, and the drawn arc method must be evaluated. Since there is considerable overlap in the stud welding capabilities of the three methods, there are many applications where more than one of them can be used. Conversely, there are many instances where one method is better suited for the application. Setting up specific guidelines for selection of the best capacitor discharge method is rather difficult. However, general guidelines for the selection of the three different methods are described in the following sections.

Initial Contact Method

The initial contact method is used only with portable equipment, principally for welding mild steel studs. The simple and relatively lightweight equipment makes this method ideal for welding mild steel insulation pins to galvanized ductwork.

Initial Gap Method

The initial gap mode is used with portable and fixed equipment for welding mild steel, stainless steel, and aluminum. It is generally superior to both the drawn arc and initial contact methods for welding dissimilar metals and aluminum. Inert gas shielding is not needed for aluminum welding.

Drawn Arc Method

The types of equipment and materials welded are the same as those used in the initial gap method. The stud does not require a special tip. The method is ideally suited for high-speed production applications involving automatic feed systems with either portable or fixed equipment. Inert gas is required for aluminum welding.

APPLICATION CONSIDERATIONS

Studs can be welded with the workpiece in any position, i.e., flat, vertical, and overhead. The use of the gravity drop-head principle, of course, is limited to the flat position. Stud welding in the vertical position is limited to studs with a maximum diameter of 19 mm (3/4 in.).

Studs can be welded to curved or angled surfaces, plate edges, and plate in the vertical, sloping, and overhead positions, and in many other special applications. However, because the basic arc stud welding process melts considerably more metal than capacitor discharge or short-cycle welding, the ceramic ferrule must be designed to fit the contour and position of the workpiece surface. It should be noted that stud welding in positions other than the flat position must be qualified for the application according to the latest edition of *Structural Welding Code—Steel*, AWS D1.1;[16] *Bridge Welding Code*, AWS D1.5;[17] *Structural Welding Code—Stainless Steel*, AWS D1.6;[18] or *Structural Welding Code—Aluminum*, AWS D1.2.[19]

The arc stud welding process is much more tolerant of workpiece surface contaminants, such as light coatings of paint, scale, rust, or oil, than the capacitor discharge stud welding mode. The long arc duration with the arc stud welding process assists in burning the contaminants away. In addition, the molten metal expulsion tends to consume and expel any residue out of the joint. The arc stud welding process is suitable for welding through thick galvanized coatings using special welding procedures, provided the base material is thick enough to withstand the long arc time.

In contrast, the capacitor discharge welding process is suitable for welding small-diameter studs to thin-gauge galvanized sheet metal. The percussive nature of

16. See Reference 4.
17. American Welding Society (AWS) Committee on Structural Welding, 2002, Bridge Welding Code, AASHTO/AWS D1.5M/D1.5:2002, Miami: American Welding Society.
18. See Reference 9.
19. American Welding Society (AWS) Committee on Structural Welding, 2003, *Structural Welding Code—Aluminum*, AWS D1.2/D1.2M, 2003, Miami: American Welding Society.

the capacitor discharge arc tends to expel thin metallic coatings, such as those applied by electroplating and galvanizing, out of the joint.

WELD QUALITY, INSPECTION, AND TESTING

Arc stud weld quality assurance requires the proper materials, equipment, setup, operating procedures, and operators trained in the process. Proper setup includes such variables as gun retraction (lift), stud extension beyond the ferrule (plunge), welding current, and welding time. This section covers weld quality concerns for basic arc stud welding, followed by similar information on capacitor discharge stud welding.

BASIC ARC STUD WELDING

Weld quality is maintained by close attention to the factors that may produce variations in the weld. To maintain weld quality and consistency, the stud welder or welding operator must implement the following conditions and techniques:

1. Ensure sufficient welding power for the size and type of stud being welded;
2. Use direct current electrode negative for steels, and direct current electrode positive for aluminum and magnesium;
3. Ensure a good workpiece connection;
4. Use welding cables of sufficient size with good connections;
5. Use correct accessories and ferrules (ceramic arc shields) appropriate for the stud configuration and welding position as required;
6. Clean the workpiece surface where the stud is to be welded;
7. Adjust the gun so that the stud extends the recommended distance beyond the ferrule and retracts it the proper distance for good arc characteristics (Stud extension or plunge should be about equal to the length reductions shown in Table 9.6 or the setting for plunge shown in Table 9.1);
8. Hold the gun steady at the proper angle to the workpiece. Generally, it is perpendicular. Accidental movement of the gun during the weld cycle may cause a defective weld;
9. Keep stud welding equipment properly cleaned and maintained; and
10. Make test welds before starting and at selected intervals during the job.

STEEL STUDS

The AWS document, *Structural Welding Code—Steel*, AWS D1.1, contains provisions for the installation and inspection of steel studs welded to steel components.[20] Quality control and inspection requirements for stud welding are also included. *Recommended Practices for Stud Welding*, ANSI/AWS C5.4-93, briefly covers inspection and testing of both steel and aluminum stud welds.[21]

Visual Inspection

Welded studs can be inspected visually for weld appearance and consistency and also tested mechanically. Threaded production studs can be proof-tested by applying a specified load (force) on them. If they do not fail, the studs are considered acceptable. Production studs should not be bent or twisted for proof-testing.

The weld flash around the stud base should be inspected for consistency and uniformity. Lack of flash may indicate a faulty weld. Figure 9.28(A) indicates a satisfactory stud weld with a good weld flash formation. In contrast, Figure 9.28(B) shows a stud weld in which the plunge was too short. (This type of defect may also be caused by arc blow.) Prior to welding, the stud should always project the proper length beyond the bottom of the ferrule. Figure 9.28(C) illustrates *hang-up*, a condition in which the stud did not plunge into the weld pool. A hang-up may be corrected by realigning the gun components to ensure completely free movement of the stud during lift and plunge.

Figure 9.28(D) shows poor alignment, which may be corrected by positioning the stud gun perpendicular to the workpiece. Figure 9.28(E) shows the results of low weld current. To correct this problem, the workpiece lead and all connections should be checked. The current setting or the time setting, or both, should also be increased. The effect of too much weld current is shown in Figure 9.28(F). Decreasing the current setting or the welding time setting, or both, will lower the weld power.

Mechanical Testing

Mechanical tests should be included as part of the procedure and performance qualifications before production welding is initiated to ensure that the welding schedule is satisfactory. Mechanical tests can also be made during the production run or at the beginning of a shift to ensure that the welding conditions have not changed.

20. See Reference 4.
21. American Welding Society (AWS) Subcommittee on Stud Welding, 1993, *Recommended Practices for Stud Welding*, ANSI AWS C5.4-93, Miami: American Welding Society.

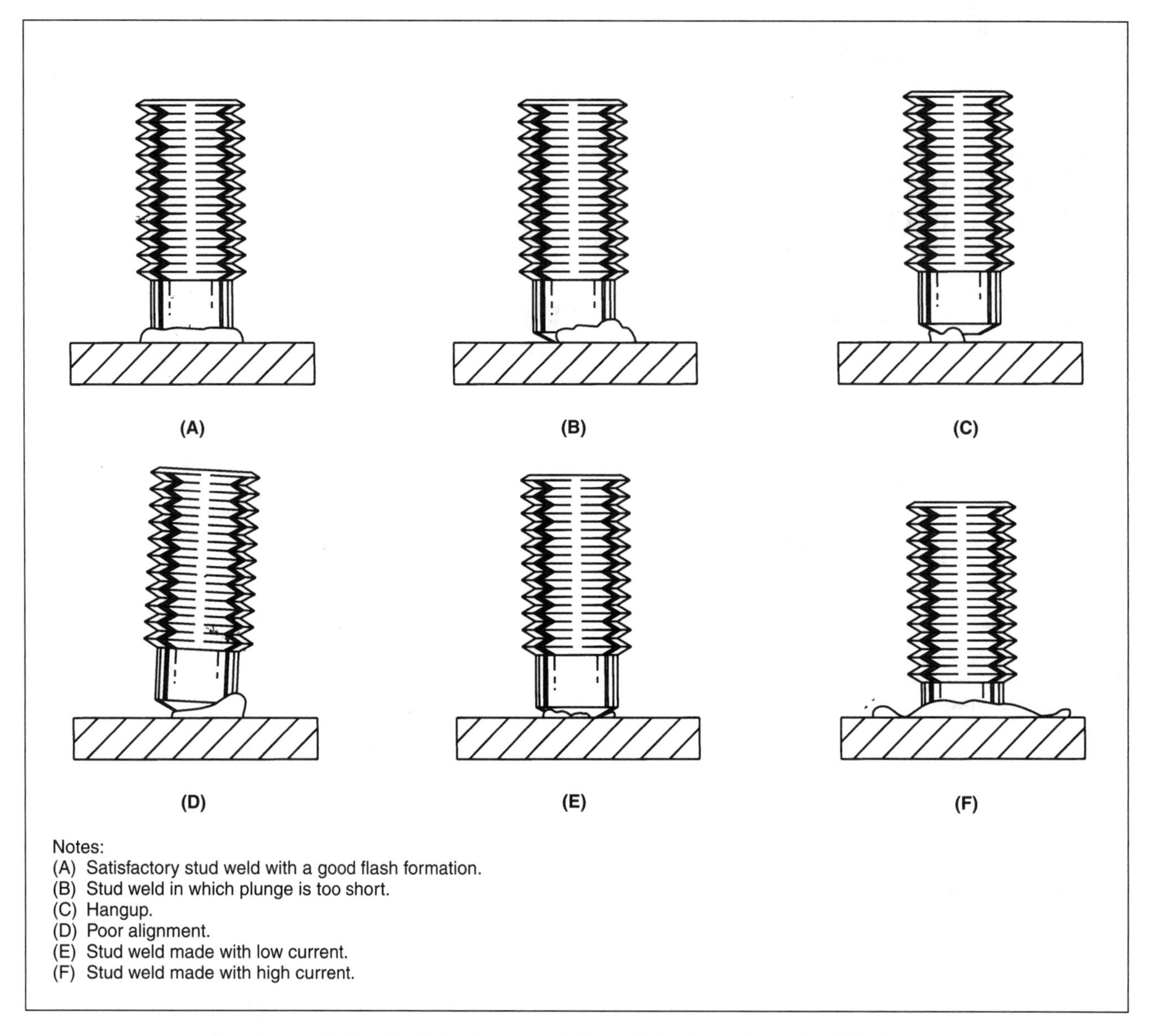

Figure 9.28—Satisfactory and Unsatisfactory Arc Stud Welds

Arc stud welding in positions other than the flat position (also called *downhand*) must be qualified to the application in accordance with *Structural Welding Code—Steel*, AWS D1.1M/AWS D1.1;[22] *Structural Welding Code—Stainless Steel*, AWS D1.6/D1.6M;[23] *Structural Welding Code—Aluminum*, AWSD1.2;[24] or *Bridge Welding Code,* AASTO/AWS D1.5M/D1.5.[25] This test requires a 10-stud application qualification when using any ferrule other than the flat-position ferrule and welding to any curved or angled surface or position other than the flat position, or when welding studs to non-qualified base materials or through metal deck. All 10 studs must be welded to the material to be used in production and tested to failure, using bend, torque or tensile test methods, or a combination of two of these. The failure must be in the stud shank (not the weld). This test qualifies the operator, the application detail being qualified, and the stud welding process being used.

22. See Reference 4.
23. See Reference 9.
24. See Reference 18.
25. See Reference 16.

Bend testing can be accomplished by striking the stud with a hammer or by using a bending tool such as a length of a tube or pipe, as shown in Figure 9.29. The angle through which the stud will bend without weld failure depends on the stud and base metal compositions and conditions (i.e., cold worked, heat-treated) and the stud design. The application of bend testing should be determined when the welding procedure specification is established, or by consulting the applicable welding code. As bend testing will damage the stud, it should be performed on qualification samples only. For some applications, however, unthreaded studs can remain in the bent condition.

The method used to apply a tensile load on an arc welded stud often depends on the stud design. Special tooling may be required to grip the stud properly without damage, and a special tensile loading device may be needed. A simple method of applying tensile load to straight-threaded studs is shown in Figure 9.30. A hardened sleeve, washer and nut of the appropriate size and same material as the stud is placed over the stud. The nut is tightened with a torque wrench against a washer bearing on the sleeve. This applies a tensile load (and some shear) on the stud.

The relationship between nut torque, *T*, and tensile load, *F*, can be estimated using the following equation:

$$T = kFd \tag{9.3}$$

where

T = Torque in appropriate units, ft. lbs., in. lbs., or joules.

d = Nominal thread diameter, mm (in.); and

k = Torque constant. Constant related to such factors as thread angle, helix angle, thread diameters, and coefficients of friction between the nut and thread, and the nut and washer. It can vary from 0.06–0.35.[26]

For clean, non-plated mild steel studs, *k* is approximately 0.2 for all thread sizes and for both coarse and fine thread. However, the many factors that influence friction will influence the value of *k*. These include the stud, nut, and washer materials, surface finishes, and lubrication. For other materials, *k* may have some other value because of the differences in friction between the parts. The value of *k* should be established by experiment on the stud being tested and should be based on the actual yield and tensile strength of the stud as received and the sleeve, washer, and nut combination used in the torque test setup.

26. Industrial Fasteners Institute, *Fastener Standards,* Sixth Ed., 1988. *Design of Bolted Joints—An Introduction,* pp M64-M70, Cleveland, Ohio: Industrial Fasteners Institute.

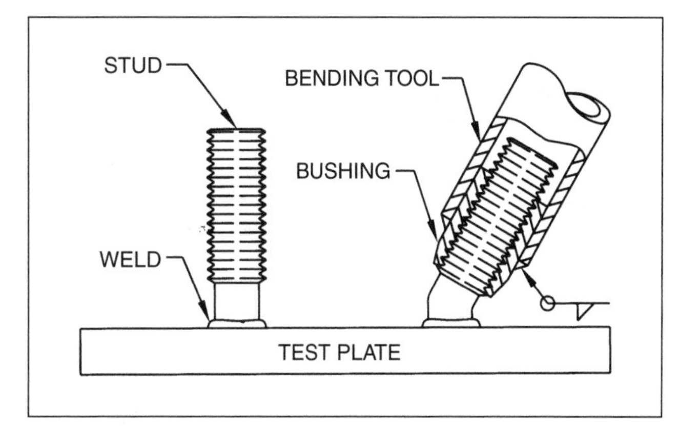

Figure 9.29—Bend Test for Welded Studs to Determine Acceptable Welding Procedures

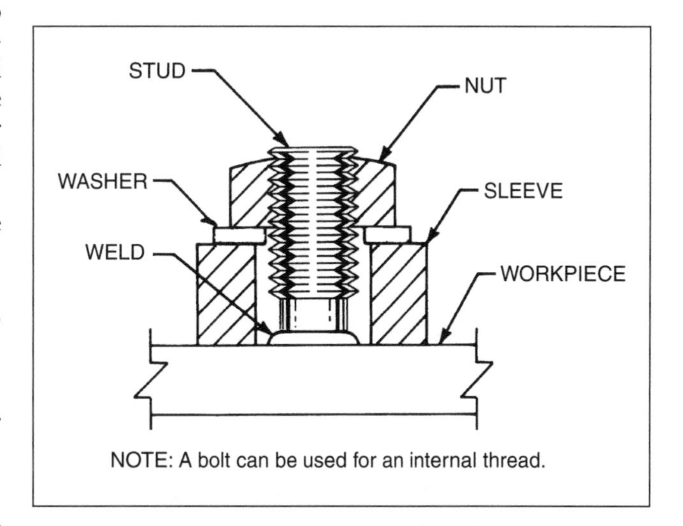

Figure 9.30—Method of Applying a Tensile Load to a Welded Stud Using Torque

ALUMINUM STUDS

Visual inspection of aluminum stud welds for acceptance is limited because the appearance of the weld flash does not necessarily indicate quality. Therefore, visual inspection of aluminum stud welds is recommended only to determine complete fusion and absence of undercut around the periphery of the weld.

Aluminum studs can be tested to establish acceptable welding procedures using the bend test shown in Figure 9.29. If the stud bends about 15 degrees or more from the original axis without breaking the stud or weld, the welding procedures should be considered satisfactory. Production studs should not be bent and then straightened because of possible damage to them. In this case,

Table 9.11
Typical Nut Torques Causing Failure of Aluminum Alloy Studs

Thread Size	Failure Load	
	N · m	ft-lb · in.
M6 (1/4-20 in.)	7	60
M8 (5/16-18 in.)	13	115
M10 (3/8-16 in.)	22	195
M11 (7/16-14 in.)	33	290
M12 (1/2-13 in.)	49	435

Source: Structural Welding Code—Aluminum, AWS D1.2.

the torque test or separate qualification test plates may be substituted.

The method used for torque testing of threaded aluminum studs is the same as that used for threaded steel studs. Torque is applied to a predetermined value or until the stud fails. Typical torque tests rendered the failure loads shown in Table 9.11. For a particular application, the acceptable proof load should be established by suitable laboratory tests that relate applied torque to tensile loading.

CAPACITOR DISCHARGE WELD QUALITY AND INSPECTION

Some aspects of capacitor discharge stud welding evaluation are more difficult than those for arc stud welding because of the absence of weld flash. The operator does not hear and see the welding arc, nor can the operator use the characteristics of a weld flash to evaluate weld quality. However, there should be some flash at the weld joint.

The best method of quality control for capacitor discharge stud welding is to perform destructive tests on studs that have been welded to base metal similar to the metal that will be used in actual production. The destructive test should be a bend, torque, or tensile test. Once a satisfactory welding schedule is established, the production run can begin. It is best to check weld quality at regular intervals during production and especially after maintenance to ascertain that the welding results have not changed.

Following are some requirements for producing good capacitor discharge stud welds and maintaining consistent weld quality:

1. Size of the power source must be appropriate to the stud size being welded;
2. Proper operation and maintenance of equipment must be assured;
3. Cable connections should be tight;
4. Studs and stud gun must be properly handled during welding;
5. Weld surface must be clean and free from oils, grease, and other lubricants and from rust, mill scale, plating, and oxides that may contribute to high electrical resistance in areas of welding and grounding;
6. Weld surface imperfections, such as extreme roughness, should be avoided or remedied to assure complete fusion in the weld area;
7. The stud axis to the workpiece surface should be maintained in a position perpendicular to the workpiece to assure complete fusion; and
8. Design of the weld end of the stud, including tip size, face angle, and weld base diameter, must be correct for the application.

Capacitor discharge stud welds may be inspected both visually and mechanically. The success of visual inspection depends on the interpretation of the appearance of the weld. Figure 9.31 illustrates good and bad capacitor discharge stud welds. If a questionable weld is evident after the welds have been visually inspected, the weld should be mechanically tested.

Mechanical Testing

Mechanical testing of capacitor discharge stud welds should be performed with the same methods described above for arc stud welding. Bend testing and proof tensile loading tests are used to establish welding conditions and to qualify studs to be used in production.

Maximum nut torque values for proof testing studs by this method are listed in Table 9.12 for various stud materials and sizes. The torque values listed will produce tensile stresses in studs that are slightly below the material yield strengths.

Table 9.12 also lists the tensile loads that will develop the approximate nominal tensile strength of the stud material in the various diameters and the maximum shear load that the studs can carry. The user should follow appropriate safety precautions during proof testing or stud selection.

ECONOMICS

Arc stud welding provides opportunities for cost-effective welding because of the following inherent attributes:

(A) Good Weld

(B) Weld Power Too High

(C) Weld Power Too Low

Figure 9.31—Examples of Satisfactory and Unsatisfactory Capacitor Discharge Stud Welds

Table 9.12
Torque, Tensile, and Shear Loads for Capacitor Discharge Welded Studs of Various Materials and Sizes

		Maximum Fastening Torque[a]		Maximum Tensile Load[b]		Maximum Shear Load	
Stud Material	Stud Size	N·m	lb f·in.	kN	lb	kN	lb
Low-carbon, copper-flashed steel	6–32	0.7	6	2.2	500	1.7	375
	8–32	1.4	12	3.4	765	2.6	575
	10–24	1.6	14	4.3	960	3.2	720
	1/4–20	4.9	43	7.8	1750	5.8	1300
	5/16–18	8.1	72	13	2900	9.8	2200
	3/8–16	12	106	19	4300	14	3250
Stainless steel 304 or 305	6–32	1.1	10	3.5	790	2.6	590
	8–32	2.3	20	5.6	1260	4.2	940
	10–24	2.6	23	6.8	1530	5.1	1150
	1/4–20	8.5	75	13	2880	9.6	2160
	5/16–18	14	126	17	3750	14	3100
	3/8–16	21	186	22	4850	20	4550
Aluminum alloy 1100	6–32	0.3	2.5	0.9	200	0.6	125
	8–32	0.6	5	1.3	295	0.8	185
	10–24	0.7	6.5	1.7	380	1.0	235
	1/4–20	2.4	21.5	3.0	670	1.9	415
	5/16–18	4.1	36	5.0	1125	3.1	695
	3/8–16	6.0	53	7.4	1660	4.4	1000
Aluminum alloy 5086	6–32	0.4	3.5	1.7	375	1.0	235
	8–32	0.8	7.5	2.6	585	1.6	365
	10–24	1.1	10	3.3	735	2.0	460
	1/4–20	3.7	32.5	6.1	1360	3.8	850
	5/16–18	6.2	54.5	10	2300	6.2	1400
	3/8–16	9.2	81	15	3400	9.4	2100
Copper alloy (brass) 260 and 268	6–32	0.9	8	2.7	600	1.7	390
	8–32	1.8	16	3.8	860	2.5	560
	10–24	2.1	18.5	4.6	1040	3.0	680
	1/4–20	6.4	61	8.7	1950	5.7	1275
	5/16–18	12	102	15	3280	9.5	2140
	3/8–16	16	150	21	4800	14	3160

a. These values should develop stud tensile stresses to slightly below the yield strengths of the materials.
b. These values should develop the nominal tensile strengths of the materials.

1. There is little heat input to the base metal compared to conventional arc welding, keeping distortion at a minimum;
2. Access to the back side of the base member is not required, eliminating drilling, tapping, or riveting for installation;
3. Thinner base metals can be specified to get required tap depths for threaded fasteners, saving material, workpiece preparation, and welding costs;
4. The capacitor discharge variation can weld small studs with no damage to prefinished opposite sides of thin materials, therefore no subsequent cleaning or finishing is required;
5. Dissimilar metals can be joined with the capacitor discharge method, allowing the designer to choose from metals that may be less expensive; and
6. Very low weld cycle times.

Conversely, stud welding may not be as cost effective as other welding processes because only one end of a stud can be welded to the workpiece; if studs are required on both sides of the workpiece, two welds are necessary. Ceramic ferrules, flux, and shielding gas may be needed. Capacitor discharge applications require close-tolerance projections adding to the cost. These reasons may prove to be economically prohibitive for the application being considered.

SAFE PRACTICES

Some of the potential hazards associated with other arc welding processes also apply to the materials and equipment used in arc stud welding, and the process can be dangerous if the equipment is not installed and maintained regularly or if the operator does not use proper safety precautions. Operators must know and practice appropriate safety precautions, as recommended in the standard *Safety in Welding, Cutting and Allied Processes*, ANSI Z49.1.[27]

Personnel operating stud welding equipment should wear protective clothing and face and skin protection to guard against burns from weld spatter produced during welding. Long-sleeve shirts, long pants and high boots along with leather aprons, gloves and leggings should be used. The protective shirts and pants should be made from natural fibers or flameproof materials and not from flammable synthetic fabrics. When welding in the workpiece vertical or overhead positions, weld hats, helmets, and face shields should be used.

Eye protection in the form safety glasses with side shields or a face shield with a No. 3 filter lens should be worn to protect against arc radiation. See Appendix A for additional information on lens selection. Helpers or workers within 1.5 m (5 ft) should wear clear safety glasses with side shields and protective clothing similar to that mentioned in the above paragraph.

It is necessary for operators or workers in the welding area to use proper ear protection when welding with capacitor discharge equipment. Other stud welding procedures in confined areas should be evaluated for noise levels and similar protection used.

Stud welding is an electrical process. Electric shock can be fatal. Workers in the welding area must not touch live electrical parts and the welding operator must be insulated from live electrical parts. The equipment should be properly installed, grounded and maintained. All weld cables and connections should be inspected regularly for broken or frayed insulation, damaged insulators or other electrical hazards and repaired or replaced at once. Cables and electrical connections should be kept out of standing water and operators should not work in wet areas or weld when the area or their clothing and gloves are wet.

Before repairs to equipment are started, electric power should be turned off and electric switch boxes locked out. Capacitors used in capacitor discharge equipment should be completely drained or discharged before repairs begin. Equipment manuals concerning the safe operation of the equipment are furnished by the manufacturer and should be consulted. Only qualified personnel trained and experienced with the manufacturers' repair and maintenance procedures should be allowed to repair and maintain the equipment.

All combustible or volatile materials must be removed from the weld area. Stud welding weld berries or spatter are minimal, but can result in fires or explosions if they reach combustible or volatile materials. Gas cylinders must be properly stored and restrained. Hoses should be checked for leaks. The cylinders should be isolated from excessive heat or welding spatter. Operators should ensure that cylinders never become part of the electrical circuit.

Continuous welding in an enclosed area or welding and cleaning base materials with paint, epoxy, galvanizing or other coatings may produce toxic fumes. Natural or forced ventilation in the welding area must be provided to prevent fume accumulation.

Material safety data sheets are available from stud manufacturers for the stud materials and the ceramic ferrules and should be consulted. Although ceramic ferrules are made from fired clay, which contains silica, they are fired at high temperature and fused, so there is

27. American National Standards Institute (ANSI) Accredited Standards Committee Z49, 1999, *Safety in Welding, Cutting, and Allied Processes*, ANSI Z49.1. Miami: American Welding Society

no release of particulate silica small enough to be inhaled when the ferrule is removed from the weld.

Hands, feet, clothing, and tools should be kept away from the weld stud, chuck, and other moving parts during the weld cycle to avoid pinch points or electric shock. Safe practices in lifting, moving and handling of welded assemblies must be enforced. Safety shoes are necessary when working with heavy assemblies.

The United States government document *Occupational Safety and Health Standards for General Industry* should be consulted.[28] Appendix B presents a list of additional sources of safety information and publishers of safety standards.

CONCLUSION

Conventional arc stud welding and capacitor discharge stud welding are widely used by many of the metalworking industries. Conventional arc stud welding is most commonly used in structural applications. The capacitor discharge method is used for welding ferrous and nonferrous studs to similar materials in light-gauge sheet metal applications.

Literally millions of studs of all types and configurations are welded every year in all types of industries, including construction, aviation, transportation, and consumer goods manufacturing. Arc stud welding provides one-sided, no-hole, no-leak features and leaves no marking on prefinished or painted thin metal. In addition to these benefits, the efficiency and speed of the stud welding process make it an economical choice that can be considered for many important joining applications when stringent specifications must be met.

BIBLIOGRAPHY[29]

American National Standards Institute (ANSI) Accredited Standards Committee Z49, 1999, *Safety in Welding, Cutting, and Allied Processes*, ANSI Z49.1. Miami: American Welding Society.

American Society for Testing and Materials (ASTM). 2000. *Stainless and Heat-Resisting Steel Bars and Shapes*, ASTM A276-2000a. West Conshohocken, Pennsylvania: American Society for Testing and Materials.

American Society for Testing and Materials (ASTM) Subcommittee A108. 1999. *Standard Specifications for Steel Bars, Carbon, Cold Finished, Standard Quality, Grades 1010 through 1020*, A108-99. West Conshohocken, Pennsylvania: American Society for Testing and Materials.

American Society for Testing and Materials, (ASTM). 2000. *Stainless and Heat-Resisting Steel for Cold Heading and Cold Forging–Bar and Wire*, A493-95. West Conshohocken, Pennsylvania: American Society for Testing and Materials.

American Society for Testing and Materials (ASTM) Subcommittee A496-90a. 1990. *Standard Specifications for Steel Wire, Deformed, for Concrete Reinforcement,* A496-90a. West Conshohocken, Pennsylvania: American Society for Testing and Materials.

American Welding Society (AWS) Committee on Structural Welding. 2002. *Structural Welding Code—Steel,* AWS D1.1/D1.1M:2002. Miami: American Welding Society.

American Welding Society (AWS) Committee on Structural Welding. 2002. *Structural Welding Code—Aluminum*, AWS D1.2/D1.2M:2002. Miami: American Welding Society.

American Welding Society (AWS) Committee on Structural Welding. 2002. *Structural Welding Code—Stainless Steel*, AWS D1.6/D1.6M, Miami: American Welding Society.

American Welding Society Committee on Structural Welding. 2002. *Bridge Welding Code*, D1.5M/D1.5: 2002. Miami: American Welding Society.

American Welding Society (AWS) Committee on Definitions and Symbols, 2001. *Standard Welding Terms and Definitions*, AWS A3.0. Miami: American Welding Society.

American Welding Society (AWS) Subcommittee on Stud Welding. 1993. *Recommended Practices for Stud Welding,* ANSI/AWS C5.4, Miami: American Welding Society.

American Welding Society (AWS) Committee on Structural Welding. 1999. *Structural Welding Code—Stainless Steel*, D1.6:1999. Miami: American Welding Society.

American Welding Society (AWS) Welding Handbook Committee. Jenney, C. L., and A. O'Brien, eds. 2001. *Welding Science and Technology.* Vol. 1 of *Welding Handbook,* 9th ed. Miami: American Welding Society.

American Welding Society (AWS) Welding Handbook Committee. O'Brien, R. L., ed. 1991. *Welding Processes*, Vol. 2 of *Welding Handbook*, 8th ed. Miami: American Welding Society.

28. Occupational Safety and Health Administration (OSHA). 1999. *Occupational Safety and Health Standards for General Industry,* in *Code of Federal Regulations (CFR),* Title 29 CFR 1910. Washington, D.C.: Superintendent of Documents, U.S. Government Printing Office.

29. The dates of publication given for the codes and other standards listed here were current at the time this chapter was prepared. The reader is advised to consult the latest edition.

International Organization for Standardization (ISO). 1998. *Studs and Ceramic Ferrules for Arc Stud Welding*, ISO 13918:1998(E). Geneva, Switzerland: International Organization for Standardization.

Occupational Safety and Health Administration (OSHA). 1999. *Occupational Safety and Health Standards for General Industry*, in *Code of Federal Regulations (CFR)*, Title 29 CFR 1910. Washington, D.C.: Superintendent of Documents, U.S. Government Printing Office.

Society of Automotive Engineers (SAE). Iron and Steel Technical Committee. 2001. SAE Handbook, Volume 1, *Steel Fasteners*. Warren, Pennsylvania: Society of Automotive Engineers.

SUPPLEMENTARY READING LIST

ASM International. 1993. *Welding, brazing and soldering*. Vol. 6 of *Metals handbook*. 10th ed. Materials Park, Ohio: ASM International.

Automated system welds heat transfer studs. 1974. *Welding Journal*. 53(1): 29–30.

Baeslack, W. A., G. Fayer, S. Ream, and C. E. Jackson. 1975. Quality control in arc stud welding. *Welding Journal*. 54(11): 789-s–798-s.

Chambers, Harry A. 2001. Principles and practices of stud welding. *PCI Journal*. 46(6): 2-25. Chicago: Precast/Prestressed Concrete Institute.

Hahu, O., and K. G. Schmitt. Microcomputerized quality control of capacitor discharge stud welding. 1982. In *Proceedings: 4th International JWS Symposium*, November 24–26, 1982. Vol. 2. Osaka Japan: Japan Welding Society.

Irving, B. 1992. Welding in space. *Welding Journal*. 71(1): 67–69.

Lockwood, L. F. 1967. Gas shielded steel welding of magnesium. *Welding Journal* 46(4): 168-s–174-s.

Masubuchi, K., A. Imakita and M. Miyaki. 1988. An initial study of remotely manipulated stud welding for space applications. *Welding Journal*. 67(4): 25–34.

Pease, C. C. 1969. Capability studies of capacitor discharge stud welding on aluminum alloy. *Welding Journal*. 48(6): 253-s–257-s.

Pease, C. C., and F. J. Preston. 1972. Stud welding through heavy galvanized decking. *Welding Journal*. 51(4): 241–244.

Pease, C. C., F. J. Preston, and J. Taranto. 1973. Stud welding on 5083 aluminum and 9% nickel steel for cryogenic use. *Welding Journal*. 52(4): 232–237.

Ramasamy, S. 2000. Drawn arc stud welding, crossing over from steel to aluminum. Welding Journal. 79(1): 35–39.

Shoup, T. E. 1976. Stud welding. *Welding Research Council Bulletin 214*. New York: Welding Research Council.

CHAPTER 10

ELECTROSLAG WELDING

Prepared by the Welding Handbook Chapter Committee on Electroslag Welding:

J. R. Hannahs, Chair
Edison Community College

D. R. Amos
Siemens Westinghouse Power Corporation

W. L. Bong
Arcmatic Integrated Systems

J. H. Devletian
Portland State University

S. Liu
Colorado School of Mines

J. E. Sims
Consultant

R. B. Smith
Select Arc

Welding Handbook Volume 2 Committee Member:

D. R. Amos
Siemens Westinghouse Power Corporation

Contents

CHAPTER 10

ELECTROSLAG WELDING

INTRODUCTION

Electroslag welding produces the coalescence of metals using heat generated by passing electric current through molten conductive slag, which melts the base and filler metals. Although this process is most often employed to weld metals in the vertical or near-vertical position, usually in a single pass, it can be used at angles of 45° or greater from vertical.

The single-pass welding of heavy plates has long been desired as a means of avoiding multiple-pass welding techniques. Before 1900, graphite molds were placed on each side of a space between vertical plates to contain the molten metal created by graphite electrodes, fusing the edges to form the weld. Graphite molds were subsequently replaced by copper or ceramic molds, and conventional welding arcs, gas torches, and thermite mixtures were devised to generate the molten metal with a degree of superheat sufficient to obtain uniform coalescence.

In the early 1950s, Russian scientists at the Paton Institute of Electric Welding in Kiev announced the development of machines that employed the principle of an electrically conductive slag to produce single-pass vertical welds. Subsequent work at the Bratislava Institute of Welding in Czechoslovakia was made available to engineers in Belgium in 1958 and through them to the rest of the Western world. An electroslag unit was introduced in the United States in 1959.

In the 1970s, the electroslag welding process was widely used in bridge building, shipbuilding, pressure vessel fabrication, and other applications. Various problems were perceived to be associated with the process, including difficulties with nondestructive examination. These concerns prompted research that resulted in many refinements and modifications to the process, including the development of a narrow-groove technique for bridges and other structures.

This chapter describes the electroslag welding (ESW) process and its process variation, consumable guide electroslag welding (ESW-CG). The narrow-groove electroslag welding technique is also described. The discussion addresses the fundamentals of the process, equipment, materials, process capabilities, typical applications and uses, inspection and weld quality, economics, and safe practices.

FUNDAMENTALS

Electroslag welding produces the coalescence of metals through molten slag melting the filler metal and the surfaces of the workpieces to be welded. The weld pool is shielded by this slag, which moves along the full cross section of the joint as welding progresses. The process is initiated by an arc that heats a granulated flux and melts it to form the slag. The arc is then extinguished by the conductive slag, which is kept molten by its resistance to the electric current passing between the electrode and the workpieces.

A square-groove joint is usually positioned so that the axis or length of the weld is vertical or nearly vertical. Except for circumferential welds, no manipulation of the workpiece is required once welding has started. Electroslag welding is a mechanized welding process, and once started, it continues to completion. Since no arc exists, the welding action is quiet and spatter-free. Extremely high metal deposition rates allow the welding of very thick sections in one pass. A high-quality weld deposit results from the nature of the melting and solidification during welding. No angular distortion occurs in the welded plates.

The process is initiated by starting an electric arc between the electrode and the joint bottom. Granulated welding flux is then added and melted by the heat of the arc. As soon as a sufficiently thick layer of molten slag (flux) is formed, all arc action stops, and the welding current passes from the electrode through the slag by electrical conduction. Welding is initiated in a sump or on a starting tab to allow the process to stabilize before the welding action reaches the workpiece.

The heat generated by the resistance of the molten slag to passage of the welding current is sufficient to melt the welding electrode and the edges of the workpiece. The interior temperature of the bath is approximately 1925°C (3500°F). The surface temperature is approximately 1650°C (3000°F). The melted electrode and the base metals collect in a pool beneath the molten slag bath and slowly solidify to form the weld. Progressive solidification occurs from the bottom upward, and molten metal is always present above the solidifying weld metal.

Run-off weld tabs (described in the section "Welding Procedures") are required to allow the molten slag and some weld metal to extend beyond the top of the joint. Both starting and run-off tabs are usually removed flush with the ends of the joint.

ADVANTAGES AND LIMITATIONS

The electroslag welding process offers many opportunities for reducing welding costs on specific types of joints. The advantages of the process are the following:

1. Extremely high metal deposition rates of 16 kilograms per hour (kg/h) to 20 kg/h (35 pounds per hour [lb/h] to 45 lb/h) per electrode can be achieved;
2. Very thick materials can be welded in one pass;
3. Only one equipment setup is required, with no interpass cleaning, since only one pass is involved;
4. Preheating is not normally required, even on materials of high hardenability;
5. High-quality weld deposits can be made because the weld metal stays molten for an appreciable time, allowing gases to escape and slag to float to the top of the weld;
6. Minimum joint preparation and fitup is required, as mill edges and flame-cut square edges are normally employed;
7. The process has a high duty cycle because it is mechanized and, once started, continues to completion, causing little operator fatigue;
8. A minimum of materials handling is involved, as the workpiece needs to be positioned only to place the axis of the weld in a vertical or near-vertical position. No manipulation of the workpieces is needed once the welding operation has started;
9. Weld spatter is eliminated, which results in 100% filler metal deposition efficiency;
10. Low flux consumption; 0.45 kg (1 lb) of flux is used for each 9 kg (20 lb) of weld metal;
11. No angular distortion occurs in the horizontal plane and is minimum in the vertical plane, but this is easily compensated for; and
12. Welding time is a minimum, as electroslag welding is the fastest welding process for large, thick material.

Limitations of the electroslag welding process are the following:

1. The electroslag welding process welds only carbon- and low-alloy steels and some stainless steels;
2. Joints must be positioned in the vertical or near-vertical position;
3. Once welding has started, it must be carried out to completion or a defective start-stop area is likely to result;
4. Electroslag welding cannot be used on materials thinner than about 13 millimeters (mm) (1/2 inch [in.]); and
5. Complex material shapes may be difficult or impossible to weld using electroslag welding.

CONVENTIONAL METHOD

The conventional method of electroslag welding uses a wire electrode with a nonconsumable contact guide tube to direct the electrode into the molten slag bath. A schematic of this method is shown in Figure 10.1. One or more electrodes are fed into the joint, depending on the thickness of the material being welded. The electrodes are fed through nonconsumable wire guides that are maintained 50 mm to 75 mm (2 in. to 3 in.) above the molten slag. Horizontal oscillation of the electrodes can be used to weld very thick materials.

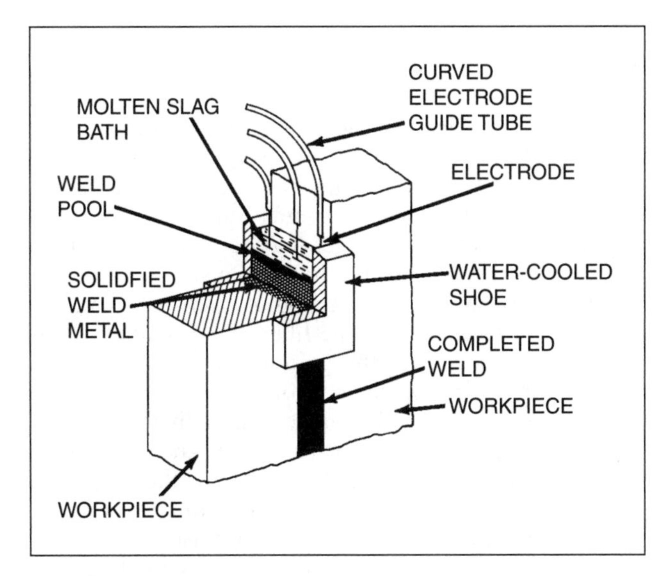

Figure 10.1—Nonconsumable Guide Method of Electroslag Welding (Three Electrodes)

Water-cooled copper shoes (dams) are normally used on both sides of the joint to contain the molten weld metal and slag bath. The shoes, which are attached to the welding machine, move vertically with the machine. The vertical movement of the welding machine is consistent with the electrode deposition rate. This movement may be either mechanized or controlled by the welding operator.

The vertical movement of the shoes exposes the weld surfaces. A slight reinforcement is normally present on the weld, which is shaped by a groove in the shoe. The weld surfaces are covered with a thin layer of slag. This slag consumption must be compensated for during welding by the addition of small amounts of flux to the molten slag bath. Fresh flux is usually added manually. Flux-cored wires can be used to supply flux to the bath.

The conventional method of electroslag welding can be used to weld plates ranging in thickness from approximately 13 mm to 500 mm (1/2 in. to 20 in.). Thicknesses from 19 mm to 460 mm (3/4 in. to 18 in.) are most commonly welded. One oscillating electrode can successfully weld thicknesses up to 120 mm (5 in.); two electrodes, thicknesses up to 230 mm (9 in.); and three electrodes, thicknesses up to 500 mm (20 in.).

With each electrode, the process deposits approximately 11 kg to 20 kg (25 lb to 45 lb) of filler metal per hour. The diameter of the electrode used is generally 3.2 mm (1/8 in.). The electrode metal-transfer efficiency is almost 100%. The normally large weld produced by electroslag welding consumes approximately 2.3 kg (5 lb) of flux for each 45 kg (100 lb) of deposited weld metal.

CONSUMABLE GUIDE METHOD

Consumable guide electroslag welding (ESW-CG) is a process variation similar to the conventional method, except that a consumable guide extends down the length of the joint. In the conventional method, the welding head moves progressively upward as the weld is deposited. However, in the consumable guide method, the welding head remains stationary at the top of the joint, and both the guide tube and the electrode are progressively melted by the molten slag.

The consumable guide method of electroslag welding is shown in Figure 10.2. The filler metal is supplied by both an electrode and the guiding member. The electrode wire is directed to the bottom of the joint by a guide tube extending the entire length (height) of the joint. The welding current is carried by the guide tube, which melts off just above the surface of the slag bath. Thus, the welding machine does not move vertically, and stationary or non-sliding shoes are used.

On short welds, the shoes may be the same length as the joint. Several sets of shoes may be required for longer joints. As the metal solidifies, a set of shoes is removed from below the weld pool and placed above the top shoes. This "leapfrog" pattern is repeated until the weld is complete. As welding proceeds and the slag bath rises, the consumable guide melts and becomes part of the weld metal. The consumable guide provides approximately 5% to 15% of the filler metal.

As with the conventional method, one or more electrodes can be used. These may oscillate horizontally in the joint. Because the guide tube carries electrical current, it may be necessary to insulate it from the groove faces of the base plate and the shoes.

A flux coating can be provided on the outside of the consumable guide to insulate it and help replenish the slag bath. Other forms of insulation include doughnut-shaped insulators, fiberglass sleeves, and tape.

The consumable guide method can be used to weld sections of virtually unlimited thicknesses. When using stationary electrodes, each electrode welds approximately 63 mm (2.5 in.) of plate thickness. One oscillating electrode successfully welds thicknesses up to 130 mm (5 in.); two oscillating electrodes, thicknesses up to 300 mm (12 in.); and three oscillating electrodes, thicknesses up to 450 mm (18 in.). Weld lengths up to 9 meters (m) (30 feet [ft]) have routinely been accomplished with a single stationary electrode. Oscillation control problems may occur on long weld lengths. Therefore, if the required amount of oscillation cannot be properly controlled, additional electrodes can be added, and oscillation can be reduced or eliminated.

Wing-type guide tubes can be used for certain applications when a round guide tube cannot properly heat the entire cross-sectional area or when the weldments are irregular in shape. When oscillation equipment is unavailable, the wing-type guide is most commonly used. Wing-type guides are also used when the joint is large enough to require two guides, but only one wire feeder is available. This application is limited to 100 mm (4 in.) joints. Joints that exceed 115 mm (4-1/2 in.) in thickness are marginal when welding with one guide.

Wing-type guide tubes are made by tack welding light-carbon-steel bar stock to the sides of a round guide tube. This procedure is illustrated in Figure 10.3. Normally, wing-type guides extend to within 6 mm (1/4 in.) of the edge of the joint. The additional cross section passes current into the slag bath and keeps the bath temperature high enough to melt the plate edges remote from the filler wire. Because of this extra current, the power demand is equal to that of a multiple-electrode setup.

Narrow Groove Electroslag Welding Technique

A narrow groove electroslag welding technique was developed during the 1980s and field-tested during the 1990s. It is an optimized version of standard consum-

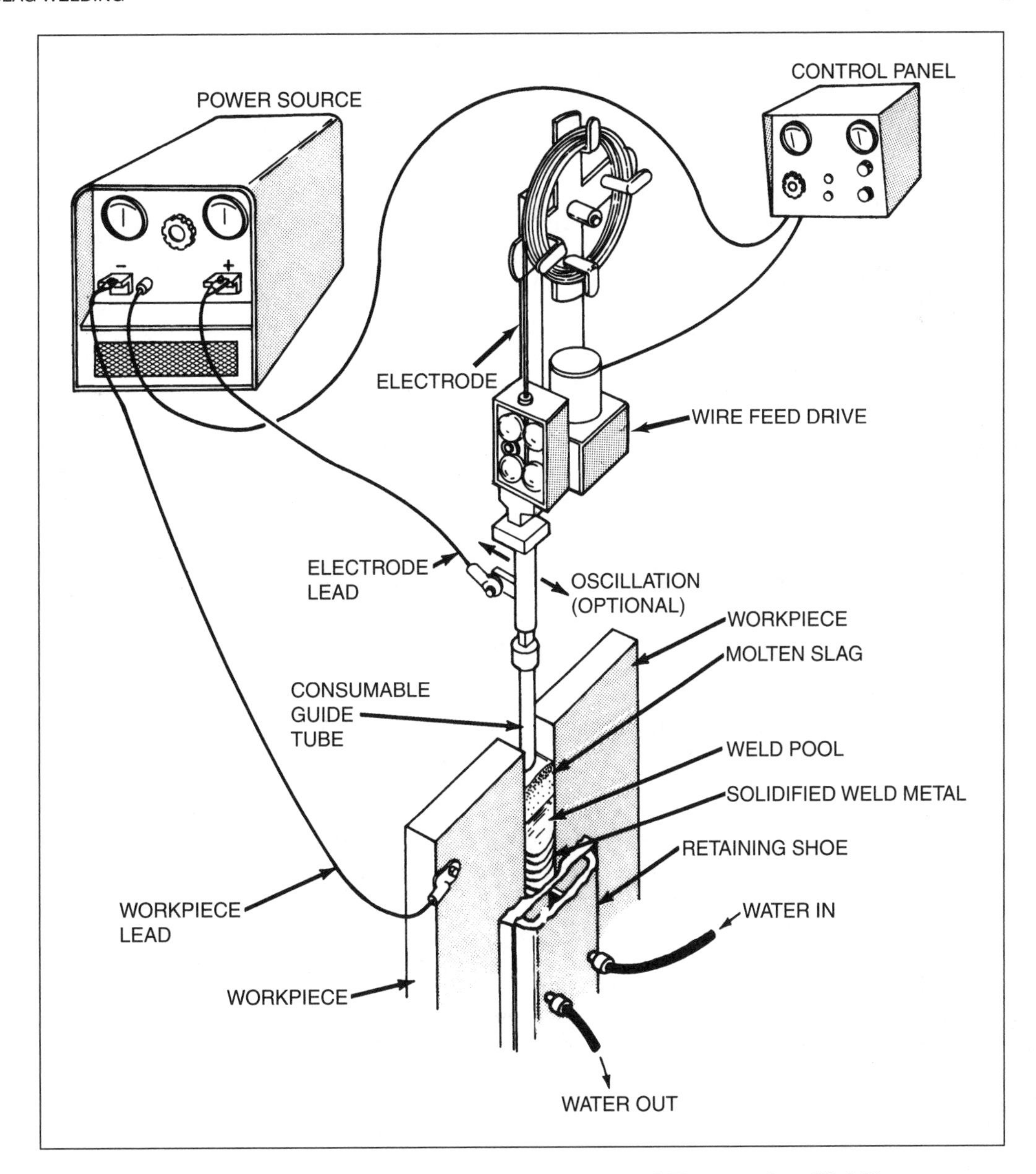

Figure 10.2—Consumable Guide Method of Electroslag Welding

able guide electroslag welding designed to weld thick-section bridge steels in a single pass in the vertical position. The developers and users of this technique refer to it as *narrow-gap improved electroslag welding.*[1]

1. This narrow groove welding technique was developed by the Oregon Graduate Institute of Science and Technology and Northwestern University in cooperation with the Federal Highway Authority. Although the developers and users of the technique refer to it as *narrow-gap improved electroslag welding,* it is not listed as such in the American Welding Society (AWS) standard A3.0: 2001, *Standard Welding Terms and Definitions.*

The steels that can be welded with this technique include A 709 Grades 36, 50, and 50W. Typically, these steels can be welded in sizes as thin as 19 mm (3/4 in.) with virtually no upper limit on thickness. A typical thick plate, however, would measure 50 mm (2 in.) or more in thickness. In addition, many transition thick-to-thin joints—32 mm (1-1/4 in.) thick to 50 mm (2 in.) thick plates, for example—can be welded with narrow groove electroslag welding.

This electroslag welding technique is characterized by improved Charpy V-notch (CVN) impact toughness

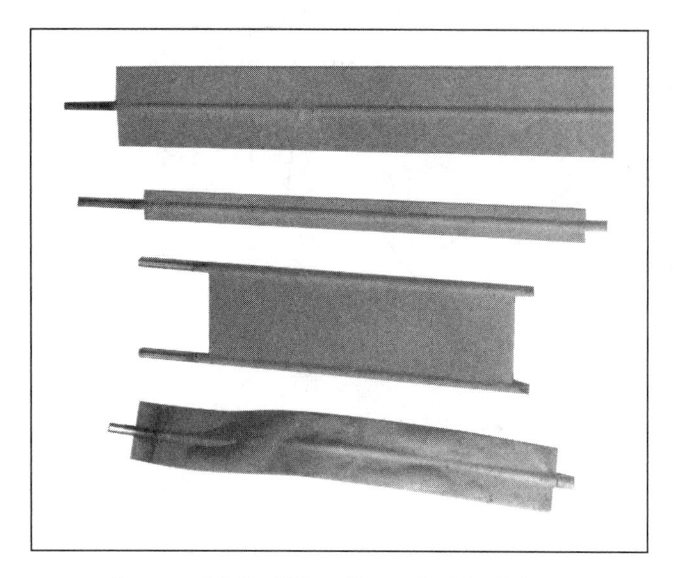

Figure 10.3—Wing-Type Guide Tubes

in both the weld metal and the heat-affected zone (HAZ), excellent fatigue strength, increased productivity, and high resistance to both solidification cracking and hydrogen-assisted cracking (also known as *ferrite vein cracking*).

Compared to consumable guide electroslag welding, the narrow groove electroslag welding technique includes the following characteristic features:

1. Reduced heat input through the use of tubular filler metal and faster travel speed;
2. Narrow root opening of only 19 mm (0.75 in.) in width;
3. Alloyed filler metal designed to be compatible with the A 709 Grades 36, 50, and 50W bridge steels; and
4. Consumable-web or wing-guide assembly.

All of these factors serve to increase weld-metal toughness. However, the CVN impact toughness of the HAZ is increased only by reducing heat input. Typically, the welding conditions to weld 50 mm (2 in.) thick A 709 Grade 50W are 1000 amperes (A), 36 volts (V) and 1 mm/s (2 in./min) travel speed for a heat input of approximately 40 kilojoules per millimeter (kJ/mm) (963 British thermal units per inch [Btu/in.]). This value of heat input is about half that produced by conventional consumable guide electroslag welding.

The filler metal used for narrow groove electroslag welding is metal-powder cored AWS A.5.25 FES70-EWT-G welding wire of 2.4 mm (3/32 in.) diameter. The chemical composition of the steel filler metal (in weight percent (wt %) is typically 0.04C(max)-1.2Mn-3Ni-0.2Mo-0.01Ti. Since the dilution from both the A 709 Grade 50W base metal and the 1008 steel consumable guide constitutes about 50% of the weld metal, the mechanical properties of the weld metal are dependent on the composition of the admixture of the resulting weld metal. The wing and web guides are constructed from low-carbon steel such as AISI/SAE 1010, 1008, or 1006.[2,3] The flux used for narrow groove electroslag welding is a fused neutral flux for both starting and running conditions.

The dimensions of the web and wing guides include a thickness of 6 mm (1/4 in.) and a width equal to the plate thickness minus 12 mm (1/2 in.). For example, when welding 50 mm (2 in.) thick plate, the guide would be 6 mm (1/4 in.) thick and 38 mm (1-1/2 in.) wide. The dual-wire web guide is used for plate thicknesses greater than 50 mm (2 in.), and the wing guide is used for thickness of 50 mm (2 in.) or less. By using the wing and web guides, oscillation of the guide is not normally performed in the narrow groove electroslag welding method. Long run-on and run-off tabs of 150 mm (6 in.) are used to ensure the production of sound weld metal.

The power source for narrow groove electroslag welding is typically a direct-current constant-voltage machine rated for 1500 A at 100% duty cycle. Water-cooled stationary shoes are mandatory for the deposit of full-penetration electroslag welds because moderate weld cooling rates are needed to achieve excellent toughness in both the weld metal and heat-affected zone. An automatic flux feeder is sometimes attached above the slag pool to replenish the thin layer of slag that is continuously consumed on the cold shoe surfaces during welding.

A continuous chart recorder is recommended for monitoring and recording current, voltage, and wire feed speed. Although not mandatory, the chart recorder serves three useful purposes: It provides a permanent record of the entire weld; it indicates the slag level in real time so that adjustments in slag height can be made before any problems develop; and it is a valuable aid when troubleshooting welding problems.

Because narrow groove electroslag welding utilizes higher travel speeds and faster weld cooling rates than conventional electroslag welding, greater resistance to cracking is required of the resulting weld. Susceptibility to both solidification cracking and hydrogen-assisted cracking increase with increasing travel speed. However, increased travel speed is necessary to achieve high CVN impact toughness in the weld metal and particularly in the heat-affected zone. Thus, the following

2. American Iron and Steel Institute, 1101 17th St. N.W., Suite 1300, Washington, DC 20036.
3. Society of Automotive Engineers International, 400 Commonwealth Drive, Warrendale, PA 15096-0930.

modifications are needed to prevent cracking while still maintaining high impact toughness:

1. Metal-powder tubular wire is used to increase the ratio of weld width to weld depth (W/D);
2. Web and wing-guide plates made from grade 1010 or 1008 steel are also used to increase the W/D ratio; and
3. C-Mn-Ni-Mo tubular filler metal very low in carbon is used to optimize impact toughness and provide resistance to solidification and hydrogen-assisted cracking.

The benefits of the reduced heat input for the weld metal include the production of a finer overall grain size, a greater amount of acicular ferrite, finer acicular ferrite, and reduced grain-boundary ferrite. The heat-affected zone benefits from a reduced heat input by developing a finer microstructure and decreased amounts of grain-boundary ferrite, Widmanstatten ferrite, and blocky ferrite. The alloys in the filler metal are selected to provide as much resistance to solidification cracking and hydrogen-assisted cracking as possible while still utilizing nickel, in particular, for toughness.

EQUIPMENT

The equipment for both the conventional and consumable guide electroslag welding processes is the same except for the design of the electrode guide tubes and the requirements for vertical travel. The major components of an electroslag welding installation include the following:

1. Power source,
2. Wire feeder and oscillator,
3. Electrode guide tube,
4. Welding controls,
5. Welding head, and
6. Retaining shoes.

POWER SOURCE

Power sources are typically of the constant-voltage transformer-rectifier type with ratings of 750 A to 1000 A, direct current (dc), at 100% duty cycle. The power sources are similar to those used for submerged arc welding. Load voltages generally range from 30 V to 55 V; therefore, the minimum open-circuit voltage of the power source should be 60 V. Constant-voltage alternating-current power sources of similar ratings are used for some applications. A separate power source is required for each electrode.

The power source is generally equipped with a contactor, a means for remote control of output voltage, a means for balancing multiple electrode installations, a main power switch, a range control, an ammeter and a voltmeter.

WIRE FEEDER AND OSCILLATOR

The function of a wire feeder is to deliver the wire electrode at a constant speed from the wire supply through the guide tube to the molten slag bath. The wire feeder is usually mounted on the welding head.

Each wire electrode is generally driven by its own drive motor and feed rolls. A dual gearbox to drive two electrodes from one motor can be used, but this does not provide redundancy in the event of a feeding problem. In the case of multiple-electrode welding, the failure of one wire drive unit need not shut down the welding operation if a corrective measure can be accomplished quickly. It should be emphasized, however, that for successful electroslag welding, it is vital to avoid a shutdown because weld repair at the restart can be costly. At times, fifty or more continuous operating hours are demanded of these wire drives for long, heavy weldments.

The motor-driven wire feeders used in electroslag welding are similar in design and operation to those used for other continuous electrode welding processes such as gas metal arc welding and submerged arc welding. A typical wire feed unit is illustrated in Figure 10.4. The feed rolls normally consist of a geared pair, the driving force thus being applied by both rolls. Configuration of the roll gap may vary depending on whether a solid or cored electrode is used. With solid wire, care must be taken to ensure that the wire feeds without slipping but is not squeezed so tightly that it becomes knurled. Knurled wire can create the effect of a file and abrade the components between the feed rolls and the weld. Rolls with an oval groove configuration have been found to perform best for both types of wire without danger of crushing metal-cored electrodes.

A wire straightener, either the simple three-roll design or the more complex revolving design, should be used to remove the cast in the wire electrode. Cast causes the electrode to wander as it emerges from the guide. This, in turn, can cause changes in the position of the weld pool, which may cause discontinuities such as incomplete fusion. Cast in the electrode causes a more severe problem on large, heavy weldments.

Electrode speed depends on the current required for the desired deposition rate as well as on the diameter and type of electrode being used. Generally, a speed range of 17 mm/s to 150 mm/s (40 in./min to 350 in./min) is entirely suitable for use with 2.4 mm (3/32 in.) or 3.2 mm (1/8 in.) diameter metal-cored or solid electrodes.

Electrodes should be packaged for uniform, uninterrupted feeding. The wire supply should allow feeding with minimal driving torque requirements so that binding or wire stoppage does not take place. The wire

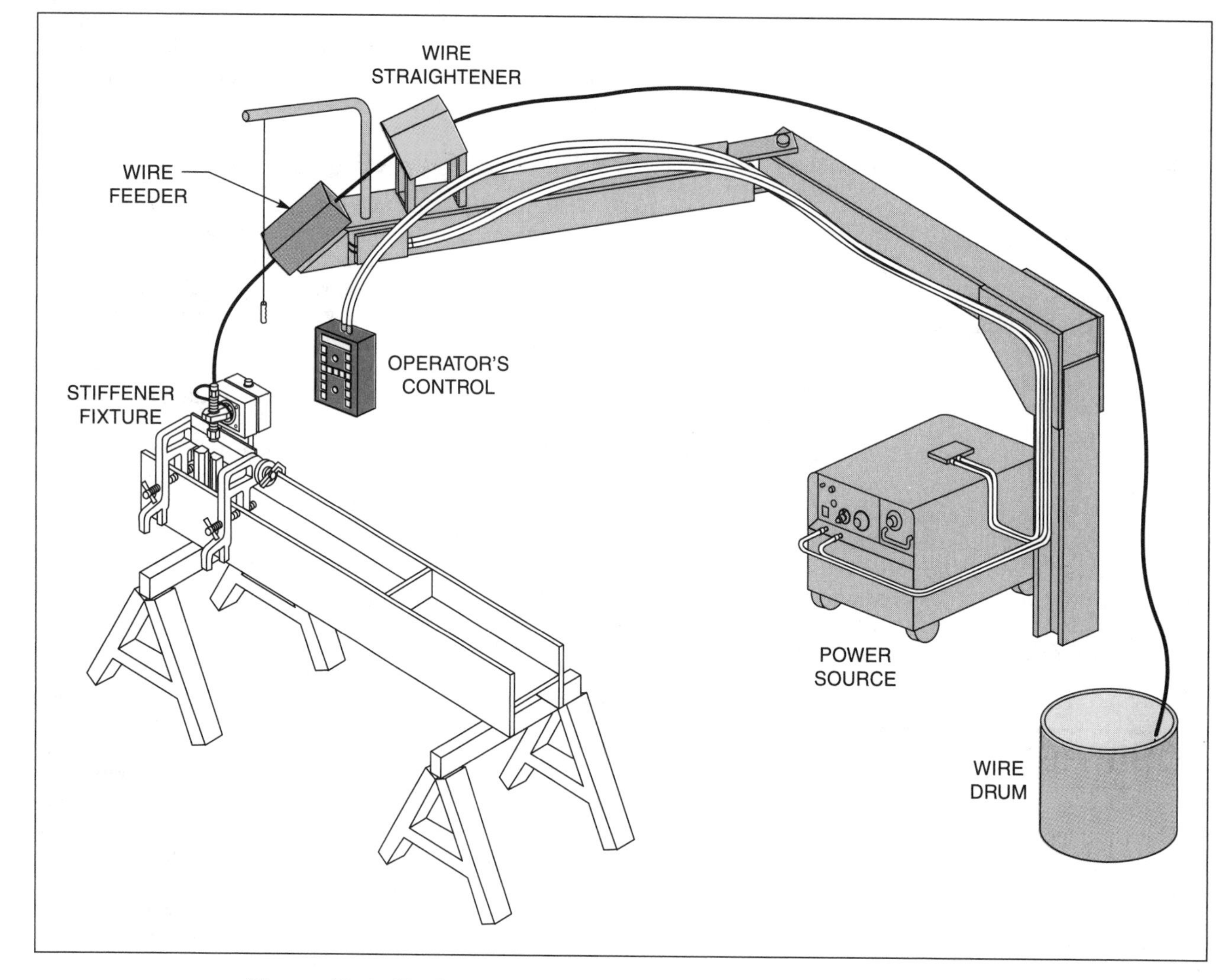

Figure 10.4—Typical Wire Feed Unit (Consumable Guide Method)

supply must be adequate to complete the entire weldment without stopping.

Electrode oscillation devices are required when the joint thickness exceeds approximately 57 mm (2-1/4 in.) per electrode. Oscillation of electrode guide tubes can be provided by motor-operated mechanical drive devices such as a lead screw or a rack and pinion. The drive must be adjustable for travel distance, travel speed, and variable delay at the end of each stroke. Electronic circuitry is generally used to control the oscillation movement.

ELECTRODE GUIDE TUBE

In conventional electroslag welding, the nonconsumable wire guide tube (often referred to colloquially as the *snorkel*) guides the electrode from the wire feed rolls into the molten slag bath. It also functions as an electrical contact to energize the electrode. Because the exit end of the tube is positioned close to the molten slag, it deteriorates with time.

Guide tubes are generally made of beryllium copper alloy and supported by two narrow, rectangular bars brazed to the tubes. Beryllium copper is used because it retains reasonable strength at elevated temperatures. The tubes are wrapped with insulating tape to prevent short-circuiting to the workpiece.

To feed the electrodes vertically into the molten slag, the guides must be curved and narrow enough to fit into the root opening of the joint. They are generally less than 13 mm (1/2 in.) in diameter. To overcome cast in the electrode, integral wire straightening can be designed into the tubes.

Consumable guide tubes are made of steel that is compatible with the base metal. They are slightly longer than the joint to be welded and commonly have an outside diameter of 12 mm to 16 mm (1/2 in. to 5/8 in.) and an inside diameter of 3.2 mm to 4.8 mm (1/8 to 3/16 in.). Smaller diameters are necessary for the welding of sections less than 19 mm (3/4 in.) thick. The guide tube is attached to a copper-alloy support tube, which is mounted on the welding head. Welding current is transmitted from the copper tube to the steel tube and subsequently to the electrode.

For welds more than 600 mm to 900 mm (2 ft to 3 ft) long, it is necessary to insulate the guide tubes to prevent short-circuiting with the workpiece. The entire tube length can be coated with flux. As an alternative, insulator rings spaced 300 mm to 450 mm (12 in. to 18 in.) apart can be slipped over the tube and held in place with small weld buttons on the tube. The flux covering or insulator rings melt and help replenish the slag bath as the guide tube is consumed.

WELDING CONTROLS

Electroslag welding controls are contained in a console mounted near the welding head. The console may provide the following component control groups:

1. Alarm systems for equipment or system malfunctions;
2. Ammeters, voltmeters, and remote contactor controls for each of the power sources and remote voltage control in the form of either a manual rheostat or a motorized rheostat, which is triggered by a reversing switch;
3. A speed control for each of the wire drive motors that also jogs and reverses the wire drive. On some consoles, the power contactor and wire drive can be activated by a common switch;
4. Oscillator controls to move the guides back and forth in the weld joint. Adjustable limit switches mounted on the welding head control the established length of stroke; timers control dwell duration at each end of the stroke; and
5. Control for the vertical rise of the welding head (conventional method only). The type of control depends on whether the rise is activated manually or automatically.

A sensor can be used with the rise device for automatic control. The sensor is aimed at a point below the top of the containment shoe and adjusted to detect the top of the molten slag bath. When the bath rises above the designated point, the rise drive is activated. It moves the welding head and shoe up until the bath is no longer detectable. A continuous incremental rise is obtained automatically in this manner.

In a given joint, if the plate thickness, root opening, number of electrodes, and electrode feed speed are known, the approximate vertical rate of rise can be computed. This speed can be set on a variable-speed motor by the welding operator. As welding progresses, minor adjustments can be made to keep the molten slag bath and liquid-metal pool within the containment shoes.

WELDING HEAD

The normal weldments in electroslag welding are relatively large and heavy. Therefore, a single convenient location should be established for the power source, with long cables to the portable welding head at the weld joint location. A remote control system is used. The control boxes are generally lightweight and contain a minimum of components. The controls interconnect between the power source and wire feeders on the welding head.

The welding head includes the wire feeder, electrode supply, wire guide tube(s), electrical connections to the guide tubes, and a means of attaching the welding head assembly to the workpiece. It may also contain provisions for multiple-electrode operation and an electrode oscillation drive unit. When portability is required, the wire feeder and electrode supply can be located a short distance away from the welding head, as is the case in semiautomatic gas metal arc welding.

An electroslag plate crawler is shown schematically in Figure 10.5. A single electrode is fed into the joint through a guide tube. Once welding is initiated, the crawler assembly is propelled upward by a serrated drive wheel that tracks the weld joint. The vertical speed of the plate crawler is set by remote control. Two water-cooled copper shoes slide along the joint to contain the molten slag and weld metal. The copper shoes are held tightly against the plate by spring tension between the front and rear sections of the plate crawler.

The plate crawler can be used to weld plates ranging in thickness from 13 mm to 50 mm (1/2 in. to 2 in.). Normally, either a square-groove or single-V-groove joint is used. Vertical travel speeds of up to 3 mm/s (7 in./min) are possible.

RETAINING SHOES (DAMS)

The retaining shoes and the associated water circulation system are included in this category. The function of the shoes is to maintain the molten metal and slag bath within the weld cavity. The shoes, which are fabricated of copper, generally include water passages at critical heat build-up points to prevent overheating or

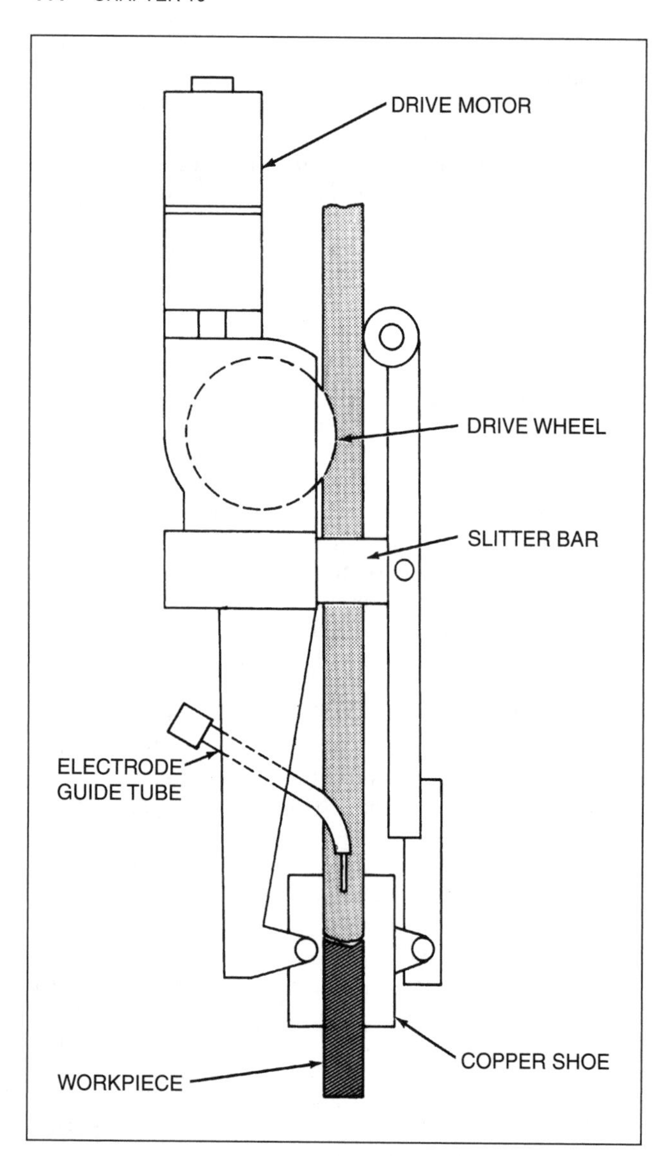

Figure 10.5— Plate Crawler Used with Conventional Electroslag Welding

melting. To provide for a slight reinforcement of weld metal, each shoe typically has a cavity machined in the side toward the weld, as shown in Figure 10.2.

The shoe can be cooled by a water circulation system or by tap water. Water circulators must have a heat removal capacity of 32 kilojoules per hour (kJ/h) to 42 kJ/h (30,000 British thermal units per hour [Btu/h] to 40,000 Btu/h). When using a recirculating system, condensation does not normally occur on the shoes. However, the use of tap water may cause condensation since tap water is frequently at a lower temperature than the ambient air. If the condensation runs down the shoes and collects in the starting weld tabs prior to starting the weld, weld porosity is likely to occur. Condensation on the inside of the shoes during welding evaporates ahead of the advancing slag bath. Thus, it is best to turn on the tap water just before starting the weld.

With conventional electroslag welding, water-cooled shoes mounted on the welding head travel upward as the welding progresses. With the consumable guide method, the shoes do not move. However, they can be repositioned in leapfrog fashion as welding progresses upward. The shoes are sometimes not water-cooled, in which case the shoes must be massive to avoid melting. The shoes are clamped in place by means of wedges against U-shaped bridges (strongbacks) across the joint or by means of large C-clamps on short welds made with the consumable guide method.

MATERIALS

The composition of electroslag weld metal is determined by the compositions of the base metal, the filler metal, and their relative dilution. The filler metal and the consumables used in electroslag welding include the electrode, the flux, and, in the case of consumable guide welding, a consumable guide and its insulation. The welding consumables can effectively control the final chemical composition and mechanical properties of the weld metal.

ELECTRODES

Two types of electrodes, solid and metal-cored, are used with the electroslag welding process. Solid electrodes are more widely used. Various chemical compositions are available with each type of electrode to produce the desired mechanical properties in the weld metal.

For the welding of alloy steels, metal cored electrodes permit the adjustment of the composition of the filler metal through alloy additions in the core. Metal cored electrodes also provide a means for replenishing flux in the molten bath. The metal tube is made of low-carbon steel. The use of flux cored electrodes may result in excessive buildup of slag in the bath when the core is composed entirely of flux.

In the electroslag welding of carbon steels and high-strength low-alloy steels, the electrodes usually contain less carbon than the base metal. Weld metal strength and toughness are achieved by alloying with a variety of elements. This approach reduces the tendency of weld-metal cracking in steels containing up to 0.35% carbon.

The electrode wire compositions used to weld higher-alloy steels normally match the compositions of the base metal. Higher-alloy steels generally develop mechanical properties by means of a combination of chemical composition and heat treatment. It is usually necessary to heat treat an electroslag weld in a higher-alloy steel to develop the desired properties of the weld metal and the heat-affected zone. Thus, the best approach is to select a weld-metal composition and a base metal composition designed so that both respond to heat treatment to approximately the same degree.

When the electrode wire for electroslag welding is selected, dilution with the base metal must be considered. In a typical electroslag weld, the dilution of the base metal generally runs from 30% to 50%. The amount of dilution from base metal melting is dependent on the welding procedure. The filler metal and melted base metal mix thoroughly to provide the weld an almost uniform chemical composition throughout.

The most popular electrode diameters are 2.4 mm (3/32 in.) and 3.2 mm (1/8 in.). However, 1.6 mm to 4.0 mm (1/16 in. to 5/32 in.) diameter electrodes have been successfully used. Smaller-diameter electrodes provide a higher deposition rate than larger electrodes at the same welding amperage. Based on practical applications, it has been determined that either 2.4 mm (3/32 in.) or 3.2 mm (1/8 in.) diameter electrodes can provide the optimum combination of deposition rate, feeding ability, welding amperage ranges, and the ability to be straightened.

The electrodes used in electroslag welding are typically supplied in 27 kg (60 lb) coils, large spools weighing 270 kg (600 lb), or large drums. Since it is essential that enough electrode wire remains available to complete the entire weld joint, spools or drums of up to 340 kg (750 lb) are generally used because they are practical and economical.

FLUX

The flux is a major part of the successful operation of the electroslag welding process. Flux composition is of the utmost importance since its characteristics determine how well the electroslag process operates. During the welding operation, the flux is melted into slag, which transforms the electrical energy into thermal energy to melt the filler metal and base metal. The slag (flux) must conduct the welding current, protect the molten weld metal from the atmosphere, and provide for stable operation.

The fluxes used for electroslag welding are usually combinations of complex oxides of silicon, manganese, titanium, calcium, magnesium, and aluminum, with some calcium fluoride always present to adjust the electrical characteristics. Special characteristics can be achieved by varying the composition of the flux.

An electroslag welding flux must have several important characteristics. When molten, it must be electrically conductive and yet have adequate electrical resistance to generate sufficient heat for welding. However, if its resistance is too low, arcing may occur between the electrode and the surface of the slag bath. To ensure the even distribution of heat in the joint, the viscosity of the molten slag must be fluid enough for good circulation. A slag that is too viscous may cause slag inclusions in the weld metal, whereas one that is too fluid may leak out of small openings between the workpiece and the retaining shoes.

The melting point of the flux must be well below that of the metal being welded, and the boiling point must be well above the operating temperature to avoid losses that could change operating characteristics. The molten slag is generally chosen to be relatively inert to reactions with the metal being welded. It should be stable over a wide range of welding conditions and slag bath sizes. However, it is sometimes advantageous to select a flux that produces a reactive slag. Reactive slags can be used to refine the weld metal or to adjust the level of impurities such as oxygen.

Solidified slag on the weld surfaces should be easy to remove, although some commercial fluxes may not have this characteristic.

Only a relatively small amount of flux is used during electroslag welding. An initial quantity of flux is required to establish the process. Flux solidifies as slag in a thin layer on the cold surfaces of the retaining shoes and on both weld faces. It is necessary to add flux to the molten bath during welding to maintain the required depth. Not including losses incurred through leakage, the total flux used is approximately 0.5 kg (1 lb) for each 9 kg (20 lb) of deposited metal. However, as the plate thickness or weld length increases, the flux consumption drops to 0.5 kg (1 lb) for each 36 kg (80 lb) of deposited metal.

Under normal conditions, electroslag flux in the original unopened packages is usually protected against moisture pickup by the packaging. Flux reconditioning may be necessary if the flux has been exposed to high humidity.

CONSUMABLE GUIDE TUBE

The primary functions of the consumable guide tube are to provide support to the electrode wire from the welding head to the molten slag bath and to act as the primary current path. The consumable guide melts periodically just at the top of the rising molten slag bath. This permits the welding head to be fixed in position at the top of the vertical joint. The electrode cable is attached to the guide tube. The welding current is conducted to the electrode as it passes through the end of the guide tube and then into the molten slag bath.

The outside diameter of most consumable guide tubing measures 13 mm (1/2 in.) or 16 mm (5/8 in.). The inside diameter of the tubing is normally determined by the size of the electrode wire being used. The amount of metal contributed to the weld by the melted guide tube is generally small, except when welding thin sections. Since the guide tube becomes part of the weld, the composition of the guide tube should be compatible with the desired weld metal composition.

When short welds are to be made, a bare consumable guide tube can be used. However, for long welds, the tube must be insulated to prevent electrical contact with the base metal. A flux coating can be used to provide electrical insulation and add flux to the slag bath. Other forms of insulation include doughnut-shaped insulators, fiberglass sleeves, and tape. Because the insulation becomes part of the slag pool, a type of insulation that does not affect the deposited weld metal or the operating characteristics of the flux should be selected.

SPECIFICATION FOR ELECTRODES AND FLUXES

The standard *Specification for Carbon and Low Alloy Steel Electrodes and Fluxes for Electroslag Welding*, ANSI/AWS A5.25, classifies the electrodes and fluxes used for electroslag welding.[4,5] Metal cored electrodes are classified on the basis of a chemical analysis of weld metal taken from an undiluted ingot. Solid electrodes are classified on the basis of their chemical composition as manufactured. Since the consumable guide tube usually contributes only a small amount of filler metal to the joint, it does not change the flux-electrode classification. However, the guide tubes must conform to AISI specifications for 1008 to 1020 carbon-steel tubing.

Metal cored electrodes deposit weld metals of low-carbon steel and low-alloy steels. The low-alloy steel deposits contain small amounts of nickel and chromium in addition to either copper or molybdenum. The carbon content is less than 0.15%. Solid electrodes are divided into three classes—medium manganese (approximately 1%), high manganese (approximately 2%), and a special class.

In the flux-electrode classification system, both metal cored and solid electrodes are used in any combination with nine fluxes. The fluxes are classified on the basis of the mechanical properties of a weld deposit made with a particular electrode and a specified base metal, resulting in nine potential classes. The compositions of the fluxes are left to the discretion of the manufacturers. Three levels of tensile strength for flux-electrode combinations are specified: 430 megapascals (MPa) to 550 MPa (60 thousand pounds per square inch [ksi] to 80 ksi); 480 MPa to 650 MPa (70 ksi to 95 ksi); and 550 to 700 MPa (80 to 100 ksi). For each level of strength, two of the three flux-electrode classifications must meet minimum toughness requirements as determined by the Charpy V-notch impact test.

WELDING VARIABLES

Welding variables are the factors that affect the operation of the process, weld quality, and the economics of the process. A smooth-running process and a quality weld deposit result when all of the variables are in proper balance. In electroslag welding, it is essential that the effects of each variable be fully understood, since they differ from those associated with the conventional arc welding processes.

FORM FACTOR

The slow cooling rate and solidification patterns of an electroslag weld are similar to those of metal cooling in a mold. In electroslag welding, heat is removed from the molten weld metal by the cool base metal and the water-cooled retaining shoes. Solidification begins at these cooler areas and progresses toward the center of the weld, as shown in Figure 10.6(A). However, since filler metal is added continuously and the joint fills during welding, solidification progresses from the bottom of the joint, as indicated by the solidified grain structure shown in Figure 10.6(B).

The angle at which the grains meet in the center is determined by the shape of the weld pool. The weld pool shape can be expressed by the form factor. The *form factor* is the ratio of the width of the weld pool to the maximum depth. The width is comprised of the root opening plus the total penetration into the base metal. The depth is the distance from the top of the weld pool to the lowest level of the liquid-solid interface. The dendrites grow competitively into the weld pool at an angle of approximately 90° to the solid-liquid interface.

A line drawn perpendicular to the dendritic grains approximates the shape of the solid-liquid interface. This is illustrated in Figure 10.6(B). This boundary defines the width and depth of the weld pool, and the

4. American Welding Society (AWS) Committee on Filler Metals and Allied Materials, 1997, *Specification for Carbon and Low Alloy Steel Electrodes and Fluxes for Electroslag Welding*, ANSI/AWS A5.25/AS.25M-97, Miami: American Welding Society.

5. At the time of the preparation of this chapter, the referenced codes and other standards were valid. If a code or other standard is cited without a date of publication, it is understood that the latest edition of the document referred to applies. If a code or other standard is cited with the date of publication, the citation refers to that edition only, and it is understood that any future revisions or amendments to the code or standard are not included; however, as codes and standards undergo frequent revision, the reader is advised to consult the most recent edition.

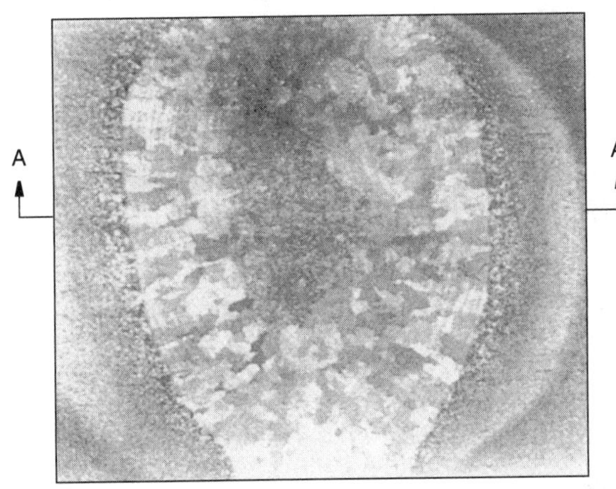

(A) Transverse Section

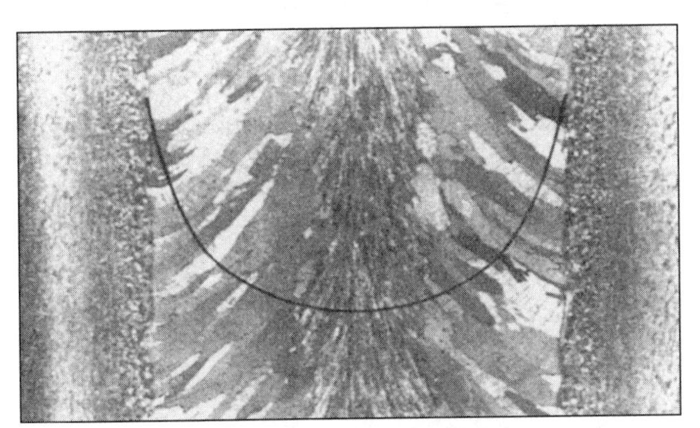

(B) Longitudinal Section at A-A

Figure 10.6—Transverse and Longitudinal Section through a 100 mm (4 in.) Thick Electroslag Weld

form factor can be easily determined. Welds having a high form factor (i.e., a wide width and a shallow weld pool) tend to solidify with the grains meeting at an acute angle. Welds having a low form factor (i.e., a narrow width and a deep weld pool) tend to solidify with the grains meeting at an obtuse angle. Thus, the form factor indicates how the grains from opposite sides of the weld meet at the center.

The angle at which the grains meet in the center determines whether the weld has high or low resistance to hot centerline cracking. If the dendritic grains meet head on at an obtuse (large) included angle, the cracking resistance will be low. However, if the angle is acute (small), the cracking resistance will be high. Therefore, maximum resistance to cracking is obtained with a high form factor.

The shape of the weld pool and the resultant form factor are controlled by the welding variables. However, the form factor alone does not control cracking. The base-metal composition (especially the carbon content), the filler-metal composition, and joint restraint have a significant effect on the propensity for cracking. Some studies have indicated that the manganese content, or the manganese-to-silicon ratio, is also important with respect to cracking.

WELDING AMPERAGE

Welding amperage and electrode feed rate are directly proportional to one another and therefore can be treated as one variable. Increasing the electrode feed speed increases the welding amperage and the deposition rate when a constant-voltage power source is used.

As the welding amperage is increased, so is the depth of the weld pool. When welding with a 3.2 mm (1/8 in.) diameter electrode below approximately 400 A, an increase in amperage also increases weld width. The net result is a slight increase in form factor. However, when operating a 3.2 mm (1/8 in.) diameter electrode above 400 A, an increase in amperage reduces weld width. Thus, the net effect of increasing the welding amperage is to decrease the form factor and thus lower the resistance to cracking.

Welding amperages of 500 A to 700 A are commonly used with 3.2 mm (1/8 in.) electrodes. Metals prone to cracking or conditions that promote cracking may require a high form factor associated with welding amperages below 500 A.

WELDING VOLTAGE

Welding voltage is an extremely important variable. It has a major effect on the depth of fusion into the base metal and on the stable operation of the process. Welding voltage is the primary means for controlling the depth of fusion. Increasing the voltage increases both the depth of fusion and the width of the weld. The depth of fusion must be somewhat greater in the center of the weld than at the edges to ensure complete fusion at the outside edges, where the chilling effect of the water-cooled shoes must be overcome.

Since an increase in welding voltage increases the weld width, it also increases the form factor and thereby increases cracking resistance. The voltage must also be maintained within limits to ensure stable operation of the process. If the voltage is low, short-circuiting or arcing to the weld pool may occur. Too high a voltage may produce unstable operation because of slag spatter and arcing on the top of the slag bath. Welding voltages of 32 V to 55 V per electrode are used. Higher voltages are used with thicker sections.

ELECTRODE EXTENSION

When using conventional electroslag welding, the distance between the surface of the slag bath and the end of the guide tube is referred to as the *dry electrode extension*. No dry extension occurs in the consumable guide method since the guide tube melts primarily by means of heat conduction from the molten slag. However, at high-heat input ranges, radiation heat transfer from the slag bath may be sufficient to melt the guide above the slag pool. Resistance is increased by using constant-voltage power and constant electrode feed speed and by increasing the dry electrode extension. This causes the power source to reduce the current output, which slightly increases the form factor.

Electrode extensions of 50 mm to 75 mm (2 in. to 3 in.) are generally used. Extensions of less than 50 mm (2 in.) usually cause overheating of the guide tube. Those greater than 75 mm (3 in.) cause overheating of the electrode because of the increased electrical resistance. Hence, at long extensions the electrode melts at the surface of the slag bath instead of in the bath. This results in instability and improper slag bath heating.

ELECTRODE OSCILLATION

Plates up to 75 mm (3 in.) thick can be welded with a stationary electrode and high voltage. However, the electrode is usually oscillated horizontally across the plate thickness when the material exceeds 50 mm (2 in.). The oscillation pattern distributes the heat and helps to obtain better edge fusion. Oscillation speeds vary from 8 mm/s to 40 mm/s (20 in./min to 100 in./min), with the speed increasing to match the plate thickness.

Generally, oscillation speeds are based on a traverse time of 3 seconds to 5 seconds. Increasing the oscillation speed reduces the weld width and hence the form factor. Thus, the oscillation speed must be balanced with the other variables. A dwell period is used at each end of the oscillation travel to obtain complete fusion with the base metal and overcome the chilling effect of the retaining shoes. The dwell time may vary from 2 seconds to 7 seconds.

DEPTH OF THE SLAG BATH

A minimum slag bath depth is necessary so that the electrode enters the bath and melts beneath the surface. Too shallow a bath causes slag expulsion (spitting) and arcing on the surface. Excessive bath depth provides excessive area for heat transfer into the retaining shoes and the base metal. This reduces the overall temperature of the slag bath, which reduces the weld width and hence the form factor. Slag bath circulation is poor with excessive depth, and the cooler slag may tend to solidify on the surface of the base metal, resulting in slag inclusions. A bath depth of 38 mm (1-1/2 in.) is optimum, but it can be as low as 25 mm (1 in.) or as high as 50 mm (2 in.) without significant effects.

NUMBER AND SPACING OF ELECTRODES

As the metal thickness per electrode increases, the weld width decreases slightly, but the depth of the weld pool decreases greatly. Thus, the form factor improves as the thickness of material increases for a given number of electrodes. However, a point is reached at which the weld width at the cool retaining shoes is smaller than the root opening, and incomplete edge fusion results. At this point, the number of electrodes must be increased.

In general, one oscillating electrode can be used for sections up to 130 mm (5 in.) thick, and two oscillating electrodes can be used for sections up to 300 mm (12 in.) thick. Each additional oscillating electrode accommodates approximately 150 mm (6 in.) of additional thickness. This applies to both the conventional and consumable guide methods. If non-oscillating electrodes are used, each electrode handles approximately 65 mm (2-1/2 in.) of plate thickness.

ROOT OPENING

A minimum root opening is needed for sufficient slag bath size, good slag circulation, and in the case of the consumable guide method, clearance for the guide tube and its insulation. Increasing the root opening does not affect the depth of the weld pool. However, it increases the weld width and hence the form factor.

Root openings are generally in the range of 20 mm to 40 mm (3/4 in. to 1-1/2 in.), depending on the thickness of the base metal, the number of electrodes, and the use of electrode oscillation. Excessive root openings require extra amounts of filler metal, which may not be economical. In addition, they may cause incomplete edge fusion.

WELDING PROCEDURES

The properties and quality of welded joints are determined by the specific welding procedure used and by the ability of the operator to apply the procedure. Predictable properties and soundness of a joint can be secured only by adherence to a welding procedure that has been properly qualified. It cannot be expected that predictable and repeatable results will be obtained, even by careful and painstaking welding operators, if poor material or inadequate or worn-out equipment is used. Also, a properly qualified welding procedure will not result in the desired welded joint if the operator has not been properly trained or supervised to ensure that all essential details of the specific welding procedures are followed. In addition, preparation of the coupons for testing must be done correctly to obtain proper and consistent test results.

JOINT PREPARATION

One of the major advantages of the electroslag process is the relatively simple joint preparation. It is basically a square-groove joint, so the only preparation required is a flat, straight edge on each groove face. This can be produced by thermal cutting, machining, or a similar method. If sliding retaining shoes are used, the plate surfaces on both sides of the groove must be reasonably smooth to prevent slag leakage and the jamming of the shoes.

The joint should be free of oil, heavy mill scale, or moisture, just as with any welding process. However, the joint need not be as clean as would normally be required for other welding processes. Oxygen-cut surfaces should be free of adhering slag, but a slightly oxidized surface is not detrimental.

Care should be exercised to protect the joint prior to initiating the weld. Porosity may be caused by moisture-bearing materials, referred to colloquially as *mud,* packed around the shoes. The joint must be completely dry before welding. Leaking water-cooled shoes may also cause porosity and other discontinuities at the weld face.

JOINT FITUP

Prior to welding, components should be set up with the proper joint alignment and root opening. Rigid fixturing or strongbacks that bridge the joint should be used. Strongbacks are bridge-shaped plates that are welded to each component along the joint so that alignment during welding is maintained. They are designed with clearance to span fixed-position and movable retaining shoes. After welding is completed, they are removed.

For the consumable guide method, imperfect joint alignment can be accommodated to some degree. Plates with large misalignments can be welded with special retaining shoes that are adaptable to the fitup. Alternatively, the space between the shoes and the workpiece can be packed with refractory material or steel strips (the steel must be similar in composition to the base metal). After the weld is complete, the steel strips can be removed in a way that will blend the weld faces smoothly with the adjacent base metal.

Experience dictates the proper root opening for each application. As welding progresses up the joint, the workpieces are drawn together by weld shrinkage. Therefore, for long welds, the root opening at the top of the joint should be approximately 3 mm to 6 mm (1/8 in. to 1/4 in.) more than at the bottom to allow for this shrinkage. The factors influencing shrinkage allowance include material type, joint thickness, joint length, and vertical travel speed.

Using the proper root opening and shrinkage allowance is important for maintaining the dimensions of the weldment. However, if the root opening proves incorrect, the welding conditions can be varied, within limits, to compensate for it. For example, if the initial root opening is too small, the wire feed speed can be lowered to reduce deposition rate and increase penetration. If the root opening is too large, the wire feed speed can be increased within good operating limits. Regardless of the type of base metal, the voltage should be increased to account for a wider root opening.

If the root opening is greater than the capability of the normal number of electrodes used, an additional electrode can be added, if space is available. In some cases, electrode oscillation can compensate for excessive root opening

When the root opening is too small, the joint may fill too fast, causing weld cracks or incomplete edge fusion. It is also possible that a small root opening may close up because of weld shrinkage and thus stop an oscillating guide tube from traversing.

INCLINATION OF THE WORKPIECE

The axis of the weld joint is generally in the vertical or near-vertical position. If it deviates by more than 10° to 15° from vertical, special welding procedures must be used. With greater deviations, it becomes increasingly difficult to weld without slag inclusions and incomplete edge fusion. However, acceptable electroslag welds have been produced at 45° and more from the vertical position.

The alignment of the guide tube can be a problem with inclined welds when using the consumable guide method. Large insulators (flux rings) or flux-coated guide tubes with spring clips are often required to keep the tube aligned in the joint.

ELECTRICAL CONNECTION TO THE WORKPIECE

A good workpiece connection, or electrical return, is important because of the relatively high welding currents used for the electroslag welding process. Two 4/0 welding leads are normally sufficient for each electrode. It is best to attach the workpiece lead directly under the sump, that is, below the electrode. In this location, the effect of any strong magnetic field in the weldment on filler-metal transfer is minimized.

Spring-type ground clamps are not recommended because they tend to overheat. A more positive connection is best, such as one made with a C- clamp.

STARTING WELD TABS AND RUN-OFF WELD TABS

When full penetration is required for the full length of the joint, starting weld tabs (often called *run-on tabs)* and run-off weld tabs are required. The starting weld tabs are located at the bottom of the joint. They are used in conjunction with a starting plate to initiate the welding process. Generally, the tabs and starting plate are made from metal that is the same as or similar to the base metal. The starting weld tabs and plate form the sump in which the weld is started. In this case, the sump is removed and discarded after welding. The tabs and starting plate are the same thickness as the base plates.

When the starting weld tabs and starting plate are disposable, a tab is welded to the bottom of each base plate. The starting plate is welded across the bottom of the tabs to form the sump. The faces of the sump are flush with the base plate surfaces.

Copper sumps can be used, in which case water-cooling is usually necessary. The arc is not initiated on the copper sump because it would melt through the water jacket. Normally, the arc is started on one or two small blocks of base metal placed in the bottom of the copper sump.

Disposable run-off tabs should also be of the same metal or a metal similar to the base metal. Copper tabs can be used, but they must be water-cooled. The run-off tabs should be the same thickness as the base metal and should be securely attached to both plates at the end of the joint. The weld is completed in the cavity that they form above the base plates.

ELECTRODE POSITION

The position of the electrode determines where the greatest amount of heat is generated. The electrode should normally be centered in the joint. However, if the electrode is cast toward one side, the wire guide may be displaced in the opposite direction to compensate for the cast. When welding corner and T-joints or any joint where a fillet is to be formed, the electrode may need to be offset to produce the required weld metal geometry.

INITIATION AND TERMINATION OF WELDING

One method of initiating electroslag welding is to pour molten slag into the joint. The usual starting method is to strike an arc between the electrode and the starting plate. This may be done by inserting a steel wool ball between the electrode and the base metal and turning the power on, or by advancing the electrode toward the starting plate with the power on. The latter method requires a chisel point on the end of the electrode. Once an arc is struck, flux is slowly added until the arc is extinguished. At this point, the process is in the electroslag mode.

It is extremely important that the welding operation proceed without interruption. The equipment should be checked and the supplies of electrode and flux should be verified and determined to be adequate before the weld is started. The weld must not be interrupted to replenish the electrode supply. The welding equipment must be capable of operating continuously until the weld is finished.

When the weld has reached the run-off tabs, termination should follow a procedure that fills the crater; otherwise, crater cracking may occur. In this procedure, the electrode feed is gradually reduced when the slag reaches the top of the run-off tabs and the crater fills. The welding amperage decreases simultaneously. When the electrode feed stops, the power source is turned off. The starting weld tabs, the run-off tabs, and the excess weld metal are then cut off flush with the top and bottom edges of the weldment.

SLAG REMOVAL

A chipping gun is effective in slag removal, although a slag hammer or pick also does the job. Slag also adheres to the copper shoes and the copper sump, if used. These must be cleaned before another weld is made. Eye protection should be worn during slag removal operations.

CIRCUMFERENTIAL WELDS

A square-groove butt joint is used to join two cylindrical components end to end. The weld is made by rotating the workpieces and allowing the slag and molten weld metal to remain at the three o'clock position with clockwise rotation. The welding head is stationary until the finish of the weld. Special starting and run-off techniques are required to complete the joint. Small-diameter weldments may not be economically feasible. However, as the wall thickness and diameter increase, the cost savings may make the process advantageous.

METALLURGICAL CONSIDERATIONS

During electroslag welding, heat is generated by the resistance to current flow as it passes from the electrode(s) through the molten slag into the weld pool. The molten slag, being conductive, is electromagnetically stirred in a vigorous rotating action. Heat diffuses throughout the entire cross section being welded.

The temperatures attained in the electroslag welding process are considerably lower than those achieved with the arc welding processes. However, the temperature of the molten slag bath must be higher than the melting range of the base metal for satisfactory welding. The molten zone of both slag and weld metal advances relatively slowly, usually approximately 13 mm/min to 50 mm/min (1/2 in./min to 2 in./min), and the weld is generally completed in one pass.

Conventional electroslag welds differ from consumable electrode arc welds in several ways, including the following:

1. In the as-deposited condition, a generally favorable residual stress pattern is developed in joints produced with electroslag welding. The weld surfaces and heat-affected zones are normally in compression, and the center of the weld is in tension.
2. Because of the symmetry of most vertical electroslag butt welds (square-groove joints welded in a single pass), no angular distortion occurs in the weldment. A slight distortion may occur in the vertical plane caused by the contraction of the weld metal, but compensation can be made for this during joint fitup.
3. The weld metal stays molten long enough to permit some slag-refining action, as in electroslag remelting. The progressive solidification allows gases in the weld metal to escape and nonmetallic inclusions to float up and mix with the slag bath. High-quality, sound weld deposits are generally produced with electroslag welding.
4. The circulating slag bath washes the groove faces and melts them in the lower portion of the bath. The weld deposit contains up to 50% admixed base metal, depending on welding conditions. Therefore, the composition of steel being welded and the amount melted significantly affect both the chemical composition of the weld metal and the mechanical properties of the resulting welded joint.
5. The prolonged thermal cycle results in a weld-metal structure that consists of large prior austenite grains that typically follow a columnar solidification pattern. The grains are oriented horizontally at the weld-metal edges and turn to a vertical orientation at the center of the weld, as shown in Figure 10.6(B). The microstructure of electroslag welds in low-carbon steels generally consists of acicular ferrite and pearlite grains with proeutectoid ferrite outlining the prior austenite grains. It is very common to observe coarse, prior austenite grains at the periphery of the weld and a much finer-grained region near the center of the weld. This fine-grained region appears equiaxed in a transverse cross section; however, longitudinal sections reveal its columnar nature, as shown in Figure 10.6(B). Changes in the composition of the weld metal and, to a lesser extent, the welding procedure can markedly change the relative proportions of the coarse- and fine-grained regions to the extent that only one may be present.
6. The relatively long time at high temperatures and the slow cooling rate after welding result in wide heat-affected zones with a relatively coarse grain structure. The cooling rate is slow enough to form only relatively soft, high-temperature transformation products. For most steels, this is an advantage, particularly if stress-corrosion cracking might be a problem.

PREHEAT AND POSTHEAT

Preheating is not required and generally not used in electroslag welding. The absence of preheating is a significant advantage over arc welding for many types of steel. The electroslag welding process is self-preheating in that a significant amount of heat is conducted into the workpieces, preheating them ahead of the weld. Moreover, because of the very slow cooling rate after welding, postheating is usually unnecessary.

POSTWELD HEAT TREATMENT

Most applications of electroslag welding, particularly the welding of structural steel, require no postweld heat treatment. As discussed previously, as-deposited electroslag welds have a favorable residual stress pattern that would be negated by a postweld heat treatment. Subcritical postweld heat treatments such as stress relief can be either detrimental or beneficial to mechanical properties, particularly notch toughness. Such heat treatments are generally not employed after electroslag welding.

The properties of carbon and low-alloy steel welds can be greatly altered by heat treatment. Normalizing removes nearly all traces of the cast structure of the weld and nearly equalizes the properties of the weld metal and base metal. This may improve the resistance to brittle fracture initiation and propagation above certain temperatures as measured by the Charpy V-notch impact test.

Quenched-and-tempered steels are not usually joined with the electroslag welding process. They must be heat-treated after welding to obtain adequate mechanical strength properties in the weld and heat-affected zones. Heat treatment is very difficult to apply to large, thick structures.

The data shown in Tables 10.1 to 10.9 are based on butt welds in carbon steels using water-cooled shoes. Adjustments may need to be made depending on the flux, electrode diameter, and joint design used. The welding voltages shown are measured directly across the slag bath, not at the power source, as a voltage drop occurs in the welding cables and consumable guide tube(s).

WELDING SCHEDULES

Typical welding conditions that produce sound butt welds in normal situations are listed in Tables 10.1 to 10.9. However, the typical welding conditions shown are not necessarily the only conditions that can be used for a particular weld. It is possible that as the particular requirements of a repetitive weld become better known, the settings can be adjusted to obtain optimum welding results. Qualification tests often required for code work may provide conditions different from those listed in the tables.

Table 10.1
Typical Electroslag Welding Conditions: One Non-Oscillating Electrode

Plate Thickness		Joint Opening		Current	Voltage
mm	in.	mm	in.	A	V
19	3/4	25	1	500	35
25	1	25	1	600	38
50	2	25	1	700	39
76	3	25	1	700	52

Table 10.2
Typical Electroslag Welding Conditions: One Oscillating Electrode

Plate Thickness		Joint Opening		Oscillation Distance		Oscillation Speed*		Current	Voltage
mm	in.	mm	in.	mm	in.	mm/s	in./min	A	V
50	2	32	1-1/4	32	1-1/4	11	25	700	39
76	3	32	1-1/4	57	2-1/4	19	45	700	40
102	4	32	1-1/4	83	3-1/4	27	65	700	43
127	5	32	1-1/4	108	4-1/4	36	85	700	46

*A dwell time of 2 seconds is used at each shoe.

Table 10.3
Typical Electroslag Welding Conditions: Two Non-Oscillating Electrodes

Plate Thickness		Joint Opening		Electrode Spacing		Current	Voltage
mm	in.	mm	in.	mm	in.	A	V
76	3	25	1	64	2-1/2	425/wire	40
102	4	25	1	64	2-1/2	425/wire	43
127	5	25	1	64	2-1/2	425/wire	46

Table 10.4
Typical Electroslag Welding Conditions: Two Oscillating Electrodes[a]

Plate Thickness		Joint Opening		Oscillation Distance		Oscillation Speed[b]		Current	Voltage
mm	in.	mm	in.	mm	in.	mm/s	in./min	A	V
127	5	32	1-1/4	25	1	8	20	700/wire	41
152	6	32	1-1/4	50	2	17	40	700/wire	42
203	8	32	1-1/4	102	4	34	80	700/wire	45
254	10	32	1-1/4	152	6	50	120	700/wire	48
305	12	32	1-1/4	203	8	50	120	700/wire	51

a. Electrode spacing is 83 mm (3-1/4 in.) for two oscillating electrodes.
b. A dwell time of 2 seconds is used at each shoe.

Table 10.5
Typical Electroslag Welding Conditions: Three Non-Oscillating Electrodes

Plate Thickness		Joint Opening		Electrode Spacing		Current	Voltage
mm	in.	mm	in.	mm	in.	A	V
152	6	25	1	64	2-1/2	500/wire	41
178	7	25	1	64	2-1/2	550/wire	45
203	8	25	1	70	2-3/4	600/wire	49
229	9	25	1	76	3	625/wire	53

Table 10.6
Typical Electroslag Welding Conditions: Three Oscillating Electrodes[a]

Plate Thickness		Joint Opening		Electrode Spacing		Oscillation Distance		Oscillation Speed[b]	
mm	in.	mm	in.	mm	in.	mm	in.	mm/s	in./min
305	12	38	1-1/2	114	4-1/2	50	2	25	60
330	13	38	1-1/2	127	5	64	2-1/2	32	75
356	14	38	1-1/2	133	5-1/4	70	2-3/4	35	82.5
381	15	38	1-1/2	140	5-1/2	76	3	38	90
406	16	38	1-1/2	152	6	89	3-1/2	44	105
432	17	38	1-1/2	159	6-1/4	95	3-3/4	48	112.5
457	18	38	1-1/2	165	6-1/2	102	4	50	120

a. Initially, 600 A and 55 V should be used on each electrode. Once the welder has gained experience, the welding current and the voltage can be adjusted in accordance with the section titled "Welding Variables."
b. A dwell time of 4 seconds is used at each shoe.

Table 10.7
Typical Electroslag Welding Conditions: Four Oscillating Electrodes[a]

Plate Thickness		Joint Opening		Electrode Spacing		Oscillation Distance		Oscillation Speed[b]	
mm	in.	mm	in.	mm	in.	mm	in.	mm/s	in./min
457	18	38	1-1/2	127	5	50	2	32	75
483	19	38	1-1/2	133	5-1/4	57	2-1/4	35	82.5
508	20	38	1-1/2	140	5-1/2	64	2-1/2	38	90
533	21	38	1-1/2	146	5-3/4	70	2-3/4	41	97.5
559	22	38	1-1/2	152	6	76	3	44	105
584	23	38	1-1/2	159	6-1/4	83	3-1/4	48	112.5
610	24	38	1-1/2	165	6-1/2	89	3-1/2	51	120

a. Initially, 600 A and 55 V should be used on each electrode. Once the welder has gained experience, the welding current and the voltage can be adjusted in accordance with the section titled "Welding Variables."
b. A dwell time of 4 seconds is used at each shoe.

Table 10.8
Typical Electroslag Welding Conditions: Five Oscillating Electrodes[a]

Plate Thickness		Joint Opening		Electrode Spacing		Oscillation Distance		Oscillation Speed[b]	
mm	in.	mm	in.	mm	in.	mm	in.	mm/s	in./min
609	24	38	1-1/2	127	5	76	3	32	75
635	25	38	1-1/2	133	5-1/4	76	3	35	82.5
660	26	38	1-1/2	140	5-1/2	76	3	38	90
686	27	38	1-1/2	146	5-3/4	76	3	41	97.5
711	28	38	1-1/2	152	6	76	3	44	105
737	29	38	1-1/2	152	6	102	4	48	112.5
762	30	38	1-1/2	159	6-1/4	102	4	50	120

a. Initially, 600 A and 55 V should be used on each electrode. Once the welder has gained experience, the welding current and the voltage can be adjusted in accordance with the section titled "Welding Variables."
b. A dwell time of 4 seconds is used at each shoe.

Table 10.9
Typical Electroslag Welding Conditions: Six Oscillating Electrodes[a]

Plate Thickness		Joint Opening		Electrode Spacing		Oscillation Distance		Oscillation Speed[b]	
mm	in.	mm	in.	mm	in.	mm	in.	mm/s	in./min
762	30	38	1-1/2	133	5-1/4	70	2-3/4	35	82.5
787	31	38	1-1/2	140	5-1/2	64	2-1/2	38	90
813	32	38	1-1/2	143	5-5/8	73	2-7/8	40	94
838	33	38	1-1/2	146	5-3/4	83	3-1/4	42	99.5
864	34	38	1-1/2	152	6	76	3	44	105
889	35	38	1-1/2	156	6-1/8	86	3-3/8	46	109
914	36	38	1-1/2	159	6-1/4	95	3-3/4	48	112.5

a. Initially, 600 A and 55 V should be used on each electrode. Once the welder has gained experience, the welding current and the voltage can be adjusted in accordance with the section titled "Welding Variables."
b. A dwell time of 4 seconds is used at each shoe.

APPLICATIONS

Many types of carbon steels can be joined using electroslag welding. These include AISI 1020, AISI 1045, ASTM A36, ASTM A441, and ASTM A515. They can generally be welded without postweld heat treatment.

In addition to carbon steels, other steels can be successfully welded using this process. They include AISI 4130, AISI 8620, ASTM A302, HY80, austenitic stainless steels, ASTM A514, ingot iron, and ASTM A387. Most of these steels require special electrodes and a grain-refining postweld heat treatment to develop the required weld or weld heat-affected-zone properties.

JOINT DESIGN

The basic joint design used with electroslag welding is the square-groove butt joint. Square-edge plate preparations can be used to produce other types of joints such as corner, T, and edge joints. It is also possible to make transition joints, fillet welds, cross-shaped joints, overlays, and weld pads with the electroslag welding process. Specially designed retaining shoes are needed for joints other than butt, corner, and T-joints.

Typical electroslag weld joint designs and schematic drawings of the final welds are shown in Figure 10.7. The depth of fusion is also shown.

TYPICAL USES

Electroslag welding is recognized by important national and international codes. Several of the codes have requirements that differ from those for other welding processes. For example, in *Structural Welding Code—Steel*, AWS D1.1,[6] electroslag welding is not permitted as a prequalified welding process. Thus, the contractor must prepare a welding procedure qualification test plate and submit the joint to destructive testing. The test must demonstrate that the contractor is capable of successfully implementing the process. Electroslag welds in pressure vessels that are fabricated in accordance with the ASME *Boiler and Pressure Vessel Code*[7] must be normalized after welding. As indicated previously, this refines the grain structure of the weld metal and the heat-affected zone.

Other fabrication and construction codes such *Welding of Pipelines and Related Facilities,* ANSI/API 1104,[8] and the American Bureau of Shipping (ABS) *Rules for Building and Classing: Steel Vessels*[9] do not require any special testing or heat-treating of electroslag welds. However, purchase contracts may permit owners, owners' representatives, and regulatory agencies to require special tests prior to approval of weld procedures, and these prerogatives may be exercised prior to approving electroslag welding.

Structural Applications

Electroslag welding has many unique advantages that make it a highly desirable and widely used welding process for structural applications. A common structural application is the welding of the transition joint (essentially a type of butt joint) between different flange thicknesses. The varying thicknesses present no problem when copper shoes designed for this type of joint configuration are used. Electroslag welding is often used to weld stiffeners in box columns and wide flanges. In all cases, the stiffener weld would be a T-joint. A typical application is shown in Figure 10.8.

The high weld metal deposition rates, the low percentage of weld discontinuities, and the fact that it is a mechanized process are strong reasons for the use of electroslag welding. For thick sections, electroslag welding is a low-cost process if the weldments meet the design requirements and service conditions. However, if the welding process is stopped during the welding of a joint for any reason, the restart area must be carefully inspected for discontinuities. Those considered unacceptable for the application must be repaired using another welding process.

Machinery

The manufacturers of large presses and machine tools work with large, heavy plates. The design often requires plates that are larger than the mill can produce in one piece. Electroslag welding is used to splice together two or more plates. A schematic illustration of the electroslag welding setup for this application is shown in Figure 10.9.

Other machinery applications include kilns, gear blanks, motor frames, press frames, turbine rings, shrink rings, crusher bodies, rebuilding metal mill rolls, rims for road rollers, and products that are formed from plates and welded along longitudinal joints.

6. American Welding Society (AWS) Committee on Structural Welding, *Structural Welding Code—Steel*, ANSI/AWS D1.1, Miami: American Welding Society.
7. American Society of Mechanical Engineers (ASME) International, *Boiler and Pressure Vessel Code*, New York, American Society of Mechanical Engineers.
8. American Petroleum Institute (API), *Welding of Pipelines and Related Facilities*, ANSI/API 1104, Washington. D.C.: American Petroleum Institute.
9. American Bureau of Shipping (ABS), *Rules for Building and Classing: Steel Vessels*, Houston, Tex.: American Bureau of Shipping.

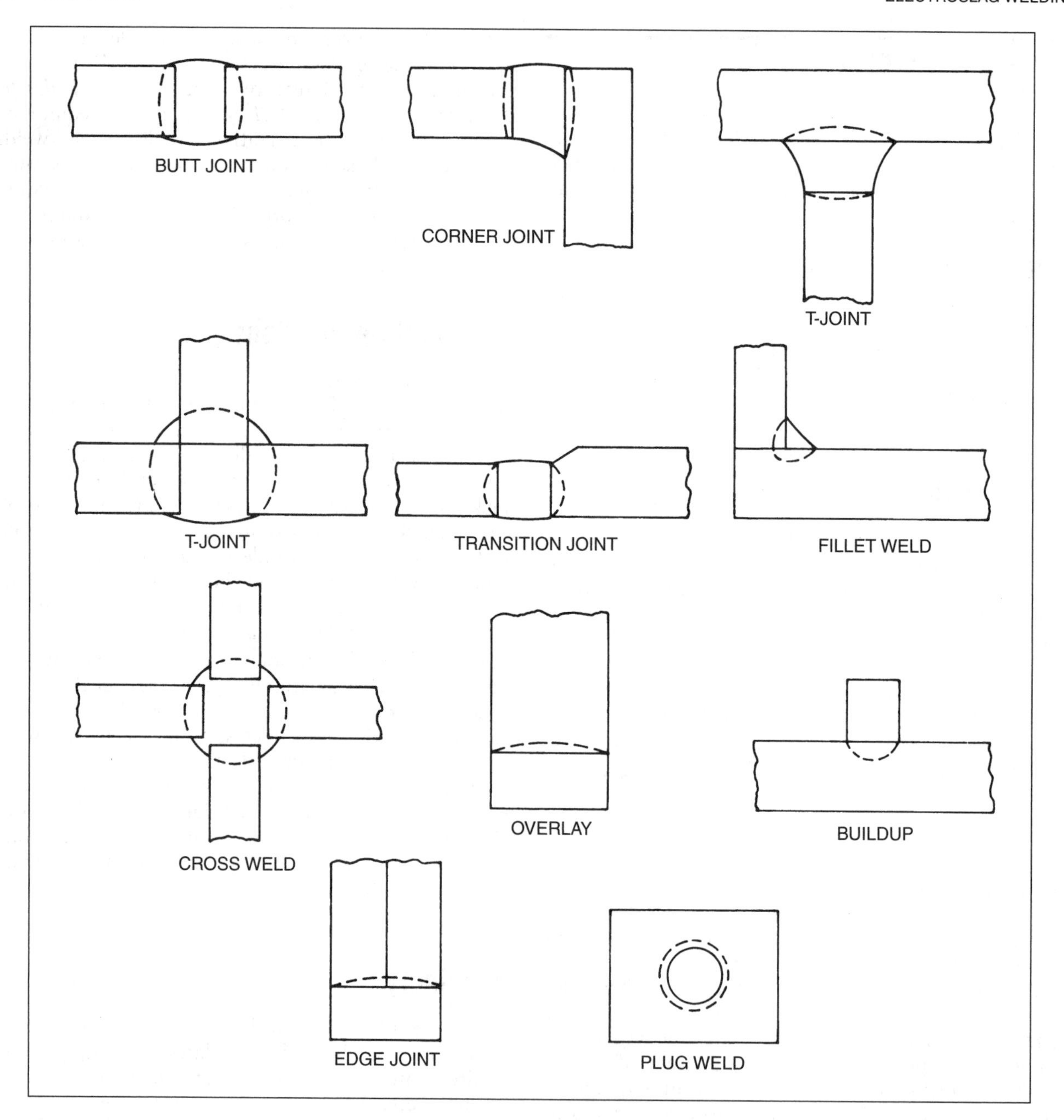

Figure 10.7—Joint Designs for Electroslag Welding Showing Depth of Fusion into the Base Metal

Pressure Vessels

Pressure vessels for the chemical, petroleum, marine, and power-generating industries are made in all shapes and sizes and with wall thicknesses from less than 13 mm (1/2 in.) to greater than 400 mm (16 in.). In current practice, plate can be rolled to form the shell of the vessel, and the longitudinal joint must be welded. In very large or thick-walled vessels, the shell can be fabricated from two or more curved plates and joined using several longitudinal electroslag welds.

The steels used in pressure-vessel construction are generally heat-treated. Consequently, when these steels are welded with a high heat input, as occurs with electroslag welding, the weld heat-affected zone does not have adequate mechanical properties. To improve the mechanical properties, weldments are heat treated as required.

Photo courtesy of Arcmatic Integrated Systems

Figure 10.8—Welding a Stiffener Plate Using the Electroslag Process

Ships

Electroslag welding is used in the shipbuilding industry for both in-shop and on-ship applications. Main hull section joining is performed with the conventional electroslag welding method. Vertical welding of the side shell from the bilge area to, but not including, the sheer strake can be performed with an electroslag plate crawler. Plate thicknesses of 13 mm to 32 mm (1/2 in. to 1-1/4 in.) are commonly found in marine side shells. The weld length may be 12 m to 21 m (40 ft to 70 ft) depending on the size of the ship.

Castings

Electroslag welding is often used to weld cast components. The metallurgical characteristics of a casting and an electroslag weld are similar, and both respond to postweld heat treatment in a similar manner. Many large, difficult-to-cast components can be produced in smaller, higher-quality units and then joined by electroslag welding. Costs are reduced, and the quality is usually improved. Compatible weld metal produces a homogeneous structure that can be machined. Color match and other desirable properties are thus produced.

A typical application is shown in Figure 10.10, in which two castings are set up for electroslag welding.

INSPECTION AND QUALITY CONTROL

Because of the nature of the electroslag welding process, the weld discontinuities and defects that may occur do not always resemble those of a multiple-pass arc weld. Weld metal discontinuities that may be

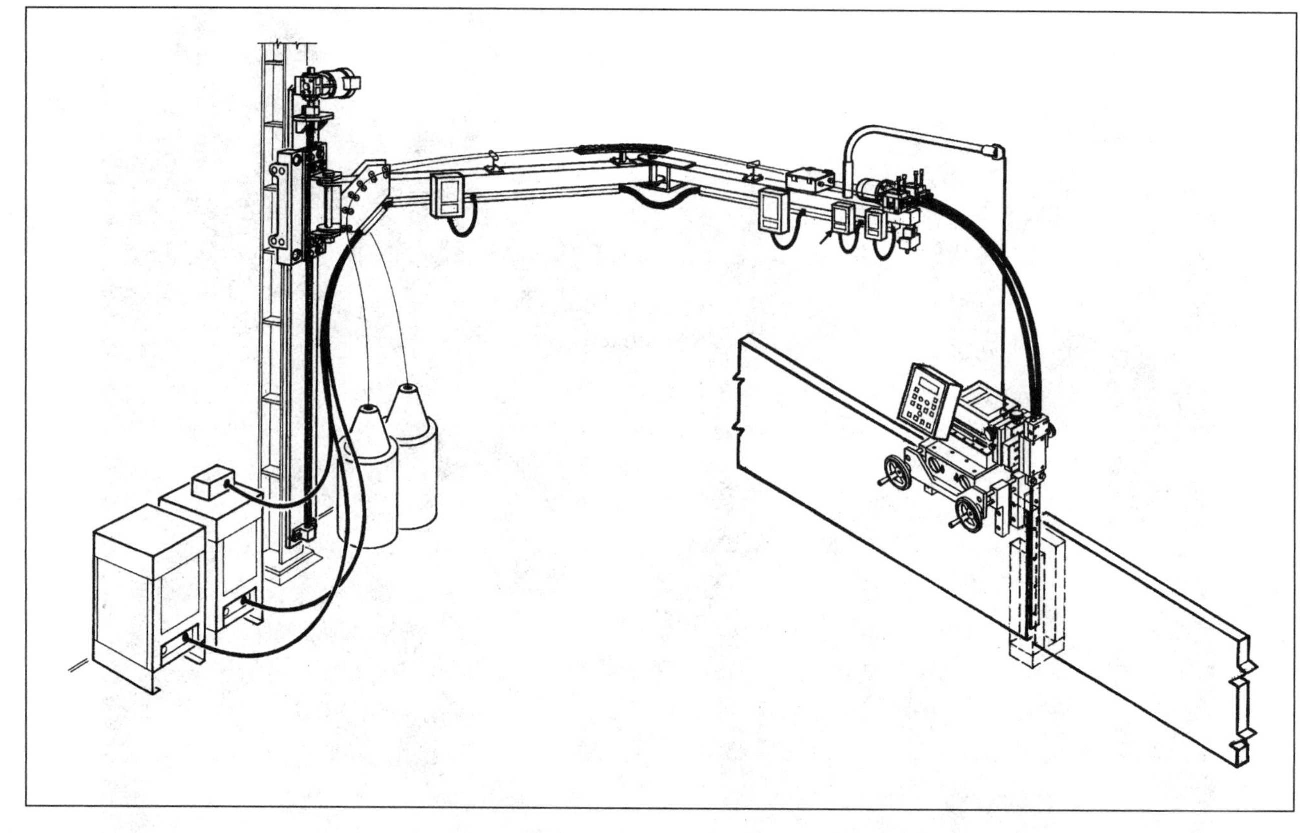

Figure 10.9—Electroslag Welding Setup for Splicing Large Plates

encountered in the electroslag process are discussed in the section titled "Weld Quality."

Conventional nondestructive examination can be used to determine the soundness of the electroslag-welded joint. Nondestructive examination techniques normally used include the following:

1. Visual examination (VT),
2. Liquid penetrant inspection (PT),
3. Magnetic particle inspection (MT),
4. Radiographic (isotope or X-ray) inspection (RT), and
5. Ultrasonic testing (UT). It should be noted that in UT the large grain size may require special techniques.

All nondestructive examination should be performed in accordance with qualified procedures by qualified technicians.

METHODS OF INSPECTION

The inspection method used depends not only on the requirements of the code or standard pertinent to the weldment but also on the contract specifications of the owner or purchaser.

Every welded joint should undergo a thorough visual examination for incomplete edge fusion, undercut, and surface cracks. When the sump and run-off tabs are removed, discontinuities such as inclusions, cracking, and porosity may be seen. However, for internal discontinuities such as porosity, inclusions, cracking, and in rare instances, incomplete internal fusion, radiographic and ultrasonic testing are the most effective means of examination. Magnetic particle testing may also be used in searching for cracks and incomplete fusion, but it is limited to the surface and immediate subsurface.

ACCEPTANCE

Most welding acceptance criteria are established by the customer or a regulatory agency, or both. Electroslag applications are usually welded to various code requirements. Occasionally, customer requirements supplement or alter these requirements.

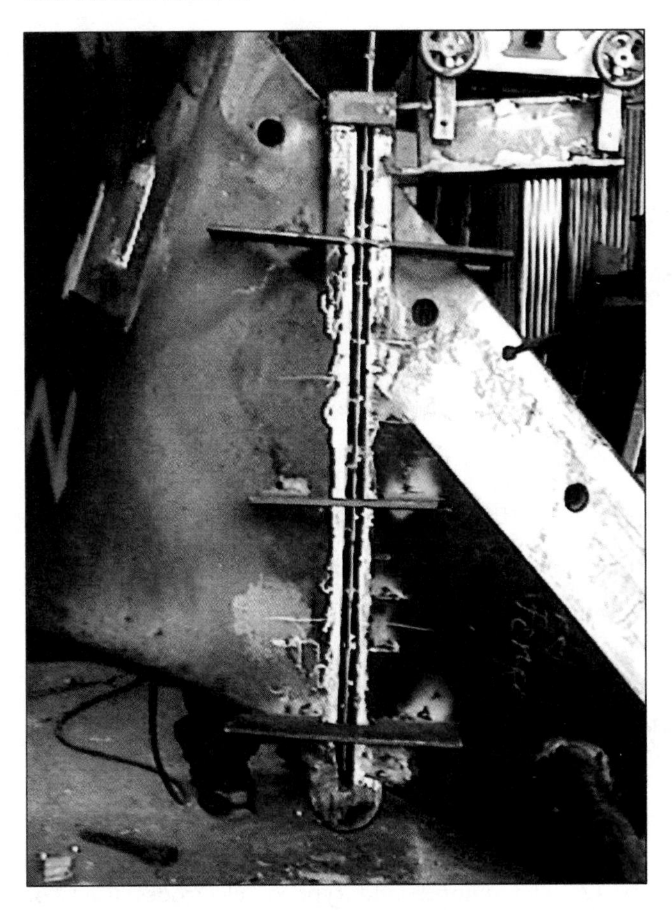

Figure 10.10—Large Castings Positioned for Electroslag Welding

QUALITY CONTROL

A written procedure that identifies the essential variables to be used in electroslag welding should be prepared. This procedure should be reviewed to determine its adequacy on the basis of comparison with applicable code requirements. Only an approved welding procedure should be issued to the welding supervisor or welding operator, and any additional instructions considered necessary should be included.

A checklist can be used by the welding operator to ensure that the equipment is properly set up and that all of the required operating adjustments have been made. The operator may need an assistant when welding joints that are not readily visible from both sides. Proper quality-control procedures should be used to ensure that the qualified welding procedure is being followed.

REWORK

In general, the repair of defective joints can be minimized by establishing a good preventive maintenance program for the equipment using well qualified and properly trained welding operators and implementing a sound welding procedure.

A discontinuity such as undercut can often be repaired by rewelding with the shielded metal arc welding or gas metal arc welding process without gouging or grinding. Discontinuities such as incomplete fusion at the joint surface, overlap, copper pickup on the weld face, and metal spillage around the shoes are visible defects that can be repaired by gouging or grinding to sound metal and rewelding. Discontinuities such as porosity, cracking, and internal incomplete fusion are usually detected by radiographic or ultrasonic testing and can be repaired by gouging to sound metal and rewelding. Economic considerations should dictate whether electroslag welding or some other process is used for repair.

The restarting of electroslag welds should be avoided if possible. When such restarts are made in a partially completed electroslag weld, some repairs in the starting area are almost inevitable. One method of preparing the plate for a restart is to use an arc-gouged start area as illustrated in Figure 10.11. The restart area should be closely inspected for discontinuities.

TESTING OF WELDS

All codes and specifications have definite rules for the testing of qualification welds to determine compliance with requirements. The most frequently required tests for groove welds are mechanical tests, such as tensile and bend tests, using specimens cut from specific locations in the welds. Fillet welds do not readily lend themselves to mechanical bend tests. In such cases, fillet weld break tests or macro-etch tests, or both, may be required. Test procedures and methods of determining the mechanical properties are detailed in *Standard Methods for Mechanical Testing of Welds*, AWS B4.0.[10]

Radiographic testing is sometimes allowed as an alternative to mechanical testing when qualifying operators.

WELD QUALITY

Welds made with the electroslag welding process under proper operating conditions are normally high in

10. American Welding Society (AWS) Committee on Mechanical Testing of Welds, *Standard Methods for Mechanical Testing of Welds*, AWS B4.0, Miami: American Welding Society.

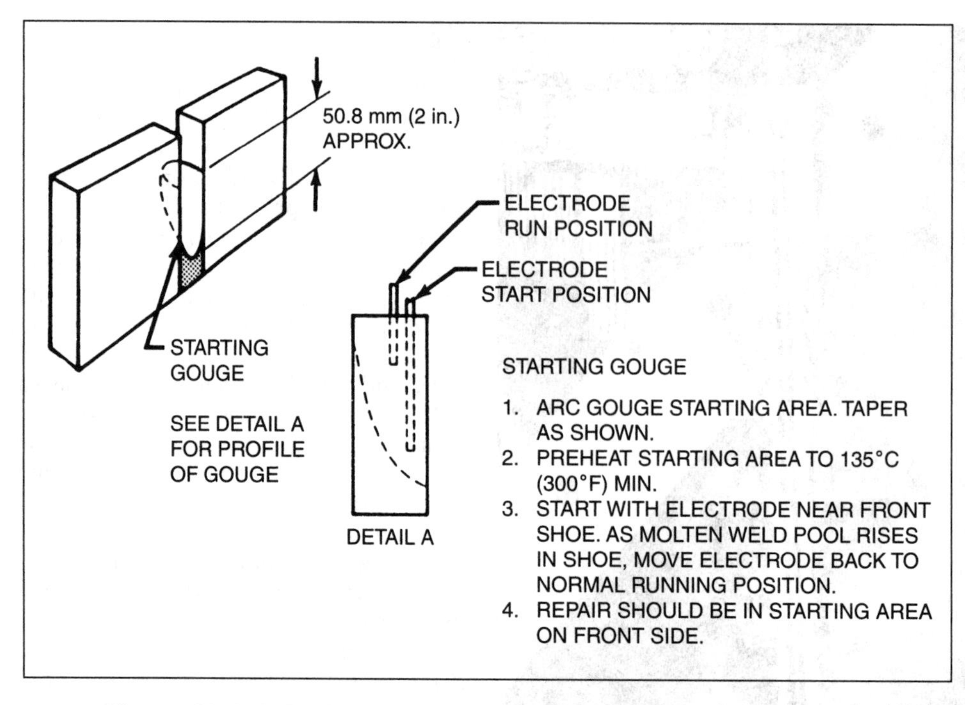

Figure 10.11—Typical Electroslag Welding Restart Procedure

quality and free from harmful discontinuities. In any welding process, however, abnormal conditions may occur during welding and cause discontinuities in the weld. The most common of these discontinuities, possible causes, and remedies are shown in Table 10.10. The information is primarily applicable to electroslag-welded joints in carbon steels and low-alloy steels.

MECHANICAL PROPERTIES

The mechanical properties of electroslag welds depend on the type and thickness of base metal, the composition of the electrode, electrode-flux combination, and welding conditions. All of these influence the chemical composition, metallurgical structure, and mechanical properties of the weldment. In general, electroslag welds are used in structures that are to be loaded under static or fluctuating load conditions. One major concern is the notch toughness of the weld metal and heat-affected zones under service conditions, particularly at low temperatures. This must be carefully evaluated for each particular application so that design requirements are met and the weldments perform satisfactorily.

Typical mechanical properties of deposited weld metal in selected structural steels using the consumable guide method are shown in Table 10.11. Those for various carbon and low-alloy steels are presented in Table 10.12. The number and type of electrodes used for welding are reported in the table.

ECONOMICS

The advantages associated with electroslag welding result in considerable cost savings, particularly when thicker materials are to be joined. Savings have been achieved when components are joined to make larger units instead of initially producing massive castings or forgings. In thicker-section weldments, electroslag welding is often less expensive than the more conventional joining methods such as submerged arc welding. Even in some applications involving thinner base metals, electroslag welding has resulted in cost savings because of its efficiency and simple joint preparation.

To appreciate the true overall economy of electroslag welding, the first consideration is the cost of joint preparation. A square oxygen-cut joint is suitable preparation for the process in carbon steels. Elaborate joint matching or close fitup is not required. In welds with a thickness of 75 mm (3 in.) or more, a weld produced with electroslag welding requires much less weld metal and up to 90% less flux than a comparable weld fabricated with submerged arc welding.

Table 10.10
Electroslag Weld Discontinuities: Causes and Remedies

Location	Discontinuity	Causes	Remedies
Weld	Porosity	1. Insufficient slag depth	1. Increase flux additions.
		2. Moisture, oil, or rust	2. Dry or clean workpiece.
		3. Contaminated or wet flux	3. Dry or replace flux.
	Cracking	1. Excessive welding speed	1. Slow electrode feed rate.
		2. Poor form factor	2. Reduce current; raise voltage; decrease oscillation speed.
		3. Excessive center-to-center distance between electrodes or guide tubes	3. Decrease spacing between electrodes or guide tubes.
	Nonmetallic inclusions	1. Rough plate surface	1. Grind plate surfaces.
		2. Unfused nonmetallics from plate laminations	2. Use better quality plate.
Fusion line	Incomplete fusion	1. Low voltage	1. Increase voltage.
		2. Excessive welding speed	2. Decrease electrode feed rate.
		3. Excessive slag depth	3. Decrease flux additions; allow slag to overflow.
		4. Misaligned electrodes or guide tubes	4. Realign electrodes or guide tubes.
		5. Inadequate dwell time	5. Increase dwell time.
		6. Excessive oscillation speed	6. Slow oscillation speed.
		7. Excessive electrode-to-shoe distance	7. Increase oscillation width or add another electrode.
		8. Excessive center-to-center distance between electrodes	8. Decrease spacing between electrodes.
	Undercut	1. Welding speed too slow	1. Increase electrode feed rate.
		2. Excessive voltage	2. Decrease voltage.
		3. Excessive dwell time	3. Decrease dwell time.
		4. Inadequate cooling of shoes	4. Increase cooling water flow to shoes or use larger shoe.
		5. Poor shoe design	5. Redesign groove in shoe.
		6. Poor shoe fitup	6. Improve fitup; seal gap with refractory cement dam.
Heat-affected zone	Cracking	1. High restraint	1. Modify fixturing.
		2. Crack-sensitive material	2. Determine cause of cracking to ascertain remedy.
		3. Excessive inclusions in plate	3. Use better quality plate.

Another cost consideration is welding time. Once the workpieces are in place for welding, the electroslag weld is completed without stopping if the process remains under control. In contrast, the downtime or nonproductive time encountered in most arc welding processes can range from 30% to 75%. Nevertheless, it must be clearly recognized that stoppage of a heavy electroslag weld in progress can be very costly. Moreover, to restart without producing a defect is difficult, if not impossible. Electroslag welding speeds for various plate thicknesses are shown in Figure 10.12.

The deposition rate for electroslag welding is approximately 16 kg/h to 20 kg/h (35 lb/h to 45 lb/h) per electrode. In very heavy plates, 47 kg/h to 61 kg/h (105 lb/h to 135 lb/h) of weld metal can be deposited using three electrodes. Figure 10.12 shows welding speeds for various plate thicknesses when using a root opening of 29 mm (1-1/8 in.). Heavy plates ranging from 76 mm to 305 mm (3 in. to 12 in.) in thickness are welded at speeds between 610 mm/h to 1220 mm/h (24 in. to 48 in.).

Significant savings are achieved by the elimination of angular distortion and subsequent rework. Angular distortion can become a major factor in heavy multiple-pass welding, whether the welding is done from one side or from both sides.

Table 10.11
Typical Mechanical Properties of Weld Metal from Electroslag Welds in Structural Steels, As-Welded (Consumable Guide Method)

	Thickness		Electrode		Yield Strength		Tensile Strength		Elongation	Reduction of Area	Impact Strength*	
Base Metal, ASTM	mm	in.	Type (AWS)	No.	MPa	ksi	MPa	ksi	% in 51 mm (2 in.)	%	J	Ft-lb
A 441	25	1	EM13K-EW	1	344	49.9	523	75.8	28.0	59.0	23	17
A 441	64	2-1/2	EM13K-EW	1	316	45.8	505	73.3	26.5	66.0	37	27
A 36	152	6	EM13K-EW	2	317	46.0	548	79.5	28.5	52.8	—	—
A 36	305	12	EM13K-EW	2	255	37.0	464	67.3	33.5	71.0	—	—
A 572 Gr. 42	203	8	EM13K-EW	2	400	58.2	585	84.8	25.0	67.6	38	28
A 572 Gr. 60	57	2-1/4	EH10Mo-EW	1	423	61.5	680	98.5	18.0	35.6	—	—

*Impacts at -18°C (0°F)

Table 10.12
Typical Mechanical Properties of Weld Metal from Electroslag Welds in Carbon and Alloy Steels

											Charpy Impact Tests, –12.2°C (10°F) Notch Location					
	Thickness		Electrode			Yield Strength		Tensile Strength		Elongation	WMFG[a]		WMCG[b]		BM HAZ[c]	
Base Metal, ASTM	mm	in.	Type	No.	Heat Treatment	MPa	ksi	MPa	ksi	% in 51 mm (2 in.)	J	ft-lb	J	ft-lb	J	ft-lb
A 204-A	89	3-1/2	Mn-Mo	1	NT[d]	382	55.5	562	81.5	27	46	34	39	29	18	13
A 515 Gr. 70	38	1-1/2	Mn-Mo	1	SR[e]	358	52.0	579	84.1	26	61	45	35	26	10	7
A 515 Gr. 70	51	2	Mn-Mo	1	SR	469	68.1	587	85.2	23	63	46	29	21	7	5
A 515 Gr. 70	86	3-3/8	Mn-Mo	2	NT	396	57.5	537	78.0	29	46	34	45	33	30	22
A 515 Gr. 70	171	6-3/4	Mn-Mo	2	NT	313	45.5	512	74.3	31	33	24	29	21	16	12
A 302-B	76	3	Mn-Mo-Ni	1	NT	393	57.0	565	82.0	28	72	53	71	52	86	63
A 387-C	76	3	1-1/4 Cr-1/2 Mo	2	NT	320	46.5	503	73.0	29	95	70	103	76	78	57
A 387-D	83	3-1/4	2-1/4 Cr-1 Mo	1	SR	396	57.5	565	82.0	25	63	46	68	50	65	48
A 387-D	191	7-1/2	2-1/4 Cr-1Mo	2	SR	551	80.0	658	95.5	20	84[f]	62	102[f]	75	113[f]	83

a. WMFG = Weld metal, fine grain
b. WMCG= Weld metal, coarse grain
c. BM HAZ = Base metal, heat-affected zone
d. NT = Normalized and tempered
e. SR = Stress relieved
f. Impacts at 10°C (50°F)

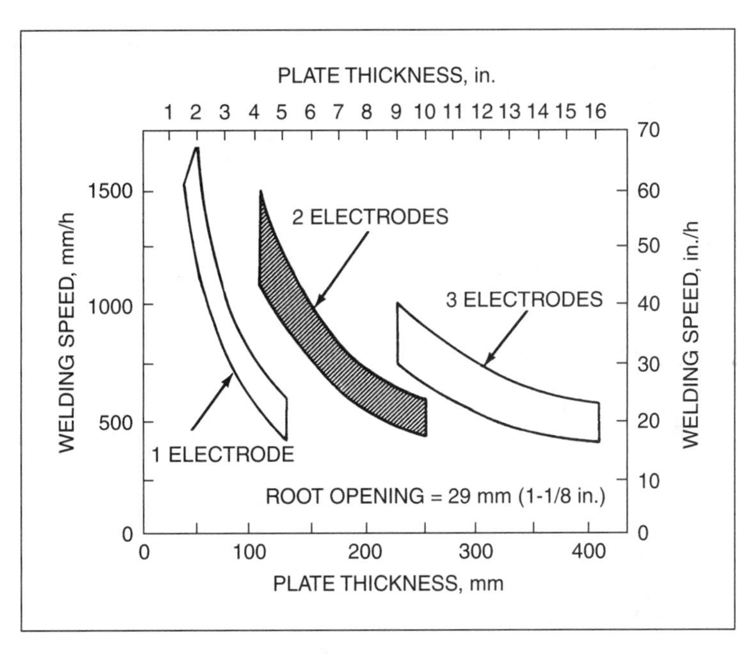

Figure 10.12—Electroslag Welding Speeds for Various Plate Thickness

Electroslag welding normally produces a high percentage of weldments that are free of discontinuities and defects, thereby minimizing repair costs. Slag entrapment, porosity, and incomplete fusion can be avoided in most cases.

SAFE PRACTICES

As in any type of welding, reasonable care must be exercised in the setup, welding, and postwelding procedures for electroslag welding. A number of potential hazards exist—some minor and others serious—but all can be minimized. Failure to use safety protection equipment or follow safe practices can result in physical hazards to personnel and damage to production parts, equipment, and facilities.

Detailed safety information on equipment and materials is available in the manufacturers' instructions and in the documents *Safety in Welding, Cutting, and Allied Processes,* ANSI Z49.1;[11] *Practice for Occupational and Educational Eye and Face Protection,* ANSI Z87.1;[12] and Chapter 17, "Safe Practices," in *Welding Science and Technology,* Volume 1 of the *Welding Handbook,* ninth edition.[13] Mandatory federal safety regulations established by the U.S. Department of Labor, Occupational Safety and Health Administration (OSHA), are available in *Code of Federal Regulations,* Title 29 Part 1910,[14] and the National Safety Council publication *Accident Prevention Manual: Engineering and Technology.*[15]

GENERAL SAFETY GUIDELINES

The workpieces welded with the electroslag process are generally large, and because they are welded vertically they are sometimes tall and awkwardly posi-

11. American National Standards Institute (ANSI) Accredited Standards Committee Z49, *Safety in Welding, Cutting, and Allied Processes*, ANSI Z49.1, Miami: American Welding Society.

12. *Practice for Occupational and Educational Eye and Face Protection*, ANSI Z87.1, New York: American National Standards Institute.

13. American Welding Society (AWS) Welding Handbook Committee, Jenney, C. L., and A. O'Brien, eds., 2001, *Welding Science and Technology*, Vol. 1 of *Welding Handbook*, 9th ed., Miami: American Welding Society.

14. Occupational Safety and Health Administration (OSHA), *Occupational Safety and Health Standards for General Industry*, in *Code of Federal Regulations (CFR)*, Title 29 CFR 1910, Subpart Q, Washington D.C.: Superintendent of Documents, U.S. Government Printing Office.

15. National Safety Council, 2000, *Accident Prevention Manual: Engineering and Technology*, 12th ed., Itasca, Illinois: National Safety Council.

tioned. Caution should be used in positioning these components, and qualified welders should be used to weld needed braces, clamps, strongbacks, and other restraints or fixtures. The same care should be exercised when removing the fixtures by cutting or air carbon arc gouging. When removing sumps and run-off blocks, welders and welding operators must ensure that the sumps and blocks fall in an appropriate place, not onto cutting-gas lines or electrical cables.

PERSONAL PROTECTIVE EQUIPMENT

Appropriate eye protection and suitable clothing must be worn to protect against arc radiation from this and other welding operations, radiant heat from hot workpieces or weldments, and the hazards of hot molten slag. Insulating gloves should be worn at all times.

Safety glasses with side shields are recommended to provide protection against the hot slag or spatter balls that may fly from the joint during welding. Shade No. 12 is recommended if the arc must be observed before the slag bath is established. Shade No. 4 is recommended for observing the slag bath. Considerable caution should be used when removing the cooling shoes. As solidified slag is essentially glass, it breaks up and splinters erratically.

ELECTRICAL HAZARDS

The hazard of electrical shock is possible with any electrical equipment. However, with electroslag welding, the operator does not touch the equipment except for occasional adjustments. The electrode wire, and in effect, everything that comes into contact with it, is "electrically hot." The electrode wire and other "live" parts of the welding system should not be touched by the operator or other personnel in the area.

As a normal procedure, the operator should make a preliminary check of the equipment to spot potential problems. Electrical hazards such as loose or worn connections, frayed insulation, or cables in or around water should be repaired and the hazard should be removed. Special care should be taken to prevent the drainage of water from cooling shoes or circulators around electroslag welding equipment. Manufacturers' recommendations for cable size, and for the installation, use, and maintenance of the equipment should always be followed.

GASES AND FUMES

Fumes containing fluoride compounds may be generated during electroslag welding, but it is not necessary for the electroslag welding operator to be constantly in or around any fumes produced. Electroslag welding is essentially a mechanized process, although the operator monitors the welding. If welding is to be performed in a poorly ventilated area, exhaust fans or smoke extractors must be used.

FIRE AND BURN PREVENTION

Welding operations should not be carried out in the presence of combustible materials. Solvents used to clean the weld joint area must be completely removed from the work area. Aerosol cans must not be placed near the weld joint, as they may explode.

Hot slag may spill out the top of the joint or leak around the cooling shoe. Poor fitup of the cooling shoe or misalignment of sumps and runoffs may also open gaps where leaks can occur. Slag leakage can be especially dangerous because of the large volume of slag bath associated with the electroslag welding process.

For the same reason, special attention should be given to shoe fitup and alignment, and the cooling shoes should be left in place after completion of the weld until the slag bath has solidified. Water-cooled shoes are much safer to handle than solid shoes, but both types can cause severe burns.

CONCLUSION

The electroslag welding process offers many opportunities for reducing welding costs, and where it can be applied it is very often the least costly welding process. The advantages include extremely high metal deposition rates and the ability to weld very thick material in one pass. There is only one setup and no interpass cleaning since welds are usually made in one pass. Preheating is normally not required, even on materials of high hardenability.

Electroslag welding requires minimal joint preparation and fitup. The process is mechanized, and once started, it continues to completion. There is no weld spatter, therefore resulting in 100% filler-metal transfer efficiency. Flux consumption is low, and there is no angular distortion in the horizontal plane. Minimum distortion may occur in the vertical plane but it is easy to compensate for it. Electroslag welding is the fastest welding process for large weldments in thick material.

BIBLIOGRAPHY[16]

American Bureau of Shipping (ABS). 2001. *Rules for building and classing: steel vessels 2001.* Houston: American Bureau of Shipping.

16. The dates of publication given for the codes and other standards listed here were current at the time this chapter was prepared. The reader is advised to consult the latest edition.

American National Standards Institute (ANSI) Accredited Standards Committee Z49. *Safety in welding, cutting, and allied processes,* ANSI Z49.1. Miami: American Welding Society.

American National Standards Institute (ANSI) Accredited Standards Committee Z87. *Practice for occupational and educational eye and face protection,* ANSI Z87.1. New York: American National Standards Institute.

American Petroleum Institute (API). 1999. *Welding of pipelines and related facilities.* ANSI/API 1104. Washington, D.C.: American Petroleum Institute.

American Society of Mechanical Engineers (ASME) International. *Boiler and pressure vessel code.* New York: American Society of Mechanical Engineers.

American Welding Society (AWS) Committee on Filler Metals and Allied Materials. 1997. *Specification for carbon and low-alloy steel electrodes and fluxes for electroslag welding.* ANSI/AWS A5.25/A5.25M-97. Miami: American Welding Society.

American Welding Society (AWS) Committee on Structural Welding. 2003. *Structural welding code—steel,* ANSI/AWS D1.1, Miami: American Welding Society.

American Welding Society (AWS) Committee on Mechanical Testing of Welds. 2000. *Standard methods for mechanical testing of welds,* AWS B4.0M:2000, Miami: American Welding Society.

American Welding Society, (AWS) Welding Handbook Committee, Jenney, C. L., and A. O'Brien, eds. 2001. *Welding science and technology.* Vol. 1 of *Welding handbook.* 9th ed. Miami: American Welding Society.

National Safety Council. 2000. *Accident prevention manual: engineering and technology.* 12th ed. Itasca, Illinois: National Safety Council.

Occupational Safety and Health Administration (OSHA). 1999. *Occupational safety and health standards for general industry.* In *Code of Federal Regulations (CFR),* Title 29 CFR 1910, *Subpart Q.* Washington D.C.: Superintendent of Documents, U.S. Government Printing Office.

SUPPLEMENTARY READING LIST

Brosholen, A., E. Skaug, and J. J. Visser. 1977. Electroslag welding of large castings for ship construction. *Welding Journal* 56(8): 26–30.

Campbell, H. C. 1970. *Electroslag, electrogas, and related welding processes.* Welding Research Council Bulletin No. 154. New York: Welding Research Council.

des Ramos, J. B., A. W. Pense, and R. D. Stout. 1976. Fracture toughness of electroslag welded A537G steel. *Welding Journal* 55(12): 1-s to 4-s.

Davenport, J. A., B. N. Qian, A. W. Pense, and R. D. Stout. 1981. Ferrite vein cracking in electroslag welds. *Welding Journal* 60(12): 237-s to 2243-s.

Devletian, J. H., and S. J. Chen. 1989. Joining of thick-section titanium alloys by electroslag welding. *Welding Journal*: 68(9): 37–42.

Delevetian, J. H., D. Singh, and R. B. Turpin. 1997. Electroslag welding of an advanced double-hull design ship. *Welding Journal* 76(8): 49–52.

Dilawari, A. H., T. W. Eager, and J. Szekely. 1978. An analysis of heat and fluid flow phenomena in electroslag welding. *Welding Journal* 57(1): 24-s to 30-s.

Dorschu, K. E., J. E. Norcross, and C. C. Gage. 1973. Unusual electroslag welding applications. *Welding Journal* 52(11): 710–716.

Eichhorn, E., J. Remmel, and B. Wubbels. 1984. High-speed electroslag welding. *Welding Journal* 63(1): 37–41.

Forsberg, S. G. 1985. Resistance electroslag (RES) surfacing. *Welding Journal* 64(8): 41–48.

Frost, R. H., G. R. Edwards, and M. D. Rheinlander. 1981. A constitutive equation for the critical energy input during electroslag welding. *Welding Journal.* 60(1): 1-s to 6-s.

Frost, R. H., D. L. Olson, and G. R. Edwards. 1984. In *Modelling of Casting and Welding Processes II. Proceedings 1983 Engineering Foundation Conference. Henniker, New Hampshire: 31 July-5 Aug. 1983.* J. A. Dantzig, and J. T. Berry, eds. Warrendale, Pennsylvania: The Metallurgical Society of AIME.

Hannahs, J. R., and L. Daniel. 1970. Where to consider electroslag welding. *Metal Progress* 98(5): 62–64.

Johnsen, M. R. 2000. Electroslag welding stands poised for a comeback in bridge construction. *Welding Journal* 79(2): 39–41.

Jones, J. E., D. L. Olson, and G. P. Martins. 1980. Metallurgical and thermal characteristics of non-vertical electroslag welds. *Welding Journal.* 59(9): 245-s to 254-s.

Kenyon, N., G. A. Redfern, and R. R. Richardson. 1977. Electroslag welding of high nickel alloys. *Welding Journal* 54(7): 235-s to 239-s.

Konkol, P. J. 1983. Effects of electrode composition, flux basicity, and slag depth on grain-boundary cracking in electroslag weld metals. *Welding Journal* 62(3): 63-s to 71-s.

Konkol, P. J., and W. F. Domis. 1979. Causes of grain-boundary separations in electroslag weld metals. *Welding Journal.* 58(6): 161-s to 167-s.

Lawrence, B. D. 1973. Electroslag welding curved and tapered cross-sections. *Welding Journal* 52(4): 240–246.

Liu, S., and C. T. Su. 1989. Grain refinement in electroslag weldments by metal powder addition. *Welding Journal* 68(4): 132-s.

Lowe, G., S. R. Bala, and L. Malik. 1981. Hydrogen in consumable guide electroslag welds—its sources and significance. *Welding Journal.* 60(12): 258-s to 268-s.

Malin, V. Y. 1985. Electroslag welding of titanium and its alloys. *Welding Journal* 64(2): 42–49.

Myers, R. D. 1980. Electroslag welding eliminates costly field machining on large mining shovel. *Welding Journal* 59(4): 17–22.

Noruk, J. S. 1982. Electroslag welding used to fabricate world's largest crawler driven dragline. *Welding Journal* 61(8): 15–19.

Oh, Y. K., J. H. Devletian, and S. J. Chen. 1990. Low-dilution electroslag cladding for shipbuilding. *Welding Journal* 69(8): 27–44.

Okumura, M., M. Kumagai, N. Nakamura, and K. Kohira. 1976. Electroslag welding of heavy section 2 1/4 Cr-1Mo steel. *Welding Journal* 55(12): 389-s to 399-s.

Parrott, R. S., S. W. Ward, and G. D. Uttrachi. 1974. Electroslag welding speeds shipbuilding. *Welding Journal* 53(4): 218–222.

Patchett, B. M., and D. R. Milner. 1972. Slag-metal reactions in the electroslag process. *Welding Journal* 51(10): 491-s to 505-s.

Paton, B. E. 1962. Electroslag welding of very thick material. *Welding Journal* 41(12): 1115–1123.

Paton, B. E., ed. 1962. *Electroslag welding*. 2nd ed. Miami: American Welding Society.

Pense, A., J. D. Wood, and J. W. Fisher. 1981. Recent experiences with electroslag welded bridges. *Welding Journal* 60(12): 33–42.

Pussegoda, L. N., and W. R. Tyson. 1981. Sensitivity of electroslag weld metal to hydrogen. *Welding Journal.* 60(12): 252-s to 257-s.

Raman, A. 1981. Electroslag welds: problems and cures. *Welding Journal* 60(12): 17–21.

Ricci, W. S., and T. W. Eagar. 1982. A parametric study of the electroslag welding process. *Welding Journal* 61(12): 397-s to 400-s.

Ritter, J. C., B. F. Dixon, and R. H. Phillips. 1987. Electroslag welding of ship propeller support frames. *Welding Journal* 66(10): 29–39.

Schilling, L. G., and K. H. Klippstein. 1981. Tests of electroslag-welded bridge girders. *Welding Journal* 60(12): 23–30.

Scholl, M. R., R. B. Turpin, J. H. Devletian, and W. E. Wood. 1982. *Consumable guide tube electroslag welding of high carbon steel of irregular cross-section.* ASM Paper 8201-072. Metals Park, Ohio: American Society for Metals.

Shackleton, D. N. 1982. Fabricating steel safely using the electroslag welding process. Part 2. *Welding Journal* 61(1): 23-s to 32-s.

Shackleton, D. N. 1981. Fabricating steel safely using the electroslag welding process. Part 1. *Welding Journal* 60(12): 244-s to 251-s.

Solari, M., and H. Biloni. 1977. Effect of wire feed speed on the structure in electroslag welding of low-carbon steel. *Welding Journal* 56(9): 274-s to 280-s.

Souak, J. F. 1981. Fracture resistance of 4 in. thick A36 and A588 grade A electroslag weldments. *Welding Journal* 60(12): 269-s to 272-s.

Tribau, R., and S. R. Balo. 1983. Influence of electroslag weld metal composition on hydrogen cracking. *Welding Journal* 62(4): 97-s to 104-s.

Yu, D., H. S. Ann, J. H. Devletian, and W. E. Wood. 1986. Solidification study of narrow-gap electroslag welding. In *Welding research: The state of the art. Proceedings: Joining Division Council, University Research Symposium, Toronto, Canada, 15–17 Oct. 1985.* E. F. Nippes and D. J. Ball, eds. Materials Park, Ohio: American Society for Metals.

CHAPTER 11

OXYFUEL GAS WELDING

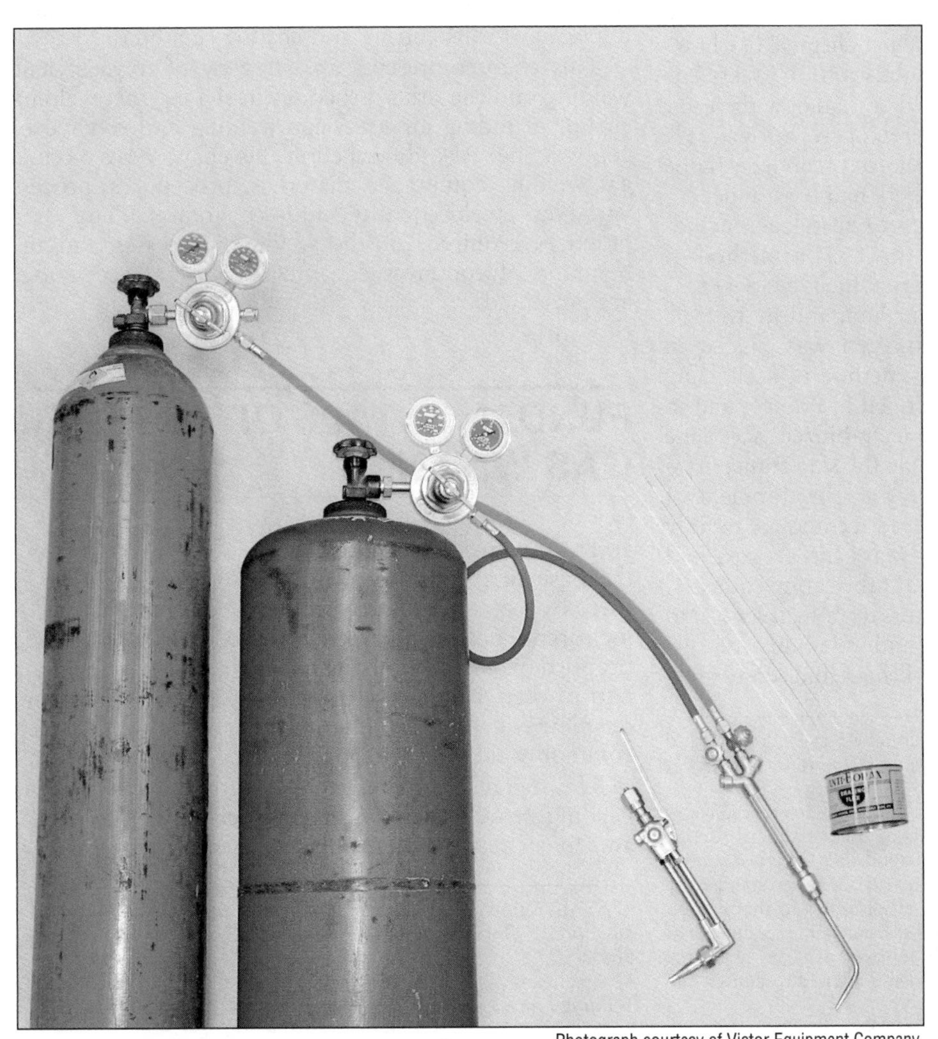

Photograph courtesy of Victor Equipment Company

Prepared by the Welding Handbook Chapter Committee on Oxyfuel Gas Welding:

G. R. Meyer, Chair
Consultant

J. J. Jones
Victor Equipment Company

J. D. Compton
College of the Canyons

Welding Handbook Volume 2 Committee Member:

D. W. Dickinson
The Ohio State University

Contents

CHAPTER 11

OXYFUEL GAS WELDING

INTRODUCTION

The term *oxyfuel gas welding* (OFW) refers to the group of welding processes that achieve the coalescence of metals by heating the workpieces with an oxyfuel gas flame. These processes are implemented with or without the application of pressure and with or without a filler metal.[1,2] Of this group, oxyacetylene welding (OAW) has the widest range of applications.

Oxyacetylene welding is one of the oldest welding processes. Experiments by the French chemist LeChatelier demonstrated that the combustion of acetylene combined with oxygen produced a flame with a far higher temperature than any previously known. He described the characteristics of the oxyacetylene flame in a paper delivered in 1895. Others had been independently investigating similar aspects of acetylene, including its commercial possibilities. In 1892, a method of producing calcium carbide, from which acetylene is made, was invented in North Carolina, and in 1895, a machine for producing liquid oxygen was placed in operation in Indiana. In 1897, a method of stabilizing acetylene with acetone was developed in France, and in 1904, a pressurized container for stabilized acetylene was introduced. This made possible the safe transportation and handling of this volatile gas. Development of oxyacetylene welding apparatus and techniques rapidly followed, and in fewer than ten years the process had become very valuable to the metal fabricating industry. It was especially well known for its effectiveness in repair welding. The usefulness and reliability of this new process were solidly established and confirmed many times over during the war years of 1914 through 1918.[3,4]

Although the use of oxyfuel gas welding has been surpassed by the arc welding processes for most applications, oxyfuel gas welding remains popular because it is versatile, portable, and inexpensive. The process is capable of making quality welds in many materials, and when used with modified basic equipment, is also important for its cutting and heating capabilities.

This chapter presents an overview of oxyacetylene welding and the other processes in the oxyfuel welding group, including air acetylene welding and oxyhydrogen welding. Also described in this chapter are oxyfuel gas welding equipment, materials, procedures, process variables, weld quality and economics. The safe practices required for the welding processes and the handling of compressed gas cylinders and accessories are presented.

FUNDAMENTALS OF OXYFUEL GAS WELDING

The oxyfuel gas welding processes involve melting the base metal and applying a filler metal (if one is used) by means of a flame produced at the tip of a welding torch. Fuel gas and oxygen are combined in specific proportions inside a mixing chamber, which is usually a part of the welding torch assembly. Molten metal from the plate edges and the filler metal, if used, intermix in a common weld pool and coalesce on cooling.

The metals normally welded with the oxyfuel gas welding process include carbon steels, low-alloy steels, and most nonferrous metals, but generally not refrac-

1. American Welding Society, (AWS) Committee on Definitions and Symbols, 2001, *Standard Welding Terms and Definitions*, A3.0:2001, Miami: American Welding Society.
2. At the time of the preparation of this chapter, the referenced codes and other standards were valid. If a code or other standard is cited without a date of publication, it is understood that the latest edition of the document referred to applies. If a code or other standard is cited with the date of publication, the citation refers to that edition only, and it is understood that any future revisions or amendments to the code or standard are not included; however, as codes and standards undergo frequent revision, the reader is advised to consult the most recent edition.

3. ESAB Welding and Cutting Products, 1995, *The Oxy-Acetylene Handbook*, Florence, South Carolina: ESAB Welding and Cutting Products.
4. American Welding Society, 1997, O'Brien, R. L., ed., *Jefferson's Welding Encyclopedia*, Miami: American Welding Society.

tory or reactive metals. The process is used to weld thin sheet, tubes, and small-diameter pipe, and is ideally suited for repair welding. Thick-section welds, except those associated with repair work, are not cost effective when compared to many of those produced with the available arc welding processes.

Oxyfuel gas welding is also used for many surfacing operations, some of which are not possible with arc welding processes. For example, the application of surfacing materials high in zinc content, such as admiralty metal, can be accomplished with oxyfuel gas welding. Automated operations are often used to apply surfacing to products such as tube sheets or heat exchangers.

ADVANTAGES AND LIMITATIONS

An important advantage offered by oxyfuel gas welding is the control the welder can exercise over the heat input and temperature, independent of the addition of filler metal. The welder can also control weld bead size, shape, and weld pool viscosity. The equipment used in oxyfuel gas welding is low in cost, usually portable, and versatile enough to be used for a variety of related operations, including bending and straightening, preheating, postheating, surfacing, brazing, braze welding, and soldering. Cutting attachments, multiflame heating nozzles, and a variety of special application accessories add greatly to the overall versatility of the basic oxyfuel gas welding equipment. This versatility makes the oxyacetylene process particularly attractive from the viewpoint of initial investment.

Oxyacetylene welding is usually not recommended for high-strength heat-treatable steels, especially when they are being fabricated in the heat-treated condition. When welding quenched and tempered steels, the slow rate of heat input of oxyacetylene welding may cause metallurgical changes in the heat-affected zone and so may destroy the heat-treated base metal properties. One of the arc welding processes should be selected for these metals.

OXYFUEL GAS WELDING PROCESS VARIATIONS

Variations of the oxyfuel gas welding processes are air acetylene welding, oxyacetylene welding, oxyhydrogen welding and pressure gas welding. These techniques are described briefly in this section.

Air Acetylene Welding

Air acetylene welding (AAW) is an oxyfuel gas welding process that uses an air-acetylene flame to achieve the coalescence of materials. This obsolete or seldom-used process is performed (without the application of pressure) only on metals with very low melting points that are not affected by the flame chemistry. However, the process is frequently adapted to various soldering and brazing applications in which only the filler metal is melted.

Oxyacetylene Welding

Oxyacetylene welding (OAW), which uses acetylene as the fuel gas, is the most widely used of the oxyfuel gas welding processes. Acetylene is the fuel gas of choice because of its high combustion intensity. The acetylene flame, enhanced with the addition of varying amounts of oxygen, provides the "tool" for welding and can be adjusted according to the needs of the application.

Oxyacetylene welding equipment consists of a welding torch, regulators, an oxygen cylinder, and an acetylene cylinder, typically on a two-wheeled cart. With the addition of a cutting attachment, the unit is a complete and relatively inexpensive welding and cutting outfit. This self-contained unit can be readily moved about in a shop or plant. It can be moved easily into the field on a small truck to repair a breakdown wherever it may have occurred. This equipment is illustrated in Figure 11.1, shown with welding rods and flux, which are used in some applications.

Oxyacetylene welding is almost universally used for maintenance and repair, where its flexibility and mobility result in significant savings of time and labor. The process is also well suited for use in machine and automobile repair shops, and is equally useful in shops devoted entirely to welding, where the repair of industrial, agricultural, and household equipment may be the main business.

Oxyacetylene welding is not recommended for the repair of high-strength heat-treatable steels. These should be welded with an arc welding process.

Oxyacetylene welding is used in the fabrication of sheet metal, tubing, pipe, and other metal shapes. Pipelines up to 51 millimeters (mm) (2 inches [in.]) in diameter are welded with this process. Applications include projects such as the installation of industrial piping systems and the fabrication of many products used in automotive manufacturing. The process is widely used by artists and metal sculptors. Figure 11.2 shows an oxyfuel gas welding application, in which a neutral flame is used as the heat source and the forehand technique is used.

Oxyhydrogen Welding

Oxyhydrogen welding (OHW) is an oxyfuel gas welding process that uses hydrogen as the fuel gas. This process is almost exclusively used for the melting and welding of low-melting metals, such as aluminum, magnesium, lead, and their alloys. The low-temperature, nearly invisible oxyhydrogen flame is capable of

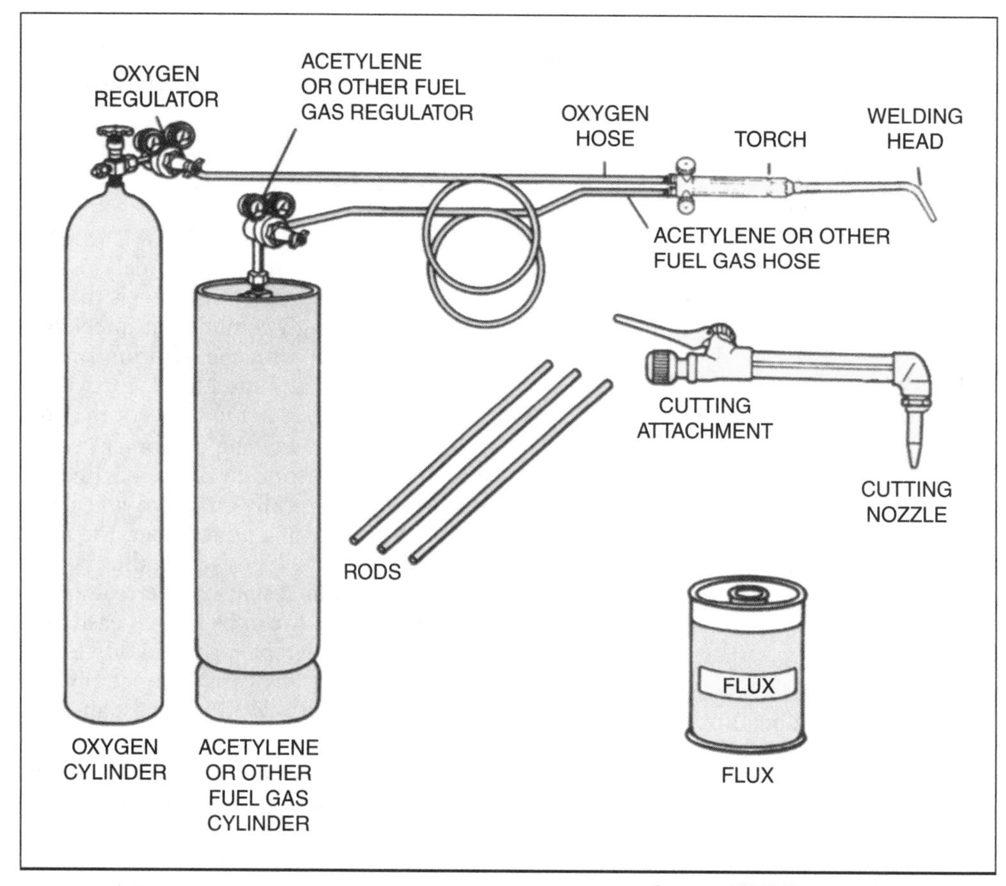

Courtesy of ESAB Welding and Cutting Products

Figure 11.1—Oxyfuel Gas Equipment

Photograph courtesy of Victor Equipment Company

Figure 11.2—Oxyacetylene Welding on Mild Steel Using a Neutral Flame and the Forehand Technique

maintaining its very small size without producing carbon soot, thus allowing the process to be used for very fine, precise work, such as joining the intricate components of jewelry and electronic assemblies.

Oxyhydrogen welding is an especially convenient process for intricate applications when hydrogen and oxygen can be generated from a water electrolysis process.

Pressure Gas Welding

Pressure gas welding (PGW) is an oxyfuel gas welding process in which the abutting surfaces of the workpieces are heated to the welding temperature, and with the application of pressure, a weld is produced simultaneously over the entire faying surfaces of the workpieces. A filler metal is not used.

Pressure gas welding is used to make upset welds in butt joints. The joints are heated with the oxyfuel flame and forced together to obtain the forging action needed to produce sound welds in plain carbon steel, low-carbon and high-carbon steels, low-alloy and high-alloy steels and some nonferrous metals. The process is useful in joining dissimilar metals. Pressure gas welding adapts well to mechanized operations such as pipe welding, although the process has largely been replaced by gas metal arc welding.[5]

MATERIALS

The materials used in oxyfuel gas welding are gases; filler metal in the form of welding rods, and flux, although rods and fluxes are not used in all applications. The principal gases are oxygen, which supports the combustion of the fuel gas, and acetylene, which provides the heat and the atmosphere required for welding. Hydrogen, methane (natural gas), and various proprietary fuel gases are less commonly used.

OXYGEN

Oxygen in the gaseous state is colorless, odorless, and tasteless. A chief source of oxygen is the earth's atmosphere, which contains approximately 21% oxygen by volume. Although air contains sufficient oxygen to support fuel gas combustion, the use of pure oxygen speeds up burning reactions and increases flame temperatures. The use of the term *oxygen* throughout this chapter refers to pure oxygen.

Most oxygen used in the welding industry is extracted from the atmosphere by liquefaction techniques employed by gas manufacturing companies. In the extraction process, air may be compressed to approximately 20 megapascals (MPa) (3000 pounds per square inch gauge [psig]), although some types of equipment operate at much lower pressures. Carbon dioxide and any impurities in the air are removed. The air passes through coils, where it is allowed to expand to a rather low pressure. During the expansion, the air becomes substantially cooled. It is then conveyed back over the coils, inducing further cooling until liquefaction occurs.

The liquid air is sprayed on a series of evaporating trays or plates in a rectifying tower. The nitrogen and other gases that boil at lower temperatures than the oxygen escape from the top of the tower, and high-purity liquid oxygen collects in a receiving chamber at the base. Some plants are designed to produce liquid oxygen for bulk delivery; in other plants, gaseous oxygen is withdrawn for compression into cylinders.

FUEL GASES

Commercial fuel gases have one common property—they all require oxygen to support combustion. To be suitable for welding operations, a fuel gas, when burned with oxygen, must have the following characteristics:

1. High flame temperature,
2. High rate of flame propagation,
3. Adequate heat value, and
4. Minimum chemical reaction of the flame with base and filler metals.

Among the commercially available fuel gases, acetylene most closely meets all of these requirements. Other fuel gases such as methylacetylene-propadiene (stabilized) (MPS), propylene, propane, methane (natural gas), and proprietary gases based on these gases, provide sufficiently high flame temperatures but exhibit lower flame propagation rates.

The gas flames produced by these gases (other than acetylene) are excessively oxidizing at oxygen-to-fuel gas ratios that are high enough to produce usable heat transfer rates required for welding. To ensure stable operation and optimal heat transfer, flame-holding devices such as counter bores on the tips are usually necessary. However, these commercial fuel gases are frequently used for oxygen cutting. They are also used for torch heating, brazing, soldering, and other operations in which the demands on the flame characteristics and heat transfer rates are not as stringent as those for welding.

Safety precautions recommended throughout this chapter and in the "Safe Practices" section should be

5. See Reference 4, p. 401.

Table 11.1
Characteristics of Common Fuel Gases

Fuel Gas	Chemical Formula	Specific Gravity*	Volume-to-Weight Ratio*		Oxygen-to-Fuel Combustion Ratio[a]	Flame Temperature with Oxygen[b]		Heat of Combustion					
								Primary		Secondary		Total	
		Air = 1	m^3/kg	ft^3/lb		°C	°F	MJ/m^3	Btu/ft^3	MJ/m^3	Btu/ft^3	MJ/m^3	Btu/ft^3
Acetylene	C_2H_2	0.906	0.91	14.6	2.5	3087	5589	19	507	36	963	55	1470
Propane	C_2H_8	1.52	0.54	8.7	5.0	2526	4579	10	255	94	2243	104	2498
Methylacetylene-propadiene (stabilized) (MPS)[c]	C_3H_8	1.48	0.55	8.9	4.0	2927	5301	21	571	70	1889	91	2460
Propylene	C_3H_4	1.48	0.55	8.9	4.5	2900	5250	16	438	73	1962	89	2400
Methane (natural gas)	CH_4	0.62	1.44	23.6	2.0	2538	4600	0.4	11	37	989	37	1000
Hydrogen	H_2	0.07	11.77	188.7	0.5	2660	4820						325

*At 15.6°C (60°F)

a. The volume units of oxygen required to completely burn a unit volume of fuel gas according to the formulas shown in Table 11.2. A portion of the oxygen is obtained from the atmosphere.
b. The temperature of the neutral flame.
c. May contain significant amounts of saturated hydrocarbons.

observed when using fuel gases. Storage and distribution systems should be installed according to applicable national, state, or local codes. Recommendations listed in Material Safety Data Sheets (MSDS) from gas suppliers should be observed and the manufacturers' instructions for the safe installation and use of oxyfuel gas apparatus should be followed. Additional safety resources are listed in Appendix B, "Safety and Health Codes and Other Standards."

Gas-Related Terminology

Table 11.1 summarizes some of the pertinent characteristics of common fuel gases. In order to appreciate the significance of the information in this table, it is necessary to understand the following terms and concepts used to describe fuel gases.

Specific Gravity. Specific gravity is the density of a gas compared to the density of air (as a ratio). The value indicates how the gas may accumulate in the event of a leak. For example, gases with a specific gravity less than that of air tend to rise. They may collect in the corners of rooms, in lofts, and in ceiling spaces. Gases with a specific gravity greater than that of air tend to accumulate in low, still areas.

Volume-to-Weight Ratio. A quantity of gas at a standard temperature and pressure can be described by a known volume or weight. The values shown in Table 11.1 provide the volume per unit weight of gases at 15.6°C (60°F) under atmospheric pressure. Multiplying these figures by the known weight yields the volume. If the volume is known, multiplying the volume by the reciprocal of the figures shown by the volume yields the weight.

Combustion Ratio. The oxygen-to-fuel combustion ratio indicates the volume of oxygen theoretically required for complete combustion of a fuel gas as a multiple of the fuel gas volume. These oxygen-to-fuel gas ratios, termed *stoichiometric mixtures*, are obtained from the balanced chemical equations given in Table 11.2. The values shown for complete combustion are useful in calculations. They do not represent the oxygen-to-fuel gas ratios actually delivered by an operating torch because, as explained with the next term, complete combustion is partly supported by oxygen in the surrounding air.

Table 11.2
Chemical Equations for the Complete Combustion of the Common Fuel Gases

Fuel Gas	Reaction with Oxygen
Acetylene	$C_2H_2 + 2.5O_2 \rightarrow 2CO_2 + H_2O$
Methylacetylene-propadiene (stabilized) (MPS)	$C_3H_4 + 4O_2 \rightarrow 3CO_2 + 2H_2O$
Propylene	$C_3H_6 + 4.5O_2 \rightarrow 3CO_2 + 3H_2O$
Propane	$C_3H_8 + 5O_2 \rightarrow 3CO_2 + 4H_2O$
Methane (natural gas)	$CH_4 + 2O_2 \rightarrow CO_2 + 2H_2O$
Hydrogen	$H_2 + 0.5O_2 \rightarrow H_2O$

Heat of Combustion. The total heat of combustion (heat value) of a hydrocarbon fuel gas is the sum of the heat generated in the primary and secondary reactions that take place in the overall flame. The combustion of hydrogen takes place in a single reaction. The theoretical basis for these chemical reactions and their heat effects are discussed in *Welding Science and Technology*, Volume 1 of the *Welding Handbook*, 9th edition, Chapter 2.[6]

Typically, the heat content of the primary reaction is generated in an inner, or primary, flame, where combustion is supported by the oxygen supplied by the torch. The secondary reaction takes place in an outer, or secondary, flame envelope in which the combustion of the primary reaction products is supported by oxygen from the air.

Although the heat of the secondary flame is important in most applications, the more concentrated heat of the primary flame is the major contributor to the welding capability of an oxyfuel gas system. The primary flame is said to be *neutral* when the chemical equation for the primary reaction is exactly balanced, yielding only carbon monoxide and hydrogen. Under these conditions, the primary flame atmosphere is neither carburizing nor oxidizing. Since the secondary reaction is necessarily dependent on the products of the primary reaction, the term *neutral* serves as a convenient reference point for describing combustion ratios and comparing the various heat characteristics of different fuel gases.

Flame Temperature. The flame temperature of a fuel gas varies according to the oxygen-to-fuel ratio. Although the flame temperature provides an indication of the heating ability of the fuel gas, it is only one of the many physical properties to be considered in an overall evaluation. Flame temperatures are usually calculated mathematically, as no practical method of accurately measuring these values is available.

The flame temperatures listed in Table 11.1 are for the neutral flame, that is, the primary flame that is neither oxidizing nor carburizing in character. Flame temperatures higher than those listed may be achieved, but these flames are oxidizing in nature, an undesirable condition in the welding of most metals.

Combustion Velocity. One characteristic property of a fuel gas, the combustion velocity (flame propagation rate), is an important factor in the heat output of the oxyfuel gas flame. This is the velocity at which a flame front travels through the adjacent unburned gas. It influences the size and temperature of the primary flame.

The combustion velocity also affects the velocity at which gases may flow from the torch tip without causing a flame standoff or backfire. A flame standoff occurs when combustion takes place some distance away from the torch tip rather than right at the torch tip. A backfire is the momentary recession of the flame into the welding tip or mixer, followed by the reappearance or complete extinction of the flame.

As shown in Figure 11.3, the combustion velocity of a fuel gas varies in a characteristic manner according to the proportions of oxygen and fuel gas in the mixture.

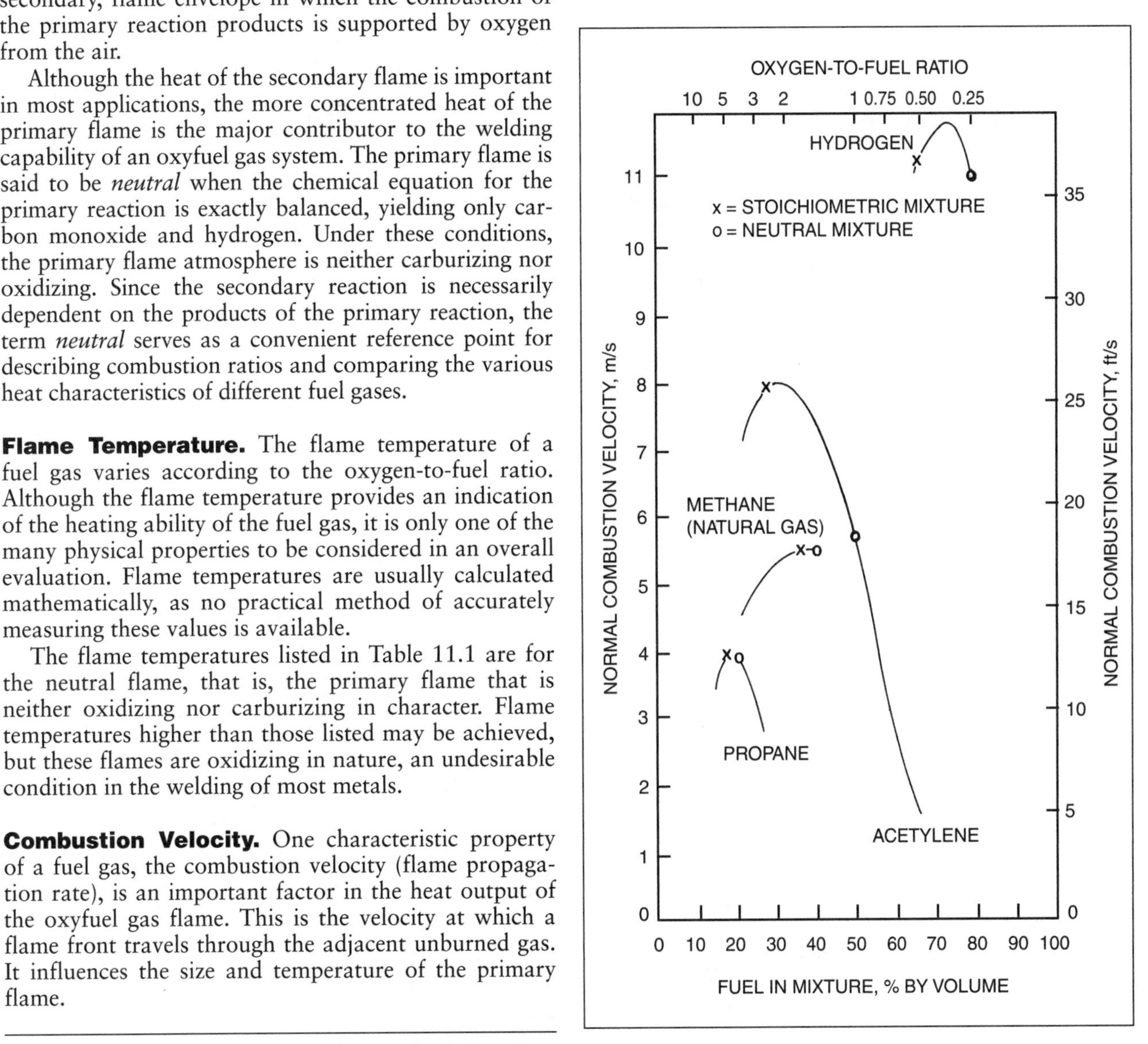

Figure 11.3—Normal Combustion Velocity (Flame Propagation Rate) of Various Fuel Gas-Oxygen Mixtures

6. American Welding Society (AWS) Welding Handbook Committee, Jenney, C. L. and A. O'Brien, eds., 2001, *Welding Science and Technology*, Vol. 1 of Welding Handbook, 9th ed., Vol. 1, Miami: American Welding Society, pp. 60–61.

Combustion Intensity. The flame temperatures and heating values of fuels have been used almost exclusively as the criteria for evaluating fuel gases. However, these two factors alone do not provide sufficient information for a complete appraisal of fuel gases for heating purposes. A concept known as *combustion intensity,* or *specific flame output,* is used to evaluate different oxygen-fuel gas combinations. The combustion intensity takes into account the burning velocity of the flame, the heating value of the mixture of oxygen and fuel gas, and the area of the flame cone issuing from the tip. The combustion intensity, C_i, is expressed as follows:

$$C_i = C_v \times C_h \tag{11.1}$$

where

C_i = Combustion intensity, joules per square meter (J/m^2) × seconds (s) (British thermal units per square foot [Btu/ft^2] × s);
C_v = Normal combustion velocity of flame, meters per second (m/s) (feet per second [ft/s]; and
C_h = Heating value of the gas mixture under consideration, joules per cubic meter (J/m^3) (British thermal units per cubic foot [Btu/ft^3]).

Therefore, the combustion intensity, C_i, is maximum when the product of the normal burning velocity of the flame, C_v, and the heating value of the gas mixture C_h, is maximum.

Like the heat of combustion, the combustion intensity of a gas can be expressed as the sum of the combustion intensities of the primary and secondary reactions. However, the combustion intensity of the primary flame, located near the torch tip where it can be concentrated on the workpiece, is of major importance in welding. The secondary combustion intensity influences the thermal gradient in the vicinity of the weld.

Figures 11.4 and 11.5 show the typical rise and fall of the primary and secondary combustion intensities of several fuels with varying proportions of oxygen and fuel gas. Figure 11.6 provides the total combustion intensities for the same gases. These curves illustrate that acetylene produces the highest combustion intensities of the gases plotted.

Acetylene (C_2H_2) is a hydrocarbon compound that contains a larger percentage of carbon by weight than any of the other hydrocarbon fuel gases. Colorless and lighter than air, it has a distinctive odor resembling garlic. Because the acetylene contained in cylinders is dissolved in acetone, it has a slightly different odor from that of pure acetylene.

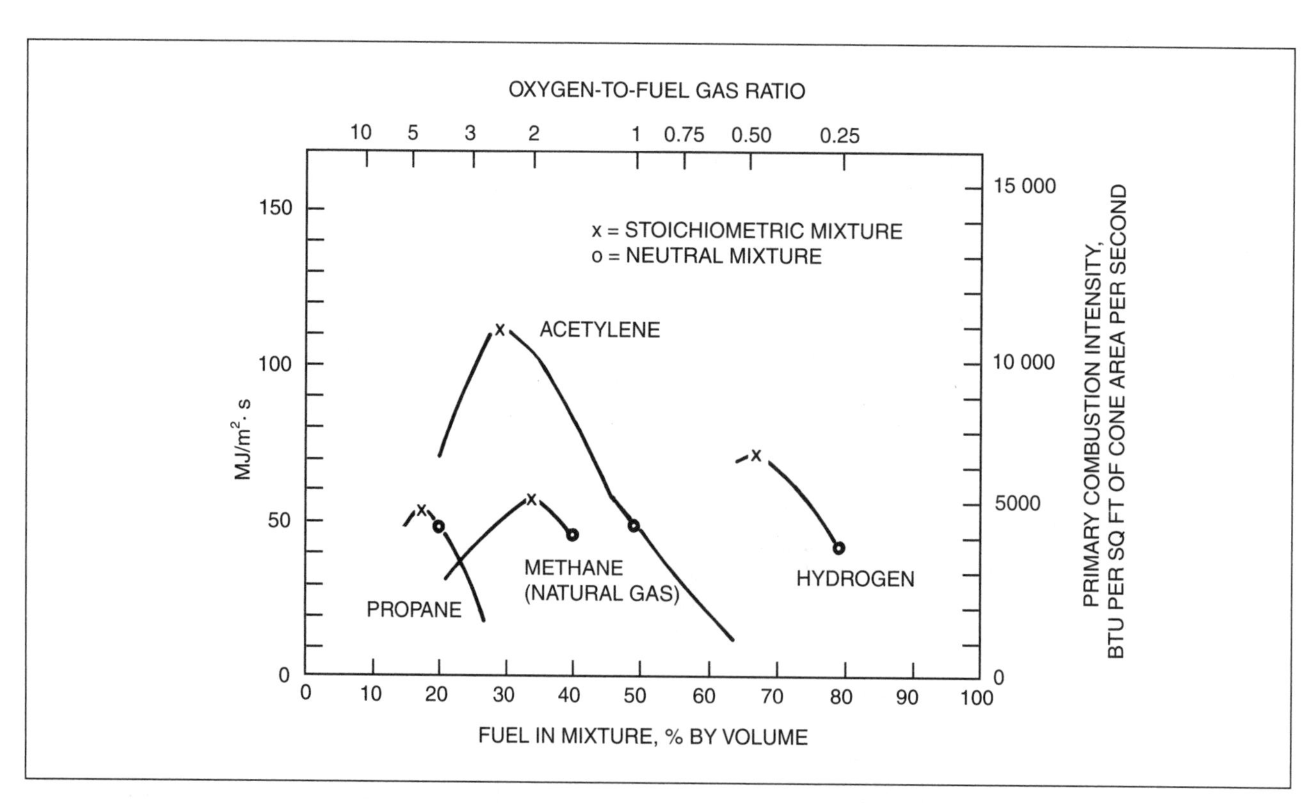

Figure 11.4—Primary Combustion Intensities of Various Fuel Gas-Oxygen Mixtures

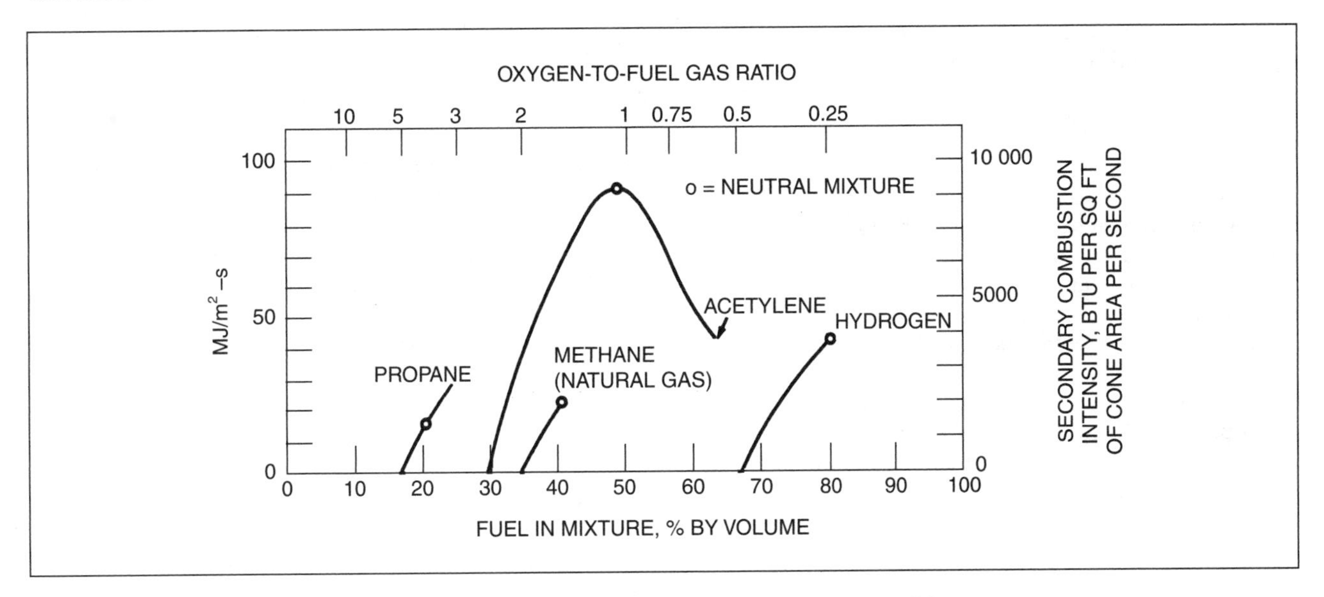

Figure 11.5—Secondary Combustion Intensities of Various Fuel Gas-Oxygen Mixtures

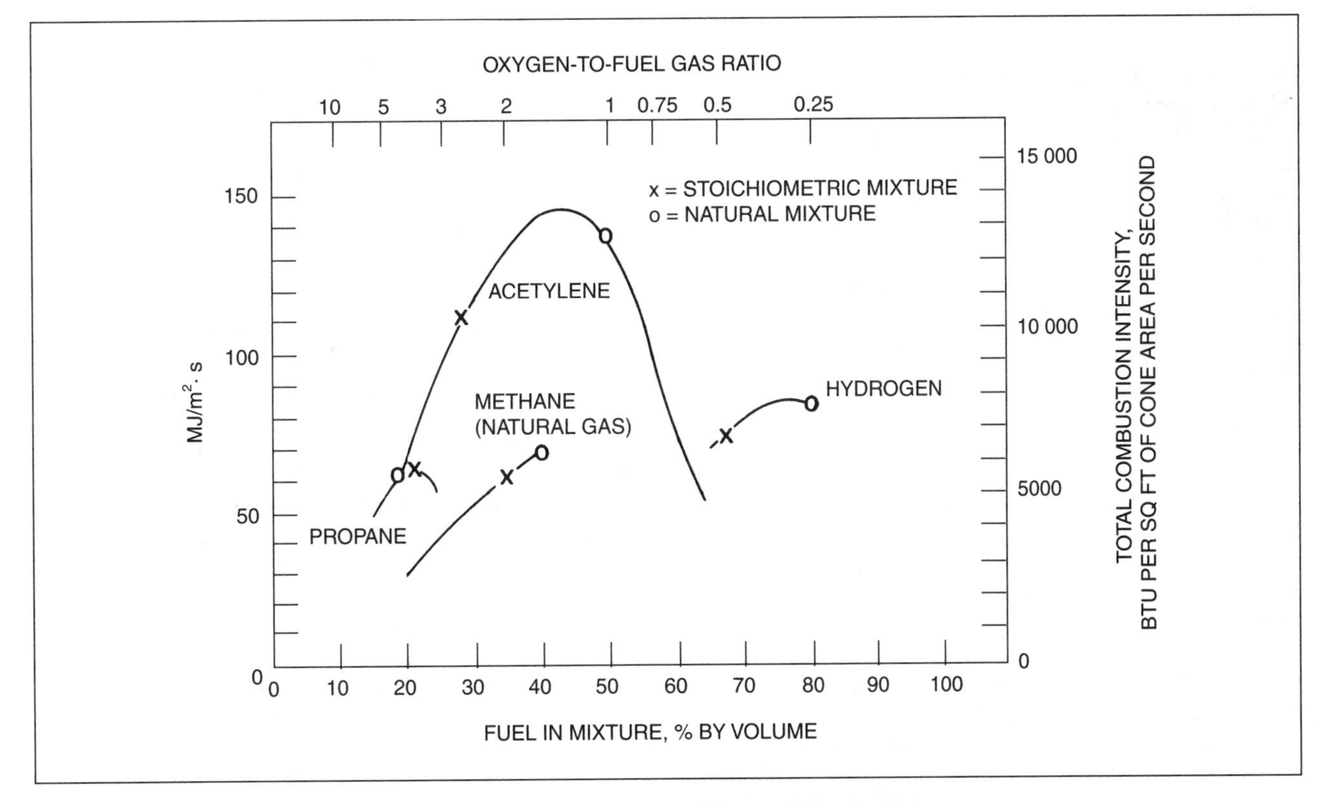

Figure 11.6—Total Combustion Intensities of Various Fuel Gas-Oxygen Mixtures

The complete combustion of acetylene is theoretically represented by the following chemical equation:

$$C_2H_2 + 2.5O_2 \rightarrow 2CO_2 + H_2O \quad (11.2)$$

This equation indicates that one volume of acetylene (C_2H_2) and 2.5 volumes of oxygen (O_2) react to produce two volumes of carbon dioxide (CO_2) and one volume of water vapor (H_2O). The volumetric ratio of oxygen to acetylene is 2.5 to 1.

As previously noted, the reaction of Equation 11.2 does not proceed directly to the products shown. Combustion takes place in two stages. The primary reaction takes place in the inner zone of the flame (referred to as the *inner cone*) and is represented by the following chemical equation:

$$C_2H_2 + O_2 \rightarrow 2CO + H_2 \quad (11.3)$$

In this case, one volume of acetylene and one volume of oxygen react to form two volumes of carbon monoxide and one volume of hydrogen. The heat content and high temperature of this reaction (see Table 11.1) result from the decomposition of the acetylene and the partial oxidation of the carbon resulting from that decomposition.

When the gases issuing from the torch tip are in the one-to-one ratio indicated in Equation 11.3, the reaction produces the typical brilliant blue inner cone. This relatively small flame creates the combustion intensity needed for welding steel. The flame is considered neutral because no excess carbon or oxygen exists to carburize or oxidize the metal. The products are actually in a reducing status, a benefit when welding steel.

In the outer envelope of the flame, the carbon monoxide and hydrogen produced by the primary reaction burn with oxygen from the surrounding air. This forms carbon dioxide and water vapor, respectively, as shown in the following chemical equation for the secondary reaction:

$$2CO + H_2 + 1.5O_2 \rightarrow 2CO_2 + H_2O \quad (11.4)$$

Although the heat of combustion of this outer flame is greater than that of the inner flame, its combustion intensity and temperature are lower because of its large cross-sectional area. The final products are produced in the outer flame because they cannot exist in the high temperature of the inner cone.

The oxyacetylene flame is easily controlled by valves on the welding torch. By a slight change in the proportions of oxygen and acetylene flowing through the torch, the chemical characteristics in the inner zone of the flame and the resulting action of the inner cone on the molten metal can be varied over a wide range. Thus, by adjusting the torch valves, it is possible to produce an oxidizing, neutral, or carburizing flame.

Acetylene Production. Acetylene is the product of a chemical reaction between calcium carbide (CaC_2) and water. In that reaction, the carbon in the calcium carbide combines with the hydrogen in the water, forming gaseous acetylene. At the same time, the calcium combines with oxygen and hydrogen to form a calcium hydroxide residue. Following is the chemical expression:

$$2CaC_2 + 2H_2O \rightarrow C_2H_2 + Ca(OH)_2 \quad (11.5)$$

The calcium carbide used in this process is produced by smelting lime and coke in an electric furnace. When removed from the furnace and cooled, the carbide is crushed, screened, and packed in airtight containers. The most common of these holds 45 kg (100 lb) of the hard, grayish solid. Approximately 0.28 m^3 (10 ft^3) of acetylene can be generated from 1 kg (2.2 lb) of calcium carbide.

Acetylene is also produced in petrochemical plants and is used for a variety of purposes other than oxyfuel gas welding and cutting.

Safe Handling of Acetylene. Free acetylene under certain pressure and temperature conditions may dissociate explosively into its hydrogen and carbon constituents; therefore, cylinders to be filled with acetylene are initially packed with a porous filler. Acetone, a solvent capable of absorbing 25 times its own volume of acetylene per atmosphere of pressure, is then added to the filler to stabilize the acetylene. By dissolving the acetylene and dividing the interior of the cylinder into small, partly separated cells within the porous filler in this manner, a safe acetylene-filled container is produced.

Gaseous acetylene is unstable at temperatures above 1435°F (780°C) or at pressures above 207 kPa (30 psig), and decomposition may result even in the absence of oxygen. This characteristic has been taken into consideration in the preparation of a code of safe practices for the generation, distribution, and use of acetylene gas.[7] It is a requirement of safe practice that acetylene must never be used in generators, pipelines, or hoses at pressures exceeding 103 kPa (15 psig).

Cylinders of acetylene are available in sizes containing from 0.28 m^3 to 12 m^3 (10 ft^3 to 420 ft^3) of the gas. The cylinders are equipped with fusible safety plugs made of a metal with an approximate melting point of 100°C (212°F). This allows the gas to escape if the

7. See, for example, Compressed Gas Association (CGA) *Acetylene*, CGA-G1, Arlington, Virginia: Compressed Gas Association.

cylinder is subjected to excessive heat, resulting in a relatively controlled burn rather than a rupture of the cylinder.

Hydrogen

The relatively low heat value of the oxyhydrogen flame restricts the use of hydrogen to the welding of aluminum, magnesium, lead, and similar metals that melt at lower temperatures than steel and to certain torch brazing operations. Other welding processes, however, have largely supplanted all forms of oxyfuel gas welding for many of these metals.

Hydrogen is available in seamless drawn steel cylinders charged to a pressure of about 14 MPa (2000 psig) at a temperature of 70°F (21°C). It can also be supplied as a liquid, either in individual cylinders or in bulk. Liquid hydrogen is vaporized into gas at the point of use.

Methylacetylene-Propadiene (Stabilized) Gas

Several commercially prepared fuel gas mixtures are available for welding, but they are not generally used for this purpose. They are more extensively used for cutting, torch brazing, and other heating operations. The compositions of one group of mixed fuel gases approximate methylacetylene-propadiene (stabilized) and contain mixtures of propadiene, propane, butane, butadiene, and methylacetylene.

A characteristic of these mixed fuel gases is that the heat distribution within the flame is more even than that of acetylene, thus requiring less manipulation of the gas torch for controlling the heat input. The flame temperatures of these gases are lower than those achieved with acetylene and neutral gas mixtures. Adjusting the flames to oxidizing can increase temperatures. These gases are popular because they may cost less than acetylene, and the cylinders contain a greater volume of fuel for a given size and weight.

Methane (Natural Gas)

Natural gas is obtained from wells and distributed by pipelines. Its chemical composition varies widely, depending on the locality from which it is obtained. The principal constituents of most natural gases are methane (CH_4) and ethane (C_2H_6). The volumetric requirement of natural gas is, as a rule, about 1-1/2 times that of acetylene to provide an equivalent amount of heat. The principal use of natural gas in the welding industry is for oxygen cutting and heating operations.

Propane

Propane (C_3H_8) is used in heating operations related to welding and metal fabrication, but it is primarily used for oxygen cutting. Propane is derived from crude oil and gas mixtures obtained from active oil and natural gas wells. It is sold and transported in steel cylinders containing up to 45 kg (100 lb) of the liquefied gas. Large-volume consumers are supplied by rail-car tanks or other methods of bulk delivery.

Propylene

A single-component fuel gas, propylene (C_3H_6), is a petroleum product with performance characteristics similar to those of the MP gases. Although not suitable for welding, propylene is used in related operations for oxygen cutting, brazing, heating, and flame hardening. Propylene gas delivery equipment is similar in design to that used with the MPS gases.

FILLER METALS

The properties of the weld metal must closely match those of the base metal. Because of this requirement, filler metal in the form of welding rods with various chemical compositions are available for welding most ferrous and nonferrous materials. The welding process itself influences the filler metal composition, since certain elements may be lost or added during welding. Welding rods are available for welding almost all of the commonly used base metals with the oxyfuel gas welding process. The standard diameters of rods vary from 1.6 mm to 10 mm (1/16 in. to 3/8 in.); the standard rod length is 6.10 m and 9.14 m (24 in. and 36 in.).

The chemical composition of a filler metal should be within the limits specified for the particular metal to be welded. Filler metal must be free from porosity, nonmetallic inclusions, and any other foreign matter and should deposit smoothly. Many proprietary filler metals are available for specific applications; manufacturers of filler metals have designed welding rods that allow for the metallurgical changes that take place during welding so that the deposited metal will have the correct chemical composition. The deposited metal should be free flowing and should unite readily with the base metal to produce sound, clean welds.

In general, an effort should be made to match the filler metal to the base metal in repair welding with oxyfuel gas welding. In maintenance and repair work, however, the composition of the base metal may not be obvious. Fortunately, the composition of the welding rod need not always match that of the base metal. A steel welding rod of nominal strength can be used to repair broken parts made of alloy steels. When it is necessary to heat-treat a steel workpiece after welding, carbon can be added to a deposit of mild steel by the judicious use of the carburizing flame. It is preferable, however, to use a welding rod of low-alloy steel.

The American Welding Society (AWS) Committee on Filler Metals and Allied Materials has prepared a number of specifications for filler metals for various welding processes. Many of the filler metals used in oxyfuel gas welding meet these specifications for use with carbon steel, low-alloy steel, cast iron and some of the nonferrous metals, as well as for surfacing alloys. Information on welding aluminum and aluminum alloys and the selection of filler materials will be found in the chapters on the various metals and alloys in the *Welding Handbook*, 8th edition, Volumes 3 and 4, *Materials and Applications, Parts 1 and 2.*[8,9]

The welding rods for use with steel are listed in *Specification for Carbon and Low Alloy Steel Rods for Oxyfuel Gas Welding,* ANSI/AWS A5.2.[10] The rods are classified on the basis of strength. The most commonly used is RG60 (414 MPa [60 ksi] minimum tensile strength), which has properties compatible with most low-carbon steels.

For the oxyfuel gas braze welding of cast iron, both cast iron and copper-base welding rods are used, according to *Specification for Welding Electrodes and Rods for Cast Iron,* ANSI/AWS A5.15.[11] These filler metals are classified on the basis of chemical composition.

For welding aluminum and aluminum alloys, *Specification for Bare Aluminum and Aluminum Alloy Welding Electrodes and Rods,* ANSI/AWS A5.10, should be consulted.[12]

FLUXES

Oxides that fail to flow from the weld zone become entrapped in the solidifying metal, interfering with the addition of filler metal. These conditions may occur when the oxides have a higher melting point than that of the base metal. One of the most important ways to control weld quality is to remove oxides and other impurities from the surface of the metal to be welded. Unless the oxides are removed, fusion may be difficult to attain, the joint may lack strength, and inclusions may be present. Fluxes are applied as an aid to removing oxides and other impurities.

Steel and its oxides, as well as slag that forms during welding, do not fall into the above category and thus require no fluxing. Aluminum, however, forms an oxide with a very high melting point. This oxide must be removed from the welding zone before satisfactory results can be obtained.

Certain substances react chemically with the oxides of most metals, forming fusible slags at the welding temperature. These substances, either singly or in combination, make efficient fluxes. Gas tungsten arc welding (GTAW) and gas metal arc welding (GMAW) are generally used to weld aluminum in order to avoid these slag problems.

A good flux should assist in removing the oxides during welding by forming fusible slags that float to the top of the weld pool and do not interfere with the deposition and fusion of filler metal. A flux should protect the weld pool from the atmosphere and prevent the weld pool from absorbing or reacting with gases in the flame. This must be done without obscuring the welder's vision or hampering the manipulation of the weld pool.

During the preheating and welding periods, the flux should clean and protect the surfaces of the base metal and, in some cases, the welding rod. Fluxes are excellent metal cleaners, but they should not be used as a substitute for the cleaning of the base metal during joint preparation.

Flux is commercially available as a dry powder, a paste or thick solution, or as a preplaced coating on the welding rod. Some fluxes produce much more favorable results if they are used dry. Braze welding fluxes and the fluxes for use with cast iron are usually in this class. These fluxes are applied by heating the end of the rod and dipping it into the powdered flux. Enough adheres to the rod to ensure adequate fluxing. Dipping the hot rod into the flux again will coat another portion. Dropping some of the dry powder on the base metal ahead of the welding zone may sometimes help, especially in the repair of "dirty" castings—those that cannot be completely cleaned before welding.

Fluxes in paste form are usually painted on the base metal with a brush, and the rod can either be painted or dipped. Commercially precoated rods can be used without further preparation, and when required, additional flux can be placed on the base metal. A precoated rod may sometimes need to be dipped in powdered flux if the flux melts off too far from the end of the rod during welding.

8. American Welding Society (AWS) Welding Handbook Committee, Oates, W. R., ed. 1996, *Materials and Applications—Part 1*, Vol. 3 of *Welding Handbook*, 8th ed., Miami: American Welding Society. pp. 81–83.
9. American Welding Society (AWS) Welding Handbook Committee, Oates, W. R., and A. M. Saitta, eds., 1998, Materials and Applications—Part 2, Vol. 4 of *Welding Handbook*, 8th ed. Miami: American Welding Society.
10. American Welding Society (AWS) Committee on Filler Metals and Allied Materials, 1990, *Specification for Carbon and Low-Alloy Steel Rods for Oxyfuel Gas Welding*, ANSI/AWS A-5.2-92R, Miami: American Welding Society.
11. American Welding Society (AWS) Committee on Filler Metals and Allied Materials, 1990, *Specification for Welding Electrodes and Rods for Cast Iron,* ANSI/AWS A5.15-90, Miami: American Welding Society.
12. American Welding Society (AWS) Committee on Filler Metals and Allied Materials, 1999, *Specification for Bare Aluminum and Aluminum Alloy Welding Electrodes and Rods,* AWS A5.10/A5.10M:1999, Miami: American Welding Society.

OXYFUEL GAS WELDING EQUIPMENT

The basic equipment in an oxyfuel gas welding setup consists of fuel gas and oxygen cylinders, each with a gas regulator to reduce cylinder pressure, hoses to convey the gases to the torch, gas control valves to regulate flow, and a torch and tip combination to produce the desired flame.

Each component of the oxyfuel gas unit has an essential function in the application and control of the heat necessary for welding. The same basic equipment is used for torch brazing and for many heating operations. By the simple substitution of a cutting attachment and tip combination, the equipment can be converted to manual or mechanized oxygen cutting.

A variety of general purpose and specialized oxyfuel gas welding equipment provides options for selecting the most suitable equipment for the particular application. The operator controlling the use of this equipment must be thoroughly familiar with the capabilities and limitations of the equipment and the rules of safe operation. The manufacturers' operating instructions and the gas suppliers' Material Safety Data Sheets (MSDS) for the equipment and gas should be consulted.

Important standards and recommended practices are published by the Occupational Safety and Health Administration (OSHA),[13] the American National Standards Institute (ANSI),[14] the American Welding Society (AWS),[15,16] and the Compressed Gas Association (CGA),[17] and other organizations to guide users in the correct installation, operation and maintenance of equipment and apparatus used for oxyfuel gas welding, including torches, cylinders, regulators and hoses. These documents and the publishers are listed in Appendix B. Some of the standards applicable to oxyfuel gas welding equipment are listed in Table 11.3.

Table 11.3
Standards and Recommended Practices for Oxyfuel Gas Welding

Designation	Document Title
ANSI Z49.1	*Safety in Welding, Cutting, and Allied Processes*
AWS C4.3	*Operator's Manual for Oxyfuel Gas Heating Torch Operation*
AWS C4.5M	*Uniform Designation System for Oxyfuel Nozzles*
CGA B-57-1	*Compressed Gas Cylinder Valve Outlet and Inlet Connections*
CGA E-5	*Torch Standard for Welding and Cutting*
CGA TB-3	*Hose Line Flashback Arrestors*
CGA E-1	*Standard Connections for Regulator Outlets, Torches, and Fitted Hose for Welding and Cutting Equipment*
CGA E-4	*Standard for Gas Pressure Regulators*
CGA S-1.1	*Pressure-Relief Device Standards-Part I: Cylinders for Compressed Gases*
CGA V-6	*Standard for Cryogenic Liquid Transfer Connections*
OSHA	*Occupational Safety and Health Standards for General Industry, Code of Federal Regulations (CFR), Title 29, CFR1910, Subpart Q*

Designations:
ANSI = American National Standards Institute
AWS = American Welding Society
CGA = Compressed Gas Association
OSHA = Occupational Safety and Health Administration

WELDING TORCHES

A typical welding torch consists of a torch handle, a mixer, and a tip assembly. The torch provides a means for independent control of the flow of each gas, a method of attaching a variety of welding tips or other apparatus, and a convenient handle for controlling the movement and direction of the flame. Figure 11.7 shows the components of a welding torch.

The gases pass through the control valves to separate passages in the torch handle and then to the head (in the upper end of the handle). They are subsequently delivered into the mixer assembly, where the oxygen and fuel gas are mixed and transported out through an orifice at the end of the welding tip. The tip is a simple tube, narrowed at the front end to produce a flame with

13. Occupational Safety and Health Administration (OSHA), 1999, *Occupational Safety and Health Standards for General Industry,* in *Code of Federal Regulations (CFR),* Title 29 CFR 1910, *Subpart Q,* Washington D.C.: Superintendent of Documents, U.S. Government Printing Office.
14. American National Standards Institute (ANSI) Accredited Standards Committee, ANSI/CGA B-57-1, *Compressed Gas Cylinder Valve Outlet and Inlet Connections,* Arlington Virginia: American National Standards Institute.
15. American Welding Society (AWS) C-4 Committee on Oxyfuel Gas Welding and Cutting, *Operator's Manual for Oxyfuel Gas Heating Torch Operation*; AWS C4.3, and *Uniform Designation System for Oxyfuel Nozzles,* C4.5M; and ANSI Z49.1, *Safety in Welding, Cutting and Allied Processes,* Miami: American Welding Society.
16. American Welding Society (AWS) Committee on Oxyfuel Gas Welding and Cutting, 2000, *Uniform Designation System for Oxyfuel Nozzles,* AWS C4.5M: 2000, Miami: American Welding Society.
17. Compressed Gas Association (CGA), *Torch Standard for Welding and Cutting* CGA E-5; *Hose Line Flashback Arrestors* CGA TB-3; *Standard Connections for Regulator Outlets, Torches, and Fitted Hose for Welding and Cutting Equipment,* CGA E-1; *Standard for Gas Pressure Regulators,* CGA E-4; *Pressure-Relief Device Standards,* Part 1: *Cylinders for Compressed Gases,* CGA S-11; and *Standard for Cryogenic Liquid Transfer Connections,* CGA V-6, Compressed Gas Cylinder Valve Outlet and Inlet Connections, CGA B-576-1, Arlington, Virginia: Compressed Gas Association.

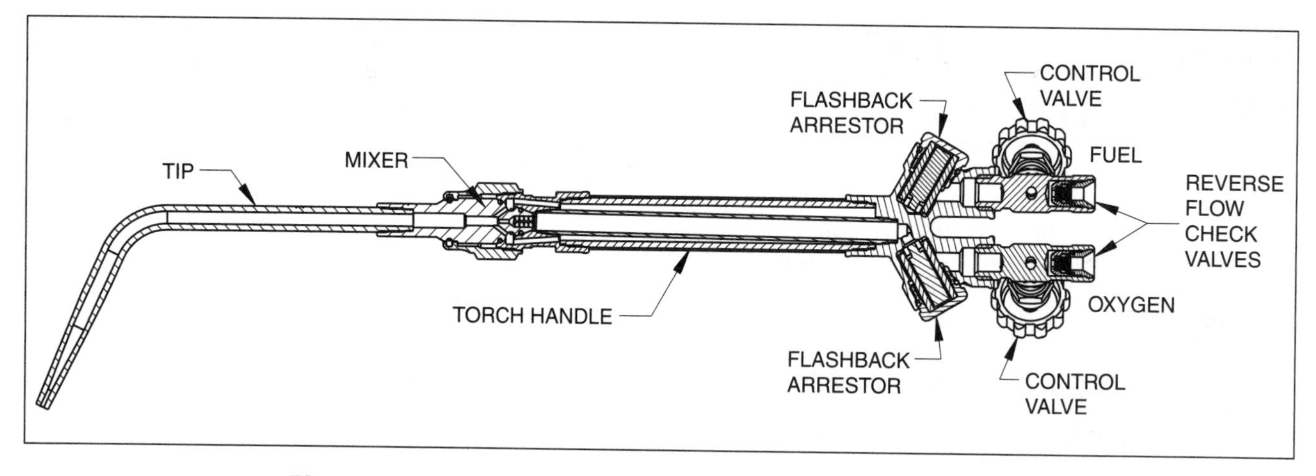

Figure 11.7—Components of an Oxyfuel Gas Welding Torch

a suitable welding cone. Sealing rings or surfaces are provided in the torch head or on the mixer seats to facilitate leak-tight assembly.

Torch Handle

The welding torch handle encases the controls for the mixing and flow of oxygen and fuel gas. Torch handles are manufactured in a variety of styles and sizes range from the small size for extremely light (low-volume gas flow) applications to the extra heavy (high-volume gas flow) types generally used for localized heating operations.

A typical small welding torch used for the oxyacetylene welding of sheet metal delivers acetylene at volumetric rates ranging from about 0.007 cubic meters per hour (m^3/h) to 1.0 m^3/h (0.25 cubic feet per hour [ft^3/h] to 35 ft^3/h). Medium-sized torches are designed to provide acetylene flows from about 0.028 m^3/h to 2.8 m^3/h (1 ft^3/h to 100 ft^3/h). Heavy-duty heating torches may permit acetylene flows as high as 11 m^3/h (400 ft^3/h). Fuel gases other than acetylene may be used with even larger torches at higher pressures. These large torches are designed to deliver fuel gas flow rates as high as 17 m^3/h (600 ft^3/h).

Torch handles can be used with a variety of mixer and welding tip designs as well as with special purpose nozzles, cutting attachments, and heating nozzles. A typical torch handle and accessories are shown in Figure 11.8.

GAS MIXERS

Manufacturers of gas mixers have designed styles and sizes to accommodate a large variety of OFW applications. The chief function of the gas mixer is to combine the fuel gas and oxygen thoroughly to assure complete combustion. Mixers are constructed to serve as a heat sink to help prevent the flame from flashing back into the tip or mixer.

The most commonly used oxygen-fuel gas mixers are the positive-pressure type (also referred to as *equal-pressure* or *medium-pressure*) and the injector, or low-pressure type. The positive-pressure mixer requires the gases to be delivered to the torch at pressures above 14 (kPa) (2 psig). For acetylene, the pressure should be between 14 kPa and 103 kPa (2 psig and 15 psig). Oxygen is generally supplied at approximately the same pressure. However, no restrictive limit on the oxygen pressure exists. Thus, it can range up to 172 kPa (25 psig) with the larger tips.

A typical mixer for a positive-pressure torch is shown in Figure 11.9(A). The oxygen enters through a center duct, and the fuel gas enters through several angled ducts, creating a swirling flow to accomplish the mixing. Mixing turbulence decreases to a laminar flow as the gas passes through the tip.

The purpose of the injector-type mixer is to increase the effective use of fuel gases supplied at pressures of 14 kPa (2 psig) or lower. In this torch, oxygen is supplied at pressures ranging from 70 kPa to 275 kPa (10 psig to 40 psig), the pressure increasing to match the tip size. The relatively high velocity of the oxygen flow is used to aspirate or draw in more fuel gas than would normally flow at the low-supply pressures.

The gas mixers designed for injector-type torches employ the principle of the venturi tube to increase the fuel gas flow. As shown in Figure 11.9(B), high-pressure oxygen passes through the small central duct, creating a high-velocity jet. The oxygen jet crosses the openings of the angled fuel gas ducts at the point where the venturi

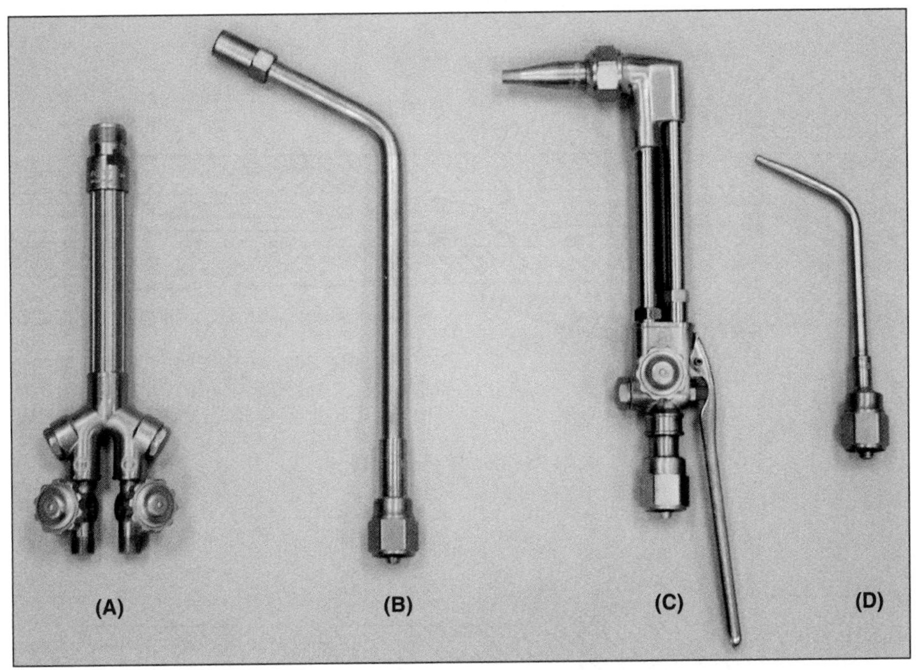

Photograph courtesy of Victor Equipment Company

Figure 11.8—Typical Torch Handle (A) to be Used with Welding Tip (B), Cutting Tip (C), and Heating Tip (D)

tube is restricted. This action produces a pressure drop at the fuel gas openings, causing the low-pressure fuel gas flow to increase as the mixing gases pass into the enlarged portion of the venturi.

WELDING TIPS

The welding tip is the portion of the torch through which the gases pass just prior to ignition and burning. The tip enables the welder to guide the flame and direct it to the work with maximum efficiency.

The welding tips used in oxyfuel gas welding are generally made of a nonferrous metal with high thermal conductivity, such as a copper alloy, to reduce the risk of overheating. Tips are generally manufactured by drilling bar stock to the proper orifice size or by swaging tubing to the proper diameter over a mandrel. The bore in both types must be smooth in order to produce the required flame cone. The front end of the tip should also be shaped to facilitate easy use and assure a clear view of the welding operation.

Welding tips are available in a variety of sizes, shapes, and construction designs. Two types of tip and mixer combinations are employed. Special tips may be used for each size of mixer, or one or more mixers can cover the entire range of tip sizes. In the latter approach, the tip unscrews from the mixer, and each size of mixer has a particular thread size to prevent the improper coupling of a tip and a mixer. A single mixer with a "gooseneck" into which the various sizes of tips may be threaded is used for some types of welding.

During welding, the flame must be carefully adjusted relative to the size of torch, mixer, and tip. Controlling the flow rate to obtain the desired flame characteristics is discussed in the next section, "Volumetric Flow Rate." When a series of welding tips are selected to accommodate a variety of metal thicknesses, the thickness range covered by one tip should slightly overlap that covered by the subsequent tip. As no single standard for tip size designations exists, manufacturers' recommendations should be followed.

Volumetric Flow Rate

The most important factor in determining the usefulness of a welding torch tip is the action of the flame on the metal. If it is too violent, it may blow the metal out of the weld pool. Under such conditions, the volumetric flow rate should be reduced to a velocity at which the metal can be welded. The resulting rate represents the

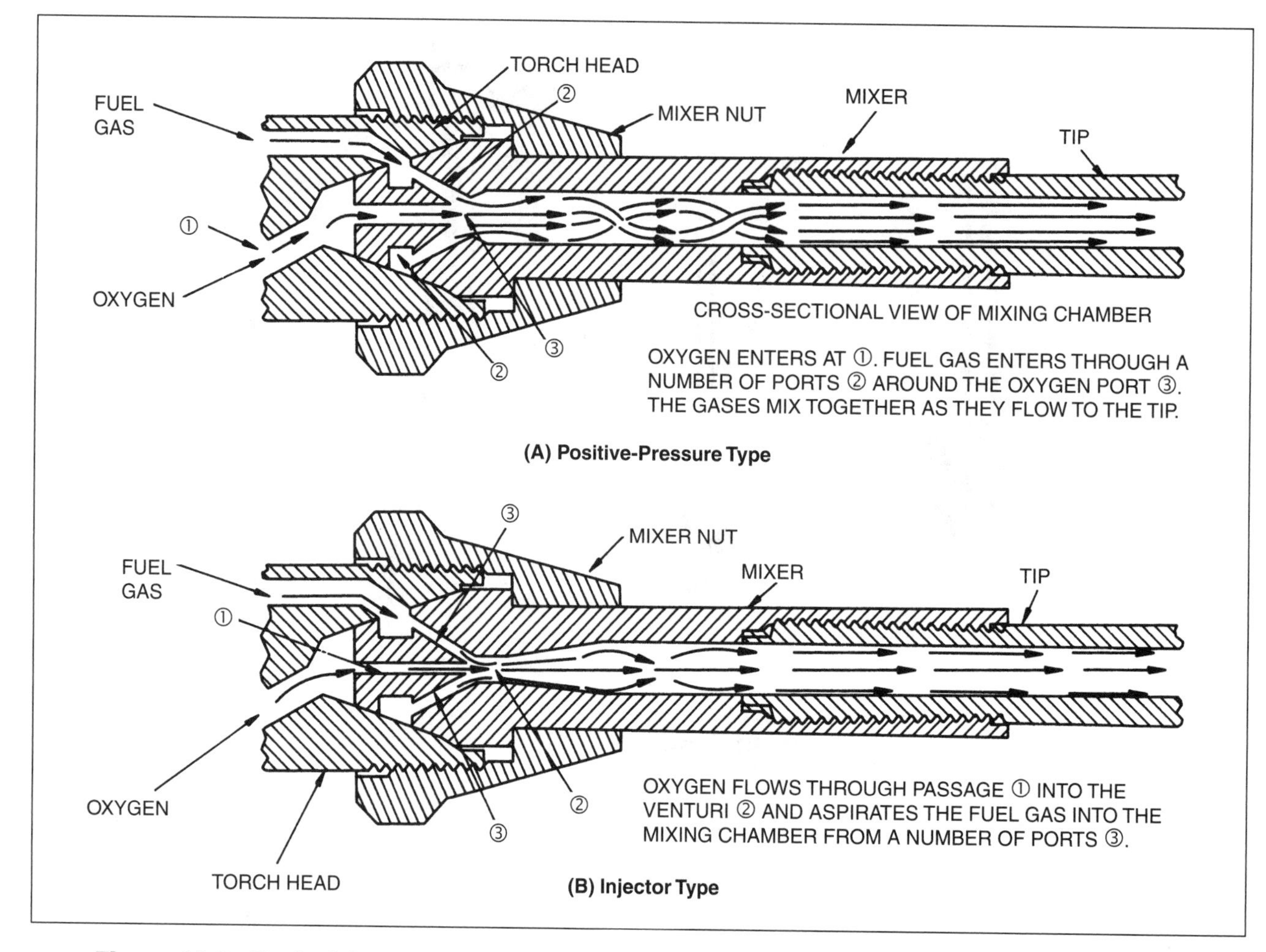

Figure 11.9—Typical Gas Mixer Designs: (A) Positive-Pressure Torch and (B) Injector Torch

maximum volumetric flow that can be handled by a given size of welding tip.

A flame may also be too "soft" for effective welding. When the flame is too soft, the volumetric flow rate must be increased to prevent overheating the tip, which may cause a backfire or flashback.

Backfire and Flashback. A backfire occurs if the flame momentarily recedes into the welding tip, where the flame immediately reappears or is extinguished—usually followed by a loud report. A flashback occurs when the flame recedes into or back of the mixing chamber of the welding torch. The flame can travel through the torch, the hoses and even into the regulator or cylinder, often with serious consequences. Approximately 20% of operator injuries are caused by flashback. Although preventive measures can control most instances of flashback, a flashback arrester can be installed to prevent propagation of the flame beyond the location of the arrester and limit damage.

Welding tips with a hooded or cup-shaped end are available for uses with gases that have a low combustion velocity, such as propane. These tips are usually used for heating, brazing, and soldering.

Flame Cones. The purpose of the welding flame is to raise the temperature of metal to the point of fusion. This can best be accomplished when the welding flame, or cone, permits the heat to be directed easily. Consequently, the cone characteristics become important. Laminar or streamlined gas flow throughout the length of the tip is of paramount importance, especially during passage through the front portion of the tip.

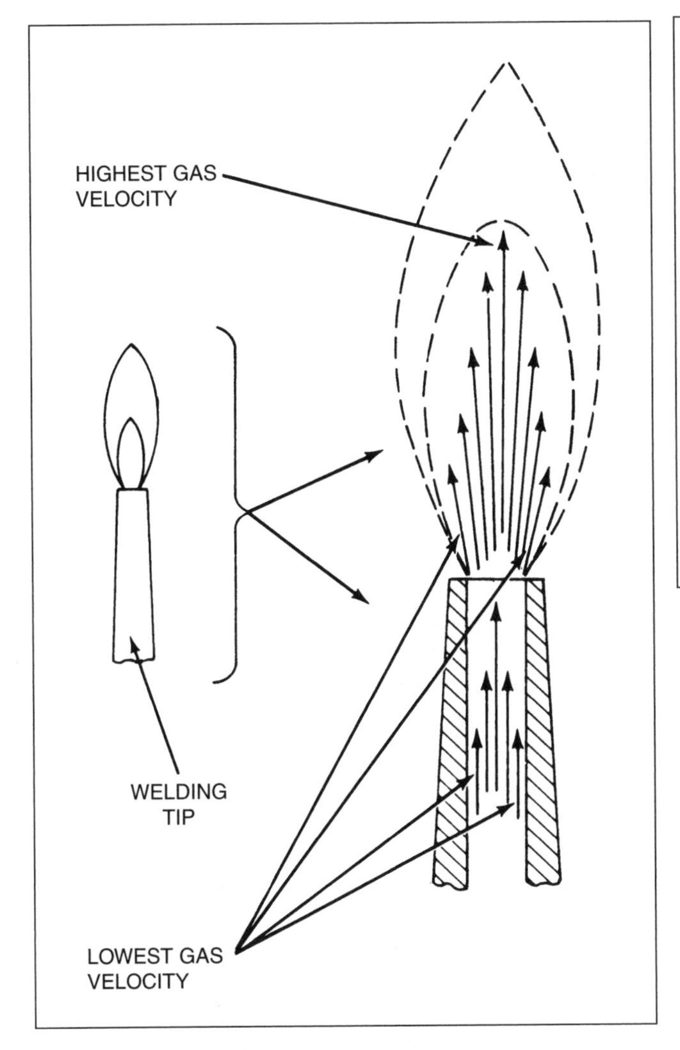

Figure 11.10—Vector Representation of Laminar Flow Velocity in a Welding Tip and in the Formation of a Uniform Flame Cone

The high-velocity flame cone shown in Figure 11.10 is an illustration of the velocity gradient extending across a circular orifice when the exiting flow is laminar. Since the greatest velocity exists at the center of the stream, the flame at the center is the longest. Similarly, since the velocity of the gas stream is lowest at the wall of the tip (bore) where the flowing friction is greatest, the portion of the flame bordering the wall is the shortest.

From the analysis of the principles that underlie the formation of a flame cone, it is possible to understand the flow conditions that exist along the upper portion of the gas passageway in a tip. The shape of the flame cone depends on a number of factors, including the smoothness of the bore, the ratio of the lead-in to the final run diameter, and the cleanliness of the tip orifice.

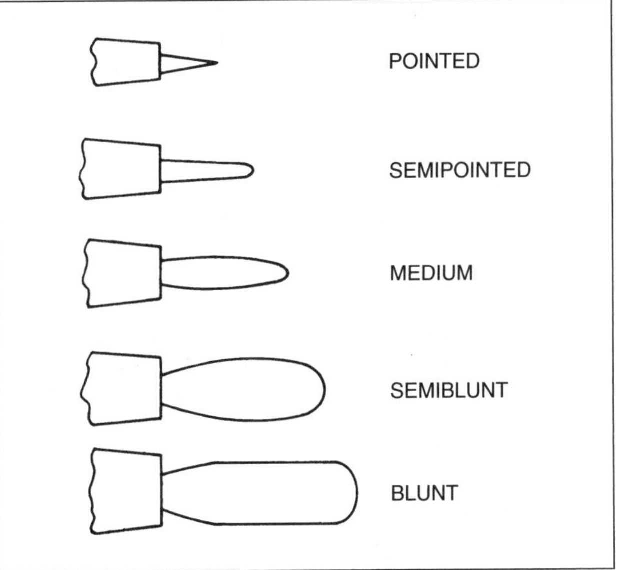

Figure 11.11—Representative Flame Cone Shapes Produced by Welding Tips

A small tip produces a cone that varies from a pointed to a blunt shape. A medium tip produces a cone that varies from a semi-pointed to a medium shape, and cones produced by a large tip vary from a semi-blunt to a blunt shape. Typical flame cone shapes produced by welding tips are shown in Figure 11.11.

Welding tips are generally made of a relatively soft copper alloy and must be handled with care to guard against damage. They should be cleaned using tip cleaners especially designed for this purpose. Tip and mixer threads and all sealing surfaces must be kept clean and in good condition. A poor seal can result in leaks, which may cause a backfire or flashback. Welding tips should never be used to move or hold the workpiece.

HOSES

The hoses used in oxyfuel gas welding and allied operations are manufactured specifically to meet the utility and safety requirements for this service. Hoses must be flexible to assure mobility and easy manipulation in welding and must also withstand high line pressures at moderate temperatures.

Reverse-flow check valves and a flashback arrestor should be located at the torch inlet. An additional valve may be used at the regulator outlet. The purpose of a check valve is to help prevent the reverse flow of gases into the hose, regulator, or cylinder. A reverse flow of gas may cause a flashback, fire, or explosion in any part of the apparatus. A flashback arrestor at the torch inlet offers additional protection to the operator and the hose.

For rapid identification, fuel-gas hoses are colored red in the United States. As a further precaution, the swivel nuts used to make fuel-gas hose connections are identified by a groove cut into the outside of the nut. The nuts also have left-hand threads to match the fuel-gas regulator outlet and the fuel-gas inlet fitting on the torch. In contrast, oxygen hoses are colored green, and the connections each have a plain nut with right-hand threads matching the oxygen regulator outlet and the oxygen inlet fitting on the torch. These methods of identification are standard in the United States, but they may not be used in some other countries; therefore these aspects of identification should be reviewed before using hoses.

The standard means of specifying hose is by inside diameter and working pressure. The nominal inside diameters most commonly used are 3.2 mm, 4.8 mm, 6.4 mm, 7.9 mm, 9.5 mm, and 12.7 mm (1/8 in., 3/16 in., 1/4 in., 5/16 in., 3/8 in., and 1/2 in.), although larger sizes are available. Standard industrial welding hoses and fittings have a maximum working pressure of 1400 kPa (200 psig).

Hoses should be as short as is practical. Lengths of hose over 8 m (25 ft) or small-diameter hose may restrict the flow of gas to the torch. In some cases, this restriction may be overcome by using a higher regulator pressure, but usually a larger-diameter hose is recommended.

Whenever possible, hoses should be supported in an elevated position to avoid damage by falling objects, falling sparks, or hot metal. Damaged hose must be replaced; if repairs are made, replacement fittings intended for this purpose must be used. As noted in Table 11.3, specifications covering the various types of welding hose are published in CGA E-1-1994.[18]

REGULATORS

A regulator is a mechanical device that maintains the delivery of a gas at a substantially constant reduced pressure even though the pressure at the source may vary. The regulators used in oxyfuel gas welding and allied applications are adjustable pressure reducers that are designed to operate automatically after an initial setting.

Except for minor differences, all regulators operate on the same basic principle. They fall into different application categories according to their designed capabilities for handling specific gases, different pressure ranges, and varying volumetric flow rates. Regulators must be used only for their intended purpose. In oxyfuel gas welding, the requirements for cylinder regulators are considerably different from those of station regulators used on pipeline systems. The safe use of regulators is discussed in this chapter in the section, "Safe Practices."

In the commonly used one-torch setup (shown in Figure 11.1), oxygen and acetylene are supplied from single cylinders. Each cylinder is connected to a regulator, which may be either the single-stage or the two-stage type. A regulator is equipped with two pressure gauges, one indicating the inlet or cylinder pressure and the other indicating the outlet (working) pressure to the torch. Regulators and cylinder pressure gauges are built to withstand high pressures with a safe overload margin. Pressure gauges are built to deliver graduated working pressures to accommodate the intended service application.

In shops or fabricating plants where large volumes of oxygen and fuel gases are needed, it is often practical to pipe the gases to various work areas. A station regulator is used to control piped gas pressure.

Pipeline pressures for oxygen seldom exceed 1400 kPa (200 psig). For acetylene, the pressure must not exceed 103 kPa (15 psig). Station regulators are therefore built for low-pressure operation, although they usually have high volumetric flow capacity. Single-stage types equipped only with a working pressure gauge adequately meet station regulator requirements. Due to their pressure limitations, station regulators must never be substituted for cylinder regulators because of the possibility of a serious accident. Cylinder regulators must not be used on station outlets because the low inlet pressures to these regulators may not allow adequate flow capacity.

Regulators are generally classed as single-stage or two-stage, depending on whether the pressure is reduced in one or two steps. The output pressure of the single-stage type exhibits a characteristic known as *rise* or *drift*. This is a slight rise or drop in the delivery pressure that occurs as the volume of gas in the cylinder diminishes and pressure is depleted. This characteristic is usually detrimental only when a large quantity of the gas is withdrawn from a high-pressure cylinder during a single usage. Periodic readjustment of regulator pressure will correct any detrimental effects.

Two-stage regulators essentially consist of two single-stage regulators operating in series within one housing. They provide a constant delivery pressure as cylinder pressure is depleted. The principal components of pressure-reducing regulators are the following:

1. An adjusting screw that controls the thrust of a bonnet spring;
2. The bonnet spring that transmits the thrust to a diaphragm;

18. Compressed Gas Association (CGA), 1994, Standard *Connections for Regulator Outlets, Torches, and Fitted Hose for Welding and Cutting Equipment*, Arlington, Virginia: Compressed Gas Association.

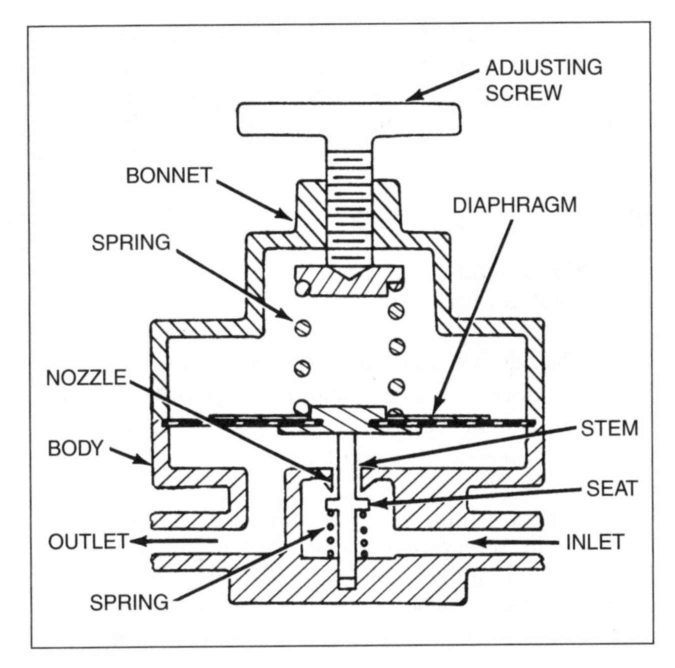

Figure 11.12—Cross Section of the Major Components of a Single-Stage Stem-Type Regulator

3. The diaphragm that contacts a stem on a movable valve seat;
4. A valve consisting of a nozzle, the movable valve seat; and
5. A small spring located under the moveable valve seat.

These components, shown schematically in Figures 11.12, 11.13, and 11.14, permit the gas to be delivered at the specified pressure.

The bonnet spring force tends to hold the seat open while the forces on the underside of the diaphragm tend to cause the seat to close. When gas is withdrawn at the outlet, the pressure under the diaphragm is reduced, thus further opening the seat and admitting more gas until the forces on either side of the diaphragm are equal.

A given set of conditions—such as constant inlet pressure, constant volumetric flow, and constant outlet pressure—will produce a balanced condition that will maintain a fixed relationship between the nozzle and the seat member. As previously noted, the inlet pressure from a cylinder drops as gas is used, causing a gradual

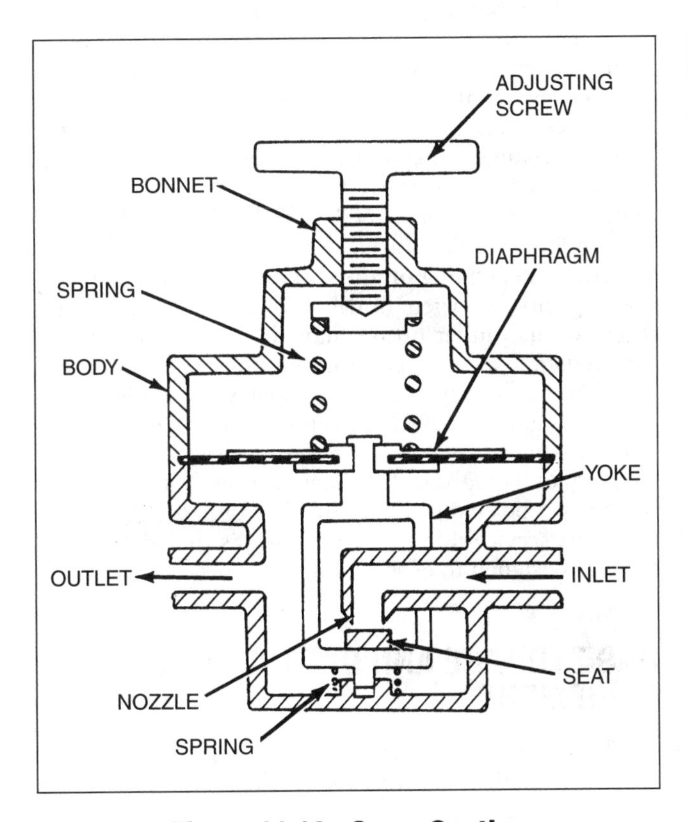

Figure 11.13—Cross Section of the Major Components of a Typical Single-Stage Nozzle-Type Regulator

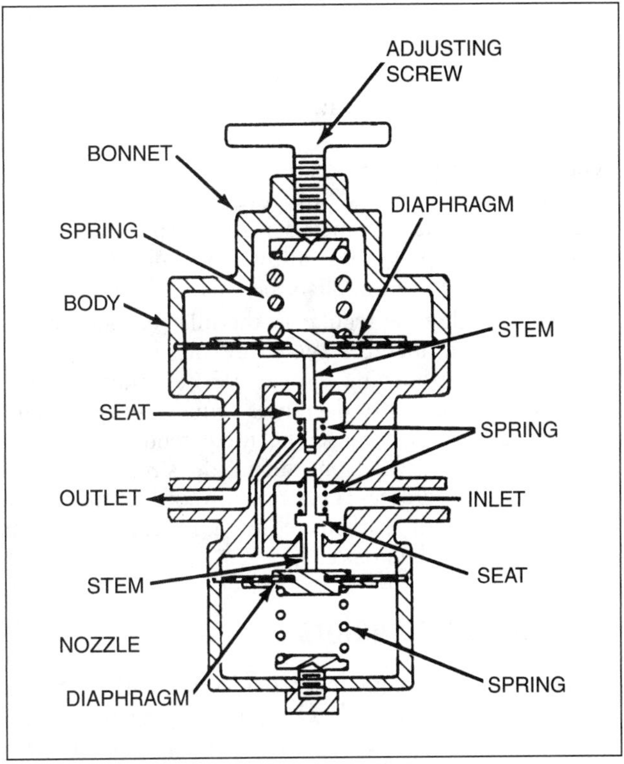

Figure 11.14–Cross Section of the Major Components of a Typical Two-Stage Regulator

change in regulator outlet pressure. The factors affecting the extent of this change depend on the type of single-stage regulator used.

Single-Stage Regulators

The two basic pressure-reducing single-stage regulators are the stem and nozzle types. The stem type (sometimes referred to as the *inverse* or *negative type*), is closed by inlet pressure, as illustrated in Figure 11.12. The nozzle type (sometimes referred to as the *direct-acting* or *positive type*) is opened by inlet pressure, as illustrated in Figure 11.13.

In the stem-type regulator, inlet pressure tends to close the seat member (pressure closing) against the nozzle. The outlet pressure has a tendency to increase somewhat as the inlet pressure decreases. This increase is caused by a reduction in the force produced by the inlet gas pressure against the seating area as the inlet pressure decreases.

The gas outlet pressure for any particular setting of the adjusting screw is regulated by the following:

1. A balance of forces between the bonnet spring thrust and the opposing forces created by the gas pressure against the underside of the diaphragm;
2. The force created by the inlet pressure against the valve seat; and
3. The force exerted by the small spring located under the valve seat.

When the inlet pressure decreases, its force against the seat member decreases, allowing the bonnet spring force to move the seat member away from the nozzle. Thus, more gas pressure is allowed to build up against the diaphragm to re-establish the balanced condition.

In the nozzle-type regulator, the inlet pressure tends to move the seat member away (pressure opening) from the nozzle, thus opening the valve. The outlet pressure of this type of regulator decreases somewhat as the inlet pressure decreases because the force tending to move the seat member away from the nozzle is reduced as the inlet pressure decreases. Less pressure on the underside of the diaphragm is then required to close the seat member against the nozzle.

Two-Stage Regulators

The two-stage regulator provides regulation that is more precise over a wide range of varying inlet pressures. Two-stage regulators are suggested for precise work, such as precision welding or continuous mechanized cutting, in order to maintain a constant working pressure and therefore a controlled volumetric flow at the welding or cutting torch.

A two-stage regulator, as illustrated in Figure 11.14, incorporates two single-stage regulators in series as one unit. The outlet pressure from the first stage is usually preset to deliver a specified inlet pressure to the second stage. In this way, a nearly constant delivery pressure can be obtained from the outlet of the second stage of the regulator as the supply pressure to the first stage decreases.

The combinations that can be incorporated in a two-stage regulator are as follows:

1. Nozzle-type first stage and stem-type second stage;
2. Stem-type first stage and nozzle-type second stage;
3. Two stem-type regulators, (Figure 11.14); and
4. Two nozzle-type regulators.

Regardless of the configuration used, the increase or decrease in outlet pressure is usually so slight (and apparent only at very low inlet supply pressures) that for all practical purposes the variation in delivery pressure is disregarded in welding and cutting operations.

Regulator Inlet and Outlet Connections

By design, cylinder outlet connections are of different sizes and shapes to preclude the possibility of connecting a regulator to the wrong cylinder. Therefore, regulators must be manufactured with different inlet connections to fit the various gas cylinders. The Compressed Gas Association (CGA) establishes standardized non-interchangeable cylinder valve outlet connections in the standard *Compressed Gas Cylinder Valve Outlet and Inlet Connections*, ANSI/CGA B-57-1.[19]

Regulator outlet fittings also differ in size and threading, depending on the specific gas and the regulator capacity. Oxygen outlet fittings have right-hand threads. Fuel outlet fittings have left-hand threads and grooved nuts. Inlet connections for regulators are designed to fit only specific cylinder valve outlets. Different designs are coordinated for almost all gases. Pipeline station valves for the various gases are different from cylinder valves. Inlet connections on regulators must never be changed to allow the regulator to be used for a gas service different from the gas that the regulator was designed to accommodate.

GAS STORAGE AND DISTRIBUTION EQUIPMENT

The selection of a method of distribution of gases to the work facility depends on the location, size of the

19. Compressed Gas Association (CGA), *Compressed Gas Cylinder Valve Outlet and Inlet Connections*, ANSI/CGA B-5-1, Arlington, Virginia: Compressed Gas Association.

project, gas consumption requirements, and specific oxyfuel gas process to be used. Gases can be delivered in single cylinders, portable or stationary cylinder manifolds, bulk supply systems, and pipelines.

Individual cylinders of gaseous oxygen and fuel gas provide an adequate supply of gas for welding and cutting operations that consume limited quantities of gas. A hand truck or cylinder cart provides a convenient, safe, and portable support for a cylinder of oxygen and a cylinder of fuel gas.

Unlike acetylene cylinders, which contain a porous filler material, the cylinders used for the storage of liquefied fuel gas do not contain a filler material. These welded steel or aluminum cylinders hold the liquefied fuel gas under pressure. The pressure in the cylinder is a function of the temperature. Cylinders containing liquefied fuel gas have relief valves that are set at predetermined pressures to prevent over-pressurization in the event that the cylinder is exposed to temperatures approaching 93°C (200°F).

If these high temperatures are reached, the rapid discharge of fuel gas through the relief valve causes the cylinder to cool down, the cylinder pressure then drops and the relief valve closes. In a fire, the cylinder relief valve opens and shuts intermittently until all the fuel in the cylinder has been discharged or the source of the extreme heat has been removed.

The withdrawal rate of liquefied fuel gases from cylinders is a function of the temperature, the amount of fuel in the cylinder, and the specified operating pressure. Information pertaining to the withdrawal rate should be obtained from the gas supplier.

Oxygen can be delivered to the user in individual cylinders as a compressed gas or as a liquid. Cylinder manifolds and bulk distribution systems, described in the following sections, are also available. Gaseous oxygen in cylinders is usually under a pressure of approximately 15 170 kPa (2200 psig). Cylinders of various capacities are used, holding approximately 2 cubic meters (m^3), 2.3 m^3, 3.5 m^3, 6.9 m^3, and 9.4 m^3 (70 cubic feet [ft^3], 80 ft^3, 122 ft^3, 244 ft^3, and 330 ft^3) of oxygen. Cylinders containing liquid oxygen yield the equivalent of over 85 m^3 (3000 ft^3) of gaseous oxygen. These cylinders are used for applications that do not warrant a bulk oxygen supply system but are too large to be supplied conveniently by gaseous oxygen in cylinders. The cylinders containing liquid oxygen are equipped with liquid-to-gas converters or may use external gas vaporizers.

The withdrawal rate for cylinders containing acetylene must be carefully monitored so that the specified withdrawal rate from a given size cylinder is not exceeded. If the volumetric demand is too high, acetone may be drawn from the cylinder along with the acetylene. Therefore, it has become standard practice to limit the withdrawal of acetylene from a single cylinder to an hourly rate not exceeding one seventh of the cylinder's volumetric contents. Two or more acetylene cylinders should be used in manifold to provide high flow rates.

Cylinder Manifolds

Cylinder manifolds can be implemented when individual cylinders lack the capacity to supply high rates of gas flow, particularly when needed for continuous operation over long periods of time. A reasonably large volume of gas can be provided by this means, and it can be discharged at a moderately rapid rate. Manifolds must be obtained from reliable manufacturers and must be installed by personnel familiar with the proper construction and installation of oxygen and fuel manifolds and pipelines.

Manifolds can be either portable or stationary. Portable manifolds like that shown in Figure 11.15 can be installed with a minimum of effort. They are useful when moderate volumes of gas are required for jobs of a nonrepetitive nature, either in the shop or the field.

Stationary manifolds such as the arrangement shown in Figure 11.16 are installed in shops where large volumes of gas are required. These manifolds feed a pipeline system that distributes the gas to various stations throughout the plant. This arrangement enables a number of operators to work from a common pipeline system without interruption, or the pipeline may supply large automated torch-brazing or oxygen cutting operations.

An important protective device for the stationary fuel system is a hydraulic seal or hydraulic flashback arrestor. This device prevents flashbacks originating at a torch station from passing back into the system. It consists of a small pressure vessel partially filled with water through which the fuel gas flows. The gas continues through the piping system to the station regulator. A flashback or high-pressure backup causes a relief valve in the vessel head to vent the excess pressure to the atmosphere outside. A check valve prevents the water from backing up into the supply line.

Bulk Systems

To satisfy the large consumption required by many industries, gaseous oxygen can be transported from the producing plant to the user in portable multiple-cylinder banks or in long, high-pressure tubes mounted on truck trailers. The trailers hold from 850 m^3 to 1420 m^3 (30,000 ft^3 to 50,000 ft^3) of gaseous oxygen in the larger units and typically 10,000 ft^3 (285 m^3) in the smaller units.

Bulk oxygen can also be delivered as a liquid in large insulated containers mounted on trucks, trailers, or rail cars. The liquid oxygen is transferred to an insulated storage tank on the consumer's property. The oxygen is

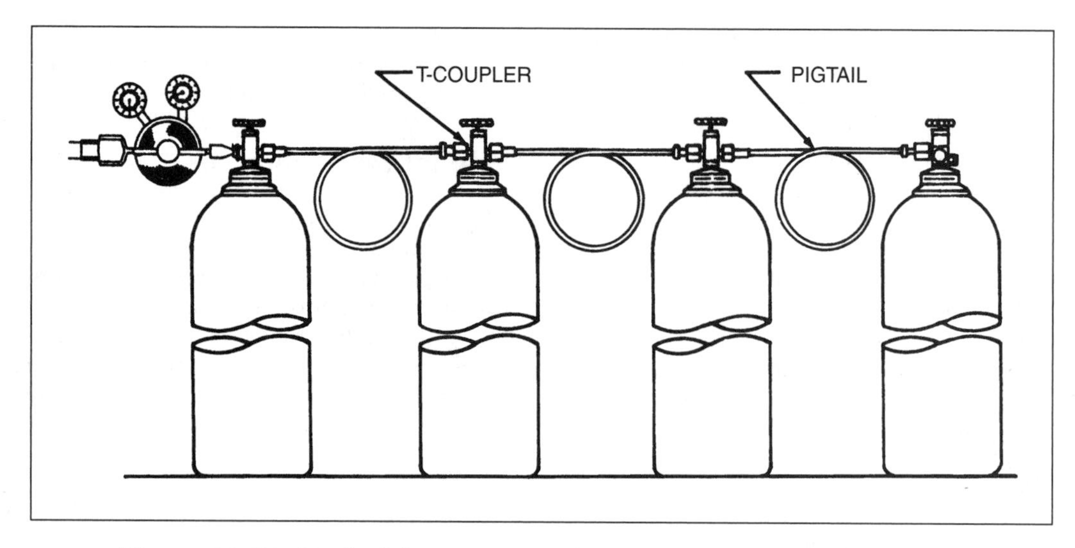

Figure 11.15—Typical Arrangement for a Portable Oxygen Manifold

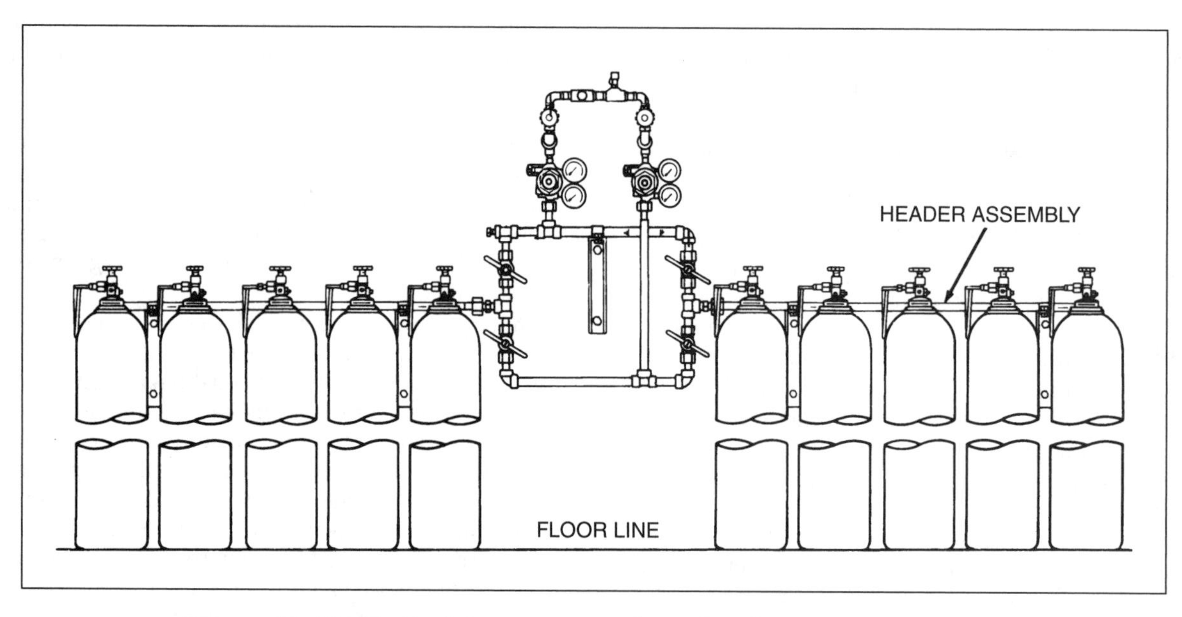

Figure 11.16—Typical Arrangement for a Stationary Gas Manifold

withdrawn from the storage tank, converted to gas, and transported into distribution pipelines, as needed, by means of automatic regulating equipment. The standard *Bulk Oxygen Systems at Consumer Sites,* NFPA 50-96, establishes the rules and regulations for bulk oxygen systems located on the consumer's premises.[20]

20. National Fire Protection Association (NFPA), *Bulk Oxygen Systems at Consumer Sites,* NFPA 50-96, Quincy, Massachusetts: National Fire Protection Association.

Liquefied fuel gases can be distributed from on-site bulk tanks with a capacities of 2000 liters (L) to 45 000 L (500 gallons [gal] to 12,000 gal). The tanks are filled periodically by truck delivery.

Other Supply Systems

When stationary manifolds, trailer trucks, or bulk supply systems are used as sources of oxygen or fuel gases or when the gases are distributed by pipelines to the points of use, the pipelines must be designed to handle and distribute the gases in sufficient volume without an undue pressure drop. They must also incorporate all necessary safety devices. It is important to note that while copper tubing is satisfactory for use as oxygen pipelines, it must not be used for acetylene pipelines. When acetylene is in proximity to copper, copper acetylide may form and explode spontaneously.

The rules and regulations set forth in *Design and Installation of Oxygen-Fuel Gas Systems for Welding, Cutting, and Allied Processes,* NFPA 51, govern the installation of oxygen and fuel gas manifolds and pipelines.[21] Compliance with local regulations and ordinances must also be ensured.

PROCESS VARIABLES AND OPERATING PROCEDURES

Variable conditions that should be considered when using the oxyfuel gas welding process include joint design and preparation, base metal selection and preparation, metallurgical considerations, flame adjustment and welding techniques. Planning the appropriate joint design, correctly preparing the edges to be welded, meticulously cleaning the base metal, and using the correct welding technique are all important to producing a good weld.

JOINT DESIGN AND BASE METAL PREPARATION

Cleanliness along the joint and on the sides of the base metal is of the utmost importance. Dirt, oil, and oxides may cause incomplete fusion, inclusions, and porosity in the weld.

The spacing between the workpieces to be joined (root opening) should be considered carefully. The root opening for a given thickness of metal should be narrow enough to permit the opening to be bridged without difficulty, yet it should be large enough to permit full penetration. The specifications for root openings relative to base metal and thickness variations should be closely followed.

The thickness of the base metal at the joint determines the type of edge preparation for welding. Thin sheet metal can easily be melted completely by the flame. Thus, edges with a square face can be butted together and welded. This type of joint is limited to material under 3.2 mm (1/8 in.) in thickness. For thicknesses of 3.2 mm to 4.8 mm (1/8 in. to 3/16 in.), a slight root opening or groove is necessary for complete penetration, but filler metal must be added to compensate for the opening.

Joint edges 6.4 mm (1/4 in.) and greater in thickness must be beveled. Beveled edges at the joint provide a groove for better penetration and fusion at the sides. The angle of bevel for oxyacetylene welding varies from 35° to 45°, which is equivalent to a variation of the included angle of the joint from 70° to 90° degrees, depending on the application. A root face 1.6 mm (1/16 in.) wide is normal, but feathered edges are sometimes used. Plate thicknesses 12.8 mm (1/2 in.) and above are double-beveled when welding can be done from both sides. The root face can vary from 0 mm to 3.2 mm (0 in. to 1/8 in.). Beveling both sides reduces the amount of filler metal required by approximately one half. The gas consumption per unit length of weld is also reduced. Thicknesses of 12.8 mm (1/2 in.) and above are usually not welded by the oxyfuel gas process.

A square edge is the easiest to prepare. This edge can be sheared, machined, chipped, ground, or cut with an oxygen cutting process. The thin oxide coating on an oxygen-cut steel surface need not be removed because it is not detrimental to the welding operation or to the quality of the joint. A bevel angle can be cut with an oxygen cutting process.

TORCH OPERATION

The oxyfuel gas welding torch is designed to mix the combustible and combustion-supporting gases and to provide the means for applying the flame at the desired location. Welding tips in a range of sizes are available to produce the required volume or size of welding flame. Tip flames vary from short, small-diameter needle flames to flames of 4.8 mm (3/16 in.) or greater in diameter and 50 mm (2 in.) or greater in length.

The inner cone, or the vivid blue flame of the burning gas mixture issuing from the tip is termed the *working flame.* The closer the end of the inner cone is brought to the surface of the metal being welded or heated, the more effective the heat transfer is from flame to metal. The flame can be made soft or harsh by varying the gas flow. A gas flow that is too low for a given tip size results in a soft, ineffective flame that is sensitive to backfiring. A gas flow that is too high may result in a harsh, high-velocity flame that is hard to handle and blows the molten metal out of the weld pool.

The chemical action of the flame on the weld pool can be altered by changing the ratio of the volume of oxygen to the volume of fuel gas issuing from the tip.

21. National Fire Protection Association (NFPA), *Standard for the Installation and Operation of Oxygen-Fuel Gas Systems for Welding, Cutting, and Allied Processes,* NFPA 51, Quincy, Massachusetts: National Fire Protection Association.

Most oxyacetylene welding is done with a neutral flame with approximately a 1-to-1 ratio of fuel gas to oxygen. An oxidizing action can be obtained by increasing the oxygen flow, and a reducing action can be obtained by increasing the acetylene flow. Both adjustments are valuable aids in welding.

Flame Adjustment

Welding torches should be lighted with a friction lighter or a pilot flame. The instructions of the equipment manufacturer must be followed when adjusting operating pressures at the gas regulators and torch valves before the gases issuing from the tip are ignited. A pure acetylene flame is shown in Figure 11.17(A), with three oxyacetylene flame adjustments also shown: (B) the carburizing flame, (C) the neutral flame, and (D) the oxidizing flame.

The neutral flame (C) is obtained most easily by an adjustment from an excess acetylene flame (A), recognized by the feather extension of the inner cone. The feather diminishes as the acetylene flow is decreased and the flow of oxygen is introduced or increased. The flame is neutral just at the point of disappearance of the feather extension of the inner cone. This neutral flame, while neither carburizing nor oxidizing within itself, may have a slight reducing effect on the metal being welded.

To determine the amount of excess acetylene in a reducing flame, the length of the feather can be compared with the length of the inner cone, measuring both from the torch tip. An excess acetylene flame with twice the acetylene required for a neutral flame has an acetylene feather that is twice the length of the inner cone. Starting with a neutral flame adjustment, the welder can produce the desired acetylene feather by increasing the acetylene flow.

The oxidizing flame adjustment (D) is sometimes specified as the amount by which the length of a neutral inner cone should be reduced—for example, by one tenth. Starting with the neutral flame, the welder can increase the oxygen until the length of the inner cone is decreased to the desired amount.

Oxidation and Reduction

Certain metals have such a high affinity for oxygen that oxides form on the surface almost as rapidly as they can be removed. This affinity for oxygen can be a useful characteristic in certain welding operations. For example, the manganese and silicon contained in plain carbon steel are important in oxyfuel gas welding because they deoxidize the weld pool. Therefore, the selection of the correct manganese and silicon content of steel welding rods is important when welding plain carbon steels.

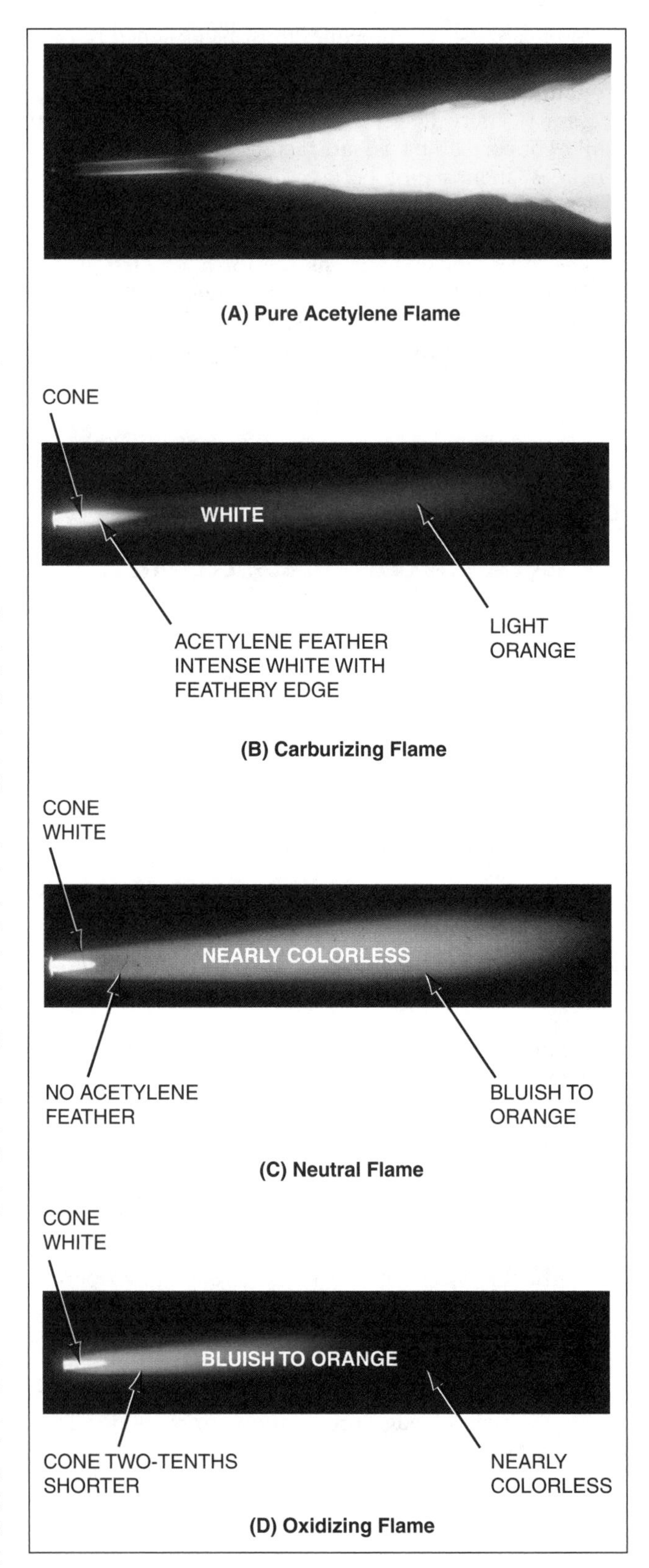

Figure 11.17—Oxyacetylene Flame Adjustments

The type of flame used in welding various metals contributes to securing the most desirable weld metal deposit. The proper type of flame in combination with the appropriate welding technique can be used to achieve shielding, which reduces the effect of oxygen and nitrogen (found in the atmosphere) on the molten metal. Such a flame also has the effect of stabilizing the molten weld metal and preventing the loss of carbon, manganese, and other alloying elements. In oxyacetylene operations, oxides are usually removed by applying a flux.

FOREHAND AND BACKHAND WELDING

Oxyfuel gas welding that is performed with the torch tip pointed forward in the direction of the weld progression is called *forehand welding*. Welding that progresses in the opposite direction, with the torch pointing toward the completed weld, is known as *backhand welding*. Each method has its advantages, depending on the application, and each method imposes some variations in deposition technique.

The forehand method is recommended for welding materials up to 4.8 mm (3/16 in.) thick because it provides good control of the weld pool, resulting in a smoother weld at both top and bottom. The pool of molten metal is small and easily controlled. A great deal of pipe welding is performed using the forehand technique, even in 9.5 mm (3/8 in.) wall thicknesses.

However, increased speeds and better control of the weld pool may be possible with the backhand technique when metal 4.8 mm (3/16 in.) or thicker is welded. This recommendation is based on careful study of the speeds normally achieved with the backhand technique and on the greater ease with which fusion at the root of the weld can be obtained.

Backhand welding may be used with a slightly carburizing flame (slight acetylene feather) in steel welding, for example, when it is desirable to melt a minimum amount of steel in making the joint. The increased carbon content obtained from this flame lowers the melting point of a thin layer of steel and increases welding speed. The backhand technique increases the speed of making pipe joints in which the wall thickness is 6.4 mm to 7.9 mm (1/4 in. to 5/16 in.) and the groove angle is less than normal. Backhand welding is sometimes used in surfacing operations.

MULTIPLE-LAYER WELDING

Multiple-layer welding is used to achieve the maximum ductility of a steel weld in the as-welded or stress-relieved condition or when several layers are required to weld thick metal. It is accomplished by depositing filler metal in successive passes along the joint until it is filled. Since the area covered with each pass is small, the weld pool is reduced in size. This procedure enables the welder to obtain complete joint penetration without overheating while the first few passes are being deposited. The smaller weld pool is more easily controlled, and the welder can thus avoid oxides, slag inclusions, and incomplete fusion.An increase in ductility in the deposited steel results from grain refinement in the underlying passes when they are reheated. The final layer does not possess this refinement unless an extra pass is added and removed or unless the torch is passed over the joint to bring the last deposit up to normalizing temperature.

APPLICATIONS

Oxyfuel gas welding can be used on a wide range of commercial ferrous and nonferrous metals and alloys. As in any welding process, however, physical dimensions and chemical composition may limit the weldability of certain materials and workpieces. During welding, the metal is taken through a temperature range that almost replicates the original manufacturing or casting procedure. The base metal in the weld area loses the properties that resulted from prior heat treatment or cold working.

METALS WELDED

Carbon steels, low-alloy steels, and most nonferrous metals can be welded by the oxyfuel gas welding process. The welding procedure for plain carbon steels is usually straightforward and poses little difficulty to the welder. Welding such materials as high-carbon and high-alloy steels is limited by the equipment available for heat-treating after welding. These metals are successfully welded when the size or nature of the workpiece permits postheat treating operations. Sound welds are produced in other materials by variations in preheating, welding technique, heat-treating, and fluxing.

The oxyfuel gas welding process can be used for repairing the usual assemblies encountered in maintenance work, and can be used to repair metals of considerable thickness. Very thick cast iron machinery frames are often repaired by welding with a cast iron filler rod or by braze welding. The oxyfuel gas welding process is not recommended for refractory or reactive metals.

Steels and Cast Irons

Low-carbon, low-alloy steels, cast steels and cast irons are the metals most easily welded by the oxyacetylene welding process. Fluxes are usually not required when welding these materials. In oxyfuel gas welding, steels having more than 0.35% carbon are considered high-carbon steels. They require special care to maintain their particular properties. Alloy steels of the air-hardening type also require extra precautions to maintain their properties, even though the carbon content may be 0.35% or less. The joint area is usually preheated to retard the cooling of the weld by means of conducting heat into the surrounding base metal. This slows cooling and prevents the hardness and brittleness associated with rapid cooling. A full-furnace anneal or heat treatment may be required immediately after welding air-hardening steels.

Procedures. The welder should use a neutral or slightly carburizing flame and take care not to overheat and decarburize the base metal. The preheat temperature required depends on the composition of the material to be welded. Temperatures ranging from 150°C to 540°C (300°F to 1000°F) are often used.Once the correct preheat temperature has been determined, it is important that this temperature be maintained uniformly during welding. A uniform temperature can be maintained by protecting the workpiece with a heat-retaining covering. Other means of shielding can also be used to retain the temperature in the workpiece. The interpass temperature should normally be maintained within 65°C (150°F) of the preheat temperature. Lower interpass temperatures cause excessive shrinking forces that can result in either distortion or cracking at the weld or other sections. This type of cracking frequently occurs in the welding of circular structures made of brittle metals such as cast iron.

Modifications in procedures are required for stainless and similar steels. Because of their high chromium-nickel content, these steels have relatively low thermal conductivities. A flame smaller than that used for equal thicknesses of plain carbon steel is recommended. Because chromium oxidizes easily, a neutral flame is employed to minimize oxidation. A flux is typically used to dissolve the oxides and protect the weld metal. Filler metal with a high percentage of chromium or nickel-chromium steel is used. Even with these precautions, when a weld with a critical level of quality is required, a process other than oxyfuel gas welding is generally recommended. Table 11.4 summarizes the general welding conditions for welding ferrous metals with the oxyacetylene process.

Cast iron, malleable iron, and galvanized iron all present particular problems in welding by any method; however, the microstructure of gray cast iron can be maintained throughout the weld area by the use of preheat, a flux, and an appropriate cast-iron welding rod.

Table 11.4
General Conditions for the Oxyacetylene Welding of Various Ferrous Metals

Metal	Flame Adjustment	Flux	Welding Rod
Cast steel	Neutral	No	Steel
Steel pipe	Neutral	No	Steel
Steel plate	Neutral	No	Steel
Steel sheet	Neutral, slightly oxidizing	No Yes	Steel Bronze
High-carbon steel	Slightly carburizing	No	Steel
Wrought iron	Neutral	No	Steel
Galvanized iron	Neutral, slightly oxidizing	No Yes	Steel Bronze
Cast iron, gray	Neutral, slightly oxidizing	Yes Yes	Cast iron Bronze
Cast iron, malleable	Slightly oxidizing	Yes	Bronze
Cast-iron pipe, gray	Neutral, slightly oxidizing	Yes Yes	Cast iron Bronze
Cast-iron pipe	Neutral	Yes	Cast iron or base metal composition
Chromium-nickel steel castings	Neutral	Yes	Base metal composition or 25-12 chromium-nickel steel
Chromium-nickel steel (18-8 and 25-12)	Neutral	Yes	Niobium stainless steel or base metal composition
Chromium steel	Neutral	Yes	Niobium stainless steel or base metal composition
Chromium iron	Neutral	Yes	Niobium stainless steel or base metal composition

Figure 11.18—The Oxyacetylene Process Used for the Heat-Treating of Castings

Nodular iron requires materials in the welding filler metal that will assist in promoting agglomeration of the free graphite in order to maintain ductility and shock resistance in the heat-affected zone. The filler metal manufacturer should be consulted for information on preheat and interpass temperature control for the filler metal under consideration.

A heat-treating application is shown in Figure 11.18, in which the heating of castings was accomplished with acetylene as the fuel gas.

Of course, instances occur in which cast irons are welded without preheat, particularly in salvage work. In most applications, however, preheat from 200°C to 320°C (400°F to 600°F), with control of interpass temperature and provision for slow cooling, assures more consistent results. Protection such as a heat-resistant covering can be used to ensure slow and uniform cooling. Care should be taken to prevent localized cooling. It should also be stressed that in the salvage of cast iron, removal of all foundry sand and slag is necessary for consistently good repair results.

Nonferrous Metals

The particular properties of each nonferrous alloy should be considered when selecting the most suitable oxyfuel gas welding process to use. When the necessary precautions are taken, acceptable results can be obtained.

For example, aluminum, unlike some other metals, gives no warning by changing color prior to melting and will collapse suddenly at the melting point. Consequently, welders require practice to learn to control the rate of heat input. Aluminum and its alloys are prone to hot shortness (loss of strength at high temperatures), and welds should be supported adequately in all areas during welding. Any exposed aluminum surface is always covered with a layer of oxide that, when combined with the flux, forms a fusible slag which floats on top of the weld pool.

When copper is welded, allowances are necessary for the chilling of the welds because of the very high thermal conductivity of the metal. Preheating is often required. Considerable distortion can be expected in copper because the thermal expansion is higher than in other commercial metals. These characteristics pose obvious difficulties that must be overcome for satisfactory welding.

Procedures. The workpieces should be fixtured or tack welded securely in place. The section least subject to distortion should be welded first so that it forms a rigid structure for the balance of the welding. When the design of the structure permits welding from both sides, welding alternately on each side of the joint can minimize distortion. Strongbacks or braces can be applied to sections most likely to distort. The welds may be peened to reduce distortion. If properly performed, peening can overcome severe warping.

A backstep sequence of welding may be used to help control distortion. This method consists primarily of making short weld increments in the direction opposite to the progress of welding. The weldment should be designed so that distortion during welding is minimized.

WELD QUALITY

Although the appearance of a weld does not necessarily indicate its quality, visual examination of an oxyfuel gas weld is a primary method of inspection. Discontinuities can be grouped into two broad classifications—those that are revealed by visual inspection and those that are not.

Visual examination of the underside of a weld reveals whether penetration is complete and whether excessive globules of metal are present. Incomplete joint penetration may be due to insufficiently beveled edges, thick root faces, high welding speeds, or poor torch and welding rod manipulation.

Oversized and undersized welds can be readily observed. Weld gauges can be used to determine whether a weld has excessive or insufficient reinforcement. Undercut or overlap at the sides of the welds can usually be detected by visual examination.

Although other discontinuities such as incomplete fusion, porosity, and cracking may or may not be apparent externally, excessive grain growth and the presence of hard spots cannot be determined visually. Incomplete fusion may be caused by insufficient heating of the base metal, rapid travel, or gas or dirt inclusions. Porosity is a result of entrapped gases, usually carbon monoxide, which can be avoided by more careful flame manipulation and adequate fluxing where needed. Hard spots and cracking are a result of metallurgical characteristics of the weldment. Further details on this topic are presented in Chapter 4, "Welding Metallurgy," in *Welding Science and Technology*, Volume 1 of the *Welding Handbook*, 9th edition.[22]

22. American Welding Society (AWS) Welding Handbook Committee, 2001, C.L. Jenney and A. O'Brien, eds., 2001, *Welding Science and Technology*, Vol. 1 of *Welding Handbook*, 9th ed., Miami: American Welding Society, Chapter 4.

METALLURGICAL EFFECTS

During oxyfuel gas welding, the temperature of the base metal varies from that of the elevated temperatures of the weld pool to room temperature in areas of the workpieces most remote from the weld. When steels are involved, the weld and the adjacent heat-affected zones are heated considerably above the transformation temperature of the steel. This results in a coarse grain structure in the weld and adjacent base metal. The coarse grain structure can be refined by a normalizing heat treatment, such as heating to the austenitizing temperature range (approximately 900°C [1650°F]) and cooling in air after welding.

When the steel is heated to above the transformation temperature, hardening of the heat-affected zone of the base metal can occur if the steel contains sufficient carbon and the cooling rate is high enough. Hardening can be avoided in most hardenable steels by using the torch to keep heat on the weld for a short time after the weld has been completed. If air-hardening steels are welded, the best heat treatment is a full-furnace anneal of the weldment.

The oxyfuel flame allows a degree of control to be maintained over the carbon content of the deposited metal and over the portion of the base metal that is heated to its melting temperature. When an oxidizing flame is used, a rapid reaction results between the oxygen and the carbon in the metal. Some of the carbon is lost in the form of carbon monoxide, and the steel and the other constituents are oxidized. When the torch is used with an excess acetylene flame, carbon is introduced into the molten weld metal.

When heated to a temperature range between 430°C and 870°C (800°F and 1600°F), carbide precipitation occurs in unstabilized austenitic stainless steel. Chromium carbides gather at the grain boundaries and lower the corrosion resistance of the heat-affected zone. If this occurs, a heat treatment after welding is required, unless the steel is an alloy stabilized by the addition of niobium or titanium and welded with the aid of a niobium-bearing stainless steel welding rod. The niobium combines with the carbon, thus minimizing the formation of chromium carbide. The chromium that remains is dissolved in the austenitic matrix, the form in which it can best resist corrosion.

Another factor to be considered in welding is the possible tendency toward hot shortness of the metal, which is a marked loss in metal strength at high temperatures. Some copper-based alloys are highly prone to this tendency. If the base metal has this tendency, it should be welded with care to prevent hot cracking in the weld zone. Allowances should be made in the welding technique used with these metals, and fixturing or clamping should be done with caution. The proper welding sequence and multiple-layer welding with narrow stringer beads help to reduce hot cracking.

WELDING WITH OTHER FUEL GASES

Propane, natural gas, butane, methylacetylene-propylene (stabilized) gas, propylene, and other similar gases are not suitable as fuel gases for welding ferrous materials due to their oxidizing characteristics. However, many nonferrous and some ferrous metals can be braze welded with these gases, depending on the careful adjustment of the flame and the use of flux. These gases are used extensively for both manual and mechanized brazing and soldering operations.

These fuel gases have relatively lower flame propagation rates than acetylene, therefore it is important to use welding tips designed for the fuel gas being used. When standard oxyacetylene welding tips are used, the maximum flame velocity may interfere with heat transfer from the flame to the workpiece. The highest flame temperatures of these gases are obtained at high ratios of oxygen to fuel gas. These ratios produce highly oxidizing flames, which prevent the satisfactory welding of most metals. Tips with flame-holding devices, such as skirts, counter bores, and holder flames, should be used to permit higher gas velocities before the gas leaves the tip. This excellent heat transfer efficiency makes it possible to use these fuel gases for many heating applications.

Air contains approximately 80% nitrogen by volume. Since nitrogen does not support combustion, fuel gases burned with air produce lower flame temperatures than those burned with oxygen. The total heat content is also lower. The air-fuel gas flame is suitable only for welding light sections of lead and for light brazing and soldering operations. The torches for use with air and fuel gas generally are designed to aspirate the proper quantity of air from the atmosphere to support combustion. The fuel gas flows through the torch at a supply pressure of 70 to 275 kPa (10 psig to 40 psig) and aspirates the needed air. For light work, the fuel gas usually is supplied from a small cylinder that is easily transportable.

The plumbing, refrigeration, and electrical trades use propane in small cylinders for many heating and soldering applications. The propane flows through the torch at a supply pressure from 20 kPa to 415 kPa (3 psig to 60 psig). The gas flow serves to aspirate the air. The torches are used for soldering electrical connections, sealing the joints in copper pipelines, and light brazing jobs. Small self-contained propane torch sets are available for incidental heating operations and home workshop use.

Standard oxyacetylene equipment, with the important exception of torch tips and fuel gas regulators, can usually be used to distribute and burn these gases. Special regulators are designed for the various gases and specialized heating and cutting tips are available. Connections on regulators must not be changed to allow the regulator to be used for a gas service different from the gas for which the regulator was designed. The manufacturer's recommendations should be followed regarding the use of regulators and torch tips.

ECONOMICS

The oxyfuel gas welding process is one of the least expensive of all the welding processes. However, the economic factors applied to welding are stated not only in dollars and cents, but also in terms of efficiency and convenience. The OFW process is simple and the equipment is portable and versatile. It is so widely used that the equipment and materials are readily available everywhere.

High-quality OFW welds that meet code requirements can often be completed at speeds competitive with arc welding processes, for example, in circumferential welds in pipe or tubing. The two economic advantages of OFW are lower equipment costs and shorter setup time.

Repair welds in thick sections are cost-effective with oxyfuel gas welding. Oxyfuel gas welding can augment mechanized operations of arc processes when a limited number of nonrepetitive welds are required. The level of skill required to use the OFW process is usually not as high as that required to effectively use the arc welding processes.

The process converts easily to heating and cutting applications, again at very low cost. A modest investment in oxyacetylene apparatus provides entry into metalworking applications beyond welding—bending and straightening, preheating, surfacing, brazing, braze welding, soldering, and cutting.

SAFE PRACTICES

No attempt should be made to operate oxyfuel gas welding apparatus until the welder or operator has been thoroughly trained in its proper use. The manufacturer's operating instructions and recommendations for safe use must be followed at all times. The reader is advised to consult the recommended safe practices which are detailed in the American National Standards Institute document *Safety in Welding, Cutting, and Allied Processes, ANSI Z49.1*[23] and in the Occupational Safety

23. See Reference 14.

and Health Administration document *Occupational Safety and Health Standards for General Industry.*[24] Other safety standards and recommended practices are listed in Appendix B of this volume.

FIRE AND EXPLOSION PREVENTION

Oxygen by itself does not burn or explode, but it supports combustion. Under high pressure, oxygen may react violently with oil, grease, or other combustible materials. Cylinders, fittings, and all equipment to be used with oxygen should be kept away from oil, grease, and other contaminants at all times. Oxygen cylinders must never be stored near highly combustible materials. Oxygen must never be used to operate pneumatic tools, to start internal combustion engines, to blow out pipelines, to dust clothing, or for any other potentially unsafe use.

Acetylene is a fuel gas that burns readily. Therefore, it must be kept away from open flames or other sources of ignition. Acetylene cylinder and manifold pressures must always be reduced through pressure-reducing regulators. Cylinders must always be protected against excessive temperatures and should be stored in well-ventilated, clean, dry locations at a safe distance from other combustibles. They should be stored and used with the valve end up.

When gas is taken from an acetylene cylinder that is lying on its side, acetone may be withdrawn along with the acetylene. This can cause damage to the apparatus or contaminate the flame, resulting in welds of inferior quality.

Acetylene that comes into contact with copper, mercury, or silver may form acetylides, especially if impurities are present. These compounds are violently explosive and can be detonated by a slight shock or by the application of heat. Thus, alloys containing more than 67% copper, other than tips and nozzles, should not be used in any acetylene system.

Cylinders containing liquefied fuel gases must always be used in an upright position. If these cylinders are placed on their side, liquid rather than vapor may be withdrawn from the cylinder. This could cause damage to the apparatus and produce a large, uncontrollable flame.

PERSONAL PROTECTIVE EQUIPMENT

Welders must use goggles or eye shields as a protection against sparks and the intense glare and heat radiated from the oxyfuel flame and molten metal. Appendix A is a guide designed to help in the selection of protective lenses for the eyes. Safety glasses, hearing protection and a hard hat should be worn; along with flame-retardant gloves, sleeves, leather aprons, leggings and safety shoes. Woolen clothing is preferred, as cotton burns readily and synthetic materials often melt. Flame-extinguishing equipment should be at hand. In closed or semiclosed areas, a forced-air ventilation or supplemental breathing system may be required. Other safety equipment should be in place for both the operator and others in the work environment.

STORAGE AND HANDLING OF GASES IN CYLINDERS

Cylinders containing gases should be stored in the upright position in clean, dry, well-ventilated locations. Cylinders can become a hazard if tipped over, and care should be exercised to avoid this possibility. If the cylinder valve is ruptured as the result of a fall, the escaping gas can cause the cylinder to become a hazardous projectile. It is standard practice to fasten the cylinders on a cylinder truck or to secure them against a rigid support.

Refrigerated oxygen cylinders are of double-wall construction, like a thermos bottle, with a vacuum between the inner and outer shell. They must be handled with extreme care to prevent damage to the internal piping and the loss of vacuum. Such cylinders must always be transported and used in an upright position. Cylinders containing fuel gases are also pressurized and must be handled with care. Provisions for the safe use, handling, and storage of gas cylinders are stated in the American Nation Standards Institute document *Safety in Welding, Cutting, and Allied Processes* ANSI Z49.[25]

REGULATOR SAFETY

To assure the safe use of regulators, the manufacturers' recommended operating procedures must always be carefully followed. The following safety precautions should always be observed to prevent accidents when using regulators:

1. The operator must be trained in the proper use of regulators or be under competent supervision;
2. The regulator must always be clean and in good working condition;
3. Cylinders must be secured to a wall, post, or cart so they will not tip or fall;
4. Cylinder valves must be inspected for damaged threads, dirt, dust, oil, or grease. Dust and dirt must be removed with a clean cloth. The regulator must not be attached to a gas cylinder if oil, grease, or damage is present, and the gas sup-

24. See Reference 13.

25. See Reference 14, pp. 26–27.

plier should be informed of this condition. Regulators should only be repaired or serviced by qualified technicians;

5. The cleaned cylinder value should be cracked open for an instant and then closed quickly. This will blow out any foreign matter that may be inside the valve port. However, if the cylinder valve is opened too much, the cylinder may tip over due to the force of escaping gas. The welder must not stand in front of the valve port;
6. The regulator should be attached to the cylinder valve and tightened securely with a wrench;
7. Before opening the cylinder valve, the regulator adjusting screw must be turned counterclockwise until the adjusting spring pressure is released;
8. The welder must stand to the side of the regulator when opening the cylinder valve, never in front of or behind the regulator. The cylinder valve must be opened carefully and slowly until the cylinder pressure is indicated on the high-pressure gauge. Acetylene cylinder valves must never be opened more than one and one-half turns. All other cylinder valves should be opened completely to seal the valve packing;
9. The regulator adjusting screw should be turned clockwise to attain the desired delivery pressure for the apparatus being used;
10. The system should be tested for leaks using the methods recommended by the manufacturer. Before lighting the torch, the system must be purged by opening only one valve on the torch at a time. The manufacturer's recommendations for purging and lighting the torch must be followed; and
11. The cylinder valves must be kept closed at all times except when the cylinder is in use. When the welder has finished using the apparatus, both cylinder valves should be closed. To bleed the system, the fuel-gas valve on the torch is opened and the gas is allowed to escape at a safe location. The torch fuel valve is then closed and the regulator-pressure adjusting screw is turned counterclockwise until the screw turns freely. This operation is repeated on the oxygen system. It is important never to bleed both systems at the same time. A reverse flow or mixing of gases may result, which may be hazardous.

CARE OF TORCHES

Welding torch handles, mixers, and tips are designed to withstand the tough operating conditions to which they are exposed. In order to provide safe and efficient usage, they must be properly maintained in good working condition at all times. They should be used only with the proper fuel gas and for the purpose for which they were designed.

It is important to follow manufacturers' operating instructions and recommendations for safe practices. If torches require service, only qualified repair technicians should do this work. The ANSI standard *Safe Practices in Welding, Cutting, and Allied Processes* should be consulted.[26]

HOSES AND HOSE CONNECTIONS

Hose connections must comply with the standard hose connection specification in the Compressed Gas Association document *Standard Connections for Regulator Outlets, Torches, and Fitted Hose for Welding and Cutting Equipment* (CGA E-1).[27] Hose connection size, shape and threading must not be compatible with connections for ordinary breathing air. To prevent error, the generally recognized color of an oxygen hose is green, and a fuel gas hose is red. This color-coding system is used in the United States, but may not be used in other countries.

The torch and hoses should be purged before lighting the torch for the first time each day and also after every cylinder change. The purge should not take place in confined spaces or near sources of ignition.

AUXILIARY EQUIPMENT

In addition to the equipment and materials described above, a variety of auxiliary equipment is used with oxyfuel gas welding. Two of the most universally required safety items are the friction lighter, which should always be used to ignite the gas, and check valves at the torch inlet. A check valve may also be used at the regulator outlet. Torch-mounted flashback arrestors offer additional protection to the operator and the apparatus. Other accessories, such as tip cleaners, cylinder trucks, clamps, and holding fixtures are also important auxiliary aids for safe oxyfuel gas welding.

FUMES AND GASES

The fumes and gases produced during oxyfuel gas welding may be hazardous. This is especially true with certain coated or clad metals, such as galvanized or plated surfaces, or metals with a high-alloy content. The workplace must be adequately ventilated by either natural air or forced-air methods to keep fumes and gases at safe levels. The American National Standards

26. See Reference 14
27. See Reference 18.

document, *Safety in Welding, Cutting, and Allied Processes*, should be consulted.[28]

The potential hazards presented by welding fumes and gases and the safe practices that must be followed to protect the welder are also outlined in a standard published by the American Conference of Governmental Industrial Hygienists (ACGIH) *TLVs® and BEIs®: Threshold Limit Values for Chemical Substances and Physical Agents in the Workroom Environment*.[29]

CONCLUSION

Historically, acetylene welding was an important first step toward the development of modern welding processes. The development and application of this process demonstrated to fabricators of an earlier time that welding was a practical and reliable means of joining metals and had the capability of replacing other methods, specifically riveting. Since those early days, oxyfuel gas welding has served as an introduction to the field of welding for countless students and has remained the preferred welding process of many for metal fabricating and repairs.

The initial education welders receive in oxyfuel gas welding has been a well-used building block for learning the more sophisticated arc welding and solid-state welding processes. In spite of the incredibly advanced technology of modern welding processes, oxyfuel gas welding, for its simplicity, reliability, and convenience, still has its practitioners—and will continue to have them for years to come.

BIBLIOGRAPHY[30]

American Conference of Governmental Industrial Hygienists (ACGIH) *TLVs® and BEIs®: Threshold limit values for chemical substances and physical agents in the workroom environment.* Cincinnati, Ohio: American Conference of Governmental Industrial Hygienists.

American National Standards Institute (ANSI) Accredited Standards Committee Z49.1. 1999. *Safety in welding, cutting, and allied processes.* ANSI Z49.1: 1999. Miami: American Welding Society.

American Welding Society (AWS) Committee on Definitions and Symbols. 2001. *Standard welding terms and definitions,* A3.0:2001. Miami: American Welding Society.

American Welding Society (AWS) Committee on Filler Metals and Allied Materials. 1992. *Specification for carbon and low-alloy steel rods for oxyfuel gas welding.* ANSI/AWS A5.2-1992. Miami: American Welding Society.

American Welding Society (AWS) Committee on Filler Metals and Allied Materials. 1990. *Specification for welding electrodes and rods for cast iron,* ANSI/AWS A5.15-90R. Miami: American Welding Society.

American Welding Society (AWS) Committee on Filler Metals and Allied Materials. 1999. *Specification for bare aluminum and aluminum-alloy welding electrodes and rods,* ANSI/AWS A5.10/A5.10M:1999. Miami: American Welding Society.

American Welding Society (AWS) Welding Handbook Committee. 2001. Jenney, C. L. and A. O'Brien, eds. *Welding science and technology,* Vol. 1, *Welding handbook*, 9th ed. Miami: American Welding Society.

American Welding Society (AWS) Welding Handbook Committee. 1996. Oates, W. R., ed. *Materials and applications—Part 1*, Vol. 3. *Welding handbook*, 8th ed. Miami: American Welding Society.

American Welding Society (AWS) Welding Handbook Committee. 1998. Oates, W. R., and A. M. Saitta, eds. *Materials and applications—Part 2*, Vol. 4. *Welding handbook*, 8th ed. Miami: American Welding Society.

Compressed Gas Association (CGA). 1994. *Compressed gas cylinder valve outlet and inlet connections*, ANSI/CGA B-57-1. Arlington, Virginia: Compressed Gas Association.

Compressed Gas Association (CGA). 1994. *Standard for gas pressure regulators.* CGA E-4, 1994. Arlington, Virginia: Compressed Gas Association.

Compressed Gas Association (CGA). *Standard connections for regulator outlets, torches, and fitted hose for welding and cutting equipment,* CGA E-1-1994. Arlington, Virginia: Compressed Gas Association.

ESAB Welding and Cutting Products. 1995. *Oxy-Acetylene Handbook*. Florence, South Carolina: ESAB Welding and Cutting Products.

National Fire Protection Association (NFPA). 1997. *Standard for the installation and operation of oxygen-fuel gas systems for welding, cutting, and allied processes.* NFPA 51-97. Quincy, Massachusetts: National Fire Protection Association.

28. See Reference 20.
29. American Conference of Governmental Industrial Hygienists (ACGIH), 2003, *TLVs® and BEIs:® Threshold Limit Values for Chemical Substances and Physical Agents in the Workroom Environment*, Cincinnati, Ohio,: American Conference of Governmental Industrial Hygienists.
30. The dates of publication given for the codes and other standards listed here were current at the time the chapter was prepared. The reader is advised to consult the latest edition of the code or other standard.

National Fire Protection Association (NFPA). 1996. *Bulk oxygen systems at consumer sites.* NFPA 50-96. Quincy, Massachusetts: National Fire Protection Association.

Occupational Safety and Health Administration (OSHA). 1999. *Occupational Safety and Health Standards for General Industry,* in *Code of Federal Regulations (CFR),* Title 29, CFR 1910, *Subpart Q.* Washington D.C.: Superintendent of Documents, U.S. Government Printing Office.

SUPPLEMENTARY READING LIST

Ballis, W. L., C. DeHaven and R. Baker. 1977. Training of oxyacetylene welding to weld mild steel pipe. *Welding Journal* 56(4): 15–19.

National Fire Protection Association (NFPA). 1999. *Standard for fire prevention during welding, cutting, and other hot work.* NFPA 51B-1999. Quincy, Massachusetts: National Fire Protection Association.

National Fire Protection Association (NFPA). 1998. *Storage, use, and handling of compressed and liquefied gases in portable cylinders.* NFPA 55–98. Quincy, Massachusetts: National Fire Protection Association.

The National Training Fund for the Sheet Metal and Air Conditioning Industry. 1979. *Welding book I.* Alexandria, Virginia: The National Training Fund for the Sheet Metal and Air Conditioning Industry.

Occupational Safety and Health Administration (OSHA). *Oxygen-fuel gas welding and cutting.* In *Code of Federal Regulations (CFR),* Title 29, CFR 1910.253, *Subpart Q.* Washington D.C.: Superintendent of Documents, U.S. Government Printing Office.

Sosnin, H. A. 1982. Efficiency and economy of the oxyacetylene process. *Welding Journal* 61(10): 46–48.

CHAPTER 12

BRAZING

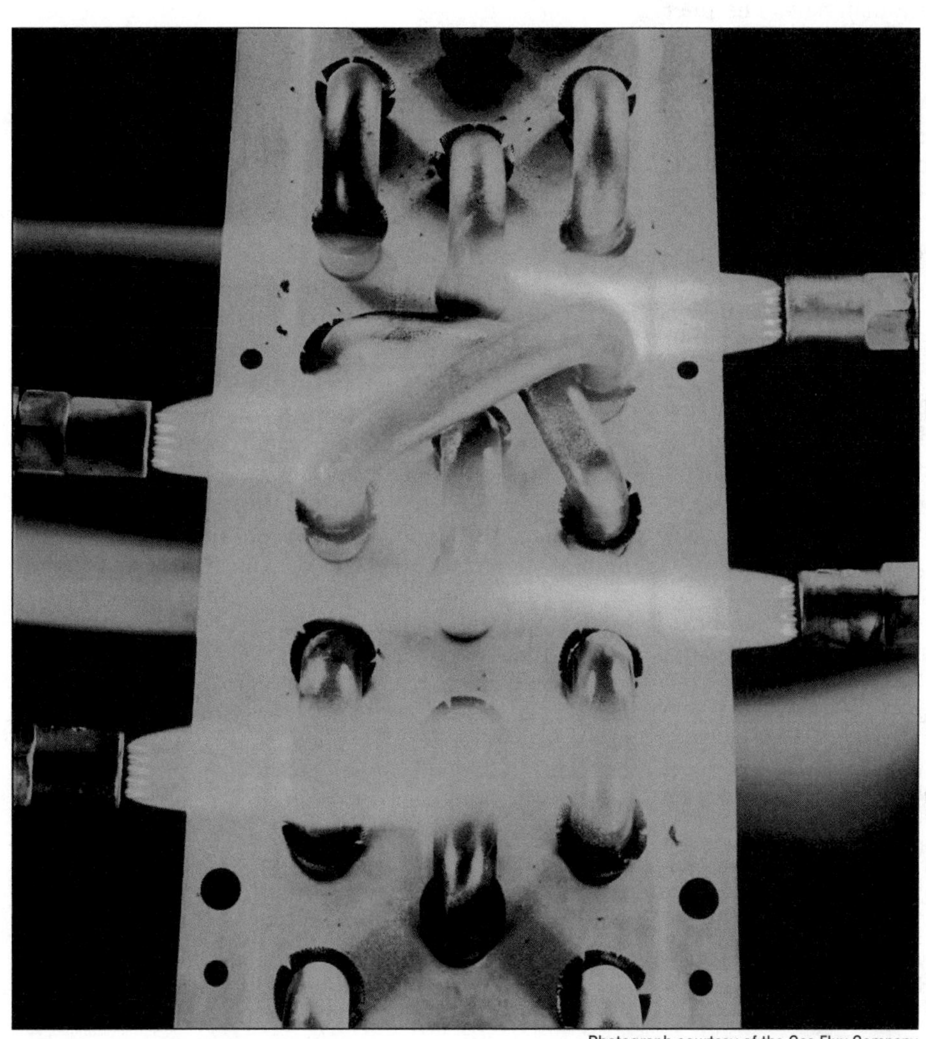

Photograph courtesy of the Gas Flux Company

Prepared by the Welding Handbook Chapter Committee on Brazing:

N. C. Cole, Chair
NCC Engineering

D. W. Bucholz
Colorado School of Mines

F. M. Hosking
Sandia National Laboratories

M. J. Lucas, Jr.
G.E. Aircraft Engines

R. L. Peaslee
Wall Colmonoy Corporation

Welding Handbook Volume 2 Committee Member:

D. W. Dickinson
The Ohio State University

Contents

CHAPTER 12

BRAZING

INTRODUCTION

The term *brazing* refers to a group of joining processes that includes torch brazing, furnace brazing, induction brazing, dip brazing, diffusion brazing, resistance brazing, and infrared brazing. Brazed joints are formed when an assembly of materials is heated to the brazing temperature in the presence of a filler metal with a liquidus above 450°C (840°F) but below the solidus, or melting point, of the base materials. The filler metal distributes itself between the closely fitted surfaces of the joint by capillary action.[1,2] Heat is provided by a system appropriate to the process and application.

Brazing is economically attractive for the production of high-strength metallurgical bonds while preserving desired base material properties. The various brazing processes discussed in this chapter serve many industrial and commercial joining needs. Examples of brazing applications throughout the chapter illustrate when to select brazing as the joining process, how to design the joint, and what materials are best suited for an individual application.

Some of the techniques, equipment, and physical mechanisms of both brazing and welding are integrated in a process known as braze welding. Although classified as a stand-alone welding process by the American Welding Society, braze welding is discussed in this chapter because of these close relationships.

The brazing processes discussed in this chapter include torch brazing, furnace brazing, induction brazing, resistance brazing, dip brazing, infrared brazing, and diffusion brazing. Several seldom-used brazing processes—blanket brazing, block brazing, exothermic brazing, and twin carbon arc brazing—are addressed briefly.

Other topics included in this chapter are brazing equipment and materials, techniques, braze quality, inspection, economics, and safe practices.

1. Definitions throughout this chapter are adapted from American Welding Society, Committee on Definitions and Symbols, 2001, *Standard Welding Terms and Definitions,* Miami: American Welding Society. AWS A3.0: 2001. Miami: American Welding Society.
2. At the time of preparation of this chapter, the referenced codes and other standards were valid. If a code or other standard is cited without a date of publication, it is understood that the latest edition of the referenced document applies. If a code or other standard is cited with the date of publication, the citation refers to that edition only, and it is understood that any future revisions or amendments to the code or standard are not included. As codes and standards undergo frequent revision, the reader is advised to consult the most recent edition.

FUNDAMENTALS

According to its definition, brazing must meet each of three criteria:

1. The components of the assembly must be joined without melting the base materials.
2. The filler metal must have a liquidus temperature above 450°C (840°F).
3. The filler metal must wet the surfaces of the base materials and be drawn into or held in the joint by capillary action.

To achieve a good joint using any of the various brazing processes, the components must be properly cleaned and must be protected by either flux or a protective atmosphere during the heating process to prevent excessive oxidation or other surface modifications that would affect the brazing process. The components must be designed to afford a capillary for the filler metal when properly aligned, and a heating process must be selected that will provide the proper brazing temperature and heat distribution.

Braze welding is a specialized method of welding with a brazing filler metal. In braze welding, the filler metal (usually in the form of a welding rod or electrode) is melted and deposited in grooves and fillets exactly at the points where it is to be used. Capillary action is not a factor in distribution of the brazing filler metal; limited base metal fusion may occur in braze welding.

APPLICATIONS

The brazing process is used for a variety of reasons to join various materials. In many instances, the motivation is to join dissimilar materials to obtain the maximum benefit of both materials, or it may be to produce the most cost- or weight-effective joint. By using the proper joint design, the resulting braze can function better than the either of the base materials being joined. Applications of brazing cover the entire range of manufactured products, from inexpensive toys to the highest quality aircraft engines, automotive assemblies, nuclear systems, electronic equipment, food processing equipment, pressure vessels, biomedical components, and aerospace vehicles.

ADVANTAGES

Like any joining process, brazing has both advantages and limitations. The advantages vary with the heating method employed, but in general brazing is very economical when done in large batches. The brazing process can join dissimilar metals without melting the base materials, which distinguishes it from other joining methods. In many instances, several hundred assembled parts with many feet of braze joints can be brazed at one time. When protective-atmosphere brazing is used, the assemblies are kept clean and the heat treatment process can be facilitated as part of the brazing cycle.

Brazing is used because it can produce results that are not always achievable with other joining processes. Brazing is often the preferred joining process because it provides the following advantages:

1. Economical for production of complex assemblies,
2. Simple way to join large joint areas,
3. Excellent stress and heat distribution,
4. Ability to preserve coatings and claddings,
5. Ability to join dissimilar materials,
6. Ability to join metals to nonmetals,
7. Ability to join widely different thicknesses,
8. Capability of joining precision parts,
9. Production of joints requiring little or no finishing,
10. Capability of joining many brazements simultaneously (batch processing) or on a continuous basis, and
11. Sequential fabrication of multiple joints at progressively lower brazing temperatures.

LIMITATIONS

Since the brazing process uses a molten metal flowing between the materials to be joined, there is the possibility of liquid-metal interactions on some materials, which may be unfavorable. Depending on the material combinations involved and the thickness of the base metal, erosion of the base metal may occur. In many cases, the erosion may be of little consequence. However, when brazing joints for heavily loaded applications or when using thin materials, the erosion can weaken the joint and make it unsatisfactory for its intended application. Also, the formation of brittle intermetallics or other phases can make the resulting joint too brittle to be acceptable.

A disadvantage of some of the manual brazing processes is that highly skilled technicians or operators are required to perform the operation. This is especially true for torch brazing when using a filler metal with a melting point close to the melting point of the base metal.

Nevertheless, with the proper joint design, brazing filler metal, and process selection, a satisfactory brazing technique can be developed for joining most materials for which a fusion welding process is not feasible because of strength requirements or economic considerations.

PRINCIPLES OF OPERATION

Capillary flow is the dominant physical principle that assures good brazements. Capillary flow into the joint clearance is a result of surface tension between the base metal and filler metal promoted by the contact angle between the two. The joint is protected by a flux or protective atmosphere. Capillary flow begins as the faying surfaces are wetted by the molten filler metal. Brazing filler metal flow is influenced by the dynamics of fluidity, viscosity, vapor pressure, gravity, and especially by the effects of metallurgical reactions between the filler metal and base metal. The joint clearance must be spaced to permit efficient capillary action and coalescence. Torch brazing and furnace brazing are two major methods of producing brazements.

The typical torch braze has a relatively large area and a very small joint clearance. In the simplest brazing application, the surfaces to be joined are cleaned to remove contaminants and oxides. Next, the surfaces are coated with flux (a chemical material capable of dissolving solid metal oxides and preventing reoxidation). The joint area is heated until the flux melts and cleans the remaining oxides on the base metal surface, which is protected against further oxidation by the layer of liquid flux adjacent to the joint. This occurs before the filler metal melts. As the joint area continues to be heated and the filler metal is melted, it is drawn into the joint clearance. Capillary attraction between the base metal and the filler metal is much higher than that between the base metal and the flux. Accordingly, the flux is displaced by the filler metal. On cooling to room temperature, the resulting joint will be filled with solid filler metal and the solidified flux will be found at the joint periphery, although some flux may be trapped and left in the joint.

Joints to be brazed are usually designed with clearances of 0.025 to 0.25 millimeters (mm) (0.001 to 0.008 inches [in.]). Therefore, fluidity of the filler metal is an important factor. High fluidity is a desirable characteristic of a brazing filler metal because capillary action may be insufficient to draw a viscous filler metal into closely fitted joints.

Furnace brazing typically takes place in the atmosphere of a reducing gas, such as hydrogen, or in an inert gas or vacuum. Brazing in this protective atmosphere eliminates the necessity for postbraze cleaning and ensures the absence of corrosive mineral flux residue. Carbon steels, stainless steels, and superalloy components are most often processed in atmospheres of reacted gases, dry hydrogen, dissociated ammonia, argon, or a vacuum. Vacuum furnaces are generally used to braze zirconium, titanium, stainless steels, and the refractory metals. With good processing procedures, aluminum alloys can also be brazed in a vacuum furnace with excellent results.

The American Welding Society publishes detailed information on brazing and braze welding in the specifications, standards, and recommended practices listed in Table 12.1.

Table 12.1
Standards, Specifications and Recommended Practices for Brazing and Braze Welding

A2.4	Standard Symbols for Welding, Brazing, and Nondestructive Examination
A5.8	Specification for Filler Metals for Brazing and Braze Welding
A5.31	Specification for Fluxes for Brazing and Braze Welding
B2.2	Standard for Brazing Procedure and Performance Qualification
C3.2	Standard Method for Evaluating the Strength of Brazed Joints
C3.3	Recommended Practices for Design, Manufacture, and Inspection of Critical Brazed Components
C3.4	Specification for Torch Brazing
C3.5	Specification for Induction Brazing
C3.6	Specification for Furnace Brazing
C3.7	Specification for Aluminum Brazing
C3.8	Recommended Practices for Ultrasonic Inspection of Brazed Joints
D10.13M	Recommended Practices for the Brazing of Copper Pipe and Tubing for Medical Gas Systems

PROCESSES, EQUIPMENT, AND TECHNIQUES

Brazing processes are customarily designated according to the sources or methods of heating. The methods significant for industrial use are the following:

1. Torch brazing (TB),
2. Furnace brazing (FB),
3. Induction brazing (IB),
4. Resistance brazing (RB),
5. Dip brazing (DB),
6. Infrared brazing (IRB), and
7. Diffusion brazing (DFB).

Other processes, mainly used for special applications, are blanket brazing, block brazing, exothermic brazing, and twin carbon arc brazing. Whichever process is used, the characteristics of brazing remain the same: the filler metal has a melting point above 450°C (840°F) but below that of the base material, and it spreads within the joint by capillary action.

TORCH BRAZING

Torch brazing (TB) is accomplished by heating the assembly with a fuel-gas flame, using one or more gas torches. Depending on the temperature and the amount of heat required, the fuel gas, such as acetylene, propane, or natural gas, can be burned with air, compressed air, or oxygen. Additional information on torch brazing is presented in the American Welding Society publication *Specification for Torch Brazing*, ANSI/AWS C3.4.[3]

Manual torch brazing, totally without automation, represents the simplest brazing technique. Manual torch brazing of copper tubing is illustrated in Figure 12.1. While torch brazing is labor intensive and low in productivity, it has some practical and economic justification. First, the braze joint is visible to the operator, who adjusts the process based on observation. Second, heat is directed only to the joint area. This is an important

3. American Welding Society (AWS) Committee on Brazing and Soldering, *Specification for Torch Brazing*, ANSI/AWS C3.4. Miami: American Welding Society.

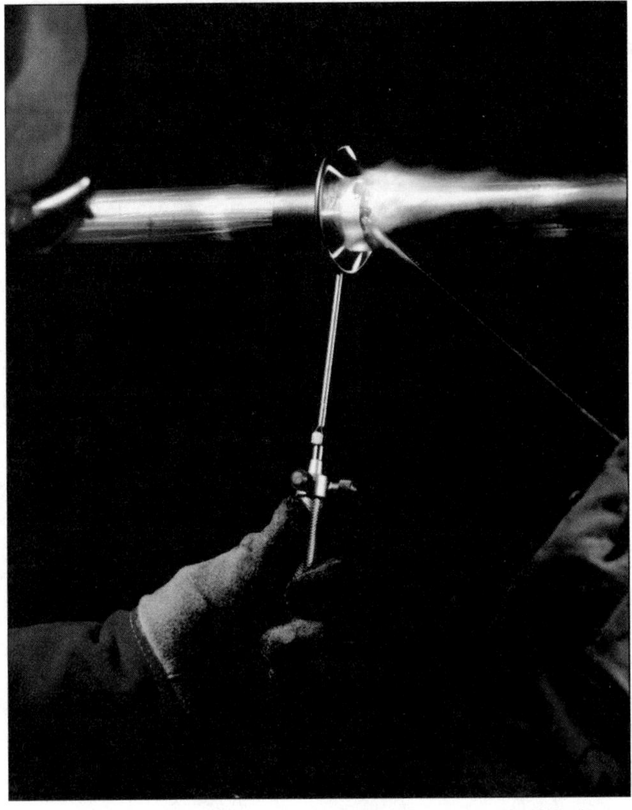

Photograph courtesy of Smith Equipment

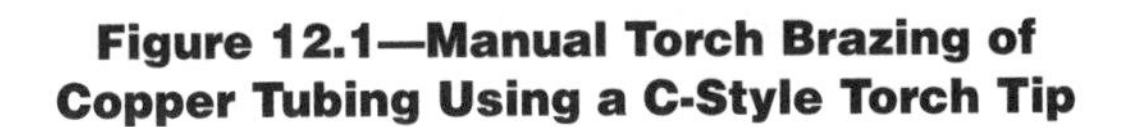

Figure 12.1—Manual Torch Brazing of Copper Tubing Using a C-Style Torch Tip

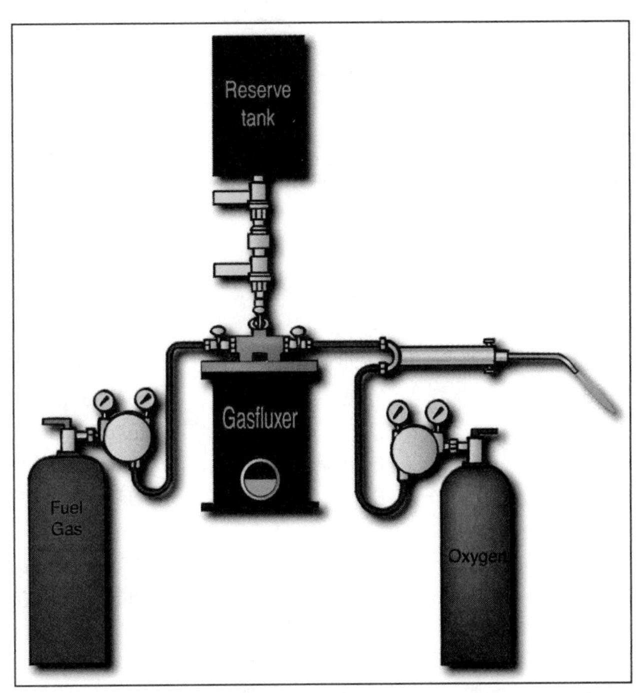

Photograph courtesy of Fusion, Incorporated

Figure 12.2—Generic Setup for Manual Torch Brazing

consideration when energy costs represent a large fraction of the cost of a brazed joint.

Equipment

Of the torches used in brazing, air-natural gas torches provide the lowest flame temperature and the least heat. Acetylene under pressure is used in the air-acetylene torch with air at atmospheric pressure. Both air-natural gas and air-acetylene torches can be used to advantage on small components and thin sections.

Torches that use oxygen with natural gas or other gases (such as propane or butane) produce higher flame temperatures. When properly applied as a neutral flame or a slightly reducing flame, excellent results are obtainable for many brazing applications. A generic setup for manual torch brazing is shown in Figure 12.2.

Oxyhydrogen torches are often used for brazing aluminum and nonferrous alloys because they operate at lower temperatures, which reduces the possibility of overheating the assembly during brazing. An excess of hydrogen provides the joint with additional cleaning and protection.

Specially designed torches with multiple tips or multiple flames can be used to an advantage to increase the rate of heat input; however, local overheating must be carefully avoided. This can be accomplished by constantly moving the torch relative to the assembly to disperse the heat.

For manual torch brazing, the torch can be equipped with a single tip that provides either a single flame or multiple flames. Manual torch brazing is particularly useful on assemblies involving sections of unequal mass. When the volume and rate of production warrants, mechanized operations can be set up using torches. One or more torches equipped with single- or multiple-flame tips may be an acceptable option. Machines can be designed to move either the workpieces or the torches, or both. For premixed natural gas-air flames, a refractory type of nozzle or tip is used.

Torch heating for brazing with filler metal rods can be accomplished with flux-covered rods or flux-cored rods. Only copper-phosphorus filler metals are self-fluxing; they are used for brazing in air without flux. With the exception of these copper-phosphorus filler

Table 12.2
Classification of Brazing Fluxes with Brazing or Braze Welding Filler Metals

Classification*	Form	Filler Metal Type	Activity Temperature Range °C	Activity Temperature Range °F
FB1-A	Powder	BAlSi	580–615	1080–1140
FB1-B	Powder	BAlSi	560–615	1040–1140
FB1-C	Powder	BAlSi	540–615	1000–1140
FB2-A	Powder	BMg	480–620	900–1150
FB3-A	Paste	BAg and BCuP	565–870	1050–1600
FB3-C	Paste	BAg and BCuP	565–925	1050–1700
FB3-D	Paste	BAg, BCu, BNi, BAu, and RBCuZn	760–1205	1400–2200
FB3-E	Liquid	BAg and BCuP	565–870	1050–1600
FB3-F	Powder	BAg and BCuP	650–870	1200–1600
FB3-G	Slurry	BAg and BCuP	565–870	1050–1600
FB3-H	Slurry	BAg	565–925	1050–1700
FB3-I	Slurry	BAg, BCu, BNi, BAu, and RBCuZn	760–1205	1400–2200
FB3-J	Powder	BAg, BCu, BNi, BAu, and RBCuZn	760–1205	1400–2200
FB3-K	Liquid	BAg and RBCuZn	760–1205	1400–2200
FB3-L	Dispensable paste	BAg and BCuP	565–870	1050–1600
FB3-M	Dispensable paste	BAg and BCuP	565–925	1050–1700
FB3-N	Dispensable paste	BAg, BCu, BNi, BAu, and RBCuZn	760–1205	1600–2200
FB4-A	Paste	BAg and BCuP	595–870	1100–1600

*Note: The selection of a flux designation for a specific type of work can be based on the form, the filler metal type, and the description above, but more information than is presented in this table is generally needed for flux selection.

metals, all other filler metals for torch brazing require fluxes. For certain applications, even the self-fluxing copper-phosphorus filler metals require added flux. This is shown in Table 12.2,[4] which lists flux classifications used with various filler metals.

The brazing filler metal can be preplaced in the joint as the joint is prepared for brazing and fluxed before heating, or it can be face-fed into the joint. Heat is applied to the joint first to melt the flux and then continuing until the filler metal melts and flows into the joint. Overheating of the base metal and brazing filler metal should be carefully avoided, as it may cause drop-through, an undesirable sagging or surface irregularity. Natural gas is well suited for torch brazing because its relatively low flame temperature reduces the risk of overheating.

Brazing filler metal can be preplaced at the joint in the form of rings, washers, strips, slugs, or powder, or it can be fed by hand, typically in the form of wire or rod. It should be noted that silver filler metals do not diffuse in iron. In any case, proper cleaning and fluxing are essential.

Techniques

Torch brazing techniques with oxyfuel gas welding equipment differ from those used for oxyfuel gas welding. Operators experienced only in welding techniques may require instruction in brazing techniques. It is good practice, for example, to prevent the inner cone of the flame from coming in contact with the joint, potentially causing overheating. Overheating could result in the melting of the base metal and dilution with the filler metal. This may increase its liquidus temperature and cause the flow to stop or be more sluggish. In addition, the flux may become overheated and thus lose its ability to promote capillary flow, and the low-melting constituents of the filler metal may evaporate.

FURNACE BRAZING

Furnace brazing (FB) is a process in which the assembly is placed in a furnace and heated to the brazing tem-

4. American Welding Society (AWS) Committee on Brazing and Soldering, Jenney, C. L., ed., (forthcoming), *Brazing Handbook*, 5th Ed., Miami: American Welding Society.

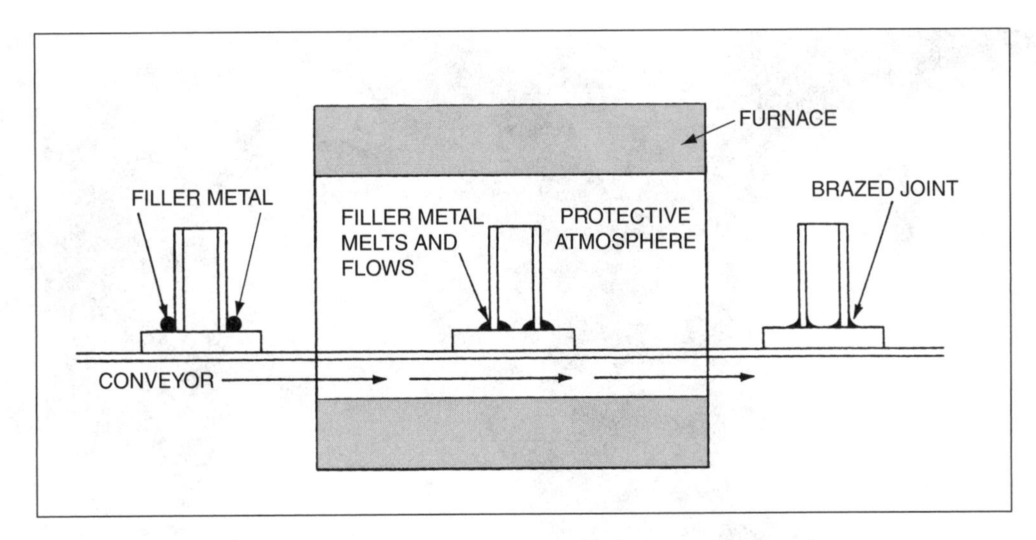

Figure 12.3—Sequence of Steps in Furnace Brazing

perature. Furnace brazing is used extensively when the following conditions can be met:

1. The components to be brazed can be preassembled or placed in a fixture to hold them in the correct position,
2. The brazing filler metal can be placed in contact with the joint,
3. Multiple brazed joints are to be formed simultaneously on complete assemblies,
4. Many similar assemblies are to be joined, and
5. Complex components can be heated uniformly to prevent the distortion that would result from local heating of the joint area.

Electric, gas, or oil-heated furnaces with automatic temperature controls capable of holding the temperature within ± 6°C (± 10°F) should be used for furnace brazing. Specially controlled atmospheres that perform fluxing functions must be provided. Brazing in a controlled atmosphere requires no flux; it is the rare exception that a furnace-brazing application requires flux. Additional information on furnace brazing is presented in the latest edition of the American Welding Society document, *Specification for Furnace Brazing*, ANSI/AWS C3.6.[5]

The components to be brazed should be assembled with the filler metal located in or around the joints. The preplaced filler metal can be in the form of wire, foil, filings, slugs, powder, paste, or tape. The assembly is placed on the conveyor, heated in the furnace until it reaches brazing temperature and brazing takes place. The assembly is then removed. These steps are illustrated in Figure 12.3.

Brazing time depends somewhat on the thickness of the pieces to be brazed and the mass of fixturing necessary to position them. The brazing time should be restricted to the time necessary for the filler metal to flow through the joint without allowing excessive interaction between the filler metal and base metal. Normally, one or two minutes at the brazing temperature is sufficient to complete the braze. A longer time at the brazing temperature is beneficial when the filler metal remelt temperature is to be increased and when diffusion will improve joint ductility and strength. Times ranging from 30 to 60 minutes at the brazing temperature are often used to increase the braze remelt temperature. Care must be exercised in selecting the appropriate filler metal to produce the increase in remelt temperature.

Furnace Brazing Equipment

Brazing furnaces are classified into the following four types: batch furnaces with either air or controlled atmospheres, continuous furnaces with controlled atmospheres, retort furnaces with controlled atmospheres, and vacuum furnaces with controlled temperatures.

Figure 12.4 shows a high-vacuum side-loading brazing furnace in which a load of 304 stainless steel assemblies were brazed using a nickel filler metal.

5. American Welding Society (AWS) Committee on Brazing and Soldering, *Specification for Furnace Brazing*, ANSI/AWS C3.6. Miami: American Welding Society.

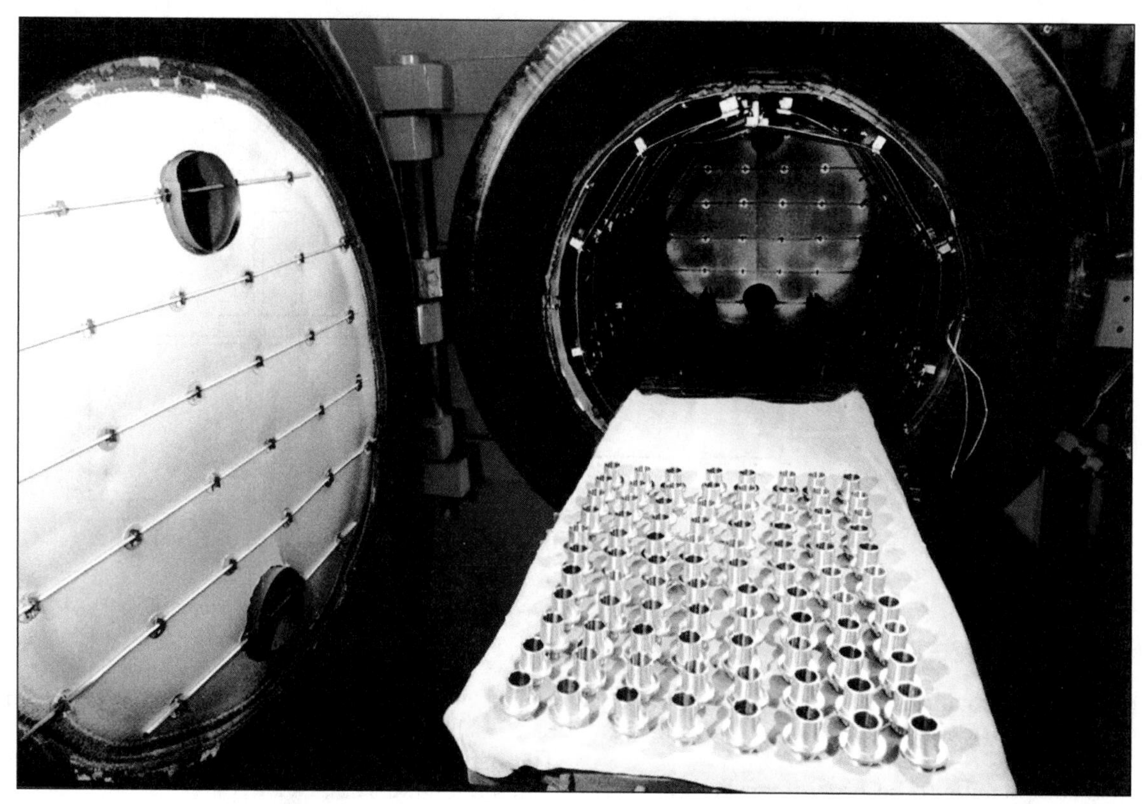

Photograph courtesy of HiTecMetal Group

Figure 12.4—A Load of Stainless Steel Assemblies Nickel-Brazed in a Vacuum Furnace

Most brazing furnaces have a potentiometer-type temperature control connected to thermocouples, gas controls, and resistance heating elements using silicon-carbide, nickel-chromium, or refractory metal (molybdenum, tantalum, tungsten). When a gas or oil flame is used for heating, the flame must not impinge directly on the parts to be brazed.

Controlled-Atmosphere Furnace. Controlled atmosphere furnaces maintain a continuous flow of the atmosphere gas in the work zone to avoid contamination from the outgassing of the metal components of the assembly and dissociation of oxides. If the controlled atmosphere is flammable or toxic, adequate venting of the work area and protection against explosion are necessary.

Batch Furnace. Batch furnaces heat each workload separately. They may be top loading (pit type), side loading, or bottom loading. Some furnaces, called *bell furnaces,* are designed to be lowered over the assemblies to be brazed. Gas- or oil-fired batch furnaces without retorts require that a protective atmosphere be used. Electrically heated batch furnaces are usually equipped for controlled-atmosphere brazing, since the heating elements can usually be operated in the controlled atmosphere, or the atmosphere can be contained in a retort.

Continuous Furnace. Continuous furnaces receive a steady flow of incoming assemblies. The heat source may be gas or oil flames or electrical heating elements. The assemblies are conveyed through the furnace either singly or in trays or baskets. Continuous furnaces (conveyor, mesh belt, or roller hearth types) and shaker-hearth, pusher, or slot types are commonly used for high-production brazing. Continuous furnaces usually contain a preheating or purging area through which the parts or assemblies enter. In this area, the assemblies are slowly brought to a temperature below the brazing temperature. If an atmosphere gas is used in the brazing zone, it also flows over and around the assemblies in the preheat zone under positive pressure. The gas flow removes any entrapped air and starts the reduction of

Photograph courtesy of HiTecMetal Group

Figure 12.5—Aircraft Components Copper-Brazed in a Continuous-Belt Hydrogen Furnace

surface oxides. The assemblies are heated to the brazing temperature in the last heating zone immediately prior to entering the cooling zone. Atmosphere gas trails the brazements into the cooling zone.

Figure 12.5 shows the copper brazing of aircraft components in a continuous mesh belt hydrogen furnace.

Retort Furnace. Retort furnaces are batch furnaces in which the assemblies are placed in a sealed retort for brazing. The air in the retort is purged and replaced by a controlled-atmosphere gas, and the retort is placed in the furnace. After the assemblies have been brazed, the retort is removed from the furnace, cooled, and the controlled atmosphere is purged. The retort is opened, and the brazed assemblies are removed. A protective atmosphere is sometimes used within a high-temperature furnace but outside of the retort to reduce external scaling of the retort.

Vacuum Furnace. Vacuum furnace brazing is widely used in the aerospace, nuclear, and other specialized fields for applications in which reactive metals are joined or where entrapped fluxes would be intolerable. If the vacuum atmosphere is maintained by continuous pumping, it removes the volatile constituents liberated during heating.

Vacuum brazing equipment is generally used to braze stainless steels, superalloys, aluminum alloys, titanium alloys and metals containing refractory or reactive elements. A vacuum is a relatively economical atmosphere that prevents oxidation by removing air from around the assembly. Prior surface cleanliness is nevertheless required for good base-metal wetting and filler-metal flow. Base metals containing chromium and silicon can be vacuum brazed. Base metals that can generally be brazed only in vacuum are those that contain low percentages of aluminum, titanium, zirconium, or other elements with particularly stable oxides. However, the use of a nickel-plate barrier may still be preferred to obtain optimum quality.

The most commonly used vacuum-brazing furnace is the cold-walled vacuum furnace. Furnace configurations may be side loading (horizontal), top loading (pit type), or bottom loading. Work zones are usually rectangular for side-loading furnaces, and circular for top- and bottom-loading types.

Various combinations of vacuum pumps and blowers can be employed to evacuate brazing furnaces. They are frequently used in the following combinations:

1. Roughing pump (RP), oil sealed (133 pascals [Pa] to 67 Pa [1 torr to 0.5 torr]),
2. Roughing pump with roots blower (RB) (1.33 Pa to 0.13 Pa [10^{-2} torr to 10^{-3} torr]),
3. Roughing pump with roots blower and diffusion pump (13.3 Pa to 1.33×10^{-4} Pa [10^{-1} torr to 10^{-6} torr]), and
4. Roughing pump and roots blower with turbo-molecular pump (1.33 Pa to 0.13 Pa [10^{-2} torr to 10^{-3} torr]).

The oil-sealed roughing pump is used for pressures from 13 to 1300 Pa (0.1 to 10 torr). Pressures of 1.3 Pa to 0.13 Pa (10^{-2} torr to 10^{-3} torr) are best obtained with the high-speed, dry roots, or turbo-molecular type of pump. The brazing of base metals containing chromium, silicon, or other elements that promote rather strong oxide formation usually requires a turbo-molecular type of pump. The turbo-molecular vacuum pump is not capable of exhausting directly to the atmosphere and requires a roughing back-up pump.

Base materials containing more than a low percentage of aluminum, titanium, or zirconium, which form very stable oxides during brazing, require a vacuum of

0.13 Pa (10^{-3} torr) or lower. The vacuum-furnace brazing of these materials usually requires a diffusion pump that will obtain pressures of 1.3 Pa to 0.0001 Pa (10^{-2} torr to 10^{-6} torr). The diffusion pump is backed by a mechanical vacuum pump or by a combination of a roots-type of pump and a mechanical pump.

Vacuum Retort Furnace. Vacuum retort furnaces have a sealed retort, usually constructed of fairly thick metal. The retort with the assembled brazements loaded inside is sealed, evacuated, and heated from the outside by a gas-fired or electric furnace. Most brazing work requires vacuum pumping continuously throughout the heating cycle to remove gases that outgas from the workload. The retort size and its maximum operating temperature are limited by the ability of the retort to withstand the collapsing force of atmospheric pressure at the brazing temperature. The maximum temperature for vacuum brazing furnaces of this type is about 1150°C (2100°F). Argon, nitrogen, or another gas is often introduced into the retort to accelerate cooling after brazing.

Double-Pumped, Double-Walled Retort Furnace. The typical double-pumped, double-walled retort furnace has an inner retort that contains the assemblies to be brazed within an outer wall or vacuum chamber. The vacuum chamber also contains the thermal insulation and electrical heating elements. A moderately reduced pressure, typically 1.33 to 13.3 Pa (1.0 to 0.1 torr), is maintained within the outer wall, and a much lower pressure, below 1.3 Pa (10^{-2} torr), within the inner retort. Brazing requires continuous vacuum pumping of the inner retort throughout the heat cycle to remove gases given off by the workload. In this type of furnace, the heating elements and the thermal insulation are not subjected to the high vacuum. Heating elements are typically made of nickel-chromium alloy, graphite, stainless steel, or silicon carbide materials. Thermal insulation is usually made of silica or alumina brick, castable material, or fibrous material.

Cold-Wall Vacuum Furnace. A typical cold-wall vacuum furnace has a single vacuum chamber with thermal insulation and electrical heating elements located inside the chamber. The vacuum chamber is usually water-cooled. The maximum operating temperature is determined by the materials used for the thermal insulation (the heat shield) and the heating elements, which are subjected to both the high vacuum and the operating temperature of the furnace.

Heating elements for cold-wall furnaces are usually constructed of high-temperature, low-vapor-pressure materials, such as molybdenum, tungsten, graphite, or tantalum. Heat shields are typically made of multiple layers of molybdenum, tantalum, nickel, or stainless steel. Thermal insulation is usually made of high-purity alumina brick, graphite, or alumina fibers sheathed in stainless steel. Temperatures up to 2200°C (4000°F) and pressures as low as 1.33×10^{-4} Pa (10^{-6} torr) are obtainable.

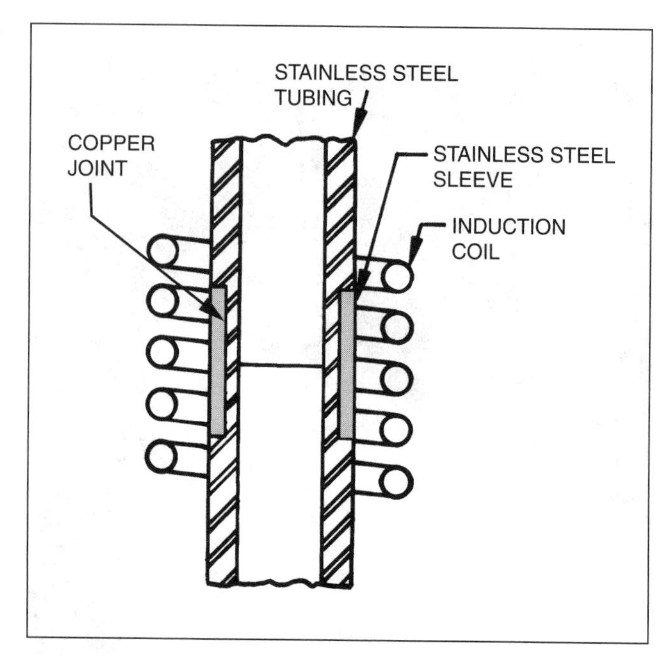

Figure 12.6—Joint in Stainless Steel Tubing Induction Brazed in a Controlled Atmosphere

INDUCTION BRAZING

The heat for induction brazing (IB) is obtained from an electric current induced in the components to be brazed, hence the name *induction brazing*. For induction brazing, the assemblies to be brazed are placed in or near a water-cooled coil carrying alternating current. The assemblies do not form a part of the electrical circuit. The components to be heated function as the short-circuited secondary of a transformer in which the work coil, when connected to the power source, functions as the primary. On both magnetic and nonmagnetic brazements, heating is obtained from the resistance of the components to currents induced in them by the transformer action. The latest edition of the American Welding Society document *Specification for Induction Brazing*, ANSI/AWS C3.5, provides additional information on induction brazing.[6]

Figure 12.6 provides a schematic view of a joint in stainless steel tubing, illustrating the placement of the joint in the induction coil.

6. American Welding Society (AWS), Committee on Brazing and Soldering, *Specification for Induction Brazing*, ANSI/C3.5. Miami: American Welding Society.

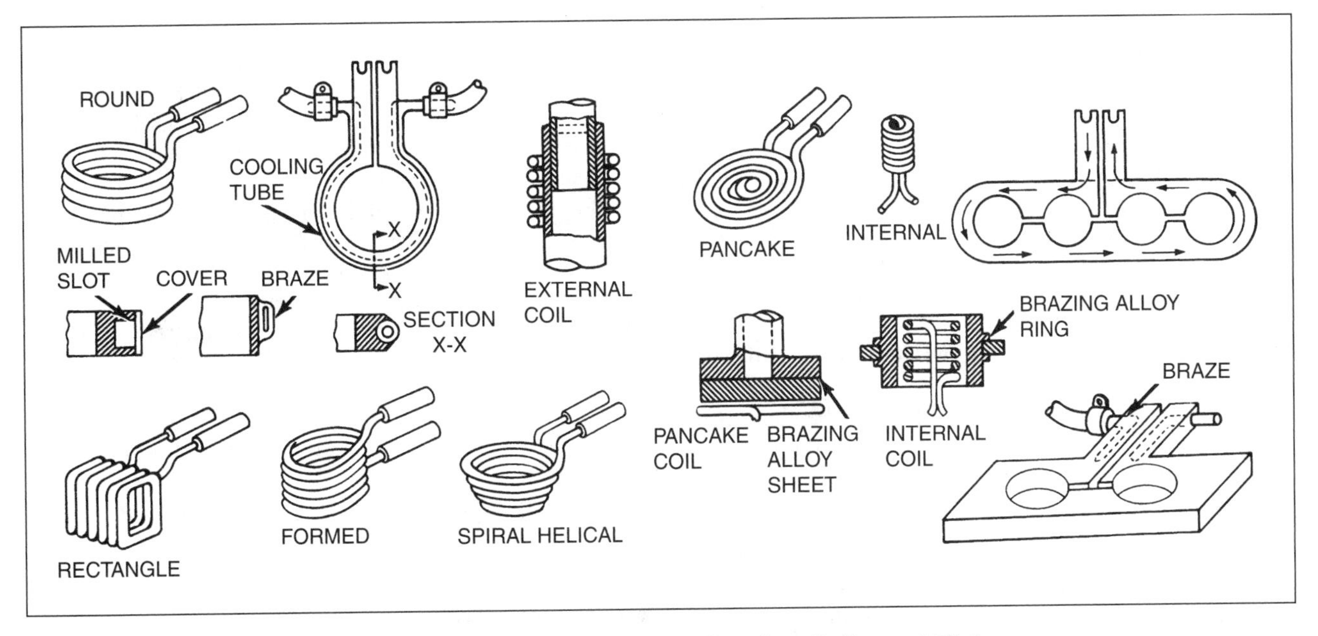

Figure 12.7—Typical Induction Brazing Coils and Plates

The brazing filler metal is preplaced. Careful design of the joint and the coil setup are necessary to assure that the surfaces of all members of the joint reach the brazing temperature at the same time. Flux is employed except when a protective atmosphere is specifically introduced to perform the same function as the flux.

Frequencies for induction brazing generally vary from 10 kilohertz (kHz) to 450 kHz. The lower frequencies are obtained with solid-state generators and the higher frequencies with vacuum-tube oscillators. Induction generators are manufactured in output sizes from one kilowatt (kW) to several hundred kilowatts. Various induction brazing coil designs are illustrated in Figure 12.7. To heat the assembly to the brazing temperature, one generator with a transfer switch can be used to energize several individual workstations in sequence, or assemblies in holding fixtures can be indexed or continuously processed through by means of a conveyor coil.

Induction brazing is used when very rapid heating is required. Processing time is usually in the range of a couple of seconds to a few minutes when large numbers of assemblies are handled automatically. Induction brazing has been used to produce an extensive array of consumer and industrial products, structural assemblies, electrical and electronic products, mining equipment, machine and hand tools, military and ordnance equipment, and aerospace assemblies.

Assemblies can be induction brazed in a controlled atmosphere by placing the components and coil in a nonmetallic chamber, or by placing the chamber and assemblies inside the coil. The chamber can be quartz, tempered glass, or other suitable material such as a high-temperature silica glass.

RESISTANCE BRAZING

Resistance brazing (RB) is a process in which the heat necessary for brazing is obtained from the resistance to the flow of an electric current through electrodes and the joint to be brazed. The components of the joint become part of the electric circuit. The equipment consists of tongs or clamps with the electrodes attached at the end of each arm. The arms are current-carrying conductors with leads attached to a transformer. The tongs or clamps should preferably be water-cooled to avoid overheating. A spot welding machine can be adapted for resistance brazing.

The assembly is held between two electrodes, and the proper pressure and current are applied. The pressure should be maintained until the joint has solidified. In some cases, both electrodes can be located on the same side of the joint using a suitable backing to maintain the required pressure. Fluxing is done with due attention to the conductivity of the fluxes. (Most fluxes act as insulators when dry.) Flux is employed except when an atmosphere is specifically introduced to perform the same function.

Brazing filler metal in the form of preplaced wire, shims, washers, rings, powder, or paste is used. In some instances, face-feeding is possible. For copper and copper alloys, the copper-phosphorus filler metals are most satisfactory because they are self-fluxing. Silver-based filler metals may be used, but a flux or atmosphere is necessary. A wet flux is usually applied as a very thin mixture immediately before the assembly is placed in the brazing fixture. Dry fluxes are not used because they function as insulators and will not permit sufficient current to flow.

The components of the assembly must be clean prior to brazing. The components, brazing filler metal, and flux are assembled and placed in a fixture. Pressure is then applied. As current flows, the electrodes become heated, frequently to incandescence, and the flux and filler metal melt and flow. The current should be adjusted to obtain uniform rapid heating in the assemblies. Too much current risks overheating and the oxidizing or melting of the assembly and causes the electrodes to deteriorate. Too little current lengthens the time of brazing. Experimenting with electrode compositions, geometry, and voltage will determine the best combination of rapid heating with reasonable electrode life.

Quenching the brazement from an elevated temperature will help flux removal. The assembly must first cool sufficiently to permit the braze to hold the components together. When brazing insulated conductors, it may be advisable to quench the brazement rapidly while it is still in the electrodes to prevent overheating of the adjacent insulation. The use of water-cooled clamps can prevent damage to the insulation.

Resistance brazing is most successfully applied to joints with a relatively simple configuration. It is difficult to obtain uniform current distribution, and therefore uniform heating, if the area to be brazed is large or discontinuous or is much longer in one dimension. Components to be resistance brazed should be designed so that pressure can be applied to them without causing distortion at brazing temperatures. When possible, the components should be designed to be self-nesting, which eliminates the need for dimensional features in the fixtures. The components should also be free to move as the filler metal melts and flows in the joint.

One common source of current for resistance brazing is a step-down transformer power source with a secondary circuit that can furnish sufficient current at low voltage (2 to 25 volts [V]). The current ranges from about 50 amperes (A) for small, delicate jobs to thousands of amperes for larger jobs. A varied selection of commercial equipment is available for resistance brazing.

Electrodes for resistance brazing are made of high-resistance electrical conductors such as carbon or graphite blocks, tungsten or molybdenum rods, or in some instances, even steel. The heat for brazing is mainly generated in the electrodes and flows into the work by conduction. It is usually unsatisfactory to attempt to use the resistance of the brazing assembly alone as a source of heat.

The pressure applied by a spot welding machine, clamps, pliers, or other means must be sufficient to maintain good electrical contact and to hold the pieces firmly together as the filler metal melts. The pressure must be maintained during the time of current flow and after the current is shut off until the joint solidifies. The time of current flow will vary from about one second for small, delicate work to several minutes for larger assemblies. This time is usually controlled manually by the operator, who determines when brazing has occurred by observing the temperature and the extent of filler metal flow.

DIP BRAZING

Dip brazing (DB) is a process that uses heat from a molten chemical or a molten metal bath to bring the assemblies to the brazing temperature. When a chemical bath is used, the bath may function as a flux; when molten metal is used, the bath provides the filler metal.

Molten Metal Bath Method

This method is usually limited to the brazing of very small assemblies, such as wire connections or metal strips. A crucible, usually made of graphite, is heated externally to the required temperature to maintain the brazing filler metal in fluid form. A cover of flux is maintained over the molten filler metal. The size of the crucible and the heating method of the molten bath must assure that the immersion of the assembly will not lower the bath temperature below the brazing temperature. The assembly should be clean and protected with flux prior to introduction into the bath. The ends of the wires or other components forming the joint must be held firmly together when they are removed from the bath until the brazing filler metal has fully solidified.

Molten Chemical (Flux) Bath Method

This brazing method requires either a metal or ceramic container for the flux and a method of heating the flux to the brazing temperature. Heat can be applied externally with a torch or internally with an electrical resistance heating unit. A third method involves the electric resistance heating of the flux itself. In this case, external heating must initially melt the flux. Suitable controls are necessary to maintain the flux within the brazing temperature range. The bath

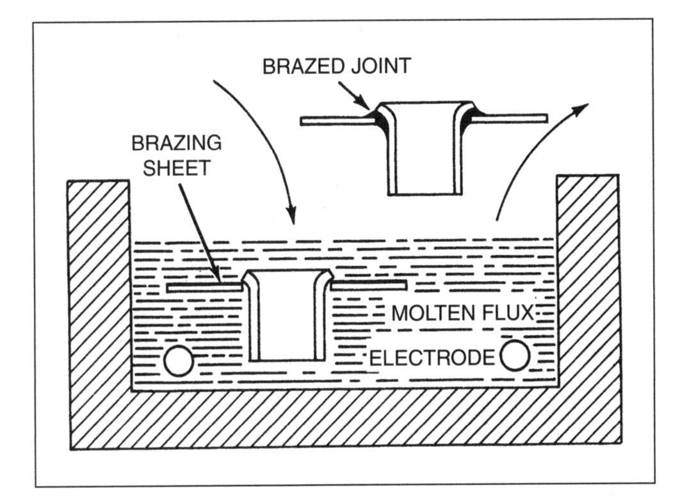

Figure 12.8—Illustration of Chemical Bath Dip Brazing

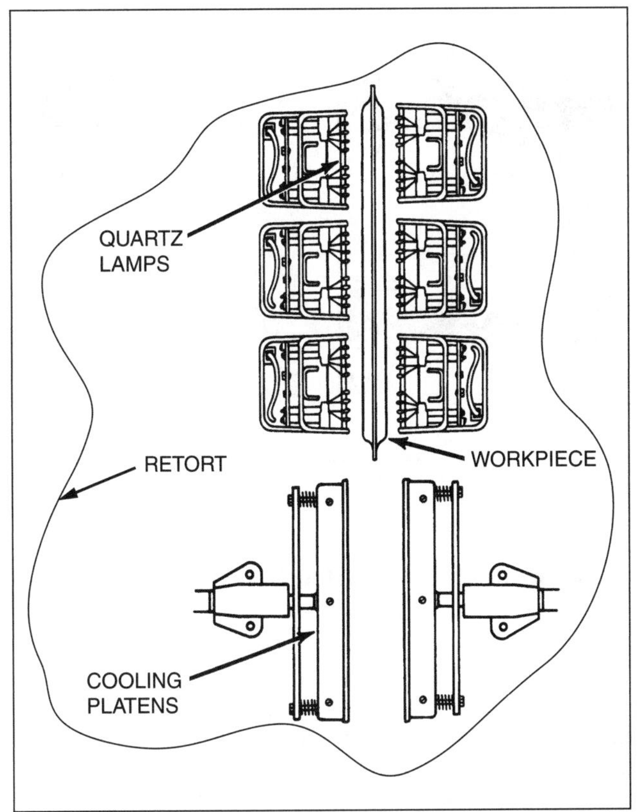

Figure 12.9—Infrared Brazing Apparatus, Usually Employed in an Argon Atmosphere Below Atmospheric Pressure

must be of an appropriate size to assure that immersion of components for brazing will not cool the flux below the brazing temperature. Chemical bath dip brazing is illustrated in Figure 12.8.

Workpieces should be cleaned, assembled, and preferably held in fixtures prior to immersion into the bath. Brazing filler metal is preplaced as rings, washers, slugs, paste, or as a cladding on the base metal. Preheat may be necessary to assure dryness of the assemblies to prevent explosion when placed in the molten flux. Preheat also prevents the freezing of flux on the components of the assembly, which may cause the selective melting of the flux and brazing filler metal. Preheat temperatures are usually close to the melting temperature of the flux. A certain amount of flux adheres to the brazement after brazing. Molten flux must be drained off while the brazements are hot. Flux remaining on cold brazements must be removed by washing with water or by chemical means.

INFRARED BRAZING

Infrared brazing (IRB) uses heat from radiation to bring the assembly to the brazing temperature. This process could be considered a form of furnace brazing, with heat supplied by long-wavelength light radiation. Heating is obtained by invisible radiation from high-intensity quartz lamps capable of delivering up to 5000 watts (W) of radiant energy. Heat input varies inversely as the square of the distance from the source to the assembly to be heated, but the lamps are not usually shaped to follow the contour of the assembly. Concentrating reflectors are used to focus the radiation on the assembly.

For vacuum brazing with vacuum or inert gas protection, the assembly and the lamps are placed in a bell jar or retort that can be evacuated or filled with inert gas. The assembly is then heated to a controlled temperature, as indicated by thermocouples. Figure 12.9 is a schematic illustration of an infrared brazing arrangement, which usually employs an argon atmosphere below atmospheric pressure. The brazement is moved to the cooling platens after brazing.

DIFFUSION BRAZING

Diffusion brazing (DFB) is a furnace brazing process. Coalescence is produced by heating the components to the brazing temperature and by using a filler metal or an *in situ* liquid phase. The filler metal is distributed by capillary action, or it can be preplaced or formed at the faying surfaces. The base metals and filler metals must

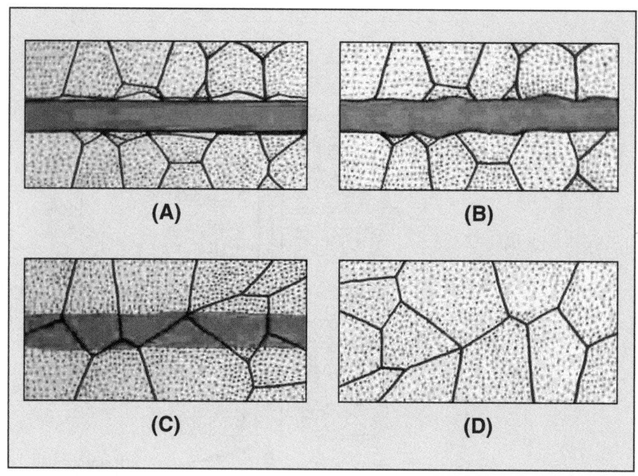

Source: Adapted from American Welding Society (AWS) Committee on Brazing and Soldering, Jenney, C.L., ed., (forthcoming), *Brazing Handbook*, 5th Ed., Miami: American Welding Society, Figure 17.1.

Figure 12.10—Diffusion Brazing Sequence

have high mutual solubility. At the appropriate temperature and with sufficient time in the furnace, the filler metal diffuses with the base metal and isothermally solidifies during the brazing cycle. A substantial change takes place in the microstructure of the joint: when there is considerable interdiffusion between base and filler metals, the properties of the brazed joint change and become very similar to those of the base metal. Diffusion brazing can result in joints with increased strength and ductility, and with a significantly increased joint remelt temperature.

Figure 12.10 is metallographic illustration of the diffusion brazing sequence. In this case, in (A), a filler metal containing boron is either preplaced in the joint or it flows into the joint by capillary action. In (B) brazing begins as the filler metal is distributed in the joint by capillary action. In Figure 12.10(C), as diffusion time or temperature, or both, are increased, boron from the filler metal diffuses into the base metal. Full diffusion occurs, as shown in Figure 12.10(D), when the grains of the base metal and filler metal interdiffuse and the base metal grains align across the joint. When this grain modification is complete, the resulting joint is the equivalent of a solid-state joint and the joint area is difficult to distinguish from the base metal.[7]

Base Metal-Filler Metal Combinations

Diffusion brazing can be used to join any metal or material if the brazing filler metal is mutually soluble with the base metal. The following base metal-filler metal combinations are commonly used in diffusion brazing:

1. Copper base metal brazed with a silver filler metal,
2. Low-carbon steels and low-alloy steels brazed with copper filler metal,
3. AISI 410 stainless steel brazed with copper or nickel filler metals,
4. AISI 300-series stainless steel brazed with copper or nickel filler metals, and
5. Nickel and nickel-alloy base metals brazed with nickel filler metals.

Nickel base metals are highly soluble with the many commercially available nickel filler metals. Carbon steels and low-alloy steels can be brazed with copper filler metals. Although copper has very limited solubility in iron at the usual brazing temperature of 1121°C (2050°F), in a press-fitted clearance with sufficient roughness to allow the copper to fill the joint, the copper diffuses into the iron and the iron back-diffuses into the remaining copper, thus creating a strengthened joint. When the correct diffusion time and temperature are applied, the joint seemingly disappears and the brazement appears as a single unit.

Aluminum base metals and aluminum filler metals are mutually soluble to a high level, but the high brazing temperature used with aluminum and the strength of the filler metal make it unnecessary to use full diffusion as a means to increase the strength of the joint or to raise the melting point of the joint.

The American Welding Society publication, *Specification for Filler Metals for Brazing and Braze Welding,* ANSI/AWS A5.8, lists the most commonly used filler metals for brazing.[8] Some of the filler metals listed may be used for diffusion brazing, depending on the base metals involved and the requirements of the specific application.

VARIABLES

The major variables for the diffusion brazing of an assembly are brazing temperature, holding time at brazing temperature, quantity of filler metal required in the joint clearance, and the level of mutual solubility of the base metals and the filler metal. The rate of diffusion is directly proportional to the holding time, quantity of filler metal, and level of solubility. The diffusion rate is

7. See Reference 4, Chapter 17.

8. American Welding Society (AWS) Committee on Filler Metals and Allied Materials, 1992 (Reaffirmed 2003) *Specification for Filler Metals for Brazing and Braze Welding*, ANSI/AWS A5.8, Miami: American Welding Society.

inversely proportional to the joint clearance and the quantity of filler metal used.

The desired diffusion can be established and controlled by maintaining the assembly at a temperature near the liquidus temperature of the filler metal for an extended time, and also by maintaining a much higher temperature for a much shorter time.

Desired changes in joint properties of the brazement can be achieved by varying the brazing procedure and filler metal selection for the base metal used in the brazement. The diffusion portion of the cycle is usually accomplished at the brazing temperature. However, brazing at a high temperature and diffusing near the solidus temperature or below can also be successful. This particularly applies when using filler metal BNi-5 with a solidus of approximately 1079°C (1975°F), a liquidus of 1135°C (2075°F), and a brazing range of 1149°C to 1204°C (2100°F to 2200°F).

Diffusion Brazing Equipment

The batch furnace is the most suitable type for diffusion brazing because the heating rate, hold times during heating, length of brazing time, cooling cycle, and hold times during cooling can be controlled. The vacuum furnace and the hydrogen retort furnace are well suited for the diffusion brazing of high-temperature heat- and corrosion-resistant base metals.

The continuous-belt furnace is not appropriate for most diffusion brazing cycles, although this type of furnace can be used for some brazements if the belt speed is lowered and the temperature of the furnace is increased. For example, carbon steels and alloy steels can be diffusion brazed using a copper filler metal. The joint should be a press-fit joint with a reasonably smooth surface.

Less sophisticated equipment and atmospheres can be used for the diffusion brazing of copper base metals using a silver filler metal. For example, a very simple retort furnace can be used to diffusion braze copper base metal using the ANSI/AWS designated filler metal BAg-8 in a high dew point atmosphere.

Torch and induction brazing equipment is not used for diffusion brazing, as it does not allow sufficient time for diffusion to take place.

OTHER PROCESSES

The brazing methods included in this section are not common, but are they important in some special situations. They vary according to the specific need and application.

Blanket Brazing

Blanket brazing uses a blanket made of insulating material that is resistance-heated; the heat is transferred to the components by conduction and radiation, but mostly by radiation.

Block Brazing

Block brazing (BB) is a process that uses heat from heated metal or graphite blocks applied to the joint. This technique is obsolete or seldom used.

Exothermic Brazing

Exothermic brazing (EXB) is a special process that heats commercial filler metal by means of a solid-state exothermic chemical reaction between two metals, a metal oxide, and a metal or an inorganic nonmetal. An exothermic chemical reaction generates heat released as the free energy of the reactants. Nature has provided countless numbers of solid-state or nearly solid-state metal-metal oxide reactions that are suitable for use in exothermic brazing units. In special circumstances, exothermic brazing has been used in microelectronic packaging, petrochemical equipment, and aerospace components.

Exothermic brazing uses simplified tooling and equipment. The reaction heat brings adjoining metal interfaces to a temperature at which preplaced brazing filler metal melts and wets the base metal interface surfaces. Several commercially available brazing filler metals have suitable flow temperatures. The process is limited only by the thickness of the base metal and the effect of brazing heat, or any previous heat treatment, on the metal properties.

Twin Carbon Arc Brazing

Twin carbon arc brazing (TCAB) uses the heat from an arc between two carbon electrodes to bring the assembly to the brazing temperature. It is an obsolete or little-used process. Artists sometimes choose this process for special effects.

AUTOMATION

The important variables involved in brazing are the temperature, time at temperature, filler metal, and brazing atmosphere. Other variables are joint fit-up, amount of filler metal, and the rate and mode of heating. All of these features can be automated and applied to the commonly used brazing processes, specifically torch brazing, furnace brazing (vacuum and atmosphere), resistance brazing, induction brazing, dip brazing, and infrared brazing.

Generally, the amount of heat supplied to the joint is automated by controlling the temperature and time at temperature. Filler metal and flux can be preplaced at the joints during assembly of the components or automatically fed into the joints while at the brazing temperature. Further automation may include in-line inspection and cleaning (flux removal), simultaneous brazing of multiple joints in an assembly, and continuous brazing operations. Figure 12.11 shows an automatic brazing machine with a multiple gas-air torch heating system capable joining 100 to 200 assemblies per hour.

In general, the more automated a process becomes, the more rigorous its economic justification must be. Usually the high cost of automation is justified by increased productivity. In the case of brazing, further justification may well be found in the energy saved with efficient joint heating.

Photograph courtesy of Fusion Incorporated

Figure 12.11—Rotary Index Brazing Machine with Multiple-Torch Heating System

Brazing processes that can accommodate robotic manipulation and meet standard processing requirements are the most suitable for automation. Torch brazing is routinely automated to fabricate tubular joints.

Automated brazing offers the following major advantages:

1. High production rates,
2. Consistency of results,
3. Filler metal savings,
4. High productivity per worker,
5. Energy savings, and
6. Adaptability and flexibility.

A continuous-belt furnace increases production but eliminates the possibility of in-line inspection and lowers energy efficiency because the entire assembly is heated.

Automatic brazing machines improve torch brazing. Typically, heat is directed only to the joint area by one or more torches. An example is shown in Figure 12.12, which shows a multiple-torch heating system used to bring silver-alloy brazing paste to brazing temperature to braze a three-piece steel assembly. Similar effects can be obtained by induction heating. A typical machine has provisions for assembly and fixturing, automatic fluxing, preheating (if needed), brazing, air- or water-quenching, removal of the brazement, and inspection.

The rotary index brazing machine shown in Figure 12.13 is an example of automated brazing. The machine is capable of brazing from 300 to 800 assemblies per hour.

MATERIALS

The materials used in brazing—the base metals or materials, filler metals, atmospheres, and fluxes—are interdependent and should be selected on the basis of their ability to form brazed joints with the mechanical and physical characteristics required for the intended service application.

FILLER METALS

Brazing filler metals must have the following properties:

Photograph courtesy of Fusion Incorporated

Figure 12.12—Multiple Gas/Air Torch Heating Pattern Brings Silver Brazing Paste Alloy to Brazing Temperature on a Three-Piece Steel Assembly

Photograph courtesy of Fusion Incorporated

Figure 12.13—Rotary Index Brazing Machine Joins 300–800 Assemblies per Hour

1. A melting point or melting range compatible with the base metals being joined, and sufficient fluidity at brazing temperature to flow and distribute the melt into properly prepared joints by capillary action;
2. A composition of sufficient homogeneity and stability to minimize liquation (the separation of constituents during brazing);
3. The ability to wet surfaces of base metals and form a strong, sound bond; and
4. The ability to produce or avoid filler-metal interactions with the base metal, depending on the requirements. Filler metals are manufactured in innumerable forms and shapes to accommodate many diverse applications. A few examples are shown in Figure 12.14.

Melting and Fluidity

Pure metals melt at a single temperature and are generally very fluid. Binary compositions have differing characteristics, depending on the relative contents of the two metals. Figure 12.15 is the equilibrium diagram for the silver-copper binary system. The solidus line, ADCEB, traces the temperature of the alloys at the start of melting. The liquidus line, ACB, shows the temperatures at which the alloys become completely liquid. At point C the two lines meet (72% silver-28% copper), indicating that this particular alloy melts at this fixed temperature (or eutectic temperature). This alloy is the eutectic composition; it is as fluid as a pure metal, whereas the other alloy combinations are mushy (typical of liquation, discussed in the next section) between their solidus and liquidus temperatures. The wider the temperature spread is, the more sluggishly the alloys will flow in a capillary joint.

The α region is a solid solution of copper in silver; the β region is a solid solution of silver in copper. The central solid zone consists of an intimate mixture of α and β solid solutions. Above the liquidus line, the silver and copper atoms are thoroughly interspersed as a liquid solution.

Liquation

Liquation—the separation of the filler metal constituents during brazing—can be caused by partial melting of the constituents. Because the solid and liquid alloy phases of a brazing filler metal generally differ, the composition of the melt gradually changes as the temperature increases from the solidus to the liquidus. If the portion that melts first is allowed to flow out, the

Photographs courtesy of Lucas-Milhaupt, Inc.

Figure 12.14—Assortments of Filler Metals in Various Shapes and Forms

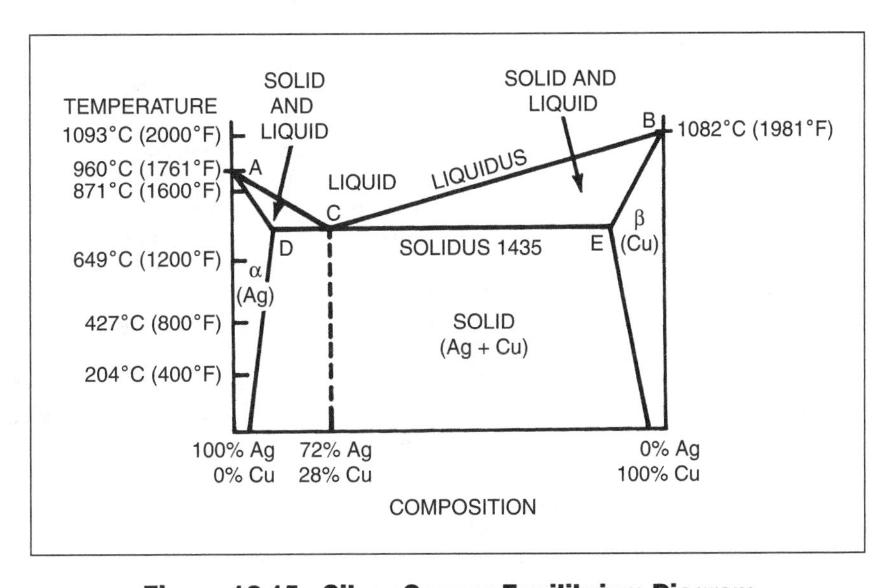

Figure 12.15—Silver-Copper Equilibrium Diagram

remaining solid may not melt and therefore may remain behind as a residue, or *skull*. Filler metals with narrow melting ranges tend not to separate, so they flow quite freely into joints with extremely narrow clearances. Filler metals with wide melting ranges need rapid heating to minimize separation. Heating should be accomplished quickly or the application of the filler metal to the joint should be delayed until the base metal reaches the brazing temperature. Filler metals subject to liquation have a sluggish flow, require wide joint clearances, and form large fillets at joint extremities.

Wetting and Bonding

The term *wetting* describes the phenomenon that takes place when a liquid filler metal spreads and adheres to the base metal in a thin continuous layer. This action promotes capillary flow and enhances the effectiveness of the brazing filler metal.

The function of the filler metal is to alloy with the surface of the base metal without causing detrimental effects such as the following:

1. Undesirable diffusion into the base metal,

2. Dilution with the base metal,
3. Base metal erosion, and
4. Formation of brittle compounds.

The first three of these, which affect the base metal, depend on the mutual solubility between the brazing filler metal and the base metal, the amount of brazing filler metal present, and the temperature and duration of the brazing cycle.

Some filler metals diffuse excessively, which changes the base metal diffusion and dilution characteristics. To control diffusion, the brazing operator should select a suitable filler metal, apply the minimum quantity of filler metal, and follow the appropriate brazing cycle. The more efficiently the filler metal wets the base metal, the better the capillary flow will be. In long capillaries between the metal surfaces, mutual solubility may change the filler metal composition by alloying. This will usually raise the liquidus temperature and cause it to solidify before completely filling the joint.

Base metal erosion occurs if the base metal and the brazing filler metal are mutually soluble. Sometimes such alloying produces brittle intermetallic compounds that reduce the joint ductility.

Compositions of brazing filler metals are adjusted to control these factors and to provide desirable characteristics, such as corrosion resistance in specific media, favorable brazing temperatures, or material economies. Thus, to overcome the limited wettability of the silver-copper alloys when brazing iron and steel, the composition of these filler metals should include either zinc or cadmium, and tin or nickel to lower the liquidus and solidus temperatures. Tin is added in place of zinc or cadmium if these constituents with high vapor pressures are objectionable. (See the "Safe Practices" section in this chapter.)

Similarly, silicon is used to lower the liquidus and solidus temperatures of aluminum and nickel-based brazing filler metals. Other brazing filler metals contain elements such as lithium, phosphorus, or boron, which reduce surface oxides on base metals and form compounds with melting temperatures below the brazing temperature. These molten oxides then flow out of the joint, leaving a clean metal surface for brazing. These filler metals are essentially self-fluxing. Boron and phosphorus are used to lower the melting temperature of nickel filler metals.

Filler Metal Selection

Four factors that must be considered when selecting a brazing filler metal are compatibility with the base metal and joint design, service requirements of the brazement, brazing temperature, and heating method.

Compatibility with Base Metal and Joint Design. The brazed product and its intended service are the first considerations in the selection of base metals and joint design; then a compatible filler metal can be selected. Compatible base metals and filler metals are listed in Table 12.3. The American Welding Society publication *Specification for Filler Metals for Brazing and Braze Welding,* ANSI/AWS A5.8 can be consulted.[9] Additional information is published in *Standard Method for Evaluating the Strength of Brazed Joints,* AWS C3.2M/AWS C3.2[10] and *Recommended Practices for Design, Manufacture, and Inspection of Critical Brazed Components,* ANSI/AWS C3.3.[11]

Service Requirements. Filler-metal compositions must be selected to accommodate the operating requirements of the application, such as the temperature in service, thermal cycling, life expectancy, stress loading, corrosive conditions, radiation stability, and vacuum operation. Service environments may range from extreme high-temperature environments such as 560°C to 654°C (1900°F to 2200°F) used for jet engine brazements and 1504°C (3500°F) used in X-ray tube applications to the very low temperatures related to cryogenic applications.

Brazing Temperature. The filler metal selected must also be compatible with the brazing temperature, the protective environment, and the method of heating required. Low brazing temperatures are usually preferred to economize on energy, minimize the effects of the heat on the base metal (annealing and grain growth), minimize base metal-filler metal interaction, and increase the life of fixtures and other tools.

The reasons for using high brazing temperatures include taking advantage of a more economical brazing filler metal with a higher melting point; to accomplish the annealing, stress relief, or heat treatment of the base metal during brazing; to permit subsequent processing at elevated temperatures; to promote base metal-filler metal interactions to increase the joint remelt temperature; or to promote the removal of certain refractory oxides by a vacuum or gas-atmosphere.

Heating Method. Any heating method can be used with filler metals that have narrow melting ranges—less than 28°C (50°F) between the solidus and liquidus. The brazing filler metal can be preplaced in the joint area in

9. See Reference 8.
10. American Welding Society (AWS) Committee on Soldering and Brazing, 2002, *Standard Method for Evaluating the Strength of Brazed Joints*, AWS C3.2M/AWS C3.2:2002, Miami: American Welding Society.
11. American Welding Society (AWS) Committee on Soldering and Brazing, 2002, *Recommended Practices for Design, Manufacture, and Inspection of Critical Brazed Components*, AWS C3.3:2002, Miami: American Welding Society.

Table 12.3
Base Metal–Filler Metal Combinations

Base Metals	Al & Al Alloys	Mg & Mg Alloys	Cu & Cu Alloys	Carbon & Low-Alloy Steels	Cast Irons	Stainless Steels	Ni & Ni Alloys	Ti & Ti Alloys	Be, Zr, & Alloys (Reactive Metals)	W, Mo, Ta, Nb & Alloys (Refractory Metals)	Tool Steels
Al & Al alloys	BAlSi										
Mg & Mg alloys	X	BMg									
Cu & Cu alloys	BAlSi	X	BAg, BAu, BCuP, RBCuZn, BNi								
Carbon & low-alloy steels	BAlSi**	X	BAg, BAu, RBCuZn, BNi	BAg, BAu, BCu, RBCuZN, BNi							
Cast irons	X	X	BAg, BAu, RBCuZn, BNi	BAg, RBCuZn, BNi	BAg, RBCuZN, BNi						
Stainless steels	BAlSi**	X	BAg, BAu, BNi	BAg, BAu, BCu, BNi, RBCuZn	BAg, BAu, BCu, BNi, RBCuZn	BAg, BAu, BCu, BNi					
Ni & Ni alloys	BAlSi	X	BAg, BAu, RBCuZn, BNi	BAg, BAu, BCu, RBCuZn, BNi	BAg, BCu, RBCuZn, BNi	BAg, BAu, BCu, BNi	BAg, BAu, BCu, BNi				
Ti & Ti alloys	BAlSi	X	BAg	BAg	BAg	BAg	BAg	BAg, BAlSi			
Be, Zi & alloys (reactive metals)	Y	X	BAg	BAg	BAg	BAg	BAg	BAg	Y		
W, Mo, Ta, Nb, & alloys (refractory metals)	X	X	BAg	BAg, BCu, BNi,* BAu	BAg, BCu, BNi*	BAg, BCu, BNi,* BAu	BAg, BCu, BNi,* BAu	Y	Y	BCu, BAg, BNi, BAu	
Tool steels	X	X	BAg, BAu, RBCuZn, BNi	BAg, BAu, BCu, RBCuZn, BNi	BAg, BAu, RBCuZn, BNi, BCu	BAg, BAu, BCu, BNi	BAg, BAu, BCu, RBCuZn, BNi	X	X	BCu, BAg, BNi, BAu	BAg, BAu, BCu, RBCuZn, BNi

Notes:
Refer to ANSI/AWS A5.8, *Specification for Filler Metals for Brazing*, for information on the specific compositions within each classification (See Reference 8).
X—Not recommended; however, special techniques may be practicable for certain dissimilar metal combinations.
Y—Generalizations on these combinations cannot be made. Refer to American Welding Society (AWS) Committee on Brazing and Soldering, (forthcoming), *Brazing Handbook*, 5th ed., Miami: American Welding Society, for usable filler metals.
*Special brazing filler metals are available and are used successfully for specific metal combinations.
**Recommended only if the nonaluminum is nickel plated first.

Key:
BAlSi—Aluminum
BAg—Silver base
BAu—Gold base
BCu—Copper base
BCuP—Copper phosphorus
RBCuZn—Copper zinc
BMg—Magnesium base
BNi—Nickel base

the form of rings, washers, formed wires, shims, powder, or paste. In some applications, the filler metals can be manually or automatically face-fed into the joint after the base metal is heated.

Filler metals that tend to liquate should be used with heating methods that bring the joint to brazing temperature quickly, or the brazing filler metal should be introduced after the base metal reaches the brazing temperature. Heating the workpiece too rapidly should be carefully avoided, or high thermal stresses, misalignment, or warping could occur.

FILLER METAL CLASSIFICATIONS

To simplify filler metal selection, the American Welding Society published the *Specification for Filler Metals for Brazing and Braze Welding,* ANSI/AWS A5.8,[12] which divides filler metals into seven categories and into various classifications within each category. The specification lists filler metals that are commonly used and are commercially available. Other brazing filler metals not currently covered by the specification are available for special applications. Suggested base metal and filler metal combinations are listed in Table 12.3. Commonly used filler metals are described below.

Aluminum-Silicon Filler Metals

Aluminum-silicon filler metals are used for joining aluminum grades 1060, 1100, 1350, 3003, 3004, 3005, 5005, 5050, 6053, 6061, 6951, and cast alloys A712.0 and C711.0. While all types of aluminum-silicon filler metals are suitable for furnace and dip brazing, some types are also suitable for torch brazing using lap joints rather than butt joints.

Brazing sheet or tubing is a convenient source of aluminum filler metal. It consists of a core of aluminum alloy and a coating of lower-melting filler metal. The coatings are aluminum-silicon alloys applied to one or both sides of the sheet. Brazing sheet is frequently used as one member of an assembly with the other member made of an unclad brazeable alloy. The coating on the brazing sheet or tubing melts at the brazing temperature and flows by capillary action and gravity to fill the joints.

Magnesium Filler Metals

Magnesium filler metal (BMg-1) is used to join AZ10A, K1A, and M1A magnesium alloys by the torch brazing, dip brazing, or furnace brazing processes. Heating must be closely controlled to prevent the melting of the base metal. Joint clearances of 0.10 to 0.25 mm (0.004 to 0.010 in.) are best for most applications. Corrosion resistance is good if the flux is completely removed after brazing. Brazed assemblies are generally suitable for continuous service up to 120°C (250°F) or intermittent service to 150°C (300°F), subject to the usual limitations of the actual operating environment.

Copper and Copper-Zinc Filler Metals

Copper and copper-zinc brazing filler metals are used to join ferrous metals and nonferrous metals. The corrosion resistance of the copper-zinc alloy filler metals is generally inadequate for joining copper, silicon bronze, copper-nickel alloys, or stainless steel.

The essentially pure copper brazing filler metals are used to join ferrous metals, nickel-base alloys, and copper-nickel alloys. Figure 12.16 shows aircraft components that were copper brazed in a continuous mesh belt hydrogen furnace. The component on the left is stainless steel and the one on the right is stainless steel brazed to carbon steel using copper filler metal.

Figure 12.17 shows low-carbon steel industrial gears in a mesh basket. Copper paste filler metal was applied to the joint areas to prepare for brazing in a continuous mesh belt hydrogen furnace. After brazing, the components will be carburized during the continuous belt furnace process.

The copper brazing filler metals are free flowing and are often used in furnace brazing with a combusted gas, hydrogen, or a dissociated ammonia atmosphere without flux. Copper filler metals are available in wrought and powder forms. One specialized copper filler metal is a copper oxide designed to be suspended in an organic vehicle.

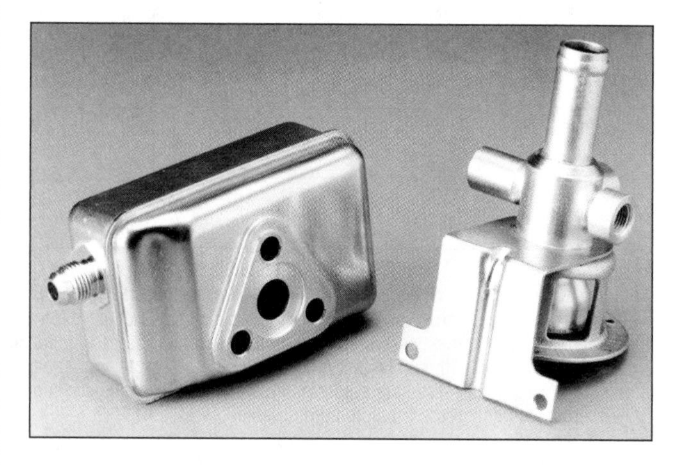

Figure 12.16—Aircraft Components Brazed with Copper Filler Metal: Stainless Steel (Left) and Stainless Steel Brazed to Carbon Steel (Right)

12. See Reference 8.

Figure 12.17—Copper Paste Filler Metal Applied to Low-Carbon Steel Industrial Gears for Brazing and Carburizing During a Continuous Belt Furnace Operation

Copper-zinc filler metals are used on steel, copper, copper alloys, nickel and nickel-base alloys, and stainless steel for applications that do not require corrosion resistance. They are used with torch, furnace, and induction brazing processes. Fluxing is required; a borax-boric acid flux is commonly used.

Copper-Phosphorus Filler Metals

Copper-phosphorus filler metals are primarily used to join copper and higher-melting copper alloys. Highly stressed copper alloys may be susceptible to cracking during brazing unless stress-relieved prior to the flow of the filler metals. They have some limited use for joining silver, tungsten, and molybdenum. They should not be used on ferrous or nickel-base alloys, or on copper-nickel alloys with more than 10% nickel. These filler metals are suited for all brazing processes and have self-fluxing properties when used on copper. They tend to liquate if heated slowly. These filler metals are widely used in the manufacturing of refrigeration heat exchangers, using torch brazing in air without flux.

Silver Filler Metals

Silver filler metals are used to join most ferrous and nonferrous metals, except aluminum and magnesium, and are compatible with all methods of heating. Silver filler metals can be preplaced in the joint or fed into the joint area after heating.

Silver-copper alloys high in silver do not wet steel well when brazing is done in air with a flux. Copper will alloy with cobalt and nickel much more readily than silver. Thus, copper wets many of these metals and their alloys satisfactorily, whereas silver does not. When brazing in certain protective atmospheres without flux, silver-copper alloys will wet and flow freely on most steels at the proper temperature.

Zinc is commonly used to lower the melting and flowing temperatures of silver-copper alloys. It is by far the most helpful wetting agent when joining alloys based on iron, cobalt, or nickel. Used alone or in combination with cadmium or tin, zinc produces alloys that wet the iron-group metals but do not alloy with them to any appreciable depth.

Cadmium is incorporated in some silver-copper-zinc filler metal alloys to further lower the melting and flow temperatures and to increase the fluidity and wetting action on a variety of base metals. Since cadmium oxide fumes are a health hazard, cadmium-bearing filler metals should be used with caution. See the "Safe Practices" section in this chapter.

Tin has a low vapor pressure at normal brazing temperatures. Tin is added to silver brazing filler metals in place of zinc or cadmium when volatile constituents are objectionable, such as when brazing is done without flux in atmosphere furnaces or vacuum furnaces, or when the brazed assemblies will be used in high vacuum at elevated temperatures. Silver-copper filler metals with tin additions have wide melting ranges. The wetting action of fillers containing zinc is more effective for ferrous metals than those containing tin, and when zinc is tolerable, they are preferred over fillers with tin.

Stellite®, cemented carbides, and molybdenum-rich and tungsten-rich refractory alloys are brazed with filler metals with added manganese, nickel, and, infrequently, cobalt to increase wettability.

When stainless steels and base metals that form refractory oxides are brazed in reducing or inert atmospheres without flux, silver brazing filler metals that contain lithium as the wetting agent are quite effective. The heat of formation of lithium oxide is such that the lithium metal tends to reduce adherent oxides on the base metal during brazing. The resulting lithium oxide is readily displaced by the liquid brazing filler metal.

Gold Filler Metals

Gold filler metals are used to join components in applications requiring higher melting temperatures or corrosion resistance, or both, and applications where volatile components are undesirable. They are used to braze iron-, nickel-, and cobalt-base metals when the application requires resistance to higher temperatures, oxidation, or corrosion. They are commonly used on thin sections because of their low rate of interaction with the base metal.

Nickel Filler Metals

Nickel filler metals for brazing are generally used on AISI 300- and 400-series stainless steels, nickel- and cobalt-base metals, and even carbon steel, low-alloy steels, and copper when specific properties are desired. They exhibit good corrosion and heat-resistance properties. They are normally applied as powders, pastes, rod, foil, or in the form of sheet or rope with plastic binders.

Nickel filler metals have the very low vapor pressures that vacuum systems require at elevated temperatures.

The filler metals containing phosphorus are low in ductility because they form nickel phosphides; however, ductility is improved by diffusion brazing and maintaining small joint clearances (≤ 0.002 in. [≤ 0.05 mm]). The brazing process must be carefully controlled when brazing thin sections with filler metals containing boron. This is to prevent potentially undesirable properties that arise due to the rapid diffusion of boron into the base metal. Nickel filler metals containing boron are used in diffusion brazing because the boron atom is very small and is extremely mobile, which improves the properties of the brazed joint. However, it can also cause weld cracks if welds are subsequently made near the brazed joint. The presence of boron may cause the weld to become brittle.

Cobalt Filler Metal

Cobalt filler metal is used for its high-temperature properties and its compatibility with cobalt-base metals. Diffusion brazing in a high-purity atmosphere with low partial-pressure oxygen and other contaminants will produce optimum brazing results. Special high-temperature fluxes are available for torch brazing.

Filler Metals for Refractory Metals

Brazing is an excellent joining process for fabricating assemblies of refractory metals, particularly those involving thin sections. However, only a few filler metals have been specifically designed for both high-temperature and high-corrosion applications.

The filler-metal compositions and pure metals used to braze refractory metals are listed in Table 12.4. Low-melting filler metals, such as silver-copper-zinc, copper-phosphorus, and copper, are used to join tungsten for electrical contact applications, but these filler metals cannot operate at high temperatures. The use of less-available filler metals with higher melting points, such as tantalum and niobium, is warranted in those cases.

Nickel-base and precious-metal-base filler metals can also be used to join tungsten. Various brazing filler metals can be used to join molybdenum. The effect of the brazing temperature on base-metal recrystallization must be considered. When brazing above the recrystallization temperature, the brazing time must be kept short. If high-temperature service is not required, copper-base and silver-base filler metals can be used.

Niobium and tantalum can be brazed with a number of filler metals based on refractory or reactive metals. Platinum, palladium, platinum-iridium, platinum-rhodium, titanium, and nickel-base filler metals (such as nickel-chromium-silicon alloys) are typically used for brazing niobium and tantalum. Copper-gold alloys containing gold in amounts between 46% and 90% (considered a gold filler metal) form brittle, age-hardening compounds. Silver-base filler metals are not recommended because they may cause embrittlement of the base metals.

FLUXES AND ATMOSPHERES

Metals and alloys may react with the air atmosphere to which they are exposed, with increased reaction as the temperature is raised. The common reaction is oxidation, but nitrides, silicides, and carbides are sometimes formed. Fluxes and protective atmospheres (including vacuum) are used to prevent undesirable reactions during brazing. Some fluxes and protective atmospheres may also reduce oxides already present. The American Welding Society publication *Specification for Fluxes for Brazing and Braze Welding,* ANSI/AWS A5.31, provides further information.[13] Chapter 4 of the *Brazing Handbook*, 5th edition, provides detailed information on brazing fluxes and atmospheres.[14]

Titanium, zirconium, niobium, and tantalum become permanently embrittled when brazed in any atmosphere containing hydrogen, oxygen, or nitrogen. Hydrogen will cause embrittlement in copper if the copper has not been thoroughly deoxidized.

The use of flux or a protective atmosphere does not eliminate the need to clean the components prior to brazing. Recommended cleaning procedures are discussed in the section "Joint Design and Preparation" in

13. American Welding Society (AWS) Committee on Filler Metals and Allied Materials, *Specification for Fluxes for Brazing and Braze Welding*, ANSI/AWS A5.31, Miami: American Welding Society.
14. See Reference 4, Chapter 4.

Table 12.4
Brazing Filler Metals for Refractory Metals[a]

Brazing Filler Metal	Liquidus Temperature °C	Liquidus Temperature °F	Brazing Filler Metal	Liquidus Temperature °C	Liquidus Temperature °F
Nb	2416	4380	Mn-Ni-Co	ß1021	1870
Ta	2997	5425			
Ag	960	1760	Co-Cr-Si-Ni	1899	3450
Cu	1082	1980	Co-Cr-W-Ni	1427	2600
Ni	1454	2650	Mo-Ru	1899	3450
Ti	1816	3300	Mo-B	1899	3450
Pd-Mo	2860	1571	Cu-Mn	871	1600
Pt-Mo	3225	1774	Cb-Ni	1190	2175
Pt-30W	2299	4170			
Pt-50Rh	2049	3720	Pd-Ag-Mo	1306	2400
			Pd-Al	1177	2150
Ag-Cu-Zn-Cd-Mo	619–701	1145–1295	Pd-Ni	1205	2200
Ag-Cu-Zn-Mo	718–788	1324–1450	Pd-Cu	1205	2200
Ag-Cu-Mo	780	1435	Pd-Ag	1306	2400
Ag-Mn	971	1780	Pd-Fe	1306	2400
			Au-Cu	885	1625
Ni-Cr-B	1066	1950	Au-Ni	949	1740
Ni-Cr-Fe-Si-C-B	1066	1950	Au-Ni-Cr	1038	1900
Ni-Cr-Mo-Mn-Si	1149	2100	Ta-Ti-Zr	2094	3800
Ni-Ti	1288	2350			
Ni-Cr-Mo-Fe-W	1305	2380	Ti-V-Cr-Al	1649	3000
Ni-Cu	1349	2460	Ti-Cr	1481	2700
Ni-Cr-Fe	1427	2600	Ti-Si	1427	2600
Ni-Cr-Si	1121	2050	Ti-Zr-Be[b]	999	1830
			Zr-Cb-Be[b]	1049	1920
			T i-V-Be[b]	1249	2280
			Ta-V-Cb[b]	1816–1927	3300–3500
			Ta-V-Ti[b]	1760–1843	3200–3350

a. Not all of the filler metals listed are commercially available.
b. Depends on the specific composition.

this chapter and also in Chapter 7 of the American Welding Society's *Brazing Handbook*, 5th edition.[15]

The purpose of brazing filler metal is to flow into the capillary and make a joint, but it may also flow over portions of the base metal surface. This may be undesirable from a cosmetic viewpoint, or there may be holes or features on the brazement that must not be filled or plugged so that the function of the device is not impeded. When extraneous flow must be prevented, the brazing operator applies a *stopoff*, a compound that retards the flow of the filler material. Great care must be exercised to prevent the stopoff material from getting into the actual braze joint because this would prevent bonding.

Stopoff materials are generally oxides of titanium, calcium, aluminum, or magnesium. They are applied by brush, tape, spray, or a hypodermic-needle system. These materials work quite well when furnace brazing

15. See Reference 4, Chapter 7.

without flux; however, when flux is used, the cleaning action of the flux may counteract the stopoff effect. After brazing, the stopoff material can be removed by washing with hot water or by chemical or mechanical stripping.

BASE METALS

The effect of brazing on the mechanical properties of the metal in a brazement and the final joint strength must be considered. Base metals strengthened by cold working will be heat-treated by the brazing process temperatures and times, usually in the annealing range of the base metal being processed. Brazed, hot-worked or cold-worked, heat-resistant base metals will normally exhibit only the annealed physical properties after brazing. The brazing cycle, by its very nature, will usually anneal cold-worked base metal unless the brazing temperature is very low and the time at temperature is very short. It is not practical to cold-work the base metal after the brazing operation.

A heat-treatable base metal should be selected when the strength of the brazement after brazing must be above the annealed properties of the base metal. The base metal can be an oil-quench type, an air-quench type that can be brazed and hardened in the same or a separate operation, or a precipitation-hardening type that can be brazed and solution-treated in a combined cycle. Components already hardened can be brazed with a low-temperature filler metal using short times at temperature to maintain the mechanical properties.

Aluminum and Aluminum Alloys

The non-heat-treatable wrought aluminum alloys that are most successfully brazed are the ASTM 1000 and 3000 series. Also, the low-magnesium alloys of the ASTM 5000 series may be brazed successfully. Available filler metals melt below the solidus temperatures of all commercial wrought, nonheat-treatable alloys.

The heat-treatable wrought alloys most commonly brazed are the ASTM 6000 series. The ASTM 2000 and 7000 series of aluminum alloys are low-melting alloys and therefore not normally brazeable, with the exception of the 7072 and 7005 alloys.

The aluminum sand-cast alloys and permanent-mold-cast alloys that are most commonly brazed are the ASTM 443.0, 356.0, and 712.0 alloys. Aluminum die castings are generally not brazed because of blistering caused by the high gas content. Table 12.5 lists the common aluminum base metals that can be brazed.

Most brazing of aluminum is accomplished by the torch brazing, dip brazing, or furnace brazing processes. Furnace brazing takes place in air or in controlled atmospheres, including a vacuum. Additional information on brazing aluminum and aluminum alloys is contained in the latest edition of the American Welding Society document *Specification for Aluminum Brazing,* ANSI/AWS C3.7,[16] and also in Chapter 12 of the *Brazing Handbook,* 5th Edition.[17]

Magnesium and Magnesium Alloys

Brazing techniques similar to those used for aluminum are used for magnesium alloys. Furnace brazing, torch brazing, and dip brazing can be employed, although the latter process is the most widely used.

Magnesium alloys that are considered brazeable are listed in Table 12.6. Furnace and torch brazing experience is limited to the M1A alloy. Dip brazing can be used for AZ10A, AZ31B, AZ61A, K1A, M1A, ZE10A, ZK21A, and ZK60A alloys.

Referring to the filler metals used for brazing magnesium listed in Table 12.6, BMg-1 brazing filler metal is suitable for the torch brazing, dip brazing, or furnace brazing processes. Other commercial filler metals provide even lower-melting composition and is suitable for use with dip brazing only.

Beryllium

Brazing is the preferred method for achieving a metallurgical bond in beryllium.[18] Suitable brazing filler metal systems and temperature ranges include the following:

1. Zinc: 427°C to 454°C (800°F to 850°F);
2. Aluminum-silicon: 566°C to 677°C (1050°F to 1250°F);
3. Silver-copper: 640°C to 904°C (1200°F to 1660°F); and
4. Silver: 882°C to 954°C (1620°F to 1750°F).

Zinc melts below 449°C (840°F), the temperature defined by AWS for brazing filler metal. Nevertheless, it is generally accepted as the lowest-melting filler metal for brazing beryllium.

Aluminum-silicon filler metals can be used in high-strength, wrought beryllium assemblies because the brazing temperature is well below the base metal recrystallization temperature. The BA1Si-4 type of filler metal brazes well with fluxes. Brazing without flux in a pro-

16. American Welding Society (AWS) Committee on Brazing and Soldering, *Specification for Aluminum Brazing,* AWS C3.7. Miami: American Welding Society.

17. See Reference 4, Chapter 12.

18. Beryllium and its compounds are toxic. Proper identification and handling of beryllium metal are required by federal regulations. See Appendix B for sources of safety information.

Table 12.5
Nominal Composition and Melting Range of Common Brazeable Aluminum Alloys

Aluminum Association Designation	ASTM Alloy	UNS Number	Brazeability Rating[b]	Nominal Composition[a]						Approximate Melting Range	
				Cu	Si	Mn	Mg	Zn	Cr	°C	°F
1100	1100	A91100	A		Al 99% min					643–657	1190–1215
1350	1350	A91350	A		Al 99.5% min					646–657	1195–1215
3003	3003	A93003	A	0.12	—	1.2	—	—	—	643–654	1190–1210
3004	3004	A93004	B	—	—	1.2	1.0	—	—	629–651	1165–1205
3005	3005	A93005	A	0.3	0.6	1.2	0.4	0.25	0.1	638–657	1180–1215
5005	5005	A95005	B	—	—	—	0.8	—	—	632–654	1170–1210
5050	5050	A95050	B	—	—	—	1.4	—	—	624–657	1155–1210
5052	5052	A95052	C	—	—	—	2.5	0.25	—	607–650	1125–1210
6151	6151	A96151	C	—	0.9	—	0.6	—	0.25	588–650	1190–1200
6951	6951	A96951	A	0.29	0.35	—	0.6	—	—	615–654	1140–1210
6053	6053	A96053	A	0.25	0.7	—	1.2	—	0.25	557–651	1070–1205
6063	6063	A96063	A	—	0.4	—	0.7	—	—	615–654	1140–1210
7005	7005	A97005	B	—	—	0.45	1.4	4.5	0.13	641–657	1185–1215
7072	7072	A97072	A	—	—	—	—	1.0	—	607–646	1125–1195
Cast 433.0	Cast 443.0	A04430	A	—	5.0	—	—	—	—	629–632	1065–1170
Cast 356	Cast 356.0	A03560	C	—	7.0	—	0.3	—	—	557–613	1035–1135
Cast 406	Cast 406.0		A	—	—	Al 99% min		—	—	643–657	1190–1215
Cast 710.0			B	0.5	—	—	0.7	6.5	—	596–646	1105–1195
Cast 712.0	Cast 712.0		A	—	—	—	0.35	6.5	—	604–643	1120–1190

a. Percent of alloying elements: aluminum and normal impurities constitute remainder.
b. Brazeability ratings:
A = Alloys readily brazed by all commercial methods and procedures.
B = Alloys that can be brazed by all techniques with moderate care.
C = Alloys that require special care to braze.

tective atmosphere requires stringent control. Aluminum-base filler metals have less metallurgical interaction with the base metal than silver-base fillers. This is a significant advantage when thin beryllium sections or foils are to be joined.

Silver and silver-base brazing filler metals are used in structures exposed to elevated temperatures. Controlled-atmosphere brazing with these alloy systems is straightforward and can be performed in purified controlled atmospheres, including vacuum.

Copper and Copper Alloys

The copper-alloy base metals include copper-zinc alloys (brass), copper-silicon alloys (silicon bronze), copper-aluminum alloys (aluminum bronze), copper-tin alloys (phosphor bronze), copper-nickel alloys, dispersion-strengthened copper alloys, and several others. The brazing of copper and copper alloys and appropriate filler metals are discussed in detail in Chapter 18 of the *Brazing Handbook*, 5th edition.[19] Figure 12.18 shows copper tubing being torch brazed with a silver-base filler metal using a special torch design. The silver filler metal is in wire form and is fed into the joint as it is melted.

Low-Carbon and Low-Alloy Steels

Low-carbon steels and low-alloy steels are brazed without difficulty. They are frequently brazed at temperatures above 1080°C (1980°F) with copper filler

19. See Reference 4, Chapter 18.

Table 12.6
Brazeable Magnesium Alloys and Filler Metals

ANSI/AWS A5.8 Classification	ASTM Alloy Designation	UNS Number	Available Forms*	Solidus Temperature		Liquidus Temperature		Brazing Temperature Range		Suitable Filler Metal	
				°C	°F	°C	°F	°C	°F	BMg-1	BMg-2A
Base Metal											
—	AZ10A	—	E	632	1170	643	1190	582–616	1080–1140	X	X
—	AZ31B	M11311	E, S	566	1050	627	1160	582–593	1080–1100	—	X
—	AZ61A	—	E	510	950	615	1140	495–505	925–940	—	—
—	K1A	—	C	649	1200	650	1202	582–616	1080–1140	X	X
—	M1A	—	E, S	648	1198	650	1202	582–616	1080–1140	X	X
—	ZE10A	—	S	593	1100	646	1195	582–593	1080–1100	—	X
—	ZK21A	M16210	E	626	1159	642	1187	582–616	1080–1140	X	X
—	ZK60A	—	E, S	520	970	635	1175	495–505	925–940		
Filler Metal											
BMg-1	AZ92A	M11920	W, R	443	830	599	1110	604–616	1120–1140	—	—

Key to available forms:
E = Extruded shapes and structural sections
S = Sheet and plate
C = Castings
W = Wire
R = Rod

metal in a controlled atmosphere or at lower temperatures with silver-base filler metals and flux.

For alloy steels, the filler metal should have a solidus well above any heat-treating temperature to avoid damage to joints that will be heat-treated after brazing. In some cases, air-hardening steels can be brazed and then hardened by quenching from the brazing temperature.

A filler metal with a brazing temperature lower than the critical temperature of the steel can be used when no change in the metallurgical properties of the base metal is specified.

High-Carbon Steels and High-Carbon Tool Steels

High-carbon steels contain more than 0.45% carbon. High-carbon tool steels usually contain 0.60% to 1.40% carbon. The brazing of high-carbon steels is best accomplished prior to or during the hardening process. Hardening temperatures for carbon steels range from 760°C to 820°C (1400°F to 1500°F). Filler metals with brazing temperatures above 820°C (1500°F) should be used. When brazing and hardening are done in one operation, the filler metal should have a solidus at or below the austenitizing temperature.

Tempering and brazing can be combined for high-carbon tool steels and high-carbon, high-chromium alloy tool steels that have tempering temperatures in the range of 540°C to 650°C (1000°F to 1200°F). Filler metals with brazing temperatures in that range are used. The component is removed from the tempering furnace, brazed by localized heating methods, and then returned to the furnace for completion of the tempering cycle.

Cast Irons

Cast irons generally require special considerations for brazing. The types of cast iron brazed include gray, malleable, and ductile. White cast iron is seldom brazed.

Prior to brazing, the faying surfaces are generally cleaned electrochemically or chemically, seared with an oxidizing flame, or grit blasted. When low-melting silver filler metals are used, wetting by the brazing filler metal is easiest. Ductile and malleable cast irons should be brazed below 760°C (1400°F).

When high-carbon cast iron is brazed with copper, a low brazing temperature should used to avoid the melting of localized areas of the cast iron, particularly in thin sections.

Stainless Steels

All of the stainless steel alloys, formerly considered difficult to braze because of their high chromium con-

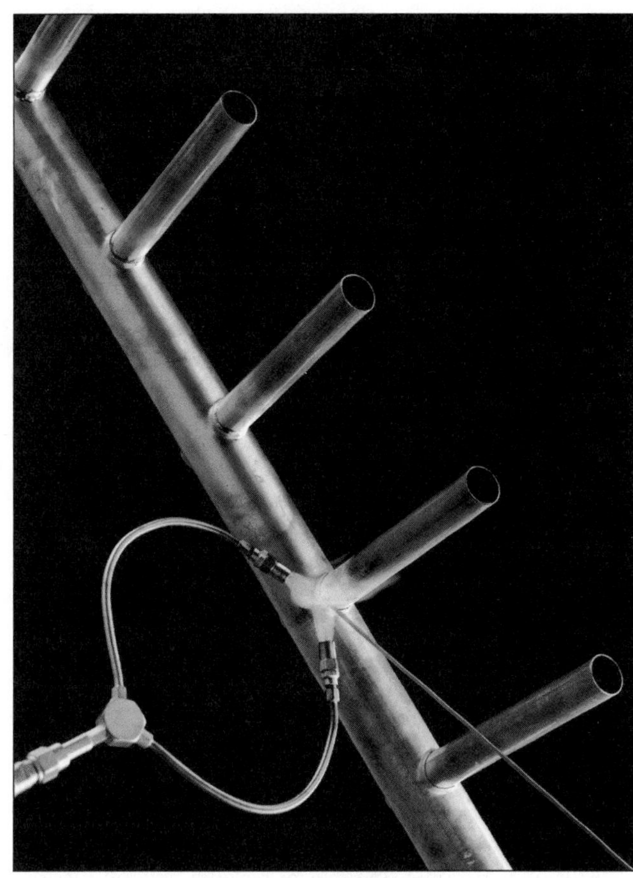

Photograph courtesy of Gas Flux Company

Figure 12.18–Silver Brazing of Copper Tubing with Special Torch Design

tent, can be brazed using equipment with precise atmosphere controls. Brazing these alloys is best accomplished in a purified (dry) hydrogen atmosphere or in a vacuum. Dew points below –51°C (–60°F) must be maintained because wetting becomes difficult following the formation of chromium oxide. Torch brazing requires the application of flux to reduce any chromium oxides present.

Most of the silver-alloy, copper, and copper-zinc filler metals are used for brazing stainless steels. Silver filler metals containing nickel are generally best for corrosion resistance. Filler metals containing copper-phosphorus should not be used on highly stressed brazements because brittle nickel and iron phosphides may be formed at the joint interface. Diffusion-brazing nickel-phosphorus and nickel-base filler metals are generally suitable.

Nickel filler metals containing boron are generally best for the diffusion brazing of stainless steels and heat-resistant base metals because boron has a mild fluxing action, which aids in wetting these base metals. Care should be taken to avoid potential embrittlement if welding is to be performed later near the brazed joint containing boron. The presence of boron may make the weld susceptible to cracking.

Diffusion brazing produces joints with improved physical properties. The brazing of austenitic chromium-nickel stainless steels is discussed further in Chapter 24 of the *Brazing Handbook*, 5th edition.[20]

Chromium Irons and Steels. The martensitic stainless steels (403, 410, 414, 416, 420, and 431) air-harden on cooling from the brazing temperature, which occurs at or above the austenitizing temperature ranges. Therefore, they must be annealed after brazing or isothermally annealed during the brazing operation. These steels are also subject to cracking during brazing with certain brazing filler metals, unless properly stress-relieved prior to or during brazing.

The ferritic stainless steels (405, 406, and 430) cannot be hardened and their grain structure cannot be refined by heat treatment. The properties of these alloys degrade when brazed at temperatures above 980°C (1800°F) because of excessive grain growth. They lose ductility after long heating times between 340°C and 600°C (650°F and 1100°F). However, some of the ductility can be recovered by heating the brazement to approximately 790°C (1450°F) for a suitable time.

Precipitation-Hardening Stainless Steels. Precipitation-hardening stainless steels are basically stainless steels with additions of one or more of the elements copper, molybdenum, aluminum, and titanium. Such alloying additions make it possible to strengthen the alloys by precipitation-hardening heat treatments. When alloys of this type are brazed, the brazing cycle and temperature must match the heat-treatment cycle of the alloy. Manufacturers of these alloys have developed recommended brazing procedures for their particular steels.

Nickel and Nickel Alloys

Nickel and the high-nickel alloys are embrittled by sulfur and low-melting metals present in brazing alloys, such as zinc, lead, bismuth, and antimony. Base-metal surfaces must be thoroughly cleaned prior to brazing to remove any substances that may contain these elements. Sulfur and sulfur compounds must also be excluded from the brazing atmosphere.

Highly stressed or cold-worked nickel and its alloys are subject to stress cracking in the presence of molten brazing filler metals. The base metal should be annealed

20. See Reference 4, Chapter 24.

prior to brazing to remove residual stresses or carefully stress-relieved during the braze cycle.

Silver brazing filler metals are commonly used. In corrosive environments, brazing filler metals with a high silver content are preferred. Cadmium-free brazing filler metals are chosen to avoid stress corrosion cracking.

Nickel-base brazing filler metals offer the greatest corrosion and oxidation resistance and elevated temperature strength. Some high-temperature and corrosive environments may require gold-based filler metals.

Brazing is a preferred method for joining dispersion-strengthened nickel alloys in applications that must function at elevated temperatures. High-strength brazements have been made with special nickel-base brazing filler metals and then tested in temperatures up to 1300°C (2400°F).

Heat-Resistant Alloys

Heat-resistant alloys (often called *superalloys*) are generally brazed in a dry hydrogen atmosphere or in a vacuum furnace using nickel-base, cobalt-base, or special filler metals.

The cobalt-base metals are the easiest of the superalloys to braze because most of them do not contain titanium or aluminum. Base metals that are high in titanium or aluminum are difficult to braze in dry hydrogen or vacuum atmospheres because titanium and aluminum oxides are not reduced at brazing temperatures. Electrolytic nickel plating can be used as a barrier coating to achieve excellent brazing.

Titanium and Zirconium

Titanium and zirconium combine readily with oxygen, hydrogen, and nitrogen and react with many metals by forming brittle intermetallic compounds. Base metals must be thoroughly cleaned before brazing and brazed immediately after cleaning. Reoxidation of the surface occurs with time at room temperature.

Silver and silver-base filler metals were formerly used in the brazing of titanium, but it was found that brittle intermetallics formed and crevice corrosion resulted. Type 3003 aluminum foil filler metal is used to join thin, lightweight structures such as complex honeycomb structures and sandwich panels. Electroplating various elements on the faying surfaces of the base metal will allow the elements to react *in situ* with the titanium during brazing to form a titanium alloy eutectic. This transient liquid phase flows well and forms fillets and then solidifies by means of interdiffusion.

Other brazing filler metals with high service capability and corrosion resistance include Ti-Zr-Ni-Be, Ti-Zr-Ni-Cu, and Ti-Ni-Cu alloys. The best braze processing is obtained in high-vacuum atmosphere furnaces using closely controlled temperatures in the range of 900°C to 955°C (1650°F to 1750°F) to maintain the desired titanium alloy properties after brazing.

Carbides and Cermets

Carbides of the titanium and refractory metals that are bonded with cobalt are used in the manufacturing of cutting tools and dies. Closely related materials called *cermets* consist of ceramic particles bonded with various metals.

The brazing of carbides and cermets is more difficult than the brazing of metals. Torch, induction, or furnace brazing is used, often with a sandwich brazing technique: a layer of weak, ductile metal (pure nickel or pure copper) is brazed between the carbide or cermet and a hard metal support. The cooling stresses cause the soft metal to deform instead of cracking the ceramic.

Silver-base brazing alloys, copper-zinc alloys, and copper are often used in the manufacture of carbide tools. Silver alloys containing nickel are preferred for better wettability. The nickel-base alloys containing boron and a 60% Pd-40% Ni alloy may be satisfactory for brazing nickel- and cobalt-bonded cermets of tungsten carbide, titanium carbide, and niobium carbide.

Ceramics

Alumina, zirconia, magnesia, beryllia, silicon nitride, silicon carbide, aluminum nitride, and thoria are ceramic materials that can be joined by brazing. They are inherently difficult to wet with conventional filler metals. Differences in thermal expansion, heat conduction, and ductility result in cracking and crack propagation at relatively low stresses.

If the ceramic is premetallized to facilitate wetting, copper, silver-copper, and gold-nickel filler metals are used. Titanium or zirconium hydride can be decomposed at the ceramic-metal interface to form an intimate bond. Filler metals containing titanium and zirconium are available and are used for the direct brazing of ceramics and ceramic-to-metal brazing. These filler metals are referred to as active brazing alloys (ABA).

Nonmetallized ceramics are brazed with silver-copper-clad or nickel-clad titanium wires. Useful titanium and zirconium filler metals are Ti-Zr-Be, Ti-V-Zr, Zr-V-Cb, Ti-V-Be, and Ti-V-Cr.

Precious Metals

The precious metals—silver, gold, platinum, and palladium—present few brazing difficulties. The thin oxide films on these metals are readily removed by fluxes and protective atmospheres.

A typical application for these metals is the resistance brazing or furnace brazing of electrical contacts to holders using silver and gold filler metals.

Refractory Metals

Four refractory metals—tungsten, molybdenum, niobium and tantalum—are discussed in this section. They are primarily important for their superior strength, corrosion resistance, and the ability to withstand high-temperature service. They are frequently used by the aerospace industry in structures and components for airframes and rocket engines. Metallurgical researchers continue to investigate and improve the materials and methods for brazing the refractory metals.

The properties of tungsten, molybdenum, niobium and tantalum include very high melting temperatures, high recrystallization temperatures, high to very high densities, low specific heats, and low coefficients of thermal expansion. The body-centered cubic (BCC) crystal structure of these metals presents a significant characteristic—a well-defined ductile-to-brittle transition behavior. The characteristics of refractory metals that must be considered when brazing procedures are established are the ductile-to-brittle transition behavior, the recrystallization temperature, and the reaction with gases and interstitial elements such as carbon.

The transition temperature range for these metals is not a fixed property of the metal but is influenced by the strain rate, alloying additions, impurities, heat treatment, and the fabrication process. Thus, all aspects of the brazing process must be carefully established and managed. Chapter 31 of the AWS *Brazing Handbook*, 5th edition,[21] provides detailed information on the brazing of the refractory metals.

Tungsten. Tungsten can be brazed to tungsten and to other metals and nonmetals with nickel-base filler metals, but interaction between tungsten and nickel will cause recrystallization of the base metal. The assembly should be designed to accommodate the loss of properties associated with recrystallization and diffusion of impurities. The tungsten should be stress-relieved by heat treatment prior to brazing, and the brazing cycle should be short to limit interaction with the filler metal.

Molybdenum. A wide variety of brazing filler metals can be used to join molybdenum and its alloys. If possible, the brazing temperature should be kept below the recrystallation temperature of the base metals for better properties. Copper- and silver-base filler metals are good candidates. Gold-copper, gold-nickel, and copper-nickel filler metals can be used for electronic and nonstructural applications. Higher-melting metals and alloys can be used as filler metals if higher-temperature service is necessary.

21. See Reference 4, Chapter 31.

Tantalum and Niobium. Tantalum and niobium require special techniques to be satisfactorily brazed. All reactive gases must be removed from the brazing atmosphere. These include oxygen, nitrogen, carbon monoxide, ammonia, and hydrogen. Niobium becomes embrittled by small amounts of oxygen. Tantalum forms oxides, nitrides, carbides, and hydrides very readily, leading to a loss of ductility. For oxidation protection at high temperatures, tantalum and niobium are often electroplated with copper or nickel. The brazing filler metal must be compatible with the specific plating used. These metals are usually brazed in a vacuum atmosphere.

Dissimilar Metal Combinations

Many dissimilar metal combinations can be brazed, even those with metallurgical incompatibilities that preclude welding.

Important criteria to be considered start with differences in thermal expansion. If a metal with high thermal expansion surrounds a low-expansion metal, clearances at room temperature that are satisfactory for capillary flow will be too great at brazing temperature. Conversely, if a low-expansion metal surrounds a high-expansion metal, no clearance may exist at brazing temperature. For example, when brazing a molybdenum plug in a copper block, the assembly must be a press fit at room temperature; if a copper plug is to be brazed in a molybdenum block, a properly centered loose fit at room temperature is required.

When brazing tube-and-socket joints as shown in Figure 12.33(A)–(C) between dissimilar base metals, the tube should be the low-expansion metal and the socket the high-expansion metal. Thus, at brazing temperature the clearance will be maximum and the resulting capillary will fill with the brazing filler metal. When the joint cools to room temperature, the brazed joint and the tube will be in compression.

With a tongue-in-groove joint, the groove should be in the low-expansion material. The fit at room temperature should be designed to give joint clearances for capillary flow on both sides of the tongue at brazing temperature. Longitudinal shear stresses in the braze metal are limited by reducing the size of the overlap distances.

"Sandwich brazing" is commonly used to manufacture carbide-tipped metal cutting tools. A relatively ductile metal is coated on both sides with brazing filler metal and preplaced in the joint. This places a third material in the joint; the ductile material will deform during cooling and reduce the stresses caused by differential contraction of the components of the brazement.

The filler metal used to braze dissimilar metals must be compatible with both base metals. The filler metal should have corrosion or oxidation resistance at least equal to the resistance of the poorest of the two metals being brazed. It should not form galvanic couples, which could promote crevice corrosion in the braze area. Brazing filler metals form low-melting phases with many base metals, requiring adaptation of the joint design and the brazing cycle, and adjustments to the quantity and placement of filler metal.

Metallurgical reactions between the brazing filler metal and dissimilar base metals may produce an unacceptable brazed joint. An example is the brazing of aluminum to copper. Copper reacts with aluminum to form a brittle compound. Such problems can be overcome by coating one of the base metals with a metal that is compatible with the brazing filler metal. Therefore, to braze aluminum to copper, the copper is plated with a silver or high-silver alloy. The joint is then brazed at 816°C (1500°F) with a standard aluminum brazing filler metal. Nickel plating also forms a suitable diffusion barrier.

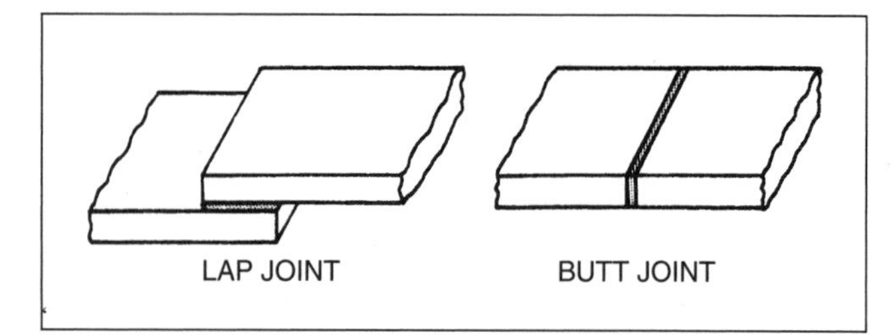

Figure 12.19—Basic Lap and Butt Joints for Brazing

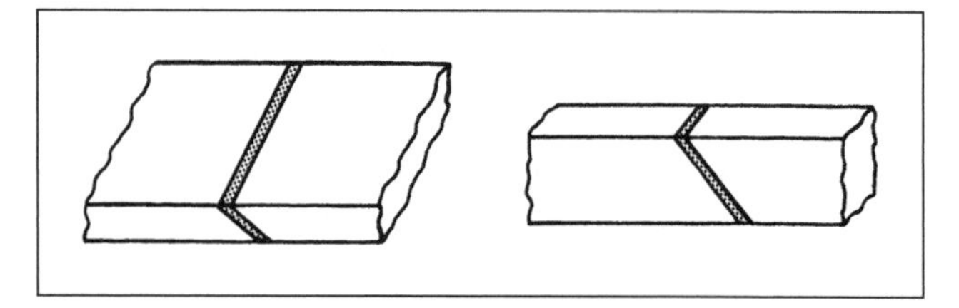

Figure 12.20—Typical Scarf Groove Designs for Brazing

JOINT DESIGN

Two basic joint designs are used in brazing: the lap joint and the butt joint. These joints are shown in Figure 12.19.

The lap joint can be made as strong as the weaker member (even when using a high-strength filler metal or in the presence of small discontinuities in the joint) by specifying an overlap at least three times the thickness of the thinner member. Lap joints are highly efficient and are easy to fabricate; however, they have the disadvantage that the increased metal thickness at the joint creates a stress concentration at the abrupt changes in the cross section. The stress concentration can be corrected by machining a base metal fillet at the end of each component.

Butt joints are used when the lap-joint thickness would be objectionable or when the strength of a brazed butt joint will satisfactorily meet service requirements. The joint strength depends only partly on the filler metal strength. The strength of these joints increases as the clearance decreases, the base metal hardness increases, and diffusion brazing is incorporated.

The scarf groove (bevel joint) is a variation of the butt joint. As shown in Figure 12.20, the beveled cross-sectional area of this joint is increased without an increase in metal thickness. Two disadvantages limit its use: the sections are difficult to align, and the joint is difficult to prepare, particularly in thin base metals. Fixturing is often necessary to hold components in place during brazing. Since the joint is at an angle to the axis of tensile loading, the load-carrying capacity is that of a lap joint.

For further information, the latest edition of the American Welding Society document *Recommended Practices for Design, Manufacture, and Inspection of Critical Brazed Components*, ANSI/AWS C3.3, can be consulted.[22]

Many assemblies can be designed to be self-locating and self-supporting. Examples of some self-fixturing joint designs are shown in Figure 12.21.

JOINT CLEARANCE

Joint clearance is defined as the distance between the faying surfaces of a brazement. Joint clearance has a major effect on the mechanical performance of a brazed joint. This applies to all types of loading, such as static, fatigue, and impact loading, and to all joint designs. Several effects of joint clearance on mechanical performance are the following:

1. The purely mechanical effect of restraint to plastic flow of the filler metal by a higher-strength base metal;
2. The possibility of slag entrapment;
3. The possibility of voids;

22. See Reference 11.

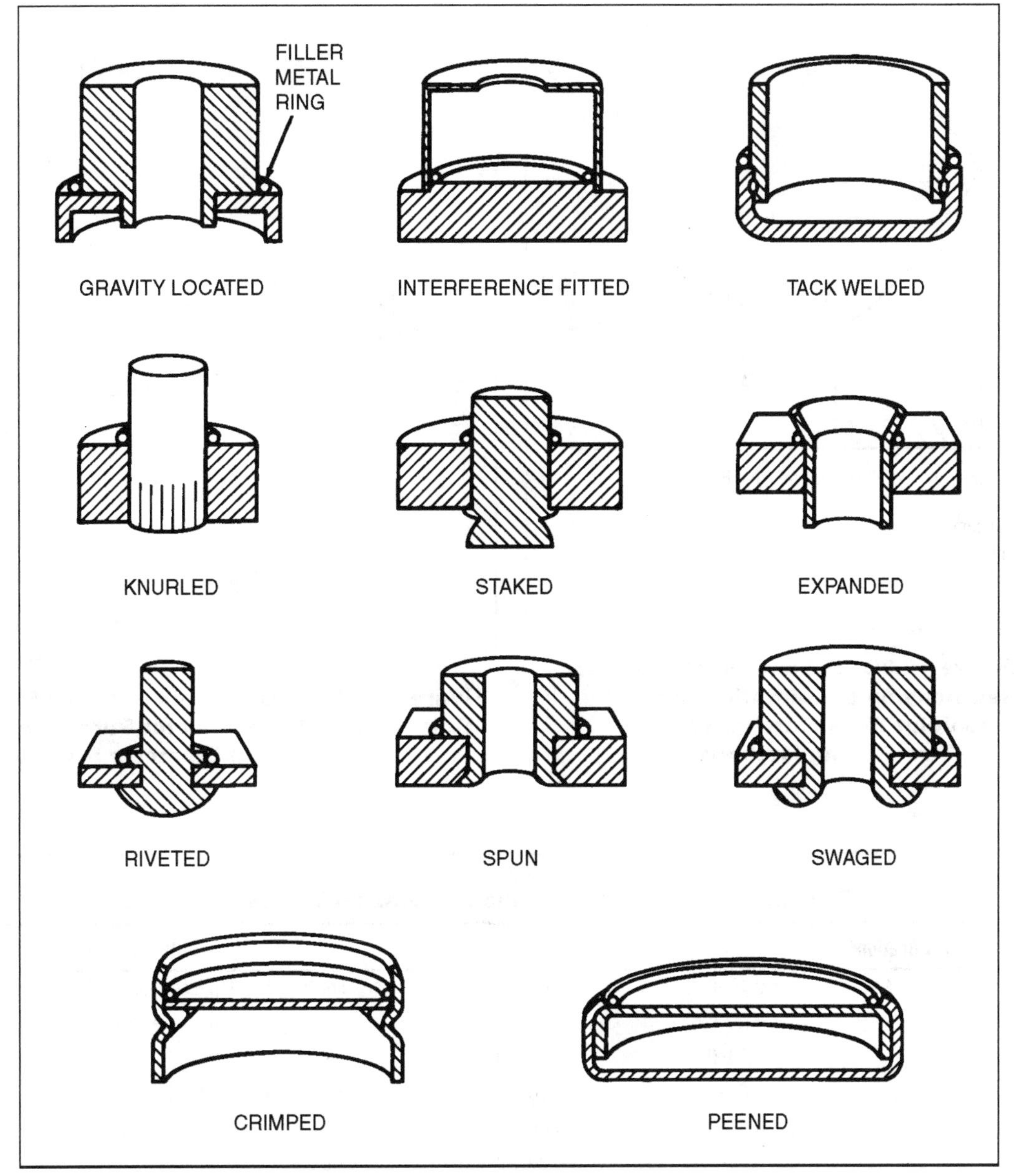

Source: Adapted from American Welding Society (AWS) Committee on Brazing and Soldering, C. L. Jenney, ed., Forthcoming, *Brazing Handbook*, 5th ed., Miami: American Welding Society.

Figure 12.21—Typical Self-Fixturing Methods for Brazed Assemblies

4. The relationship between joint clearance and capillary force, which accounts for filler-metal distribution; and
5. The amount of filler metal that must be diffused with the base metal when the diffusion-brazing process is used.

If the brazed joint is free of defects (no flux inclusions, voids, unbrazed areas, pores, or porosity), its strength in shear depends on the joint thickness, as illustrated in Figure 12.22. This figure indicates the change in joint shear strength relative to joint clearance. The information listed in Table 12.7 can be used as a nominal guide for joint clearance size at brazing temperature when designing brazed joints for maximum strength.

Some specific data on the relationship of joint clearance to joint strength for silver-brazed butt joints in steel are shown in Figures 12.23 and 12.24. Figure

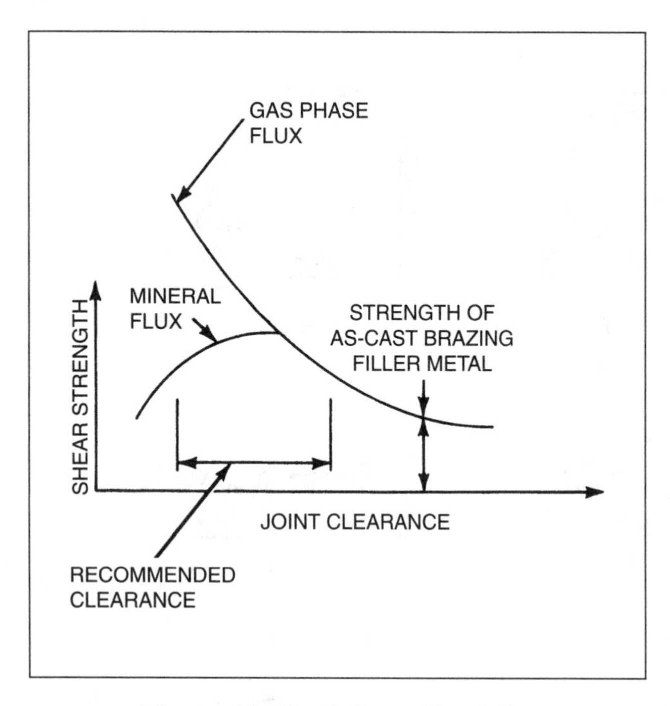

Figure 12.22—Schematic of the Relationship of Joint Clearance to Joint Shear Strength for Two Fluxing Methods

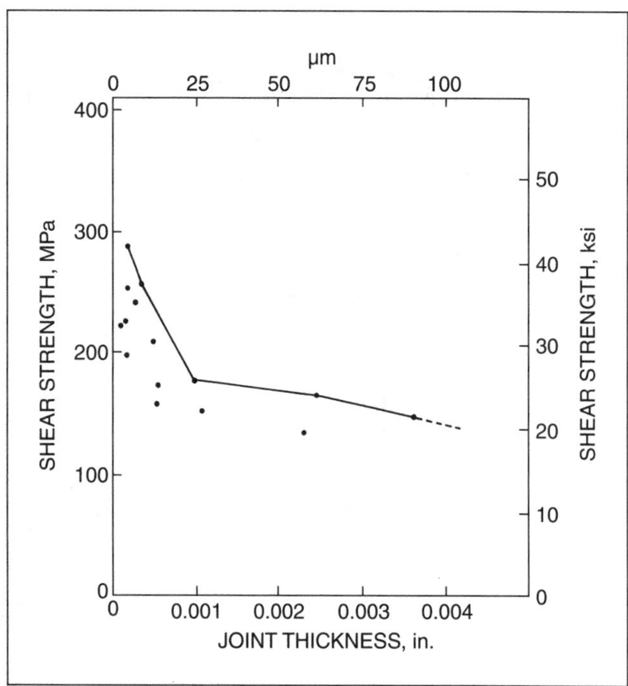

Figure 12.23—Relationship of Shear Strength to Brazed Joint Thickness for Pure Silver Joints in 12.7 mm (0.5 in.) Diameter Steel Drill Rod

Table 12.7
Recommended Joint Clearance at Brazing Temperature

Filler Metal AWS Classification[a]	mm	in.	Joint Clearance[b]
BAlSi Group	0.15–0.25	0.006–0.010	For length at lap less than 6.35 mm (1/4 in.)
	0.25–0.6	0.010–0.025	For length at lap greater than 6.35 mm (1/4 in.) mm)
BCuP Group	0.03–0.12	0.001–0.005	—
BAg Group	0.05–0.12	0.002–0.005	Flux brazing (mineral fluxes)
	0.03–0.05	0.001–0.002[c]	Atmosphere brazing (gas-phase fluxes)
BAu Group	0.05–0.12	0.002–0.005	Flux brazing (mineral fluxes)
	0.00–0.05	0.000–0.002[c]	Atmosphere brazing (gas-phase fluxes)
BCu Group	0.00–0.05	0.000–0.002[c]	Atmosphere brazing (gas-phase fluxes)
BCuZn Group	0.05–0.12	0.002–0.005	Flux brazing (mineral fluxes)
BMg Group	0.10–0.25	0.004–0.010	Flux brazing (mineral fluxes)
BNi Group	0.05–0.12	0.002–0.005	General applications (flux or atmosphere)
	0.00–0.05	0.000–0.002	Free-flowing types (atmosphere brazing)

a. Key:
BAlSi —Aluminum
BAg—Silver base
BAu—Gold base
BCu—Copper
BCuP—Copper phosphorus
RBCuZn—Copper zinc
BMg—Magnesium base
BNi—Nickel base

b. Clearance on the radius when rings, plugs, or tubular members are involved. On some applications it may be necessary to use the recommended clearance on the diameter to assure not having excessive clearance when all the clearance is on one side. An excessive clearance will produce voids. This is particularly true when brazing is accomplished in a high-quality atmosphere (gas-phase fluxing).

c. For maximum strength, a press fit of 0.03 to 0.05 mm/mm (0.001 to 0.002 in./in.) of diameter should be used.

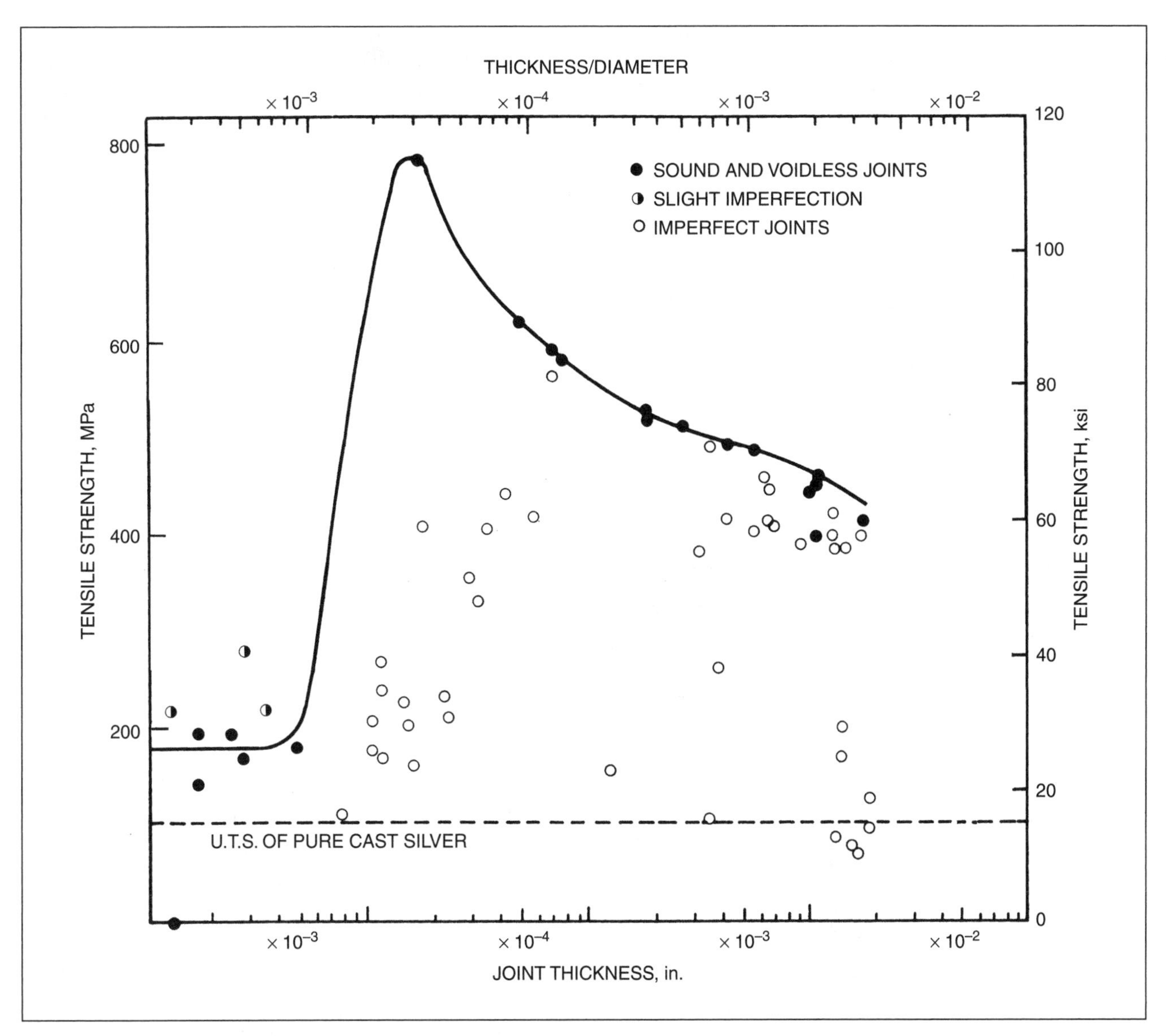

Figure 12.24—Relationship of Tensile Strength to Brazed Joint Thickness of 12.7 mm (0.5 in.) Diameter Silver Brazed Butt Joints in 4340 Steel Rod

12.23 shows the optimum shear values obtained with joints in drill rod 12.7 mm (0.5 in.) in diameter using pure silver. The butt joints in the rods were brazed by induction heating in a dry 10% hydrogen-90% nitrogen atmosphere. Figure 12.24 relates tensile strength to joint thickness for brazed butt joints of the same size. The strength is noticeably decreased at extremely small clearances. It should be noted that the data in Figures 12.23 and 12.24 were obtained with nonstandard test specimen designs. The AWS publication *Standard Method for Evaluating the Strength of Brazed Joints*, ANSI/AWS C3.2, should be consulted.[23]

Preplaced filler is brazing filler metal placed in the joint as the joint is prepared for brazing; for example, foil placed between two plates. The clearances noted in Table 12.7 generally do not apply to this application. In applications using preplaced filler metal, the brazement should be preloaded so that the joint clearance will

23. See Reference 10.

decrease during the brazing operation. This forces the filler metal into voids created by the normal roughness of the faying surfaces. In some applications, additional filler metal is made available by extending the filler metal shim out beyond the joint edges.

The type of flux used has an important effect on brazeability. A mineral flux must melt at a temperature below the melting range of the brazing filler metal, and it must flow into the joint ahead of the filler metal. When the joint clearance is too small, the mineral flux may be held in the joint and cannot be displaced by the molten filler metal. This will produce joints that will have little or no strength. When the clearance is too large, the molten filler metal will flow around pockets of flux, causing excessive flux inclusions or entrapment.

For a joint between dissimilar base metals, the joint clearance at the brazing temperature must be calculated from thermal expansion data. Figure 12.25 shows thermal expansion data for some materials. The data in Figure 12.26 can be used to find the diametral clearance at the brazing temperature between dissimilar metals.

To withstand high differential thermal expansion of two metals being brazed, the brazing filler metal must be strong enough to resist fracture, and the base metal must

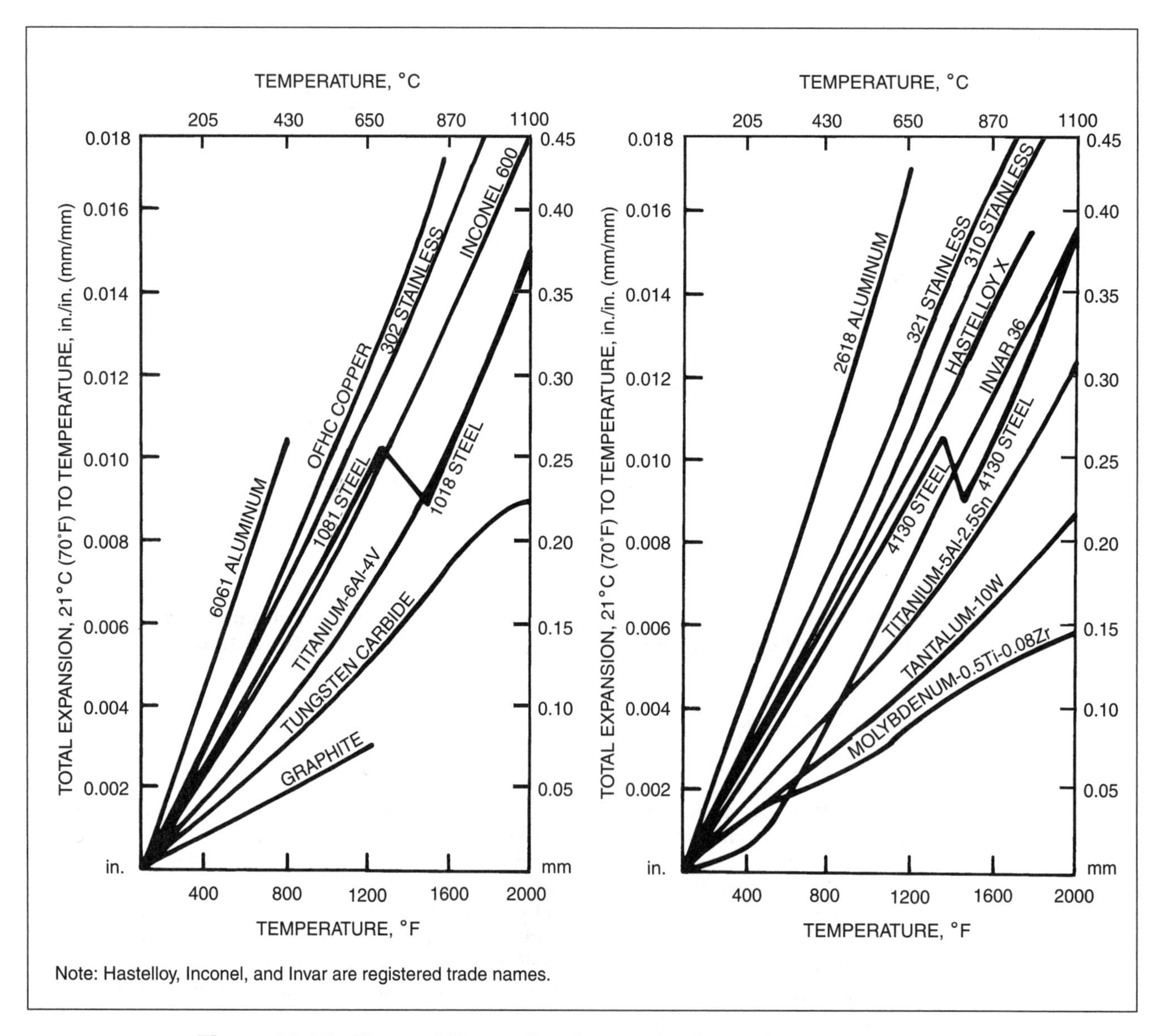

Figure 12.25—Thermal Expansion Curves for Some Common Materials

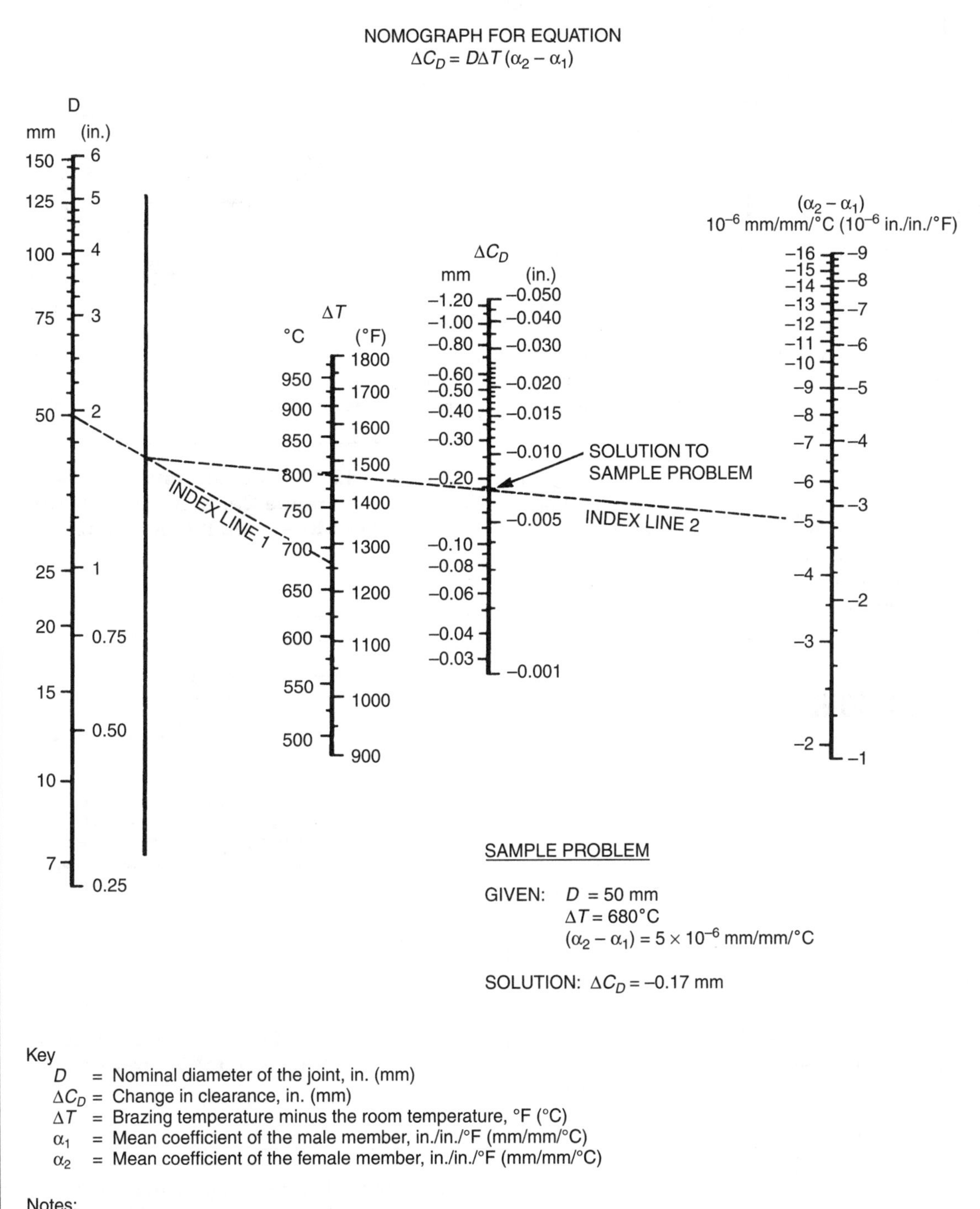

Notes:

1. This nomograph provides the change in diameter caused by heating. The clearance to promote the flow of the brazing filler metal must be provided at the brazing temperature.
2. This nomograph assumes a case where α_1 exceeds α_2; thus, the scale value for $(\alpha_1 - \alpha_2)$ is negative. The resultant values for ΔC_D are also negative, signifying that the joint gap reduces upon heating. Where $(\alpha_1 - \alpha_2)$ is positive, the values of ΔC_D are read as positive, signifying an enlargement of the joint gap upon heating.

Figure 12.26—Nomograph for Finding the Change in Diametral Clearance in Dissimilar Metal Joints

yield during cooling. Some residual stress will remain in the final brazement. Thermal cycling of such a brazement during its service life will repeatedly stress the joint area, which may shorten the service life. Dissimilar metal brazements should be designed so that residual stresses do not add to the stresses imposed during service.

STRESS DISTRIBUTION

A good brazement design should incorporate joints that will avoid high stress concentration at the edges of the braze and will distribute the stresses uniformly into the base metal. High-strength brazements should be designed to fail in the base metal rather than in the braze filler metal. A smaller overlap should be used only under light service loads. For applications in which joints will be lightly loaded, it is economical to use the simplest joint design that will not break in service. Typical designs are shown in Figures 12.27 through 12.30.

Using a fillet of brazing filler metal is not good brazing design because it is seldom possible to make the brazing filler metal consistently form a desired fillet size and contour. When the fillets become too large, shrinkage or piping porosity will occur and act as a stress concentration.

ELECTRICAL CONDUCTIVITY

Brazing filler metals in general have low electrical conductivity compared to copper. However, when brazing a properly designed joint in an electrical circuit, the filler metal will not add appreciable resistance to the circuit.

With butt joints, the brazed joint thickness (resistance) is very small compared to the lengthwise resistance of the conductor, even though the unit resistivity of the filler metal is much higher than that of the base metal. Nevertheless, a filler metal with low resistivity should be used provided that it meets all other requirements of the application.

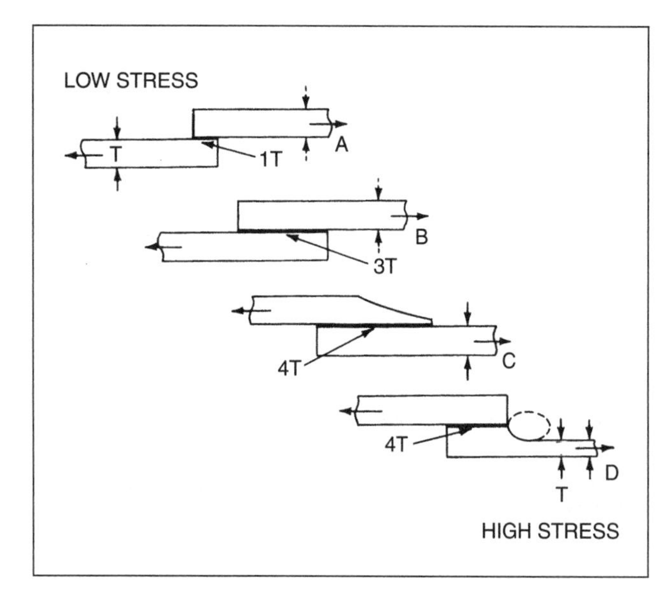

Figure 12.27—Brazed Lap Joint Designs for Use at Low and High Stresses

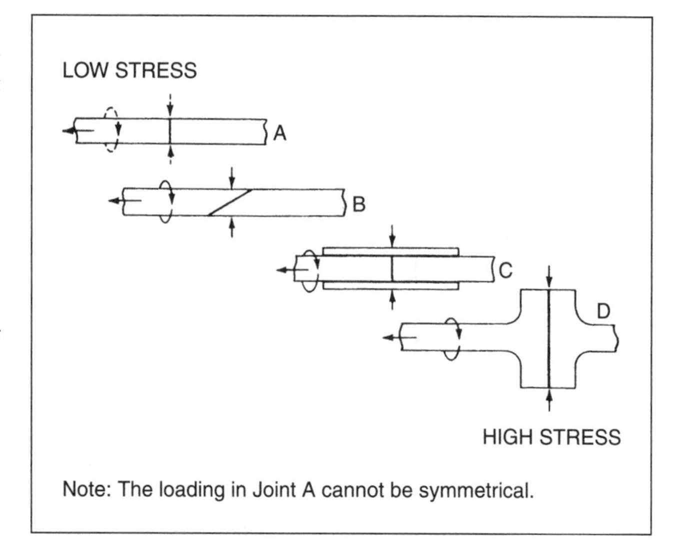

Figure 12.28—Brazed Butt Joint Designs to Increase Capacity of Joint for High Stress and Dynamic Loading

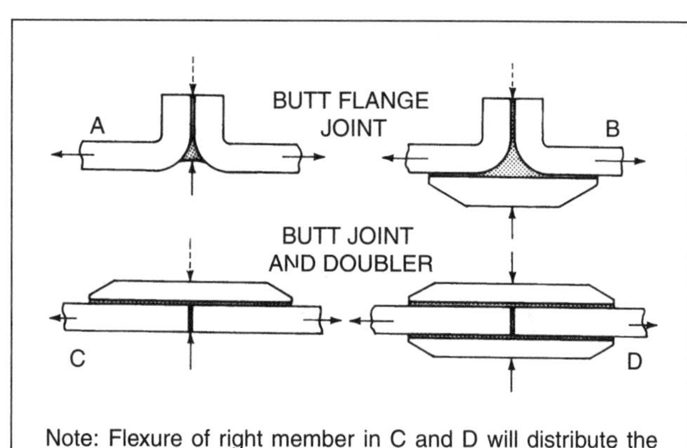

Figure 12.29—Butt Joint Designs for Sheet-Metal Brazements

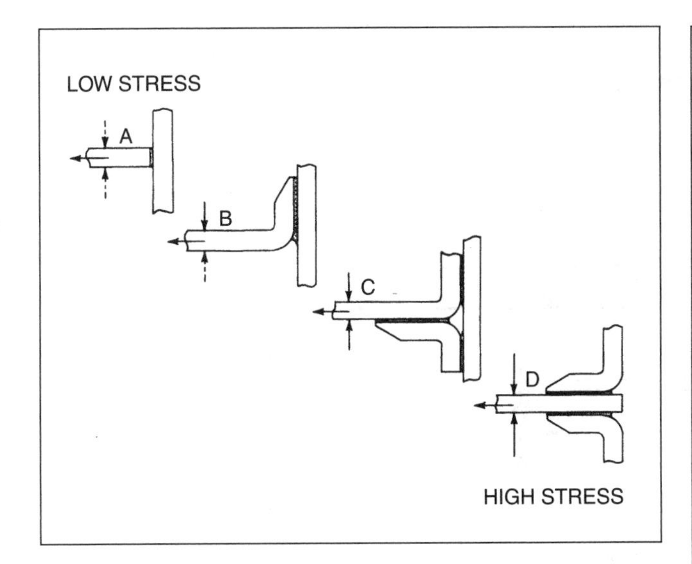

Figure 12.30—T-Joint Designs for Sheet-Metal Brazements

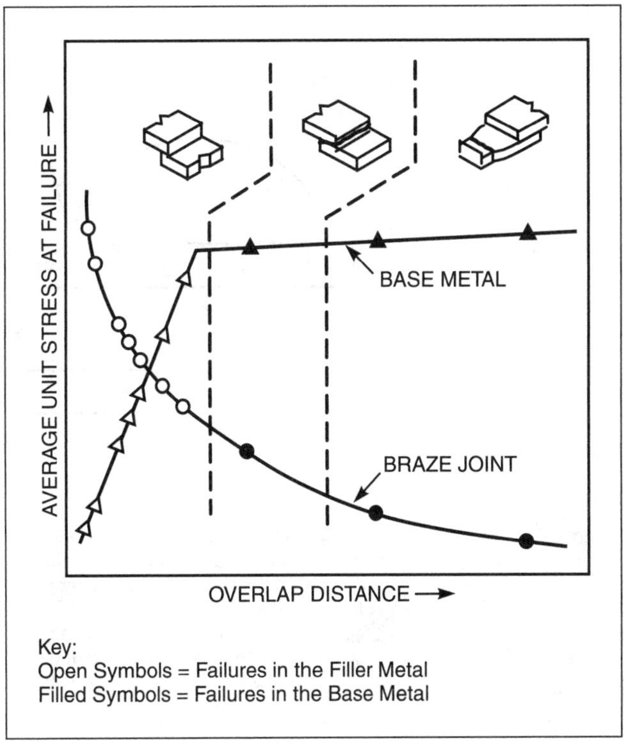

Since voids in the brazed joint will reduce the effective area of the electrical path, lap joints are recommended. A lap length at least 1-1/2 times the thickness of the thinner member will have a joint resistance approximately equal to the same length in solid copper.

AVERAGE UNIT STRESS AT FAILURE →
BASE METAL
BRAZE JOINT
OVERLAP DISTANCE →

Key:
Open Symbols = Failures in the Filler Metal
Filled Symbols = Failures in the Base Metal

Figure 12.31—Average Unit Shear Stress in the Brazed Lap Joint in the Base Metal as Functions of Overlap Distance

TESTING OF BRAZED JOINTS

Standardized testing must be adopted to evaluate the strength of brazed joints. Different designs of test specimens yield different results. Note in Figure 12.31 that the apparent joint strength of the brazed joint measured for a low overlap distance is high in comparison to the long overlap strength. Each of two laboratories that test only one overlap distance may test at opposite ends of the braze joint curve, and may reach widely different conclusions. The entire usable overlap range of the curve must be sampled to obtain adequate data.

The load-carrying capacity of the joint is best revealed in the right-hand portion of the base metal curve in Figure 12.31. The brazement should be designed to fail in the base metal without an excessive overlap. Further information is available in a standard published by the American Welding Society, *Standard Method for Evaluating the Strength of Brazed Joints*, ANSI/AWS C3.2.[24]

24. See Reference 10.

METALLURGICAL CONSIDERATIONS

Brazing is accomplished with temperatures that are below the solidus of the metals being joined. Metallurgical changes that accompany brazing are restricted to solid-state reactions in the base metal, solidification and interface reactions between the brazing filler metal and base metal, and reactions within the solid filler metal.

Capillary Flow

The capillary flow of brazing metal depends on its surface tension, wetting characteristics, and physical and metallurgical reactions with the base metal, flux or atmosphere, and oxides on the base-metal surface. The flow is further controlled by hydrostatic pressure within the joint. Figure 12.32 is an idealized presentation of the wetting behavior.

Contact angles greater than 90° indicate that no wetting (or dewetting) has occurred (see Figure 12.32(A). The angle for ideal wetting ($\theta = 0°$) is shown in (C). A contact angle less than 90° measured between the solid

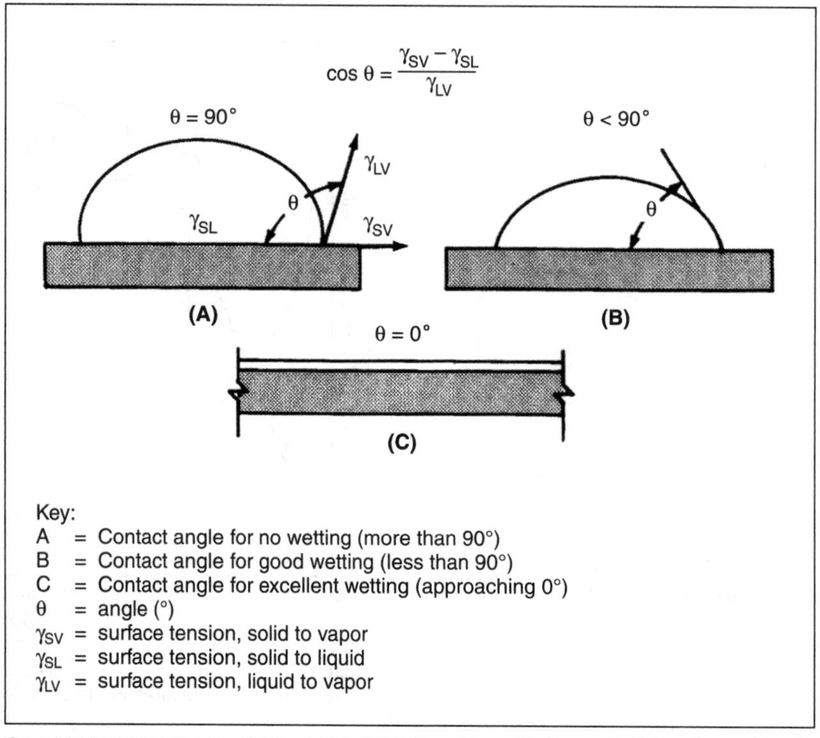

Source: Adapted from American Welding Society (AWS) Committee on Brazing and Soldering, Jenney, C.L., ed., (forthcoming), *Brazing Handbook*, 5th Ed., Miami: American Welding Society, Figure 1.3.

Figure 12.32—Contact Angle for a Liquid Droplet on a Solid Surface: Wetting Angles of Brazing Filler Metals

and liquid usually identifies a positive wetting characteristic (B).

In some brazing processes, wetting and spreading are assisted by the addition of flux. In protective-atmosphere brazing, wetting and flow depend entirely on surface interactions between the liquid metal and base metal. Most oxides cannot form in protective atmospheres or are readily displaced or removed by flux. Oxides of chromium, aluminum, titanium, and manganese require special treatments.

Interfacial Reactions

When liquid filler metal is present in the joint at the peak temperature in the brazing cycle, erosion can occur in some base metals. The rate of dissolution of the base metal by the filler metal depends on the mutual solubility limits, the quantity of brazing filler metal available in the joint, the brazing temperature, and the potential formation of lower-temperature eutectics or compounds.

Sometimes an interlayer of intermetallic compound may form between the filler metal and the base metal during the joining operation. Phase diagrams are used to predict intermetallic compound formation.

Once the filler metal has solidified to form the joint, subsequent effects may be controlled by diffusion phenomena. As an illustration of this metallurgical joining mechanism, when superalloys are brazed with a nickel-base filler metal containing boron and held at the brazing temperature, the boron diffuses into the base metal. Grain boundaries grow across the joint, sometimes obscuring the location of the joint. Liquid filler metal penetration between base metal grain boundaries may also occur. This process is called *diffusion brazing*.

Liquid-Metal Embrittlement

Base metals in a stressed state are particularly susceptible to liquid-metal penetration and subsequent embrittlement. This is sometimes called *stress corrosion cracking*. The molten metal penetrates and weakens the grain boundaries of the base metal. Copper-base filler metals used on high iron-nickel alloys under stress fail rapidly. Alloying elements diffuse more rapidly into grain boundaries than into the crystal lattice of a grain. Any filler metal that attacks the grain boundary of the base metal during brazing increases the risk of embrittling the brazed joint.

In a few cases, low-melting alloy elements may fill grain-boundary cracks as they separate. This helps to mitigate damage and is occasionally referred to as an *intrusion*.

Service Conditions

The dynamic metallurgical characteristics of the brazing process may extend into the service environment of the brazement. Careful consideration should be given to the subsequent diffusion and metallurgical changes that can occur in service. At elevated temperatures, changes may occur in the solid state as a direct result of diffusion, oxidation, or corrosion. This means that the metallurgical and mechanical properties of these joints may change in service and must be evaluated as part of the joint qualification procedure.

PROCEDURES

Successful brazing results depend on careful management of procedures from surface preparation through brazing, cleanup, and inspection. The latest edition of the American Welding Society publication *Standard for Brazing Procedure and Performance Qualification,* ANSI/AWS B2.2, should be consulted.[25]

PRECLEANING AND SURFACE PREPARATION

Metals to be brazed must be free of oxides and other contaminants. Clean, oxide-free surfaces are essential to ensure sound brazed joints of uniform quality. Grease, oil, dirt, and oxides prevent the uniform flow and bonding of the brazing filler metal and also impair fluxing action, resulting in discontinuities (voids and inclusions). With the metals that form refractory oxides or with critical-atmosphere brazing applications, precleaning must be more thorough and the cleaned components must be preserved and protected from contamination.

The length of time that cleaning remains effective depends on the metals involved, the atmospheric conditions, the amount of handling of the components, the manner of storage, and similar factors. It is recommended that brazing be done as soon as possible after the components have been cleaned.

Degreasing is usually the first procedure. The following degreasing methods are commonly used:

1. Solvent cleaning with petroleum solvents or chlorinated hydrocarbons;
2. Vapor degreasing with stabilized perchloroethylene;
3. Alkaline cleaning with commercial mixtures of silicates, phosphates, carbonates, detergents, soaps, wetting agents, and, in some cases, hydroxides;
4. Emulsion cleaning with mixtures of hydrocarbons, fatty acids, wetting agents, and surface activators; and
5. Electrolytic cleaning, both anodic and cathodic.

The effectiveness of these methods can be enhanced by mechanical agitation or by applying ultrasonic vibrations to the bath if a bath method is used.

Scale and oxide removal can be accomplished mechanically or in a chemical bath, often with a pickling solution. Prior removal of oils and greases allows intimate contact of the pickling solution with the components. Vibration aids in descaling with the solutions in the following cleaning methods:

1. Acid cleaning with phosphate-type acid cleaners;
2. Acid pickling with sulfuric, nitric, and hydrochloric acid; or
3. Electrolytic pickling in a salt bath.

The selection of a chemical cleaning agent depends on the nature of the contaminant, the base metal, the surface condition, and the joint design. For example, base metals containing copper and silver should not be pickled with nitric acid. In all cases, the chemical residue must be removed by thorough rinsing to prevent formation of other equally undesirable films on the faying surfaces or subsequent chemical attack of the base metal.

Mechanical cleaning removes oxide and scale and also roughens the faying surfaces to enhance capillary flow and wetting by the brazing filler metal. Mechanical methods include grinding, filing, machining, wire brushing, and grit blasting. Grit blasting should be done with clean blasting grit made of iron, stainless steel, modified nickel-base grit, or silicon carbide. The material must not leave any deposit on the surfaces that would impair wetting by the filler metal. Grit blasting should never be done with blasting materials such as

25. American Welding Society (AWS) Committee on Welding Qualification, 1991, *Standard for Brazing Procedure and Performance Qualification,* ANSI/AWS B2.2-91, Miami: American Welding Society.

silica sand, alumina, and other nonmetallic materials, as they cause brazing problems.

FLUXING

When a flux is selected for use, it must be applied as an even coating, completely covering the joint surfaces of the assembly. Fluxes are most commonly applied in the form of pastes or liquids. Dry powdered flux can be sprinkled on the joint or applied by dipping the heated end of the filler metal rod into the flux container. The particles should be small and thoroughly mixed to improve metal coverage and fluxing action. The areas surrounding the joints can be kept free from discoloration and oxidation by applying flux to a wide area on each side of the joint.

Paste and liquid flux should adhere to the clean metal surfaces. If the metal surfaces are not clean, the flux will ball up and leave bare spots. Thick paste fluxes can be applied by brushing. Less viscous formulations can be applied by dipping, manual squirting, or automatic dispensing. The proper consistency depends on the types of oxides present as well as the heating cycle. For example, ferrous oxides formed during rapid heating of the base metal are soft and easy to remove, and only limited fluxing action is required. However, when joining copper or stainless steel or when a long heating cycle is used, a concentrated flux is required. Flux reacts with oxygen, and once it becomes saturated, it loses all its effectiveness. The viscosity of the flux can be reduced without dilution by heating it to 50°C to 60°C (120°F to 140°F), preferably in a ceramic-lined flux or glue pot with a thermostat control. Warm flux has low surface tension and adheres to the metal more readily. Recommendations for fluxes are published in the American Welding Society standard *Specification for Fluxes for Brazing and Braze Welding,* ANSI/AWS A5.31.[26]

Stopoff

When filler metal flow must be restricted to defined areas, a *stopoff* is applied. Stopoffs are commercially available preparations that outline and protect the areas that are not to be brazed. Stopoffs are produced in several forms. One form is a slurry that includes water or an organic binder of oxides of aluminum, chromium, titanium, or magnesium. Others are called *parting compounds* and *surface reaction stopoffs.* They are in the form of powders that are mixed with a binder or paints and viscous materials that can be brushed, sprayed or extruded.

Brazing Filler Metal Preplacement

When designing a brazed joint, the brazing process to be used and the manner in which the filler metal will be placed in the joint should be established. In most manually fabricated brazed joints, the filler metal is simply fed into the face of the joint. For furnace brazing and high-production brazing, the filler metal is preplaced at the joint. Manual or automatic dispensing equipment is available to perform this operation.

Brazing filler metal is manufactured in several forms—wire, shims, strip, powder, and paste. Figures 12.33 and 12.34 illustrate methods of preplacing brazing filler metal in wire and sheet forms. When the base metal is grooved to accept preplaced filler metal, the groove should be cut in the heavier section. When computing the strength of the intended joint, the groove area should be subtracted from the joint area, since the brazing filler metal will flow out of the groove and into the joint interfaces, as shown in Figure 12.35.

Powdered filler metal can be applied in any of the locations indicated in Figures 12.33 and 12.34. It can be applied dry to the joint area and then moistened with

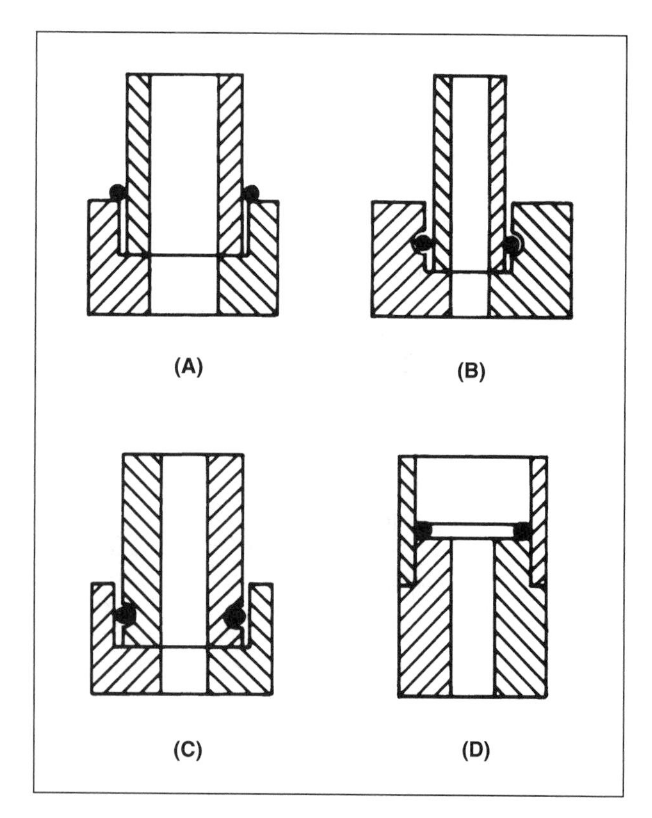

Figure 12.33—Methods of Preplacing Brazing Filler Wire

26. See Reference 13.

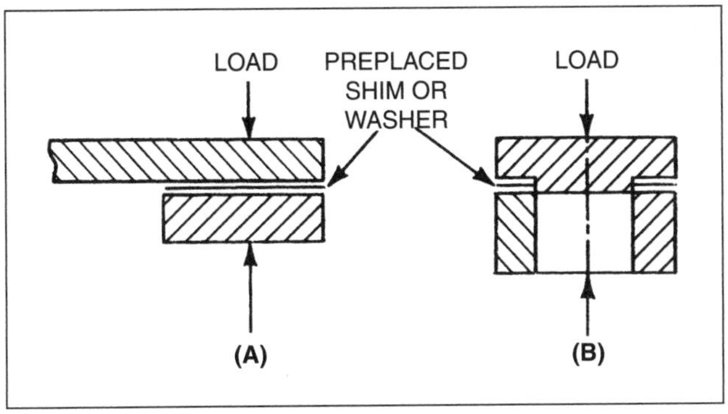

Figure 12.34—Preplacement of Brazing Filler Shims

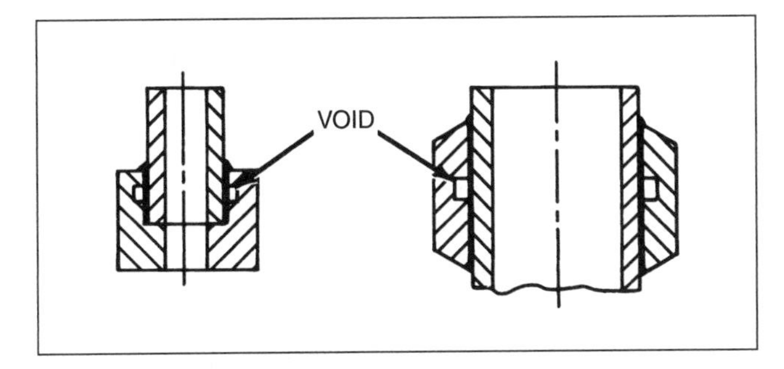

Figure 12.35—Brazed Joints with Grooves for Preplacement of Filler Metal; After the Brazing Cycle the Grooves are Void of Filler Metal

a binder, or it can be premixed with the binder and applied to the joint. The density of powder is usually only 50% to 70% of a solid metal, so the groove volume must be larger for powder.

When preplaced shims are used, the sections being brazed should be free to move together when the shims melt. Some type of loading may be necessary to accommodate this movement and force excess filler metal and flux (when flux is used) out of the joint.

Assembly and Fixturing

When flux is used, the components to be brazed should be assembled immediately after fluxing, before the flux has time to dry and flake off.

When fixtures are needed to maintain alignment or dimensions, the mass of a fixture should be minimized. It should have pinpoint or knife-edge contact with the part to be brazed, but should be away from the joint area to minimize heat loss through conduction to the fixture. The fixture material must have adequate strength at brazing temperature to support the brazement. It must not readily sinter at elevated temperatures with the components at the points of contact. For torch brazing, extra clearance is needed to access the joint and the brazing filler metal with the torch flame. For induction brazing, fixtures are generally made of ceramic materials to avoid having extraneous metal in the field of the induction coil. Ceramic fixtures can be designed to serve as a heat shield or a heat absorber.

Some components to be brazed can be designed to be self-fixturing, which represents an economical advantage. (Refer to Figure 12.21.)

Flux Removal

For all processes, all traces of flux residue should be removed from the brazement after brazing is completed because of the potential for corrosion in certain environments while in service for extended periods of time. Flux residues can usually be removed by rinsing with hot water. Oxide-saturated flux is glasslike and more difficult to remove. If the metal and joint design can withstand quenching, saturated flux can be removed by quenching the brazement from an elevated temperature. This treatment cracks off the flux coating. In stubborn cases, it may be necessary to use a warm acid solution, such as 10% sulfuric acid or one of the proprietary cleaning compounds that are available commercially. Nitric acid should not be used on alloys containing copper or silver.

Oxidized areas adjacent to the joint can be restored by chemical cleaning or by mechanical methods such as wire brushing or blast cleaning.

Fluxes used for brazing aluminum are not readily soluble in cold water. The brazements are usually rinsed in very hot water, above 82°C (180°F), with a subsequent immersion in nitric acid, hydrofluoric acid, or a combination of those acids. A thorough after-rinse with water is then necessary.

Stopoff Removal

Stopoff materials of the parting-compound type can be easily removed mechanically, by wire brushing, air blasting, or water flushing. The surface-reaction type of stopoff used on corrosion- and heat-resistant base metals can best be removed by pickling in hot nitric acid-hydrofluoric acid, except in assemblies containing copper and silver. Sodium hydroxide (caustic soda) or ammonium bifluoride solutions can be used in all applications, including copper and silver, because they will not attack base metals or filler metals. A few stopoff materials can readily be removed by dipping in 5% to 10% nitric or hydrochloric acid or by grit blasting.

INSPECTION

Inspection of brazements should always be required to assure the proper quality of the brazement, and inspections are often specified by regulatory codes and by the fabricator. Inspection of brazed joints can be conducted on test specimens or on the finished brazed assembly. The tests may be nondestructive or destructive.

Brazing discontinuities can be grouped in the following general classes: those associated with drawing or dimensional requirements, those associated with structural discontinuities in the brazed joint, and those associated with the braze metal or the brazed joint.

NONDESTRUCTIVE EXAMINATION METHODS

The two objectives of nondestructive examination (NDE) of brazed joints should be to find discontinuities defined in quality standards or codes and to obtain clues to the causes of irregularities in the fabrication process.

Visual Inspection

Every brazed joint should be examined visually. It is also a convenient preliminary test when other test methods are to be used.

The joint can be visually examined to assure that it is free from materials such as oxide film, flux, and stopoff residue. Visual examination can reveal flaws due to damage, misalignment and poor fit-up of assemblies, dimensional inaccuracies, inadequate flow of brazing filler metals, exposed voids in the joint, surface flaws such as cracks or porosity, and heat damage to the base metal. Visual inspection will not detect internal flaws such as flux entrapment in the joint or incomplete filler metal flow between the faying surfaces.

Proof Testing

Proof testing is a method of inspection that subjects the completed joint to loads slightly in excess of those that will be experienced during its subsequent service life. These loads can be applied by hydrostatic methods, by tensile loading, by spin testing, or by numerous other methods. Occasionally, it is not possible to assure that a brazement is serviceable by any of the other nondestructive methods of inspection, and proof testing then becomes the most satisfactory method.

Leak Testing

Pressure testing determines the gas or liquid tightness of a closed vessel. It can be used as a screening method to find gross leaks before adopting more sensitive test methods. A pressure test with air or liquid can be performed by one of the following methods (and is sometimes used in conjunction with a pneumatic proof test):

1. Submerging the pressurized vessel in water and noting signs of leakage by rising air bubbles;
2. Pressurizing the assembly, closing the gas (including air) inlet source, and then noting any change in internal pressure over a period of time (corrections for temperature may be necessary); or
3. Pressurizing the assembly and checking for leaks by brushing the joint area with a soap solution or a commercially available liquid and noting any bubbles and their source.

A method sometimes used in conjunction with a hydrostatic proof test is to examine the brazed joints visually for indications that the hydrostatic fluid is escaping through the joint.

The leak testing of brazed assemblies with a refrigerant gas is extremely sensitive. The brazed vessel under test is pressurized using either pure refrigerant gas or a gas such as nitrogen containing a tracer. Areas are probed with a sampling device that is sensitive to the halide ion. An audible alarm indicates the detection of a leak. A leak can be measured quantitatively with this method. Precautions must be exercised to avoid contaminating the surrounding air with the refrigerant gas, which will decrease the sensitivity of the method.

The mass spectrometer leak test is the most sensitive and accurate way of detecting extremely small leaks in a brazed vessel. The mass spectrometer is a sensing device that detects gas and converts it to an electrical signal. The mass spectrometer is coupled to the interior and the brazed assembly is evacuated. A tracer gas, such as helium, is used in conjunction with this device in one of two ways. The area to be tested is surrounded with the tracer gas, or the brazed assembly is pressurized with the tracer gas and the exterior is monitored with the mass spectrometer probe, which presents a readout.

Liquid Penetrant Inspection

Liquid penetrant inspection is a nondestructive method capable of finding cracks, porosity, incomplete flow, and similar flaws open to the surface of a brazed joint. Commercially available colored or fluorescent penetrants are applied to the surface and allowed to flow into surface openings by capillary action. After the surface penetrant has been removed, a white developer is applied that will draw out any colored penetrant in a flaw to the surface. Colored penetrant is visible under ordinary light. When fluorescent penetrant is used, flaw indications will glow when placed under an ultraviolet light source. Penetrant inspection is not applicable to brazed fillets, as it often leads to misinterpretation. It should be used only after machining the joint.

Detailed information on liquid penetrant inspection is published in three AWS documents: *Specification for Torch Brazing*, ANSI/AWS C3.4;[27] *Specification for Induction Brazing*, ANSI/AWS C3.5;[28] and *Specification for Furnace Brazing*, ANSI/AWS C3.6.[29]

Radiographic Examination

Radiographic examination of brazements can reveal incomplete bonding or incomplete flow of the filler metal. The joints should be uniform in thickness and the exposure made straight through the joint. The sensitivity of the method is generally limited to 2% of the joint thickness. X-ray absorption by certain filler metals, such as gold and silver, is greater than absorption by most base metals. Therefore, areas in the joint that are devoid of braze metal are revealed by a much darker image than the brazed area on the film or viewing screen.

Ultrasonic Inspection

The ultrasonic examination method uses low-energy, high-frequency mechanical vibration (sound waves) and readily detects, locates, or identifies discontinuities in brazed joints. Whether ultrasonic examination is applicable to brazements depends largely on the design of the joint, surface condition, material grain size, and the configuration of adjacent areas. The latest edition of the American Welding Society document *Recommended Practices for Ultrasonic Inspection of Brazed Joints*, ANSI/AWS C3.8, provides detailed information on this method of inspection.[30]

Thermal Heat Transfer Examination

Examination by heat transfer is a method used to detect incomplete bonding in brazed assemblies such as honeycomb structures and covered skin panel surfaces. With one technique, the surfaces are coated with a developer that is a melting-point powder. The developer melts and migrates to cool areas when the heat from an infrared lamp is applied. The bonded areas act as heat sinks, resulting in a thermal gradient to which the developer will react. Sophisticated techniques use phosphors, liquid crystals, and temperature-sensitive materials.

Infrared-sensitive electronic devices that provide some form of readout are available to monitor temperature differences less than 1°C (2°F), which indicate variations in braze quality.

DESTRUCTIVE TESTING METHODS

Destructive methods of inspection clearly demonstrate whether a brazement design will meet the requirements of intended service conditions. Destructive methods must be restricted to partial sampling. It is used to verify the nondestructive methods of inspection by sampling production specimens at suitable intervals.

METALLOGRAPHIC EXAMINATION

Metallographic examination requires the removal of sections from the brazed joints and preparation for macroscopic or microscopic examination. This method reveals the microstructure of the brazed joint, enabling the inspector to detect porosity and other flaws such as poor flow of brazing filler metal, excessive base metal erosion, the diffusion of brazing filler metal, or improper fitup of the joint.

Peel Tests

Peel tests are frequently employed to evaluate lap joints. In a peel test, a member of the brazed specimen is clamped rigidly in a vise, and the free member is peeled away from the joint. The broken members reveal the general quality of the bond and the presence of voids and flux inclusions in the joint. The permissible number, size, and distribution of these discontinuities should be defined in the job contract, specification, or code.

Tension and Shear Tests

Carefully controlled tension and shear tests can be used to quantitatively determine the strength of the brazed joint or to verify the relative strengths of the joint and base metal. These tests are widely used when developing a brazing procedure. Random sampling of brazed joints is used for quality control and verification of brazing performance.

Torsion Tests

Torsion tests are performed to evaluate brazed joints by using a stud, screw, or tubular member that has been

27. See Reference 3.
28. See Reference 6.
29. See Reference 5.
30. American Welding Society (AWS) Committee on Brazing and Soldering, *Recommended Practices for Ultrasonic Inspection of Brazed Joints,* ANSI/AWS C3.8, Miami: American Welding Society.

brazed to a base member. In a torsion test, the base member is clamped rigidly; then the stud, screw, or tube is rotated to failure, which will occur in either the base metal or the brazing alloy.

BRAZE QUALITY

Nondestructive and destructive examinations are used to identify several types of brazing discontinuities. The limits of acceptability should be defined specifically in the job contract or by reference to codes or other standards. Some common discontinuities in brazed joints are discussed in this section.

Voids, Porosity, and Incomplete Fill

Voids, porosity, or incomplete fill can be the result of improper cleaning, excessive joint clearances, insufficient filler metal, entrapped gas, solidification shrinkage, and movement of the faying surfaces of the brazement caused by improper fixturing. The filler metal is vulnerable (i.e., susceptible to oxidation and cannot support loads) when in the liquid or partially liquid state. Incomplete fill reduces the strength of the joint by reducing the load-carrying area, and it may provide a path for leakage.

Flux Entrapment

Entrapped flux is a possible source of discontinuities that may be found in any brazing operation where a flux is added to prevent and remove oxidation during the heating cycle. Flux trapped in the joint prevents the flow of the filler metal into that area, thus reducing joint strength. It may also falsify leak- and proof-test indications. If open to the surface, entrapped corrosive flux may reduce service life. When sealed, entrapped flux is not normally corrosive.

Discontinuous Fillets

Discontinuous fillets are usually noted during visual inspection. Whether this condition can be waived depends on the job contract.

Base Metal Erosion

Erosion occurs by the excessive dissolution of the base metal by the brazing filler metal. It may result in the thinning of the base metal. Erosion reduces the strength of the base metal by changing the composition of the materials and by reducing the base metal cross-sectional area.

Unsatisfactory Surface Appearance

Unsatisfactory appearance of the brazing filler metal, including excessive spreading and roughness, is objectionable for more than aesthetic reasons. These defects may act as stress concentrations, corrosion sites, or may interfere with inspection of the brazement.

Cracks

Cracks reduce both the strength and service life of the brazement. Cracks act as stress raisers, lowering the mechanical strength of the brazement and causing premature fatigue failure.

TROUBLESHOOTING

Unsatisfactory brazing is usually the result of one or more of the following failures:

1. No wetting—no capillary flow resulting in discontinuities or voids;
2. Excessive wetting—too much filler metal where it is not desired, e.g., in holes or on machined surfaces; and
3. Erosion—attack on the base metal by the brazing filler metal, which reduces the thickness of base metal areas.

If the basic cause of each of these failures can be identified, the solution to the brazing problem will be found. Table 12.8 lists items for consideration for each of these problems.

BRAZE WELDING

Braze welding (BW) is accomplished with a brazing filler metal that has a liquidus above 450°C (840°F) but below the solidus of the base metals to be braze welded. As noted on the first page of this chapter, braze welding differs from brazing in that the filler metal is not distributed in the joint by capillary action. The filler metal in the form of brazing rod is face-fed into the joint groove or is deposited from a gas metal arc welding electrode. The base metals are not melted; only the filler metal melts. Bonding takes place between the deposited filler metal and the hot unmelted base metals in the same manner as conventional brazing, but without intentional capillary flow. Joint designs for braze welding are similar to those used for oxyacetylene welding.

Table 12.8
Solutions to Typical Brazing Problems

PROBLEM	CAUSES
No Flow, No Wetting	Braze filler—different lot or wrong selection
	Low temperature—poor technique, thermocouple/controller error
	Time—too short
	Dirty parts—not cleaned properly
	Poor atmosphere—too little flux, wrong flux, bad gas or vacuum
	No nickel-plate or other coatings (when required)—allowing oxidation of base metal
	Joint clearance too large—poor fitup control
Excess Flow or Wetting Causing Hole Plugging or Brazing Wrong Joints	Temperature too high—poor technique, furnace error
	Time—too long
	Too much filler metal—poor technique, different joint clearance size
	Braze filler—different lot or wrong selection
	No stopoff used
Erosion-Braze Filler Metal Eats Away Base Metal	Temperature too high—poor technique, furnace error
	Time at temperature too long—poor technique, controller error
	Excessive braze filler metal—poor technique, change in joint clearance, components in different attitude
	Cold-worked components highly susceptible—change in manufacturer—not stress relieved
	Braze-filler metals temperatures too high above liquidus, or high concentration of melting point depressants

Braze welding was originally developed to repair cracked or broken cast-iron components. Fusion welding of cast iron requires extensive preheating and slow cooling. This minimizes the development of cracks and the formation of hard cementite. With braze welding, cracks and cementite are easier to avoid, and fewer expansion and contraction problems are encountered.

COMMONLY USED BRAZE WELDING PROCESSES

At least two processes are used for braze welding. The most common are braze welding with an oxyfuel gas welding torch or using the gas metal arc welding process.

Torch Braze Welding

Although the shielded metal arc welding process is sometimes used for the braze welding of cast iron, most braze welding is done with an oxyfuel gas welding torch, a copper alloy brazing rod, and a suitable flux. Braze welding also is done with gas tungsten arc and plasma arc torches without flux. The gas tungsten arc welding and plasma arc welding torches, which use inert gas shielding, accomplish braze welding with filler metals that have relatively high melting temperatures.

Gas Metal Arc Braze Welding

The gas metal arc welding (GMAW) process, when adapted to brazing, uses filler metals with relatively low melting points that minimize the melting of the base metal. A copper-based electrode (e.g., aluminum bronze or silicon bronze) is used to deposit the weld bead. CuAl and CuSi gas metal arc welding electrodes are examples. This mode is very successful for joining thin sheet. It is used extensively in the auto body manufacturing industry and to join heat-sensitive materials such as cast iron.

Gas metal arc braze welding is also used to join galvanized steels because the low heat input minimizes the amount of galvanized coating that is melted or removed. The copper-based weld bead is more corrosion resistant than welds made with a carbon-steel electrode.

ADVANTAGES AND LIMITATIONS OF BRAZE WELDING

Braze welding equipment is simple and easy to use. The process can produce joints with adequate strength for many applications. Braze welding has the following advantages over conventional fusion welding processes:

1. Less heat is required to accomplish bonding, which permits faster joining and lower fuel consumption;
2. The process produces little distortion from thermal expansion and contraction;
3. The deposited filler metal is relatively soft and ductile, can be machined, and is under low residual stress;
4. Brittle metals, such as gray cast iron, can be braze welded without extensive preheat; and
5. The process provides a convenient way to join dissimilar metals such as copper to steel or cast iron and nickel-copper alloys to cast iron and steel.

Braze welding has these limitations:

1. Weld strength is limited to that of the filler metal;
2. Permissible performance temperatures of brazements are lower than those of fusion welds because of the lower melting temperature of the filler metal, e.g., with copper-alloy filler metal, service temperatures are limited to 260°C (500°F) or lower;
3. The braze welded joint may be subject to galvanic corrosion and differential chemical attack; and
4. The brazing filler metal color may not match the base metal color.

BRAZE WELDING EQUIPMENT

Conventional braze welding performed with an oxyfuel gas welding torch and the associated equipment is described in Chapter 11 of this volume. In some applications, an oxyfuel preheating torch may be needed. Special applications use the gas metal arc, gas tungsten arc, or plasma arc welding equipment described in Chapter 3, Chapter 4, and Chapter 7, respectively. Clamping and fixturing equipment may be needed to hold the components in place and to align the joint.

BASE METALS

Braze welding is employed primarily on base metals such as carbon steels, low-alloy steels, and cast-iron alloys. It can also be used to join copper, nickel, and nickel alloys. Braze weldments of dissimilar metals between many of these metals are possible if suitable filler metals are used. Other metals can be braze welded with suitable filler metals that wet and form a strong metallurgical bond with them.

FILLER METALS

Commercial braze welding filler metals are of the BCuZn type—the brasses containing approximately 60% copper and 40% zinc. Brazing alloys with small additions of tin, iron, manganese, and silicon improve flow characteristics and decrease the volatilization of the zinc. These alloys scavenge oxygen and increase the weld strength and hardness. Filler metal with added nickel (10%) has a whiter color and produces higher weld-metal strength.

Chemical compositions and properties of four standard copper-zinc welding rods used for braze welding are listed in Table 12.9. The minimum joint tensile strength achievable is approximately 275 to 413 MPa

Table 12.9
Copper-Zinc Welding Rods for Braze Welding

AWS Classification*	Approximate Chemical Composition, %					Minimum Tensile Strength		Liquidus Temperature	
	Copper	Zinc	Tin	Iron	Nickel	MPa	ksi	°C	°F
RBCuZn-A	60	39	1	—	—	275	40	900	1650
RBCuZn-B	60	39	1	—	—	477	65	882	1620
RBCuZn-C	60	38	1	1	—	344	50	890	1630
RBCuZn-D	50	40	—	—	10	413	60	935	1714

*See American Welding Society (AWS) Committee on Filler Metals and Allied Materials, *Specification for Filler Metals for Brazing and Braze Welding*, ANSI/AWS A5.8, Miami: American Welding Society.

(40 to 60 ksi). The joint strength decreases rapidly when the weldment is exposed to temperatures above 260°C (500°F).

Because a braze weld is a bimetal joint, the susceptibility to corrosion of both the base metal and filler metal must be considered in the application of braze welding to a given product. The completed joint is subject to galvanic corrosion in certain environments, and the filler metal may be less resistant to certain solutions than the base metal.

BRAZE WELDING FLUXES

Fluxes for braze welding are proprietary compounds developed for the braze welding of specific base metals with brass filler metal rods. They are designed for use at temperatures higher than those encountered in brazing operations, and so they remain active for longer times at temperature than similar fluxes used for capillary brazing. The following types of flux are in general use for the braze welding of iron and steels:

1. A basic flux that cleans the base metal and weld beads and assists in the precoating (wetting) of the base metal, used for steel and malleable iron;
2. A flux that performs the same functions as the basic flux and also suppresses the formation of zinc oxide fumes; and
3. A flux containing iron oxide or manganese dioxide that combines with free carbon on the cast iron surface and removes the carbon, formulated specifically for braze welding of gray or malleable cast iron.

Further information is presented in *Specification for Fluxes for Brazing and Braze Welding*, AWS A5.31, published by the American Welding Society.[31]

Flux can be applied by one of the following methods:

1. The heated filler rod can be dipped into the flux and transferred to the joint during braze welding,
2. The filler rod can be precoated with flux,
3. The flux can be brushed on the joint prior to brazing, or
4. The flux can be introduced through the oxyfuel gas flame.

METALLURGICAL CONSIDERATIONS FOR BRAZE WELDING

The bonding mechanism between filler metal and base metal in braze welding is the same bond that occurs with conventional brazing. Clean base metal components are heated to a temperature at which the faying surfaces are wetted by the molten filler metal, producing a metallurgical bond between the two components. Cleanliness is a prerequisite. The presence of dirt, oil, grease, oxide film, or carbon will inhibit wetting.

Following wetting, atomic diffusion takes place between the brazing filler metal and the base metal in a narrow zone at the interface. Some base metals allow the brazing filler metal to slightly penetrate the grain boundaries of the base metal, further contributing to bond strength.

Filler materials for braze welding are alloys that have sufficient ductility as cast to let them flow plastically during solidification and subsequent cooling. The alloys thereby accommodate shrinkage stresses. It should be noted that two-phase alloys that have a low-melting grain boundary constituent are not useable—the boundaries crack during solidification and cooling.

BRAZE WELDING APPLICATIONS

The most common use of braze welding is the repair of broken or defective steel and cast iron components of machinery and tools. Since large parts can be repaired in place, significant cost savings result. Braze welding also rapidly joins thin-gauge mild steel sheet and tubing for applications in which fusion welding would be difficult.

A major application of braze welding is the joining of galvanized steel ducting for air conditioning systems. The brazing temperature is held below the vaporization temperature of the zinc overcoat. This minimizes the loss of the protective zinc coating from the steel surfaces. Exhaust ventilation is required for this application because it exposes the welder to a significant amount of zinc fumes.

The thicknesses of metals that can be braze welded range from thin-gauge sheet to very thick cast iron sections. Fillet and groove welds are used to make butt, corner, lap, and T-joints.

Braze Welding Procedures

Groove, fillet and edge welds are used to braze weld assemblies made from sheet and plate, pipe, tubing, rods, bars, castings, and forgings. To obtain good joint strength, an adequate bond area between the brazing filler metal and the base metal is required. Weld groove geometry should provide an adequate groove face area so that the joint will not fail along the interfaces.

Fixturing. Fixturing is usually required to hold the components in the proper location and alignment for

31. See Reference 13.

braze welding. When repairing cracks and defects in cast-iron components, fixturing may not be necessary unless the component is broken into pieces.

Joint Design and Preparation. Joint designs for braze welds are similar to those used for oxyacetylene welding. For thicknesses over 2 mm (3/32 in.), single- or double-V-grooves are prepared with included angles of 90° to 120° to provide large bond areas between base metal and filler metal. Square grooves can be used for thickness less than 2 mm (3/32 in.).

The prepared joint faces and adjacent surfaces of the base metal must be cleaned to remove all oxide, dirt, grease, oil, and other foreign material. The joint faces of cast iron must also be free of graphite smears caused by prior machining. Graphite smears can be removed by quickly heating the cast iron to a dull red color and then wire brushing it after it cools to black heat. If the casting has been heavily soaked with oil, it should be heated in the range of 320°C to 650°C (600°F to 1200°F) to burn off the oil. The surfaces should be wire brushed to remove any residue.

In production braze welding of cast iron components, the surfaces to be joined are usually cleaned by immersion in an electrolytic salt bath.

Preheating

Preheating may be required to prevent cracking from thermally induced stresses in large cast iron braze welds. When braze welding copper, preheating reduces the amount of heat required from the brazing torch and the time required to complete the joint.

Local or general preheating can be applied. The temperature should be in the range of 425°C to 480°C (800°F to 900°F) for cast iron. Higher temperatures can be used for copper. When braze welding is completed on cast-iron components, they should be thermally insulated for slow cooling to room temperature to minimize the development of thermally induced stresses.

Technique

When using the oxyfuel gas torch for braze welding, the joint must be aligned and fixtured in position. Braze welding flux, when required, is applied to preheated filler rod (unless precoated) and also sprinkled on thick joints during heating with the torch. The base metal is heated until the filler metal melts, wets the base metal, and flows onto the joint faces (precoating). The braze welding operator then progresses along the joint, precoating the faces and filling the groove with one or more passes, using operating techniques similar to oxyfuel gas welding. The inner cone of an oxyacetylene flame should not be directed on copper-zinc alloy filler metals or on an iron or steel base metal.

With electric arc torches, the technique is similar to oxyfuel gas braze welding, except that flux is not generally used.

ECONOMICS

Cost considerations for brazed products—materials, labor and overhead—are the same as those associated with any manufactured product and very similar to those for welded products. Selecting brazing as the joining process for a given application is often a good economic choice, primarily because brazing of large numbers of identical parts can be done in batches.

Automated brazing operations are highly economical. The high cost of automation is usually justified by increased productivity. However, manual torch brazing, although labor intensive, is the better economic choice when a small number of brazements are needed or when the configuration of the product is too complex to be practical for automation.

Dissimilar metals and materials can be joined by brazing; this often provides an opportunity to choose less expensive base metals or filler metals. Energy cost savings can be achieved because the brazing heat is directed only to the joint area. Conversely, a continuous belt furnace increases production but lowers energy efficiency because the entire assembly is heated.

The following are several other cost considerations:

1. Some components can be designed to be self-fixturing, which is an economical advantage;
2. Large castings, or broken or defective steel and cast-iron components of machinery and tools can be repaired in place, resulting in significant cost savings;
3. The brazed joint often blends in with the base metal and does not need to be finished by grinding or other mechanical means.
4. The annealing, stress relief, or heat treatment of the base metal can often be combined with the brazing operation, eliminating separate thermal operations for each; and
5. Simplified joint designs can be used in service applications in which joints will be lightly loaded.

SAFE PRACTICES

Hazards encountered during brazing operations are similar to those associated with welding and cutting. At brazing temperatures some elements vaporize, producing toxic gases. Personnel and property must be protected against hot materials, gases, fumes, electrical shock, radiation, and chemicals. Before a brazing operation is started, Material Safety Data Sheets (MSDS) and equipment operating manuals from the manufacturers should be consulted.

Minimum safety requirements are specified in the latest edition of the American National Standard, *Safety in Welding, Cutting and Allied Processes,* ANSI Z49.1.[32] This standard applies to brazing, braze welding, and soldering, as well as other welding and cutting processes. A number of organizations publish standards that provide detailed information on safety issues associated with the use of these processes. These and other safety and health documents and their publishers are listed in Appendix B of this volume. Some of the safety concerns are discussed briefly in this section.

WORK AREA AND PERSONNEL SAFETY

Brazing equipment, machines, cables, and other apparatus should be placed so that they present no hazard to personnel in work areas, in passageways, on ladders, or on stairways. Good housekeeping should be maintained.

Precautionary signs conforming to the requirements of *Environment and Facility Safety Signs*, ANSI Z535.2, should be posted designating the applicable hazards and safety requirements.[33]

Ventilation

It is essential that adequate ventilation be provided so that personnel will not inhale gases and fumes generated while brazing. Some filler metals and base metals contain toxic materials such as cadmium, beryllium, zinc, mercury, or lead, which are vaporized during brazing. Fluxes contain chemical compounds of fluorine, chlorine, and boron, which are harmful if they are inhaled or come in contact with the eyes or skin.

Solvents and cleaning compounds containing chlorinated hydrocarbons, acids, and alkalis may be toxic or flammable or cause chemical burns when present in the brazing environment.

To avoid suffocation of personnel, atmosphere furnaces must be carefully checked to ensure that they are purged with air before personnel enter them.

Eye and Face Protection

Eye and face protection must comply with *Practices for Occupational and Educational Eye and Face Protection,* ANSI Z87.1.[34]

For torch brazing, operators and associated workers should wear goggles or spectacles with filter lens of shade number four or five. A lens selector guide is presented in Appendix A. Operators and workers associated with resistance, induction, or salt-bath dip-brazing equipment should use appropriate face shields, spectacles, or goggles for face and eye protection.

Protective Clothing

Appropriate protective clothing for brazing should provide sufficient coverage and be made of suitable materials to minimize skin burns that could be caused by spatter or radiation. Heavier materials such as wool or heavy cotton clothing are preferable to lighter materials because they are more difficult to ignite. All clothing must be free from oil, grease, and combustible solvents. Brazers and brazing operators should wear protective heat-resistant gloves made of leather or other suitable materials.

Respiratory Protective Equipment

When controls such as ventilation of the area fail to reduce air contaminants to allowable levels and where the implementation of such controls is not feasible, respiratory protective equipment should be used to protect personnel from hazardous concentrations of airborne contaminants. Only approved respiratory protection equipment should be used. Approvals of respiratory equipment are issued by the National Institute of Occupational Safety and Health (NIOSH) and the Mine Safety and Health Administration (MSHA).[35] Selection of the proper equipment should be in accordance with ANSI Z88.2, *Practices for Respiratory Protection.*[36]

PRECAUTIONARY INFORMATION

Two important sources of precautionary information are safety recommendations from suppliers of equipment and materials, and safety signs and labels issued by safety standards publishers for posting in the work area.

32. American National Standards Institute (ANSI) Accredited Standards Committee, 2002, *Safety in Welding, Cutting and Allied Processes.* ANSI Z49.1, Miami: American Welding Society.
33. American National Standards Institute (ANSI) Accredited Standards Committee, *Environment and Facility Safety Signs*, ANSI Z535.2. New York: American National Standards Institute.
34. American National Standards Institute (ANSI) Accredited Standards Committee, *Practices for Occupational and Educational Eye and Face Protection*, ANSI Z87.1, New York: American National Standards Institute.
35. See Appendix B for contact information for NIOSH, MSHA, and ANSI.
36. American National Standards Institute (ANSI) Accredited Standards Committee, *Practices for Respiratory Protection*, ANSI Z88.2, New York: American National Standards Institute.

Safe practices are also addressed in *Safety and Health Fact Sheets* written by the American Welding Society Project Committee on Labeling and Safe Practices.[37]

Material Safety Data Sheets

The suppliers of brazing and braze welding materials provide Material Safety Data Sheets (MSDS) and distribute them to users. In accordance with *Hazard Communications Standard*, OSHA 29CFR 1910.1200, the MSDS identifies hazardous materials if any are present in the products.[38]

A number of potentially hazardous materials may be present in fluxes, filler metals, coatings, and atmospheres used in brazing processes. When the fumes or gases from a product contain a component whose individual limiting value will be exceeded before the general brazing fume limit of 5 mg/m^3 is reached, the component must be identified on the MSDS. These include, but are not limited to, materials with low permissible exposure limits (PEL®).[39]

Precautionary Labels and Signs

Brazing and braze welding operators should be warned against the potential hazards from fumes, gases, electric shock, heat, and radiation as applicable. Precautionary information is defined in ANSI Z49.1.[40] Examples of precautionary information are shown in Figures 12.36 through 12.49.

Resistance and Induction Brazing Processes

As a minimum, the precautionary information shown in Figure 12.36, Figure 12.38, and Figure 12.39, or equivalents, should be placed on stock containers of consumable materials and on major equipment such as power sources, wire feeders, and controls used in electrical resistance or induction brazing processes. The information should be readily visible to the worker and may be on a label, tag, or other printed form as defined in the latest editions of the American National Standards Institute documents *Environmental and Facility Safety Signs*, ANSI Z535.2,[41] and *Product Safety Signs and Labels*, ANSI Z535.4.[42]

37. American Welding Society (AWS) Project Committee on Labeling and Safe Practices, 1998, *Safety and Health Fact Sheets*, Miami: American Welding Society.
38. Occupational Safety and Health Administration (OSHA), *Hazard Communications Standard*, OSHA 29CFR 1910.1200, Superintendent of Documents, Washington, D.C.: U.S. Government Printing Office.
39. See Reference 36.
40. See Reference 32.
41. See Reference 33.

WARNING:

PROTECT yourself and others. Read and understand this information.

FUMES AND GASES can be hazardous to your health.

ARC RAYS can injure eyes and burn skin.

ELECTRIC SHOCK can KILL.

- Before use, read and understand the manufacturer's instructions, Material Safety Data Sheets (MSDSs), and your employer's safety practices.
- Keep your head out of the fumes.
- Use enough ventilation, exhaust at the arc, or both, to keep fumes and gases from your breathing zone and the general area.
- Wear correct eye, ear, and body protection.
- Do not touch live electrical parts.
- See American National Standard ANSI Z49.1, *Safety in Welding, Cutting, and Allied Processes,* published by the American Welding Society, 550 N.W. LeJeune Rd., Miami, Florida 33126; OSHA *Safety and Health Standards*, available from the U.S. Government Printing Office, Superintendent of Documents, P.O. Box 371954, Pittsburgh, PA 15250-7954.

DO NOT REMOVE THIS INFORMATION

Source: Adapted from American National Standards Institute (ANSI) Accredited Standards Committee, 1999, *Safety in Welding, Cutting, and Allied Processes*, ANSI Z-49.1:1999, Miami: American Welding Society, Figure 1.

Figure 12.36— Precautionary Information for Brazing and Braze Welding Processes and Equipment

Oxyfuel Gas, Furnace, and Dip Brazing Processes

As a minimum, the information shown in Figure 12.37, or its equivalent, should be placed on stock containers of consumable materials and on major equipment used in oxyfuel gas, furnace (except vacuum), and dip brazing processes. The information should be readily visible to the worker and can be posted as a

42. American National Standards Institute (ANSI) Accredited Standards Committee, *Product Safety Signs and Labels,* ANSI Z535.4, Washington D.C.: National Electrical Manufacturer's Association.

WARNING:

PROTECT yourself and others. Read and understand this information.

FUMES AND GASES can be hazardous to your health.

HEAT RAYS (INFRARED RADIATION) from flame or hot metal can injure eyes.

- Before use, read and understand the manufacturer's instructions, Material Safety Data Sheets (MSDSs), and your employer's safety practices.
- Keep your head out of the fumes.
- Use enough ventilation, exhaust at the flame, or both, to keep fumes and gases from your breathing zone and the general area.
- Wear correct eye, ear, and body protection.
- See American National Standard ANSI Z49.1, *Safety in Welding, Cutting, and Allied Processes*, published by the American Welding Society, 550 N.W. LeJeune Rd., Miami, Florida 33126; OSHA *Safety and Health Standards*, available from the U.S. Government Printing Office, Superintendent of Documents, P.O. Box 371954, Pittsburgh, PA 15250-7954.

DO NOT REMOVE THIS INFORMATION

Source: Adapted from American National Standards Institute (ANSI) Accredited Standards Committee, 1999, *Safety in Welding, Cutting, and Allied Processes*, ANSI Z-49.1:1999, Miami: American Welding Society, Figure 2.

Figure 12.37—Precautionary Information for Oxyfuel Gas Processes and Equipment

DANGER: CONTAINS CADMIUM.

PROTECT yourself and others. Read and understand this information.

FUMES ARE POISONOUS AND CAN KILL.

- Before use, read and understand the manufacturer's 'instructions, Material Safety Data Sheets (MSDSs), and your employer's safety practices.
- Do not breathe fumes. Even brief exposure to high concentrations should be avoided.
- Use enough ventilation, exhaust at the work, or both, to keep fumes and gases from your breathing zone and the general area. If this cannot be done, use air supplied respirators.
- Keep children away when using.
- See American National Standard Z49.1, *Safety in Welding, Cutting, and Allied Processes*, published by the American Welding Society, 550 N.W. LeJeune Rd., Miami, Florida 33126; OSHA *Safety and Health Standards*, available from the U.S. Government Printing Office, Superintendent of Documents, P.O. Box 371954, Pittsburgh, PA 15250-7954.
- *First Aid:* If chest pain, shortness of breath, cough, or fever develop after use, obtain medical help immediately.

DO NOT REMOVE THIS INFORMATION

Source: Adapted from American National Standards Institute (ANSI) Accredited Standards Committee, 1999, *Safety in Welding, Cutting, and Allied Processes*, ANSI Z-49.1:1999, Miami: American Welding Society, Figure 3.

Figure 12.38—Precautionary Information for Brazing Filler Metals Containing Cadmium

label, tag, or other printed form as defined in ANSI Z535.2[43] and ANSI Z535.4.[44]

Filler Metals Containing Cadmium

As a minimum, brazing filler metals containing more cadmium than 0.1% by weight should carry the information shown in Figure 12.38, or its equivalent, on tags, boxes, or other containers, and on any coils of wire or strip not supplied to the user in a labeled container. Precautionary information requirements should also conform to ANSI Z535.4.[45]

43. See Reference 33.
44. See Reference 42
45. See Reference 42

Brazing Fluxes Containing Fluorides

As a minimum, brazing fluxes and aluminum salt bath dip brazing salts containing fluorine compounds should have precautionary information as shown in Figure 12.39 or its equivalent, on tags, boxes, jars, or other containers. Labels for other fluxes should conform to the requirements of *Hazardous Industrial Chemicals—Precautionary Labeling*, ANSI Z129.1.[46]

46. American National Standards Institute (ANSI) Accredited Standards Committee, *Hazardous Industrial Chemicals—Precautionary Labeling*, ANSI Z129.1, New York: American National Standards Institute.

WARNING: CONTAINS FLUORIDES.

PROTECT yourself and others. Read and understand this information.

FUMES AND GASES CAN BE HAZARDOUS TO YOUR HEALTH. BURNS EYES AND SKIN ON CONTACT. CAN BE FATAL IF SWALLOWED.

- Before use, read and understand the manufacturer's instructions, Material Safety Data Sheets (MSDSs), and your employer's safety practices.
- Keep your head out of the fume.
- Use enough ventilation, exhaust at the work, or both, to keep fumes and gases from your breathing zone and the general area.
- Avoid contact of flux with eyes and skin.
- Do not take internally.
- Keep children away when using.
- See American National Standard ANSI Z49.1, *Safety in Welding, Cutting, and Allied Processes*, published by the American Welding Society, 550 N.W. LeJeune Rd., Miami, Florida 33126; OSHA *Safety and Health Standards*, available from the U.S. Government Printing Office, Superintendent of Documents, P.O. Box 371954, Pittsburgh, PA 15250-7954.
- *First Aid:* If contact in eyes, flush immediately with water for at least 15 minutes. If swallowed, induce vomiting. Never give anything by mouth to an unconscious person. Call a physician.

DO NOT REMOVE THIS INFORMATION

Source: Adapted from American National Standards Institute (ANSI) Accredited Standards Committee, 1999, *Safety in Welding, Cutting, and Allied Processes*, ANSI Z-49.1:1999, Miami: American Welding Society, Figure 4.

Figure 12.39–Precautionary Information for Brazing Fluxes Containing Fluorides

FIRE PREVENTION AND PROTECTION

For detailed information on fire prevention and protection in brazing processes, the NFPA document *Fire Prevention During Welding, Cutting and Other Hot Work,* NFPA 51B, should be consulted.[47]

Brazing should preferably be done in specially designated areas that have been designed and constructed to minimize fire risk. No brazing should be done unless the atmosphere is either nonflammable or unless gases (such as hydrogen), which can become flammable when mixed with air, are confined and prevented from being released into the atmosphere.

Fire extinguishing equipment should be available and ready for use where brazing work is being done. The fire extinguishing equipment may be pails of water or a water hose, buckets of sand, portable extinguishers, or an automatic sprinkler system, depending on the nature and quantity of combustible material in the area. The method appropriate to the type of fire should be used to extinguish combustible material, chemical, electrical, and metal fires as per NFPA recommendations.

Before brazing begins in a location not specifically designated for such purposes, inspection and authorization by a responsible person should be performed.

When repairing containers that have held flammable or other hazardous materials, there is the possibility of explosions, fires, and the release of toxic vapors. Brazers and brazing operators must be fully familiar with the American Welding Society standard *Recommended Safe Practices for the Preparation for Welding and Cutting of Containers and Piping,* F4.1: 1999.[48] The applicable state, local, and federal specifications should also be consulted.

Brazing Atmospheres

Flammable gases are sometimes used as atmospheres for furnace brazing operations. These include combusted fuel gas, hydrogen (2% hydrogen is used routinely in the furnace brazing of carbon steel), dissociated ammonia, and some nitrogen-hydrogen mixtures. Prior to introducing such atmospheres, the furnace or retort must be purged of air according to the safe procedures recommended by the furnace manufacturer.

Adequate area ventilation must be provided that will exhaust and discharge to a safe place the explosive or toxic gases that may emanate from furnace purging and brazing operations. Local environmental regulations should be consulted when designing the exhaust system.

In dip brazing, the assembly to be immersed in the bath must be completely dry. Moisture on the components will cause an instantaneous generation of steam that may explosively expel the contents of the dip pot. Predrying the assembly prevents this hazard. If supplementary flux must be added, it must be dried to remove the surface moisture and also the water of hydration.

47. National Fire Protection Association (NFPA), 1999, *Fire Prevention During Welding, Cutting and Other Hot Work,* NFPA 51B, Quincy, Massachusetts: National Fire Protection Association.

48. American Welding Society (AWS) Safety and Health Committee, 1999, *Recommended Safe Practices for the Preparation for Welding and Cutting of Containers and Piping,* F4.1, Miami: American Welding Society.

ELECTRICAL HAZARDS

All electrical equipment used for brazing should conform to the latest edition of *National Electric Code* ANSI/NFPA 70.[49] The equipment should be installed by qualified personnel under the direction of a competent technical supervisor. Prior to production use, the equipment should be inspected by competent safety personnel to ensure that it is safe to operate.

CONCLUSION

Brazing is used for many applications and under many circumstances. This chapter has offered the reader a brief overview of the materials and processes involved in this remarkable process. It is the best joining process available when the following are parameters of a joining project:

1. Dissimilar metals or materials must be joined, e.g., metals-to-ceramics or ceramics-to-ceramics;
2. It is desirable not to melt a portion of the base metal;
3. A base metal that is not weldable must be joined;
4. The joint design is complicated; and
5. Multiple joints can be made at once or in sequence.

Many resources are available to the users of brazing processes. When choosing filler metals, the latest edition of *Specification for Filler Metals for Brazing and Braze Welding,* AWS A5.8[50] should be consulted. For detailed information on testing brazed joints to determine optimum strength, *Standard Method for Evaluating the Strength of Brazed Joints*, AWS C3.2M/C3.2:2002[51] should be consulted. If the application is a critical one in which hazard to life or significant cost is involved, *Recommended Practices for the Design, Manufacture, and Examination of Critical Brazed Components*, AWS C3.3:2002[52] should be consulted.

Additional specifications include the following:[53]

Specification for Torch Brazing, ANSI/AWS C3.4;

Specification for Induction Brazing, ANSI/AWS C3.5;

Specification for Furnace Brazing, AWS/AWS C3.6;

Specifications for Aluminum Brazing, AWS/AWS C3.7.

49. See Reference 7.
50. See Reference 9.
51. See Reference 21.
52. See Table 12.1.
53. See Reference 7.

The publications listed in the Supplementary Reading List are sources of specialized information, as is the *Brazing Handbook*.[54]

Many new brazing processes are being developed and used under appropriate conditions. Examples are laser, electron beam, resistance, and microwave brazing. Brazement behavior is also being extensively examined through finite-element analysis and modeling of thermal, fluid flow, and mechanical characteristics that map the probable residual stresses and predict braze-joint reliability and failures. This continuing research illustrates that although brazing has its place in history as one of the earliest metal joining processes and is of major importance to present-day industry, it has not yet reached its ultimate potential for joining metals and materials for the progressively complex applications of the future.

BIBLIOGRAPHY[55]

American National Standards Institute (ANSI) Accredited Standards Committee. *Safety in welding, cutting, and allied processes.* ANSI Z49.1:1999. Miami: American Welding Society.

American National Standards Institute (ANSI) Accredited Standards Committee. *Environment and facility safety signs*, ANSI Z535.2. New York: American National Standards Institute.

American National Standards Institute (ANSI) Accredited Standards Committee. *Practices for occupational and educational eye and face protection*, ANSI Z87.1. New York: American National Standards Institute.

American National Standards Institute (ANSI) Accredited Standards Committee. *Practices for respiratory protection*, ANSI Z88.2. New York: American National Standards Institute.

American National Standards Institute (ANSI) Accredited Standards Committee. *Product safety signs and labels*, ANSI Z535.4. Washington, D.C.: National Electrical Manufacturers Association.

American National Standards Institute (ANSI) Accredited Standards Committee. *Precautionary labeling for hazardous industrial chemicals*, ANSI Z129.1. New York: American National Standards Institute.

American Welding Society AWS Committee on Safety and Health. 1998. *Safety and Health Fact Sheets.* Miami: American Welding Society.

54. See Reference 47.
55. The dates of publication given for the codes and other standards listed here were current at the time the chapter was prepared. The reader is advised to consult the latest edition.

American Welding Society (AWS) Committee on Brazing and Soldering. 2002. *Recommended practices for design, manufacture and inspection of critical brazed components*, AWS C3.3:2002. Miami: American Welding Society.

American Welding Society (AWS) Committee on Brazing and Soldering. 1999. *Specification for torch brazing*, ANSI/AWS C3.4:1999. Miami: American Welding Society.

American Welding Society (AWS) Committee on Brazing and Soldering. 1999. *Specification for induction brazing,* ANSI/AWS C3.5:1999. Miami: American Welding Society.

American Welding Society (AWS) Committee on Brazing and Soldering. 1999. *Specification for furnace brazing*. ANSI/AWS C3.6:1999. Miami: American Welding Society.

American Welding Society (AWS) Committee on Brazing and Soldering. 1999. *Specification for aluminum brazing,* ANSI/AWS C3.7:1999. Miami: American Welding Society.

American Welding Society (AWS) Committee on Filler Metals and Allied Materials, 1992 (Reaffirmed 2003) *Specification for Filler Metals for Brazing and Braze Welding*, ANSI/AWS A5.8, Miami: American Welding Society.

American Welding Society (AWS) Committee on Brazing and Soldering. (Forthcoming). *Brazing Handbook*, 5th ed. Miami: American Welding Society.

American Welding Society (AWS) Committee on Brazing and Soldering. 2002. *Standard method for evaluating the strength of brazed joints*, ANSI/AWS C3.2M/C3.2:2002. Miami: American Welding Society.

American Welding Society (AWS) Committee on Brazing and Soldering. 1998. *Recommended practices for ultrasonic inspection of brazed joints,* ANSI/AWS C3.8-98R. Miami: American Welding Society.

American Welding Society (AWS) Committee on Filler Metals and Allied Materials. 1992. *Specification for fluxes for brazing and braze welding*, ANSI/AWS A5.31-92. Miami: American Welding Society.

American Welding Society (AWS) Committee on Fumes and Gases. 1999. *Recommended safe practices for preparation for welding and cutting of containers and piping*, F4.1:1999. Miami: American Welding Society.

ASM International. 1993. *Welding, brazing, and soldering*. Vol. 6 of ASM handbook. Materials Park, Ohio: ASM International.

National Fire Protection Association (NFPA). 1999. *Fire prevention during welding, cutting and other hot work,* NFPA 51B. Quincy, Massachusetts: National Fire Protection Association.

National Fire Protection Association (NFPA). 1999. *National Electric Code*, NFPA 70. Quincy, Massachusetts: National Fire Protection Association.

Occupational Safety and Health Administration (OSHA). *Hazard communications standard*. Washington, D.C.: Superintendent of Documents, U.S. Government Printing Office.

SUPPLEMENTARY READING LIST

Cadden, C. H., N. Y. C Yang, and T. H. Headley. 1997. Microstructural evolution and mechanical properties of braze joints in Ti-13.4Al-21.2Nb. *Welding Journal* 76(8): 316-s to 325-s.

Cole, N. C. 1979. Corrosion resistance of brazed joints. Bulletin 247. New York: Welding Research Council.

Cole, N. C. 1980. Corrosion resistance of brazed joints. *Source book on brazing and brazing technology.* Metals Park, Ohio: American Society for Metals. 365.

Cole, N. C. 1981. Corrosion behavior of brazed joints. *DVS Ber. 69 (Hart-Hochtemperatureloeten Diffusionschweissen)*. 118–120 (English).

Cole, N. C., R. W Gunkel, and J. W. Koger. 1973. Development of corrosion-resistant filler metals for brazing molybdenum. *Welding Journal* 52(10): 446-s to 473-s.

Davé, V. R., R. W. Carpenter, D. T. Christensen, and J. O. Milewski. 2001. Precision laser brazing utilizing nonimaging optical concentration. *Welding Journal* 80(6): 142-s to 147-s.

Gale, W. F., and E. R. Wallach. 1991. Wettability of nickel alloys by boron-containing brazes. *Welding Journal*. 71(1): 25-s to 33-s.

Gilliland, R. G., and G. M. Slaughter. 1969. The development of brazing filler metals for high-temperature service. *Welding Journal* 48(10), 463-s to 469-s.

Hammond, J. P., S. A. David, and M. L. Santella. 1988. Brazing ceramic oxides to metals at low temperature. *Welding Journal* 67(10): 227.

Helgesson, C. I. 1968. *Ceramic-to-metal bonding*. Cambridge, Massachusetts: Boston Technical Publishers, Inc.

Hosking, F. M. 1985. Sodium compatibility of refractory metal alloy—type 304L stainless steel joints. *Welding Journal* 64(7): 181-s to 190-s.

Hosking, F. M., and J. A Koski. 1992. Graphite–metal brazing for thermal applications. *The Metal Science of Joining*. Warrendale, Pa.: TMS Publications. 307–314.

Hosking, F. M., J. J. Stephens, and J. A. Rejent. 1999. Intermediate temperature joining of dissimilar metals. *Welding Journal* 78(4): 127-s to 136-s.

Hosking, F. M., C. H. Cadden, N. Y. C. Yang, S. J. Glass, J. J. Stephens, P. T. Vianco, and C. A. Walker.

2000. Microstructural and mechanical characterization of actively brazed alumina tensile specimens. *Welding Journal* 79(8): 222-s to 230-s.

Hosking, F. M., S. E. Gianoulakis, R. C. Givler, and R. P. Schunk. 2000. Thermal and fluid flow brazing simulations. *Advanced brazing and soldering technologies: International Brazing and Soldering Conference 2000 proceedings*. Miami, Fla.: American Welding Society. 389–397.

Jones, T. A., and C. E. Albright. 1984. Laser beam brazing of small-diameter copper wires to laminated copper circuit boards. *Welding Journal* 63(12): 34–37.

Kawakatsu, I. 1973. Corrosion of BAg brazed joints in stainless steel. *Welding Journal* 52(6): 223-s to 239-s.

Lugscheider, E., and T. Cosack. 1988. High temperature brazing of stainless steel with low-phosphorus nickel-based filler metal. *Welding Journal* 67(11): 215-s to 219-s.

Lugscheider, E., and H. Krappitz. 1986. The influence of brazing conditions on the impact strength of high-temperature brazed joints. *Welding Journal* 65(10): 261-s.

Lugscheider, E., K. D. Partz, and R. Lison. 1982. Thermal and metallurgical influences on AISI 316 and Inconel® 625 by high temperature brazing with nickel base filler metals. *Welding Journal* 61(10): 329-s to 333-s.

Lugscheider, E., H. Zhuang, and M. Maier. 1983. Surface reactions and welding mechanisms of titanium- and aluminum-containing nickel-base and iron-base alloys during brazing under vacuum. *Welding Journal* 62(10): 295-s to 300-s.

Lugscheider, E., T. Schittny, and E. Halmoy. 1989. Metallurgical aspects of additive-aided wide-clearance brazing with nickel-based filler metals. *Welding Journal* 68(1): 9-s to13-s.

McDonald, M. M., D. L. Keller, C. R. Heiple, and W. E. Hofman. 1989. Wettability of brazing filler metals on molybdenums and TMZ *Welding Journal*: 68: 389-s to 393-s.

Mizuhara, H., and K. Mally. 1985. Ceramic-to-metal joining with active brazing filler metal. *Welding Journal* 63(10): 27–32.

Moorhead, A. J., and P. F. Becher. 1987. Development of a test for determining fracture toughness of brazed joints in ceramic materials. *Welding Journal* 66(1): 26-s to 31-s.

Neilsen, M. K., and J. J. Stephens. 2000. Residual stress in metal-to-ceramic braze joints: advanced braze alloy constitutive model. *Advanced brazing and soldering technologies: international brazing and soldering conference 2000 proceedings*. Miami: American Welding Society. 411–418.

Onzawa, T., A. Suzumura, and M. W. Ko. 1990. Brazing of titanium using low-melting point Ti-based filler metals. *Welding Journal* 69(12): 462-s to 467-s.

Patrick, E. P. 1975. Vacuum brazing of aluminum. *Welding Journal* 54(6): 159–163.

Pattee, H. E. 1972. Joining ceramics to metals and other materials. Bulletin 178. New York: Welding Research Council.

Pattee, H. E. 1973. High-temperature brazing. Bulletin 187. New York: Welding Research Council.

Roulin, M., J. W. Luster, G. Karadeniz, and A. Mortensen. 1999. Strength and structure of furnace-brazed joints between aluminum and stainless steel. *Welding Journal* 78(5): 151-s to 155-s.

Rugal, V., N. Lehka, and J. K. Malik. 1974. Oxidation resistance of brazed joints in stainless steel. *Metal Construction and British Welding Journal*. 183–176.

Sakamoto, A., C. Fujiwara, T. Hattori, and S. Sakai. 1989. Optimizing processing variables in high-temperature brazing with nickel-based filler metals. *Welding Journal* 68(3): 63–67.

Santella, M. 1992. Fundamental metallurgical considerations in brazing and soldering. *The metal science of joining*. Warrendale, Pa.: TMS Publications. 61–65.

Schmatz, D. J. 1983. Grain boundary penetration during brazing of aluminum. *Welding Journal* 62(10): 267-s to 271-s.

Schultze, W., and H. Schoer. 1973. Fluxless brazing of aluminum using protective gas. *Welding Journal* 52(10): 644–651.

Schwartz, M. M. 1975. The fabrication of dissimilar metal joints containing reactive and refractory metals. Bulletin 210. New York: Welding Research Council.

Schwartz, M. M. 1973. Brazed honeycomb structures. Bulletin 182. New York: Welding Research Council.

Schwartz, M. M. 1969. *Modern metal joining techniques*. New York: John Wiley & Sons.

Selverian, J. H., and S. Kang. 1992. Ceramic-to-metal joints brazed with palladium alloys. *Welding Journal* 71(1): 25-s to 33-s.

Stephens, J. J., et al. 1992. High-temperature creep properties of eutectic and near-eutectic silver-copper alloys: application to metal/ceramic joining. *The metal science of joining*. Warrendale, Pa.: TMS Publications. 285–294.

Swaney, O. D., D. E. Trace, and W. L Winterbottom. 1986. Brazing aluminum automotive heat exchangers in vacuum. *Welding Journal* 65(5): 49–57.

Takemoto, T., T. Ujie, H. Chaki, and A. Matsunawa. 1996. Influence of oxygen content on brazeability of a powder aluminum braze filler metal. *Welding Journal* 75(11): 372-s to 378-s.

Terrill, J. R., C. N. Cochran, J. J. Stokes, and W. E. Hanpin. 1971. Understanding the mechanisms of aluminum brazing. *Welding Journal* 50(12): 833-s to 839-s.

The Aluminum Association. 1971. *Aluminum brazing handbook*. New York: The Aluminum Association.

Vianco, P. T., and M. Singh, eds. 2000. *Advanced brazing and soldering technologies: International Brazing and Soldering Conference 2000 proceedings*. Miami, Fla.: American Welding Society.

Vianco, P. T., F. M. Hosking, J. J. Stephens, C. A. Walker, M. K. Nielsen, S. J. Glass, and S. L. Monroe. 2002. Aging of brazed joints—interface reactions in base metal/filler metal couples—part 1: low-temperature Ag-Cu-Ti filler metal. *Welding Journal* 81(10): 201-s to 210-s.

Vianco, P.T., F. M. Hosking, J. J. Stephens, C. A. Walker, M. K. Nielsen, S. J. Glass, and S. L. Monroe. 2002. Aging of brazed joints—interface reactions in base metal/filler metal couples—part 2: high-temperature Au-Ni-Ti braze alloy. *Welding Journal* 81(11): 256-s to 264-s.

Winterbottom, W. L. 1984. Process control criteria for brazing under vacuum. *Welding Journal* 63(10): 33–39.

Witherell, C. E., and T. J. Ramos, 1980. Laser brazing. *Welding Journal* 59(10): 267-s to 277-s.

CHAPTER 13

SOLDERING

Photograph courtesy of Sandia National Laboratories

Prepared by the Welding Handbook Chapter Committee on Soldering:

F. M. Hosking, Chair
Sandia National Laboratories

P. T. Vianco, Co-Chair
Sandia National Laboratories

W. D. Rupert
Wolverine Joining Technologies

R. W. Smith
Materials Resources International

Welding Handbook Volume 2 Committee Member:

D. W. Dickinson
The Ohio State University

Contents

CHAPTER 13

SOLDERING

INTRODUCTION

Soldering is a joining technique that has been used since ancient times. Many artifacts discovered in archeological excavations were found to have been joined by soldering. The technology has existed for several thousand years, with innovative changes occurring as metallurgical knowledge and new metals were developed.

Copper and lead alloys were the first base metals to be joined. Early metallurgists learned to identify low-melting eutectics in binary systems, which permitted the soldering of complex shapes for jewelry and utensil applications. The industrial revolution promoted widespread use of soldered joints.

It is interesting to note that this ancient joining method is used today at the highest levels of technology. Advancements in solder compositions, processing techniques, and applications have continued in many industrial applications such as microelectronics, telecommunications, automotive, aerospace, plumbing, and decorative packaging. Many soldering processes, procedures, filler metals, and types of equipment are used to join numerous base materials. Specific applications require consideration of all of these factors to obtain the optimum manufacturing and service results.

This chapter presents an overview of soldering technology. It covers soldering principles and practices in some detail. Topics include the fundamentals of soldering, the various soldering processes, equipment, materials and consumables, basic operating procedures, inspection and quality, economics, and safe practices. Sources of additional information are suggested in the Supplementary Reading List at the end of the chapter.

FUNDAMENTALS

The term *soldering* refers to the group of joining processes that bond base materials together with a filler metal at a soldering temperature below the solidus temperature of the base material. The solder alloy has a liquidus temperature not exceeding 450°C (840°F). If the filler metal temperature exceeds 450°C (840°F), it is considered a brazing alloy.[1, 2]

A soldered joint is the result of a metallurgical reaction that occurs between the solder filler metal and the base materials being joined. The solder alloy wets and flows along the joint clearance and generally forms an intermetallic reaction layer with the base material. On solidification, the bond between adjacent atoms at the reaction interface is equivalent to that found in a continuous solid metal or compound. However, some solder joints do not rely on a metallurgical bond but are held together mechanically by an interlocking interface. The strength of the joint can be enhanced by the joint design. The selection and use of the proper materials and processes can achieve reliable solder joints.

Temperature ranges of commonly used soldering alloys are compared with base-metal melting points in Figure 13.1.

Soldering is preferred by many fabricators because of the relative ease with which high-quality joints can be made. The major reasons for the popularity of the process include relatively low processing temperatures, low energy input, and precise process controls that have minimum effect on the properties of the base materials. A wide range of heating methods can be used, giving flexibility to design and manufacturing procedures. Modern automation permits the simultaneous fabrication of many joints, simple or complex, such as those

1. American Welding Society (AWS) Committee on Definitions and Symbols, 2001, *Standard Welding Terms and Definitions,* AWS A3.0:2001, Miami: American Welding Society, p. 35.

2. At the time of preparation of this chapter, the referenced codes and other standards were valid. If a code or other standard is cited without a date of publication, it is understood that the latest edition of the document referred to applies. If a code or other standard is cited with the date of publication, the citation refers to that edition only, and it is understood that any future revisions or amendments to the code or standard are not included. As codes and standards undergo frequent revision, the reader is advised to consult the most recent edition.

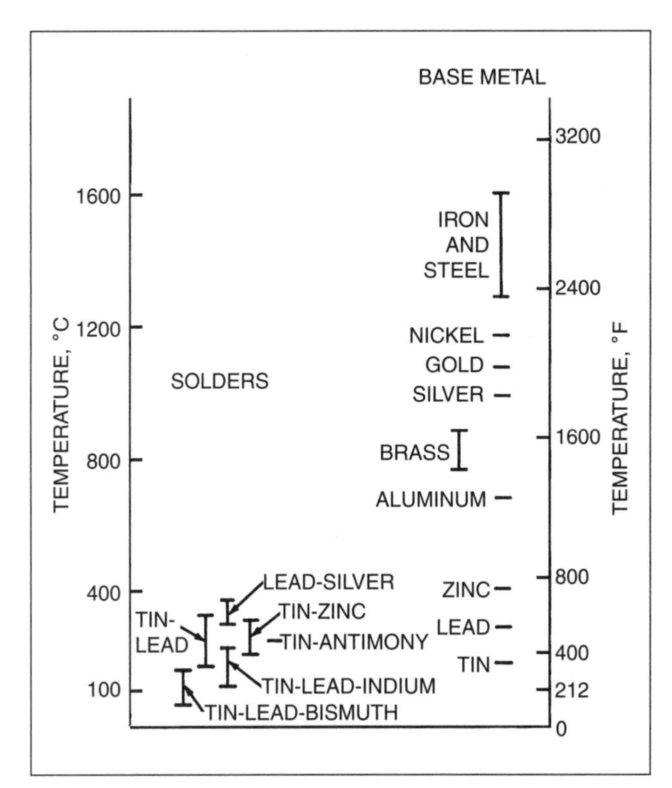

Figure 13.1—Solder Temperature Ranges Compared with Base Metal Melting Points

used in microelectronics. Highly reliable joints can be obtained with carefully controlled processing. The occasional solder joint that contains imperfections can easily be reworked or repaired by reprocessing.

PHYSICAL, CHEMICAL AND METALLURGICAL ASPECTS OF SOLDERING

Many principles of physics, chemistry, and metallurgy are involved in the preparation and fabrication of a soldered joint. Essential to the process is the phenomenon of solder wetting and the adherence of liquid solder on a solid metallic or nonmetallic surface. Wetting is affected by the surface tension properties of the materials involved and the degree of alloying that occurs during the soldering action. Soldering normally requires the presence of a flux. The flux removes oxides from the base materials and solder alloy. It also lowers the surface tension between the molten metal and solid substrate. This flux behavior improves the wetting and spreading of the solder alloy.

Wetting takes place when the solder reacts with the base-metal surface. Alloying depends on the solubility of the base metal in the molten solder metal. A high level of alloying between the base metal and solder alloy can raise the melting temperature and retard spreading. Therefore, good solder filler metals usually dissolve only a moderate amount of base metal. Intermetallic compounds may form at the interface, depending on the metal systems involved.

Solder joints are often designed with joint clearances that require capillary flow of the molten solder between the fixtured base metals. The capillary action is improved by lowering the surface tension of the solder, narrowing the joint clearance, and using a compatible chemical flux.

The principles at work during soldering require that the surfaces of the workpieces (the materials to be joined) are clean and free of dirt, oxides, or other contaminants that normally inhibit the wetting and flow of solder. One important function of a flux is to produce a pristine base-metal surface that is free of any film that would be detrimental to wetting. This chemical reaction should be minimal but effective; however, it should not be considered a substitute for prior cleaning.

Flux is activated when heated. It cleans contacted surfaces and protects the cleaned areas from oxidation or recontamination during soldering. The solder filler metal is applied when the joint has been heated to the soldering temperature. The surfaces of the base material are protected by the activated flux during the soldering action. After the solder joint is cooled, some residual flux may be present. Depending on the level of activity of the residual flux, it should be removed to prevent potential corrosion-related failures of the joint while in service.

Physical conditions that have an adverse effect on wetting, spreading, and capillary action can result in unsatisfactory joints. Unsatisfactory joints generally result from poor surface conditions, improper joint fit-up, and incorrect flux selection. Some base metals, such as chromium, cannot be readily wet by most conventional solder filler metals. The phenomenon known as *dewetting* is typical for alloys that contain these metals. Dewetting occurs when the molten solder retracts from its original position on the wetted surface and leaves areas of incomplete coverage. Inadequate cleaning, incorrect flux selection, and inappropriate solder alloy compositions are the main causes of dewetting.

APPLICATIONS

Soldering is extensively used in many industrial sectors for fabricating low-temperature joints. Products vary from simple structural assemblies to high-end microelectronic components. The common benefit soldering brings to its users is the generally easy, inexpensive setup needed to fabricate reliable assemblies with a broad range of different materials.

The automotive, aerospace, communications and microelectronics industries, the military sector, manufacturers of consumer products and structures, the food processing industry, and many others rely on soldering. Products include automotive radiators, electronic packages and chip attachment, sheet metal, piping and plumbing applications, instruments and electronic assemblies, cryogenic containers, jewelry and art glass. Soldering is also used to make vacuum seals and other types of seals, and it is used in various coating and surfacing operations. A typical structural soldering application is shown in Figure 13.2, a soldered bracket-and-pin static charge dissipater assembly. Figure 13.2(A) shows the top fillet side of the joint and (B) shows the bottom fillet side of the joint.

Although most soldering is done at temperatures below 250°C (482°F), many applications require solder alloys with higher melting temperatures (between 300°C and 450°C [572°F and 842°F]) to meet strength or other thermal specifications. Step soldering is one such example, in which an assembly is soldered through a sequence of soldering operations, each one using a lower-melting solder alloy to avoid remelting the solder joints already fabricated. There is continued interest, especially in the electronics industry, in assembling components for use in relatively high-temperature environments. Typical solder alloys include compositions rich in lead or gold to achieve the higher melting temperatures and operating conditions, which normally must sustain temperature exposures as high as 300°C (572°F).

A typical range of different interconnects for an electronic assembly that can require high-to-low-melting solder alloys during fabrication is shown in Figure 13.3.

Another application is the fabrication of photovoltaic solar panels that convert sunlight to electrical energy. The electrical connection between individual silicon cells is made by soldering copper conductors to the metallized contact points on the conductive grid of each solar cell. Soldering is done either manually, with hot air or irons, or with automated equipment. The automated systems use hot bars or quartz lamps to heat the areas to be soldered. Figure 13.4(A) shows a cell, module, and array unit of a photovoltaic system. The cell design and interconnects are shown in Figure 13.4(B).

These high-temperature conditions are also encountered in down-hole applications such as oil, gas, and geothermal well exploration and monitoring operations. The 95% zinc, 5% aluminum solder shows the most promise for use in these applications because of its melting properties and commercial availability. Hand soldering with an iron or microtorch is the preferred soldering process.

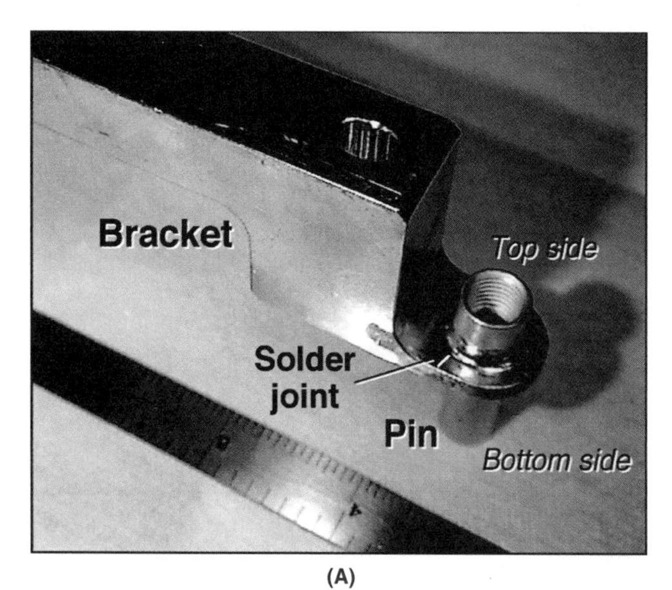

(A)

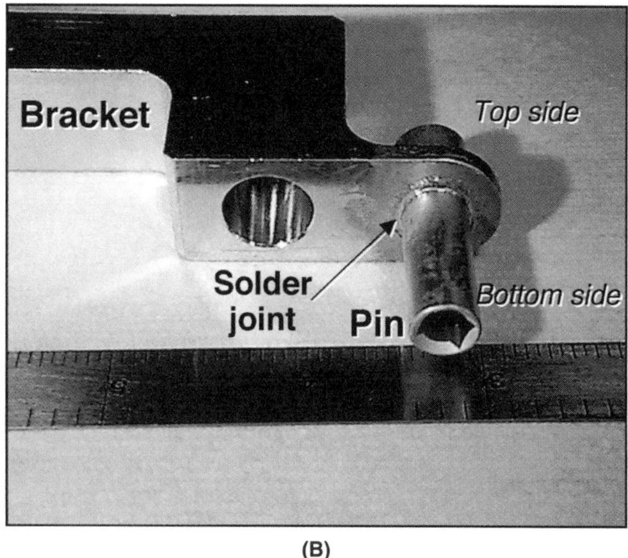

(B)

Source: Vianco, P. T., and J. A. Regent, 2001, A Furnace Process for Structural Soldering, *Welding Journal* 80(4), pp. 217–218, Figure 3.

Figure 13.2—Soldered Bracket-and-Pin Static Charge Dissipater Assembly: (A) Top Fillet Side of a Soldered Joint in a Static Charge Dissipater Assembly and (B) Bottom Fillet Side

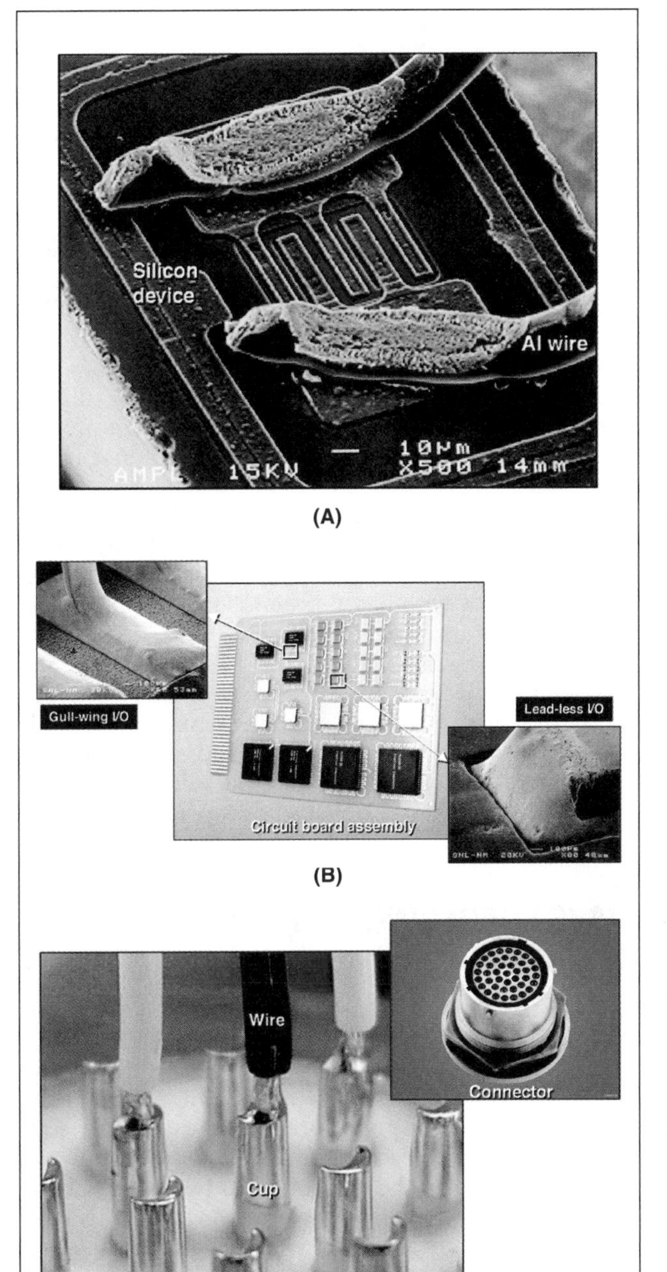

(C)

Notes:

(A) Level 1 interconnect represented by a wire bond to an integrated circuit pad.

(B) Level 2 interconnects represented by the circuit-board assemblies—laminate and ceramic (hybrid microcircuit technologies).

(C) Level 3 interconnects such as those used for connectors.

Source: Adapted from Vianco, P. T., 2002. Solder technology for ultrahigh temperatures. *Welding Journal,* 81(10): 52. Research performed by Sandia National Laboratories, Albuquerque, N. Mex.

Figure 13.3—Common Solder Interconnects in Commercial and Military-Grade Electronic Products

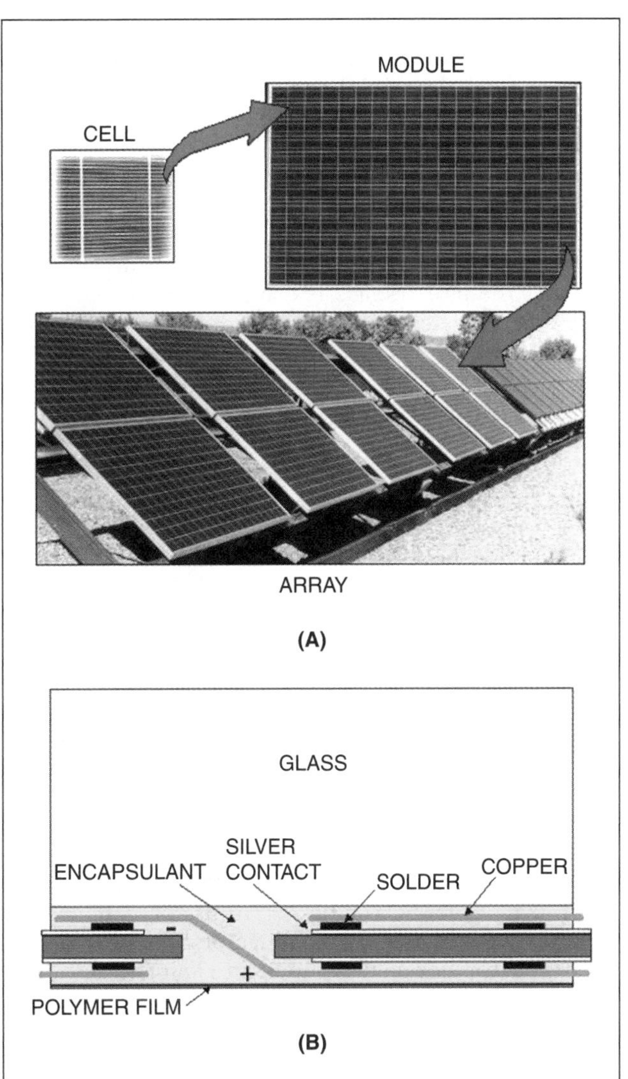

Source: Adapted from Hosking, F. M., and M. A. Quintana, 2001. Interconnect soldering of silicon photovoltaic module cells. *Welding Journal,* 80(10), p. 57. Research performed by Sandia National Laboratories, Albuquerque, N. Mex.

Figure 13.4—(A) Photovoltaic Cell, Module, and Array of Solar Panels and (B) Photovoltaic Cell Design

PROCESS VARIATIONS

Soldering can be accomplished by using one of several process variations. When selecting a method for a particular job, however, the solderer should carefully consider the conditions and principles that are common to all soldering practices. Among these considerations are proper soldering temperature, heat distribution, and the rates of heating and cooling. Other requirements are joint preparation, cleaning, fluxing, preheating, soldering, and final cleaning. The soldering process that best provides the properties required to meet product specifications should be selected.

Soldering is generally a low-temperature joining method that takes advantage of the reactions that occur during flux activation and solder wetting. Each choice of a solder and flux has a corresponding process that provides the best results. The application of the solder and flux is determined by the requirements of the selected soldering method.

Soldering often must be done in proximity to other heat-sensitive materials or metals that have been given a specific thermal or mechanical treatment. Cold-worked metals can become softened or relaxed during the soldering process. Other alloys could be overaged. These factors should be taken into consideration in the design of the finished product. Soldering requires the maintenance of close tolerances to ensure quality joints. It is often advisable to make sample components with the soldering method under consideration and then to subject the samples to destructive testing to be sure that production results will be satisfactory.

The amount of operator interaction depends on the soldering method. Soldering processes can be highly automated when all the material and process variables have been evaluated and are carefully controlled. In contrast, the successful and efficient soldering of individual workpieces or small lots can be carried out using hand-held soldering torches or irons, or can be performed on hot plates. An example of manual soldering on a table-top hot plate is shown in Figure 13.5.

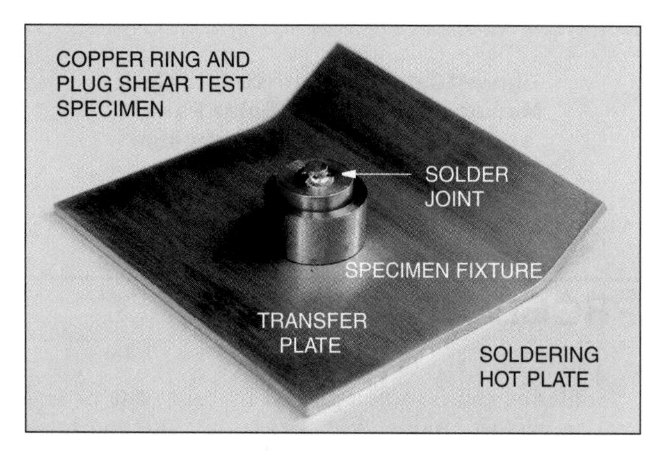

Figure 13.5—Copper Ring and Plug Shear Test Specimen Soldered Manually on a Table-Top Hot Plate, Including Fixturing

DIP SOLDERING

Dip soldering (DS) is a process that uses a molten bath of solder to supply both the heat and the solder necessary to join the workpieces. Solder is drawn into the fixtured joint clearance and forms the joint as it is removed from the bath and solidifies. Dipping is done manually or by automated means. The bath is generally resistively heated in a solder pot. The solder pot should be large enough to maintain the required production rate. Workpieces being dipped should not appreciably lower the temperature of the solder bath. Pots of adequate size can be held at lower operating temperatures and still supply sufficient heat to solder the dipped joints. Solder composition should be monitored to maintain the recommended alloy chemistry and impurity levels.

Dip soldering is illustrated in Figure 13.6(A). The workpieces are immersed in the solder bath, where solder is drawn into the fixtured joint clearance. As the workpieces are withdrawn from the bath, the solder solidifies and forms the joint.

Dip soldering is a useful and economical method because an entire unit comprising any number of joints can be soldered in one operation. Fixtures are usually required to hold the workpieces and maintain joint clearances during solidification of the solder.

WAVE SOLDERING AND CASCADE SOLDERING

Wave soldering (WS) is an automated soldering process in which the workpieces are passed through a bath of molten solder. In this process, illustrated in Figure 13.6(B), the solder is pumped out of a narrow slot above the solder pot to produce a wave or series of waves. A workpiece conveyor can pass over the waves at a small angle to the horizontal to assist in draining the solder. Double waves or special waveforms also can be used for this purpose.

Integrated wave soldering systems, often used for producing printed electronic circuit boards, are units that can apply the flux, dry and preheat the circuit board, solder the components, and clean the completed assembly. Some of these soldering systems have special features in which the flux is applied by passing the workpiece through a wave or by spraying, rolling, or dipping. Several systems employ a mixture of oil with the solder to aid in the elimination of defects such as "bridges" and "icicles" between conductor paths. Bridges and icicles are soldering defects on printed circuit boards or similar interconnects that result from solder breakaway as the workpieces are withdrawn from the solder bath. Bridges are excess solder that

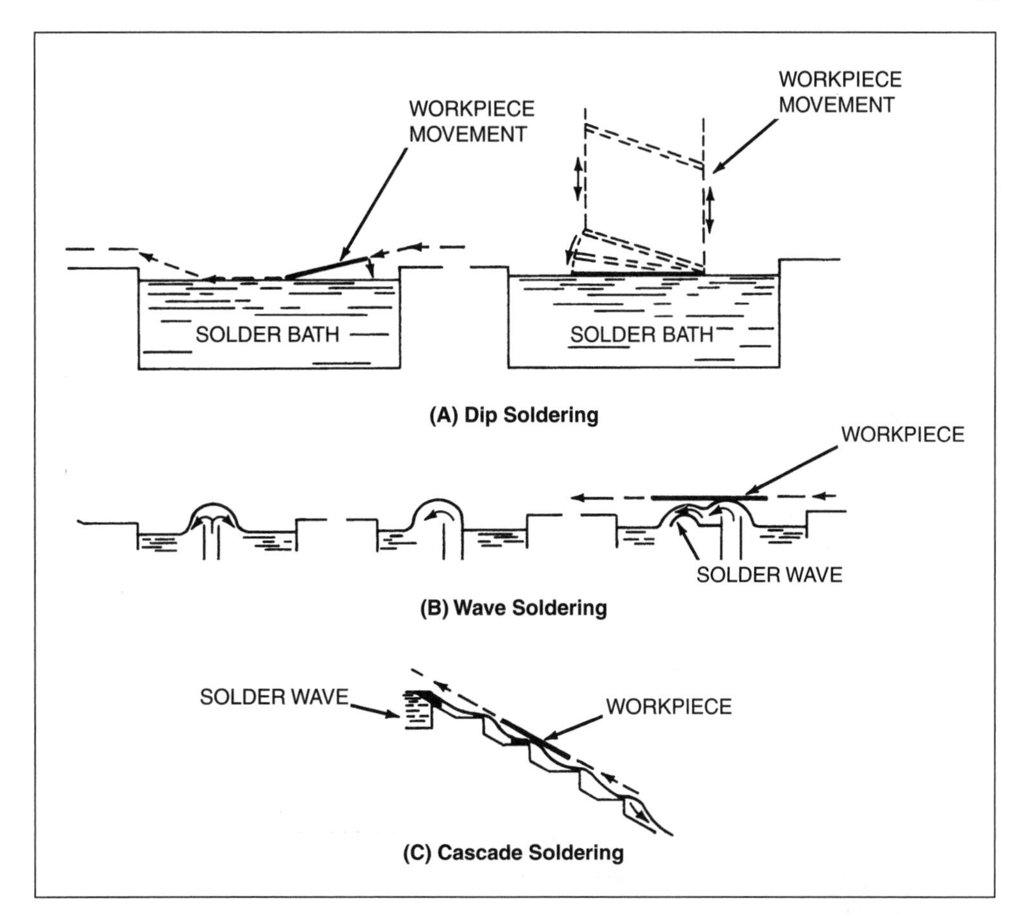

Figure 13.6—Soldering Techniques Used for Large Production Runs: (A) Dip Soldering; (B) Wave Soldering; (C) Cascade Soldering

form undesired conductive paths across the electrical circuit and are potential sources for short circuits. Icicles are formed by excess solder that has solidified in long, drawn-out features on circuit lands, terminations or other conductor surfaces. These can range from simple cosmetic defects to potential short circuits. Some soldering systems are made inert by flowing technical-grade nitrogen through the work area of the hot zone, thus avoiding bridging and icicling. Figure 13.7 shows how these defects may form during solder breakaway as the workpiece is withdrawn from the solder bath. Icicle formation is shown in Figure 13.7(A), (B), and (C); bridge formation is shown in Figure 13.7(D), (E), and (F).

Another wave-soldering system uses dual waves with the solder alloy flowing in the direction opposite to the circuit board travel. Wave-solder systems are excellent when the product calls for oxide-free solder surfaces.

Cascade soldering, an alternate technique of wave soldering, is illustrated in Figure 13.6(C). With cascade soldering, the solder flows by gravity down a trough and is returned by pump to the upper reservoir. The workpiece travels over the solder stream, which facilitates solder-joint formation.

FURNACE SOLDERING

Furnace soldering (FS) is a soldering process in which the workpieces are placed in a furnace or oven and heated to the soldering temperature. This process produces consistent and satisfactory results for many applications, especially in high-volume reflow soldering. Furnace soldering, shown in Figure 13.8, should be considered when the following conditions apply:

1. Entire assemblies can be brought to the soldering temperature without damage to any of the components;

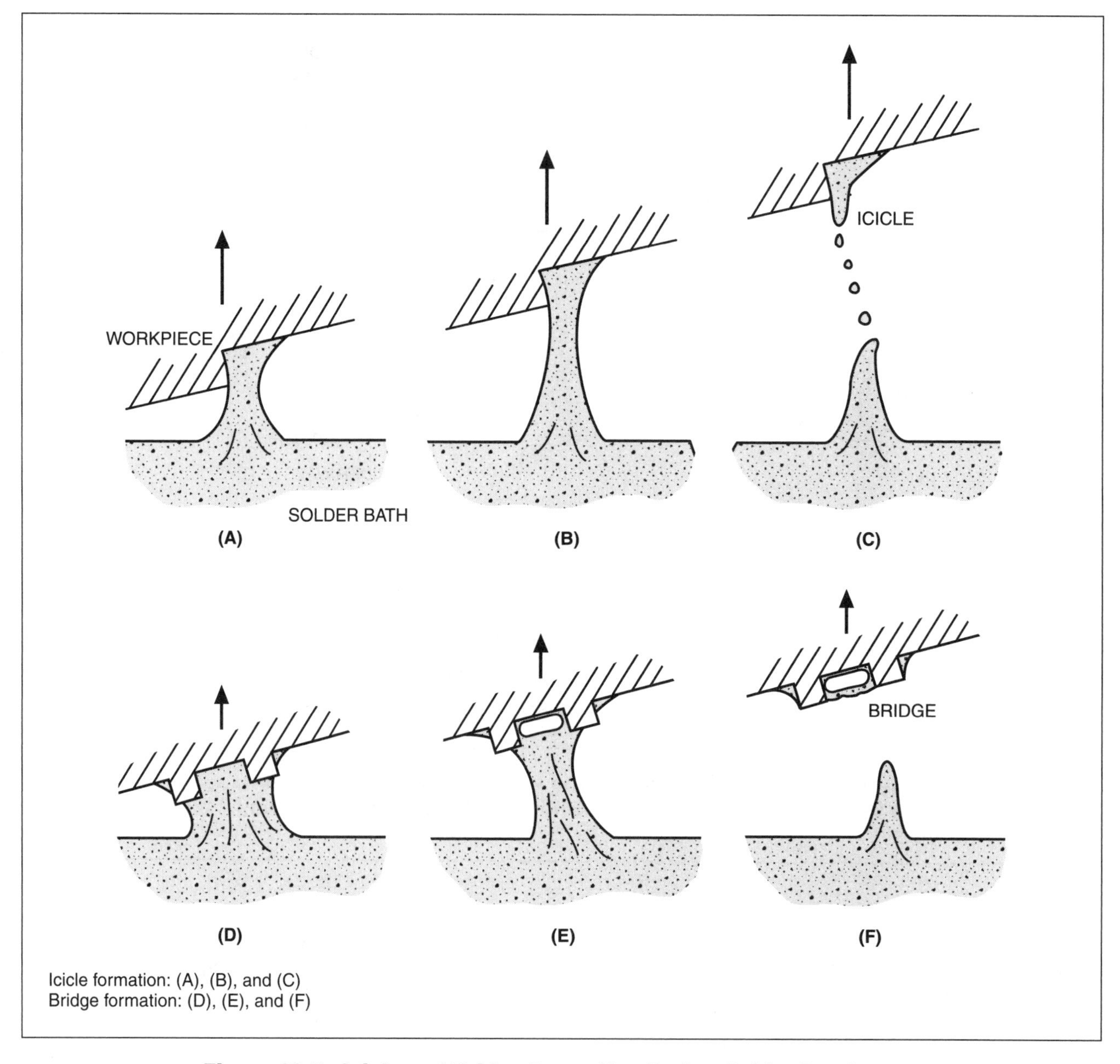

Figure 13.7—Icicle and Bridge Formation During Solder Breakaway as Workpiece Is Withdrawn from Solder Bath

2. Production is sufficiently large to allow expenditure for fixtures to hold the components during soldering;
3. The assembly is complex, making other heating methods impractical. The process is based on standard batch or continuous furnace operating principles. Heating can be performed either in air or in a protective (or reactive) furnace atmosphere.

The solder filler metal is generally preplaced as a solid preform (previously shaped solder manufactured in a range of sizes and shapes) or solder paste. If flux is used, it also is pre-applied. Application of the solder or flux in the furnace during heating is usually not recommended.

Most standard fluxes can be used in furnace soldering. The use of a reducing atmosphere in the furnace or

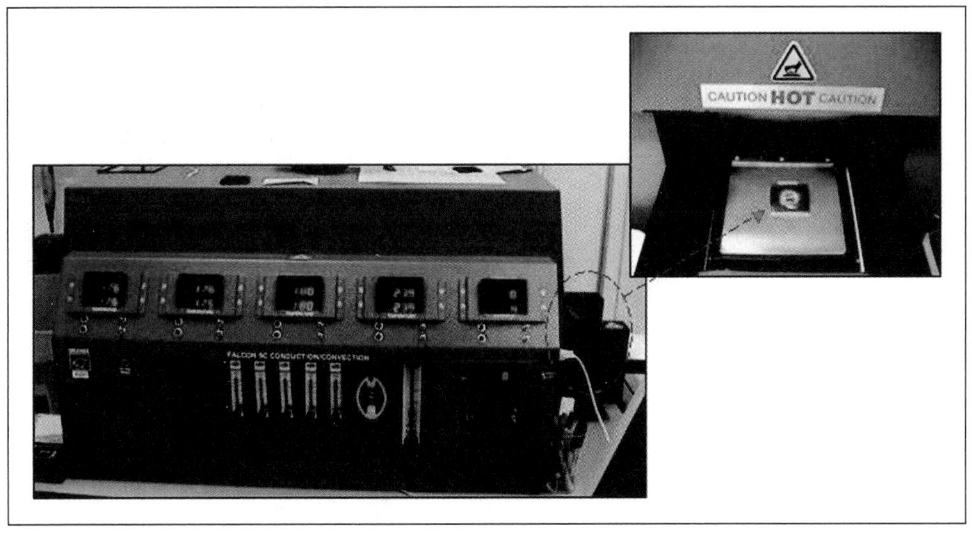

Photograph courtesy of Sandia National Laboratories

Figure 13.8—Furnace Reflow Soldering Operation in Progress (Front Oven Control Panel and Hood Exit with Soldered Assembly)

oven allows joints to be made with less aggressive types of fluxes, depending on the metal and solder combination. The use of an inert-gas atmosphere will prevent further oxidation of the workpieces, but adequate and appropriate fluxing is still required.

Proper clamping fixtures are important during furnace soldering. Any movement of the joint during solidification of the solder may result in a poor joint.

Furnaces should be equipped with adequate temperature controls because the flow of solder has an optimum temperature range, which depends on the flux used. The optimum heating condition exists when the heating capacity of the oven is sufficient to heat the workpieces rapidly under controlled flux application. Heating is usually accomplished by conduction, convection, or radiation in air or in a protective atmosphere such as nitrogen. It is often advantageous to accelerate the cooling of the workpieces on their removal from the oven to minimize soldering reactions. Forced air or nitrogen has been found to be satisfactory for cooling.

INDUCTION SOLDERING

Induction soldering (IS) is a soldering process in which the heat required to effect coalescence is obtained from the resistance of the workpieces to induced electric current.

The material that is to be induction soldered must be an electrical conductor or heated with a conductive susceptor. The rate of heating depends on the induced current flow. The distribution of heat obtained with induction heating is a function of the induced wave frequency. Higher frequencies concentrate the heat at the surface. The types of equipment available for induction heating include the vacuum-tube or solid-state oscillators, resonant spark-gap systems, motor-generator units, and solid-state electrical power sources.

The induction technique requires that components being joined have clean surfaces and accurate joint clearances. High-grade solders spread rapidly and produce good capillary flow. Preforms are often the best means of supplying the correct amount of solder and flux to the joint. Application during heating is generally difficult and not recommended.

When soldering dissimilar metals by induction, particularly joints composed of both magnetic and nonmagnetic components, attention must be given to the design of the induction coil in order to bring both workpieces to approximately the same temperature.

Induction soldering is generally chosen for large-scale production runs. The simple joint design required by induction soldering lends itself to automation. It is ideal when minimum oxidation of the surface adjacent to the joint is desired and when good appearance and consistent high-quality joints are necessary. Induction soldering can also be used for the application of heat to a localized area. Typical joint and coil designs for induction soldering are illustrated in Figure 13.9.

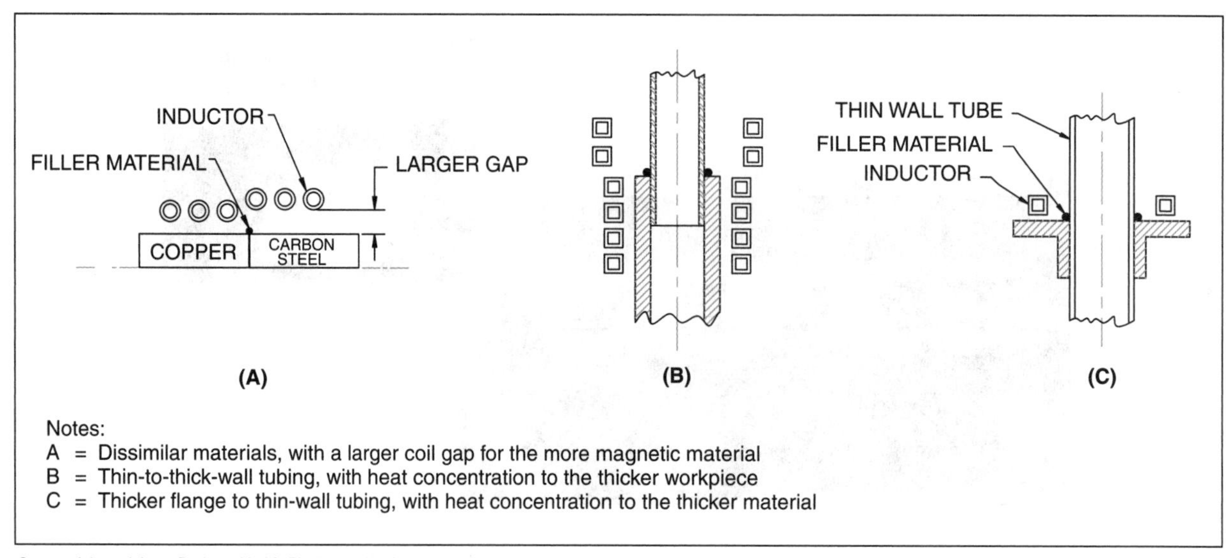

Source: Adapted from Rudnev, V., M. Black, and W. Albert, 2003. "Intricacies of Induction Joining," in *International Brazing and Soldering Conference Proceedings*, San Diego: American Society of Metals and American Welding Society, Figures 2–4.

Figure 13.9—Typical Joint and Coil Designs for Induction Soldering

INFRARED SOLDERING

Infrared soldering (IRS) is a soldering process in which the heat required for coalescence is supplied by infrared radiation. Heating depends on the type of material and its radiation absorption properties. Fixtured workpieces, with the solder alloy in place, are passed under the infrared radiation source and heated quickly to the soldering temperature. Typical applications involve high-volume and throughput manufacturing. A comparison of the different reflow soldering methods, including infrared heating, is shown in Figure 13.10.

Infra-optical soldering systems are also available. Heat is obtained by focusing infrared light (radiant energy) on the joint by means of a lens. Lamps ranging from 45 to 1500 watts (W) can be used for different applications. The devices can be programmed through a solid-state controlled power source with an internal timer. This heating method is very well suited to reflow soldering.

IRON SOLDERING

Iron soldering (INS) is a process that uses the tip of a resistance-heated soldering iron to conductively heat the workpieces and solder alloy to the soldering temperature. The process is typically a manual operation and is the most common method of fabricating simple solder joints.

LASER BEAM SOLDERING

Laser beam soldering is a process in which a laser light provides the heat source for soldering. It is a specialized soldering process used when accurate localized heating of the joint area is required, particularly with heat-sensitive materials; however, larger areas can also be heated with this process.

Soldering is usually performed in a protective gas cover with or without flux. Since coupling of the laser beam is usually materials-dependent, this must be carefully considered when designing and fabricating the solder joint. Several joints can be processed simultaneously by multiplexing the beam. The process is used in high-end military and commercial soldering applications where volume is relatively low and the level of joint reliability is high. Solder and flux are normally applied to the joint area as preforms.[3]

RESISTANCE SOLDERING

Resistance soldering (RS) is a process that uses heat generated by resistance to the flow of electric current in a circuit that includes the workpiece. Resistance soldering requires the workpiece to be placed either between a ground and a movable electrode or between two mov-

3. Vianco, P. T., *Soldering Handbook*, 3rd ed., 1999, Miami: American Welding Society, pp. 423–427.

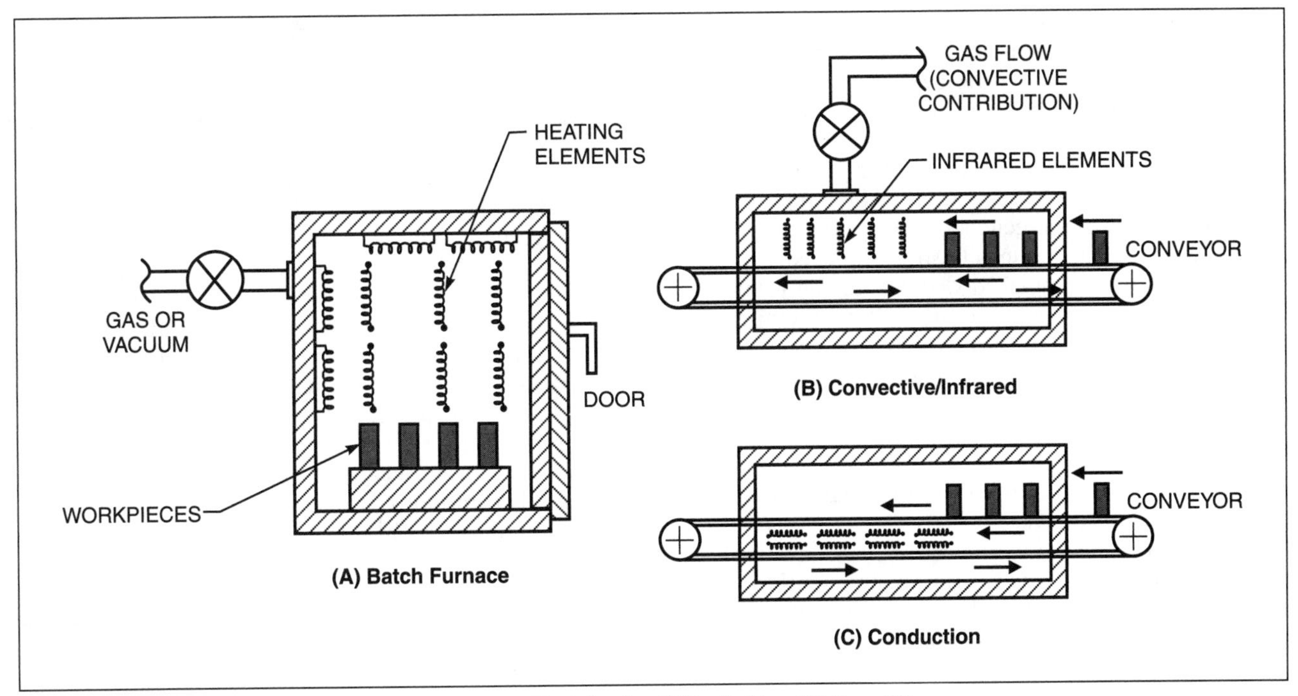

Source: Adapted from Vianco, P. T., 1999, *Soldering Handbook*, 3rd ed., Miami: American Welding Society, p. 351, Figure 6.25.

Figure 13.10—Schematic Diagrams of (A) a Batch Furnace, (B and C) In-Line Furnace Concepts Using Radiation or Convection Heat Modes, or (C) Thermal Conduction

able electrodes to complete an electrical circuit. Heat is applied to the joint by both the electrical resistance of the metal being soldered and by conduction from the moving electrode, which is usually carbon. The process is similar to other resistance joining processes. The required equipment includes a power source, process controller, and electrodes.

Production assemblies may use multiple electrodes, rolling electrodes, or custom-designed electrodes, whichever will be advantageous to soldering speed, localized heating, and power consumption.

Resistance soldering electrode tips cannot be pre-coated (tinned), and the solder must be fed into the joint or supplied by preforms or solder coatings on the work-pieces. Typical applications include the fabrication of "spot" joints that require localized heating or placement.

SPRAY SOLDERING

Spray soldering is a process that uses a thermal spray gun to apply the solder alloy to the workpieces. This method is generally selected when the contour of the workpiece is difficult to handle with other soldering techniques.

Gas-fired or electrically heated guns are used in spray soldering. Both types are designed to spray molten or semimolten solder onto the workpiece from a continuously fed solid solder wire.

Gas-fired guns use propane with oxygen or natural gas with air to heat and spray a continuously fed solid solder wire, approximately 3.2 millimeters (mm) (0.125 inches [in.]) in diameter. About 90% of the solder wire is melted by the flame of the gun. The solder strikes the workpiece in a semiliquid form. The work-piece, also heated by the flame, then supplies the balance of the heat required to melt and flow the solder. Adjustments can be made to the spray gun to control the solder spray.

Electrically heated guns are similar to gas-fired guns except that they use a heating element to melt the solder. Compressed air is then used to spray the molten solder onto the workpieces.

TORCH SOLDERING

Torch soldering (TS) is a process that uses the heat from a fuel-gas flame for soldering. Propane and oxy-acetylene are typical fuel gases used.

The flame temperature is controlled by the nature of the gas or gas mixtures used. The highest flame temperatures are attained with acetylene. Lower temperatures are produced with propane, butane, natural gas, and manufactured gas, roughly in that order. Fuel gas burned with oxygen generates higher flame temperatures than fuel gas burned with air. The flame of a fuel gas burned with oxygen is sharply defined; when burned with air, the flame is bushy and flared.

Solder is either preplaced or fed into the joint area. The workpieces are heated to the soldering temperature by the directed flame from the torch. Equipment is similar to that used for torch brazing. Typical applications involve both manual and automated soldering on tubular assemblies.

The gas torch selected for soldering should be appropriate to the size, mass, and configuration of the assembly to be soldered.

Multiple-flame tips, or burners, are available in various shapes and designs and are frequently used to solder complex assemblies. The tips may be designed to operate on torches for oxygen and fuel gas, compressed air and fuel gas, or on Bunsen-type torches. Adjustments of tips or torches should be made with care to avoid adjustments that result in a sooty flame, which deposits carbon on the workpiece and prevents the flow of solder.

HOT-GAS SOLDERING

A variation of torch soldering is hot-gas soldering. Hot gas soldering uses a fine jet of inert gas heated to above the liquidus of the solder. The gas acts as a heat transfer medium and as a shield to reduce the access of air to the joint. The process is usually automated to make tubular solder joints.

ULTRASONIC SOLDERING

Ultrasonic soldering (USS) is a process variation in which high-frequency vibratory energy is transmitted through molten solder to remove undesirable surface films and thereby promotes the wetting of the base metal. The process depends on the direct line-of-sight to the ultrasonic probe or horns and is usually performed without the use of flux. Equipment is available for ultrasonic dip-soldering and hand-soldering operations.

An ultrasonic transducer produces the high-frequency vibrations that break up tenacious oxide films on base metals. The freshly exposed base metal is readily wetted without the use of flux, or with a less aggressive flux. Ultrasonic soldering units are useful in soldering return bends to the sockets of aluminum air-conditioner coils. Ultrasonic soldering is also used to apply solderable coatings on difficult-to-solder metals.

VAPOR-PHASE SOLDERING (CONDENSATION)

Vapor-phase soldering uses the latent heat of vaporization of a condensing saturated liquid to provide the heat required for soldering. This process uses preplaced flux and solder. Typical condensing fluids are fluorinated organic compounds with boiling points between 215°C and 255°C (420°F and 490°F). A reservoir of saturated vapor over a boiling liquid provides a constant-controlled temperature with rapid heat transfer that is useful for soldering large assemblies as well as temperature-sensitive parts. Commercial equipment is available with conveyors that provide an in-line, continuous process for electronics manufacturing. The process is best suited for applications in which uniform heating is required to minimize thermal warping in assemblies with dissimilar materials.

EQUIPMENT

The proper application of heat is of paramount importance in any soldering operation. Heat should be applied in such a manner that the faying surface becomes sufficiently hot to melt the solder and facilitate the wetting and flow of the molten solder along the base metal. A variety of equipment, tools, and methods are available as heat sources.

Depending on the application, the soldering process can be manual, semi-automated, or robotic. The equipment setup may vary from simple fixturing to computer-controlled processing equipment, depending on the soldering method chosen. More detailed information is included in the *Soldering Handbook*, 3rd edition, published by the American Welding Society.[4]

SOLDERING IRONS

Soldering irons are the most commonly used tools, and they are available for a variety of joining applications. The selection of soldering irons may be simplified by classifying them into the following groups: conventional soldering irons; transformer-type, low-voltage pencil irons; special quick-heating and pliers-type irons; and heavy-duty industrial irons.

Several types of soldering irons are shown in Figure 13. 11.

The conventional soldering iron has a copper tip that is heated electrically or by oil, coke, or gas burners. To lengthen the usable life of a copper tip, a coating of

4. See Reference 3.

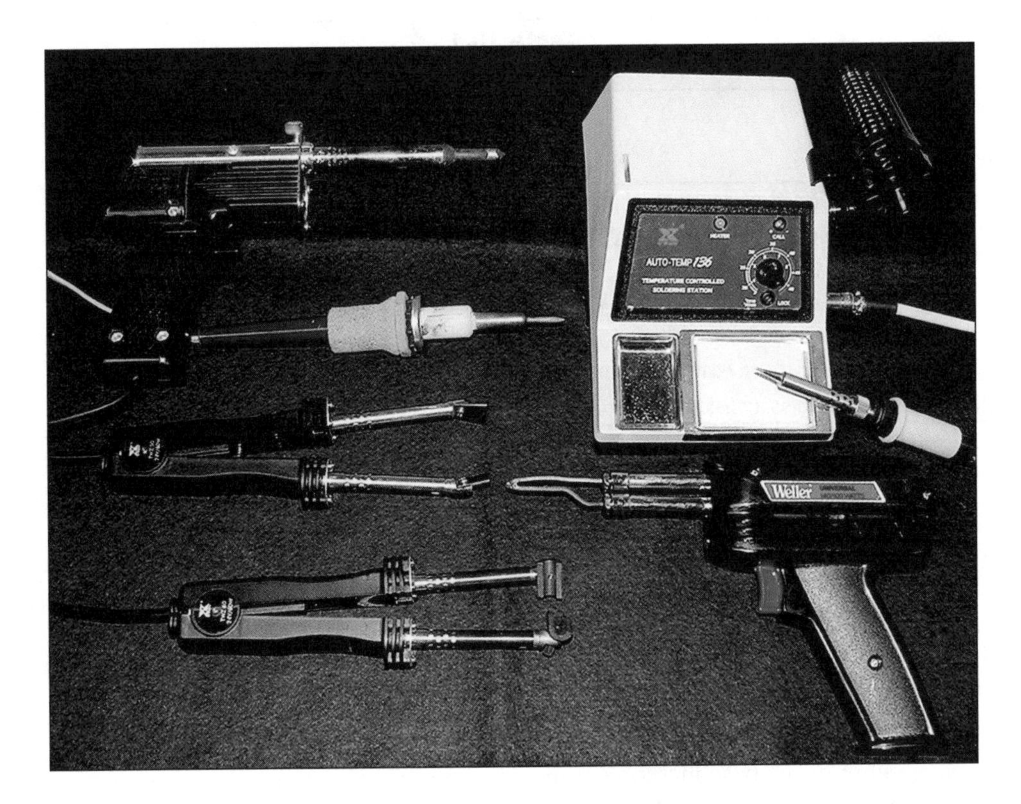

Figure 13.11—Typical Soldering Irons

solderable metal such as iron, with or without additional coatings, is applied to the surface of the copper. The rate of dissolution of the iron coating in molten solder is substantially less than the rate for copper. The iron coating also protects better than uncoated copper against wear, oxidation, and pitting.

Modern hand-soldering irons are manufactured with tips that closely control temperatures at the tip. They are available in a wide range of tip sizes especially designed to work with specific solder-wire diameters and to maintain the required soldering temperatures. These irons are usually selected on the basis of the application, tip diameter, and power range.

Other irons heated by a gas flame are commonly used in plumbing and roofing applications. A guide for the selection of some commonly used soldering irons is presented in Table 13.1.

Regardless of the heating method used, the tip of the soldering iron performs four important functions. It stores and conducts heat from the heat source to the workpieces, stores molten solder, conveys molten solder, and withdraws surplus molten solder.

The angle at which the copper tip is applied to the workpiece is important in delivering maximum heat to the workpiece. The flat side of the tip should be applied to the workpiece to obtain the maximum area of contact. Flux cored solders should not be melted on the soldering tip because this destroys the effectiveness of the flux. Cored solder should be touched to the soldering tip to initiate good heat transfer, and then the solder should be melted on the workpieces to complete the solder joint.

The performance of electric industrial soldering irons cannot be measured solely by the wattage rating of the heating element. The material of manufacture and the design affect the heat reserve and temperature recovery of the copper tip.

MATERIALS

Base metals and nonmetallic materials are usually selected for the specific properties required by the product design. The solder alloy is usually selected to meet specific melting requirements of the base material to provide good flow, penetration, and wettability in the soldering operation and the desired joint properties in the finished product.

Table 13.1
Guide for the Selection of Soldering Irons

Soldering Application	Tip-Diameter Range		Power Range
	mm	in.	Watts
Miniature printed circuits, thin substrates, temperature-sensitive components	1 to 3	1/32 to 1/8	10 to 20
Intermittent light assembly work, printed circuits, instruments, jewelry	3 to 5	1/8 to 3/16	20 to 35
Repetitive assembly work, telephones and appliances, art glass	5 to 6	3/16 to 1/4	40 to 60
High-speed production soldering, light tinware, general duty, medium electrical, light plumbing	6 to 13	1/4 to 1/2	70 to 150
Medium tinware, light roofing, shipboard repair, heavy electrical, heavy plumbing	13 to 38	1/2 to 1-1/2	170 to 350
Heavy tinware, roofing, radiators, armatures, transformer cans	38 to 53	1-1/2 to 2	350 to 1250

BASE METALS AND MATERIALS

Typical base-metal properties include strength, ductility, electrical and thermal conductivity, density, and corrosion resistance. The solderability of the base materials is an important consideration. The selection of solders, fluxes, and surface-preparation will be affected by the solderability of the base materials to be joined. Solderability tests are widely used on materials to be soldered. These tests give important information, but the tests may not cover effects such as those resulting from storage and variation in materials. Also, the tests will not cover the ability to clean previously prepared components unless they are designed into the testing.

In general, the metals most suited to soldering are platinum, gold, copper, silver, cadmium plate, tin plate and solder plate (e.g., Sn-Pb). Somewhat less solderable are lead, nickel plate, brass, bronze, rhodium, and beryllium-copper. Metals that are difficult to solder include galvanized iron, tin-nickel, nickel-iron, and low-carbon steel. Increased difficulty can be expected when soldering chromium, nickel-chromium, nickel-copper and stainless steel. Aluminum and aluminum-bronze materials are the least solderable. Beryllium and titanium cannot be soldered with conventional technology; however, joints in titanium and other materials, such as ceramics and composites, have been successfully soldered using active solder and mechanical agitation of the molten alloys or applying solderable coatings.

SOLDERS

Solder alloys typically have melting ranges below 450°C (842°F). A wide variety of commercially available solder filler metals have been designed to work with most industrial metals and alloys. These generally flow satisfactorily with the appropriate fluxes and produce good surface wetting and solder joints with satisfactory properties.

Specifications for general solder alloys are published by the American Society for Testing and Materials (ASTM) in the *Annual Book of ASTM Standards*,[5] which includes *Standard Specification for Solder Metal*, ASTM B 32-00.[6] These standards have replaced the U.S. Federal Solder Specification, *Electronic Solder (96–485°C)*, QQ-S-571E.[7]

The IPC–Association Connecting Electronics Industries (formerly the Institute for Interconnection and Packaging Electronic Circuits) and the Electronic Industries Alliance (EIA) work with the electronics industry to develop internationally recognized solder specifications such as *General Requirements for Solder Electronic Interconnections*, IPC-S-815A.[8] The IPC-S-815A standard was superseded by IPC/EIA J-STD-001C, *Requirements for Soldered Electrical and Electronic Assemblies*;[9] and IPC/EIA J-STD-006A, *Requirements for Electronic Grade Solder Alloys and Fluxed and*

5. American Society for Testing and Materials (ASTM). 2000, Annual *Book of ASTM Standards*, West Conshohocken, Pennsylvania: American Society for Testing and Materials.
6. American Society for Testing and Materials (ASTM), 2000, *Standard Specification for Solder Metal*, B 32-00, in *Annual Book of ASTM Standards*, West Conshohocken, Pa.: American Society for Testing and Materials.
7. General Services Administration (GSA), 1986, *Electronic Solder (96–485°C)*, QQ-S-571E, Washington, D.C.: General Services Administration.
8. IPC–Association Connecting Electronics Industries, 1987, *General Requirements for Solder Electronic Interconnections*, IPC-S-815A, Northbrook, Ill.: IPC–Association Connecting Electronics Industries.
9. IPC–Association Connecting Electronics Industries/EIA, 2000, *Requirements for Soldered Electrical and Electronic Assemblies*. IPC/EIA J-STD-001C, Northbrook, Ill.: IPC–Association Connecting Electronics Industries.

Table 13.2
Typical Commercial Solder Product Forms

Solder Product	Specifications
Pig	Available in 25 and 45 kg (50 and 100 lb)
Ingots	Rectangular or circular in shape, weighing 1.4, 2.3, and 4.5 kg (3, 5, and 10 lb)
Bars	Available in numerous cross sections, weights, and lengths
Paste or cream	Available as a mixture of powdered solder and flux
Foil, sheet, or ribbon	Available in various thickness and widths
Segment or drop	Triangular bar or wire cut into any desired number of pieces or lengths
Wire, solid	Diameters of 0.25 to 6.35 mm (0.010 to 0.250 in.) on spools
Wire, flux cored	Solder cored with rosin, organic, or inorganic fluxes. Diameters of 0.25 to 6.35 mm (0.010 to 0.250 in.)
Preforms	Unlimited range of sizes and shapes to meet special requirements

Non-Fluxed Solid Solders for Electronic Soldering Applications.[10]

The International Organization for Standardization (ISO), Geneva, Switzerland provides another source of solder specifications with the publication *Soft Solder Alloys—Chemical Composition and Forms,* ISO/DIS 9453).[11] Solder alloy specifications that are developed for electronic applications are generally suitable for structural applications.

Solders are commercially available in various forms and products that can be grouped into about a dozen classifications. The major groups of solder product forms are listed in Table 13.2. This list is by no means complete, since solder in any desired size, weight, or shape in any product form is available by special order from solder manufacturers.

Tin-Lead Solders

Tin-lead solder alloys are the most widely used solder filler metals. Consequently, a large variety of commercial fluxes, cleaning methods, and soldering processes have been developed to be used with tin-lead solders. In describing solders, it is customary to identify the tin content first. As an example, 60/40 solder is 60% tin and 40% lead, by weight percent.

10. IPC–Association Connecting Electronics Industries/EIA, 2001, *Requirements for Electronic Grade Solder Alloys and Fluxed and Non-Fluxed Solid Solders for Electronic Soldering Applications,* IPC/EIA J-STD-006A, Northbrook, Ill.: IPC–Association Connecting Electronics Industries.
11. International Organization for Standardization (ISO), 1990, *Soft Solder Alloys–Chemical Composition and Forms,* SIS-ISO 9453, Geneva, Switzerland: International Organization for Standardization.

The behavior of the various tin-lead alloys can be illustrated best by reference to a constitutional diagram, as shown in Figure 13.12. The following terms are used to describe this diagram:

1. *Solidus temperature*—The highest temperature at which a metal or alloy is completely solid. The solidus temperature is represented by Curve ACEDB in Figure 13.12;
2. *Liquidus temperature*—The lowest temperature at which a metal or alloy is completely liquid. The liquidus temperature is represented by Curve AEB in Figure 13.12;
3. *Eutectic alloy*—An alloy that melts at one temperature and not over a range of temperatures. The eutectic temperature is the solidus temperature represented by Curve CED in Figure 13.12. The eutectic alloy is the composition noted at Point E in Figure 13.12. This alloy is approximately 63% tin by weight; and
4. *Melting range*—The temperature between the solidus, ACEDB, and the liquidus, AEB, where the solder alloy is partially melted.

As shown in Figure 13.12, pure lead melts at 327°C (621°F), (Point A), and pure tin melts at 232°C (450°F) (Point B). Solders containing 19.5% tin (Point C) up to 97.5% tin (Point D) have the same solidus temperature, namely the eutectic temperature, which is 183°C (361°F). The eutectic composition is completely liquid above 183°C (361°F). Any other composition will contain solid and liquid components that are in thermodynamic equilibrium. The primary phase constituents are rich in either tin (β) or lead (α). These compositions do not melt completely until above the liquidus temperature.

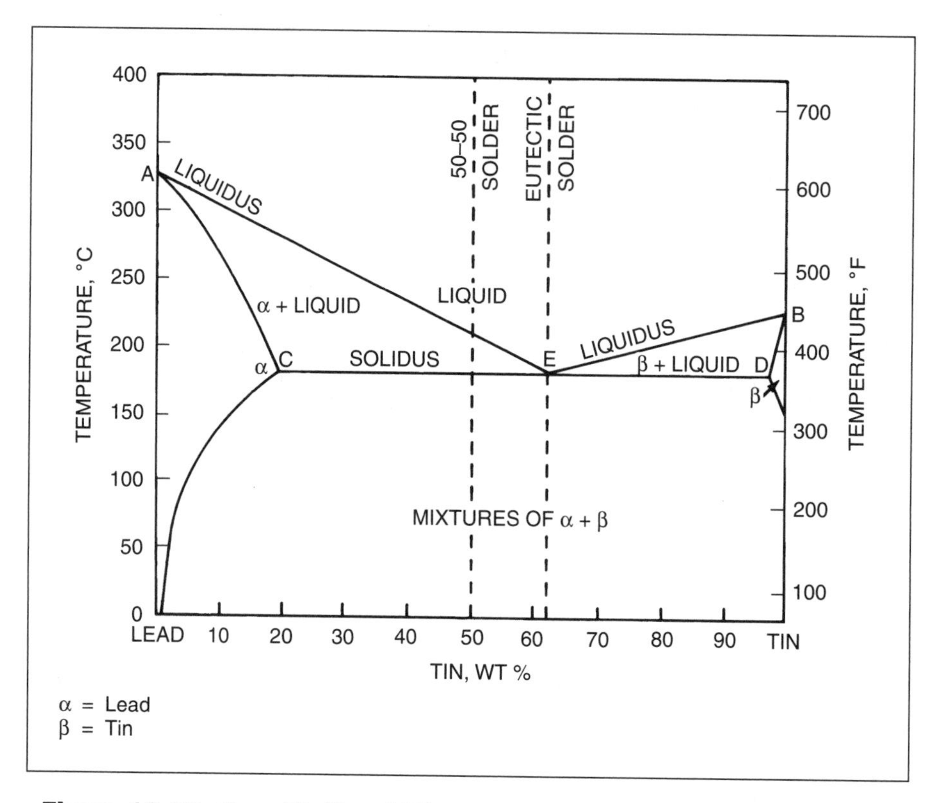

Figure 13.12—Constitutional Diagram for the Tin-Lead Alloy System

For example, 50/50 tin-lead solder has a solidus temperature of 183°C (361°F) and a liquidus temperature of 214°C (417°F), which is a melting range covering 31°C (56°F). The melting range is the temperature difference between the solidus and liquidus. The melting characteristics of specific tin-lead solders are shown in Table 13.3.

The 5/95 tin-lead solder has a relatively high melting temperature within a narrow melting range. Wetting and flow characteristics of 5/95 tin-lead solder are less effective compared to solders with higher tin content. Proper wetting and flow of 5/95 tin-lead solder requires extra care in surface preparation. Solders with a high lead content have better creep properties at 150°C (302°F) than solders containing more tin. The high soldering temperature limits the use of organic-based fluxes such as rosin or those of the intermediate type. The 5/95 tin-lead solder is particularly adaptable to torch, dip, induction, or furnace soldering. It is used for sealing precoated containers, electronic packages, automotive radiators, and other moderately elevated temperature applications. This solder is also used in chip attachment applications for the electronics industry.

The 10/90, 15/85, and 20/80 tin-lead solders have lower liquidus and solidus temperatures, but they have wider melting ranges than 5/95 tin-lead solder. They also provide better wetting and flow characteristics with most fluxes. However, extreme care must be taken to avoid movement of these solders during solidification to prevent hot tearing and cracking during cooling. These solders are typically used for sealing cellular automobile heater cores and radiators. They also are used for general joining and surfacing applications.

The 25/75 and 30/70 tin-lead solders have lower liquidus temperatures than all previously mentioned alloys but have the same solidus temperature as the 20/80 tin-lead solder. Therefore, their melting ranges are narrower than that of the 20/80 solder. All standard cleaning, fluxing, and soldering techniques can be used with these solders. They are widely used for torch and dip soldering. These alloys are used in radiator fabrication, radiator repair, and a variety of other structural applications.

The 35/65, 40/60, and 50/50 tin-lead solders have progressively lower liquidus temperatures. The solidus temperature is the same as the 20% to 30% tin solders, and the melting ranges are narrower. Solders of this group have very good wetting properties, satisfactory strength, and are economical for electronic applications. These solders are also extensively used in sheet-metal work. The 50/50 combination is used for non-potable

Table 13.3
Tin-Lead Solder Classifications and Melting Properties

ASTM Solder Classification*	Composition, Weight %		Solidus		Liquidus		Melting Range	
	Tin	Lead	°C	°F	°C	°F	°C	°F
5	5	95	300	572	314	596	14	24
10	10	90	268	514	301	573	33	59
15	15	85	225	437	290	553	65	116
20	20	80	183	361	280	535	97	174
25	25	75	183	361	265	117	84	150
30	30	70	183	361	255	491	72	130
35	35	65	183	361	247	477	64	116
40	40	60	183	361	235	455	52	94
45	45	55	183	361	228	441	65	80
50	50	50	183	361	217	421	34	60
60	60	40	183	361	190	374	7	13
70	70	30	183	361	192	378	9	17

*See ASTM Specification B 32-00, *Standard Specification for Solder Metal*, in Annual Book of ASTM Standards, West Conshohocken, Pa.: American Society for Testing and Materials.

water plumbing and piping. These solders are typically in rosin-cored wire form and are also used in industrial wave-soldering applications.

The 60/40 alloy and the 63/37 eutectic tin-lead solder are used when the application requires tight control of the soldering temperature and the reaction of the solder alloy with the base metal. Tight control is particularly necessary for delicate instruments and electronic assemblies. The 60/40 composition is close enough to that of the eutectic tin-lead alloy to have an extremely narrow melting range. All methods of cleaning, fluxing, and heating may be used with this solder. These alloys are widely used, often in the form of a solder paste, in the soldering of electronics by means of wave, furnace, and vapor-phase soldering.

The 70/30 tin-lead solder is a special-purpose alloy used when a higher tin content is required for improved wetting with a slightly higher liquidus temperature. All soldering techniques are applicable.

Impurities in Tin-Lead Solders

Impurities in tin-lead solders can occur during manufacture of the alloys or may result from contamination during use. Specifications for solder alloys usually limit the maximum total impurity content, with specific limitations for certain metals. Solder metals covered by *Standard Specification for Solder Metals*, ASTM B 32[12] are shown in Table 13.4. The reported values are in weight percent. Individual users sometimes require additional restrictions on impurities for particular applications.

Impurities can cause a reduction in wetting properties, sluggishness of flow within solder joints, an increase in oxidation rate, and changes in melting-temperature ranges. Strength properties of joints can be adversely affected with increased tendencies to cracking of the solder or difficulties with adhesion to the base materials. Impurity elements also affect the appearance and quality of the molten solder.

The combined effects of impurities can be disastrous in soldered joints. These concerns are recognized in manufacturing specifications and take into account the quality of solder needed for different applications. Solder materials should be purchased with care to assure that the appropriate alloy and grade is obtained.

Aluminum. Aluminum (Al) as an impurity in tin-lead solder at levels of more than 0.005% can cause grittiness in the solder. A noticeable deterioration in oxidation of a solder bath surface or an increase in surface dross can be an indication of aluminum contamination.

Antimony. Antimony (Sb) is often used in solders as a desirable addition; however, this metal tends to reduce wetting and spreading qualities and can cause adhesion problems when present at higher levels than required.

12. See Reference 6.

Table 13.4
Solder Compositions and Melting Properties

Alloy Grade	Composition, Weight %[a, b]											Melting Range[c]			
												Solidus		Liquidus	
	Sn	Pb	Sb	Ag	Cu	Cd	Al	Bi	As	Fe	Zn	°C	°F	°C	°F
Sn96	balance	0.10	0.12 max	3.4–3.8	0.08	0.005	0.005	0.15	0.01 max	0.02	0.005	221	430	221	430
Sn95	balance	0.10	0.12	4.4–4.8	0.08	0.005	0.005	0.15	0.01	0.02	0.005	221	430	245	473
Sn94	balance	0.10	0.12	5.4–5.8	0.08	0.005	0.005	0.15	0.01	0.02	0.005	221	430	280	536
Sn70	69.5–71.5	bal.	0.50	0.015	0.08	0.001	0.005	0.25	0.03	0.02	0.005	183	361	193	377
Sn63	62.5–63.5	bal.	0.50	0.015	0.08	0.001	0.005	0.25	0.03	0.02	0.005	183	361	183	361
Sn62	61.5–62.5	bal.	0.50	1.75–2.25	0.08	0.001	0.005	0.25	0.03	0.02	0.005	179	354	189	372
Sn60	59.5–61.5	bal.	0.50	0.015	0.08	0.001	0.005	0.25	0.03	0.02	0.005	183	361	190	374
Sn50	49.5–51.5	bal.	0.50	0.015	0.08	0.001	0.005	0.25	0.025	0.02	0.005	183	361	216	421
Sn45	44.5–46.5	bal.	0.50	0.015	0.08	0.001	0.005	0.25	0.025	0.02	0.005	183	361	227	441
Sn40A	39.5–41.5	bal.	0.50	0.015	0.08	0.001	0.005	0.25	0.02	0.02	0.005	183	361	238	460
Sn40B	39.5–41.5	bal.	0.50	0.015	0.08	0.001	0.005	0.25	0.02	0.02	0.005	185	365	231	448
Sn35A	34.5–36.5	bal.	1.8–2.4	0.015	0.08	0.001	0.005	0.25	0.02	0.02	0.005	183	361	247	447
Sn35B	34.5–36.5	bal.	0.50	0.015	0.08	0.001	0.005	0.25	0.02	0.02	0.005	185	365	243	470
Sn30A	29.5–31.5	bal.	1.6–2.0	0.015	0.08	0.001	0.005	0.25	0.02	0.02	0.005	183	361	255	491
Sn30B	29.5–31.5	bal.	0.50	0.015	0.08	0.001	0.005	0.25	0.02	0.02	0.005	185	365	250	482
Sn25A	24.5–26.5	bal.	1.1–1.5	0.015	0.08	0.001	0.005	0.25	0.02	0.02	0.005	183	361	266	511
Sn25B	24.5–26.5	bal.	0.50	0.015	0.08	0.001	0.005	0.25	0.02	0.02	0.005	185	365	263	504
Sn20A	19.5–21.5	bal.	0.8–1.2	0.015	0.08	0.001	0.005	0.25	0.02	0.02	0.005	183	361	277	531
Sn20B	19.5–21.5	bal.	0.50	0.015	0.08	0.001	0.005	0.25	0.02	0.02	0.005	184	363	270	517
Sn15	14.5–16.5	bal.	0.50	0.015	0.08	0.001	0.005	0.25	0.02	0.02	0.005	225	437	290	554
Sn10A	9.0–11.0	bal.	0.20	0.015	0.08	0.001	0.005	0.25	0.02	0.02	0.005	268	514	302	576
Sn10B	9.0–11.0	bal.	0.50	1.7–2.4	0.08	0.001	0.005	0.03	0.02	0.02	0.005	268	514	299	570
Sn5	4.5–5.5	bal.	0.50	0.015	0.08	0.001	0.005	0.25	0.02	0.02	0.005	308	586	312	594
Sn2	1.5–2.5	bal.	0.50	0.015	0.08	0.001	0.005	0.25	0.02	0.02	0.005	316	601	322	611
Sb5	94.0 min	0.20	4.5–5.5	0.015	0.08	0.03	0.005	0.15	0.05	0.04	0.005	233	450	240	464
Ag1.5	0.75–1.25	bal.	0.40	1.3–1.7	0.30	0.001	0.005	0.25	0.02	0.02	0.005	309	588	309	588
Ag2.5	0.25	bal.	0.40	2.3–2.7	0.30	0.001	0.005	0.25	0.02	0.02	0.005	304	580	304	580
Ag5.5	0.25	bal.	040	5.0–6.0	0.30	0.001	0.005	0.25	0.02	0.02	0.005	304	580	380	716

a. Limits are % maximum unless shown as a range or stated otherwise.

b. For purposes of determining conformance to these limits, an observed value or calculated value obtained from analysis shall be rounded to the nearest unit in the last right-hand place of figures used in expressing the specified limit, in accordance with rounding method of Recommended Practice E 29 in American Society for Testing and Materials (ASTM). 2002. *Standard Practice for Using Significant Digits in Test Data to Determine Conformance with Specifications*, ASTM E 29-02. West Conshohocken, Pennsylvania: American Society for Testing and Materials.

c. Temperatures given are approximations and for information only.

Arsenic. Arsenic (As) is a cause of dewetting when as little as 0.005% is present. The problem becomes more severe at higher levels, and therefore less than 0.002% arsenic in the solder metal is desirable.

Bismuth. Bismuth (Bi) is often used as an additive to solder alloys. It also can exist as a trace element. Low levels of bismuth can be tolerated, although it can change metallurgical characteristics of the joint and produce low-melting phases.

Cadmium. Cadmium (Cd) modifies the surface tension of solders and can cause deleterious effects, such as bridging and solder icicles, on printed circuit boards.

Copper. The amount of copper (Cu) that can be present in a solder without causing problems depends largely on the application. ASTM specifications limit the copper content of tin-lead solders to 0.08%. However, copper can be present up to 0.3% without any observable reduction in soldering properties. Copper pickup should be closely monitored in dip-soldering operations, particularly with high-volume and automated manufacturing conditions.

Iron and Nickel. Iron (Fe) and nickel (Ni) are not normally present in solder alloys. Specifications usually limit the iron and nickel content to a maximum of 0.02%. Severe reductions in wetting properties have been observed with higher levels. Iron and nickel contamination often occurs as pickup from the process fixtures or the walls of the solder-bath container.

Phosphorus and Sulfur. Phosphorus (P) and sulfur (S) should be kept at the absolute minimum levels to prevent oxidation and grittiness, which can result from the formation of phosphides and sulfides.

Zinc. Zinc (Zn) affects the wetting and surface-tension properties of molten solder. Thus tin-lead solders should contain less than 0.005% zinc. Dewetting on copper surfaces has been ascribed to zinc at levels of only 0.01% in the alloy.

Tin-Antimony Solder

The 95% tin, 5% antimony solder has melting characteristics as shown in Table 13.5. This solder provides a narrow melting range at a temperature higher than the tin-lead eutectic. The solder is used in many

Table 13.5
Melting Properties for Various Solder Compositions

Composition, Weight %		Solidus		Liquidus		Melting Range	
		°C	°F	°C	°F	°C	°F
Tin	**Antimony**						
95	5	232	450	240	464	8	14
Tin	**Zinc**						
91	9	199	390	199	390	0	0
80	20	199	390	269	518	70	128
70	30	199	390	531	192	112	202
60	40	199	390	340	645	141	255
30	70	199	390	375	708	176	318
Cadmium	**Zinc**						
82.5	17.5	265	509	265	509	0	0
40	60	265	509	335	635	70	126
10	90	265	509	339	750	134	241
Cadmium	**Silver**						
95	5	338	640	393	740	55	100
Zinc	**Aluminum**						
95	5	382	720	382	720	0	0

plumbing, refrigeration, and air-conditioning applications because it has good creep properties and does not contain lead, an increasingly important environmental consideration. The tin-antimony solder should not be used to join zinc-containing base metals such as brass because of the potential for the formation of brittle intermetallic compounds.

Tin-Antimony-Lead Solders

Antimony may be added to a tin-lead solder as a substitute for some of the tin. The addition of antimony improves the mechanical properties of the joint with only slight impairment to the soldering characteristics. All standard methods of cleaning and heating may be used. Specialized fluxes are needed for best results with these alloys.

Tin-Silver, Tin-Copper-Silver, and Tin-Lead-Silver Solders

The melting characteristics of solders containing silver are listed in Table 13.6. The 96% tin-4% silver solder is free of lead and is often used to join stainless steel for food-handling equipment. It has good shear and creep strengths and excellent flow characteristics.

The tin-silver and tin-copper-silver solders are the standard lead-free alloys used with copper pipe and tubes in potable water systems that must meet strict environmental, safety, and health regulations. The most commonly used tin-copper-silver alloy is the 95.5% tin-4% copper-0.5% silver solder, which melts near 220°C (430°F).

The 62% tin-36% lead-2% silver solder is used when soldering to silver-coated surfaces in electronic applications. The silver addition retards the dissolution of the silver coating during the soldering operation. The addition of silver also increases creep resistance due to the presence of Ag_3Sn precipitates that form in the solder during solidification.

High-lead solders containing tin and silver provide higher temperature solders for many applications, including automobile radiators. They exhibit good tensile, shear, and creep strengths, and they are recommended for cryogenic applications. Inorganic fluxes are generally recommended for use with these solders.

Tin-Zinc Solders

A large number of tin-zinc solders, some of which are listed in Table 13.5, are used for joining aluminum. Galvanic corrosion of soldered aluminum joints is minimized if the solder and the base metal are galvanically close in the electrochemical series table. Alloys containing 70% to 80% tin with the balance zinc are recommended for soldering aluminum. The addition of 1% to 2% aluminum, or an increase of the zinc content as high as 40%, improves corrosion resistance. However, the liquidus temperature rises correspondingly, and those solders are therefore more difficult to apply. The 91/9 and 60/40 tin-zinc solders may be used at temperatures above 140°C (284°F). The 80/20 and 70/30 tin-zinc solders are more widely used for coating the workpieces before the soldering operation.

Cadmium-Silver Solder

The 95% cadmium-5% silver solder has melting characteristics as shown in Table 13.5. Its primary use is for applications in which service temperatures will be higher than permissible with lower melting solders. Butt joints in copper can be made to produce room-temperature tensile strengths of 170 Megapascals (Mpa) (25 thousand pounds per square inch [ksi]). At 220°C (428°F), the tensile strength drops to 18 MPa (2.6 ksi).

The joining of aluminum to aluminum or to other metals is possible with 95% cadmium-silver solder. Improper use of solders containing cadmium may lead to health hazards. Therefore, care should be taken in

Table 13.6
Tin-Silver and Tin-Lead-Silver Solder Melting Properties

Composition, wt %			Solidus		Liquidus		Melting Range	
Tin	Lead	Silver	°C	°F	°C	°F	°C	°F
96	—	4	221	430	221	430	0	0
62	36	2	180	354	190	372	10	18
5	94.5	0.5	294	561	301	574	7	13
2.5	97	0.5	303	577	310	590	7	13
1	97.5	1.5	530	988	309	588	0	0

their application and during soldering, particularly with respect to fume inhalation. (Refer to the "Safe Practices" section of this chapter for information on safety standards covering fumes and Appendix B for a list of standards and publishers.) The material safety data sheet (MSDS) should be consulted for specific handling requirements and exposure limits.

Cadmium-Zinc Solder

Cadmium-zinc solders are also useful for soldering aluminum. The melting characteristics of these solders are presented in Table 13.5. The cadmium-zinc solders develop joints with intermediate strength and corrosion resistance. One application for the 40% cadmium-60% zinc solder is the soldering of aluminum lamp bases. This solder alloy is generally not used in high-volume applications. Improper use of this solder may lead to health hazards, particularly with respect to fume inhalation. The MSDS supplied by the manufacturer and pertinent safety standards should be consulted.

Zinc-Aluminum Solders

Zinc-aluminum solder is specifically used to join aluminum. It develops high-strength joints with good corrosion resistance. The solidus temperature is high, which limits its use to applications where soldering temperatures in excess of 370°C (698°F) can be tolerated. A major application is the dip soldering of return bends in aluminum air-conditioner coils. These coils are also made by flame brazing with fluxes. Ultrasonic solder pots that do not require the use of flux also are employed. In manual soldering operations, the heated aluminum surface is rubbed with the solder stick to promote wetting without a flux. The melting properties of zinc-aluminum solder are shown in Table 13.5.

Zinc-Base Solders

Zinc solders, with 95% zinc and other additions to restrict copper dissolution, were developed specifically for automotive radiator applications to improve wetting, joint strength, and corrosion resistance. Melting temperatures in the range of 425°C (797°F) are typical. The solders are generally heated with a torch or in a furnace or oven. A series of inorganic fluxes is available for use with these solders.

Fusible Alloys

Fusible or low-melting solder alloys make substantial use of bismuth. They are used in soldering operations where soldering temperatures below 183°C (361°F) are required. The melting characteristics and compositions of a representative group of fusible alloys are shown in Table 13.7.

Solders with low melting temperatures should be used when the following conditions must be addressed:

1. Heat-treated base metals are soldered, and higher soldering temperatures would result in softening the workpiece;
2. Materials adjacent to soldered joints are sensitive to temperature and would deteriorate at higher soldering temperatures;
3. Step-soldering operations are used to avoid remelting a nearby joint that has been made with a solder with a higher melting temperature; and
4. Temperature-sensing devices, such as fire sprinkler systems, are activated when the fusible alloy melts at a relatively low temperature and performs as the sensing weak link.

Table 13.7
Melting Properties of Typical Fusible Alloys

Composition, wt %				Solidus		Liquidus		Melting Range	
Lead	Bismuth	Tin	Other	°C	°F	°C	°F	°C	°F
26.7	50	13.3	10 Cd	70	158	70	158	0	0
25	50	12.5	12.5 Cd	70	158	74	165	4	7
40	52	—	8 Cd	91	197	91	197	0	0
32	52.5	15.5	—	95	203	95	203	0	0
28	50	22	—	96	204	107	225	11	25
28.5	48	14.5	9 Sb	102	217	227	440	125	223
44.5	55.5	—	—	124	255	124	255	0	0

Many of these solders, particularly those containing a high percentage of bismuth, are very difficult to use successfully in high-speed soldering operations. Particular attention must be paid to the cleanliness of metal surfaces. Strong, potentially corrosive fluxes must be used to make satisfactory joints directly to untreated metals such as aluminum, copper, and steel. If the surface can be plated for soldering with finishes such as tin or tin-lead, noncorrosive rosin fluxes may be satisfactory. However, most fluxes are not effective below 175°C (347°F) since full activation of the flux does not occur.

Indium Solders

Indium (In) solders have properties that make them valuable for many electronic and special applications. Melting characteristics and compositions of a representative group of these solders are shown in Table 13.8.

A 50% tin-50% indium alloy adheres to glass readily and may be used for glass-to-metal and glass-to-glass soldering. The low vapor pressure of this alloy makes it useful for seals in vacuum systems. It also offers the added value of a lead-free solder.

High fatigue resistance, especially to thermal cycling, has resulted in the increased use of indium alloys for the soldering of electronic systems, particularly indium-lead and indium-lead-silver solders.

Indium solders do not require special handling techniques. All of the soldering methods, fluxes, and processes used with the tin-lead solders are applicable to indium solders; however, indium solders are sensitive to corrosion in the presence of chlorides. Joints should be cleaned after soldering to remove all flux residues. The indium alloys perform best when covered by a protective organic conformal coating or under hermetically sealed conditions.

Special Solders

A class of specialized solder alloys has been developed to bond directly to nonmetallic materials without the need to metallize the faying surface. The alloys use conventional solder compositions, primarily tin-based, with small additions of active elements such as titanium or other oxide-forming constituents. These additions are generally less than 1% to 2% by weight. During soldering, the active element diffuses to the solder and nonmetallic interface and chemically reacts with the nonmetallic material. The key to the process is that the resulting reaction product is more stable than the initial nonmetallic surface. The alloys are used for joining ceramics to metals in automotive, aerospace, and microelectronic applications. Alumina, silicon nitride, and aluminum nitride are some of the nonmetallic base materials that have been soldered.

Figure 13.13(A) shows the microstructure of a copper-alumina joint. Figure 13.13(B) shows the microstructure of a titanium-to-titanium joint. In these applications, mechanical agitation was used to permit the wetting and flowing of the solder. Both were soldered with an active solder (Sn-Ag-Ti) filler metal. Figure 13.14 is a schematic illustration of active soldering under mechanical agitation.

FLUX

Flux is formulated to enhance solder wetting of base materials by removing nonwettable films from precleaned base-metal surfaces and by preventing oxidation during the soldering operation. The flux must protect those surfaces during heating and be available to protect the molten solder at the proper processing temperature. The flux must have sufficient activity to continue these functions until the joint has completely solidified.

When heated, the flux promotes the wetting and flow of molten solder on the base-metal surface. The flux should remove thin oxide or similar nonwettable layers from the base metal surfaces. An efficient flux also prevents reoxidation of the surfaces during the soldering process and is readily displaced by the molten solder.

Table 13.8
Typical Indium Solder Melting Properties

Composition, wt %			Solidus		Liquidus		Melting Range	
Tin	Indium	Lead	°C	°F	°C	°F	°C	°F
50	50	—	117	243	125	257	8	14
37.5	25	37.5	138	230	138	230	0	0
—	50	50	180	356	209	408	29	52

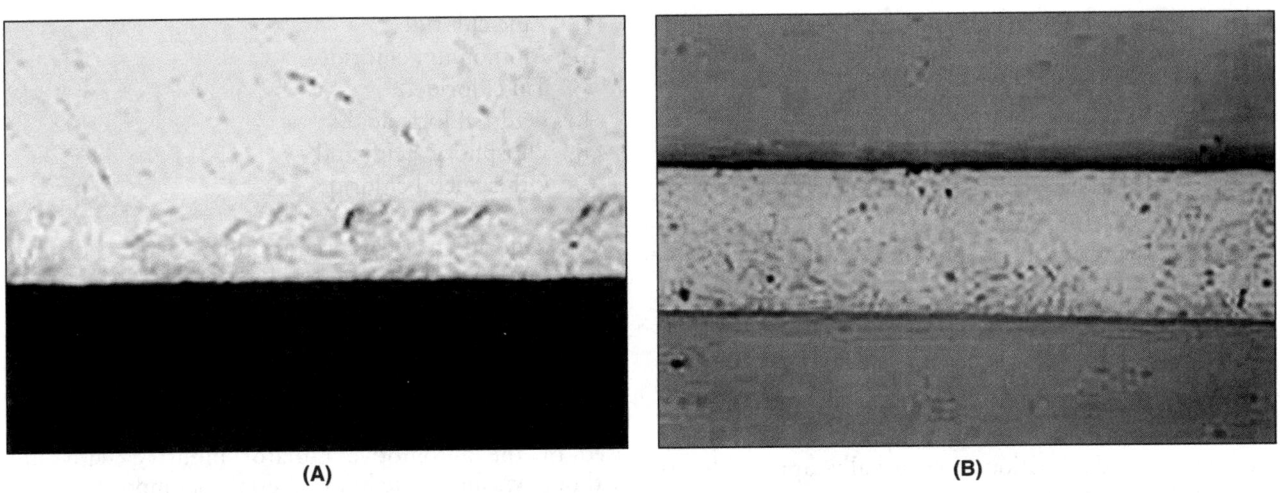

Source: Adapted from Smith, R. W., 2001, Active Solder Joining of Metals, Ceramics and Composites, *Welding Journal*, 80(10): 30–35.

Figure 13.13—(A) Microstructure of a Copper-Alumina Joint and (B) Titanium-Titanium Joint Soldered with an Active Solder (Sn-Ag-Ti)

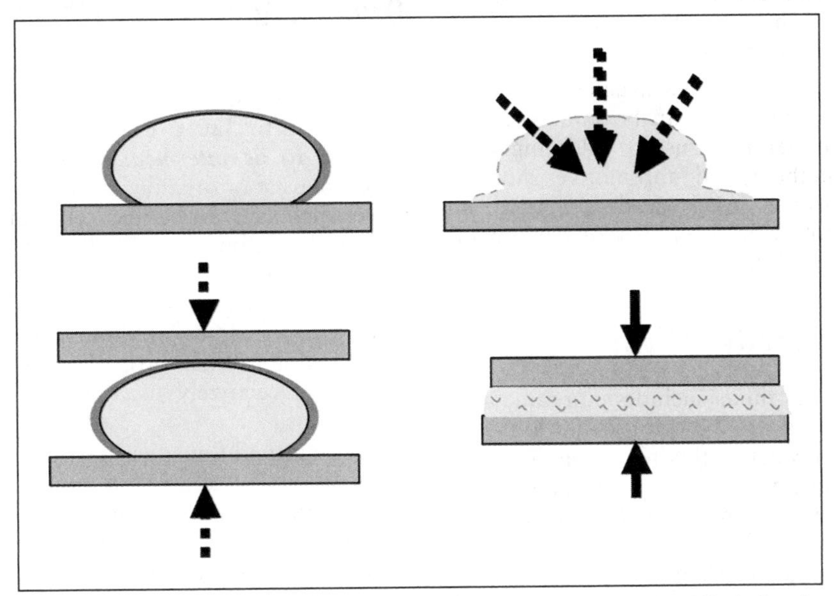

Source: Adapted from American Welding Society, Smith, R.W., 2001, Active Solder Joining of Metals, Ceramics and Composites, *Welding Journal*, 80(10): 30–35.

Figure 13.14—Active Solder under Mechanical Agitation

Soldering flux is available in liquid, solid, or gaseous form, as well as in single or multiple cores in wire solder. Not all fluxes are available in each form.

Flux Specifications

Flux specifications are available as standard references for handling, application, testing, processing, and cleaning. With the phasing out of military standards,[13] the use of industrial specifications is increasing in both commercial and military applications. The most frequently referenced industrial flux specification is the

13. U.S. Department of Defense, *General Specification for Flux, Soldering, Liquid, Paste Flux, Solder Paste and Solder-Paste Flux (for Electronic/Electrical Use)*, MIL-F-14256F.

Joint Industry Standard, *Requirements for Soldering Fluxes* ANSI/J-STD 004,[14] from the Electronic Industries Association and IPC. Flux activity and corrosiveness classifications are based on a series of tests specified by *Test Methods Manual,* IPC-TM-650,[15] and *General Requirements for Soldering Electronic Interconnections,* IPC-S-815A.[16] ASTM and ISO publish comparable specifications on flux designations and flux property testing, such as *Standard Specification for Liquid and Paste Fluxes for Soldering Applications of Copper and Copper Alloy Tube,* ASTM's B 813.[17]

ANSI/J-STD, ASTM, and ISO specifications are also available for flux cored solder wire and solder pastes. Although most of these specifications and documents were developed for electronics applications, the test methods and designations are usually applicable to structural applications.

A functional method of classifying fluxes is based on the acid activity of the flux, i.e., its ability to remove metal tarnishes, oxides, sulfides, and so forth. Thus, fluxes can be classified according to the level of acidity, with inorganic fluxes the most active, organic fluxes moderately active, and rosin fluxes the least active.

The selection of the type of flux usually depends on the ease with which a material can be soldered. Inorganic fluxes are often used in industrial soldering applications such as plumbing and automotive radiators where a more aggressive flux is needed. The recommended fluxes for soldering a number of base metals are presented in Table 13.9.

Inorganic Fluxes (Fully Active)

The inorganic class of fluxes includes inorganic acids and salts. These fluxes are best for conditions that require rapid and highly active fluxing action. They can be applied as solutions, pastes, or dry salts. They function equally well with torch, furnace, resistance, or induction soldering methods since they do not readily char or burn at elevated soldering temperatures. These fluxes can be formulated to provide stability over a wide range of soldering temperatures.

The following are typical inorganic flux constituents:

1. Zinc chloride,
2. Ammonium chloride,
3. Tin chloride,
4. Hydrochloric acid,
5. Phosphoric acid, and
6. Other metal chlorides.

The chloride-based inorganic fluxes have a distinct disadvantage in that the residue remains chemically active after soldering. This residue, if not removed, may cause severe corrosion at the joint. Residues from the spraying of flux and flux vapors may also attack adjoining areas.

The bromide group of inorganic fluxes is widely used by the automotive radiator industry, with and without washing facilities. Certain compositions of these fluxes can be used without washing, and those residues will not cause corrosion of the parts that have been soldered.

Organic Acid Fluxes (Moderately Active)

Organic acid fluxes, while less active than the inorganic materials, are effective at soldering temperatures from 90°C to 320°C (200°F to 600°F). They are also referred to as *intermediate fluxes.* The organic acid fluxes consist of organic acids, bases, and similar derivatives such as hydrohalides. They are active at soldering temperatures, but the period of activity is short because of their susceptibility to thermal decomposition. The tendency of organic acid fluxes to volatilize, char, or burn when heated can limit their use with torch or flame heating. When properly used, residues from these fluxes are relatively inert and can be removed with water.

Organic acid fluxes are particularly useful in applications in which controlled quantities of flux can be applied and sufficient heat can be used to fully decompose or volatilize the corrosive constituents. Precaution is necessary to prevent excess flux from wicking into the insulation sleeving and causing corrosive reactions after soldering. Care must also be exercised when soldering in closed systems where corrosive fumes may condense on critical parts of the assembly.

The following are typical organic acid flux constituents:

1. Abietic acid,
2. Ethylene diamine,
3. Glutamic acid,
4. Hydrazine hydrobromide,
5. Oleic acid,
6. Stearic acid, and
7. A wide range of other acid-based or acid-forming organic chemicals.

14. American National Standards Institute, Joint Industry Standard, *Requirements for Soldering Fluxes*, ANSI/J-STD 004, West Conshohocken, Pa.: American Society for Testing and Materials.
15. IPC–Association Connecting Electronics Industries, 2000, *Test Methods Manual,* IPC-TM-650, Northbrook, Ill.: IPC-Association Connecting Electronics Industries.
16. See Reference 8.
17. American Society for Testing and Materials (ASTM) Subcommittee B05.04, *Standard Specification for Liquid and Paste Fluxes for Soldering Applications of Copper and Copper Alloy Tube,* ASTM B 813, West Conshohocken, Pa.: American Society for Testing and Materials.

Table 13.9
Fluxes Recommended for Soldering to Base Metals and Surface Finishes

Base Metal or Applied Finish	Rosin	Organic	Inorganic	Special Flux and/or Solder	Soldering Not Recommended*
Aluminum				X	
Aluminum-bronze				X	
Beryllium					X
Beryllium-copper		X	X		
Brass	X	X	X		
Cadmium	X	X	X		
Cast iron				X	
Chromium					X
Copper	X	X	X		
Copper-chromium			X		
Copper-nickel	X	X	X		
Copper-silicon			X		
Gold	X	X	X		
Inconel™				X	
Lead	X	X	X		
Magnesium					X
Manganese-bronze					X
Monel™		X	X		
Nickel		X	X		
Nickeliron		X	X		
Nichrome™				X	
Palladium	X	X	X		
Platinum	X	X	X		
Rhodium			X		
Silver	X	X	X		
Stainless steel			X		
Steel			X		
Tin	X	X	X		
Tin-bronze	X	X	X		
Tin-lead	X	X	X		
Tin-nickel		X	X		
Tinzinc	X	X	X		
Titanium					X
Zinc		X	X		
Zinc die castings					X

*With proper procedures, such as precoating, most metals can be soldered.

Rosin Fluxes

Rosin fluxes are normally used with base metals in electrical and electronic applications or with metals that are precoated with a solderable finish. The following three conventional types of rosin fluxes are recognized by the soldering industry and categorized by their level of reactivity are designated as:

1. (Non-activated) rosin (R),
2. Rosin mildly activated (RMA),
3. Rosin activated (RA).

Non-Activated Rosin. The least reactive rosin-based fluxes contain resins dissolved in a combination of aliphatic alcohols that are relatively stable with low-activity or non-activated chemistries. These fluxes, designated *R*, are primarily comprised of pure (water-white) rosin or colophony. They have physical and chemical properties that make them particularly suitable for use in the electronics industry. The active constituent, abietic acid, becomes mildly active at soldering temperatures between 175°C (347°F) and 315°C (599°F). The residue is hard, nonhygroscopic, electrically nonconductive, and noncorrosive.

Rosin Mildly Activated. Because of the low activity of rosin, mildly activated rosin fluxes have been developed to increase the fluxing action without significantly altering the noncorrosive nature of the residue. They are referred to as *RMA fluxes*, and are the preferred fluxes for military, telecommunication, and electronic products, and other products requiring high reliability.

Rosin Activated. A third and still more active type of rosin-base flux is called *activated rosin*. These fluxes, commonly referred to as *RA*, are widely used in commercial electronics and in high-reliability applications where the residue must be completely removed after soldering. The activating material can be an organic compound that reacts to release chlorides, other halides, or a low level of organic acid.

Low-Solids Fluxes

A useful development in flux chemistry has been the formulation of low-solids or "no-clean" fluxes. These fluxes were designed to minimize flux residues and subsequent solvent cleaning after soldering. They are just as effective as conventional fluxes but are more sensitive to the processing conditions because they contain fewer active constituents.

Special Fluxes

Reaction fluxes are a special group of fluxes that are useful when soldering aluminum. These fluxes can also be used with other metals. In practice, the decomposition of the flux cleans and displaces oxides and deposits a metallic film on the metal surface that allows for wetting, spreading, and capillary flow. Other synthetic fluxes were developed to solder difficult-to-wet metals such as stainless steel, super alloys, and other heat-resistant metals.

FLUX SELECTION

The selection of the appropriate flux for a specific soldering project can be a complex task considering the many fluxes that are available. The correct selection is very important to the success of the operation. Soldering fluxes are designed specifically for the metals or materials and the applications involved. For example, specific fluxes are available for electronics, plumbing, radiators, and a wide variety of industrial components and products.

Desirable general properties of fluxes include the ability to remove oxides, protect metal surfaces, and activate below the soldering temperature. Desirable post-soldering properties are flux residues that are electrically nonconductive and corrosion resistant. Each flux is designed for a specific heating process and has an optimum processing temperature range.

There is no universal test that can identify all the necessary properties of a flux for a specific application. Therefore, a number of tests have been developed that relate to flux characteristics and values in the fabrication of particular components. Users should perform a thorough review before selecting a flux. They should not rely entirely on commercially available data that may not be relevant to a particular application under consideration.

Table 13.10 presents a general metal solderability chart and flux selection guide for various soldering applications, including the above classifications of materials. Table 13.11 lists typical fluxing agents.

PROCEDURES

The basic stages of soldering are joint preparation, fluxing, preheating, soldering, and final cleaning to remove flux residues or other debris resulting from the soldering process. These procedures, with the exception of the soldering operation itself, are discussed in this section.

JOINT PREPARATION

All metal surfaces to be soldered should be chemically or mechanically cleaned before assembly to facili-

Table 13.10
Metal Solderability and Flux Selection Guide

Metals	Solderability	Rosin Fluxes: Non-Activated (R)	Rosin Fluxes: Mildly Activated (RMA)	Rosin Fluxes: Activated (RA)	Organic Acid Flux, Water Soluble	Inorganic Flux, Water Soluble	Special Flux and/or Solder
Platinum, gold, copper, silver, cadmium plate, tin (hot dipped), tin plate, solder plate	Easy to solder	Suitable	Suitable	Suitable	Suitable	Not recommended for electrical soldering	
Lead, nickel plate, brass, bronze, rhodium, beryllium copper	Less easy to solder	Not suitable	Not suitable	Not suitable	Suitable	Suitable	
Galvanized iron, tin-nickel, nickel-iron, low-carbon steel	Difficult to solder	Not suitable	Not suitable	Not suitable	Suitable	Suitable	
Chromium, nickel-chromium, nickel-copper, stainless steel	Very difficult to solder	Not suitable	Not suitable	Not suitable	Not suitable	Suitable	
Aluminum, aluminum-bronze	Most difficult to solder	Not suitable	Not suitable	Not suitable	Not suitable		Suitable
Beryllium, titanium	Not solderable (except with special solder and processes)						

tate wetting of the base metal by the solder. Meticulous precleaning of the joint surfaces is the first step. The use of flux should not be considered as a substitute for precleaning. An applied solderable surface finish may be necessary on base materials that are difficult to solder directly to one another.

PRECLEANING

An unclean surface prevents the solder from flowing and makes soldering difficult or impossible. Materials such as oil, grease, paint, pencil markings, drawing and cutting lubricants, general atmospheric dirt, rust, and oxide films should be removed before soldering. The importance of cleanliness to ensure sound soldered joints cannot be overemphasized. Because cleaning methods are often designed for a specific soldering operation, their suitability for a critical application should be investigated thoroughly.

Degreasing

Solvent or alkaline degreasing is recommended for cleaning oily or greasy surfaces. Of the solvent degreasing methods, vapor-condensation solvents leave the least residual film on the surface. In the absence of a vapor-degreasing apparatus, immersion in liquid solvents or in detergent solutions is a suitable cleaning option. Hot alkali detergents are widely used for degreasing. All cleaning solutions must be thoroughly removed before soldering. Residues from hard-water rinses may later interfere with soldering.

Ultrasonic Cleaning

Workpieces, fixtures, and solder preforms can be ultrasonically cleaned. Most ultrasonic cleaners use a solvent medium to couple the ultrasonic energy to the surfaces to be cleaned. The usual method is to load the workpieces into a separate container, which is then placed in the ultrasonic bath. The ultrasonic energy removes contaminants that adhere to the surfaces. Ultrasonic cleaning does not remove oxides or other nonmetallic films that are chemically bonded at the surface.

Acid Cleaning

The purpose of acid cleaning, or pickling, is to remove rust, scale, and oxides or sulfides from the metal to provide a clean surface for soldering. Most inorganic acids, including hydrochloric, sulfuric, phosphoric, nitric, and hydrofluoric acids (either full strength or diluted with deionized water), fulfill this function. Hydrochloric and sulfuric acids are the most

Table 13.11
Typical Fluxing Agents

Type	Composition	Carrier	Uses	Temperature Stability	Ability to Remove Tarnish	Corrosiveness	Recommended Cleaning After Soldering
			Inorganic				
Acids	Hydrochloric, hydrofluoric orthophosphoric	Water, petrolatum paste	Structural	Good	Very good	High	Hot water rinse and neutralize; organic solvents
Salts	Zinc chloride, ammonium chloride, tin chloride	Water, petrolatum paste polyethylene glycol	Structural	Excellent	Very good	High	Hot water rinse and neutralize; polyethylene glycol 2% HCl solution; hot water rinse and neutralize; organic solvents.
			Organic				
Acids	Lactic, oleic, stearic glutamic, phthalic	Water, organic solvents, petrolatum paste, polyethylene glycol	Structural, electrical	Fairly good	Fairly good	Moderate	Hot water rinse and neutralize; organic solvents
Halogens	Aniline hydrochloride, hydrochloride, bromide, derivative of palmitic acid, hydrazine hydrochloride (or hydrobromide)	Same as organic acids	Structural, electrical	Fairly good	Fairly good	Moderate	Same as organic acids
Amines and amides	Urea, ethylene diamine	Water, organic solvents, petrolatum paste, polyethylene glycol	Structural, electrical	Fair	Fair	Noncorrosive	Hot water rinse and neutralize; organic solvents
Activated rosin	Water-white rosin	Isopropyl alcohol, organic solvents, polyethylene glycol	Electrical	Poor	Fair	Noncorrosive (normally)	Water-based detergents; isopropyl alcohol; organic solvents
Water-white rosin	Rosin only	Same as activated	Electrical	Poor	Poor	None	Same as activated water-white rosin but does not normally require postcleaning

widely used. The workpieces should be washed thoroughly in hot water after pickling and should be dried as quickly as possible.

Mechanical Cleaning

Mechanical cleaning methods include the following:

1. Mechanical sanding or grinding,
2. Hand filing or sanding,
3. Cleaning with steel wool,
4. Wire brushing or scraping, and
5. Grit or shot blasting.

It is best to clean soft metals such as copper by gentle wire brushing or sanding. Steel wool is generally used on plumbing materials. Care should be taken to avoid embedding abrasive media in the base metal. It is best to avoid mechanical cleaning when preparing electronic components for soldering.

Aluminum can be soldered better after oxides are removed by mechanical means; wire brushing or scraping is best. Steel or stainless steel can be brushed or blasted. Shot blasting is preferable to sanding because it avoids embedding silica particles. Stainless steel shot should be used for stainless steel surfaces.

For best results, cleaning should extend beyond the joint area. It is important to follow these abrasive treatments with a surface etch to remove any embedded particles that could interfere with solder wetting and flow.

APPLIED SURFACE FINISHES

Applied surface finishes may be necessary for base materials that are difficult to solder. The metallization of base metal surfaces with a more solderable metal or alloy prior to the soldering operation is sometimes desirable to facilitate soldering. Coatings of tin, copper, gold, silver, cadmium, iron, nickel, and alloys of tin-lead, tin-zinc, tin-copper, and tin-nickel are applied for this purpose.

The advantages of precoating are twofold. First, soldering becomes more rapid and uniform. Second, the use of strong acidic fluxes can be avoided during soldering. The precoating of metals that have tenacious oxide films—aluminum, aluminum bronzes, highly alloyed steels, and cast iron—is almost mandatory. Precoating of steel, brass, and copper can sometimes be useful to avoid the need for using highly active but corrosive fluxes.

There are several methods of applying precoating to the metal surfaces. Solder or tin may be applied by means of immersion in molten metal, electrodeposition (plating), or chemical displacement.

Molten-Metal Immersion

Immersion in molten metal, or hot dipping, can be accomplished by fluxing and dipping the parts in molten tin or solder. Small workpieces are often placed in wire baskets, cleaned, fluxed, dipped in the molten metal, and centrifuged to remove excess metal. Coating by hot dipping is applicable to carbon steel, alloy steel, cast iron, copper, and certain copper alloys. Prolonged immersion in molten tin or solder should be avoided to prevent excessive base metal dissolution or the formation of intermetallic compounds at the interface between the coating and the base metal, potentially degrading solderability and joint strength.

Electrodeposition

Precoating by means of electrodeposition can be accomplished in stationary tanks, in plating units with conveyors, or in barrels. These methods are applicable to all steels, copper alloys, and nickel alloys. The coating metals are not limited to tin and solder. Copper, cadmium, silver, gold, palladium, nickel, and iron are also commonly used, as are alloy systems such as tin-copper, tin-zinc, and tin-nickel.

Electroplating of Bimetallic Layers. Duplex coatings can be applied by using certain combinations of metals to electroplate one metal over another. This technique is often used as an aid to soldering. A coating of 0.005 mm (0.0002 in.) copper and 0.008 mm (0.0003 in.) tin is particularly useful for brass. The solderability of aluminum is assisted by a coating of 0.013 mm (0.0005 in.) nickel, followed by 0.008 mm (0.0003 in.) tin, or by a combination of zincate (zinc), 0.005 mm (0.0002 in.) copper, and tin. Iron plating followed by tin plating is extremely useful over a cast-iron surface. Iron- and nickel-base alloys typically are electroplated with 0.005 mm (0.0002 in.) nickel, followed by 0.0013 mm (0.00005 in.) gold.

Chemical Displacement

Chemical displacement coatings or immersion coatings of tin, silver, gold, palladium, or nickel may be applied to some of the common base metals. These coatings are usually very thin and generally have a poor shelf life. The shelf life of a coating is defined as the ability of the coating to withstand storage conditions without impairment of solderability. The manufacturer of the coating should be consulted regarding shelf life limitations. Hot-tinned and flow-brightened electrotinned coatings have an excellent shelf life. Inadequate thickness of electrotinned or immersion-tinned coatings can yield a limited shelf life. A tin or solder coating thickness of 0.003 mm to 0.008 mm (0.0001 in. to 0.0003 in.) is recommended to assure maximum solderability after prolonged storage. Similarly, a gold coating of 0.0013 mm (0.00005 in.) is recommended.

FLUXING

Flux can be applied by dipping, spraying, or brushing the surfaces to be soldered. For consistent soldering results, the amount and chemistry of the applied flux should be maintained. Most fluxes have a specified or recommended elapsed time between when they are applied and when soldering occurs.

PREHEATING

Preheating is normally required before actual soldering. A temperature of 80°C to 100°C (176°F to 212°F) is necessary to volatize the solvent or water carrier in the flux; this is followed by a temperature of 130°C to 145°C (266°F to 325°F) to activate the remaining reactive constituents. Otherwise, extensive porosity or poor wetting can occur.

FLUX RESIDUE TREATMENT

Flux residues should be removed after soldering unless the flux is specifically designed to be consumed during the process or will not affect product performance. Flux residues that may corrode the base metal or otherwise prove harmful to the effectiveness of the joint must be removed. The removal of flux residues is especially important where joints are subjected to humid, elevated-temperature environments.

Zinc-chloride-based fluxes leave a fused residue that will absorb water from the atmosphere. Removal of the residue is accomplished best by thorough washing in hot water containing 2% hydrochloric acid, followed by a hot-water rinse. The acidified water removes the white crust of zinc oxychloride, which is insoluble in water. Complete removal can also be accomplished by additional washing in hot water that contains some washing soda (sodium carbonate), followed by a clear water rinse. Occasionally some mechanical scrubbing may also be required.

Inorganic flux residues containing inorganic salts and acids should be removed completely. Residues from the organic fluxes that are composed of very mild organic acids—stearic acid, oleic acid, ordinary tallow, or the highly corrosive combinations of urea plus various organic hydrochlorides—should also be removed. To determine whether all of the salts have been removed, the joint should be washed with warm water containing a few drops of silver nitrate. If any chloride salts are present, the wash will turn milky with the precipitation of silver chloride.

Residues from the organic acid fluxes are usually soluble in hot water. Double rinsing in warm water is always advisable.

Rosin flux residues can normally be left on the joint unless appearance is a prime factor or if the joint area is to be painted or subsequently coated. Activated rosin fluxes can be treated in the same manner, but they should be removed for critical electronic applications. If rosin residues must be removed, alcohol or chlorinated hydrocarbons can be used. Certain rosin activators are insoluble in water but soluble in organic solvents. These flux residues require removal by organic solvents, followed by a water rinse.

The residues from activated-type fluxes used on aluminum are usually removed with a rinse in warm water. If this does not remove all traces of residue, the joint can be scrubbed with a brush and then immersed in 2% sulfuric acid, followed by immersion in 1% nitric acid. A final warm-water rinse is then required.

Soldering pastes for plumbing systems are usually emulsions of petroleum jelly and a water solution of zinc ammonium chloride. Because of the corrosive nature of the acid salts contained in the flux, residues must be removed to prevent corrosion of the soldered joints and the copper pipes. Oily or greasy flux-paste residues are generally removed with an organic solvent. The handling of these chemicals should follow recommended safety procedures.

Final Cleaning

Most applications require the removal of flux residues. This is especially critical when the soldered assemblies are intended for service in corrosive environments. The flux residues should be cleaned with recommended or approved solvent cleaners. Cleaning procedures recommended by the flux manufacturer should be followed.

PROCESS VARIABLES

Most soldering techniques have common process variables that require different levels of control to yield the desired soldering results. The type of materials, surface conditions, time and temperature profiles, process atmosphere, heating method, and joint design all affect how the solder joint is made and how it will perform in service. Each application requires a thorough understanding of how these parameters may interact. This requires knowing what the process space is and performing the necessary process characterization to assure meeting product specifications.

JOINT DESIGN

Joints should be designed to fulfill the requirements of the finished assembly, to permit application of the flux and solder by the soldering process to be used, and to maintain the proper clearance during heating. Special fixtures may be necessary, or the components can be staked, crimped, clinched, wrapped, or otherwise held together.

The selection of a joint design for a specific application depends largely on the service requirements of the assembly. It may also depend on factors such as the heating method to be used, the fabrication techniques prior to soldering, the number of items to be soldered, and the method of applying the solder.

When service requirements of a joint are severe, it is generally necessary to design the joint so that it does not limit the function of the assembly. Solders have low strength compared to the metals usually soldered; therefore, the soldered joint should be designed to avoid dependence on the strength of the solder alloy. For example, an interlocking joint can provide the necessary service strength, with the solder serving as only the seal or stiffener in the assembly.

In some industrial applications, the solder joint itself must carry the load. A typical example is a pipe joint in a plumbing system, where a lap joint is used with no additional mechanical support. In this case, the properties of the solder alloy and the fabricated joint are important to the response of the soldered assembly while in service.

Two joint designs, the lap joint and the lock-seam joint, are basic designs used for soldering. Other joint designs frequently used are illustrated in Figure 13.15. It should be noted that butt joints are not often used because of their small load-bearing surface areas.

The lap or lock-seam joint should be employed whenever possible, since they offer the best possibilities of obtaining maximum joint area and strength.

An important factor in joint design is the manner in which the solder is to be applied to the joint. The designer must consider the number of joints per assembly and the number of assemblies to be manufactured. For limited production using a manual soldering process, the solder may be face-fed into the joint with little or no problem. However, for large numbers of assemblies containing multiple joints, an automated process such as wave soldering may be advantageous. In this case, the design must provide accessible joints that are suitable for automated fluxing, soldering, and cleaning.

Clearance between the workpieces should allow the solder to be drawn into the space between them by capillary action, but the joint clearance should not be so wide that the solder cannot fill it. Joint clearances up to 0.125 mm (0.005 in.) are preferred for optimum strength, but variations are allowable in specific instances. For example, when soldering precoated metals, a clearance as small as 0.025 mm (0.001 in.) is possible.

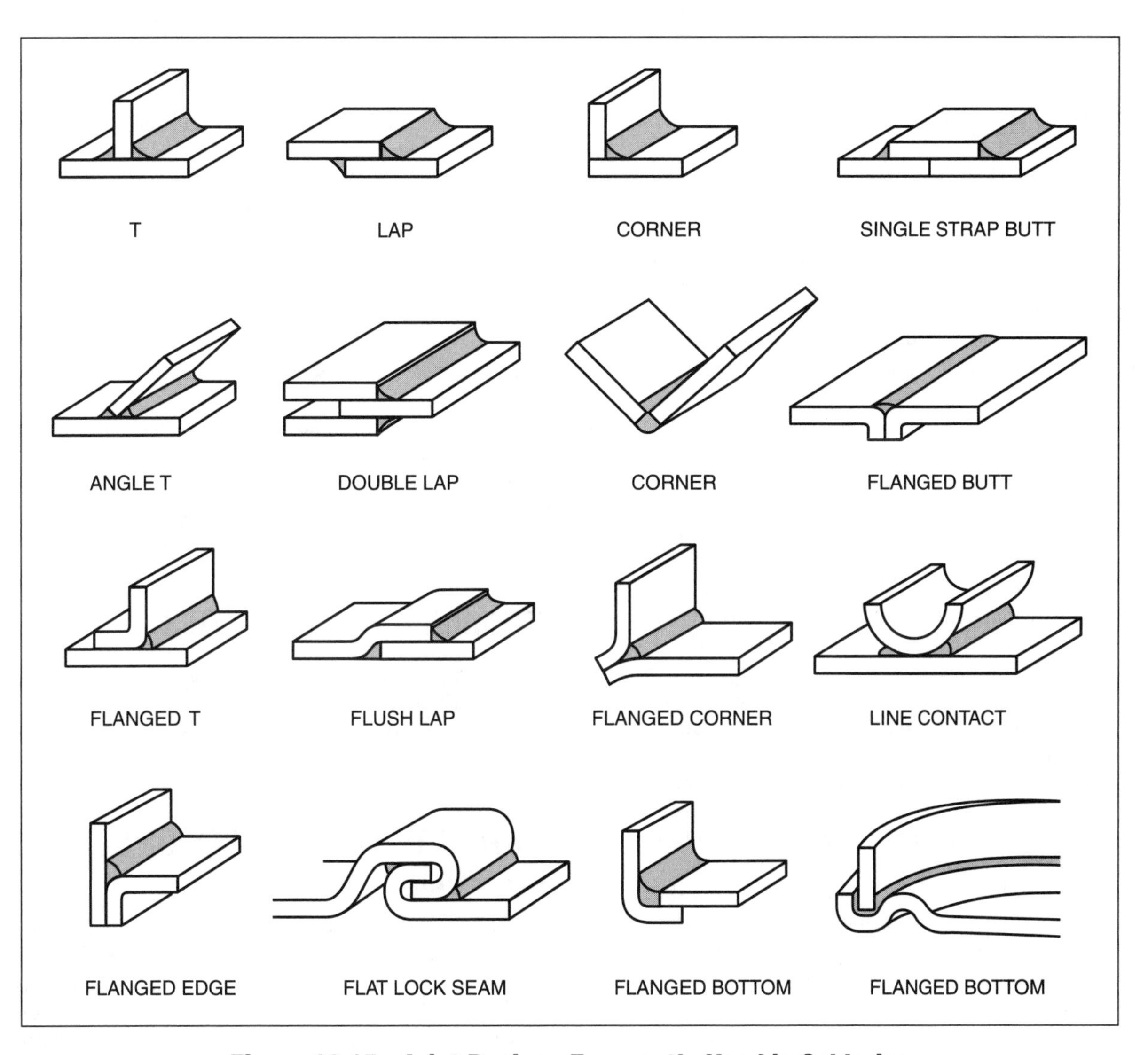

Figure 13.15—Joint Designs Frequently Used in Soldering

Self-Fixtured Joints

Twenty-one designs for self-fixturing solder joints are illustrated in Figure 13.16. Several means of improving the strength of soldered joints in printed circuits are shown in Figure 13.17. They include increasing the area and length of the soldered seals.

INSPECTION AND TESTING

The inspection and testing of soldered joints can be performed using standard inspection techniques. These techniques are common to most joining processes and take advantage of both nondestructive and destructive testing methods. Examination methods specific to soldered joints are discussed in this chapter and in further detail in Chapter 7 of the *Soldering Handbook*.[18]

NONDESTRUCTIVE EXAMINATION

Visual inspection is normally adequate for soldered joints. The inspector ensures that soldered joints are smooth and free of obvious voids, holes, or porosity. The profile between the soldered joint and the material being joined should show a smooth transition with a relatively low angle of contact between the solder and the base metal. The joint should be examined for any areas that have not been successfully wetted. Poor wetting or dewetting can be observed where the metal retains its original color. Dewetting occurs where solder has originally flowed across the joining surfaces and then pulled back into globules, leaving a discolored, dirty-looking surface. These discontinuities are usually related to poor surface precleaning or the use of an inappropriate flux.

Solder joints can readily be overheated or underheated. Overheated joints can be detected by the presence of burned or charred fluxes and oxides on the solder joint. Underheated joints generally show poor flow characteristics with uneven solder distribution over the base metal surface. These features usually indicate poor metallurgical bonding.

Soldered printed circuit boards may produce a set of discontinuities unique to the product. The alloy composition or processing conditions can cause bridging of the solder. Bridging may occur between electrical connections that are closely spaced. The connections should be insulated from each other. Another discontinuity unique to circuit boards is called *icicling*, which produces spikes of solder beneath the board. (Refer to Figure 13.7). This may cause electrical interference in the finished product. Icicling is promoted by impurities such as cadmium or zinc and by the lack of flux activity. The design of the printed circuit board materials can also cause porosity in the joint. All of these defects can be detected by visual inspection.

Other Nondestructive Examination Methods

In addition to visual inspection, other methods of nondestructive examination are used to inspect some soldered products. Pressure-vacuum fluid-seal testing and leakage-rate testing can be used on closed systems. Examples include plumbing systems, which are inspected by water pressure tests; vehicle radiators, which are inspected with air pressure tests; food cans, which are inspected using vacuum tests; and gas-filled systems, which are inspected with halogen-leak testing.

Radiography can be used for pipe joints or other applications where large surface areas of lead solder joints are present.

Laser and infrared inspection techniques are used in electronic fabrications. Heat generated by the laser provides an indication of solder joint quality. Surface dimensions may also be checked.

Acoustic emission testing is useful, but this process may affect the joint quality if the acoustic couplant is not completely removed after testing.

MECHANICAL TESTING

Conventional destructive testing techniques including mechanical tests, corrosion evaluation, and metallurgical analysis are also used to assess solder joint properties. Tests can be conducted with standard test specimens or actual production parts.

COMMON DISCONTINUITIES

Common discontinuities include voids, porosity, incomplete fill, run-out, discolored surfaces, bridging, and icicles. Any evidence of cracking is considered a serious defect.

TESTING THE PROPERTIES OF SOLDERS AND SOLDER JOINTS

Details of the physical and mechanical properties of solders are usually provided by the supplier. This information is used in specifications to ensure the consistent quality of filler metals. Typical properties reported by filler-metal manufacturers might not apply to commercial products and applications. The reported properties

18. See Reference 3, pp. 447–489.

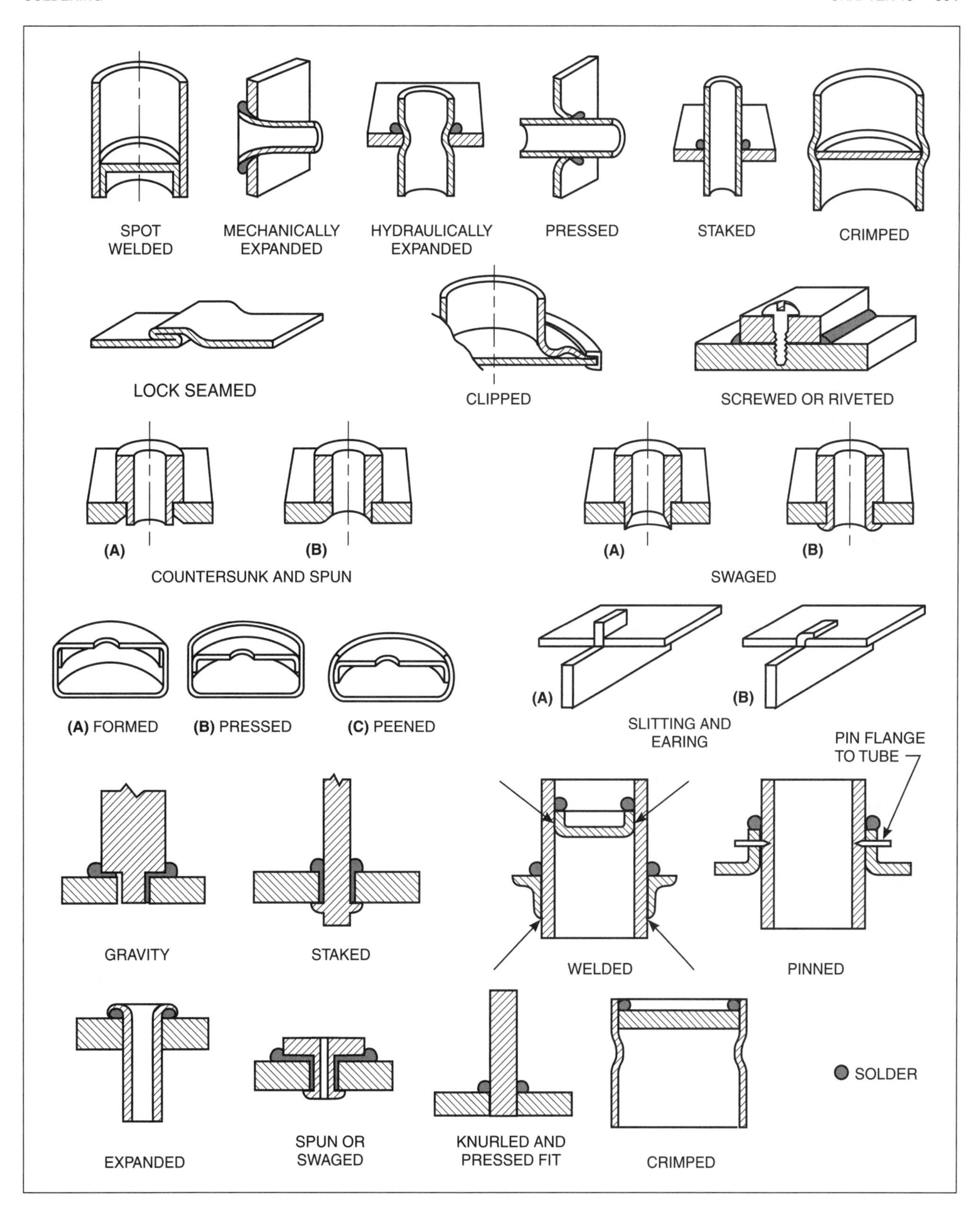

Figure 13.16—Self-Fixturing Solder Joint Designs

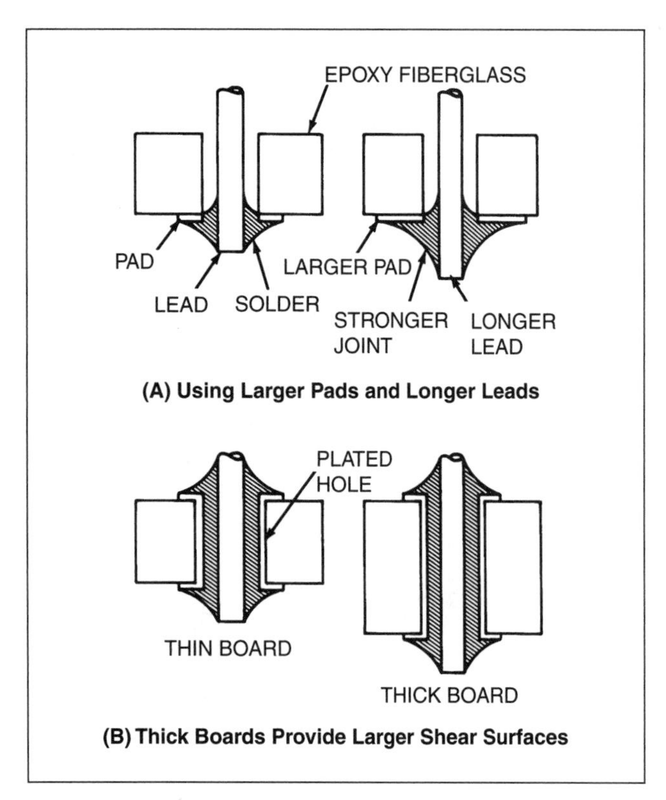

Figure 13.17—Methods of Improving Joint Strength in Feedthrough Components

of these alloys serve only to provide a basis of selection from the various available solder filler metals. Therefore, users should conduct tests on their own manufactured products to determine the suitability of the filler metal and the solder process.

For soldered joints in shear, lap joints are mainly used. For joints in peel, lock-seam or base-material-supported joints are used. The test method must be appropriate to the product for mechanical property evaluation. Short-time tensile tests are good for manufacturing quality control and for comparisons. Because most solder joints are subject to some stress in service, the results of creep, stress-rupture, and fatigue tests are important indicators of product performance.

Ultimately, the total soldered product must be tested to closely simulate actual service, otherwise serious deficiencies can occur as a result of premature joint failure. The mechanical properties of soldered products are highly dependent on the product design, alloy selection, manufacturing process, and service conditions. Each individual product should be studied for all of these factors so that an optimum balance can be obtained between costs and utility. Additional sources of information are listed in the Supplementary Reading List at the end of the chapter.

ECONOMICS

Many of the concepts involved in estimating and controlling costs of manufactured products can be applied to soldering costs. The topics discussed in this section apply more specifically to soldering.[19]

LABOR AND MATERIALS COSTS

Labor costs are essentially based on the number of worker hours required to make the soldered joint. The most economical joint is designed for maximum solderability. The joint design should reflect not only fitness for purpose, but also should facilitate the fastest rate of soldering with the fewest defects. The configurations of successful solder joints, shown in Figure 13.15, can save time and reduce design costs because these designs can be adapted for joints required by a particular product.

Self-fixturing designs save assembly time and often preclude the need to purchase fixturing equipment. Figure 13.16 shows examples of self-fixturing designs that can be adapted to a particular product.

Labor costs may or may not be reduced by the selection of an automated process; this must be evaluated in terms of total cost of manufacturing, which would include the purchase, installation, and maintenance of the automated system.

Material costs include solder alloys and the costs associated with delivering the solder to the joint. These include wire, preforms, paste and other consumables such as fluxing agents, soldering-iron tips, and materials required for surface preparation and final cleaning of the workpieces.

The amount of solder and flux used should be carefully controlled. Too much solder is expensive because it is wasted material and may also affect joint performance as a result of metallurgical reactions. If an inadequate amount of solder is used, it may lessen joint strength and create voids that are subject to premature failures and corrosion, thus adding to the number of rejected joints that must be reworked or discarded. Using too much flux results in wasted material, causes defects, and adds time and material costs to the cleanup process. Too little flux results in incomplete joint information.

Cost control of a soldering project begins with the development of the soldering process best suited to the product. This involves the following three steps:

19. See Reference 3, pp. 107–109, 319–321.

1. Performing an analysis of materials to be soldered and selecting the appropriate solder alloy by conducting laboratory tests on specimens with standardized joint geometries.
2. Fabricating and testing a prototype with joint configurations similar to the soldered product to determine whether the joints can be soldered in the actual manufacturing process.
3. Fabricating the actual product and evaluating it in terms of defects related to the inherent properties of the materials and the parameters of the soldering process.

The objective of these evaluations is to document the parameters and assess the incidence of defects. Performing these steps in the above order has cost advantages. If the first two steps are omitted to save time and costs, the result is often even more expensive because there may be production delays, increased scrap, and the need for failure analysis and reworking.

The first step will eliminate combinations of base metals, solders, fluxes and surface preparations that do not perform well.

The second step will verify the parameters of the soldering and manufacturing processes, including assembly, method of introducing solder and flux, and the means of heating. The results of solderability tests will assure that the solder wets and spreads on the selected base material surfaces. These tests can determine whether the solder is compatible with the fluxes being considered for the application, whether alternative substrate materials, finishes, and solder temperatures can be considered, and whether precleaning methods are effective. Mechanical tests appropriate to the product's intended service can be conducted to evaluate various solder joints, and corrosion tests can be performed.

The third step will examine the finished product to determine the incidence and types of defects that may result from the soldering process parameters and identify any damage to the joint that may have occurred during postsoldering fabrication. This evaluation will determine if the product meets established specifications.

HAND SOLDERING

While hand soldering may be viewed as inefficient when compared to modern manufacturing processes, hand soldering is flexible and versatile and is sometimes the most economical option. Hand soldering should be considered when the product quantity is too small to justify the purchase of automated equipment. Several other factors may also favor hand soldering. Quality can be more consistent because operators inspect as they work, identifying and correcting problems as they occur. In production runs, a large number of defective joints might be soldered before the problem is discovered, resulting in either costly repair or excessive scrap. When dissimilar metals or dissimilar thicknesses of metal are involved, the manual operation allows good temperature control and heating of both base pieces. Hand soldering may be more cost effective for unusual workpiece geometries that would require expensive fixturing for production lines.

SAFE PRACTICES

The effects of manufacturing on the safety and health of the worker and those nearby and the impact of the operation on the environment are of particular concern to both management and employees. Soldering operators are exposed to many common hazardous materials and processes in the course of performing most soldering operations. It is important to develop a preliminary hazard assessment (PHA) plan prior to conducting potentially hazardous soldering processes, as this can help avoid major safety infractions, serious injuries, damaged equipment and lengthy operational downtime.

Numerous resources are available that address general hazards such as fumes and gases, workplace ventilation, corrosives and poisonous chemicals, fire hazards and heat sources (such as hot metal, tools, flames, metal and chemical splatter), electrical shock, and falls. General safe practices information is readily available in the American Welding Society publication, *Safety in Welding, Cutting, and Allied Processes*, ANSI Z49.1:1999;[20] Chapter 17 of the *Welding Handbook,* Vol. 1, 9th ed. of *Welding Science and Technology;*[21] and in Chapter 8 of the *Soldering Handbook,* 3rd ed.[22] *Appendix B* of this volume lists many applicable safety codes and other standards. *Safety and Health Fact Sheets,* 3rd ed., published by the American Welding Society, is available electronically.[23]

Management must assure that soldering operations are carried out under safe conditions. Employees must do the following:

1. Read all labels on base metals or nonmetallic materials, solder filler metals, fluxes, and cleaning agents before use to ensure safe handling;

20. American Welding Society (AWS) Committee on Safety and Health, 1999, *Safety in Welding, Cutting, and Allied Processes*, ANSI Z49.1, Miami: American Welding Society.
21. American Welding Society (AWS) Welding Handbook Committee, Jenney, C. L. and A. O'Brien, 2001, *Welding Science and Technology*, Vol. 2 of *Welding Handbook*, 9th ed., Miami: American Welding Society, Chapter 17.
22. Vianco, P. T., 1999, *Soldering Handbook*, 3rd ed., Miami: American Welding Society, Chapter 8.
23. http://www.aws.org/.

2. Recognize any potentially toxic metals or chemicals;
3. Use the materials only for the purposes intended; and
4. Wear adequate protective clothing and eye protection.

Material safety data sheets supplied by the manufacturers of the materials used in soldering are excellent sources of information and must always be consulted. In addition to safety recommendations issued by the manufacturers of soldering products and publishers of industry standards, federal, state, and local governments also issue standards for the protection of workers and the environment.

All hand-soldering operations should be carried out in a ventilated area with working surfaces kept free of solder droplets, particles, and residual fluxes. Many of the base metals or materials, solders, fluxes and cleaning agents produce fumes during soldering operations. Overheated solder pots can give off toxic metal vapors and fumes. Ventilation systems should be installed to eliminate these fumes. The American Conference of Governmental Industrial Hygienists (ACGIH) publishes recommended practices for the workroom environment in *Industrial Ventilation: A Manual of Recommended Practice*;[24] and *Threshold Limit Values for Chemical Substances and Physical Agents and Biological Exposure Indices*.[25]

Industrial soldering often requires electrical equipment that operates at relatively high levels of power. The *National Fire Protection Association (NFPA) document, Electrical Standard for Industrial Machinery,* NFPA 79, should be consulted.[26] This organization also provides safety procedures for furnaces and oven processes in *Standard for Ovens and Furnaces, NFPA 86, and Industrial Furnaces using Vacuum as an Atmosphere, NFPA 86D.*[27]

All soldering irons and equipment should be properly grounded. When electrical heaters are used for dip soldering operations, current-leakage safety devices should be used for worker protection.

Employees should be aware of all factors involved in soldering that could have an influence on their health and safety. Standard procedures for identifying toxic or hazardous materials and assuring employee awareness are published by the Occupational Health and Safety Administration (OSHA) in *OSHA Fact Sheet 93-26, Hazard Communication Standard.*[28]

CONCLUSION

Soldering is a relatively mature technology with many comprehensive sources of information that address most of the critical material and processing issues. Even with this vast source of available technical knowledge, there is a continual need to expand the development of soldering science and engineering. Whether the application involves a simple throwaway design or a high-reliability product, the fundamental metallurgical understanding of what comprises the fabrication and performance of a good solder joint is the same. Attention to the details of soldering technology will assure consistently good results that meet product requirements.

BIBLIOGRAPHY[29]

American Conference of Governmental Industrial Hygienists (ACGIH). 2003. *TLVs® and BEIs® Threshold limit values for chemical substances and physical agents and biological exposure indices.* Cincinnati, Ohio: American Conference of Governmental Industrial Hygienists.

American Conference of Governmental Industrial Hygienists (ACGIH). 1998. *Industrial ventilation: a manual of recommended practice,* 23rd ed., Publication 2092, Cincinnati, Ohio: American Conference of Governmental Industrial Hygienists.

24. American Conference of Governmental Industrial Hygienists (ACGIH), 1998, *Industrial Ventilation: A Manual of Recommended Practice,* 23rd ed. Publication 2092, Cincinnati, Ohio: American Conference of Governmental Industrial Hygienists.

25. American Conference of Governmental Industrial Hygienists (ACGIH), *TLVs® and BEIs® Threshold Limit Values for Chemical Substances and Physical Agents and Biological Exposure Indices,* Cincinnati, Ohio: American Conference of Governmental Industrial Hygienists. (Published annually; the latest edition should be consulted.)

26. *National Fire Protection Association (NFPA), Electrical Standard for Industrial Machinery, NFPA 79, Quincy, Massachusetts: National Fire Protection Association.*

27. *National Fire Protection Association (NFPA), Standard for Ovens and Furnaces, NFPA 86, Quincy, Massachusetts: National Fire Protection Association.* and *National Fire Protection Association (NFPA), Industrial Furnaces using Vacuum as an Atmosphere, NFPA 86D, Quincy, Massachusetts: National Fire Protection Association.*

28. Occupational Health and Safety Administration (OSHA), 1993, *OSHA Fact Sheet 93-26, Hazard Communication Standard, 29 CFR 1910.1200* (available on line at http://www.osha-slc.gov/OshDoc/Fact_data/FSNO93-26.html).

29. The dates of publication given for the codes and other standards listed here were current at the time this chapter was prepared. The reader is advised to consult the latest edition.

American National Standards Institute (ANSI) Accredited Standards Committee Z49. 1999. *Safety in welding, cutting, and allied processes,* ANSI Z49.1, Miami: American Welding Society.

American Society for Testing and Materials (ASTM). 2002. *Annual book of ASTM standards.* West Conshohocken, Pa.: American Society for Testing and Materials.

American Society for Testing and Materials (ASTM). 2002. *Standard practice for using significant digits in test data to determine conformance with specifications,* ASTM E 29-02. West Conshohocken, Pennsylvania: American Society for Testing and Materials.

American Society for Testing and Materials (ASTM) Subcommittee B05.04. 2000. *Standard specification for liquid and paste fluxes for soldering applications of copper and copper alloy tube,* ASTM B 813, West Conshohocken, Pa.: American Society for Testing and Materials.

American Society for Testing and Materials (ASTM). 2000. *Standard specification for solder metal, ASTM B 32.* West Conshohocken, Pa.: American Society for Testing and Materials.

American Welding Society (AWS) Welding Handbook Committee. Jenney C. L. and A. O'Brien, eds. 2001. *Welding Science and Technology.* Volume 1 of *Welding handbook.* 9th ed. Miami: American Welding Society.

American Welding Society (AWS) Safety and Health Committee. 1999. *Safety in welding, cutting, and allied processes, ANSI* Z49.1:1999. Miami: American Welding Society.

American Welding Society (AWS) Committee on Definitions and Symbols. 2001. *Standard welding terms and definitions,* AWS A3.0:2001. Miami: American Welding Society.

American Welding Society (AWS) Committee on Safety and Health. 1998. *Safety and health fact sheets.* Miami: American Welding Society. Fact sheet link: http://www.aws.org/technical/FACT-PDF.EXE/SNH.HTM.

General Services Administration (GSA). 1986. *Electronic solder (96 to 485°C),* QQ-S-571E. Washington, D.C.: General Services Administration.

Hosking, F. M., and M. A. Quintana. 2001. Interconnect soldering of silicon photovoltaic module cells. *Welding Journal* 80(10): 57–58.

International Organization for Standardization (ISO). 1990. *Soft solder alloys—chemical composition and forms,* SIS-ISO 9453. Geneva, Switzerland: International Organization for Standardization.

IPC–Association Connecting Electronics Industries. 2002. *General Requirements for Electrical and Electronic Assemblies,* J-STD-001C. Lincolnwood, Ill.: IPC–Association Connecting Electronics Industries.

IPC–Association Connecting Electronics Industries. 2001. *Requirements for electronic grade solder alloys and fluxed and non-fluxed solid solders for electronic soldering applications,* IPC/EIA J-STD-006A. Northbrook, Ill.: IPC–Association Connecting Electronics Industries.

IPC–Association Connecting Electronics Industries. 2000. *Test methods manual,* IPC-TM-650. Northbrook, Ill.: IPC–Association Connecting Electronics Industries.

National Fire Protection Association (NFPA). 2002. Electrical standard for industrial machinery, NFPA 79. Quincy, Mass.: National Fire Protection Association.

National Fire Protection Association (NFPA). Standard for ovens and furnaces. 1999. NFPA 86. Quincy, Mass.: National Fire Protection Association.

National Fire Protection Association (NFPA), Industrial furnaces using vacuum as an atmosphere, NFPA 86D. Quincy, Mass.: National Fire Protection Association.

Occupational Health and Safety Administration (OSHA), 1993, *OSHA Fact Sheet 93-26, Hazard communication standard, 29 CFR 1910.1200.* Washington D.C.: Superintendent of Documents. Government Printing Office.

Rudnev, V., M. Black, and W. Albert. 2003. *Intricacies of induction joining.* International brazing and soldering conference proceedings. San Diego, California. ASM International/American Welding Society. Miami: American Welding Society.

Smith, R. W. 2001. Active solder joining of metals, ceramics and composites. *Welding Journal* 80(10): 30–35.

Vianco, P. T. 2002. Solder technology for ultrahigh temperatures. *Welding Journal* 81(10): 51–54.

Vianco, P. T., and J. A. Regent. 2001. A furnace process for structural soldering. *Welding Journal* 80(4): 217–218.

Vianco, P.T. 1999. *Soldering handbook,* 3rd edition. Miami: American Welding Society.

SUPPLEMENTARY READING LIST

Aluminum Association. 1996. *Aluminum soldering handbook.* Publication ASH-22-516116. Washington D.C.: Aluminum Association.

ASM International. 1993. *Welding, brazing, and soldering.* Vol. 6 of *ASM Handbook.* Materials Park, Ohio: ASM International.

Copper Development Association. 1998. *Soldering and brazing copper tube and fittings*. CDA Publication A1143-00/98. New York: Copper Development Association.

Frear, D. R., S. N. Burchett, H. S. Morgan, and J. H. Lau. 1994. *The mechanics of solder alloy interconnects*. New York: Van Nostrand Reinhold/International Thomson Publishing.

Frear, D. R., W. B. Jones, and K. R. Kinsman. 1991. Solder mechanics. Warrendale, Pa.: TMS Publications.

Humpston, G. and D. M. Jacobson. 1993. *Principles of soldering and brazing*. Materials Park, Ohio: ASM International.

Hwang, J. S. 1996. *Modern solder technology for competitive electronics manufacturing*. New York: McGraw-Hill.

Kireta, A. 2002. Tips for soldering and brazing copper tubing. *Welding Journal* 81(8): 36–42.

Klein Wassink, R. J. 1989. *Soldering in electronics*. 2nd ed. Ayr, Scotland: Electrochemical Publications.

Lau, J. H. 1991. *Solder joint reliability*. New York: Van Nostrand Reinhold/International Thompson Press.

Manko, H. H. 1979. *Solders and soldering*, 2nd ed. New York: McGraw-Hill.

National Center for Manufacturing Sciences. 1998. *Lead-free solder project*. NCMS Final Report, Ann Arbor, Mich.: National Center for Manufacturing Science.

Pecht, M. G. 1993. *Soldering processes and equipment*. New York: John Wiley and Sons.

Rahn, A. 1993. *The basics of soldering*. New York: John Wiley and Sons.

Smith, R. W., P. Vianco, C. Hernandez, E. Lugscheider, I. Rass, and F. Hiller. 2000. *A new active solder for joining electronic components*. Albuquerque, NM: International Brazing and Soldering Conference Proceedings. ASM International/American Welding Society.

Thwaites, C. J. 1982. *Capillary joining—Brazing and soft-soldering*. New York: Research Studies Press, John Wiley and Sons.

Vianco, P. T. and J. A. Rejent. 2001. A furnace process for structural soldering. *Welding Journal* 80(4): 217s–219s.

Vianco, P. T. 1999. Corrosion issues in solder joint design and service. *Welding Journal* 78(10): 39–46.

Vianco, P. T. 1998. Determining the mechanical strength of soldered joints. *Welding Journal* 77(10): 49–52.

Vianco, P. T. 1993. The present triumphs and future problems with wave soldering. *Welding Journal* 72(10): 49–52.

Winstanley, A. 1997. *The basic soldering and desoldering guide*. Ferndown, Dorset, UK: Wimborne Publishing, Ltd.

Woodgate, R. W. 1993. *A guide to defect-free soldering*. Atlanta, Ga.: Miller Freeman (Up Media Group, Inc.).

Yost, F. G., F. M. Hosking, and D. R. Frear. 1993. *The mechanics of solder alloy wetting and spreading*. New York: Van Nostrand Reinhold/International Thomson Publishing.

CHAPTER 14

OXYGEN CUTTING

Prepared by the Welding Handbook Chapter Committee on Oxygen Cutting:

G. R. Meyer, Chair
Consultant

J. D. Compton
College of the Canyons

B. F. Jezek
TREGASKISS Welding Products

Welding Handbook Volume 2 Committee Member:

D. W. Dickinson
The Ohio State University

Contents

Photograph courtesy of ESAB Welding and Cutting Products

CHAPTER 14

OXYGEN CUTTING

INTRODUCTION

Oxygen cutting (OC) refers to the group of cutting processes used to sever, gouge, pierce or remove metals by means of the high-temperature exothermic reaction of oxygen with a base metal. With some oxidation-resistant metals, the reaction may be aided by the use of a chemical flux or a metal powder.[1,2]

The oxygen cutting processes include oxyacetylene cutting (OFC-A); oxyhydrogen cutting (OFC-H); oxynatural gas cutting (OFC-N); oxypropane cutting (OFC-P), which includes propadiene and methylacetylene-propadiene (stabilized) gas; oxygen arc cutting (OAC); oxygen gouging (OG); oxygen lance cutting (OLC); metal powder cutting (OC-P); and flux cutting (OC-F).

Oxyfuel gas cutting and its variations are important processes commonly used in structural fabrication, shipbuilding, the manufacture of heavy machinery, and in the fabrication, repair and maintenance of pressure vessels and storage tanks. This group of processes can be used to cut structural shapes, pipe, rod, and similar materials for construction and maintenance, and to cut scrap in salvage operations. In a steel mill or foundry, extraneous projections can be quickly severed from billets and castings. In disassembly operations, bolts, rivets, and pins can be rapidly severed.

Mechanized and automated oxyfuel gas cutting processes are used in steel warehouses and many other industrial settings to cut steel to size, to pierce or cut holes, and to cut various shapes from plate. An important application is preparing plate edges for welding. The processes are also used in the fabrication of machine parts such as gears, clevises, frames and tools.

The fundamentals of the oxygen cutting process, oxyfuel gas cutting, and several commonly used variations are described in this chapter. A brief discussion of the economics of oxygen cutting also is included. Safety recommendations and references to additional safety standards are presented in the "Safe Practices" section near the end of the chapter.

FUNDAMENTALS OF OXYGEN CUTTING

The oxygen cutting processes are based on the ability of high-purity oxygen to combine rapidly with iron when it is heated to its ignition temperature, which is above 870°C (1600°F). The iron is quickly oxidized by the high-purity oxygen, and heat is liberated by several reactions. The balanced chemical equations for these reactions are as follows:

First reaction: $Fe + O \rightarrow FeO$ + heat (267 kilojoules [kJ])

Second reaction: $3Fe + 2O_2 \rightarrow Fe_3O_4$ + heat (1120 kJ); and

Third reaction: $2Fe + 1.5O_2 \rightarrow Fe_2O_3$ + heat (825 kJ).

The tremendous heat release that occurs during the second reaction predominates over that of the first reaction, which is supplementary in most cutting applications. The third reaction is usually limited to heavy cutting applications in which metals from 300 millimeters (mm) to 500 mm (12 inches [in.] to 20 in.) and up to 2 meters (m) (7 feet [ft]) thick are cut.

Stoichiometrically, 0.29 m^3 (104 ft^3) of oxygen will oxidize 1 kilogram (kg) (2.2 pounds [lb]) of iron to iron oxide. In actual operations, the consumption of cutting

1. American Welding Society (AWS) Committee on Definitions and Symbols, *Standard Welding Terms and Definitions*, AWS A3.0:2001, Miami: American Welding Society.
2. At the time this chapter was prepared, the referenced codes and other standards were valid. If a code or other standard is cited without a date of publication, it is understood that the latest edition of the document referred to applies. If a code or other standard is cited with the date of publication, the citation refers to that edition only, and it is understood that any future revisions or amendments to the code or standard are not included; however, as codes and standards undergo frequent revision, the reader is advised to consult the most recent edition.

oxygen varies with the thickness of the metal. Oxygen consumption is higher than the ideal stoichiometric reaction for thicknesses under approximately 40 mm (1-1/2 in.), and it is lower for greater thicknesses. The oxygen consumption is lower than the ideal stoichiometric reaction in thick sections because only part of the iron is completely oxidized to iron oxide. Some unoxidized or partly oxidized iron is removed as slag by the kinetic energy of the rapidly moving oxygen. Chemical analysis has shown that over 30% of the slag is unoxidized metal in some instances. The heat generated by the rapid oxidation of iron melts some of the iron adjacent to the reaction surface. This molten iron is usually swept away with the iron oxide by the motion of the oxygen stream. The concurrent oxidizing reaction heats the layer of iron at the active cutting front. Part of the unoxidized metal (slag) may adhere to the lower surface of the metal being cut.

The heat generated by the iron-oxygen reaction at the focal point of the cutting reaction (the *hot spot*) must be sufficient to preheat the material to the ignition temperature. Although some heat is lost by radiation and conduction, ample heat remains to sustain the reaction. In actual practice, the top surface of the material is frequently covered by mill scale or rust. This layer must be melted away by the preheating flames to expose a clean metal surface to the oxygen stream. Preheating flames help sustain the cutting reaction by providing heat to the surface. The flames also shield the oxygen stream from turbulent interaction with air.

The alloying elements normally found in carbon steels are oxidized or dissolved in the slag without markedly interfering with the cutting process. When appreciable amounts of alloying elements are present in steel, the effect of these elements on the cutting process must be considered. Steels containing minor additions of oxidation-resistant elements such as nickel and chromium can be cut with the oxygen cutting process. However, when oxidation-resistant elements are present in large quantities, modifications to the cutting technique are required to sustain the cutting action. This is the case for stainless steels, for example.

OXYFUEL GAS CUTTING

The oxyfuel gas cutting (OFC) processes cut or remove metal by means of the chemical reaction of oxygen with the metal at elevated temperatures. The necessary heat is provided by a flame of fuel gas burning with oxygen. These cutting processes are often referred to by various other names, including *burning*, *flame cutting*, and *flame machining*, but the standard term designated by the American Welding Society (AWS) is *oxyfuel gas cutting*.

The cutting operation begins by using a fuel gas flame to heat the area of the workpiece to be cut to the ignition temperature of approximately 870°C (1600°F). The actual cutting is performed by an oxygen stream that is directed at the preheated location on the workpiece, resulting in the oxidation of the heated metal and the generation of heat. This heat sustains the continued oxidation of the metal throughout the cut. The oxygen stream and combusted gas transport the molten oxide away, and the metal in the path of the stream burns, producing a narrow cut known as a *kerf*.

The torch directs the preheat flame produced by the controlled combustion of fuel gases and controls the cutting oxygen. Because the cutting oxygen jet has a 360° cutting edge, it provides a rapid means of cutting straight edges or curved shapes to required dimensions without expensive handling equipment. The cutting direction can be continuously changed during operation. Oxyfuel gas cutting is illustrated schematically in Figure 14.1. Figure 14.2 shows the oxyfuel gas cutting of a 57 mm (2 1/4 in.) low-carbon steel beam.

A comparison of oxyfuel gas cutting to other metal cutting methods such as sawing, milling, and arc cutting reveals a number of advantages and disadvantages of the process. The following are several advantages of oxyfuel gas cutting:

1. Steels can generally be cut faster with oxyfuel gas cutting than by mechanical chip removal processes;
2. Section shapes and thicknesses that are difficult to cut by mechanical means can be severed economically by oxyfuel cutting;
3. Basic manual oxyfuel cutting equipment costs are low compared to machine tools or the equipment used in other cutting processes;
4. Manual oxyfuel cutting equipment is portable and easily deployed in the field;
5. Cutting direction can be changed rapidly on a small radius during operation;
6. Large metal plate sections can be rapidly cut in place by moving the oxyfuel cutting torch rather than the plate; and
7. Oxyfuel cutting is an economical method of plate edge preparation for bevel and groove weld joint designs.

The following limitations are inherent to the oxyfuel gas cutting of metals:

1. Dimensional tolerances achieved may be greater than those attained with mechanical cutting;
2. The process is essentially limited to the cutting of steels and cast iron, although other readily oxidized metals such as titanium can be cut;

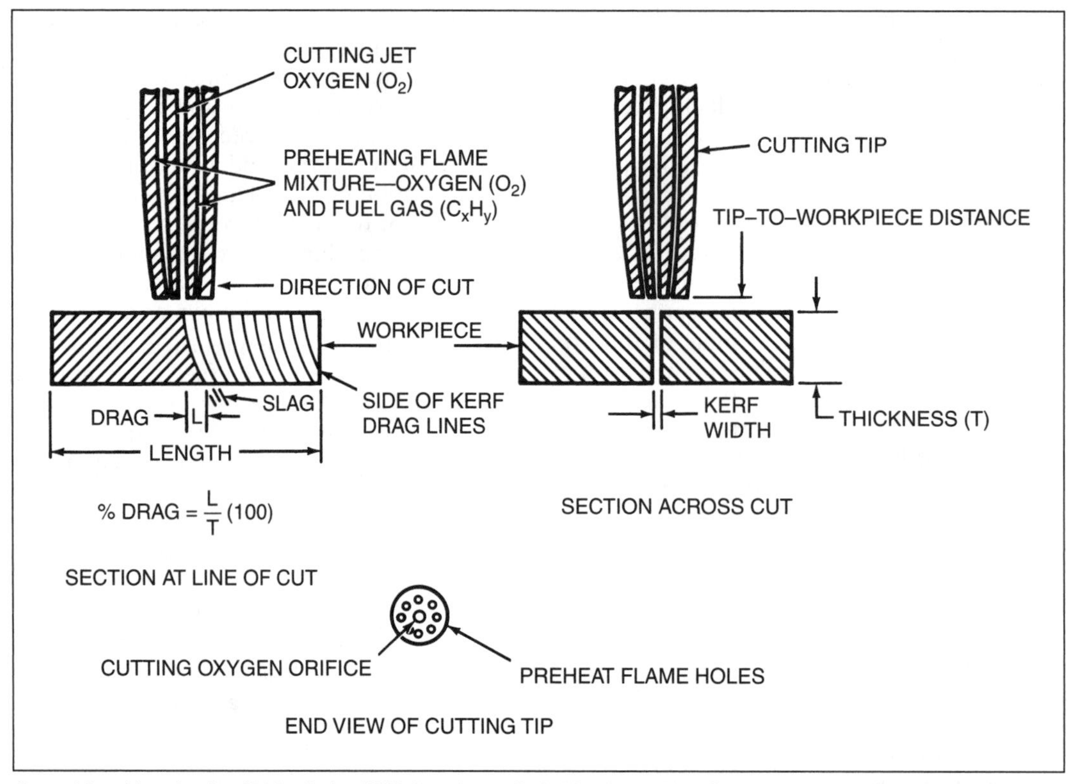

Source: Adapted from American Welding Society (AWS) Committee on Oxyfuel Gas Cutting, 2002, Operator's Manual for Oxyfuel Gas Cutting, Miami: American Welding Society, p. 1.

Figure 14.1—Schematic of Oxyfuel Gas Cutting

Photograph courtesy of Messer-MG Systems & Welding, Inc.

Figure 14.2—Oxyfuel Gas Torch Completes a Cut on 57 mm (2-1/4 in.) Low-Carbon Steel Beam

3. Preheat flames and expelled red-hot slag constitute fire and burn hazards;
4. Fuel combustion and oxidation of the metal require proper fume control and adequate ventilation;
5. Hardenable steels may require preheat or postheat, or both, to control their metallurgical structures and mechanical properties adjacent to the cut edges; and
6. Special process modifications are required for the cutting of high-alloy steels and cast irons.

PRINCIPLES OF OPERATION

The oxyfuel gas cutting process employs a torch with a tip (nozzle). The torch cutting tip is designed with a number of preheat flame ports and a center passage for the cutting oxygen. The functions of the torch and tip are to mix the fuel gas and the oxygen in the correct proportions, to produce preheat flames, and to supply a concentrated stream of high-purity oxygen to the reaction zone. The oxygen oxidizes the hot metal and blows the molten reaction products away from the cut.

The preheat flames serve to heat the metal to a temperature at which the metal will react with the cutting oxygen. The oxygen jet rapidly oxidizes most of the metal in a narrow section to make the cut. Metal oxides and molten metal are expelled from the cut by the kinetic energy of the oxygen stream. Moving the torch across the workpiece at the proper rate produces a continuous cutting action. The torch may be moved manually or by means of a mechanized or automated carriage.

The accuracy of a manual operation depends largely on the skill of the operator.The accuracy and speed of the cut and the finish of the cut surfaces are usually improved when mechanized or automated cutting operations are used. Several torches may be used on one machine. Automated operations can be programmed to perform a series of complex cuts with a high level of accuracy.

Kerf Control

When a workpiece is cut using an oxyfuel cutting process, a narrow width of metal is progressively removed, producing the cut, or kerf. Control of the kerf is important in cutting operations in which the dimensional accuracy of the workpiece and the squareness of the cut edges are significant factors in quality control. Kerf width is a function of the size and type of tip used, the oxygen port, the cutting speed, and the flow rates of cutting oxygen and preheating gases. Oxygen flow rates must usually be adjusted in proportion to the thickness of the workpiece. Thicker metals require increased oxygen flow. Cutting tips with larger cutting oxygen ports are required to handle the higher flow rates. Consequently, the width of the kerf increases in proportion to the thickness of the material being cut. The width of a kerf is shown in Figure 14.3.

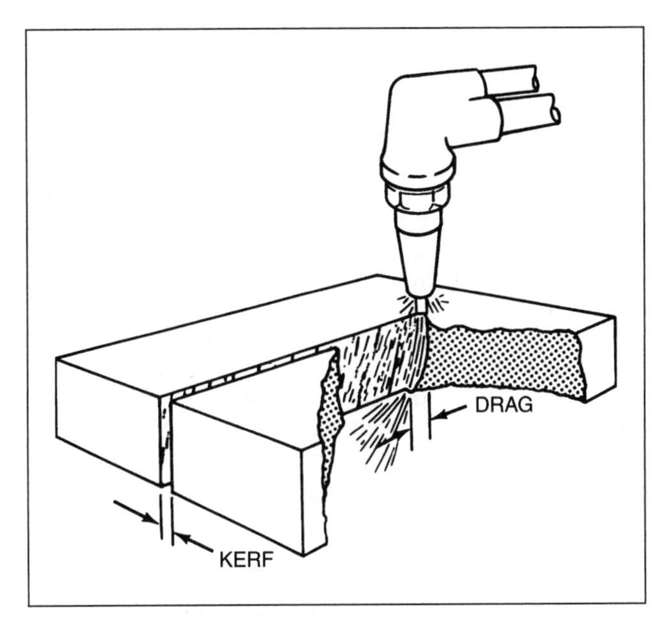

Figure 14.3—Kerf and Drag in Oxyfuel Gas Cutting

Cutting speeds below those recommended for optimal-quality cuts usually result in irregularities in the kerf. In this case the oxygen stream may inconsistently oxidize and wash away additional material from each side of the cut. Excessive preheat flame results in undesirable melting and the widening of the kerf at the top.

Kerf width is especially important in shape cutting. Compensation must be made for kerf width in the layout of the work or the design of the template. Kerf width can generally be maintained to within +0.4 mm (+1/64 in.) on materials up to 50 mm (2 in.) thick.

Drag

If the cutting speed is increased or if the oxygen flow is decreased, the oxygen available in the lower regions of the cut decreases. With less oxygen available, the oxidation reaction rate decreases, and the oxygen jet has less energy to carry the reaction products out of the kerf. As a result, the most distant part of the cutting stream may lag behind the portion closest to the torch tip. The length of this lag, measured along the line of cut, is referred to as the *drag*. This is shown in Figure 14.3. When the speed of the cutting torch is adjusted so that the oxygen stream enters the top of the kerf and exits from the bottom of the kerf along the axis of the tip, the cut is said to have zero drag.

Drag may also be expressed as a percentage of the cut thickness. A drag of 10% signifies that the far side of the cut lags the near side of the cut by a distance equal to 10% of the material thickness. An increase in cutting speed with no increase in oxygen flow usually results in a larger drag. This may cause a deterioration of cut quality. A strong possibility of losing the cut exists at excessive speeds. Reverse drag may occur when the oxygen flow is too high or the travel speed is too slow. Under these conditions, poor-quality cuts usually result. It should be noted that the cutting stream lag caused by improper torch alignment is not considered drag.

EQUIPMENT

Oxyfuel gas cutting equipment is available commercially for manual, mechanized, and automated operations. All types of equipment operate on the same principle. Manual equipment is used primarily for operations that do not require a high degree of accuracy or a high-quality cut surface. Mechanized or automated cutting equipment is used for accurate, high-quality work and for large-volume cutting such as that required in steel fabricating shops.

Operators of both manual and mechanized equipment must not attempt to operate any oxyfuel apparatus until they are trained in its proper use or under competent supervision. They must adhere to the manufacturer's recommendations and operating instructions and all applicable standards to ensure safe and efficient operation.

MANUAL EQUIPMENT

A typical installation for manual oxyfuel gas cutting requires the following equipment, accessories, and supplies:

1. One or more cutting torches suitable for the preheat fuel gas to be used and the range of material thicknesses to be cut;
2. Torch cutting tips for a range of material thicknesses;
3. Oxygen and fuel gas hoses;
4. Oxygen and fuel gas pressure regulators;
5. Sources of oxygen and fuel gases to be used;
6. Flame strikers, eye protection, flame-resistant and heat-resistant gloves and clothing, safety devices; and
7. Operating instructions from the equipment manufacturer and the supplier of oxygen and fuel gas.

The functions of an oxyfuel gas cutting torch are to control the mixture and flow of preheat fuel gas and oxygen, control the flow of cutting oxygen, and discharge the gases through the cutting tip at the proper velocities and volumetric flow rates for preheating and cutting. These functions are controlled by the operator, by the pressures of incoming gases, and by the design of the torch and cutting tips. Oxyfuel gas cutting torches are versatile tools that can be readily transported to the work site. Oxyfuel gas cutting torches can be used to cut metal up to 2 m (7 ft) thick. The latest edition of the Compressed Gas Association's publication *Torch Standard for Welding and Cutting*, CGA E-5, should be consulted.[3]

Manual oxyfuel cutting torches are available in various sizes, and the size chosen should be one that is easily manipulated by the operator. However, torch selection generally depends on the thickness range of the material to be cut.

Oxyfuel gas cutting torches are classified according to the manner in which they mix the fuel gas and the oxygen. In the tip-mix torch, the fuel and oxygen for the preheat flames are mixed in the tip. In the torch-mix type, the mixing takes place within the torch. Torch-mix torches are further categorized as positive-pressure (equal pressure) designs or injector (low-pressure) designs. Positive-pressure torches are used when sufficient fuel gas pressure exists to supply the torch mixer with the required volume of gas. The injector-type torches are used when the fuel gas pressure (usually natural gas at less than 14 kilopascals (kPa) (2 pounds per square inch, gas pressure [psig]) is such that the fuel gas must be drawn into the torch by the venturi action of the injector mixer. In addition, some manufacturers offer a mixer design that operates effectively at both low and high fuel pressures. This torch design is referred to as a *universal pressure mixer.*

Diagrams of these two types of torches are shown in Figures 14.4 and 14.5. Figure 14.6 shows a torch and two interchangeable accessories: a multiple-port heating tip and a cutting tip.

Manual Cutting Tips

Precision-machined copper-alloy or swaged cutting tips of various designs and sizes are used in cutting

3. Compressed Gas Association (CGA), *Torch Standard for Welding and Cutting*. CGA E-5. Arlington, Virginia: Compressed Gas Association.

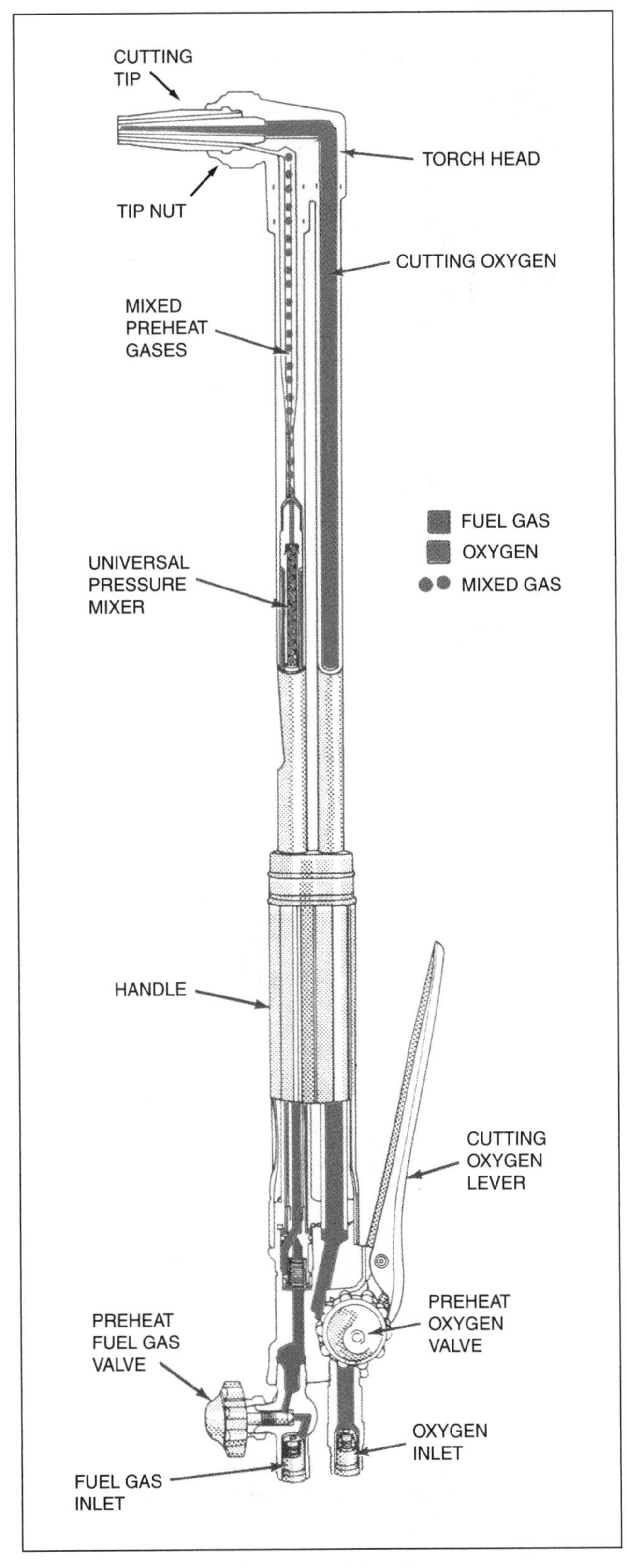

Figure 14.4—Positive-Pressure Oxyfuel Gas Cutting Torch

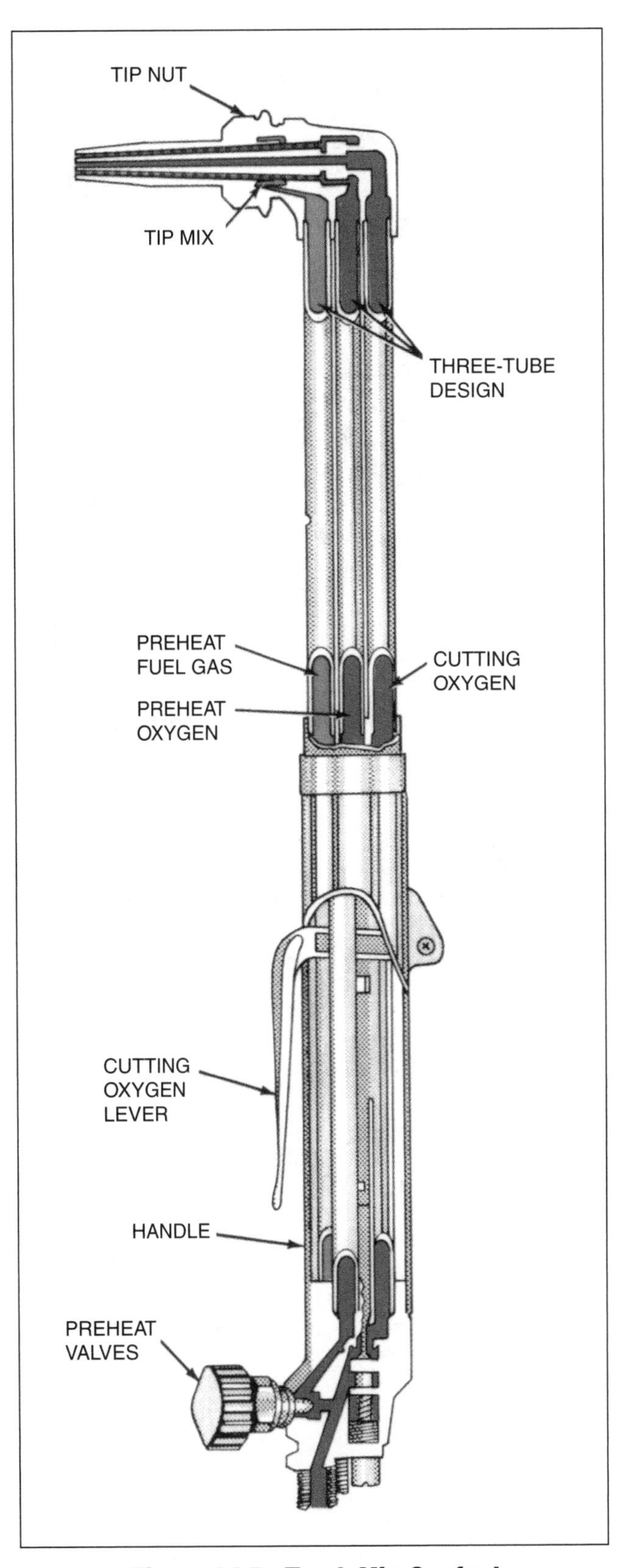

Figure 14.5—Torch-Mix Oxyfuel Gas Cutting Torch

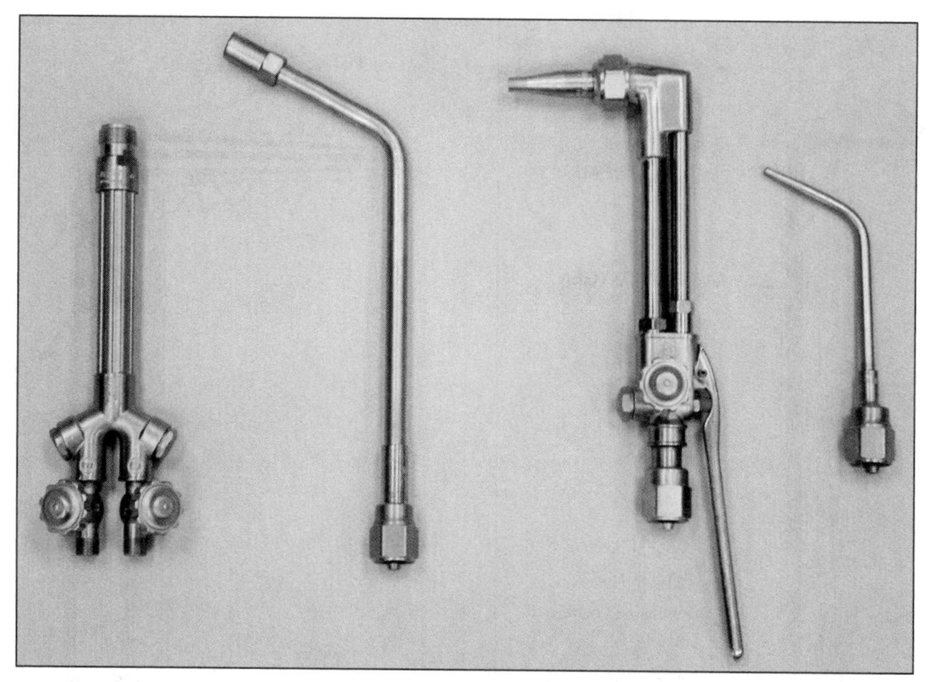

Photograph courtesy of Victor Equipment Company

Figure 14.6—Oxyfuel Gas Torch with Interchangeable Heating and Cutting Heads

torches. They are held in the torch by a tip nut. The design selected depends on the fuel gas used and the type of work to be performed. For example, when cutting rusty or scaly steel, a tip that furnishes a great amount of preheat should be selected. All cutting tips have preheat flame ports, usually arranged in a circle around a central cutting-oxygen orifice. The preheat flame ports and the cutting-oxygen orifice are sized for the thickness range of the metal to be cut.

Cutting tips are designated as standard or high speed. Standard tips have a straight-bore oxygen port. They are typically used with oxygen pressures ranging from 205 kPa to 415 kPa (30 psig to 60 psig). High-speed tips differ from standard tips in that the exit end of the oxygen orifice flares out or diverges. The divergence allows the use of higher oxygen pressures, typically 415 kPa to 690 kPa (60 psig to 100 psig), while maintaining a uniform oxygen jet at supersonic velocities. High-speed tips are ordinarily used only for mechanized cutting. They usually permit cutting at speeds approximately 20% greater than speeds available with standard tips. Even though higher pressures are used, the high-speed orifices are usually smaller than standard tip orifices. Thus, the smaller size and faster speed do not necessarily result in greater oxygen consumption. Standard and high-speed tips are shown schematically in Figure 14.7.

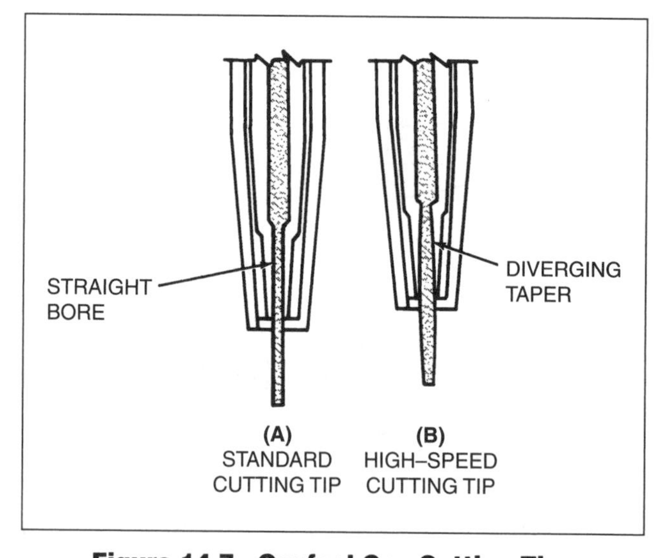

Figure 14.7—Oxyfuel Gas Cutting Tips

The size and design of the cutting-oxygen orifice are not usually affected by the type of fuel gas used. However, the design of the preheat flame port is dependent on the specific fuel gas to be used. Various fuel gases require different volumes of oxygen and fuel, and they burn at different velocities and temperatures. Therefore,

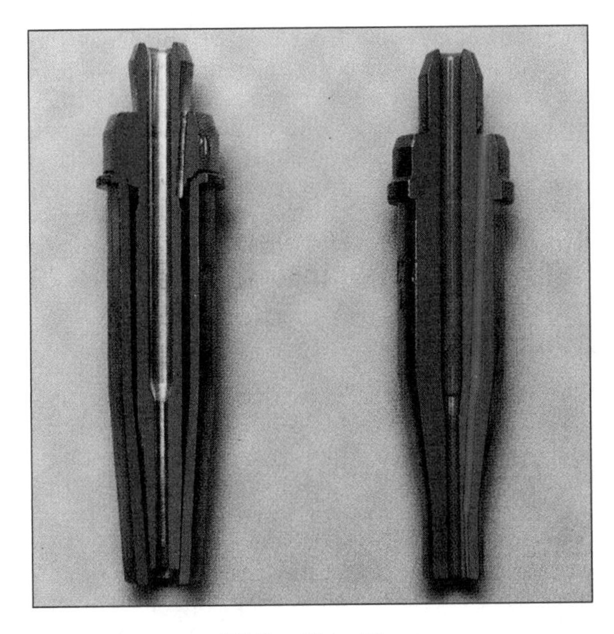

(A) One-Piece Tips

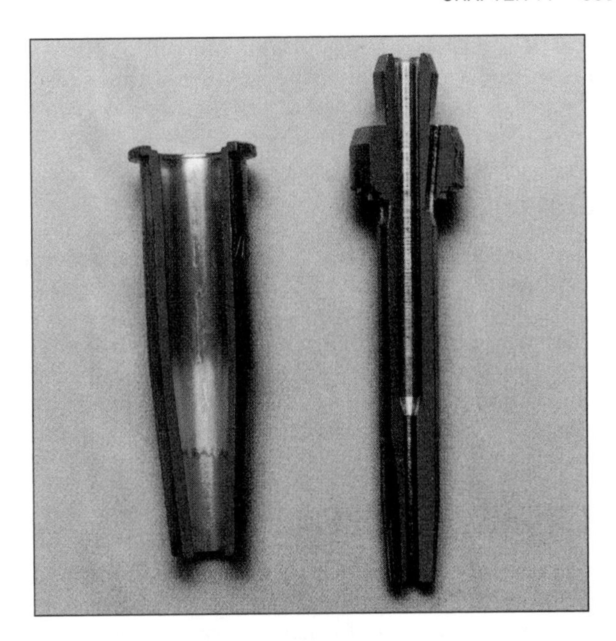

(B) Two-Piece Tips

Figure 14.8—Cross Sections of One-Piece and Two-Piece Tips Used for Fuel Gases Other Than Acetylene

the preheat flame port size and number are designed to provide both a stable flame and adequate preheat for applications with the particular fuel gas being used.

Tips for use with acetylene are usually one-piece units with drilled or swaged flame ports. They are flat on the flame end. Tips for use with other fuel gases are one-piece, similar to acetylene tips, or two-piece with milled splines on the inner member, as illustrated in Figure 14.8. Tips designed for use with methylacetylene-propadiene (stabilized) (MPS) gas have a flat surface on the flame end. Most tips intended for use with propylene have a slight recess, whereas tips for natural gas and propane have a deeper recess or a cupped end.

Cutting tips, although considered consumable items, are precision tools. The tip is considered to have the greatest influence on cutting performance. The proper maintenance of tips can greatly extend their useful life and provide for continued high-quality performance. For example, the accumulation of slag in and around the preheat and cutting oxygen passages disturbs the preheat flame and oxygen stream characteristics. This can result in an obvious reduction in performance and quality of cut. When this occurs, the tip should be restored to a good working condition or replaced.

Gas Pressure Regulators

The ability to make a successful cut requires not only the proper choice of cutting torch and tip for the fuel gas selected but also requires a means of precisely regulating the gas pressures and volumes. Regulators are pressure control devices used to reduce high source pressures to the required working pressures by means of adjustable pressure valves. Regulators vary in design, performance, and convenience features. Gas pressure regulators are designed for use with specific types of gases and for established pressure ranges.

The gas pressure regulators used in oxyfuel gas cutting operations are generally similar in design to those used for oxyfuel gas welding (OFW). However, regulators with higher capacities and delivery pressure ranges than those used for oxyfuel gas welding may be required for multiple torch operations and heavy cutting. The regulators for most other fuel gases are similar in design to those used for acetylene. The Compressed Gas Association's *Standard for Gas Pressure Regulators*, CGA E-4, should be consulted.[4]

Hoses

Oxygen and fuel gas hoses used for oxyfuel gas cutting are the same as those used for oxyfuel gas welding, as discussed in Chapter 11, "Oxyfuel Gas Welding."

4. Compressed Gas Association (CGA), 1994, *Standard for Gas Pressure Regulators*, CGA E-4-1994. Arlington, Virginia: Compressed Gas Association. See also Chapter 11, "Oxyfuel Gas Welding."

Hoses for oxygen cutting should conform to specifications in the standard published by the Rubber Manufacturers Association (RMA), *Specification for Rubber Welding Hose*, ANSI/RMA IP-7.[5]

Other Equipment

Tip cleaners, wrenches, strikers, and appropriate safety devices, including protective clothing, must be used. Tinted goggles or other eye protection devices are available in a number of different shades. Further information on personal protective equipment is presented in the section titled "Safe Practices." Recommended lens shades for oxygen cutting are presented in Appendix 1.

MECHANIZED EQUIPMENT

Mechanized oxyfuel gas cutting equipment can vary in complexity from simple hand-guided cutting machines to very sophisticated computer-controlled units. Although mechanized equipment is similar to the manual equipment in principle, it differs in design to accommodate higher pressures, faster cutting speeds, and the means for starting the cut. Many machines are designed for special purposes, such as those for making vertical cuts, beveling or preparing edges for welding, and pipe cutting and beveling. A variety of mechanized cutting systems are commercially available.

Mechanized oxyfuel cutting may require additional equipment, including one or more of the following, depending on the application:

1. A machine to move one or more torches in the required cutting pattern,
2. Torch mounting and adjusting arrangements on the machine,
3. A cutting table to support the workpiece,
4. Means for loading and unloading the cutting table, and
5. Automatic preheat ignition devices for multiple-torch machines.

Torches

A typical mechanized cutting torch consists of a cutting tip and a barrel similar to that of a manual torch, but of heavier construction. The torch body and barrel encase the oxygen and fuel-gas tubes that carry the gases to the cutting tip. The cutting tip is secured by a tip nut. The body of the torch may have a rack for indexing the tip to a desired position from the work surface.

5. Rubber Manufacturers Association (RMA), 1999, *Specification for Rubber Welding Hose*, ANSI/RMA IP-7, 1999, Washington, DC: Rubber Manufacturers Association.

Mechanized torches have two or three gas (hose) inlets. Torches with two gas inlet fittings have a fuel-line connection and one oxygen connection with two valves. Torches with three inlet fittings have separate connections for fuel gas, preheat oxygen, and cutting oxygen and permit separate regulation of preheat and cutting oxygen. They are recommended for remote-control operations. Figure 14.9 shows the components of a torch for mechanized oxyfuel gas cutting.

Cutting Tips

Cutting tips for mechanized cutting are designed to operate at higher oxygen and fuel pressures than those normally used for manual cutting. The cutting tip shown in Figure 14.7(B) is one type used for operation at high cutting speeds. Divergent cutting tips are based on the principle of gas flow through a venturi. High velocities are reached as the gas emerges from the venturi nozzle. Divergent cutting tips are precision machined to minimize any distortion of the gases when they exit the nozzle. They are used for the majority of mechanized cutting applications because of their superior cutting characteristics for materials up to 150 mm (6 in.) thick. They are not recommended for cutting materials over 250 mm (10 in.) thick.

Regulator Systems

When natural gas, propane, or similar fuels are used as preheat fuels in mechanized cutting, fuel and oxygen can be conserved by using a combination high-low pressure regulating system. Because these fuels burn at lower heat transfer intensities than acetylene, high flow rates of fuel and preheat oxygen are required to heat the metal to ignition temperature in a reasonable time. Once the cut has been started, less heat is needed to maintain cutting action, yielding savings in gas costs and an increase in cut quality.

High-low pressure regulating systems permit the starting gas flow rates to be reduced to a predetermined level when the flow of cutting oxygen is initiated. This reduction may be performed manually or automatically, depending on the design of the regulators and the control system.

Cutting Machines

Oxyfuel gas cutting machines are designed for portable or stationary use. Portable cutting machines are primarily used for straight-line cutting; however, some are designed to cut circles and other shapes. Portable machines typically consist of a motor-driven carriage with an adjustable mounting device for the cutting torch, as shown in Figure 14.10. In most cases,

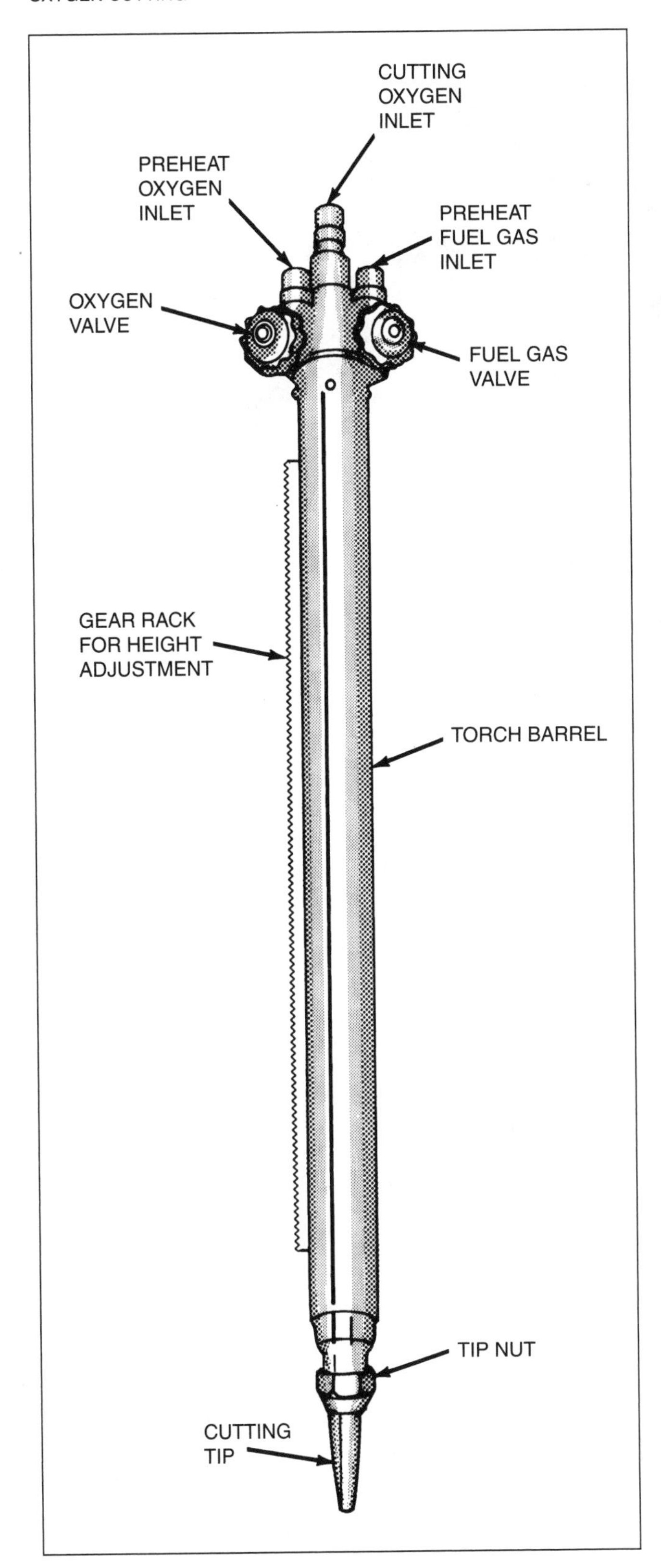

Figure 14.9—Three-Hose Torch for Mechanized Oxyfuel Gas Cutting

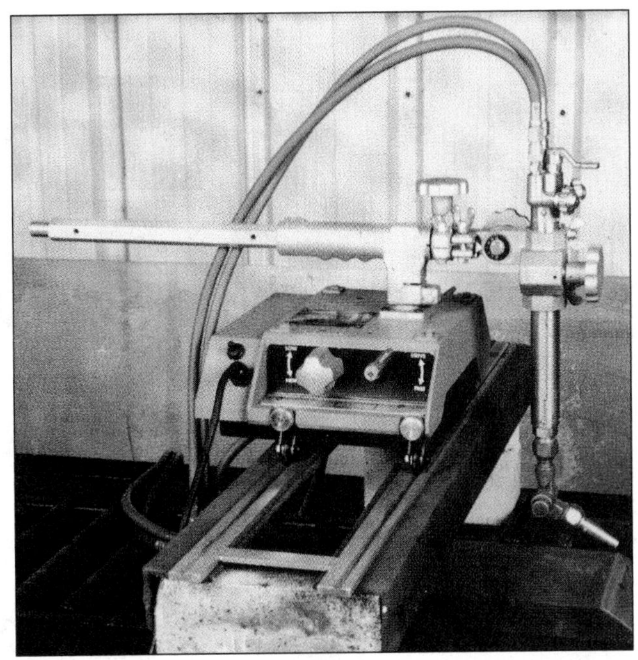

Photograph courtesy of Victor Equipment Company

Figure 14.10—Mechanized Cutting Torch Mounted on a Portable Carriage

the machine travels on a track, which performs the function of guiding the torch. The carriage speed is adjustable over a wide range. The degree of cutting precision depends upon the accuracy of the track and the fit between the track and the driving wheels of the carriage.

Portable machines are available in various weights and sizes, depending on the type of work to be performed. The smallest machines weigh only a few pounds. They are limited to carrying light-duty torches used to cut thin materials. Large portable cutting machines are heavy and rugged. They have the capacity to carry one or more heavy-duty torches and the necessary auxiliary equipment required for the cutting of thick sections.

To begin operation, the operator ignites the torch, positions it at the starting point, and initiates the cutting oxygen flow and the carriage travel. He or she then adjusts the torch height to maintain the preheat flames at the correct distance from the work surface. To ensure that quality cuts are made, the operator may initiate a trial cut and follow the carriage to make any necessary adjustments. When the cut is completed, the operator turns off the cutting torch and the carriage.

Stationary cutting machines are designed to remain in a fixed location. The raw material is moved to the machine, and the completed cut shapes are transported

Photograph courtesy of Messer-MG Systems and Welding, Inc.

Figure 14.11—Gantry-Type Cutting Machine with Computerized Numerical Control Drive

away. The workstation is composed of the cutting machine, a system to supply the oxygen and the preheat fuel to the machine, and a material handling system.

The torch-support carriage runs on tracks. The support structure may span the workpiece with a gantry-type bridge across the tracks, as shown in Figure 14.11, or it may be cantilevered off to one side of the tracks as shown in Figure 14.12. Torch carriage designs are usually classified according to the width of plate that can be cut, which is equivalent to the travel distance on the tracks (the transverse motion). The maximum cutting length is dictated by physical limitations of the gas and electric power supply lines. An operator station with consolidated controls for gas flow, torch movement, and machine travel is generally incorporated in the machine.

Multiple torches can be mounted on a shape-cutting machine, depending on the size of the machine. The machine has the capacity to cut shapes of nearly any complexity and size. In multiple-torch operations, several identical shapes can be cut simultaneously. The number depends on the shape, size, plate size, and the number of available torches.

A rectilinear or coordinate-drive machine often incorporates a sine-cosine potentiometer to coordinate separate drive motors for longitudinal and transverse motion of the torch. The carriage and the cross arm, each with its own driving motor, are driven in predetermined directions, while the linear speed of the torch remains at a constant pre-selected value. This type of construction permits the design and manufacture of cutting machines with sufficient rigidity to carry all modern monitoring and control equipment.

Suitable control systems are designed to feed information to the electric drive motors of the carriage and cross arm. One control system uses a photoelectric cell tracer that can follow line drawings or silhouettes. Computerized numerical controls (CNC) direct the motion using coordinates supplied by pre-programmed shapes stored in local memory, punched paper tape, disk drives, or a remote computer via direct, or distributed, numerical control (DNC).

GASES

In addition to the oxygen used to support combustion, various fuel gases are used in oxyfuel gas cutting, depending on the process variation. These process variations include OFC-A when acetylene is used as a fuel gas, OFC-H when hydrogen is used; OFC-N when natural gas is used, and OFC-P when propane, propylene or methylacetylene-propadiene (stabilized) gas is used as the fuel gas.

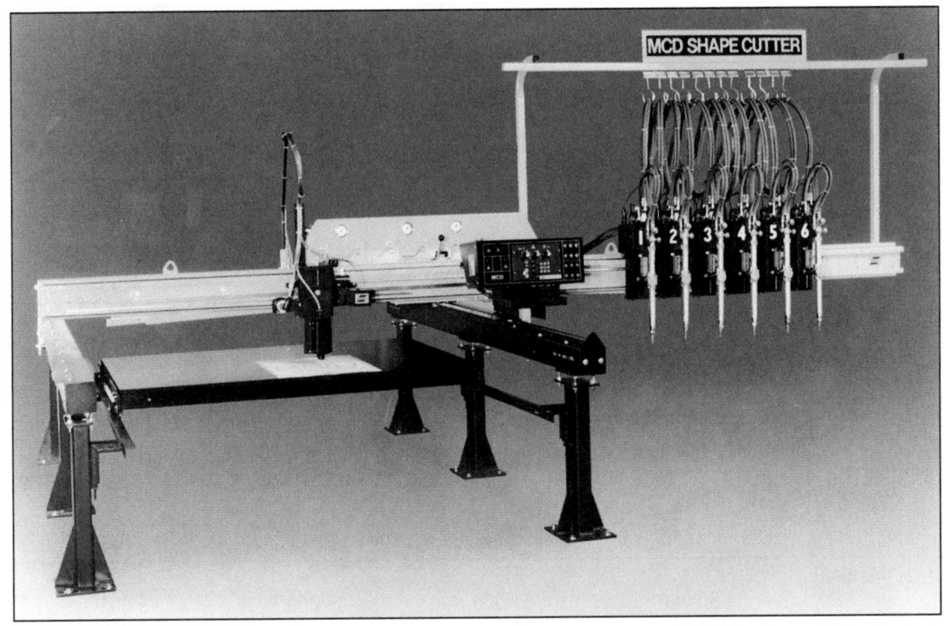

Photo courtesy of ESAB Welding and Cutting Products

Figure 14.12—Cantilever-Type Mechanized Shape-Cutting Machine Equipped with Photocell Tracer and Six Oxyfuel Gas Cutting Torches

OXYGEN

The oxygen used for cutting must have a purity of 99.5% or higher. Lower purity reduces the efficiency of the cutting operation. For example, a 1% decrease in oxygen purity to 98.5% results in a decrease in cutting speed of approximately 15% and an increase of about 25% in the consumption of cutting oxygen. The quality of the cut is impaired, and the amount and tenacity of the adhering slag increases. When oxygen purity is below 95%, the familiar cutting action disappears, degenerating into a melt-and-wash action that is usually unacceptable.

PREHEATING GASES

A number of commercially available fuel gases are used with oxygen to provide the preheating flames required to carry out the cutting operation. Some of these are proprietary compositions. The preheat flames perform the following functions in the cutting of steel:

1. Raises the temperature of the steel to the ignition point;
2. Adds heat energy to the steel to maintain the cutting reaction;
3. Provides a protective shield between the cutting oxygen stream and the atmosphere; and
4. Dislodges any rust, scale, paint, or other foreign substance from the upper surface of the steel that would stop or retard the normal forward progress of the cutting action.

A preheat intensity that rapidly raises the temperature of steel to the ignition temperature is usually adequate to maintain cutting action at high travel speeds. However, the quality of the cut may not be optimum. High-quality cutting can be carried out at considerably lower preheat intensities than those normally required for rapid heating. Most large cutting machines are equipped with dual-range gas controls that limit high-intensity preheating to the starting operation. The preheat flames are reduced to lower intensity during the cutting operation to save fuel and oxygen and provide a better cut surface.

Fuel gases are generally selected on the basis of availability, performance, and cost. The following factors should be considered when selecting a preheat fuel:

1. Time required for preheating when starting cuts on square edges or rounded corners, and when piercing holes for cut starts;

Table 14.1
Properties of Common Fuel Gases

Fuel Gas	Acetylene	Propane	Propylene	Methylacetylene-Propadiene (Stabilized)	Natural Gas
Chemical Formula	C_2H_2	C_8H	C_3H_6	C_3H_4	Methane (CH_4)
Neutral flame temperature					
°C	3100	2520	2870	2870	2540
°F	5600	4580	5200	5200	4600
Primary flame heat emission					
btu/ft^3	507	255	433	517	11
MJ/m^3	19	10	16	20	0.4
Secondary flame heat emission					
MJ/m^3	36	94	72	70	37
btu/ft^3	963	2243	1938	1889	989
Total heat value (after vaporization)					
MJ/m^3	55	104	88	90	37
btu/ft^3	1470	2498	2371	2406	1000
Total heat value (after vaporization)					
kJ/kg	50 000	51 000	49 000	49 000	56 000
btu/lb	21 500	21 800	21 100	21 100	23 900
Total oxygen required (neutral flame)					
volume O_2/volume fuel	2.5	5.0	4.5	4.0	2.0
Oxygen supplied through torch (neutral flame)					
volume O_2/volume fuel	1.1	3.5	2.6	2.5	1.5
m^3 oxygen/kg (15.6°C) fuel	1.0	1.9	1.4	1.4	2.2
ft^3 oxygen/lb fuel (60°F)	16.0	30.3	23.0	22.1	35.4
Maximum allowable regulator pressure					
kPa	103	1030	1030	1030	
psi	15	150	150	150	Line
Explosive limits in air, %	2.5–8.0	2.3–9.5	2.0–10	3.4–10.8	5.3–14
Volume-to-weight ratio					
m^3/kg (15.6°C)	0.91	0.54	0.55	0.55	1.4
ft^3/lb (60°F)	14.6	8.66	8.9	8.85	23.6
Specific gravity of gas (15.6°C [60°F])					
Air = 1	0.906	1.52	1.48	1.48	0.62

Note: To evaluate the significance of the information in this table, including combustion intensity and specific flame output for various fuel gases, refer to the terms and concepts involved in the burning of fuel gas in Chapter 11, "Oxyfuel Gas Welding."

2. Effect on cutting speeds for straight-line, shape, and bevel cutting;
3. Cost and availability of the fuel in cylinder, bulk, and pipeline volumes;
4. Cost of the preheat oxygen required to burn the fuel gas efficiently;
5. The effect of these combined factors on work output for a specific cutting operation;
6. The ability to use the fuel efficiently for other operations such as welding, heating, and brazing, if required; and
7. Safety in transporting and handling the fuel gas containers.

For optimal performance and safety, the torches and tips used should always be designated specifically for use with the particular fuel gas selected.

The properties of various commonly used fuel gases are listed in Table 14.1.[6]

6. To evaluate the significance of the information in this table, refer to the terms and concepts involved in the burning of fuel gas as presented in Chapter 11, "Oxyfuel Gas Welding." Combustion intensity or specific flame output for various fuel gases is also covered in that chapter. These properties are important considerations in fuel gas selection.

ACETYLENE

Acetylene is the fuel gas most often used for manual cutting processes. The chief advantages of acetylene for oxyfuel gas cutting (OFC-A) and the reasons for its preference are its availability, the high flame temperature it produces, and users' widespread familiarity with its flame characteristics. The combustion of acetylene with oxygen produces a hot, short flame with a bright inner cone at each preheat port. The hottest point is at the end of this inner cone. Combustion is completed in the long outer flame. The sharp distinction between the two flames facilitates the adjustment of the ratio of oxygen to acetylene to achieve the desired flame characteristics.

Depending on this ratio, the flame may to be adjusted to exhibit the characteristics of reducing (carburizing), neutral, or oxidizing flame types, as shown in Figure 14.13. The neutral flame, obtained with a ratio of approximately one part oxygen to one part acetylene, is generally used for manual cutting. A neutral flame is shown in Figure 14.13(A). As the oxygen flow is decreased, a bright streamer begins to appear, like that shown in Figure 14.13(B). This streamer indicates a reducing flame, which is sometimes used to rough-cut cast iron. When excess oxygen is supplied, the inner flame cone shortens and becomes more intense. The flame temperature increases to a maximum at an oxygen-to-acetylene ratio of about 1.5 to 1. An oxidizing flame like that pictured in Figure 14.13(C) is used for short preheating times and for cutting very thick sections.

The high flame temperature and heat transfer characteristics of the oxyacetylene flame are particularly important for bevel cutting. They are also an advantage for operations in which the preheat time is an appreciable fraction of the total time for cutting, for example, when making short cuts.

Acetylene in the free state must not be used at pressures higher than 103 kPa (15 psig) or 207 kPa (30 psi) absolute pressure. At higher pressures, it may decompose with explosive force when exposed to heat or shock. Chapter 11, "Oxyfuel Gas Welding," contains additional information on acetylene, its production and storage, and on the oxyacetylene flame.

HYDROGEN

Hydrogen mixed with oxygen burns with a very hot, almost colorless flame. The flame does not have a well-defined inner cone, which makes it difficult to determine whether or not the flame is neutral. Hydrogen is seldom used as a cutting fuel gas except for special applications. It is sometimes used in underwater cutting if it presents an economic advantage, but natural gas and MPS gas are generally used.

METHYLACETYLENE-PROPADIENE (STABILIZED) GAS

One of the gases used in the OFC-P process is methylacetylene-propadiene (stabilized) (MPS), a liquefied, stabilized acetylene-like fuel that can be stored and handled similarly to liquid propane. Methylacetylene-propadiene (stabilized) is a mixture of several hydrocarbons, including propadiene (allene), propane, butane, butadiene, and methylacetylene. Methylacetylene, like acetylene, is an unstable, high-energy, triple-bond compound. The other compounds in MPS dilute

(A) Neutral Flame

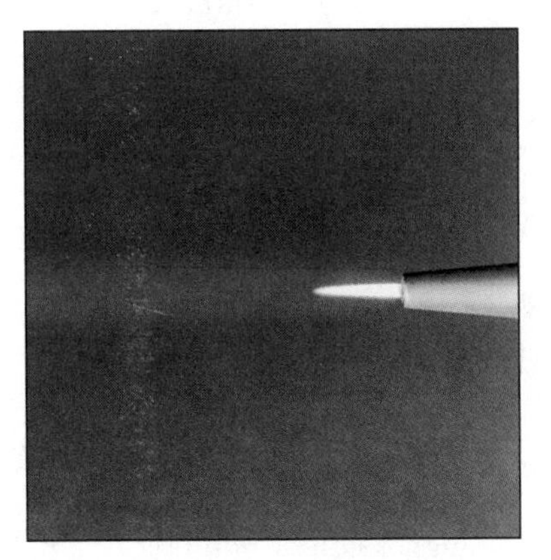

(B) Reducing Carburizing Flame

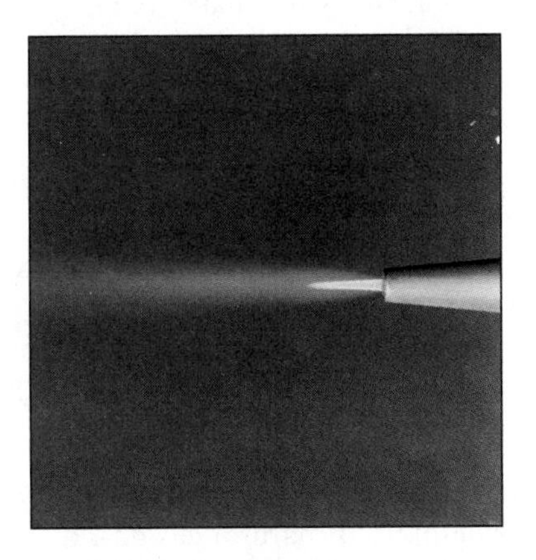

(C) Oxidizing Flame

Figure 14.13—Types of Oxyacetylene Flames

the methylacetylene sufficiently to make the mixture safe for handling.

Methylacetylene-propadiene (stabilized) burns hotter than either propane or natural gas. Like acetylene, it affords a high release of energy in the primary flame cone. The outer flame produces a relatively high release of heat, similar to that of propane and propylene. The overall heat distribution in the flame is the most even of any of the gases.

A neutral flame is achieved at a ratio of 2.5 parts of torch-supplied oxygen to 1 part MPS. Its maximum flame temperature is reached at a ratio of 3.5 parts of oxygen to 1 part of MPS. These ratios are used for the same applications as acetylene.

Although many of the characteristics of MPS gas are similar to those of acetylene, MPS requires about twice the volume of oxygen per volume of fuel for a neutral preheat flame. Thus, oxygen costs will be higher when MPS gas is used instead of acetylene for a specific job. To be competitive, the cost of MPS gas must be lower than that of acetylene for the job.

Methylacetylene-propadiene (stabilized) gas provides an advantage over acetylene for underwater cutting in deep water or when higher operating pressures are required. Because the outlet pressure of acetylene is limited to 207 kPa (30 psi) absolute, oxyacetylene cutting is usually not applicable at depths below 6 m (20 ft) of water. Methylacetylene-propadiene (stabilized) can be used at these and greater depths, as can hydrogen. For underwater applications, MPS, acetylene, and hydrogen should all be evaluated for selection as the preheat fuel gas.

NATURAL GAS

The composition of the natural gas used in oxynatural gas cutting varies, depending on its source. Its main component is methane (CH_4). The ratio of torch-supplied oxygen to natural gas required to produce a neutral flame is 1.5 parts oxygen to 1 part natural gas. The flame temperature achieved with natural gas is lower than that of acetylene. The natural gas flame is also more diffuse and less intense. The characteristics of the flame for carburizing, neutral, or oxidizing conditions are not as distinct as those exhibited by the oxyacetylene flame.

Because of the lower flame temperature and the resulting lower heating efficiency, significantly greater quantities of natural gas and oxygen are required to produce heating rates equivalent to those of oxygen and acetylene. To compete with acetylene, the cost and availability of natural gas and oxygen, the higher consumptions, and the longer preheat times of these gases must be considered. The use of tips designed to provide a heavy preheat flame for cutting machines that allow a high-low preheat setting may compensate for deficiencies of the lower heat output of natural gas.

PROPANE

Propane is used for oxyfuel gas cutting in many industrial plants because of its availability and because it has a much higher total heat value than natural gas (see Table 14.1). For proper combustion during oxypropane cutting, propane requires 4 to 4-1/2 times its volume of preheat oxygen. This requirement is offset somewhat by the higher heat value it provides.

Propane is stored in liquid form in pressurized tanks and may be piped to individual workstations. Small tanks are easily transported to the work site.

PROPYLENE

Propylene, marketed under many different brand names, is commonly used as the fuel gas for many types of oxyfuel gas cutting. One volume of propylene requires 2.6 volumes of torch-supplied oxygen for a neutral flame and 3.6 volumes for maximum flame temperature. The cutting tips used for propylene are similar to those used for MPS. The characteristics of propylene are considered to be similar to natural gas and acetylene; thus, the comparative properties of propylene as the fuel gas for cutting would place it about mid-way between those of oxyacetylene cutting and oxynatural gas cutting.

OPERATING PROCEDURES

The recommendations of the equipment manufacturer in assembling and using the equipment must always be closely followed in the operation of oxyfuel gas cutting equipment—whether the fuel gas used is acetylene, hydrogen, MPS, natural gas, propane, or propylene. Strict adherence to the manufacturers' and suppliers' instructions will ensure proper and safe use of the equipment and will help prevent damage. Also, oxyfuel gas cutting operations must conform to all applicable industry standards.

MAINTENANCE OF GAS REGULATORS

Oxygen and fuel-gas regulators must be clean and in good working condition. If oil, grease, or foreign material is present on a regulator or other equipment, or if the equipment is damaged, it must not be used prior to being properly cleaned or serviced by a qualified repair technician. Hoses must be in good condition and of the appropriate size to provide adequate volume and pressure of both oxygen and fuel gas to the cutting torch.

PROCEDURES FOR FLASHBACK AND BACKFIRE

Two concerns that the operators of oxyfuel gas cutting equipment must deal with are flashback and backfire. A flashback occurs when the flame recedes into or behind the mixing chamber of the oxyfuel gas torch. When the flashback occurs at the mixer, it is often accompanied by a whistling or humming sound. This phenomenon represents a serious burn hazard, and corrective action must be taken immediately to extinguish it. The oxygen valve on the torch must be quickly turned off, and then the fuel gas valve should be turned off as quickly as possible.

One cause of flashback is the failure to purge the hose lines before lighting the torch; another cause is the overheating of the torch tip. While a torch-mounted flashback arrestor is designed to stop the flashback flame from entering the hose, the torch can still be damaged. A regulator-mounted flashback arrestor protects the regulator, but not the torch or hose. A reverse-flow check valve will not stop a flashback.

A backfire is the momentary recession of the flame into the torch tip or mixer, followed by the immediate reappearance or complete extinguishing of the flame, and accompanied by a loud pop. After this condition, the torch is still workable. If backfiring continues, the torch or tip, or both, should be removed from service for cleaning and possible repair and the cause of the backfire should be determined and corrected.

TORCH OPERATION

The manufacturer's recommendations for opening the valves, lighting, testing, and using the torch must always be followed. Only a spark lighter or other recommended lighting device specifically designed for lighting oxyfuel gas torches should be used. Shaded or tinted eye protection and appropriate clothes must be worn.

When using acetylene or MPS gas, the most widely accepted method of lighting the torch is to open the fuel-gas valve slightly, about 1/8 to 1/4 turn, and light the gas with a spark lighter or other recommended device. The operator should continue to open the fuel valve until the smoke and soot has cleared. The oxygen valve should then be opened until the flame is stable, after which the flow of fuel and oxygen may be increased alternately until the desired flame is obtained.

For other fuel gases, the operator's manual for the specific torch being used must be followed. The intensity of the flame may be adjusted by slightly increasing or decreasing the volumes of both gases.

Flame Adjustment

Flame adjustment is a critical factor in attaining satisfactory torch operation. The amount of heat produced by the flame depends on the intensity and type of flame required. As shown in Figure 14.13, three types of flames—carburizing, neutral, or oxidizing—can be set by properly adjusting the torch valves.

A carburizing flame with acetylene, MPS, or propylene is indicated by trailing feathers on the primary flame cone or by long yellow-orange streamers in the secondary flame envelope. Propylene-based fuels, propane, and natural gas have a long, rounded primary flame cone. A carburizing flame is often used to achieve the best cut finish and is used for the stack cutting of thin material.

A neutral flame with acetylene, MPS, or propylene is indicated by a sharply defined, dark primary flame cone and a pale blue secondary flame envelope. Propane- and propylene-base fuels and natural gas have a short, sharply defined cone. This flame is obtained by adding oxygen to a carburizing flame. It is the flame most frequently used for cutting.

An oxidizing flame for acetylene or MPS has a light-colored primary cone and a smaller secondary flame shroud. It also generally burns with a harsh whistling sound. With propane- and propylene-base fuels and natural gas, the primary flame cones are longer, less sharply defined, and have a lighter color. This type of flame is obtained by adding oxygen to the neutral flame. This flame is frequently used for fast, low-quality cutting. It is used selectively for piercing and high-quality beveling.

MANUAL CUTTING PROCEDURES

Several methods can be used to start a cut on an edge. The most common method is to place the preheat flames halfway over the edge, holding the end of the flame cones 1.5 mm to 3 mm (1/16 in. to 1/8 in.) above the surface of the material to be cut. The tip axis should be aligned with the plate edge. When the top corner reaches a reddish-yellow color, the cutting-oxygen valve is opened and the cutting process starts. Torch movement is started after the cutting action reaches the far side of the edge.

Another starting method is to hold the torch halfway over the edge, with the cutting oxygen turned on, but not touching the edge of the material. When the metal reaches a reddish-yellow color, the torch is moved onto the material and cutting starts. This method wastes oxygen, and starting is more difficult than with the first method. It should be used only for cutting thin material when preheat times are very short.

A third method is to put the tip entirely over the edge of the material to be cut. The preheat flame is held there until the metal reaches its kindling temperature. The tip is then moved to the edge of the plate so the oxygen stream will slightly clear the metal. After turning the cutting oxygen on, the cut is initiated. This method has the advantage of producing sharper corners at the beginning of the cut.

Once the cut has been started, the torch is moved along the line of cut with a smooth, steady motion. The operator should maintain as constant a tip-to-workpiece distance as possible. The torch should be moved at a speed that produces a light ripping sound and a smooth spark stream.

For plate thicknesses of 13 mm (1/2 in.) or more, the cutting tip should be held perpendicular to the plate. For thin plate, the tip can be tilted in the direction of the cut. Tilting increases the cutting speed and helps prevent slag from freezing across the kerf. When cutting material in the vertical position, the cut should be started on the lower edge of the material and proceed upward.

It is often necessary to start a cut at some point other than on the edge of a piece of metal. This technique is known as *piercing*. Piercing usually requires a somewhat larger preheat flame than the one used for an edge start. In addition, the flame should be adjusted to slightly oxidizing to increase the heat energy. The area where the pierce cut is to begin should be located in scrap area of the material. The torch tip should be held in one spot until the steel surface turns a yellowish red color and a few sparks appear from the surface of the metal. The tip should be angled and lifted up as the cutting oxygen valve is opened. The torch should be held stationary until the cutting jet pierces through the plate.

Torch motion is then initiated along the cut line. If the cutting oxygen is turned on too quickly and the torch is not lifted, slag may be blown into the tip and may plug the gas ports.

MECHANIZED AND AUTOMATED CUTTING PROCEDURES

Operating conditions for mechanized and automated oxygen cutting vary depending on the fuel gas and the style of cutting torch being used. Tip size designations, tip design, and operating data can be obtained from the torch manufacturer.

Startup and shutdown procedures for mechanized and automated oxyfuel cutting are essentially the same as those used for manual torch operation. However, proper evaluation and adjustment of operating conditions are more critical for obtaining high-speed, high-quality cuts. The cutting charts furnished by the equipment manufacturer or gas supplier should be used to select the proper tip size for the material thickness to be cut. In addition to the tip size, initial fuel and oxygen pressure settings and travel speeds should be selected from the chart. The chart also frequently lists gas flow rates, the drill size of the oxygen orifice, preheat cone lengths, and kerf width. Operating conditions should then be adjusted to give the desired cut quality.

Proper tip size and cutting-oxygen pressure are important in making a high-quality mechanized cut. If the proper tip size is not used, maximum cutting speed and the best quality of cut will not be achieved. The cutting-oxygen pressure setting is an essential condition; deviations from the recommended setting will greatly affect cut quality. For this reason, some manufacturers specify setting the pressure at the regulator and operating with a specific length of hose. When longer or shorter hoses are used, an adjustment in pressure should be made. A preferred method is to measure oxygen pressure at the torch inlet. Pressure settings for cutting oxygen are then adjusted to obtain the recommended pressure at the torch inlet, rather than at the regulator outlet.

Other important adjustments include the preheat fuel and oxygen pressure settings and the travel speed. Once the regulators have been adjusted, the torch valves may be used to throttle gas flows to provide the desired preheat flame. If sufficient flow rates are not obtained, pressure settings at the regulator can be increased to compensate. Cleanliness of the nozzle, type of base metal, purity of cutting oxygen, and other factors have a direct effect on performance.

Manufacturers differ in their recommendations regarding travel speed. Some suggest a range of speeds for specific thicknesses, while others list a single speed. In either case, the recommended settings are intended only as guides. To determine the correct speed for an application, the cut should be initiated at a slower speed than that recommended and gradually increased until the cut quality falls below the required level. At this point, the speed should be reduced until the cut quality is restored, and the cutting operation is then continued at this speed.

Stack Cutting

The stacking of thin plate and cutting several pieces at one time can be more economical than cutting individual pieces, particularly when the material thickness is under 6 mm (1/4 in.). When data on mechanized oxyfuel cutting speeds and gas requirements are plotted against the material thickness, it can be observed that the requirements are not directly proportional to material thickness. Gas consumption per unit of thickness decreases as the thickness increases. Consequently, stack cutting may decrease the cost of cutting thin plate. Stack cutting is limited to sheet and plate up to 13 mm (1/2 in.) thick because of the difficulty in clamping heavier material in a tight stack.

The plasma arc cutting (PAC) and laser beam cutting (LBC) processes have largely replaced the oxygen cutting process for stack cutting. Both plasma arc and laser beam cutting offer high-speed, high-quality cuts on thin material. Regardless of the procedure employed, the economics of a stack cutting operation must be carefully evaluated with respect to the total costs involved, including such items as material preparation, stack makeup, clamping devices, and the increased skill and care required of the operator.

Preheating and Postheating

The material being cut may be preheated to provide desired mechanical and metallurgical characteristics or to improve the cutting operation. Preheating the workpiece can accomplish several useful purposes, including the following:

1. Increase the efficiency of the cutting operation by permitting higher travel speed, which reduces the total amount of oxygen and fuel gas required to make the cut;
2. Reduce the temperature gradient in the steel during the cutting operation, which reduces distortion and also lowers thermally induced stresses or results in their more favorable distribution, preventing the formation of quenching or cooling cracks.
3. Prevent hardening of the cut surface by reducing the cooling rate; and
4. Decrease migration of carbon toward the cut face by lowering the temperature gradient in the metal adjacent to the cut.

The temperatures used for preheating steel generally range from 90°C to 700°C (200°F to 1300°F), depending on the size and the type of steel to be cut. The majority of carbon and alloy steels can be cut with the steel heated to the 200°C to 315°C (400°F to 600°F) temperature range. The higher the preheat temperature is, the more rapidly the oxygen reacts with the iron. This facilitates higher cutting speeds.

It is essential that the preheat temperature be uniform throughout the section in the areas to be cut. If the metal near the surfaces is at a lower temperature than the interior metal, the oxidation reaction will proceed more rapidly in the interior. Large pockets may form in the interior, resulting in unsatisfactory cut surfaces or slag entrapment, which may interrupt the cutting action. If the material is preheated in a furnace, cutting should be initiated as soon as possible after the material is removed from the furnace to take advantage of the heat in the plate.

If furnace capacity is not available for preheating the entire piece, local preheating in the vicinity of the cut may be beneficial. For light cutting, preheating may be accomplished by passing the cutting torch emitting the preheating flames slowly over the line of the cut until the desired preheat temperature is reached. Another method that may yield better results involves preheating the location to be cut with a multiple-flame heating torch mounted ahead of the cutting torch. Temperatures can be measured in several ways. The most common method is using a type of crayon that melts at specific temperatures.

To reduce thermally induced internal stresses in the cut workpieces, they may be annealed, normalized, or stress-relieved. With proper postheat treatment, most metallurgical changes caused by the cutting heat can be eliminated. If a furnace of the required size is not available for postheat treatment, the cut surface may be reheated to the proper temperature with multiple-flame heating torches.

Plate Edge Preparation

Bevel, V-groove, and U-groove joint designs may be specified for the welding of steel components. Oxygen cutting or gouging can be used to prepare the edges to be welded. Single and double bevels for straight-line beveling are readily produced (usually in mechanized cutting operations) using standard cutting tips and torches. U-grooves are produced by means of oxygen gouging with specially designed cutting tips.

Plate Beveling

The beveling of plate edges in preparation for welding is necessary in many applications to ensure the proper dimensions and fit and to accommodate standard welding techniques. Plate edges can be beveled with the use of a single torch or multiple torches operating simultaneously. Although single-torch beveling can be accomplished manually, beveling is best done by machine for accurate control of the cutting variables. When cutting bevels with two or three torches, plate-riding devices should be used to ensure a constant tip position above the plate, as shown in Figure 14.14.

In single-torch beveling, the amount and type of torch preheat is a dominant factor. With bevel angles below 15°, the loss of preheat efficiency is minimal. When the bevel angle is greater than 15°, the heat transferred from the preheat flames to the plate decreases rapidly as the bevel angle increases. Considerably greater preheat input is required, particularly for thicknesses up to

Figure 14.14—Mechanized Cutting Arrangement for Beveling a Plate Edge

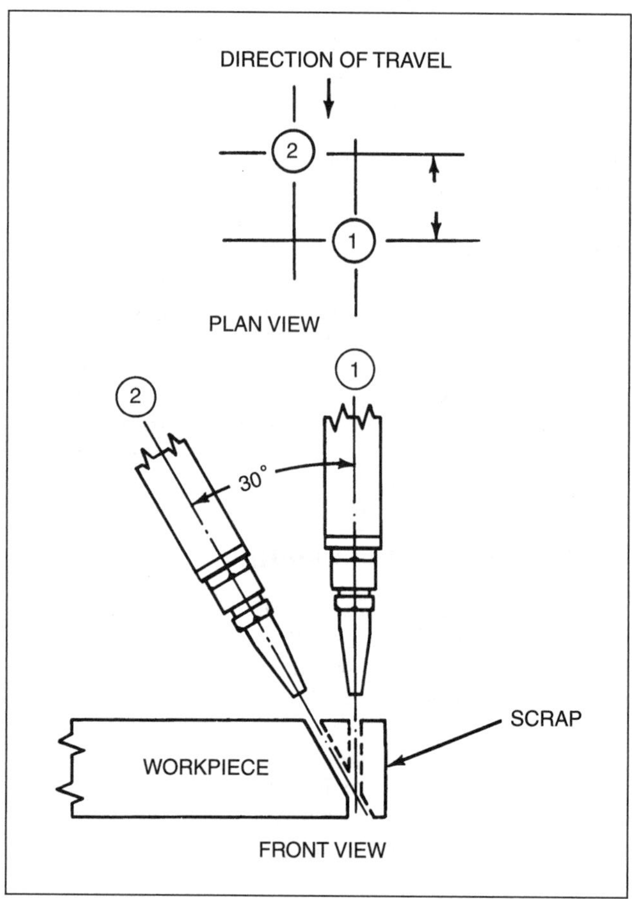

Figure 14.15—Cutting a Single-Bevel Edge Preparation With a Root Face

25 mm (1 in.). Best results are obtained by using high oxygen-to-fuel ratios and by positioning the tip very close to the workpiece. When cutting bevel angles greater than 30° or when working with heavy plate, special bevel tips are used to provide the additional preheat capacity required.

Faster beveling speeds may be obtained with one of two techniques. The first uses an auxiliary torch (with only preheat flames burning) mounted perpendicular to the workpiece. The second uses an auxiliary adapter that divides the preheat and applies a portion of it at right angles to the workpiece. Both of these methods consume less total preheat gas than a single-angled tip.

The best quality of cut face is usually not obtained at the highest cutting speed. In fact, the cut face finish can usually be improved by operating at lower speeds. When the speed is reduced to obtain improved surface finish of the cut, the preheat flames should be decreased to prevent excessive meltdown of the top edge of the faces.

Figures 14.15, 14.16, and 14.17 illustrate the torch positions used to cut the three basic types of beveled edges. In each case, the torch position spacing is governed by plate thickness, tip size, and the speed of cutting. The cutting torches are spaced in positions that are practical and do not interrupt the cutting action of any of the cutting-oxygen streams. When the distance between the torches is too great, the cutting action of the trailing torch may not span the kerf of the leading torch. This causes the oxygen stream to be deflected into the kerf of the leading torch, resulting in gouges in the cut face. This produces a rough surface, usually with a light slag that adheres to the underside of the prepared edge.

The positioning of the torches in a lateral direction for multiple-bevel cutting is usually accomplished by trial and error. However, this can be costly and can result in lengthy reworking or possible scrapping of portions of the workpiece. A simple machined template that is typical of the desired edge geometry is useful for torch alignment. A kerf-centering device is attached to each cutting tip, as shown in Figure 14.18. The torches are then properly angled and adjusted to the edge template. The multiple-torch cutting head is now ready to duplicate the template profile.

Precise torch-conveying equipment is necessary to obtain close dimensional tolerances when preparing plate edges. Reproducibility, accuracy, and maximum efficiency are achieved when using large gantry-and-rail machines. This type of apparatus may be classified as a machine tool. A plate is placed on a flat cutting table between the rails of a three-gantry cutting machine, as

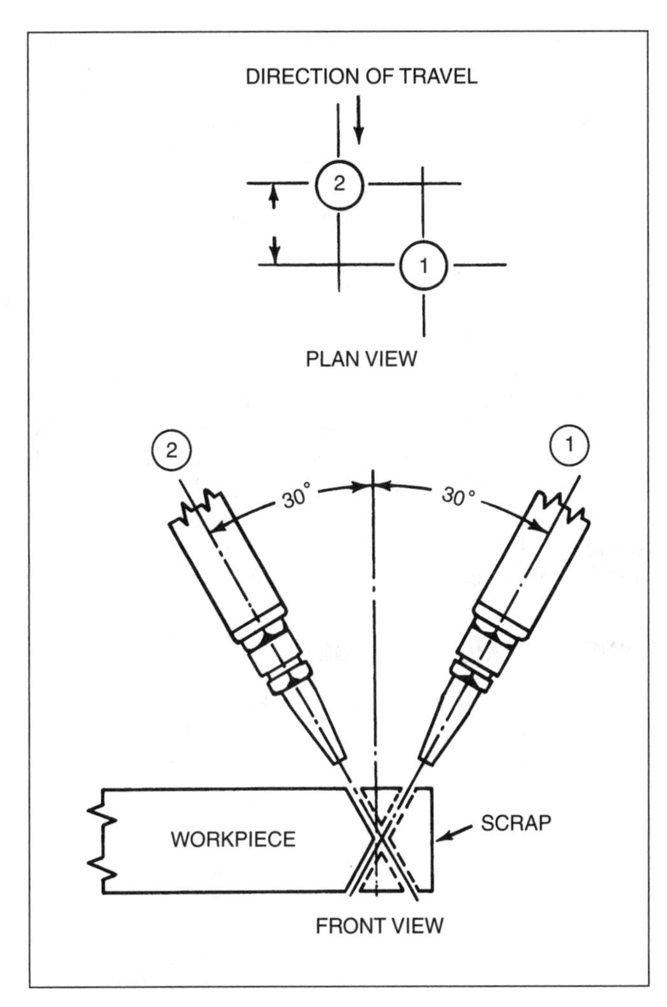

Figure 14.16—Cutting a Double-Bevel Edge Preparation with No Root Face

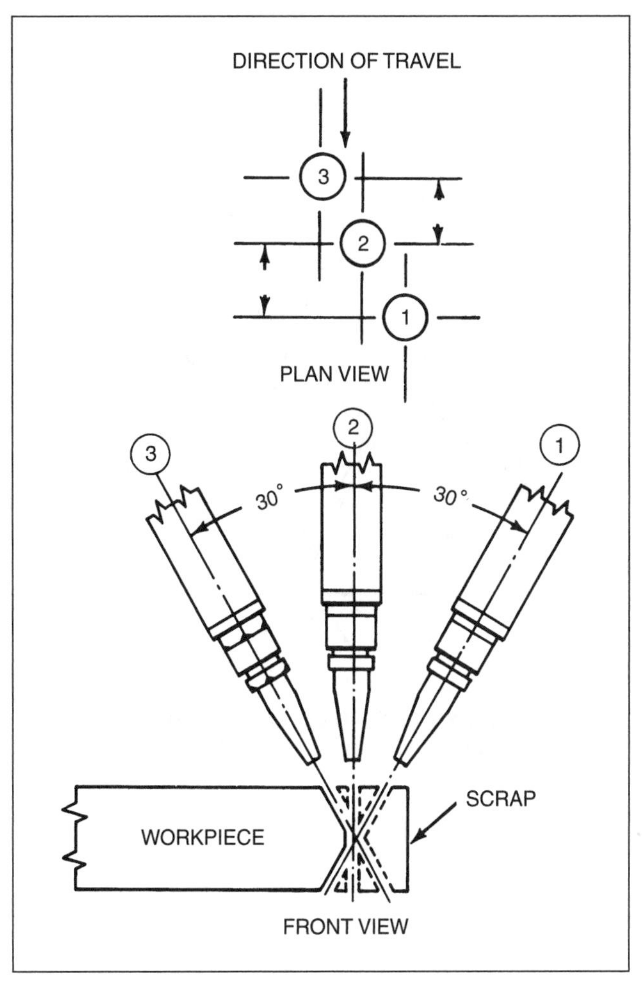

Figure 14.17—Cutting a Double-Bevel Edge with a Root Face

shown in Figure 14.19. The machine can prepare all four edges of the plate without repositioning. It can also cut the plate into smaller segments at the same time.

PROCESS VARIATIONS

Oxyfuel gas cutting process variations categorized by the American Welding Society include oxyacetylene cutting (OFC-A), oxyhydrogen cutting (OFC-H), oxy-natural gas cutting (OFC-N), and oxypropane cutting (OFC-P), each designated by the gas used with oxygen.[7]

7. American Welding Society (AWS) Committee on Definitions and Symbols, 2001, *Standard Welding Terms and Definitions*, AWS A3.0:2001, Miami: American Welding Society, p. 114.

It should be noted that propylene, a liquid petroleum gas used as the fuel gas in many cutting operations, is included in the designation OFC-P, as is MPS gas.

TECHNIQUES FOR SPECIAL APPLICATIONS

Several techniques have been devised for the oxyfuel gas cutting of oxidation-resistant steels. These are also applicable to cast irons. The most important techniques are torch oscillation, the waster-plate technique, and the wire-feed technique.

The quality of the cut surface is impaired when traditional techniques are used to cut oxidation-resistant metals. Scale and slag may adhere to the cut faces. Pickup of carbon or iron, or both, usually appears on the cut surfaces of stainless steels and nickel-alloy steels.

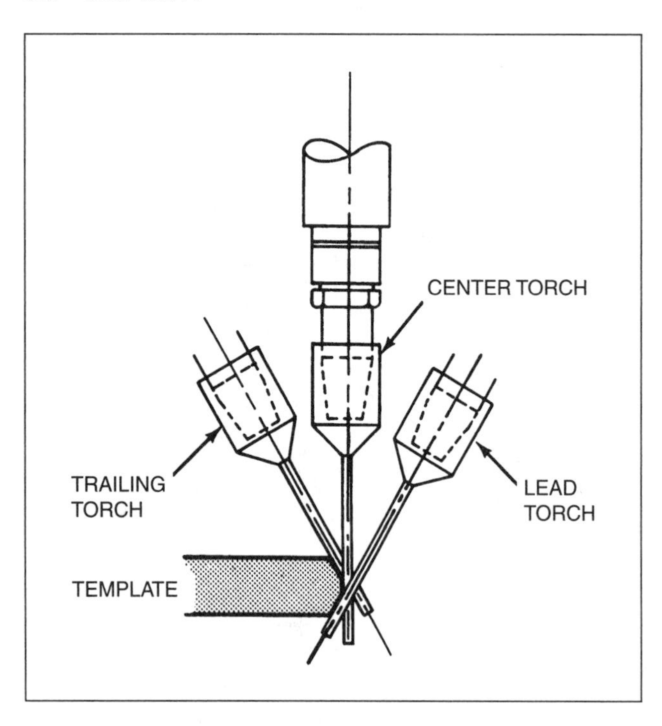

Figure 14.18—Kerf Centering and Bevel-Angle Setting Method

This may affect the corrosion resistance and magnetic properties of the metal. If the corrosion resistance or magnetic properties of the material are critical, approximately 3 mm (1/8 in.) of metal should be machined from the cut edges. For critical cuts in oxidation-resistant metals, the plasma arc cutting process usually proves more successful than the oxyfuel cutting processes.

Torch Oscillation

The torch oscillation technique for the cutting of cast iron, described in the "Cast Iron" section under "Process Applications," can be applied to stainless steels. Low-alloy stainless steels up to 100 mm (4 in.) thick can sometimes be severed with a standard cutting torch and oscillation. The entire thickness of the starting edge must be preheated to a bright red color before the cut is started. This technique should be combined with another of the cutting methods listed.

Waster-Plate Technique

Oxidation-resistant steels can be cut by clamping a low-carbon steel waster plate on the upper surface of the material to be cut. The cut is started in the waster

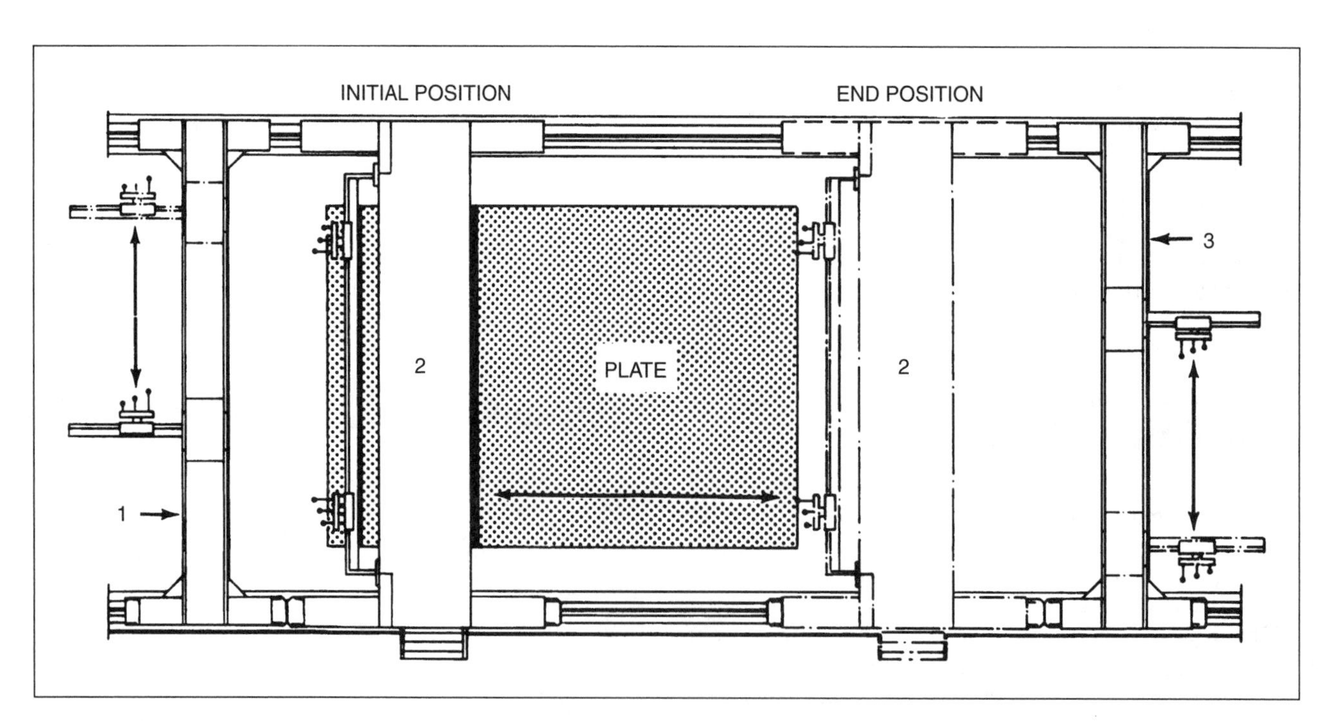

Figure 14.19—Plan View of a Three-Gantry Cutting

plate. The heat liberated by the oxidation of the low-carbon steel provides additional heat at the cutting face to sustain the oxidation reaction. The iron oxide from the low-carbon steel helps to wash away the refractory oxides from the stainless steel. The required thickness of the waster plate varies in proportion to the thickness and the oxidation resistance of the material being cut.

The limitations of this technique include the cost of the waster-plate material, additional set-up time, slow cutting speeds, and roughness in the quality of the cut.

Wire-Feed Technique

With the appropriate equipment for the wire-feed technique, a small-diameter low-carbon steel wire is fed continuously into the torch preheat flames ahead of the cut. The end of the wire should melt rapidly into the surface of the alloy-steel plate. The effect of the wire addition on the cutting action is the same as that of the waster plate. Wire- and powder-feed techniques are often used in steel mills in continuous casting slab cut-off lines for high-alloy steels.

The deposition rate of the low-carbon steel wire must be adequate to maintain the oxygen cutting action. The deposition rate should be determined by trial cuts. The thickness of the alloy plate and the cutting speed are also factors that must be considered in selecting this process. A motor-driven wire feeder and wire guide mounted on the cutting torch are required accessory equipment.

GOUGING

Using the oxyfuel gas cutting processes to gouge steel plate is usually limited to thicknesses up to 25 mm (1 in.). Oxyfuel gas gouging is frequently used on the underside of a welded joint to remove discontinuities in the original root pass. The process is also used to remove defective weld joints or cracks when repairing previously fabricated metal.

Gouging usually requires a special gouging tip with an extra-heavy preheat capacity and a central oxygen orifice that produces a high level of turbulence in the oxygen stream. This turbulence causes a wide flow of oxygen that can be controlled by the operator to achieve the desired width and depth of the gouge. Other factors that determine the shape of the gouge are speed, tip angle, pressure, amount of preheat, and tip size. One of the significant advantages of oxyfuel gouging is that no additional equipment is required other than that already used in the oxyfuel gas cutting process.

UNDERWATER CUTTING

Underwater cutting is used for salvage work and for cutting below the water line on piers, dry docks, and ships. The techniques for underwater cutting with the oxyfuel gas process are not materially different from the techniques used in the cutting of steel in open air. An underwater oxyfuel gas cutting torch has the same features as a standard oxyfuel gas cutting torch, but it has the additional feature of supplying its own ambient atmosphere in the form of an air bubble.

As in a standard torch, fuel and oxygen are mixed together in the underwater cutting torch and burned to produce the preheat flame. Cutting oxygen is provided through the tip to sever the steel. The underwater torch provides an air bubble around the cutting tip that is maintained by a flow of compressed air around the tip, as shown in Figure 14.20. The air shield stabilizes the preheat flame and at the same time displaces the water from the cutting area.

The underwater cutting torch has connections for three hoses to supply compressed air, fuel gas, and oxygen. A combination shield and spacer device is attached at the cutting end of the torch. The adjustable shield controls the formation of the air bubble. The shield is adjusted so that the preheat flame is positioned at the correct distance from the work. This feature is essential for underwater work because of poor visibility and reduced operator mobility caused by cumbersome diving suits. Slots in the shield allow the burned gases to escape. A short torch is used to reduce the reaction force produced by the compressed air and cutting oxygen pushing against the surrounding water.

As the depth at which the cutting is being done increases, the gas pressures must be increased to overcome both the added water pressure and the frictional losses inherent in the use of longer hoses. Approximately 3.5 kPa (0.5 psig) for each 300 mm (12 in.) of depth must be added to the basic gas pressure requirements used in air for the thickness being cut.

Methylacetylene-propadiene (stabilized) gas and propylene are the best all-purpose preheat gases because they can be used at any depths to which divers can descend and perform satisfactorily. Acetylene must not be used at depths greater than approximately 20 ft (6 m) because its maximum safe operating pressure is 100 kPa (15 psig).

The underwater oxyfuel gas cutting torch will sever steel plate in thicknesses from 13 mm (1/2 in.) to approximately 101 mm (4 in.) with no great difficulty. However, the constant quenching effect of the surrounding water lowers the efficiency of preheating in metal less than 13 mm (1/2 in.) thick. Thin materials require much larger preheating flames and preheating gas flows. The orifice size for the cutting oxygen is considerably larger for underwater cutting than for cutting in air. A special apparatus for lighting the preheat flames under water is also needed.

Some manufacturers have developed a spacing sleeve to be used for underwater cutting with a standard

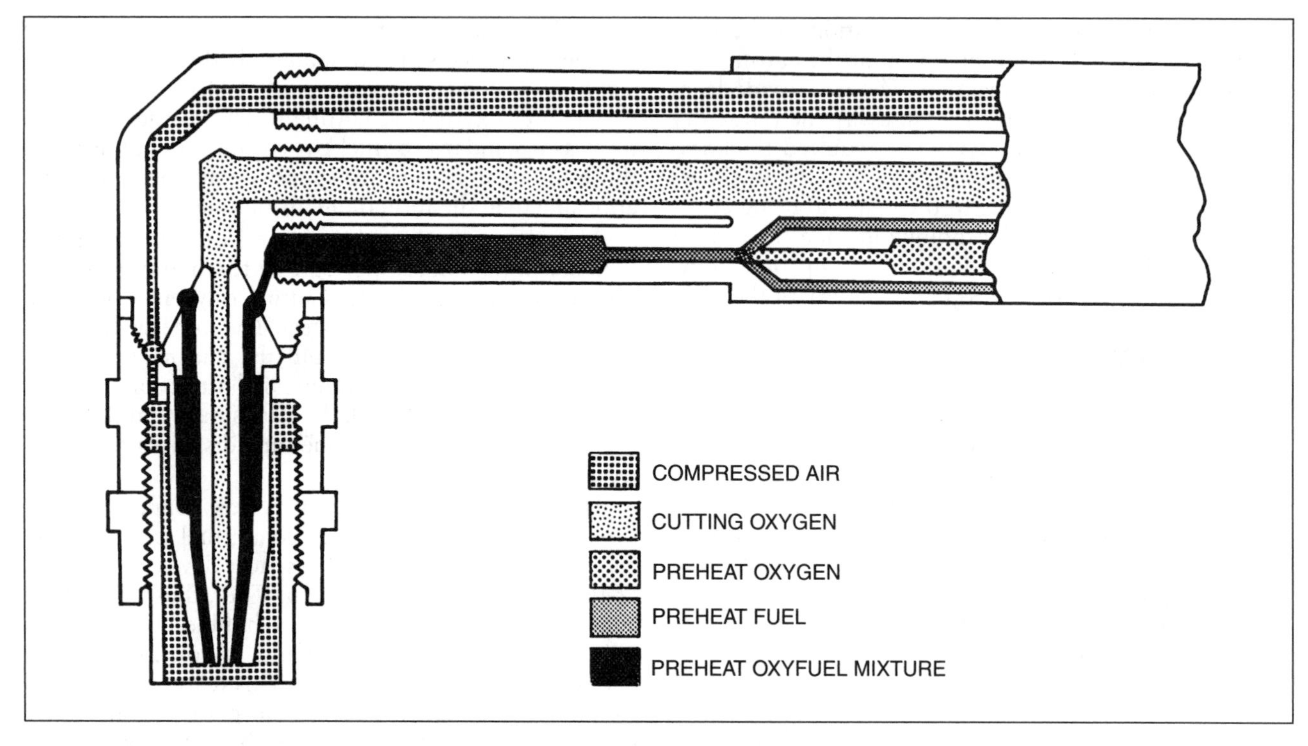

Figure 14.20—Basic Design of an Underwater Oxyfuel Gas Cutting Torch

cutting torch. This device clamps over the cutting tip and provides a guide for the proper tip-to-workpiece distance. A source of compressed air is not required for this unit.

The recommendations of the manufacturer must be followed when setting up and operating underwater oxyfuel gas cutting equipment.

APPLICATIONS

Oxyfuel gas cutting is widely used for the severing of carbon and low-alloy steel as well as other iron alloys and some nonferrous metals such as titanium. The process is not generally applicable to stainless steels, high-alloy steels, nickel, and cast iron because these metals do not readily oxidize and provide the degree of heat needed to sustain the cutting operation. For the cutting of high-alloy and stainless steel, it is necessary to use one of the oxyfuel gas cutting special techniques described previously, another OC process such as flux cutting or metal powder cutting, or one of the arc cutting processes. For the same reason, preheating or postheating, or both, are often necessary to accommodate increased percentages of carbon and alloys in the metal to be cut. The cutting of carbon steel and low-alloy steels, cast irons, and oxidation-resistant steels are common applications.

Figure 14.21 is an example of automatic oxyfuel gas cutting, showing a shape-cutting machine with six torches cutting precision parts from steel plate. Oxyfuel gas cutting is commonly used in steel mills. One application is the mechanized cutting of billets, as shown in Figure 14.22; another is the oxygen lance cutting of cast steel, shown in Figure 14.23. Figure 14.24 shows a manual cut on pipe.

CARBON AND LOW-ALLOY STEELS

Carbon steels are readily cut with the oxyfuel gas cutting processes. Low-carbon steels are cut without difficulty using standard procedures. Typical data for the cutting of low-carbon steel using commonly available fuel gases are presented in Table 14.2. The gas flow rates and cutting speeds listed are to be considered as guides for determining more precise settings for a particular job. When a new material is to be cut, a few trial cuts should be made to obtain the most efficient operating conditions. It should be noted that Table 14.2 pre-

Photograph courtesy of Messer-MG Systems & Welding

Figure 14.21—Precision Plate Cutting with Six-Torch Automatic Oxyfuel Gas Cutting Machine

Figure 14.22—Mechanized Oxyfuel Gas Cutting of Billets in a Steel Mill

Figure 14.23—Oxygen Lance Cutting of Cast Steel

Photograph courtesy of Victor Equipment Company

Figure 14.24—Manual Oxyfuel Gas Cutting of Pipe

sents data on thicknesses up to at 300 mm (12 in.), the maximum thickness normally encountered when shape cutting in production shops. The cutting of steel plate over approximately 300 mm (12 in.) thick is considered heavy cutting, discussed in the section "Range of Thicknesses Cut."

Alloying elements have two possible effects on the oxyfuel gas cutting of steel. They may make the steel more difficult to cut, or they may give rise to hardened or heat-checked cut surfaces, or both. The effects of alloying elements are summarized in Table 14.3.

A large quantity of heat energy is liberated in the kerf when steel is cut with the oxygen jet. Much of this energy is transferred to the sides of the kerf, where it raises the temperature of the steel adjacent to the kerf above its critical temperature. Since the torch is moving forward, the source of heat quickly moves along. The mass of cold metal near the kerf acts as a quenching medium, rapidly cooling the hot steel. This quenching action may harden the cut surfaces of high-carbon and alloy steels.

The depth of the heat-affected zone depends on the carbon and alloy contents, the thickness of the base metal, and the cutting speed employed. The hardening of the heat-affected zones of steels containing up to 0.25% carbon is not critical in the thicknesses usually cut. Higher carbon steels and some alloy steels are hardened to a degree that the thickness may become critical.

Typical depths of the heat-affected zones in oxygen-cut steel are shown in Table 14.4. For most applications of oxyfuel gas cutting, the affected metal need not be removed. However, if it is removed, removal should be by mechanical means.

CAST IRONS

The high-carbon content of cast irons resists the ordinary oxyfuel gas cutting techniques used to cut low-carbon steels. Cast irons contain some of the carbon in the form of graphite flakes or nodules and some in the form of iron carbide (Fe_3C). Both of these constituents hinder the oxidation of the iron. High-quality production cuts typically made in steels cannot be obtained in cast irons, therefore most cutting involving cast irons is carried out to remove risers, gates, or defects, to repair or alter castings, or for scrapping.

Cast irons can usually be cut manually by using an oscillating motion of the cutting torch, as shown in Figure 14.25. The degree of motion depends on the section thickness and carbon content. Torch oscillation helps the oxygen jet to blow the slag and molten metal out of the kerf. The kerf is normally wide and rough.

A larger cutting tip with greater preheat than that used for steel is required for cutting the same thickness of cast iron. A hot carburizing flame is used, with the streamer extending to the far side of the cast iron section. The burning of the excess fuel gas helps to maintain preheat in the kerf.

Sometimes cast irons can be cut using the special techniques for cutting oxidation-resistant steels. However, cast irons are readily cut with the air carbon arc cutting (CAC-A) and plasma arc cutting (PAC) processes, and these processes are frequently preferred over oxyfuel gas cutting.

OXIDATION-RESISTANT STEELS

The absence of alloying materials permits the oxidation reaction to proceed rapidly in pure iron. As the quantity and number of alloying elements in iron increase, the oxidation rate decreases from that of pure iron. Cutting thus becomes more difficult.

The oxidation of the iron in any alloy steel liberates a considerable amount of heat. The iron oxides produced have melting points near that of iron. However, the oxides of many of the alloying elements in steels (e.g., aluminum and chromium) have melting points higher than those of iron oxides. These high-melting-point oxides, which are refractory in nature, may shield the material in the kerf so that fresh iron is not continuously exposed to the cutting-oxygen stream. Thus, the quality and speed of cutting decreases in proportion to increased amounts of refractory oxide-forming elements in the iron.

For ferrous metals with high alloy content, such as stainless steel, the use of plasma arc cutting and in some

Table 14.2
Data for the Manual and Mechanized Cutting of Clean Low-Carbon Steel without Preheat

Inch-Pound Units

Thickness of Steel, in.	Diameter of Cutting Orifice, in.	Cutting Speed, in./min.	Gas Flow, ft^3/h				
			Oxygen	Acetylene	MPS	Natural Gas	Propane
1/8	0.020–0.040	16–32	15–45	3–9	2–10	9–25	3–10
1/4	0.030–0.060	16–26	30–55	3–9	4–10	9–25	5–12
3/8	0.030–0.060	15–24	40–70	6–12	4–10	10–25	5–15
1/2	0.040–0.060	12–23	55–85	6–12	6–10	15–30	5–15
3/4	0.045–0.060	12–21	100–150	7–14	8–15	15–30	6–18
1	0.045–0.060	9–18	110–160	7–14	8–15	18–35	6–18
1-1/2	0.060–0.080	6–14	110–175	8–16	8–15	18–35	8–20
2	0.060–0.080	6–13	130–190	8–16	8–20	20–40	8–20
3	0.065–0.085	4–11	190–300	9–20	8–20	20–40	9–22
4	0.080–0.090	4–10	240–360	9–20	10–20	20–40	9–24
5	0.080–0.095	4–8	270–360	10–25	10–20	25–50	10–25
6	0.095–0.105	3–7	260–500	10–25	20–40	25–50	10–30
8	0.095–0.110	3–5	460–620	15–30	20–40	30–55	15–32
10	0.095–0.110	2–4	580–700	15–35	30–60	35–70	15–35
12	0.110–0.130	2–4	720–850	20–40	30–60	45–95	20–45

SI Units

Thickness of Steel, mm	Diameter of Cutting Orifice, mm	Cutting Speed, mm/s	Gas Flow, L/min				
			Cutting Oxygen	Acetylene	MPS	Natural Gas	Propane
3.2	0.51–1.02	6.8–13.5	7.2–21.2	2–4	2–4	4–12	2–5
6.4	0.76–1.52	6.8–11.0	14.2–26.0	2–4	2–5	4–12	2–6
9.5	0.76–1.52	6.4–10.1	18.9–33.0	3–5	2–5	5–12	3–7
13	1.02–1.52	5.1–9.7	26.0–40.0	3–5	2–5	7–14	3–8
19	1.14–1.52	5.1–8.9	47.2–70.9	3–6	3–5	7–14	3–9
25	1.14–1.52	3.8–7.6	51.9–75.5	4–7	4–7	8–17	4–9
38	1.52–2.03	2.5–5.9	51.9–82.6	4–8	4–8	9–17	4–10
51	1.52–2.03	2.5–5.5	61.4–89.6	4–8	4–8	9–19	4–10
76	1.65–2.16	1.7–4.7	89.6–142	4–9	4–10	10–19	5–11
102	2.03–2.29	1.7–4.2	113–170	5–10	4–10	10–19	5–11
127	2.03–2.41	1.7–3.4	127–170	5–10	5–10	12–24	5–12
152	2.41–2.67	1.3–3.0	123–236	5–12	5–12	12–24	6–19
203	2.41–2.79	1.3–2.1	217–293	7–14	10–19	14–30	7–15
254	2.41–2.79	0.85–1.7	274–331	7–17	10–19	16–33	7–15
305	2.79–3.30	0.85–1.7	340–401	9–19	15–29	20–75	10–22

Notes:

1. Preheat oxygen consumptions: Preheat oxygen for acetylene = 1.1 to 1.25 × acetylene flow m^3/h (ft^3/h); preheat oxygen for natural gas = 1.5 to 2.5 × natural gas flow m^3/h (ft^3/h); preheat oxygen for propane = 3.5 to 5 × propane flow m^3/h (ft^3/h).
2. Operating notes: Higher gas flows and lower speeds are generally associated with manual cutting, whereas lower gas flows and higher speeds apply to mechanized cutting. When cutting heavily scaled or rusted plate, high gas flow and low speeds should be used. Maximum indicated speeds apply to straight-line cutting; for intricate shape cutting and best quality, lower speeds are required.

Table 14.3
Effect of Alloying Elements on the Resistance of Steel to Oxyfuel Gas Cutting

Element	Effect
Carbon	Steels up to 0.25% carbon can be cut without difficulty. Steels with a higher carbon content should be preheated to prevent hardening and cracking. Graphite and cementite (Fe3C) are detrimental, but cast irons containing up to 4% carbon can be cut using special techniques.
Manganese	Steels containing approximately 14% manganese and 1.5% carbon are difficult to cut and should be preheated for best results.
Silicon	Silicon, in the amounts usually present, has no effect. Transformer irons containing as much as 4% silicon are cut. Silicon steel containing large amounts of carbon and manganese must be carefully preheated and post-annealed to avoid air hardening and possible surface fissures.
Chromium	Steels with up to 5% chromium are cut without much difficulty when the surface is clean. Higher chromium steels, such as 10% chromium steels, require special techniques (see the section "Oxidation-Resistant Steels"), and the cuts are rough when the usual oxyacetylene cutting process is used. In general, carburizing preheat flames are desirable when cutting this type of steel. In addition, the flux cutting and metal powder cutting processes enable cuts to be made in the common straight chromium irons and steels as well as in stainless steel.
Nickel	Steels containing up to 3% nickel may be cut using the conventional oxyfuel gas cutting processes; cuts are satisfactory in steels up to about 7% nickel content.
Molybdenum	This element affects cutting about the same as chromium. Aircraft-quality chrome-molybdenum steel presents no difficulty. High molybdenum-tungsten steels, however, may be cut only with special techniques.
Tungsten	The usual alloys with up to 14% tungsten may be readily cut, but cutting is difficult with a higher percentage of tungsten. The limit is approximately 20% tungsten.
Copper	Copper has no effect in amounts up to about 2%.
Aluminum	The effect of aluminum is not appreciable unless present in large amounts (on the order of 10%).
Phosphorus	This element has no effect in the amounts usually tolerated in steel.
Sulfur	Small amounts such as are present in steels have no effect. With higher percentages of sulfur, the rate of cutting is reduced and sulfur dioxide fumes are noticeable.
Vanadium	In the amounts usually found in steels, this alloy may improve rather than interfere with cutting.

Table 14.4
Approximate Depths of Heat-Affected Zones in Steels Cut with Oxyfuel Gas Cutting*

Steel Thickness		Depth of Heat-Affected Zone			
		Low-Carbon Steels		High-Carbon Steels	
mm	in.	mm	in.	mm	in.
Under 13	Under 1/2	Under 0.8	Under 1/32	0.8	1/32
13	1/2	0.8	1/32	0.8 to 1.6	1/32 to 1/16
152	6	3.2	1/8	3.2 to 6.4	1/8 to 1/4

*The depth of the fully hardened zone is considerably less than the depth of the heat-affected zone.

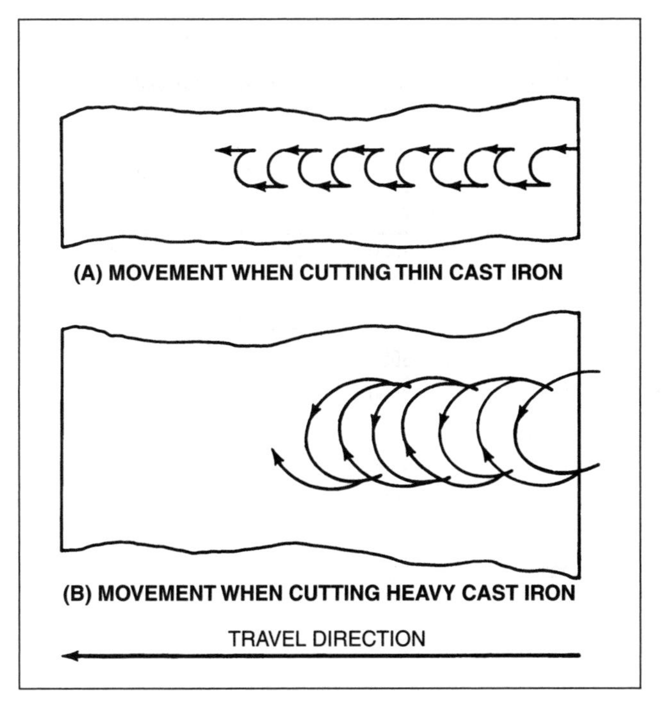

Figure 14.25—Typical Cutting-Torch Manipulation for the Cutting of Cast Irons

cases air carbon arc cutting should be considered. If these options are not available or practical, special oxygen cutting techniques should be considered.

RANGE OF THICKNESSES CUT

Oxyfuel gas cutting is used to cut a wide range of steel thicknesses from approximately 3 mm to 2100 mm (1/8 in. to 84 in.). Thicknesses over approximately 500 mm (20 in.) are not generally cut except in steel mill operations, where the steel is cut while still at high temperatures. Cutting machines capable of cutting to tolerances of 0.8 mm to 1.6 mm (1/32 in. to 1/16 in.) are used to produce components that can be assembled into final product form without intermediate machining. These cutting machines can also be used for rapid material removal prior to machining to close tolerances.

Light Cutting

The cutting of metal less than 3.2 mm (1/8 in.) is difficult with the OFC processes. There is an insufficient mass of metal adjacent to the kerf to absorb the heat from the preheat flame and subsequent oxidation to prevent the melting of the kerf edges. Thicknesses up to 6.4 mm (1/4 in.) can be cut by using the lowest preheat allowable, and where possible, by angling the torch forward by as much as 45° in the direction to be cut.

Stack cutting, plasma arc cutting, or laser beam cutting are often the preferred methods for these thin materials.

Medium Cutting

The oxyfuel gas cutting process works very well on medium material thicknesses, from 6.4 mm (1/4 in.) to 225 mm (10 in.). Clean, sharp, top and bottom corners with few surface defects can be readily obtained. Quality cuts up to about 102 mm (4 in.) can be obtained using manual, mechanized, or automated torches. Manual torches are usually used for scrapping or material removal at thicknesses greater than these.

The number of torches incorporated in mechanized or automated systems may range from two to twelve on a single machine. The speed and quality of the cuts made with multiple torches may equal or exceed that of plasma arc cutting or laser beam cutting, usually with less investment.

Heavy Cutting

Heavy cutting, the cutting of steel over approximately 300 mm (12 in.) thick, is performed in a wide variety of operations such as ingot cropping, scrap cutting, and riser cutting. The basic reactions that permit the oxygen cutting of thick steel are the same as those for cutting thinner sections. Thicknesses ranging from 300 mm to 1525 mm (12 in. to 60 in.) may be cut using heavy-duty torches. In heavy cutting, preheat and cutting oxygen flows increase and cutting speed decreases in direct relation to the thickness of the material to be cut.

The most important factor in heavy cutting is oxygen flow. The flow of cutting oxygen is controlled by tip size and operating pressure, which must be adjusted to provide the flow required for the thickness being cut. Oxygen cutting pressures in the range of 70 kPa to 380 kPa (10 psig to 55 psig), measured at the cutting torch, are usually adequate for the heaviest cutting when using the appropriate tip size and equipment. The oxygen flow at the torch entry is of paramount importance.

It is possible by experimentation to arrive at an approximate oxygen flow requirement that can be useful as a guide in selecting equipment suitable for a given job. These requirements may vary, but in terms of thickness, they usually fall within the approximate range of 89 liters per millimeter (L/mm) to 139 L/mm of oxygen (80 ft^3/in. to 125 ft^3/in.) per mm (in.) of thickness.

Table 14.5 presents the range of operating conditions that cover normal heavy-cutting operations. It should

Table 14.5
Typical Cutting Action Data for the Oxyfuel Gas Cutting of Thick Low-Carbon Steel

Material Thickness		Cutting Oxygen					
		Orifice Diameter		Flow Rate		Pressure at Torch	
mm	in.	mm	in.	L/min	ft³/h	kPa	psig
305	12	3.74–5.61	0.147–0.221	472–708	1000–1500	386–228	56–33
406	16	4.32–7.36	0.170–0.290	614–944	1300–2000	372–172	54–25
508	20	4.93–8.44	0.194–0.332	803–1180	1700–2500	359–152	52–22
610	24	5.61–8.44	0.221–0.332	944–1416	2000–3000	331–200	48–29
711	28	6.35–9.53	0.250–0.375	1087–1652	2300–3500	283–179	45–26
813	32	6.35–9.53	0.250–0.375	1274–1888	2700–4000	352–207	40–30
914	36	7.37–10.72	0.290–0.422	1416–2120	3000–4500	276–179	40–26
1016	40	7.37–10.72	0.290–0.422	1605–2360	3400–5000	317–207	40–30
1118	44	7.37–11.90	0.290–0.468	1792–2600	3800–5500	352–179	40–26
1219	48	8.44–11.90	0.332–0.468	1888–2830	4000–6000	276–193	40–20

be noted that the data presented in this table may not be entirely suitable for all heavy-cutting operations, but can be used as a guide in selecting the correct equipment and operating conditions. Although these values have been used successfully, the actual values for the most efficient operation of a specific cutting application are always best established by trial cuts. When heavy cutting is performed with the torch in a horizontal position, it may be necessary to increase the cutting-oxygen pressure to aid in removing slag from the kerf.

Although recommended travel speeds are not included in Table 14.5, speeds from 0.85 mm/s to 2.5 mm/s (2 in./min to 6 in./min) are used in the range of thicknesses covered. A speed of 1.3 mm/s (3 in./min) is possible for thicknesses up to at least 910 mm (36 in.). The correct speed is obtained by observing the operating conditions carefully and making suitable adjustments while the cutting operation is in progress.

Because heavy workpieces usually have a scale-covered surface, the technique used to start the cut differs from that used on clean, thin material. The start is made more slowly on the rougher edges. Figure 14.26 indicates correct and incorrect starting procedures. Figure 14.26(A) shows the desirable starting position with the preheat flames on the top corner and extending down the face of the material. The cutting reaction begins at the top corner. It proceeds down the face of the material to the bottom as the torch moves forward. Figures 14.26(B), (C), (D), (E), and (F) illustrate difficulties that arise if incorrect procedures are used.

When the cut proceeds properly with the correct oxygen flow and forward speed, the reaction proceeds to the end of the cut without leaving a skipped corner. Figure 14.27 illustrates various correct and incorrect terminating conditions and the proper drag conditions. The correct conditions that produce a drop cut (a complete severing of the workpiece) are depicted in Figure 14.27(A).

Overall, the following conditions are required for successful heavy cutting on a production basis:

1. Adequate gas supply to complete the cut, considering that a lost cut on heavy materials is extremely difficult, if not impossible, to restart;
2. Equipment of sufficient structural size to maintain rigidity and carry the equipment;
3. Equipment of sufficient capacity to handle the range of speeds and gas flows required; and
4. Skilled personnel trained in proper heavy cutting techniques.

QUALITY

The determination of acceptable quality depends on the requirements of the specific job. Salvage operations and severing members for scrap do not require high-quality cutting. For these applications, the oxygen cutting processes are used to complete the operation rapidly with little regard to the quality of the cut surfaces. When the cut materials are used in fabrications

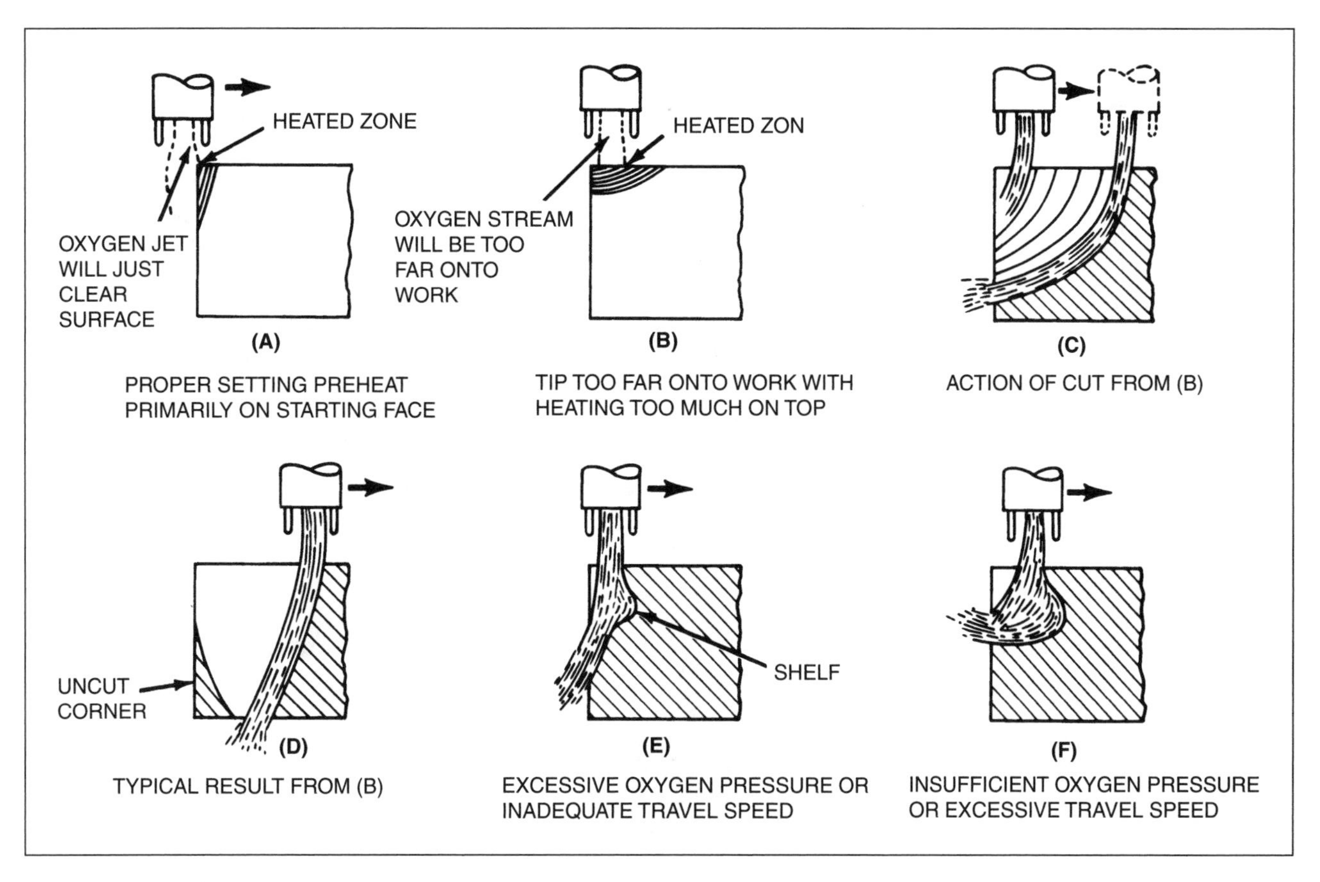

Figure 14.26—Correct and Incorrect Starting Procedures for Heavy Cutting

with no other processing of the cut surfaces, the quality requirements of the surfaces may be significant.

Close control of these factors is generally confined to mechanized and automated oxyfuel gas cutting. Good control of torch position, initiation of the cut, travel speed, and template stability are required for high-quality cutting, as are consistent cleanliness and maintenance of the equipment.

Given the proper equipment in good condition and a reasonably clean and well-supported workpiece, a well-trained operator can cut shapes to tolerances of 0.8 mm to 1.6 mm (1/32 in. to 1/16 in.) from material not more than 51 mm (2 in.) thick. The correct cutting tip, preheat flame adjustment, cutting oxygen pressure and flow, and travel speed must be used.

Regardless of operating conditions, the phenomena known as *drag lines* are inherent in oxygen cutting. These are the lines that appear on the cut surface, as shown in Figure 14.28, resulting from the manner in which the iron oxidizes in the kerf. The amount of drag is important. Light drag lines on the cut surface are not considered detrimental. If drag is too great, the corner at the end of the cut may not be completely severed, and the cut part will not drop.

Cut surface quality is dependent on many variables. The most significant are following:

1. Type of steel;
2. Thickness of the material;
3. Quality of steel (lack of segregations, inclusions, and other discontinuities);
4. Condition of the steel surface;
5. Intensity of the preheat flames and the preheat oxyfuel gas ratio;
6. Size and shape of the cutting-oxygen orifice;
7. Purity of the cutting oxygen;
8. Cutting-oxygen flow rate;
9. Cleanliness and flatness of the exit end of the nozzle; and
10. Cutting speed.

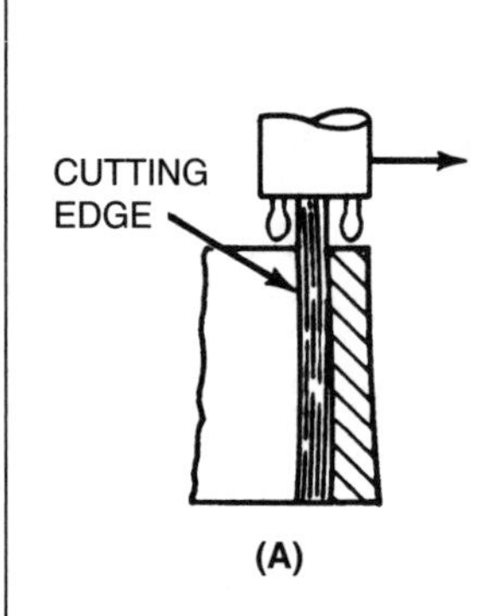

NO DRAG PERMITS STREAM TO BREAK THROUGH FACE UNIFORMLY AT ALL POINTS. TYPICAL OF BALANCED CONDITIONS

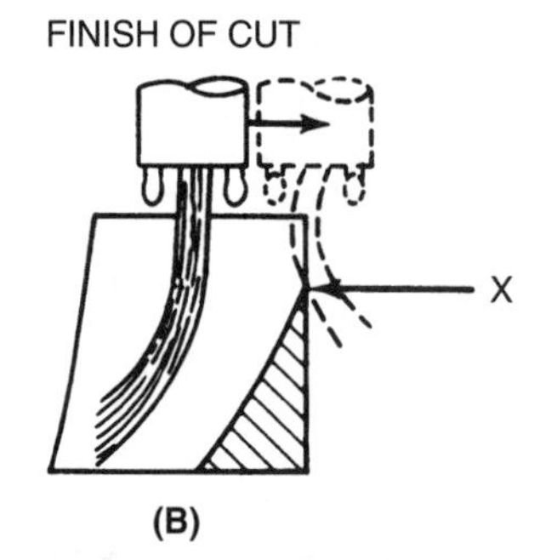

DRAG CAUSES ACTION TO CARRY THROUGH AT X AND TO PASS BEYOND MATERIAL, LEAVING UNCUT CORNER. TYPICAL OF INSUFFICIENT OXYGEN OR EXCESSIVE SPEED

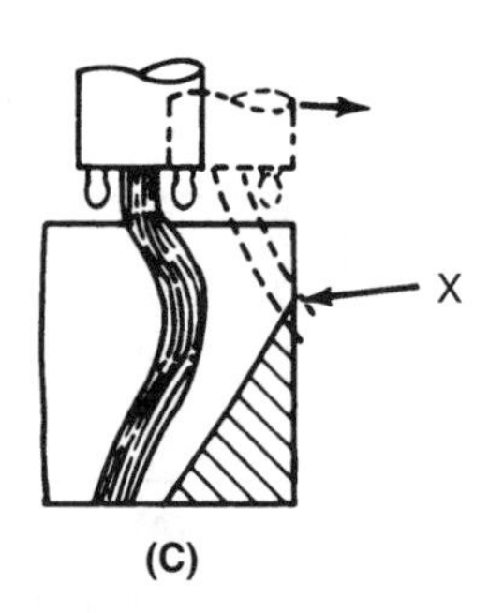

FORWARD DRAG CAUSES STREAM TO BREAK THROUGH AT X AND BECOME DEFLECTED, LEAVING UNCUT CORNER. TYPICAL OF HIGH CUTTING OXYGEN PRESSURE OR TOO LITTLE SPEED

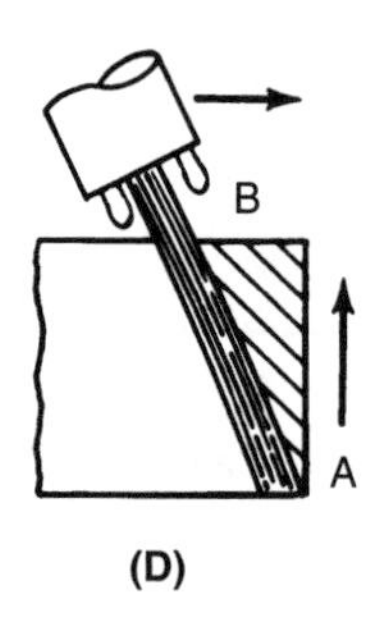

IF CUTTING FACE SUCH THAT BREAKTHROUGHT AT BOTTOM, A, LIES AHEAD OF ENTRY POINT, B, AND AT NO POINT DOES FACE EXTEND BEYOND A, ACTION WILL SEVER FROM A UPWARD

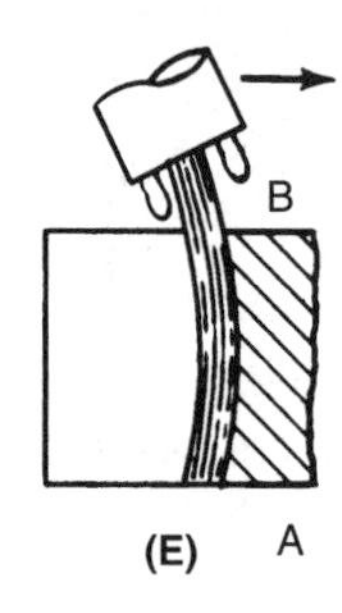

ANGULAR TIP DISPOSITION SIMILAR TO (D) SHOWING LIMIT OF EFFECTIVENESS, SIMILAR TO (A)

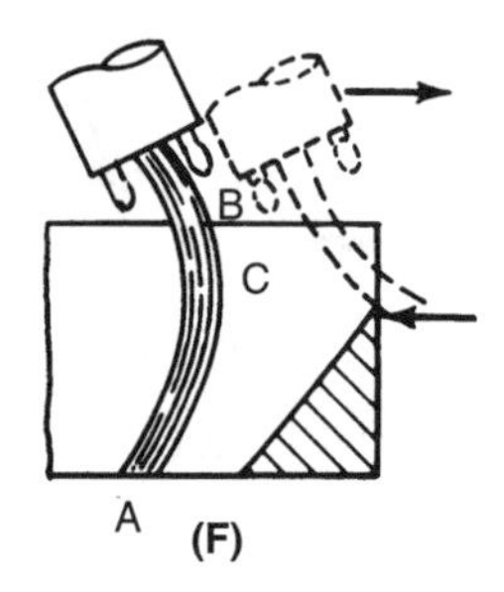

IF CONDITIONS ARE SUCH THAT A AND B ARE IN LINE OR OTHERWISE DISPOSED, BUT C LIES AHEAD OF A. STREAM WILL BREAK AT X, LEAVING UNCUT CORNER, SIMILAR TO (C)

Notes:
1. Torch held vertically in (A), (B), and (C).
2. Torch angled in direction of cutting in (D), (E), and (F).

Figure 14.27—Correct and Incorrect Terminating Conditions for Heavy Cutting

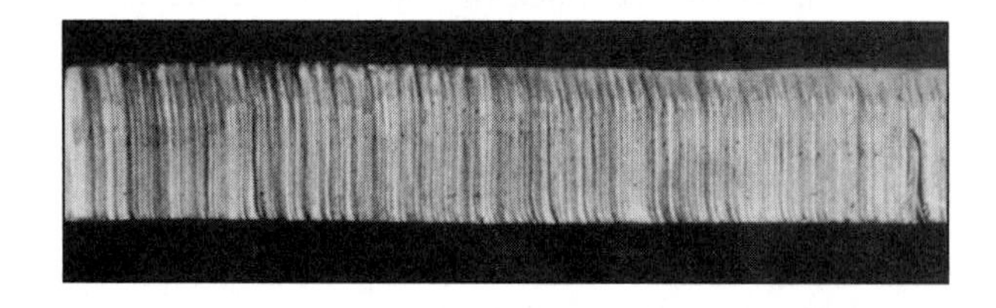

Figure 14.28—Drag Lines on the Kerf Wall Resulting From Oxygen Cutting

For any given cut, these variables should be evaluated so that the required cut quality is obtained with the minimum aggregate cost incurred in oxygen, fuel gas, labor, and overhead.

Figure 14.29 illustrates typical edge conditions that result from variations in the cutting procedure for material of uniform type and thickness.

As dimensional tolerance and surface roughness are somewhat interdependent, they must be considered together when judging the quality of a cut. Most speci-

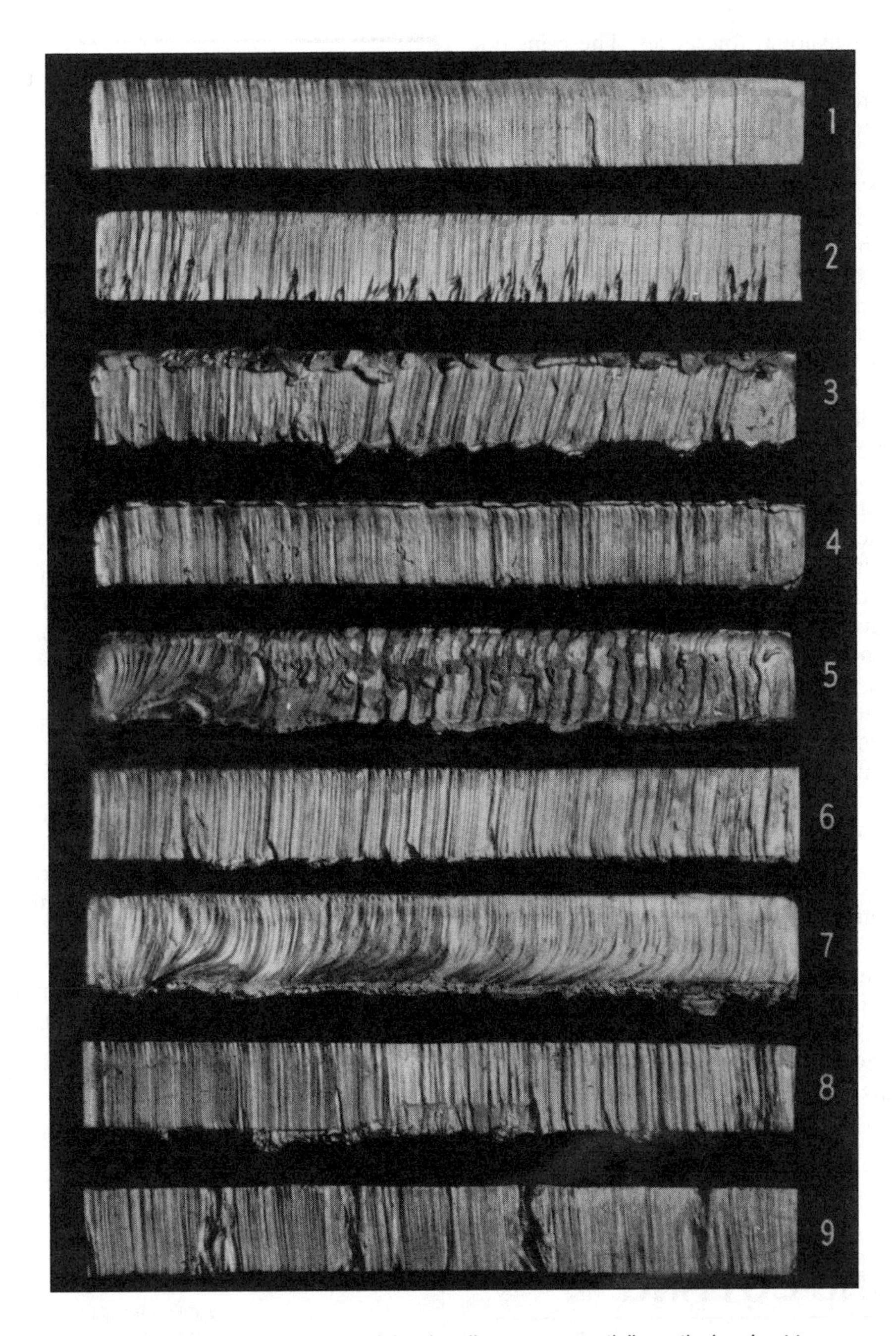

Notes:

1. Example of a good cut—the edge is square and the drag lines are essentially vertical and not too pronounced.
2. Preheat flames were too small and the cutting speed was too slow for this cut, causing unacceptable gouging at the bottom.
3. Preheating flames were too long, causing melting of the top surface and an irregular cut edge with an excessive amount of adhering slag.
4. Oxygen pressure was too low and cutting speed too slow, causing melting of the top edge.
5. Oxygen pressure was too high and the nozzle size too small, resulting in loss of control of the cut.
6. Cutting speed was too slow, increasing the irregularties of the drag lines.
7. Cutting speed was too fast, resulting in a pronounced break in the dragline and an irregular cut edge.
8. Torch travel speed was unsteady, resulting in a wavy and irregular cut edge.
9. Cut was lost and not carefully restarted, causing unacceptable gouges at the restarting point.

Figure 14.29—Typical Edge Conditions Resulting From Technique Variations in Manual Oxyfuel Gas Cutting of 25 mm (1 in.) Steel Plate

fications include dimensional tolerances. These include straightness of edge, squareness of edge, and permissible variation in plate width. All of these are primarily a function of the cutting equipment and its mechanical operation.

When the torch is held rigidly and advanced at a constant speed, as in mechanized oxyfuel gas cutting, dimensional tolerances can be maintained within reasonable limits. The degree of longitudinal precision of a mechanized cut depends primarily on such factors as the condition of the equipment, trueness of the guide rails, clearances in the operating mechanism, and the uniformity of control of the drive speed. In addition to equipment, dimensional accuracy is dependent on the control of thermal expansion of the material being cut. A lack of dimensional tolerance may result from the buckling of the material (thin plate or sheet). Warping may result from the heat being applied to one edge or the workpiece shifting while it is being cut.

The oxyfuel gas cutting operation should be planned carefully to minimize the effect of the variables on dimensional accuracy. For instance, when trimming opposite edges of a plate, warping is minimized if both cuts are made simultaneously in the same direction. Distortion can often be controlled when cutting irregular shapes from plates by inserting wedges in the kerf, following the cutting torch, to limit movement of the metal during thermal expansion and contraction. When cutting openings in the middle of a plate, distortion may be limited by making a series of unconnected cuts. The section is left attached to the plate in a number of places until cutting is almost completed, then connecting locations are finally cut though. This intermittent cutting results in somewhat reduced cut quality.

Thin material may be cut in stacks to eliminate warping and buckling. Another technique is to cut thin plate while it is partially submerged in water to remove the heat.

OXYGEN LANCE CUTTING

Oxygen lance cutting (OLC) is an oxygen cutting process that uses oxygen supplied through a consumable steel pipe or lance to produce the cut. The preheat required to start the cutting operation is obtained by other means. Oxygen lance cutting is used to cut refractory brick and mortar and remove slag. It has also been used to open furnace tap holes and to remove solidified material from vessels, ladles, and molds.

The earliest version of oxygen lance cutting employed a plain black iron pipe as a lance through which oxygen was blown. In a typical contemporary system, an oxyfuel gas cutting or welding torch is used to heat the cutting end of an iron pipe lance to a cherry red, at which point the oxygen flow is initiated. The pipe burns in a self-sustaining exothermic reaction, and the heating torch is removed. When the burning end of the lance is brought close to the workpiece, the heat of the flame melts the work. Oxygen lance cutting is illustrated in Figure 14.30.

Arc-started oxygen lancing, a variation of the oxygen lance cutting process, uses an arc to start the iron-oxygen reaction. This equipment utilizes tubes that are typically 450 mm (18 in.) long and either 6.4 mm or 9.5 mm (0.25 in. or 0.375 in.) in diameter. A 12-volt battery can be used as a power source, with the cutting tube connected to one battery terminal and a copper striker plate connected to the other. To start the burning operation, the operator initiates the oxygen flow and draws the steel tube across the copper plate at a 45° angle. Sparking at the copper plate ignites the tube. The burning rod can then be used for cutting, piercing, or beveling steel. It can also be used to remove pins, rivets, and bolts.

OXYGEN ARC CUTTING

Oxygen arc cutting (OAC) is an oxygen cutting process that produces a cut by means of an arc struck between the workpiece and a consumable tubular electrode through which oxygen is directed to the workpiece.[8] Oxygen arc cutting, although classified by the American Welding Society as an oxygen cutting process, is discussed in Chapter 15, "Arc Cutting and Gouging," because of its equally close relationship to the arc cutting processes.

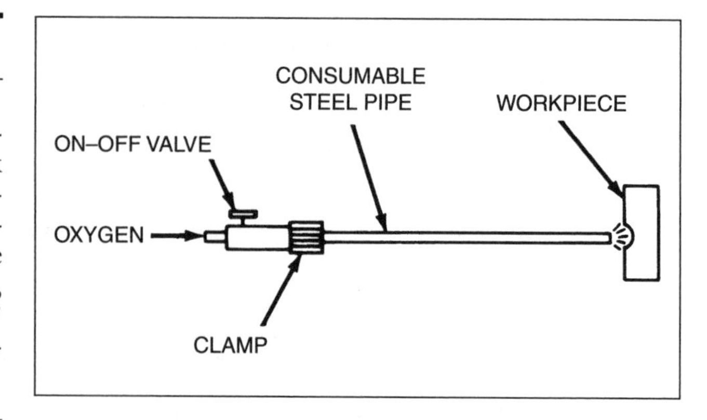

Figure 14.30—Oxygen Lance Cutting

8. See Reference 2, p. 27.

Figure 14.31—Multiple Holes Pierced by an Oxygen Lance to Sever a Cast Iron Roll 1 m (40 in.) in Diameter

Figure 14.32—Holes Pierced In a Cast Iron Roll Using an Oxygen Lance

APPLICATIONS

Oxygen lance cutting can be used to pierce virtually all materials. It has been used successfully on aluminum, cast iron, steel, and reinforced concrete.

The oxygen lancing of a 1 m (40 in.) diameter cast iron roll used in a paper mill is shown in Figure 14.31. The variable angle bracket shown in Figure 14.31 helped in guiding the lance. The cutting oxygen was supplied at 550 kPa to 870 kPa (80 psig to 120 psig). The holes pierced in the roll are shown in Figure 14.32.

A more sophisticated version of the lance involves a number of low-carbon steel wires packed into the steel tube. This increases the cutting life and capability of the lance. Commercially available tubes are typically 3.2 m (10-1/2 ft) long and 16 mm (0.625 in.) in diameter. Using this variation, a hole 63 mm (2-1/2 in.) in diameter can be made in 600 mm (24 in.) of reinforced concrete at a rate of about 100 mm/min (4 in./min). This operation would consume approximately 1.7 m^3 (60 ft^3) of oxygen.

The oxygen lance cutting process can be employed underwater. The lance must be lighted before it is placed underwater, but then piercing proceeds essentially as in air. One limitation of underwater oxygen lance cutting is that the violent bubbling action produced restricts visibility.

METAL POWDER CUTTING

Metal powder cutting is an oxygen cutting process that uses the heat produced by an oxyfuel gas flame along with iron powder or another metal powder to aid in cutting. The powdered metal accelerates and propagates the oxidation reaction and also the melting and spalling action of hard-to-cut and oxidation-resistant materials.

In this process, the metal powder is directed into the kerf through either the cutting tip or single or multiple jets external to the tip. In the first method, gas-conveyed powder is introduced into the kerf through special orifices in the cutting tip. In the second method, powder is introduced via the external jets; the gas conveying the powder imparts sufficient velocity to the powder particles to carry them through the preheat envelope into the cutting-oxygen stream. The short time required for the powder particles to pass through the preheat envelope is sufficient to produce the desired reaction in the cutting zone.

The metal powder reacts chemically with the refractory oxides produced in the kerf and increases their fluidity. The resultant molten slags are washed out of the reaction zone by the oxygen jet. Fresh metal surfaces are continuously exposed to the oxygen jet and powder. Iron powder and mixtures of metallic powders, such as iron and aluminum, are used.

The cutting of oxidation-resistant steels using the metal powder cutting method can be performed at approximately the same speeds as oxyfuel gas cutting of carbon steel of equivalent thicknesses. The cutting oxygen flow must be slightly higher with the metal powder process. This process has been largely replaced by the plasma arc cutting process and is only used in specialized applications.

EQUIPMENT

Two general types of powder dispensers are used in the metal powder cutting process. One type of dispenser is a vibratory device in which the quantity of powder dispensed from a hopper is governed by a vibrator. Desired amounts of powder can be obtained by adjusting the amplitude of vibration. The vibratory-type dispenser is generally used when uniform and accurate powder flow is required. The other type of dispenser is a pneumatic device. An ejector or fluidizing unit is located in the bottom of a low-pressure vessel. The powder-conveying gas is brought into the dispenser in a manner that fluidizes the powder to make it flow uniformly into an ejector unit, where it is picked up by a gas stream that serves as the transporting medium to the torch.

In addition to the fuel and oxygen hoses, another hose is used to convey the powder to the torch. A special manual powder cutting torch mixes the oxygen and fuel gas and then discharges this mixture through multiple orifices in the cutting tip. The powder valve is an integral part of the torch. The cutting-oxygen lever on the torch also opens the powder valve in proper sequence. The powder carried by the conveying gas is brought through a separate tube into a chamber forward of the preheat gas chamber in the torch head. The powder then enters a separate group of passages in a two-piece cutting tip. From there, it discharges at the mouth of the tip in a conical pattern. The powder emerges with sufficient velocity to pass through the burning preheat gas and surrounds the central cutting-oxygen stream.

FLUX CUTTING

Flux cutting is an oxygen cutting process that uses the heat from an oxyfuel gas flame with a flux added to the flame to aid in making the cut. This process is primarily intended for the cutting of stainless steels. The flux is designed to react with oxides of alloying elements, such as chromium and nickel, to produce compounds with melting points approximating those of iron oxides.

Special apparatus is required to introduce the flux into the kerf. With the addition of flux, stainless steels can be cut at a uniform linear speed without torch oscillation. Cutting speeds approaching those for equivalent thicknesses of carbon steel can be attained. The tip sizes are larger and the oxygen flow is somewhat greater than those used in the cutting of carbon steels.

This process is limited to specialized applications, as it has largely been replaced by the plasma arc cutting process.

EQUIPMENT

A flux-feeding unit consisting of a dispenser designed to operate at normal oxygen pressures is required for the flux cutting process. The oxygen transports the flux through a hose from the dispenser to a conventional three-hose cutting torch. A mixture of oxygen and flux flows from the cutting-oxygen orifice of the torch tip. Special operating procedures are used to prevent the buildup of flux in the cutting-oxygen hose and the cutting torch.

ECONOMICS

The manual oxygen cutting process is often a sound economical choice, particularly when a limited number of cuts are to be made or the application does not require cuts of consistent high quality. Manual oxyfuel gas cutting equipment costs are much lower than those of machine tools or other cutting process equipment. Some shapes and material thicknesses that are difficult to cut by mechanical methods can be readily cut by an oxyfuel gas cutting process. Material handling equipment is usually not necessary.

Operator training is simple and inexpensive. Oxyfuel gas cutting operators usually start with manual oxyfuel gas cutting, which provides basic training, and with this skill they may progress to the more sophisticated mechanized oxyfuel gas cutting or the electric arc cutting processes.

Mechanized oxyfuel gas cutting is an economical method for straight-line, shape and bevel cutting. The process is particularly cost effective for beveling plate edges in preparation for bevel and groove welds.

The stack cutting of sheet or plate up to 13 mm (1/2 in.) can be a cost-saving technique when it is an advantage to cut several pieces at one time. However, the cost of material preparation, stack makeup, waster plates (if used), clamping devices, and operator skill requirements are pertinent factors in determining the economical advantages or limitations of stack cutting.

For high-volume production cutting, the use of multiple-torch cutting can offer significant economic advantages. The average cost per piece is primarily affected by labor. When two or more pieces are cut simultaneously, the incremental labor cost is spread over the total numbers of pieces cut. This economic benefit may be sacrificed at some point, however, as the number of pieces exceed the ability to efficiently remove them from the cutting table. At some point the material handling time will reduce the average torch-on time.

The most expensive supplies used in oxyfuel gas cutting are high-purity oxygen and acetylene, the fuel gas of choice; therefore, techniques that require less oxygen and the use of a less expensive fuel gas instead of acetylene are cost-saving options. These options must be carefully evaluated, however. For example, when MPS gas is used instead of acetylene, approximately twice as much oxygen is required to maintain a neutral preheat flame. When oxygen and natural gas are used, larger volumes of both are required for heating rates equivalent to oxygen and acetylene.

In mechanized cutting, natural gas, propane or similar fuels can be used as the preheating fuel in conjunction with a system of high-low pressure regulators, which results in less gas consumption.

Costs of preheating fuels can be evaluated by observing the time required for starting cuts, or piecing holes to start cuts, and the cutting speeds that can be maintained. Other variables that affect costs are the required quality of the cut, the quality and thickness of the material to be cut, the condition of the surface, the size and shape of the cutting-oxygen orifice, the purity of the oxygen, the intensity of preheat flames and the oxygen-to-fuel gas ratio. Gas consumption per unit of thickness can be calculated.

SAFE PRACTICES

As many of the safe practices that apply to oxyfuel gas welding systems also apply to oxygen cutting, the reader should refer to the "Safe Practices" section in Chapter 11 of this volume. Detailed recommendations are prescribed in the American National Standard *Safe Practices in Welding, Cutting, and Allied Processes*, ANSI Z49.1.[9]

Regulations established by the Occupational Health and Safety Administration (OSHA), U.S. Department of Labor, for the oxygen cutting processes are published in the standard *Oxygen-Fuel Gas Welding and Cutting*, Title 29, *Code of Federal Regulations (CFR)*, Part 1910.253, Subpart Q, and other sections referred to in the standard.[10]

Operators, supervisors and management personnel must ensure that these practices are followed, as well as the recommendations in the operating and safety manuals from equipment manufacturers and safety instructions in the Material Safety Data Sheets provided by gas suppliers.

SAFE HANDLING, USE, AND STORAGE OF CYLINDERS

Oxygen, acetylene and other fuel gas cylinders, including refrigerated gas cylinders, should be transported according to the U.S. Department of Transportation (DOT) specifications. Provisions for the safe use, handling, and storage of gas cylinders are stated in the American National Standards Institute document *Safety in Welding, Cutting and Allied Processes,* ANSI Z49.[11]

It is important to call oxygen and fuel gases by the correct names, to determine the specific hazards associated with each of the gases, and to follow appropriate procedures for handling and use. Oxygen vigorously accelerates combustion; improper use may result in fire or explosion. Oxygen should be used only for its intended purpose. It must never be used in place of compressed air in pneumatic tools, to clean equipment, or to provide ventilation.

Cylinders should be secured to prevent falling and should be stored in the vertical position, valve end up, at a location with reduced likelihood of being accidentally knocked over or hit by moving or falling objects. If the cylinder valve is ruptured as the result of a fall, the escaping gas can cause the cylinder to become a hazardous projectile. Oxygen and acetylene or other fuel gas cylinders should not be stored together. Empty cylinders should be marked as such and should not be stored with full cylinders. Cylinder caps should be in place when the cylinders are not in use.

Gas suppliers' Material Safety Data Sheets should be consulted, and their recommendations should be strictly followed. Safety and operating instructions from the manufacturer regarding attaching regulators, opening oxygen and fuel-gas valves, adjusting pressure, and lighting the torch should be carefully followed. Procedural details and instructions are also provided in

9. American National Standards Institute (ANSI) Accredited Standards Committee Z49, *Safety in Welding, Cutting, and Allied Processes*, ANSI Z49.1, Miami: American Welding Society.

10. Occupational Safety and Health Administration (OSHA), *Oxygen-Fuel Gas Welding and Cutting*, in *Code of Federal Regulations (CFR)*, Title 29, Part 1910.253, *Subpart Q*, Washington D.C.: Superintendent of Documents, U.S. Government Printing Office.

11. See Reference 8, pp. 26–27.

Operator's Manual for Oxyfuel Gas Cutting, AWS C4.2.[12]

It is important to store fuel gas cylinders in the upright position. When gas is taken from an acetylene cylinder that is lying on its side, acetone may be withdrawn along with the acetylene. This can contaminate the flame, resulting in loss of control of the cut or unacceptable cut quality. It can cause damage to the apparatus. If cylinders containing liquefied fuel gases are placed on their sides, liquid instead of vapor may be withdrawn from the cylinder. This could cause damage to the apparatus and produce a large, uncontrollable flame.

REGULATORS

Regulators should be maintained in good operating condition. Before attaching an oxygen regulator to a cylinder or pipe valve, the operator should assure that all connections are free of dust, dirt, oil, or grease.

Using regulators for purposes for which they are not designed can result in injury to personnel or damage to the equipment. Regulators should be used only the purpose for which they are designed and only for the specific gases for which they are designed. Station regulators should not be used on high-pressure cylinders. Regulators should never be modified and should never be forced onto a cylinder or pipeline valve.

HOSES AND FITTINGS

In the United States, oxygen hoses are usually green, and fuel gas hoses are red. However, manufacturers of hoses made for use outside the United States may not use these colors; therefore the labels printed on the hoses should be checked to verify that the correct hose is selected for use. Oxygen hoses must never be used for fuel gas, and fuel gas hoses must never be used for oxygen. The Rubber Manufacturers Association in their standard, *Specifications for Rubber Welding Hose*, RMA IP-7, provides specifications for hoses for oxygen cutting.[13]

PERSONAL PROTECTIVE EQUIPMENT

The appropriate protective clothing and equipment to be used for cutting operations vary with the nature and location of the work to be performed. Some or all of the following personal protective equipment may be required for oxygen cutting operations:

1. Tinted goggles or face shields with filter lens. The recommended filter lenses for various cutting operations are Shade No. 3 or 4 for light cutting (i.e., thicknesses up to 25 mm [1 in.]), Shade No. 4 or 5 for medium cutting (i.e., thicknesses of 25 mm to 150 mm [1 in. to 6 in.]), and Shade No. 5 or 6 for heavy cutting (i.e., thicknesses of over 150 mm [6 in.]);[14]
2. Flame resistant gloves;
3. Safety glasses;
4. Hearing protection;
5. Flame-resistant jackets, coats, hoods, aprons, and similar apparel;
   (a) Woolen clothing, preferably, or cotton, but no synthetic materials;
   (b) Sleeves, collars, and pockets kept buttoned;
   (c) Cuffs eliminated;
6. Hard hat;
7. Leggings and spats;
8. Safety shoes;
9. Flame-extinguishing equipment;
10. Supplemental breathing equipment; and
11. Other safety equipment as required for both the operator and the work environment.

GASES AND FUMES

In the oxygen cutting processes, gases and fumes are a potential health hazard. When oxygen cutting is performed in an enclosed or semi-enclosed area, exhaust ventilation should be provided, and the operator should be equipped with a respirator if needed. The American National Standards Institute (ANSI) standard *Safety in Welding, Cutting, and Allied Processes*, ANSI Z49, should be consulted.[15]

NOISE

Noise from cutting operations may exceed safe levels in some circumstances. When necessary, ear protection should be worn by the operator. Allowable noise levels are specified in the Occupational Safety and Health Administration document *Occupational Safety and*

12. American Welding Society (AWS) Committee on Oxyfuel Gas Cutting, 2002, *Operator's Manual for Oxyfuel Gas Cutting*, AWS C4.2:2002, Miami: American Welding Society.
13. See Reference 5.
14. American Welding Society (AWS) Safety and Health Committee, 2001, *Lens Shade Selector* (Chart), F2.2:2001, Miami: American Welding Society.
15. See Reference 8, pp 10–13.

Health Standards for General Industry, CFR 1910, Subpart Q.[16]

FIRE AND EXPLOSION

Oxyfuel gas cutting procedures inherently produce a variety of hazardous conditions that must be closely monitored and controlled. The use of combustible gases and oxygen, cutting near combustible materials, gas leaks from torches, hoses, cylinders and related equipment are only a few of the conditions that can cause fire and explosions. However, these conditions may be avoided with careful attention to safe practices. Fire and explosions may occur if carelessness exists in the workplace.

Safe practices in oxygen cutting are published by the National Fire Protection Association's *Standard for Fire Prevention during Welding, Cutting, and Other Hot Work*, NFPA 51B.[17]

CONCLUSION

Oxyfuel gas cutting is solidly established in the metal cutting field and is here to stay. The reader may find it rewarding to explore all the values and capabilities of this old, yet modern, process.

The oxyfuel cutting process was first used in the beginning of the 20th century. Equipment failures, both in the production and supply of gas and the cutting apparatus, were not uncommon. In only a few years, however, cutting torches and related equipment reached stabilized design parameters that, to a large extent, remain with us today.

While improvements in manufacturing methods and materials have allowed the equipment to last longer and perform more efficiently and safely, the basic process has not changed. The kindling temperature of steel is identical to what it was 100 years ago. Oxygen and iron still react in the same manner, and the end result is no different. The process is not expected to change in the foreseeable future.

In the early days of oxyfuel gas cutting, the process was sometimes used in applications for which it was not well suited, for example, for the cutting of stainless steel and cast iron. The results were marginal, but there were few alternatives. Processes such as plasma arc cutting, laser beam cutting, and air carbon arc cutting were not yet available. Although these newer processes have made inroads into the field of metal cutting, none are expected to replace the efficiency and convenience of oxyfuel gas cutting when it is used within its capabilities. Mechanized oxyfuel gas cutting is competitive with some of the most sophisticated cutting processes. No other process is comparable in terms of low initial investment, portability, versatility, ease of training, and low operating costs. No other process is as widely used, whether calculated by the number of torches in use, areas of use, or perhaps even by equipment durability. Torches over eighty years old are still in use today.

Because there have been few innovations in oxyfuel gas cutting, its advantages are sometimes overlooked. Large amounts of time and money are often spent to install new equipment and processes to replace existing oxyfuel gas cutting, when possibly a minor tune-up and a little training would accomplish equal results at far less cost.

16. Occupational Safety and Health Administration (OSHA) document *Occupational Safety and Health Standards for General Industry*, CFR 1910, Subpart Q, Washington, D.C., Superintendent of Documents, Government Printing Office.

17. National Fire Protection Association, *Standard for Fire Prevention During Welding, Cutting, and Other Hot Work*, NFPA 51B, Quincy, Massachusetts: National Fire Protection Association.

BIBLIOGRAPHY[18]

American National Standards Institute (ANSI) Accredited Standards Committee Z49. 1999. *Safety in welding, cutting, and allied processes.* ANSI Z49.1:1999. Miami: American Welding Society.

American Welding Society (AWS) Committee on Definitions and Symbols. 2001. *Standard welding terms and definitions.* AWS A3.0:2001. Miami: American Welding Society.

American Welding Society (AWS) Committee on Oxyfuel Gas Welding and Cutting. 2000. *Uniform designation system for oxyfuel gas nozzles.* AWS C4.5M:2000. Miami: American Welding Society.

American Welding Society (AWS) Committee on Oxyfuel Gas Welding and Cutting. 2002. *Operator's manual for oxyfuel gas cutting.* ANSI/AWS C4.2:2002. Miami: American Welding Society.

Compressed Gas Association (CGA). 1994. *Standard for gas pressure regulators.* CGA E-4-1994. Arlington, Virginia: Compressed Gas Association.

Compressed Gas Association (CGA). 1998. *Torch standard for welding and cutting.* CGA E-5-1998. Arlington, Virginia: Compressed Gas Association

National Fire Protection Association (NFPA). 1999. *Standard for fire prevention during welding, cutting,*

18. The dates of publication given for the codes and other standards listed here were current at the time this chapter was prepared. The reader is advised to consult the latest edition.

and other hot work. NFPA 51B-1999. Quincy, Massachusetts: National Fire Protection Association.

ESAB Welding and Cutting Products. 1994. *The oxy-acetylene handbook*. Florence, S. C.: ESAB Welding and Cutting Products.

O'Brien, R. L., ed. 1992. *Jefferson's welding encyclopedia*. Miami: American Welding Society.

Rubber Manufacturers Association, 1999. *Specification for rubber welding hose*, ANSI/RMA IP-7. Washington, D.C.: Rubber Manufacturers Association.

Occupational Safety and Health Administration (OSHA). *Oxygen-fuel gas welding and cutting*. In *Code of Federal Regulations (CFR)*, Title 29 CFR 1910.253, *Subpart Q*. Washington D.C.: Superintendent of Documents, U.S. Government Printing Office.

Victor Equipment Company. 2002. *Welding, cutting and heating guide*. Denton, Texas: Victor Equipment Company.

SUPPLEMENTARY READING LIST

American Conference of Governmental Industrial Hygienists (ACGIH). 1999. *1999 TLVs® and BEIs®: Threshold limit values for chemical substances and physical agents in the workroom environment*. Cincinnati: American Conference of Governmental Industrial Hygienists. (Editions of this publication are also available in Greek, Italian, and Spanish).

American Conference of Governmental Industrial Hygienists (ACGIH). 1998. *Industrial ventilation: A manual of recommended practice*. 23rd ed. Publication 2092. Cincinnati: American Conference of Governmental Industrial Hygienists.

American National Standards Institute (ANSI)/American Welding Society (AWS) Safety and Health Committee. 2001. *Lens shade selector.* ANSI/AWS F2.2:2001. Miami: American Welding Society.

American Welding Society (AWS) Committee on Labeling and Safe Practices. 1998. *Safety and health fact sheets*. 2nd ed. Miami: American Welding Society. (Also available online at http://www.aws.org.)

American Welding Society (AWS) Committee on Labeling and Safe Practices. 1994. *Recommended safe practices for the preparation for welding and cutting of containers and piping*. ANSI/AWS F4.1-94. Miami: American Welding Society.

ASM International, 1993. Vol. 6, *ASM handbook. Welding, brazing and soldering*. Materials Park, Ohio: ASM International.

Compressed Gas Association (CGA). 1999. *Safe handling of gas in containers*. CGA P-1-1999. Arlington, Virginia: Compressed Gas Association.

Compressed Gas Association (CGA). 1998. *Hose line flashback arrestors*. Technical Bulletin TB-3. Arlington, Virginia: Compressed Gas Association.

Compressed Gas Association (CGA). 1998. *Acetylene*. CGA G-1-1998. Arlington, Virginia: Compressed Gas Association.

Compressed Gas Association (CGA). 1996. *Oxygen*. CGA G-4. Arlington, Virginia: Compressed Gas Association.

National Fire Prevention Association (NFPA). 1997. *Design and installation of oxygen-fuel gas systems for welding, cutting, and allied processes*. NFPA 51. Quincy, Massachusetts: National Fire Prevention Association.

Occupational Safety and Health Administration (OSHA). 1999. *Occupational safety and health standards for general industry.* In *Code of Federal Regulations (CFR),* Title 29, CFR 1910, *Subpart Q*. Washington, D.C.: Superintendent of Documents, U.S. Government Printing Office.

CHAPTER 15

ARC CUTTING AND GOUGING

Photograph courtesy of ESAB Welding and Cutting Products

Prepared by the Welding Handbook Chapter Committee on Arc Cutting and Gouging:

I. D. Harris, Chair
Edison Welding Institute

J. D. Colt
Hypertherm, Incorporated

R. C. Fernicola
ESAB Welding and Cutting Products, Incorporated

C. A. Landry
Centricut, Incorporated

C. B. Wilsoncroft
Tweco Arcair (Thermadyne)

Welding Handbook Committee Member:

J. H. Myers
Weld Inspection and Consulting Services

Contents

CHAPTER 15

ARC CUTTING AND GOUGING

INTRODUCTION

Arc cutting (AC) comprises a group of thermal cutting (TC) processes that sever or remove metal by melting it with the heat of an arc between an electrode and the workpiece. Gouging is a thermal cutting process variation that removes metal by melting or burning the entire removed portion to form a bevel or groove. The category *arc cutting and gouging* includes the following processes that are or have been used for cutting or gouging metals:

Plasma arc cutting and gouging (PAC),
Air carbon arc cutting and gouging (CAC-A),
Shielded metal arc cutting (SMAC),
Gas metal arc cutting (GMAC),
Gas tungsten arc cutting (GTAC),
Oxygen arc cutting (AOC), and
Carbon arc cutting (CAC).

This chapter describes these arc cutting and gouging processes and provides information on fundamentals, required equipment, materials and consumables, process variables, applications, quality, economics, and safe practices. The advantages and limitations of the processes and factors affecting process selection are discussed.

Plasma arc cutting and air carbon arc cutting are addressed in detail in this chapter because of the wide range of applications for these processes. The remaining processes, used less frequently, are discussed briefly in a subsequent section of the chapter.

PLASMA ARC CUTTING

The plasma arc cutting (PAC) process, shown in Figure 15.1, is widely used in a variety of industries to produce rapid, clean cuts in ferrous and nonferrous metals. Users of plasma arc cutting systems include manufacturers of ships and offshore platforms; general fabricators; producers of structural beams, pipe, and ducts; processors of copper, steel and aluminum; and contractors and maintenance personnel.

The plasma arc cutting process severs metal by using a constricted arc to melt a localized area of a workpiece and removes the molten material with a high-velocity jet of ionized gas issuing from the constricting orifice. The ionized gas is plasma, defined as a gas that has been heated by an arc to at least a partially ionized condition, enabling it to conduct an electric current.[1,2] Plasma arcs typically operate at temperatures of 10 000°C to 14 000°C (18,000°F to 25,000°F).

Invented in the mid-1950s, conventional plasma arc cutting became commercially successful shortly after its introduction to industry. The ability of the process to sever any electrically conductive material provided special advantages for the cutting of nonferrous metals that could not be cut with the oxyfuel gas cutting (OFC) process. Plasma arc cutting was initially used to cut stainless steels and aluminum.

1. American Welding Society (AWS) Committee on Definitions and Symbols, 2001, *Standard Welding Terms and Definitions,* AWS A3.0:2001, Miami: American Welding Society, p. 2.
2. At the time of the preparation of this chapter, the referenced codes and other standards were valid. If a code or other standard is cited without a date of publication, it is understood that the latest edition of the document referred to applies. If a code or other standard is cited with the date of publication, the citation refers to that edition only, and it is understood that any future revisions or amendments to the code or standard are not included; however, as codes and standards undergo frequent revision, the reader is advised to consult the most recent edition.

Figure 15.1—Manual Plasma Arc Cutting

ADVANTAGES AND LIMITATIONS

As the plasma arc cutting process developed, it continued to add new advantages over other cutting processes such as oxyfuel gas cutting and mechanical cutting, including the cutting of carbon steel as well as nonferrous metals. The non-contact plasma arc cutting process requires much less force to hold the workpiece in place and move the torch (or vice versa) than in mechanical cutting processes such as shearing and nibbling.

The process offers enhanced capabilities in cutting shaped parts. Compared to oxyfuel gas cutting, the plasma arc cutting process operates at a much higher energy level, resulting in considerably faster cutting speeds. The reduction in cutting time translates into significant cost savings, thus making the process highly cost-effective. In addition to higher speeds, plasma arc cutting has the advantage of instantaneous startup without preheating. Instant startup is particularly advantageous for applications involving interrupted cutting such as the severing of mesh, expanded metal, and grates. Very high-quality cuts can be made, as the narrow heat-affected zone produced by this cutting process has a favorable effect on the strength of the workpiece and provides resistance to corrosion and cracking. A major advantage of plasma arc cutting is its versatility; the plasma arc cutting process can cut all commercial metals, whereas oxyfuel gas cutting is mainly limited to the cutting of carbon steel.

However, plasma arc cutting has several limitations when compared to oxyfuel gas cutting. Plasma arc cutting has limited thickness capability. It produces a wider kerf than OFC, and the edge of a PAC kerf may not be square. Also, torch components deteriorate with repeated arc initiations.

Plasma arc cutting introduces hazards such as fire, electric shock, intense light, fumes and gases, and noise levels that are generally not associated with mechanical processes and may be associated with other arc processes to a lesser extent. When performing close-tolerance work, it is difficult to control plasma arc cutting as precisely as some mechanical processes.

Plasma arc cutting equipment tends to be more expensive than that required for oxyfuel gas cutting and requires a large amount of electrical power. However, when all factors are considered, the process has many economic benefits. These and other aspects are discussed in the "Economics" section near the end of the chapter.

PRINCIPLES OF OPERATION

As previously defined, plasma is a gas that has been heated by an arc to an ionized condition, or partially ionized to the point that it conducts an electric current. Plasma exists in all electric arcs, but the term *plasma arc* is associated exclusively with torches that produce a constricted arc. The principal features that distinguish plasma arc torches from other arc torches are the constricting nozzle, the use of a second gas for shielding, and the higher arc voltage in the constricted arc for a given current and gas flow rate.

The basic components of a plasma arc cutting torch are shown in Figure 15.2. The arc is constricted as it is forced through a small nozzle orifice downstream of the electrode. As plasma gas passes through the arc, it is rapidly heated to a high temperature. It expands and accelerates as it passes through the constricting orifice

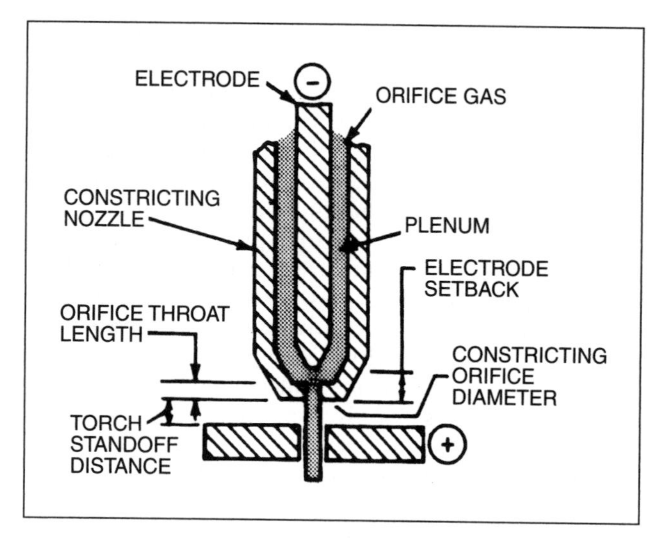

Figure 15.2—Cross Section of a Plasma Arc Cutting Torch

toward the workpiece. The intensity and velocity of the plasma is determined by several variables including the type of gas used, the pressure and flow pattern of the gas, the electric current, the size and shape of the orifice, and the distance to the workpiece.

The electric circuitry for plasma arc cutting is shown in Figure 15.3. The process operates on direct current electrode negative (DCEN) polarity. The orifice directs the superheated plasma stream from the electrode toward the workpiece. When the arc melts the workpiece, the high-velocity jet blows away the molten metal to form the cut, referred to as the *kerf*. The cutting arc, known as a *transferred arc,* attaches or transfers to the workpiece.

Arc Initiation

One of three starting methods can be used to initiate the cutting arc: the pilot arc, direct high-frequency, and electrode (tip) retract starting.

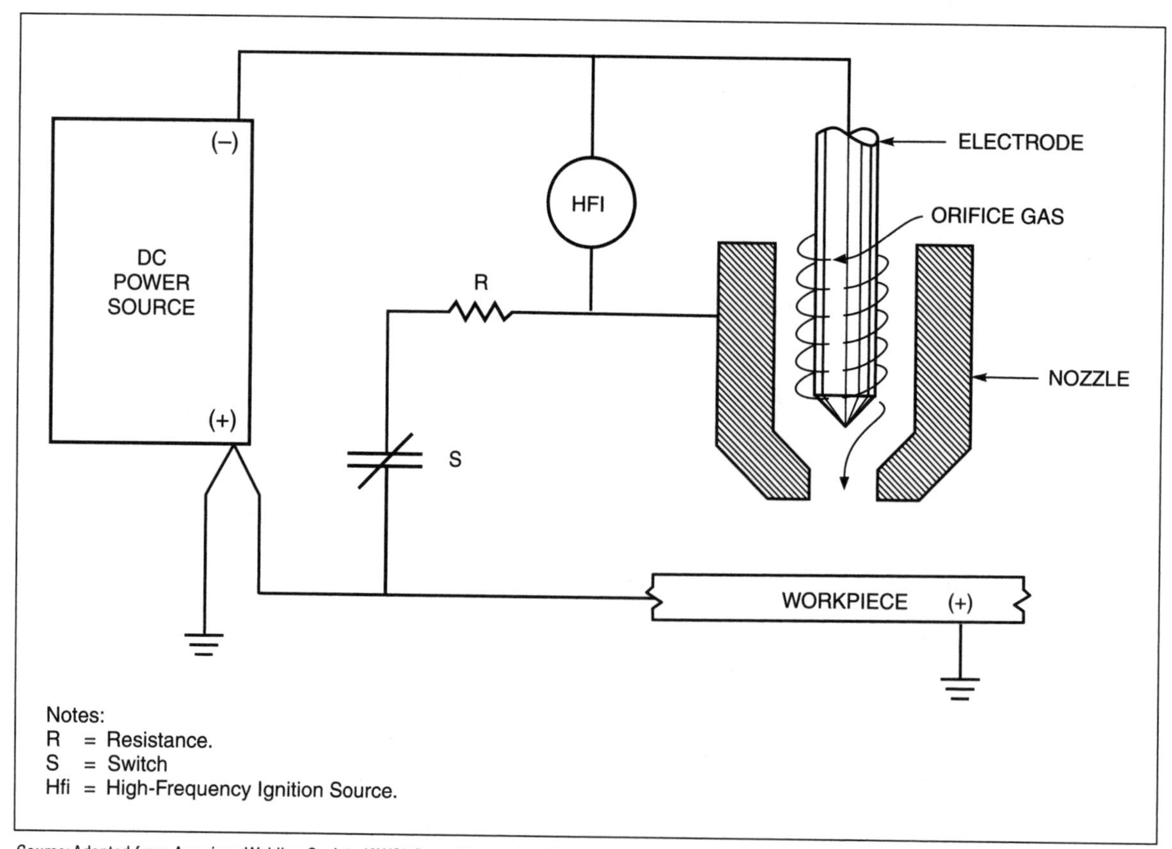

Source: Adapted from American Welding Society (AWS) Committee on Arc Welding and Arc Cutting, 2001, *Recommended Practices for Arc Cutting and Gouging*, AWS C5.2:2001, Miami: American Welding Society, Figure 3.

Figure 15.3—Basic Plasma Arc Cutting Circuitry

A pilot arc is an arc between the electrode and the torch nozzle. This arc is sometimes referred to as a *nontransferred arc* because, unlike the transferred arc, it does not attach or transfer to the workpiece. The pilot arc provides an electrically conductive path from the direct-current power source through the electrode in the torch, the nozzle, the resistor, the relay contacts, and back so that the main cutting arc can be initiated. The most common pilot arc starting technique is to strike a high-frequency spark between the electrode and the torch nozzle. A pilot arc is established across the resulting ionized path. When the torch is close enough to the workpiece so that the plume of the pilot arc touches the workpiece, an electrically conductive path from the electrode to the workpiece is established. The cutting arc follows this path to the workpiece.

Direct high-frequency starting is similar to pilot arc starting, except in this case, the high-frequency spark strikes directly between the electrode and the workpiece, establishing the necessary ionized path.

The third method, the electrode (or tip) retract starting technique, involves the use of electrode retract starting torches. These torches have a moveable nozzle or electrode so that the nozzle and electrode can be momentarily shorted together and then separated or retracted to establish the cutting arc.

EQUIPMENT

Plasma arc cutting and gouging equipment includes torches and accessories, a power source, controls, and fixtures or positioners.

Torches

The plasma arc cutting process is applied with a torch designed for manual or mechanized operations. Manual torches are designed for hand-held use. Mechanized torches can be mounted either on a tractor or on a computer-controlled cutting machine or robot. Several types and sizes of manual or mechanized torches are available. The choice depends on the thickness of the metal to be cut.

Manual cutting torches, especially those designed for operation at over 100 amperes (A), are typically water-cooled to protect the torch and cool the torch components that might be consumed by the intense heat. Mechanized torches are universally water-cooled because of the high power used in mechanized cutting. Recirculating water coolers similar to those used for high-current welding operations are typically built into the power source to provide torch cooling. A schematic illustration of a typical plasma arc cutting torch is shown in Figure 15.4.

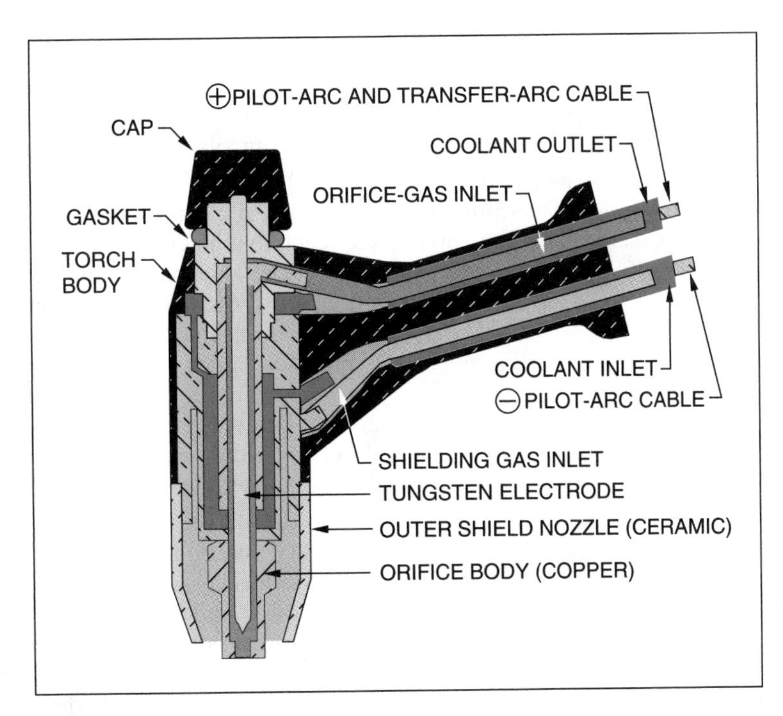

Figure 15.4— A Typical Plasma Arc Cutting Torch Design

The diameter of the orifice determines the maximum amount of current that can be used; the larger the diameter of the orifice is, the higher the current that can be passed through it. Plasma arc cutting torches are available in various current ranges. Torches generally categorized as low-power operate at 30 A or less. Medium-power torches operate at 30 A to 100 A, and high-power torches operate at 100 A to 1000 A. Different power levels are appropriate for different applications, with the higher power levels used to cut thicker metal at higher speeds.

Although some torches can be dragged along in direct contact with the workpiece, most require maintaining a standoff distance (a distance between the torch nozzle and the workpiece) to obtain the best cut quality. The standoff distance must be maintained within relatively close tolerances to achieve uniform results. Many mechanized torches are equipped with an automatic standoff controlling device, based on arc voltage controls, to maintain a fixed distance between the torch and workpiece. In other cases, mechanical followers are used to accomplish this.

Plasma arc cutting torches operate at such extremely high temperatures that various parts of the torch must be considered consumable. The nozzle and the electrode are particularly vulnerable to wear during cutting. Therefore, cutting performance usually deteriorates as they wear. The timely replacement of consumable torch parts is required to maintain high-quality cuts.

Modern plasma torches have self-aligning and self-adjusting consumable parts. As long as they are assembled in accordance with the manufacturer's instructions, they should require no further adjustment for proper operation.

Other torch parts—for example, shielding gas nozzles, insulators, and seals—may also require periodic inspection and replacement if they are worn or damaged. The manufacturer's instructions concerning the frequency of inspection and replacement of the torch parts should be followed.

Power Source

Plasma arc cutting requires a specially designed power source to accommodate the electrical characteristics of the constricted arc. These relatively high-voltage direct-current power sources are of the constant-current (drooping) volt-ampere (V-A) curve type. To achieve satisfactory arc starting performance, the open-circuit voltage of the power source is generally about twice the operating voltage of the torch. Operating voltages range from 50 volts (V) or 60 V to over 200 V; thus, plasma arc cutting power sources have open-circuit voltages ranging from about 150 V to over 400 V.

Several types of plasma arc cutting power sources are available. The simplest is the fixed-output type, which consists of a transformer and rectifier. The transformer of the fixed-output power source is wound with a drooping characteristic so that the output voltage drops as the cutting current increases. In some cases, several outputs are available from a single power source through a switching arrangement. The switching arrangement can select various taps provided on the transformer or reactor of the power source. Variable-output power sources are also available. The most widely used types have a saturable reactor and current feedback circuit so that the output can be stabilized at the desired current level.

Other types of controls are available on plasma arc cutting power sources, including electronic phase control and various types of switch-mode power sources. Switch-mode power sources use high-speed, high-current semiconductors to control the output. These controls can regulate the output of a standard direct-current power source, the high-performance *chopper power source*, or they can be incorporated in an inverter-type power source. With their high efficiency and small size, switch-mode power sources are especially appropriate for applications in which portability is a consideration.

Cutting Process Controls

Most manual torches are controlled by a trigger switch. The switch is pressed to start the cutting arc and released to stop the cut. For mechanized cutting, starting and stopping the cutting arc can be accomplished manually by pushbutton or automatically using the motion controls, e.g., computerized numerical controls (CNC) within the system.

Plasma arc cutting controls provide the proper sequence for the cutting operation as needed, including varying the gas flow, water supply, and power level. If the plasma torch is run without an adequate supply of gas, the torch may be damaged by internal arcing. For this reason, a gas pressure switch is usually included in the circuit to ensure that adequate gas pressure is present before the torch can operate. This interlock also shuts down the torch in the event of a gas supply failure during cutting.

For high-current torches, which are liquid-cooled, an additional interlock is included in the coolant system. This interlock prevents operation of the torch without coolant flow and shuts the power source off to prevent damage if the coolant flow is interrupted during operation. Some systems have component interlocks that prevent the activation of the cutting operation if the parts are not correctly installed in the torch.

Motion Equipment

A variety of motion equipment is available for use with plasma arc cutting torches. These devices range from straight-line tractors to numerically controlled or direct computer-controlled shape cutting machines with parts-nesting capabilities. Plasma arc cutting equipment can also be adapted to robotic actuators for cutting three-dimensional parts.

Environmental Protection Controls

The plasma arc cutting process is inherently a noisy, fume-generating process. Several devices and techniques are available to control and contain these hazards. One commonly used approach to reducing noise and fume emissions in mechanized applications is to perform the cutting over a water table and surround the arc with a water shroud. This method requires a cutting table filled with water up to the workpiece-supporting surface, a water shroud attachment to go around the torch, and a recirculating pump to draw water from the cutting table and pump it through the shroud. A relatively high water flow (55 liters per minute [L/min] to 75 L/min [15 to 20 gallons per minute (gal/min)]) is used.

Another process variation, underwater plasma arc cutting, is also in common use. With this method, the working end of the torch and the plate to be cut are submerged under approximately 75 millimeters (mm) (3 inches [in.]) of water. While the torch is underwater but not cutting, a constant flow of plasma gas is maintained throughout the torch to keep the water out.

The primary requirements in water-table design are adequate strength for supporting the workpiece, sufficient scrap capacity to hold the *dross* (the cutting debris or metal removed in the cutting process), a means for removing the dross, and the ability to maintain the water level in contact with the work. Cutting tables for manual or mechanized plasma arc cutting are usually equipped with a downdraft exhaust system. Fume removal or filtering devices are generally required to meet air pollution regulations. When the table is used for underwater cutting, it is necessary to provide a means of rapidly raising and lowering the water level. This can be achieved by pumping the water in and out of a holding tank or by displacing it with air in an enclosure under the surface of the water.

MATERIALS

Materials used in plasma arc cutting include gases, electrodes, and consumable torch components. The electrode and torch nozzle are subject to intense heat and are subject to melting or deformation; thus they are considered to be consumables. They must be monitored for wear and replaced when necessary.

Gases

The most frequently used gases for manual plasma arc cutting are air, nitrogen, and mixtures of argon and hydrogen. If the equipment requires a cooling gas, argon, nitrogen, carbon dioxide, air, or oxygen is used.

Mechanized cutting uses nitrogen, air, or oxygen as the plasma gas, with water injection to further restrict the arc. An argon-hydrogen mixture is frequently used for cutting base metals over 75 mm (3 in.) thick.[3]

Electrodes

Single and dual gas torches using inert or reducing gases use tungsten alloy electrodes similar to those used for gas tungsten arc welding (GTAW) and plasma arc welding (PAW). Torches designed for operation in oxidizing gases such as air or oxygen use an air- or water-cooled hafnium insert electrode set into a copper alloy body. The electrode is a one-piece consumable type.

APPLICATIONS

Manual plasma arc cutting is widely used in automobile body repair, in sheet metal fabrication, and for general fabrication jobs requiring relatively short cuts on plate and other products made from most commercial alloys, including carbon steels, high strength and low alloy steels, stainless steels, aluminum alloys, nickel alloys, and copper alloys. Instant arc initiation and high travel speeds reduce heat input and help maintain base metal properties.

The first commercial application of plasma arc cutting involved the mechanized cutting of access holes in aluminum railroad tank cars. The process has since become the standard choice for use on a wide variety of aluminum applications. Table 15.1 shows typical conditions for the mechanized cutting of aluminum-alloy plate. Typical conditions for the mechanized cutting of stainless steel plate are shown in Table 15.2.

The majority of applications for the mechanized plasma arc cutting of carbon steel are for thicknesses up to 16 mm (5/8 in.), reflecting the use of plate in industry in general. However, using oxygen plasma arc cutting (described in the section "Process Variations") at 260 A to 340 A, the plasma process is more economical than the oxyfuel gas cutting process for thicknesses up to 32 mm (1-1/4 in.). The higher cost of plasma arc equipment compared to oxyfuel gas cutting equipment can be justified by the higher cutting speeds achieved by

3. American Welding Society (AWS) Committee on Arc Welding and Cutting, *Recommended Practices for Shielding Gases for Welding and Plasma Arc Cutting,* AWS C5.10-1994, Miami: American Welding Society, pp. 22–23.

Table 15.1
Typical Conditions for the Plasma Arc Cutting of Aluminum Alloys

Thickness		Speed		Orifice Diameter*		Current (DCEN)	Power
mm	in.	mm/s	in./min	mm	in.	A	kW
6	1/4	127	300	3.2	1/8	300	45
13	1/2	86	200	3.2	1/8	250	37
25	1	38	90	4.0	5/32	400	60
51	2	9	20	4.0	5/32	400	60
76	3	6	15	4.8	3/16	450	67
102	4	5	12	4.8	3/16	450	67
152	6	3	8	6.4	1/4	750	150

*Depending on the orifice diameter and the gas used, plasma gas flow rates vary from about 47 L/min (100 ft^3/h) for a 3.2 mm (1/8 in.) orifice to about 120 L/min (250 ft^3/h) for a 6.4 mm (1/4 in.) orifice. The gases used are nitrogen or argon with hydrogen additions up to 35%. The equipment manufacturer should be consulted for each application.

Table 15.2
Typical Conditions for the Plasma Arc Cutting of Stainless Steels

Thickness		Speed		Orifice Diameter*		Current (DCEN)	Power
mm	in.	mm/s	in./min	mm	in.	A	kW
6	1/4	86	200	3.2	1/8	300	45
13	1/2	42	100	3.2	1/8	300	45
25	1	21	50	4.0	5/32	400	60
51	2	9	20	4.8	3/16	500	100
76	3	7	16	4.8	3/16	500	100
102	4	3	8	4.8	3/16	500	100

*Depending on the orifice diameter and the gas used, plasma gas flow rates vary from about 47 L/min (100 ft^3/h) for a 3.2 mm (1/8 in.) orifice to about 94 L/min (200 ft^3/h) for a 4.8 mm (3/16 in.) orifice. The gases used are nitrogen or argon with hydrogen additions up to 35%. The equipment manufacturer should be consulted for each application.

the plasma arc cutting process. The conditions for the mechanized plasma arc cutting of carbon steel plate are provided in Table 15.3.

The plasma arc process has also been used for the stack cutting of carbon steel, stainless steel, and aluminum. Although the plasma arc cutting process can tolerate wider gaps between plates than oxyfuel gas cutting, the plates to be stack-cut should preferably be clamped together.

The beveling of plate and pipe edges for weld joint preparation is performed with techniques similar to those used in oxyfuel gas cutting. One to three plasma arc cutting torches can be used, depending on the joint preparation required. A single-torch application is shown in Figure 15.5.

PROCESS VARIATIONS

In addition to single-gas (conventional) plasma arc cutting systems for manual and mechanized applications, the plasma arc cutting process has several variations, including dual-flow plasma arc cutting, air plasma arc cutting, water-shield plasma arc cutting, water-injection plasma arc cutting, underwater cutting, oxygen plasma arc cutting, and high-density plasma arc cutting.[4] Each variation requires specialized equipment.

4. American Welding Society (AWS) Committee on Arc Welding and Cutting, 2001, *Recommended Practices for Plasma Arc Cutting and Gouging*, AWS C5.2:2001, Miami: American Welding Society, pp. 8–10.

Table 15.3
Typical Conditions for the Mechanized Plasma Arc Cutting of Carbon Steel

Thickness		Speed		Orifice Diameter*		Current (DCEN)	Power
mm	in.	mm/s	in./min	mm	in.	A	kW
6	1/4	86	200	3.2	1/8	275	40
13	1/2	42	100	3.2	1/8	275	40
25	1	21	50	4.0	5/32	425	64
51	2	11	25	4.8	3/16	550	110

*Depending on the orifice diameter and the gas used, plasma gas flow rates vary from about 94 L/min (200 ft^3/h) for a 3.2 mm (1/8 in.) orifice to about 104 L/min (300 ft3/h) for a 4.8 mm (3/16 in.) orifice. The gases used are usually nitrogen, oxygen, or air. The equipment manufacturer should be consulted for each application.

Figure 15.5—Bevel Cutting of Steel Plate Using Plasma Arc Cutting

Equipment manufacturers should be consulted regarding the installation, use, maintenance, and safe practices appropriate to the equipment.

Dual Flow (Shielded) Plasma Arc Cutting

The dual-flow technique requires a plasma arc cutting torch with an outer nozzle that provides for the concentric flow of a secondary shielding gas around the plasma gas. The shielding gas is usually nitrogen, air, carbon dioxide, argon, oxygen, or mixtures of argon and hydrogen. The advantages of this technique are that the secondary gas shields the plasma and the cutting zone, which reduces or eliminates cut-surface contamination. Also, the nozzle is recessed within an outer cap, which prevents the nozzle from contacting the workpiece and reduces *double arcing*. (Double arcing damages the nozzle. It occurs when the metallic torch nozzle forms a part of the current path from the electrode back to the power source and forms two arcs, one from the electrode to the nozzle and the second from the nozzle to the workpieces).

The dual-flow technique is selected on the basis of the type of metal to be cut or gouged, the metallurgical properties required for the finished edge, and the physical properties of the edge such as the presence of dross, the accuracy of the square or bevel cut, smoothness, and flatness.

Water Shield Plasma Arc Cutting

Water shield plasma arc cutting is a variation of the dual-flow technique of mechanized plasma arc cutting in which water is used instead of a shielding gas. The technique is mainly used to cut stainless steels. The cooling effect of water results in longer service life of the torch nozzles and better appearance of the cut surfaces. However, this technique is not generally used when cutting speed, squareness of the cut edge, and elimination of dross along the kerf are critical.

Air Plasma Arc Cutting

Air plasma arc cutting uses the oxygen content of air to increase the speed of cutting mild steel by about 25% over conventional plasma arc cutting. The use of an oxidizing gas requires a hafnium insert electrode rather than a tungsten electrode. Improvements in electrodes have normalized the service life of the electrode, and cutting with lower currents has resulted in higher quality of the cut surfaces. Thus, the air plasma arc cutting technique, formerly limited in use, has gained wide usage.

Water Injection Plasma Arc Cutting

Water injection plasma arc cutting is a mechanized process, generally using from 250 A to 750 A. Water is injected around the arc, either as a high-velocity radial or tangential spray. The water imparts a swirling action and impinges on the arc to create a more constricted arc than that of the conventional process. The advantages of this technique are improved accuracy of the square or bevel cut, increased cutting speeds, and minimized dross formation. The cooling action of the injected water allows the use of a ceramic section, which nearly eliminates double arcing. The technique facilitates piercing operations.

Underwater Cutting

Plasma arc cutting systems can be modified for underwater cutting performed in water no deeper than 14.2 meters (m) (50 feet [ft]). The primary modification made to manual plasma arc cutting systems is to increase the electrical isolation of the hand torch and the leads from the power source to the torch.

The most important application of underwater plasma arc cutting has been in the repair or dismantling of radioactive components of nuclear reactors. The manual plasma arc cutting equipment is sometimes submerged into the protective water shield used by the nuclear industry to protect the environment from radioactive emissions and the cutting work is carried on in this water shield. Divers perform the cuts in protective "dry" diving suits.

Mechanized plasma arc cutting systems have been modified to cut radioactive stainless steel sections 75 mm to 100 mm (3 in. to 4 in.) thick in depths of 8.5 m (30 ft). Safety concerns for divers using high-voltage equipment in water have limited the applications to the nuclear industry.

Oxygen Plasma Arc Cutting

Oxygen is widely used as the plasma gas for cutting all grades of steel because of the high cutting speeds and excellent cut quality that can be achieved. Less dross is produced with oxygen, and the level of nitrides in the cut edges is very low. The low level of nitrides results in less porosity in the weld metal when the edges prepared by oxygen plasma arc cutting are welded.

Pure oxygen reacts chemically with the iron in the steel in a reaction that liberates a great deal of heat that facilitates very high cutting speeds. Numerically or computer-controlled cutting machines capable of 6.35 m/min (250 in./min) should be implemented to take full advantage of the high cutting speeds. Oxygen plasma arc cutting systems are available for both water-injected and dry oxygen for cutting at current levels up to 400 A.

Electrode life is shortened when oxygen is used as the plasma gas. Arc starting and stopping techniques can be modified to preserve the hafnium electrode, and special electrodes designed to control erosion of the hafnium insert can be used to increase the life of the electrode.

High-Current-Density Plasma Arc Cutting

High-current-density plasma arc cutting is generally a dual-flow technique using air or oxygen as the plasma gas with a variety of shielding gases. Any metal up to 13 mm (1/2 in.) can be cut, and cut-edge quality is excellent. This technique uses a high-current-density plasma arc cutting torch designed to produce a superconstricted arc. Cuts made with this torch have a narrow kerf width, and the quality of the cut can be compared to a laser beam cut in some applications.

CUT QUALITY

Factors to consider in evaluating the quality of a cut made with the plasma arc cutting process include surface smoothness, kerf width and angle, dross adherence, and sharpness of the top edge. These factors are affected by the type of material being cut, the equipment being used, and the cutting conditions. Metallurgical effects should also be considered when evaluating the quality of cuts obtained using plasma arc cutting.

Surface Smoothness

Plasma arc cuts in plate up to approximately 75 mm (3 in.) thick may have a surface smoothness very similar to that produced by oxyfuel gas cutting. On thicker plate, low travel speeds produce discoloration and a somewhat rougher surface. Surface oxidation is practically nonexistent with mechanized equipment that uses water injection or water shielding. It should be noted this might not be the case when oxygen is used as the plasma gas on carbon steel. Oxygen typically produces a very thin oxide film.

Kerf Width and Angle

The kerf widths of plasma arc cuts are 1-1/2 to 2 times the width of oxyfuel gas cuts in plate up to 50 mm (2 in.) thick. For example, a typical kerf width in 25 mm (1 in.) steel is approximately 5 mm (3/16 in.). Kerf width increases with plate thickness.

The plasma jet tends to remove more metal from the upper part of the kerf than from the lower portion. This results in beveled cuts that are wider at the top than at the bottom. A typical included angle of a cut in 25 mm (1 in.) steel is 4° to 6°. This bevel occurs on the left-hand side of the cut when clockwise orifice gas swirl is used. The right-hand side (production-part side of the cut) has a typical bevel of 2° for a good quality cut. The bevel angle on both sides of the cut tends to increase with cutting speed.

Dross

Dross is the material that melts during cutting and adheres to the bottom edge of the cut face. With modern mechanized equipment, cuts with little or no dross can be produced in aluminum and stainless steel up to approximately 75 mm (3 in.) thick, and in carbon steel up to approximately 40 mm (1-1/2 in.) thick. With carbon steel, the selection of speed and current are more important. A dross-free range of cutting speeds exists for each particular combination of material type, thickness, cutting gas, and current. This range is wide for oxygen and air plasma arc cutting, but narrow for nitrogen plasma arc cutting. Dross is usually present on thick materials.

Sharpness of the Top Edge

A good plasma cut should not have a rounded top edge, i.e., there should be an angle of about 90° between the top surface of the workpiece and the cut face. Rounding of the top edge may result when excessive power is used to cut a given plate thickness or when the torch standoff distance is too great. It may also occur in the high-speed cutting of materials less than 6 mm (1/4 in.) thick.

Metallurgical Effects

During plasma arc cutting, the material at the cut surface is heated to its melting temperature and ejected by the force of the plasma jet. This produces a heat-affected zone along the cut surface similar to fusion welding operations. The heat not only alters the structure of the metal in this zone but also introduces internal tensile stresses from the rapid expansion and contraction of the metal at the cut surface.

The depth to which the arc heat penetrates the workpiece is inversely proportional to the cutting speed. The heat-affected zone on the cut face of a 25 mm (1 in.) stainless steel plate severed at 21 mm/s (50 in./min) is approximately 0.08 mm to 0.13 mm (0.003 in. to 0.005 in.) deep, as determined from microscopic examination of the grain structure at the cut edge of a plate.

Because of the high cutting speed used when cutting stainless steel and the quenching effect of the base plate, the cut face passes through the critical 650°C (1200°F) temperature very rapidly. Thus, the chromium carbide constituent of the steel has virtually no opportunity to precipitate along the grain boundaries, and corrosion resistance is maintained. Measurements of the magnetic properties of Type 304 stainless steel made on base metal and on plasma arc cut samples have indicated that magnetic permeability is unaffected by arc cutting.

The metallographic examination of cuts in aluminum plates indicates that the heat-affected zones in aluminum plate are deeper than those in stainless-steel plate of the same thickness. This results from the higher thermal conductivity of aluminum. Microhardness surveys indicate that the heat penetrates about 5 mm (3/16 in.) into a 25 mm (1 in.) thick plate. Age-hardenable aluminum alloys of the 2000 and 7000 series are crack-sensitive at the cut surface. Cracking appears to result when a grain boundary eutectic film melts and separates under stress. Machining to remove the cracks may be necessary on the edges that will not be welded.

Hardening occurs in the heat-affected zone of a plasma arc cut in high-carbon steel if the steel has a very fast cooling rate. The degree of hardening is also affected by the type of plasma gas. Nitrogen produces the highest surface hardness and oxygen produces the lowest, based on the amount of nitrogen absorbed into the cut edge.

Various metallurgical reactions may occur when cutting long, narrow, or tapered workpieces or outside corners. The heat generated during a preceding cut may adversely affect the quality of a following cut.

Safety

Safety recommendations for arc cutting are included in the last section of this chapter, "Safe Practices."

PLASMA ARC GOUGING

Plasma arc gouging is a variation of the plasma arc cutting process. Although the process had been used for a number of years it was often replaced with other processes. However, interest in plasma arc gouging as a low-fume process was renewed during the late 1990s because of increased emphasis on environmental and quality issues, and it is again a widely used process. A plasma arc gouging application is shown in Figure 15.6.

The concept of plasma arc gouging can be described as taking a small step backward from plasma arc cutting. In cutting, the goal is to constrict the plasma arc and focus it as tightly as possible to give the greatest energy concentration and cutting power possible. In plasma arc gouging, the arc is slightly defocused by increasing the size of the constricting orifice to produce an arc that is much larger than the arc used for cutting. A cutting arc is directed downward through the metal to blow the molten metal down and out through the kerf to cause the two pieces of metal to separate, while the less focused gouging arc is inclined at an angle to the workpiece to plow out a groove on the surface of the material and blow the molten metal off to the side. If the intense, tightly focused cutting arc were used in this manner, the groove created would be a deep and very narrow shape that would be useless for most applications.

ADVANTAGES

Plasma arc gouging is often the preferred method for accomplishing gouging tasks for several reasons. It is a low-fume process; it produces a high-quality gouge in

Photograph courtesy of ESAB Welding and Cutting Products

Figure 15.6—Plasma Arc Gouging

many applications, which is particularly important when gouged parts are to be rewelded; and it may be the lowest cost alternative when the overall cost of the gouging operation is considered.

Some of the alternate gouging techniques produce large quantities of vaporized metal fumes with the attendant problems. When plasma arc gouging is used to accomplish the same task, the fume generation issue is quite different. Like air carbon arc gouging, plasma arc gouging uses an electric arc to melt the metal to be gouged out, but in plasma arc gouging, the plasma gas itself is used to push the molten metal out of the groove rather than a blast of air. This is done with considerably less violence than air carbon arc gouging, resulting in considerably less vaporization of the molten metal and a much lower occurrence of metallic vapor and reaction with the surrounding atmosphere.

When air is used as the plasma gas, there is still some reaction with the air in the plasma stream and some fume is produced; however the volume of fumes is considerably lower than that produced by air carbon arc gouging. When plasma arc gouging is accomplished with an inert gas in the plasma gas mixture, the fume reduction is very dramatic. The molten metal in the gouge is protected from the surrounding atmosphere by the inert gas of the plasma stream and has little chance to react with the surrounding atmosphere. The result is a very low fume level in most cases. One exception is aluminum, where the lightness and the strong affinity of oxygen for aluminum can cause some fume as the molten aluminum is ejected from the gouge and reacts with the surrounding air.

Because of the strong ultraviolet content of the radiation from the plasma arc, there is some increase in the amount of carbon monoxide, ozone, and oxides of nitrogen generated. However, the amount of these gases is generally below threshold limits.[5]

The gas used for plasma arc gouging primarily determines the condition of the final groove. When gouging carbon steel, the oxidation left when using air as the plasma gas is usually of little concern. However, when gouging stainless steel, aluminum, and other corrosion-resistant alloys, an inert gas should be used as the plasma gas. The inert gas shields the groove from the contaminating atmosphere and the groove is generally free of oxidation and other contamination. In most cases the groove can be welded with almost any welding process without additional cleanup.

The arc is less constricted for gouging, which results in a lower arc velocity. The temperature of the arc and the velocity of the gas stream combine to melt and expel metal in a manner similar to other gouging processes. A major difference is that the gouge is bright and clean, particularly on materials such as aluminum and stainless steel. An argon-hydrogen plasma gas is usually employed to gouge these materials. Virtually no post-cleaning is required when the plasma-gouged surface is to be welded.

EQUIPMENT

Creating an effective plasma arc gouging system begins with the selection of a power source, a gouging torch, and accessory equipment. Most plasma arc equipment manufacturers offer gouging nozzles that fit existing plasma arc cutting torches. However, more is involved than simply putting a different nozzle on the torch. The power source and the design of the torch are important to effective plasma arc gouging.

Power Source

Most plasma arc gouging is done at currents of 100 A and above. Plasma arc gouging can be accomplished at lower currents, but this would result in low metal removal rates and small gouge sizes. Plasma arc gouging generally requires a much higher arc voltage than plasma arc cutting for several reasons. Since the torch is operated at an angle to the plate, the nozzle-to-work distance is inherently longer than in plasma arc cutting, requiring a higher arc voltage. In addition, access problems are common when using plasma arc gouging to do repair work. This often causes the nozzle-to-workpiece distance to be even longer, further increasing the arc voltage demands on the power source. Arc voltages of 200 V or more may be required. Most power sources originally designed for plasma arc cutting cannot supply the magnitude of arc voltage required for plasma arc gouging without significant current drop-off or arc outages. To be a practical production tool, the plasma arc power source used for gouging must be able to supply full output current at high arc voltages.

Gouging Torches

The design of the torch is very important. In normal plasma arc cutting, the heat and molten metal is directed downward and away from the torch, so the torch body is usually not heavily spattered with molten metal and is not affected by most of the radiated heat from the plasma arc. In plasma arc gouging, the arc is actually on top of the plate and much of the heat is radiated directly into the plasma torch head. This puts an external heat load on the torch head that many plasma arc torches were not designed to withstand. As a result, the service life of the torch head may be shortened. In addition, in many plasma arc gouging

5. American Conference of Governmental Industrial Hygienists (ACGIH), TLVs and BEIs —*Threshold Limit Values for Chemical Substances and Physical Agents and Biological Exposure Indices*, Cincinnati, Ohio: American Conference of Governmental Industrial Hygienists.

applications, obstructions in the gouging area deflect some of the molten metal being ejected from the groove back onto the torch head, further accelerating torch-head deterioration. The best approach to this problem is to use a torch head molded out of special high-temperature, glass-filled polymers. Auxiliary protection devices such as Fiberglas®-silicon insulators, or "mittens," that wrap around the torch head can be used for additional protection.

Air plasma arc torches can be used for carbon steel gouging, but will create somewhat higher fume levels. Inert gas plasma torches reduce the fume levels. Inert gas torches are required for stainless steel and aluminum gouging. Dual-gas torch construction is best because it provides for a secondary shielding gas to further shield the gouging zone from atmospheric contamination. This type of torch construction also helps clear the molten metal from the groove.

OPERATING PROCEDURES

The conditions and parameters for plasma arc gouging are very important to ensuring success. Gas selection and current amperage are primary considerations. Other parameters are torch angle, travel speed, groove size, and groove configuration.

Gas and Gas Mixtures

The recommended plasma gas for all gouging is a mixture of argon with 35% to 40% hydrogen. Premixed gases may be supplied from cylinders or prepared using a gas-mixing device. Helium can be substituted for the argon-hydrogen mixture, but the resulting gouge will be shallower.

The secondary or cooling gas, when used, is argon, nitrogen, or air. The selection is based on the desired brightness of the gouge, fume generation, and cost.

Air is sometimes used for the plasma gas with air-operating systems, but this is an option only for the gouging of carbon steel. Air produces a somewhat higher fume level than an inert gas, but considerably less than that produced with the air carbon arc gouging process. Manual air cutting systems are available with output up to 300 A, but most are limited to 200 A.

An inert plasma gas is the best choice for aluminum, stainless steel, and other highly alloyed metals. Nitrogen or an argon-hydrogen mixture should be used. In general, nitrogen lacks the intense arc heat of an argon-hydrogen mix and, as a result, the metal removal rate is lower. An argon-hydrogen mixture is usually the best selection, as the high-temperature plasma arc it produces maximizes the metal removal rate and does not produce surface contamination.

When a dual-gas torch is used, the secondary gas should also be inert. Argon is the best choice, but nitrogen can also be used with results that are almost as good. When gouging aluminum, a dual-gas torch must be used if the current is over 200 A because, without a secondary shield at this higher current, atmospheric contamination causes oxidation of the exposed molten metal and results are unsatisfactory.

Current

The current selection is usually a function of the amount of metal removal desired. The minimum current is usually 100 A for most plasma arc gouging operations; however small repairs and defect removal can be done at lower currents.

Groove Size and Configuration

The groove size of a plasma arc gouge is a function of current level, travel speed, and torch angle. Groove configuration is influenced by several factors. The width of a gouge increases as the current is increased or the travel speed slows down. However, the groove depth is slightly affected by travel speed but is very strongly affected by the magnitude of the current. In manual applications, the torch can be held at a shallow angle and moved from side to side in a washing motion to skim off metal or remove hardfacing.

The best torch angle to start the gouge is from 35° to 40° with an arc length of 13 mm to 25 mm (1/2 in. to 1 in.) The amount of metal removed in one pass when gouging in the flat position is subject to an upper limit. Since plasma arc gouging does not eject the metal from the gouge with as much force as air carbon arc gouging, it may be difficult to remove the metal from deep grooves in one pass. For instance, the maximum groove depth recommended at 150 A with one pass is 8 mm (5/16 in.). Deeper grooves can be obtained with multiple passes.

APPLICATIONS

Many industries use the plasma arc gouging process to great advantage. As an example, plasma arc gouging systems were installed in a railcar and locomotive repair facility, where as many as 30 or 40 air carbon arc and oxyfuel gas gouging arcs were operating simultaneously. Plasma arc gouging substantially reduced the amount of fumes and smoke and greatly improved the control of the metal removed. Access into tight places was substantially easier. In addition, the process worked particularly well on aluminum and stainless steel tank and passenger cars.

Plasma arc gouging is also an excellent tool for the maintenance and repair of trucks and off-road equipment. The rebuilding or repair of truck bodies, many of which are made of stainless steel or aluminum, is similar to the railcar repair industry. Typical users of plasma arc gouging are garbage truck repair facilities and aluminum flatbed trailer repair facilities.

Roll repair represents another significant use of plasma arc gouging. The process is used in the repair and rebuilding of a variety of turning rolls. Coal crush rolls, steel mill rolls, and paper calendaring rolls must have the worn hardfacing layer removed prior to repair and rebuilding. Plasma arc gouging works very well in a lathe-type arrangement to skim off the residual layer quickly with minimal damage to the underlying roll. Previously used methods were carbon arc gouging or grinding with diamond wheels.

Plasma arc gouging is an excellent process for fin and riser removal in foundries. Using a pad washing technique (described in Figure 15.11 in the next section), the plasma process produces considerably less smoke than air carbon arc gouging.

Plasma arc gouging is recognized as an effective process for use in the repair, rebuilding, and maintenance of manufacturing and processing plants, where large amounts of stainless steel, aluminum, or corrosion-resistant alloys are used. Power plants, chemical plants, refineries, paper mills, and food processing plants are examples.

Plasma arc gouging is an important process used in weld backgouging in a variety of fabrication situations in which high-integrity, two-sided welds are required, such as in pressure vessels and cryogenic tanks. The high-quality results are the major advantage. The elimination of carbon pick-up problems and the need for additional grinding and cleanup (associated with air carbon arc gouging) result in improved quality and drastically reduced cost with plasma arc gouging.

Another typical application is in the retrofit of armored personnel carriers. These vehicles are made of aluminum. Repairing or retrofitting them with items such as new gas tanks is difficult using weld removal techniques such as carbon arc gouging, requiring extensive cleanup with mechanical methods, which are slow and costly.

Safety

Safety recommendations for plasma arc gouging are included in the last section of this chapter, "Safe Practices."

AIR CARBON ARC CUTTING

Air carbon arc cutting (CAC-A) was developed in the 1940s as an extension of carbon arc cutting. Unlike carbon arc cutting, which must be performed in the vertical or overhead position to permit gravity to remove the melted metal, the air carbon arc cutting process enables the operator to cut or remove metal in all positions.

The first attempts at achieving an air-blast version of carbon arc cutting involved two operators. One operator applied a carbon arc cutting torch to melt the metal, and the other directed a nozzle with an air jet at the molten pool. A single torch combining the air blast with the carbon electrode holder soon evolved as the forerunner to modern air carbon arc cutting torches. The first commercial air carbon arc cutting torch was introduced in 1948.

Air carbon arc cutting is widely used in the metal fabrication, construction, petrochemical, and mining industries as well as in general maintenance and repair operations. It is also used in foundries to prepare castings for shipment. The process is an important tool used extensively by the welding industry to prepare grooves for welding butt joints. When the process is performed properly, the grooves require little additional cleaning or grinding. The air carbon arc cutting process is also used to backgouge joints to ensure complete joint penetration. Applicable to both ferrous and nonferrous metals, this versatile process is efficient and cost-effective.

The process for air carbon arc cutting and gouging physically removes molten metal from the workpiece with a jet of compressed air. The intense heat of the arc between a carbon-graphite electrode and the workpiece melts a portion of the workpiece. Simultaneously, a jet of air of sufficient volume and velocity is passed through the arc to blow away the molten material. The newly exposed solid metal is then melted by the heat of the arc, and the sequence continues.

Air carbon arc cutting does not depend on oxidation to maintain the cut; therefore it is capable of cutting metals that cannot be cut by the oxyfuel gas cutting process. The process is used successfully on aluminum, carbon steel, stainless steel, many copper alloys, cast iron, and nickel. The melting rate is a function of the electrical current. The metal removal rate depends on the melting rate and the efficiency of the air jet in removing the molten metal. The air must be capable of blowing the molten metal out and clear of the arc region before it can solidify. The process is shown schematically in Figure 15.7.

EQUIPMENT

The equipment for the air carbon arc cutting process includes an electrode holder, cutting electrodes, a power source, and an air supply. For mechanized and automated cutting, a control and a carriage are also required. Figure 15.8 illustrates a typical configuration for air carbon arc cutting equipment.

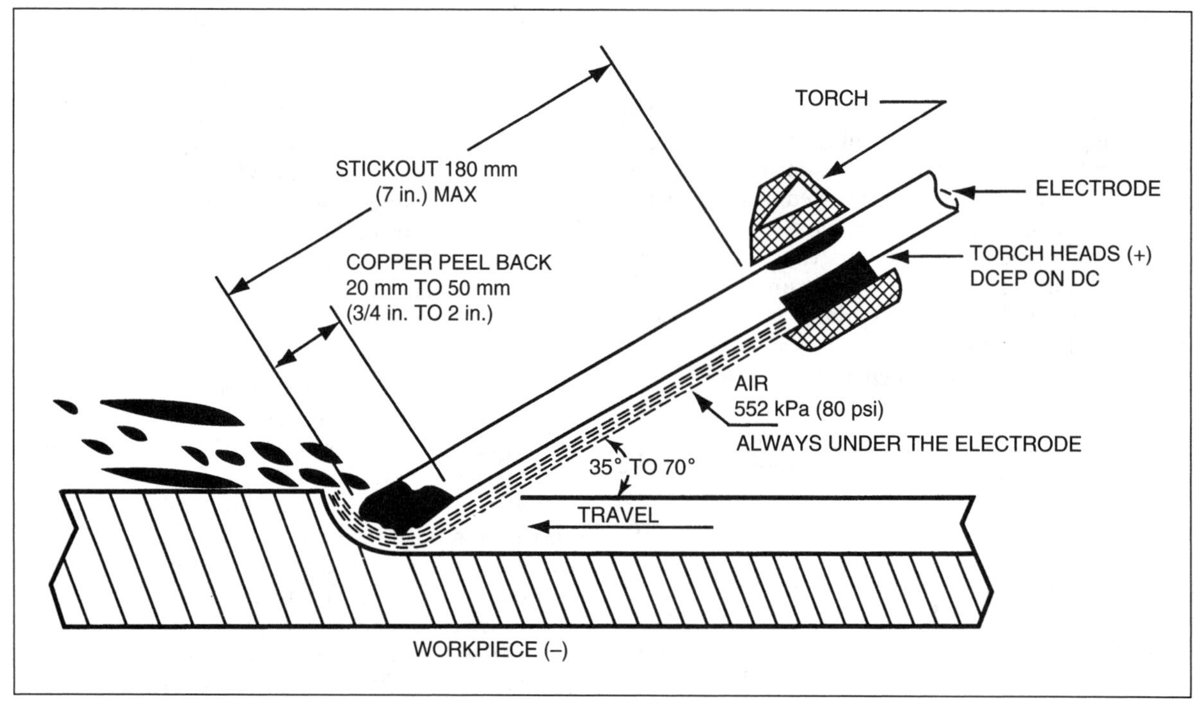

Source: Adapted from American Welding Society (AWS) Committee on Arc Welding and Arc Cutting, 2000, *Recommended Practices for Arc Gouging and Cutting*, AWS C5.3:2000, Miami: American Welding Society, Figure 2.

Figure 15.7—Schematic Illustration of the Air Carbon Arc Cutting Process

Cutting Torches

Manual electrode holders, or torches, for air carbon arc cutting and gouging are similar to conventional heavy-duty shielded metal arc welding holders, as shown in Figure 15.9. The electrode is held in a rotating head that contains one or more air orifices so that regardless of the angle at which the electrode is set relative to the cutting torch, the air jet remains in alignment with the electrode. A valve is provided for turning the cutting air on and off.

Commercially available torches range in size from light-duty to extra heavy-duty foundry torches. Following is a guide for air carbon arc cutting and gouging torch selection.

Light-Duty Torches. Light-duty torches are recommended for small shops, farms, and maintenance operations with a limited compressed air supply. The maximum current is approximately 450 A direct current (dc).

General Purpose Torches. These torches are used for general-purpose applications in shipyards, fabrication shops, and general maintenance. They are limited to a maximum of 1000 A.

Heavy-Duty Torches. Heavy-duty torches are used for general foundry work, pad washing and cutoff, and high-amperage work in shipyards and fabrication shops. They are limited to 1600 A with air-cooled cables and 2000 A with water-cooled cables.

Mechanized Torches. Mechanized electrode holders are used for edge preparation, backgouging, and high-production applications. They are used with 8 mm to 19 mm (5/16 in. to 3/4 in.) jointed carbon electrodes.

Controls

Three types of control systems—amperage-controlled, voltage-controlled, and a dual system—are available for use with mechanized and automated air carbon arc cutting. All of these systems are capable of making grooves of consistent depth to a tolerance of 0.6 mm (0.025 in.) and are used when high-quality, high-production cuts and gouges are required.

The amperage-controlled air carbon arc cutting control maintains the arc current by amperage signals

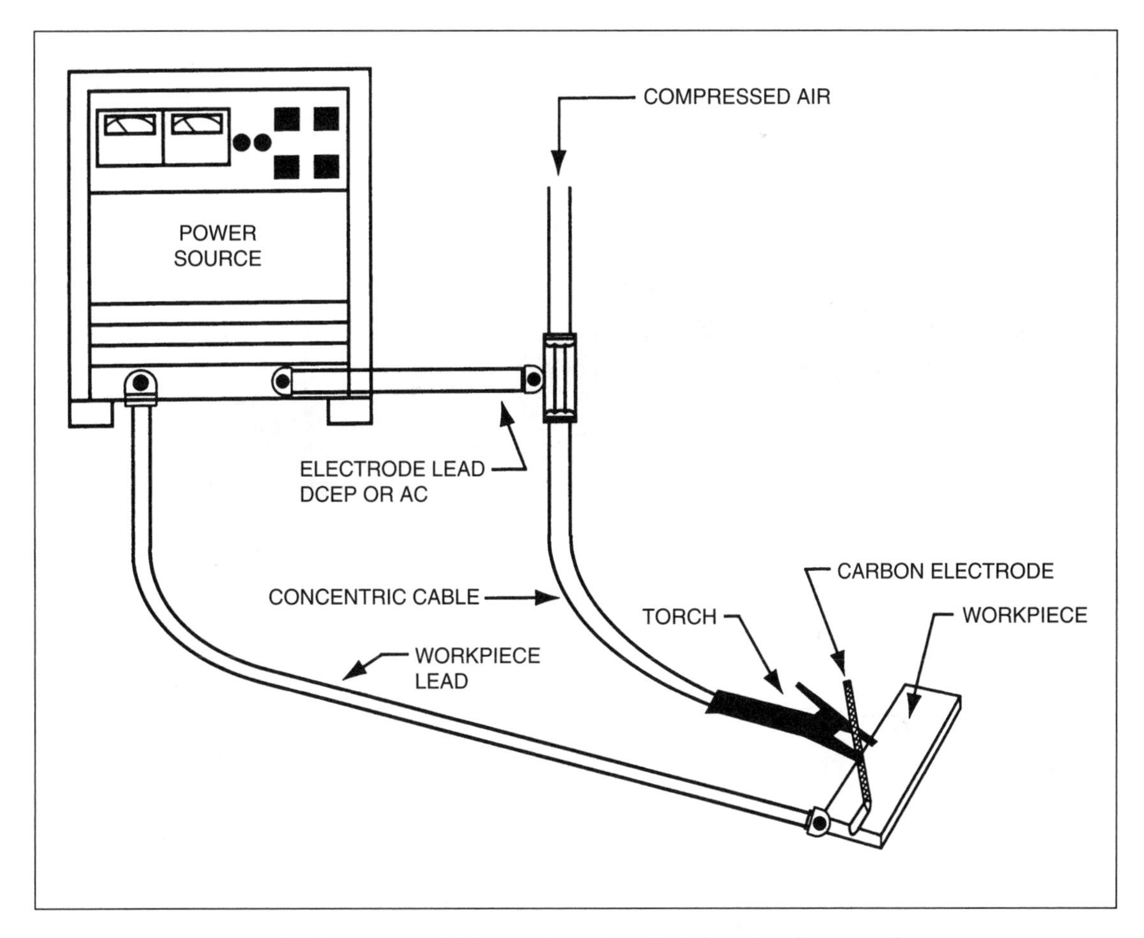

Figure 15.8—Typical Air Carbon Arc Cutting Equipment Setup

through solid-state controls. This system controls the electrode feed speed, which maintains the preset amperage. It can be operated with a constant-voltage power source only.

The voltage-controlled type maintains arc length by voltage signals through solid-state electronic controls. This type controls the arc length determined by the preset voltage and can be used with a constant-current power source only.

The dual-system control used for automated systems operates on a signal from the arc, and can be adjusted for amperage and voltage control by means of a selector switch located in the control. The dual system control can be used with either constant-current or constant-voltage power sources, with the choice dependent on whether voltage or amperage is being controlled.

Electrodes

The three types of electrodes used for air carbon arc cutting are direct-current copper-coated, direct-current plain, and alternating-current copper-coated electrodes. The shape most frequently used for electrodes is round. Flat and half-round electrodes are available for the creation of rectangular grooves and the removal of weld reinforcement.

Direct-current copper-coated electrodes are the most widely used because of they have a comparatively long service life, stable arc characteristics, and groove uniformity. These electrodes are manufactured from a special mixture of carbon and graphite with a suitable binder. The mixture is extruded and baked to produce dense, homogeneous graphite electrodes with low electrical resistance. The electrodes are then coated with a controlled thickness of copper. They are available in diameters ranging from 3.2 mm to 19 mm (1/8 in. to 3/4 in.).

Jointed direct-current copper-coated electrodes that prevent stub loss can be used. Designed with a socket and tenon to form a tight joint, they are available in diameters ranging from 8 mm to 25.4 mm (5/16 in. to 1 in.).

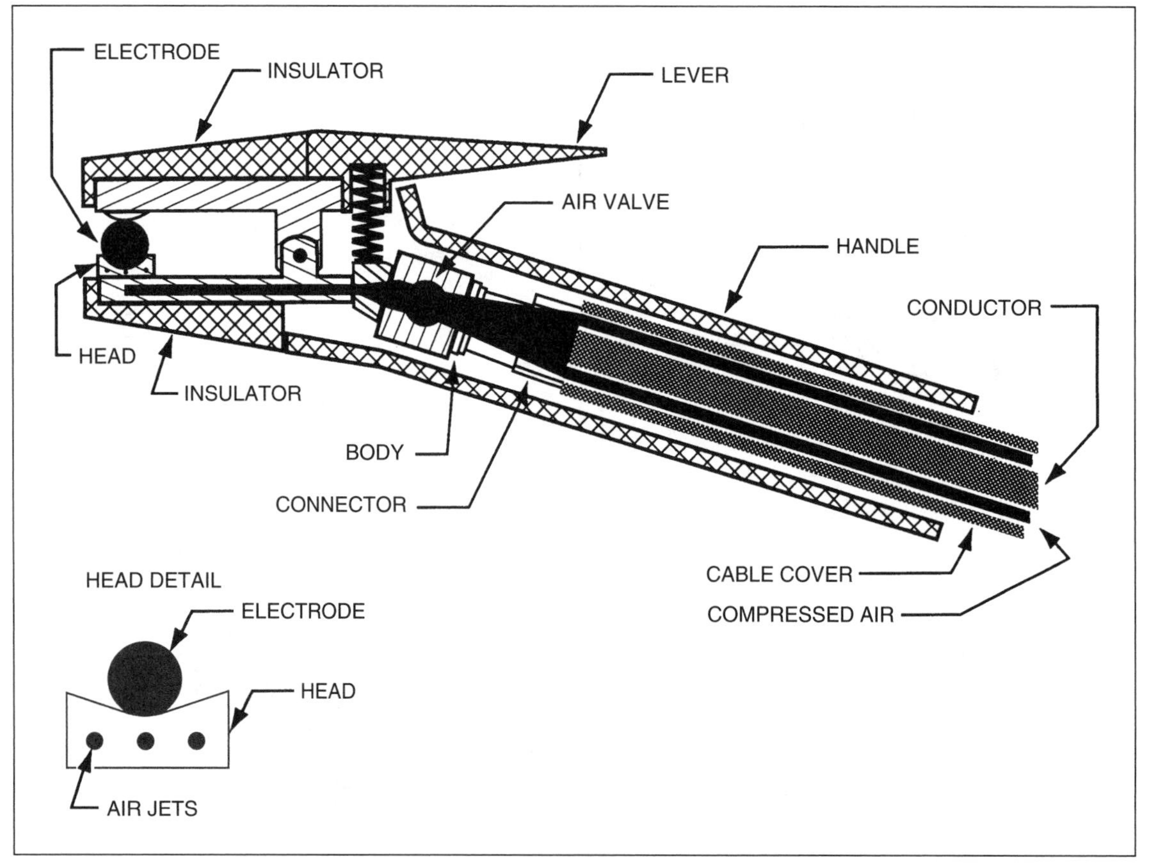

Source: Adapted from American Welding Society (AWS) Committee on Arc Welding and Cutting, *Recommended Practices for Air Carbon Arc Gouging and Cutting*, AWS C5.3:2000, Miami: American Welding Society, Figure 3.

Figure 15.9—Cross Section of a Torch for Manual Air Carbon Arc Cutting and Gouging

Of limited use, direct-current plain electrodes have no copper coating. During cutting, they are consumed more rapidly than coated electrodes. Plain electrodes are available in sizes ranging from 3.2 mm to 25.4 mm (1/8 in. to 1 in.) diameter, but their principal use involves diameters of less than 9.5 mm (3/8 in.).

Alternating-current coated electrodes are made from a mixture of carbon and graphite with rare-earth materials added to provide arc stabilization to permit cutting with an alternating current. These electrodes are coated with a controlled thickness of copper and are available in diameters ranging from 4.8 mm to 12.7 mm (3/16 in. to 1/2 in.).

Power Sources

Most standard welding power sources can be used for the air carbon arc cutting process. The open-circuit voltage should be sufficiently higher than the required arc voltage to allow for the voltage drop in the circuit. The arc voltages used in air carbon arc gouging and cutting range from 35 V to 55 V. An open-circuit voltage of at least 60 V is adequate. The actual arc voltage in air carbon arc gouging and cutting is determined to a large extent by the size of the electrode and the application.

Recommendations for the selection of power sources are summarized in Table 15.4.

The manufacturer of the power source should be consulted concerning its use for air carbon arc cutting because some types of power sources that are satisfactory for use with welding are not applicable to air carbon arc cutting.

Electrical leads in the cutting circuit should consist of standard welding cables recommended for arc welding. Cable size is determined by the maximum cutting current to be used.

Table 15.4
Power Sources for Air Carbon Arc Gouging and Cutting

Type of Current	Type of Power Source	Remarks
Direct current	Constant-current motor generator, rectifier, or resistor grid unit	Recommended for all electrode sizes.
Direct current	Constant-potential motor generator or rectifier	Recommended for 6.4 mm (1/4 in.) and larger diameter electrodes only. May cause carbon deposit with small electrodes. Not suitable for automatic torches with voltage control.
Alternating current	Constant-current transformer	Recommended for ac electrodes only.
Alternating current or direct current	Constant-current power source	Direct current supplied from three-phase transformer-rectifier sources is satisfactory, but direct current from single-phase sources provides unsatisfactory arc characteristics. Alternating-current output from ac/dc units is satisfactory provided ac electrodes are used.

Air Supply

Compressed air with pressure ranging from 560 kilopascals (kPa) to 700 kPa (80 pounds per square inch gauge [psig] to 100 psig) is normally required for air carbon arc gouging. Light-duty manual electrode holders allow for gouging and cutting with as little pressure as 280 kPa (40 psig) at 38.5 L/min (81.6 cubic feet per hour [ft^3/h]). Compressed nitrogen or inert gas can be used when compressed air is not available. Oxygen should not be used for air carbon arc cutting or gouging.

The air stream must be of sufficient volume and velocity to properly remove the melted slag from the kerf. The orifices in air carbon arc torches are designed to provide an adequate air stream for gouging and cutting. However, poor-quality cuts and gouges may result if the air pressure falls below the minimum specified by the torch manufacturer, or if the volume of air is restricted by hoses or fittings that are too small.

While gouges or cuts made with insufficient air may not always look particularly bad, they may be laden with slag and carbon deposits. For this reason, it is important that the air pressure be at or above the minimum specified for the type of torch being used. The inside diameter of all hoses and fittings must be large enough to allow the intended volume of air to reach the electrode holder.

Hoses and fittings with an inside diameter of 6.4 mm (1/4 in.) are sufficient for light-duty electrode holders. A minimum inside diameter of 9.5 mm (3/8 in.) is required for general purpose and heavy-duty electrode holders. Automated gouging electrode holders should be equipped with hoses and fittings with a minimum inside diameter of 12.7 mm (1/2 in.).

Operating Procedures and Techniques

In air carbon arc cutting, the electrode is held at an angle of between 70° and 80° to the surface of the workpiece. To cut thick nonferrous metals, the electrode should be held perpendicular to the workpiece surface, with the air jet in front of the electrode in the direction of travel. With the electrode in this position, the metal can then be severed by moving the arc upward and downward through the metal with a sawing motion.

Air carbon arc cutting electrodes are designed to operate with alternating current or direct current, or both, depending on the material to be cut. Table 15.5 lists the recommended electrodes and types of current for cutting several common alloys.

AIR CARBON ARC GOUGING

The variables that require attention in a given application of the air carbon arc gouging process are electrode diameter and type, amperage, voltage, air pressure and flow rate, travel speed, electrode push angle, electrode extension, and the base metal used. The functions of these variables are summarized in Table 15.6.

For gouging, the electrode should be held so that a maximum of 178 mm (7 in.) extends from the cutting torch, as shown in Figure 15.10. For nonferrous materials, this extension should be reduced to 76.5 mm (3 in.).

The air jet should be turned on before striking the arc, and the cutting torch should be gripped as shown in Figure 15.10. The electrode slopes back from the direction of travel with the air jet behind the electrode.

Table 15.5
Electrode and Current Recommendations for the Air Carbon Arc Cutting of Several Alloys

Alloy	Electrode Type	Current Type	Remarks
Carbon, low-alloy, and stainless steels	Direct current	DCEP	
	Alternating current	AC	Only 50% as efficient as DCEP
Cast irons	Alternating current	DCEN	At the middle of the electrode current range
	Alternating current	AC	
	Direct current	DCEP	At maximum current only
Copper alloys			
Copper 60% or less	Direct current	DCEP	At maximum current
Copper over 60%	Alternating current	AC	
Nickel alloys	Alternating current	AC	
	Alternating current	DCEN	
Magnesium alloys	Direct current	DCEP	Before welding, the surface must be cleaned.
Aluminum alloys	Direct current	DCEP	Electrode extension should not exceed 100 mm (4 in.). Before welding, surface must be cleaned.

Table 15.6
Primary Variables for Air Carbon Arc Gouging

Variable	Function
Electrode diameter	Determines the size of the groove.
Amperage	Determined by the diameter of the electrode being used. It is the current flow that melts the base metal.
Voltage	The electric potential behind the amperage, or the arc force. Voltage is set on constant-voltage power sources and is determined by arc length on constant-current power sources.
Air pressure and flow rate	The means for removal of the molten metal.
Travel speed	Determines the depth and quality of finished grooves.
Electrode travel and work angle	Can determine groove shape.
Electrode extension	Affects metal removal rates and quality of groove.
Base metal	Determines selection of parameters for other variables.

Source: Adapted from American Welding Society (AWS) Committee on Arc Welding and Cutting, *Recommended Practices for Air Carbon Arc Gouging and Cutting,* AWS C5.3:2000, Miami: American Welding Society, Table 6.

Under the proper operating conditions, the air jet sweeps beneath the electrode end and removes all molten metal. The arc can be struck by lightly touching the electrode to the workpiece. The electrode should not be drawn back once the arc is struck. A short arc should be maintained by progressing in the direction of the cut fast enough to keep up with metal removal. The steadiness of progression controls the smoothness of the resulting cut surface.

The current ranges for commonly used air carbon arc gouging electrodes are presented in Table 15.7. The actual current used for a given electrode size depends on the operating conditions such as the material being cut, type of cut, cutting speed, cutting position, and required cut quality. The recommendations of the manufacturer should be followed for the operation and maintenance of the equipment and the use of consumable materials.

When using jointed carbon electrodes, it is important to strike the arc with the open or blunt end of the electrode. The reason for this becomes apparent when the electrode has been almost completely consumed and is approaching the jointed section. If the arc had been struck on the tapered end of the electrode, the jointed

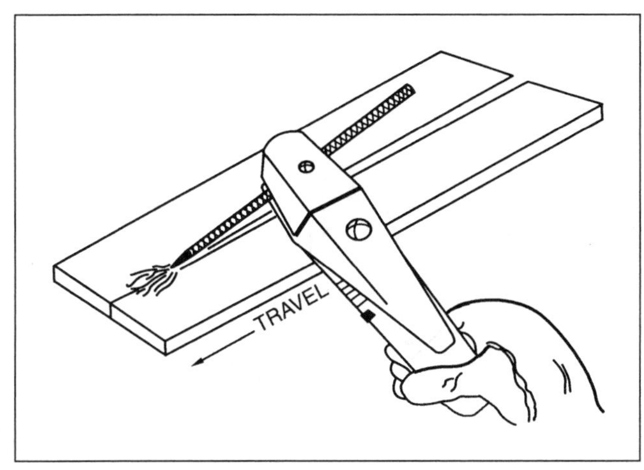

Source: Adapted from American Welding Society (AWS) Committee on Arc Welding and Arc Cutting, 2000, *Recommended Practices for Arc Gouging and Cutting*, AWS C5.3:2000, Miami: American Welding Society, Figure 5.

Figure 15.10—Manual Air Carbon Arc Gouging in the Flat Position

section would consist of a tapered end surrounded by a loose red-hot sleeve of carbon. This hot sleeve tends to be ejected violently from the gouging arc and, like weld spatter, can cause burns or set combustibles on fire. When the arc is struck with the open end of the electrode and the electrode is consumed to the jointed section, the sleeve forms part of the incoming electrode and is restrained from violent ejection.

When gouging a workpiece in the vertical position, the operation should be performed downhill to allow gravity to assist in removing the molten metal. Gouging in the horizontal position can be performed either to the right or to the left, but it should always be done in the forehand direction.

In gouging to the left, the cutting torch should be held as shown in Figure 15.10. In gouging to the right, the cutting torch is reversed to locate the air jet behind the electrode. When gouging overhead, the electrode and torch should be held at an angle that will prevent molten metal from falling on the operator.

The depth of the groove is controlled by the travel speed. Slow travel speeds produce a deep groove; fast speeds produce a shallow groove. Grooves up to 25 mm (1 in.) deep can be made. However, the deeper the groove, the more experience is required on the part of the operator for successful results.

The width of the groove is determined by the size of the electrode used. The groove width is usually about 3.2 mm (1/8 in.) wider than the electrode diameter. A wider groove can be produced by oscillating the electrode using a circular or weaving motion.

When gouging, a push angle of 35° to 70° from the surface of the workpiece is used for most applications. A steady rest is recommended in gouging to ensure a smoothly gouged surface. This is particularly advantageous for use in the overhead position. The proper travel speed depends on the size of the electrode, type of base metal, cutting amperage, and air pressure. An indication of the proper speed, and thus good gouge quality, is a smooth, hissing sound in the arc.

Table 15.7
Suggested Current Ranges for Commonly Used Air Carbon Arc Gouging Electrodes

Electrode Diameter		Direct-Current Electrode with DCEP,* A		Alternating-Current Electrode with AC, A		Alternating-Current Electrode with DCEN,** A	
mm	in.	Minimum	Maximum	Minimum	Maximum	Minimum	Maximum
4.0	5/32	90	150	—	—	—	—
4.8	3/16	150	200	150	200	150	180
6.4	1/4	200	400	200	300	200	250
7.9	5/16	250	450	—	—	—	—
9.5	3/8	350	600	300	500	300	400
12.7	1/2	600	1000	400	600	400	500
15.9	5/8	800	1200	—	—	—	—
19.1	3/4	1200	1600	—	—	—	—
25.4	1	1800	2200	—	—	—	—

*Direct Current Electrode Positive
**Direct Current Electrode Negative

Source: Adapted from American Welding Society (AWS) Committee on Arc Welding and Arc Cutting, 2000, *Recommended Practices for Arc Gouging and Cutting*, AWS C5.3:2000, Miami: American Welding Society, Table 3.

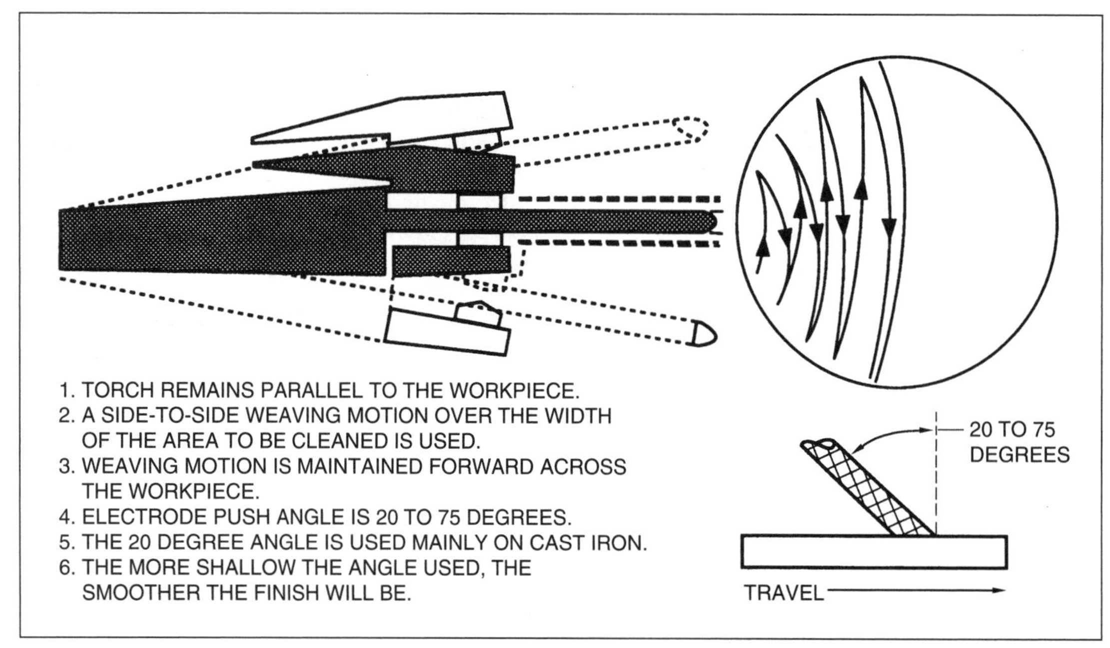

Source: Adapted from American Welding Society (AWS) Committee on Arc Welding and Arc Cutting, 2000, Recommended Practices for Arc Gouging and Cutting, AWS C5.3:2000, Miami: American Welding Society, Figure 10.

Figure 15.11—Pad Washing Technique with Air Carbon Arc Electrode Holder

Washing

When using the air carbon arc gouging process for washing, or the removal of metal from large areas as is done when removing surfacing metal or riser pads from castings, the proper position of the electrode should be the position shown in the lower right corner of Figure 15.11. The electrode should be oscillated from side to side while pushing forward at the depth desired. In pad washing operations, an angle of 20° to 75° to the workpiece surface is used. The 20° angle is used for cast iron or light finishing passes, while the steeper angles allow deeper rough gouging to be done with greater ease on most materials.

Cutting torches with fixed-angle heads that hold the electrode at the correct angle are particularly well suited for this application. With other types of torches, care should be exercised in keeping the air behind the electrode. The steadiness of the operator's hand determines the smoothness of the surface produced.

APPLICATIONS

The air carbon arc gouging process can be used to sever, gouge, wash, and bevel most metals including carbon, low-alloy, and stainless steels; cast iron; and alloys of aluminum, magnesium, copper, and nickel. Gouging is often used to prepare plate and pipe edges for welding. Two edges can be butted together and a U-groove gouged along the joint, as shown in Figure 15.10. The root of a weld can be gouged out to sound metal before completing the weld on the second side. Similarly, defective weld metal can be gouged out for repair. Another application is the removal of old surfacing material before a workpiece is resurfaced.

Metallurgical Effects

To avoid difficulties with carburized metal, users of the air carbon arc process should be aware of the metallurgical events that occur during gouging and cutting. When the carbon electrode is positive, the current flow carries ionized carbon atoms from the electrode to the base metal. The free carbon particles are rapidly absorbed by the melted base metal. Since this absorption cannot be avoided, it is important that all carburized molten metal is removed from the kerf, preferably by the air jet.

When the air carbon arc cutting process is used under improper conditions, the carburized molten metal left on the surface can usually be recognized by its dull, gray-black color in contrast to the bright blue color of a properly made groove. Inadequate air flow may leave small pools of carburized metal in the bottom of the groove. Irregular electrode travel, particularly in a manual operation, may produce ripples in the groove wall that tend to trap the carburized metal. Finally, an improper electrode angle may cause small beads of carburized metal to remain along the edge of the groove.

The effect of carburized metal that remains in the kerf or groove through a subsequent welding operation depends on many factors. These include the amount of carburized metal present, the welding process to be employed, the kind of base metal, and the weld quality required. Although it may seem that filler metal deposited during welding would dissolve small pools or beads of carburized metal, experience with steel base metals has shown that traces of metal containing approximately 1% carbon may remain along the weld fusion line. Carbon pickup in the weld metal becomes significant with demands for increasing weld strength and toughness. Increased carbon content can decrease weld toughness, especially in quenched-and-tempered steels. No evidence suggests that the copper from copper-coated electrodes is transferred to the cut surface in the base metal.

Carburized metal on the cut surface can be removed by grinding, but it is much more efficient to completely prevent the retention of undesirable metal by performing air carbon arc gouging and cutting properly within prescribed conditions.

Comparisons of corrosion rates typical for Type 304L stainless steel welds were made to determine whether air carbon arc gouging carried out in the prescribed manner would adversely affect corrosion resistance. No significant difference was observed in the corrosion rates of welds prepared by air carbon arc cutting and those prepared by grinding. If any appreciable carbon absorption had occurred, the corrosion rates for welds prepared by air carbon arc cutting would have been significantly higher. However, surfaces prepared using air carbon arc cutting may be more susceptible to stress corrosion cracking, depending on the service environment. If a concern exists, the surfaces should be mechanically dressed following air carbon arc cutting.

Air carbon arc cutting requires a lower heat input than oxyfuel gas cutting. For this reason, a workpiece gouged or cut by the air carbon arc process has less distortion than one prepared by oxyfuel gas cutting.

Safety

Safety recommendations for air carbon arc cutting and gouging are included in the last section of this chapter, "Safe Practices."

OTHER ARC CUTTING PROCESSES

This section provides a brief description of five additional arc cutting processes—shielded metal arc cutting, oxygen arc cutting, gas tungsten arc cutting, gas metal arc cutting, and carbon arc cutting. These processes are not widely used in industry because of economic considerations. However, the reader should be aware of them because they can be used when other processes are not available. Additional sources of information about these processes are cited in the supplementary reading list.

SHIELDED METAL ARC CUTTING

Shielded metal arc cutting (SMAC) is an arc cutting process that uses a covered electrode. A constant-current power source operating on DCEN is preferred. The principal functions of the electrode covering during cutting are to serve as electrical insulation, permitting the insertion of the electrode into the groove of the cut without short-circuiting the sides of the electrode, and to act as an arc stabilizer, concentrating and intensifying the action of the arc.

The effectiveness of this procedure in cutting thick base metals is a function of the manipulation of the electrode. Electrodes E6010, E6012, and E6020 are usually employed, but cutting can be achieved using virtually any shielded metal arc welding electrode. Electrodes with coverings especially made for cutting are also available.

The principles of operation, power sources, and other equipment are substantially similar to those for shielded metal arc welding.

Although a dc constant-current welding machine is preferred for shielded metal arc cutting, an ac constant-current power source can also be used. For shielded metal arc cutting in air, heavy-duty electrode holders should be used with 4.8 mm (3/16 in.) diameter and larger electrodes. For shielded metal arc cutting under water, specially constructed, fully insulated electrode holders are mandatory. A DCEN power source must be used to protect the electrode holder and the metal parts of the diver's outfit from electrolytic corrosion.

Applications

Shielded metal arc cutting has been used for the cutting of risers and gates in nonferrous foundries and to cut nonferrous scrap for remelting. The workpiece should be positioned so that gravity assists in removing the molten metal. Generally, the process does not provide satisfactory edge preparation for welding without considerable cleanup by chipping or grinding.

OXYGEN ARC CUTTING

Oxygen arc cutting (OAC) is an oxygen cutting process that uses an arc between the workpiece and a consumable tubular electrode through which oxygen is directed to the workpiece. Mild steel is cut by using the arc to raise the temperature of the material to its kindling point in the presence of oxygen. The combustion reaction that occurs is self-sustaining, liberating sufficient heat to maintain the kindling temperature on all sides of the cut. The necessary preheat at the start of the cutting operation is provided by the electric arc. A schematic illustration of the process is shown in Figure 15.12.

For oxidation-resistant metals, the cutting mechanism is more of a melting action. In these instances, the covering on the electrode provides flux that helps the molten metal flow from the cut.

Operating Techniques

In the oxygen arc method of cutting, piercing, and gouging, the coating is kept in contact with the base metal at all times. The coating insulates the core from the workpiece and automatically maintains the proper arc length.

The technique for starting cutting and piercing operations is much the same as for arc welding. The tip of the electrode is tapped on the workpiece at the desired location and the arc is maintained for a moment while the oxygen valve is opened. Piercing action begins immediately, and the electrode is pushed through the plate as the hole is formed. The coating insulates the electrode core from shorting against the sides of the hole.

In cutting, the electrode is dragged along the plate surface at the speed of travel dictated by the progress of the cut. The inclination of the electrode and the speed of motion are adjusted to render the most efficient and highest quality cut.

Template-guided cutting is common. The electrode is pressed against the template and is insulated from it by the coating. For straight-line cuts, any straight edge can be clamped along the line to be cut. The cut is accomplished by holding the electrode against the guide and the plate at the same time. Circular openings in tanks are sometimes cut by using the circumference of a suitably sized pipe as the guiding template.

When cutting in air (up to 75 mm [3 in.] of mild steel or 12 mm [1/2 in.] of certain nonferrous alloys), the technique involves dragging the electrode along the intended line of cut while applying slight pressure. In underwater cutting, positive pressure must be maintained against the metal being cut, regardless of the thickness of the metal.

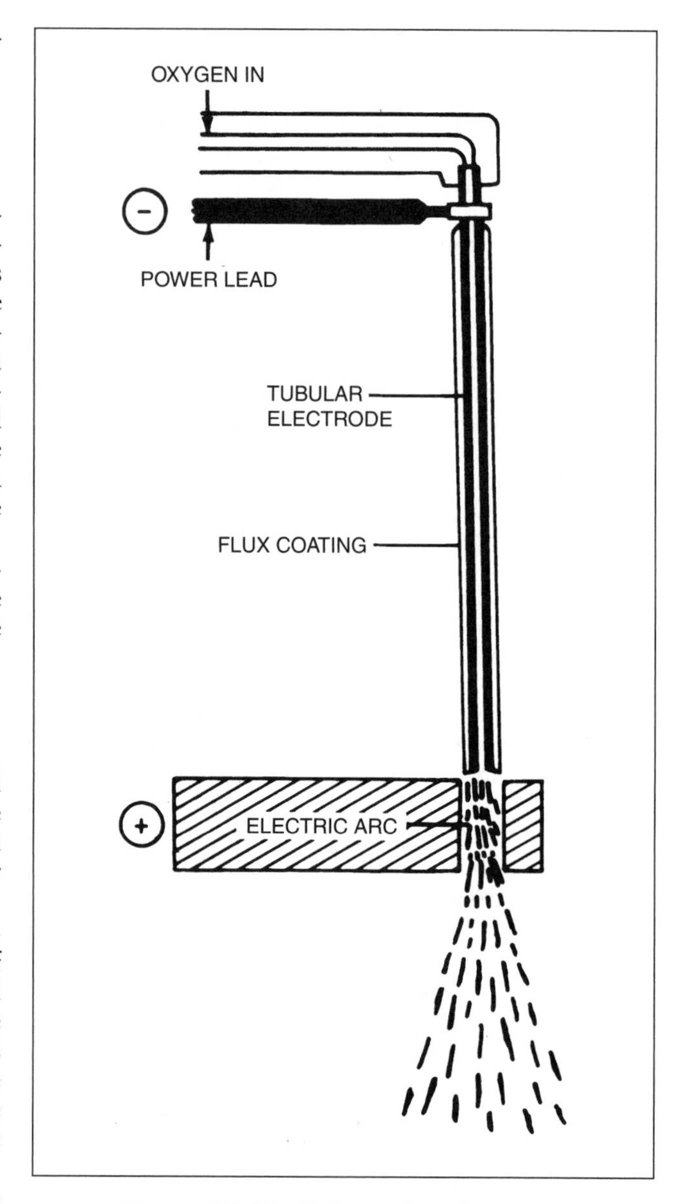

Figure 15.12—Schematic of Oxygen Arc Cutting Electrode in Operation

Gouging is performed by striking the arc, releasing the oxygen stream, and inclining the electrode until it is almost parallel to the plate surface and pointed away from the operator along the line of the prospective gouge. The arc and the oxygen melt the plate surface, and the molten metal is blown away by the force of the oxygen jet.

Equipment

Oxygen arc cutting can be performed using alternating current or direct-current power sources. Direct current electrode negative is preferred for rapid cutting.

The electrode holder used for oxygen arc cutting is of a special design. It must convey both electric current to the electrode and oxygen to the cut. This is accomplished by bringing oxygen to the electrode holder and passing it through the bore of the electrode into the arc. For cutting in air, a fully insulated electrode holder is desirable. When the process is used for underwater cutting, a fully insulated holder equipped with a suitable flashback arrester is mandatory.

Tubular steel electrodes are available in 5 mm to 8 mm (3/16 in. to 5/16 in.) diameters, 0.46 m (18 in.) long, with bore diameter approximately 1.6 mm (1/16 in.). The extruded covering is comparable to a mild steel electrode of AWS classification E6013. Underwater electrodes are steel tubes with waterproof coatings.

Applications

Oxygen arc cutting electrodes were developed primarily for use in underwater cutting and later applied to cutting in air. In both applications, the electrodes can cut ferrous and nonferrous metals in any position.

Oxygen arc cutting has been used effectively by foundries and scrap yards for cutting mild and low-alloy steels, stainless steel, cast iron, and nonferrous metals in any position. The usefulness of the process varies with the thickness and composition of the material being cut.

The edges of metal cut by the oxygen arc torch are somewhat uneven. Thus, they usually require a light surface preparation to make them suitable for welding.

Cut Quality

The oxygen arc method of cutting produces metallurgical effects in the heat-affected zone comparable to those that occur in shielded metal arc welding. The power input approaches that of shielded metal arc welding, but the heat penetration is generally not as deep in oxygen arc cutting because of the faster speed of travel. This produces a somewhat more pronounced quench effect. Metals that do not require a postheat treatment after welding can be severed by this process without detrimental effects. Grades of austenitic stainless steels that are sensitive to corrosion attack when subjected to shielded metal arc welding are sensitized along the cut when severed by this process.

Oxygen arc cuts in cast iron and medium-carbon, low-alloy steels are apt to develop cracks on the face of the cut. The extent and frequency of cracking depend on the composition and hardenability of the steel.

GAS TUNGSTEN ARC CUTTING

Gas tungsten arc cutting (GTAC) can be used to sever nonferrous metals and stainless steels in thicknesses up to 12 mm (1/2 in.) with standard gas tungsten arc welding equipment. Among the metals cut with GTAC include aluminum, magnesium, copper, silicon-bronze, nickel, copper-nickel, and various types of stainless steels. This cutting process can be used in either a manual or mechanized mode. The same electric circuit is used for cutting as for welding. Higher current is required to cut a given thickness of plate than to weld it. An increased gas flow is also required to remove the molten metal and sever the plate.

Fundamentals of Operation

In GTAC, a 4 mm (5/32 in.) diameter 2% thoriated tungsten electrode is extended approximately 6.4 mm (1/4 in.) beyond the end of a 9.5 mm (3/8 in.) diameter metallic or ceramic gas nozzle. A mixture of approximately 65% argon and 35% hydrogen is delivered to the torch at a flow rate of 17 L/min (60 ft^3/h). Nitrogen can also be used, but the quality of the cut is not as good as that obtained with an argon-hydrogen mixture. Best cutting results are obtained using DCEN, but alternating current with superimposed high frequency has produced satisfactory cuts on material up to 6.4 mm (1/4 in.) thick.

Arc starting can be accomplished with a high-frequency spark or by scratching the electrode on the workpiece. An electrode-to-workpiece distance of 1.6 mm to 3.2 mm (1/16 in. to 1/8 in.) is used, but this is not a critical factor. As the torch is moved over the plate, a small section of the plate is melted by the heat of the arc, and the molten metal is blown away by the gas stream to form the kerf. At the end of the cut, the torch is raised from the workpiece to break the arc.

Equipment

Standard gas tungsten arc welding torches can be used for cutting. Cutting currents up to 600 A are used, as noted in Table 15.8. Welding torches can be used for cutting at currents up to 175% of their nominal ratings because little heat is reflected from the cutting operation. For example, a 300-A torch can be used to perform cutting operations with 500 A for short periods of time.

A constant-current, direct-current power source (either a rectifier or motor-generator) with a minimum open-circuit voltage of 70 V is recommended for cutting. Cuts made with alternating current are limited

Table 15.8
Conditions for Gas Tungsten Arc Cutting

	Thickness		Travel Speed		DCEN Current	
Material	mm	in.	mm/min	In./min	A	Type of Gas
Stainless Steel	3	1/8	510	20	350	80% Ar + 20% H_2
	6	1/4	510	20	500	65% Ar + 35% H_2
	13	1/2	380	15	600	65% Ar + 35% H_2
Aluminum	3	1/8	760	30	200	80% Ar + 20% H_2
	6	1/4	510	20	300	65% Ar + 35% H_2
	13	1/2	510	20	450	65% Ar + 35% H_2

to plate thicknesses of 6.4 mm (1/4 in.). The major difficulty encountered when using alternating current is the loss of tungsten from the electrode at the high currents required.

One face of the gas tungsten arc cut is usually dross-free, with dross adhering to the side of the workpiece farthest from the workpiece lead. The cut quality on the dross-free side is usually acceptable, while the other side requires considerable cleanup.

Table 15.8 provides conditions for the gas tungsten arc cutting of commonly used thicknesses of stainless steels and aluminum.

GAS METAL ARC CUTTING

Gas metal arc cutting (GMAC) is an arc cutting process that uses a continuous consumable electrode and a shielding gas. The process was developed soon after the commercial introduction of the gas metal arc welding process. Gas metal arc cutting first occurred accidentally during a welding operation when it was found that if the electrode feed rate were set too high, the plate would be penetrated. When the torch was moved, a cut was made.

Gas metal arc cutting is used to cut shapes in stainless steel and aluminum. Using normal welding equipment and a 2.4 mm (3/32 in.) diameter carbon-steel electrode, stainless steel up to 38 mm (1-1/2 in.) thick and aluminum up to 76 mm (3 in.) thick can be cut.

The main drawbacks to the use of gas metal arc cutting are the high consumption rate of welding electrodes and the high current (up to 2000 A) required for cutting.

CARBON ARC CUTTING

Carbon arc cutting, the oldest arc cutting process, is rarely used and is considered obsolete. The process used an arc between a carbon (graphite) electrode and the base metal to melt the surface of the workpiece. Since the process depended on gravity to remove the molten metal, it could only be used in the vertical and overhead positions.

One variation used the arc force to assist in pushing metal out of the kerf by using higher amperages. The cuts produced required extensive cleanup of dross and slag. Prior to welding, the cut edges required grinding to remove the melted area remaining on the metal, which was high in carbon picked up from the carbon electrode.

SAFETY

Safety recommendations for other cutting and gouging processes are included in the last section of this chapter, "Safe Practices."

ECONOMICS

Severing mild steel less than 50 mm (2 in.) thick is the most common industrial cutting application. Plasma arc cutting is the preferred process because it provides high-quality cuts at faster cutting speeds at a lower cost, not only in mild steel, but also in stainless steels and aluminum alloys. The cost of plasma arc cutting is compared with other cutting processes in this section. The relative costs of plasma arc gouging and air carbon arc gouging are also discussed.

The main categories of costs associated with metal cutting are the cost of the cutting system, worktables, ventilation equipment and other accessories, and operating expenses.

PLASMA ARC CUTTING

The range of costs of a plasma arc cutting system depends on the intended use. Small, hand-held cutting torches are inexpensive and are adequate for general use. Precision, high-speed plasma arc cutting systems are more expensive, and the purchase price increases for more advanced systems that produce higher quality cuts or higher cutting speeds.

For the plasma arc process, cutting tables with CNCs and ventilation equipment often account for the largest portion of the cost of a complete cutting system. Other equipment purchases may include material handling systems, water chilling equipment, and gas and water supply lines.

Operating expenses include the cost of the consumable torch parts, gas, electricity, labor, and overhead associated with operating and maintaining the cutting system. Costs vary according to the type and size of the system and the material being cut. Air can be used as the plasma and shield gases in some operations, but in others a significantly more expensive gas is needed. A typical allocation of costs for an oxygen plasma arc cutting operation is shown in Figure 15.13.

Economies can be achieved with higher cutting speeds, which reduce the per-unit cost, and by cutting square, clean edges that reduce the overall cost of the cut parts. The cost of secondary operations such as dross removal must be considered if required by the application.

The selection of the cutting process is often based on the characteristics of the process. Conventional plasma arc, precision plasma arc, oxyfuel gas and laser cutting systems each excel at some aspect of cutting, such as cut edge characteristics or type or thickness of metal. If there is not a clear choice, a decision can be based on cut quality and the cost per unit length of cutting.

A comparison of the costs of air plasma arc cutting, oxygen plasma arc cutting, and oxyfuel gas cutting of mild steel up to 50 mm (2 in.) is shown in Figure 15.14.

PLASMA ARC GOUGING

In some applications, the critical issue is simply to optimize the economics of a gouging application, and the advantages of plasma arc gouging are irrelevant. Obviously, the gouging job can be done by other means, but the overall cost of another process can often be much higher than that of plasma arc gouging when the requirement and cost of post-gouge cleanup is considered. The actual cost per unit length of gouge, which includes gouging materials, gas, electric power, and labor, must be calculated for a particular application.

Plasma Arc Gouging and Air Carbon Arc Cost Comparisons

Plasma arc gouging with inert gas is often considered more expensive than air carbon arc gouging, which uses compressed air. Compressed air is often perceived to be

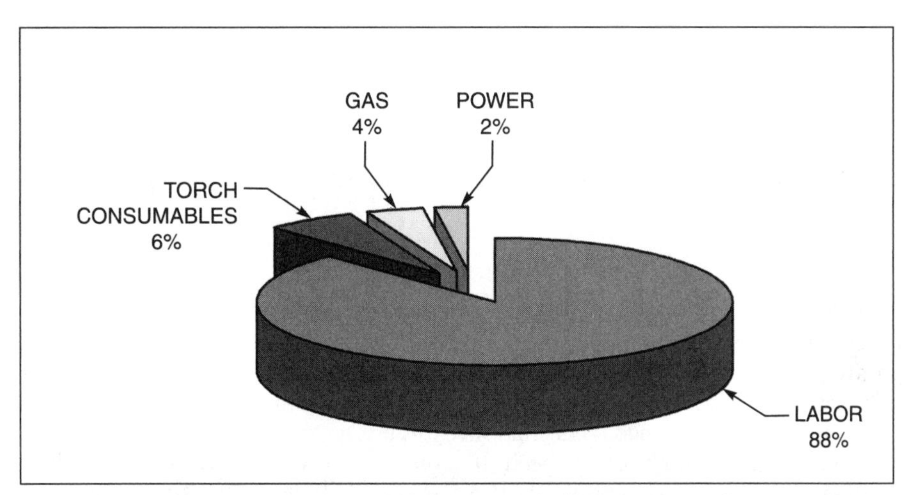

Source: Adapted from American Welding Society (AWS) Committee on Arc Welding and Cutting, *Recommended Practices for Plasma Arc Cutting and Gouging*, AWS C5.2:2001, Miami: American Welding Society, Figure 21.

Figure 15.13–Typical Allocation of Costs of an Oxygen Plasma Arc Cutting Operation at 200 A

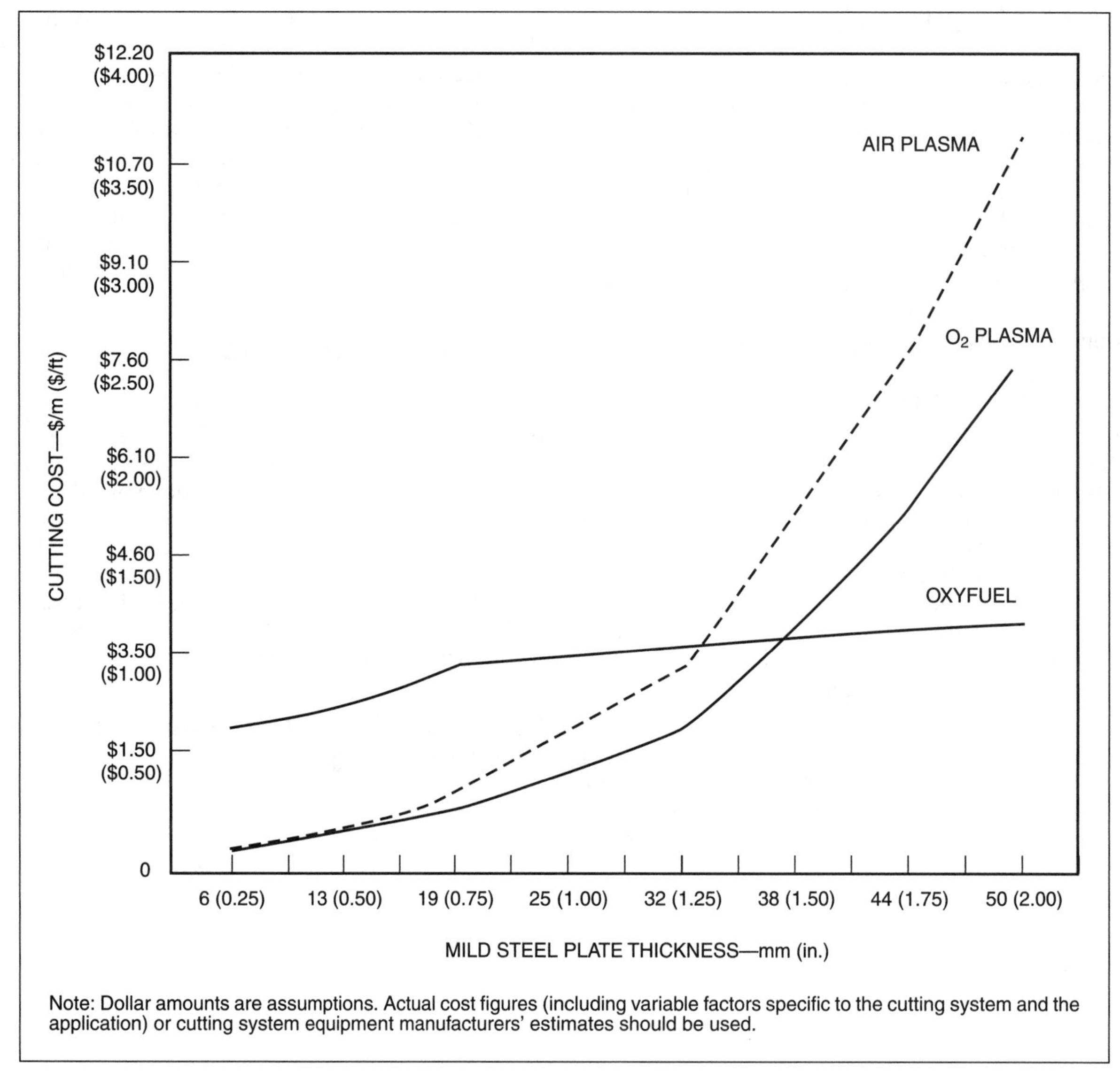

Source: Adapted from American Welding Society (AWS) Committee on Arc Welding and Cutting, 2001, Recommended Practices for Plasma Arc Cutting and Gouging, AWS C5.2:2001, Miami: American Welding Society, Figure 22.

Figure 15.14–Cost Comparison of Plasma Arc Cutting at 200 A and Oxyfuel Gas Cutting of Mild Steel

free of cost, but it actually is not. However, since many users of air carbon arc gouging do not associate a cost with compressed air, for the purpose of the comparison shown in Table 15.9, no cost is assigned. The major advantage of plasma arc gouging is high travel speed. It is four times faster than air carbon arc gouging. At the assumed labor and overhead rate of $30 per hour in the cost comparison, plasma arc gouging results in a much lower cost per unit length of gouge than air carbon arc gouging.

The carbon electrode is consumed, so there is a cost per unit length of gouge for the electrode. With plasma arc gouging there is no consumable rod; however, the torch electrode and gas nozzle have to be replaced periodically. In plasma arc gouging, the electrode and nozzle typically last more than 6 hours, longer than typical plasma arc cutting torch parts.

Although the cost of a plasma arc gouging electrode is higher than a carbon arc gouging electrode, the cost is insignificant when balanced against other advantages of plasma arc gouging. Because the process uses a more efficient heat source, less power is consumed by plasma arc gouging than air carbon arc gouging. This further contributes to cost savings. On the basis of cost per cm (foot) of gouge, the direct cost of plasma arc gouging is considerably lower than air carbon arc gouging.

In addition to direct cost savings, plasma arc gouging provides indirect cost savings that vary from one application to another. The cost of grinding and cleanup is an example. This may be negligible on carbon steel, but when air carbon arc gouging is used with stainless steel or aluminum, the secondary cleanup costs include grinding wheels and labor, which can be substantial. This is avoided with plasma arc gouging, particularly if an inert gas system is used.

Coping with fume generation with air carbon arc gouging is a significant cost consideration. Depending on the duty cycle, location (inside or outside), local environmental laws, size of the shop, and so forth, fume generation may not be a problem or it may present a major problem. The installation of fume removal equipment can be expensive to the point of being prohibitive. If fumes are a significant problem, plasma arc gouging may be the only practical choice.

Conversely, a much higher initial investment is required for plasma arc gouging than is required for an air carbon arc setup. For air carbon arc gouging, users often make use of existing welding power sources and a compressed air supply and merely add a gouging torch for an investment of a few hundred dollars. A plasma arc gouging system for high-production gouging may cost from ten to twenty times more than an air carbon arc gouging setup. However, the payback is usually quite rapid.

Table 15.9 presents a cost comparison of plasma arc gouging at 150 A with an inert gas versus air carbon arc gouging with a 5/16 in. electrode that gives a similar metal removal rate and groove size. Cost figures are assumptions.

SAFE PRACTICES

The general subjects of safety and safe practices in welding and thermal cutting processes are addressed in the standards *Safety in Welding, Cutting, and Allied Processes,* ANSI Z49.1,[6] and NFPA 51B, *Fire Prevention during Welding, Cutting, and Other Hot Work,* NFPA 51B.[7] All personnel engaged in thermal cutting should be familiar with the practices discussed in these documents. As a rule, equipment should not be operated until the manufacturer's instructions have been read and understood. In addition, potential physical hazards such as those due to the high-pressure gas and water systems must be considered.

ARC CUTTING AND GOUGING

Many of the potential hazards and safety recommendations to prevent accidents apply to all arc cutting and arc gouging processes. Among the hazards are fumes and gases, noise, radiant energy, fire, and electric shock. General precautions are described in this section, which includes separate segments for potential hazards of plasma arc cutting and gouging and air carbon arc cutting and gouging.

Fumes and Gases

Arc cutting produces fumes and gases that may be hazardous to health. The composition and generation rate of fumes and gases depend on many factors including arc current, cutting speed, the material being cut, and the gases used. The fume and gas byproducts usually consist of the oxides of the metal being cut, ozone, and oxides of nitrogen, including nitrogen dioxide. Phosgene gas could be present as a result of the thermal or ultraviolet decomposition of chlorinated hydrocarbon cleaning agents or suspension agents used in some aerosol antispatter agents or paints. Degreasing or other operations involving chlorinated hydrocarbons should be located so that vapors from these operations are not exposed to radiation from the arc.

The metal fumes generated by the arc cutting processes can be controlled by natural ventilation, local exhaust ventilation, or the respiratory protective equipment described in *Safe Practices in Welding, Cutting, and Allied Practices,* ANSI Z49.1.[8] The method of ventilation required to keep the level of toxic substances that may be in the cutter's breathing zone within acceptable concentrations is directly dependent on several factors including the metal being cut, the size of the work area, and the degree of confinement or obstruction to normal air movement where the cutting is taking place. Each operation should be evaluated on an individual basis to determine requirements.

The acceptable levels of the toxic substances associated with cutting, designated as time-weighted average threshold limit values (TLVs®) and ceiling values, have been established by the American Conference of Governmental Industrial Hygienists (ACGIH)[9] and by the

6. American National Standards Institute (ANSI) Accredited Standards Committee Z49, 1999, *Safety in Welding, Cutting and Allied Processes,* ANSI Z49.1. Miami: American Welding Society

7. National Fire Protection Association, *Standard for Fire Prevention During Welding, Cutting, and Other Hot Work,* NFPA 51B, Quincy, Massachusetts: National Fire Protection Association.

8. See Reference 6.

9. See Reference 5.

Table 15.9
Typical Cost Comparison: Plasma Arc Gouging versus Air Carbon Arc Gouging (U.S. Customary Units Only)

	Plasma Arc Gouging	Air Carbon Arc Gouging
I. Gas Cost		
A = Cooling gas	\$0.075 ft^3 (argon)	0 (air)
B = Cooling gas flow rate	280 ft^3/hr	3600 ft^3/hr
C = Plasma gas cost	\$0.21 ft^3 (H-35)	\$0
D = Plasma gas flow rate	130 ft^3/hr	\$0
E = Travel speed	100 in./min	25 in./min
$\dfrac{(A \times B) + (C \times D)}{E \times 60 \text{ min/hr}}$ 12 in./ft	\$0.097/ft	\$0/ft
II. Labor and Overhead Cost		
F = Labor and overhead	\$30./hr	\$30./hr
$\dfrac{I \times J \times 12 \text{ in./ft} \times 60 \text{ sec/min}}{E \times 60 \text{ min/hr} \times 60 \text{ sec/min}}$ 12 in./ft	\$0.06/ft	\$0.24/ft
III. Electrode Cost per Foot		
G = Electrode cost		
Electrode cost per foot = 12.5 ft × 5/16 in.	\$0	\$0.563
H = Gouge length per electrode	N/A	42 in.
$\dfrac{G \times 12 \text{ in./ft}}{H}$	\$0	\$0.161/ft
I = Power consumption per electrode	28.5 kW	24.8 kW
IV. Power Cost		
J = Power cost		
$\dfrac{I \times J \times 12 \text{ in./ft} \times 60 \text{ sec/min}}{E \times 60 \text{ min/hr} \times 60 \text{ sec/min}}$	\$0.0057	\$0.0198
Total (I + II + III + IV)	**\$0.1623/ft**	**\$0.4208/ft**
Savings Using Plasma Gouging	**\$0.2585/ft**	

Key:
A = Cooling gas
B = Cooling gas flow rate
C = Plasma gas
D = Plasma gas flow rate
E = Gouging speed
F = Labor and overhead
G = Electrode cost
H = Gouge length per electrode
I = Power consumption per electrode
J = Power cost

Occupational Safety and Health Administration (OSHA). Compliance with these levels can be verified by sampling the atmosphere under the cutter's helmet or in the immediate vicinity of the cutter's breathing zone. Sampling should be in accordance with *Method for Sampling Airborne Particulates Generated by Welding and Allied Processes,* ANSI/AWS F1.1.[10]

The ultraviolet light emitted from the arc acts on the oxygen in the surrounding atmosphere to produce ozone. The amount of ozone produced depends on the intensity of the ultraviolet energy, the humidity, and the amount of screening afforded by the fume, among other factors. The concentration of ozone generally increases with an increase in current and when cutting or gouging aluminum. Fume control methods are described in American National Standards Institute Z49 Committee, *Safety In Welding, Cutting and Allied Processes,* ANSI Z49.1.[11]

With respect to nitrogen dioxide, tests have shown that high concentrations are found only close to the arc. Natural ventilation quickly reduces these concentrations to safe levels in the typical breathing zone of cutters; however, they must keep their heads out of the cutting fumes. With some applications this may not be possible. Fumes must be removed from the work area or eliminated at the source by using an exhaust system or other engineering controls. Codes generally require that the exhaust be filtered before being vented to the atmosphere.

Plasma Arc Cutting. Several alternative fume-removal systems are available for mechanized plasma arc cutting. One system consists of two parts, an annular nozzle that generates a water shroud around the arc and a cutting table that maintains a bed of water that contacts the bottom surface of the workpiece. The water shroud is predominantly used to reduce arc noise.

Underwater plasma arc cutting also uses a waterbed, but instead of having the level of the water contact only the bottom surface of the workpiece, it is totally submerged in the water. This system does not require the use of a water-shroud nozzle. However, it requires that the level of the water be lowered periodically for loading and unloading the plate and for the positioning of the torch and plate. Since the operator cannot see the plate during cutting with this system, it is intended for use only with CNC systems.

10. *American Welding Society (AWS)* Committee on Fumes and Gases, *1992, Methods for Sampling Airborne Particulates Generated by Welding and Allied Processes,* ANSI/AWS F1.1-92, Miami: American Welding Society.
11. See Reference 6.

Table 15.10
Recommended Eye Protection for Plasma Arc Cutting and Air Carbon Arc Cutting

	Minimum Protective Lens Shade	Suggested Lens Shade Number (Comfort)
Plasma Arc Cutting Current, A		
Less than 300	8	9
300 to 400	9	12
400 to 800	10	14
Air Carbon Arc Cutting, A		
Less than 500	10	12
500 to 1000	11	14

Source: Adapted from American National Standards Institute (ANSI), Accredited Standards Committee Z49, 1999, *Safety in Welding, Cutting, and Allied Processes,* ANSI Z49.1:1999, Miami: American Welding Society, p. 7.

Radiation

It is necessary for the cutter to wear eye and skin protection when exposure to radiation is unavoidable. The recommended eye protection for plasma arc cutting and air carbon arc cutting is shown in Table 15.10. As a rule, the operator should start the lens shade selection process by choosing a shade that is too dark to see the cutting zone, then successively go to the lighter shade that gives a sufficient view of the cutting zone without going below the recommended minimum. The likelihood of radiation exposure can be reduced by the use of mechanical barriers such as walls and welding curtains.

Plasma Arc Cutting and Gouging. The plasma arc emits intense visible and invisible (ultraviolet and infrared) radiation that is potentially harmful to the eyes and skin. This radiation may also produce ozone, oxides of nitrogen, or other toxic fumes in the surrounding atmosphere.

In addition to baffling the sound of plasma arc cutting, the water shroud also serves as a light-absorbing shield, especially when dye is added to the water in the table. When the use of dye is contemplated, the manufacturer should be contacted for information on the type and concentration of dye to use. It is advisable that the operator use eye protection even when these dyes are used because of the possibility of unexpected interruption of water flow through the water shroud. Underwater plasma arc cutting reduces the amount of radiation because of the greater depth of the water. Additional dye is not generally required.

Operators should wear adequate eye and face protection. A standard welding helmet fitted with a filter glass appropriate for the current used by plasma arc cutting should be sufficient. Heavy gloves should be worn, as plasma gouging radiates a significant amount of heat. At higher currents, aluminized gloves or a heat-resistant backpad are appropriate. Exposed areas of skin should be covered to protect against ultraviolet rays, and heavy flame retardant clothing should be worn to protect against molten metal and sparks. In extreme situations in confined areas, welding jackets or coveralls may be appropriate.

Since there is some spray of molten metal with plasma arc gouging, flammable materials should be removed from the work area and a fire extinguisher kept at hand. Although plasma arc gouging produces less fume than the air carbon arc process, some fume is produced, so good ventilation is required. If gouging a material that can produce dangerous or carcinogenic fumes, a respirator may be required.

Since high voltages are used in plasma arc gouging, operators should exercise care in wet or damp areas. The equipment should always be turned off before changing any torch parts.

Plasma gouging is quieter than air carbon arc gouging, but the noise level produced can still damage hearing if proper ear protection is not used.

If inert gases are used for gouging, operation in a confined space can displace oxygen and create an asphyxiation hazard. Proper ventilation or a fresh-air respirator must be used in these situations. If argon-hydrogen is used as the plasma gas in a confined space, it is important that precautions be taken to assure that hydrogen does not build up and create an explosion hazard. Hydrogen is lighter than air, so the potential for buildup exists when it is prevented from rising and dissipating. Proper ventilation is required in these cases.

Above all, the safety instructions the manufacturer provides with the product should be read and understood by all personnel involved in the cutting or gouging operation.

Air Carbon Arc Cutting. Any person within the immediate vicinity of the cutting arc should have adequate eye and skin protection from the radiation produced by the cutting arc. The filter shade recommended for air carbon arc cutting is shown in Table 15.10. Leather or wool clothing that is dark in color is recommended to reduce reflection, which could cause ultraviolet burns to the neck and face inside the helmet.

Noise

Plasma arc cutting and air carbon arc cutting are frequently noisy processes. Exposure to excessive noise can affect hearing capability. The condition is usually temporary if exposure is limited, but it can be permanent if exposure to the same noise takes place for a longer time.

Plasma Arc Cutting. The amount of noise generated by a plasma arc cutting torch operated in the open depends primarily on the cutting current. A torch operating at 400 A typically generates approximately 100 decibels (dBA) measured at about 1.8 m (6 ft). At 750 A, the noise level is approximately 110 dBA. Much of the noise is in the frequency range of 5000 to 20,000 hertz (Hz). These noise levels can cause hearing damage. Hearing protection must be worn when the noise level exceeds the specified limits. These values, which may vary locally, are specified by OSHA[12] for most industrial environments.

The water-shroud technique previously described is commonly used to reduce noise in mechanized cutting applications. The water effectively acts as a sound-absorbing enclosure around the torch nozzle. The water directly below the plate keeps noise from coming through the kerf opening. Noise reduction is typically about 20 dBA. This reduction is usually sufficient to bring the operation within safe limits that comply with OSHA standards.

The water-shroud technique should not be confused with water injection or water shielding, since neither of these process variations uses sufficient water to significantly reduce noise. Underwater plasma arc cutting provides greater noise reduction than the water shroud because the nozzle end of the torch and the arc are totally submerged.

Fire Hazards

Fire prevention measures are necessary for all cutting and gouging operations. Arc cutting and gouging processes produce significant heat, metal sparks, and hot spatter, which create a fire hazard. Personnel should wear fire-retardant clothing, and fire extinguishers should be readily available near the cutting operation.

Air carbon arc cutting requires special fire prevention precautions because of the metal removal process. All combustibles within 11 m (35 ft) of the work area should be removed. Protection such as metal screens should be placed in the line of hot metal ejected by the compressed air stream if ample room for dissipation is not available. Additional information on this topic is presented in the document *Fire Prevention during Welding, Cutting, and Other Hot Work,* NFPA 51B.[13]

12. Occupational Safety and Health Administration (OSHA), *Occupational Safety and Health Standards for General Industry,* in *Code of Federal Regulations (CFR),* Title 29 CFR 1910, *Subpart Q,* Washington, D.C.: Superintendent of Documents, U.S. Government Printing Office.
13. See Reference 7.

Explosion Prevention

Some cutting-gas mixtures used with plasma arc cutting and gouging contain hydrogen. The inadvertent release of such mixtures can result in explosion and fire hazards. When gas leaks are suspected, equipment must not be operated. The equipment manufacturer should be contacted regarding questions about operation with certain gases.

When cutting aluminum or magnesium plate on a water table, a possibility of hydrogen detonation beneath the workpiece exists. This is believed to be due to hydrogen released by the interaction of molten aluminum or magnesium and water. The hydrogen can accumulate in pockets under the workpiece and ignite when the cutting arc is near the accumulation. Aeration of the water in the water table to disperse the hydrogen is an effective solution. Before cutting aluminum or magnesium on a water table, the equipment manufacturer should be contacted for recommended practices.

Electrical Hazards

Electric shock can be fatal. Safety precautions for all arc cutting and gouging processes include the following:

1. All electrical circuits must be kept dry, as moisture can provide an unexpected path for current flow. Equipment cabinets that contain water, as well as gas lines and electrical circuits, should be checked periodically for leaks.
2. Equipment must be properly grounded and connected as recommended by the manufacturer.
3. All electrical connections should be kept mechanically tight, as poor electrical connections can generate heat and start fires.
4. High-voltage cable must be used, and cables and wires must be kept in good repair. The manufacturer's instructions should be consulted for the proper cable and wire sizes.
5. Personnel must not touch live circuits. Equipment panels must be kept in place, and access doors must be kept closed.
6. Only trained personnel should be permitted to operate or maintain the equipment.

The risk of electrical shock is probably the greatest when replacing used torch parts. Operators must make sure that the primary power to the power sources and the power to the control circuitry is turned off when replacing torch parts. Operators and maintenance personnel should be aware that plasma arc cutting equipment presents a greater hazard than conventional welding equipment because it uses higher voltages. The voltages used in plasma arc cutting equipment range from 150 V to 400 V dc.

SAFETY RESOURCES

Emergency first aid should be readily available. Prompt, trained emergency response may reduce the extent of injury in the event of accidental electrical shock. Additional resources are included in the following codes and standards.[14]

1. *National Electrical Safety Code,* ANSI C2;[15]
2. *Safety in Welding, Cutting, and Allied Processes,* ANSI Z49.1; [16]
3. OSHA General Industry Standards, 29 CFR 1910;[17] and
4. *Fire Prevention during Welding, Cutting, and Other Hot Work,* NFPA 51B.[18]

Appendix B of this volume, "Safety and Health Codes and Other Standards," lists these and other health and safety standards, codes, specifications, and other publications.

CONCLUSION

Arc cutting and gouging is dominated by the plasma arc cutting, plasma arc gouging, and the air carbon arc gouging processes. For most mechanized cutting operations for sheet and plate up to 38 mm (1-1/2-in.) thick, the plasma arc cutting process dominates the majority of applications based on economics when compared to oxyfuel gas cutting and laser beam cutting. Manual plasma arc cutting is extensively used for sheet metal and thinner plate. Robotic cutting with the plasma arc cutting process is used in an increasing number of applications.

With the trend toward the use of lighter weight and higher performance materials, the plasma arc gouging process is often selected because of its flexibility and its ability to gouge all metals. It is easier to adapt to mechanization or automation than air carbon arc cutting.

Labor is the highest cost element in any cutting or gouging operation, so any ability to mechanize, automate, or apply robotic operation results in reduced production costs. This will lead to further use of these processes in the future. While laser beam cutting is superior for cutting very intricate parts with fine detail and can cut thinner (less than 6 mm [1/4-in.]) materials with high productivity, the cost of laser

14. See Reference 2.
15. Institute of Electrical and Electronic Engineers (IEEE), *National Electrical Safety Code,* ANSI C2-1997, New York: American National Standards Institute.
16. See Reference 6.
17. See Reference 12.
18. See Reference 7.

systems continues to be prohibitive for many applications. The advent of high-precision plasma arc cutting and its general use in production operations highlights the healthy state of cutting technology. The process gives the user increasingly sophisticated production tools to reduce manufacturing costs while providing faster, higher quality cut products.

BIBLIOGRAPHY[19]

American Conference of Governmental Industrial Hygienists (ACGIH). 2003. *2003 TLVs® and BEIs®: Threshold limit values for chemical substances and physical agents and biological exposure indices.* Cincinnati: American Conference of Governmental Industrial Hygienists. (Editions of this publication are also available in Greek, Italian, and Spanish.)

American National Standards Institute (ANSI) Accredited Standards Committee Z49. 1999. *Safety in welding, cutting, and allied processes.* ANSI Z49.1:1999. Miami: American Welding Society.

American Welding Society (AWS) Committee on Arc Welding and Arc Cutting. 2001. *Recommended Practices for Arc Cutting and Gouging.* AWS C5.2:2001. Miami: American Welding Society.

American Welding Society (AWS) Committee on Arc Welding and Arc Cutting. 2000. *Recommended practices for arc gouging and cutting.* AWS C5.3:2000. Miami: American Welding Society.

American Welding Society (AWS) Committee on Arc Welding and Cutting. 1994. *Recommended practices for shielding gases for welding and plasma arc cutting.* ANSI/AWS C5.10-94. Miami: American Welding Society.

American Welding Society (AWS) Committee on Definitions and Symbols. 2001. *Standard welding terms and definitions.* AWS A3.0:2001. Miami: American Welding Society.

American Welding Society (AWS) Committee on Fumes and Gases. *1992. Methods for sampling airborne particulates generated by welding and allied processes.* ANSI/AWS F1.1-92. Miami: American Welding Society.

Institute of Electrical and Electronic Engineers (IEEE). *National electrical safety code.* ANSI C2-1997. New York: American National Standards Institute.

National Fire Protection Association (NFPA). 1999. *Fire prevention during welding, cutting, and other hot work.* NFPA 51B. Quincy, Mass.: National Fire Protection Association.

Occupational Safety and Health Administration (OSHA). 1999. *Occupational safety and health standards for general industry.* In *Code of Federal Regulations (CFR),* Title 29 CFR 1910, *Subpart Q.* Washington, D.C.: Superintendent of Documents, U.S. Government Printing Office.

Sibley, C. R. September 29, 1959. *Electric arc cutting.* U.S. Patent 2,906,853.

19. The dates of publication given for the codes and other standards listed here were current at the time this chapter was prepared. The reader is advised to consult the latest edition.

SUPPLEMENTARY READING LIST

Fernicola, R. C. 1998. A guide to manual plasma arc cutting. *Welding Journal* 77(3): 53–55.

Fernicola, R. C. 1994. New oxygen plasma process rivals laser cutting methods. *Welding Journal* 72(6): 65–69.

Frappier, M. B. 1988. Plasma arc cutting supplies explained. *Welding Journal* 67(2):48–50.

Hebble, C. M., Jr. 1973. Cutting with low current broadens application of plasma process. *Welding Journal* 52(9): 587-s–589-s.

Heflin, R. L. 1985. Plasma arc gouging of aluminum. *Welding Journal* 64(5): 16–19.

Hughey, H. G. 1947. Stainless steel cutting. *Welding Journal* 26(5): 393-s–400-s.

Kandel, C. 1946. Underwater cutting and welding. *Welding Journal* 25(3): 209-s–212-s.

Marshall, W. J., D. H. Sliner, M. Hoekkala, and C.E. Moss. 1980. Optical radiation levels produced by air carbon arc cutting processes. *Welding Journal* 59(3): 43–46.

McGough. M. S., W. E. Austin, and G. J. Kretl. 1989. Underwater plasma arc cutting in Three Mile Island's reactor. *Welding Journal* 68(7): 22–26.

Na, S. J., S. W. Park, S. H. Cho, and T. J. Ho. 1988. A microprocessor-based shape and velocity control system for plasma arc cutting. *Welding Journal* 67(2): 27–33.

Panter, D. 1977. Air carbon arc gouging. *Welding Journal* 56(5): 32–37.

Sasse, F. H. 1991. Oxygen plasma process increases quality in carbon steel cutting. *Welding Journal* 70(2): 64–66.

Severance, W. S., and D. G. Anderson. 1984. How plasma arc cutting gases affect productivity. *Welding Journal* 73(2): 35–38.

Skinner, G. M., and R. J. Wickham. 1967. High quality plasma arc cutting and piercing. *Welding Journal* 46(8): 657-s –664-s.

Thielsch, H., and J. Quass. 1954. Shielded-metal-arc cutting and grooving. *Welding Journal* 33(5): 438-s–446-s.

U.S. Government Printing Office. *Underwater cutting and welding manual.* NAVSHIPS 250-692-9. Washington, D. C.: Superintendent of Documents, U. S. Government Printing Office.

Wait, J. D., and S. H. Resh. 1959. Tungsten arc cutting of stainless steel shapes in steel warehousing operations. *Welding Journal* 38(6): 576-s–581-s.

Wodtke, C. H., W. A. Plunkett, and D. R. Firzzell. 1976. Development of underwater plasma arc cutting. *Welding Journal* 55(1): 15–24.

Appendix A

LENS SHADE SELECTOR

Shade numbers are given as a guide only and may be varied to suit individual needs.

Process	Electrode Size in. (mm)	Arc Current (Amperes)	Minimum Protective Shade	Suggested* Shade No. (Comfort)
Shielded Metal Arc Welding (SMAW)	Less than 2.4 (3/32)	Less than 60	7	—
	2.4–4.0 (3/32–5/32)	60–160	8	10
	4.0–6.4 (5/32–1/4)	160–250	10	12
	More than 6.4 (1/4)	250–550	11	14
Gas Metal Arc Welding (GMAW) and **Flux Cored Arc Welding (FCAW)**		Less than 60	7	—
		60–160	10	11
		160–250	10	12
		250–500	10	14
Gas Tungsten Arc Welding (GTAW)		Less than 50	8	10
		50–150	8	12
		150–500	10	14
Air Carbon Arc Cutting (CAC-A)				
(Light)		Less than 500	10	12
(Heavy)		500–1000	11	14
Plasma Arc Welding (PAW)		Less than 20	6	6–8
		20–100	8	10
		100–400	10	12
		400–800	11	14
Plasma Arc Cutting (PAC)		Less than 20	4	4
		20–40	5	5
		40–60	6	6
		60–80	8	8
		80–300	8	9
		300–400	9	12
		400–800	10	14
Torch Brazing (TB)		—	—	3 or 4
Torch Soldering (TS)		—	—	2
Carbon Arc Welding (CAW)		—	—	14

	Plate Thickness		Suggested* Shade No. (Comfort)
	mm	in.	
Oxyfuel Gas Welding (OFW)			
Light	Under 3	Under 1/8	4 or 5
Medium	3 to 13	1/8 to 1/2	5 or 6
Heavy	Over 13	Over 1/2	6 or 8
Oxygen Cutting (OC)			
Light	Under 25	Under 1	3 or 4
Medium	25 to 150	1 to 6	4 or 5
Heavy	Over 150	Over 6	5 or 6

*As a general rule for selecting the correct lens shade, the welder should start with a shade that is too dark to see the weld zone, then go to a lighter shade which gives sufficient view of the weld zone without going below the minimum. In oxyfuel gas welding, cutting, or brazing where the torch or the flux, or both, produces a high yellow light, it is desirable to use a filter lens that absorbs the yellow or sodium line of the visible light spectrum.

Source: Adapted from American Welding Society (AWS) Committee on Safety and Health and Subcommittee on Labeling and Safe Practices, F2.2:2001 *Lens Shade Selector* (Chart), Miami: American Welding Society.

Appendix B

HEALTH AND SAFETY CODES AND OTHER STANDARDS

Appendix B is excerpted from Annex B of *Safety in Welding, Cutting, and Allied Processes*, ANSI Z49.1:1999.[1,2] It is not a part of that standard, but is included for information.

The following codes, standards, specifications, pamphlets, and books contain information that may be useful in meeting the safety and health requirements of welding, cutting, and allied processes. The names of the standards organizations and their letter designations are also listed in this appendix. Inquiries as to the availability and cost of any of these publications should be addressed directly to the publishers.

ACGIH		Threshold Limit Values (TLV®) for Chemical Substances and Physical Agents in the Workroom Environment Industrial Ventilation Manual
AGA		Purging Principles and Practices
ANSI	A13.1	Scheme for the Identification of Piping Systems
	B11.1	Safety Requirements for Construction, Care and Use of Mechanical Power Presses
	B15.1	Safety Standard for Mechanical Power Transmission Apparatus (with ASME)
	B31.1	Power Piping (with ASME)
	Z535.4	Standard for Product Safety Signs and Labels
	Z87.1	Practice for Occupational and Educational Eye and Face Protection
	Z88.2	Respiratory Protection
	Z89.1	Protective Headwear for Industrial Workers
API	1104	Standard for Welding Pipelines and Related Facilities
	PUBL 2009	Safe Welding and Cutting Practices in Refineries, Gasoline Plants, and Petrochemical Plants
	PUBL 2013	Cleaning Mobile Tanks in Flammable or Combustible Liquid Service
	STD 2015	Safe Entry and Cleaning of Petroleum Storage Tanks, Planning and Managing Tank Entry from Decommissioning through Recommissioning
	PUBL 2201	Procedures for Welding or Hot Tapping on Equipment in Service
AVS		Vacuum Hazards Manual

1. American National Standards Institute (ANSI) Committee Z49, 1999, *Safety in Welding, Cutting, and Allied Processes*, ANSI Z49.1:1999, American Welding Society.
2. At the time of preparation of this appendix, the codes and other standards were valid, however, as codes and standards undergo frequent revision, the reader is advised to consult the most recent edition.

CGA	C 4	Method of Marking Portable Compressed Gas Cylinders to Identify the Material Contained
	E 1	Regulator Connection Standards
	E 2	Standard Hose Connection Specification
	G 7.1	Commodity Specification for Air
	P 1	Safe Handling of Compressed Gas Cylinders
	V 1	Compressed Gas Cylinder Valve Outlet and Inlet Connections
NEMA	EW1	Electric Arc Welding Power Sources
NFPA	50	Bulk Oxygen Systems at Consumer Sites
	51	Standard for the Design of Oxygen-Fuel Gas Systems for Welding and Cutting, and Allied Processes
	51B	Standard for Fire Prevention in Use of Cutting and Welding Processes
	70	National Electrical Code®
	79	Electrical Standard for Industrial Machinery
	306	Control of Gas Hazards on Vessels
	327	Standard Procedures for Cleaning or Safeguarding Small Tanks and Containers Without Entry
	701	Standard for Flame-Resistant Textiles and Films
NIOSH	78 138	Safety and Health in Arc Welding and Gas Welding and Cutting
	80 144	Certified Equipment List (with Supplements)
NSC		Accident Prevention Manual for Industrial Operations
		Fundamentals of Industrial Hygiene
OSHA		Occupational Safety and Health Standards for General Industry (29 CFR Part 1910, Subpart Q)
		Occupational Safety and Health Standards for Construction (29 CFR Part 1926, Subpart J)
RMA	IP 7	Specification for Rubber Welding Hose
RWMA		Resistance Welding Machine Standards
UL	252	Regulators
	551	Transformer-Type Arc Welding Machines

PUBLISHERS OF SAFETY CODES AND OTHER STANDARDS

ACGIH **American Conference of Governmental Industrial Hygienists**
6500 Glenway Avenue, Building D 7
Cincinnati, OH 45211-4438
www.acgih.org

AGA **American Gas Association**
1515 Wilson Boulevard
Arlington, VA 22209
www.aga.com

ANSI **American National Standards Institute**
11 West 42nd Street, 13th Floor
New York, NY 10036-8002
www.ansi.org

API **American Petroleum Institute**
1220 L Street NW
Washington, DC 20005
www.api.org

ASME **American Society of Mechanical Engineers**
345 East 47th Street
New York, NY 10017-2392
www.asme.org

ASTM **ASTM**
1916 Race Street
Philadelphia, PA 19103
www.astm.org

AVS **American Vacuum Society**
120 Wall Street, 32nd Floor
New York, NY 10005
www.avs.org

AWS **American Welding Society**
550 N.W. LeJeune Road
Miami, FL 33126
www.aws.org

CGA **Compressed Gas Association**
1725 Jefferson Davis Highway, Suite 1004
Arlington, VA 22202-4102
www.cganet.com

MSHA **Mine Safety and Health Administration**
4015 Wilson Boulevard
Arlington, VA 22203
www.msha.gov

NEMA **National Electrical Manufacturers Association**
2101 L Street NW
Washington, DC 20037
www.nema.org

NFPA **National Fire Protection Association**
One Batterymarch Park
Quincy, MA 02269-9101
www.nfpa.org
www.sparky.org

NIOSH **National Institute for Occupational Safety and Health**
4676 Columbia Parkway
Cincinnati, OH 45226
www.cdc.gov/niosh/homepage/html

NSC **National Safety Council**
1121 Spring Lake Drive
Itasca, IL 60143-3201
www.nsc.org

OSHA **Occupational Safety and Health Administration**
200 Constitution Avenue NW
Washington, DC 20210
www.osha.gov

RWMA **Resistance Welder Manufacturers Association**
1900 Arch Street
Philadelphia, PA 19103
www.rwma.org

RMA **Rubber Manufacturers Association**
1400 K Street NW
Washington, DC 20005
www.rma.org

UL **Underwriters Laboratories, Incorporated**
333 Pfingsten Road
Northbrook, IL 60062
www.ul.com

U.S. Government Printing Office
Superintendent of Documents
P.O. Box 371954
Pittsburgh, PA 15250-7954
www/access/gpo.gov

Appendix C

FILLER METAL SPECIFICATIONS

Compiled by the American Welding Society Committee on Filler Metals and Allied Products.[1,2]

AWS Filler Metal Specifications by Material and Welding Process

	OFW	SMAW	GTAW GMAW PAW	FCAW	SAW	ESW	EGW	Brazing
Carbon Steel	A5.2	A5.1	A5.18	A5.20	A5.17	A5.25	A5.26	A5.8, A5.31
Low-Alloy Steel	A5.2	A5.5	A5.28	A5.29	A5.23	A5.25	A5.26	A5.8, A5.31
Stainless Steel		A5.4	A5.9, A5.22	A5.22	A5.9	A5.9	A5.9	A5.8, A5.31
Cast Iron	A5.15	A5.15	A5.15	A5.15				A5.8, A5.31
Nickel Alloys		A5.11	A5.14		A5.14			A5.8, A5.31
Aluminum Alloys		A5.3	A5.10					A5.8, A5.31
Copper Alloys		A5.6	A5.7					A5.8, A5.31
Titanium Alloys			A5.16					A5.8, A5.31
Zirconium Alloys			A5.24					A5.8, A5.31
Magnesium Alloys			A5.19					A5.8, A5.31
Tungsten Electrodes			A5.12					
Brazing Alloys and Fluxes								A5.8, A5.31
Surfacing Alloys	A5.21	A5.13	A5.21	A5.21	A5.21			
Consumable Inserts			A5.30					
Shielding Gases			A5.32	A5.32			A5.32	

1. American Welding Society (AWS) Committee on Filler Metals and Allied Products.
2. At the time of preparation of this appendix the referenced codes and standards were valid. However, as codes and standards undergo frequent revision, the reader is advised to consult the most recent edition.

AWS Filler Metal Specifications and Related Documents

Designation	Title
FMC	*Filler Metal Comparison Charts*
UGFM	*User's Guide to Filler Metals*
A4.2M/A4.2	*Standard Procedures for Calibrating Magnetic Instruments to Measure the Delta Ferrite Content Austenitic and Duplex Ferritic-Austenitic Stainless Steel Weld Metal*
A4.3	*Standard Methods for Determination of the Diffusible Hydrogen Content of Martensitic, Bainitic, and Ferritic Steel Weld Metal Produced by Arc Welding*
A4.4M	*Standard Procedures for Determination of Moisture Content of Welding Fluxes and Welding Electrode Flux Coverings*
A5.01	*Filler Metal Procurement Guidelines*
A5.1	*Specification for Carbon Steel Electrodes for Shielded Metal Arc Welding*
A5.2	*Specification for Carbon and Low Alloy Steel Rods for Oxyfuel Gas Welding*
A5.3/A5.3M	*Specification for Aluminum and Aluminum-Alloy Electrodes for Shielded Metal Arc Welding*
A5.4	*Specification for Stainless Steel Electrodes for Shielded Metal Arc Welding*
A5.5	*Specification for Low-Alloy Steel Electrodes for Shielded Metal Arc Welding*
A5.6	*Specification for Covered Copper and Copper Alloy Arc Welding Electrodes*
A5.7	*Specification for Copper and Copper Alloy Bare Welding Rods and Electrodes*
A5.8	*Specification for Filler Metals for Brazing and Braze Welding*
A5.9	*Specification for Bare Stainless Steel Welding Electrodes and Rods*
A5.10/A5.10M	*Specification for Bare Aluminum and Aluminum-Alloy Welding Electrodes and Rods*
A5.11/A5.11M	*Specification for Nickel and Nickel-Alloy Welding Electrodes for Shielded Metal Arc Welding*
A5.12/A5.12M	*Specification for Tungsten and Tungsten-Alloy Electrodes for Arc Welding and Cutting*
A5.13	*Specification for Surfacing Electrodes for Shielded Metal Arc Welding*
A5.14/A5.14M	*Specification for Nickel and Nickel-Alloy Bare Welding Electrodes and Rods*
A5.15	*Specification for Welding Electrodes and Rods for Cast Iron*
A5.16/A5.16M	*Specification for Titanium and Titanium Alloy Welding Electrodes and Rods*
A5.17/A5.17M	*Specification for Carbon Steel Electrodes and Fluxes for Submerged Arc Welding*
A5.18/A5.18M	*Specification for Carbon Steel Electrodes and Rods for Gas Shielded Arc Welding*
A5.19	*Specification for Magnesium Alloy Welding Electrodes and Rods*
A5.20	*Specification for Carbon Steel Electrodes for Flux Cored Arc Welding*
A5.21	*Specification for Bare Electrodes and Rods for Surfacing*
A5.22	*Specification for Stainless Steel Electrodes for Flux Cored Arc Welding and Stainless Steel Flux Cored Rods for Gas Tungsten Arc Welding*
A5.23/A5.23M	*Specification for Low-Alloy Steel Electrodes and Fluxes for Submerged Arc Welding*
A5.24	*Specification for Zirconium and Zirconium Alloy Welding Electrodes and Rods*
A5.25/A5.25M	*Specification for Carbon and Low-Alloy Steel Electrodes and Fluxes for Electroslag Welding*
A5.26/A5.26M	*Specification for Carbon and Low-Alloy Steel Electrodes for Electrogas Welding*
A5.28	*Specification for Low-Alloy Steel Electrodes and Rods for Gas Shielded Arc Welding*
A5.29	*Specification for Low-Alloy Steel Electrodes for Flux Cored Arc Welding*
A5.30	*Specification for Consumable Inserts*
A5.31	*Specification for Fluxes for Brazing and Braze Welding*
A5.32/A5.32M	*Specification for Welding Shielding Gases*

INDEX OF MAJOR SUBJECTS

A

D

E

F

G

N

O

P

T

U

V

Z

INDEX

A

B

C

D

E

F

G

H

I

J

K

L

M

N

O

P

Q

R

S

T

U

V

W

Z